AF517736

AGROFORESTRY
THEORY AND PRACTICES

ANTONY JOSEPH RAJ ***Ph.D.***
Associate Dean
College of Forestry and Environment
Sam Higginbottom Institute of Agriculture, Technology and Sciences
(Formerly Allahabad Agricultural Institute) Deemed University
Allahabad, Uttar Pradesh, India

S.B. LAL ***Ph.D.***
Pro-Vice Chancellor
Sam Higginbottom Institute of Agriculture, Technology and Sciences
(Formerly Allahabad Agricultural Institute) Deemed University
Allahabad, Uttar Pradesh, India

Published by:

SCIENTIFIC PUBLISHERS (INDIA)
5A, New Pali Road, P.O. Box 91
Jodhpur 342 001, India
E-mail: info@scientificpub.com
Website: www.scientificpub.com

Print: 2025

ISBN: 978-81-72338-66-4
eISBN: 978-93-86102-96-6

Printed in India

FOREWORD

World's total forest area is 4.03 billion hectares, corresponding to 31 per cent of the total land area or an average of 0.6 ha of per capita, whereas India's forest and tree cover is 78.29 million hectares which is only 23.81 per cent of the geographical area. We are continuously putting efforts to achieve the national goal of 33 per cent geographic area of the country under the forest and tree cover as enshrined in the National Forest Policy, 1988. India has only 2.4 per cent of the world's geographical area and 0.5 per cent of the world's grazing area but supports over 16 per cent of the world's human population and over 18 per cent of world's cattle population. This ever increasing population places enormous demands and pressures on the land resources and forest resources. Agro-forestry is the potential option for increasing the tree cover outside the notified forest areas.

India is endowed with a variety of soils, climate, biodiversity and ecological regions. An area of 46.70 million ha has been estimated under wastelands which is 14.75 per cent of the total geographical area of the country. The land degradation issue threatens country's food security and the quality of the environment which assumes a major significance nowadays. Agroforestry practices are considered as most vital technology and potential farming system for minimizing the land degradation. Agroforestry practices increase farm productivity, diversify income sources for farmers and provide environ-mental services. Agroforestry improves soil, water and air quality, and biodiversity while supporting sustainable production of food, feed, fibre and energy. IPCC's prediction of temperature increase between 1.1°C and 6.2°C by the end of the century due to excessive carbon dioxide emission will most likely create extreme changes in temperature and precipitation. Agroforestry represents a significant opportunity for sequestering more carbon per unit area on agricultural lands and it can be better climate change mitigation option than ocean and other terrrestrial options because of vast production and protective benefits.

The new comprehensive editorial textbook "Agroforestry: Theory and Practices" by Dr.Antony Joseph Raj and Prof.S.B.Lal is a rich source of knowledge and practical information on agroforestry drawn from the scientific literature, databases and field experiences from all over world. This textbook provides thoroughly up-to-date principles and methods on agroforestry and excellently covers the latest and modern technologies in the agroforestry field. This book, with its high standards, will enormously benefit the students for their preparation of competitive exams like UPSC-Civil Services, UPSC-Indian Forest Service, ICAR-ARS Scientist Exam, ICFRE Forestry Scientist Exam, NET Exam, State Public Service Commission Exams etc. I am confident that scientists, University and college teachers, and foresters from all over world should find this resource book useful in creating effective and innovative training programmes and manpower in agroforestry.

I welcome this latest Agroforestry textbook which will be most valuable to the students of agriculture, forestry, horticulture, soil science, water science, ecology, environment science and other plant sciences.

I would like to congratulate the authors, Dr. Antony Joseph Raj and Prof. S.B. Lal, for their tireless efforts in bringing this high quality textbook for the benefit of students, teachers, scientists and agricultural community.

[Rev.(Prof. Dr.) RAJENDRA B. LAL]
Vice-Chancellor
Sam Higginbottom Institute of Agriculture, Technology and Sciences
(Formerly Allahabad Agricultural Institute) Deemed University
Allahabad, Uttar Pradesh, India

July 2013

PREFACE

Agroforestry provides a different land use option, compared with traditional agricultural and forestry systems. Agroforestry combines trees, shrubs, forages, grasses, livestock, and crops in innovative, flexible combinations tailored to the needs of farmers and landowners. This intensive integration in agroforestry ensures sustained availability of multiple products as direct benefits such as food, vegetables, fruits, fodder, fuel, manure, medicine, timber, etc. It makes use of the complementarity between trees and crops, so that the available resources can be more effectively exploited. It is a practice that respects the environment and has an obvious landscape benefit. The integration of trees, agricultural crops, and/or animals into an agroforestry system has the potential to enhance soil fertility, reduce erosion, improve water quality, enhance biodiversity, increase aesthetics and sequester carbon. Efficient, modern versions of agroforestry have been developed around the world which can be adapted to different agroclimatic conditions.

The agroforestry always remains productive for the farmer and generates continuous revenue. Agroforestry allows for the diversification of farm activity and makes better use of environmental resources. The goal of agroforestry is to optimize productivity and conservation benefits within a set of integrated land use practices. Agroforestry has interesting advantages from three different perspectives viz. agriculture, forestry and environment. The agroforestry practices enhance biodiversity, sequester more carbon dioxide from atmosphere, diversify farmers' income sources, generate greater profits than annual crops, and create a more integrated, interesting and visually appealing land use system that may be more environmentally, economically and socially sustainable than the original farmland. With the shrinking per capita land availability, agroforestry system with the integration of perennial woody trees with crops/ pastures is most suitable technology for increasing total productivity of food, feed and fuel and thereby reducing the risk of farming.

This editorial textbook "Agroforestry: Theory and Practices" is one of the finest books on agroforestry that offers a global review of the basic approaches, tools and technologies, research innovations and real-world practices in agroforestry. The book offers a comprehensive guide to basic principles, techniques and applications, integrative strategies, economic and environmental concerns, and future trends in agroforestry in different regions of the world. This textbook is an effort to create a coherent and wide-ranging guide to the practice of agroforestry. This book covers key areas in agroforestry, namely agroforestry practices and its distribution, agroforestry systems classification, agroforestry trees, agroforestry management, technologies and modern concepts in agroforestry, production benefits of agroforestry, environmental services of agroforestry, agroforestry education, research & extension, etc.

In many aspects, the topics and structure of this textbook is highly meritorious and unique than other agroforestry books. This textbook is intended for university and college students, professors, scientists, researchers, foresters, farmers, policy makers and professionals in the field of agriculture, forestry, horticulture, other agricultural sciences and biological sciences. The main intention of this textbook is to provide a state-of-the-art and up-to-date knowledge of recent developments in agroforestry as a potential future land use system. This textbook on agroforestry will enormously benefit the students for their preparation of competitive exams like UPSC-Civil Services, UPSC-Indian Forest Service, ICAR-ARS Scientist/NET Exam, ICFRE Forestry Scientist Exam, State Public Service Commission Exams and University Entrance Exam for admission to M.Sc. and Ph.D. programmes

This agroforestry textbook will contribute significantly to academic teaching and scientific research. Additional information or suggestions are invited from experienced researchers and experts for improving the quality of the book in future editions. We thank Dr.(Mrs.) Roselin Antony, Assistant Professor of Mekelle University (Ethiopia) for her editorial help and for enhancing the English language & technical writing of manuscript. We are confident that this agroforestry textbook will become a huge success just like our earlier book "Forestry: Principles and Applications".

ANTONY JOSEPH RAJ
S.B. LAL

July 2013

LIST OF CONTRIBUTORS

Dr. Antony Joseph Raj
Associate Professor (Forestry), LaRMEP, Mekelle University, Mekelle, Ethiopia & Associate Dean (On Leave), College of Forestry and Environment, SHIATS Deemed University, Allahabad

Dr. M.M. Roy
Director, Central Arid Zone Research Institute (CAZRI), Jodhpur, Rajasthan

Shri. N. Satheesh, IFS
District Forest Officer, Tiruchirappalli, Tamil Nadu

Dr. K. Sasikumar, IFS
District Forest Officer, North Tripura District, Kumarghat, Tripura

Dr. T. Mohan Raj, IFS
Divisional Forest Officer, Tonk Forest Division, Tonk, Rajasthan

Shri. N. Gurunathan, IFS
Scientist in National Research Center for Agroforestry, Jhansi (Earlier) and Presently Probationary Officer, Indira Gandhi National Forest Academy, Dehradun, Uttarakhand

Dr. Suresh Kumar
Head & Principal Scientist, Department of Integrated Land Use Management, Central Arid Zone Research Institute (CAZRI), Jodhpur, Rajasthan

Prof. S.B. Lal
Pro-Vice Chancellor and Dean, College of Forestry and Environment, SHIATS Deemed University, (Formerly Allahabad Agricultural Institute), Allahabad, Uttar Pradesh

Shri. P. Durairasu, IFS
Dean (Forestry), Forest College and Research Institute (FC&RI), Mettupalayam, Tamil Nadu

Prof. S.L. Madiwalar
Dean (Forestry) & Professor (Agroforestry), College of Forestry, Sirsi, (UAS-Dharwad), Karnataka

Prof. S.J. Patil
Professor and Head, Dept. of Silviculture and Agroforestry & Officer In-charge for AICRP Agroforestry, University of Agricultural Sciences, (UAS-Dharwad), Karnataka

Prof. K.T. Parthiban
Professor and Head, Dept. of Tree Breeding, Forest College and Research Institute (FC&RI), Mettupalayam, Tamil Nadu

Prof. M.P. Divya
Professor (Agroforestry), Officer In-charge for AICRP Agroforestry, Forest College and Research Institute (FC&RI), Mettupalayam, Tamil Nadu

Prof. C.T. Ashok Kumar
Professor (Entomology), Department of Entomology, College of Agriculture, U.A.S., G.K.V.K., Bangalore, Karnataka

Dr. Ramakrishna Hegde
Associate Professor & Head,
Department of Silviculture and Agroforestry,
College of Forestry, Ponnampet
(UAS, Bangalore), Karnataka

Dr. T.K. Kunhamu
Associate Professor (Agroforestry),
Dept. of Silviculture and Agroforestry and Officer In-charge for AICRP Agroforestry, College of Forestry, Kerala Agricultural University, Thrissur, Kerala

Dr. K.S. Pant
Senior Scientist (Agroforestry),
Department of Silviculture and Agroforestry,
College of Forestry, Dr.Y.S.Parmar University of Horticulture and Forestry, Nauni, Solan, H.P.

Dr. C. Buvaneswaran
Scientist E, Division of Forestry, Land Use and Climate Change, Institute of Forest Genetics and Tree Breeding, Coimbatore, Tamil Nadu

Dr. Sanjeev K. Chauhan
Associate Professor (Forestry),
Dept. of Forestry & Natural Resources,
Punjab Agricultural University,
Ludhiana, Punjab

Dr. Ajit
Senior Scientist, National Research Center for Agroforestry (NRCAF),
Jhansi, Uttar Pradesh

Dr. Munish Kumar
Associate Professor, Department of Soil Conservation and Water Management/Forestry, C.S. Azad University of Agriculture and Technology, Kanpur, Uttar Pradesh

Dr. Sharad Nema
Associate professor (Forestry) & Head,
School of Studies (Forestry & Wildlife),
Bastar University, Jagdalpur, Chhattisgarh

Dr. Naresh Kaushik
Senior Scientist (Agroforestry),
CCS Haryana Agricultural University,
Hissar, Haryana

Prof. M.S. Malik
Professor and Head, Dept. of Silviculture and Agroforestry and Officer In-charge for AICRP Agroforestry, Faculty of Forestry, Birsa Agriculture University, Kanke, Ranchi, Jharkhand

Dr. O.P. Chaturvedi
Head, Department of Plant Science & Principal Scientist, Central Soil and Water Conservation Research and Training Institute (ICAR), Dehradun

Dr. A. Venkatesh
Principal Scientist (Forestry),
ICAR Research Complex for NEH Region,
Umiam (Barapani), Meghalaya

Dr. J.C. Tewari
Principal Scientist (Forestry),
Central Arid Zone Research Institute (CAZRI),
Jodhpur, Rajasthan

Dr. Harsh Mehta
Principal Scientist, Central Soil and Water Conservation Research and Training Institute (ICAR Institute),
Dehradun, Uttarakhand

Dr. J.M.S. Tomar
Senior Scientist, Central Soil and Water Conservation Research and Training Institute (ICAR), Dehradun, Uttarakhand

Dr. P. Ratha Krishnan
Senior Scientist (Forestry),
Central Arid Zone Research Institute (CAZRI),
Jodhpur, Rajasthan

Dr. Rajesh Kaushal
Senior Scientist, Central Soil and Water Conservation Research and Training Institute (ICAR), Dehradun, Uttarakhand

Dr. D.R. Palsaniya
Senior Scientist,
Indian Grassland and Fodder Research Institute (ICAR), Jhansi, Uttar Pradesh

Dr. P.J. Mishra
Officer In-charge for AICRP Agroforestry, College of Agriculture, Orissa University of Agriculture and Technology, Bhubaneswar, Odisha

Dr. Lalit M. Gupta
Associate Professor (Agroforestry), Division of Agroforestry, Sher-E-Kashmir University of Agriculture, Science and Technology, Jammu, J&K

Dr. Manmohan J.R. Dobriyal
Associate Professor (Silviculture), Department of Silviculture and Agroforestry, Aspee College of Horticulture and Forestry/ College of Forestry, Navsari Agricultural University, Navsari, Gujarat

Dr. Emiru Birhane
Assistant Professor, Department of Land Resources Management and Environmental Protection, Mekelle University, Mekelle, Ethiopia

Mr. Etefa Guyassa
Department of Land Resources Management and Environmental Protection, Mekelle University, Mekelle, Ethiopia

Dr. Ajay Thakur
Scientist, Botany Division, Forest Research Institute (ICFRE Institute), Dehradun, Uttarakhand

Mrs. Smitha G Nair
Jr. Scientist, AICRP on Agroforestry, Orissa University of Agriculture and Technology, Bhubaneswar, Odisha

Dr. S. Kala
Scientist (Forestry), ICAR-CSWCRTI Research Centre, Agra, Uttar Pradesh

Dr. A. Sudha
Assistant professor (Forest Pathology) Forest college and Research Institute Mettupalayam, Tamil Nadu

Dr. J. Jayaprakash
Scientist, Central Soil and Water Conservation Research and Training Institute (ICAR), Dehradun, Uttarakhand

Dr. S. Varadha Raj
Assistant Professor, Department of Agricultural Economics, Tamil Nadu Agricultural University, Coimbatore, Tamil Nadu

Dr. S. Umesh Kanna
Assistant Professor (Forestry), Forest College and Research Institute (FC&RI), Mettupalayam, Tamil Nadu

Mr. Mengisteab Hailu
Head, Department of Land Resources Management and Environmental Protection, Mekelle University, Mekelle, Ethiopia

Mr. Samuale Tesfaye
Department of Land Resources Management and Environmental Protection, Mekelle University, Mekelle, Ethiopia

Mr. Hailemariam Gebrewahid
Institute of Geo-Information and Earth Observation Sciences (I-GEOS), Mekelle University, Mekelle, Ethiopia

Dr. Afaq Majid Wani
Assistant Professor, College of Forestry and Environment, SHIATS Deemed University, (Formerly Allahabad Agricultural Institute), Allahabad, Uttar Pradesh

Dr. Munesh Kumar
Assistant Professor, Department of Forestry, HNB Garhwal University, Srinagar Garhwal, Uttarakhand

Dr. Amol Vasishth
Assistant Professor (Forestry), College of Horticulture and Forestry, MPUAT, Jhalawar, Rajasthan

Shri. V. Mahewarappa
Asst Professor, Department of Silviculture and Agroforestry, College of Forestry, Ponnampet, University of Agricultural Sciences (UAS-Bangalore), Karnataka

Dr. Vipan Guleria
Dr.Y.S.Parmar University of Horticulture and Forestry,
Nauni, Solan, Himachal Pradesh

Ms. Meenakshi Gupta
Assistant Professor (Agroforestry),
Division of Agroforestry,
Sher-E-Kashmir University of Agriculture, Science and Technology,
Jammu, J&K

Ms. S. Vennila
Forest College and Research Institute
Mettupalayam, Tamil Nadu

Ms. V. Priyanka
Forest college and Research Institute,
Mettupalayam, Tamil Nadu

Dr. Raj Kumar
Scientist, Central Soil and Water Conservation Research and Training Institute (ICAR),
Dehradun, Uttarakhand

Dr. S.K. Uttam
Assistant Professor, Department of Soil Conservation and Water Management/Forestry,
C.S. Azad University of Agricultural & Technology,
Kanpur, Uttar Pradesh

Shri. M.N. Ramesha
Scientist,
ICAR-CSWCRTI Research Centre,
Datia, Madhya Pradesh

B.B. Behra
Jr. Scientist, AICRP on Agroforestry,
Orissa University of Agriculture and Technology, Bhubaneswar, Odisha

Ms. B. Vinothini
Forest college and Research Institute
Mettupalayam, Tamil Nadu

CONTENTS

PART 7: AGROFORESTRY PRACTICES FOR EXTREME SITE CONDITIONS

PART 8: AGROFORESTRY PRACTICES FOR SOCIO-ECONOMIC DEVELOPMENT

LIST OF TABLES

LIST OF FIGURES

ABBREVIATIONS AND ACRONYMS

A&R : Afforestation and Reforestation

ACIAR : Australian Centre for International Agricultural Research

AEZs : Agro-Ecological Zones

AF : Agroforestry

AFLP : Amplified Fragment Length Polymorphism

AFNETA : Alley Farming Network for Tropical Africa

AFRENA : African Research Network on Agroforestry

AFS : Agroforestry System

AFTA : Association for Temperate Agroforestry

AFTPs : Agroforestry Tree Products

AGB : Aboveground Biomass

AICRP : All India Coordinated Research Project

AICRP-Agroforestry : All India Coordinated Research Project on Agroforestry

AIDAB : Australian International Development Assistance Bureau

ANAFE : African Network for Agriculture, Agroforestry and Natural Resources Education

APAN : Asia-Pacific Agroforestry Network

ASB : Alternatives to Slash-and-Burn

B/C Ratio : Benefit/Cost Ratio

BCTIO : British Colombia Trade and Investment Office

BNF : Biological Nitrogen Fixation

CA : Conservation Agriculture

CAF : Complex Agroforestry Systems

CAMCORE : Central American and Mexico Coniferous Resources Cooperative

CAPS : Cleaved Amplified Polymorphic Sequences

CBD : Convention on Biological Diversity

CDIAC : Carbon Dioxide Information Analysis Center

CEC : Cation Exchange Capacity

CFA : Commonwealth Forestry Association

CFCs : Chlorofluorocarbons

CGIAR : Consultative Group on International Agricultural Research

CIAT : International Center for Tropical Agriculture

CIDA : Canadian International Development Agency

CIFOR : Centre for International Forestry Research

CIRAD-Forêt : Centre de coopération internationale en recherche agronomique pour le développement

COP : Conference of the Parties

CPR : Common Property Resources

CRD : Completely Randomized Design

CSIRO : Commonwealth Scientific and Industrial Research Organization

CSSRI : Central Soil Salinity Research Institute

CSTAF : Center for Subtropical Agroforestry

CSWCRTI : Central Soil and Water Conservation Research and Training Institute

D&D : Diagnosis and Design

DANIDA : Danish International Development Agency

DBH : Diameter at Breast Height

DoLR : Department of Land Resources

EAI : Equivalent Annual Income

EFI : European Forest Institute

ESP : Exchangeable Sodium Percent

ESTs : Expressed Sequence Tags

ET : Evapotranspiration

ETFRN : European Tropical Forest Research Network

EURAF : European Agroforestry Federation

FACT Net : Forest, Farm, and Community Tree Network

FAO : Food and Agricultural Organization of the United Nations

FARA : Forum for Agricultural Research in Africa

FDC : Forest Development Corporation

FGNFTs : Fast Growing Nitrogen Fixing Trees
FINNIDA : Finnish International Development Agency
FIO : Forest Industries Organization
FRIM : Forest Research Institute Malaysia
FSR/E : Farming Systems Research/Extension
FYM : Farm Yard Manure
GDP : Gross Domestic Product
GEF : Global Environment Facility
GHGs : Greenhouse Gases
GIS : Geographic Information System
GPP : Gross Primary Production
Gt : Gigaton
GTZ : Deutsche Gesellschaft für Technische Zusammenarbeit (German Agency for Technical Cooperation)
GWP : Global Warming Potential
HCFCs : Hydrochlorofluorocarbons
HDSR : High Density Short Rotation
HFCs : Hydrofluorocarbons
IAASTD : International Assessment of Agricultural Knowledge, Science and Technology for Development
IARCs : International Agricultural Research Centers
IBPGR : International Board for Plant Genetic Resources
ICAR : Indian Council of Agricultural Research
ICARDA : International Center for Agricultural Research in the Dry Area
ICEF : India-Canada Environment Facility
ICFRE : Indian Council of Forestry Research and Education
ICRAF : International Centre for Research in Agroforestry (World Agroforestry Centre)
ICRISAT : International Crops Research Institute for the Semi-Arid Tropics
IDRC : International Development Research Centre
IFPRI : International Food Policy Research Institute
IFS : Integrated Farming System
IGNFA : Indira Gandhi National Forest Academy
IIED : International Institute for Environment and Development
IIFS : Intensive Integrated Farming System
IIRR : International Institute of Rural Reconstruction
IISD : International Institute for Sustainable Development
IITA : International Institute of Tropical Agriculture
IITF : International Institute of Tropical Forestry
ILCA : International Livestock Centre for Africa
ILRI : International Livestock Research Institute
INBAR : International Network for Bamboo and Rattan
INM : Integrated Nutrient Management
IPCC : Intergovernmental Panel on Climate Change
IPM : Integrated Pest Management
IPNS : Integrated Plant Nutrition Systems
IRR : Internal Rate of Return
ISTC : International Standard Trade Classification
IUCN : International Union for Conservation of Nature
IUFRO : International Union of Forest Research Organizations
JFMC : Joint Forest Management Committees
KVK : Krishi Vigyan Kendra
LAMPS : Large-Area Multipurpose Societies
LEV : Land Expectation Value
LKTS : Lesser Known Tree Species
LPG : Length of Growing Period
LSD : Latin Square Design
LULUCF : Land-Use, Land-Use Change and Forestry
LUS : Land Use System
MAF : Multistoried Agroforestry System
MAPs : Medicinal and Aromatic Plants
MAS : Marker Assisted Selection
MDGs : Millennium Development Goals
MFPF : Minor Forest Produce Federation
Mha : Million Hectare
MPTs : Multipurpose Trees
MPTS : Multipurpose Trees and Shrubs
MT : Million Tons
MTS : Modified Taungya System
MUSLE : Modified Universal Soil Loss Equation
MW : Mega Watt
NAPCD : National Action Programme to Combat Desertification
NASA : National Aeronautics and Space Administration
NBM : National Bamboo Mission
NBP : Net Biome Production
NBPGR : National Bureau of Plant Genetic Resources
NEP : Net Ecosystem Production

NEPED : Nagaland Environment Protection and Economic Development
NFAP : National Forestry Action Programme
NFP : National Forest Policy
NFTA : Nitrogen Fixing Tree Association
NFTs : Nitrogen Fixing Trees
NPP : Net Primary Production
NPV : Nuclear Polyhedrosis Virus
NPV : Net Present Value
NRCAF : National Research Centre for Agroforestry
NRM : Natural Resource Management
NRSA : National Remote Sensing Agency
NRSC : National Remote Sensing Centre
NTFPs : Non-Timber Forest Products
NWDB : National Wasteland Development Board
NWFPs : Non-Wood Forest Products
NWTPs : Non-Wood Tree Products
OFI : Oxford Forestry Institute
PAGR : Plant and Animal Genetic Resources
PFCs : Perfluorocarbons
Pg : Pentagram
PGR : Plant Genetic Resources
ppm : Parts Per Million
QTL : Quantitative Trait Loci
RaCSA : Rapid Carbon Stock Appraisal
RAPD : Random Amplified Polymorphic DNA
RBD : Randomized Block Design
RCBD : Randomized Complete Block Design
REDD : Reducing Emissions from Deforestation and forest Degradation
RFLP : Restriction Fragment Length Polymorphism
RH : Relative Humidity
ROR : Rate of Return
SAC : Space Application Centre
SAF : Simple Agroforestry Systems
SALT : Sloping Agricultural Land Technology
SALWA : Semi-Arid Lowlands of West Africa
SAR : Sodium Adsorption Ratio
SAT : Semi-Arid Tropics
SEANAFE : Southeast Asian Network for Agroforestry Education
SFCs : State Forest Corporations
SFR : Short Rotation Forestry
SFRC : Southern Forest Rangers College
SIDA : Swedish International Development Agency
SLM : Sustainable Land Management
SNPs : Single Nucleotide Polymorphisms
SOC : Soil Organic Carbon
SOM : Soil Organic Matter
SPWD : Society for Promotion Wasteland Development
SSRs : Simple Sequence Repeats
SWEET : Sloping Watershed and Environmental Engineering Technology
TAFCORN : Tamil Nadu Forest Plantation Corporation
TAI : Tree-Animal Interface or Tree-Animal Interaction
TBOs : Tree Borne Oil Seeds
TCI : Tree-Crop Interface or Tree-Crop Interaction
TDC : Tribal Development Corporation
TDP : Tree Domestication Program
TLC : Total Literacy Campaign
TOF : Trees Outside Forests
UMCA : University of Missouri Center for Agroforestry
UNCBD : United Nations Convention on Biological Diversity
UNCCD : United Nations Convention to Combat Desertification
UNCED : United Nations Conference on Environment and Development
UNDP : United Nations Development Programme
UNEP : United Nations Environment Programme
UNFCCC : United Nations Framework Convention on Climate Change
UNICEF : United Nations Children's Fund
USDA-NAC : United States Department of Agriculture – National Agroforestry Center
USLE : Universal Soil Loss Equation
VFC : Village Forest Committees
VOCs : Volatile Organic Compounds
VVK : Van Vigyan Kendra
WBG : World Bank Group
WEHAB : Water, Energy, Health, Agriculture and Biodiversity Initiative
WMO : World Meteorological Organization

1 INTRODUCTION TO AGROFORESTRY

INTRODUCTION

The existence of large human population in the fragile ecosystems of the developing world, which occupy the greater proportion of the land of the developing economies, underlines the need for devising strategies to assist in increasing the productivity of these ecosystems while at the same time either rehabilitating them or halting the process of degradation. Agroforestry is a system of land management which seems to be suitable for these ecologically brittle areas. It combines the protective characteristics of forestry with the productive attributes of both forestry and agriculture. Agroforestry is a new name for the old practice of growing trees on farmland. Farmers have been practising agroforestry for thousands of years by combining trees with crops or animals. When land is scarce or when soil has a low fertility or is sensitive to erosion, agroforestry techniques offer considerable benefits for long term agricultural sustainability. Agroforestry can help to make small farms more productive by increasing family income.

Achieving the Millennium Development Goals (MDGs) has been the prime importance of all international agenda. The goals embody the world's aspirations to eliminate desperate hunger and poverty, ensure decent health, enable universal education and elevate the status of women, while conserving and regenerating the global environment. Attaining these goals is the greatest challenge of our generation and it will bring benefits to everyone, including greater economic abundance, peace and security to all people on the earth. A clear vision is evolving that articulates how agroforestry research and development can contribute materially to achieve these goals and aspirations (Garrity *et al.*, 2006).

The economic and environmental challenges currently facing the agricultural and forestry sectors are prompting stakeholders in these sectors to consider new ways of operating. By combining the strengths of agriculture and forestry, agroforestry opens up promising new prospects for the future of rural communities. Agroforestry is gaining momentum at a time when global community is increasingly demanding sustainable natural resource management and environmental friendly development approaches. Landscape enhancement, integrated land use management and sustainable development are all areas in which agroforestry can make a positive contribution. Agroforestry takes several forms and is attracting growing interest in global context from local, regional and international stakeholders, who are incorporating agroforestry activities in their land management and development projects, viewing it as a concrete tool for the sustainable development of rural land resources.

PARADIGM SHIFTS IN AGRICULTURAL DEVELOPMENT

TRADITIONAL FARMING SYSTEMS

Farmers in large parts of the tropics have no tradition of aiming at maximum production per hectare and their main concern was to reduce the risk of crop failure. The principal reason is that there were no convenient markets for food surpluses. The traditional farming systems were integrated, based on self-sufficiency and hence on internal supplies and services between different farm components: mixed cropping with a legume supplying nitrogen to a cereal, field crops supplying fodder for livestock in exchange for manure, etc. However, in the last century these systems proved to be not sufficiently productive to feed the rapidly growing population. These traditional systems were not touched by agricultural science, which was the driving force for ever-increasing yield levels in the temperate zones of the world. Agronomists neglected the traditional mixed farming systems for two reasons: (1) the prevailing notion that the market economy demands specialization, and (2) the lack of suitable research methods to study the intricately interwoven farm components.

GREEN REVOLUTION

Modern agricultural science, in spite of its roots in systems research, gradually became very much preoccupied with the improvement of single crops. When called upon to combat hunger in tropical regions, agronomists concentrated on raising the yield of the principal annual staple food crops such as rice and wheat. These crops were already grown as monocrops for the market and lent themselves to the approach which had been so successful in the temperate zones. Agronomists indeed succeeded with these crops, as for instance shown by the 'green revolution' in Asia.

LOSS OF NATURAL VEGETATION

Unfortunately, green revolution did not benefit poor farmers and this approach failed to alleviate hunger, in particular in tropical regions with rainfed farming systems and underdeveloped markets for food crops. In those regions the situation continued to deteriorate, also because the prices of non-food commodities (e.g. cotton, coffee, spices, fibres) kept falling worldwide, strapping small-holders of cash income and reducing them to subsistence levels of farming. Mounting population pressure resulted in ever-smaller farm size and need to sacrifice unproductive trees in favour of planting food crops. Moreover, the inability to raise yield levels forced the growing population to bring more land (often marginal land) under cultivation at the expense of the natural vegetation.

DETERIORATION OF LAND USE SYSTEMS

Before long these trends led to alarming reports about deteriorating land use systems: (1) expanding dust bowls and deserts (because of wind erosion), (2) degraded land following loss of topsoil and silting up of irrigation systems (because of water erosion), and (3) declining soil fertility and yield levels (because of inadequate inputs of manure and fertilizer and the opening up of marginal lands). And then it was realized that these precarious situations had something in common: *"Trees were vanishing from the Landscape"*. Deforestation to open more arable fields, trees and shrubs killed by overgrazing, cutting of trees for fuelwood, etc. all added up to a landscape being denuded of its permanent vegetation (mainly forests and grazing land for cattle).

NEED FOR TREES IN AGRICULTURAL DEVELOPMENT

Thus it became clear that trees not only yield useful products but also play a vital role as more permanent elements in the landscape, sustaining the capacity of the land to feed people. This brought trees onto the agenda of agricultural development. Although trees are the largest perennial plants and they exhibit the qualities which are important in sustaining the productive capacity of the land, the most important factor is that permanent vegetation or woody plants cover the land, whether it consists of trees, shrubs, vines or perennial herbs (grasses, plantain, yams, etc.).

BIRTH OF AGRFORESTRY SCIENCE

Unfortunately the benefits of a permanent vegetation cover of trees and other perennials only become obvious when the land has been denuded under excessive popu-lation pressure, overgrazing and deforestation. Once the elements have free play over the bare land, 'regreening' becomes very difficult because only the hardiest of plants can reclaim the area and these tend to produce little in the way of food for man and animals. It is therefore of the utmost importance to reverse this overexploitation process before land degradation has impoverished the people who live in the area. Annual crops cannot provide permanent cover and in dryland farming the fields lie a large part of the year unprotected. The realization that these annual crops should benefit from suitable combinations with tree crops in mixed cropping systems, gave rise to *AGROFORESTRY* as a distinct discipline in agricultural science in the 1970s. Further consideration of the role of trees, shrubs and vines in mixed cropping systems resulted in the inclusion of crop mixtures of woody plants as well as cropping systems combining woody plants and animal husbandry.

PREFERENCE FOR AGROFORESTRY OVER MIXED FARMING

Meanwhile agricultural science had rediscovered its origins - dating back to the early 19th century - in farming systems in the tropics. Crop science had already extended its reach from single crops to the study of mixed crops. This research confirmed the farmer's claim that mixed cropping reduces the risk of crop failure, but it also dampens the response to crop care. The outcome of the research on comparison of simple crop mixture (e.g. maize and beans grown together) over two plots of maize and beans grown separately, is not spectacular at all. The mixture yields only slightly more than the sum of the two crops on their own and the total yield of the mixture is more stable from year to year.

Compared to the interactions between maize and beans in mixed cropping, the interactions in agroforestry systems are a good deal more complex. Moreover, trees require years before they reach an effective size. During these years their interactions with companion crops and/or livestock keep changing. The result of agroforestry technology is the reversal of a downward trend in land use into an upward trend, putting land-use back on a sustainable basis. Farmers adopt agroforestry practices for two reasons. They want to increase their economic stability and they want to improve the management of natural resources under their care.

WHAT IS AGROFORESTRY?

Farmers have integrated trees in their farming systems for centuries. Agroforestry is an ancient land use practice and modern science involving the deliberate management of trees on farms and in surrounding landscapes. Agroforestry systems vary greatly in tree species mix, complexity, configu-ration and input requirements, producing a wide range of products and services. With appropriate technical and institutional support, the practice of agroforestry can contribute to rural food and health systems and help to buffer households against health and nutrition shocks. As a science, agroforestry integrates perspectives from agriculture, ecology and rural development.

Agroforestry is the production of trees and non-tree crops or animals on the same piece of land. It employs integrated concepts which make this system of land management possibly the most self-sustaining and ecologically sound of any agricultural system. It is the integration of trees, plants and animals in conservative and long-term productive systems. Agroforestry can be considered more as an approach than a single finished technology. Although several finished systems have been devised and tested, such technology may require adjustment for particular situations. The flexibility of the agroforestry approach is one of its advantages.

OBJECTIVES OF AGROFORESTRY

- To increase awareness of the benefits of including trees with growing crops as part of sustainable agriculture.
- To raise awareness of the value of agroforestry in improving soil fertility and preventing soil erosion.
- To raise awareness of the nutritional benefits of various products obtained from agroforestry.
- To gain understanding of the various benefits provided by recommended tree species.

WHY AGROFORESTRY?

Agroforestry systems make maximum use of the land. Every part of the land is considered suitable for plants that are useful. Emphasis is placed on perennial, multiple purpose crops that are planted once but yield benefits over a long period of time. Furthermore, systems of agroforestry are designed for beneficial interactions of the crop plants and to reduce unfavourable interactions. They are designed to reduce the risks associated with agriculture, small scale or large and to increase the sustainability of agriculture. Agroforestry practices normally help to conserve and even improve the soil. Agroforestry includes recognition of the interaction of crops, both favourable and unfavourable. The most common interaction is competition, which may be for light, water, or soil nutrients. Competition invariably reduces the growth and yield of any crop. Yet competition occurs in monoculture as well and this need not be more deleterious in agro-forestry systems. Interactions may be complementary, as in the case of trees, pasture and foraging animals, where trees provide shade and/or forage and animals provide manure.

Agroforestry systems are designed to produce a range of benefits including food, feed, fuels, fibers and usually improving soil fertility. Agroforestry systems take advantage of trees for many uses, to hold the soil, to increase fertility through nitrogen fixation, or through bringing minerals from deep in the soil and depositing them by leaf-fall, to provide shade, construction materials, foods and fuel. Agroforestry systems may be thought of as principle parts of the farm system itself, which contains many other sub-systems that together define a way of life.

ANTICIPATED OUTCOMES OF AGROFORESTRY

- Communities are encouraged to regard agroforestry and tree planting as a normal activity on farmland.
- The building and maintenance of tree nurseries in local communities and the introduction of a variety of recommended tree species.
- Farmers are encouraged to experiment and compare various techniques for tree planting.
- People learn the value of local trees and gain confidence in their local knowledge.
- The improvement of soil fertility and conservation.
- The improvement of family nutrition.

- Fuelwood becomes more available, benefiting the local environment.
- The provision of opportunities to improve small holder income.

HISTORY OF AGROFORESTRY

ERA OF TRADITONAL AGROFORSTRY

Agroforestry is the traditional practice of growing trees on farms for the benefit of the farm family. The scientific rediscovery of agroforestry dates back to the mid-1970's but the roots of the agroforestry tradition go much deeper. The concept of agroforestry lies embedded in the traditional agroforestry practices of farmers all over the world. In some areas, traditional agroforestry practices have declined but in others they are still going strong. Systematic study and improvement of viable traditional practices is still the best route of entry for those seeking to make a modest contribution to rural development in the short term. Scientific agroforestry holds promise of major improvements in the medium to long term.

Trees have been used in cropping systems since the beginning of agriculture. Throughout the world, at one period or another in its history, it has been the practice to cultivate tree species and agricultural crops in intimate combination. In much of the tropics, human beings underwent a transition from hunting/gathering to the use of domesticated plants and livestock. As a part of the process they cut down trees, cleared the debris by burning, and sowed crops in the ash-enriched soil. Thus was born slash-and burn agriculture, a primary forerunner of the present day agroforestry and a practice that may have originated in the Neolithic period, around 7000 BC.

It was the general custom in Europe, at least until the Middle Ages, to clear-fell degraded forest, burn the slash, cultivate food crops for varying periods on the cleared areas, and plant or sow tree species before, along with, or after the sowing of the agricultural crops. This "farming system" is no longer popular in Europe. But it was still widely followed in Finland up to the end of the last century, and was being practised in few areas in Germany as late as the 1920s (King 1968).

In tropical America, many societies have traditionally simulated forest conditions in their farms in order to obtain beneficial effects of forest structures. Farmers in Central America, for example, have long imitated the structure and species diversity of tropical forests by planting a variety of crops with different growth habits. Plots of no more than one-tenth of a hectare contained an average of two dozen different species of plants each with a different form, together corresponding to the layered configuration of mixed tropical forests. A farmer would plant coconut or papaya with a lower layer of bananas or citrus, a shrub layer of coffee or cacao, tall and low annuals such as maize, and finally spreading ground cover of plants such as squash (Wilken 1977).

In Asia, the Hanunoo of the Philippines practised a complex and somewhat sophisticated type of shifting cultivation. In clearing the forest for agricultural use, they deliberately left certain selected trees which, by the end of the rice-growing season, would provide a partial canopy to the new foliage to prevent excessive exposure of soil to the sun. Trees were an indispensable part of the Hanunoo farming system and were either planted or conserved from the original forests to provide food, medicines, construction wood and cosmetics, in addition to their protective services (Conklin 1957). Similar farming systems have also been common in many other parts of the humid lowland tropics of Asia.

This situation was little different in Africa. In southern Nigeria, yams, maize, pumpkins and beans were typically grown together under a cover of scattered trees (Forde 1937). The Yoruba of western Nigeria, who have long practised an intensive system of mixing herbaceous shrub and tree crops, claim that the system is a means of conserving human energy by making full use of the limited space won from the dense forest. The Yoruba also claim that this system is an inexpensive means of maintaining the soil's fertility, as well as combating erosion and nutrient leaching (Ojo 1966). There are innumerable examples of traditional land-use practices involving combined production of trees and agricultural species on the same piece of land in many parts of the world. These are some examples of what is now known as agroforestry. Trees were an integral part of a farming system and they were kept on established farmland to support agriculture. The ultimate objective of these practices was not tree production but food production.

ERA OF TAUNGYA SYSTEM

By the end of the 19th century, however, the establishment of forest plantations had become the dominant objective wherever agroforestry was being utilized as a system of land management. In the beginning, the change of emphasis was not deliberate. At an outpost of the British Empire in 1806, U Pan Hle, a Karen in the Tonze forests of Thararrawaddy Division in Myanmar (Burma), established a plantation of teak (*Tectona grandis*) by using a method he called "Taungya", and presented it to Sir Dietrich Brandis, the Governor. Brandis is reported to have said, "this, if the people can ever be brought to do it, is likely to become the most efficient way of planting teak" (Blanford 1958). From this beginning, the practice became increasingly widespread. It was introduced into South Africa as early as 1887 (Hailey 1957) and was taken, from what was then Burma, to the Chittagong and Bengal areas in colonial India in 1890 (Raghavan 1960).

The ruling philosophy of the taungya system was to establish forest plantations whenever possible using available unemployed or landless labourers. In return for performing forestry tasks, the labourers would be allowed to cultivate the land between the rows of tree seedlings to grow agricultural produce. This is a simplification of a system whose details varied depending on the country and locality. The wider implications and potentials of the Taungya approach were first recognized by Kenneth King (1968) and others at the University of Ibadan in Nigeria. King coined the name "Agrisilviculture" to refer to what later came to be called agroforestry.

As a result of forester's preoccupations with the forests and the forest estate, the research was undertaken and designed to ensure that: (1) little or no damage occurred to the forest tree species, (2) rates of growth of the forest tree species were not unduly inhibited by competition from the agricultural crop, (3) optimum time and sequence of planting of either the tree or agricultural crop be ascertained in order to ensure the survival and rapid growth of the tree crop, (4) forest species that were capable of withstanding competition from agricultural species be identified, and (5) optimum planting-out espacements for the subsequent growth of the tree crop be ascertained. In short the research, which was conducted and undertaken for forestry by foresters, never envisioned the system as being capable of making a significant contribution to agricultural development, or its potential as a land-management system (King 1987).

TRANSFORMATION OF LAND USE SYSTEM

Many factors and developments in the 1970s contributed to the general acceptance of agroforestry as a system of land management that is applicable to both farm and forest. These factors are: (1) re-assessment of the development policies of the World Bank by its President, Robert McNamara, (2) re-examination by the Food and Agricultural Organization of the United Nations of its policies pertaining to forestry, (3) re-awakening of scientific interest in both intercropping and farming systems, (4) deteriorating food situation in many areas of the developing world, (5) increasing spread of tropical deforestation and ecological degradation, (6) energy crisis of the 1970s and consequence price escalation and shortage of fertilizers, and (7) establishment of a project by the International

Development Research Centre (IDRC) of Canada for the identification of tropical forestry research priorities.

With the concern for the rural poor, World Bank actively considered the possibility of supporting nationally oriented forestry programmes. As a result, it formulated a new Forestry Sector Policy paper which is still being used as the basis for much of its lending in the forestry sub-sector. Indeed, its social forestry programme, which has been extended considerably over last few decades, not only contains many elements of agroforestry but is designed to assist the peasant and the ordinary farmer to increase food production and to conserve the environment as much as it helps the traditional forest services to produce and convert wood (Spears 1987).

At the same time, FAO also redirected its thrust and assistance in the direction of the rural poor. Its new policies, while not abandoning the traditioinal areas of forestry development, emphasized the importance of forestry for rural development. It also focused on the benefits that could accrue to both the farmer and the nation if greater attention were paid to the beneficial effects of trees and forests on food and agricultural prodcution, and advised land managers in the tropics to incorporate both agriculture and forestry into their farming system and "eschew the false dichotomy between agriculture and forestry" (King 1979). FAO also stressed the necessity of devising systems which could provide food and fuel and yet conserve the environemnt.

The two strands of forest policy reforms by World Bank and FAO, which evolved independently, were added the simultaneous efforts of a large number of tropical land-use experts and institutions. Faced with the problems of deforestation and environmental degradation, these experts and institutions intensified their search for appropriate land-use approaches that would be socially acceptable, ensure the sustainability of the production base and meet the need for production of multiple outputs. Efforts to design major programmes which would allow local communities to benefit directly from forests paved the way for new forestry concepts, such as social forestry and farm forestry, which were implemented in many countries (Nair 1993).

Under the auspices of the Consultative Group on International Agricultural Research (CGIAR), several International Agricultural Research Centers (IARCs) were established in different parts of the world to undertake research with the objective of enhancing the productivity of major agricultural crops (or animals) of the tropics. The development of high-yielding varieties of cereals and related technologies through the joint efforts of some of these centers and relevant national programmes paved the way for '*Green Revolution*' (Borlaug and Dowswell 1988). However, it was soon realized that many of the green revolution technologies that placed a heavy demand on increased use of fertilizers and other costly inputs, were beyond the reach of a large number of resource-poor farmers in the developing countries. Most of the IARCs and the national programmes were focusing on individual crops such as rice, wheat, maize and potato, and production technologies for monocultural or sole-crop production systems of these crops. However, the farmers, especially poor farmers, often cultivated their crops in mixed stands of more than one crop and sometimes crop and trees. In such circumstances the production technologies developed for individual crops would seldom be applicable. As a consequence, there was renewed and heightened interest in the concepts of intercropping in plantations and integrated farming systems.

DEVELOPMENT OF AGROFORESTRY AND ICRAF

The most significant single initiative that contributed to the development of agroforestry came from the International Development Research Centre (IDRC) of Canada. In July 1975, the IDRC commissioned John Bene, a philanthropist and prominent figure in the Canadian forest industry, to lead a study to (1) identify the critical priorities for world forestry research and training, (2) assess the interdependence between forestry and agriculture in low-income tropical countries and propose research leading to the optimization of land use, (3) formulate forestry research programmes which promise to yield results of considerable economic and social impact on developing countries, (4) recommend institutional arrangements to carry out such research effectively and expeditiously, and (5) prepare a plan of action to obtain international donor support.

John Bene appointed an advisory committee and regional consultants to make recommendations on the forest research needs of the tropics. The culmination of these various activities, including extensive travel by Bene, was the publication in 1977 of a report entitled *Trees, Food and People* (Bene et al 1977). The Bene Commission concluded that the top priority for tropical forestry lay in "agroforestry", the term coined by the commission to refer to the new field of interdisciplinary endeavour. In short, there was a shift in emphasis from forestry to combined production systems which would integrate forestry, agriculture and/or animal husbandry in order to optimize tropical land use (Bene et al 1977). The IDRC Project Report recommended the establishment of an international organization which would support, plan and coordinate, on a world-wide basis, research works combining the land-management systems of agriculture and forestry. IDRC met with other potential donors (Canada, Switzerland and the Netherlands) in 1976 to discuss the report. This proposal was generally well received by international and bilateral agencies and an executing agency was set up, which prepared a charter, elected a Board of Trustees (BoT) and established the International Council for Research in Agroforestry (ICRAF) in 1977 and ICRAF started its work in The Hague.

In July 1978, ICRAF moved to its present headquarters in Nairobi, Kenya and Dr.Kenneth King, a former Assistant Director General in charge of Forestry at FAO, was appointed as Director General. In 1980, the donors asked the Board to restructure ICRAF's work and to develop a realistic strategy for its future programme. Dr.Howard Steppler, Board Chair and Interim Director drafted a new strategy for ICRAF in 1981. The new strategy set the Council on a path to develop a diagnostic methodology to determine relevance of agroforestry interventions in particular situations. Following the new strategy, ICRAF concentrated on analyzing land-use systems and designing relevant agroforestry technologies to help overcome constraints. In-house capability and methods, a knowledge base, and information acquisition and distribution were developed and ICRAF adapted farming systems research techniques into the now well known "Diagnosis & Design" (D&D) methodology. One of the projects launched in the early 1980s was systematically to inventory agroforestry systems and Nair (1985) published a first approximation of a classification of described agroforestry systems.

ICRAF joined the Consultative Group on International Agricultural Research (CGIAR) in 1991 to conduct strategic research on agroforestry at a global scale and was renamed as the International Centre for Research in Agroforestry (ICRAF). In 2002, the Centre acquired the brand name the 'World Agroforestry Centre'. However, the 'International Centre for Research in Agroforestry' remains as legal name. World Agroforestry Centre reflects the fact that Centre is now recognized as the international leader in agroforestry research and development.

DEFINITIONS OF AGROFORESTRY

- It was the Bene Commission that proposed the first widely accepted definition: "Agroforestry is a sustainable management system for land that increases overall production, combines agricultural crops, tree crops and forest plants and/or animals simultaneously or sequentially, and applies management practices that are compatible with the cultural patterns of the local population" (Bene et al 1977).
- Another definition was given in the ICRAF's first publication *The Wasted Lands*: "Agroforestry has been defined as a sustainable land management system which increases

the overall yield of the land, combines the production of crops (including tree crops) and forest plants and/or animals simultaneously or sequentially, on the same unit of land, and applies management practices that are compatible with the cultural practices of the local population" (King and Chandler 1978).

- Nair (1979) defined agroforestry as "a land use system that integrates trees, crops and animal in a way that is scientifically sound, ecologically desirable, practically feasible and socially acceptable to the farmers".
- A strictly scientific definition of agroforestry should stress two characteristics common to all forms of agroforestry and separate them from the other forms of land use, namely:
  - the deliberate growing of woody perennials on the same unit of land as agricultural crops and/or animals, either in some form of spatial mixture or sequence
  - there must be a significant interaction (positive and/or negative) between the woody and non-woody components of the system, either ecological and/or economical

  When promoting agroforestry one should then stress the potential of it to achieve certain aims, not only by making theoretical and qualitative remarks about the benefits of trees, but also, and more importantly, by providing quantitative information (Lundgren 1982).
- The following definition was suggested by ICRAF and was increasingly used in ICRAF publications and thus achieved wide acceptability. "Agroforestry is a collective name for land-use systems and technologies where woody perennials (trees, shrubs, bamboos, palms, etc.) are deliberately used on the same land-management units as agricultural crops and/or animals, in some form of spatial arrangement or temporal sequence. In agroforestry systems there are both ecological and economical interactions between the different components" (Lundgren and Raintree 1982). This definition implies that:
  - agroforestry normally involves two or more species of plants (or plants and animals), at least one of which is a woody perennial
  - an agroforestry system always has two or more outputs
  - the cycle of an agroforestry system is always more than one year
  - even the simplest agroforestry system is more complex, ecologically (structurally and functionally) and economically, than a monocropping system
- "Agroforestry is a collective name for all land-use systems and practices where woody perennials are deliberately grown on the same land management unit as agricultural crops and/or animals, either in spatial mixture or in temporal sequence. There must be significant ecological and economic interactions between the woody and non-woody components" (Lundgren 1987).
- Young (1989a) proposed and defined agroforestry as "a collective name for land use system in which woody perennials (trees, shrubs etc.) are grown in association with herbaceous plants (crops, pastures) and /or livestock in spatial arrangement or rotation of both and in which there are both ecological and economic interaction between the trees and non tree components of the system".
- The general definition of Somarriba (1992): "Agroforestry are diversified technical practices that have in common the following:
  - there is at least two different plants in biological interaction
  - one of these two plants is a perennial

  - one of these two plants is a forage, a food crop or a tree crop"

- "The deliberate growing of woody perennials on the same unit of land as agricultural crops and/or animals, either in some form of spatial mixture or sequence. There must be a significant interaction (positive and/or negative) between the woody and non-woody components of the system, either ecological and/or economical" (Nair 1993).
- "An intensive land management system that optimizes the benefits (physical, biological, ecological, economic, social) arising from biophysical interactions created when trees and/or shrubs are deliberately combined with crops and/or livestock" (Garrett et al 1994).
- Sanchez (1995) defined agroforestry as "a collective name for land use system and practices in which woody perennials are deliberately integrated with crops and /or animals on the same land management unit".
- Agroforestry is generally practised with the intention of developing a more sustainable form of land use that can improve the farm productivity and the welfare of the rural community. It has been argued that the above definitions are not sufficiently dynamic and the following alternative has been suggested by Leakey (1996): "Agroforestry should be considered as a dynamic, ecologically based, natural resource management system that, through the integration of trees in farm and rangeland, diversifies and sustains smallholder production for increased social, economic and environmental benefits"
- Another definition has been suggested by the "Laboratoire de Botanique Tropicale" of Montpellier: "Agroforestry is a land use system, controlled by local population where perennial trees are associated to agriculture and/or stock farming on the same piece of land in such a way that the consequent ecosystem tend to mimic the natural forest ecosystem according to aerial and soil biomass, vegetation structure and specific richness".

COMPONENTS OF AGROFORESTRY

The agroforestry systems comprise the following four components: (1) land, (2) trees, (3) non-tree crops, and (4) farm animals.

LAND

Land, a non-renewable resource, is central to all primary production system. Land is considered as the most important basic natural resource. It is a dynamic and complex combination of geology, topography, hydrology, soil, flora and fauna and influences every sphere of human activity. Different sectors including agriculture, industries, infrastructure and power projects put forth competing demand for land. Subsistence farming practices, accelerated soil and water erosion, erratic rainfall, increasing population and high density of livestock population have all contributed to unsustainable land use that has led to degradation of this valuable resource. Agroforestry is a system to manage the agricultural resource and land for the benefits of the farmer and the long-term welfare of society. While this is appropriate for all land, it is especially important in the case of hillside farming where agriculture may lead to rapid loss of soil. The farmers must think conservatively to maintain the land over long periods of time.

TREES

In agroforestry, particular attention is placed on multipurpose trees and perennial shrubs. The most important of these trees are legumes because of their ability to fix nitrogen and thus make it available to other plants. Considerable knowledge on properties of trees is necessary to incorporate them in agroforestry systems such as size and form of the canopy, root system, climatic adaptation of the species, adaptation to various soils and stresses and suitability for various agroforestry practices. The trees can be planted in agroforestry in different ways viz. individual trees, scattered trees, lines of trees with crops (alley cropping), strips of trees along contour and waterways, boundary lines, living fences, windbreaks, shelterbelts, terrace planting on hills, wood lots, etc. The role of trees in agroforestry includes:

- Source of fruits, nuts and edible leaves
- Source of non-edible materials including resin, tannin, insecticides and medicinal products
- Source of construction materials, posts, lumber and thatching
- Source of fuel
- Beautification and shade
- Soil conservation and improvement of soil fertility
- Water quality improvement
- Climate amelioration
- Carbon sequestration and climate change mitigation

NON-TREE CROPS

Any crop plants can be used in agroforestry systems. The choice of crop plants in designing agroforestry systems should be based on those crops already produced in a particular region for home consumption, marketing, feeding animals, or that have great promise for production in the region. Choice of crops in agroforestry is based on certain values of crops viz. food crops, vegetable crops, forage crops, oil crops, cash crops and medicinal crops. The crops are selected based on the following criteria:

- Crops for making money
- Crops for feeding the farmer
- Crops for good nutrition
- Crops for self sufficiency
- Crops for feeding the farm animals
- Crops for protecting the soil

ANIMALS

Any farm animals can be used in agroforestry systems viz. cow, bull, buffalo, goat, sheep, poultry, piggery and fish. The choice of animal will be based on the following values:

- Animals for making money
- Animals for feeding the farmer
- Animals for supplying labour
- Animals for non-food products
- Animals for using crop residues
- Animals for producing manure

Fig. 1.1: Agroforestry – A Potential Land Use System (Source: Xu et al 2011)

SCOPE OF AGROFORESTRY

In more common terms, agroforestry is the art of planting trees in fields. The expression "farm forestry" is also used sometimes, but it can be misleading since pure forestry can sometimes be called farm forestry if it is implemented by farmers. The same holds true for social forestry. One can also give a "default definition" of agroforestry as any tree planting which is not forestry or arboriculture with the classical meaning of these terms, is agroforestry. Any tree under farmer's management, and for the benefit of the farm, is agroforestry. ICRAF has acknowledged and stated that the definition of agroforestry has been developed for scientific purposes and that, for development purposes, there is no reason to draw sharp distinctions between agroforestry, horticulture, farm forestry and other land uses of land involving trees.

The term "agroforestry" itself has been subject of debate. The main reason is that the majority of agroforestry technologies do not have similarities with forests (or, if they do, are not managed like forests e.g. homegardens) but rather with agricultural fields. Many scientists have suggested that a term like "agro-arboriculture" or "agro-silviculture" could have been better. However, the term agroforestry is universally accepted and now in common use. Agroforestry is a science that distinguishes itself from both agriculture and forestry. Its objective is to optimize positive interactions between woody and non-woody components, so that the production system may be more sustainable and diversified than with a conventional approach under given agroecological and socioeconomic conditions. The interactions between crops and animals are not enough to make up an agroforestry land use system, but the presence of woody perennials is necessary for the system to be considered as agroforestry. Mixed cropping is not agroforestry, although part of the wide knowledge existing on mixed cropping could be applied to agroforestry.

The objective of agroforestry is the reconciliation of two types of land use systems that have deeply affected during the last centuries countrysides of both tropical and temperate countries: agriculture and forestry. The main feature which characterizes agroforestry is the combination, or association of several plants, annual and perennial, on the same field. Agroforestry systems are a kind of cropping systems and it means that the field is homogeneously managed, through a particular technical pathway (technological pattern). The agroforestry strategies can be defined mainly through three main features: (1) the minimization of risk or crop failure, (2) the optimization of labour efficiency, and (3) the possible use of improved planting material and inputs

Agroforestry has been in use for at least 1300 years, although tree domestication probably started much earlier. Agroforestry was brought from the realm of indigenous knowledge into the forefront of agricultural research less than four decades ago and was promoted widely as a sustainability-enhancing practice that combines the best attributes of forestry and agriculture. Growing trees along with crops and livestock was postulated to enhance crop yields, conserve soil and recycle nutrients while producing fuelwood, fodder, fruit and timber. Mixing trees, crops and/or animals can take on many shapes and forms. Agroforestry is essentially meant for smallholder farmers who live partially from subsistence agriculture, but there are also agroforestry practices that can be used in intensive land use systems on large-scale farms.

The potential of agroforestry was initially based on largely descriptive and information, usually collected from good quality sites and later applied research was vigorously pursued. While such efforts raised awareness and helped to institutionalize agroforestry in national programmes throughout the world, the need for a critical, science-based approach gradually emerged. Agroforestry is a field of study which involves the combined application of ecology, economics, agronomy, forestry, soil science, animal science, tree genetics, biometrics, anthropology and other applied sciences. One of the challenges of agroforestry is to create conditions for an efficient multidisciplinarity. Agroforestry involves all the above disciplines to include trees in agricultural lands. Hence agricultural and forestry institutions should be closely involved in conducting coordinated research works to develop agroforestry technologies. In short, foresters and agronomists must work together to develop promising agroforestry innovations.

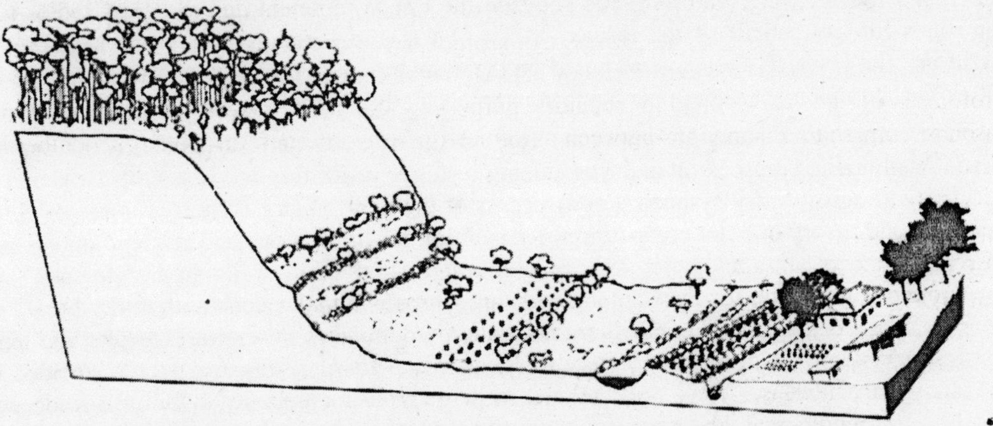

Fig. 1.2: Agroforestry Practice – An Effective Method for Sloping Land Management

CONCEPTS OF AGROFORESTRY

Agroforestry practices are *intentional* combinations of trees with crops and/or livestock which involve *intensive* management of the *interactions* between the components as an *integrated* agroecosystem. These four key criteria - intentional, intensive, interactive and integrated - are the essence of agroforestry and are what distinguish it from other farming or forestry practices. To be called agroforestry, a land use practice must satisfy all of the following four criteria/ four "I" words (Gold and Garrett 2009):

Intentional: Combinations of trees, crops and/or animals are intentionally designed and managed as a whole unit, rather than as individual elements which may occur in close proximity but are controlled separately.

Intensive: Agroforestry practices are intensively managed to maintain their productive and protective functions, and often involve annual operations such as cultivation, fertilization and irrigation.

Interactive: Agroforestry management seeks to actively manipulate the biological and physical interactions between the tree, crop and animal components. The goal is to enhance the production of more than one harvestable component at a time, while also providing conservation benefits such as non-point source water pollution control or wildlife habitat.

Integrated: The tree, crop and/or animal components are structurally and functionally combined into a single, integrated management unit. Integration may be horizontal or vertical, and above-ground or below-ground. Such integration utilizes more of the productive capacity of the land and helps to balance economic production with resource conservation.

Two key principles which set agroforestry systems apart from agricultural or forestry systems are competition and complexity. They in turn determine two desirable properties viz. profitability and sustainability. *Agroforestry* as a science should be based on these four key features: (1) competition, (2) complexity, (3) profitability, and (4) sustainability.

COMPETITION

When plants grow in proximity to each other they interact in positive ways (complementarity) or in negative ways (competition). The biophysical bottom-line of agroforestry is how to manage the interaction for light, water and nutrients between the tree component and the crop and/or livestock components for the benefit of the farmer. Much of the initial enthusiasm, when agroforestry was brought into the research agenda, was based on the assumption that when trees and crops are brought together they complement rather than compete with each other in their capture of growth resources.

Plant competition plays an important role in the success and sustainability of agroforestry systems. Minimizing competition and maximizing resource utilization increase crop yield and overall productivity of agroforestry systems. For a success of the agroforestry system, it is essential to study the concept and theory of plant competition, competition for light, water and nutrient among trees and crops, and concept of allelopathy and competition related plant traits in agroforestry systems. Aboveground competition for light among trees and crops is a major constraint under humid climate conditions, while belowground competition for water predominates in semi-arid tropics and temperate agroforestry systems. In agroforestry systems, plant competitive ability for light is decided by leaf area, canopy structure, etc. Root traits related to belowground competitive ability include occupied soil space, morphological and physical elasticity and spatial and temporal soil partitions. It is suggested that future research on agroforestry systems focus on comparisons among different components in different climatic conditions, competition processes and physiological mechanisms, plant allelopathy and the role and mechanism of mycorrhizae in plant competition, interactions between above-ground and below-ground competition and the effect of global climate change on plant competition of the agroforestry systems.

The anecdotal evidence about the potential of agroforestry raised some suspicions when it was realized that most of the successful examples of agroforestry came from high-potential environments, where water or nutrients were not major limiting factors. A soil-agroforestry hypothesis was formulated out of this inquiry (Sanchez 1987) and Antony Young transformed it into ten hypotheses (Young 1989b) and the number keeps growing. Process-oriented research works have been conducted globally to test these hypotheses and some remain to be tested or rejected in a sufficiently rigorous manner and under a wide enough range of conditions to be accepted as broadly applicable principles. All of them deal with competition or complementarity. Many of these hypotheses are very general and they should be tested in specific agroforestry systems under specific biophysical and socioeconomic conditions, such as differing climate, soil, slope, cultural conditions and income levels.

COMPLEXITY

Agroforestry practice is full of complexity and there are several biophysical interactions. In addition, there are major socioeconomic, ecological and methodological complexities.

SOCIOECONOMIC COMPLEXITY: The socioeconomic dimension of agroforestry is complex because of aspects related to temporal and spatial variability; scale factors; the multiplicity of products and services; the economic, social and ecological processes involved; the methods of characterization and diagnosis; and the diversity of institutions involved. Agroforestry systems vary in space and time. Trees are seldom used uniformly across the farmer's land and typically occupy the discrete areas such as homegardens, contours, boundaries and fallow plots. Agroforestry frequently coexists with pure agricultural systems and pure forestry. Some sequential systems closely resemble agriculture in their early stages and forestry in their later stages (Sanchez 1995).

The socioeconomic front is complex also because of the scale factor. At one end is the farm household and at the other are the international pressures on national governments to 'internalize externalities' (assigning monetary value to positive or negative consequences beyond farm, such as siltation, loss of biodiversity and carbon sequestration). In between are two critical scales and one is the community or watershed scale with influences by local elders or their equivalents in the control of communal land. Community or watersheds are seldom the same and one is a social demarcation while the other is geographical. The other critical scale is the national government where, for example, policies favouring timber exports may obliterate traditional community rights. The interactions between the different scales are indeed complex.

Another level of complexity is the multiplicity of products and services. Besides providing food, fruit, fodder, fuelwood, timber and medicines, agroforestry also provides services such as conserving soil and biodiversity and delineation of boundaries. Most of the products have economic value, while most of the services have non-monetary environmental value. The need to develop a predictive understanding on how farm households make decisions regarding land use is as essential as developing a predictive understanding of competition between tree and crop roots. There is great deal of site specificity such as cultural values, policies and income levels, tree crop species, climate and soil properties. Examples of socioeconomic processes are how households react to a diminishing resource base caused by nutrient depletion, how communities react to a different policy environment caused by increasing environmental consciousness by governments and how farmers intensify their land use in response to demographic pressure.

An operational dimension of additional complexity is the diversity of institutions involved in agroforestry research. Such institutions are attached to several public sectors such as agriculture, forestry, environment and soil conservation. Within a research institution or university, agroforestry cuts across many disciplinary or commodity-oriented departments. This is an integral part of natural resource management research.

ECOLOGICAL COMPLEXITY: Agroforestry systems exert two main ecological functions, productive and protective. The high total biomass of most agroforestry systems is responsible for the protective or buffering function. The scale and time factors of agroforestry add complexity from the ecological perspective. For example, the effects of a few trees on one farmer's field are different from those of thousands of trees on many farmers fields at the watershed scale. Ecological externalities such as stream sedimentation or, on the positive side, increase in plant species diversity, are seldom felt at the farm scale but they are certainly felt at higher spatial scales.

Ecological parameters interact with socioeconomic ones building complexity upon complexity. This, however, is part of all natural resource management research, of which agroforestry is one subject. Scientifically sound approaches to natural resource management are to be developed with integrated research. In particular, there is need to study the effects of different agroforestry systems on the dynamics of pest-predator-disease interactions affecting food chains and life cycles that eventually determine sustainability and risk. The pantropical spread of the leucaena psyllid (*Heteropsylla cubana*) form its center of origin in the Caribbean to Hawaii in 1984, then to Asia and in Africa since 1992 has caused major damage to *Leucaena leucocephala* (Vandenbeldt and Napompeth 1992). Exotic species leave their predators behind when they are moved away from their center of origin, but eventually predators catch up with them or new ones arise. Multipurpose tree species should not be expected to be pest or disease proof. Scientists especially entomologists and plant pathologists should be prepared for such outbreaks by having a diverse genetic base available (Sanchez 1995).

METHODOLOGICAL COMPLEXITY: These complexities are also part and parcel of agroforestry research. The challenge is to make research methods simple while remaining accurate and reproducible, experimental design and statistical analyses require extra care in agroforestry. Over simplification will lead to major short comings. One example is the inappropriate design of many alley-cropping trials which has rendered them invalid as a result of root interference between plots. There have been major methodological advances in measuring competition for light and water. Water-use measurement in agroforestry systems has been greatly simplified by straightforward methodologies that can follow water uptake of individual roots, branches and whole trees or crop plants.

The main methodological challenge is below ground, where straightforward ways to measure soil processes and root interactions are not well developed. Conventional soil fertility methods for assessing the availability of nitrogen and phosphorus are not sensitive enough to detect changes in low external input systems. New methods and approaches need to be developed. Relevant changes in soil organic matter content, for example, are seldom detected by conventional methods such as total soil carbon and total nitrogen determinations. Conventional soil tests often fail to detect important changes in phosphorus availability. Methods for evaluating soil biological processes have advanced much but several parameters such as microbial biomass and labile organic phosphorus pools still lack replicability and threshold or critical levels. Simplifying root interactions is taking on a new dimension and methods for assessing the resilience and sustainability of agroforestry systems in economic, social and ecological terms remain to be strengthened. Modelling in agroforestry needs further research strength, in terms of both mechanistic and simulation models. High priority should be given to developing robust models based on hard data, now that such data are becoming increasingly available.

PROFITABILITY

Agroforestry is also about securing profits for the farmers. To be regarded sustainable, improved agroforestry systems must increase profitability in the medium term. Profitability is a necessary but not a sufficient condition of sustainability. In fact, many profitable systems are not sustainable in agronomic, social or ecological terms. The most sustainable agroforestry systems are those that combine high-value marketable products with sound biophysical practices. The complex agroforestry systems are perhaps the epitome of sustainability, with biodiversity levels approaching the natural forests. They provide farmers with a steady flow of income from high-value products that have assured markets and that are grown with little nutrient removal from the soil. Three research strategies to increase profitability in agroforestry are domestication, marketing and policy innovation.

SUSTAINABILITY

A rural production system's sustainability corresponds to its ability to meet the ever-increasing needs of mankind, without affecting and if possible improving, the resource base on which the system depends. A sustainable rural production system is only one of the elements in the global concept of sustainable development that includes a series of conditions outside the rural system that are classified as economic, social, ecological, political and institutional. Agroforestry fulfils many requirements for sustainability by including trees in agricultural production systems, by utilizing existing resources and management practices that optimize the combined production of several products instead of maximizing the production of only one product, and through its numerous service roles.

The current challenge

After Agroforestry Development

Fig. 1.3: Agroforestry System – A Successful Approach for Land Rehabilitation

The concept of sustainability has profitability, productivity and environmental components, with a strong time dimension. Agroforestry systems that are profitable make farming systems more robust and less subject to perturbations. The diversity of outputs and the positive environmental attributes of trees on farms make them potential. Agroforestry interventions can make many agricultural systems more robust. Agroforestry is more promising in terms of environmental sustainability, even with low use of external inputs. Sustainability can be described in terms of conserving the soil, enhancing biodiversity, conserving carbon in terrestrial ecosystems and enhancing nutrient capture and retention.

BENEFITS OF AGROFORESTRY

The integration of trees and shrubs with other plants and/or livestock in agroforestry can provide the following benefits:

ENVIRONMENTAL BENEFITS

- Reduction of pressure on forest
- Efficient recycling of nutrients by deep rooted trees on farms
- Reduction of surface run-off, nutrient leaching and soil erosion through impending effect of trees
- Improvement of soil fertility status through addition and decomposition of leaf fall
- Improvement of soil structure through constant addition of organic matter from decomposed litter
- Improvement of microclimate, such as lowering of soil surface temperature and reduction of evaporation of soil moisture through a combination of mulching and shading
- Improvement of biodiversity
- Reduction in the incidence of pest and disease outbreaks that are prevalent in monocropping
- Increased protection of soil, water and wildlife
- Better protection of ecological systems

ECONOMIC BENEFITS

- Increment in farm productivity and maintenance of outputs of food, fuelwood, fodder, fertilizer and timber
- Increase in farm yield per unit of agricultural land compared to other cropping methods
- Reduction in incidence of total crop failure, common to monocropping or cultivation of single crops
- Increment in levels of farm incomes due to improved and sustained productivity

SOCIAL BENEFITS

- Creation of additional sources of income for farmers
- Increment in farm job stability by spreading farm labour throughout the year
- Improvement in rural living standards from sustained employment and higher incomes
- Improvement in nutrition and health due to increased quality and diversity of food outputs
- Increment in health care and medicine by increasing access to pharmacological trees and shrubs
- Stabilization and improvement of upland communities through elimination of the need to shift sites of farm activities

NATIONAL BENEFITS

- More efficient and sustainable land management system than other land uses
- Reduction in hunger and poverty by increasing production of farms for both consumption and sale
- Sustained supply of raw materials to forest based industries
- Reduction in the speed of climate change by more carbon sequestration and storage by trees

LIMITATIONS OF AGROFORESTRY

Besides numerous advantages, agroforestry is considered to possess certain negative aspects which are furnished below:

- Possible competition between trees and food crops for space, sunlight, moisture and nutrients which may reduce the yield of food crops
- Damage to food crops during tree harvest operation
- Potential of trees to serve as hosts to insect pests that may be harmful to food crops
- Rapid regeneration of prolific trees which may displace food crops and take over entire fields
- Requirement for more labour inputs which may be a problem in labour scarcity areas
- Longer rotation period of trees to attain maturity and acquire an economic value
- Unwillingness of farmers to displace food crops with trees, especially where land is scarce
- The fact that agroforestry is more complex, less well understood and more difficult to apply, compared to monocropping practice

AGROFORESTRY TERMINOLOGIES

- **Agrisilviculture:** The conscious and deliberate use of land for the concurrent production of agricultural crops and forest trees.
- **Agrodeforestation:** Promotion of single-species agricultural systems and forestry plantations (usually of exotic species), at the expense of the traditonal agroforestry systems. It is the removal of agroforestry systems for large-scale cash crops.
- **Agroforestry Components or Elements:** Trees or other woody perennials, seasonal plants, and animals are called the components or elements of agroforestry.
- **Agroforestry D&D:** With the objective of analyzing farming systems, identifying possible constraints and proposing agroforestry solutions that should alleviate these constraints, agroforestry scientists use an approach known as D& D (Diagnosis and Design).
- **Agroforestry Intervention:** An agroforestry technology proposed to alleviate some of the constraints identified by a D & D exercise.
- **Agroforestry Management:** It implies the methods used in agroforestry practice to turn inputs into outputs. In more simple terms, this is what farmers must do for their seedlings to become trees that provide good fuelwood or fodder, fix nitrogen and for their seeds to become an either edible or saleable product.
- **Agroforestry Performance:** An agroforestry system's performance is the quantified relationship between the inputs and outputs or the input/output ratio. It is a system's measure of efficiency. For economic performance of agroforestry system, one must take into account

the market value of the inputs and outputs, including subsistence goods or inputs that are reinvested in the system and must be quantified in monetary terms. Economic performance is expressed in output units (or their monetary equivalent) per unit of time and/or space or of another input. Social performance of agroforestry system is a matter of evaluating the system's impact on the well-being of the populations concerned. The satisfaction of households, or village communities, or the behaviour of a group of the population after adopting a new agroforestry system are all facts belonging to social performance.

- **Agroforestry Practice:** It denotes a specific land-management operation of an agroforestry nature. Agroforestry practice is a distinctive arrangement of agroforestry components in time and space.
- **Agroforestry Sustainability:** The ability to combine production and resource conservation gives agroforestry its undeniable quality of sustainability. Agroforestry fulfils many requirements for sustainability by including trees in agricultural production systems, by utilizing existing resources and management practices that optimize the combined production of several products instead of maximizing the production of only one product, and through its numerous service roles.
- **Agroforestry System (AFS):** A set of interdependent agroforestry components (trees with crops and/or animals) representing a current type of land use in a given region. An agroforestry practice or agroforestry technology become an agroforestry system once they are well developed and commonly used in a given region in such a way that they form a well-defined land use system for that region.
- **Agroforestry Technology:** It is an innovation or improvement, usually through scientific intervention, to either modify an existing system or practice, or develop a new one. It is a set of specifications for the roles, arrangement and management of multipurpose trees and associated components. These specifications must include the technology related tree characteristics, referred to as the ideotype of the tree. An agroforestry technology is generally adapted to a particular site, but can be described with more or less specifications, so that its application domain is flexible e.g. agroforestry land management proposal, on-station or on-farm agroforestry experiment.
- **Agroforestry:** The deliberate association of woody perennials (trees) with crops and/or animals on the same unit of land.
- **Agroforests:** They are particular kind of agroforestry land use and created as the end-result of all agroforestry systems with a typical structure and composition.
- **Agrosilviculture:** It can encompass all forms of agriculture (including animal husbandry) with trees, and would thus be another word for agroforestry.
- **Agrosilvopastoral System:** The land management systems for the concurrent production of agricultural crops, tree crops and forage crops for the rearing of domesticated animals.
- **Alley Cropping/ Hedgerow Intercropping/Alley Farming:** Practice of growing annual crops between rows (alleys) of closely planted trees (mainly leguminous). The trees are pruned periodically to give a hedge appearance and the resulting mulch is placed on the alleys to provide nurtrients and control weed.
- **Annuals:** It means the plant-growing season is annual.
- **Apisilviculture:** Cultivation of trees and shrubs valuable as a source of nectar for honey bees (apiculture).

- **Aquaforestry or Aquasilviculture:** It is the integration of trees and shrubs with fish production. Trees are planted around ponds and tree leave being used as 'forage' for fish.
- **Arid:** A climate in which potential evaporation exceeds rainfall in all months of the year. A condition that makes cropping possible only with the support of water harvesting or irrigation. It refers to an area with an average of less than about 200 mm annual rainfall.
- **Bamboo:** Although they do not contain lignin, they are considered to be the same class as woody perennials in agroforestry.
- **Beautification:** Planting trees for ornamental purposes.
- **Bench Terrace:** An embankment constructed across sloping fields with a steep drop on the downslope side.
- **Biomass:** The weight of material produced by a living organism or collection of organisms. Biomass is expressed in terms of fresh weight or dry weight.
- **Boundary Plantings:** Trees planted along boundaries or property lines to mark them well and for protection.
- **Browsing:** The feeding on buds, shoots and leaves of woody plants by livestock or wild animals. Browse is the material consumed.
- **Bush:** It is a low, densely branched shrub.
- **Cinderella Species:** Indigenous fruit and nut trees in the tropics have been des-cribed as "Cinderella Species" as their importance has been largely overlooked.
- **Commercial Farm Forestry:** Commercial farm forestry is defined as the process under which farmers grow trees on commercial basis on farm lands.
- **Community Forestry/ Rural Forestry:** Community forestry is defined as raising of trees on public or community land rather than on privately owned lands as in case of farm forestry.
- **Companion Cropping:** The practice of planting a long and short season crops together with the intent of harvesting the short season crops to make room for long-season crops to grow and take over fields.
- **Complex Agroforestry Systems (CAF):** They are tree-based systems with a forest-like configuration. They associate a high number of agroforestry components, among which trees are dominant and they mimic natural forest structures with a complex multistrata structure and a closed or almost closed canopy that is usually dominated by a few tree species.
- **Contour Buffering:** They are created by planting trees and shrubs in rows on a contour or cross-slopes. Rows are often planted along with crops and leaf litter provides soil cover and fertilizer. Most importantly for contour areas, buffers reduce sheet and rill water erosion, and even increase soil levels by trapping sediment within the rows.
- **Contour Hedgerows/ Hedgerow Barriers:** They are rows of trees or shrubs closely planted along the contours of sloping land and pruned to form hedges.They are planted to reduce soil erosion by run-off water and maintain the soil fertility.
- **Contour Line:** An imaginary line on a field, joining all points which are at the same height above sea level.
- **Contour Vegetation Strips:** They are living barriers of trees and shrubs which are planted along the contour lines of a slope to control water and soil erosion. These lines of vegetation can also provide useful products such as food, fodder, fuel, poles, etc.

- **Crops/Annual Crops:** They are seasonal plants which are geneally herbaceous.
- **Crown:** The canopy of a tree or other woody plant, which rises above the trunk or stem.
- **Deciduous Tree:** A tree which stands virtually leafless for a while after shedding its leaves and before a new flush leaves out. By contrast an evergreen tree changes its leaves gradually.
- **Desertification:** It is a form of land degradation with the process of continuous decline in the biological productivity of arid or semi-arid land resulting in a skeletal soil that is difficult to revitalize.
- **Dispersed Trees:** Trees planted alone or in small numbers on patures or otherwise treeless areas.
- **Domestication:** It refers the selection, breeding and adaptation of plant and tree germplasm to increase production and quality during cultivation.
- **Earthworks:** Construction made of earth, usually to conserve or control water.
- **Eco-agroforestry:** Eco-agroforestry integrates agroforestry and biodiversity - realizing the full potential of agroforestry as a self-sustaining diverse production system - that works within ecological cycles and supports ecosystem health.
- **Entomoforestry or Entomosilviculture:** Cultivation of trees and shrubs valuable as a source of feed for beneficial insects.
- **Erosion:** The process of wearing away soil by wind and/or rain. Soil erosion is harmful because of the loss of fertile top soil. Where the soil is deposited, it may also cause problems, e.g. silting up of waterways.
- **Evaporation:** Transition of water to vapour. Usually the water evaporates from the soil or vegetation and as a result the soil gradually dries out.
- **Extension Forestry:** Planting of trees on the sides of roads, canals and railways, along with planting on wastelands is known as extension forestry for increasing the boundaries of forests.
- **Fallow:** Land resting from cropping, which may be grazed or left unused and often leads to colonization by natural vegetation.
- **Farm Forestry:** Farm forestry is defined as the practice of forestry in all its aspects on farm or village lands, generally integrated with other farm operations. (or) It is the incorporation and management of tree growing into the farming system by farmers alone or in partnerships and take many forms: plantations on farms, woodlots, timber belts, alleys, wide-spaced tree plantings, native forest plantations, etc.
- **Farming Enterprises:** They are all activities undertaken to produce an output that contributes to total production or income of the farm family, while the household is a social organization in which members normally live and sleep in the same place and share their meals.
- **Farming System:** A unique and reasonably stable arrangement of farming enterprises that the household manages according to well-defined practices in response to the physical, biological and socio-economic environments, and in accordance to the household's goals, preferences and resources.
- **Foliage:** The mass of leaves on plants and trees.
- **Forest Farming:** Forest farming is a specific form of agroforestry that involves the cultivation of high-value non-timber crops under the protection of a forest canopy that has

been modified to provide the shade level appropriate for a specific crop. It is the system of managing natural forests in order to produce food crops or foliage. The system yields wood or tree products in addition to agronomic crops or forage, improves crop or forage quality and improves soil nutrient recycling for crop and forage use. The diversity created with forest farming can also attract a variety of wildlife species. Trees should be planted or native forests thinned (5 to 40% tree crown cover) to allow adequate light into the understorey crops or forage.

- **Green Manure:** Green plant material used as fertilizer.
- **Homegarden:** It is the intimate, multistorey combination of various trees, crops and animals around homesteads. It is the system in which the perennial crops and annual crops are grown side by side in a complex and layered structure (upper storey with tall trees, middle storey with small trees, understroey with small plants and root crops and ground cover with grasses). Animals are also usually included in the system. The home garden is able to provide an extra and continuous flow of products for daily use. They are common in the humid tropics and are characterized by the intensive use of multipurpose trees, shrubs, food crops and animals.
- **Humid:** A climate in which rainfall exceeds potential evaporation during at least 9 months of the year. It refers to tropical areas that receive more than about 1500 mm annual rainfall.
- **Improved Fallows:** It is the replacement or enhancement of natural fallow vegetation by the introdcution of selected trees or shrubs which can help to restore nutrients to the soils and to suppress weeds as well as providing useful by-prodcuts. Areas left to grow up with selected trees as part of a tree-crop rotation system. The purpose of improved fallow is to shorten the fallow period and/or increase the yield of subsequent crops. The trees therefore enrich the fallow both biologically and economically.
- **Individual Trees:** Trees occurring alone, whether spontaneously or planted.
- **Infiltration Rate:** The rate at which water can move through a soil.
- **Intercropping:** This is generally associated with small farms and is achieved by planting short crops in between rows of tall ones. It is a highly efficient use of land, but makes cultivation difficult due to a lack of space.
- **Land Cover:** It refers to vegetation types that cover the earth's surface; it is the interpretation of a satellite (digital) image of different land cover. In simple terms, it is what can be seen on a map, including water, vegetation, bare soil, and/or artificial structures.
- **Land Use Systems:** Land use systems combine land cover and land use with the addition of the cycle of vegetation changes and management activities (planting and harvesting, among others); this needs more on-ground information.
- **Land Use:** It refers to human activities (such as agriculture, forestry and building construction) at a particular location that alter land surface processes including biogeochemistry, hydrology and biodiversity; of course, the uses interact strongly with land cover, however they are not always identical: the same land cover can be used differently and the same uses can be applied to different land cover.
- **Leaching:** The process by which nutrients in the soil are washed down through rain or irrigation water to a depth at which plant roots can no longer reach them. After leaching, the nutrients may be carried away by ground water movement.
- **Leeward:** The side of an object and its surrounding area that is sheltered from the wind.

- **Legume:** Member of a large family of trees, shrubs and herbs, the Leguminosae. On the roots of these plants, there are small nodules which contain bacteria. These bacteria convert inert nitrogen from the air into a form which the bacteria and plants can use for their growth.
- **Litter:** Organic material on the soil surface, including leaves, twigs and flowers, freshly fallen or slightly decomposed.
- **Living Fences:** They are lines of trees or shrubs planted in close spacings on farm boundaries or on the borders of farmyards, pasture plots, animal enclosures or around agricultural fields. They have been used by farmers in areas where materials for fenceposts are scarce or expensive. They can be made of single or multiple, densely planted rows consisting of a mixture of plant species.
- **Lopping:** A form of pruning in which some of the branches of a tree are removed. Usually the lower branches are cut, while the upper part of the crown is allowed to continue to grow. Lopping should lead to the sprouting of new shoots near the cuts.
- **Microclimate:** The temperature, sunlight, humidity and other climatic conditions in a small localized area, e.g. in a field, under a tree or in the topsoil.
- **Mulch:** Protective covering of the soil surface by various substances, such as green or dry matter, sand or stones, applied to prevent evaporation of moisture, moderate soil temperature and control weeds.
- **Multilayer Tree Gardens:** Multispecies, multilayer dense plant associations with no organized planting arrangements.
- **Multipurpose Forestry/ Multipurpose Forest Tree Production System:** Forest tree species are regenerated and managed for their ability to produce not only wood, but also fodder, fruit, fuel, manure, etc.
- **Multipurpose Trees on Crop Lands:** Trees scattered haphazardly or according to some systematic patterns on bunds, terraces or plot/field boundaries.
- **Multipurpose Woodlot:** An area planted with trees for wood, fodder, fuel, soil protection, soil reclamation, etc.
- **Nectar Crop:** Trees valuable as a source of nectar for honey bees.
- **Nitrogen Fixation:** The processing of inert nitrogen in the air into a form that can be used by plants. The process is performed by organisms that live in association with the roots of certain plants, e.g. legumes.
- **Non-Commercial Farm Forestry:** Non-commercial farm forestry is defined as increasing the number of trees raised by individual farmers for their own family uses.
- **Nutrients:** Mineral substances and nitrogen, which are absorbed by the roots to enable plants to grow.
- **Orchard:** A field planted with fruit trees.
- **Parklands (Scattered Trees):** Parklands are a very common type of agroforestry system in tropics and characterized by well-grown scattered trees on cultivated and recently fallowed land. These parklands develop when crop cultivation on a piece of land becomes more permanent. Canopy cover of the trees in parklands averages from 5 to 10 per cent, with variations mainly due to farmer's attitude towards trees in cultivated fields. As crop cultivation intensifies, there are usually fewer trees. Parklands are best developed near the villages, as here they can be well protected and managed.

- **Partnership Cropping:** This is achieved when two crops are partnered, enhancing the growth and productivity of one or both.
- **Perennials:** It means the plants last several growing seasons and live for longer than one or two years.
- **Permeability:** The capacity to allow air, water or other material to pass through. A desirable property of soils.
- **Pollarding:** Removing all the branches, including the top of the tree, leaving only the trunk. Shoots are allowed to sprout to form a new crown.
- **Polyculture:** The practice of planting two or more crops in order to create competition or complementation between crop species.
- **Protein Bank:** Production of protein-rich tree fodder on farm/range lands for cut-and-carry fodder production.
- **Pruning:** The cutting of parts of a woody plant, often to stimulate better-placed new growth.
- **Range Lands:** An extensive area of land on which livestock can graze.
- **Relay Cropping:** The practice of planning crops in succession with up to three or four crops in a single year.
- **Riparian Buffer:** This buffer is used specifically to protect wetlands and to filter water. It is created by planting a corridor of trees and/or shrubs adjacenet to or parallel to the river bank, stream bank, wetland or body of water. Plants must be of sufficient width and should be placed up-gradient near the water. Riparian buffers protect near-stream soils from over-bank flows, trap harmful chemicals or sediment transported by surface and subsurface flows from adjacent land uses or provide shade, detritus and increase biodiversity by harbouring wildlife.
- **Root Collar:** The point near ground level where the root system merges with the stem.
- **Rotation:** Cultivation of a succession of crops, possibly including a fallow period, on the same land. One rotation cycle usually takes several years to complete.
- **Run-off:** Rain or other water that flows over the soil surface and does not infiltrate into the soil.
- **Sapling:** A young tree, no longer a seedling, but not yet big enough to be called as s pole. Usually few meters high and at most 2.5 cm in diameter at breast height.
- **Semi-arid:** A climate with an average rainfall of about 200-900 mm and large variations from year to year.
- **Sequential Agoforestry:** In this agroforestry systems, the maximum growth rates of the crop and the tree components occur at different times even though both components may have been planted at the same time and are in close proximity e.g. shifting cultivation, improved fallows, taungya and some multistrata systems. Interactions between the crop and the tree components are minimized with time in sequential agroforestry.
- **Shelterbelt:** A shelterbelt is a barrier formed by trees and shrubs strategically planted to reduce wind speed in order to protect agricultural lands, people, animals and buildings. They can also be used to support sand dune stabilization. Shelterbelts are made up of strips of trees, shrubs and grasses planted in multiple rows to create typical triangle shape. Shelterbelts are most successfully introduced in areas where there are high wind speeds

and/or prevailing wind for long periods, or where the soil is dry for long period and has loose soil structures.

- **Shifting Cultivation/ Slash-and-Burn Agriculture/ Jhum Cultivation:** The term refers to farming of agricultural systems in which land under vegetaion is cleared, trees are cut, debris burnt, ash incorported into the soil for cropping with agricultural crops for few years and then the land is left unattended while the natural vegetation regenerates. The cultivation phase is usually short (2-3 years) but the regeneration phase is much longer (10-20 years).
- **Shrubs:** They are small, usually multi-stemmed woody plants.
- **Silvi-Sericulture:** Cultivation of trees and shrubs valuable as a source of host and feed for silkworms.
- **Silvopastoral System:** A form of agroforestry system consisting of the trees (woody perennial) and pasture/animal component.
- **Simple Agroforestry Systems (SAF):** It refers to associations involving a small number of agroforestry components arranged with obvious, usually well-arranged pattern: one or a couple of tree species, either as a continuous canopy, in equally distant lines or in edges, and some annual species for ground cover. The tree component can be of major economic importance or have a more qualitative role (fodder, soil fertility, etc.), whereas the annual species is usually always economically important (paddy, wheat, maize, vegetables, forage herbs, etc.) and this ground species can also be a semi-perennial (banana, cacao, coffee, or coffee)
- **Simultaneous Agroforestry:** It is where the tree and the crop components grow at the same time and in close enough proximity for interactions to occur e.g. alley cropping (hedgerow intercropping), contour hedges, boundary plantings, homegardens and several silvopastoral systems.
- **Small Trees:** Single-stemmed woody perennials which are less than 7 m high.
- **Social Forestry:** Social forestry is defined as forests "of the people, by the people and for the people". It means the management and protection of forests and afforestation on barren lands with the purpose of helping in the environmental, social and rural development, as against the traditional objective of securing revenue.
- **Splash Erosion:** Raindrops that fall on soil aggregates, causing small soil particles to splash in all directions.
- **Stake:** It refers to a wooden pole used to support climbers. The live stakes strike roots easily and are in fact very large cuttings.
- **Sub-humid:** In the tropics, a climate with an average annual rainfall of roughly 900-1500 mm.
- **Sustainability:** It refers to management of resources in agriculture to satisfy changing human needs, while maintaining or enhancing the quality of the environment and productivity of the land.
- **Taproot:** The first root to emerge from the seed and usually vigorous, persistent and growing down-wards.
- **Taungya System**: This system consists of growing annual agricultural crops along with forest trees during the early years of establishement of the forest plantation.

- **Terrace Planting:** Terraces are usually put in place as soil and water conser-vation measures on slopes and hills. They provide flat areas of land that can be planted with crops. Building terraces involves digging ditches and making ridges along the contours of a slope. Grasses, trees and shrubs can be planted on the ridges to stabilize the ground, provide leaf mulch and protection from wind for crops and provide other useful products such as food, fodder, fuel, poles, etc. Trees can be planted on the ridge of the terrace or at the back of the terrace.
- **Terraces:** Level areas constructed along the contours of hills, often but not necessarily planted with trees.
- **Traditional Agroforestry Practice:** Once an agroforestry technology has been in existence for a certain amount of time and is practised regularly by farmers, it can be called a traditional agroforestry practice. In other words, a traditional agroforestry practice is always an agroforestry technology but an agroforestry technology is not always a traditional agroforestry practice.
- **Transpiration:** Loss of water in the form of vapour from stomata of the plants.
- **Tree-Crop Interaction:** When trees and crops are grown together on the same pieces of land there will be interactions between the two components, which may have positive or negative results. Tree-crop interaction is defined as the effect of one component of an agroforestry system on the performance of another component and/or the overall systems.
- **Tree-Crop Interface (TCI):** The area where the tree interacts directly with the crops.
- **Trees:** They are woody perennial plants having single main stem, constitute the majority of the ligneous plants. It is usually accepted that, to be called a tree, the plant needs to be of a mimimum size of about 7 m.
- **Turbulence:** Whirling of wind after it has passed an object.
- **Urban Forestry:** It is the practice of growing trees on non-forest land in urban areas for recreational purposes and aesthetic value.
- **Vegetative Strips:** Long, narrow areas of any type of vegetation, usually planted along contours for erosion control and it may include trees.
- **Windbreak:** Windbreak are created by strategically planting of several rows trees and shrubs along property lines, or in fields to protect the crops and livestocks from strong winds and hard rain. Windbreaks are most effective when planted perpendicular to strong winds and generally more than 200 feet long.
- **Wind-ward:** The side of an object and its surrounding area that is exposed to the wind.
- **Woodlot:** An area planted with trees for timber or fuel.
- **Woody Hedgerows:** Woody hedges for browse, mulch, green manure, soil conservation, etc.
- **Woody Plants/Ligneous Plants:** They are plants that contain lignin, an organic substance that impregnates and unites the cells and fibres of the plant, whose tissues are arranged in such a way that they take on the characteristics of wood.
- **Woody Vines:** They belong to ligneous plants and generally require a support.

■■■■

2 AGROFORESTRY AND FOOD SECURITY

INTRODUCTION

In many developing countries, there is an inextricable link between the tree resources and the livelihoods of rural communities. Many poor rural populations traditionally rely on forest trees for most of their livelihoods including fuel wood and timber as well as non-wood forest products (NWFPs). The majority of rural communities are engaged in agriculture for their livelihoods. Additionally, they exploit natural resources as a means of supplementing their cash income. Many authors have highlighted the importance and use of trees and woody shrubs for the availability of food over critical periods. Such periods include the beginning of the rainy season when food storage is most acute as households have usually exhausted previous years harvest but the harvest for the new season has not commenced. The availability of NWFPs especially indigenous fruits serves as an important gap-filler when food stocks are low and also as a source of income.

Agroforestry, by its very nature, creates complex systems with impacts ranging from the site or practice level up to the landscape and beyond. Successful design of agroforestry practices hinges on the ability to pull together very diverse and sometimes large sets of information (i.e., biophysical, economic and social factors), and then implementing the synthesis of this information across several spatial scales from site to landscape. In South Asia, the tree-based farming systems have been integral part of human civilization and hence huge traditional ecological knowledge base exists on the practice *per se*. In the global climate change scenario as well, the developing world in the humid tropics, particularly in South Asia are experiencing a global challenge of mitigating greenhouse gas emission, where agroforestry practices could help in manuring carbon sequestration process. Likewise, the agroforestry or any tree-based farming system has also intangible benefits such as soil and water conservation, carbon sequestration, etc. that has been classified as environmental services. Thus, the systems of agroforestry are having ecological linkages with human ethics and economic benefits. One of the best examples have been cited is home gardens that is a complex integration of a group of trees and other understorey crops that is manifested by human tradition. In this chapter, the paradigm shift in the agroforestry trends have been attempted to address the issues of food security.

GLOBAL DEMOGRAPHY

Skewed access of the rural poor to the resource base and food insecurity has been a major challenge to sustainable development of most of the developing countries of the world. Therefore, improvement of the well-being of the masses while maintaining a viable and healthy environment is the most daunting task before the policy makers across the globe. About 870 million people (12.5 per cent) of the global population are estimated to have been undernourished (in terms of dietary energy supply) during the period 2010-12. Majority of these (852 million) live in the developing countries, where the prevalence of undernourishment is now estimated at 14.9 per cent of the population. The undernourishment in the developing countries is slightly more than 23.2 per cent of the population in 1990-92 (substantially higher than previously estimated), thus implying the Millennium Development Goal (MDG) target of 11.6 percent for 2015 (FAO 2012).

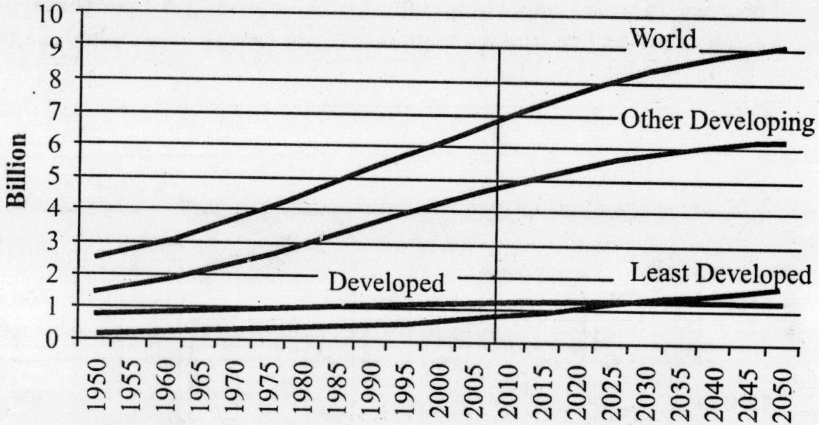

Fig. 2.1: Population Growth in the Globe (Source: UN Population Division-2009)

Global population is increasing steadily and the rate of increase was higher in the developing nations when compared to developed and least developed nations (Fig. 2.1). However, there is a marginal increase in the arable land in the developing nations (Fig. 2.2), which shall give some opportunity to moderate the enormity of food insecurity in the developing world. Significant progress in global agricultural production driven by green revolution in the Asian subcontinent in the second half of the last century helped to break cycles of crop failures, food deficits and wide spread famines. However, in recent past agricultural production has been hampered by changing climate, land and soil degradation, shortage of quality irrigation water. The problem is further magnified by the burgeoning population in the Asian and African countries.

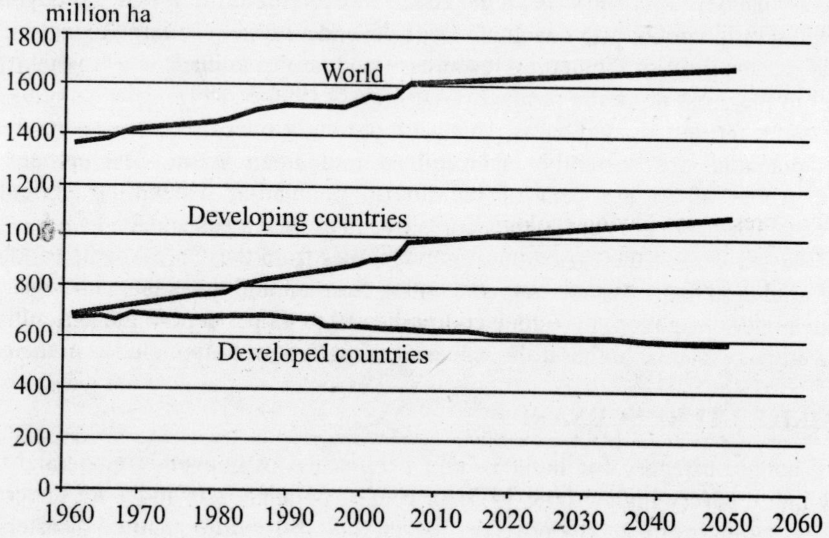

Fig. 2.2: Scenario of Arable Land in the Globe (Source: Bruinsma 2009)

GLOBAL FOOD PRODUCTION

World cereal production in 2012 was reduced marginally (2 million tons) and figure to 2,282 million tons (including rice in milled terms), 2.8 per cent down from the previous year's record (Fig. 2.3). The overall decrease in world cereal output this year comprises a 5.7 per cent reduction in wheat production and a 2.6 percent decline for coarse grains, which largely outweighed the marginal 0.9 per cent growth expected in the global rice crop.

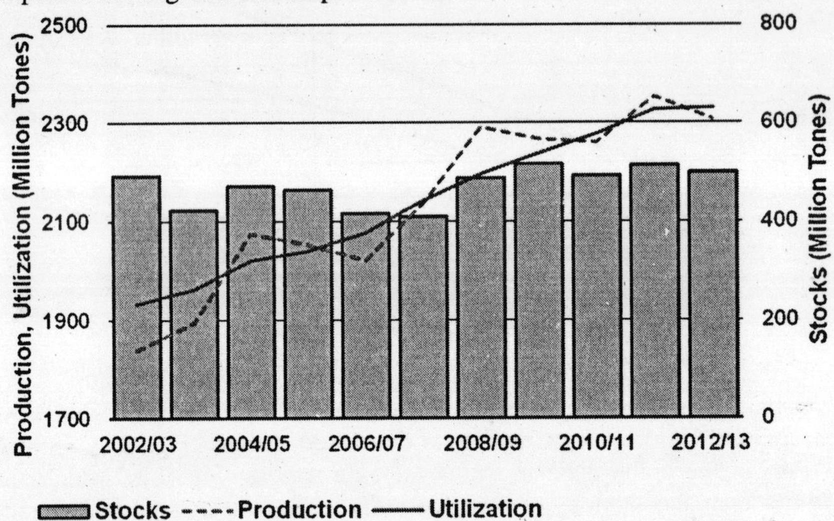

Fig. 2.3: Status of World Food (Cereal) Production (Source: FAO 2013)

ASIAN AGRICULTURE

Agriculture is a major sector in Asia, and an important determinant of economic growth. The successful economic development through the Green Revolution of the 1960s and beyond is testimony to this fact. Success in agricultural development is dependent to a very large extent on the efficiency in the use and management of the natural resources (land, crops, animals and water). It contributes 25 to 43 per cent to the gross domestic product (GDP) (FAO 1996 & 1997). Much of this contribution is made by the more fertile irrigated areas which are presently over used. However, the rainfed areas which have potential are currently underutilized and merit more development attention for development. About 43-88 per cent of the human population depend on agriculture for their livelihoods, of which 12-93 per cent of the people live in rainfed areas and 26-84 per cent in the arable land. Some 5-41 per cent of the agricultural output comes from these areas. Due to low productivity, the shares of total crop and livestock outputs coming from rainfed areas is much lower than the share of the total area under irrigation. Livestock contributes 10 to 45 per cent to the agricultural GDP in the developing world, and can be higher if the values of draught power are included in the calculation.

INDIAN AGRICULTURE SCENARIO

The score of hunger category for India is 3 i.e., the status of hunger is moderately high. Climate vulnerability index is very high. About 17.5 per cent of population in India are undernourished. The food security situation is affected by poverty, uneven food distribution, natural disasters, deforestation and low education levels.

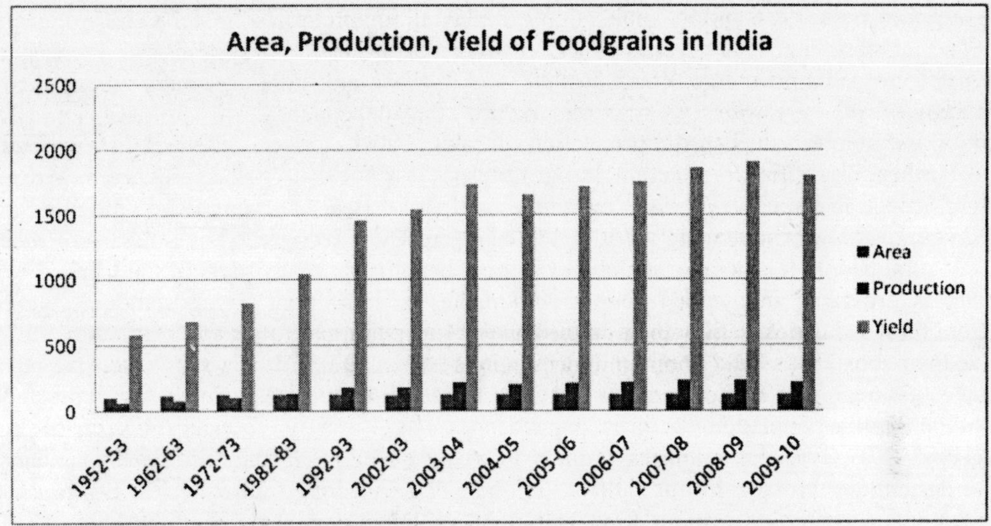

Fig. 2.4: Area, Production, Yield of Food Grains in India
(Source: (a) Ministry of Agriculture, Government of India, 2008 & 2010;
(b) Economic Survey, 2009-10)

During 1952-53, 102.02 Mha were covered under food grain with the production of 59.2 million tons and productivity of 580 kg/ha. Between 1952-53 and 2009-10 only 98.28 Mha were added to the existing area under food grain cultivation. However, due to impact of green revolution and the use of modern agro-services, the total production increased from 59.2 MT in 1952-53 to 218.20 MT in 2009-10 and yield increased from 580 kg/ha to 1,798 kg/ha in the intervening period. While 18.1 percent area under food grains had assured irrigation in 1952-53, the irrigation coverage increased to 46.8 percent by 2007-08. It is evident from the graph that since 2003-04, the area under food grain has remained more or less stagnant with relatively stagnated yield rate. These stagnant trends have been shown by their growth rates from the year 2003-04 to 2009-10 (Fig. 2.4).

MYRIAD OF AGROFORESTRY SYSTEM

Agriculture is the primary livelihood strategy for 85 per cent of the rural population in developing regions (Dixon et al 2001) and the main constraint to agricultural productivity is soil nutrient deficiency, especially nitrogen and phosphorous due to continuous cultivation practices. Increased use of inorganic fertilizers will improve the yields over the short term; however it will mask the soil nutrient depletion, rather than correcting it, especially soil micronutrients (Beedy et al 2012). Moreover, nutrient depletion will reduce the productivity and thereby generates food crisis and threats the food security. Soil fertility depletion has been recognized as the fundamental biophysical root course responsible for declining food security in Africa (Pieri 1998). Food security means to guarantee sufficient quality and quantity of foods to population; this is a main problem over large areas of the world (Massolino and Pardini 2009). A key challenge facing the international community is, therefore, to ensure food security for present and future generations, while protecting the natural resource base on which we all depend. In the coming years, improved livelihoods in many of the developing countries in Asia and Africa will depend on ability of the stakeholders to arrest degradation increasing food production.

The agroforestry science and its application can play an important role in achieving greater food security. The incorporation of a diverse variety of trees into agricultural systems can increase crop productivity, increase the income of small farmers, and improve nutrition, especially among the rural poor. Moreover, the agroforestry practices reduce the dependency on primary agricultural commodities and help to establish the production of value added products based on raw agricultural materials. Further, diversified production in agroforestry is a form of risk avoidance, which is of special significance in the context of the current agricultural crises that many developing and under developed countries are experiencing. Worldwide, agroforestry is recognized as a means of meeting the needs of rural people for income, social and cultural benefits. In many parts of the tropical world, agroforestry is providing indigenous people with means to raise their living standards, generate income from rural industries, satisfy their cultural needs for traditional foods and medicines, etc., and to improve their social status and community aspirations (Anon. 2003). It is a very extensive practice in the tropics although it is found from the Arctic to the south temperate regions. Approximately 24 per cent of the world's population has direct dependence on agroforestry products and services (Leaky and Sanchez 1997). Over the countries farmers in many developing countries have maintained a productive agriculture through shifting cultivation combined with long fallow cycles. But in the past few decades the need to grow more food and improve livelihood has forced farmers to go for maximum productions, much beyond the capacity of the fragile ecosystem using limited input and soil management practices. This often resulted in breakdown of the traditional farming system and resource degradation.

AGROFORESTRY IN INDIA

Trees have a special role in the ethos of the people of India. There are several sacred trees and sacred groves valued by the people. India also has a long historical tradition of tree-growing on farms and around homes. Such traditions and indigenous ethics had and continue to have an impact and implications for tree-growing and ecological, economic and social well-being of the people. Sacred elements and traditional practices in cultural landscape of India also have a substantial livelihood and conservation value (Bawa et al 2004).

Agroforestry systems in India include trees in farms, community forestry and a variety of local forest management and ethnoforestry practices (Pandey 1998). A wider definition of agroforestry encompasses a variety of practices, including trees on farm boundaries, trees grown in close association with village rainwater collection ponds, crop-fallow rotations, and a variety of agroforests, silvopastoral systems, and trees within settlements. These systems have been presented as a solution to rising fuelwood prices in India resulting from increase in demand and decrease in supply of fuelwood due to forest degradation (Bowonder et al 1988). Overall, India is estimated to have between 14,224 million (Ravindranath and Hall 1995) and 24,602 million (Prasad et al 2000) trees outside forests, spread over an equivalent area of 17 million ha (NFAP 1999), supplying 49 per cent of the 201 million tons of fuelwood and 48 per cent of the 64 million m^3 of timber consumed annually by the country (Rai and Chakrabarti 2001).

POTENTIAL OF AGROFORESTRY

Globally forestry is one of the options to mitigate climate change. It is estimated that the total global technical potential for afforestation and reforestation activities for the period 1995-2050 is between 1.1–1.6 GtC/yr of which 70 per cent will be in the tropics (IPCC 2000). Agroforestry is an attractive option for carbon mitigation for the reasons: (i) it sequesters carbon in vegetation and soil depending on the pre-conversion vegetation and soil carbon; (ii) the wood products produced serve as substitute for similar products unsustainably harvested from natural forests; and (iii) it increases the income to farmers (Makundi and Sathaye 2004). Approximately 1.2 billion people, making 20 per cent of the world's population depend directly on agroforestry products and services in rural and urban areas of developing countries (Leakey and Sanchez 1997).

The potential land area suitable for agroforestry in Africa, Asia and the Americas is 585-1215 Mha (Dixon 1995). It is estimated that an additional 630 Mha of current crop lands and grass lands could be converted into agroforestry, primarily in the tropics. Agroforestry activities could be of two types; converting fallow and marginal croplands to agroforestry and adopting agroforestry practices into existing cropping system.

TIMBER

Currently the gap between supply and demand is met through imports. The deficit of 16 million cum (7.23 mt) as on 2008 could be met by promoting about 3,38,068 ha of Poplars (@ 160 t/ha productivity at 7 years rotation) or a combination of other species like *Eucalyptus*, Subabul, *Acacia mangium,* etc. and restricting imports. At a modest rate of Rs. 1000/t, this additional contribution will be Rs. 7,230 million. The value addition in processing and the employment generation are other additional contributions. The major contribution of agroforestry has been in the supply of small timber, where more than half of the demand is being met from agroforestry (Table 2.1).

FOOD

By virtue of diversity of the components of the agroforestry systems like food grains, vegetables, fruits, nutrition security to the communities could be ensured. Induction of fodder cultivation under agroforestry land use will ensure production of milk, meat and animal products. Wide range of food crops, pulses and oil seeds can meet diverse needs of society. The estimated demand for food by 2025 will be 320 million ton and food crops grown under agroforestry alone are likely to contribute about 25 MT/annum (Table 2.1).

FRUITS

In recent past when growth rate of agriculture is hovering around less than 4 per cent, the growth in horticulture has been recorded 7 per cent. Fruit production in the country during 2002 was 48.57 million ton with almost 54 per cent contribution from Mango and Banana alone (Kalloo 2003). Under agroforestry system, fruit owe special significance on account of short juvenile phase than timber trees and assure regular production for long period (up to 40 years). Experiences have indicated that agrihorti-culture, obviously due to assured returns and notional security (Kareemulla et al 2002), is preferred by the farmers than agrisilviculture. In recent years, aonla (*Emblica officinalis*) has been promoted under agrihorticulture system and already more than 50,000 ha area is under improved cultivars of the species. A number of industries utilizing its products are providing value addition as well as employment generation. Further, on wastelands there is plenty of scope for introducing fruit trees as hortipasture components for immediate monetary gains and higher biomass yield without hampering forage production. An area of 14 Mha of rainfed lands can be developed under hortipasture land use.

FUELWOOD

Fuelwood production from various sources was around 169 million tons in 1990, which increased to 201 million tons by 2001. Forests account for over 50 per cent of this supply. The rest comes obviously from the farmlands and community lands. About 10 million rural people are estimated to be engaged in collection of fuelwood (Mahapatra 2003). Besides this many more are engaged in the marketing of fuelwood in rural and urban areas. The fuelwood scarcity has been obviously felt in areas where forests are absent. Therefore, in such areas, agroforestry has played its role to fulfill the firewood demand. Further, as the scope for extraction of firewood from the forests is decreasing gradually, the only alternative left for the people is to patronize agroforestry on farm and community lands. By promoting either silvipasture or pure woodlots supply from wastelands could be enhanced (Table 2.1).

Table 2.1: Contributions of Agroforestry Produce to National Demand by 2025 (NRCAF 2007)

Product Category	Total Demand	Contribution from Agroforestry			Contribution from Agroforestry (%)
		Traditional	Improved	Total	
Timber (million cubic m)	116	52	48	100	86
Fuel wood (million ton)	330	60	13	73	22
Fodder (million ton dry)	1,040	100	12	112	11
Fruits (million ton)	43	1.5	4.4	5.9	14
TBOs (million of bio-diesel)	10	0.5	5.4	5.9	59
Food (million ton)	308	7	15	22	7
Bioenergy (MW)	16,000	4,000	1,000	5,000	3
Forest cover (million ha)	100	10	15	25	25

FODDER

Trees and shrubs often contribute substantial amount of leaf fodder in arid, semi arid and hill regions during lean period through lopping/pruning of trees, popularly known as top feed. The fodder scarcity could be reduced through fodder cultivation on farmlands and through silvipasture on farm/ community wastelands. Assuming availability of another 10 per cent of community wastelands for silvipasture, the additional fodder production (@ 2.0 t/ha/yr at 40% of research yield) would be 12 million ton.

LIVESTOCK

Promotion of silvipasture provides additional fodder supply that ensures proper rearing of livestock and environmental protection. Small ruminant farming based on goats and sheep is most ideal under agroforestry. The average net income from rearing of small ruminants in one ha of developed silvipasture is Rs. 7868 per year (Kareemulla et al 2002). Given the potential for silvipasture-based livestock farming at 10 per cent of total silvipasture potential, the annual contribution would be about Rs. 439 crores.

BIOENERGY SECTOR

The main biomass energy sources in rural areas, which are being used in the households, include wood from forest, croplands and homesteads, cowdung and crop biomass. Bioenergy from tree biomass is one significant contribution from agroforestry. Species like *Prosopis juliflora* are being used for power generation in states like Karnataka and Andhra Pradesh. *P. juliflora* due to high calorific value of over 5000 Kcal is preferred in these power plants. An estimated 0.51 million ha area is under *P. juliflora*. Even if 25 per cent of the area is meant for power generation leaving the rest 75 percent for fuel and charcoal, the bioenergy potential works out to 1,000 MW. Similarly other tree species can be used for running the biomass based power plants. Thus a total of 5000 MW power could be produced from the biomass sources from trees under moderate conditions, which meets almost one third of the ultimate potential of 16,000 MW from biomass. Establishing power plants of such a capacity attract an investment of Rs.90,830 crores. R & D can help in enhancing productivity and assisting the power plants in captive plantation management on degraded lands. Promoting bioenergy through *P. juliflora* also encourages tremendous employment generation to the tune of 6.34 million mandays and 703 million woman days for fuel making in Tamil Nadu alone.

EMPLOYMENT OPPORTUNITIES

Increased supply of wood in the market had triggered a substantial increase in the number of small scale industries dealing with wood and wood based products in the near past. Such industries have promoted agroforestry and contributed significantly to increasing area under farm forestry. Recognizing agroforestry as a viable venture, many business corporations, limited companies such as ITC, WIMCO, West Coast Paper Mills Ltd, Hindustan Paper Mills Ltd., financial institutes such as IFFCO have entered into the business and initiated agroforestry activities in collaboration with farmers on large scale. One of the major contributions of agroforestry for the economy is the livelihood impact, both in terms of income and employment generation with improved agroforestry systems to the tune of 943 million person days annually from the 25.4 million ha of agroforestry area (Table 2.2). The maximum potential is in the Tree Borne Oil Seeds (TBOs) followed by silvipasture. Dhyani and Sharda (2005) have indicated the potential of agroforestry for rural development and employment generation to the tune of 5.763 million human days yr^{-1} from Indian Himalayas alone.

Table 2.2: Employment Generation Potential through Agroforestry (NRCAF 2007)

AF System	Area (million ha)	Additional Employment Generation/ha/yr (man days)	Total Employment/yr (million man days)
Silviculture	1.8	30	53.3
Agrisilviculture (Irri)	2.3	40	91.3
Agrisilviculture (RF)	1.3	30	38.0
Agrihorticulture (Irri)	1.5	50	76.1
Agrihorticulture (RF)	0.5	40	20.3
Silvipasture	5.6	30	167.4
TBOs	12.4	40	497.1
Total	25.4	-	943.4

AGROFORESTRY VIS-A-VIS FOOD SECURITY

Agroforestry research and development has emerged as a new and vibrant discipline over the past 25 years. The services provided by agroforestry practices to rural livelihoods as well as towards conservation of biodiversity have attracted wide attention among agroforestry and conservation scientists. It applies scientific principles to find practical solutions towards natural resource management and agricultural production problems (Jama et al 2006) and subsequently, supports in improving food security of the vast rural population. Generally the farmers have little knowledge of crop management, soil fertilization, complexity of the farming systems in the era of high input, mechanized, intensive agricultural scenario, due to which they may be passed from self-subsistence to high risk of food security. Therefore, the farmers should be getting aware of the knowledge of the crop management, which should be aimed to both high yields and soil conservation, instead of the excessive forest clearing that favours soil erosion and reduction of soil fertility. There are many agroforestry practices that not only secure the livelihood of the rural populace but also enhanced the income, preserve soil condition as well as conserve the biodiversity.

There are various agroforestry practices made up of combination of several types of products, which are both subsistence and income generating, help the farmers to meet their basic needs and minimize the risk of the production systems and total failure (ICRAF 1993). Agrisilvicultural system is one of them in which there is a concurrent production of agricultural crops and trees in same piece of land either simultaneously or alternately. The agrisilviculture systems aim at production of enough foodgrains, timber, fodder, firewood and other products. Moreover, agrisilviculture practice helps in improvement of shifting cultivation practices through improving soil quality and reducing soil erosions. The system may include multipurpose trees, homestead plantation, alley cropping or hedgerow cropping. The multipurpose forest tree production is the system in which trees/shrubs are deliberately kept and managed for more than one preferred use, product, and/or service along with the agricultural crops. The retention or cultivation of these trees is usually economically, but also sometimes ecologically motivated, in a multiple-output land use system. By integrating multipurpose trees/shrubs in the agricultural lands, farmers can reduce their dependency on a single staple crop. For example, if a drought destroys the annual crop, trees will still provide fruits, fodder, firewood, timber and other products that often achieve high commercial value.

The multipurpose trees can be planted in the farm in several types of plantation geometry such as scattered, on the field bund or border, row plantation or strip plantation. Moreover, in the homesteads diverse products (fruits, vegetables, spices etc.), which are available year-round not only contribute to food security during the "lean" seasons but also ensure food diversity. The mixed tree-gardens represent a substantial unexploited potential for enhancing productivity and profitability (Kumar 2006). A study in eleven types of homegardens across Guinean Highland Savannahs revealed that the total yield can be as high as 112.43 tons per year. It was not only contributing to food security but nutritional of the farmers also (Mopongmetsem et al 2011). Alley cropping and hedgerow cultivation is very useful for sloping land. The system is claimed to be culturally appropriate, ecologically fit, economically sound and technically astute development tool (Dwivedi 1992). Moreover, green manuring by leaves of trees in the hedgerow may add organic matter to the extent of 2.0 to 5.0 t/ha and 150 kg of N, 15 kg of P and 75 kg K per hectare annually (Dwivedi 1992). This increased in soil fertility, subsequently enhanced the crop yield and improved the food security. A study of 1,000 farmers from 15 districts in Kenya found that tree fruits contributed 18 per cent of crop revenue, and tea and coffee contributed an additional 29 per cent of the revenue (Place and Wanjiku 2006). Mithoefer and Waibel (2003) reported that in Zimbabwe indigenous fruits provided higher returns to labour than annual crop production. Similar reports of alternate sources of income and employment to the rural poor population through agroforestry are also highlighted by various workers from different parts of the world (Balooni 2003; Puri and Nair 2004; Samra et al 2005).

Land degradation and desertification are two cardinal processes, which render agricultural lands unproductive and threaten food security in several parts of Asia (Kumar 2006). Although trees are expected to improve soil fertility, the extent to which different agroforestry practices accomplish this depends on tree species, stocking level, growth rate and the input of litter (Kumar 2006). It will be greatest where fast growing trees are integrated with a high density and adding the pruned materials into the soil thereby increasing the soil organic materials. Higher soil organic matter and available nutrients in tree-based agroecosystems can also increase yields in smallholder farming systems (Neufeldt et al 2009). A study from Nepal on the impact of agroforestry on soil fertility and farm income showed that agroforestry intervention nearly doubled farm income per hectare from USD 800-1580 (Neupane and Thapa 2001). Similar report of increased crop production and generating additional income to the rural people is reported in Malawi of South Africa by improving soil condition through the intervention of agroforestry practice. Fertilizer tree technologies of intercropping, relay cropping, improved fallows and biomass transfer, have been promoted as sustainable, low-input alternative or complimentary inputs to inorganic fertilizers in Malawi. The choice of the technologies is driven by the size of the land holdings and more benefits are associated with large land holdings (Quinion et al 2010). Moreover, Kalaba et al (2010) have highlighted the impact of agroforestry adoption on livelihoods of farmers in Malawi, Mozambique and Zambia includes increase in crop yields, increase in income, increased savings resulting in change of wealth and soil improvement.

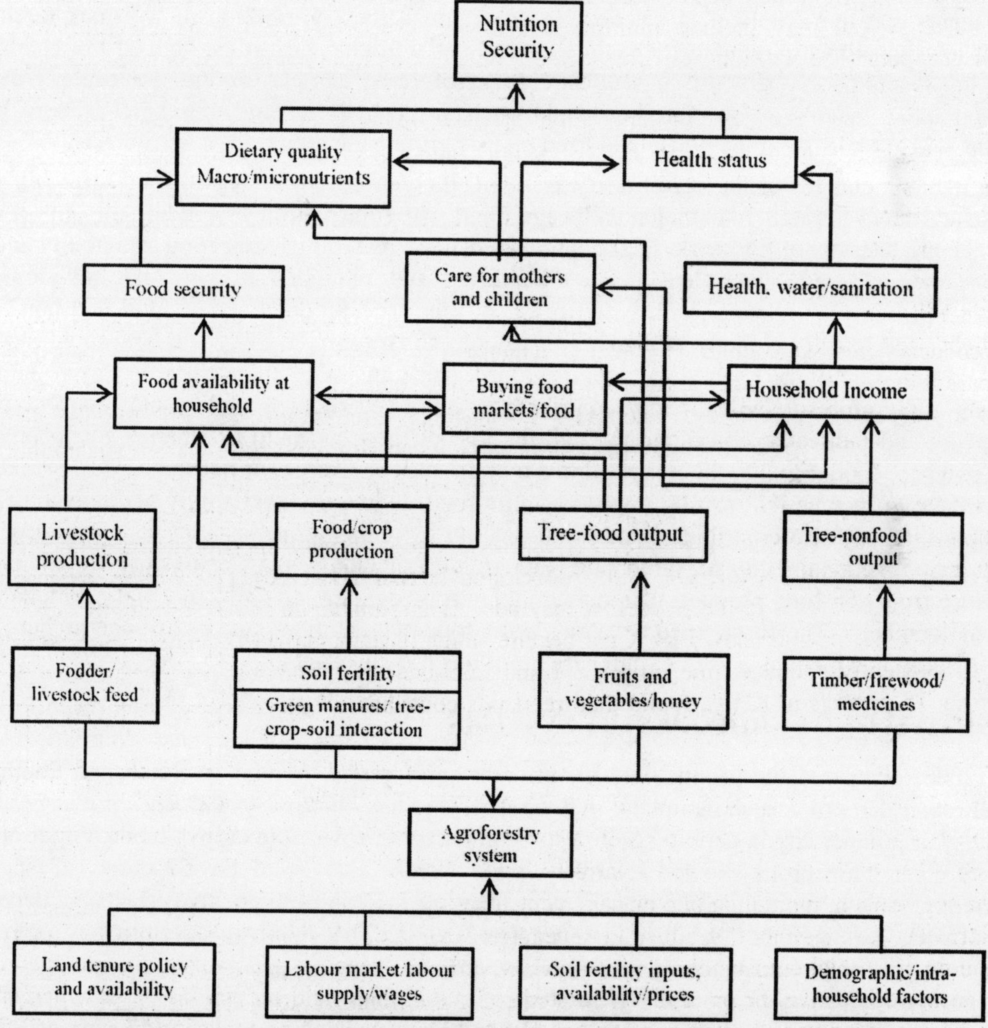

Fig. 2.5: Agroforestry and Food Nutrition Security Linkages (Babu and Rhoe 2002)

Understanding the agroforestry-income-consumption-food-nutrition security linkages is important for identifying, analyzing, and rectifying the constraints and challenges in maximizing the benefits of agroforestry systems for food and nutrition security. Figure 2.5 presents a conceptual framework for studying the agroforestry-nutrition security linkages. Adoption of an agroforestry system as a food-based strategy depends on several underlying factors. First, land availability to grow tree crops is an important determinant of agroforestry farming system. Second, land tenure policies that do not facilitate private ownership of land may reduce the investment on land improvement because the benefits may not accrue to the occupant in the long run. Third, the opportunity cost of growing trees in place of annual food crops could be higher, particularly in irrigated and high-potential farming areas. Fourth, labour market factors such as high rural wages and low supply of agricultural labourers may encourage the transfer to tree crops thus facilitating the shift from monocropping to agroforestry (Babu and Rhoe 2002).

Farming systems that face poor soil fertility provide incentives for adoption of agroforestry for green manures and tree-crop-soil interaction. In areas where the contribution of livestock to cattle manure is low, agroforestry remains the main source of soil nutrients. Given the high cost of chemical fertilizer inputs and poor availability in remote areas, agroforestry can play an important role. Finally, the demographic factors within the household such as availability of household labour, land ownership, and knowledge of the household head may determine the adoption of agroforestry.

The tree components of the agroforestry system can contribute to increased food availability at the household level through four major pathways. First, the contribution of fodder supplement can increase the productivity of livestock. Higher productivity can result in greater food availability and/or higher incomes. Second, green leaf manure harvested and ploughed *in situ* and tree-crop-soil interaction can increase the soil fertility and result in improved productivity of food and non-food crops. Higher soil fertility is likely to increase the food/cash crop productivity and household income that will, in turn, contribute to food availability. Third, fruits and vegetables harvested from the tree component of agroforestry will directly contribute to improved dietary quality and micro-nutrient consumption, and indirectly, contribute through the sale of these items to the market for income in order to purchase food. Finally, the non-food tree outputs such as timber for housing and firewood for fuel contribute to income that can be used to acquire food. Indirectly, availability of fuelwood from tree components may also contribute to nutrition security by reducing the search for fuelwood which could increase time availability for child care (Filmer and Pritchett 2002). Presumably, the increased cash income from non-food products can increase the resources spent on child care and health care of household members. This cash income along with food availability, can contribute to nutrition security if additional income is used to purchase nutritious food products.

PRODUCTIVITY OF AGROFORESTRY SYSTEMS

Kumar (2006) compared the productivity of different agroforestry systems in South and Southeast Asia and found that crops such as upland rice, ginger (*Zingiber officinale*), and *Kaempferia galanga* showed higher productivity in certain agroforestry combinations (over sole crops), while fodder plants and many other grain crops showed relatively lower yields. That is, of the 67 cases (48 species combinations, some in more than one management situations), 16 showed positive effects, 12 depicted neutral effects, and another 39 illustrated negative trends. The relative superiority is probably dependent on species/circumstances, and is not amenable to sweeping generalizations, i.e., the effect may be positive, negative, or neutral. These studies reveals that yield reductions occur when shade intolerant crops [e.g., many fodder species, cereals, and legumes such as soybean (*Glycine max*)] are grown in association with tree species especially after canopy closure (Table 2.3).

Asymmetric competition (i.e., resource acquisition at differential rates) and thereby resource pre-emption by the dominant component is a major cause of production decrease in competing mixtures (Kumar et al 1999). Differences in resource acquisition capabilities (e.g., crown spread and rooting characteristics) are also magnified during the course of competitive interactions. Consequently, understorey yield declines are more probable in denser and older stands of trees compared to poorly stocked young stands. Likewise, nutrient-rich sites generally hasten tree canopy closure and aggravate interspecific competition. It is, therefore, hypothesised that reductions in understorey crop yield in tree-crop mixtures may be more probable on good sites, especially with high input usage. Conversely, degraded sites and shade tolerant crops (e.g., ginger) may show better subcanopy productivity or that productivity may be at par with that of open grown crops (e.g., *Kaempferia galanga*; Table 2.3).

Table 2.3: Agroforestry Systems and Productivity in South and Southeast Asia (Kumar 2006)

Systems/Practices and Parameters Evaluated	System Description	Productivity (kg ha^{-1})	Source	Effects on Productivity
Coconut (*Cocos nucifera*) + intercrops (occupying 70 to 75% of the net area) in Kerala, India (inter crop yield)	*Manihot esculenta*	60 to 75% of open area yield	Nair (1983)	0
	Amorphophallus campanulatus			0
	Ipomoea batatas			0
	Dioscorea alata			0
	Dioscorea esculenta			0
	Coleus parviflorus			0
	Zingiber officinale			0
	Curcuma longa			0
Acacia tortilis–silvopastoral in Rajasthan, India (fodder yield)	*Cenchrus ciliaris* + trees 10 x 10 m spacing	5580	Shankarnarayan et al (1987)	+
	Cenchrus ciliaris + trees 5 x 10 m spacing	5290		+
	Cenchrus ciliaris alone	4600		
Sorghum–nitrogen fixing tree mixtures in semiarid central India (grain yield; tree age = 8 years)	Sole crop	1154	Suresh and Rao (1999)	–
	Faidherbia albida	1013*		–
	Acacia ferruginea	890*		–
	Albizia lebbeck	720*		–
Acacia nilotica + rice, Chhattisgarh, India (rice grain yield)	Year 1	2000	Viswanath et al (2000)	+
	Year 10	3400		
Subsistence farming systems in the mid-hills of Nepal (pooled rice, maize, wheat and millet grain yields)	uplands: Agroforestry[1]	5686[a]	Neupane and Thapa (2001)	+
	no Agroforestry	3036[b]		
	lowlands: Agroforestry[1]	6853[a]		+
	no Agroforestry	4002[b]		
Forage grasses in association with fast growing multipurpose trees in Kerala, India (Cumulative annual biomass yield for tree+grass combinations; dry weight comparisons at age year 6 yr)	*Pennisetum purpureum* (sole crop)[2]	1257[b]	Kumar et al (2001a)	
	Leucaena leucocephala + *P.purpureum*	540[a]		–
	Casuarina equisetifolia + *P.purpureum*	355[a]		–
	Ailanthus triphysa + *P. purpureum*	630[a]		–
	Acacia auriculiformis + *P. purpureum*	510[a]		–
	Panicum maximum (sole crop)	1020[a]		
	Leucaena leucocephala + *P. maximum*	583[a]		–
	Casuarina equisetifolia + *P. maximum*	1350[b]		+
	Ailanthus triphysa + *P. maximum*	973[a]		0
	Acacia auriculiformis + *P. maximum*	535[a]		–
	Brachiaria ruziziensis (sole crop)	830[b]		

Systems/Practices and Parameters Evaluated	System Description	Productivity (kg ha^{-1})	Source	Effects on Productivity
	Leucaena leucocephala + *B. ruziziensis*	480[a]		–
	Casuarina equisetifolia + *B. ruziziensis*	645[a]		–
	Ailanthus triphysa + *B. ruziziensis*	410[a]		–
	Acacia auriculiformis + *B. ruziziensis*	393[a]		–
	Zea mexicana (sole crop)	507[a]		
	Leucaena leucocephala + *Z. mexicana*	710[a]		+
	Casuarina equisetifolia + *Z. mexicana*	663[a]		+
	Ailanthus triphysa + *Z. mexicana*	417[a]		–
	Acacia auriculiformis + *Z. mexicana*	305[a]		–
Ailanthus triphysa + *Zingiber officinale* at different densities in Kerala, India (ginger rhizome dry weight at five years of tree age).	Sole ginger	3500	Kumar et al (2001b)	
	Ailanthus (3333 trees ha^{-1})	3700[a]		+
	Ailanthus (2500 trees ha^{-1})	5000[b]		+
	Ailanthus (1660 trees ha^{-1})	3600[a]		+
	Ailanthus (1111 trees ha^{-1})	4000[a]		+
Morus alba–Phaseolus mungo production in subtropical India (grain yield; 5 year old trees spaced at 5 x 8 m)	Open	435	Thakur and Singh (2002)	
	No crown removal	251*		–
	25% crown removal	299*		–
	50% crown removal	337*		–
	75% crown removal	354*		–
Alley cropping (upland rice with *Gliricidia sepium* and *Senna spectabilis*), northern Mindanao, Philippines[3] (upland rice grain yield)	Control (no inputs)	90 to 830	MacLean et al (2003)	
	Mulched (10 Mg ha^{-1} of *S. spectabilis* fresh materials)	310 to 1040		–
	Incorporation of 10 Mg ha^{-1} of *G. sepium* fresh materials	910 to 1510		–
	Incorporation of 10 Mg ha^{-1} *G. sepium* fresh materials + mulching (5 Mg ha^{-1} *G. sepium* green manure + 5 Mg ha^{-1} of *S. spectabilis* mulch)	1270 to 1480		+
	Farmer's practice + hedgerows	230 to 1150		+
Poplar (*Populus deltoides*)– soybean (*Glycine max*) agrisilviculture systems in Chattisgarh, India (grain yield)	Sole crop	1450	Mishra et al (2004)	
	Mixed with poplar at 4 x 5 m spacing (year 6)	970 to 1420*		–
Agrisilviculture involving rice (*Oryza sativa*) *Casuarina equisetifolia* and *Acacia auriculiformis* at different spacing in Konkan region, Maharashtra, India (grain yield)	Sole crop	4900	Thaware et al (2004)	
	C. equisetifolia (year 6) 5 x 2 m	3300*		–
	10 x 2 m	3500*		–
	15 x 2 m	4000*		–
	A. auriculiformis (year 6) 5 x 2 m	3600*		–
	10 x 2 m	3900*		–
	15 x 2 m	4200*		–

Systems/Practices and Parameters Evaluated	System Description		Productivity (kg ha^{-1})	Source	Effects on Productivity
Poplar–mungbean (*Vigna radiata*) agroforestry system in Uttaranchal, India (grain yield)	Sole crop		1054	Pandey and Tewari (2004)	
	In association with 6 year-old trees		864*		–
Kaempferia galanga in multistrata systems involving *Cocos nucifera, Vateria indica, Ailanthus triphysa* or *Grevillea robusta* in Kerala, India (dry weight of rhizomes)[4] when coconut palms were 17 years and dicot trees three–year–old)	No over canopy		1619[a]	Kumar et al (2005)	
	Single strata (coconut canopy; palms at 7.5 x 7.5 m spacing)		1696[a]		0
	Multistrata (coconut+one row of multi–purpose trees in the middle of two rows of coconut in both directions)		1477[a]		0
	Multistrata (coconut+two rows of multipurpose trees in the middle of two rows of coconut in one direction only)		1641[a]		0
Tea (*Camellia sinensis*) – hedgerow system in the sloping lands of Sri Lanka (made tea yield for 36 months)	Control		7404	De Costa and Surenthran (2005)	
	Calliandra calothyrsus	Mulched	5540*		–
		Unmulched	4949*		–
	Senna spectabilis	Mulched	5178*		–
		Unmulched	4681*		–
	Eupatorium inulifolium	Mulched	9092*		+
		Unmulched	7576		+
	Flemingia congesta	Mulched	5764*		–
		Unmulched	5113*		–
	Gliricidia sepium	Mulched	5290*		–
		Unmulched	4482*		–
	Tithonia diversifolia	Mulched	5096*		–
		Unmulched	4432*		–

* significant at 0.05 level compared to the control. Values with the same superscripts under a source category do not differ significantly.

[1]Agroforestry with exotic fodder and grass species such as *Leucaena leućocephala, Calliandra calothyrsus, Flemingia congesta, Morus alba, Guazuma ulmifolia, Pennisetum* spp. and *Stylosanthes guianensis.*

[2]Difference between tree grass combinations and year after planting were significant ($p<0.01$).

[3]The range of values represents grain yield at two sites over two consecutive years for which the treatment differences were statistically significant ($p<0.01$).

[4]Differences not statistically significant. '+' indicates strong positive effect, '0' means neutral effect and '–' signifies strong negative effect (comparison is made with respect to sole crops, farmer's practices, wherever relevant).

Another approach is the application of intensive integrated farming system (IIFS) which refers to the farms where in two or more components are integrated with the farm resources with an objective of optimum utilization of available resources to realize maximum profits and also to stabilize returns. It provides an opportunity to utilize the land, labour, water, manure and fertilizers more efficiently. Farming system approach examines the full range of farm activities closely related to one another by the common use of farmers land, labour capital and management factors (Sharma et al 1991). The farming components include crop, livestock, poultry, fish, trees, plantation crops, forestry sericulture, etc. A combination of one or more components with cropping, when carefully chosen, planned and executed, gives greater surpluses than single component especially for small and marginal farmers. Farm as a unit is to be considered and planned for effective integration of the components to be combined with crop production activity. The IIFS improves the space utilization and increase productivity per unit area. Moreover, IIFS provides diversified products, enhanced utilization of crop residues and livestock wastes and less reliance to outside inputs such as fertilizers, agrochemicals, feeds, energy, etc. Further, the system improves higher net returns to land and labour resources of the farming family (Table 2.4). Besides the various integration as mentioned above, the farmers can also increase their income through some secondary agricultural activities such as beekeeping, mushroom culture, vermicomposting, silk rearing, etc. Thus, integrated farming system might be the only potential sustainable alternative to agricultural development as it offers unique opportunities for food security as well as income generation at household level.

Table 2.4: IIFS Models and Productivity in Meghalaya, India

IIFS	Productivity/yr	Stocking Density of Animals/Birds and Fishes	Net Income (Rs/ha/yr)
Chicken-crop-fish-duck horticulture along with hedgerow on contour bunds.	Duck: 1.21 q Eggs: 3,410 Nos. Fish: 6.43 q/ha	Duck- 500 nos./ha Fish- 9,000 fingerlings/ha	65,306
Crop-fish-poultry multipurpose trees	Live chicken: 6.43q Fish: 4.25 q/ha	Broiler birds- 200/batch Fish- 9,000 fingerlings/ha	31,573
Crop- fish-goat-multipurpose trees	Meat: 116 kg Fish: 4.29 q/ha	Goat- 12 nos./ha Fish- 9,000 fingerlings/ha	33,735
Crop-fish-pig- vermicompost-bamboo-multipurpose trees hedgerow-broom	Pork: 0 .78 q/ pig	Pigs- 16 nos./ha Fish- 9,000 fingerlings/ha	34,276
Crop-fish-dairy-mushroom vermicompost - horticulture hedgerow	Milk: 1634 lit Fish: 16.22 q/ha	Cattle- 2+2 nos./ha Fish- 9,000 fingerlings/ha	1,21,634
Crop and fishery (without integration)	-	-	13,965

BREAKING POVERTY AND FOOD INSECURITY CYCLE

Agroforestry could contribute to livelihood improvement, where people have a long history and accumulated local knowledge. India is particularly notable for ethnoforestry practices and indigenous knowledge systems on tree-growing. In terms of household income, Central Indian upland rice fields provide an illuminating economics (Viswanath et al 2000). The farms often have an average of 20 *Acacia nilotica* trees per ha, of 1 to 12 years of age. Small farms have more tree density. At a ten-year rotation, these trees provide a variety of products, including fuelwood (30 kg/tree), brushwood for fencing (4 kg/tree), small timber for farm implements and furniture (0.2 m^3), and non-timber forest products such as gum and seeds. Thus, trees account for nearly 10 per cent of the annual farm income – distributed uniformly throughout the year than in rice monoculture – of smallholder farmers with less than 2 ha farm holding. A combination of *Acacia* and rice traditional agroforestry system has a benefit/cost (B/C) ratio of 1.47 and an internal rate of return (IRR) of 33 per cent at 12 per cent annual discount rate during a ten-year period.

In the northeast Indian State of Meghalaya, guava-based and in Assam, lemon-based agrihorticultural agroforestry systems (i.e. farming systems that combine domesticated fruit trees and forest trees) gave 2.96 and 1.98-fold higher net return respectively, in comparison to farmlands without trees. Average net monetary benefit to guava-based agroforestry systems was Rs.20,610/ha (US$ 448.00) and to Assam lemon-based agroforestry systems, Rs.13,787.60/ha (US$ 300.00). Such systems are most useful livelihood improvement strategies in the rainfed agriculture of Meghalaya (Bhatt and Misra 2003). Similarly, the net present value for the different agroforestry models on six-year rotation in Haryana varied from Rs. 26,626 to Rs. 72,705 ha/yr, whereas the B/C ratio and IRR varied from 2.35 to 3.73 and 94 to 389 per cent respectively. Thus, agroforestry has not only uplifted the socio-economic status of farmers, but also contributed towards the overall development of the region (Kumar et al 2004).

There are numerous non-timber forest products collected from wilderness for subsistence and cash income. Often, harvesting is unsustainable because of lack of knowledge about silviculture of trees and destructive exploitation strategies driven by market forces. Domestication of such species aimed at commercialization and production of valued products can reduce the pressure on natural ecosystems (Belcher et al 2005). Domestication of forest fruit trees and other species grown in agroforestry systems offers significant opportunity for livelihood improvement through nutritional and economic security of the poor in the tropics (Milne et al 2006). The wild edible plants form an important constituent of traditional diet in Sikkim Himalaya, where about 190 species are eaten and almost 47 species are traded in local market. Wild edible fruit species have high carbohydrate content (Sundriyal and Sundriyal 2001) ranging between 32 and 88 per cent. Such fruit trees can be taken up for domestication in agroecosystems on priority action. Trees in agroforestry systems can provide host to globally valued products and thus support livelihoods locally. A study of the 8 year old agroforestry intervention in Palamau District, Jharkhand found that the community depended solely on rainfed farming and animal husbandry definitely gains positively by agroforestry interventions (Minj and Quli 2001). Suitable community plantations of non-timber forest products in tribal areas such as Jharkhand can potentially serve the dual purpose of conserving useful species as well as livelihood improvement of local people (Quli 2001). Such programmes in tribal areas have enhanced likelihood of success as communities are dependent on the wild resources for livelihood. In Jharkhand, trees in agroecosystems are particularly valued as host to insects that yield marketable products such as silk (Singh et al 1994), lac products (Jaiswal et al 2002) and honey (Dwivedi 2001).

Woodcarving industry is emerging as an important source of income to local artisans worldwide (Cunningham et al 2005). Promotion of species used in woodcarving industry facilitates long-term locking-up of carbon in carved wood and supports local knowledge. It therefore, strengthens livelihoods. For example, Jodhpur, Rajasthan has emerged as a major centre of woodcarving, exporting woodcraft worth Rs.60 million annually, facilitated by traditional knowledge and skill, and growing tourism. Suitable agroforestry programmes may enhance the availability of wood in agroecosystems, thereby improving the ability of developing countries to participate in the growing global economy.

CAVEATS AND CLARIFICATIONS

Trees in a variety of ethnoforestry and agroforestry systems contribute to food security, rural income generation through diversity of products and services, and can enhance nutrient cycling, improve soil productivity, soil conservation and soil faunal activities. Nonetheless, trees in agroforestry systems can also cause competition with the associated food crops. Agroforestry may, thus, reduce the yield of the agricultural produce in farmlands. For instance, in Haryana, *Azadirachta indica* and *Prosopis cineraria* did not produce any significant difference in the wheat yield, while *Dalbergia sissoo* and *Acacia nilotica* gave a reduction in yield. *A. nilotica* had a more prominent effect with a reduction of 40 to 60 per cent wheat yield and *D. sissoo* reduced yield by 4 to 30 per cent, but the reduction effect was only upto a distance of 3 m (Puri et al 1995). Interestingly, species that did not negatively affect the yield are indigenous trees occurring in traditional agroforestry systems, and they are economically more useful for providing multiple benefits. Selection of such species to enrich agroforestry systems shall be useful for local and national food security. Not all species desirable for livelihood improvement can be grown without designing an optimum species combination. Many fruit-yielding species that are suitable to tolerate highly alkali soil (pH > 10) become susceptible to waterlogging. The desirability for agroforestry systems due to high potential for livelihood improvement requires special techniques for planting. For example, pomegranate (*Punica granatum*) trees are unable to tolerate water stagnation. To avoid mortality due to water stagnation during the monsoon, the raised and sunken bed technique may be necessary for agroforestry practices on highly alkali soil (Dagar et al 2001).

Designing a sustainable tree mixture for agroforestry systems is another challenge. In agroforestry, differences in functional group composition do have a larger effect on ecosystem processes than does functional group richness alone. Thus, much time and expense need to be invested in finding species or genetic varieties that combine in more diverse agroecosystems to improve total yield. For instance, a five-year field experiment of tree mixtures for agroforestry system in tropical alfisol of southern India involving mango (*Mangifera indica*), sapota (*Achras sapota*), eucalyptus (*Eucalyptus tereticornis*), casuarina (*Casuarina equisetifolia*) and leucaena (*Leucaena leucocephala*) found that growth of sapota can be enhanced by 17 per cent when grown in mixture with leucaena. But a reduction of 12 per cent in the growth of mango may occur when co-planted with casuarina or leucaena (Swaminathan 2001). Eucalyptus is incompatible with mango and sapota. Many species suffer from root competition and thus selection of tree species with either low root competitiveness or trees with complementary root interaction is of strategic importance in agroforestry systems (Kumar et al 1999).

CONCLUSION

Although agroforestry options for carbon sequestration are attractive, they present critical challenges for carbon and cost accounting due to dispersed nature of farmlands and dependence of people on the multiple benefits from agroforestry. Additionally, important concerns regarding monitoring, verification, leakage and the establishment of credible baselines also need to be addressed. Another challenge is the incentives that promote tree growing by rural people. Not everyone is willing to adopt agroforestry. We shall need effective communication strategy to extend innovations among people to adopt and maintain agroforestry to supply fuelwood and other products. The likelihood of adoption depends on the availability of lands, progressive attitude of farmers, supportive village institutions, their wealth status and their perceived risk concerning agricultural production (Mahapatra and Mitchell 1999).

In order to use agroforestry systems as an important option for livelihood improvement, climate change mitigation and sustainable development in India, research, policy and practice will have to progress towards: (i) effective communication with people in order to enhance agroforestry practices with primacy to multifunctional values; (ii) maintenance of the traditional agroforestry systems and strategic creation of new systems; (iii) enhancing the size and diversity of agroforestry systems by selectively growing trees more useful for livelihood improvement; (iv) designing context-specific silvicultural and farming systems to optimize food production, carbon sequestration, biodiversity conservation; (v) maintaining a continuous cycle of regeneration-harvest-regeneration as well as locking the wood in non-emitting uses such as wood carving and durable furniture; (vi) participatory domestication of useful fruit tree species currently growing in wilderness to provide more options for livelihood improvement; and (vii) strengthening the markets for non-wood forest products. Prevalence of a variety of traditional agroforestry systems in India offers opportunity worth reconsidering for carbon sequestration, livelihood improvement, biodiversity conservation, soil fertility enhancement and poverty reduction.

■■■■

3 AGROFORESTRY AND ECOSYSTEM SERVICES

INTRODUCTION

Rates of deforestation and land degradation, and losses of biodiversity and ecosystem services, continue to rise globally and the international community is faced with the challenge of finding land use interventions that can alleviate or reduce the impact of these environmental concerns. Cultivation of trees in the agricultural systems is one of the major primeval practices in the tropical regions of the world. It is characterized by an intensive integration of forest trees, agricultural and horticultural crops, and shrubs with a basic objective to ensure sustained availability of multiple products as direct benefits such as food, vegetables, fruits, fodder, fuel, manure, medicine and raw materials for agricultural implements. Other indirect benefits and services such as ornamentals, shading, live-fencing and shelterbelt or windbreaks are also derived. Such intermixing of species of agriculture and forestry, often termed as 'agroforestry' came into international prominence as a potential source of solutions of many interrelated problems of production and conservation disturbing land use systems in the tropical and subtropical regions.

Researchers all over the world realized that this age old conventional practice of integrating trees on farm land has the potential for providing rural livelihoods and habitats for species outside formal forest lands, connecting nature reserves, and easing resource use pressure on conservation areas. In the recent few decades, there has been a growing interest in agroforestry because of its deliverable goods and services. A recent global assessment of tree cover found that 43 per cent of the world's agricultural land had at least 10 per cent of tree cover which indicates that trees are increasingly being planted as part of farming systems throughout the world (Zomer et al 2009). The integration of trees, agricultural crops, and/or animals into an agroforestry system has the potential to enhance soil fertility, reduce erosion, improve water quality, enhance biodiversity, increase aesthetics and sequester carbon (Garrett and McGraw 2000; Garrity 2004; Williams-Guille et al 2008; Nair et al 2009). It has been well recognized that these services and benefits provided by agroforestry practices occur over a range of spatial and temporal scales (Jose 2009). Many of these ecosystem services originated at farm level are being enjoyed by the society at local or regional or total landscapes.

Important realized ecosystem services of agroforestry are well documented and the major ecosystem services delivered by agroforestry systems are categorized into:

- Agroforestry for Microclimate Modification
- Agroforestry for Soil Enrichment
- Agroforestry for Soil Stabilization and Erosion Control

- Agroforestry for Flood Mitigation
- Agroforestry for Water Yield and Quality
- Agroforestry and Pest Control
- Agroforestry for Pollination Improvement
- Agroforestry for Carbon Sequestration and Air Quality
- Agroforestry for Climate Change Mitigation
- Agroforestry for Biodiversity Conservation
- Agroforestry for Aesthetics and Cultural Services

AGROFORESTRY FOR MICROCLIMATE MODIFICATION

Simultaneous growing of trees on farm land along with agricultural crops has the ability to modify the microclimatic conditions including temperature, water vapour content and wind speed, improvement of crop-water efficiency and energy balance which can have beneficial effects on crop growth and animal welfare (Brenner 1995; Jose et al 2004). Microclimatic modifications in agroforestry is manifested through soil moisture and soil temperature relations resulting primarily from the use of trees for shade, or as live supports and live fences, or windbreaks and shelterbelts. The shade of trees on farm land in tropics causes a net effect of complex interactions which extend far beyond the simple reduction of heat and light (Willey 1975). In crop production system, temperature, humidity and movement of air, as well as temperature and moisture of the soil, directly affect photosynthesis, transpiration and the energy balance of associated crops (Rosenberg et al 1983) and the net effect of which may translate into increased yields. Presence of trees on farm land would have blanket effect on soil surface. In general, shading causes a reduction of temperature and temperature fluctuations as well as the vapour pressure deficit under tropical conditions.

Barradas and Fanjul (1986) compared shaded and open grown coffee plantations in Mexico and found that, in a coffee plantation under the shade of *Inga* spp. (205 trees/ha; average tree height 14 m), the average maximum temperature was 5.4°C lower and the minimum temperature 1.5°C higher, and that both vapour pressure deficit and evaporation were substantially reduced as compared to open-grown coffee. The minor temperature fluctuations under shade could be ascribed to reduced radiation load on the coffee plants during the day and reduced heat loss during the night. The lower vapour pressure deficit was probably caused by a higher water input through the trees' transpiration stream in combination with the lower temperatures. Similar results, indicating a buffering effect of the trees on the microclimate beneath them, were also reported for a combination of coconut and cacao in India (Nair and Balakrishnan 1977) and for an alley cropping system of millet and *Leucaena* in India (Corlett et al 1989). A reduction of vapour pressure deficit will cause a consequent reduction in transpiration and, hence, less probability of water stress for the under storey or shaded crop (Rosenberg et al 1983). This is highly useful during short periods of drought and would result least reduction in yield due to water stress which was found in the study of Neumann and Pietrowicz (1989) wherein the bean plants associated with *Grevillea robusta* trees in Rwanda showed no signs of wilting in hot afternoons, whereas those grown on a field without trees did.

In wheat based farming system of Punjab (India), yield reduction due to higher temperatures during reproductive stage is often mitigated through introduction of trees on farm land. Tree row orientation and distance influence the growth behaviour of the crop but the effect of sun angle (which changes with season) can modify this influence. Lower heat load during grain filling and subsequent increased duration of grain filling can mitigate the effect of quantitative reductions in radiant energy and of poor quality light during the initial stages of crop growth (Kohli and Saini 2003). In this context, the use of deciduous trees such as *Dalbergia sissoo*, *Trewia nudiflora*, *Populus deltoides* and *Salix alba* with leaf fall coinciding with the vegetative stage of the crop could be a good proposition for agroforestry. Tree canopy pruning as a management tool can also be advocated to provide a favourable microclimate to the intercropped wheat.

Reduction of wind on farm land has significant effect on farm production. Wind speed reductions can extend to 30 times the height of tree belts on the leeward side (Tamang et al 2010). The resultant reduction in wind erosion effects can have multiple benefits for crops including increased growth rate and quality, protection from windblown soil, moisture management and soil protection. Furthermore, higher air and soil temperatures in the leeward of a shelterbelt can extend the growing season, with earlier germination and improved growth at the start of the season (Brandle et al 2004). In hot and dry environments, the primary effect of windbreaks may be on modifying crop water use efficiency by reducing the turbulent transfer of heat and water vapour and therefore reducing evapotranspiration. The primary effect of trees on crop energy balance is through the interception of radiation and the reduction of wind speed. The specific elements that may be affected by modifying the crop energy balance are: radiation and photosynthesis and its rate and duration; temperature and plant development; transpiration and its consequences on soil water use and water use efficiency (Messing and Noureddine 1991).

Thus agroforestry, under any given set of environmental conditions, can improve the microclimatic conditions when proper selection of tree species, optimizing the tree spacing and proper root pruning as well as shoot pruning are made a part of the technology for crop growing in agroforestry.

AGROFORESTRY FOR SOIL ENRICHMENT

Land degradation and declining soil fertility pose a major peril to agricultural production. Use of artificial or synthetic fertilizers to augment soil nutrient status could not succeed to afford satisfactory result because majority of the commercial fertilizers are expensive or unavailable in majority of poor developing countries in tropics. Incorporation of trees in the farm land can facilitate the sustenance of the nutrient pool and enhance soil fertility both under sequential and simultaneous agroforestry (Young 1997). Intermixing of trees that are able to fix nitrogen biologically along with crops is fairly common in tropical agroforestry systems. Non N-fixing trees can also enhance soil physical, chemical and biological properties by adding significant amount of above and belowground organic matter and releasing and recycling nutrients in agroforestry systems (Jose 2009). Influence of trees in enhancing the soil fertility status has been described by several researchers (Young 1997; Nair and Latt 1997; Schoth and Sinclair 2003; Jose et al 2004). Agroforestry promotes closed system with internal recycling of nutrients, whereby nutrients are accessed from lower soil horizons by tree roots and returned to the soil through leaf fall, agroforestry systems enhance soil nutrient pools and turnover and reduce reliance on external inputs.

Trees on farm land play a crucial role in improving soil physical and biological properties (Rao et al 1998). One of the important roles of trees in agroforestry is the enrichment of organic matter in the system. Role of organic matter in enhancement of soil physical properties such as soil structure, aggregation, porosity, moisture retention, and erosion resistance is well documented. Apart from enrichment of soil nutrient status, trees on farm land can reclaim the degraded soils. Tejwani (1994) reported that salt affected farm lands could be reclaimed through agroforestry. Introduction of trees on farm lands significantly enhanced diversity of birds, insects, and earthworms; increased soil organic carbon content and N cycling; and improved soil health (Thevathasan and Gordon 2004).

Incorporation of litter and pruned materials of trees in the farm land improves soil fertility through release of nutrients and addition of organic materials to soil. The amount of nutrient released and organic matter added to soil is governed by quantity and quality of litter or pruned material added, climatic conditions, field conditions and soil type (Hegde 1995; Kumar 2008). Introduction of nitrogen fixing trees like *Dalbergia sissoo* in agroforestry significantly increased nutrient pool, organic biomass, and activities of enzymes-hydrogenases and alkaline phosphatise in the soil (Chander et al 1998). Seiter et al (1995) reported that in alley cropping system of maize and red alder (*Alnus rubra*), 32-58 per cent of total N in maize was obtained from N fixed by red alder and interestingly the nitrogen transfer increased with proximity between trees and crops. Trees can also significantly manipulate nutrient embellishments through intercepting rainfall, via throughfall and stemflow. Zhang (1999) reported that stemflow and throughfall together contributed 10.99 and 15.22 kg N ha^{-1} yr^{-1} in hybrid poplar and silver maple systems respectively.

Arbuscular mycorrhizal (AM) fungi enhance plant nutrient uptake and growth, soil stability and soil aggregation, litter decomposition rates, and could potentially enhance crop yields while reducing the need for chemical fertilizer input. It has been reported that agroforestry systems might enhance AM fungal richness compared to monocropped systems (Chifflot et al 2009). In nut shell, the mechanisms by which trees in agroforestry improve soil physiochemical and biological properties are as follows:

- Release of nutrients from tree litter and prunings
- Nitrogen input through biological nitrogen fixation (through N-fixing trees)
- Phosphorus input through mycorrhizal associations
- Reduced soil erosion and nutrient leaching
- Nutrient capture from the subsoil through deep-rooted trees
- Redistribution of nutrients through lateral roots of some trees

AGROFORESTRY FOR SOIL STABILIZATION AND EROSION CONTROL

Soil management is a key feature of agroforestry systems, and in both tropical and temperate climates, agroforestry systems are designed and implemented to counter soil erosion and degradation, and improve soil quality and health. Trees on farm land play a crucial role in soil stabilization and erosion control. The probable role of trees in soil conservation can be perceived in the following way

- Increase of soil cover, by litter and prunings
- Provision of partly permeable hedgerow barriers
- Lead to the progressive development of terraces, through soil accumulation upslope of hedgerows
- Increase soil resistance to erosion, by maintenance of organic matter
- Stabilization of earth structures by root systems
- Make productive use of the land occupied by conservation works

The replacement of natural forest and scrublands by croplands and grasslands devoid of trees on susceptible soils has resulted in increased run-off and accelerated erosion in many agricultural areas. Whereas most agroforestry systems are capable of maintaining soil organic matter at levels higher than under pure agriculture, and organic matter is the major variable factor controlling resistance to erosion. Presence of trees on farm land is useful in increasing structural stability of the soil, tree roots can enhance water infiltration and improve water storage by increasing the number of soil pores. Macropores rapidly channel surplus surface water flow and allow air and moisture to move into the soil. In this way the risk of soil erosion is reduced. Tree roots and trunks also act as physical barriers to reduce surface flow of water and sediment (Udawatta et al 2008). In New Zealand, widely-spaced poplars reduced pasture production losses due to landslides during a cyclonic storm by 13.8 per cent with, on average, each tree saving 8.4 m^2 from failure (Hawley and Dymond 1988). Mature willow and poplar trees at 12 m spacing can reduce mass movement by 10-20 per cent (Hicks 1995).

AGROFORESTRY FOR FLOOD MITIGATION

Changes in climate and weather patterns are having a dramatic impact on ecosystem-based livelihoods, especially in agriculture, a key source of income for majority of the countries in tropics. One factor contributing to household vulnerability is the use of monoculture production systems, which lead to reliance on one crop for income. Agricultural diversification, particularly when undertaken in relation to environmental or economic risks, has the potential to increase household adaptive capacity under climate change (Smit and Skinner 2002) and contribute to agrobiodiversity and the sustenance of vital functions, structures, and processes in agroecosystems (Mijatovic et al 2011).

The integration of trees into cultivated land, or agroforestry, is one approach to agricultural diversification that has been shown to provide a range of potential benefits. Agus et al (2003) found that tree-based farming systems maintain most of the flood mitigation capacity exerted by forest. Trees, undergrowth and woody debris on farm land would increase "hydraulic roughness" of the flood plain slowing down the passage of flood flow. On farm land, trees and woody debris could direct or concentrate flows forming multiple channels and back water floods enhancing flood storage that could result in net effect of delayed and reduced size of flood peak. Thus, risk of flash flooding following periods of heavy rainfall could be reduced in agroforestry systems, with the tree roots and trunks acting as permeable barriers to reduce sediment and debris loading into rivers following floods.

AGROFORESTRY FOR WATER YIELD AND QUALITY

Civilization on the earth began by felling trees, further development and advancement in civilization resulted in clearing natural forests to grow crops or to build infrastructure to human habitation. In the recent past, most of the developmental activities in tropical countries are being undertaken at the cost of natural forests. Clearance of forests for expansion of agriculture, development of infrastructure for industry and human settlements, and urbanization resulted in reduction of water that enters the soil and increase in the overland mud stream flow. Scarcity of land in certain regions lead to building on floodplains and wetlands which can reduce water storage and buffer capacity and put the new developments at risk of flooding. Additionally, human habitation and industry can lead to streams becoming polluted while increasing the demand for clean water. Thus, the need to increase water productivity and quality is a growing global concern as the demand for water will increase every year.

Many developing countries in tropics are experiencing unstable water supplies, wherein small decreases in rainfall induce much larger reductions in stream flow (UNEP 2006). Rapid degradation of water catchments and climate change are major factors responsible for decreasing water supplies in many parts of the world. Vital water catchments have been lost or are being degraded, creating concerns about the loss of hydrological functions and increasing competition for scarce water resources between agriculture, urban centres, industry and wildlife. Apart from these, the quality of the water is also a major concern in several developing countries including India. The challenge is to ensure the availability of adequate quality water at the landscape or river basin level, especially for the rapidly growing populations in many developing countries.

Trees on farm land increase the water-holding capacity of the soil, reduce soil evaporation, increase water infiltration into the soil (Nair 1993), and efficiently capture rainwater compared with traditional agricultural practices (Lott et al 2003). Jackson et al (2000) reported that the initial recharge of the soil profile following heavy rain was greater in agroforestry treatments than in sole maize or sole tree treatments at Machakos in Kenya, possibly because percolation of water down through the soil profile was facilitated by channels created by tree roots which had died and decomposed (Chirwa et al 2006). In Midwestern USA, a multispecies buffer that included woody perennials increased infiltration rates to five times that of cultivated and grazed fields (Seobi et al 2005). Agroforestry strips in Missouri, USA, reduced surface water runoff by 9 per cent after just two years of establishment, compared with a control watershed (Udawatta et al 2002). Agroforestry can reduce soil water content during critical times such as fallow periods and increase water infiltration and water storage. Furthermore on aboveground, stems, leaf litter and pruning debris in agroforestry systems can reduce runoff flow rates, thereby enhancing sedimentation within the agroforestry strip and increasing infiltration (Seobi et al 2005).

Trees on farm land are useful in reducing the water requirement of the production system. The shade provided by the trees requires the plants being grown beneath them to be watered less. Also, the presence of tree on farm land reduces loss of water through evapotranspiration from the farm land. Trees can increase the quantity of water used on-farm for tree or crop transpiration and may also improve the productivity of the water that is used by increasing the biomass of trees or crops produced per unit of water used. Recently, it has been proposed that trees can efficiently increase water productivity, particularly under semiarid regions (Ong and Swallow 2003; Ong et al 2007).

Trees also help to clean fine particulates from the water through their root systems and hold water for keeping the soil moist. Because root systems are helping to keep the soil in place and maintaining soil water levels, farming can be done in areas that would be at risk of erosion otherwise. Streams are then not as susceptible to being overrun with sediment or by flash flooding. Agroforestry can also help in improving water quality by reducing levels of pollution and soil erosion and thus landscape amelioration (Nair and Graetz 2004; Schultz et al 2004). For example, riparian buffer zones, if well designed and properly located, can be very helpful in this direction (Dosskey 2001). These buffers help in reducing the transport of polluted runoffs to the rivers and streams. Agroforestry also improves water-use efficiency and increases environmental sustainability.

Combination of trees with agricultural crops in agricultural production system can contribute considerable hydrological services wherein the hydrological function is between that of natural forests and agricultural croplands. It is clear that agroforestry can make solid contributions in maintenance of watershed functions and productive agriculture. However, it is important to remind that one agroforestry system cannot be sufficient unless it is large enough to benefit from its own water purification and related hydrological services. Most of the benefits of water quality are passed along downstream. It is important to strategically use agroforestry systems in agricultural areas that border water sources to maintain good water quality over all.

AGROFORESTRY AND PEST CONTROL

In simple word, agroforestry is a combination of trees and shrubs with crops and livestock in time and space on a land management unit to achieve optimum benefits from biological interactions between soils, plants, and associated animals including arthropods. Agroforestry systems create a complex landscape structure than monoculture that is important for biological pest control. In structurally complex landscapes, parasitism is higher and crop damage lower than in simple landscapes with a high percentage of agricultural use (Thies and Tscharntke 2000). In small-scale, subsistence agriculture in the tropics, traditional farming practices have evolved that provides a sustainable means of reducing the incidence and damage caused by pests, including nematodes. It is believed that agroforestry could reduce pest outbreak usually associated with monoculture. However, pest control depends on microclimatic conditions, nutrition, nature of agroforestry components, prevalence of natural enemies, etc. and synergetic effects of these factors. Complementary interactions between the various biodiversity components can also be of a multiple nature and some of these interactions can be used to induce positive and direct effects on the biological control of specific crop pests, soil fertility regeneration, and enhancement and soil conservation. The exploitation of these interactions is of greater significance in real situations which involves agroforestry design and management that requires an understanding of the numerous relationships between soils, microorganisms, plants, insect herbivores, and natural enemies.

The conscious incorporation of tree components with agronomic crops can result in insect management benefits because of the structural complexity and permanence of trees and to their modification of microclimates within the production area. Individual plants in annual cropping systems are usually highly synchronized in their phenology and short lived. In such systems, the lack of temporal continuity is a problem for natural enemies because prey availability is limited to short periods of time and other resources, such as pollen, nectar, and neutral insects, are not consistently available. The addition of trees of variable phenologies or diverse age structure through staggered planting can provide refuge and a more constant nutritional supply to natural enemies because resource availability through time is increased (Rao et al 2000). Akbulut et al (2003) found that useful arthropods reached notably higher numbers in maize, bean, and zucchini grown between alleys of hybrid poplar than in monocultures. Trees provided a more favourable habitat for useful insects, and therefore agroforestry contributed to increased arthropod diversity. Stamps and Linit (1997) opined that agroforestry could ensure increased insect diversity and reduced pest problems because the combination of trees and crops provides greater niche diversity and complexity in both time and space than the polyculture of annual crops.

Shade from trees may markedly reduce pest density in understory intercrops. Hedgerows or windbreaks of trees have a dramatic influence on microclimate; almost all microclimate variables (heat input, wind speed, soil desiccation, and temperature) are modified downwind of a hedgerow. Tall intercrops or thick groundcovers can also alter the reflectivity, temperature, and evapotranspiration of shaded plants or at the soil surface, which in turn could affect insects that colonize according to ''background'' colour or those that are adapted to specific microclimatological ranges (Cromartie 1991). Trees can also provide alternate hosts to natural enemies, as in the case of the planting of prune trees adjacent to grape vineyards to support overwintering populations of the parasitoid *Anagrus epos*, which later migrate into adjacent vineyards and regulate populations of the grape leafhopper (Murphy et al 1996). The biodiversity, inherent in multiple cropping and multiple cultivar traditional farming systems, increases the available resistance or tolerance to nematodes (Bridge 1996).

Thus, it can be concluded that well designed agroforestry practice can reduce the crop stress by micro-site enrichment, microclimatic modification and other services which in turn increases the crop tolerance to pests and diseases apart from providing food and habitat for birds, spiders, parasites, and other natural enemies of pests.

AGROFORESTRY FOR POLLINATION IMPROVEMENT

More than 100 crop plants or varieties require pollinators to be most productive. In the recent past, due to lack of nearby habitat for pollinators in and around the farm lead to decline in the population of pollinators and crop yield. Normally, agricultural crops are seasonal and insect pollinators and other pollinators which are useful in pollination activity might not find habitat after the harvest of the crop. In such situation, trees on farm land can improve and provide habitat for important pollinators. In agroforestry, trees may be present in hedges or wind breaks or as shade trees or in blocks increase the overall diversity of plants and physical structure in a landscape and, as a result, provide habitat for pollinators. It is prudent to mention that wide variation in flowering of trees and understory crops provide forage resources for pollinators such as honey bee. Honey bees are very active and effective pollinators for many kinds of crops.

In an cardamom and coffee based agroforestry practice in south India, canopy trees play an important role in maximizing cardamom and coffee production by providing the shade needed by cardamom and coffee plants, nesting sites for pollinating bees and floral resources for pollinators at times of the year when cardamom or coffee is not blooming thus creating sequential blooms in coffee and cardamom plantations (Mayfield and Belavadi 2008).

AGROFORESTRY FOR CARBON SEQUESTRATION AND AIR QUALITY

Agroforestry is one of the important land use systems which has both annuals and woody perennials. The role of land use systems, such as agroforestry as a climate change mitigation and adaptation strategy, has gained considerable importance lately following the realization of the ability of these systems to capture atmospheric carbon dioxide and store the carbon in plant parts and soil (Nair 2012). Agroforestry systems have attracted particular consideration in view of their perceived advantages of the large volume of aboveground biomass and deep root systems of trees. The potential of agroforestry seems to be substantial, but it has not been even adequately recognized and exploited. Proper design and management of agroforestry practices can make them effective carbon sinks. As in other land use systems, the extent of carbon (C) sequestered will depend mainly on the amounts of C in standing biomass and recalcitrant C remaining in the soil. The potential of agroforestry systems to sequester carbon varies depending upon the type of the system, species composition, age of component species, geographic location, environmental factors, and management practices (Jose 2009).

Nair et al (2009) reviewed carbon sequestration potential of different agroforestry land uses and reported that the carbon sequestration potential of the vegetation component (above and belowground) varied from 0.29 Mg ha^{-1} yr^{-1} in a fodder bank agroforestry system of West African Sahel to 15.21 Mg ha^{-1} yr^{-1} in mixed species stands of Puerto Rico. Soil carbon estimates ranged from 1.25 Mg ha^{-1} in a Canadian alley cropping system to 173 Mg ha^{-1} in an Atlantic Coast silvopastoral system in Costa Rica. They concluded that in general, agroforests on arid, semiarid, and degraded sites had a lower carbon sequestration potential than those on fertile humid sites. Similarly temperate agroforestry systems had relatively lower rates compared to tropical systems.

Another indirect avenue of C sequestration is through the use of agroforestry technologies for soil conservation, which could enhance C storage in trees and soils. Agroforestry systems with perennial crops may be important carbon sinks, while intensively managed agroforestry systems with annual crops are more similar to conventional agriculture. In order to exploit this vastly unrealized potential of C sequestration through agroforestry in both subsistence and commercial enterprises in the tropics and the temperate region, innovative policies, based on rigorous research results, hav[illegible] put in place (Montagnini and Nair 2004).

Some of the important agroforestry practices, such as windbreaks and shelterbelts, are [illegible] have numerous benefits such as protecting buildings, roads from drifting sand in [illegible] protecting livestock animals from cold wind, protecting crops from desiccating dry wi[illegible] shelter to birds and scavengers, removing atmospheric carbon dioxide and prod[illegible] reducing wind velocity and particulate matters in the air, mitigating odour from the ne[illegible]

It is said that majority of odour causing agents are in the form of aerosols and vegetative barriers in form of trees either in windbreak or shelterbelt can filter airstreams or remove dust, gas and microbial contaminants (Tyndall and Colletti 2007).

AGROFORESTRY FOR CLIMATE CHANGE MITIGATION

It is explicit that the earth's climate is changing at an unprecedented rate which is the result of increased emission of green house gases to the atmosphere both directly and indirectly by the anthropogenic activity. Climate change will have consequences for farmers all over the developing world. These effects will play out over long periods of time, while farmers and policy makers tend to have relatively short planning horizons. The effects of climate change are geographically inequitable, varied and unpredictable with potentially devastative for the existence of life on earth chiefly human survival. Plants are of particular importance as they are major regulators of global climate and are the keystone of the carbon cycle. The uptake of carbon dioxide (CO_2), one of the principle greenhouse gases, during photosynthesis is the major pathway by which carbon is removed from the atmosphere and made available to animals and humans for growth and development. Agroforestry which combines woody perennials along with agricultural crops has potential role in climate change mitigation.

Agroforestry offers a set of innovative practices that are designed to enhance productivity that often could contribute to climate change mitigation through enhanced carbon sequestration, avoided emission from degradation and deforestation, reduce non CO_2 emission in the agricultural sector, biomass production to substitute emission from fossil fuels, manage vulnerable carbon stock under climate change, manage biomass products and biogenic waste streams, etc. An extensive array of studies (Albrecht and Kandji 2003; Nair et al 2009; Jose 2009; Nair 2012) have substantiated the fact that agroforestry systems, even if not primarily designed for carbon sequestration, present a unique opportunity to increase carbon stocks in the terrestrial biosphere. The quantitative importance of agroforestry as carbon sink derives from wide applicability in existing agricultural systems. Worldwide it is estimated that 630×10^6 hectares are suitable for agroforestry. Role of agroforestry in carbon dioxide mitigation is well documented by the several researchers.

Agroforestry practices in the humid tropics are part of a continuum of landscapes ranging from primary forests and managed forests to row crops or grasslands. In agroforestry systems, there is a reduced need for supplementary nitrogen applications, and recycled nitrogen from leaf litter provides a quantifiable contribution to adjacent crops that can replace inorganic N additions and thus reduce N_2O emissions. A decrease in nitrogen leaching out of the rooting zone will reduce NO_x emissions as a result of denitrification in surface water resources. Models estimate that nitrates leaving a tree-based intercropping system can be reduced by 50 per cent compared to a monoculture control (Thevathasan and Gordon 2004).

Agroforestry offers a potential as biomass energy provider and has the potential to reduce reliance on fossil fuel consumption in a number of ways. Production of firewood from arable or grazing land or by product of timber production presents interesting opportunities in CO_2 mitigation through the substitution of fossil energy consumption by using wood as energy sources and the protection of existing forests and other natural landscapes. Biomass energy from agroforestry is a carbon-neutral source of energy that doesn't contribute to CO_2 enrichment of the atmosphere. Trees planted around homesteads can also contribute to energy savings in farm buildings and they can reduce the amount of energy needed to heat or cool a house by up to 30 per cent (DeWalle and Heisler 1988). Furthermore, internal cycling of nutrients, and enhanced pest and disease control, can reduce the need for oil-based agrochemicals and localised production of multiple outputs can avoid the need for long-distance transportation of goods and therefore reduce fuel use which indirectly contributes for climate change mitigation. However, adequate understanding of these secondary effects of agroforestry with regards to climate change mitigation would require more research.

AGROFORESTRY FOR BIODIVERSITY CONSERVATION

Agroforestry is increasingly being acknowledged as an integrated land use that can directly enhance biodiversity and contribute to the conservation of landscape biodiversity. Biodiversity plays a vital role in sustaining human life and the health of our planet. Biodiversity is defined as the totality of genetic, species and ecosystem diversity that constitutes life on Earth. Biodiversity which contributes immeasurably to the sustainable production of many goods and services is continued to decline at an alarming rate. Some of the important responsible factors for the current affairs of biodiversity are overexploitation of species, invasion by alien species, environmental pollution and contamination, global climate change, alteration of ecosystems, and degradation and loss of habitats (Rands et al 2010). Alarming rate of important ecosystems and species disappearance in sustaining human life and the health of our planet called for immediate action to conserve biodiversity worldwide. Several researchers documented and elucidated the mechanism of agroforestry in enhancing the biodiversity (Shastri et al 2002; Schroth et al 2004; McNeely 2004; McNeely and Schroth 2006; Harvey and Gonzalez-Villalobos 2007; Jose 2012).

Jose (2009) portrayed the five major roles of agroforestry in conserving biodiversity: (1) agroforestry provides habitat for species that can tolerate a certain level of disturbance; (2) agroforestry helps to preserve germplasm of sensitive species; (3) agroforestry helps to reduce the rates of conversion of natural habitat by providing a more productive, sustainable alternative to traditional agricultural systems that may involve clearing natural habitats; (4) agroforestry provides connectivity by creating corridors between habitat remnants which may support the integrity of these remnants and the conservation of area-sensitive floral and faunal species; and (5) agroforestry helps to conserve biological diversity by providing other ecosystem services such as erosion control and water recharge, thereby preventing the degradation and loss of surrounding habitat.

Agroforestry systems by their very nature are more diverse than monocultures of crops and livestock; this increase in 'planned' biodiversity, i.e. the components chosen by the farmer, increases the 'associated' biodiversity i.e. the wild plants and animals occurring on the farm land. Role of agroforestry in conserving tropical biodiversity has been documented by the various researchers of the world. These studies demonstrate that agroforestry systems support floral and faunal assemblages that can be as species-rich, abundant and diverse as forests, but often with modified species compositions that include non-forest species (Harvey et al 2007). Thus, in agroforestry integration of trees and livestock on farm land increases biodiversity, this varies with land management practices, cropping pattern and arrangement of different components. These components in agroforestry introduce another kind of biodiversity in soil flora and fauna, herbivores, carnivores, decomposers, etc.

AGROFORESTRY FOR AESTHETICS AND CULTURAL SERVICES

The visual impact of monocultures of crops or trees is unattractive for many people and in agroforestry integration of trees into agricultural landscapes can increase the diversity and attractiveness of the landscape (McAdam et al 2008). Traditional agroforestry systems such as traditional homegardens, tree based coffee farming, grazed orchards, parkland and wood pastures are valued for their visual appeal. However, establishing modern agroforestry systems which tend to be more artificial, geometric and rigid in appearance than traditional systems causes aesthetic changes at a landscape scale, and such changes must be carefully considered in the design and location of such systems (Bell 2000).

Agroforestry is the traditional practice of cultivating trees on farm land and it is said that in many countries in the world, trees are indispensible part of their culture. People believe that forests or trees are the links between the sky and earth often symbolise links between the spiritual world of ancestors and people. The variety of cultural values and symbolic functions ascribed to the trees or forests are as numerous and diverse as the communities and cultures of the region. Tangibly and intangibly, trees or forests feature in all aspects of culture: language, history, art, religion, medicine, politics, and even social structure itself. Trees may house the spirits of ancestors as well as those of the newborn. The tree features in many myths and tales. It consistently reflects a few important symbolic images. The tree stands between heaven and earth and is associated with creation as well as the underworld ("cosmic" tree). The tree is a motherly symbol: a protector and provider who gives fruit, other foods and medicines, provides a reservoir for water, protects against the elements and evil spirits. The tree often symbolizes human productiveness. It may also be a phallic and fatherly symbol, symbolically linking people with their ancestors. Owing to the associated cultural values trees are being introduced either on farm land or in homesteads in many tropical countries.

Thus, agroforestry can sustain human life through production of food, wood and in addition to environmental services such as climate change mitigation, enhanced water quality and quantity, and biodiversity conservation. Today, the provision of ecosystem services generally, and the subset of environmental services in particular, is being challenged as never before by the combined effects of expanding populations, rapid economic growth and greater global integration. A mechanism should be developed to generate growing interest in agroforestry through payment for such ecosystem services.

■■■■

4 AGROFORESTRY SYSTEMS CLASSIFICATION

INTRODUCTION

Agroforestry is a new name for old practices. The term now is usually used to include all land use systems in which woody perennials form a part of or are grown deliberately on the same piece of land as agricultural crops and/or animals either in the form of a spatial arrangements or in temporal sequence. Agroforestry systems refers to such distinct agroforestry practices in which agriculture, forestry and pastures uses of land are combined either temporally or spatially. There may not be distinct pathways of inputs and outputs as required in typical systems. However, the arrangements of different types of crops and the level of interaction between different components are distinct for individual agroforestry systems. Several agroforestry systems are old and some of the systems retain their form to considerable extent. These agroforestry systems vary depending upon the region, terrain, climate and soil.

In order to understand and evaluate the existing agroforestry systems and to develop action plans for their improvement, it is necessary to classify them according to some common criteria. The most organized effort to understand the systems has been a global inventory of agroforestry systems and practices in developing countries undertaken by ICRAF between 1982 and 1987. This activity involved systematically collecting, collating and evaluating data pertaining to large number of land use systems around the world (Nair 1987).

The primary purpose of a general classification of agroforestry is to identify the different types of agroforestry that occur and to group those that are similar, thereby facilitating communication and the organized storage of information about them. While different classifications may be suitable for different purposes, it is useful to have a generally accepted way of classifying and describing the major types that occur frequently to be able to communicate without having to perpetually clarify definitions (Sinclair 1999).

The classification scheme predominantly used in agroforestry since the mid 1980s, based on an inventory of agroforestry systems in developing countries (Nair 1987 and 1989a), has remained essentially unchanged over the past decades (Nair 1985, 1987, 1989b, 1990 and 1993) despite major shifts in how agroforestry is perceived both as a development imperative in temperate and tropical regions and as an area of scientific enquiry (Sanchez 1995; Leakey 1996; Gordon et al 1997).

CLASSIFICATION OF AGROFORESTRY SYSTEMS

The most obvious and easy-to-use criteria for classifying agroforestry systems are the spatial and temporal arrangement of components, importance and role of components, production aims or outputs from the system and the social and economic features. They correspond to the system's structure, function (output), socioeconomic nature, or ecological (environmental) spread. These characteristics also represent the main purpose of a classification scheme. Objectives for classification of agroforestry systems should:

- Include a logical grouping according to the major factors on which production of the system will depend
- Indicate how the system is managed
- Offer flexibility in regrouping the information
- Should be easily understood

Agroforestry systems can be generally categorized according to the following sets of criteria (Tejwani 1994):

- *Structural basis*: It refers to the composition of the components, including spatial arrangement of the woody component, vertical stratification of all the components and temporal arrangement of the different components.
- *Functional basis*: It refers to the major function or role of the system, usually delivered by the woody components (these can be of a productive, service or protective nature, e.g., soil conservation, windbreak, shelterbelt, etc.).
- *Ecological basis*: It refers to the environmental condition and ecological suitability of systems based on the assumption that certain types of systems can be more appropriate for certain ecological conditions i.e., there can be separate sets of agroforestry systems for arid and semiarid lands, tropical highlands, lowland humid tropics, etc.
- *Socioeconomic basis*: It refers to the level of inputs of management (low input, high input) or intensity of management and commercial goals (subsistence, commercial and intermediate).
- *Physiognomic basis*: If refers to the general characters and appearance of vegetation.

Several agroforestry systems have been evolved for different areas and one system differs from the other in respect of structure, function, arrangement of components and composition. Nair (1985) classified the agroforestry systems based on nature of components, structure, function, socio-economic and ecological factors, etc.

STRUCTURAL CLASSIFICATION OF AGROFORESTRY SYSTEMS

The structure of the system can be defined in terms of its components and the expected role or function. However, it is important to consider the arrangement of components in addition to their type. In this structural classification, the type of component and their arrangement both are important consideration. Hence, on the basis of structure, Agroforestry systems can be grouped into two categories: classification based on nature of components; classification based on arrangement of components. The most common and generally accepted agroforestry systems classification based on nature of components (Nair 1985 and 1993) is presented below.

CLASSIFICATION BASED ON NATURE OF COMPONENTS

Based on the nature of components, AF systems can be classified into the following categories:

(1) Agrisilvicultural systems

(2) Silvopastoral systems

(3) Agrosilvopastoral systems

(4) Other systems

AGRISILVICULTURAL SYSTEMS

It refers to combined production of agricultural crops and trees including shrubs/vines and forest crops. This system involves the conscious and deliberate use of land for the concurrent production of agricultural and forest crops both. Based on the nature of the components this system can be categorized into various forms:

- Improved fallow species in shifting cultivation
- Taungya system
- Alley cropping (Hedgerow intercropping)
- Shelterbelts and Windbreaks
- Soil conservation hedges
- Multispecies/Multilayer tree gardens
- Multipurpose trees and shrubs on farmlands
- Plantation crop combinations
- Agroforestry fuelwood production

SILVOPASTORAL SYSTEMS

It refers to combined production of trees and fodder crops/grasses and/or animals. This system deals with combined cultivation of timber yielding tree and fodder yielding tree plus grasses and /or animals. This system is further classified into three groups:

- Protein bank
- Livefence of fodder trees and hedges
- Trees and shrubs in pasture

AGROSILVOPASTORAL SYSTEMS

It refers to combined production of trees, agricultural crops and pasture and/or animals for production of food, fodder, timber and also grasses/grazing. Under these systems, production of woody perennials is combined with annuals and pastures. This system is grouped into two categories:

- Homegardens
- Woody hedgerows for browse, mulch, green manure and soil conservation

OTHER SYSTEMS

The following systems can be included in this category:

- Apiculture with trees
- Aquaforestry
- Sericulture with trees
- Mixed woodlots

AGRISILVICULTURAL SYSTEMS

If refers to the use of land for the concurrent production of agricultural crops and forest crops. It covers all systems in which land is used to produce both forest trees and agricultural crops either simultaneously or alternately on sustainable basis. The system is used for intercropping of forest plantation with agricultural crops in the initial years until the canopy of the forest trees closed. This system can solve the problem of shortage of food, fuelwood, timber; conserve soil and water; and ameliorate the harsh climatic condition.

Fig. 4.1: Wheat + Poplar Agrisilvicultural System in NRCAF, Jhansi

In agrisilvicultural system, agricultural crops are intercropped with tree crops in the interspace between the trees. Under this system, agricultural crops can be grown up to two years under protective irrigated condition and up to four years under rainfed farming. The crops can be grown profitably upto the above said period beyond which it is uneconomical to grow grain crops. However fodder crops, shade loving crops and shallow rooted crops can be grown economically. Wider spacing is adopted without sacrificing tree population for easy cultural operation and to get more sunlight to the intercrop. Performance of the tree crops is better in this system when compared to monoculture.

Fig. 4.2: Soybean + *Melia azedarach* Agrisilvicultural System in AICRP, UAS Dharwad

IMPROVED FALLOW SPECIES IN SHIFTING CULTIVATION

Shifting cultivation is a pattern of land use and a system of production of crops under which plots of land are cleared, cultivated for a short period for raising one, two or three crops, after which the land is allowed to rest longer than the period of cultivation. However, during the period of rest the land reverts to some modified form of its original cover. Fallows are cropland left without crops for periods ranging from one season to several years. The objective of improved fallow species in shifting cultivation is to recover depleted soil nutrients. Once the soil has recovered, crops are reintroduced for one or more seasons. This system is practised extensively in the north-eastern hill region comprising the states of Assam, Meghalaya, Manipur, Nagaland and Tripura and the two Union territories of Arunachal Pradesh and Mizoram and to some extent Andhra Pradesh, Bihar, Madhya Pradesh, Odisha and Karnataka states. It is called 'Jhum' in the north-eastern hill region and 'Podu' in AP and Odisha states and considered most destructive for forest areas.

The main feature of the improved fallow system of agroforestry is that trees and shrubs are not grown with crops on the same plot at the same time. The fallow periods vary from region to region but are presently becoming shorter due to an increasingly acute land shortage. The best species for the fallow system should induce good nitrogen fixation in the soil. While the main function of the fallow is to maintain or restore soil fertility and reduce erosion, some plants can be introduced primarily for their economic value. Species choice should not be exclusively confined to 'soil improvers', but plants with marketable products should also be considered. Plants included in improved fallows should be compatible with future crops, free of any negative physical or chemical effects on the soil and not in competition with the crops to be planted later on the same site. The exact techniques in this system vary with the previous land use, value of the fallow vegetation, condition of the land and expected duration of the fallow. Improved fallows can be established in a variety of ways and at various stages of the fallow. Methods might include:

- Direct seeding of clean tilled, harvested plots
- Selective cutting of bush, followed by enrichment planting with tall seedlings
- Introducing tall seedlings and cuttings into poor-quality fallows on degraded land
- Planting tree seedlings in closely spaced, deep planting holes or furrows within blocks of cleared cropland

TAUNGYA SYSTEM

The taungya (taung = hill, ya = cultivation) is a Burmese word coined in Burma in 1850s (Blanford 1958). The taungya system was introduced into India by Brandis in 1890 and the first taungya plantations were raised in 1896 in North Bengal. It is practised in the states of Kerala, West Bengal and Uttar Pradesh and to a lesser extent in Tamil Nadu, Andhra Pradesh, Odisha, Karnataka and the north-eastern hill region. In southern India, the system is called 'Kumri'. It is practised in areas with an assured annual rainfall of over 1,200-1,500 mm. This is a modified form of shifting cultivation in which the labour is permitted to raise crops in an area but only side by side with the forest species planted. This labour is responsible for the upkeep of a plantation. The practice consists of land preparation, tree planting, growing agricultural crops for 1-3 years, until shade becomes too dense, and then moving on to repeat the cycle in a different area. In some cases, crops may be grown one year before the trees are planted. A large variety of crops and trees are cultivated depending on the soil and climatic conditions:

- Trees: *Tectona grandis, Bamboo, Santalum album, Tamarindus indica, Acacia nilotica, Acacia mearnsii, Ceiba pentandra*, Cashew, Rubber, etc.

- Crops: Millet, pulses, groundnut, cotton, etc.

Three Types of Taungya systems are:

- *Departmental Taungya*: Under this, agricultural crops and plantation are raised by the forest department by employing a number of labourers on daily wages. The main aim of raising crops along with the plantation is to keep down weed growth.
- *Leased Taungya*: The plantation land is given on lease to the person who offers the highest money for raising crops for a specified number of years and ensures care of tree plantation.
- *Village Taungya*: This is the most successful of the three taungya systems. In this, crops are raised by the people who have settled down in a village inside the forest for this purpose. Usually each family has about 0.8 to 1.7 ha of land to raise trees and cultivate crops for 3 to 4 years.

Advantages offered by the taungya system are:

(i) Artificial regeneration of the forest is obtained cheaply
(ii) Problems of unemployment are solved
(iii) Helps towards maximum utilization of the site
(iv) Low cost method of forest plantation establishment
(v) In every case highly remunerative to the forest departments
(vi) Provision of food crops from forest land
(vii) Weed, climber growth etc. is eliminated

Disadvantages of the taungya system are:

(i) Loss of soil fertility and exposure of soil
(ii) Danger of epidemics
(iii) Legal problems created
(iv) Susceptibility of land to accelerated erosion increases
(v) It is a form of exploitation of human labour.

The taungya farmers are given the following concessions as a part and parcel of success of the system:

- Free grazing for animals
- Free timber for house construction and agricultural implements
- Schooling facilities for children
- Monitory loan at nominal interest
- Water supply through excavation of wells and construction of ponds

ALLEY CROPPING (HEDGEROW INTERCROPPING)

Alley cropping, also known as hedgerow intercropping, involves managing rows of closely planted (within row) woody plants with annual crops planted in alleys in between hedges. The woody plants are cut regularly and leaves and twigs are used as mulch on the cropped alleys in order to reduce evaporation from the soil surface, suppress weeds and/or add nutrients and organic matter to the top soil. Where nitrogen is required for crop production, nitrogen-fixing plants are the main components of the hedgerows. The primary purpose of alley cropping is to maintain or increase crop yields by improvement of the soil and microclimate, and weed control. Farmers may also obtain tree products from the hedgerows, including fuelwood, building poles, food, medicine and fodder. On sloping land, the hedgerows and pruning's may help to control erosion. Alley cropping usually works best in places where people feel a need to intensify crop production but face soil fertility problems. This situation is often characteristic of crowded, densely populated areas, but may also occur wherever some farmers wish or forced to increase production on a plot of limited size.

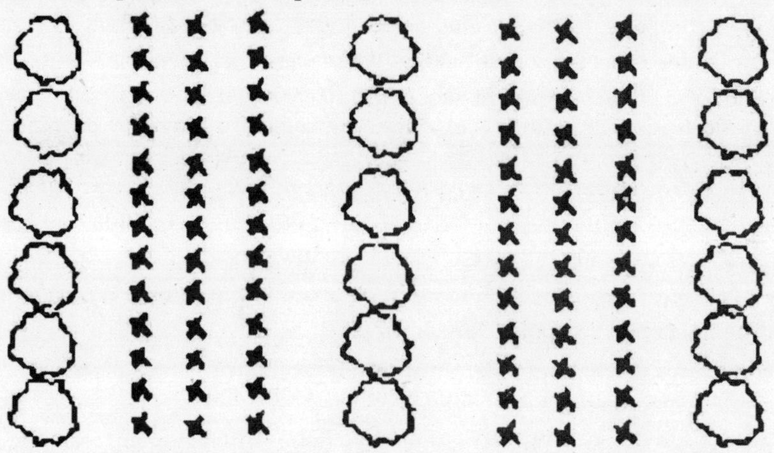

Fig. 4.3: Arrangement of Trees and Crops in Alley Cropping

Alley cropping with *Leucaena* is considered of prime importance in moisture conservation and efficient utilization of resources and Kanaujia et al (2002) found that moisture content was high in closer inter-row (4 m) and intra- row spacing (0.5 m) as compared to wider inter-row (6 m) and intra-row spacing (1 and 1.5 m). Woody plants are introduced as hedgerows in farm fields to maximize the positive and minimize the negative effects of trees on crop management and yields. Without doubt, trees compete with farm crops for soil nutrients, soil moisture and light. However, the right kind of trees at the right spacing, with proper management, may actually produce a net increase in yields from croplands. Trees may also provide new products such as fuelwood, fodder or food, in addition to the annual crops.

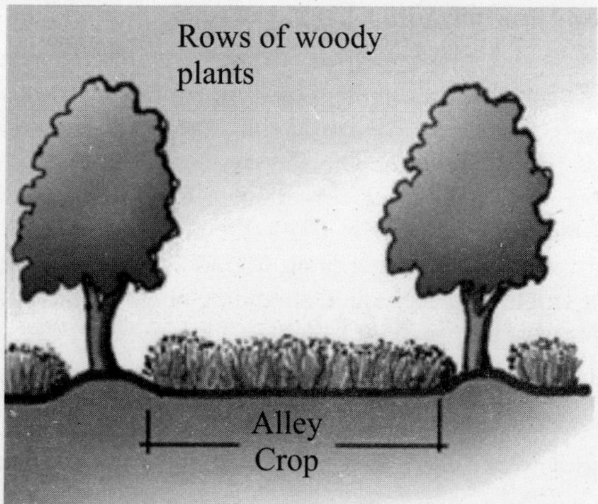

Fig. 4.4: Pattern of Alley Cropping System (Source: USDA-NAC)

The position and spacing of hedgerow and crop plants in an alley-cropping system depend on plant species, climate, slope, soil conditions and the space required for the movement of people and tillage equipment. Ideally, hedgerows should be positioned in an East-West direction so that plants on both sides receive full sunlight during the day. The spacing used in fields is usually 4 to 20 m between rows and 25 cm to 2 m between trees within rows. The closer spacing is generally used in humid areas and the wider spacing in sub-humid or semi-arid regions.

The position and spacing of hedgerows may also be affected by slope and the placement and design of soil and water conservation structures when these are combined with alley cropping. On sloping land, hedgerows should always be placed on the contour. If this means that they do not have the desirable east-west orientation, then they may need regular thinning to prevent excessive shading of adjacent crops. *Cassia siamea, Leucaena leucocephala*, *Gliricidia sepium*, *Callianda calothyrsus* and *Sesbania sesban* are commonly used tree species for alley cropping. Alley cropping usually includes leguminous trees to improve soil fertility through nitrogen fixation and hence an ideal alley cropping tree or shrub species should have following characteristics.

- It should have a sparse, small crown to permit sunlight penetration or should resprout rapidly after pruning, coppicing, pollarding or lopping.
- It should form a deep taproot system with few lateral root branches near the surface so as not to compete with crop roots.
- It should have shallow lateral roots that are easily 'pruned' by ploughing along the hedgerow, without serious damage to the plants.
- The leaf litter or some portion of it, should decompose at a rate that makes nutrients available when they are needed in the cropping cycle.
- Ideally, trees and shrubs used for alley cropping should fix nitrogen and should also produce wood, fuelwood, food, fodder, medicine or other products used by farmers or other local community.
- The species selected should grow well under the specific limitations of the site such as saline or acid soils, drought, flooding, heavy winds, insect pests or other hazards.

SHELTERBELTS AND WINDBREAKS

SHELTERBELTS: These are belts/blocks consisting of several rows of trees established at right angles to the prevailing wind. Shelter-belts have a typical triangular shape. This can usually be brought about by raising tall trees in the centre. A certain degree of penetration by winds is planned as by raising a solid wall, the protection decreases very fast on the leeward side. Shelterbelts up to 50 m width are considered ideal under Indian conditions. The ratio of height and width should be roughly 1:10. However, this figure may vary with local conditions. Orientation of shelterbelts depends on the direction and velocity of the prevailing winds. Shelterbelts may be raised in quadrangles if the wind direction tends to change very often. Length is an important consideration as far as shelterbelts are concerned. The minimum length of a shelterbelt should be about 25 times its height (Negi 1986). The purposes of planting shelterbelts are:

- To deflect air currents
- To reduce the velocity of prevailing winds
- To provide general protection to the leeward areas against the effects of wind erosion
- To protect the leeward areas from the desiccating effects of hot wind
- To provide food, fodder, timber etc.

The tree species recommended for shelterbelts are *Acacia arabica, Acacia leucopholea, Dalbergia sissoo, Lannea coromandelica, Eucalyptus* spp., *Tamarix articulata, Parkinsonia aculeata, Prosopis juliflora, Prosopis spicigera, Casuarina equisetifolia*, etc. The shrubs recommended include *Calotropis procera, Clerodendron phlomoides, Cassia auriculata, Dodonaea viscosa*, etc. The grass species include *Saccharum spontaneum, Saccharum munja, Panicum antidotale, Cenchrus* spp., etc.

WINDBREAKS: Windbreaks are strips of trees and/or shrubs planted to protect fields, homes, canals or other areas from wind and blowing soil or sand. When properly designed and maintained, windbreaks reduce the speed of the wind and thus its ability to carry and deposit soil and sand. They also improve growing conditions by decreasing water evaporation from soil and plants and can be used to reduce evaporation from water surfaces, such as irrigation ponds, canals or streams. In addition, windbreaks can provide wide range of useful products viz. poles, fuelwood, fruit, fodder, fibre, mulch, etc. The important reasons for which windbreaks are planted include:

- To protect livestock from cold winds
- To protect crops and pastures from hot, drying winds
- To reduce/prevent soil erosion
- To provide habitat for wildlife
- To reduce evaporation from farmlands
- To improve the microclimate for growing crops and to shelter people and livestock,
- To retard grass fire
- For fencing and boundary demarcation

From a side view, the windbreak should have the shape of an "A". Shorter bushes and grasses should be on the front and back, and taller trees should be in the idle. Windbreaks protect fields for a distance of 25 times the height of the tallest trees. Windbreaks must not be too dense and when you look through the windbreak, you should be able to see 60-80 per cent foliage. In general, trees with narrow, vertical growth are ideal for windbreaks to minimize the land removed from crop production. For best results, windbreaks should be planted at right angles to winds from which protection is needed. Windbreaks planted north-south are a good compromises as they provide protection from winds coming from the western quarter. They also give better shading of adjacent crops and pastures than windbreaks planted east-west.

The windbreak height determines the size of the sheltered area. The taller the windbreak, the greater the area it protects. On level ground, a well designed windbreak will reduce the speed of wind for about 25 times the tree height on downwind side. Maximum reduction of wind speed is in the area 5 to 15 times the tree height away from the windbreak. On the upwind side, some protection is gained up to a distance of 5 times the tree height away from the windbreak. Thus a wind break of 20 m tall will give some protection for 100 m on the upwind side to 500 m on the downwind side. Wind breaks are most effective when they stretch (length) without major gaps for distances exceeding 12 times the mature height of the trees. The wind velocity in the protected area may be reduced between 25 to 75 per cent of the wind speed.

Some fast growing species should be used to establish the desired effect as rapidly as possible. Some of the tree species used for windbreaks are *Eucalyptus*, *Cassia*, *Prosopis*, *Leucaena*, *Casuarina*, *Acacia*, *Grevillea*, *Syzygium*, *Dalbergia*, etc. A single row wind break should be used only where the land is so valuable that only a small amount of space can be spared for tree planting. If a single row windbreak is to be planted, tree species that retain their foliage and give a fairly dense growth should be selected. Windbreaks of three to five rows are more effective for most farm situations and are less affected by gaps caused by missing trees.

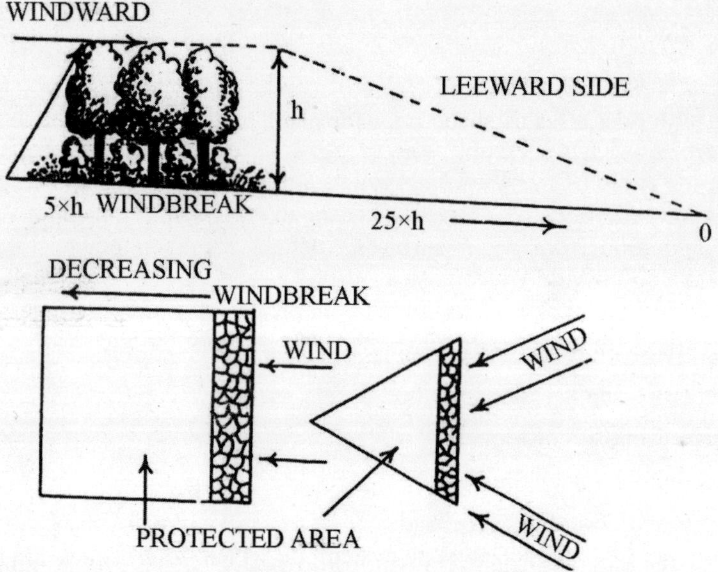

Fig. 4.5: Windbreak and Protected Area

SOIL CONSERVATION HEDGES

Trees can be planted for physical soil conservation works (grass strips, bunds, risers and terraces) wherein they play two roles, viz. stabilizing the structure and making productive use of the land they occupy. Stabilization is through the root system. In some of sloping landscapes of the country, the risers or terraces are densely planted with trees, with multiple use being made of them for fruit, fodder and fuelwood. In this system, the major groups of components are multipurpose trees and common agricultural species. The primary role of multipurpose tree species is soil conservation and provision of various tree products. The following tree species are used for soil conservation: *Grevillea robusta*, *Acacia catechu*, *Pinus roxburghii*, *Acacia modesta*, *Prosopis juliflora*, *Alnus nepalensis*, *Leucaena leucocephala*, etc.

MULTISPECIES/MULTILAYER TREE GARDENS

Multispecies, multilayer dense plant associations with no organized planting arrangements. Different woody components of varying form and growth habits form the overstorey. Herbaceous plants are usually absent in understorey or shade tolerant plants are sometimes present. This system is practised in areas with fertile soils, good availability of labour and high human population pressure.

MULTIPURPOSE TREES AND SHRUBS ON FARMLANDS

Trees are scattered haphazardly or according to some systematic patterns on bunds, terraces or plot/field boundaries. Multipurpose trees and other fruit trees are generally grown in farmlands along with common agricultural crops or in animal production fields. Trees provide fruit, timber, fuelwood, fodder etc. to the farmers' home consumption and for sale in local market. This agroforestry system is practised in all ecological regions especially in subsistence farming and also commonly integrated with animals.

PLANTATION CROP COMBINATIONS

This agroforestry system includes the integrated and intercropped multistory (mixed, dense) mixtures of plantation crops; mixtures of plantation crops in alternate or other regular arrangement; shade trees for plantation crops with scattered shade trees; and intercropping with agricultural crops. Woody plants consist of plantation crops like coffee, cacao, coconut, etc.; fruit trees in integrated multistory mixtures; fuelwood or fodder trees as shade trees for plantation crops. Herbaceous plants are usually present in 'intercropping with agricultural crops system' and shade tolerant herbaceous plants are present to some extent in 'integrated multistory mixtures'. Plantation crop combinations system is practised in humid lowlands or tropical humid/sub-humid highlands (depending on the plantation crops concerned) usually as smallholder subsistence system.

AGROFORESTRY FUELWOOD PRODUCTION

This agroforestry system comprises interplanting firewood species on or around agricultural lands/ animal production fields. In this system, firewood tree species form the woody component and agricultural crops of the locality form the herbaceous plant component. This system is practised in all ecological regions.

SILVOPASTORAL SYSTEMS

Silvopastoral system is a land management system in which forest trees are managed for the production of wood as well as for rearing domesticated animals. The land is managed for the concurrent production of forest crops and pasture grasses for grazing by domestic animals. Under Indian conditions, silvopastoral system includes management of fodder grasses in natural forests or in plantation or in grasslands with a view to obtaining the maximum yields of wood, fodder and other products. This system is classified into three categories: protein bank; livefence of fodder trees and hedges; trees and shrubs in pasture.

Fig. 4.6: Goat Browsing in Silvopastoral System

PROTEIN BANK

In this Silvopastoral system, various multipurpose trees (protein rich trees) are planted on or around farmlands and rangelands for cut and carry fodder production to meet the feed requirement of livesto-ck during the fodder deficit period in winter. Leguminous fodder trees or shrubs are established as small stands on certain portions of the farm or pasture area serving as a supplementary source of protein-rich fodder for livestock. They also serve as fence. They are regularly pruned and the top and branch prunings are then fed to animals. The example for fodder trees under this system includes *Acacia nilotica, Albizia lebbeck, Azadirachta indica, Leucaena leucocephala, Gliricidia sepium, Sesbania grandiflora, Sesbania sesban, Erythrina* sp., *Flemingia* sp., *Desmodium* sp., *Calliandra* sp. etc.

LIVEFENCE OF FODDER TREES AND HEDGES

In this system, various fodder trees and hedges are planted as livefence to protect the property from stray animals or other biotic influences. The Example for fodder trees for livefence includes *Gliricidia sepium, Sesbania grandiflora, Erythrina* spp., *Acacia* spp., *Euphorbia* spp. etc.

TREES AND SHRUBS ON PASTURE

In this system, various tree and shrub species are scattered irregularly or arranged according to some systemic pattern in pastures and rangelands of all ecological regions to supplement forage production. Example includes *Acacia nilotica, Acacia leucophloea, Tamarindus indica, Azadirachta indica*, etc.

AGROSILVOPASTORAL SYSTEMS

The production of woody perennials combined with annuals and pastures is referred as agrosilvopastoral systems. This system is grouped into: home gardens; woody hedgerows for browse, mulch, green manure and soil conservation.

HOMEGARDENS

This system is found extensively in high rainfall areas in tropical South and South East Asia. Home gardens represent land use systems involving deliberate management of multipurpose trees and shrubs in intimate association with annual and perennial agricultural crops and livestock within the compounds of individual houses. The whole tree-crop-animal units are being intensively managed by family labour (Fernandes and Nair 1986). Home gardens are highly productive, sustainable and very practicable. Food production is primary function of most home gardens.

Homegardens are characterized by high species diversity and usually 3-4 vertical canopy strata. The layered configuration and compatible species admixture are the most conspicuous characteristics of all home gardens. Generally all home gardens consist of an herbaceous layer near the ground, a tree layer at the upper levels and an intermediate layer. The lower layer can be partitioned into two, the lowermost being at less than 1 m in height, dominated by different vegetables and the second layer of 1-3 m height comprising food crops such as banana, papaya, etc. The upper tree layer can also be divided into two, consisting of emergent, full grown timber and fruit trees occupying the upper most layer of 25m height and medium size trees of 10-20m occupying the next lower layer. The intermediate layer of 5-10 m height is dominated by various fruit trees.

This practice finds expression in the states of Kerala and Tamil Nadu in India with humid tropical climates and where coconut is the main crop. Many species of trees, bushes, vegetables and other herbaceous plants are grown in dense and in random or spatial and temporal arrangements. Most home gardens also support a variety of animals. Fodder grass and legumes are also grown to meet the fodder requirement of cattle. In India, every homestead has around 0.20 to 0.50 ha land for personal production.

Woody species in the homegarden include coconut (*Cocos nucifera*), jackfruit (*Artocarpus heterophyllus*), mango (*Mangifera indica*), guava (*Psidium guajava*), cashew nut (*Anarcardium occidentale*), citrus (*Citrus* spp.), papaya (*Carica papaya*), passionfruit (*Passiflora edulis*), pineapple (*Ananas comosus*), custard apple (*Annona squamosa*), tamarind (*Tamarindus indicus*), breadfruit (*Artocarpus altilis*), cacao (*Theobroma cacao*), coffee (*Coffea arabica*, *C. robusta*), *Calliandra calothyrsus*, *Moringa oleifera*, *Flemingia* spp., *Gliricidia sepium*, subabul (*Leucaena leucocephala*), *Sesbania grandiflora*.

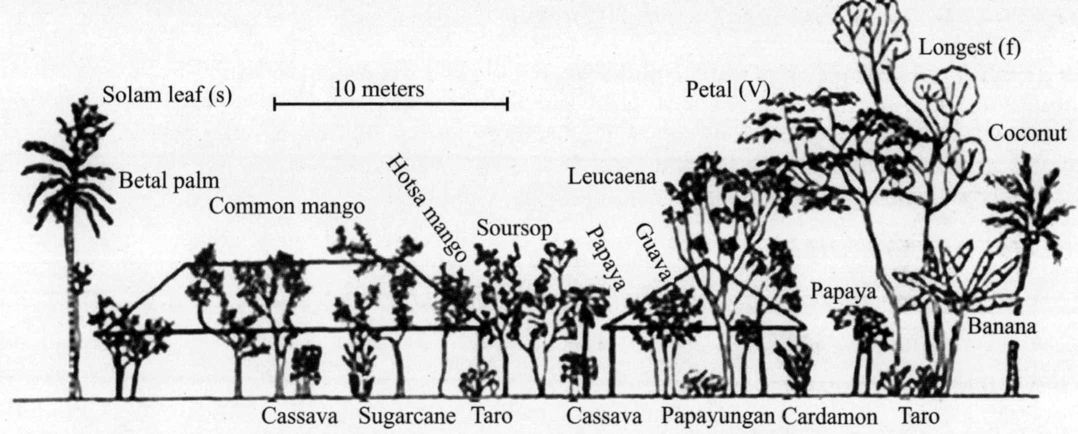

Fig. 4.7: Typical Homegarden in the Tropics

Herbaceous species are grain crops, oilseed crops, sweet potato, cassava or tapioca, taro or cocoyam, yam, potato, carrot, peanut, garlic, onion, turmeric, black pepper, chilli, cabbage, pumpkin, melon, okra, tomato, beans, pea, banana/plantain, etc.

WOODY HEDGEROWS

In this system, various woody hedges, especially fast growing and coppicing fodder shrubs and trees are planted for the purpose of browse, mulch, green manure, soil conservation, etc. The following species viz., *Erythrina* spp., *Leucaena leucocephala, Sesbania grandiflora* are generally used. They are very common in different forms especially in arid and semi-arid regions to provide browse or serve as mulch or litter to conserve the available moisture.

OTHER SYSTEMS

The following systems can be included in this category:

APICULTURE WITH TREES

In this system, various honey (nectar) producing tree species frequently visited by honeybees are planted on the boundary, mixed with an agricultural crop. The main purpose of this system is the production of honey. Species preferred are *Eucalyptus* spp., Litchi, *Mangifera indica, Madhuca indica,* etc.

AQUAFORESTRY

In this system, various trees and shrubs preferred by fish are planted on the boundary and around fish-ponds. Tree leaves are used as forage for fish. The main or primary role of this system is fish production and bund stabilization around fish-ponds. Species preferred are *Salix* spp., *Eucalyptus* spp., *Syzygium cuminii, Terminalia arjuna*, etc.

SERICULTURE WITH TREES

Cultivation of trees and shrubs valuable as a source of host and feed for silkworms.

MIXED WOODLOTS

In this system, special location-specific MPTs are grown mixed or separately planted for various purposes such as wood, fodder, soil protection, soil reclamation, etc.

CLASSIFICATION BASED ON ARRANGEMENT OF COMPONENTS

The arrangement of components i.e. trees, shrubs, crops and/or animals or multispecies combinations involve in the dimensions of space and time. Nair (1985) classified agroforestry systems also on the basis of arrangement of components in respect of space and time.

SPATIAL ARRANGEMENT

- Dense mixed stand
- Sparse mixed stand
- Strip/Zonal/Boundary

TEMPORAL ARRANGEMENT

- Coincident
- Concomitant
- Intermittent
- Interpolated
- Separate

STRATIFICATION OF COMPONENTS

- Single strata
- Double strata
- Multilayered

***SPATIAL ARRANGEMENT*:** Spatial arrangements of plants in an agroforestry mixture may result in dense mixed stands (as in homegardens) or in sparse mixed stands (as in most systems of trees in pastures). The species (or species mixtures) may be laid out in strips of varying widths either on edge (boundary) or inside the fields (zonal).

***TEMPORAL ARRANGEMENT*:** Temporal arrangements of plants in agroforestry may also take various forms. An extreme example is the conventional shifting cultivation cycles involving 2-4 years of cropping and more than 15 years of fallow cycle. Similarly, some silvopastoral systems may involve grass leys in rotation with some species of grass remaining on the land for several years. On the basis of the arrangement of components in respect of time (temporal), the agroforestry systems are classified into coincident, concomitant, overlapping (relay cropping), separate and interpolated.

- *Coincident*: when different crops occupy the land together as coffee under shade trees, and pasture under
- *Concomitant*: when components stay together for some part of life as in taungya plantation

- *Intermittent* (space dominated): when annual crops are grown with perennial crops e.g. paddy with coconut
- *Interpolated* (space and time dominated): when different components occupy space during different times e.g. home gardens
- *Separate*: when components occupy space during separate time e.g. shifting cultivation

STRATIFICATION OF COMPONENTS: On the basis of vertical stratification, agroforestry systems may be classified into:

- *Single layered* or *Single stratum*: The crops are usually in one layer or one storey e.g. MPTs woodlot, tree gardens
- *Double layered* or *Double stratum*: The crops are usually in two layers e.g. silvipasture, agrisilviculture
- *Multiple layered* or *Multistratum*: The crops form several layers e.g. homegardens

FUNCTIONAL CLASSIFICATION OF AGROFORESTRY SYSTEMS

Agroforestry systems yield various types of products and perform various functions. The production function and protection function are theoretically two fundamental attributes of all agroforestry systems. This implies that agroforestry systems have a productive function yielding one or more products that usually meet basic needs, as well as a service role. They yield food, fruits, vegetables, etc. and also moderate climate, improve soil productivity, and conserve soil and moisture. In view of these functions, AF systems can be grouped into three main classes as under.

PRODUCTIVE AGROFORESTRY SYSTEM

Productive functions of agroforestry systems are producing one or more products for food, fodder, manure, timber, fuelwood, shelter, etc. These systems are predominantly producing the goods required for meeting the basic needs of society. The objective of the land use is to maximize the production of annual and perennial crops. These systems are practised in most fertile lands e.g. homegardens, trees on fields boundaries, plantation crops, fuelwood and pulpwood plantations, etc.

PROTECTIVE AGROFORESTRY SYSTEM

Protective functions of agroforestry systems are protecting and maintaining production systems by the ecosystem services like climate amelioration, reducing wind and water erosion, improving soil fertility, providing shelter and shade (for crop, animal and man). However, the emphasis on production of outputs should not diminish the importance of sustainability. These systems aim at providing protection to land and bring gradual improvement in productivity e.g. windbreak, shelterbelt, livefence, soil conservation hedges, etc.

MULTIPURPOSE AGROFORESTRY SYSTEM

These Agroforestry systems aim at optimizing the both of the above functions. It should have productivity in sustainable manner.

CLASSIFICATION BASED ON ECOLOGICAL GROUPING

Agroforestry system characterizations pertain to specific ecological conditions of different geographical regions. The most famous and intensive effort in describing several existing agroforestry systems were made by Nair (1985). Based on the major agroecological zones, agroforestry systems are grouped into the following categories:

HUMID/ SUB-HUMID LOWLANDS

This region is characterized by hot humid climate for all or most of the year and evergreen or semi-evergreen vegetation. The lowland humid and sub-humid tropics (commonly referred as the humid tropics) are by far the most important ecological regions. Because of climatic conditions that favour rapid growth of a large number of plant species, various types of agroforestry plant associations can be found in areas with a high human population e.g. plantation of crops with combinations, multilayer tree gardens, trees on rangelands and pastures, improved fallow in shifting cultivation and multipurpose woodlots.

SEMI-ARID/ARID LANDS

This region is characterized by less rainfall confined to 9-21 days in July-September, 2-4 wet months, solar radiation incidence (400-500 cal/cm^2/day), high wind velocity (20 km/hour), high potential evapotranspiration (6 mm/day) and high mean aridity index (70-74.8 per cent). Various forms of silvopastoral systems, windbreaks and shelterbelts are typical examples of these ecological regions.

HIGHLANDS

Variable rainfall, degraded and shallow lands at high altitude to deep rich soils in valleys and great climatic variations are the features of highlands. This area is a storehouse of great biological diversity. The Himalayan region is an excellent example of this type of area. Agroforestry has long encompassed many well known land use systems practices. Typical example includes production systems involving plantation crops such as coffee, tea, use of woody perennials in soil conservation and improved fallow.

Table 4.1: Common Agroforestry Systems in the Tropics (Nair 1990)

Humid lowlands	Semi-arid lowlands	Highlands
Shifting cultivation	Silvopastoral systems	Soil conservation hedges
Taungya	Windbreaks/shelterbelts	Silvopastoral combinations
Homegardens	Multipurpose trees for fuel/fodder	Plantation crop systems
Plantation crop combinations	Multipurpose trees for farmlands	
Multilayer tree gardens		
Intercropping systems		

A system based on the nature of the components and their major functional characteristics for specific purpose appears more logical, simple and practical purpose oriented approach to classification of agroforestry systems. Again, the choice for a system may be based on many interacting considerations - social, ecological and economical. The right choice for the right situation is necessary.

CLASSIFICATION BASED ON SOCIO-ECONOMIC CONDITIONS

Socioeconomic criteria such as scale of production, level of technology, input and management have been used as a basis for classifying agroforestry systems. Based on such socioeconomic criteria as scale of production and level of technology input and management, agroforestry systems have been grouped into three categories: commercial, intermediate and subsistence.

COMMERCIAL AGROFORESTRY SYSTEM

The major aim of the system is production of the output (usually a single commodity) for sale. The typical example is commercial production of plantation crops such as rubber, oilpalm, and coconut with permanent underplanting of food crops, pastures, etc. Another example is commercial production of shade tolerating plantation crops such as coffee, tea and cacao under overstorey of shade trees.

INTERMEDIATE AGROFORESTRY SYSTEM

Intermediate agroforestry systems are those that are intermediate between commercial and subsistence scales of production and management. The example is production of perennial cash crops and subsistence food crops undertaken on medium to small sized farms wherein the cash crops fulfill the cash needs and the food crops meet the family's food needs.

Table 4.2: Classification of Agroforestry Systems and Practices (Nair 1985)

Categorization of Systems (Based on their structure and functions)			**Grouping of Systems (According to their spread and management)**	
Structure (nature and arrangement of components, especially woody ones)		**Function (role and/or output of components, especially woody ones)**	**Agro-ecological environmental adaptability**	**Socio-economic and Management level**
Nature of components	**Arrangement of components**			
Agrisilviculture (crops and trees incl. woody shrubs)	*In space (spatial)* *Mixed dense* (e.g., home garden)	*Productive function* Food Fodder	*Lowland humid tropics*	*Based on level of technology input* Low input (marginal)
Silvopastoral (pasture/animals and trees)	*Mixed sparse* (e.g. most systems of trees in pastures)	Fuelwood Other woods Other products	*Highland humid tropics* (above 1,200 msl, Malaysia)	Medium input
Agrosilvopastoral (crops, pasture/ animal and trees)	*Strip* (width of strip to be more than one tree)	*Protective function* Windbreak Shelterbelt Soil conservation	*Lowland subhumid* tropics (savanna zone of Africa, Cerrado of South America)	High input
Others (multipurpose tree lots, apiculture with trees, aquaculture with trees, etc.)	*Boundary* (trees on edges of plots/fields) *In time (temporal)* Coincident Concomitant Overlapping Sequential (separate) Interpolated	Moisture conservation Soil improvement Shade (for crop, animal and man)	*Highland subhumid* tropics (tropical highlands) (e.g. in Kenya, Ethiopia)	*Based on cost/benefit relations* Commercial Intermediate Subsistence

SUBSISTENCE AGROFORESTRY SYSTEM

Subsistence AF systems are those wherein the use of land is directed towards satisfying basic needs and is managed mostly by the owner and his family. Subsistence farmers produce most of what they consume, or consume most of what they produce.

PHYSIOGNOMIC CLASSIFICATION OF AGROFORESTRY

Agroforestry systems are classified based on the general appearance, character or aspect of vegetation without reference to its scientific characteristics. As the visible structure or external morphology of the plants is the basis of classification and starting point toward further knowledge. In physiognomic classification, the classes are explicitly analysed in terms of climate, soils, landforms and vegetation of particular ecosystem. The agroforestry systems are classified into three types based on the general characters and appearance of vegetation: xeromorphic AF system, mesomorphic AF system and hydromorphic AF system.

XEROMORPHIC AGROFORESTRY SYSTEM

Xerophytes or xerophytic plants have adapted to survive in an environment that lacks average or sufficient amounts of water, such as a desert and dry regions. Water is a limiting factor for crop and plant growth in this agroforestry. Example for this AF system includes dry farming and agroforestry in arid and semi-arid regions.

MESOMORPHIC AGROFORESTRY SYSTEM

Mesophytes are terrestrial plants which are adapted to neither a particularly dry nor particularly wet environment. Most of the agroforestry systems of tropical and temperate regions fall under this catergory.

HYDROMORPHIC AGROFORESTRY SYSTEM

Hydrophytes are plants that grow only in water or very moist soil. Water is available in sufficient quantity for crop and plant growth in agroforestry. Example includes wetland agroforestry, paddy cultivation with trees, fish culture and aqua-silviculture.

OTHER WAYS OF AGROFORESTRY SYSTEMS CLASSIFICATION

Agroforestry systems can be classified in many other ways using structural and functional considerations. Avila (1992) classified different land use systems in a most simple way: Homogenous Systems and Mixed Systems.

Table 4.3: Types of Land Use Systems

Type of System	Examples of Components
Homogenous Systems	
Monocropping systems	Maize, wheat, rice
Mono-animal systems	Cattle, sheep, poultry
Mono-tree systems	Timber plantations, woodlots
Mixed Systems	
Crop-crop	Maize/cassava, maize/beans
Animal-animal	Cattle/goat, cattle/poultry
Crop-animal	Maize/cattle, cereals/poultry/household waste
Crop-tree	Alley farming, mixed intercropping, boundary tree planting
Animal-tree	Alley grazing, fodder tree banks
Crop-animal-tree	Homegardens, alley farming with Livestock

Another classification from Young (1987) divides agroforestry systems into:

(1) Mainly Agrosilvicultural (i.e., trees with crops)

(2) Mainly or Partly Silvopastoral (i.e., trees with pasture and livestock)

(3) Tree-Component Predominant

(4) Other Components Present and Special Aspects

This scheme recognizes further subdivision according to structural or functional considerations. This particular classification is probably best suited for analysis of the potentials of agroforestry.

(1) Mainly Agrosilvicultural (Trees with Crops)

Rotational

- Planted tree fallow
- Taungya

Spatial Mixed

- Trees on cropland
- Plantation crop combination
  - o With upper-storey trees
  - o With lower-storey
  - o Tree/shrubs crops
  - o With herbaceous crops
- Tree gardens
  - o Multistorey tree gardens
  - o Home gardens

Spatial Zoned

- Alley farming
- Boundary planting
- Windbreaks and shelterbelts
- Biomass transfer
- Trees for soil conservation
  - o Barrier hedges
  - o On grass barrier steps
  - o On bunds, etc.
  - o On terraces

(2) Mainly or Partly Silvopastoral (Trees with Pastures and Livestock)

Spatial Mixed

- Trees on rangeland or pastures
- Plantation crops with pastures

Spatial Zoned

- Fodder banks
- Live fences

- o Mainly barrier function
- o Multipurpose

(3) Trees Component Predominant

- Woodlots with multipurpose management
- Reclamation forestry leading to production
  - o On eroded land
  - o On salinized land
  - o On moving sands

(4) Other Components Present and Special Aspects

- Apiculture with forestry
- Aquaforestry (trees with fisheries)
- Trees in water management
- Irrigated agroforestry

With a view to reviewing and synthesizing the state-of-the-art in agroforestry research and development Torquebiau (1989) adopted the following classification:

(1) Alley Farming (Hedgerow intercropping)

(2) Crops under tree cover

(3) Pastures and animals under tree cover

(4) Agroforests (live fencing, boundary planting, windbreaks, shelterbelts)

(5) Sequential technologies (shifting cultivation, taungya, improved fallow)

(6) Other technologies (aquaculture and apiculture with trees).

■■■■

5 MULTIPURPOSE TREES IN AGROFORESTRY

INTRODUCTION

Agroforestry, social forestry, community forestry, village forestry and farm forestry are all terms used to describe tree growing that is undertaken mainly outside the gazetted/notified forest areas. Agroforestry is a land use system in which trees are grown in association with agricultural crops, shrubs, pastures or livestock. This integration of trees and shrubs in the land use system can be either a spatial arrangement or in a time sequence. Sometimes the phrase 'woody perennials' is used instead of trees and shrubs. Of course, woody perennials include all trees and shrubs with a lifespan of more than a year, but they also cover bamboos and palms.

The trees in an agroforestry system are not necessarily planted. Instead natural regeneration of trees may be protected or mature trees may be deliberately left in the fields or pastures. Woody perennials are sometimes referred to as "multipurpose trees" (MPTs) or "multipurpose trees and shrubs" (MPTS). Almost all trees can be said to be MPTs, but the concept was introduced to distinguish the multiple roles often played by trees in an agroforestry system from the single purpose of wood production in pure forest plantations. Tree growing in such forest areas normally aims at meeting demands for wood for industrial purposes, and is often called industrial forestry. The growing of MPTs in agroforestry has the potential to alleviate many land use problems and uplift the people's standard of living.

Agroforestry research, however, now reached a stage to recommend suitable tree-crop combinations, their package of practices and genetically improved planting material for a given situation. In addition, continuous research advancements are being achieved by international and national agroforestry research organizations to benefit the farmers. Availability of MPTs information for many species has been an advantage for the adoption of agroforestry as a land use system. But information on certain MPTs for agroforestry is still limited by problems of lack of basic data on growth rates, breeding system, resilience to environmental stress and response to management. Multipurpose trees (MPTs) well suited to an agroecosystem and meeting the needs of local people and/or industrial needs of the area will encourage its adoption under agroforestry systems. Further, genetically improved high yielding clones within such tree species which may accomplish specific agroforestry system-based objectives will enhance the popularity of agroforestry among farmers.

WHAT IS A MULTIPURPOSE TREE?

One of the key components in any agroforestry system is the multipurpose trees (MPTs). MPTs are woody perennials which are deliberately kept or grown in a land use system to produce multipurpose products and benefits. Many authors have given many definitions of MPTs and MPTS. Burley and von Carlowitz (1984) reviewed many definitions and gave a synthesized definition as "multipurpose trees and shrubs (MPTS) are those which are deliberately grown or kept and managed for preferably more than one intended use, usually economically and/or ecologically motivated major products and/or services in any multipurpose land use systems, especially agroforestry systems". Forestry /Fuelwood Research Development Project (1992) defined precisely "multipurpose trees" (MPTs) to mean "tree species that are grown to provide more than one significant crops or functions on the farm. On small farms, this can often mean, for instance, that the farmer uses both wood and leaves from the same tree"

There are large numbers of MPTs belonging to a wide taxonomic range. Farmers have been growing trees for different purposes for thousands of years. Certainly, all trees provide shade and protection from soil erosion. In this sense, all trees can be said to have at least two purposes. More precisely, the term 'multipurpose tree species denote tree species that are grown to provide more than one significant crop of function or form. These may include fuelwood, timber, fibre, fodder, food, medicine and environmental services such as soil conservation and fertility improvement, climatic amelioration, flood control, carbon sequestration, etc.

Tree species can be multipurpose in two ways:

- **A single tree can yield more than one crop**. For example, farmers in Central America grow *Gliricidia sepium* as living fences that provide fuelwood, fodder and green manure for agricultural crops- all at the same time. Similarly farmers in some tropical regions grow jackfruit (*Artocarpus heterophyllus*) that provides food, fodder and timber.
- **Trees of the same species, when managed differently, can yield different crops**. In tropics, for example, *Leucaena leucocephala* is managed so that some trees will principally yield wood while others principally produce leaf fodder.

Farmers can grow multipurpose trees in various combinations with other crops, as in agroforestry, in block plantations of trees or in naturally regenerating tree farms. In certain settings, multipurpose trees are grown and managed for only one purpose. In the Solomon Islands, *Gliricidia sepium* is grown only to provide shade in coffee plantations. The same species is planted in Central America and the Philippines for multiple uses, yet the way it is managed results in a very different shape and use. Farmers facing changing conditions in their environment or market can also change the way they manage a tree. For example, market changes may persuade farmers, who previously grew trees for fodder and fuel, to cut their trees for sale as roundwood for construction material.

WHY GROW MULTIPURPOSE TREES?

REDUCING RISK OF TOTAL CROP FAILURE

Growing multipurpose trees can reduce the risk of total crop failure. For example, if farmers usually grow *Leucaena leucocephala* for animal fodder and the tree's leaves are destroyed by pests, they will still have wood that can be used for fuel, pulp, or light-weight construction material. Having a variety of plants on a farm as in agroforestry systems can also reduce the risk of total crop failure. Evidence shows that combining several types of plants provides a type of insurance. If the economic benefit of one crop is reduced by pest damage or market failure, the farmer can make up for it by harvesting another crop. Also, growing a variety of species makes the farm less vulnerable to any one pest.

INCOME GENERATION AND DISTRIBUTION

Agroforestry practices can increase farmers' annual income. Some increases in revenue come from harvesting different tree crops in different seasons. For example, in many Asian countries farmers use the leaves of *Artocarpus heterophyllus* (jackfruit) as fodder during the driest season. The same trees produce fruits providing income and food to the owner at other times. Ultimately it gives timber when tree is harvested. The result is that income and employment are distributed more evenly throughout the year (Alam 1993).

PRODUCTION SERVICES

Dozens of different products and benefits obtainable from woody perennials have been identified and listed by different scientists and practioners. In the early days of agroforestry when trees and shrubs were discovered, or rather rediscovered, as useful components of land use systems, it was believed that the longer the list of uses the more convincing would be the value and usefulness of trees and shrubs. In principle, this option still holds true but it requires some interpretation, qualification and systematization. Benefits obtainable from trees and shrubs can be divided into production and service uses. From systems point of view and its economic performance, a further distinction can be made between those services that directly promote agricultural production. MPTs on croplands/farmlands bring significant economic benefits to the farmer, the community, the region or the nation. Such benefits may include increment in maintenance of outputs of food, fuelwood, fodder, fertilizer, timber, medicines, etc.

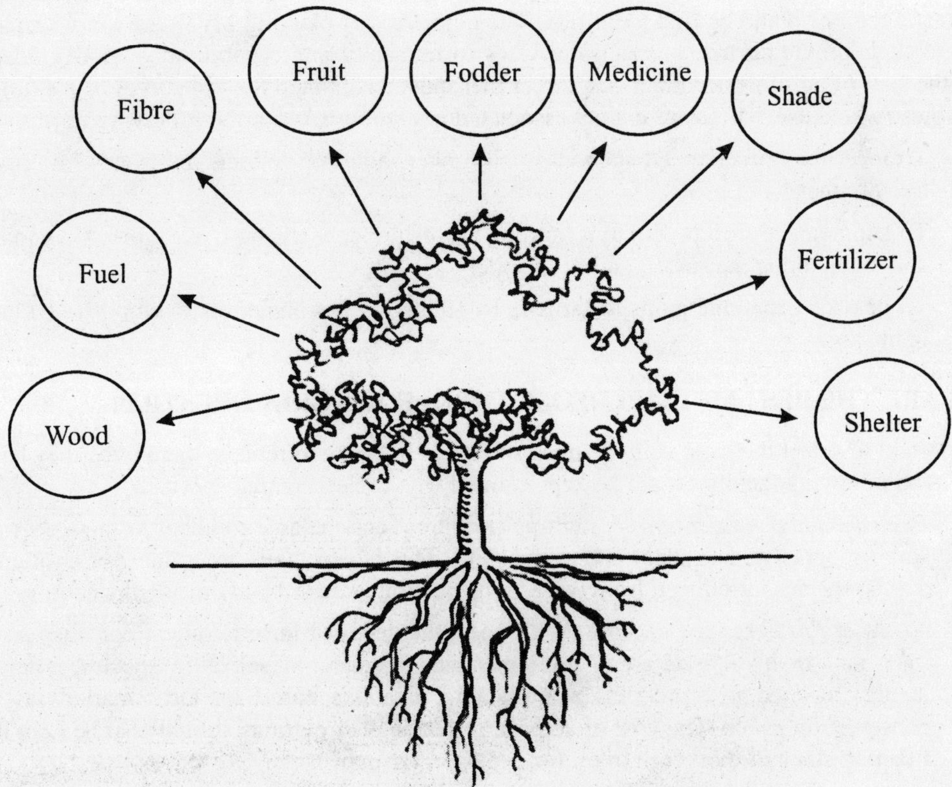

Fig. 5.1: Benefits of MPTs - '6Fs' - (Food, Fruit, Fodder, Fuel, Fibre and Fertilizer)

ECOSYSTEM SERVICES

Of course there are many other reasons for growing trees on farms. The integration of trees into an agroforestry system has the potential to improve ecosystem. Trees provide shade for a cooler soil temperature, reduce loss of soil moisture and also protect the soil from wind and water erosion. Other Examples for ecosystem services include carbon sequestration and climate change mitigation, nitrogen fixation and soil fertility improvement, soil stabilization and conservation, windbreaks, live fences, biodiversity improvement, increment of aesthetic value, flood control, clean water and air, pest control, etc.

OBJECTIVES OF PLANTING MULTIPURPOSE TREES

All trees are multipurpose. They bring subsoil nutrients to the surface, provide shade and slow erosion. Many trees provide fodder, living fenceposts, fruit and other edible parts, shade, insecticides, and wood; they all have some role in soil stabilization and offer quality-of-life benefits like beauty and a shelter for informal gatherings. Working with trees is an important investment which can be significant to the future of farming and rural community. Developing agroforestry systems, tree nurseries, and fruit and nut tree species is most appropriate for those with a long-term commitment in an area. Learning the valued qualities of the trees already present in and native to local area is very important in agroforestry.

For each agroforestry system, tree species should be selected based on the site, agroclimate, products and services to be obtained. The MPTs with major economic value suitable for particular agroclimatic region should be first identified. The objectives of planting MPTs are quire variable with respect to their providing tree products, services to humanity and rehabilitation of degraded lands. Before the start of agroforestry practices, the farmer must understand the objectives of planting MPTs in agroforestry land use. MPTs are mainly associated in agroforestry for the following objectives:

- To meet the farmers need such as small timber, housing, fodder, fuelwood, fruit and minor tree products.
- To supplement crop productivity through moisture conservation, addition of organic matter and maintaining the sustainability of soil.
- To provide economic gains to farmers by supplying raw materials to wood-based industries in the area.

WHAT ARE THE BEST MULTIPURPOSE TREES FOR AGROFORESTRY?

It is important to select the most suitable trees since it is not easy to replace them once they have been planted. The following factors should be kept in mind when selecting tree species:

- *Environmental Adaptation*: A multipurpose tree must be able to adapt to the area's climate, soil, topography, and plant and animal life. This is especially important for exotic species, that is, species introduced from outside the planting area or from outside the country.
- *Needs of Farmers*: The species should meet the needs of farm families. For this reason, it is important to involve farmers, both men and women, in selecting species. They should identify desired tree products and planting locations based on local markets as well as conservation needs. The cost of acquiring seedlings or cuttings should also be kept in mind. Planting stock of most fruit trees, for example, are expensive.
- *Ease of Maintenance*: Some species are more difficult to look after than others. Farmers should consider beforehand how much time they have to care for the trees. If they require

additional skills and knowledge to grow a particular species, training or demonstration programs should be organized.

- *Availability of Genetic Materials*: Seeds or seedlings of the species being considered must be easy to obtain. If vegetative propagation is required, farmers should receive training in how to do this. Seeds of *Gliricidia sepium* and *Sesbania* spp., for example, might not be readily available. Farmers may need help in how to propagate the trees using root cuttings. They may not even have enough time to harvest the cuttings.

SELECTION CRITERIA FOR MULTIPURPOSE TREES

Leguminous trees and shrubs that can fix atmospheric nitrogen are preferred over non-legumes. Multipurpose species that have additional uses are generally preferable because they give the agroforestry system more flexibility. Ideally, trees and shrubs suitable for agroforestry should meet the following criteria:

- establish easily
- grow rapidly
- easy to propagate
- easy to manage
- ability to survive in any type of soil
- have a deep root system with few lateral branches near the surface
- have a suitable branching pattern, including high branch and leaf productivity, both quantitatively (biomass production) and qualitatively (mulch quality, decomposition, etc.)
- regenerate readily after pruning
- have good coppicing ability
- provide useful by-products such as fuelwood, stakes, food, etc.
- be free from pest and diseases
- all the above qualities should not be impeded as the tree matures

Few tree species meet all of the above criteria and some have serious disadvantages that must be overcome. For example, *Leucaena* has slow early growth and its seedlings must be protected against weeds during early establishment; once established, however, the seedlings grow vigorously. Occasionally, it may be desirable to choose a species that is excellent for one specific purpose, for example, *Acioa barteri,* for its slow decomposing mulch, or the fast-growing *Calliandra calothyrsus* for its ability to produce large quantities of biomass within a short time.

When livestock production will be incorporated into an agroforestry system, the tree or shrub must provide forage in addition to its other functions. The following characteristics are desirable:

- high forage productivity that is unimpeded by maturity, (good juvenile - mature correlation)
- good feeding value and high palatability
- legumes are preferred because of their high protein value

A species should grow well under the specific limitations of the site, such as drought, flooding, heavy winds, insect pests, or other hazards. Species should perform well in spite of a site's limiting climatic and soil factors. The ongoing research programme, and other research efforts, should eventually yield a good deal of information on the performance of numerous MPTs under agroforestry across a very broad spectrum of conditions in tropical regions.

METHODS FOR MULTIPURPOSE TREES SCREENING AND EVALUATION

Screening of multipurpose tree species for agroforestry practices is an important step in agroforestry research.

STAGES OF EXPERIMENTATION

In theory, the number of multipurpose trees and shrubs (MPTS) with potential for agroforestry is high. For example, the number of potentially useful multipurpose trees and shrubs currently entered in the MPTS database at the World Agroforestry Centre (ICRAF) stands at nearly 1,100. Most of these MPTS have not been subjected to scientific study the world over. The task of comprehensive evaluation of MPTS for agroforestry development is large, presenting a major challenge to ICRAF and its partner institutions all over the world. In practice, of course, agroforestry researchers in tropical regions do not have to wait for the completion of such a comprehensive review. They are able to design manageable, effective MPT screening experiments by working with a limited number of promising species. The initial evaluation of MPTs should be carried out in three stages:

1. A list of promising MPTs is developed based on previous research worldwide experience with local MPTs, and local research priorities. The criteria for suitable agroforestry MPTs should be applied. Both local and exotic species should be included on the list.
2. Field trials for continuous growth assessment are conducted. Continuous growth trials permit general evaluation of each species for adaptation to local conditions, relative growth performance, and freedom from pests and diseases.
3. In a separate set of field trials, each species is subjected to agroforestry management. This permits evaluation of design establishment, biomass production, and pruning recovery.

The field trials (stages 2 and 3) may be conducted simultaneously for improved efficiency. In the agroforestry research, experimental design is kept simple: the continuous growth and management trials both employ the Randomized Complete Block design with three or four replications. The end result should be a short list of "best-bet" MPTs that are promising enough to merit further experimentation (e.g., spacing and fertilization trials, on-farm testing). If very few or no species perform satisfactorily, or if there is a need for MPT improvement, the experimentation process should be repeated with a new list of species and accessions.

ADDITIONAL GUIDELINES FOR MPT SCREENING EXPERIMENTS

The experimental design issues in species and provenance research are generally well known to researchers in agriculture and forestry. For example, comprehensive coverage is provided in Wood and Burley (1991). Six issues worth stressing in connection with screening MPTs for agroforestry development are:

1. In cases where the numbers of species and provenances are likely to be large, the simpler designs such as the Randomized Complete Block design may not be the most efficient. Incomplete Block Designs such as lattice designs could be more efficient in such experiments.
2. When screening mixtures of trees and shrubs, care should be taken to subdivide the species into near-homogenous sets with respect to their growth forms and rates. Such an

arrangement will safeguard against undue competition for light in the later stages of field experiments.

3. The issues of plot size and shape should be given special consideration. Small rectangular plots (including single and two-row plots) are used in screening MPTs, instead of the traditional large, square plots.
4. As compared with the cultivars and provenances which are currently used in intensive forestry and agricultural production systems, the MPTs of potential value in agroforestry systems are at much lower levels of domestication. At these lower levels, great opportunities exist for exploiting within-species genetic variation. The traditional research path in forestry has been to screen for species and subsequently to screen for the best provenances within top priority species. This time-consuming path may not be necessary for MPT screening, because information on general growth performance of numerous species is now more readily available to researchers. Efforts should be made to combine both species and provenance screening in appropriately designed experiments.
5. For ease of comparison of data among scientists, it is important that measurements of trees and shrubs be standardized. Standardization becomes even more important in experiments which are established on a network basis.
6. The Diagnosis and Design (D&D) methodology developed at ICRAF can assist in the selection and assessment of MPTs. D&D provide a rationalized approach for diagnosing land use system needs and designing suitable agroforestry interventions.

CONSIDERATIONS FOR CHOICE OF MULTIPURPOSE TREES

In order to select appropriate MPT species, the required tree attributes are carefully specified based on agroforestry technology recommended during Diagnosis-and-Design analysis. This is followed by searching MPT data bases, the literature and selecting seed sources that match the ecological zone under consideration. Due to high number of accessions considered in initial screening experiments, appropriate field designs for handling large numbers of germplasm are recommended. The designs should allow a broad range of data to be collected to provide the initial prognosis for integrating promising MPTs in agroforestry systems. The process which leads to the final decision of what species to choose is divided into the following basic stages (von Carlowitz 1989):

Diagnostic Stage:

- Inventory, assessment and analysis of prevailing land use situations, characteristics and components
- Identification and prioritization of land use problems
- Determination of the point of intervention

Design Stage:

- Decision on the type of agroforestry technology to effect the intervention
- Identification of MPTs associated with the technology and its intended role and function

Technical Stage:

- MPT specification related to the expected functions and outputs of the technology they are associated with

Once the priority land use problems have been identified, agroforestry technologies that have the capability to overcome successfully or ease the cause and effects of these problems are chosen. The problem solving effectiveness and efficiency of a chosen agroforestry technology depend on:

- Choosing the right place for the technology in the default land use system
- The appropriate arrangement of the components, mainly MPTs, within the technology
- The adequate management of the technology and its components
- On the capability of the MPTs to provide the outputs and services expected as part of the function of the technology, and to respond to the management that is characteristic of the technology

The actual selection of species, for technologies chosen for their capability of solving or easing identified land use constraints, has to be based on the following criteria:

- Biophysical conditions (climate and soils) at the site of the intended tree that has to be matched with species-specific environmental requirements and tolerances
- Technology specific tree characteristics (ideotype) and ideotype tree is a plant model designed to yield products of greater quality and quantity than a conventional cultivar or wild plant (Dickmann 1985)
- Positive response of candidate species to management and tree manipulation required as part of the technology concept and function.

MULTIPURPOSE TREES GERMPLASM DOCUMENTATION AND IMPROVEMENT

Many potentially useful MPTs are presently unknown to science. Valuable MPT germplasm remains to be collected and properly documented. This activity requires particularly urgent attention in tropical countries experiencing rapid rates of devegetation threatening total loss of some species and varieties. National, regional, and international research centers could accord high priority to this activity. ICRAF and the International Board for Plant Genetic Resources (IBPGR) have been doing MPT germplasm collection and documentation projects in tropical countries. The International Livestock Centre for Africa (ILCA) maintains a germplasm collection of fodder trees and shrubs. IBPGR and the Royal Botanic Gardens, Kew, have published a valuable reference book on forage and browse plants for arid and semi-arid Africa.

However, much works still remains to be done in strengthening national institutions in the tropics for collection, documentation, and storage of MPT germplasm. Great gains could also be derived from breeding MPT for agroforestry development. This would require the initiation of well planned MPT germplasm improvement programs. ICRAF is currently developing cooperative breeding strategies with national research systems.

GREATER ROLE OF MULTIPURPOSE AGROFORESTRY TREE SPECIES

The multipurpose trees are woody perennials that can provide multiple products and services in agroforestry systems. They may supply food in the form of fruit, fodder, leaves and at same time supplying wood, fuelwood, fiber, flosses, fertilizer (add nutrients to the soil) and supply some other products (silk, gum, lac etc.). In short it can be said that multipurpose trees are those trees which provide multiple products and perform various functions and the trees have a greater impact on a farmer's well being because they fulfill more than one basic need. The MPTs can be integrated with agroforestry or any other form of tree farming system to improve total biomass and yields, to diversify products, to increase economic resiliency and to improve farm viability and sustainability in the long-term. The multiple products and functions of multipurpose trees are the added advantage. The most important benefit of multipurpose trees planting and integrating in other forms of tree farming system is that they can provide an assured amount of insurance in the event of a agricultural crop failure or market fluctuation. Various multipurpose trees and shrubs are planted in the agroforestry systems of different agroclimatic zones to provide various benefits and services.

Table 5.1: Multipurpose Tree Species suitable for Agroforestry (Sharad Nema 2012)

Species	Climate	Uses
Acacia nilotica	Arid, semiarid tropics	fuelwood
Acacia tortillis	Semiarid tropics	fuelwood
Albizia lebbek	Humid tropics, semiarid tropics	fuelwood, timber
Artocarpus heterophyllus	Humid tropics, semiarid tropics	fruit, fodder, bark dye, windbreak
Azadirachta indica	Semiarid tropics	timber, oil, insecticidal and medicinal, fuelwood, windbreaks
Casuarina equisetifolia	Humid tropics	fuelwood, timber
Cocos nucifera	Humid tropics	nut, nutritional drink, fuel, crafts, housing thatch
Dalbergia sissoo	Semiarid tropics	timber, fuelwood
Eucalyptus camaldulensis	Humid tropics	fuelwood, timber
Eucalyptus tereticornins	Humid tropics, semiarid tropics	fuelwood, pole timber, wind break
Gliricidia sepium	Humid tropics	food, fuelwood, poles, fodder, Nitrogen fixing, living fence
Gmelina arborea	Humid tropics	timber, fuelwood, furniture
Grevillea robusta	Subhumid tropics, humid tropics	timber, fuelwood, building materials
Leucaena leucocephala	Humid subtropics, humid tropics	fuelwood, pole, fodder, windbreak, soil improver
Mangifera indica	Semiarid tropics	fruit, medicine, leaves as mulch, timber, windbreak
Moringa oleifera	Semiarid tropics	fruit, medicine, fodder,
Prosopis cineraria	Arid and semiarid tropics	windbreak
Sesbania grandiflora	Humid tropics	fodder, fuelwood, food
Tamarindus indica	Semiarid tropics	shade, fruit, light timber, medicinal uses
Ziziphus mauritiana	Semiarid tropics, subhumid tropics	food, shade

ICRAF's MULTIPURPOSE TREE AND SHRUB (MPTS) DATABASE

The MPTS Database was created under the Multipurpose Tree Improvement Programme at ICRAF. The Multipurpose Tree and Shrub (MPTS) Database held by the International Council for Research in Agroforestry (ICRAF), is a user-friendly computing tool developed for the retrieval of detailed information on multipurpose tree and shrub species. The database comes with software that allows the user to browse species or create parameters for the selection of species based on any number of physical and social variables. The MPTS User Manual describes it as, "a computer-based information and decision-support system founded on an inventory of existing information about multipurpose tree and shrub species in the tropics and sub-tropics. In the context of the database, multipurpose trees and shrubs (MPTS) are those that are deliberately grown or kept in integrated land-use systems and are often managed for more than one output. They are expected to make a significant economic or ecological impact, or both. This definition also includes forest plantation species that are grown primarily for lumber, provided that they are used in integrated land-use systems. The main objective of the database is to provide detailed information on a large number of species to field workers and researchers who are engaged in activities involving woody perennials suitable for agroforestry systems and technologies. It is designed to help them make rational decisions regarding the choice of candidate species for specific sites and defined purposes".

The database currently consists of some 1,093 species and 2,340 descriptions (records) of tropical and subtropical woody perennials. The database is built under a graphical interface that allows the user to select species based on any number of criteria. MPTS software allows the user to create complex site specific parameters (e.g., biophysical characteristics, cultivation problems, intended uses, etc.), and combine these with Boolean expressions, or simply select species by name. Information regarding the selected species can then be retrieved and sent to a file or printer for display. Information on individual/appropriate species can be retrieved by stating any part of a species name (taxonomic/binome or vernacular), or by constructing a search with site specific parameters. Many variables can be combined with Boolean expressions to form these complex searches. Additional criteria may also be entered into a query (e.g., specific climatic characteristics, agroforestry practice, etc.), while additional features for dealing with search results are also available (e.g., refining a search, sending profile descriptions to a system file, etc.).

ICRAF's AGROFORESTREE (AFT) DATABASE

The Agroforestree (AFT) Database is a species reference and selection guide for agroforestry trees developed by ICRAF. In the context of the database, agroforestry trees are those that are deliberately grown or kept in integrated land-use systems and are often managed for more than one output. They are expected to make a significant economic or ecological impact, or both. The main objective of the database is to provide detailed information on a number of species to field workers and researchers who are engaged in activities involving trees suitable for agroforestry systems and technologies. It is designed to help them make rational decisions regarding the choice of candidate species for defined purposes. Information for each species covers species identity, ecology and distribution, propagation and management, functional uses, pests and diseases and a bibliography. To date, more than 500 species have been included. The specific aims of the database are:

- To enable quick and efficient access to a consolidated pool of information on tree species that can assume useful production or service functions, or both;
- To provide a tool that will assist with the selection of species for use in agroforestry and related research using factors that are relevant to the chosen agroforestry technologies;
- To help researchers assess potential agroforestry trees for uses other than those commonly known, such as timber;
- To provide indicators for the economic assessment of species through yield information on tree products.

IMPORTANT MULTIPURPOSE TREES FOR TROPICS

WOOD YIELDING TREES: *Acacia* spp., *Dalbergia* spp., *Albizia* spp., *Tectona grandis*, *Melia azedarach*, *Azadirachta indica*, *Gmelina arborea*, *Populus deltoides*, *Eucalyptus* spp., *Casuarina equisetifolia*, *Grevillea robusta*, *Hardwickia binata*, *Artocarpus heterophyllus*, *Madhuca indica*, *Thespesia populnea*, Bamboo

FUELWOOD TREES: *Eucalyptus* spp., *Leucaena leucocephala*, *Acacia tortilis*, *Albizia lebbeck*, *Calliandra* spp., *Casuarina equisetifolia*, *Prosopis cineraria*, *Prosopis juliflora*, *Cassia siamea*, *Faidherbia albida*, *Acacia auriculiformis*, *Acacia senegal*, *Ailanthus* spp., *Pithecellobium dulce*, *Capparis decidua*, *Flemingia* spp.

FODDER TREES (Leaves and Pods): *Leucaena leucocephala*, *Leucaena diversifolia*, *Sesbania sesban*, *Sesbania grandiflora*, *Sesbania aculeata*, *Gliricidia sepium*, *Prosopis cineraria*, *Prosopis juliflora*, *Acacia nilotica*, *Acacia tortilis*, *Acacia senegal*, *Acacia auriculiformis*, *Faidherbia albida*, *Pithecellobium dulce*, *Calliandra calothyrsus*, *Albizia amara*, *Albizia lebbeck*, *Albizia procera*, *Albizia saman*, *Erythrina* spp., *Hardwickia binata*, *Parkia speciosa*, *Dalbergia sissoo*, *Grewia* spp., *Moringa oleifera*, *Bauhinia* spp., *Cassia auriculata*, *Parkia speciosa*, *Desmanthus virgatus*

FERTILIZER TREES: *Gliricidia sepium*, *Leucaena leucocephala*, *Faidherbia albida*, *Sesbania grandiflora*, *Sesbania sesban*, *Calliandra calothyrsus*, *Prosopis cineraria*, *Acacia tortilis*, *Casuarina equisetifolia*, *Erythrina* spp., *Acacia senegal*, *Adansonia digitata*

FRUIT TREES: *Ziziphus mauritiana*, *Mangifera indica*, *Tamrindus indica*, *Moringa oleifera*, *Emblica officinalis*, *Morus alba*, *Artocarpus heterophyllus*, *Cocos nucifera*, *Areca catechu*.

■■■■

6 NITROGEN FIXING TREES IN AGROFORESTRY

INTRODUCTION

Tropical agroforestry systems are characterized by great diversity among and within tree species that are found in contrasting biological, complex niches in a spectrum from managed natural forest to intensive farmland production. Different species and production systems have immense value to communities that are dependent on trees (in forests and farmlands) and to local, national and international markets. The diversity of species and their unique biological characteristics, and the wide range of landscapes in which tree species are found, pose major challenges in developing generic models and principles for managing tree resources effectively. The lack of quality tree germplasm, inappropriate farm management practices and a lack of market integration mean that a lot of the potential for improving livelihoods and the environment remain untapped.

Valuable opportunities to improve the livelihood of communities are being lost due to threats to natural and planted populations of tree stands, with inadequate integration of trees into sustainable farming practices. Under-performing trees species that are integrated in agricultural landscapes due to inadequate approaches for silvicultural/horticultural and genetic management, result in negative impact on livelihood of communities. Similarly under-investment in the improvement, prioritization, domestication and wider cultivation of more high value tree species will result in poor livelihood development of farming communities.

The introduction of trees that symbiotically fix atmospheric nitrogen is widely acknowledged as one of the most efficient means to sustain the productivity of agroecosystems through the improvement of the soil nitrogen balance, especially in the tropics. Nevertheless, very few research programmes specifically focus on this topic. Quite varied results have been obtained on fixed-nitrogen enrichment of the soil and use of this nitrogen to fertilize annual intercrops. The N_2-fixing potential of tree species is often low or inhibited by environmental constraints such as drought, salinity and excess mineral nitrogen or plant diseases. The symbiotic bacteria/host tree association could possibly be improved through plant biotechnology and genetic engineering research. Symbiotic nitrogen fixation could also be optimized by modifying current agricultural practices. Nitrogen-fixing trees can be intercropped with other annual crops in various conditions, e.g. hedges, bush fallows and long-term rotations. Such management programmes should be better integrated in rural environments through the promotion of multipurpose species producing wood and high-protein livestock fodder as well as edible fruits or seeds (Galiana et al 2004).

OVERVIEW OF BIOLOGICAL NITROGEN FIXATION

NITROGEN

Nitrogen gas comprises about 78 per cent of the earth's atmosphere. Nitrogen is present in various forms and it is also found in rocks, soils, sediment, oceans and living matter. Nitrogen is essential component of amino acids and proteins that build cell material and plant tissue. It is also necessary for the function of other essential biochemical agents including chlorophyll, many enzymes, nucleic acids (DNA, RNA). Most microorganisms and plants obtain nitrogen from the surrounding soil and water. Animals get nitrogen from the food they eat. Nitrogen is a major plant nutrient for plant growth and crop yield. Plants lacking nitrogen show stunted growth and yellowish leaves. Nitrogen is naturally available in three forms:

- Soil Humus (Organic Form): Plants cannot access and they are unavailable for uptake
- Soil Nutrients (In-organic Form): Examples include NO_3^-, NO_2^-, NH_3, NH_4^+
- Atmospheric Nitrogen (N_2)

NITROGEN CYCLE

It is the biogeochemical cycle that describes the transformations of nitrogen and nitrogen containing compounds in nature. The nitrogen cycle describes the movement of nitrogen through the lithosphere, hydrosphere, biosphere and atmosphere. It is the transformation of nitrogen from atmosphere (air) to soil, plants, animal life and returns back to air or soil through decay or denitrification.

BIOLOGICAL NITROGEN FIXATION

All plants need relatively large amounts of nitrogen (N) for proper growth and development. Biological nitrogen fixation (BNF) is the term used for a process in which nitrogen gas (N_2) from the atmosphere is incorporated into the tissue of certain plants. Only a select group of plants is able to obtain N this way, with the help of soil microorganisms. *Rhizobium* bacteria are mutualistic with certain plant species e.g. legumes and they grow in root nodules and *Frankia* actinomycete also nodulate trees such as *Alnus*, *Casurarina*, etc to fix nitrogen. *Azotobacter* and *Azospirillum* are free living nitrogen fixing bacteria associated with the rooting zone (the rhizosphere) of plants. Blue-green algae and Clostridium are non-symbiotic nitrogen fixers. Cyanobacteria are nitrogen fixers that also fix carbon (these are photosynthetic).

Only prokaryotes show nitrogen fixation. These organisms possess the 'nif gene' complex which makes the proteins, such as nitrogenase enzyme, used in nitrogen fixation. Nitrogenase is a metalloprotein, protein subunits being combined with an iron, sulphur and molybdenum complex. The reaction involves splitting nitrogen gas molecules and adding hydrogen to make ammonia.

$$N_2 \rightarrow 2N - 669\ kJ$$

$$2N + 8H^+ \rightarrow 2NH_3 + 2H_2 + 54\ kJ$$

Although the process involves a number of complex biochemical reactions, it may be summarized in a relatively simple way by the following equation:

$$N_2 + 8H^+ + 16ATP \rightarrow 2\ NH_3 + 2H_2 + 16ADP + 16\ P_i$$

The equation above indicates that one molecule of nitrogen gas (N_2) combines with eight hydrogen ions (also known as protons) ($8H^+$) to form two molecules of ammonia ($2NH_3$) and two molecules of hydrogen gas ($2H_2$). This reaction is conducted by an enzyme known as nitrogenase. The 16 molecules of ATP (ATP = Adenosine Triphosphate, an energy storing compound) represent the energy required for the BNF reaction to take place. In biochemical terms 16 ATP represents a relatively large amount of plant energy. Thus, the process of BNF is 'expensive' to the plant in terms of energy usage. The sun, via the process of photosynthesis, is the ultimate source of this energy needed for BNF. As ammonia (NH_3) is formed, it is converted to an amino acid such as glutamine. The Nitrogen in amino acids can be used by the plant to synthesize proteins for its growth and development.

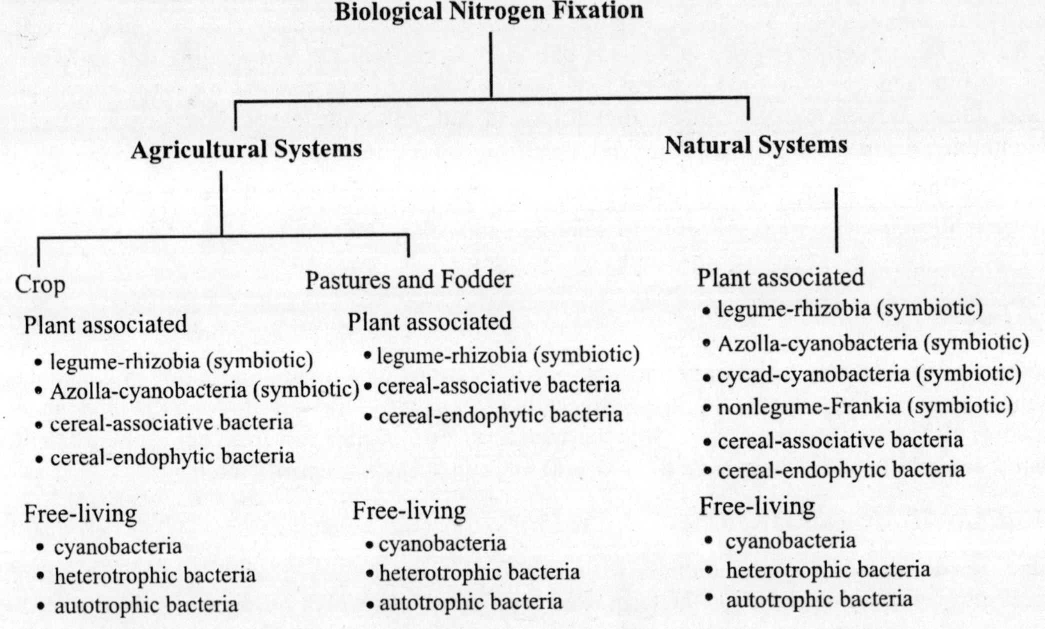

Fig. 6.1: Biological Nitrogen Fixation

NITROGEN FIXING TREES

Trees play the important and most dominant roles in Agroforestry systems. Basically, the trees ideal to be incorporated in various agroforestry systems are called the multipurpose trees (MPTs). The term multipurpose tree (MPT) refers to all woody perennials that are purposefully grown so as to provide more than one significant to the production and/or service functions (shelter shade, land sustainability) of the land use system they implement. Better emphasis is given which are indigenous in nature and have the capability to fix atmospheric nitrogen. Selection and management of MPTs influence the success of any agroforestry system. The nitrogen fixing trees (NFTs) besides their 'N' fixing ability provide protein rich foliage which can be used for fodder and green manuring. More than 650 tree species are known to fix nitrogen. Most trees used in agroforestry grow relatively quickly, whilst many also fix nitrogen i.e. many MPTs are fast growing nitrogen fixing trees (FGNFTs).

Nitrogen fixing plants are key constituents in many natural ecosystems in the world. They are the major source of all nitrogen that enters the nitrogen cycle in these ecosystems. Many nitrogen fixing plants are woody perennials, or nitrogen fixing trees (NFTs), most of these being found in the tropics. In temperate areas, the nitrogen fixers tend to be herbaceous. NFTs have been removed or reduced in most man-made ecosystems such as agricultural, forest lands and urban environments. These lands require expensive chemical fertilizer inputs in order to maintain their productivity. Man-made systems can be improved by learning and adopting from natural ecosystems. For example, the reintroduction of NFTs, with appropriate management such as agroforestry land use practices, can increase and sustain productivity. No plant grows without nitrogen, and many tropical soils have low supplies of this nutrient. NFTs do not depend solely on soil nitrogen, but "fix" nitrogen through symbiotic microorganisms that live in root nodules and convert atmospheric nitrogen into a plant usable form. This form of nitrogen is known as "green manure" and is a nutrient that helps plants, such as food crops, to grow. There are two basic types of N-fixing systems found in trees, based on two different symbiotic microorganisms.

- Bacteria of the genus *Rhizobium* inoculate trees in the families Fabaceae and Ulmaceae
- Actinomycete of the genus *Frankia* inoculates several other families like Betulaceae, Casuarinaceae, etc.

Beneficial partnerships are the way of nature. In particular, these microbes (*Frankia* and *Rhizobium*) form associations with certain plants allowing them to fix atmospheric nitrogen into a form that plants can use. These symbiotic partners can help to rehabilitate damaged landscapes, preparing the soil for a succession of more long-term plants. Among the woody nitrogen-fixers in temperate regions, actinorhizal plants are an important group. These pioneering plants are able to grow in poor soils, enriching them with nitrogen and organic matter. This makes them very valuable in repairing disturbed or damaged soils.

Table 6.1: Nitrogen Fixing Woody Actinorhizal Plants (NFTA 1989)

Family	Genera
Betulaceae	*Alnus*
Casuarinaceae	*Allocasuarina, Casuarina, Gymnostoma*
Coriariaceae	*Coriaria*
Elaeagnaceae	*Elaeagnus, Hippophae, Shepherdia*
Myricaceae	*Comptonia, Myrica*
Rhamnaceae	*Ceanothus, Colletia, Descaria, Kentrothamnus, Retanilla, Talguena, Trevoa*
Rosaceae	*Cercocarpus, Chamaebatia, Cowania, Dryas, Purshia*

The next important group of woody plants associates with the bacteria of the genus *Rhizobium*. There are 3 subfamilies of the family Fabaceae (Leguminosae). These families are Papilionaceae, Mimosae and Caesalpiniaceae. Not all the trees in these subfamilies are nitrogen-fixers. Among the Caesalpiniaceae, 23 per cent are nitrogen fixers. For Mimosae, the figure is 90 per cent, and for Papilionaceae, 97 per cent are nitrogen-fixers. The Fabaceae (Leguminosae), however, make up a vast majority of the 650 popularly known NFT species.

Table 6.2: Nitrogen Fixing Woody Leguminous Plants (NFTA 1989)

Sub-family	Species	Percentage Fixers	Representative NFT Genera
Caesalpiniaceae	1,900	23	*Chamaecrista, Cordeauxia*
Mimosae	2,800	90	*Acacia, Albizia, Calliandra, Enterolobium, Leucaena, Mimosa, Paraserianthes, Pithecellobium, Prosopis*
Papilionaceae	12,300	97	*Cajanus, Dalbergia, Erythrina, Flemingia, Gliri-cidia, Pterocarpus, Robinia, Sesbania, Tephrosia*

To rapidly revegetate a damaged landscape, it must be ensured to include plenty of these species to help quickly build up the soils. In areas of very problematic soil, such as arid, tropical and subtropical regions, it is recommended to make 90 per cent of initial planting of trees nitrogen fixing, pioneer species (associating with either *Frankia* or *Rhizobium*), and 10 per cent of species of long-term canopy overstory species. When the system reaches maturity, the proportions will be reversed with 10 per cent nitrogen-fixing, support species and 90 per cent canopy species. The same formula could be followed for humid temperate regions, but the soils in these areas are not so fragile and can stand a lower percentage of nitrogen fixers. A 70/30 or even lower may suffice in these areas, as the seasonal cycles of death and regrowth feed these soils well. As the Figure 6.2 shows, the nitrogen-fixing support trees can be pruned (coppiced, pollarded, shredded or sacrificed) to provide mulch, fodder, fuel or fibre. As this is done, the roots of the tree self-prune and release nitrogen into the soil.

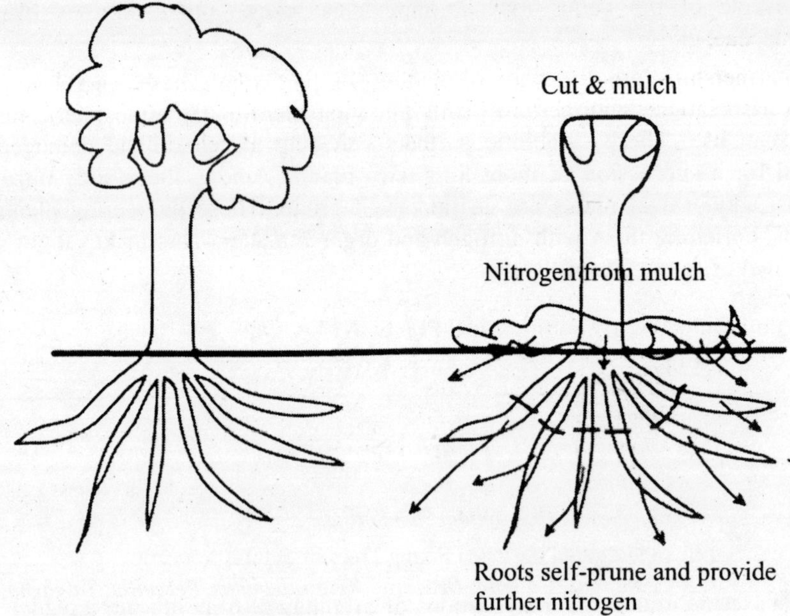

Fig. 6.2: Nitrogen Fixing Support Trees for fertilizing Long-term Canopy Trees

FERTILIZER TREES

The use of trees, especially NFTs in agroforestry and silvopastoral systems, is receiving considerable attention recent days. The use of NFTs in agroforestry systems in tropical regions is attractive as a source of N and organic matter needed in the rehabilitation of nutrient depleted soil. Nitrogen fixing fertilizer trees maintain or improve soils by the processes described below:

- Fixing atmospheric nitrogen and providing in plant available form
- Deep nutrient capture from sub-soils through roots
- Addition of organic matter, nitrogen and other nutrients to the topsoil through leaf fall
- Addition of nitrogen and other nutrients to soil by decaying roots
- Reduction in losses from soil, leading to more closed cycling of organic matter and nutrients
- Improvement of soil physical conditions
- Improvement of soil chemical conditions
- Affecting soil biological processes and conditions

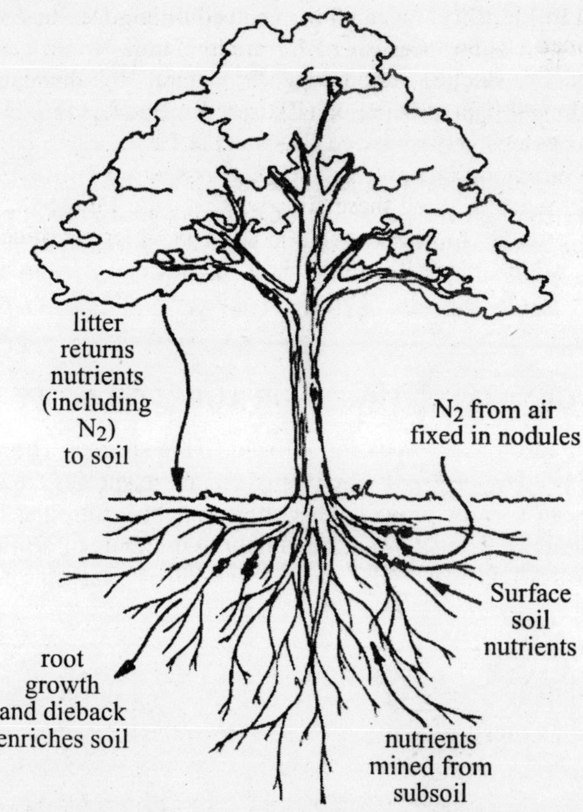

Fig. 6.3: Role of Nitrogen Fixing Trees in Soil Enrichment

Sanginga et al (1986) found that the additions of prunings of 6-month old inoculated *Leucaena leucocephala* increased subsequent maize yield from 2.2 t ha^{-1} to 4.6 t ha^{-1}. Lal (1989) reported that *L. leucocephala* planted at 2 m intervals reduced erosion from interrow maize plantings from 4.3 t ha^{-1} (plow-till) to 0.1 t ha^{-1}, i.e. same as no-till. Ladha et al (1989) found that green manure from the nitrogen-fixing *Sesbania rostrata* (a stem and root nodulating legume) was equivalent to 60-120 Kg N ha^{-1} as urea in the rice ecosystem. Analysis of soils from under *Faidherbia albida* (Charreau and Vidal 1965) indicated a remarkable fertility gradient from bare soil to soil under foliage and protein yields of millet near trees were increased 3-4 fold. Leguminous trees such as *Acacia cyanophylla* are commencing to have an important agroforestry role in areas with 250-400 mm rainfall. Significant accumulations of organic C, N, Ca, Mg, P and K have been measured in the surface soil beneath *Prospopis* canopies in natural ecosystems (Virginia 1986), due partly to transport of these elements from lower depths and to biological nitrogen fixation. *Casuarina equisetifolia* (symbiotic with the actinomycete *Frankia*) has found extensive use to consolidate sandy coastal soils.

The problems associated with biological nitrogen fixation (BNF) measurements are much more complex with nitrogen fixing trees (NFTs) than with annuals due to largely to the size of trees and their potential nature. The practical difficulties in harvesting whole trees or of realistic non-destructive sampling, and in obtaining suitable controls valid over seasons or years, are not easy to overcome, and for this reason many researchers have estimated BNF in young trees over short periods. However, questions such as – For how long do trees continue to fix N_2? What are the effects of root and litter turnover on N_2 fixation and soil fertility? – cannot be resolved through studies on N_2 fixation in young trees. The commonly used methods for measuring BNF are total nitrogen difference (TND) , acetylene reduction assay (ARA), the ^{15}N isotope methodology, the natural ^{15}N abundance method, procedure with ^{15}N fertilizer addition to soil [isotype dilution (ID) and A-value (Av)], and Uride as a measure of BNF, all of which have been extensively reviewed (Danso et al 1992).

It has been estimated the biological nitrogen fixation (BNF) of *Leucaena leucocephala* ranges from 300-550 Kg N ha^{-1} yr^{-1} through various studies, *Casuarina equisetifolia* ranges from 60-110 Kg N ha^{-1} yr^{-1}, *Sesbania rostrata* ranges from 83-109 Kg N ha^{-1} yr^{-1}, *Sesbania sesban* ranges from 43-102 Kg N ha^{-1} yr^{-1}, *Gliricidia sepium* 13-108 Kg N ha^{-1} yr^{-1}, *Inga jinicuil* 35-40 Kg N ha^{-1} yr^{-1}, *Acacia mearnsii* 200 Kg N ha^{-1} yr^{-1}, *Faidherbia albida* 20 Kg N ha^{-1} yr^{-1} and *Albizia lebbeck* 94 Kg N ha^{-1} yr^{-1} (Danso et al 1992; Nair 1993).

SELECTION OF NITROGEN FIXING TREES FOR AGROFORESTRY

Many of the most important agroforestry trees are nitrogen fixing species. Through a symbiotic relationship with *Rhizobium* soil bacteria or *Frankia* actinomycete, these species are able to fix atmospheric nitrogen into a form they can use for growth. This ability enables nitrogen fixing trees (NFTs) to tolerate infertile sites and produce diverse products without high inputs of artificial N fertilizer. There are a number of characteristics that make NFT species particularly valuable for agroforestry. These include:

- Rapid growth rate
- Nitrogen fixing ability
- High biomass production
- Coppicing ability
- Multiple utility
- Ability to survive in any type of soil (Adaptability to poor and difficult sites)
- Easy propagation
- The ability to compete with other vegetation
- Complementary tree architecture
- Deep root system to draw water from deeper soil profile
- Leaf shading in summer to conserve moisture
- Water binding mechanism
- Tolerance to salinity and saline water, and alkalinity
- Productivity under repeated harvesting
- High nutritive value fodder and acceptability to animals
- High calorific value fuelwood
- Absence of toxic substances in foliage and root exudates (Allelopathy)

In order for a NFT to be productive, it must have the ability to survive and compete in the climatic and soil conditions of the site in question. Humid zone species are not productive in arid environments, lowland species are not productive in alpine regions. When propagating NFTs, the site's environmental conditions should be matched with the environmental requirements of potential species. Elevation, mean annual precipitation, length of annual dry season, mean annual temperature and soil characteristics are of particular importance. Many NFTs grow well on infertile or difficult sites. This ability is valuable as most fertile soils are used for food crop production, leaving marginal or poor soils for tree cultivation. NFTs help to improve the sites when their nitrogen-rich leaves and branches senesce, fall to the ground as litter and are incorporated into the soil (Roshetko and Gutteridge 1996).

Seed germination and seedling growth of NFT trees should be relatively fast. This will provide a rapid supply of livestock feed and enable trees to compete with other vegetation. Competition between tree seedlings and pre-existing grasses can be intense. In some situations, tree species can be excessively competitive. Before introducing a new species to an area, a small evaluation trial has to be established to determine its potential. If the trees show signs of becoming weedy, early prolific seeding or root sprouts, these should be removed before these problems occur.

Most nitrogen fixing trees provide multiple products and services. They are planted on contours to stabilize soil, on boundaries as living fences, with grasses and shrubs to form multiple layer fodder banks, or scattered around the farm where site characteristics are too poor to support food crops. In all of these systems, NFT trees may provide fodder, fuelwood, poles, timber, fruit or other products for home consumption and sale in local markets. Simultaneously, trees may decrease erosion, rejuvenate soil, or serve as ornamentals. Many NFTs have deep roots that utilize nutrients in the subsoil below the rhizosphere of companion crops. Through leaf and branch litter, these trees recycle nutrients to the soil surface for use by other plants. Conversely, trees with shallow roots may compete heavily with other plants. Such trees are inappropriate components of most agroforestry or intercropping systems. Some NFTs have wide spreading crowns which cast light shade. These trees improve the microclimate for livestock, pasture grasses and shade loving crops. NFTs with vertical crowns cast little shade and are ideal companion plants for crops which need full sunlight. Always tree species is to be selected that provides products and services that are complementary to the landowner's objectives and management system.

A final characteristic required of NFT trees is ease of propagation. A species cannot furnish reliable quantities of products if establishment is difficult or prolonged. NFT trees must produce adequate supplies of viable seed, or be easily propagated by vegetative methods. Vegetative propagation is time consuming when establishing large areas and only practical with a limited number of trees. In most circumstances, it is best to select a species which produces viable seed in the environment in which it is planted, or a species for which dependable inexpensive supplies of seed are available on a commercial basis (Roshetko and Gutteridge 1996).

USES AND PLANTING OF NITROGEN FIXING TREES IN AGROFORESTRY

Many NFTs are important to rural households throughout the tropics providing a variety of products and services:

- *Firewood and Charcoal* are the primary energy sources for almost 50 per cent of the world's population. Fast-growing, high density NFTs make excellent fuelwood and charcoal. Many re-sprout, or coppice, vigorously after cutting, allowing repeated harvests without replanting.
- *Fodder* to feed animals is a constant concern to many farmers in developing countries. The highly nutritious and digestible leaves of some NFTs make them excellent feed for animals. The deeply penetrating roots of NFTs can reach retreating moisture and provide fresh feed during dry seasons.
- *Soil Fertility* is critical to crop production, but many resource-poor farmers cannot afford chemical fertilizers. Leaves of many NFTs are high in nitrogen and other plant nutrients and can be a renewable, free source of fertilizer.

- *Timber and Poles* are needed all over the world for house and other general construction. NFTs include both fast growing trees for rough wood, and some of the most valuable luxury timbers in the world.
- *Human Food* is harvested from several species of NFT in various parts of the world, in some instances supplying important seasonal staples.

Depending on local needs and preferences, a variety of different planting schemes with NFTs can be utilized yielding a wide variety of products and staples (NFTA 1989).

- *Living Fences and Hedges* protect crops from large pests such as wildlife, domestic animals, and man, and are often managed for fuelwood and fodder production. Trees are arranged densely, or planted as fence posts, and trimmed frequently to attain the desired form.
- *Windbreaks* are single or multiple rows of trees planted on windward field boundaries. Windbreaks help prevent soil desiccation and yield secondary tree products. Consistent foliage closure is achieved by choosing trees with dense canopies and by managing the canopy to encourage lateral branching.
- *Alley Farming,* or *Hedgerow Intercropping*, is a labour intensive management system which leads to major crop yield increases through alternating rows of tree hedges and crops. Cut leaves and green twigs are incorporated into or laid on top of the soil for multiple benefits of green manuring, soil and water conservation, and weed control. This practice is successfully being introduced on steep sloped hillsides using several NFT species. *Gliricidia, Leucaena* and *Calliandra* are important tree species for alley cropping.
- *Shade and Support* are attained quickly from fast-growing NFTs. Shade is an important benefit in hot climates for some crops, such as cocao, coffee, and tea, as well as for humans and animals. Living, soil-enriching support is quickly established for vining crops such as yarns, vanilla, and black pepper. *Gliricidia sepium* has been used for all these purposes.
- *Fodder Banks* are intensive plantings of fodder plants, spaced to maximize production, and provide a source of "cut and carry" fodder. Many NFTs have leaves or pods that are very high in protein. *Leucaena leucocephala* is called the "alfalfa of the tropics" because of its extensive use as fodder.
- *Pasture Improvement* is achieved through increased grass production, tree fodder browsed directly by animals, and shade; livestock digest more efficiently when shade is available. *Acacia* species are found throughout the tropical grazing lands.
- *Homegardens* utilize NFTs for soil fertility, as well as for edible fruits, leaves, or flowers, and as medicinals. *Parkia* species are important seasonal food sources in both Southeast Asia and West Africa.
- *Woodlots* planted with fast growing NFTs can yield quick returns, especially in less productive areas of the farm. Coppicing (sprouting) trees are the species of choice, particularly for stands of fuelwood. *Calliandra calothyrsus* is an important woodlot tree.
- *Improved Fallow* is most useful in areas where slash and burn agriculture is practised. When a field is exhausted of its nutrients from intensive cultivation, NFTs can be planted for soil enrichment, and hasten the return of fertility. *Sesbania sesban* is one of the best species for improved fallow system.
- *Land Reclamation* using NFTs is commonly practised on eroded mountainside, exhausted grazing areas, unproductive mined areas, and for dune stabilization.

NITROGEN INPUTS AND OUTPUTS IN AGROFORESTRY

NITROGEN INPUT

Trees can provide nitrogen inputs in agroforestry systems by two processes: biological nitrogen fixation (BNF) and deep nutrient capture. Although the magnitude of BNF is methodologically difficult to quantify, overall annual estimates are in the order of 150 kg N per hectare (Giller and Wilson 1991). Empirical evidence, such as the presence of active nodules of leguminous species of the Papilionaceae and Mimosae families, indicate that BNF can supply considerable nitrogen inputs to crops via litter in soils that are sufficiently well supplied with phosphorus. This is a definite nutrient input. There is also ample evidence that non-fixing trees, including several species of *Cassia* (recently renamed *Senna*), accumulate as much or more nitrogen in their leaves than nitrogen-fixing legumes, presumably because of their greater root volume and ability to capture nutrients (Garrity and Mercado 1994). It is, however, important to note that these non-fixing trees are only cycling nitrogen, not adding inputs to the system.

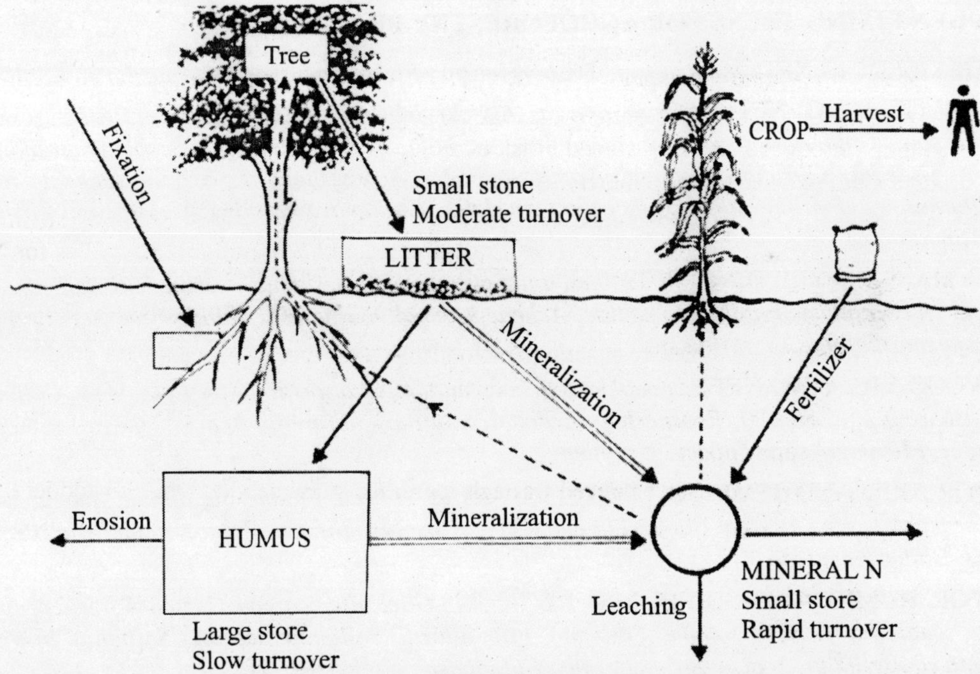

Fig. 6.4: Nitrogen Cycle under Agroforestry (Source: Young 1989)

Deep nutrient capture is the uptake of nutrients by tree roots at depths where crop roots are not present. It can be considered an additional nutrient input in agroforestry systems because such nutrients are leached as far as the crop is concerned. They become an input on being transferred to the soil via tree litter decomposition. An exciting discovery was made in western Kenya regarding subsoil nitrate in nitrogen-deficient nitisols (red alfisols and oxisols with a high iron content), where Hartemink et al (1997) detected nitrate levels in the order of 120 kg N per hectare at subsoil depths of 50 to 200 cm. The authors also found that *Sesbania sesban* fallows depleted this pool, thus capturing a resource that was unavailable to agricultural crops especially maize. The source of this nitrate pool is believed to be the result of the mineralization of topsoil organic nitrogen, which is relatively high in these soils, followed by the leaching of nitrate from topsoil layers. The nitrate anions are then held in the subsoil by positively charged clay surfaces. What the trees have done, in effect, is to expand the volume of soil used by a considerable amount.

NITROGEN OUTPUT

A typical maize crop in smallholder African farms yields less than 1 tonne per hectare of grain and requires a plant accumulation of less than 40 kg N per hectare. A 4 tonne per hectare maize crop requires 100 kg N per hectare while a 7 tonne crop requires 200 kg N (Sanchez 1976). Two-thirds of this nitrogen is accumulated in the grain and will be exported on harvest. Much of the remainder, located in the stover, may not be cycled back to the soil because it is frequently fed to livestock outside the system and manure is seldom brought back to the field where the maize was grown. Other losses through soil erosion, leaching and denitrification provide a similar nitrogen output to that of grain harvest removals (Smaling 1993). A negative annual balance of 112 kg N per hectare was calculated in maize farm by Smaling where nitrogen inputs totalled 55 kg and nitrogen outputs 167 kg per hectare.

NITROGEN FIXING TREES FOR AGROFORESTRY PRACTICES

TIMBER YIELDING NFTs: *Acacia* spp., *Dalbergia* spp., *Pterocarpus* spp., *Casuarina equisetifolia*

FORAGE YIELDING NFTs: *Albizia amara*, *Albizia lebbeck*, *Albizia procera*, *Albizia saman*, *Erythrina* spp., *Dalbergia sissoo*, *Leucaena leucocephala*, *Sesbania sesban*, *Sesbania grandiflora*, *Sesbania aculeata*, *Gliricidia sepium*, *Acacia nilotica*, *Acacia tortilis*, *Acacia senegal*, *Acacia auriculiformis*, *Faidherbia albida*, *Pithecellobium dulce*, *Calliandra calothyrsus*, *Prosopis cineraria*, *Prosopis juliflora*

GREEN MANURE YIELDING NFTs: *Sesbania grandiflora*, *Sesbania sesban*, *Gliricidia sepium*, *Leucaena leucocephala*, *Faidherbia albida*, *Acacia Senegal*, *Calliandra calothyrsus*, *Acacia tortilis*, *Casuarina equisetifolia*, *Erythrina* spp.,

FUELWOOD YIELDING NFTs: *Leucaena leucocephala*, *Acacia tortilis*, *Albizia lebbeck*, *Calliandra* spp., *Casuarina equisetifolia*, *Faidherbia albida*, *Acacia auriculiformis*, *Acacia senegal*, *Pithecellobium dulce*, *Flemingia* spp., *Gliricidia sepium*

NFTs FOR ARID AND SEMI-ARID REGIONS: *Albizia lebbeck*, *Faidherbia albida*, *Acacia tortilis*, *Acacia senegal*, *Cassia siamea*, *Pithecellobium dulce*, *Prosopis juliflora*, *Prosopis cineraria*, *Prosopis chilensis*, *Dalbergia sissoo*

NFTs FOR HUMID AND SUB-HUMID REGIONS: *Gliricidia sepium*, *Leucaena leucocephala*, *Mimosa scabrella*, *Albizia saman*, *Sesbania grandiflora*, *Sesbania sesban*, *Sesbania bispinosa*, *Casuarina equisetifolia*, *Acacia auriculiformis*, *Calliandra calothyrsus*

NFTs FOR TROPICAL HIGHLANDS: *Alnus nepalensis*, *Alnus acuminata*, *Casuarina oligodon*, *Albizia stipulata*, *Acacia mearnsii*, *Lespedeza bicolor*, *Inga jinicuil*, *Gleditsia triacanthos*.

■■■■

7 GENETIC IMPROVEMENT OF INDUSTRIAL AND NON-INDUSTRIAL AGROFORESTRY TREES

INTRODUCTION

The long history of selection and improvement of many agricultural crops is a contrast with that of forest tree species, where the shift from exploitation to domestication is much more recent. Sound management of genetic resources is critical to the optimization of reforestation efforts, for both industrial and non-industrial applications. For industrial forestry species, of which there are probably 100 million hectares of plantations worldwide, the most common approach to improvement is recurrent selection in genetically quite broad breeding populations. The use of open-pollinated general combiner seed orchards is the most common approach to the capture of genetic gain, while a few advanced programmes are characterised by the deployment of superior full-sib families or clones. Reproductive biology and patterns of variation are reasonably well documented for the established industrial species. Vigour, form and wood quality are the priority selection criteria in most programmes, with disease-resistance or insect-resistance of importance in only a few cases. The poplars are exceptional industrial species in many respects; hybridization and clonal selection in much narrower populations have been the traditional approach to improvement and, perhaps not coincidentally, disease resistance is a major selection criterion. Significant genetic gains are being achieved but, in particular for the long rotation species, there has been only a minor impact to date on the genetic quality of plantations. The major limitations to rapid improvement of most of the established industrial species are:

- Long generation intervals, related to poor juvenile-mature correlations, and the long juvenile phase before flowering.
- Low effectiveness of selection for many characters, due to low heritability or difficulty in assessment.
- Through the use of open-pollinated seed orchards, the exploitation of only a part of the genetic variation available.

In order to make use of a broader range of sites, and to supply products currently provided by harvesting of natural forests, it is likely that some of the additional plantation area will be established with tropical species which are not widely used or known at present. Selection criteria for these "new" industrial species will be similar to those for established species. Some are probably very amenable to improvement, while others present problems, e.g. flowering and seed problems, and insect susceptibility. Many potentially valuable species are simply not well known, in terms of their adaptation and biological and genetic features, and their gene pools are probably under threat. The broader implementation of good improvement programmes will be an important priority for these species. The testing of potentially useful species, characterization of mating systems, provenance collection, establishment of trials, implementation of gene conservation measures and commencement of other breeding work pose a very large task.

There is also a global requirement for several large programmes of non-industrial plantings over at least as large an area as the industrial plantations, mainly located in developing countries. Breeding programmes for non-industrial species differ markedly from those for industrial species in the wide variety, and the complexity, of selection criteria. Several of the taxa are highly variable and display features which render them particularly amenable to very rapid improvement by traditional means. Biological features of many potentially valuable non-industrial species, however, remain largely unknown. Although not well studied, some gene pools are under threat, and some species display unpredictable patterns of variation as a result of human activity. Although selection work has been undertaken in a few programmes, most non-industrial species remain at the species testing stage. The need to establish plantings on a wide range of sites, many of which are marginal or difficult, dictates that non-industrial plantings are likely to be reliant on a substantial number of the thousands of potentially useful species. The work involved in selecting the most promising species is formidable. The major emphasis of improvement for non-industrial species are likely to be exploration, collection, taxonomic studies of variation, species and provenance testing, the assessment of reproductive features, and conservation activities. Interspecific hybridization may be of considerable value for several taxa. Where regionally consistent selection criteria can be defined and the tree crop is considered sufficiently important, clonal selection programmes are likely to be of value for species easily propagated vegetatively, and seed tree selection for species propagated by seed. Sophisticated breeding programmes are unlikely to be warranted for most species. It is possible that a "once only" or "quantum leap" approach to breeding may be more appropriate than programmes offering continuous incremental gains. Both types of approaches will necessitate incorporation of genetic conservation concerns into the breeding programmes to ensure lasting gains.

All tree improvement programmes in developing countries are greatly under-resourced. Moreover, the funding provided through national, regional, bilateral and international programmes is not sufficient to conduct properly the essential activities outlined above, and where the continuity of action necessary is frequently missing. The problems are further compounded by inadequate levels of training and facilities in most areas. For the non-industrial species in particular, it is these resource limitations and the variability in user requirements, rather than biological constraints, which constitute the major impediment to rapid improvement.

In no other field of scientific investigation today are such rapid advances being made as in plant biotechnology. The potential benefits of plant biotechnology in forestry are perhaps even greater than in agriculture because of the possibility of gaining time in certain tree improvement processes. The challenges facing foresters regarding production or outturn, whether of wood or other products, are no less urgent than those facing agriculturists. Biotechnology is a collection of new technologies applied to biological systems rather than a separate academic discipline. Thus, it may be defined as the manipulation of biological systems (organisms or biological processes) via modern technologies to solve practical problems in biological science. Biotechnology may include clonal propagation via tissue culture, protoplast fusion, gene cloning, DNA recombination and genetic engineering, mutant induction and *in vitro* genetic selection. Molecular-marker-aided genome analysis and gene mapping are also sometimes included in plant biotechnology, as they are aspects of agriculture, horticulture, and forestry where genetic analysis and gene identification are used in plant breeding programmes. The large size, long regeneration cycle, and sporadic seed production of trees causes numerous problems in tree improvement programmes, such as the time in years required to reach reproductive maturity, low efficiency in producing new and improved germplasms, space limitations, time consumption and great cost.

Over the last several decades, forest genetics and tree improvement research has shown great advances. Traditionally, tree breeders have employed conventional methods, i.e., breeding and selection to make genetic combinations for creating new and improved forest stocks. Unfortunately, these traditional methods for introduction of desirable traits from one genotype to another are tedious, costly, time consuming, and not always successful. Even when potentially useful improved lines or hybrids are developed, many years of field testing are needed to prove their usefulness. However, the newly developing biotechnologies, such as tissue culture, genome analysis and gene identification, DNA recombination and genetic transformation, and RFLP/RAPD markers, provide the opportunities to solve the above problems and to accelerate tree improvement. Today geneticists are often trained to extrapolate information about what individual genes do in a model species when observed in a controlled environment and how homologous genes are likely to behave in complex genomes in a related field environment. While this approach may hold promise for simply inherited traits, the ability to understand the genetic basis of quantitatively inherited traits relies on the use of populations evaluated directly in relevant field environments. No doubt, trees are the crops for which biotechnology offers unique potential benefit towards improvement and utilization.

STATUS OF TREE IMPROVEMENT – A GLOBAL PERSPECTIVE

The general objective of a genetic improvement programme should be the sustainable management of genetic variation to produce, identify and multiply for the operational planting of well-adapted genotypes of the desired quality. Typically, this incorporates establishment of initial populations, including species and provenance testing, as well as the development of breeding and gene conservation populations; population improvement, frequently including recurrent cycles of selection and recombination and derivation and multiplication of strains to be used operationally. In principle, this applies to both industrial (i.e. biologically well-known species planted on a large scale) and non-industrial species. Tree improvement activities typically comprises of:

1. Practical tree improvement operations - species and provenance testing, establishment of gene conservation and breeding populations, recurrent cycles of selection and recombination, and propagation of selected lines (e.g. open-pollinated families or clones).
2. Supportive research - essential to run a breeding programme effectively. This includes the examination of factors such as breeding biology, mating patterns, genetic parameters (including variation) and seed problems.
3. Strategic research - aimed at the development of better breeding methods, e.g. through early selection.

While biotechnological projects currently constitute a substantial component of strategic research, practical breeding operations are largely conducted using traditional methods. Potential applications of biotechnology in the improvement of forest tree species have been well documented. While biotechnologies are expected ultimately to constitute supplements, and sometimes alternatives, to traditional methods now used in breeding programmes, clearly there still exists a wide gap between biotechnology and traditional methods. It has been estimated recently that, of a total of at least 50,000 tree species, 400 have been the subject of formal breeding or testing activities of some description, 140 with at least one generation of selection and mating, and 35 of these involving at least 20 seed parents.

IN DEVELOPED COUNTRIES

Tree improvement in developed countries mainly concerns industrial species, although agroforestry is becoming more important. Some form of genetic improvement programme is a feature of most plantation programmes for industrial species. Most are based on recurrent cycles of phenotypic selection (frequently followed by testing of progeny) and recombination among selected individuals. Vigour, stem form and wood quality are the most important selection criteria in most programmes. Resistance to disease or insects in particular, and also to cold and drought, are significant in several programmes. The importance of disease resistance in the poplars is worthy of special note. Most characters of major importance are under polygenic control. Although there are some clonal programmes, and programmes based on the propagation of full-sib families, the open-pollinated orchard is the most common approach to the capture of genetic gain. Gene pools are reasonably well preserved for most of these species, although locally adapted populations may be in danger of depletion or genetic pollution in some cases. Breeding activities (e.g. crossing programmes, progeny testing, and selection) are generally carried out under the guidance of personnel with expertise in quantitative plant breeding, supported by good systems of field experimentation and data base management.

Supportive research (e.g. assessment of reproductive behaviour, mating patterns, genetic parameters) is conducted frequently by the universities and other research organizations. Strategic research (aimed at the development of better breeding methods) is undertaken primarily by universities and other research organizations, although several of the larger private growers also conduct long term strategic research. Significant genetic gains are being achieved in these breeding programmes but, in particular for the long rotation species, there has been only a minor impact to date on the genetic quality of plantations.

Major research priorities for these species should be the broader development of methods for the propagation of full-sib families and clones, and the development of methods for early and more accurate selection and the promotion of precocious flowering. Silvicultural advances are critical for forest tree species, but it is unlikely that the intensity of horticulture given to fruit tree crops would be affordable. For some agricultural species, the genetic superiority of modern varieties requires intensive agriculture, including comprehensive pest and disease management, for its expression. Modern high yielding varieties tend to lack much of the natural resistance of old land races or wild relatives. The lower intensity of silviculture affordable renders desirable, then, the more careful management of genetic improvement for forest tree species.

IN DEVELOPING COUNTRIES

There are many examples of very successful industrial plantation programmes, including some of the world's best, in developing countries. These operations employ breeders, and in some cases hire consultants. An additional area of up to 100 million hectares of plantations will be required globally in order to meet future predicted demand for industrial timber. It is likely that the bulk of this additional area will be located in tropical countries. To some extent, this will be accomplished with already proven industrial species like tropical eucalypts, other established tropical hardwoods and conifers.

Selection criteria for "new" industrial species will be similar to those for established species. Some are probably very amenable to improvement, while others present problems, e.g. flowering and seed problems, and insect susceptibility. Many potentially valuable species are simply not well known, in terms of their adaptation and biological and genetic features, and gene pools are probably under threat. There is a global requirement also for several hundred million hectares at least of non-industrial plantings, mainly located in developing countries. It is noteworthy that while the plantation area under industrial species in the tropics doubled, the area under non-industrial species almost tripled. Non-industrial species differ markedly from industrial species in the wide variety, and the complexity, of selection criteria. Several of the taxa are highly variable and display features which render them particularly amenable to very rapid improvement by traditional means. Biological features of many potentially valuable non-industrial species, however, remain largely unknown. Although not well studied, some gene pools are under threat, and some species display unpredictable patterns of variation as a result of human disturbance. Although selection work has been undertaken in a few programmes, most non-industrial species remain at the species testing stage. There are some 2000 tree species that have been planted as "multipurpose" trees and many more species would have potential. The number of tree species of potential interest in developing countries is much larger than the number of food crops in these regions. Genetic improvement of the new industrial species and non-industrials is generally the responsibility of state, national, and in particular regional and international organizations such as FAO, agencies operating bilaterally, e.g. AIDAB, ACIAR, DANIDA, FINNIDA and GTZ, research organizations such as CAMCORE, CSIRO, CIRAD-Forêt, IBPGR, CIFOR, ICRAF, OFI, CATIE, and F/FRED, and many national institutes.

General limitations to tree improvement in developing countries are a lack of staff with the relevant skills, poor communication, particularly with forest managers and poor language skills. It is these resource limitations and the variability in user requirements, rather than biological limitations, which constitute the major impediment to the rapid improvement of non-industrial species.

GENETIC IMPROVEMENT OF INDUSTRIAL SPECIES

The term "industrial" is applied to species used predominantly in large scale plantations aimed mainly at the commercial production of timber. There are currently about 100 million hectares of industrial plantations worldwide, and it has been estimated that projected demands for industrial timber early next century might be met by a total estate of 100-200 million hectares. The industrial plantation species are grouped into broad classes though that some groups are quite heterogeneous, and programmes for many species straddle climatic and rotation length classes.

TROPICAL AND SUB-TROPICAL EUCALYPTS

These currently comprise less than 5 per cent of established industrial plantings, but this is a group of rapidly increasing importance due to the potential which exists for expanded plantings in many tropical countries. The most important taxa are *E. grandis*, *E. tereticornis* and *E. camaldulensis*, and notably hybrids between *E. grandis* and *E. urophylla* and the other two species. *E. grandis* is widely planted due to its fast growth, good silvicultural characteristics, and good pulping qualities. *E. camaldulensis* and *E. tereticornis*, more drought hardy than *E. grandis* but without the same ability to capture the site, are very widely planted, particularly in India and Africa. Other species with potential in tropical Africa, Asia and the Pacific, but much less widely used, include *E. microcorys*, *E. pilularis*, *E. cloeziana*, *E. citriodora*, *E. pellita*, and possibly *E. deglupta*. Private companies have featured prominently in the more well known plantings, but government enterprises are also important. By the standards of most industrial forest plantation species, the tropical eucalypts are relatively easy to breed for the below reasons:

- many flower abundantly and precociously
- they are generally characterised by short rotations, permitting early selection and short generations (e.g. generations of 4-7 years for *E. grandis*)
- although not as easy as for pines, controlled pollination is still quite achievable
- most coppice, and can be propagated reasonably easily by cuttings
- seeds can be stored satisfactorily.

A few very advanced breeding programmes exist, undertaken by both government institutes and private growers. Four generations of selection and open-pollinated recombination have yielded large gains in the Florida (USA) programme. Spectacular and well-publicized successes have been recorded with clonal propagation of hybrids in Brazil and the Congo, although the breeding programmes underlying these clonal selection programmes are not as advanced. Clonal programmes are underway also for *E. grandis* in other tropical countries, but are perhaps unlikely to result in the large gains reported for the hybrids. These advanced programmes represent a very small proportion of the world plantation estate in tropical eucalypts. Provenance and progeny trials have been planted in many areas, some selections made, and seedling seed orchards established, but much of the planting is still conducted with material subjected to little if any improvement. Vigour, form and wood quality are typically the most important selection criteria. Cold tolerance has been an important criterion in the *E. grandis* programme in Florida, although some freezes probably exceed the levels which can be tolerated by naturally available variants. Most programmes are characterized by open pollinated approaches to recombination, testing and capture of genetic gain. Genetic variation of the tropical eucalypts is fairly well preserved in *ex situ* and *in situ* stands.

To summarize, genetic improvement programmes offer clear benefits for industrial plantation forestry with tropical eucalypts, and no major biological impediments to the breeding of these species exist. In terms of practical priorities, there is a lot to be gained by the extension of existing technologies to additional programmes. For the pure species, open-pollinated approaches are relatively simple and give good gains. Insufficient understanding and application of sound breeding strategies is a limitation in some programmes. For the more advanced programmes, the availability of greater frost tolerance would be advantageous in some areas. The availability of better selection methods would circumvent the problem of the expense of clonal testing, and permit the imposition of higher selection intensities in the clonal programmes. Although generation intervals in these species are relatively short, large gains would be available through even earlier selection and flowering.

COLD-TOLERANT EUCALYPTS: Although comprising no more than 2-3 per cent of currently existing industrial plantations, this group is of expanding importance, particularly in plantation programmes by private companies in Chile, Africa and Europe. Around 50,000 ha of *Eucalyptus nitens* plantations have been established worldwide, while *E. globulus* plantations are much more extensive - over 10,00,000 ha in Portugal alone. Species for which plantings are more limited include *E. dunnii*, *E. smithii*, *E. radiata*, *E. fraxinoides*, *E. McArthurii* and *E. fastigata*. Although generally grown on short rotations and selectable at a similar age to the tropical eucalypts, these species are generally more difficult subjects for breeding due to:

- flowering of *E. globulus* is poor in some locations, and *E. nitens*, *E. dunnii* and *E. smithii* are all shy flowerers
- these four species are difficult to root from cuttings.

E. globulus is characterized by sophisticated and well-planned breeding programmes which have been established in recent years, e.g. the "nucleus breeding" approach in Portugal, and the "multiple population" programmes in South Africa and Chile. Clonal and seedling seed orchards have been established for these species, but are yet to produce seed in required quantities. Selection criteria are principally vigour, stem form, and wood quality. Resistance to winter freezing is of some interest in relation to possible wider planting of *E. globulus*. Genetic resources of these species are relatively secure. Addressing the flowering and rooting problems should be major priorities for more effective breeding of these species. Breeding programmes for these species would also benefit from the availability of greater frost tolerance, better selection methods, and even shorter generations.

OTHER TROPICAL HARDWOODS

Other tropical hardwoods comprising of less than 5 per cent of established industrial plantings, this heterogeneous group is of rapidly increasing importance with considerable potential in many tropical countries. Some species, in particular *Tectona grandis*, *Gmelina arborea* and the tropical Acacias, are already popular species in both private and public plantation operations. *Tectona grandis* is widely planted in Indonesia (over 10,00,000 ha), Thailand, India, several other Asian countries, South and Central America and in many parts of Africa. The species is moderately amenable to improvement - seedlings flower at age six to nine (age of first flowering is positively correlated with bole length to forking), yield of viable seed is often poor, seed can be stored, trees coppice, and juvenile cuttings root readily. Control pollination is feasible, but not easy. Long rotations and a consequent long selection interval are limitation on rapid breeding. Trials in several places demonstrate wide provenance variation. Breeding programmes are in existence in Thailand, India and elsewhere, and many clonal seed orchards have been established.

Acacia mangium has been planted extensively in Malaysia, and large plantings are continuing also in Indonesia. Other species such as *A. auriculiformis*, *A. crassicarpa* and *A. aulacocarpa* are also of interest. Seed production commences relatively early, seed can be stored satisfactorily, controlled pollination is feasible although not easy and juvenile cuttings root well. Broad provenance collections have been made and trials planted, and seedling seed orchards established at several centres. Gene pools are relatively secure.

Total plantings of *Gmelina arborea* probably stand at over 2,00,000 ha, with Brazil, West Africa, the Philippines and Malaysia the major centres. Potential exists in many other areas of the tropics, although site sensitivity is a limitation. The tree provides a very useful general purpose hardwood, and high quality pulp. With short rotations (7-12 years in some programmes), early flowering (at 3-5 years), reliable seed storage (under the right conditions), ready coppicing, and good rooting of cuttings, the species lends itself to rapid breeding, although only limited improvement has been undertaken. Controlled pollination is considerably more difficult than for the tropical eucalypts. Provenances have been quite widely tested, and seed orchards have been established in a few locations, e.g. Brazil, Malaysia, and the Philippines. Vegetative propagation of pheno-typic selections is practised in Malaysia. Selection criteria are mainly stem straightness, cylindricity and volume.

Triplochiton scleroxylon and *Terminalia superb* have been the subject of prove-nance/progeny tests and clonal trials in Africa, and plantations are being established. Selections are propagated by cuttings from stump coppice in the case of *Triplochiton*, and shoots from mature scions grafted onto young seedlings in the case of *Terminalia*. Rotation lengths and selection intervals are longer than for tropical eucalypts. *Paraserianthes falcataria* is planted in the Philippines and Malaysia, provenance/progeny trials have been established, and seedling seed orchards established. Gene conservation collections are being undertaken for *Bombacopsis quinata*, *Sterculia apetala*, *Alnus acuminata*, and *Vochysia hondurensis*. All have populations in danger of extinction or genetic erosion. Seedling seed orchards of *Bombacopsis quinata* have been established several countries in Central and South America including Honduras and Venezuela. Several members of the Meliaceae are potentially very valuable plantation species, e.g. *Swietenia macrophylla*, *S. humilis*, *S. mahogani* and *Cedrela odorata*. Damage by shoot borers, however, severely limits the extent to which plantations can be established. International provenance and progeny trials have been established of some of these species. Intraspecific variation in levels of resistance is evident, but the extent to which effective levels are available not determined. Some of these species have suffered severe genetic erosion. The use of valuable species of the Dipterocarpaceae (*Dipterocarpus alatus, Hopea* and *Shorea*) in plantations has been limited by flowering problems, seed recalcitrance, and silvicultural difficulties.

Some of these species are also threatened by erosion of gene pools. A large number of other species, as yet untested or poorly known, are of potential use, including species of *Octomeles, Cleistopholis, Pterocarpus* and *Canarium*.

Tropical hardwoods compare well with tropical eucalypts as subjects for breeding, while some are a little more difficult. In general though, breeding is not yet as advanced. Broader implementation of good improvement programmes is an important priority for these species. Some other species pose major problems: e.g. insect susceptibility in the Meliaceae, flowering and seed problems in Dipterocarpaceae. Many other potentially valuable species are simply not well known, in terms of their adaptation and biological and genetic features, and gene pools are probably under threat in parts of the species ranges. The testing of potentially useful species, characterization of mating systems, provenance collection, establishment of trials, implementation of gene conservation measures and commencement of other breeding work poses a very large and urgent task.

POPLARS AND WILLOWS

Comprising about 5-10 per cent of established industrial plantations, poplars and willows are important plantation trees in many countries, including in Europe, North America, China, Argentina, India, Africa, and Australia. Small private plantings feature strongly. The group includes a large number of species and subspecies. Relatively short rotations, early and prolific flowering and ease of vegetative propagation are features which render the poplars amenable to rapid breeding and good capture of genetic gain, although seed storage is a problem. Apart from the long history of improvement, two features of poplar breeding programmes distinguish them from those for most other forest tree species.

- The emphasis on between and within species hybridization, frequently involving well characterized clones, followed by clonal testing. Most commercial plantings comprise tested clones, frequently with cultivar names as for ornamental horticulture.
- The emphasis on disease resistance (e.g. to *Melampsora*, *Marssonina* and *Septoria*) as the major selection criterion with maintenance of growth characteristics a secondary objective. Disease resistance breeding has emerged in response to several clone-specific calamities.

The importance of the disease problem in poplars, and the long history of wide deployment of individual clones, may not be coincidental, and may provide a lesson for more considered and cautious deployment in clonal programmes with other species.An important factor limiting improvement in many poplar programmes is the availability of required levels of disease resistance. As above, large benefits would accrue through the development of better selection methods and shorter generation intervals.

OTHER TEMPERATE HARDWOODS

Although plantings of long rotation, high value species such as *Juglans* and *Prunus* are currently insignificant in total extent, it has been suggested that small private plantings of these could assume greater importance in the future on disused agricultural land. For these species, small breeding programmes are in existence. The long generation interval (related to both the long selection interval and the long delay to flowering), difficulty in controlled pollination, intermittent and light seed crops, and seed recalcitrance are limitations to the breeding of walnut. Orchards established many years ago are only now producing seed. Vigour and stem form are important selection criteria, but wood quality is also of particular importance. High values of the product have led to some use of grafted plantations, and patented clones are available. Unmanaged exploitation has led to scarcity of the species in many areas of its natural range. Breeding programmes incorporating clonal testing are underway for *Prunus* in Europe. With restricted current plantings and long generation intervals, breeding of these species beyond selection and either seed orchard establishment or propagation of superior genotypes is not readily justified. This is a group for which effective early selection would be particularly valuable.

Robinia pseudoacacia is a leguminous tree of considerable value in fast growing biomass plantations. Plantings stand at over 32,50,000 ha, with large areas in China, South Korea, and Hungary. A precocious flowerer, regular and abundant seeded, with seed readily stored, and easily propagated from shoot or root cuttings, the species is very amenable to breeding. Hungarian breeding programmes commenced several decades ago have been based on crossing (including with related species) and clonal selection, and many registered cultivars are available. Selection criteria are wood quality (in particular), frost tolerance, growth and form. The origin and genetic base of many plantings, however, are unknown. Breeding programmes incorporating provenance and progeny testing have been commenced. Greater frost resistance in the species would extend its range.

Acacia mearnsii is widely planted for bark (tannin), mining timber, poles and fuelwood especially in southern and eastern Africa, and China. Difficult to control pollinate and to propagate vegetatively, breeding programmes are based on the open-pollinated progeny trial/seeding seed orchard approach. *Alnus glutinosa* is a very precocious (as early as two to three years) and heavy seed producer. Provenance/progeny tests have been established and seed orchards have been planned. *Betula* species similarly flowerer precociously and abundantly, and greenhouse seed orchards in Finland are now producing seed. Species involved in these programmes are *B. verrucosa* and *B. pubescens*. A plus tree selection programme for *Betula verrucosa* commenced in Sweden in 1988, and progeny trials and clonal tests have been established. Small breeding programmes are in existence, and orchards have been established, for a number of other North American hardwoods, e.g. *Platanus occidentalis*, *Fraxinus pennsylvanica*, *Liquidambar styraciflua*, *Quercus rubra* and *Liriodendron tulipifera*.

MEDIUM ROTATION CONIFERS

Species of this group make up 35-40 per cent of the world plantation estate, and potential exists for further plantings in many developing countries. Included are the important *Pinus* species *P. taeda*, *P. elliottii*, *P. caribaea*, *P. radiata* and *P. pinaster*. *P. taeda* comprises about 80 per cent of some 1,00,00,000 ha of southern pine plantations in south-eastern USA, and substantial areas exist also in China (2,20,000 ha) and Brazil. *P. elliottii* makes up the major part of the remainder of the USA plantations, and is important also in China (8,80,000 ha), Brazil and South Africa (some 1,50,000 ha). Substantial plantations of *P. radiata* exist in Chile, New Zealand and Australia. *P. pinaster* is an important plantation species in France and Portugal, and is planted also in Chile and Australia. World plantings of *P. caribaea* are over 5,00,000 ha, mostly in Venezuela although the species is important also in Australia, Fiji and Brazil, and potentially so in many other tropical countries. Several other pine species are important regionally, e.g. *P. patula*, *P. rigida*, *P. palustris*, *P. echinata*, *P. virginiana* and *P. kesiya*. Some other pine species, e.g. *P. oocarpa* and *P. tecunumanii*, have potential in tropical regions. Although the seed development cycle is longer than for many hardwoods, most of these pine species have some features which facilitate breeding

- many flower reliably and well (although, for example, *P. caribaea* and *P. merkusii* flower poorly in some areas to which they have been introduced)
- controlled pollination is relatively easy
- seed can be stored readily
- grafting is easy

Cuttings can be propagated from young seedlings of many species, but propagation of cuttings from older trees generally is difficult. Well established breeding programmes are in existence, at or near the third generation of selections. A substantial proportion of the world's pine plantings are conducted with genetically improved material derived from the breeding programmes. The ease of controlled pollination has led to its frequent use in mating programmes. The large USA breeding programmes are typified by sophisticated mating designs for recombination and testing, e.g. diallels and factorials. Pair mating, polycrossing, or open pollination is more common elsewhere. Selection criteria are mainly vigour, stem form and wood quality, with disease resistance of some interest also, particularly for *P. taeda, P. elliottii* and *P. radiata.* The most advanced programmes employ controlled pollination also for the capture of genetic gain - to produce superior full-sib families, sometimes with multiplication by cuttings. The latter applies in particular to *P. radiata* in New Zealand and Australia, and to the hybrid between *P. elliottii* and *P. caribaea* in Queensland. Open pollination, however, remains the most common approach to the capture of genetic gain, and pollen contamination in seed orchards, is undoubtedly a cause of substantial leakage of this genetic gain in many programmes.

Major limitations to improvement are the long intervals to flowering and selection, inaccurate selection, and the poor rooting of cuttings from older trees. Gene pools for most species are reasonably well preserved in *ex situ* plantings. Genetic improvement offers clear benefits for planting programmes with species of this group. Broader application of good improvement programmes is an important priority, particularly in tropical countries. The broader development of technologies for the mass production of superior full-sib families would also be highly beneficial. In terms of research objectives in advanced breeding programmes, major benefits are likely to accrue from more effective control of the maturation state (particularly in relation to the rooting of cuttings), more effective selection, and shorter generation intervals.

LONG ROTATION TEMPERATE CONIFERS

Long rotation temperate confers comprises 35-40 per cent of global industrial plantations. Important species included here are *Pinus sylvestris*, *Picea abies*, other *Picea* species, e.g. *P. sitchensis*, *P. mariana*, *P. glauca* and *P. engelmannii*, *Pinus contorta*, *Pseudotsuga menziesii*, *Pinus banksiana*, *P. strobes* and *Larix* species. *Pinus contorta* and *P. banksiana* are reasonably precocious and reliable flowerers, but several of the other species mentioned are characterized by light and/or irregular flowering, e.g. *Pinus sylvestris*, *Picea abies* and *P. engelmannii*. The major obstacle to improvement, however, and the main difference from the medium rotation conifers, is the longer selection interval, rendering traditional breeding very slow.

GENETIC IMPROVEMENT OF NON-INDUSTRIAL SPECIES

There is a large requirement worldwide, mainly in developing countries, for fuel wood plantations, plantings to arrest and reverse land degradation, and, most importantly, tree plantings integrated into farming systems to provide mulch, fodder, shade, fuel wood, timber, fruit, and soil improvement and protection. Many of the species discussed in this category can be used in industrial plantations, and identification of a "non-industrial" group of species is a demarcation which is more convenient than clearly defined. Large fuel wood plantations, e.g. of tropical eucalypts or Casuarinas, located near cities are one approach to the urban fuel wood crisis. Operated on a commercial basis, these can be viewed as industrial species. Improvement beyond species and provenance selection may be difficult to justify on purely economic grounds, and may furthermore be of limited desirability in those cases where maintenance of diversity in such plantings is of major importance. Examples of integration of trees into farming systems include:

- The alley farming systems, incorporating the use of trees such as *Gliricidia* or *Leucaena* for protection and mulch production.
- Shade tree systems, e.g. *Inga* and *Erythrina* species grown as shade over coffee in Latin America.
- Trees grown principally for fodder and fuel wood in agroforestry systems, e.g. *Prosopis* and arid zone acacias.
- Trees grown in association with crops mainly for timber, e.g. *Paulownia* and *Dalbergia.*
- Windbreak plantings, e.g. of *Casuarina*.

Gliricidia is a fast growing tree used widely in the Philippines, India, Sri Lanka, Indonesia, East Africa and its native Central and South America for fodder, fuel wood, live fences, construction wood and mulch. Subjected to agricultural use for centuries, its true natural range is uncertain, and many land races have been created. Flowers are produced within the first year or two, seed production is prolific, and cuttings are easily propagated. Large collections have been made and some provenance comparisons carried out, but breeding is not very advanced. A breeding seed orchard of a particular desirable provenance will be established at the ICRAF research station in Zambia. Cuttings of very limited numbers of locally selected clones are the planting stock in some areas. Genetic variation with respect to a range of characters is substantial. Selection criteria vary with use, but timing and reliability of leaf production are important parameters for fodder production.

Leucaena is a long cultivated tree of importance in South East Asia, Nepal, India, Africa and Latin America. It is nitrogen fixer which provides good forage, green manure, firewood and small timber. Frequently used with other crops, it is also the major component of energy plantations established in the Philippines. *Leucaena leucocephala* seeds very precociously and prolifically, but is not easily propagated by cuttings. Unlike other tree species reviewed here, *L. leucocephala* is frequently self-pollinating, although this is not a feature of other species of the genus. With respect to genetic improvement, *Leucaena* has received more attention than most other non-industrial trees. Two directions have been followed in *L. leucocephala* improvement programme:

- Testing of self-pollinated lines
- Interspecific hybridization

In most areas, the psyllid (*Heteropsylla cubana)* causes serious damage and resistance to this pest is an important selection criterion. Others include biomass, cold tolerance, seedlessness, fodder quality, and dry season leaf retention. Although yield increases of well over 100 per cent have been achieved through simple selection, breeding is yet to have a major impact in planting programmes, many of which have suffered from off-site planting and a very narrow genetic base. For *L. leucocephala*, locally produced seed is what is generally used.

Prosopis is an arid zone genus of about 45 species, highly tolerant of drought and of difficult soils, grown widely on marginal land in India and countries in South America and with potential in many other countries. The trees provide fuel wood, construction timber and fodder. Coppicing is prolific. A range of provenance trials coordinated by FAO, involving several species, are in existence. Objectives of breeding include growth increases, reduction of thorniness, and improvement of fodder quality.

There are about 135 African acacias, of which *A. nilotica*, *A. tortilis*, *A. senegal* and *Faidherbia alba* are particularly important. Very drought resistant, these species provide fodder, timber, charcoal, shade, honey, gum arabic, tannins and dune stabilization in arid areas of Africa and the Indian subcontinent. Some species display salt tolerance. The species tend to be prolific seeders, seeds can be stored, and stumps produce coppice. Species are characterized by considerable genetic variation.

The genus *Paulownia* is represented by about nine species native to China, where there are reportedly 13,00,000 ha planted in mixed cultivation with crops. The species is suited to intercropping due to late leaf emergence and leaf fall, and a deep root system and produces timber within five to six years. *P. tomentosa* and *P. kawakamii* usually flower in the second year after planting, and *P. fortunei* and *P. catalpifolia* in the fifth to sixth year. Seeds can be stored satisfactorily, and cuttings can be propagated from both young seedlings and the roots of mature trees. In China, the approach to breeding envisaged is controlled pollination (in particular interspecific hybridization) and then clonal testing. Selection criteria are height (to minimize shading of crops) and freedom from the many diseases and pests. Planting stock currently used comprises selected clones of some good hybrids.

Dalbergia sissoo has long been valued in Pakistan, India and Nepal for its high value cabinet timber, fodder, fuelwood, charcoal, honey, medicinal properties, windbreak quality and capacity to fix nitrogen. It is used both with crops and as an industrial plantation species. The species flowers abundantly and regularly from age four, and good rooting of cuttings has led to their use as the most common method of propagation. The breeding system of the species is not well known whereas the selection criteria are wood yield and quality, but with consideration given also to fodder and honey properties.

Casuarina equisetifolia, *Casuarina jhunghuniana* and *C. cunninghamii* are used widely for paper and pulp industries as well as windbreak plantings and fuel wood production in areas of the tropics. *C. equisetifolia* comprises the major component of a shelterbelt covering over 10,00,000 ha in China, Coastal sand dune fixation is one of the most successful applications of this species. Both species are dioecious. For industrial plantations, provenance trial establishment, plus tree selection, and the establishment of Breeding Seed Orchards using the multiple population approach was achieved. A sterile hybrid between *C. junghuhniana* and *C. equisetifolia* is propagated vegetatively and widely planted.

Azadirachta indica is a promising species of uncertain natural distribution due to centuries of traditional use. Poor seed viability is a serious problem. Seed collections have been made and good selected provenances are available for commercial planting programme. Growth rate, insect resistance and azadirachtin content were the important selection criteria.

Moderately comprehensive data of non-industrial species on biological features influencing breeding are available for only a handful of species. Several of these taxa are highly variable and display features which render them particularly amenable to very rapid improvement by traditional means - precocious and abundant seeding, coppicing, and ease of vegetative propagation. Biological features of many potentially valuable non-industrial species remain largely unknown, and collection of data of this type will be an important component of the initial phase of any breeding programmes. Although not well studied, some gene pools are under threat, and some species display unpredictable patterns of variation as a result of human disturbance, rendering sampling strategies more difficult to design.

Major emphases of improvement for non-industrial species are likely to be taxonomic studies of variation, species and provenance testing, the assessment of some features of reproductive biology, and conservation activities. Where regionally consistent selection criteria can be defined, some clonal selection may be important for species easily propagated vegetatively, and perhaps seed tree selection for species propagated by seed. Sophisticated breeding programmes are unlikely to be warranted for most of these species. Interspecific hybridization is of proven value for some taxa, and holds promise also for others, but must be used in the context of a sound improvement programme.

BIOTECHNOLOGY AND ITS APPLICATIONS IN TREE IMPROVEMENT

Classical breeding which was practised for the past three thousand years with the advent of domestication surpassed the methods used by these ancient breeders, and classical breeders came into an era where selection is to be based on genetic characteristics involving DNA markers. Earlier breeders used phenotypic selection based on morphological characteristics which often take many cycles of breeding and backcrossing in order to place a desired characteristic from one parent into a genotype with suitable agronomic and quality characteristics. Today geneticists are often trained to extrapolate information about what individual genes do in a model species when observed in a controlled environment and how homologous genes are likely to behave in complex genomes in a related field environment. While this approach may hold promise for simply inherited traits, the ability to understand the genetic basis of quantitatively inherited traits relies on the use of populations evaluated directly in relevant field environments. Breeders currently use biotechnological approaches in order to facilitate the selection of favourable combinations of genes that occur naturally within a tree species. They also have opportunities to adjust particular traits or phenotypes via the introduction of alien genes using transgenics. In either case, it is information about genes and their functions that drives biotechnological applications in tree improvement. The introduction of biotechnological tools has revolutionized tree improvement.

MICROPROPAGATION

Over 1,000 plant species have been micropropagated, including over 100 forest tree species. It is reasonable to conclude that, with sufficient research effort, successful protocols could be developed for most tree species. In general then, a research project of this type might thus be described as short to medium term with a moderately high expectation of success.

For most industrial species, costs of planting stock and insufficient data regarding field performance remain major obstacles to be overcome before broader use of micropropagules as direct planting stock could be contemplated. Micropropagation has an immediate application in integrated clonal propagation systems featuring the commercial planting of cuttings harvested from rapidly multiplied, micropropagated stool plants of the selected clones. This approach is of value only in very advanced breeding programmes incorporating the identification of outstanding clones - currently only a few programmes with tropical eucalypts, and perhaps also some poplar programmes, but potentially also with other industrial species. Appropriate integration into breeding programmes is essential. Where clonal testing on a reasonable scale is possible and affordable, the current applicability of protocols mainly to juvenile material is not necessarily an impediment to the capture of good gains through clonal forestry. This conclusion, however, is dependent on the ability to store juvenile material for the period of a clonal test. Genetic variation in response, often substantial, is not likely to be a major problem where clonal testing can be preceded by screening for responsive genotypes, although demonstration of the absence of adverse correlations with economic traits is important. Breeding programmes with new industrial species and non-industrials are not sufficiently advanced to warrant much use of micropropagation in the short term.

Micropropagation may have wider application for the multiplication of stool plants of industrial species as breeding programmes become more advanced and other limitations to clonal forestry (e.g. maturation problems) are overcome. For some non-industrial tree species, micropropagation may ultimately have a role in the multiplication of selected varieties prior to release. Development of simple micropropagation protocols for those species for which such are not already available is therefore a useful research objective, but one which should not take priority over issues such as advancement of the breeding programme.

With considerable research, developments with somatic embryogenesis and artificial seed technology may overcome the planting stock cost limitation discussed above, and enable direct use of such propagules in plantation establishment. For industrial species then, development of these technologies is a useful long term research objective, but one which is best pursued with few appropriate model species, for which natural variation available within the taxon remains poorly investigated.

SOMACLONAL VARIATION

Assuming stability of the characteristics and the ability to regenerate from the variant cell lines, the use of somaclonal variation is likely to be of most interest where insufficient natural variation is available to provide the level sought for the trait; or Existing variation is not easily used in breeding. Many crop species in current use show greatly reduced genetic variability as a result of a long history of selection and breeding. Commercial and breeding populations of most forest tree species generally have not been subjected to the same reduction of genetic variation. Even for genetically broad populations, however, there still will be traits for which insufficient natural variation is available. A survey of populations of a number of pine species, for example, revealed no useful resistance to *Gremmeniella abietina*. Similarly, tolerance to many herbicides is not known to exist in angiosperm tree species and would thus be difficult to achieve through traditional breeding. Cold tolerance in many *Eucalyptus* species is another example. It is suggested that plants have present in their genomes a far richer array of genetic information than that expressed, that the phenotype represents a limited subset of the genes present, and that somehow cell culture gives access to some of this genetic resource.

A proven technique being available, somaclonal variation will be of valuc also where natural variation is difficult to manipulate. This approach may in some cases be less disruptive to the commercial genotype (and therefore quicker) than a backcrossing programme with a wild type. For forest tree species, breeding programmes for poplars are based on crossing desirable traits, in particular resistance, into commercial clones, and a successful somaclonal approach may offer some efficiencies, particularly if combined with *in vitro* selection.

In general though, somaclonal variation offers very little for the genetic improvement of most of the major industrial forest tree species - the quantitative traits of most interest e.g. vigour, stem form and wood properties, are not traits which have been usefully altered *in vitro*. In the longer term, the promised benefits of somaclonal variation may be more reliably delivered by genetic engineering approaches.

SOMATIC EMBRYOGENESIS/ SYNTHETIC SEEDS

The potential applications and importance of *in vitro* somatic embryogenesis and synthetic seeds in the improvement and utilization of industrial genetic resources are described below.

CLONAL PROPAGATION

Since both the growth of embryogenic cells and subsequent development of somatic embryos can be carried out in liquid medium, it is possible to combine somatic embryo-genesis with engineering technology to create large-scale mechanical or automated culture systems. Such systems are capable of producing propagules repetitively with labour input. In this process of repetitive somatic embryogenesis is initiated where by somatic embryos proliferate from the previously existing somatic embryo in order to produce clones.

RAISING SOMACLONAL VARIANTS IN TREE SPECIES

Embryos formed directly from pro-embryogenic determined cells (PEDCs) appear to produce relatively uniform clonal material, whereas the indirect pathway involving induced embryogenic determined cells (IEDCs) generates a high frequency of Somaclonal variants. Mutation during adventive embryogenesis may give rise to a mutant embryo which on germination would form a new strain of plant. For clonal propagation of tree species, somatic embryogenesis from nucellar cells may offer only rapid means or obtaining juvenile plants equivalent to seedlings with parental genotype.

SYNTHESIS OF ARTIFICIAL SEEDS

Artificial seeds are the living seeds like structures which are made experientially by a technique where somatic embryoids derived from plant tissue culture are encapsulated by a hydrogel and such encapsulated embryoids behave like a true seeds if grown in soil and can be used as a substitute for natural seeds.

SOURCE OF REGENERABLE PROTOPLAST SYSTEM

Embryogenic callus, suspension cultures, and somatic embryos have been employed as source of protoplast isolation for a range of species. Cells or tissues in these systems have demonstrated the potentiality to regeneration in culture and therefore, yield protoplast that are capable to forming whole plants.

GENETIC TRANSFORMATION

Repetitive embryos originate from single epidermal or sub epidermal cells which can readily be exposed to Agrobacterium. Thus the transformation technique applied to primary somatic embryos. Repetitive embryogenesis is also ideally suited to particle gun-mediated genetic transformation. Instead or recycling on Agrobacterium to mediate the transfer of genes into plant cells, the particle gun literally shoots DNA that has been precipitated onto particles of a heavy metals, into the plant cells. Embryogenic suspension cultures of the cotton and soybean, initiated cell lines following each firing of the gun. The transformed cell lines can then be induced to form an unlimited number of transformed somatic embryos through repetitive embryogenesis.

SYNTHETIC OF METABOLITES

The repetitive embryogenesis system is of potential use in the synthesis of metabolites such as pharmaceuticals and oils. Borage contains high level of Y-Linoleic acid, used as precursor of post-glandins or in the treatment of atopic eczema. Somatic embryos of borage also produce this metabolite but through repetitive somatic embryo-genesis a continuous supply of Y-lenolenic acid is ensured. Which otherwise would be limited to the growing season in the zygotic embryos. The same principle can be applied for production *in vitro* of industrial lubricant from jojoba and leo-palmitostearin from cacao.

IN VITRO SELECTION

Many recent publications concerning crop plants have reported useful correlations between *in vitro* responses and the expression of desirable field traits, most commonly disease resistance, although positive results are available also for tolerance to herbicides, metals, salt and low temperatures. For the selection criteria of major general importance in forest tree species, in particular vigour, stem form and wood quality, however, poor correlations with field responses will limit the usefulness of *in vitro* selection. *In vitro* selection is therefore likely to have very limited application in forest tree species.

PROTOPLAST FUSION

Many hybrids have been tested for established industrial plantation taxa such as *Eucalyptus*, *Pinus*, *Picea*, and *Larix*. A number of these hybrids are in commercial use, and it seems likely that hybrids will become of wider importance in industrial plantation forestry. Those currently in use are mostly reasonably easy to produce sexually. To be addressed then is the extent to which sexual incompatibility limits the production of other useful hybrids. This is difficult to assess. The introduction of disease resistance into commercial poplar clones has been limited by interspecific incompatibility, but other solutions to the problem have been identified. There are other instances where transfer of traits among distantly related species would be desirable - e.g. rust resistance in pines. Given the difficulties involved, however, and the fact that protoplast fusion research is empirical and likely to be long term anyway, genetic engineering may offer better hopes for the manipulation of such traits. Protoplast fusion has little apparent applicability to little known tropical hard woods and non-industrial species, for which investigations of the possibilities for even traditional hybridization remain minimal.

HAPLOID CULTURES

In *Brassica,* for example, production of haploid plantlets from anther culture has been used as a method of fixing gamete arrays of homozygous plants. An F_1 hybrid population is generated, and anther culture, with either spontaneous or colchicine induced chromosome doubling, is then used to fix true breeding homozygous diploid lines. In the absence of the anther culture technology, derivation of the homozygous lines would take several generations of inbreeding. Another culture therefore has had a direct and important application in the breeding of self-pollinated crops.

With the exception of *Leucaena leucocephala,* self-pollinating approaches are not a feature of breeding programmes for forest tree species, and the availability of haploids is not as immediately useful in these breeding programmes. Even if haploids can be obtained, the generation interval would still be an impediment in relation to consecutive testing of homozygotes and heterozygotes. This approach is not likely to be advantageous in a species which is easily propagated vegetatively. Some non-industrial species, e.g. *Gliricidia,* flower early and can be selected at an early age. Sophisticated breeding programmes are unlikely to be warranted, however, for these species.

Induction of haploid plants thus does not have immediate application in forest tree improvement programmes. Such plants may be of some use in basic genetic studies, e.g. for studies of heterosis in forest tree species. As a long-term strategic research objective for industrial species, induction of haploid plants should be of low priority until such time as methods for early selection and the promotion of early flowering are available.

SHOOT-TIP/ MERISTEM CULTURE

The multifarious applications of meristem culture in the improvement and utilization of both industrial and non-industrial wood species are:

Virus Elimination: Plants are often infected with more than one type of virus, including some even not known. A general term virus- free is used by commercial horticulturist for plants freed of any type of virus.

Micropropagation: A sexual or vegetative propagation of whole plants using tissue culture techniques referred to as micropropagation. Shoot tip or Meristem culture of many plant species can successfully be used for micro propagation.

Storage of Genetic Resources: Many plants produce seeds that are highly heterozygous in nature or that is recalcitrant. Such seeds are not accepted for storing genetic resources. So, the Meristem from such plants can be stored *in vitro*. The materials are preserved at -196°C for the long term preservation of germplasm.

Use in Plant Breeding: In many plant breeding experiments the hybrid plants produce abortive seeds or non viable seeds. As a result, it makes a barrier to crossibility in plants where non-viable seeds are unable to develop into mature plants. Shoot-tip or Meristem from such hybrid plant can be cultured to speed up breeding programme.

Propagation of Haploid Plants: Haploid plants derived from anther or pollen culture always remain sterile unless and until they are made homozygous diploid. Meristem or shoot-tip culture of haploid plants can be used for their propagation from which detailed genetic analysis can be done on the basis of morphologically character and biochemical assay.

Quarantine: There are some regulations concerning the international movement of vegetative plant material. After thoroughly checking, the plant materials may be rejected by quarantine authority if the plant material carries some diseases. Plantlets derived from shoot-tip or meristem cultures are easily accepted by the quarantine authority for international exchange without any checking. Therefore, using this technique, crop plants can be easily exchanged in crop improvement programmes that are based on materials from different parts of the world.

GAMETOPHYTE CULTURES

Although anther culture has been used for the rapid production of homozygous lines in the breeding of some self-pollinated cereals and vegetable crops, few reports exist of regeneration from gametophyte cultures of forest tree species. Research with this objective would be likely to be long-term with a relatively high risk of an unsuccessful result. Induction of haploid plants has no immediate application in forest tree improvement programmes. Such plants may be of some use in basic genetic studies, e.g. for studies of heterosis in forest tree species. As a long term strategic research objective for industrial species, induction of haploid plants should be of low priority until such time as methods for early selection and the promotion of early flowering are available. The breeding of non-industrial species is unlikely to warrant this level of sophistication.

IN VITRO EMBRYO RESCUE

This technique has been used to grow embryos that normally would abort due to incompatibility between ovule and embryo development, and also for the rescue of the zygotic embryos of apomictic species. The techniques are not difficult, and development of protocols for a new species generally would be a minor research task - short term with a high expectation of success.

Embryo rescue has been used occasionally in forest tree species, but the requirement for such technology is likely to be limited - to what probably will be a small number of hybrids which are sufficiently close to produce a normal embryo but for which embryo development *in vivo* is restricted. In the short term, research of this type is likely to have a very low priority. In the longer term, as barriers to natural hybridization become better defined, some work can be targeted at hybrids which species testing has identified as potentially of interest and for which other studies suggest that *in vitro* approaches might provide a solution.

IN VITRO CONTROL OF THE MATURATION STATE

A long history of research with *in vitro* rejuvenation has led to some successes, but little evidence that rejuvenation can be completely, permanently and reliably achieved by this method. Similarly, sporadic reports of accelerated maturation through *in vitro* manipulations are not encouraging with respect to reliable acceleration of ageing. Further empirical work with these objectives is likely to have a low probability of success. An understanding of the molecular basis of maturation is much more likely to lead to practical manipulation, but this work is in its infancy, and acceleration or reversal of maturation to precise levels remains a distant prospect.

For clonal forestry with industrial species, maintenance of juvenility as is about as useful as rejuvenation for many purposes, and probably achievable using technologies such as cryopreservation or stimulation of juvenile stump sprouts. Nevertheless, more fundamental control of the maturation state remains one of the most valuable objectives of long-term strategic research in forest tree improvement with industrial species. Rejuvenation is most applicable to programmes where good breeding programmes are in place, and where other limitations to clonal forestry do not exist. Maturation is a much less important issue with many of the non-industrials. Manipulation of the maturation state to induce early flowering and reduce generation intervals is potentially of greater interest than rejuvenation, for industrial species at least, but only of real value where active breeding programmes are in place.

CRYOPRESERVATION AND *IN VITRO* STORAGE

GERMPLASM PRESERVATION

Undoubtedly, urgent gene conservation and preservation problems exist for forest trees species, mainly among the tropical hardwoods and non-industrial species. The following features characterize these tree species:

- There is a large number of species of known or potential interest. Many may never be used commercially, and some will be important only locally.
- Biological features of many are poorly known.
- Species distributions, and those of relatives, are poorly known.
- Seeds of some species will be in the recalcitrant category, but many are orthodox.

Major impediments to the preservation of these gene pools are:

- Resources available are sufficient for only a very small fraction of the survey and collection work which would be required.
- Seed storage facilities in many tropical countries are unreliable, such that satisfactory storage of seed collected now cannot be guaranteed.

In vitro storage and cryopreservation plays a significant role and act as a complementary method for the *ex situ* conservation of long-lived perennial species. It can be observed that replacement of existing seed storage facilities with more sophisticated technology is not likely to make a positive impact on the gene conservation problem with tropical tree species. Even for the recalcitrant species, activities should favour the establishment of *ex situ* plantings which will enable evaluation of the material. Cryopreservation and *in vitro* storage techniques thus offer no solution to erosion of tropical forest gene pools. Of established industrial species, cryopreservation and *in vitro* storage of vegetative material are not advantageous approaches to gene conservation in most of the conifers and eucalypts, for which seed can be stored satisfactorily. Poplars display problem in seed storage, and breeding and commercial use based substantially on known clones. *In vitro* storage or cryopreservation may have a role as a back-up measure in large poplar breeding programmes.

Similarly, it also plays important role for other species with seed production and storage problems, e.g. *Juglans*. It should be noted though that, for most forest tree species, introduction of *in vitro* or cryopreserved material into a breeding programme will involve a delay of several years to flowering, and generally also for evaluation. Flowering genotypes in field clone banks can be much more quickly incorporated. An exception to the flowering obstacle is stored pollen, which can be rapidly incorporated into breeding programmes, and pollen storage should be more prominent in gene conservation programmes with industrial species. Cryopreservation may facilitate the broader use of pollen storage in these programmes.It should be noted also that, the use of cryopreservation and *in vitro* storage could involve selection for genotypes which tolerate the conditions imposed (e.g. freezing and thawing), and which are capable of regeneration following such treatment. The conserved material is thus not necessarily a random sample of the genetic material collected.

MAINTENANCE OF JUVENILITY

Cryopreservation is perhaps the most promising approach currently on offer for maintenance of the juvenile state and capture of genetic gains through clonal forestry with industrial species. The technology is applicable mainly in cases where good breeding programmes are in place and clonal forestry is a realistic goal. In particular, this applies to some programmes with conifers. Although some good breeding programmes are in place for the Eucalyptus, the need is not so urgent for these because rejuvenation can be achieved by other means. The approach ultimately may have some application to other tropical hardwoods, as good breeding programmes are established for these. Maturation is a much less important issue with many of the non-industrials, in particular those which flower precociously, can be selected early, and which can still be readily propagated at selection age.

EXCHANGE OF GERMPLASM

The use of plants, organs or cells cultured *in vitro* circumvents to some extent the limitations which quarantine regulations place on the import of vegetative material (in particular) into many countries. International germplasm exchanges are becoming very important in the breeding of many crops, in particular to and from international gene banks at the International Agricultural Research Centres (IARCs) and other centres. In addition, *in vitro* collection, overcoming problems of inadequate or immature seeds and deterioration of vegetative material in transit, is now being used. With the increasing emphasis on cooperative breeding of both industrial and non-industrial forest tree species in tropical countries, international movement of material will become increasingly important. In the short term, such movement is likely to be mainly in the form of seed for species and provenance testing. Movement of vegetative material, however, may become more important as breeding programmes advance, and *in vitro* approaches could find application in the longer term.

MOLECULAR MARKERS

The potential benefits of using markers linked to genes of interest in breeding programmes, thus moving from phenotype-based towards genotype based selection, have been obvious for many decades. However, realization of this potential was limited by the lack of markers. Molecular markers make it possible to accelerate the plant breeding process because it is possible to generate high density linkage maps of traits and markers and use them in many genetic backgrounds as required in a breeding programme. Having access to thousands of markers linked to traits makes it possible to select the genotype of interest based on markers and so to limit field and breeding activities to plants with the genotype of interest. Some traits and markers are conserved across related species and so comparative genomics of traits and markers between crops also helps in a breeding process. Molecular markers and marker mapping are a part of the intrusive new genetics that is pushing its way into all areas of modern biology, from genomics to breeding, from transgenics to developmental biology, from systematics to ecology, and even, perhaps especially, into plant and crop physiology.

However, because genes do not function as single entities, it is necessary to know how numerous genes function together. This, in turn, requires the knowledge of the potential and constraints of biological functions of plants. The understanding of the interaction between genes, organs and environmental factors, which include other organisms, is a major challenge for plant biologists. To obtain this information, it is important to exploit the tools of classical and molecular genetics.

RANDOM AMPLIFIED POLYMORPHIC DNA (RAPD)

RAPD technique is based on the polymerase chain reaction and has been one of the most commonly used molecular techniques to develop DNA markers. RAPD markers are amplification products of anonymous DNA sequences using single, short and arbitrary (10mers) oligonucleotide primers. RAPDs are much simpler and less expensive to work with than RFLPs because no prior knowledge of sequences is required and there is no need for radioactive probes. RAPDs produce DNA profiles of varying complexity, depending on the primer and template used. The primer amplifies pieces of DNA between 200 and 2,000 kb long, which lies between two inverted copies of itself, one copy binding to each strand of the DNA. The methodology is rapid and can be used on material without prior information on genome. A major drawback of the technique is that because of the random nature of their generation, and short primer length, they cannot be easily transferred between species. An additional drawback is that of poor reliability and reproducibility, and their sensitivity to experimental conditions.

RESTRICTION FRAGMENT LENGTH POLYMORPHISM (RFLP)

RFLP is a difference in homologous DNA sequences that can be detected by the presence of fragments of different lengths after digestion of the DNA sample into pieces with restriction enzymes and then analysing the size of the restricted fragments by gel electrophoresis. It is the sequence that makes DNA from different sources different, and RFLP analysis is a technique that can identify some differences in sequence (when they occur in a restriction site). RFLP, as a molecular marker, is specific to a single clone/ restriction enzyme combination. RFLP markers that are used for high density genomic mapping provided a new technique which overcame some of the problems associated with isozymes and proteins. RFLPs, being codominant markers, can detect the coupling phase of DNA molecules, as DNA fragments from all homologous chromosomes are detected. They are very reliable markers in linkage analysis and breeding and can easily determine if a linked trait is present in a homozygous or heterozygous state in an individual, which is information highly desirable for recessive traits. However, their utility has been hampered due to the large amount of DNA required for restriction digestion and Southern blotting. The requirement for radioactive isotope makes the analysis relatively expensive and hazardous. The assay is time-consuming and labour intensive and only one out of several markers may be polymorphic, which is highly inconvenient especially for crosses between closely-related species. Their inability to detect single base changes restricts their use in detecting point mutations occurring within the regions at which they are detecting polymorphism.

AMPLIFIED FRAGMENT LENGTH POLYMORPHISM (AFLP)

AFLP is a highly sensitive method for fingerprinting genomic DNA within any organism. Restriction endonucleases such as Mse1 and EcoR1 are used to digest the DNA before amplification. A subset of resultant fragments representing many loci is then ligated to synthetic adaptors and amplified with specified primers which are complementary to a selective sequence on the adaptors. The amplified fragments are then to be analysed using denaturing polyacrylamide gel electrophoresis which generates fingerprints to be compared as polymorphisms. The particular band differences can be cloned and sequenced to identify the source of the polymorphism. This method constitutes a rapid screening for polymorphisms. These polymorphisms are typically inherited in a Mendelian genetics fashion, enabling their use for typing, identification and mapping of genetic characteristics. AFLP analysis depicts unique fingerprints regardless of the origin and complexity of the genome. Most AFLP fragments correspond to unique positions on the genome and hence can be exploited as landmarks in genetic and physical mapping. Applications of this technique reach far beyond agricultural applications, ranging from agronomic trait analysis, diagnostics, pedigree analysis, forensics, parental heritage and may be used as a universal fingerprinting system.

MICROSATELLITES OR SIMPLE SEQUENCE REPEATS (SSRs)

These are short segments of DNA that are derived from short (usually < 6 base pairs) tandemly repeated sequences such as (GA) n, (AAT) n, (GT) n and are known as microsatellites. Many of them are located in centromeric regions, telomeric regions and in the roots of chromatin loops, whilst others are thought to play a role in pairing and synapsis of chromosomes. Though these markers are neutral and codominant, yet the high development cost is a major impediment to the routine application of SSRs in the genetic study of non-commercial species and for identifying markers located in chromosomal regions of interest. Point mutation in the primer annealing sites in some species may lead to the occurrence of null alleles, where microsatellites fail to amplify. To develop the locus-specific SSR markers, the isolation and characterization of individual loci and the construction and screening of a DNA library with microsatellite-specific probes, followed by DNA sequencing of positive clones are required. Because these regions are hyper variable, RFLP analysis with probes for micro- and minisatellites gives multilocus patterns which have resolved variation at the levels of populations and individuals. SSR markers are considered highly polymorphic as the number of repeats varies due to slippage of DNA polymerase during replication and unequal crossing over and thereby allowing us to detect many different alleles for that marker. The frequency of SSRs is higher in transcribed regions than in non-transcribed regions. SSRs are the markers of choice due to their codominant expression, multiallelism and high PIC value.

MINISATELLITES

A minisatellite (also referred as VNTR) is a section of DNA that consists of a short series of bases about 10–60 bp in length. Minisatellites consist of repetitive, generally GC-rich, variant repeats. These variant repeats are tandemly intermingled, which makes minisatellites ideal for studying DNA turnover mechanisms and have been used extensively in many areas of genetics. Minisatellites have been associated with chromosome fragile sites and are proximal to a number of recurrent translocation breakpoints. The hypervariable minisatellites are 9–24 bp long and are found mainly at the centromeric regions whereas the telomeric minisatellites are 6 bp long with repeated sequences in telomeres.

CLEAVED AMPLIFIED POLYMORPHIC SEQUENCES (CAPS)

CAPS are DNA fragments amplified using specific primers, which are afterwards digested by restriction enzymes. Sequence polymorphisms result in the cutting of products in different places, and these variants are revealed as length differences on agarose gels. The CAPS approach is sometimes known as restriction fragment length polymorphism (RFLP) PCR, and the technique bears similarities to the non-PCR-based older RFLP method. CAPS can be applied to organism-specific nuclear sequences, or to organellar DNA using universal primers. As with SSRs, sequencing is generally required in the former case in order to develop primer pairs. Similarly to SSRs, CAPS assess variation at one locus only in a particular PCR.

SINGLE NUCLEOTIDE POLYMORPHISMS (SNPs)

SNPs markers are known as the third generation markers, which are nowadays extensively used in various genomic studies for individual genotyping. Single nucleotide polymorphisms or SNPs (pronounced "snips") are DNA sequence variations that occur due to point mutations when a single nucleotide (A, T, C, or G) in the genome sequence is altered. Single nucleotide polymorphisms are single base changes in the genome that occur at a frequency of about 1 per cent and can be particularly useful in linkage mapping. They are derived from sequence information, particularly where DNA sequence (either from genomic DNA or cDNA) is available from more than one individual. Informatics tools can be used to compare the sequences and identify variations. These variable nucleotides are subsequently checked to determine whether or not the SNP is real or an artifact. The use of such informatics tools often depends on the availability of appropriate sequence data such as sequence trace files. If each polymorphism were to be inherited independently, then this would generate an exceptionally large number of haplotypes. However, this is not the observed result. Relatively few haplotypes are observed, indicating that perhaps the rate at which SNPs arise is somewhat akin to the rate at which recombination occurs across these small regions leading to a much lower number of haplotypes than would be expected from the number of polymorphic sites. Therefore, SNPs are most likely to be useful as defining haplotypes, rather than for their information individually, and so the use of SNPs is likely to involve linkage disequilib-rium studies using the haplotype rather than specific SNPs as individual molecular markers.

EXPRESSED SEQUENCE TAGS (ESTs)

An Expressed Sequence Tag or EST is a short (300–500 bp) sub-sequence of a cDNA sequence. The cDNAs used for EST generation are individual clones from a cDNA library which are complementary to mRNA, so the ESTs represent portions of expressed genes. ESTs are used to identify gene transcripts, and play an important role in gene discovery and gene sequence. Since the ESTs are often partial sequences that correspond to the same mRNA of an organism, they are assembled into contigs so as to reduce the number of expressed sequence tags for downstream gene discovery analysis.

USES OF MOLECULAR MARKERS

Modern biotechnology provides new tools such as molecular markers that can facilitate the development of improved plant breeding methods and augment our knowledge of plant genetics. Molecular markers can be used to study natural, managed and cultivated tree stands, and can measure the extent to what individuals and populations are connected to each other. They are able to establish the breeding systems of populations, to determine relationships among different taxa, to assess hybridization and other interactions between species. The primary use of molecular markers has been the generation of genetic maps. The initial purpose of these maps has been to localize the position of genes of interest (those underlying a specific identifiable trait) for either positional cloning or marker assisted selection (MAS). The various uses of molecular markers are:

CHARACTERIZATION OF BREEDING POPULATIONS: Breeding populations can be characterized by quantifying the levels and organization of genetic variation within and between different breeding groups. Marker-based systems have been used to study and compare the levels of random genetic variation throughout the different cycles of a breeding programme, thus allowing much greater flexibility and control over the rate of reduction of genetic variability. RAPD markers were successfully used to characterize a wide range of genetic variation in the germplasm bank of *E. globulus* and thereby assist in the designing of further seed collections. The application of molecular markers for directional selection is still an unfulfilled promise due to the recent domestication of tree crops resulting in the wide genetic heterogeneity of breeding populations and the inability to develop inbred lines at least on a short-term basis and also to allow a more precise understanding of genetic architecture of quantitative traits. Different markers like ISSRs and SSRs can be used to study the genetic relationship between different accessions.

BIOLOGICAL STUDIES: The pollen contamination in seed orchards and crossing programmes is widespread, and is likely to be eroding substantially genetic gains expected from breeding programmes based on open pollination or even controlled pollination. Similarly inbreeding, or distorted mating patterns, could be resulting in loss of genetic gain in open pollinated orchards. A need exists, therefore, for monitoring procedures as a routine part of tree improvement programmes with industrial species. For many tropical hardwoods and non-industrial species, species biology, in particular mating systems, is poorly known. Molecular markers are a very powerful tool for these studies. Isozymes are suitable for most of these purposes. Studies aimed at understanding gene expression and the nature of phenomena such as heterosis, perhaps the most important area of strategic research related to tree improvement, have been revolutionized by the availability of DNA markers.

GENETIC FINGERPRINTING: Misidentification of clones in seed orchard establishment or crossing programmes is probably very common. Incorrectly identified ramets in seed orchards have resulted in loss of genetic gain. The economic implications are likely to be even greater if these errors occur in clonal forestry programmes. Molecular markers will therefore have an important role in quality control in advanced breeding programmes. Although isozymes will be satisfactory for some verification problems, DNA markers are much more powerful for genotype identification, and will be necessary where identification is to be more or less exclusive. Such identification would be required for Plant Variety Rights (PVR) registration. It seems unlikely though that PVR would be of widespread importance in forestry - registration of all of the clones of interest to a programme would be expensive, and Genotype × Environment interaction (GXE) is likely to limit the extent to which clones could confidently be transferred from one programme to another without prior testing.

MARKER-ASSISTED SELECTION: Marker-assisted selection (MAS), sometimes also called marker-aided selection, is a relatively new tool for plant breeding which is primarily based on the phenotypic selection of superior individuals among segregating progenies resulting from hybridization. In its simplest form it can be applied to replace the evaluation of a trait that is difficult or expensive to evaluate. When a marker is found that co-segregates with a major gene for an important trait, it may be easier and cheaper to screen for the presence of the marker allele linked to the gene, than to evaluate the trait. In trees the long generation times along with poor juvenile-mature trait correlations have promoted interest in marker assisted selection. From time to time the linkage between the marker and the gene should then be verified. When more complex, polygenic controlled traits are concerned, the breeder is faced with the problem how to combine as many as possible beneficiary alleles for the QTLs that were detected. In this case the breeding material can be screened for markers that are linked to QTLs. Based on such an analysis, specific crosses can be devised for the creation of an optimal genotype, combining beneficiary QTL alleles from different sources. With the development and availability of an array of molecular markers and dense molecular genetic maps in crop plants, MAS has become possible for traits governed by both major genes and quantitative trait loci.

Marker-assisted selection may be used to facilitate the controlled inflow of new genetic material. 'Wild' or unadapted material often carries desired components that may be missing in cultivated material. Such components can be transferred to elite cultivated material by repeated backcrossing in the unadapted material. In a backcross programme, the presence of the desired QTL-alleles can be verified continuously by observing linked markers. At the same time, markers provide information on the origin of the remaining genome, allowing selection within the backcross material for genotypes that have lost the majority of unwanted donor DNA. Usually the application of this marker-assisted backcross procedure will also result in a reduction of the number of backcross generations that are required, thereby speeding up the breeding programme.

GENETIC LINKAGE MAPS: Segregation analysis is important for the population that is derived from a common set of ancestors. Markers that co-segregate (either are always present or absent together) must be linked, i.e. they must be located in each other's vicinity on the genome. In some cases, however, due to recombination events, the linkage between the markers may be lost. The frequency with which the linkage between co-segregating markers is broken is an indication of the genetic distance between the markers. An extensive analysis of the linkage between a large number of molecular markers yields information on their arrangement on the genome. Such an analysis can finally result in the construction of a genetic map, on which all markers are arranged in separate linkage groups or chromosomes. The main applications of genetic maps to plants are to provide the basic knowledge of genomic structure and mapping and detection of complex quantitative trait loci. Nowadays, software for the calculation of genetic maps has brought marker analyses that aim at the construction of genetic maps. Genetic linkage maps can provide a more direct method for selecting desirable genes via their linkage to easily detectable molecular markers. The initial application of the molecular marker maps was to facilitate map-based cloning, almost exclusively of Mendelian genes. Another use of these maps has been in marker-assisted selection. Having closely linked markers to the trait of interest greatly reduces the effort and number of progeny needed to introgress a gene of interest, especially if that gene has a large, measurable effect in the progeny. As the costs of marker typing decrease, the value of their use in breeding programmes will be dominated by their returns, rather than by their costs. Under the appropriate conditions, markers can replace phenotypic selection, thereby removing the need for growing or rearing of individuals.

QUANTITATIVE TRAIT LOCI

Genetic factors that are responsible for a part of the observed phenotypic variation for a quantitative trait can be called quantitative trait loci (QTLs). Although similar to a gene, a QTL merely indicates a region on the genome, and could be comprised of one or more functional genes. QTL are identified via statistical procedures that integrate genotypic and phenotypic data. QTL are assigned to chromosome locations based on the positions of markers on a linkage map. QTL are located to regions of the genome at specified levels of statistical probability. In a process called QTL-mapping the association between observed trait values and the presence/absence of alleles of markers that have been mapped onto a linkage map is analyzed. Mapping QTL in tree species is not as simple as mapping the gene that affects a qualitative trait. The major obstacle is the time necessary to complete a breeding generation. Tree breeding is made even more difficult by the changes that occur during the transition from juvenility to maturity. When it is significantly clear that the correlation that is observed did not result from some random process, it is proclaimed that a QTL is detected. Also the size of the allelic effect of the detected QTL can be estimated. A breeder can analyze QTL occurrences and use this knowledge to his advantage, for instance by using indirect selection. When selection is based on genetic information retrieved through the application of molecular markers, this is called marker-assisted selection. Once these traits are identified and mapped, map-based cloning becomes more practical to identify the gene(s) responsible for a QTL as the density of markers increases in the regions containing QTLs increases. QTL mapping in trees has been carried out using both three-generation outbred pedigrees and two-generation pedigrees involving crosses between highly heterozygous parents (pseudo-testcross strategy). Based on a detailed understanding of the molecular architecture of quantitative traits, current applications include yield-oriented advanced backcross QTL (ABQTL) systems as well as accelerated line conversion following trait introgression by marker-assisted backcrossing (MABC).

MICROARRAY-BASED MAPPING

DNA microarray technology has given rise to the study of functional genomics. The entire set of genes of an organism can be microarrayed on an area as small as a fingernail and the expression levels of thousands of genes are simultaneously studied in a single experiment. Microarrays have demonstrated significant power for genome-wide analyses of gene expression, and recently have also revolutionized the genetic analysis of segregating populations by genotyping thousands of loci in a single assay. DNA microarray technology allows comparisons of gene expression levels on a genomic scale in all kinds of combinations of samples derived from normal and diseased tissues, treated and non-treated time courses, and different stages of differentiation or development. Further computational analysis of microarray data allows the classification of known or unknown genes by their mRNA expression patterns. Global gene expression profiles in cells or tissues will provide us with a better understanding of the molecular basis of phenotype, pathology, or treatment of diseases. The markers of choice, at present, for this type of analysis are SNPs. MALDI-TOF can be used to screen large numbers of samples at many loci. This can also be used to determine allelic frequencies in phenotypically defined pools of individuals. As the number of SNPs increases and their placement on the genetic map is achieved, the number of genotypes that are needed to confirm an association will be reduced.

TAXONOMY AND PHYLOGENETIC STUDIES

Sound knowledge of the taxonomic boundaries of a species, and of its relationships to other taxa, is fundamental to tree improvement. In addition, accurate taxonomy will be required to define species for legal purposes associated with conservation programmes. The taxonomic relationships of most established plantation species have been reasonably well defined by traditional means, although markers are now being used for refinement. It is in the tropical hardwood and non-industrial groups, however, that many taxonomic relationships remain poorly defined, and molecular markers will be a valuable supplement to morphological approaches.

GENETIC ENGINEERING

Crops transformed with genes for insect and virus resistance and resistance to various types of herbicides are at or near commercial application, and plants into which these genes have been inserted include the poplars. Many projects are in progress with forest tree species, in particular for modification of lignin biosynthesis through antisense technology. Insertion of the insect or herbicide resistance genes currently available into a new species would constitute a major research undertaking, and successful application would be dependent on being able to regenerate from the transformed cells. Manipulation of more complex traits would be a much more formidable undertaking, and, although rapid progress is being made in this field, much research remains to be done. The availability of effective transformation techniques remains an obstacle, but improved techniques are being developed. Regeneration is a difficulty for some tree species, but the problem may be over-rated. The non-competence of mature material is not necessarily an obstacle to effective application of genetic engineering, provided that juvenile material responds satisfactorily. An often overlooked research component is the testing which would be required before a responsible recommendation for large scale deployment of trangenic plants could be made. Such testing could be extensive and prolonged, depending on the species and genes involved. Research projects of this type are necessarily intensive, and must be regarded as long term with only a modest expectation of success.

Insect resistance is of potential value, e.g. in the poplars and some tropical hardwoods. A single gene may be sufficient to confer resistance for very short rotation species, but a much more cautious approach should be applied with long rotation species. The work involved in introducing several different resistance genes, sufficient to ensure that insects do not acquire tolerance during a long rotation, should not be underestimated. The reduction of lignin biosynthesis is a very valuable objective for the pulp species. The introduction of herbicide tolerance gene is of some interest, but in many programmes the advantage of using unguarded herbicide application may not be sufficient to pay for the research programme. The extent to which genetic engineering permits substitution of environmentally "friendly" herbicides such as glyphosate for currently used residual herbicides could be an important factor. Cold tolerance genes are likely to be of some commercial value in many species, in particular the eucalypts. Much remains to be done, though, to establish that sufficient tolerance can be conferred using antifreeze proteins, and to extend the work to tree species. Prevention of the escape of genes into wild populations is likely to become an important concern, and sterility should be an early target of genetic engineering work with forest tree species. The major factor limiting application of genetic engineering in forest tree species is the state of knowledge of molecular control of the traits which are of most interest - those relating to growth, adaptation and stem and wood quality. Genetic engineering of these traits remains a distant prospect.

It is important that genetically engineered genotypes be of high quality with respect to other traits as well. The clonal test is the most logical basis for integration of genetic engineering into traditional tree improvement programmes. For these reasons, genetic engineering is most appropriately conducted with species where breeding programmes are advanced and clonal forestry can be realistically contemplated. Research on this subject should not assume a high priority with species

CONCLUSION

Major research priorities for these established industrial species should be the broader development of methods for the propagation of full-sib families and clones, and the development of methods for early and more accurate selection and the promotion of precocious flowering. While only a small proportion of current industrial plantations are located in developing countries, it is clear that much of the additional area of up to 100 million hectares required worldwide will be situated in tropical areas. Established, proven industrial species are likely to comprise a major proportion of this additional estate. A major practical tree improvement objective will be the establishment of sound breeding programmes for these new plantation operations.

In order to make use of a broader range of sites, and to supply products currently provided by harvesting in natural forests, it is likely that a significant proportion of the additional 100 million hectares will be established with tropical species which are not widely used or known at present. Selection criteria for "new" industrial species will be similar to those for established species. Many potentially valuable species are simply not well known, in terms of their adaptation and biological and genetic features, and gene pools are probably under threat. The broader implementation of good tree improvement programmes will be an important priority for these species. The testing of potentially useful species, characterization of mating systems, provenance collection, establishment, maintenance and assessment of trials, implementation of gene conservation measures, and commencement of other breeding work, poses a very large task.

There is a global requirement also for several hundred million hectares at least of non-industrial plantings, mainly located in developing countries. Breeding programmes for non-industrial trees differ markedly from those for industrial species in the wide variety, and the complexity, of selection criteria. Several of the taxa are highly variable and display features which render them particularly amenable to rapid improvement by traditional means. Biological features of many potentially valuable non-industrial species, however, remain largely unknown. Although not well studied, some gene pools are under threat, and some species display unpredictable patterns of variation as a result of human disturbance. Although selection work has been undertaken in a few programmes, most non-industrial species remain at the species testing stage. The need to establish plantings on a wide range of sites, many marginal or difficult, dictates that non-industrial plantings are likely to be reliant on a substantial number of the thousands of potentially useful species. The work involved in selecting the most promising species for each site is formidable. Major emphases of improvement for non-industrial species are likely to be taxonomic studies of variation, species and provenance testing, the assessment of reproductive features, and conservation activities. Interspecific hybridization is likely to be of value for several taxa. Where regionally consistent selection criteria can be defined and the tree crop is considered sufficiently important, clonal selection programmes are likely to be of value for species easily propagated vegetatively, and seed tree selection for species propagated by seed. Sophisticated breeding programmes are unlikely to be warranted for most species. It is possible that a "once only" or "quantum leap" approach to breeding may be more appropriate than breeding programmes offering continuous incremental gains.

Tree improvement programmes are greatly under-resourced. This is especially so in developing countries, where funding provided through national, regional, bilateral and international programmes is not sufficient to conduct properly the essential traditional tree improvement activities outlined above. The problems are further compounded by inadequate levels of training and facilities in most areas. For the non-industrial species in particular, it is these resource limitations and the variability in user requirements, rather than biological constraints, which constitute the major impediment to rapid improvement.

■■■■

8 DOMESTICATION AND COMMERCIALIZATION OF INDIGENOUS TREES IN AGROFORESTRY

INTRODUCTION

Agricultural developments have tended to reduce the number of crop species planted, their genetic diversity and the complexity of the agroecosystems used to produce these crops. Yields of many crops have increased dramatically, owing to improved agrotechnology and increased capital inputs. However, the erosion of crop genetic diversity is reducing our ability to develop new varieties in response to current and future agricultural requirements. The same scenario is occurring in forestry. Although a few fast growing trees are widely planted for pulp and paper, the majority of hardwood species forming the basis of timber trade have not contributed significantly to afforestation and reforestation programmes. Tropical forests contain a wealth of tree species which have traditionally been used by local people, but which have been neglected by agricultural development. These same species are cut down to clear land for development projects or to provide a subsistence living for a year or two. These overlooked and often highly desirable trees are the "Cinderella" tree species of forest ecosystem. The resources of these Cinderella tree species can be tapped by the process of domestication which has been applied so successfully to agricultural and horticultural crops (Leakey and Newton 1994b). There is a tremendous scope for the domestication and commercialization of these Cinderella trees.

The significance of trees outside forest (TOF) can be observed in several contexts. In countries with a low forest cover, tree resources outside the natural forest areas constitute the main source of wood and non-wood forest products. Growing populations, shrinking forests and degraded ecosystems all suggest that trees outside the forest are destined to play a larger local and global role in meeting the challenges of resource sustainability, poverty reduction and food security. However, the areas outside forests are characterized by overexploitation of tree resources that lead to their destruction. Tree resources in the farming areas have received little attention and are often overlooked, both in forest resources assessment and in policy and decision-making processes, despite the fact that those tree resources often play an important role in the livelihoods of the rural population. Until recently, neither foresters nor agriculturists paid much attention to the presence and roles of trees on farming lands.

Agroforestry is a delivery mechanism of multifunctional agriculture and better address the issue of declining soil fertility by rehabilitation of degraded land; restoration of above and belowground biodiversity; sequestration of carbon; and protection of soils and watersheds. Additionally, agroforestry provides unique inputs not often provided by other systems which include: (1) provision of indigenous products from underutilized species for food and markets to generate income and enhance nutritional and health status, (2) restoration of integrated agroecosystems of diverse plant species with the potential of developing into natural woodlands and forests, and (3) maintain linkages with human cultures through the availability of traditional foods and other products important to local people (Leakey 2010).

In agroforestry, the domestication of underutilized and indigenous trees was initiated in the mid 1990s by the World Agroforestry Centre (ICRAF) and its partners. This programme aimed at improving the quality and yield of products from traditionally important species that used to be gathered from forests and woodlands. In addition to meeting the everyday needs of local people, these products are widely traded in local and regional markets. Underutilized crops and trees therefore have the potential to become new cash crops for income generation and to counter malnutrition and disease by diversifying staple food/ food energy sources and dietary uptake of micro-nutrients that boost the immune system among others. These indigenous tree species also play an important role in enhancing agro-ecological functions and can help to counter climate change through carbon sequestration.

WHAT IS DOMESTICATION?

Domestication is the act/process of converting animals or plants to household use. That often means moving an organism from its natural habitat, introducing it to a modified habitat and providing the means for it to maintain itself in its new location. Domestication is a complex process akin to evolution in which human use of plant and animal species leads to morphological and physiological changes that create a distinction between domesticated taxa from their wild ancestors. Domestication can also be viewed as the selection process conducted by humans to adapt plant and animals to the needs of humans, whether as farmers or consumers.

The concept of a domesticated species is familiar to most people in terms of farm animals and household pets, but is not so common in plants. However, it is well known that modern grain cultivars have been bred over generations from wild grasses, and that our gardens and orchards are stocked with woody plants that have similarity been subjected to selection and breeding. Foresters have been slow to see the need to domesticate their crops. In the past, foresters have collected seeds from the wild and used them without selection. The use of provenance selection, an early stage in the domestication process, gained international importance in the 1960s, although techniques such as clonal selection have been used for centuries in a limited number of species such as *Salix*, *Populus*, *Cryptomeria japonica*, etc.

Broadly speaking, domestication is the process of taking a wild plant species and bringing it under management and cultivation. Humans have engaged in this process for several thousand years, mainly for food crops and ornamentals, and with varying degrees of effort. Among tree species, by far the most effort has been put into those species with edible parts, particularly to select better fruit and nut varieties. Domestication may involve genetic modification and breeding. An example may be removing a fruit-producing plant from the wild, planting it near one's house, harvesting the crop and planting the seeds for a future crop. The domesticated plant or animal may be used for one's own enjoyment or consumption, or it may be domesticated for commercial purposes. Domestication is done because it is amicable to have that particular life form around, because it is easier to deal with at home, or because through cultivation it can be improved. It is more economical to produce the product after domestication than to continually collect it from the wild. If we do not domesticate, the reasons are probably that there is no need - the organism is plentiful in the wild - or else the habitat requirements are too difficult to duplicate on a farm or in a garden. The following definitions will help to explain the processes involved in the domestication of plant species.

- To domesticate is to naturalise; make fond of home; bring under human control, tame; civilize (Concise Oxford Dictionary)
- Domestication is closely linked to the idea of selection, of fitness for purpose, of pushing nature into a higher gear and in a particular direction (Palmer 1994)
- Domestication of a plant species normally involves grower adoption for specific uses. Ultimately this process results in the careful collection, protection, characterization, evaluation and increase of germplasm of the species (Smith et al 1992)
- Domestication is a two-stage process in plants: the bringing into cultivation of wild plants or exposing them to some form of management, and subjecting these to differential production or selection (Janick et al 1982)
- Domestication is human-induced change in the genetics of a plant to conform to human desires and agroecosystems, culminating in the plant's loss of its ability to survive in natural ecosystems (Harlan 1975)
- Domestication can be defined as the production of individual genotypes or varieties exhibiting traits that are desirable to humans through the exchange of genetic material via either sexual or asexual processes (Harfouche et al 2012).

PLANT SPECIES DOMESTICATION

The word 'domestication' has had several definitions and interpretations since its first appearance in the English language in 1639 (OED 1989). With respect to plants, there is a spectrum of meaning from nurturing wild plants through to plant breeding through to genetic modification. Most commonly the word is used with reference to annual food crops that have undergone selection, breeding and adaptation in agricultural systems. Archaeologists concur that annual crop domestication began with wheat 10,000 years ago in Eurasia at a time of rising human populations and over-exploitation of local resources (Simmonds 1979). Cereal domestication surely ranks as one of the greatest technological advances in human history since not only does wheat (*Triticum aestivum*) still provide 20 per cent of food calories consumed globally but domestication of other cereal crops [e.g. rice (*Oryza sativa*), barley (*Hordeum vulgare*), maize (*Zea mays*)] has seen human population increase 1000-fold since domestication started (Diamond 2002).

The concept of plant domestication is interpreted differently by various scientists. Some have defined it in a relatively narrow sense as a human-induced change in the genetics of a plant, to adapt it to human agroecosystems. Others have defined plant domestication in a broader sense, as a process of naturalization of plant species towards specific human-induced growing conditions, during which an increased adoption for specific uses normally takes place. For instance, extraction of non-wood forest products from the forests may be considered as the initial phase of domestication of valuable tree species. During the domestication process, wild plants are first brought to some form of management. In a later stage of the process, the wild plants are actively cultivated. Only in its final phase does the process involve the selection and breeding of selected genotypes resulting in rather uniform plant populations with a narrow genetic base.

The concept of plant domestication can thus be interpreted in a restrictive sense as a process of changing of the biological characteristics of a species, or in a more extended sense as a process of change in plant exploitation practices, which brings with it changes in the plant's morphology and genetics, as well as in its growing environment. Some authors have even extended the concept of domestication from the conventional biological definition to a more inclusive one, i.e., the process of increasing human-plant interactions. According to this interpretation, domestication not only involves a change in plant characteristics and the biophysical environment, it also involves adaptations in human activities with respect to the use and manipulation of valuable natural resources. The various dimensions involved in the process of domestication can be clarified by relating it not only to plant species but also to landscapes. A domesticated landscape is one that has been modified by humans from its highly biodiverse state to a state that may still have high biodiversity, but which contains a greater concentration of resources useful to humans.

Interestingly, the process of domestication has been conducted for more than 13,000 years (i.e., since the last ice age) and occurred independently in several regions. Scientists have recognized at least six regions of domestication which include: (1) Mesoamerica, (2) the southern Andes (including the eastern piedmonts), (3) the Near East, (4) Africa (probably the Sahel and the Ethiopian highlands, (5) Southeast Asia, and (6) China. In spite of the geographic and climatic diversity in domestication centres, there is a remarkable similarity in the set of traits which were used to select completely unrelated types of crops (Table 8.1).

Table 8.1: Domestication Syndrome Traits (Harlan 1992)

Selection Stage	General Trait	Specific Trait	Examples
Seedling	Increase in seedling vigour	Loss of seed dormancy	Many crops Tomato Cassava Maize Legumes
Reproductive System	Increase selfing (multiplication) rate	Loss of seed dispersal	Fruit trees
	Amenability to vegetative propagation	More compact growth habit	Wheat, barley, maize
		Increase in number or size of inflorescences	Maize amaranth
		Increase in number of seeds per inflorescence, pulp volume, oil content/quality, active ingredients	Legumes, rice, tree crops
		Change in photoperiod sensitivity, colour, size, taste, texture	Many crops
Harvest or post-harvest	Increase in yield and shelf life	Reduction in toxic substances	Cassava, lima, bean, etc.
	Consumer appeal		

Many seed-propagated crops under domestication have been selected mainly for their traits in seed dispersal and dormancy. The domesticated varieties are characterized by a lack of seed dispersal at maturity. For example, in cereal crops, selection is focused on individual plants with a tough rachis that prevent the disarticulation of the inflorescence and the release of seeds. Whereas, in the wild relatives, an abscission layer is formed between each successive seed insertion site such that at maturity, this layer causes the rachis to break and subsequently the dispersal of seeds. Furthermore, seeds of varieties under domestication exhibit little or no dormancy compared to the wild relatives. While dormancy in the seeds from wild relatives prevent premature germination during adverse growth conditions like drought, the lack of the seed dormancy in varieties under domestication will on the contrary promote uniform germination and population, hence uniform harvest.

In perennial crops, the selection of individuals under domestication is focused towards individuals with a more compact growth habit, having fewer and shorter branches as opposed to tall single stem wild relatives. All of these changes in plant growth habit have a positive impact on the harvest index (ratio of harvested part to overall aboveground biomass) and size of the fruit or grain. Another feature that has widely been reported to have change in crop evolution is the reproductive system. The reproductive system of most species under domestication has changed towards increased selfing (as in tomatoes and peppers) or a replacement of sexual reproduction by vegetative propagation in order to maintain trueness in type of cultivated individuals faced with possibilities of outcrossing with wild relatives. The fruits of such individuals could be more tasty as the case in *Chrysophyllum cainito* in Panama where fruit from trees under domestication were found to have more pulp, less acid, larger in size and higher in sugar content as compared to fruits of wild trees (Parker et al 2010). Waruhiu et al (2004) also reported that *Dacryodes edulis* fruits from trees on farms in Cameroon and Nigeria were 66 per cent larger than those obtained from trees in the wild.

The repeated independent selection of similar 'desirable traits' in different crops under domestication that have rendered them unable to grow and reproduce successfully for more than a few seasons in natural environments, away from the care of humans who provide adequate management to reduce competition from pest and diseases. These shared traits (seed dispersal, seed dormancy, gigantism in harvest parts, determinate and synchronized growth, increased harvest index and change in flavour) have been recognized and described as "domestication syndrome" (Harlan 1992). Origins of these desirable traits are probably spontaneous mutations occurring in wild plant populations. These traits selected by man at various growth stages of wild plants, as well as after harvest could have been deleterious in the wild (Gepts 2002). Hence, fully domesticated crops can have difficulties surviving in the wild without human interventions. Partially domesticated plants show an exception to this. Among these partially domesticated species are fruit trees like apple (*Malus domestica*) in which it is sometimes difficult to distinguish between wild and natural populations.

A plant is normally considered 'wild' when it grows spontaneously in self-maintaining populations in a natural or semi-natural ecosystem and when it can exist independently of direct human action. In contrast, it is considered to be domesticated when it has been selected purposefully for specific genetic characteristics and when it is propagated and cultivated in managed agroecosystems. Crop domesticaton has been limited to less than 0.05 per cent of all plant species and about 0.5 per cent of edible species (Leakey and Tomich 1999). Out of a total of 4,00,000 flowering plant species, less than 200 have been domesticated as food and feed plants, and just 12 species provide 75 per cent of the food eaten (FAOStat 2010). Plant domestication gave rise to food surpluses; led to craft specialization; development of arts, social hierarchies, civilizations and the origin of states.

HISTORY OF TREE DOMESTICATION

Tree domestication is a far more recent phenomenon than annual crop domestication. Trees have been planted for thousands of years for food, wood, shelter and religious purposes. The first woody plants to be domesticated for cultivation were undoubtedly those yielding food or other non-timber products. The first selection and planting of a woody species may have been the olive (*Olea europea*), a thorny shrub growing near Syrian and Anatolian coasts, as early as 4000 BC. Subsequently selected varieties of olive were transported to the eastern Mediterranean e.g. the cultivation of olive has been known in Greece at least since the Minoan ear (3000 BC) for food and olive oil for clay lamps. The olive was exported to Egypt in the same period. The edible fig (*Ficus carica*) was also cultivated in

Mediterranean region as early as 4000 BC. The myrrh trees (*Commiphora myrrha*) were introduced to Egypt from Somalia in 1500 BC and planted for sources of perfume. Olives, walnuts and almonds from the Middle East and western Asia were moved into China during 208 BC-265 AD.

The apple (*Malus pumila*) was domesticated for cultivation for a few centuries BC by the Greeks and Romans and it was spread throughout Europe and Asia as a result of their travels and military conquests. Another theory states that the domestication of oranges and apples for instance, goes back about 3,000 years in China and central Asia respectively. The Romans also introduced the edible, nut-bearing, sweet chestnut (*Castanea sativa*) into Britain. In the fourth century BC, Theophrastus reported trees of frankincence (*Boswellia* spp.) and myrrh being planted on private estates in southern Arabia, and Aristotle established what was probably first arboretum in Greece. Cato (234 BC) recorded the planting of willows (*Salix* spp.) and poplars (*Populus* spp.) and noted that a conifer was planted for ship's timber. There are also several Biblical references to tree planting in the first millennium. Trees were moved around world by the Romans, Greeks, Chinese and others during military conquests. Voyages of discovery by European navigators to the Americas and Asia provided new opportunities to transfer tree germplasm between distant countries (Turnbull 2002).

Tree domestication and their subsequent cultivation in the form of plantations are not anything new. In southwestern France, for example, there are records of man-made forests as early as year 1500, as well as evidence of trading in forest products, both wood and non-wood from maritime pine (*Pinus pinaster*) plantations. These early plantations resulted mainly from the initiative of private landlords as well as smallholders and communities, primarily for commercial purposes (i.e., the trade of wood and resin products), but also for direct household needs. In this same area, there are records of zones planted with pines and other vegetation to control the advance of sand dunes before 1600 (Sargos 1997). Therefore, there is evidence of commercial, smallholder and community man-made forests, for both production and protection purposes, from as early as 400 to 500 years ago. This particular region of Southwestern France is known as the "Landes de Gascogne" which is nowadays a compact plantation forest of more than one million hectares. While ownership is more than 90 per cent private, with more than 40,000 owners of all sizes, the State plays a limited but effective role in the management of this area and so do local governments (Montane 1994).

Closer to home, teak (*Tectona grandis*) was successfully introduced in Sri Lanka as early as 1680. This same species was successfully planted in Java, India and Bangladesh during early 19th century. Additionally, eucalypts were successfully planted in India in 1790. Expanding area of planted forests, to a large extent, has resulted from the inclusion of plantation programmes in the national forest policies of many countries in recent years, especially since the mid-sixties (Evans 1992). For the most part, these man-made forests have concentrated on industrial plantations; however, they have also included plantations for protection purposes, as well as rapidly expanding agroforestry and social forestry efforts.

It has been reported that evolution has created 60,000 tree species and for thousands of years human extracted what they needed from the forest. Now population exceeds extractive capacity and most tree species are wild. In 1850, population was 1 billion and today it stands at 6 billion, whereas original global forest cover has shrunk from 70 per cent to 26 per cent now. This necessitates the importance of tree domestication in agroforestry. Domestication of trees for the provision of agroforestry tree products (AFTPs) has gained significance in recent years especially in tropical areas.

GLOBAL INITIATIVES FOR AGROFORESTRY TREE DOMESTICATION

In terms of conventional improvement, tree species are far more neglected today than agricultural crops with the exception of temperate fruit trees. In commercial forestry, fewer than 40 taxa have genetic improvement programmes underway and most of these are less than 60 years old (Barnes and Simons 1994). Attempts at improving the N_2-fixing agroforestry trees started even later, in the 1980s, coinciding with concerns about soil fertility management, tropical fuelwood crisis and renewed interest in social forestry. Domestication of other agroforestry trees has received substantial recent interest following a number of articles and conferences.

In 1992, modern agriculture's rejection of these traditionally important tree species was recognized by a most significant IUFRO conference held in 1992 "*Tropical trees: The Potential for Domestication and the Rebuilding of Forest Resources*" at Edinburgh, UK (Leakey and Newton 1994b). Out of that meeting came a global research initiative to domesticate some important tree species as new cash crops for poor, smallholder farmers in the tropics. The initiative was led by the International Centre for Research in Agroforestry (now known as the World Agroforestry Centre) and the tree domestication program (TDP) was begun in 1995 (ICRAF 1995). There appears to be no single, definitive, complete description of the objectives of the programme. However, they are stated explicitly or implicitly in the various documents describing initial activities and intentions. Essentially, the TDP was aimed at improving livelihoods of smallholder farmers by facilitating their access to high quality germplasm that would result in both higher and more stable production of tree products, particularly in silvopastoral systems, multistrata systems and improved fallows, i.e. ICRAF's priority systems at that time.

Many works had been initiated for tree domestication projects in the Pacific region: cutnut (*Barringtonia procera*) and tahitian chestnut (*Inocarpus fagifer*) in the Solomon Islands, galip nut (*Canarium indicum*) in Papua New Guinea, sandalwood (*Santalum austrocaledonicum* in Vanuatu and *S. lanceolatum* in Cape York, Queensland). Similarly domestication works have been done in relation to indigenous fruits and nuts in Africa (*Dacryodes edulis* or safou, *Irvingia gabonensis* or bush mango, and *Sclerocarya birrea* or marula). In India, the major domesticatied trees planted in agroecosystems are: ber (*Ziziphus mauritiana*), tamarind (*Tamarindus indica*), drumstick (*Moringa oleifera*), mango (*Mangifera indica*), bael (*Aegle marmelos*), jackfruit (*Artocarpus heterophyllus*), *Syzygium cumini*, *Citrus* spp., *Ficus* spp., coconut (*Cocos nucifera*), teak (*Tectona grandis*), neem (*Azadirachta indica*), etc.

The domesticated species shows 3 to 5-fold tree-to-tree variation in fruit/nut/kernel morphology. The variation is continuous, showing no indication of any distinct varieties in the wild population. Such large variation provides great potential to develop cultivars using simple horticultural techniques such as the rooting of cuttings, grafting and marcotting (marcotting is a technique of promoting root development by wounding the part of the plant to be rooted with roots induced to form on the plant stem). Interestingly, molecular studies of these tree populations have shown that the trees with superior qualities are not closely related. So, the risk of reducing genetic diversity is unlikely to be significant.

Today, about 50 of these tree species are being domesticated as new crops to improve the livelihoods of poor people, providing nutritious and tasty dietary additives to otherwise starch-heavy diets based on maize, cassava, yams and plantains. Much of the progress in tree domestication has been informed by the well-described homegarden systems of the Amazon, Southeast Asia and Africa (Kumar and Nair 2004). Because there are still active local markets for the tree products, the cultivation of these species is now also generating income and transforming the lives of marginalized rural communities. These domesticated trees in agroforestry have been regarded as "*trees of life*" which help in the transformation of tropical agriculture.

Globally, multinational companies are beginning to recognize market opportunities by partnering with communities in tropical countries. This new business paradigm has huge potential to change the fortunes of these poor and vulnerable people, who have historically been marginalized by the way in which globalized trade and industry operates. In tropical countries, local people are also starting to process products from tropical trees, making it possible to sell these products further a field, as well as 'out-of-season'. Value-adding in this way is also starting to create employment and business opportunities, which further adds to the income-generating opportunities for local people. This agroforestry approach also has important environmental benefits. These are benefits that the world needs in abundance if we are to protect our environment and overcome the scourge of poverty, malnutrition and hunger. Support for tree domestication is urgently needed to scale up these efforts to a meaningful level.

TREE DOMESTICATION IN AGROFORESTRY

Tree domestication has been defined in a number of ways. Tree domestication is the naturalization or cultivation of tree species to improve their use by people that involves selection of desired trees or genes and converting those trees or genes into the growing trees, which are harvested as a renewable resource (Libby 1973). It is a process that includes a wide range of activities: exploration and collection of natural populations, evaluation and selection of suitable species and provenances, breeding to develop superior cultivars, development of propagation techniques, multiplication and dissemination of germplasm, development of management techniques, utilization and tree product marketing, and development and dissemination of relevant technical information. Thus, any activity that results in man's control over the growing and utilization of trees can be considered as tree domestication. The full domestication of a tree species is taken to encompass the identification and characterization of its germplasm resources; the capture, selection and management of genetic resources; and the regeneration and sustainable cultivation of the species in managed ecosystems (Leakey and Newton 1994a). It is clear that the starting point can vary considerably depending on the approach taken, which in turn seems to depend on whether the domesticator is a peasant farmer or a research scientist. The process has not been completed for many tree species, thus most forest trees can at best be described as semi-domesticates.

Domesticating agroforestry trees involves accelerated and human-induced evolution to bring species into wider cultivation through a farmer-driven or market-led process. This is a science-based and iterative procedure involving the identification, production, management and adoption of high quality germplasm. High quality germplasm in agroforestry incorporates dimensions of productivity, fitness of purpose, viability and diversity. The domestication of tree species should enhance their capacity to fulfil either conservation service or production functions or both. Tree domestication in agroforestry should aim at increasing the social, economic and environmental benefits of trees to people. Strategies for domestication of individual tree species vary according to their functional uses, biology, target environments and the researcher or farmer in question.

About 200 tree species are commonly planted for timber and other purposes, such as food, shelter and beautification around the world (Evans 1992). In crops and trees, the proportions that have been domesticated out of the total number of species are small. Regarding the plant kingdom, the proportion domesticated is about 0.3 per cent in agriculture and 4.1 per cent if garden plants are included. However, if the potential for domestication is limited to higher plants (angiosperms, gymnosperms and pteridophytes) of which there are some 2,50,000 species, the proportions go upto 0.5 per cent and 6.6 per cent respectively (Leakey and Tomich 1999). This indicates that agriculture or agroforestry is not making full use of the diversity of species available. What should be done within agroforestry is the domestication of more tree species for timber and non-timber forest products as a means to reduce farmer's reliance on subsistence food production, and to promote food and nutritional security, alleviate poverty and enhance environmental resilience. In tandem with species strategies are approaches to domesticate landscapes by investigating and modifying the uses, values, interspecific diversity, ecological functions, numbers and niches of both planted and naturally regenerated trees.

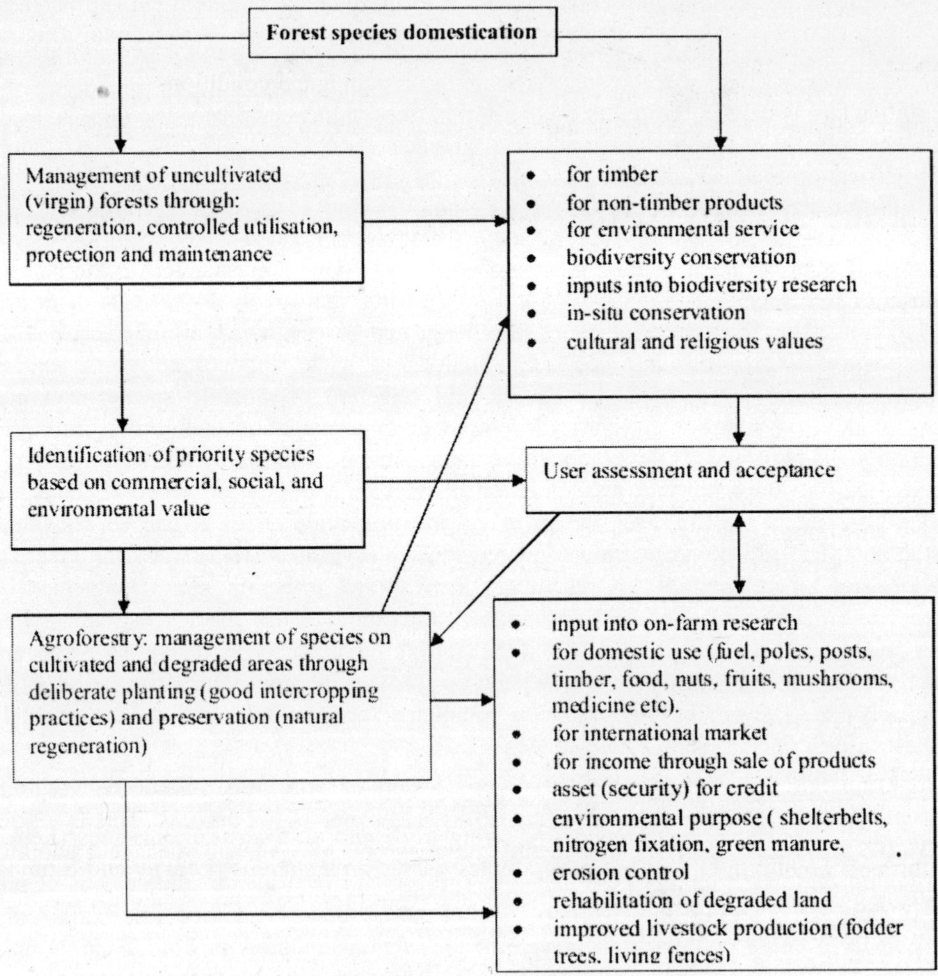

Fig. 8.1: Schematic Model of a Pathway in Agroforestry Tree Domestication Process (Source: Leakey et al 1996)

Integrated concept of domestication within the landscape seems particularly apt when we consider agroforestry. Many agroforestry systems develop as a result of the gradual modification of forests, by enriching them with useful crops. Such enrichment not only involves agricultural crops; in many cases it also concerns trees producing wood and non-wood forest products (NWFPs). The first phase of domestication of trees involves a process of concentration of naturally occurring useful tree resources in natural forests. In next phase, this will take the form of transplanting of wildlings or seedling of wild species. In subsequent phases, new species may be introduced after selection and improvement into agroecosystems. But gradually varieties will be improved through further selection. The process of domestication can therefore be considered as an evolutionary process, from gathering to breeding, during which changes at the level of both the landscape system and the plant species occur. Concomitantly, a progressively closer interaction between the tree resources and people takes place. The domestication of trees in agroforestry systems thus involves a co-evolution of human society and nature, with the combined processes of natural and cultural selection creating a great diversity of human-influenced agroforest types.

This co-evolutionary process consists of various stages along a nature-culture continuum. These stages can be associated with specific socioeconomic and ecological conditions. To understand under what conditions farmers are interested in incorporating trees in their land use system, it is necessary to understand the specific characteristics of these different stages. This will allow identification of the kind of domestication practices that offer the most scope under specific socioeconomic conditions and for given agroforestry systems. The process of domestication of trees in agroforestry systems is represented by several phases in tree management with associated ecological and institutional conditions; these phases are arranged sequentially along a gradient of increasing input of human energy per unit of exploited land.

Understanding tree domestication and good quality inputs will help farmers become successful tree growers, rather than just tree planters. Tree domestication can be characterized by four factors: identification, production, management and adoption of desirable germplasm. Key domestication activities include propagation, establishment, management and utilization. Key inputs include germplasm and the knowledge necessary for farmers to make competent decisions concerning the allocation of land and financial resources regarding to tree domestication. Tree domestication in agroforestry will improve the success of farmer's efforts and reduce deforestation pressure on local forest resources. Agroforestry tree domestication involves developing useful tree products for farmers. It leads to diversification of small-scale farms in the humid, sub-humid and drylands. It also provides farmers the opportunity to generate a cash income. It can only be successful, however, if developed with appropriate market opportunities.

STAGES OF AGROFORESTRY TREE DOMESTICATION

Two principal pathways are generally distinguished in the domestication strategy for most agroforestry species in small-holder farming systems in the tropics (Fig. 8.2). These include domestication implemented on farm by farmers, who bring trees into cultivation themselves (Leakey et al 2004) or through genetic improvement programmes on research stations (Leakey and Simons 1998). Increasingly, scientific approaches are also being introduced into on-farm domestication through the application of participatory approaches to tree improvement whereby researchers/mentors advise farmers, and sometimes implement and monitor joint on-farm research.

The different stages to the domestication of trees are: managed species, semi-domesticates and full domesticates. There are four elements of the tree domestication process in agroforestry (Libby 1973):

- The original genetic variability
- Selection of the desired trees or genes
- Packing of those genes in the plants to be used
- Converting those gene packages into the growing trees which are harvested as a renewable resource

The agroforestry tree domestication is a farmer-driven and market-led process, which matches the intraspecific diversity of local important trees of subsistence farmers, product markets and agricultural environment. The products of such domesticated trees are called Agroforestry Tree

Products (AFTPs) to distinguish them from the extractive tree resources commonly referred as non-wood forest products-NWFPs (Simons and Leakey 2004). The domestication of a species involves its manipulation and cultivation for specific uses. Through breeding and selection, this process can result in rather uniform plant populations with a narrow genetic base. The cultivation of such selected varieties is the last stage in an evolutionary continuum in people-plant interactions. In this evolutionary process, a progressively closer interaction between people and plant resources takes place with increasing inputs of human energy per unit of exploited land.

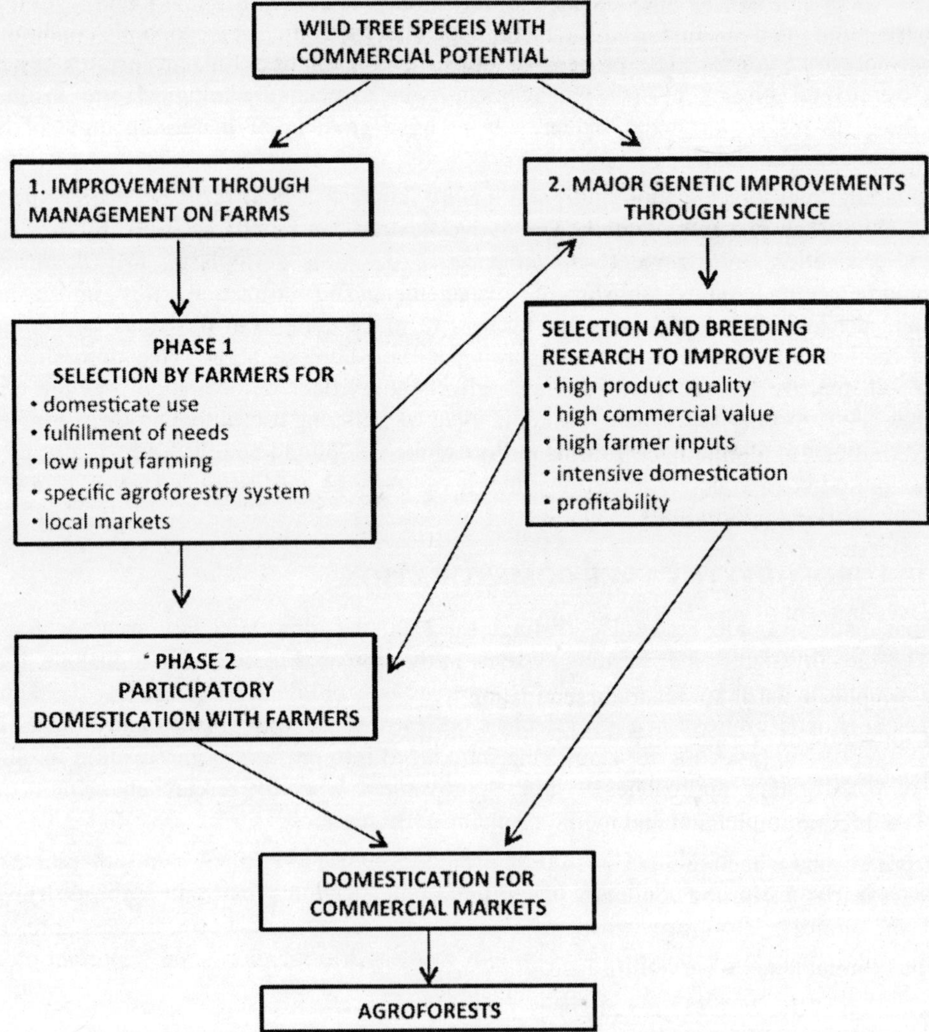

Fig. 8.2: Two Pathways for Domestication of Trees in Agroforestry
(Source: Leakey and Akinnifesi 2008)

Three dimensions of domestication may be distinguished, i.e., domestication of the biophysical as well as the human environment and the domestication of the species. The recognition of these different dimensions is of particular importance with respect to the domestication of tree species, because commercial tree utilization usually starts with the exploitation of trees from the wild in natural forests. The domestication of the forest environment involves, in the first instance, the change from uncontrolled utilization of the wild tree products to their controlled exploitation. The second phase consists of the purposeful cultivation of wild trees in farming systems. The cultivation of domesticated trees in either agroforestry systems or commercial tree-crop plantations is the last stage of this domestication process.

These three stages of domestication are associated with specific socioeconomic and ecological conditions. During the first stage of domestication of the forest environment, the management practices are, in first instance, directed at defining user rights. In the second stage, management practices with biological objectives are also undertaken. These focus at the manipulation of the growing conditions of the trees and of the tree's morphological characteristics. During the third stage of plant domestication, the genetic constitution of the tree population is also modified. The recognition of various phases of forest and tree domestication provides for a better understanding of the different options for the production of commercial tree crops in agroforestry systems, under location and time specific conditions. It also illustrates that efforts to domesticate commercial forest resources should be focused both on the manipulation of morphological characteristics of a tree (its environment) and on the modification of the tree's genetic characteristics.

The steps of agroforestry tree domestication process are:

- Selection of priority species based on their expected products or services (Prioritization)
- Definition of an appropriate domestication strategy considering the farmer, market and landscape needs
- Germplasm exploration, collection and storage (sourcing of germplasm)
- Evaluation, documentation and deployment of germplasm (seed, seedlings or clonal material)
- Tree breeding or cultivar selection pathways (genetic improvement)
- Development of propagation techniques
- Multiplication of germplasm
- Germplasm distribution and dissemination
- Cultivation of domesticated trees in agroforestry system (diffusion of germplasm)
- Development of management techniques (silviculture)
- Tree-product utilization and marketing

The research phase includes tree breeding, propagation and multiplication. The research phase may involve research institutions on their own or in participatory mode with the stakeholders such as farmers or communities. Working directly with end-users is advantageous towards economic, social and environmental goals, especially in developing countries. Prioritization is an important process in domestication of trees in agroforestry because it includes:

- Identification of community needs
- Evaluation of the species currently used
- Ranking of the primary uses (products and services) provided by the species
- Identification of a limited number of priority species
- Valuation and ranking of tentative priority species
- Selection of priority species for domestication

PRINCIPLES OF AGROFORESTRY TREE DOMESTICATION

Tree domestication is an umbrella term that is often applied erroneously to a subset of activities such as provenance testing or to a narrow application such as industrial forestry. Tree domestication refers to how humans select, manage and propagate trees where the humans involved may be scientists, civic authorities, commercial companies, forest dwellers or farmers. In the tropics, the trees involved in domestication occur in natural forest, secondary forest, communal fallow lands, plantations and farms. These trees in turn provide both products and services. Of course, human history is interwoven with forests and trees before agriculture, urbanization and commerce began. The tree domestication gains significance with largest group of people who are the rural population in developing countries and their interactions with the largest group of tree species in agricultural landscapes. In Asia, Latin America, Africa and Oceania, greatest potential exists for tree domestication to contribute sustainable development.

Recent literature on agroforestry tree domestication is dominated by research on biotechnology especially molecular genetics which, although important, is but one aspect of domestication. Much of the earlier literature dwells on propagation. Simons and Leakey (2004) offered a more comprehensive assessment of tree domestication in agroforestry addressing 14 aspects:

1. Reasons for domestication (home use, market, conservation of the species, agroecosystem diversification, improved livelihood strategies)
2. Tree uses required (products and services)
3. History and scale of cultivation (as native and exotic)
4 Natural distribution, intraspecific variation and ecogeographic survey information
5. Species biology (reproductive botany, ecology, invasiveness)
6. Scale and profile of target groups and recommendation domains (biophysical, market, cultural)
7. Collection, procurement or production of germplasm and knowledge (including ownership, attribution, benefit sharing, access and use)
8. Propagule types envisaged
9. Nursery production and multiplication
10. Tree productivity (biomass, timing, economics, risks)
11. Evaluation, both scientific and farmer participatory
12. Pests and diseases
13. Genetic gain and selection opportunities, methods and intensities
14. Dissemination, scaling up, adoption and diffusion

The prevailing problem in agroforestry tree domestication is that information is incomplete, and has led to suboptimal tree domestication strategies. While tree domestication work has increased, the documentation of the logic and the approch has been generally scant. Even when results are shared or published, it is typically the positive outcomes that are reported and not the successful process. A few case studies of tree domestication strategies have been documented and decision-frameworks have been offered for domestication of agroforestry fruit trees, but clear guidance for domestication of forest trees remains scarce (Simons and Leakey 2004).

Jamnadass et al (2009) offered a useful 'domestigram' (Figure 8.3), indicating possible pathways for domestication. A notable feature of this diagram is the central chain involving identification, production, management and adoption, which is the key to the domestication process. A 'whole of chain' approach is essential, and the success with the domestication process may depend on the weakest link in this chain. Over-emphasis on a single aspect may lead to dysfunctional outcomes. For instance, identification alone is not domestication, because there may be an inability to provide sufficient seed, and that an over-emphasis on management to the neglect of adoption, may result in guidelines that are impractical in a large-scale situation or which produce a yield far in excess of market needs. Importance of encouragement underlines "incentives must be identified which will attract investment, resolve technical problems, enhance growth and development and lead to a self-sustaining, industry-driven commercial enterprise capable of operating without direct financial input from Governments". The challenge is to ensure that such incentives can be sustained.

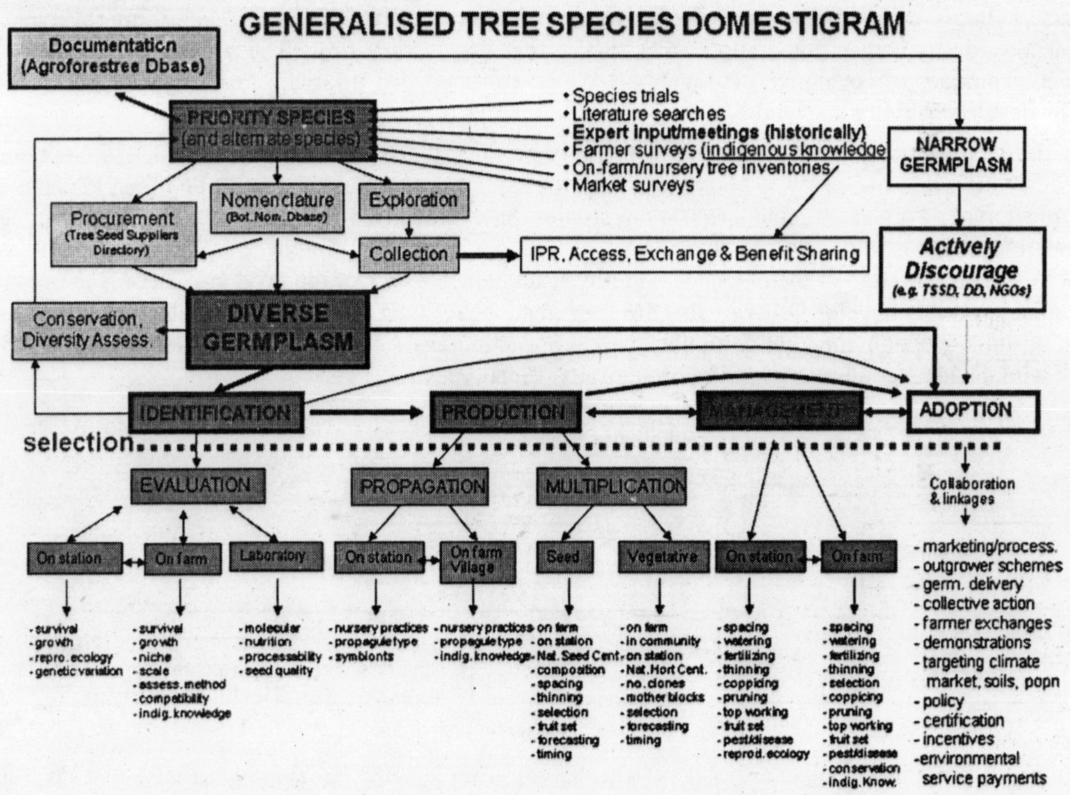

Fig. 8.3: Generalized Tree Species Domestigram (Source: Jamnadass et al 2009)

Leakey and Newton's (1994b) observations made two decades ago remain pertinent: "Opportunities are currently being lost because of a lack of awareness of the potential to domesticate forest tree species for the production of wood and non-wood products. What are the issues that have to be resolved to trigger this new revolution? From the view point of a farmer, there are:

- *Political and Social Issues*, such as how to acquire the right to won and protect a piece of land and the trees on it, and the need for incentives to plant trees.
- *Economic Issues*, such as what is the value of these trees in terms of their wood, other products and environmental services.

- *Biological Issues*, such as how to grow the trees wanted by farmers; how they can be made more desirable and productive, to the extent of satisfying the farmers' needs and even providing a surplus which could be sold to urban populations".

The main strategies for agroforestry tree domestication include priority species selection; exploration, collection and storage of germplasm; sexual propagation and breeding; vegetative propagation; genetic conservation

TREE DOMESTICATION AND VEGETATIVE PROPAGATION

The concept of domesticating crops and livestock have existed for thousands of years, however their application to tropical trees for timber and non-timber products is very recent. Basically the objective is to make them more useful to mankind by cultivation and genetic improvement. Domestication by conventional approaches to tree breeding is very slow. The advent of reliable and robust techniques of vegetative propagation have however opened up much faster approaches based on the development of clonal cultivars.

Before using vegetative propagation to mass produce a selected individual tree with superior characteristics, it is necessary to characterize the variation found within a species and then to capture the phenotype. Numerous characterization studies have identified that even within a particular population, there is very considerable (3- to 20-fold) variation in different morphological traits, as well as similar variation in the chemical and physical characteristics of the product. There is therefore potential to rapidly develop cultivars that are very much better than average. The greatest potential lies in multi-trait selection aimed at particular market opportunities. As a result, it is necessary to identify trees with the ideal combination of important traits - 'ideotypes'.

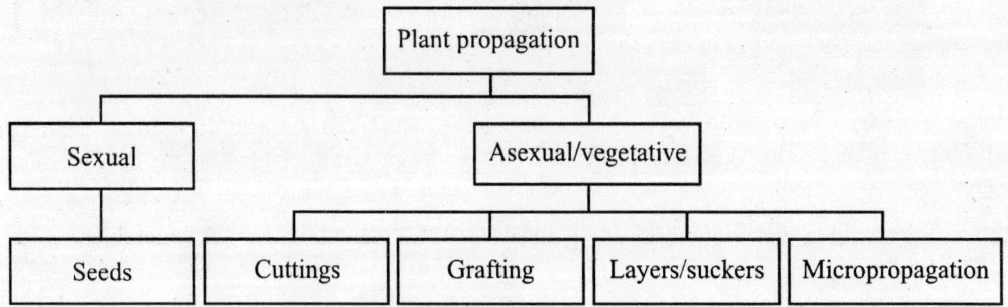

Fig. 8.4: Plant Propagation Methods

The concept of vegetative propagation is that an exact copy of the genome of a mother plant is made and continued in new individuals. This is possible because plants, unlike animals or humans, have meristematic, undifferentiated cells that can differentiate to the various organs necessary to form a whole new plant. A piece of plant shoot, root or leaf, can therefore, grow to form a new plant that contains the exact genetic information of its source tree. The different methods of vegetative propagation practices for agroforestry trees are cutting, grafting, layering/suckers, mircropropagation, etc.

In agroforestry, the producers are generally smallholder producers in developing countries where the outputs of domestication are urgently needed, the approach to domestication has been to work directly with the farmers so that they can implement the domestication programme to meet their specific needs and so that they are the beneficiaries of the work. To ensure the wise use of genetic resources and the optimum benefit flows to producers it is important to practice a wise domestication strategy that protects and conserves genetic diversity while maximizing the benefits for mankind. Because there is no point in developing domesticated cultivars if the producer is unable to market the products, it is also critical that the strategies used ensure that domestication and commercialization proceed in parallel.

DOMESTICATION OF TREES FOR WOOD AND NON-WOOD FOREST PRODUCTS

Throughout the tropics there are numerous perennial woody species that have provided indigenous people with many of their daily needs for millennia. Many of these people have now left the land for urban life, but they still demand traditional food, medi-cines, and other natural products. These traditionally important woody plants are virtually undomesticated. These neglected "Cinderella" species have great genetic diversity and also play a key role in biological, chemical, and hydrological cycles, protecting soils and providing ecological niches. The food-producing species are also important for food security, especially in the dry season, as well as a source of vitamins and minerals critical for the health and nutrition of children and pregnant woman.

The potential of tropical forests to provide a variety of non-timber forest products is at present widely acknowledged. Such products include a variety of food products as well as diverse commercial crops such as gums, resins, rattan. In the past it was often assumed that mainly tribal people who were engaged in traditional methods of forest resource use, such as gathering and swiddening, collected such products for their subsistence. But it is increasingly acknowledged that there exists a long history of commercial trade in non-timber forest products, coupled with an active management of these products. When considering the production of non-wood forest products, it is therefore not appropriate to assume a dichotomy between the gathering of NWFPs from wild trees in a natural forest and the cultivation of domesticated NWFP trees in an agroforestry system. Rather, attention should be focused on the various phases of domestication and on the creative role of human culture in regulating tree resources for human use.

There are four groups of wild trees, shrubs, and vines which could be rapidly domesticated for agroforestry and which can be viewed as potentially important sources of income for farmers. These trees produce:

- The traditionally important wild foods, mostly fruits, nuts and leaves for vegetables
- The traditionally important fibres
- Locally and industrially important pharmaceuticals and other extractives such as gums and resins
- Commercially important quality timbers and woods

The domestication of tree species is a dynamic process which develops from deciding which species to domesticate and proceeds through background socioeconomic studies, the collection of germplasm, genetic selection and improvement to the integration of domesticated species in land-use. Domestication is an ongoing process in which genetic and cultivation improvements are continuously refined. In genetic terms, domestication is accelerated and human-induced evolution. Domestication, however, is not only about selection but also integrates the four key processes of the identification, production, management, and adoption of tree resources.

Strategies for tree domestication will vary depending on the value of the products, the extent of genetic variation within a species, and many other factors. For high-value species such as those producing marketable forest products, the vegetative propagation of superior selections identified from within the existing wild populations will be appropriate. This is the approach generally followed in horticulture. Thus, an individual plant with superior yield, fruit flavour, stem form, or wood quality can be mass produced by vegetative propagation. In this way, it is possible to select those clones likely to develop above-average characteristics in any given trait. By a series of ongoing selections and an ever-increasing intensity of selection, it is also possible to achieve rapid and substantial genetic improvements.

When identifying and selecting trees for cloning by vegetative propagation, there are two important criteria: It is important, (1) to ensure that those that are highly superior for the desired traits are chosen and, (2) to ensure that the selected individuals are unrelated and as genetically diverse in other traits as possible. This calls for strategy that allows intensive selection (e.g., 1 out of 100 to 10,000 trees) from among trees from different populations, ideally from throughout the range of the species. Furthermore, this should be an ongoing rolling program of multiple trait selection, in which more and more intensive selection is imposed through the addition of selection criteria for new traits while, at the same time, new sources of genetic stock are continually added as new genetic collections enter the program. These new collections should come from further exploration and from breeding programs and so continually broaden the genetic base of the planted trees. Molecular genetics techniques can be used to ensure that genetic diversity is maintained in the clones used for commercial production.

Selection procedures vary depending on the product and the situation. For example, for indigenous fruits, rapid progress will be made if indigenous knowledge can be used. Usually rural people know which are the best individual trees in their area for yield, fruit size, or flavour. Thus, as with temperate apples, pears, etc., people can be asked to report the existence of superior trees, so reducing the task of screening large numbers of trees. On the other hand, for medicinal trees, it is more likely that a chemical screening process will be required, but the magnitude of this task can probably be reduced by starting on a population basis, since it is likely that trees from certain environments will be richer in the required metabolites. Meanwhile, for timber trees, log size and straightness are the first selection criteria. Various forms of "plus-tree" (i.e., elite selection) provenance and progeny selection are well known. To these have recently been added some "predictive tests" which can be applied in the nursery as a procedure for mass screening from genetically diverse seedling populations. This involvement of farmers, however, has to be done for their ultimate benefit, in accordance with the Convention on Biological Diversity. By capturing genetic diversity of a wide range of indigenous trees for growth in agroforestry systems, it is hoped the promises of agroforestry to alleviate poverty and to mitigate environmental degradation and the loss of biodiversity will be fulfilled (Leakey 1998).

PARTICIPATORY APPROACH IN AGROFORESTRY TREE DOMESTICATION

In agroforestry, with the greater emphasis on the social, cultural and economic needs of resource-poor subsistence farmers, there has been a recent shift towards domesticating trees producing valuable non-wood forest products (NWFPs) with the people and for the people. This requires a very different approach to tree improvement, one based more on horticultural than forestry techniques (Leakey and Jaenicke 1995); one situated on the farm rather than in a research station. In Africa, the World Agroforestry Centre has adopted a participatory approach to tree domestication with rural communities, aiming to make the communities instant beneficiaries of the research. The key strategy is to ensure that the techniques can be adopted at the village level to produce planting stock for local farmers, and improved fruit and nut products with above-average quality and greater uniformity for local marketing.

World Agroforestry Centre (ICRAF) is working in six major tropical ecoregions to help to mitigate deforestation, land depletion and rural poverty through the development of improved agroforestry systems. To achieve these goals ICRAF's research agenda is engaged in a thrust on the domestication of agroforestry trees. Since 1998 the ICRAF and its partners have developed this participatory approach to the domestication of high-value indigenous fruit and nut trees, and medicinal plants in West and Central Africa (Leakey et al 2003). Cameroon and Nigeria witnessed this approach with a number of species, especially *Dacryodes edulis* (safou) and *Irvingia gabonenis* (bush mango/ dika nut). This approach to improving agroforestry tree products (AFTPs) aims to increase significantly and stabilize income, food and health strategies of rural households and their extended families. In this way, it seeks to ensure that women, youths and other vulnerable groups in society, obtain maximum benefits from agroforestry.

Participatory tree domestication refers to the means by which rural communities select, propagate and manage trees according to their own needs, in partnership with scientists, civic authorities and commercial companies. It is usually oriented towards specific local markets and encompasses the use of both indigenous knowledge and genetic selection based on scientific principles. In such cases, scientists generally work at the community/ household level with farmers on species of their own choice. Participatory domestication thus empowers the farmers, allowing the outputs and benefits of domestication to remain with the community. Domestication activities are developed in many 'pilot villages', each creating their own set of cultivars. At least in the short term, this strategy seems to maintain genetic diversity. If this can be sustained, the risks of genetic erosion attached to any large-scale programme of vegetative propagation, can hopefully be minimized.

This model, which has been developed in Cameroon and Nigeria by ICRAF and partners, is based on involving the farmers in all stages of the tree domestication process. This starts with asking the farmers about which of the trees from natural forest they would like to cultivate on their farms, and progresses to the development of simple, low-technology plant propagators in the villages (Leakey et al 1990). These inexpensive and effective propagators, made from readily available products (wood, sand and polythene) for the rooting of stem cuttings, do not require running water or electricity. This simple and appropriate technology has many benefits for rural development projects over more complex propagation systems, as with the involvement of NGOs, villagers are trained in the basic principles of vegetative propagation so that they can themselves produce and bulk up 'cultivars' from the trees that they know and like best in their area.

This emphasis on the empowerment of the community and its use of indigenous knowledge about superior phenotypes in the forest allows rapid progress to be made, as it overcomes the need to do expensive and time consuming mass propagation and selection from populations of seedlings with unknown potential. This is particularly important, when there are a number of different fruit characteristics that together form a 'plus-tree' for ideo type (Leakey et al 2002), as the more traits for which selection is desired, the larger is the number of trees that would need to be screened in a research station approach to tree improvement.

Participatory domestication also allows farmers to be the beneficiaries and guardians of the use of their indigenous knowledge about inter and intra-specific variation in the population, and germplasm derived from it. This approach conforms to the aims of the Convention on Biological Diversity (CBD), which seeks to protect the rights of local people to their indigenous knowledge and germplasm. It is, thus, in stark contrast to the 'research station model' of tree domestication. It, however, requires that the farmers are informed about, and understand, their rights and know how to maintain and protect these rights (Leakey et al 2003).

In the early years of the African Humid Tropics (AHT) programme, action was limited to Cameroon and Nigeria, although species priorities were set in Cameroon, Ghana, Gabon and Nigeria. The development of strategies and techniques for Participatory Domestication has occurred over the last fifteen years and has been well documented (Simons 1996; Leakey and Simons 1998; Tchoundjeu et al 1998; Leakey et al 2003; Simons and Leakey 2004). This programme had been extended later into Equatorial Guinea and Gabon (2002), Democratic Republic of Congo and Ghana (2004). With this geogeographical expansion the list of species has also been extended:

- Indigenous fruit/nut trees: *Irvingia gabonensis*, *Irvingia wombolu*, *Dacryodes edulis*, *Garcinia kola*, *Ricinodendron heudelotii*, *Cola nitida*, *Chrysophyllum albidum*, *Allanblackia* spp.
- Indigenous leafy vegetable: *Gnetum africanum*
- Medicinal trees: *Prunus africana*, *Pausinystalia johimbe*, *Annickia chloranta*
- Melliferous trees and shrubs: *Vernonia*, *Alchornea*, *Dombeya*, *Myrianthus*, *Polyscias* and *Vitellaria* spp. and *Lophira lanceolata*

The participatory domestication in Africa was implemented under a Public-Private Partnership developed between ICRAF, IUCN, SNV (Netherlands Development Organization) and other Private Organizations. In addition ICRAF has also extended its tree domestication activities to the Sahel, the Miombo woodlands of southern Africa, the highlands of East Africa, the headwaters of the Amazon, Pacific region especially in the Solomon Islands and to South East Asia. Again, as in Africa, domestication of indigenous tree species has started new business and employment opportunities for rural communi-ties in the processing, value-adding and wider trade of tree products. This is an excellent example of diversified and more sustainable tropical agriculture, and of integrated rural development - a model that, with up-scaling, could alleviate poverty, hunger and malnutrition while also countering environmental degradation and climate change.

It has also demonstrated conclusively that tree-to-tree variation in fruits and nuts is not restricted to their morphology (i.e. shape, size, color) and it extends to their taste, oil content, food-thickening properties, medicinal properties and nutritional quality. Thus, the potential to develop new commercial products in the food, pharmaceutical, nutriceutical and cosmetic industries is significant.

ICRAF'S GLOBAL RESEARCH PROJECT-1 (GRP 1): DOMESTICATION, UTILIZATION AND CONSERVATION OF SUPERIOR AGROFORESTRY GERMPLASM

PROJECT GOAL

Smallholder farmers, researchers and their support agencies have increased information, knowledge and capacity needed to improve rural livelihood options through use of well conserved, characterized and developed tree genetic resources via well functioning seed and seedling systems.

PROJECT OBJECTIVE

The overall objective of GRP1 is to increase access to improved germplasm of priority tree species and ensure better functioning of tree seed and seedling supply systems. In consequence, there are opportunities to improve and optimize productivity of Agroforestry systems through improved germplasm. GRP1 aims to provide solutions to the fundamental problems that constrain the productivity of Agroforestry trees.

RATIONALE

The objectives of GRP1 are to resolve fundamental problems that constrain effective productivity of tree genetic resources tree and domestication. Such problems include but are not limited to following:

1. Constraints to characterization, tree domestication and documentation and dissemination of information
2. Constraints to tree germplasm supply and availability

TWO RESEARCH OUTPUTS

GRP1.1: Improved tree germplasm and associated information developed through appropriate methods.

GRP1.2: Sustainable tree seed and seedling supply systems developed while promoting conservation using appropriate partnerships.

ALIGNMENT TO CGIAR SYSTEM PRIORITIES (OUTPUT LEVEL)

GRP 1.1

1B Conservation/characterization of under-utilized Plant Genetic Resources

2D Genetic enhancement of selected species to increase incomes

3A High-value crops (including medicinals)

4A Integrated land, water and forest management

5D Improving R&D options to reduce rural poverty and vulnerability

GRP 1.2

3A High-value crops (including medicinals)

5D Improving R&D options to reduce rural poverty and vulnerability

COUNTRIES OF PLANNED RESEARCH (OUTPUT LEVEL)

Mali, Niger, Senegal, Burkina Faso, Ethiopia, Kenya, Tanzania, Malawi, Mozambique, Zambia, Zimbabwe, Uganda, Cameroon, DR Congo, Côte d'Ivoire, Equatorial Guinea, Gabon, Ghana, Nigeria, India, Thailand, India, Sri Lanka, Bangladesh, Indonesia, Laos, Philippines and China.

TARGET ECOREGIONS

1. Sub-Saharan Africa

RAEZ1 Warm arid and semi-arid tropics (AEZ 1): West Africa: Mali, Niger, Senegal, Burkina Faso; East Africa: Ethiopia, Kenya, Tanzania; Southern Africa: Malawi, Mozambique, Zambia and Zimbabwe.

RAEZ 2 Warm sub-humid tropics (AEZ 2): East Africa: Uganda; Southern Africa: Malawi, Mozambique, Zambia and Zimbabwe.

RAEZ 3 Warm humid tropics (AEZ 3): Cameroon, DR Congo, Côte d'Ivoire, Equatorial Guinea, Gabon, Ghana, Nigeria.

Table 8.2: Impact Pathways of ICRAF's GRP-1 by Output (ICRAF 2008)

Outputs	Outcomes	Impacts
GRP1.1: Improved tree germplasm developed for prioritized tree species using participatory tree domestication and information for these and other key agroforestry species updated and disseminated.	• Sustainability and productivity of agroforestry systems improved as farmers and other stakeholders domesticate and utilize a broader range of important tree species in more optimal ways thereby promoting current and future use. • Greater availability of updated and well-documented information on key agroforestry species will lead to increased species diversity in nurseries and on farm.	• Sustained and productive agroforestry systems contributing towards alleviation of poverty and food security and protection of the environment.
GRP1.2: Sustainable systems f tree seed and seedling supply developed and tree genetic resources conserved using appropriate strategies.	• Sustainable seed and seedling systems operating - availability of quality trees seeds and seedlings and innovative management practices. • Enhanced tree diversity on farm. • Better conserved tree germplasm.	• More buffered and productive farm environment contributing to poverty alleviation and better food security.

2. Asia and the Pacific

RAEZ 8 Warm arid and semi-arid tropics (AEZ 1): India and Thailand.

RAEZ 9 Warm sub-humid tropics (AEZ 2): India, Sri Lanka and Thailand.

RAEZ 10 Warm humid tropics (AEZ 3): Bangladesh, Indonesia, Laos, Philippines, Sri Lanka and Thailand.

RAEZ 11 Warm arid and semi-arid subtropics with summer rainfall (AEZ 5): China, India.

INTERNATIONAL PARTNERS

They primarily assist in molecular genetics work for marker assisted selection, development of strategies, databases, international policy formulation, multilateral conservation efforts, and advocacy. The international partners include: University College Dublin (UCD), ARCS Siebersdorf, Australian Tree Seed Centre, CAB International, CIFOR, CIRAD, Forest Landscape Denmark (FLD), European Forestry Institute, Food and Agricultural Organization (FAO), Ghent University, International Centre for Under-utilized Crops, IITA, Bioversity, International Society Horticultural Science, Unilever, SNV, Mars, Scottish Crop Research Institute (SCRI) and Winrock International.

DEVELOPMENTAL PARTNERS

They mainly collaborate and desire research findings linking directly into development projects. These partners include: Netherlands Development Organization SNV, Technoserve, IUCN/SECO, Mars Inc., Syngenta International AG and Unilever.

REGIONAL PARTNERS

They primarily assist in regional networking and regional conservation efforts. The regional collaborators include: CORAF, COMIFAC, APAARI, ASARECA, Centro Agronómico Tropical de

Investigación y Enseñanza (CATIE), Forum for Agricultural Research in Africa (FARA), FORNESSA, and SADC Tree Seed Centre Network.

NATIONAL PARTNERS

They primarily engage with World Agroforestry Centre in developing and testing methods, developing improved 2 germplasm and propagation methods, conservation of individual tree species, and assembling range-wide collections. The national partners include: National Tree Seed Centres; National Agricultural Research Institutes, National Horticultural Institutions, National Forestry Research Institutes, Universities, National Health Authorities (for fruit and medicinal species), National Extension Systems, and National Quarantine Authorities. In addition, nationally based NGOs and CBO partners will be engaged on scaling up research.

■■■■

9 TRADITIONALLY IMPORTANT TREES IN INDIGENOUS AGROFORESTRY SYSTEMS

INTRODUCTION

Feeding the world's population is one of the most pressing challenges facing humanity in the twenty-first century. There is need to expand by 70 per cent at the global level and to double in developing countries. This goal is made ever harder by the increasing impacts of climate change, including extreme weather events, which are putting poor households at risk. Forests as well as trees on-farms are a direct source of food and cash income for more than a billion of the world's poorest people, providing both staple foods and supplemental foods. More than 50 million people in India directly depend upon forests for subsistence, which have resulted in degradation of forest wealth as well. Instead income from on-farms trees can make a significant contribution to rural households providing both cash and subsistence benefits. Fuelwood is the main energy source for cooking and food processing in rural areas in most developing countries including India. Similarly, the rearing of animals is also an integral component of agro-ecosystem, which boosts livelihood and sustains the agriculture. The much required fodder resources are also dwindling. Fuelwood and fodder supplies thus indirectly affect the stability, quality and even quantity of food consumed. Dwindling fuelwood and fodder supplies in many rural areas have increasingly severe impacts on livelihood and poor farmers especially ladies are under much stress to arrange fodder and fuel for their livelihood security.

In comparison to other farming systems, agroforestry interventions provide diversified range of products from a given area leading to higher income and improved standards of living; increased self-sufficiency and reduced risk from climatic, biological or market impacts; better use of underutilized land; enhanced food and nutritional security; and increased fulfillment of social and cultural needs through product sharing and exchange (Kumar and Nair 2004). Basically, agroforestry has potential to offer more to the farmers than either arable crops or tree crops alone i.e., better overall productivity and economics, reduction of risk of crop failure and multiple products and services. Farmers have been raising or retaining perennial plants in a number of different arrangements for thousands of years, including mimicking forest growth to create agricultural systems that can be categorized as agroforestry systems. The number of trees on farms is increasing and interestingly, there are many examples where the rise in on-farm tree occurs in areas where population densities are high and farm size is small. The products from these on-farm trees are important to the local people and the wider population for fuelwood/fodder, as well as many food and medicinal products. Agroforestry interventions provide the opportunity for efficient nutrient cycling through species and structural diversity, constant ground cover, nitrogen fixation, symbiotic plant interactions, and long rotational time-frames.

ROLE OF AGROFORESTRY TREES IN RURAL LIVELIHOOD

Rural people most often depend heavily upon natural resources for their livelihood. The long term sustainable use of such resources, such as soils, water and biodiversity including forests, depends on local people's knowledge, management and local people's ability to maintain and utilize the available knowledge and resources. As natural forests rely on the use of vertical space for the efficient use of light, soil nutrients, and symbiotic plant interactions; similarly tree-crop interface also rely on such use of vertical space. The tallest trees combine to form a canopy, smaller tree species form an understorey, a shrub layer fills the next stratum, and an herb layer fills the lowest layer of vertical space (Kricher 1998). Similarly agroforestry systems also make use of vertical horizon due to limited scope for horizontal expansion in fast increasing population. Land resources are getting divided in the families, thus forcing farmers to adopt innovative options of intercropping to meet multiple requirements from limited land resources.

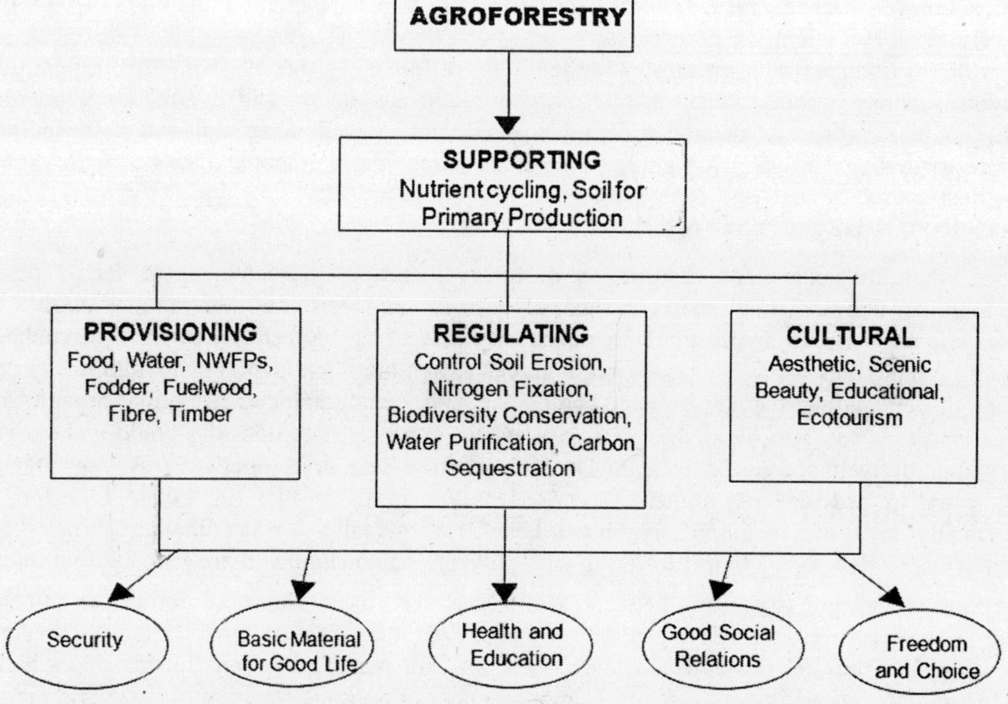

Fig. 9.1: Environmental Services and Multiple Benefits for Rural Livelihood from Agroforestry

The National Agriculture Policy-2000 emphasized the role of agroforestry for efficient nutrient cycling, nitrogen fixation, organic matter addition and improving drainage. It also underlined the need for diversification by promoting integrated and holistic development of rainfed areas on watershed basis through involvement of community to augment biomass production through agroforestry and farm forestry. The Task Force on Greening India for Livelihood Security and Sustainable Development of Planning Commission in the year 2001 has also recommended that for sustainable agriculture on rain-fed and irrigated areas, agroforestry may he introduced over an area of 18 and 10 million ha, respectively out of 43 m ha. This all will, besides ensuring ecological and economic development, provide livelihood support to about 350 million people. Gap of demand and supply of forest produce in India is widening and forests are unable to fulfill the demand. Agroforestry can play an important role in filling this gap and conservation of natural resources.

Traditional agroforestry systems in the mountains are very close to natural ecosystem as they provide ecosystem services similar to the forests such as the biodiversity conservation, provision of food and fibre, water resources and its purification, climate regulation and carbon sequestration, nutrient cycling, primary production, production of oxygen, soil formation and conservation, recreation and cultural services for the well being of the people/society and conservation.

Tree growing in combination to agriculture as well as numerous other vegetation management regimes in cultural landscape including in farms, watersheds, regional, national and global landscape, can be integrated to take advantage of services provided by adjacent natural, semi-natural or restored ecosystems, which may vary in different landscape. Increasing the livelihood security and reducing the vulnerability call for society, tree-crop interface adaptations are possible when combined with traditional resource management systems. Agroforestry as a local adaptation, therefore, is a promising area of interest for scientists, policy-makers and practitioners. The underexploited trees can play a huge role in indigenous agroforestry systems in different agro-ecosystems in respect of the biodiversity conservation, climate change mitigation and adaptation, and livelihoods improvement. Identifying the prospects of underutilized tree species, which were cultivated in indigenous agroforestry systems of India, may guide the future research, management and their adoption.

TREES IN AGROECOSYSTEMS

Understanding the extent and distribution of trees on agricultural land, at the landscape level, including the numbers and characteristics of farmers and farming communities within those landscapes, can help to assess the importance and role of agroforestry both to the livelihood of farming communities as well as to overall global agricultural production with multiple products. Further, understanding the geographic, ecological, and demographic distribution of agroforestry related land uses can also highlight those areas where increased tree densities could make a greater contribution to livelihoods or landscapes. The integration of trees in farming systems is the practice of agroforestry to provide environmental services and/or products that are either traded or used domestically to confer multiple livelihood benefits, especially for smallholder farmers in the underdeveloped/developing world suffering with poverty, malnutrition and hunger.

There is a strong association between aridity and tree cover. The more humid the climate, the higher the level of tree cover. The results from South East Asia, Central America, and South America are examples of this relationship. However, there are still many exceptions to this rule - high tree cover found in irrigated agroecosystem i.e., more number of trees are found on-farm in arid regions of Rajasthan than irrigated fertile Indo-Gangetic region. Retaining trees on farm land during clearance for agriculture or subsequent planting of trees produces a range of ecological habitats and seasonal niches ideal for wild food production. During the selective clearing of land for agriculture, farmers usually retain particular tree species. These are often fodder/fruit trees, the providers of seasonally important food for animals and human being. In other settings, trees within farming systems have been enriched by planting. Wild fruit trees may be collected from forest areas and planted in agricultural land to enrich on-farm tree species. The fruit producing potential of wild cultivars can also be upgraded by simple grafting and breeding techniques (Chauhan et al 2010).

In India, agroforestry has been a way of life. Its evolutionary process could be understood through the historical development. The process of human evolution has been from forests when man learnt the art of domesticating plant and animals after leaving hunting and gathering habit. Thus agroforestry is an age old practice whose origin is believed to be since the *Vedic* period. The period of

Indian history when sacred Ramayana was composed by Maharishi (Saint) Valmiki, there is an illustration of Ashok Vatika, a form of home garden. It comprised of plants encircling the sarovar (water tank) in three rings. Forest trees (ashok, neem, pipal, bargad) in the outer ring, horticultural trees (jackfruit, mango, banana and guava) in the middle ring and flowers (champa, chameli, juhi, bela, etc) in the inner ring. The tribal dances performed around trees are symbolic of the intimate ethnic bias to protect trees. Trees had considerable socio, ethnic and religious backing, but their use as agroforestry component was limited (Puri and Panwar 2006). The need for selection and characterization of woody perennials that can provide multipurpose products and services in agroforestry systems has been emphasized extensively in India. Toky et al (1992) listed more than 250 indigenous multipurpose trees of different eco-regions of India, including many lesser known species. Many such species (*Albizia* spp., *Acacia* spp., *Dalbergia* spp., *Gmelina arborea*, *Grewia optiva*, *Melia* spp., *Terminalia* spp., *Casuarina* spp., *Azadirachta indica*, various bamboos, etc.) are in the process of domestication and commercialization. The traditional practices and systems of agroforestry in India have been reviewed by Tejwani (1994) and complete details synthesized under diagnosis and design approach in agroforestry by Pathak et al (2000), which gives the details on regional scale to redefine the components.

IMPORTANT TREES OF INDIGENOUS AGROFORESTRY SYSTEMS/ PRACTICES IN INDIA

Farming communities have the knowledge of the available resources and their judicious utilization. This system has been working since long, which is based in indigenous knowledge and over the time further refinement resulted in current day scientific intervention. Local knowledge represents the current position of resources and their appropriate utilization. Local conditions and resources vary from place to place and farmer to farmer. Farming communities have multiple objectives and components and are characterized by their complexity to understand. Traditional agroforestry in India ranges from apparently simple forms of shifting cultivation - observed especially in the north-eastern parts of India to the complex multistoried homegardens of the humid coastal areas in the south. These traditional systems also differ for their objective for which these are maintained.

Trees which have multiple uses are preferred in traditional agroforestry systems particularly in hills, especially those yielding fruit/fodder/mulch and being suitable in providing supporting structures for the legume and cucurbit climbers. Tree fodders play an important role in traditional farming systems common across the Himalayas. A number of multipurpose tree species are conserved as scattered trees in settled croplands and terraced slopes by the traditional farmers. They are especially valuable during the dry winter season, when fodder from other sources becomes limited in quantity and quality. This diversity has also reduced risks from pests and adverse weather as they tend to affect different indigenous and traditional crops and cropping practices. The on-farm khejri (*Prosopis cineraria*) of the hot arid region, meets local needs for fodder, small timber and food, while the *Alnus nepalensis* and *Amomum subulatum* (large cardamom) combination in the eastern Himalayan region's humid sub-temperate zone is an excellent example of a commercial, but traditional, agroforestry system. Other prevalent agroforestry systems include the deliberate growing of trees on field bunds, their sporadic distribution in agricultural fields, and the systematic retention of shade trees in tea and coffee plantations. Dhyani et al (2007) have identified different agroforestry systems/practices in different agro-ecological zones of India: (1) Himalayan Region, (2) Indo-Gangetic Plains, (3) Arid and Semi-arid Region, (4) Humid and Sub-humid Region, (5) Coastal and Island Region, which reflects the area specificity and need based approach in different agroforestry components. The status and prospects of various traditionally important tree species in these agro-ecological zones are given below.

HIMALAYAN REGION

NORTH WESTERN HIMALAYAS

In Western Himalayas, 60 to 70 per cent requirement of the firewood is met from arboreal components and several multipurpose trees (MPTs) along the bunds of agricultural lands or scattered trees on the pasture lands. *Grewia optiva*, *Celtis australis*, *Bauhinia variegata*, *Albizia chinensis*, *Bombax ceiba*, *Melia azedarach*, *Toona ciliata*, etc. are common MPTs, while plum (*Prunus domestica*), apricot (*P. armeniaca*), peach (*P. persica*), almond (*P. dulcis*), pear (*Pyrus communis*), etc. are common fruit trees of agri-horti-silvicultural and agri-horticultural systems. Despite providing fodder material to livestock, trees and shrubs also provide fuelwood, poles and timber for house construction. Fuelwood is the major source of the energy needs, which is produced at the farm itself.

In Uttarakhand, Himachal Pradesh and Jammu & Kashmir traditional agroforestry systems are well established since time immemorial (Dadhwal et al 1989). Planting and harvesting of wood products, fruits, roots, leaves and cattle fodder from on-farm trees have been followed in this region since the earliest days of man's activities. Planting around on field boundaries and homesteads aims to diversify the production and increase its magnitude. Moreover, it was compulsion as well to conserve soil and water on sloping lands in addition to resource base. Economic prosperity and ecological security of farmers of hilly states has basically been due to their integration of the cropping systems. The major practice in the field is line planting on the terrace risers. The most common species is a multipurpose tree *Grewia optiva* (bhimal) and other planted tree species are *Celtis australis*, *Toona ciliata*, *Mangifera indica*, *Bauhinia variegata*, *Morus alba*, *Ficus roxburghii*, bamboo, etc. Scattered plantings of these species are also common. Generally, the density in both these types of planting is poor to medium and plants are not appropriately spaced to get full density per unit area. However, presently the trees are neither protected nor replanted/properly managed and farmers have shifted from traditional trees to the fruit plants to harness better economics.

Kala (2010) conducted a survey in middle Himalayas of India to determine the Status of an indigenous agroforestry and recorded a total of 26 herbaceous food crop species and 21 woody species that were raised by farmers in the Uttarakhand state. A total of 37 on-farm plant species have been in use for curing various ailments. Different plant parts of these species, such as bark, root, leaf, stem, flower and seed were used for medicine preparation. In general, the traditional agroforestry system in villages are raised and nurtured in such a way that the basic requirement of living for food, shelter and medicine are achieved. The indigenous system of cropping was locally called as "*Baranaja*" (growing of 12 crops) that revolved around the production of more than 12 varieties of crops. Besides food, the species grown in the agroforestry system were used for multiple purposes. Increase in population, food demand, fragmentation of land holding, water scarcity, migration of youth in search of employment, changing socio-economic and changing climatic conditions are some of the major reasons for declining agroforestry system and abandoning the agricultural land.

In subtropical and sub-temperate zone, tree like *Toona ciliata*, *Ulmus villosa*, *Melia azedarch*, *Albizia lebbeck*, *A. chinensis*, *Dalbergia sissoo*, etc. are generally retained as timber trees, whereas, *Grewia optiva*, *Morus laevigata*, *Anogeissus latifolia*, *Bauhinia variegata*, *Celtis australis*, bamboo, etc. for fodder purpose. *Acacia catechu* is retained for its multi-utility viz., katha and cutch contents, small agricultural timber, fodder, fuelwood and ecological restoration. In wet temperate high hills, *Quercus* spp., *Salix* spp., *Robinia pseudoacacia* and *Morus serrata* along with fruits viz. apple, kiwi, pear, peach, etc. are the most favoured tree species (Verma et al 2006). There is great potential for improvement of traditionally managed systems to a great extent in order to realize the real production potential of agroforestry systems/practices in this region to meet the multiple requirements.

NORTH EASTERN HIMALAYAS

The eastern Himalayan region also comprises diversity of agroforestry system as an important land use practice. Agroforestry is accepted as one of the sustainable management systems for provisioning functions adopted by the mountain communities in the region. Such systems conserve soil by improving the fertility levels and arresting erosion, provide quality water for local consumption, fodder for livestock, fuel and timber for use as energy and construction materials, and traditional crops for food security. Integration of cash crops in the system gives good economic return that promotes poverty alleviation and contributes to the human well being and good health. One good example of such a traditional agroforestry system is the cultivation of large cardamom as cash crop in the Indian state of Sikkim of the eastern Himalayan region, whereas, bamboos are lifeline of farmers in this region. Such agroforestry systems are unique examples of ecological sustenance and economic viability for the mountain people while providing goods and services to the downstream users. A large number of tree leaves are used as ruminant feed in NEH Region (Chander Datt and Niranjan 2001) and the proximate composition of promising tree/ shrubs leaves indicated that most of the tree leaves are rich in crude protein. Not only leaves, pods of some trees are also used as concentrates. The nutritive values of some of the tree fodder species are comparable with those obtained from cultivated one's including *berseem* and *lucern.* Fodder trees provide green fodder particularly during winter months from October to March when other fodders are not available. Tree leaves constitutes 25 to 60 per cent of total green fodder fed to live stock in the north-eastern Himalayan hills.

Shifting cultivation (slash and burn system) is being practised in north-eastern parts of India, which receives high rainfall. In this system, land under natural vegetation (usually forests) is cleared by slash and burn method cropped with common arable crops for a few years and then left unattended when natural vegetation regenerates. Traditionally, the fallow period was 15-20 years but in recent times, it has been reduced to 4-5 years in many areas, due to the increasing population pressure. The drastic reduction in fallow period has degenerated system thus causing serious soil erosion/ depleting soil fertility resulting to low productivity. In north-eastern India, many annual and perennial crops with diverse growth habits are being grown. At times annual crops such as potato, rice, maize and ginger are grown in monoculture or mixed culture along with *Pinus kesiya.* Another important attribute of the system is secondary succession of vegetation during fallow period. The tree species, which may be considered suitable for afforestation of fallow areas or to intercrop with arable crops must be fast growing preferably nitrogen-fixing and must efficiently recycle available nutrients within the system shortening time required to restore fertility. These may include species of *Acacia*, *Albizia*, *Alnus*, *Casuarina*, *Erythrina*, *Acacia*, *Gliricidia*, *Inga*, *Leucaena*, *Parkinsonia*, *Pithecellobium*, *Prosopis*, *Robinia*, *Sesbania*, bamboo, etc.

Thus, intercropping under or between rows of fast-growing trees in a fallow phase is one of the approaches while finds alternative to shifting cultivation. A farming-system approach based on the watershed management has been advocated as an alternative to shifting cultivation in north-eastern Himalayas. Tree + large cardamom based system is most prevalent in this region. Cardamom is a perennial cash crop grown traditionally beneath the natural forest tree cover on marginal lands and slopes. It is a shade loving plant and requires high moisture and is usually cultivated in areas where mean annual rainfall varies between 1,500-3,500 mm. Some common shade trees for this agroforestry system are *Schima wallichii*, *Engelhardtia acerifolia*, *Eurya acuminata*, *Leucosceptrum canum*, *Maesa chisia*, *Symplocos latifolia*, *Ficus nemoralis*, *Ficus hookeri*, *Nyssa sessiliflora*, *Osbeckia paniculata*, *Viburnum cordifolium*, *Litsea polyantha*, *Macaranga pustulata* and *Alnus nepalensis.* Hence, large cardamom agroforestry practice also supports conservation of tree biodiversity in the region (Sharma et al 1994). Majority of cardamom plantations have Himalayan alder (*Alnus nepalensis*) as shade trees since the combination is sympatric and has proved to be ecologically and economically viable. Biomass production as well as net productivity was much higher in alder based system than forest ecosystem. Alder though non-leguminous but enriches soil through atmospheric nitrogen fixation. In this region, interplanting tea bushes with shade trees is also traditionally followed and continued with beneficial productivity and tea quality.

INDO-GANGETIC PLAINS

In India, Indo-Gangetic plains are spread over 50.34 million ha area and form 15.31 per cent of the total geographical area of the country on which supports 32.8 per cent human population and 35.18 per cent livestock population of the country. This region contributes 51.9 per cent to the national food grain output showing the presence of good infrastructural support for high crop productivity. The Indo-Gangetic plains are formed by the sedimentation from Himalayas carried by the Indus and Ganga rivers and their tributaries. About 72 per cent of the area is under cultivation varying from 63 per cent in the lower Gangetic plain to 82 per cent in the trans Gangetic plains. The fertile soil and ample water resources attracted accelerated cultivation activity after clearing the forests. However, the on-farm trees have faced axe in this region because of commercial agriculture and very limited trees are found now in this region, though this region was quite rich in on-farm plant wealth. Tree species like *Dalbergia sissoo*, *Acacia nilotica*, *Azadirachta indica*, *Artocarpus heterophyllus*, *Madhuca latifolia*, *Melia azaderach*, Bamboo, etc. were found but now a very limited commercial species with narrow genetic base are raised in agroforestry systems i.e., *Eucalyptus tereticornis*, *Populus deltoides*, *Gmelina arborea*, *Ailanthus excelsa*, *Melia composita*, etc. Indo-gangetic region had rich resources of fruit plants (mango, guava, ber, etc.) and these fruit plants still occupy large area but under very limited number of commercial varieties.

ARID AND SEMI-ARID REGION

The life support systems in this region are constrained by bioclimatic and environmental limitations with low precipitation (100-400 mm, mean annual rainfall), high pre-monsoon temperatures (mean maximum of the hottest month: 45-47°C), high wind speed (average annual 8-10 km h^{-1} with figures up to 30-40 km h^{-1} during summers), high potential evapo-transpiration with an annual total of 1,500 to 2,000 mm. Soils are generally sandy to sandy loam in texture with poor nutrient status and low water holding capacity. *Acacia senegal*, *A. nilotica*, *A. tortilis*, *A. leucophloea*, *Ailanthus excelsa*, *Azadirachta indica*, *Lawsonia inermis*, *Salvadora persica*, *S. oleoides*, *Prosopis cineraria*, *Tecomella undulata*, *Zizyphus* spp., *Tamarix aphylla*, etc. are important tree species of the region. In the backdrop of scanty vegetation scenario, all the tree species are important in this region, though *Prosopis cineraria* (khejri) stands tall among all other tree species because of its multiple benefits and uses in the fragile desert ecosystem. Khejri is a medium sized tree, called the Royal tree of Great Indian Desert because of its role in conserving the ecosystem of the desert. It is a frost and drought resistant tree, which can withstand temperature extremes, ranging from 45-48^{0} C during the summer to less than 10°C during the winter. The tree is worshipped by a large number of people, mostly Bishnois, a great conservationist community in the Thar Desert. The medicinal value of this tree has been highlighted in ancient Ayurvedic (medical) literature. Wood is very high in energy value of about 5,000 kcal/kg. An unlopped 30-35 years old tree produces about 4-5 kg of air dried pods in normal rainfall year.

HUMID AND SUB-HUMID REGION

HUMID ZONE

This represents conditions of tropical crop cultivation, which mostly cover hot humid conditions. Due to heavy rainfall, the soils are either neutral or acidic in nature. In southern India, humid zone receives more than 200 cm of annual rainfall. The region experiences persistent high temperatures which normally do not go below 18°C even in the coolest month. The conditions for crop production are most stable in this region than in the previous three zones, however, priority is given to plantation crops.

SUB-HUMID ZONE

This climatic complex consists of temperate sub-tropical conditions. This region is potentially very productive experiencing high rains. By proper management of land, water and cropping the yields can be appreciably stepped up. The western coastal lowlands, the Western Ghats, and southern parts of Assam are included in this category. It is characterized by high temperatures throughout the year, including in the hills. The rainfall here is seasonal, but heavy and is above 78 cm in a year. The heavy rains are responsible for the tropical wet forests in this region, which consists of a large number of floral and faunal species. The temperature during the coldest months in region experiencing this climate falls between 18°C and 0°C. The summers are hot and temperatures can reach 46°C in the lowlands. May and June are the hottest months. Winter months are mostly dry with feeble winds. Frost occurs for a few weeks in winter. The foothills of the Himalayas, Punjab-Haryana plain adjacent to the Himalayas, Rajasthan east of the Aravalli range, Uttar Pradesh, Bihar and northern part of West Bengal and Assam experience this climate. On-farm tree vegetation is rich in this zone which provides multiple resources including intangible benefits.

Homegardens in Kerala, Assam, W. Bengal, etc., have a mixture of trees, shrubs herbs, etc.; Coconut, Arecanut, *Erythrina*, *Gliricidia* with black pepper, cacao, coffee, cassava and other cash crops. The important tree species in agricultural fields include *Acacia auriculiformis*, *Aegle marmelos*, *Albizia* spp., *Anogeissus latifolia*, *Anthocephalus chinensis*, *Artocarpus* spp., *Casuarina equisetifolia*, *Diospyros melanoxylon*, etc.

COASTAL AND ISLAND REGION

Coastal plains are characterized by an area of flat low lying land that is situated adjacent to a water body often a sea or ocean. Coastal plains are separated from the interior of the larger land mass by other unique features. Coastal plains have limited vegetation, however, grasses and trees are not uncommon. Coastal plains in India can be divided into two basic types, i.e., Eastern coastal plains and the Western coastal plains.

EASTERN COASTAL PLAINS

The eastern coastal plains are located on a wide stretch of land between the Eastern Ghats of India and the Bay of Bengal. This stretch of land stretches to 120 km in width at parts. The eastern coastal plains extend from Tamil Nadu in the south to West Bengal in the north. The eastern coastal plains have rivers draining into them and river deltas also occupy the valleys. The region of the eastern coastal plains is an expansive area and is divided into six regions. The six regions of the eastern coastal plains of India are the Mahanadi delta, the Southern Andhra Pradesh plain, the Krishna Godavari deltas, the Kanyakumari coast, the Coromandel and the sandy coastal regions. The eastern coastal plains are characterized by a temperature that exceeds 30°C and also experiences high levels of humidity. The rainfall of the region is also abundant (100-300 cm annually) subject to both northeast and southwest monsoon rains. Mangrove vegetation is common comprising woody, specialized types of trees of the tropics that can live on the edge, where rainforests meet oceans. They protect the coastline and prevent erosion by collecting sediment from the rivers and streams and slowing down the flow of water.

WESTERN COASTAL PLAINS

The western coastal plain of India in contrast to the eastern coastal plain is located on a narrow strip of land. The western coastal plains are located in the west of India between the Western Ghats and the Arabian Sea. They extend from Gujarat in the north down 50 km to the south in Kerala and are characterized by numerous backwaters and rivers that flow into the region. These rivers that flow into the region lead to the forming of estuaries that are found in the western coastal plains of India. The storm activity here is considerably less than on the eastern coastal plains. The western coastal plains are smaller than their eastern counterpart and are divided into the regions of Konkan, Kanara and the Malabar coast. These plains also experience a temperature of over 30ºC, characterized by high humidity level. Abundant rainfall is received by this region (100 to 300 cm annually) subject to southwest as well as northeast monsoon rains.

The important agroforestry practices include Coffee under *Erythrina lithosperma*; cacao with coconut; black pepper on *Gliricidia* spp., *Grevillea robusta*; cardamom under *Toona ciliata*, *Artocarpus*, etc., Mangrove species for shore protection are *Acanthus ilicifolius*, *Avicennia officinalis*, *Cerbera odollam*, *Rhizophora conjugata*, etc.,

The tree species planted in shelterbelts and windbreaks are *Casuarina equisetifolia*, *Azadirachta indica*, *Prosopis chilensis*, *Acacia Senegal*, etc., Trees for boundary planting and providing fodder, fuelwood and other NWFPs are *Aegle marmelos*, *Albizia* spp., *Azadirachta indica*, *Bamboos*, *Bombax malabaricum*, *Calliandra calothyrsus*, *Cassia* spp., etc.

Table 9.1: Description of Important Traditional Tree Species of Indigenous Agroforestry

S. No	Botanical Name- (Common Name)	Family	Morphological Characteristics	Salient Uses
1.	*Areca catechu* – (Arecanut)	Palmae	The areca nut is the seed of the areca palm, which grows in much of the tropical Pacific, Asia, and parts of east Africa. It is commonly referred to as betel nut, as it is often chewed wrapped in betel leaves.	Medicinal value used in pharmaceutical industry and dried nut chewed in betel leaf.
2.	*Acacia catechu* – (Khair)	Fabaceae	It is a moderate sized tree with brown or purple young shoots. Bark is dark greyish brown. Spines short and hooked from a broad triangular base. Leaves are bipinnate with glands between many of the pairs of the pinnae and a large conspicuous gland at or near the middle of the petiole; Flowers are sessile and pale yellow in auxiliary spikes. Pods are flat, thin, brown, shiny with a triangular beak at the apex and narrowed at the base into a stalk.	The timber (heartwood) is primarily used for extraction of cutch and katha, for which there is a great demand. Cutch is used for tanning, dyeing. Wood is used as fuelwood and bark has medicinal properties. As such, it is usually not available for use as timber.
3.	*Acacia nilotica* – (Kikar)	Fabaceae	A small tree, having feathery leaves, single-stemmed, stems and branchlets usually dark coloured, spines straight. Bark is grey to brownish-black, rough and longitudinally fissured. Leaves are pinnae usually in 3 to 12 pairs. Flowers are bright or golden yellow, sweetly scented, numerous, with bisexual and male flowers on the same globose heads inflorescence. Pods are grey, thick, softly	Babul is usually available only in small billets and is usually used in house construction for posts, rafters and beams; agricultural implements especially for ploughs, harrows and clod crushers and in cart making for yokes, shafts and wheels. It is also used for various

S. No	Botanical Name- (Common Name)	Family	Morphological Characteristics	Salient Uses
			tomentose, straight or slightly curved with constrictions between the seeds giving a necklace appearance.	types of tool handles.
4.	*Ailanthus excelsa* – (Maharukh)	Simaroubaceae	Indian Tree of Heaven is a large deciduous tree, trunk straight. Bark is light grey and smooth, becoming grey-brown and rough on large trees, aromatic, slightly bitter. Leaves are alternate, pinnately compound, large, leaflets in pairs, long pointed, edges coarsely toothed and often lobed. Flowers are in clusters droop at leaf bases, flowers many, mostly male and female on different trees, short stalked, greenish-yellow. Fruit is samara, flat, pointed at ends. copper red, strongly veined, twisted at the base.	Bark used as a powerful fever-cure and tonic. Leaves and bark in good repute as a tonic after labour. A good fodder yielding tree and light timber is used for industrial purpose (ship, match, plywood, etc.).
5.	*Albizia lebbeck* – (Siris)	Leguminosae	*Albizia lebbeck* is tall tree with grey-violet bark having rusty brown breathing pores, rough and fissured. Leaves are glabrous or slightly hairy, Flowers appear shortly after new leaves, are white, heavily scented, in terminal panicles. Pods are pale straw to light brown at maturity, flat and not raised or constricted between seeds; seeds brown, flat, orbicular or elliptic. Other important species i.e., *A. procera* and *A. chinensis* are also seen on cultivable land and tea gardens in India.	Siris is grown in some areas primarily as fodder for camels, water buffalo and cattle. An excellent fuelwood species. Leaves and seeds are used for eye problems.
6.	*Alnus nepalensis* – (Alder)	Betulacea	*Alder is a* deciduous tree with silver-grey bark that reaches up to 30 m in height and 60 cm in diameter. It occurs throughout the Himalaya at 500–3,000 m of elevation in Himalayas. The leaves are shallowly toothed, 7–16 cm long and 5–10 cm broad. The flowers are catkins, unusually for an alder produced in the autumn, with the seeds maturing the following year.	Wood is moderately soft. It is occasionally used for making boxes and in light construction, and as firewood. It is non-leguminous nitrogen fixing tree.
7.	*Anthocephalus cadamba* – (Kadamb)	Rubiaceae	Kadam is a large tree with a broad crown and straight cylindrical bole. The tree may reach a height of 45 m with trunk diameters of 100-160 cm. The tree sometimes has small buttresses and a broad crown. The bark is grey, smooth in young trees, rough and longitudinally fissured in old trees. Leaves are glossy green, opposite, simple more or less sessile to petiolate, ovate to elliptical. Inflorescence in clusters; terminal globose heads without bracteoles, orange or yellow flowers. Flowers are bisexual, fruitlets numerous with their upper parts containing 4 hollow or solid structures.	The fresh leaves are fed to cattle. The fragrant orange flowers attract pollinators. Sapwood white with a light yellow tinge becoming creamy yellow on exposure; not clearly differentiated from the heartwood-used in ply making.

S. No	Botanical Name- (Common Name)	Family	Morphological Characteristics	Salient Uses
8.	*Artocarpus heterophyllus* – (Kathal)	Moraceae	A medium sized important tree in tropical homegardens. The jackfruit has played a significant role in Indian agriculture for centuries. Another species of importance is *A. lakoocha* (fruits are traditionally used for making pickle)	The flesh of the jackfruit is starchy and fibrous, and is a source of dietary fibre. Raw fruit is cooked as vegetable and ripe jackfruit is naturally sweet. Leaves are used as fodder and wood for furniture and musical instruments.
9.	*Azadirachta indica* – (Neem)	Meliaceae	A medium sized tree. It has a short trunk, with wide spreading branches forming a rounded or oval crown. It has a moderately thick (1.25-2.5 cm) grey to reddish-brown bark. The outer rind is rough, woody, very much fissured, often peeling in thick slices. Leaflets are often alternate, serrate. Flowers are white, strongly honey-scented, in axillary panicles. Anthers are inserted within he tube opposite its teeth. Fruit- Drupe.	Windbreaks, shade and fodder for cattle, soil improvement and wasteland reclamation. Wood is insect repellant and used for construction, furniture, paper, pulp, chipboard and fuelwood. Azadiracthtin, an insecticidal compound can be extracted from seeds and leaves.
10.	*Bauhinia variegata* – (Kachnar)	Caeselpiniaceae	It is a small to medium-sized deciduous tree. The leaves are long and broad, rounded, and bilobed at the base and apex. The flowers are conspicuous, bright pink or white, with five petals. The fruit is a pod, containing several seeds.	The leaves, flowers and flower buds are eaten as vegetables. Leaves make good fodder and are greedily eaten by sheep, goats and cattle, The wood is used for agricultural implements.
11.	*Celtis australis* – (Khirak)	Ulmaceae	It is a deciduous tree. Leaves are simple, alternate, and sharp-toothed; rough on top, and furry underneath. Flowers are hermaphrodite (have both male and female organs) small and green without petals, either singly or in small clusters. Fruit is small, dark-purple berry-like drupes.	The fruit of this tree is sweet and edible, and can be eaten raw or cooked. The leaves and fruit are astringent, lenitive and stomachic. Decoction of both leaves and fruit is used in the treatment of amenorrhea. A yellow dye is obtained from the bark. The flexible thin shoots are used as walking sticks.
12.	*Dalbergia sissoo* – (Shisham)	Leguminosae	It is a medium to large-sized deciduous tree. Bark is thin, grey, longitudinally furrowed. Leaves are imparipinate and flowers are pale white to dull yellow, racemes. Pods are indehiscent, kidney shaped, thin and flat.	It is one of the most useful timber species of India. It is used for high-quality furniture, cabinets, decorative veneer, marine and aircraft grade plywood, ornamental turnery, carving, engraving, tool handles and sporting goods. Its root wood is used for tobacco pipes.
13.	*Dendrocalamus strictus* – (Solid Bamboo)	Poaceae	A deciduous densely tufted bamboo. Culms 8-16 m high, 2.5-8 cm diameter, pale blue green when young, dull green or yellow on maturity, much curved above half of its height; nodes somewhat swollen, basal nodes often rooting, lower nodes often with branches; internodes 30-45 cm long, thick-walled.	This species is suitable for reclamation of ravine land. It is extensively used as raw material in making bamboo furniture, paper mills and also for a variety of purposes such as construction, agricultural implements, musical instruments, furniture etc. Young shoots are commonly used as

S. No	Botanical Name- (Common Name)	Family	Morphological Characteristics	Salient Uses
				food. Leaves are used in medicines and as fodder.
14.	*Gmelina arborea* – (Gamhar)	Lamiaceae	It is moderately sized to large deciduous tree with a straight trunk. It is wide spreading with numerous branches forming a large shady crown, Bark is smooth and leaves are opposite-decussate. Flowers are abundant, scented, reddish, brown or yellow, in terminal and axillary cymes on the panicle branches. Flower is lobed and fruit is a drupe.	The wood produces good-quality pulp. Timber: Uses include the manufacture of furniture, plywood core stock, mine props, matches and timber for light construction. Tannin or dyestuff.
15.	*Grewia optiva* – (Buel/ Dhaman)	Tiliaceae	It is a small to medium-sized deciduous tree, bole clear. Branches are smooth, pale silvery-brown and bark is dark brown, thick and rough. Leaves are opposite, ovate, acuminate, closely serrate; teeth small, blunt; Flowers in penducles solitary, leaf opposed or exceptionally a few axillary; green outside, white, pale yellow or red inside. Fruit is a drupe.	The ripe fruits are edible. Raw or cooked, it has a pleasant acid taste. Fodder: The leaves are rated as good fodder and bark yields a fibre that is used for cordage and clothing.
16.	*Jatropha curcas* – (Rattanjot)	Euphorbiaceae	A small tree or bush, the leaves are green to pale green, alternate to subopposite, and three- to five-lobed. Male and female flowers are produced on the same inflorescence and the inflorescence can be formed in the leaf axil. Fruits are caspsules, the seeds are mature when the capsule changes from green to yellow. The seeds contain around 20 per cent saturated fatty acids and 80 per cent unsaturated fatty acids, and they yield 25-40 per cent oil by weight.	The oil is used as biofuel. The cake can be used for fish or animal feed and biomass feedstock to power electricity plants, or as biogas or high-quality organic fertilizer.
17.	*Madhuca latifolia* – (Mahua)	Sapotaceae	A deciduous tree, 10-15 m tall and with a spreading, dense, round, shady canopy. Bark is rough, brown in color, slightly cracked and fissured, inner bark red, exudes white, milky sap when cut. Leaves are elliptic, tip pointed, base angled, texture thick, hairy beneath and margin entire but may be wavy and stalk reddish. Flowers are in bunches at the end of the branches, sweat scented, fleshy. Fruits are ovoid, 1-4 seeded. Seeds are elongate and brown shining.	Sweet delicious flowers and consumed besides tasty fruits. Flowers are distilled for making vinegar. The seeds yield fat known as Mahua butter used in cooking, adulteration of Ghee, manufacturing chocolates and even soaps, besides treatment of rheumatism and constipation. Mahua cake is insecticidal and also used for fishing. The timber is useful variously but neglected.
18.	*Melia azedarach* – (Dek)	Meliaceae	Tall tree and deciduous in nature. Bark is brownish grey and branches are spreading. Leaves are odd-pinnate and leaflet opposite. Flowers are fragrant, petals lilac coloured, and fruit is drupe, globose.	It is used for medicinal value, industrial oil material, and for timber (furniture). Seeds have insecticidal properties.

S. No	Botanical Name- (Common Name)	Family	Morphological Characteristics	Salient Uses
19.	*Morus alba* – (Tut)	Moraceae	It is a small to moderate-sized, monoecious, deciduous tree with rough brown bark. Leaves are ovate, serrate, deeply lobed. Flowers are small and white, male and female flowers on separate stalks. Bark is smooth and light in color when young, becoming darker and fissured by age. Leaves are alternate, simple, with a serrate margin, often irregularly lobed, smooth above. Flower is normally dioecious, small, green catkins, male flowers, narrow, catkin. Fruit resembles blackberries in shape, cylindrical, fleshy.	It is mainly used for sports goods like hockey sticks, tennis and badminton rackets, and cheaper grade cricket bats. It is also used for legs of bedsteads and furniture. Leaves are main source of food for silk worms.
20.	*Pongamia pinnata* – (Sukhchain)	Leguminosae	It is a medium sized, nitrogen fixing tree. The trunk is generally short with thick branches spreading into a dense hemispherical crown of dark green leaves. The bark is thin grey to greyish- brown, and yellow on the inside. The alternate, compound pinnate leaves consist of 5 or 7 leaflets, and pointed at the tip. Flowers borne on racemes, are pink, light purple, or white. Fruit is a pod and elliptical in shape.	The wood is used for cabinet making, cart wheels, posts, agricultural implements, tool handles and combs. Oil is extracted from seeds, the leaves are eaten by cattle and readily consumed by goats, planted in homesteads as a shade or ornamental tree and in avenue plantings along roadsides and canals.
21.	*Prosopis cineraria* – (Khejri)	Fabaceae	Khejri is a flowering tree and the state tree of Rajasthan (India). It is a small tree, ranging in height from 3-5m. Leaves are bipinnate, with 7-14 leaflets on each of 1-3 pinnae. Branches are thorny along the internodes. Flowers are small and creamy-yellow, and followed by seeds in pods. The tree is found in extremely arid conditions, with rainfall as low as 150 mm annually; but is indicative of the presence of a deep water table. *Prosopis juliflora* is another important species of same genus.	Wood is used for fence posts, furniture, crafts and corrals. It is rarely used in construction, as most tree trunks are not long or straight enough, also used for fuelwood.
22.	*Terminalia arjuna* – (Arjun)	Combretaceae	Tall tree, bark grey, smooth and leaves are sub-opposite, oblong or elliptic oblong, glabrous. Flowers are small, white. Fruit is fibrous woody, glabrous with 5 hard wings, striated with numerous curved veins..	Used for timber, leaves for sericulture and fodder, bark as medicine for heart disease.
23.	*Toona ciliata* – (Toon)	Meliaceae	It is a large deciduous tree. Bark is dark grey or reddish-brown, smooth up to middle age. Leaves are usually impair-pinnate, sometimes pari-pinnate. Flowers are small, honey scented, cream coloured, in drooping or sub-erect terminal panicles. Fruit is capsule.	The wood is used for boat building, cabinet making, cigar boxes, matchboxes, decorative plywood and veneer, food containers, furniture, joinery, musical instruments, ornamental work, panelling, boxes and crates, building materials, exterior uses, millwork, mouldings.

Traditionally Important Trees in Agroecosystem

Teak based Agroforestry

Agrisilvicultural System

Fig. 9.2: Indigenous Agroforestry Systems

10 TREE ARCHITECTURE AND TREE MANAGEMENT IN AGROFORESTRY

INTRODUCTION

Agroforestry is a dynamic ecologically based natural resource management system that, through the integration of trees in farm and rangeland, diversifies and sustains smallholder production for increased social, economic and environmental benefits (Leakey 1996). In other words, agroforestry encompasses promotion of components like woody perennials, crops, livestock and other natural resources in temporal and spatial sequence which will ultimately alter livelihood options and sustainability. Rural population in many areas have long been involved in the conservation and cultivation of trees on agricultural lands and forested areas. Within the context of local spontaneous tree growing, it is necessary to review a number of indigenous tree management and conservation strategies. The relationship between rural people and the trees in their environment is generally complex. Many of the approaches used have been developed over long periods of time. Often they have emerged as responses to increasing, though sometimes subtle, pressures on the local environment. Their underlying function has been to ensure that locally-valued tree species continue to be available to rural people.

The extent to which people cultivate and manage trees varies throughout the non-industrial areas of the world. It depends largely on characteristics of local ecology, patterns of agricultural land use, cultural traditions, local demands for wood and wood products, tenure rights and economic pressures. In some societies, tree cultivation and management is a major feature of the way of life. Depending on the intensity of the management approach, different strategies will be able to withstand more environmental stress than others. Environmental degradation and the depletion of tree cover are sometimes symptomatic of a lack of comprehensive traditional tree and environmental management systems. In many cases, it has been the result of the breakdown of traditional systems because of intense and interacting pressures. Ultimately, tree planting programmes should be based on understanding any traditional or existing tree cultivation or management systems which these programmes may supplement, as well as on understanding what has made the introduction of new methods of tree management necessary in the first place (FAO 1986).

Management of trees in agroforestry (AF) system includes all operations that are required to get desired product from the unit area. In AF systems, management of tree species depends upon the choice of species, biophysical causes of land, labour and capital. It is apparent that different tree species and even the same species in different agroclimatic conditions grow at differential rate. Knowledge of these factors enables to choose appropriate silvicultural and agronomic management to get desired produce on a sustainable basis from year to year. AF systems generally involve the close interactions of trees with crops, livestock or both such a situations calls for (unique) careful tree and agricultural management. Tree selection and tree management strategies in agroforestry need thorough understanding about tree structure, tree growth, tree forms, tree crown architecture/ crown shape, arrangement of foliage in the crown and root architecture.

TREE STRUCTURE AND GROWTH

In general, trees are big plants having erect stem, branches and other parts as like plants. It is composed of an underground part: the roots and two aboveground parts: the trunk and the crown. Their size and form are extremely variable and this extends to species within genus, varieties within a species and even the same species in different environments. A tree grows upward from the tips of the branches, downward from the roots, and outward from the trunk. The roots anchor the tree to the ground, and the trunk gives support to the branches. All trees grow from actively dividing cells in apical meristem found at the tips of shoot buds and roots. Tree height growth and branch lengthening are the result of apical meristem at the growing tip whose cells divide and elongate to create upward growth in trees. This height growth in trees is termed as primary growth.

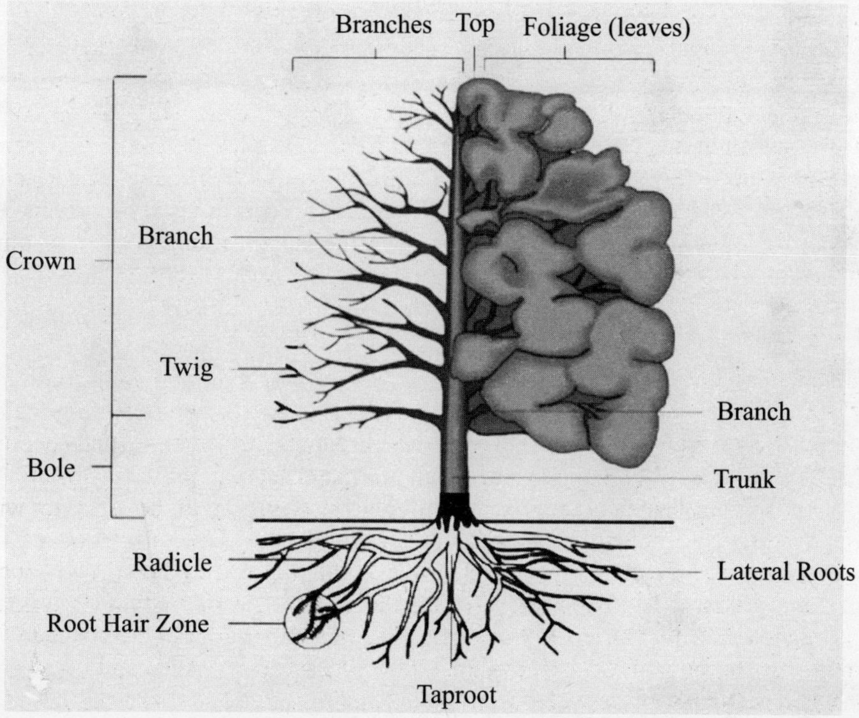

Fig. 10.1: Tree Structure (Source: www.infovisual.info)

Trees and woody plants live for many years. Trees and shrubs grow radially as well as vertically. The trees increase in diameter or girth which results from activity of cambium. During favourable seasons, new growth is made from cells just beneath the bark called cambium. The cambium cells produce new bark on the outside (phloem cells) and new wood (xylem cells) on the inside. Wood is yielded by trees, which increases in diameter by the formation and this is technically known as secondary growth. It is as a result of cell division in the vascular cambium (lateral meristem) and subsequent expansion of new cells. Secondary growth is found in most woody dicots and gymnosperms and it is absent in monocots. As the tree grows from the food it manufactures, it adds new layers of wood to its trunk. Because one is formed each year, these layers are called annual rings, and may be used to tell the age of the tree. During each growing season, a tree grows by elongation of the top shoot, side branches, roots and also adding new layer of wood and bark, all over. This increase or growth is also termed as increment. It can be expressed by measurements of height, girth, basal area, volume, quality and price over period.

TREE CROWN ARCHITECTURE

Despite the fact that most trees exhibit an indefinite growth, which produces a remarkable variability in their final size, they have a recognizable form. The many meristems of a tree are integrated into a galaxy of possible but not random morphologies. Trees exhibit an extraordinary variety of branching patterns and foliage arrangements. Luxuriance of structural details of tree canopy or the diversity of morphologies displayed by the trees can be overwhelming (Valladares and Niinemets 2007).

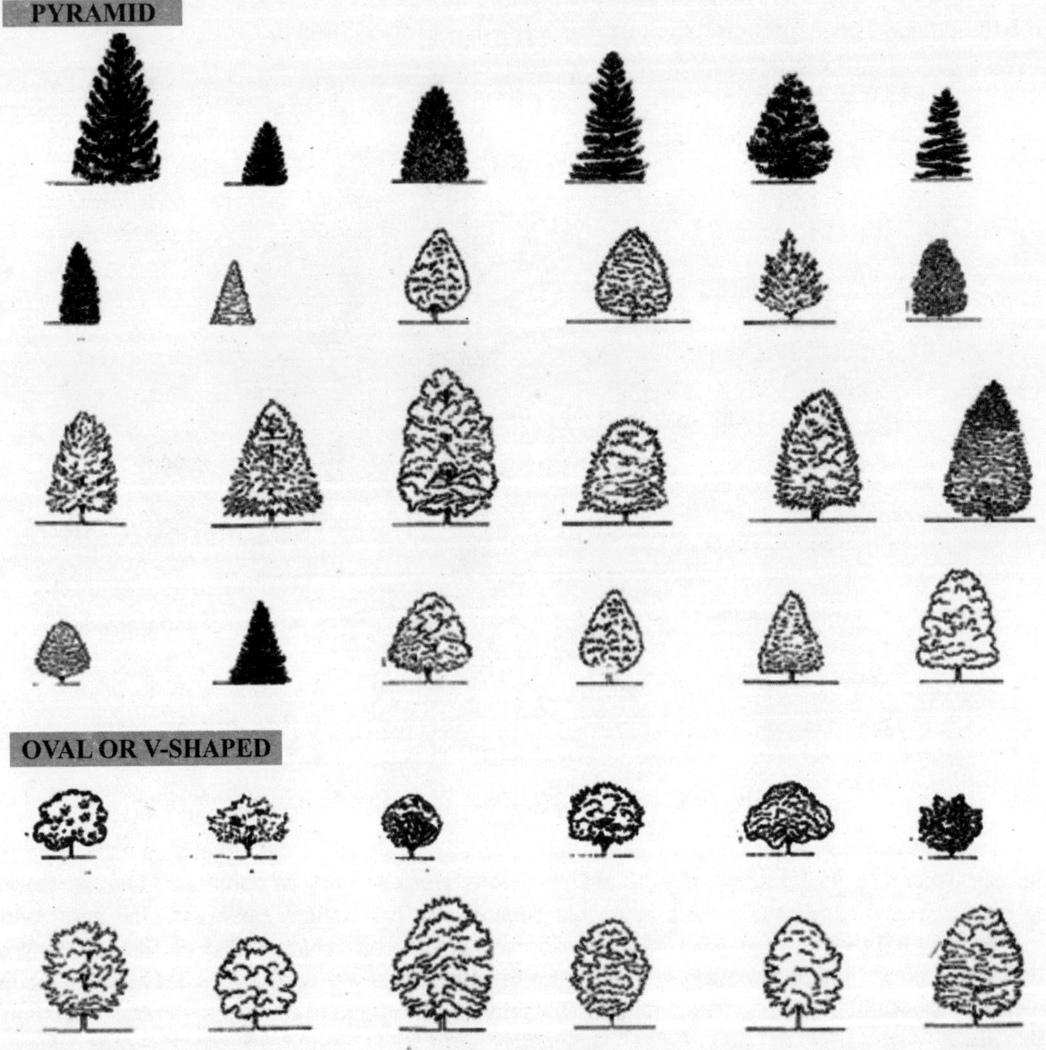

Fig.10.2: Tree Canopy Architecture

Branching complexity ranges from trees with a single axis to large trees with many orders of branching in three-dimensional space. However, the overall complex shape of a tree can be determined by surprisingly few parameters since a new branch is geometrically determined by just two parameters: branching angle and branch length (Honda et al 1997). Repetition of the branching generates the distinctive complexity of tree crowns, and the relative simplicity of the process has resulted in the generation of numerous computer models that simulate branching and growth of trees with remarkable realism (Waller and Steingraeber 1985). Although some trees have a single axis (e.g. most palms) and some have many similar branching axes, most species of trees have two or more type of axes that can be distinguished by their primary orientation, symmetry, or form. In general, leader axes are radially symmetrical whereas lateral branch axes are dorsiventrally symmetrical (Fisher 1986). The branching and consequent growth of trees and shrubs can be characterized by vertical or longitudinal, and horizontal or lateral symmetries.

Fig. 10.3: Various Shapes of Tree Crown

Vertical symmetry is characterized by growth of branches at the top (acrotony) or at the base (basitony), whereas lateral symmetry is characterized by branch growth at the upper or lower side of the lateral branch (epitony and hypotony, respectively). Logically, shrubs exhibit a basitonic branching, whereas trees are characterized by acrotonic branching. Analogously, while typical trees exhibit a hypotonic branching, most shrubs and small trees exhibit epitomic branching. However, there are many exceptions to these rules. For instance, the pyramidal shape of the crown of many conifers is due to the combination of basitonic branching (typically a shrub pattern) with a monopodial growth of the bole. The dominance of branch development when branch originates from buds on the upper side of the stems or main branches (epitomic shoots) appears to be important in shrub competition for space, since hyptonic branching confers the capacity of extending laterally but not overtopping an existing canopy (Schulze et al 1986).

Tree form can be very complex due to the combination of regular and irregular pattern formation processes. The tree canopy architecture can be classified based on dichotomic characteristics of the tree crown, such as monopodial or sympodial branching, basitonic or acrotonic branching, orthotropic or plagiotropic shoots, etc. (Valladares 1999). The shape of the crown influences important aspects of growth and survival of trees, such as light interception and competition for space. Crown shape and the arrangement of foliage within the crown are the two most basic characteristics affecting the efficiency of light capture. Various crown shapes and different dispositions of leaves within the crown result in complex diurnal and seasonal patterns of light interceptions of leaves within the crown result in complex diurnal and seasonal patterns of light interception at both the single-leaf and the whole-crown levels. Leaves at the uppermost positions of the canopy are frequently exposed to high irradiances that are often in excess for photosynthesis. Lower leaves, in turn are often heavily shaded, and the light available for these depends not only on the amount of neighbouring leaves, but is also affected by the general form of the crown and the angle and orientation of the surrounding units of the foliage (Niinemets and Valladares 2004).

The shape of the crown can be described by the absolute size, the ration of height to width, and the convexity or shape of its contour. It is further important that the crown shape can vary at any given height to width ratio. As the solar inclination angle decreases from equator to higher latitudes, crowns with differing height to width ratio have inherently varying efficiencies of light interception (Valladares and Niinemets 2007).

TREE ROOT ARCHITECTURE

In most trees, roots are underground structures that anchor the tree and provide a means to absorb the nutrients and water needed for growth of the tree. New root tips grow continuously throughout the life of the tree and provide the surfaces through which most of the nutrients and water move. Roots are used as storage organs for the food materials produced by the shoots. The major functions of roots, thus, can be summarized simply as absorption, conduction, storage, and anchorage. As cells are added to the tip by repeated cell divisions, a young root elongates and leaves behind cells that differentiate and become the primary roots of the plant. Four areas of the young root traditionally are recognized, but except for the terminal area, are not distinctly separate. Their descriptive names are only partially correct in describing the activities taking place in each area. These regions, starting at the tip and moving upwards towards the stem, are the root cap; zone of active cell division; zone of cell elongation; and zone of maturation. The first two are compacted in the first centimeter or less of the axis with the latter two no more than 4-5 centimeters from the tip. Only the root cap and the cell division regions actually move through the soil. After cells start to elongate and mature, no further extension takes place, and the root is stationary for the rest of its life.

Trees are characterized by having one main root (the taproot) from which smaller branch roots emerge. When a seed germinates, the first root to emerge is the radicle, or primary root. In conifers and most dicots, this radicle develops into the taproot. Taproots can be modified for use in storage (usually carbohydrates). Taproots are also important adaptations for searching for water, as those long taproots found in mesquite (*Prosopis juliflora*) and poison ivy (*Rhus radicans*). Some trees and most annual plants are characterized by having a mass of similarly sized roots. In this case, the radicle from a germinating seed is short lived and is replaced by adventitious roots. Most monocots have fibrous root systems. Some fibrous roots are used as storage organ. Plants with fibrous roots systems are excellent for erosion control, because the mass of roots cling to soil particles.

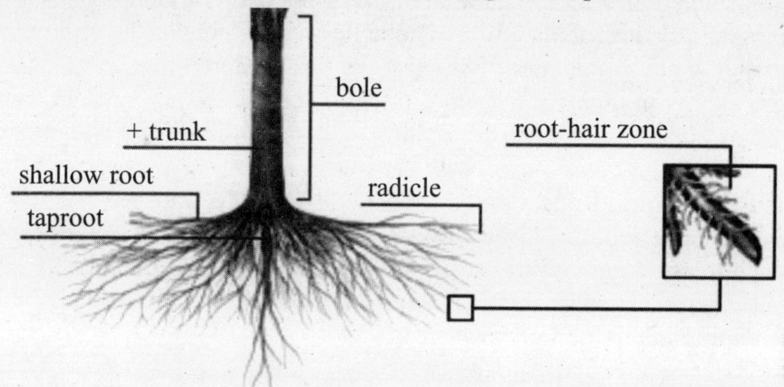

Fig. 10.4: Tree Roots Growth

Trees' root systems are made up of large, permanent roots (which mainly provide anchorage and transport), and many small, temporary feeder roots and root hairs. It is these small parts of the root system that are the primary water and nutrient absorbers. Many of these small roots function for only one or two years, and then either die or become part of the large root system. Most tree roots do not penetrate very deeply into the soil. Unless the topsoil is bare or unprotected, trees will concentrate most of their absorbing roots in the top 6 to 18 inches of soil, where water, nutrients and oxygen can be found. Tree root systems cover more area than one might expect - usually extending out in an irregular pattern 2 to 3 times larger than the crown area. However, on a dry weight basis, the "root to shoot" ratio is around 20 to 80 per cent. The type of roots formed initially is specific to a given species and with age, the initial root form is often modified by the growing environment such as soil hard-pans, water tables, soil texture and structure, and degree of compaction.

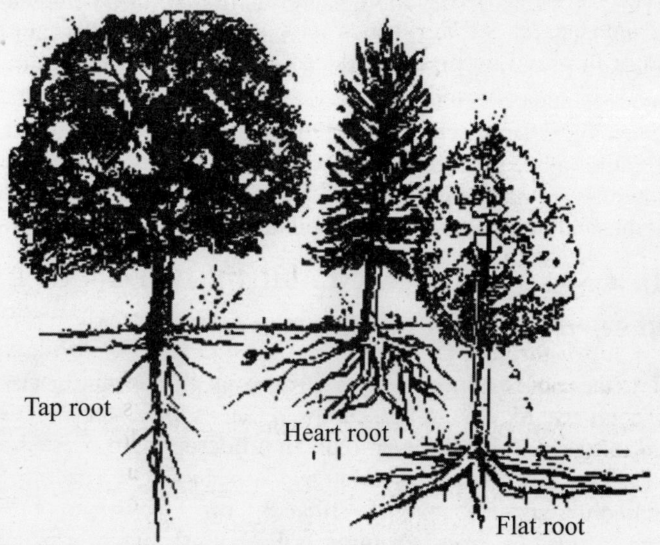

Fig.10.5: Tree Root Forms

ABOVE AND BELOW GROUND COMPETITION IN AGROFORESTRY

Low light intensity is one of the important constraints for higher yield in agroforestry. The degree of shading to annual crops increases with the increase in the proportion of land occupied by agroforestry system and size of the trees. Paddy yields will tend to be low in the conditions of low intensity of light and wheat tillering and grain production will also be less. Many tree crops are inefficient in interception of radiant energy in their early years since full canopy formation in trees may take many years. Under such circumstances, solar radiation falling on bare soil is wasted and promotes only weed growth. Intercropping of ground cover crops with the trees in their early years will therefore help in better utilization of the resource. Reduction in crop yield form second year of tree growth under agroforestry systems has been reported for many trees. The shade effects caused by full canopy of untopped and unpruned trees on agricultural crops are reported to be considerable. Compared to open field, the total and net radiation beneath the tree canopy are remarkably low. Under limitation of light, there is an inevitable depression in the intersown crop yields.

Underground competition is relatively more for moisture and nutrients. The lateral spread of tree roots is found to have more effect on utilization of moisture and nutrients which increases with higher tree density in agroforestry systems. Trees require lesser quantity of nutrients for their growth as compared to annual crop. Trees pump nutrients from the deeper layer to the surface through absorption and litter fall. However, release of nutrients through litterfall depends upon the density of trees, nature of species, season, etc. Litterfall may vary from 0.5 to 6.5 tons ha^{-1} $year^{-1}$. On the basis of dry weight, leaf litter contains 0.5 to 1.5 per cent N, 0.05 to 0.15 per cent of P, 0.25 to 0.75 per cent of K, 0.25 to 1.00 per cent Ca and 0.10 to 0.2 per cent of Mg. Growing of trees in agroforestry, markedly decreases the pH and EC and increases organic carbon, available phosphorous and potassium (Tilman 1990).

Muller (1969) emphasized that allelopathy, the direct or indirect effect of one plant upon another through the production of chemical inhibitors that are released in to the environment, should also be recognized as another factor in analyzing mechanisms of plant interactions. The species interaction due to chemical influences is also designated as allelochemistry, phytochemical ecology or ecological biochemistry and allelobiology. Most of the chemical substances involved in allelopathic reactions are secondary compounds. Though the toxic metabolites are distributed in other plant parts also, leaves are the potent source of allelochemicals. Summer materials are more toxic than those of rainy and winter season. Toxins released from tree litter are the primary causes of allelopathy. The example for allelochemicals from tree includes tannin, quercetin, azetidine, cineole, phenols, sapthulenol, resorcinol, pyrogallol, catechol, caffeic acid, chlorogenic acid, benzoic acid, hydroxy benzoic acid, dihydroxy benzoic acid, gallic acid, protocatechuic acid, vanillic acid, gentisic acid, coumaric acid, ferulic acid, sinapic acid, syringic acid, myristic acid, palmitic acid, stearic acid, oleic acid, linoleic acid, cinnamic acid, salicilic acid, tannic acid, oxalic acid, thalic acid, etc.

SILVICULTURAL OPTIONS FOR MINIMIZING THE NEGATIVE INTERACTIONS

MANIPULATION OF DENSITIES AND ARRANGEMENT OF TREES

The negative effects of trees can be minimized by increasing the spacing between the trees. The studies carried out in the semi arid regions of India showed that as the distance between the hedgerows of *Leucaena leucocephala* increased, the per cent reduction in crop yield decreased. The reduction in the yield of sorghum, cowpea and castor are relatively less in wider spacing (more than 7.0 m) between tree rows compared to narrow spacing (less than 3.5 m). Tree orientation is also important to reduce the negative effect of trees on crops. Planting in East-West direction may reduce the shading effect of trees on the crops.

MANIPULATION OF TREE CROWN AND ROOTS

The negative effects of trees can be further reduced by pruning of tree crown and roots. In alley cropping system, pruning of trees and applying the biomass to the soil will reduce the trees competitive ability and increase the growth and yield of the associated intercrops by providing green manure and by allowing more light to the intercrops. In situations of severe root competition for moisture and nutrients, root pruning operation or trenching may eliminate or reduce the negative effect of trees on the intercrops.

Certainly agroforestry has considerable potential, not as the only way to improve agricultural production, but as one important way to enhance and maintain overall productivity of the small and medium-size farms, the agricultural units that are becoming more prevalent in many parts of the world. Through skillful management practices any or all of these constraints can be controlled. For example, once it is known that trees compete with food crops and may reduce food yields, it is easy to adopt some or all of the following strategies:

- Selecting legume trees that have small or light crowns so that sufficient sunlight will reach the food crops for photosynthesis
- Selecting tree species that are deep-rooted so that they will absorb moisture and nutrients from the deeper subsoil while the food crops receive their share from the surface layer of the soil
- Spacing the trees farther apart to reduce their competitive effects on the food crops

Tree management is the method of manipulating growth of trees so as to satisfy the objectives of the grower. The main objective of tree management is to maximize the value of the tree production without greatly reducing the agricultural potential of site. The silvicultural manipulations depend upon the following conditions:

- Silvicultural requirement of the species in AF system for sustained yield and renewal of tree crop in a given set of conditions
- Characters of crops grown in the interspace
- Demand of tree produce

Some of the basic specialized techniques of tree management for achieving the aims of agroforestry systems are:

- Pruning
- Pollarding
- Thinning
- Coppicing

PRUNING

The removal of live or dead branches or multiple leaders from standing trees for the improvement of the tree or its timber is called pruning. Two kinds of pruning i.e. natural (self) and artificial pruning are prevalent in which artificial pruning is practised in agroforestry systems like agrisilviculture, where trees are grown widely spaced. Importance of pruning in agroforestry is to reduce the presence of large knots and to reduce the shade effects. The pruning operation is done with sharp tool so that knot free timber is formed. Points to keep in mind during Pruning are:

- The cut must be angled to permit the rain water to run off above the cut surface

- The branches should be cut 4-5 cm above the main stumps
- Pruning must be started from the below branches first and then gradually proceeded to top branches
- It must be done well in advance before sowing or planting the intercrops
- It is better to prune when the branches are light than to leave pruning until the branches are heavy and a saw is required

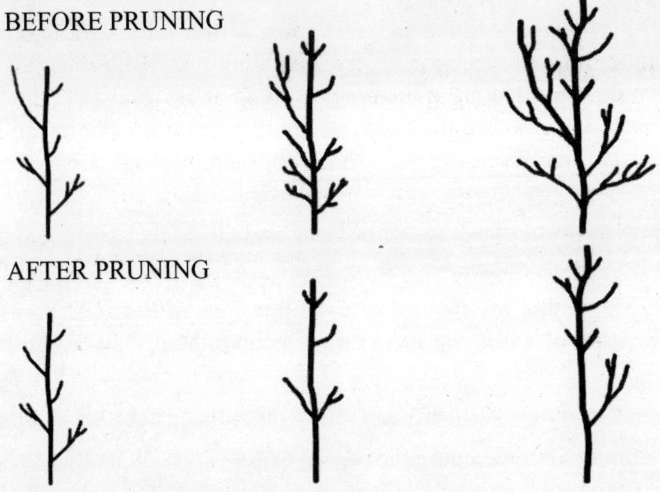

Fig. 10.6: Tree Pruning Methods

A pruning cut can be a point of infection for various diseases and this risk must be minimized. A good clean cut is essential. This will also avoid the risk of damaging the bark on the main stem below the pruned branch. Most tree species produce an obvious branch bark ridge in the junction between the main stem and the branch. This ridge should not be damaged by the cut, if it is, infection can spread rapidly. The correct pruning cut should be made as close to the stem as possible without breaking the branch bark ridge. A single cut is enough for a branch which is light enough to be pruned with a pruning knife or with hand secateurs. If the branch is heavy, it will probably require to be cut with a pruning saw and it may fall, tearing the bark on the main stem when it is weakened by cutting. In this case most of the weight of the branch can be removed by making two preliminary cuts as shown in diagram. When the stem to be pruned is above head height and up to 5 to 6 metres off the ground, a telescopic pruning saw will be required.

Pruning Techniques

1. *Fixed Lift Pruning*: It is the complete pruning of all branches below a prescribed point on the stem usually 2 metres. The lift is specified as height from ground and choice of height based on shape of development.
2. *Variable Lift Pruning*: It is complete pruning of all branches below a prescribed variable point on the stem. This point may be specified either as a proportion of height or as diameter limit. The specification is often directed towards achieving removal of constant fraction of green matter from trees regardless of size.

3. *Selective Pruning*: It is the removal of some of branches on the stem at various levels above the ground according to a particular prescription.
4. *Tip Pruning*: It is pruning of branch at a point other than at its junction with the stem. It is used to retard the development of branch.

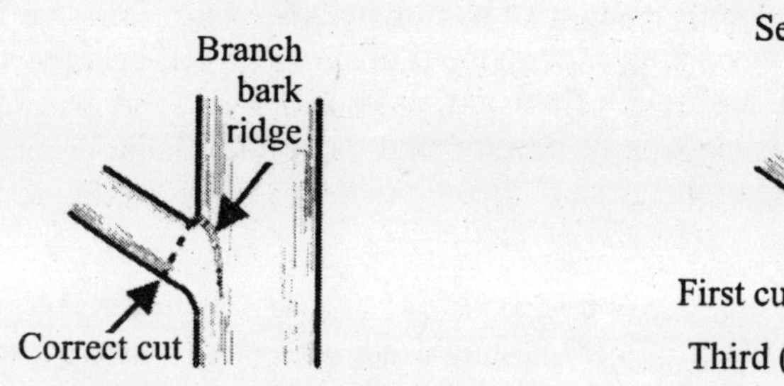

Fig. 10.7: Pruning of Light and Heavy Branch (Source: Forestry Commission 2000)

POLLARDING

Pollarding is a process in which crown of the trees are harvested at least 2 m above the ground level. Pollarding involves cutting off the crown of the tree, leaving it to send out new branches from the top of the remaining stem and this also has the advantage that the new shoots are high and therefore more apt to be protected from animal and fire damage. Vertical pollarding is the close pruning of branches along the stem. Regrowth after pollarding is vigorous because the tree's root system has already been well established. Pollarding is done to get new flush of shoots in many tree species such as:

- Arjun (*Terminalia arjun*) and mulberry (*Morus alba*) are pollarded to rear the silk worm
- Subabul (*Leucaena leucocephala*) is pollarded for forage yield
- *Gliricidium sepium* is pollarded for green manure, fodder and fuel wood

Fig. 10.8: Tree Pollarding

It has been reported that in the highlands of Kenya the pollarding of *Grevillia robusta* growing on agricultural lands is common. Intense pollarding of these trees may be carried out 15 or 20 times over a period of 50 years. The trunk will continue to widen, and the stem will increase in height unless this is deliberately prevented by pruning at the top. Whenever the farmer decides that it is large enough or that he needs the money, the trunk is felled and is sold for timber (Poulsen 1983).

THINNING

Thinning is a felling or removal of immature stands or trees so as to improve the growth and form of the trees that remain without making a permanent opening of the canopy. In agroforestry, early thinning of trees is important as it provides firewood, poles and posts, and to reduce competition with agricultural crops. Thinning at later stage of development is not advisable because late thinning results in large volume of waste covering the ground, thereby reducing agricultural production.

Objectives of thinning

- To improve the hygiene of the crop by removing dead, dying and diseases trees
- To salvage anticipated losses of the merchantable volume of wood
- To ensure the best physical condition of growth
- To obtain desired type of tree crop
- To provide poles/post that will produce some intermediate financial return from sale

The classic objective of thinning is to redistribute growth potential to fewer trees past the sapling stage, leaving a stand with a desired structure and composition. In general, five methods of thinning are recognized:

1. Low thinning or thinning from below
2. Crown thinning or thinning from above
3. Selection thinning or diameter-limit thinning
4. Free thinning
5. Mechanical thinning

COPPICING

Coppicing is the process of cutting trees down, allowing the stumps to regenerate for a number of years (usually 7-20) and then harvesting the resulting stems. This is a method of producing coppice shoots growing from a stool or living stump. In coppicing, whole tree is cutoff close to ground level. Shoots sprout from the residual stump technically referred to as the stool coppice. Coppice growth is generally much more vigorous than normal seedling growth e.g. Eucalyptus, Red sander, Indian mesquite. Coppicing helps to harvest fuelwood, poles and small sized timber within a short period. It also induces fast growth of vigourous multiple shoots. It is the method of less investment and quick returns.

Points to keep in mind during coppicing:

- Trees should be coppiced only when attains good health and shade effect is too much on intercrops
- Trees should be coppiced when it attains 3 to 5 m height
- Trees should not be coppiced in dry period

- Cutting height should be 30 to 50 cm above ground level
- Cut should be angled
- If for fuelwood, 6 shoots should be left to grow
- If for poles, only 2-3 straight shoots are allowed to grow

Fig. 10.9: Coppicing Method

11 CROP PLANNING AND MANAGEMENT IN AGROFORESTRY

CROP CULTIVATION IN AGROFORESTRY

Agroforestry is the production of trees, annual crops and animals either simultaneously or sequentially on the same piece of land under same management. The components can even be grown or raised in separate plots when materials from one are used to benefit another. The integrated approach associated with agroforestry makes it the most self-sustaining and ecologically sound land management system. In agroforestry trees, crops and animals are integrated in a long-term conservative, sustainable and productive system where greater emphasis is placed on perennial MPTs (Multi Purpose Trees) that are planted once but yield benefits over a long period of time. It reduces the risks associated with agriculture and increases sustainability of agricultural production system besides conserving and even improving the soil.

Tree-crop interaction is a very important aspect in agroforestry and its careful management is very necessary for the success of the agroforestry system (Palsaniya et al 2009; Palsaniya and Ram Newaj 2009). In agroforestry, since different components are raised together, therefore, interactions, both favourable and unfavourable do occur. Tree-crop interactions occur when more than one individual are present in a biological system and when resources are limited. The resources sharing by the components may result in complementary or competitive effects depending upon the nature of the species involved in the systems, the manner in which they are grown and the climatic factors. The positive (complementary) and negative (competitive) effects of tree in agroforestry systems have been realized by field observations. The nature and quantum of tree-crop interactions and their effects are not uniform and depends upon many factors like nature of the species grown, age and size of trees, density of components, management practices (spacing, training, pruning, irrigation, fertilization, etc) and environmental factors. More specifically, in case of erect growing and sciophyte (shade loving) plants, the extent of tree-crop interaction and the resultant effects will be less as compared to spreading type and heliophyte (sun loving) plants. Similarly, the adverse effect will be more in case of aged and large sized trees. Further, higher the density of components (tree and crop), higher will be the negative effect on the under storey crop and *vice versa*. The management practices like training, pruning, irrigation, fertilization, etc., environmental factors and availability of water, nutrients and light also affect the tree-crop complementarities or antagonism.

The most common interaction is competition, which may be for light, water, or soil nutrients. Competition invariably reduces the growth and yield of any crop. Yet competition occurs in monoculture as well and this need not be more deleterious in agroforestry systems. Interactions may be complementary, as in the case of trees, pasture, and grazing animals, where trees provide shade and/or forage, and animals provide manure. Based on the resources sharing by the components, tree-crop interaction may be complementary or beneficial or positive or synergistic in one hand and inhibitory or competitive or negative or antagonistic on the other. Tree-crop interactions may also be as aboveground and belowground interaction. Interaction may be among crop components, among tree components and between tree and crop components, the latter is more important in agroforestry.

In a legume and non-legume based intercropping system, part of the nitrogen fixed in the root nodule of the legume may become available to the non-legume component is an example of complementarity. Further, the complementarities in resource capture and sharing under mixed cropping/intercropping system may be spatial or temporal. In temporal complementarity, the leaf canopies of component crops may occupy different vertical layers with taller component tolerant of strong light and high evaporative demand and shorter component favouring shade and high relative humidity. Multistoried cropping in coconut and planting shade trees in cocoa and tea plantation uses this principle. In similar manner under agroforestry system, exploitation of different layers of soil by the root systems of trees and crops may lead better utilization of resources with much less competition. In temporal complementarity, the yield advantages provided by the mixture cannot always are explained by more effective use of growth resources at specific times. Indeed, there are substantial opportunities for temporal complementarity if species make their major demands on available resources at different times, thereby reducing the possibility of competition. For example, the usually advantageous intercropping of the slow-growing pigeonpea with fast-growing cereals such as maize or sorghum produces only limited increase in total light or water use (Natarajan and Willey 1980) but nevertheless provides intercrop advantages of 40-70 per cent.

Tree-crop interaction may be manifested in terms of effects on growth, yield and productivity, quality, intercultural operations and other agronomic practices of the components of the agroforestry (Palsaniya et al 2009). Further, the complementary effects may be manifested in terms of increased productivity, soil fertility (organic matter addition and N-fixation), nutrient recycling, soil and water conservation, microclimate improvement, anchorage to climber crops, crop protection from cold and heat waves and frost (windbreak and shelterbelt), bioamelioration, phytoremediation and sustainability while the competitive effects may be manifested in terms of competition for light, water, nutrients space and CO_2 (above and below ground), allelopathy (+ve and –ve), insects-pests and diseases incidences and birds (+ve and –ve). Therefore, effective planning and management of interactions in agroforestry systems is very much essential for taking full advantage.

CROP PLANNING AND MANAGEMENT IN AGROFORESTRY

Amongst the five inseparable components of agroforestry, viz., land, tree, crop, animals and management, crop part is most important as it affects the owner's life most. Since the performance of the agroforestry, particularly the intercrop depends largely on the type and magnitude of tree-crop interaction, management of tree and crop is very important aspect in agroforestry for its success (Palsaniya et al 2009). Management of tree-crop interaction means enhancing the complementary effect and minimizing the competitive effect by selection of species (sciophyte, erect growing), proper density, spacing and geometry, tree canopy management (training and pruning) and input management (water, nutrients, etc.).

SELECTION OF SUITABLE CROPS FOR AGROFORESTRY

In agroforestry, selection of species has a great significance. The choice of crop plants in designing such systems should be based on those crops already produced in a particular region either for marketing, feeding animals, home consumption, or that have great promise for production in the region. Thus, selection of crops requires a judgment based on knowledge of the crops, adaptations, production uses, as well as family needs, opportunities for barter and markets. Species selection should be based on local conditions, survival, adaptability, productivity, animal preference, nutritional value and digestibility. Further, while making tree and crop selections, due considerations should be given to potential markets, soil types, climatic conditions, equipment needed and species compatibility. The genetic quality of native species to withstand adverse environment becomes important for the growth and adaptability to soils with different depth and water retention capacities. The plant establishment primarily depends upon the development of good root system. Selection and use trees, crops and planting/harvesting patterns that are suitable for the site, compatible with planned practices provide desired economic and environmental returns. The following decisive factors should be kept in mind while selecting crop species in agroforestry:

- Site-specific local species
- High yielding
- Shade tolerant
- Compatibility with each other
- Silvicultural and agronomic characteristics and behaviour of the species
- Utilization potential of species
- Multispecies approach for both the agroforestry components should be followed. This would be more advantageous from the point of biotic and abiotic stresses resistance and more efficient utilization of environmental resource. This can also serve as a better cover to the soil and regeneration of soil
- Compatible with the site (soil, temperature, precipitation, planted trees)
- Productive under partial shade and moisture stress
- Highly nutritious and palatable in case of forage crops
- Enhancing soil nutrient status
- High reproductive fertility and regeneration capability
- Easy to establish
- Free from toxic principles
- Fast growing and stress tolerant
- Insect, pest and disease resistant
- Better soil covering capacity

The selection of species will have to be from the homologous zones with similar eco-climatological areas or bioclimates. The species will have to be selected habitat wise as the plant life is very selective for their habitat. Suitable MPTs and crops are listed for different regions of India in Table 11.1. After selection of suitable species, it is also very necessary to select a suitable variety for the region.

INTRODUCTION OF LEGUMINOUS INTERCROPS IN AGROFORESTRY

The importance of legumes is three fold; firstly, they constitute a high protein content in Indian dietary, secondly, they provide nutritious forage for livestock and thirdly, they enrich soil and provide nitrogen to the associated crops or trees. They constitute the cheapest source of protein to animals. They enhance soil quality by adding organic matter, and improve soil structure and water infiltration.

Properly managed legumes are nitrogen self-sufficient. These plants can achieve vigourous growth without the nitrogen fertilizers that are required for growth. This self-sufficiency is achieved through the process of biological nitrogen fixation. They act as nurse crop and supply nitrogen to the associated crops and thereby, contribute to the conservation of energy by reducing the need for N fertilization. Therefore, introduction of suitable legumes in rangelands, pastures, silvipastures and agroforestry has great significance. The important legumes which are suitable for introduction in arid and semi-arid silvipastures are *Dolichos lablab*, *Clitoria ternatea*, *Atylosia scarabaeoides*, *Macroptilium atropurpureum* and *Stylosanthes* species while red gram, green gram, black gram, chickpea, groundnut, soybean, cowpea, pea, lucerne, berseem, etc. can be introduced for cultivation under agroforestry.

Table 11.1: Suitable Crops and MPTs for Different Regions of India

S.No.	Region	Promising MPTs	Crops
1	Western Himalayan Region	*Ulmus wallichiana*, *Ailanthus excelsa*, *Morus alba*, *Robina pseudoacacia*, *Grewia optiva*, *Celtis australis*, *Pinus* spp.	Paddy, mustard, maize, wheat, rajma, berseem, etc.
2	Eastern Himalayan Region	*Acacia auriculiformis*, *Alnus nepalensis*, Bamboos, *Parasanthes falcataria*, *Gmelina arborea*, *Artocarpus heterophyllus*, *Parkia roxburghii*	Paddy, maize, mustard, colocasia, pea, potato, cabbage, turmeric
3	Indogangetic Region	Poplar, Eucalyptus, *Dalbergia sissoo*, *Acacia* spp., *Anthocephalus cadamba*, *Prosopis chilensis*, *Mangifera indica*, *Madhuca latifolia*	Paddy, wheat, oilseeds, pulses, maize, sugarcane, potato, cotton, vegetables, etc.
4	Arid and Semi-arid Region	*Dalbergia sissoo*, *Acacia tortilis*, *A. nilotica*, *Ailanthus excelsa*, *Prosopis cineraria*, *Leucaena leucocephala*, *Azadirachta indica*, Bamboo, *Hardwickia binata*, *Tectona grandis*	Pearlmillet, clusterbean, sorghum, millets, maize, pulses, oilseeds, vegetables
5	Humid and Sub-humid region	*Albizia* spp., *Erythrina*, *Gliricidia*, *Acacia auriculiformis*, *Gmelina arborea*, *Dalbergia sissoo*, Bamboo	Paddy, Jute, sugarcane, vegetables
6	Coastal and Island Region	*Casuarina equisetifolia*, *Toona ciliata*, *Grevillea robusta*, *Tamarindus indica*, *Azadirachta indica*, *Ceiba pentandra*, *Tectona grandis*, *Gliricidia* spp.	Paddy, cotton, sugarcane, pulses, oilseeds, vegetables

NUTRIENT MANAGEMENT

Agroforestry is a mix of forest and agricultural ecosystems. There exists nearly closed nutrient cycles in forest ecosystems where nutrient inputs from atmospheric deposition, biological nitrogen fixation and weathering of primary soil minerals are in balance with nutrient losses due to leaching, denitrification, runoff and erosion. Primary, secondary and micro-nutrients are absorbed by forest roots and returned to the soil via decomposition of litter and roots as well by throughfall and stemflow. On the other hand, there is a net output of nutrients from the soil via crop harvest removals in agricultural systems. This nutrient removal can result in net negative balances if nutrients are not replaced. Nutrient depletion can be offset by fertilizers, manures from outside the field and other nutrient inputs.

One fundamental principle of sustainability is to return to the soil the nutrients removed through harvests, runoff, erosion, leaching, denitrification and other loss pathways. Agroforestry is believed to promote a more efficient cycling of nutrients than agriculture. Trees can provide nitrogen inputs in agroforestry systems through biological nitrogen fixation (BNF) and deep nutrient capture. The presence of active nodules in roots of leguminous species indicates that BNF can supply considerable nitrogen inputs to crops via litter in soils. The non-fixing trees, such as *Cassia*, accumulate more nitrogen in their leaves than nitrogen-fixing legumes, presumably because of their greater root volume and ability to capture nutrients (Garrity and Mercado 1994) which can be added to the soil as green leaf manuring. *Gliricidia* and *Sesbania* are also known for their nitrogen fixation and green manuring potential. Deep nutrient capture is the uptake of nutrients by tree roots at depths where crop roots are not present. It can be considered an additional nutrient input in agroforestry systems because such nutrients are otherwise leached as far as the crop is concerned. They become an input on being transferred to the soil via tree litter decomposition.

Agroforestry can partially provide the nitrogen requirement of crops. The degree to which this nitrogen is actually taken up by the crop depends on a variety of other factors, including the decomposition rate of organic mulches, BNF, residue management, etc. The recovery of leguminous green leaf nitrogen incorporated into the soil by the crop ranges between 10 to 30 per cent and is generally lower than the recovery from nitrogen fertilizers (20 to 50 per cent). Organic inputs, however, have an important advantage over inorganic fertilizers in sustainability terms. Much of the remaining 70 to 90 per cent of the applied organic nitrogen not utilized by crops is incorporated into active and less active pools of soil organic matter because these mulches also provide a source of carbon. Soil microorganisms need a carbon substrate for growth; they also utilize the nitrogen from organic inputs, resulting in the formation of soil organic nitrogen. Furthermore, part of the nitrogen bound in the more recalcitrant organic inputs will also build up soil organic nitrogen. Inorganic fertilizers do not contain such carbon sources; therefore, most of the fertilizer nitrogen not used by crops is subject to leaching and denitrification losses, while much of the nitrogen released from organic inputs and not utilized by crops could build soil organic nitrogen capital. The potential for improved nitrogen management needs to be quantified in agroforestry systems by measuring processes such as mineralization, immobilization, denitrification, volatilization and leaching, along with changes in the soil organic nitrogen pools in systems combining organic and inorganic sources of nitrogen.

Agroforestry, however, cannot supply most of the phosphorus and other nutrients required by crops. Phosphorus is often a critical nutrient in agroforestry and other low external-input systems. Inorganic sources of phosphorus must be applied to agroforestry systems in soils depleted of this element. The strategy is to cycle all the available organic sources first, including manures, and supplement the difference with phosphorus fertilizer inputs. Combinations of organic and inorganic sources of phosphorus may result in a more efficient use of nutrients. The deep capture of phosphorus is likely to be negligible because of the very low concentrations of available phosphorus in the subsoil. Many agroforestry systems do accumulate phosphorus in their biomass and return it to the soil via litter decomposition, but this is cycling and does not constitute an input from outside the system. However, through cycling, some less available inorganic forms of phosphorus in the soil may be converted into more available organic forms.

Appropriate agroforestry systems can maintain or even restore nitrogen fertility through BNF, deep nitrate capture and several cycling mechanisms. Phosphorus fertility, however, cannot be replenished by agroforestry alone, but it can perhaps be made more available through cycling. For long-term production, agroforestry systems must include the addition of phosphorus and other nutrients and, in many cases, of nitrogen fertilizers as well in order to reverse nutrient depletion and ensure the efficient use of resources.

WATER MANAGEMENT

Within agroforestry research relatively little attention has been given to water management practices. Inclusion of trees and shrubs on the agricultural landscape, i.e., agroforestry influences adjoining microclimate (temperature, RH, light, wind), changes the surface energy balance, and alters the water requirement, actual water use, water use efficiency and productivity of adjacent crops or intercrops. Trees impact the water relations of the crop by affecting the loss of water through evaporation, transpiration, damaged leaves and mining of water from deep soil layers. In arid and semi-arid regions where wind erosion is common, wind breaks, shelterbelts or agroforestry buffers provide significant reductions in the amount of wind blown soil and subsequent abrasion of plant parts and cuticular damage. Loss of cuticular integrity or direct tearing of the leaves reduces the ability of the plant to control water loss. Moreover, wind breaks, shelterbelts and agroforestry reduce the air turbulence in the vicinity and conserve soil moisture by reducing evapotranspiration (ET) from the soil and crops. This enhances crop growth and increases its water use efficiency. But, at the same time, trees and shrubs in agroforestry themselves consume a portion of available water. So, in the area immediately adjacent to the trees, competition for water between crops and trees has a negative impact on yield.

Under irrigated condition, irrigation water supplied to intercrop will also serve the requirement of the tree component. Under rainfed condition, agroforestry or specifically silvipastures are raised without any provision of irrigation. However, irrigation may be provided during initial establishment. Further, life saving irrigations may be provided if needed. Sprinkler and drip system of irrigation may be used for higher growth and water use efficiency. *In-situ* and *ex-situ* rain water harvesting should be followed. Harvested water can be subsequently used for supplementary irrigations. Agroforestry has also been advocated as a means of managing excess water that has accumulated in the fields in canal command areas. *Eucalyptus* as a biodrainage species has widely been recognized for managing rising saline water tables.

SHADE MANAGEMENT

Shade is the most important growth limiting factor under agroforestry. Shading from trees may reduce crop yields, though the income from tree products compensates for the loss in yield. Shade in agroforestry can be managed through maintaining wider tree spacing, appropriate plantation geometry, plantation direction and training and pruning of tree component. It is desirable to have wider spacing of tree component to have lesser shade on intercrop. The row to row tree spacing should be kept relatively higher than the plant to plant. The tree row direction should be east to west to avoid shading of intercrop.

Training and pruning of plants is an integral activity under agroforestry. In case of trees, training is done during early stage to give the plant proper shape and size while pruning is done during bearing stage to maintain a balance between vegetative growth and fruiting. Pruning has the added benefit of raising the tree canopy which allows more light to hit the ground, thus providing higher production potential for the crop component of the system and producing an aesthetic open environment. Vigourous or erect type species warrant central leader system; semi vigourous, modified central leader while spreading type dwarf species warrant open centre system of training. Pruning is also done according to species. Spur pruning or cane pruning or removal of shoot to encourage new flush is in vogue in different fruit species in agrihorticulture. Removal of dead or dry shoots is a regular process of pruning.

Generally tree canopy is managed for between 25 to 45 per cent canopy for warm season crops and 40 to 60 per cent canopy for cool season crops. If canopy begins to exceed the desired amount, crop production will begin to decline. This management scheme may require thinning at intervals of approximately every 5 to 7 years depending upon species and site productivity to keep the canopy within the desired range. In general, the main trunk up to 1.5 m height is kept single stem to facilitate movement of tractor and implements in interspaces for intercultural operations. As far as possible, tree should be maintained straight to get clean bole. For fast growing species like *Acacia nilotica* and *Albizia* species, pruning up to 75 per cent tree height can be afforded without any adverse effect on tree growth. Pruning may be started from second year of plantation and go throughout rotation period either every year or alternate year. In case of timber trees like *Dalbergia sissoo*, *Tectona grandis*, etc., pruning can be done up to 50 per cent height of the tree. In arid regions of India, species like *Prosopis cineraria*, *Ailanthus excelsa* and *Acacia nilotica* are lopped peripherally while maintaining framework of tree.

In case of evergreen species, pruning or training is usually done during winter season or summer season. The best time to prune living branches is late in the dormant season (winter) or very early spring before active tree growth begins. Pruning during the dormant season may minimize sap and resin flow from cut branches. The fastest and most economical way to prune trees for maximum quality is by using a lopper or lopping shears. Pruning creates wounds to the trees and these may act as entry point for the pathogens. Therefore, wound dressing with suitable fungicide is necessary after pruning. The tree sap and other secretions like gums and resins are the natural means by which trees combat invasion by pathogens. Generally sap flow from pruning wounds is not harmful; however, excessive bleeding can weaken trees. So wound dressing should be done immediately after the wound is created to prevent introduction of pathogens and minimize bleeding.

WEED MANAGEMENT

Weeds are the most underestimated crop pest in agriculture causing higher reduction in crop yields than other pest like insects and diseases. Damages globally caused by weeds are responsible for a loss of 13.2 per cent of agriculture production or about US$ 75.6 billions per year (Oerke et al 1994). Weed is anything that competes with the crops/seedling for moisture, nutrient, space and light. Removing competition is essential to ensure survival and will certainly improve growth and yield. Generally, the weed infestation is less in agroforestry system but their management is difficult. Weeds badly affect the establishment of tree also (initial 2-3 years) in addition to crop yield. Sometimes, certain weeds act as alternative host of insect pests and diseases. For instance, presence of *Premna corymbosa* and *Vitex negundo* in teak plantation facilitate increase of *Hyblaea puera* as they are its alternative host plants. Similarly, *Cronartium himalayense*, a chir pine rust requires *Swertia* to complete its life cycle before attacking Chir. Therefore, it is necessary that such weeds are eradicated completely around the desired plants.

Generally agroforestry system is less infested with weeds as compared to sole cropping. Higher weed density is found in initial years of plantation. In subsequent years it reduces as canopy of trees restricts the amount of light reaching on ground. In addition to density, weed flora also changes and shade loving weed species dominates in agroforestry system. Sometimes, more weeds are found near tree because tree creates hindrance in cultural operation and some spaces are left over around the tree which helps in survival of perennial weeds.

In semi-arid region of India the dominant weeds of rainy season in white Siris (*Albizia procera*) based agroforestry system are *Cyperus iria*, *Digera arvensis*, *Cynodon dactylon*, *Phyllanthus niruri*, *Echinochloa crusgalli* and *Commelina benghalensis*. While in winter season *Chenopodium album*,

Fumaria parviflora, *Anagallis arvensis*, *Convolvulus arvensis*, *Melilotus* spp., *Vicia sativa*, *Phalaris minor* and *Avena fatua* are major weeds, constituting more than 90 per cent of the total weed population (Ram Newaj et al 2007). Dominant weeds of wheat are *Medicago denticulata*, *Anagallis arvensis* and *Phalaris minor*. Their intensity is greatly reduced when wheat seeded between Shisham (*Delbergia sissoo*) plantations while density of *Commelina benghalensis* increases (Tripathi et al 2001). Agroforestry systems are most often invaded by shade loving weeds. Examples of this type of weeds are blackberry, privet, wild tobacco bush, bitou bush, African boxthorn, mirror bush, cassia and vines such as honeysuckle, banana passion fruit and English ivy.

Weed control in agroforestry system is complicated due to presence of more than one species at the same time and place. Weed control need to take into account the needs of each species (trees, shrubs, herbaceous plants), the interactions between the crops and the effects each control method will have on the other crops. In most cases, trees provided with a full growing season of weed control (within 1 m of radius) will be strong enough to defend themselves. Weed control in the second and subsequent years is only justified where tree growth is very slow or there is extremely vigourous competition from deep rooted grasses or perennial weeds. There are various ways to control weeds. Single method is not sufficient to manage weeds up to desired level, integrated approach gives better results.

Cultural methods exploit the crop's competitive behaviour, growing environment and crop management practices to smothering of weed. Mulching, cover crops, crop rotation, catch crop, trap crop, appropriate time and method of sowing are some examples of cultural method of weed control. Fertilizing and watering only encourages the weeds and should not be done unless the weed control is perfect or should be applied to target crop only. Crop residues or stubbles may also be used as mulch or as a cover for a rotation crop system to minimize weed problem. Weeds are controlled by hand and used as a mulch and green manure for perennial species in agroforestry systems. Also, leguminous species are used as live cover for weed control and for reducing the weed propagules in the soil. These weed control methods are used until the trees in the agroforestry system control the weeds by shading (Carvalho and Torres 1994).

In mechanical or physical methods, physical force either manual, animal or mechanical power is used to pull out or kill weeds. Depending on weed and crop situation one or combination of these methods are used. Hand weeding, ploughing, burning, mowing and slashing are example of mechanical methods of weed control. These are oldest method of weed control and till today have relevance. Always sow the crop between tree rows after proper tillage operation to control weeds. In biological method of weed control, some organism or biological entities are employed to manage weed population below threshold level under a specific situation. It controls certain problematic and invasive weeds under non-crop situation more effectively. It requires higher initial investments but economical in long run. Chemical weed management in agroforestry system is somewhat difficult because two or more species of different nature are present. Herbicides need to be applied carefully as per recommendation in sole crop but avoid the drift of spray to other species particularly on young trees.

DISEASE MANAGEMENT

Limited information is available on the issues that are likely to influence disease incidence, etiology and epidemiology in agroforestry situations (Schroth et al 2000). Researchers are now becoming increasingly aware of the importance of disease problems in agroforestry systems (Kumar and Shukla 2009). There is a need to identify main diseases of crops and trees in various agroforestry systems and different interactions, which occur between them. Huxley and Greenland (1990) divided the above diseases interactions into primary, where component crops share diseases and secondary, where influence of one component crop on the survival and growth of diseases on the other is considered.

Alterations in disease severity of understory crops in agroforestry systems are possible due to sharing of diseases among its component crops or modification of the microclimate by the tree species. The microclimatic parameters like temperature, relative humidity (RH), light intensity and duration, presence of free water on surface of intercrop, etc., are different under open field (outside tree canopy) and under tree canopy (agroforestry). In general, temperature at outside canopy is higher during the day and it is more during night under tree canopy. RH is high during night and less during daytime under tree canopy. Further, the differences in magnitude of temperature and RH during daytime are more as compared to night hours. Usually radiation intensity is less under the canopy than outside canopy resulting in more availability of free water on leaves of under-story crops for longer duration under tree canopy, especially during morning hours. Data on disease incidence for multipurpose trees, horticultural plants and field crops in various agroforestry systems showed that pathogens common to component crops of a given agroforestry system were rare (Kumar and Shukla 2009). Only certain closely related plant species recorded common pathogens (Table 11.2).

Table 11.2: Summary of Sharing of Diseases among Crops (Kumar and Shukla 2009)

Disease	Pathogen	Recorded Hosts
Sharing among Crops		
Alternaria leaf blight	*Alternaria brassicicola, A. brassicae*	Mustard, Taramira (*Eruca sativa*)
Yellow mosaic	Yellow mosaic virus	Soybean, green gram, black gram
Cercospora leaf spot	*Cercospora cruenta*	Green gram, black gram
Sharing among Tree Crops		
Citrus canker	*Xanthomonas citri*	Lemon, Kinnow (*Citrus deliciosa*)
Ganoderma root rot	*Ganoderma lucidum*	*Leucaena leucocephala, Acacia nilotica, Dichrostachys cinerea, Dalbergia sissoo*
Sharing among Companion Crops		
Mosaic	Cucumber mosaic virus	Papaya, cucurbits

Among field crops, alternaria leaf blight was present on mustard and Taramira (*Eruca sativa*), yellow mosaic virus recorded on soybean, green gram and black gram, and cercospora leaf spot shared by green gram and black gram. Among trees, citrus canker was recorded on lemon and Kinnow (*Citrus deliciosa*) while ganoderma fruiting bodies were observed on *Leucaena leucocephala*, *Acacia nilotica*, *Dichrostachys cinerea* and *Dalbergia sissoo*. Authors mentioned that sharing of the diseases was either restricted to field crops or tree species. Pathogens, common to field crops and tree species, were not recorded, except cucumber mosaic virus (Table 11.3).

Kumar and Shukla (2009) made the following conclusions on disease scenario under agroforestry.

- Alterations in disease severity of understory crops in agroforestry systems are possible due to sharing of diseases among its component crops (primary disease interactions) or modification of the microclimate by the tree species (secondary disease interactions). Primary interactions i.e. sharing of diseases among component crops of a given agroforestry system, were uncommon because of dissimilarities of plant and pathogen species involved.

- Incidence of certain diseases viz., loose smut and alternaria leaf blight of wheat, alternaria leaf spot of mustard, cercospora leaf spot and yellow mosaic of black gram etc., did not increase under heavy shade of tree species as compared to open area.
- Incidence of some diseases was more under tree canopy during certain years as compared to open area when the climatic conditions were just suitable for the disease development, otherwise intensity of the disease remains the same. For example, incidence of white rust and downy mildew of mustard is more under tree canopy during wet winters while it remains at same level during dry winters.
- Incidence of certain diseases like powdery mildew of black gram was always more under shade as compared to open area.
- Cultivation of an intercrop in an agroforestry system also affected the disease status of the tree species in some cases. Build up of powdery mildew of ber (*Ziziphus mauritiana*) was faster in a ber-based agroforestry system with wheat as compared to ber fields without any intercrop.
- Changes in microclimatic parameters, brought about by introduction of trees in crop fields, generally influenced the disease severity of component crops if these operated in the cardinal range of the parameters for the disease development.
- Important diseases in various agroforestry systems could be kept under check by using recommended control measures for pure crops.

Table 11.3: Summary of Secondary Interactions recorded at NRCAF, Jhansi (Kumar and Shukla 2009)

Crop	**Type of Plantation**	**Associated Tree Species**	**Disease**	**Remarks**
Black gram	Block plantation	Neem	Powdery mildew	Disease severity increases under shade
			Cercospora leaf spot	No change in disease severity
			Yellow mosaic	No change in disease severity
Green gram	Windbreak	*Acacia* spp.	Cercospora leaf spot	Disease severity increases under shade
			Yellow mosaic	Masking of disease symptoms in tree shade
Mustard	Windbreak	*Dalbergia sissoo*	Downy mildew	Disease severity increases under windbreak in wet climate
			White rust	Disease severity increases under windbreak in wet climate
			Alternaria leaf blight	No change in disease severity
Wheat	Windbreak and Block plantation	*Dalbergia sissoo, Leucaena, Cassia simea*, kinnow, ber, guava, pomegranate	Alternaria leaf blight	No change in disease severity
			Loose smut	No change in disease severity
Wheat	Block plantation	Ber (*Ziziphus mauritiana*)	Powdery mildew	Irrigated crops like wheat increases disease severity

INSECT-PEST MANAGEMENT

The insect-pests of agroforestry system are essentially the pests of its components, i.e., crops and woody perennials. The mechanisms governing the insect-pest situation in agroforestry systems are yet to be investigated fully, and little information is available on comparing the insect-pest situation in monoculture or block plantation and in agroforestry combinations. There are several factors that influence the activities of insects in agroforestry. The dynamics of insect pests and their natural enemies are governed by the complexity and composition of the agroforestry system, the degree of interaction between the components, the type of agroforestry system and the composition of the plant communities in each component. Interactions among the components of the agroforestry system can be either positive (reduction in pest activity under agroforestry when compared with a monoculture or block plantation), negative (when pest problems are increased in an agroforestry system when compared with a monoculture or block plantation) or neutral (no change in pest intensities between monoculture and agroforestry) from the insect-pest management point of view.

Further, each factor governing insect-pest intensity in agroforestry may have different effect on pests at different times. The net outcome will be the sum of favourable and unfavourable effects on pests and their natural enemies. Uniformity in plant genetic material has been widely recognized as one of the main factor responsible for increased pest problems in monoculture. In monocropping, the cultivation of single plant species over large areas act as continuous and concentrated source of food supply to insects-pests in one place that supports their uninterrupted population build up. Moreover, the food plants in pure stands may be easily detected and colonized by the pests. Generally, the pest activity reduces with diverse vegetation, such as agroforestry, as compared with monoculture. However, diversity does not always result in reduced pest populations as it may be pest and site specific and also may be affected by other factors. Different degrees of pest injury may occur when a host plant is raised with different companion plants.

Plants belonging to the same or very close taxonomic group have the tendency to share common pests. In agroforestry systems, oligophagous and polyphagous insect pests are expected to thrive if both components belong to the same or a closely allied taxonomic group. Combining pulse crops with leguminous woody perennials may result in supporting pest populations common to both components because an insect feeding on a plant with a certain biochemical make-up will adapt more easily to closely related plants with similar biochemical constituents than to species that have entirely different constituents. Therefore, an agroforestry system consisting of plant species belonging to different taxonomic groups is expected to be less affected by insect pests than a system composed of closely related species. Further, the perennial plants in the agroforestry system may provide round the year food supply for the pests and thus contributes to multiplication of pest populations besides favouring maintenance of the pest population between seasons when the main food crop is absent from the field. However, it will also maintain the population of the pest's natural enemies. Therefore, it is advisable to include plants from different taxonomic groups in agroforestry to avoid sharing of common pests. Similarly, while establishing agroforestry, it should also be considered whether the component plants serve as host to any vulnerable stage of an important insect pest of the other component or as collateral or alternate or refuge hosts for the vectors of diseases. A plant believed to be a non-host for an insect pest in one stage may turn out to be a host plant for another stage as some insects utilize different host plants as food in their larval stages from those eaten in the adult stage. Exclusion of such plants from agroforestry will minimize insect-pest attack on crops or woody perennials. Moreover, the presence of component plants in agroforestry that may serve as host to a vulnerable stage of the insect pest, or that act as a collateral or alternate host for insect-pest, may be utilized to trap and kill these insects while keeping the main crop untreated. Such plants are different from trap plants that are specifically grown to attract insects from the main crop and are not otherwise a component of the agroforestry system. Complex agroforestry systems with diversified vegetation have a great potential for controlling insect-pest populations through increasing the efficiency of biological control agents especially by offering alternate prey, nectar sources and suitable micro-habitats for parasitoids and predators. At the same time, being perennial, these systems also support the natural enemy population within and between seasons, especially during the off season of the main crop. Moreover, some plants in the agroforestry may produce chemicals that enhance the efficacy of predators.

Insect damage in an agroforestry system can be minimized by manipulating crop and woody-perennial combinations. Monophagous pests can be controlled altogether by not including their host plants in the system while the host range of oligophagous and polyphagous pests can also be narrowed by eliminating palatable species from the agroforestry and replacing them with non-host plants. The modified microclimate (shade/light, temperature, RH, etc.) under agroforestry may have both direct and indirect effects on the activity of insect-pests and their natural enemies. Shade under agroforestry may provide protection to insects from direct sunlight while reduced temperature and increased RH may also affect pest activities. The damage potential of soft-bodied larval stages of most of the diurnal insects may enhance under agroforestry as they prefer to feed while avoiding direct exposure to the light. Similarly, the activity of parasites belonging to order hymenoptera may be reduced under shade (agroforestry) as they have greater host-searching capacity in bright sun. The *Bacillus thuringiensis* (entomopa-thogenic bacteria) may be more effective under agroforestry as its infection potential is greatly reduced in sunlight. Similarly, *Trichogramma* (egg parasitoid) parasitization may also be more under increased humidity in agroforestry. The increased humidity in agroforestry may be favourable for the development of disease in insect-pests.

The odoriferous plants (plants emitting volatile chemicals or odours into the surrounding environment), when raised with host plants of insect pests, can deter or enhance recognition, feeding and reproduction of the pests on their host plants. In monocropping, the odours released by the plants spread out in all directions and are perceived by the insects without any interruption. However, in agroforestry, the odours released by perennials/annuals may mask or block or reduce the effect of those released by the other component. This will make it difficult for insect to locate the hosts on which it feeds and reproduces. Moreover, such a situation will also reduce the biological-control potential of the system by affecting the performance of natural enemies. The prey-seeking behaviour of natural enemies that utilize the odours of the host plants of their prey to locate them is influenced by the presence of other vegetation in the system. Under such a scenario, identification of the plants that release pest-antagonistic and natural-enemy-attractive odours may be a big challenge and of great importance in developing pest-free agroforestry systems.

The tall woody perennial non-host plants in agroforestry may act as a physical or biological barrier to the movement of insects and their natural enemies to, from and within the system. The woody species in an agroforestry system will restrict the movement of insects through the system. The situation is advantageous if the entry of the insect pests is blocked or if outward movement of the natural enemies of the pests is hindered. The pest situation in agroforestry may also be manipulated through suitable land configuration and design of systems so that the conditions become favourable for the predatory insects and unfavourable for the pests. The type of plants, their colour, structure, height, size, density in the field, relative arrangement in time and space and the type and colour of the soil in the background will determine the sphere of activity of the insects and their natural enemies.

The proper nutrition, irrigation and other management practices also affect the insect pest situation in the agroforestry. Termites preferentially attack stressed plants (water and other stresses). Further, the same level of infestation may bring about different degrees of injury in the healthy and stressed plants. Nutrition also affects the attractiveness of plants to insects. In agroforestry, nitrogen-fixing trees make available more nitrogen to the crop components, which may enhance their attractiveness to insect pests. An undisturbed system favours the establishment of a stable pest–natural enemy equilibrium while in intensely managed systems the parasites and predators are more likely to be disturbed than the pests. Emigration of the natural enemies as a result of disturbances may lessen the benefits of biological control in the system. Mulches in an agroforestry field may serve as a source of food for many soil-dwelling insects. Severe pruning of plants in alley cropping gives rise to new vegetative growth, which is more palatable to the insects than the older growth. Removal of infested plant parts by pruning may help to get rid of sessile sucking pests. Fallowing part or the whole of a unit of land affects insect abundance in the unit, especially of the soil-dwelling stages.

■■■■

12 DISEASE MANAGEMENT IN TROPICAL AGROFORESTRY LANDSCAPES

PLANT AND TREE DISEASES

The word "pathology" is derived from Greek words 'pathos' means suffering and 'logos' means study. Thus plant pathology is the study of the suffering plant or the study of plant diseases. Forest pathology is the study of diseases in forests mainly due to pathogens. Plant/Tree disease is defined as sustained physiological disturbances and resulting structural disturbances of living tissues and organs of plants/trees caused by pathogens, ending sometimes in death. It refers to the abnormality in the condition of plant/tree caused by pathogens which may result in the death of the plant/tree. It is also defined as sustained physiological and/or structural damage to plant/tree tissues caused by biological agents (fungi, bacteria, virus, phytoplasma, nematode, parasitic plant, and protozoa) or non-biological agents ending sometimes in plant/tree death.

Tree diseases are significant of humans because they cause damage to trees and tree products. It may limit tree choice in any kind of land use system. It causes economic loss to tree growth and increases cost of production. It reduces the quantity and quality of products in the field and during storage of forest products. Tree diseases increase the prices of tree products for consumers and it also destroys the beauty of nature. Diseases can have both negative and positive impacts on forests. Negative impacts are considered to be related to timber and economic losses and positive impacts are ecological, especially related to succession, change in ecosystem biodiversity and creation of wildlife habitat.

Tree disease is any disturbance that prevents the normal development of a tree and reduces its economic or aesthetic value. Diseases of both shade and forest trees have the same pathogens, but the trees differ in value, aesthetics and utility. In forests, disease is significant only when large numbers of trees are seriously affected. Diseases with such visible symptoms as leaf spots may be alarming on shade trees but hardly noticed on forest trees. Shade trees with substantial rot may be ornamentals with high value, whereas these trees would be worthless in the forest. Emphasis on disease control for the same tree species thus requires a different approach, depending on location of the tree.

SCOPE AND SCALE OF PATHOGENS AND THEIR IMPACTS

Forest and tree health concerns in the 20th century were the result of several factors. In the beginning of the 20th century much of the forest lands were heavily logged, burned by intense fires, and cleared for agriculture. As time went by these lands were planted and forests reclaimed in abandoned agricultural fields. Nursery production of planting stock resulted in many endemic and exotic pathogens being moved across the countries and locations. Diseases of early successional species in plantations soon became important. Later in the century, land use changes, forest fragmentation and amenity planting of exotic tree species across a range of sites and habitats predisposed our forests and tree plantings to numerous damaging diseases that were not previously significant. Native pathogens often exist in equilibrium with natural forest communities. It is widely recognized that pathogens and dead trees have critical ecological roles and contribute to sustainable, healthy forest ecosystems.

Damaging disease epidemics have resulted when either non-adapted or exotic tree species were planted; trees were planted on stressful sites; management changed growing conditions or when pathogens were introduced into new ecosystems. Many forests today are being managed using different silvicultural systems than in the past in response to changing values and uses of these lands by owners and public agencies responsible for land management. Systems of intensive management, artificial regeneration, and all-aged and mixed species management can influence the incidence and severity of disease caused by native and introduced pathogens.

As productivity and yield of trees is greatly affected by diseases, the foresters should have the knowledge of common tree diseases, their causes and cures. Even if a forester is unable to cure the disease, the knowledge of the diseases and symptoms can be helpful to him in calling for help of a specialist (Forest/Plant Pathologist) in the early stages of disease.

SYMPTOMS OF PLANT/TREE DISEASES

Nature produces certain symptoms in the plant/tree to indicate the beginning of disease. Symptoms are defined as apparent expressions of presence and progress of diseases in plants/trees. Symptoms are assessed in order to identify causal pathogen, to determine amount of disease in plant and to estimate the damage (loss) caused. Symptoms can be masked, i.e. not visible because of unfavourable conditions. The symptoms may be of four kinds: (1) Necrotic Symptoms, (2) Atrophic Symptoms (3) Hypertrophic and Hyperplasia Symptoms, (4) Other Symptoms. Symptoms of disease include death and destruction of host tissue, wilting, abnormal growth and differentiation and discolouration of host tissue.

Necrotic symptoms include wilt, canker, decay, etc. Wilt is defined as drooping of foliage and ultimately the death of plant or tree. It is called pathological wilt when it is caused by some fungus and called as physiological wilt when caused temporarily due to deficiency of moisture caused by excess of transpiration during the day. Wilting occurs when water loss is greater than water intake. While pathological wilt results in the death of the plant, the physiological wilt may disappear when moisture deficiency disappears. Pathological wilt is caused in seedlings when their bases become rotted due to fungus attack and they fall to the ground. In trees, wilt causes drying of leaves followed by death of tree. Wilt results from interference with water and nutrient absorption at the roots; interference with water conduction within the plant (i.e. infection of the vascular tissue); or loss of control of transpiration. Canker is defined as a definite, relatively localized necrotic lesion primarily of the bark and cambium. Canker is slow to grow as a result of reaction of the host which produces callus tissues around the edges of the canker. It may sometimes girdle the branch or the stem thus results in the death of the part of the tree above canker. Canker is also caused by physiological causes such as frost, sun scorch, fire and drought. Some parasites, called necrotrophs, secrete enzymes that kill host tissue, extract nutrients from the cells and then live in the dead tissue. The necrotic lesions caused by pathogens can be localized or extensive. Local necrotic lesions appear as discrete necrotic areas, while extensive, or spreading lesions spread until the whole organ or plant is killed. Decay is defined as the decomposition of wood substance by fungi. The fungi may attack the living tissues, or the dead wood in trees. When fungi attack living tissues, they may kill the trees and when they attack the deadwood (heartwood), they cause rot. Typical examples include heart rot, white rot, brown rot, etc.

Atrophic Symptoms relate to slowing of growth resulting from subnormal cell division. The example includes dwarfing or stunting of entire plant especially by virus. When atrophy affects the development of chlorophyll apparatus, the symptom is called chlorosis. Discolouration of tissue is most commonly by chlorosis or mosaics of leaves, both of which can have a number of causes. Anything that interferes with the production of chlorophyll causes leaves to turn yellow, or chlorotic. Mosaicism is a symptom of many virus infections and is characterized by alternating light and dark green areas on the leaves. Hypertrophy and hyperplasia symptoms relate to abnormal growth in plants. Abnormal growth and differentiation results from deviation from the complex balance of interrelated reactions that take place in plants. Parasites can alter the hormonal balance in plants causing an abnormal increase in the cell size or number of cells, resulting in abnormal growth and differentiation. It may result from abnormal/increased cell division (hyperplasia) causing overgrowth, galls, leaf curl, witches broom, hairy roots, etc. It also results from abnormal cell enlargement (hypertrophy). Under other symptoms, plants/trees produce certain signs in addition to above described symptoms as evidence of diseases. There may be development of fruiting bodies of the pathogen that causes disease usually at or near the point of infection. Development of epicormic branches and various kinds of exudations are other examples. Some of the most common terms used to describe symptoms of plant diseases are:

- *Blight*: A disease characterized by widespread death of plant tissue.
- *Canker*: A sunken necrotic lesion often of a main stem, branch or root.
- *Damping Off*: Collapse and rot of seedlings near soil level before emergence or soon after emergence caused by *Pythium* spp., *Phytophthora* spp., *Fusarium* spp., and *Rhizoctonia* spp.
- Dieback: Partial defoliation, twig and branch death and even complete death of plants.
- *Downy Mildew*: White or grey 'bloom' on leaves and stems caused by production of sporangiophores and sporangia by members of the Peronosporales (downy mildew fungi).
- *Gall*: An abnormal growth or swelling produced as a result of pathogenic invasion.
- *Mosaic*: Patchy variation of normal green colour in leaves, usually light and dark green mosaic, symptomatic of many viral diseases.
- *Powdery Mildew*: White powdery 'bloom' on the plant surface caused by the production of fungal mycelium, conidiophores and conidia by members of the Erysiphales (powdery mildew fungi).
- *Pustule*: A blister-like spore mass breaking through a plant epidermis.
- *Rot*: Disintegration of tissue, often caused by enzymes or toxins produced by pathogens.
- *Rust*: Rust-coloured pustules formed by members of the Uredinales (rust fungi).
- *Scab*: A discrete, superficial roughened lesion.
- *Smut*: A disease characterized by black spore masses on leaves, stems or inflorescences, caused by members of the Ustilaginales (smut fungi).
- *Vascular Wilt*: A disease in which the pathogen is confined to the vascular system of the host and in which wilting is a characteristic symptom; plants lose their turgidity and become flaccid, leaves collapse.

THE DISEASE TRIANGLE CONCEPT

A disease is the physiological disturbance of the normal functioning of a plant as a result of detrimental interaction between the pathogen, the environment and the host (Agrios 1988). Diseases affect the production and utilization of trees and crops by reducing the health of the plant and directly reducing yield, quality, or storage life. The 'Disease Triangle' shows the importance of the three factors necessary for disease development:

- Pathogen or disease-causing organism (e.g. fungus, virus or bacterium)
- Susceptible plant for the pathogen to infect
- Suitable environment in which the first two will interact

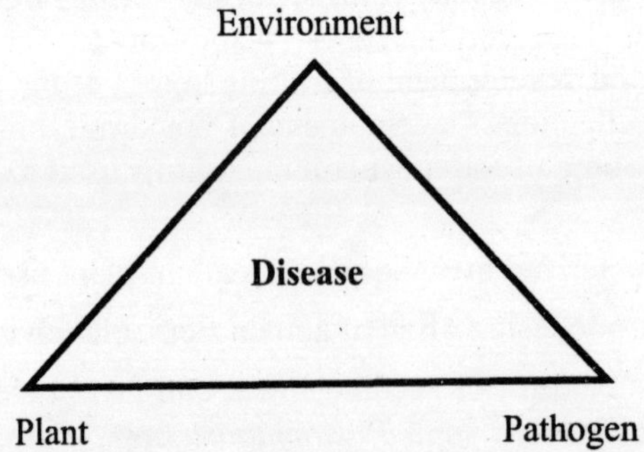

Fig. 12.1: The Disease Triangle

For example, *Thielaviopsis basicola* will attack the bedding plant *Vinca* when air and media temperatures are cool (15-17°C), but rarely at temperatures above 21-22°C. However, this pathogen will attack *Pansy* when the temperature is above 25-26°C, the media pH is high (above 6.5) and the sanitation is generally poor. The plant environment includes the soil or growing media in which the roots grow and the above-ground environment or air where the shoots, leaves and flowers are produced and developed.

GENERAL METHODS FOR PLANT DISEASE CONTROL

EXCLUSION

Preventing pathogens from becoming established in the plant propagation is a primary aspect of disease control. For example, plant propagators use this principle each year when they purchase new virus-indexed/pathogen-free cuttings for stock plant production. This minimizes the potential for fungal and bacterial infections such as rust or bacterial wilt. Growers who use soil or compost in their media pasteurize the material to eliminate soil-borne pathogens. The use of new clean pots, plug trays, root trainers, cell packs, flats and hygiene plant propagation structures prevents any potential carryover disease inoculum from infecting newly planted crops the following growing season.

AVOIDANCE

This involves using common sense practices to avoid situations that promote disease development. For example, overhead watering should be avoided late in the day in crops with a dense canopy because this creates conditions favourable for the development of *Botrytis*.

RESISTANCE

This involves the inherent ability of a host plant to remain healthy or resist attack, even though a pathogen is present and/or the environment is conducive to pathogen development. Many plant cultivars are being bred with resistance or tolerance to many specific diseases.

PROTECTION/PREVENTION

Fungicides can be applied before any disease is observed as a preventive measure to protect plants during periods when conditions are conducive to disease development. However this proactive approach should be taken judiciously and only under specific plant conditions or crop staging. For example, fungicidal treatment of seeds may protect the plants from nursery diseases. The chemicals like Cuman, Thiride, Blitox etc. are used for seed treatment for control of nursery diseases caused by fungal pathogen. Another preventive method of controlling nursery diseases of soil origin is to treat the soil with chemicals. The chemicals are broadcast and then raked into the top 15cm of the soil one day before seed sowing. The chemicals used are Captan, Thiride, Cuman, Brassicol, Blitox (Copper Oxychloride), Zinc Oxide, etc.

ERADICATION

Plant pathogens can be eliminated via several means, such as steam pasteurization of ground beds or rock-wool slabs, to eliminate root-borne diseases and nematodes. Sanitizing products containing hydrogen peroxide, chlorine bleach (sodium hypochlorite) or quaternary ammonium chloride compounds can effectively control bacteria or fungal mycelium and spores in dust or on surfaces of benches, in or on irrigation lines and emitters, and on tools used for propagation or production. Plant breeders and primary propagators routinely index and use eradication techniques to maintain vegetatively propagated plants free of plant-infecting bacteria and viruses.

THERAPY

Infected plants are treated to prevent further sporulation and spread of the disease using chemical, cultural, mechanical and biological control measures.

ACHIEVING SUCCESSFUL DISEASE MANAGEMENT

Successful disease management can be achieved if these measures are followed:

- Emphasize prevention rather than cure. This is critical in any production system, but is especially true when crops, plants and trees are being grown together in agroforestry.
- Monitor crops, plants and trees daily (or at least two times per week) by walking in the field to achieve early detection. This involves checking the foliage, stem and root systems.
- The employee responsible for monitoring must work closely with the person responsible for overall farm management. Ideally, the individual in charge of growing should be the one who monitors for disease and environmental control.
- Correct diagnosis is crucial. Without correctly identifying the pathogen, control using cultural or chemical strategies is impossible. Soil-borne fungi are the most difficult pathogens for growers to identify.
- Develop a thorough understanding of the diseases that affect the crops and plants being grown. Correct use and timing of management strategies then becomes part of an ongoing process.

- Because disease pathogens are microscopic, learn to anticipate when possible pathogen infection periods are likely to occur. By the time the problem is visible, the pathogen has usually been present for some time and is more difficult to control.
- As a grower/manager, keep detailed records that enable merging of cultural, environmental, and insect and disease data for review when a production problem arises or for doing post season summary reviews.
- Monitor the weather to help anticipate potential disease problems. For example, a cool, cloudy and wet summer with the central heating system off creates ideal conditions for many disease developments.

SEED DISEASES

Seeds are prone to attack by the fungi while still on the tree, during post harvest operations in the field, storage and subsequent handling prior to sowing. The fungi may be borne as surface contaminants or they may be present in the seed coat or may penetrate the seeds even deeper attacking the embryo and thus damage the seeds to varying extents. It has been recorded that seeds of some leguminous (pod bearing) tree species are quite susceptible to attack by fungi while they are inside the pods on the trees and they are also damaged considerably in case they are not timely collected and properly stored. Number of fungi belonging to Mucorales and Fungi Imperfecti such as Alternaria, Aspergillus, Cladosporium, Colletotrichum, Curvularia, Fusarium, Gloeosporium, Lasiodiplodia, Mucor, Penicillium, Phoma, Phomopsis, etc. are associated with seeds of various forestry species of which species of Fusarium, Lasiodiplodia, Gloeosporium and Phoma cause deep seated infection and damage the embryo.

Earlier reports show that seed mycoflora of neem during storage caused by Aspergillus spp. (Sinniah *et al* 1983). Teak seeds are attacked by a number of fungi in storage and during weathering of fruits such as *Aspergillus flavus, A. niger, Chaetomium globosum, Curvularia pallescens, C. lunata, Fusarium moniliforme, F. oxysporum, Marasmius* spp. and *Stachybotrys atra*. Singh *et al* (1989) isolated *Alternaria* spp., *Penicillium lanosum* and *Trichoderma viride* from seedcoat and *Fusarium* spp. and *Helminthosporium* spp. from embryo of teak seeds. Pods of *Dalbergia sissoo* is attacked by *Phoma* spp. Other fungi recorded on pods include *Glomerella cingulata* and *Septothyrella dalbergiae* (Pavgi and Singh 1971). *Fusarium* spp. on seeds of *Albizia lebbeck* have been found to cause both pre-emergence and post-emergence damping-off disease together with *Rhizoctonia solani*, a soil inhabitant fungus.

CONTROL MEASURES

Presence of contaminated or infected seeds in the seed lot may cause considerable spoilage of seeds particularly by molds if they are not properly stored. Timely collection, proper drying and storage of seeds under suitable temperature and humidity are necessary to protect the seeds from biodeterioration. The seeds after collection should be dried so that the seed moisture content is reduced to about 10 per cent on oven dry weight basis. Besides, it is advisable to do seed dressing with a suitable fungicide (Ceresan) @ 4-5 gm/ kg of seeds. This prophylactic treatment is easy and economical and serves the dual purpose: firstly it helps in preventing fungal deterioration of seeds in storage and secondly it minimizes the chances of seed borne pathogenic fungi attacking the seedlings in the nursery.

NURSERY DISEASES

It is essential to ensure that high quality disease free planting stock is available for planting in plantation and agroforestry systems. Disease attacks seedlings in nurseries and knowledge of the nursery diseases is essential for successful prevention and control.

DAMPING-OFF

Damping-off is probably the most serious of all the nursery diseases and is prevalent all over the world. The damage by damping off is so serious that the entire nursery stock may be destroyed. It causes large scale mortality in nursery upsetting all plantation activity (Naik 2012).

SYMPTOMS, CAUSES AND PREDISPOSING FACTORS: Mortality in seedlings may occur at three stages during their growth due to pre-emergence blight, post-emergence mortality soon after germination and root rot in older seedlings resulting in poor growth and in extreme cases even death. The most important fungi causing damping-off belong to *Pythium, Rhizoctonia, Phytophthora* and *Fusarium*. Damping-off is favoured in clayey and wet soils, because of anaerobic condition in the rooting zone and relatively high temperatures. Because of the effect of warmth damping-off tends to develop more during summer on late sowings than in the seeds sown early in spring or in autumn. Soil pH influences development of the disease, which is more serious in alkaline or neutral soils than in acid soils (Griffin 1958). Low light intensity acting both by reducing the metabolism and increasing the soil moisture favours development of damping-off. In addition, dense sowings, application of nitrogen in high dosages during early stages of growth and addition of raw undecomposed humus favour disease development (Vaartaja 1952).

CONTROL: Light textured soil and good drainage will prevent damping-off. Damping-off and other soil-borne diseases may be controlled by seed and soil treatment with chemicals and bioagents such as Bavistin (Carbandazim), Dithane M-45, Captan, Thiride, Cuman, Brassicol, Blitox (Copper Oxychloride), Zinc Oxide, Zineb, Radomil, etc. Soil acidification, steaming, fumiagation (Formalin and Methyl Bromide) and fungicidal seed and soil treatments will control damping-off.

FOLIAR DISEASES

Foliar disease occurs on most species of landscape trees, shrubs and crops. Diseases are often characterized based upon symptoms development.

1. Leaf spots are characterized by discrete lesions that discolor and become necrotic (dead). With some leaf spots, the dead tissue is compartmentalized and excised from the leaf, resulting in a "shot hole" appearance.
2. Leaf blotch, blight, or anthracnose all begin as leaf spots, but continue spreading down the leaf veins, into the petiole, and into the woody tissue of the plant.
3. Needle casts, which are simply foliar diseases in conifers, result in the death and shedding or "casting" off of dead needles.
4. Powdery mildew is diagnosed most commonly by the characteristic powdery, white coating to the leaf surface formed by the fungus that infects the flower, leaf and green stem tissue.
5. Rust forms characteristic yellow, orange leaf and stem pustules. Some rusts are solely foliar problems of broadleaved trees or conifers, whereas others can attack stems, and result in galls and cankers.
6. Leaf curl, caused by members of the fungal genus *Taphrina*, commonly deform leaves and even fruit of many hosts, including flowering plum, aspen/ cottonwood, and alder.

7. Downy mildew is a common disease that affects many landscape plants. Symptoms appear on the foliage, in addition to the stems of infected plants.
8. Mottling, mosaics and ringspots are often due to virus infection
9. Sooty Mold is not so much as a disease, but the result of a fungus feeding of the excrement of aphids.

Fungi are the primary cause of most leaf diseases, although a few exceptions of bacterial leaf spot and blight also occur. Some of these diseases are severe to plant health (most needlecasts) or aesthetics (e.g., rose black spot, scab on crabapple, anthracnose of sycamore) but most are of minor effect. For most of these diseases, the problem is an issue of aesthetics, not plant health. Recognizing this, it is advised to promote the use of resistant varieties. However, resistance does break down over time, not all shade trees have selectable resistance to incorporate into nursery production, and even resistant plants can succumb to foliar problems under unusually wet weather.

Defoliation is a dramatic event that is more traumatic for the urban dweller than the urban tree. Most broadleaved trees and shrubs will produce a second flush of later in the season, or even the following year. Key management points to keep in mind are the facts that defoliation is only problematic if the tree in question has lost more than 30 per cent of its canopy more than three times in five years. If this is the case, fungicides should be applied prophylactically prior to infection the following year, to protect new growth as it emerges, and minimize long-term impact. Most foliar pathogens infect in the spring, under cool, moist conditions. Inoculation, germination, penetration, infection and invasion are similar for both broadleaved plants and conifers. Key differences is that deciduous plant pathogens begin their infection cycle from nearby fallen leaves, whereas the needlecasts develop fruiting bodies on dead needles that usually remain attached to the host for 1-2 years.

ROOT DISEASES

The underground part of the plant, the root system, may suffer from diseases which generally remain unnoticed until manifested by symptoms on the above-ground parts. Diseases may be due to fungi, bacteria, nematodes etc. Fungi parasitic on plant roots and root infecting fungi constitute an important group of organisms causing root diseases of forest trees. Root diseases are of two types: vascular wilt and root rot (Negi 1996).

VASCULAR WILTS

Wilt is a systemic disease where the entire individual or its part exhibits wilting of the foliage in acropetal succession up the shoot. The leaves become yellow, lose turgidity and fall off. The first symptom is usually dieback of young twigs and branches. A black or brown ring can usually be seen in the vascular tissue of dead twigs. Mature trees may "seal off" the infection and live for many years. Others may die quickly. The outcome is difficult to predict. Infection occurs via the roots, then spreads up through the vascular system in the tree. The fungus carries over in the soil for several years, so only resistant species should be used for re-planting.

A growing imbalance of water economy in affected plants is common to all wilt diseases. A characteristic symptom of several wilts is the vascular discoloration in the outer layers of the sapwood, green streaks in *verlicillium* wilt in many hardwoods and pink stain in fusaral wilt. The vascular browning is ascribed as due to enzymatic action of the wilt fungus. Such discolouration may not be, however, diagnostic and cultural studies are necessary to diagnose the cause of wilt (Naik 2012).

CONTROL: Fungicides like Mancozeb, Carbendazim and Thiram are good seed dressing chemicals. Bioagents like *Trichoderma* and *Psedomonaus fluorescens* may also be used in nurseries. Application of FYM, neem cake encourages growth of bioagents present in the soil (Naik 1998). Two theories are advanced to explain the mechanism of wilt: (i) *occlusion theory*, which assumes plugging of vessels by the mycelium of the wilt fungus interfering with translocation of sap. (ii) *toxin theory*, which postulates production of plasma poison by the wilt fungus and these toxins destroy the osmotic function of the living cell (Dimond 1955).

ROOT ROTS

Some forest trees are highly susceptible to root rot. Nevertheless, root rot is almost always the result of over-watering, root damage or prior drought stress, poor soil conditions, flooding, etc. *Pythium* and *Phytophthora* can almost always be found in dead, rotted roots, so may not be the primary cause. Except for highly susceptible species, re-planting with healthy trees should not be a problem if the soil and environment are improved and good sivlicultural practices are followed.

Root rot fungi generally possess a wide host range. Once the infection has established in the roots, its further spread depends on the host resistance. In the susceptible hosts, the pathogen may spread quickly into roots and causes disease. The symptoms of root rot attack, however, do not manifest till a large percentage of roots is attacked. In the less susceptible hosts, the fungi may colonize roots, but then the progress inside the roots is slow because a large mass of healthy roots react strongly against spread of infection. Such hosts, therefore, generally do not show visible symptoms of infection till most of the roots become attacked. In the natural forests, therefore, where resistant and susceptible species are in mixture, the fungi remain in an endemic state. When natural forests are clear felled for raising plantations, the host resistance is lost and the fungi on roots of susceptible hosts are able to spread quickly (Negi 1996).

When trees are felled, the stumps and their roots remain alive for a considerable period. Gradual lowering of host resistance paves the way for colonization of roots first for specialized parasites like root rot fungi followed by weak secondary parasites and finally for saprophytes. *Armillaria mellea* which is able to spread quickly on roots of felled trees become greatly restricted or is absent if the roots are killed more quickly by ring barking the trees when green (Bakshi 1973). While the root rot fungi are considered to be specialized parasites, two groups may be distinguished: in one, the fungi cause typical root rot and do not attack the heartwood as in *Ganoderma lucidum*; in the other, the fungi may cause root and butt rot as in *Heterobacidion annosum, Armillaria mellea* and *Polyporus* spp.

CONTROL: The control of root diseases generally is more difficult because of the difficulty of combating the disease on underground plant parts which are not easily accessible as compared to diseases attacking aerial plant parts. Chemical control of root diseases with fungicides applied to soil may not be effective as these chemicals may not reach the affected parts and also they may soon become inactivated due to reaction with other chemicals present in the soil or through leaching. Difficulty in the control of root diseases of forest crop arises because of the long rotation age of such crops. In mixed plantations of susceptible and resistant species, root disease remains very much in check. However in pure plantations, once the root disease breaks out, the spread is very rapid from tree to tree and therefore, within a short time disease becomes alarming (Naik 2012).

SILVICULTURAL CONTROL OF ROOT DISEASES

- *General Methods*: Propagation of resistant tree species, raising mixed plantations and adopting measures to quarantine the host and disease are generally practised.

- *Site Selection*: The knowledge of silviculutre of species is very important for deciding which species is to be grown under what conditions. It is not advisable to raise susceptible tree species on clay soils or under conditions of waterlogging as this will make the trees prone to attack by root infecting fungi. It is also advisable to have the knowledge of root growth of a particular tree species on the site conditions.
- *Controlled Burning*: Control burning helps to keep down root diseases in some tree species.
- *Mixed Stands and Resistant Species*: In certain areas, it is advisable to raise mixed stands or plantations so as to reduce root diseases. Species prone to attack by particular root diseases should be avoided and resistant tree species should be planted. Resistant species are that which have a certain degree of resistance to a particular pathogen.
- *Site Preparation*: Removal of residual roots and stumps from the entire planting areas with simultaneous soil working is recommended.
- *Isolation Trenches*: Root disease spreads from root of one species to another. Usually this is by contact, migration of pathogens or by root grafts resulting in diseased plants occurring in groups. Isolation trenches are dug up with a view to isolate the roots which have already been infected by disease. The disease or pathogen is contained within the trenches thereby preventing further spread to other areas.
- *Thinning*: Thinning of infected trees from the site is good method to prevent the spread of root disease to other healthy trees.
- *Chemical Control*: An ideal stump treatment should be effective in preventing infection at the stump. Among chemicals tested and recommended for use are creosote, ammonium fluoride, sodium nitrite, borax, urea and ammonium sulphamate. Root feeding with Tridemorph @ 0.1% is also effective.
- *Biological Control*: Root rot due to *Fomes annosus* is severe in plantations soon after thinning because the pathogen is able to colonize freshly cut stumps. *Peniophora gigantea* is a vigorous competitor of *Fomes annosus* for initial colonization of stumps and may also replace *Fomes annosus* in stumps (Rishbeth 1963).

HEART ROTS

In the tree, decay does not set in till the heartwood is formed and this usually occurs between 15 to 30 years of age depending on the species. When the heartwood becomes exposed through openings in the bark and sapwood, decay may establish in the heartwood as heart rot which becomes progressive with age (Bakshi and Sujan Singh 1970). Tree wounds due to various causes are the principle infection courts for trunk and basal rots. When the wood of a living tree is freshly exposed following an injury, chemical and biological changes occur. Toxic natural extractives which impart resistance to decay are formed during development of the heart wood and these substances are mostly deposited in the cell wall. Heart rot causes progressive decay and therefore, the volume of decay increases in the tree and in the stand with age. Heart rots in trees may be detected either from external symptoms manifested by diseased trees or through direct probing into the heart wood. Heart rot may be detected by taking increment cores from trees suspected to contain decay and isolate the fungus in culture from the decayed wood. Advanced decay trees may give a hollow sound if struck on the bark with a hammer or back of an axe at the region of decay.

EXTERNAL INDICATORS (Boyce 1961)

- *Sporophores*: The presence of sporophores of heart rot fungi on the tree is a sure indication of decay.
- *Punk Knots*: Punk knots are cylindrical cores of soft punky tissue of the wood decayed by the fungus, embedded in the sapwood up to different depths and opening upon the bark as round or oval hard structures.
- *Swollen Knots*: Swollen knots are formed when living wood tries to grow over a knot in which a sporophore is forming.
- *Swollen Boles*: A swollen bole is a reliable indicator of decay.

Table 12.1: Diseases and Causal Pathogens of Agroforestry Tree Species (Naik 2012)

S.No	Host	Name of the Disease	Causal Pathogen
1	Shisham (*Dalbergia Sissoo*) and Rosewood (*Dalbergia latifolia*)	Wilt disease	*Fusarium solani*
		Ganoderma root rot	*Ganoderma lucidum*
		Polyporus gilvus root and butt rot	*Polyporus gilvus*
		Leaf spot	*Cercospora sissoo* *Colletogloeum sissoo* *Phyllachora dalbergiae* *Phyllachora spissa* *Phyllosticta sissoo* *Mycosphaerella dalbergiae*
		Powdery mildew	*Phyllactinia dalbergiae*
		Leaf rusts	*Maravalia achroa* *Uredo sissoo*
		Phanerogamic parasites	*Tolypanthus involucratus* *Tapinanthus dodoneifolius*
		Pod fungi	*Glomerella cingulata*
2	Teak (*Tectona grandis*)	Bacterial wilt	*Psedomonas solanacearum* *Psedomonas tectonae*
		Armillaria root rot	*Armillaria mellea* *Armillaria tabescens*
		Fomes lamaoensis root rot	*Fomes lamaoesis*
		Fomes noxius basal rot	*Fomes noxius*
		Helicobasidium compactum root rot	*Helicobasidium compactum*
		Peniophora root rot	*Peniophora rhizomorpho-sulphurea* *Polyporus zonalis*
		Polyporus zonalis root rot	*Polyporus zonalis*
		Pink disease	*Corticium salmonicolor*
		Nectria stem canker	*Nectria haematococca*

S.No	Host	Name of the Disease	Causal Pathogen
		Decay in dry coppice teak	*Fomes lividus*
			Polyporus zonalis
		Leaf spot	*Cercospora tectonae*
			Phyllosticta tectonae
			Sphaceloma tectonae
		Powdery mildews	*Phyllactinia corylea*
			Phyllactinia guttata
			Uncinula tectonae
		Leaf rust	*Olivea tectonae*
		Thread blight	*Marasmius* spp
		Phanerogamic parasites	*Dendrophthoe falcata*
			Macrosolen cochinchinensis
			Phoradendron piperoides
3	*Casuarina equisetifolia*	Algal parasites	*Stomatochroon* spp
			Cephaleuros spp
		Bacterial wilt of seedlings	*Pseudomonas solanacearum*
		Seedling blight	*Pellicularia rolfsii*
		Black root rot of seedlings	*Macrophomina phaseolina*
		Ganoderma root rot	*Ganoderma lucidum*
		Phytophthora root rot	*Phytophthora cambivora*
		Armillaria root rot	*Armillaria mellea*
		Stem wilt	*Trichosporium vesiculosum*
		Pink disease	*Coticium salmonicolor*
		Heart rot	*Polyporus glomdratus*
			Fomes fastuosus
			Fomes senex
		Leaf spot	*Phoma casuaranae*
		Phanerogamic parasites	*Cuscuta campestris*
			Dendrophthoe falcata
			Dendrophthoe lanosa
4	*Albizia* spp.	Root rots	*Clitocybe tabescens*
			Ganoderma lucidum
		Pit canker	*Fusarium solani*
		Nectria canker	*Nectria ditissima*
		Gibberellia die-back	*Gibberellia pulicaris*
		Heart rots	*Fomes fatuosus*
			Fomes robiniae
			Ganoderma applanatum
			Polyporus gilvus

S.No	Host	Name of the Disease	Causal Pathogen
		Leaf spots	*Cercospora albizziae* *Endodothella albizziae* *Phyllosticta albizzinae*
		Leaf rusts	*Revenelia japonica* *Revenelia sessilis* *Revenelia clemensiae* *Revenelia indica*
		Sooty mold	*Meliola albizziae*
5	*Eucalyptus* spp.	Cylindrocladium seedling blight	*Cylindrocladium scoparium* *C. quinqueseptatum*
		Other seedling diseases	*Cercospora eucalyptii* *Botrytis cinerea*
		Phytophthora cinnamomi root rot	*Phytophthora cinnamomi*
		Ganoderma root rot	*Ganoderma lucidum*
		Crown gall	*Agrobacterium tumefaciens*
		Pink disease	*Corticium salmonicolor*
		Diaporthe canker	*Diaporthe cubensis*
		Other canker diseases	*Hypoxylon annulatum*
6	Poplar	Ganoderma root rot	*Ganoderma lucidum*
		Stem cankers	*Corticium salmonicolor* *Botryodiplodia palmarum*
		Rust, mildew and leaf spots	*Melampsora rostrupii* *Uncinula salicis* *Myrothecium roridum* *Septoria populi* *Phyllosticta adjunct*
		Leaf spot	*Botryodiplodia palmarum*
7	Silver Oak (*Grevillea robusta*)	Root rots	*Rosellinia bunodes*
		Pink disease	*Corticium salmonicolor*
		Leaf spots	*Botryodiplodia theobromae* *Mycospharella grevilleae* *Cytospora grevilleae* *Seimatosporium grevilleae* *Phomopsis bakeri*
		Phanerogamic parasites	*Dendrophthoe falcata* *D. neelgherensis*
8	*Acacia nilotica*	Heart rot	*Fomes badius* *Ravenelia accaiae-arabicae* *Cylindrosporium accaiae*
9	*Ailanthus excelsa*	Leaf spot	*Cercospora glandulosa*

S.No	Host	Name of the Disease	Causal Pathogen
10	*Azadirachta indica*	Root rot	*Ganoderma lucidum*
		Pink disease	*Corticium salmonicolor*
		Leaf spot	*Cercospora subsessilis Cercospora azadirachtae*
		Bacterial leaf spot	*Xanthomonas azadirachtii*
			Pseudomonas azadirachtae
		Phanerogamic parasite	*Dendrophthoe falcata*
			Tapinanthus spp
11	*Bambusa* spp.	Decay fungi	*Polyporus durus*
			P. friabilis
			Ganoderma lucidum
			Trametes corrugate
			Merulius similis
		Rust	*Puccinia gracilenta*
			Puccinia xanthosperma
			Puccinia longicornis
12	*Dendrocalamus strictus*	Rust	*Dasturella divina*
			Tunicopsora bagchii
13	*Gmelina arborea*	Heart rot	*Poria rhizomorpha*
			Phyllactinia suffulta
		Powdery mildew	*Cercospora ranjita*
			Armillaria mellea
		Leaf spot	*Sclerotinia rolfsii*
		Root rot	*Fomes roseus*
			Polyporus baudoni
			Trametes straminea
			Tapinanthus spp
14	*Madhuca latifolia*	Heart rot	*Fomes caryophylli*
		Rust	*Scopella echinulata*
		Phanerogarmic parasite	*Dendroth® falcata*
15	*Pongamia pinnata*	Root rot	*Ganodema lucidum*
		Heart rot	*Fomes merilli*
		Rust	*Ravenelia hobsoni*
			R. stictica
		Leaf spot	*Microstroma pongamiae*
			Sphaceloma pongamiae
			Fusicladium pongamiae
		Tar spot	*Phyllachora pongamiae*
			P. henn
		Phanerogarmic parasite	*Cuscuta reflexa*

DISEASE MANAGEMENT IN AGROFORESTRY TREES

DISEASES OF NEEM (*Azadirachta indica*)

Damping Off

Among the nursery diseases, damping-off is the most prevalent and highly destructive disease and cause heavy loss of seedlings. It is referred to a group of disease namely pre-emergence and post-emergence damping off depending on the stage of growth of seedlings when they are attacked. This disease is caused by various soil fungi such as *Pythium*, *Phytophthora*, *Fusarium* and *Rhizoctonia* of which the last two fungi are quiet prevalent in forest nurseries in India. High soil temperature, excessive soil moisture, high soil pH (alkaline), high nitrogen content, low light intensity due to shading, stiffy or clayey soil with poor drainage, dense sowing are the conditions which favour the disease development in serious proportion. In case of neem seedlings, the disease is mostly caused by *Fusarium oxysporum*.

Control Measures

The disease has been managed through: (a) cultural practices aimed at favouring plant growth and discouraging the growth of plant pathogen; (b) use of chemicals like formalin and suitable fungicides. Formalin is applied as soil fumigant and the fungicides can be applied either as soil drench or soil mix. Also, seed dressing with fungicide (Bavistin) in certain cases is found very effective. Potting media should be properly sterilized by solar heating before use.

DISEASES OF ALBIZIA (*Albizia* spp.)

Leaf Spot and Blight

Leaf spot and blight are the major nursery diseases of *Albizia* spp. at young stage. Different pathogens such as *Cercospora albizziae*, *Colletotrichum* spp., *Alternaria alternata*, *Camptomeris albizzia*, *Pleiochaeta setosa* and *Epicoccum* spp. have recorded to attack of leaves of *Albizia lebbeck*. The disease manifests on older leaves of seedlings as a small water soaked yellowish spot which later develops into a circular light brown lesion with a distinct yellowish margin. The spot hole develops in the advanced stages of infection causing premature defoliation. The disease is aggravated under high humid conditions. The tender shoots of seedlings are also infested and damaged.

Control Measures

Application of Captaf (0.2%) fungicide solution as foliar spray is found effective in controlling the disease problem.

Rhizoctonia Leaf Web Blight

It is caused by *Rhizoctonia solani*. The disease has been reported for the first time in Assam. This disease was also reported on other broad leaved trees. It first appears on leaves close to the ground, as water soaked silvery grey blotches. The infected adjoining leaves become joined together by the fungal hyphae as if caught in a spider's web hence the name web blight. The fungal invasion of foliage is rapid over the wet leaf surface during rains.

Control Measures

It can be managed through an integrated approach involving measures such as sanitation, cultural practices and use of fungicides. Sanitary measures recommended are disposal of leaf litter by burning and segregation of diseased seedlings soon after such seedlings are spotted. This will help in preventing lateral spread of the disease through contact of the overlapping foliage of the adjoining seedlings thereby minimizing the disease incidence. Cultural practices include raising of seedlings in poly bags instead of beds; keeping seedlings in lots of 250-300 seedlings instead of 1,000 seedlings in nurseries. Application of Bayleton (0.1%) as foliar spray is found very effective in controlling the disease.

DISEASES OF GAMHAR (*Gmelina arborea*)

Poria Root-rot

Poria rhizomorpha has been recorded as a root parasite of *Gmelina arborea* in part of India in north Bengal and Assam (Bagchee 1952). The severity of the disease has been noticed more on clayey sites where the soil becomes periodically waterlogged. The fungus grows on woody debris in the soil and humus from which it spreads to the host through white coloured cord-like rhizomorphs often to a distance of 30 meters, disseminating the disease from one tree to another. It causes brown cuboidal rot in the root resulting in death of root cambium followed by die back and death of affected trees. The sporophores of the fungus effused thin, brittle, poroid, crust, inseparable from the substrate, or white to pinkish cinnamon, pores round to angular, basidiospores hyaline, ellipsoid in shape.

Root Rot and Collar Rot

These diseases are caused by *Sclerotium rolfsii* and recorded in young seedlings in Kerala (Florence et al 1987). Paling of foliage, leaf shedding and subsequent death of the seedlings are observed. In Kerala, the disease causes premature defoliation and is reported to be in mild form.

Control Measures

Soil drenching with 0.2 per cent Bavistin or Dithane M-45 at monthly intervals effectively controls the root rot and collar rot diseases in Gmelina arborea.

DISEASES OF KARANJ (*Pongamia pinnata*)

Leaf Spot and Blight

Leaf spot and blight diseases caused by *Fusicladium pongamiae* on *Pongamia pinnata*. The pathogen causes severe leaf deformities. *Microstroma pongamiae* causes white to cream-coloured spots giving a yellowish appearance to the leaves. Other fungi such as *Phyllachora pongamiae*, *Robillarda makatii* and *Urohendersonia pongamiae* cause leaf spot disease. *Cercospora pongamiae* and *Sphaceloma pongamiae* cause anthracnose spots on leaves, tender shoots and pods resulting in severe damage and early defoliating in young seedlings and trees.

Control Measures

Foliar spray of Bavistin fungicidal solution (0.1%) is found to be effective in minimizing the disease.

DISEASES OF TEAK (*Tectona grandis*)

Leaf Blight

Leaf blight in teak is caused by Rhizoctonia solani. The infected plants show water soaked grayish brown patches that enlarge rapidly and cover a large part or the entire lamina. The blighted leaves often show holes in the infected portion as a result of shedding of infected tissues during heavy rains. The infected leaves dry up and are eventually shed. The disease spreads laterally in the nursery through overlapping foliage of the adjoining seedlings often resulting in group blighting of seedlings. In each case of severe infection, defoliation is high.

Control Measures

Immediate removal of infected plants helps to prevent the disease spread. Application of Dithane M-45 (0.1%) is found effective in controlling disease.

Leaf Rust

This disease is caused by Olivea tectonae. The infected leaves are almost plastered with yellowish brown fruit bodies of the fungus. The upper leaf surface presents a grey appearance due to the formation of fleeks, which correspond to the position of sori on the lower surface. Infected leaves fall off prematurely resulting in retardation of plant growth. The disease is common in nursery and young plantations.

Control Measures

The infected seedlings can be segregated and kept in isolation. Severely infected and dead seedlings can be burnt away from the nursery to prevent the spread of the disease. The disease may be controlled in the nursery by the application of sulphur based fungicide (Sulfax) on both sides of the leaves.

Powdery Mildew

Members of the family Erysiphaceae have been recorded to cause mildews in teak. Phyllactinia corylea is recorded to attack teak leaves (Bagchee 1952). Phyllactinia guttata is also recorded to attack teak from other countries. Uncinula tectonae widely occurs in nurseries, plantations and forests. The fungus forms white powdery coating on the undersurface of teak leaves and later develops dark coloured cleistothesia over the white fungus weft (Spaulding 1961). Uncinula tectonae is restricted to the upper leaf surface and the infected leaves are coated with a dull white mycelium and conidia borne on conidiophores. Conidia are air-borne which are produced abundantly and cause fresh infection. The metabolic changes in plants take place which lead to drying of infected leaves.

Control Measures

Sulphur dust has been found as most effective in controlling powdery mildew in younger seedlings followed by Baycor, Mortesan and Calixin.

DISEASES OF SHISHAM (*Dalbergia sissoo*)

Powdery Mildew

Phyllactinia dalbergiae causes powdery mildew on sissoo seedlings and recorded from Dehra Dun and Allahabad (Uttar Pradesh), Pusa (Bihar), Poona, Bombay and Nagpur (Maharashtra) and Chichrauli and Seonti (Haryana). The fungus produces yellowish, persistent, dense mycelium on the lower surface of sissoo leaves.

Control Measures

Application of sulphur based fungicide has been found most effective followed by Baycor, Mortesan and Calixin in controlling powdery mildew disease on *D. sissoo* seedlings in nursery.

Wilt Disease

This disease is caused by the soil inhabiting fungus *Fusarium solani.* The adverse effects are yellowing and death of leaves and subsequent defoliation. Young affected trees die off after an intense attack. Waterlogging favours the incidence of this disease. Silty and clayey soils favour growth and infection by this pathogen. Selection of the proper site for raising shisham plantations is the most important preventive measure. The soil should be light textured, with adequate soil moisture and good drainage. Soil drenching with carbendazim, around the infected trees is recommended.

Ganoderama Root Rot

Disease is caused by *Ganoderma lucidum* both in natural forests and plantations. The pathogen spreads rapidly in the system of trees growing on light textured soils. The trees may die within a short span. On the other hand, the spread is relatively slow in heavy textured soils. It spreads through root contact and hence it damages pure plantations very quickly, resulting in the death of trees. The diseased trees present a *stag headed* appearance and die after a few years. Selection of suitable site, removal of the affected individuals and sanitations are the recommended measures. This disease can be prevented by removing residual roots and stumps before raising a new crop. Isolation trenches also help to prevent the spread of this pathogen.

Ganodrma lucidum also causes root rot in *Dalbergia latifolia* natural forests as well as plantations. The symptoms and control measures are same as that of *Dalbergia sissoo.*

DISEASE OF CASUARINA (*Casuarina equisetifolia*)

Stem Wilt

Trichosporium vesiculosum causes stem wilt in casuarina. Numerous spores are produced by the pathogen which is dispersed by wind. Wounds caused to the stem or branches serve as channels for infection. The spores carried by wind attack the trees through these wounds or injuries. The pathogen spreads to the shoot and root system of the tree after infection. After this, there is secondary spread of the pathogen in the plantation through root contact and root grafting. *Casuarina* trees have a well developed network of roots and thus the disease spreads very rapidly in the plantation.

Symptoms of wilting like paling of needles which dry up completely after some time. Formation of spore mass beneath the bark results in the development of blisters on the bark. After some time, these blisters burst or rupture, exposing the black powdery mass of spores on the upper surface of the bark and sporulation also occurs on the roots. Diseased trees may be found in groups as the secondary spread of this pathogen is by root contact. Prevention of primary infection by minimizing all forms of injuries to the trees is important. Isolating the diseased trees by means of trenches and groups of affected trees may thus be isolated. All the infected trees must be removed as early as possible (Naik 2012).

DISEASE OF AILANTHUS (*Ailanthus excelsa*)

Plant Parasite

Incidence of plant parasite Helicanthus elastica has been observed in many places which severely affects the growth and survival of A. excelsa trees. H. elastica belongs to the family Loranthaceae is a parasitic plant species infesting various crops. It is light green in colour and has thread like stem twining around the host plant. They are leafy and bear white coloured flowers which occur in cluster. As soon as the parasite gets in touch with the host stem, small structures like haustoria penetrate the host's tissue and derive nutrients and water from the host. This infestation kills the entire tree and spreads very rapidly to other trees through birds.

Control Measures

Mechanical removal or pruning the affected branches can control *H. elastica* to a great extent. Regular monitoring, timely location of incidence and removal of infested braches of trees from the plantations, as soon as it is noticed, can prevent spread of this plant parasite. If incidence of this parasite is observed at an early stage i.e. soon after attachment to Ailanthus tree branches, the infested portion of the branches may be cut below the point of attachment of H. elastica to host plant to avoid regeneration of this parasite from the embedded haustoria left in the host plant part. Severely infested trees should be uprooted and kept in isolation or even burnt in order to prevent further spread of the plant parasite to other trees. Flowering and fruiting in H. elastica should be prevented and thus seed production.

DISEASE OF JACKFRUIT (*Artocarpus* spp.)

Alternaria Leaf Blight

Incidence of leaf blight disease has been recorded in young seedlings of different species of *Artocarpus* and the pathogen responsible for the disease is *Alternaria* spp. Affected leaves show light yellow to dark yellow colouration on the leaf tip and spreads all over lamina at the initial stage of infection. Later the infected leaves become brown to dark brown and severely infected ones shed off. In case of *A. heterophyllus* and *A. hirsuta,* the affected leaves show greyish white colour on the tip and spread slowly to all over the lamina. At the advanced stage, the affected leaves are fully whitish to greyish brown and dark brown in colour. Severely infected leaves fall off and the whole plant gets dried from tip downwards later on.

Control Measures

Infected seedlings along with the containers should be shifted and kept in isolation in order to prevent the spread of the disease in nursery. At the initial stage, the affected portions of the leaves can be trimmed off and foliar spray with Dithane M-45 (0.01%) can be given at weekly intervals. Severely infected seedlings can be treated with both foliar spray and soil drenching with Dithane M-45 (0.1%) or Fytolan (0.2%) at weekly intervals. Application of Biocontrol agents such as *Trichoderma viride* and *Pseudomonas fluorescens* (5 gm/ 1 litre of water) is found very effective in controlling the said disease.

DISEASES OF EUCALYPTUS (*Eucalyptus* spp.)

PINK DISEASE

Eucalyptus pink disease is caused by the pathogen *Corticium salmonicolor*. Repea-ted die back in young plants occurs which lead to their death. Infection occurs, in trees which may have escaped being affected by this pathogen initially, at a later stage in the form of cankers. There occurs girdling of branches and the main stem of young plants towards the later part of the rainy season. The parts lying above girdled portion may die, causing many epicormic branches to develop. One of these becomes the leading shoot which is again infected and killed in the next season. High summer rainfall (over 250cm) and warm tropical conditions may cause epidemics. Individuals showing a fair degree of resistance to this disease in an infected plantation could be identified and selected for bre-eding programme for introduction on a large scale. *E. torelliana* is found resistant to this disease and may be used in crossing programme. Spraying of systemic fungicides like Trdemorph @ 0.1% may be done on young plantations before and during the rainy season. Diseased individuals must be removed from the plantation as early as possible.

CANKER

Diaporthe cubensis causes canker in various species of Eucalyptus, though some species are resistant to this pathogen. Young plants are attacked at the root collar causing discolouration, girdling and even death. There is development of attacks around the base of the dead branch stubs. Initially there develops longitudinal cracks in the bark, while open cankers form later on having a length of more than 1 m. There is exudation of gum from these cankers. In tracts prone to this pathogen, only resistant species should be planted. The diseased individuals must be identified, isolated and removed from the plantation as early as possible (Naik 2012).

DISEASE MANAGEMENT OF CROPS UNDER AGROFORESTRY

DISEASE OF RICE (Oryza sativa)

Blast

The blast disease in rice is caused by *Pyricularia grisea* (*P. oryzae*). It is the earliest known plant disease and it is also known as rotten neck or rice fever. It is reported from 80 rice-growing countries. It is first recorded in India during 1918.The expected grain loss is70-80 per cent. Disease can infect paddy at all growth stages and all aerial parts of plant (leaf, neck and node). Among the three, leaves and neck infections are more severe. Small specks originate on leaves and subsequently enlarge into spindle shaped spots (0.5 to 1.5 cm length, 0.3 to 0.5 cm width) with ashy center. Leaf blast is the severe cases of infection and entire crops give a blasted or burnt appearance and hence the name "blast" is given. In severe cases, lodging of crop (after ear emergence) is reported. Neck blast affects the neck region of panicle and develops a black colour and shrivels completely/ partially, grain set inhibited, panicle breaks at the neck and hangs. In nodal blast, nodes become black and break up.

Management

The excess N - fertilizer application should be avoided and application of nitrogen in three split doses is recommended. Weed hosts should be removed from bunds. Use of tolerant varieties of rice is very important. Burning of straw and stubbles after harvest is necessary. Dry seed treatment with *Pseudomonas fluorescens* talc formulation @10g/kg of seed is recommended. Water need to be stagnated to a depth of 2.5 cm over an area of 25m^2 in the nursery and 2.5 kg of *P. fluorescens* (talc) should be sprinkled and mixed with stagnated water. The root system of seedlings should be soaked for 30 min before transplant. Similarly *P. fluorescens* talc formulation must be sprayed @ 0.5% from 45 days after transplanting at 10 day intervals, three times. Seed treatment at 2.0 g/kg seed with Captan or Carbendazim or Thiram or Tricyclazole is found to be effective. It is also advisable for Spraying of Tricyclazole at 1g/litre of water or Edifenphos at 1 ml/litre of water or Carbendazim at 1.0 gm/litre. Three to four sprays each at nursery, tillering stage and panicle emergence stage may be required for complete control.

In the nursery stage for light infection, spraying of Carbendazim or Edifenphos @ 0.1% is recommended. In pre-tillering to mid-tillering stage for light infection at 2 to 5 per cent disease severities, application of Edifenphos or Carbendazim @ 0.1% is recommended. Top dressing of N fertilizers should be delayed when infection is seen. At panicle initiation to booting stage and at 2 to 5 per cent leaf area damage, Edifenphos or Carbendazim or Tricyclazole @ 0.1% should be sprayed. At flowering and after with 5 per cent leaf area damage or 1 to 2 per cent neck infection, spraying Edifenphos or Carbendazim or Tricyclazole @ 1g /litre of water is recommended.

DISEASE OF SORGHUM/JOWAR (Sorghum *bicolor*)

Anthracnose and Red Rot

Anthracnose and red rot are caused by *Colletotrichum graminicola.* Many lesions may develop close together and coalesce to kill large portions of the leaf. Mid-rib infection often occurs and is seen as elongate-elliptical red or purple lesions on which the black acervuli can be clearly seen. Infected stems when split open show discoloration (depending on cultivar) which may be continuous over a large area, or more marbled appearance.

Management

Seed treatment with Captan or Thiram @ 4g/kg of seed is recommended for effective control. Spraying with Mancozeb 1 kg/ha is also recommened.

DISEASE OF MAIZE (*Zea mays*)

Downy Mildew

Downy mildew in maize is caused by *Peronosclerospora sorghi.* Infected plant appears chlorotic and sometimes stunted, white-stripped leaves & abnormal seed set are noticed. The chlorotic area of leaf always includes the base of the blade, and transverse margins usually sharply defined between the diseased and healthy tissues. A white, downy growth may appear on both surfaces of infected leaves. Sometimes tassels of diseased plant may exhibit phyllody. Tolerant plants may show symptoms of systemic infection but have normal seed production.

Management

Seed treatment with Acylalanine fungicide metalaxyl @ 6.0 g/kg is recommended. Infected plants should be rogued out at early stage. Application of Apron 35 WP @ 2.5 g/kg as seed treatment provides effective control. Resistant varieties need to be grown to avoid the disease incidence. Spraying with Metalaxyl 1g/litre or Metalaxyl + Mancozeb @ 2.5 g/litre is also recommended.

DISEASES OF WHEAT (*Triticum aestivum*)

Leaf Rust/Brown Rust

Leaf rust/brown rust in wheat is caused by *Puccinia recondita.* The postules are circular or slightly elliptical, smaller than those of stem rust, usually do not coalesce, and contain masses of orange to orange-brown Urediospores. Infection sites primarily are found on the upper surfaces of leaves and leaf sheaths and occasionally on the neck and awns. It survives on stubbles and volunteer crops and the alternate host is *Thalictrum* spp.

Stem Rust/Black Rust

Stem rust/black Rust is caused by *Puccinia graminis* f.sp. *tritici.* Pustules (containing masses of urediospores) are dark reddish brown - occur on both sides of the leaves, on the stems, and on the spikes. Pustules are usually separate and scattered, heavy infections -coalesce. Prior to pustule formation, "flecks" may appear. Before the spore masses break through the epidermis, the infection sites feel rough to the touch. As the spore masses break through, the surface tissues take on a ragged and torn appearance. It survives on stubbles and volunteer crops and the alternate host is *Berberis* spp.

Stripe Rust/Yellow Rust

Stripe rust/yellow rust is caused by *Puccinia striiformis.* The pustules of stripe rust, which, contain yellow to orange-yellow urediospores, usually form narrow stripes on the leaves. Pustules also can be found on leaf sheaths, necks, and glumes. It survives on stubbles and volunteer crops and alternate host is unknown.

Management of All Rust Diseases

Mixed cropping and crop rotation are preferred. The excess N - fertilizer application should be avoided. Sulphur dusting @ 35-40 kg/ha is recommended. Application of Mancozeb @ 2g/litre is also recommended. Resistant varieties need to be grown to prevent infection.

DISEASE OF BAJRA/ PEARL MILLET (*Pennisetum glaucum*)

Rust

Rust in bajra is caused by *Puccinia substriata.* Small reddish-brown to reddish orange, round to elliptical uredinia develops mainly on foliage. As severity of infection increases, leaf tissue will wilt and become necrotic from the leaf apex to base. In infection sites developing late in the season, uredinia are replaced by telia which are black, elliptical, and subepidermal.

Management

Spray any one of the following fungicides when the initial symptoms of the diseases are noticed viz. Wettable Sulphur 2.5 kg or Mancozeb 1 kg/ha. The application of any one of the fungicides needs to be repeated 10 days after. Sowing during December - May results in less incidence of rust disease. Control measures are to be adopted when there is rust incidence in the early stages as spread of infection to top leaves results in poor grain filling.

DISEASE OF RED GRAM/ PIGEON PEA (*Cajanus cajan*)

Dry Root Rot

Dry root rot in red gram is caused by *Macrophomina phaseolina* (*Rhizoctonia bataticola*). Typical symptoms include root and basal stem rot with a large number of minute, fungal sclerotia visible under the bark. Plants dry prematurely, particularly when they face drought stress. Infection of seedlings and leaf infection has been reported from India. Disease incidence severe in off-season, irrigated, summer crops in several parts of India, and it is a minor one in the normal-season crop. The pathogen is both soil- and seed borne.

Management

Disease incidence can be reduced if the previous crop stubble is buried deep, and is allowed to decompose well before pigeonpea is sown. Seed dressing with biofungicides such as *Trichoderma viride* 4g/kg of seed or Carbendazim 2g/kg of seed or *Pseudomonas fluorescens* 4g/kg of seed should also reduce seedling mortality. Soil application of *T. viride* @ 2.5 kg/ha mixed with 50 kg FYM is recommended. Spot drenching with Carbendazim 1g /kg is also recommended (*T. viride* & fungicide should not be mixed).

DISEASE OF GREEN GRAM/ MUNG BEAN (*Vigna radiata*)

Yellow Mosaic

Yellow mosaic in green gram is caused by mungbean yellow mosaic virus (MYMV). Initially mild scattered yellow spots appear on young leaves. Spots gradually increase in size and ultimately some leaves turn completely yellow. Infected leaves also show necrotic symptoms. Diseased plants are stunted, mature late and produce very few flowers and pods. Pods of infected plants are reduced in size and turn yellow in colour. The disease is transmitted by the whitefly *Bemisia tabaci*.

Management

It is advisable to grow resistant varieties to yellow mosaic. It is also recommended to grow seven rows of sorghum as border crop. Seeds should be treated with Imidacloprid 70 WS @ 5ml/kg to control vector. MYMV infected plants should be rogued out early in the season to eliminate the source of inoculum. One foliar spray of systemic insecticide (Dimethoate @ 750 ml/ha) on 30 days after sowing is recommended.

■■■■

13 INSECT PEST MANAGEMENT IN TROPICAL AGROFORESTRY SYSTEMS

INTRODUCTION

Agroforestry is an intensive land management system that combines trees and shrubs with crops and livestock in time and space on a landscape level to achieve optimum benefits from biological interactions between soils, plants and arthropods. Agroforestry systems (AFS) aim at balancing ecosystem demands to sustain diversity and productivity, while meeting multiple-use and sustained yield needs of agriculture (Nair 1993; Sanchez 1995). Indigenous farmers in the developing world who usually understand land use interactions in their local ecosystems often apply the systems successfully. Examples include the multistoried coffee- and cacao-based agroforests in Latin America and the complex home gardens in Asia. Many of the benefits of AFS are derived from the increased diversity of these systems compared with corresponding monocultures of crops or trees. Although little research has been conducted on pest interactions within AFS, agroforestry has been assumed to reduce pest outbreaks usually associated with monocultures. Although the effects of various agroforestry designs on pest populations can be of a varied nature (microclimatic, nutritional, natural enemies, etc.), regulating factors do not act in isolation from each other. The few information on pest management in agroforestry (Rao et al 2000; Schroth et al 2000) stipulate that the high plant diversity associated with AFS provide some level of protection from pest and disease outbreaks. To explain such regulation, these scientists use the same theories advanced by agroecologists to explain lower pest levels in annual polycultural agroecosystems (Andow 1991; Altieri and Nicholls 2004). Some scientists caution that the use of high plant diversity as a strategy to reduce pest and disease risks in AFS meets considerable technical difficulties as the design and management of complex systems is cumbersome. Similar to orchard situations, AFS can be considered semi-permanent, relatively undisturbed systems, with no fallow or crop rotation, thus exhibiting particular biological situations affecting insects. Insect populations are more stable in complex AFS because a diverse and more permanent habitat can maintain an adequate population of the pest and its enemies at critical times (Van den Bosch and Telford 1964). For most Entomologists, the relative permanency of AFS affords the opportunity of manipulating the components of the habitat to the benefits of ecologically sound pest management practices (Prokopy 1994). Such practices include the manipulation of ground cover vegetation and of shade trees to either directly stress arthropod pests or enhance their mortality through biological control.

Among invertebrates, insects form the most dominant group of herbivores that attacks plants. A long history of co-evolution has led to the development of diverse types of associations between plants and herbivores. Trees are susceptible to insect damage just like crops, although the extent of damage depends on the stage at which they are infested and nature of the insect. They are attacked by leaf eaters in the nursery, termites and root grubs during the early establishment stage in the field, defoliators, sap suckers, gall formers, stem borers, pod or fruit borers and seed eaters perpetually during the later stages. Satya Vir and Verma (1996) reported a long list of insect pests on 17 leguminous tree species widely used in dryland agroforestry in India. A compilation of insects observed feeding on 20 agroforestry trees in sub-Saharan Africa by Singh Rathore (1995) indicates that some of the species such as eucalypts (*Eucalyptus* spp.) and pigeon pea (*Cajanus cajan*) harbour as many as 160 insect species in different parts of the tropics.

The magnitude of pest problems in agroforestry is determined by whether the herbivore interacts with (1) individual components of a system, (2) two or more components of a system, and (3) components across systems in the landscape. The herbivore interactions with multiple species over different spatial and temporal scales will have implications for pest management strategies to be followed. The species-specific pests have to be managed through host-plant resistance, biological control and/or strategic use of chemicals depending upon the economic importance of the damage.

FACTORS CONTRIBUTING TO INSECT PEST PROBLEMS IN AGROFORESTRY

EXOTIC SPECIES

Many newly introduced tree species are widely grown in tropical agroforestry. Some examples are the use of leucaena (*Leucaena leucocephala*) as barrier hedgerows for soil conservation, fodder banks and for improving soil fertility (Asia and Africa), grevillea (*Grevillea robusta*) as a shade tree in tea and coffee (east Africa and India), eucalypts (*Eucalyptus* spp.) in boundary plantings around croplands (throughout the tropics), and neem (*Azadirachta indica*) in windbreaks and boundary plantings (west Africa). Exotic species may succumb to local pests in their new environment or introduced pests from their original home, just as exotic pests accidentally introduced may pose a major threat to both indigenous and exotic trees. An exotic pest in new environment can assume epidemic proportion and cause severe damage to its host because of the absence of its natural enemies. The damage could be far greater than in the pest's native environment until populations of its natural enemies build up. That is what happened with the leucaena psyllid (*Heteropsylla cubana*) when it first spread to Southeast Asia and South Asia in the eighties and later to east and southern Africa in the nineties.

INTENSIVE USE OF SPECIES

Many agroforestry trees experience fewer pests and diseases compared with annual crops, probably because of their less susceptibility to pests, limited domestication, less intensive culture and existence of a balance between insect herbivores and their natural enemies. The inherent tolerance of a tree species to pests may be eroded in the course of its genetic improvement for faster growth, higher yield and/or better quality, as these characters often are negatively correlated with tolerance to pests.

PEST CONTROL, PEST MANAGEMENT AND INTEGRATED PEST MANAGEMENT

'Pest control' was the term commonly used in the past for the attempts to limit the damage caused by pests. We tried to kill the pest insects using chemical or other means. In spite of initial success, it was soon realized that it was not easy to kill off the insects; they reappeared when the effect of the insecticide waned or developed resistance to the chemicals. Humbled by the success of the pests, it was also realized that pests need to be controlled only if they cause economic damage. The term 'pest management' was therefore coined to indicate management of the pest population to limit it to a tolerable level. The emphasis was on regulating the population size, not killing all the pest insects, which was impracticable anyway. The concept of 'integrated pest management' (IPM) emerged in the 1970s. It envisaged the use of all the available techniques in an integrated manner to reduce the economic damage caused by pests, with the least ill effects on the environment. In the strict sense, IPM aims at regulating all the pest species including insects, pathogens and weeds in a crop production system but it is often understood in a limited sense. It marks a change in our attitude or philosophy, from supremacy over nature to acceptance of an ecologically compatible strategy to contain the pests.

Sustainable agroforestry systems should perform with fewer pest problems and minimum external inputs for protection against pests. Habitat management offers a low cost technique for small-scale farmers to minimize by creating conditions that hinder the pest populations but favour the build-up of natural enemies. Table 13.1 summarizes the broad range of effects of management factors on insect pests in both simultaneous and sequential systems. While these general effects of management may hold true in many situations, exceptions do occur with some systems and insects, which need to be determined. Removing the source of pest inoculums by producing healthy planting material and pruning the parts of trees infested by insects such as scales and mealy bugs, clean weeding at critical stages, and adequate mineral nutrition are some of the tree-related management practices that help reduce pests. Clean weeding of *Sesbania* fallows early in the season reduced *Mesoplatys* beetle populations, as the larvae succumbed to death in the course of migrating to weedy vegetation or litter cover for pupation (Sileshi et al 2000). Regularly pruned trees can be more vulnerable to insect attack as they are weakened over time and they also produce young coppice shoots that are more palatable to insects. Frequent pruning increased the stem borer (*Xyleutes capensis*) infestation of *Senna siamea* (Opondo-Mbai 1995).

Table 13.1: Management Variables and their Possible Range of Effects on Pest Infestations in Agroforestry Systems

Management Variable	**Conditions of Management Variable favouring**	
	High Pest Incidence	**Low Pest Incidence**
Simultaneous Systems		
Species diversity	Low diversity with closely related species	High diversity with widely differing species
Species number	Fewer	Many
Proportion of host to non-host species	High	Low
Spatial arrangement of trees and crops	Wide with limited interaction among species	Close with strong interaction among species
Row orientation	Along wind	Across wind
Tree density	Low density	High density
Sequential Systems		
Nature of tree fallows	Monospecies fallows	Multispecies fallows
Planting and harvesting of trees and/or crops in the landscape	Asynchronous	Synchronous
Residue management	Residues used as mulch	Residues burnt
Diversity during cropping phase	Limited diversity (e.g. monocropping)	High diversity (e.g. crop rotation and mixed cropping)
Common to both Systems		
Field size	Large fields	Small fields
Distribution of fields practising the same agroforestry system	Aggregated	Scattered

Integrated pest management in agroforestry should take into account some or all of the following strategies:

(1) Substituting susceptible tree species with alternative species that provide similar products and services
(2) Strengthening species tolerance or resistance to insects
(3) Creating conditions that reduce populations of insect pests but favour natural enemies
(4) Strategically using environmentally safe chemicals and botanical pesticides
(5) Exploiting biological control through introduction of effective natural enemies

SUBSTITUTION OF TREE SPECIES

In some instances, a susceptible tree species can be discarded for a quite different species that nevertheless performs the same function. A number of tree species can be considered in any ecoregion for some tree products such as fodder, poles and firewood, and services such as nutrient cycling, sheltering and soil conservation. Thus opportunities exist to substitute one unsuitable tree species because of pests and diseases with another free from maladies. This strategy probably works only if the alternative tree species gives similar yields and services without sacrificing quality. The scope for substitution of tree species will be limited where the trees are considered for specific products such as fruits, nuts, gums, and medicines. So, alternative strategies have to be employed to combat pests of tree species that produce such high value products.

CHEMICAL CONTROL

Chemical control becomes inevitable and justifiable where all other approaches fail, if the returns are economical and the chemicals used are environmentally safe and appropriate for the targeted pest.

HOST PLANT RESISTANCE

Host plant resistance (HPR) as a pest management tactic has particular appealing low input systems, as once the correct planting material is used, little or no further intervention is required by the grower. This is also the most appropriate strategy for agroforestry trees that produce low value products (e.g. fodder, firewood) and used for service functions (e.g. soil conservation, improving soil fertility). In some cases, different provenances may show tolerance rather than resistance by antibiosis, allowing good growth despite the presence of pests. For identifying pest resistant material in any tree species, germplasm screening should be done not only within that species but also across species in the same genus. This approach was used for *Heteropsylla cubana* through range wide collection and screening of leucaena germplasm, which lead to identifying many psyllid-resistant or tolerant *Leucaena leucocephala* provenances and accessions with good yield potential. Much work has been undertaken on resistance in eucalyptus to various pests, The other genus in which species have been compared for pest resistance is Sesbania. A general conclusion from the literature is that phenotypic assessment of resistance either within or between species should be repeated spatially and temporally to give reliable results. Ideally, heritability of the observed resistance should be determined, though there are relatively few studies of this nature.

BIOLOGICAL CONTROL

The classical biological control, or the introduction of a natural enemy from the pest's area of origin, has been implemented for some agroforestry insect pests, leucaena psyllid probably being the commonest target. Day et al (1995) listed 17 countries where the coccinellid *Curinus coeruleus* had either been introduced or arrived, seven countries for the coccinellid *Olla v-nigrum*, and seven for the encyrtid parasitoid *Psyllaephagus yaseeni*.

MANAGING VEGETATION DIVERSITY

The diversity inherent in agroforestry systems allows considerable opportunity for manipulations designed to reduce pests by directly restricting buildup of pest populations and indirectly by enhancing the effects of natural enemies. Whether increasing plant diversity in agroforestry helps reduce pests is difficult to answer. Diversity in agroforestry varies from a few trees in parkland systems of the Sahel to well over 100 species in Javanese homegardens.

MIXED CROPPING

Mixed cropping of host species with non-host or antagonistic species can help restrict infestations of some monophagous and sedentary insects. In contrast to the effect of pure *Sesbania sesban* fallows, mixed fallows of *Sesbania sesban* and *Crotalaria grahamiana* did not increase the soil populations of root-knot nematodes, and yields of nematode-susceptible beans following mixed fallows were not reduced. Many *Crotalaria* species are known to be antagonistic to *Meloidogyne* (Rodriguez-Kabana 1992). However, cropping systems may have little effect on the infestation and damage potential of species-specific insects that rapidly multiply, and that are mobile and carried by wind with limited flight control. For example, intercropping leucaena with maize, compared to sole leucaena, did not reduce the leucaena psyllid populations and damage caused by the insect (Ogol and Spence 1997). The population build up of such insects largely depends on weather conditions. Coconut planted in cacao can provide nest sites for *Dolichoderus* and *Oecophylla*, which reduce *Helopeltis* damage to cacao. Unfortunately, *Oecophylla* also bites people, so this approach has its drawbacks. Mwaiko (1998) increased *Oecophylla* populations and improved control of *Pseudotheraptus wayi* by using ant bait, as well as intercropping coconut with citrus, mango or guava. At Machakos, Kenya, it was observed that the ant *Pheidole megacephala* attacked the shoot scale *Coccus longulus* on pigeonpea and the mealy bug *Spilococcus* spp. on calliandra. The parasitoids *Metaphycus stanleyi*, *Eupelmus* spp., *Cheiloneurus carinatus* and *Tremblaya minor* on *Coccus longulus*; predators *Leucopsis* spp. and *Nephus* spp.; and the parasitoids *Anagrys nigrescens*, *C. carinatus*, *Psuedectroma* spp., *Aprostocetus* spp. and *Pachyneuron* spp. on *Spilococcus* spp. established only in the absence of the ants. Coccinellids are another important group of predators particularly for homopteran pests. When Xu and Wu (1989) intercropped bamboo with rape, aphids on the rape attracted coccinellids, which then moved to the bamboo and reduced scale populations by 97 to 99 per cent.

OVERVIEW OF PEST MANAGEMENT OPTIONS

The principles and methods of pest management are common to agricultural and forestry pests. A brief overview, with particular reference to forestry applications, is given below. Two approaches are available for pest management – prevention, where the buildup of pests is prevented by appropriate means and remedial action, where control measures are applied after the infestation has occurred. The success of preventive measures depends on our ability to identify the causes of pest build up. The preventive measure is removing the cause; the remedial measure is treating the symptom. When we cannot identify the cause, only remedial action is possible.

PREVENTIVE MEASURES

Preventive measures aim to keep pest populations at low densities and not allow them to develop into outbreaks. They rely on an understanding of the causes of pest build up. A large number of interacting factors are involved in determining the population size of a pest and it is often difficult to identify which factor is responsible for precipitating large-scale build up. Preventive measures are possible in some cases where the causes of population build up are known. These measures usually consist of silvicultural interventions aimed at tree health improvement in order to 'tune up' the tree's innate defense mechanisms. They are effective where pest build up is caused by poor tree health. Thus, preventing injury to trees by lopping can prevent infestation by the teak trunk borer, *Alcterogystia cadambae* and prompt removal of over mature trees and regular thinning of stands of *Shorea robusta*, as well as trapping and killing of moderately high populations of adult beetles, can prevent outbreak of the sal borer *Hoplocerambyx spinicornis*. Similarly, improvement of tree health and removal of dead and unhealthy trees in a pine stand can prevent pine bark beetle outbreak. Prompt removal of tree-felling refuse from a plantation site can prevent the build-up of pests like bark beetles which infest and breed on freshly felled trees and eventually attack healthy standing trees. In the case of teak defoliator outbreaks, destroying the early epicenter populations during the pre-monsoon period can prevent at least part of the subsequent large-scale outbreaks.

Silvicultural practices such as retention of plant species that support alternative hosts of pest insects or raising mixed-species plantations can also reduce pest build-up by enhancing natural enemy action. In the case of pests introduced from other countries, quarantine measures, where potential pests are intercepted at the ports of entry of commodities such as wood or planting material, is also a preventive measure. Use of pest-resistant trees can also be considered a preventive measure. Resistance refers to the genetic capability of trees to prevent, restrict or withstand pest infestation. There are not many instances of trees showing useful resistance to pests. When present, tree resistance to insects is usually polygenic. It may also be based on physical factors such as resin system characteristics. Conventional breeding for resistance is constrained by the long reproductive cycle of trees. For pests of stored timber, preventive measures include immersing the logs in water and debarking newly felled logs to prevent some groups of borers from laying eggs beneath the bark.

REMEDIAL MEASURES

Remedial measures aim to reduce the pest population level by killing the insects by one means or other. A large variety of remedial measures has been developed and tried against insects. Historically insecticides are the most common and effective means of killing insects with the use of chemical poisons. Insecticides are used either prophylactically or remedially. Prophylactic use involves application of the insecticide before the insects appear, as in the case of control of root-feeding termites of eucalypt saplings, where the insecticide is mixed with the soil to kill the termites that might attempt to penetrate to the tap root. Other examples are insecticidal treatment of nursery soil to control ants and whitegrubs or mixing of insecticide with seeds while in storage. In remedial application, insecticides are applied to the insects and the trees after the infestation is noticed. Although inorganic poisons such as lead arsenate, calcium arsenate and sulphur were used in the early days, organochlorines have been used extensively since World War II, when DDT became popular because of its effectiveness against mosquito vectors of malaria. Most major outbreaks of forest insects including defoliators were sprayed with DDT until it was withdrawn from the western market in 1973.

Organochlorine insecticides were also used in fairly large quantities to control bark beetle and termites. In the developing countries of the tropics, use of organochlorines was continued for a longer time. Other classes of insecticides that are less persistent in the environment such as organophosphates, carbamates, synthetic pyrethroids, chitin inhibitors, botanicals (like nicotine, rotenone, pyrethrum and neem products) and insect growth regulators have since been developed. Aerial application of insecticides has continued into the 1970s and 80s in many industrialized countries although over a much reduced area and with less persistent insecticides. The US Department of Agriculture guidelines for 1980 contained recommendations for use of the following insecticides against various forest pests: acephate, aldrin, azin-phosmethyl, carbaryl, carbophenothion, chlordane, chlorpyrifos, diazinon, dieldrin, diflubenzuron, dimethoate, disulphoton, ethyl dibromide, fenitrothion, methoxyclor, methyl bromide, sulfuryl fluoride and trichorfon. The turmoil created in the USA and the entire world with the publication of Rachel Carson's book in 1962 entitled 'Silent Spring' in which she vividly described the adverse impact of indiscriminate spraying of insecticides from the air over the vast stretches of forest is now part of history. Although bordering on poetic exaggeration at times, her criticism of the excessive use of pesticides, particularly in the forests, and the consequent disruption of ecological processes leading to aggravation of pest problems, resurgence of secondary pests and the accumulation of toxic residues in the human food chain hastened the development of IPM practices.

IPM involves the use of various methods such as biological control, habitat management, plant varieties resistant to pests, cultural practices and selective pesticides in a harmonious manner, as appropriate to each pest situation. It aims at reducing the pest population below economic injury level, and not at complete 'control'. The main advantages of pesticides are: (1) dramatic effectiveness by killing the insects in a short period of time, (2) broad spectrum of effectiveness and (3) commercial availability. The main disadvantages are: (1) unintended effect on non-target organisms, particularly parasitoids, predators and pollinators, (2) development of resistance by pests and (3) the temporary nature of the effect, necessitating repeated applications. Some problems like long persistence and bio-concentration in the human food chain have been overcome by the development of newer, more easily degradable pesticides. The drift of pesticides in the environment has also been reduced by improvements in application technology. Yet substantial portions of insecticides applied over the forest canopy find their way into other components of the ecosystem through drift, rain washing, leaching, etc. Its effect on non-target natural enemies is of serious concern in the forest environment where many potential pests are kept in check by their natural enemies. Application methods are still primitive in the developing countries of the tropics and entail large wastage as well as contamination of the environment. Experience in the industrialized countries has shown that unanticipated pest problems can arise as a result of widespread application of broad-spectrum insecticides as some potential pests are released from the influence of their natural enemies when these are destroyed by the insecticides.

NATURAL FOREST PEST CONTROL

CLIMATIC CONDITION

Climatic condition is expressed in terms of average physical conditions occurring throughout the year in a locality experienced over a long period of time. It also determines the length of the life cycle, successive stages and number of generation of insect species in a year. The more important climatic factors are meteorological factors, temperature, rainfall, atmospheric humidity, wind, sunshine, moon light, topographic features, mountains, plains and zoogeographical distribution. Each factor affects the insect activities in one way or the other, but in nature they seldom operate independently.

TEMPERATURE: It is the single most important physical factor which regulates the insect activity in the environment. Each species of insect has a definite range of temperature, within which the insect is active and its metabolism or growth is apparent. Any deviation from this point on either side retards metabolism and which in turn influences its activities. Near the upper and the lower limits of effective temperature zone lies the dormant zones in which the metabolism is greatly retarded and the insect becomes completely inactive and its movement ceases. The fatal temperature at which practically all the species of insects are killed instantly is 60°C.

ATMOSPHERIC HUMIDITY (*MOISTURE*): Like temperature, atmospheric humidity (moisture) is an important factor of environment, which profoundly influences the development, behaviour and distribution of insects. Each species has an optimum favourable zone and any deviation from this zone leads to a state of dormancy and beyond the effective range, the insect dies. In other words, "with increase in temperature, there is a corresponding decrease in relative humidity". Under normal conditions, both excess and deficient moisture conditions do not result in immediate death of the insect, though its activities are definitely disturbed. However, certain wood boring forest insects are profoundly influenced by variations in the moisture contents within the microclimatic conditions of its host plant. This is best explained by studying the ecological succession of wood borer fauna of a felled wood. While the coleopterous borers of families Buprestidae and most Cerambycidae require higher moisture conditions, the powder-post beetles families Bostrychidae and Lyctidae and Hymenopterans wood borers (*Xylocopa* spp., family: Xylocopidae) require dry and seasoned wood and cannot survive in moist wood.

HEAVY PRECIPITATION: Heavy precipitation (snowfall, hailstorm, rainfall or frost, etc.) may also sometimes cause direct mortality of different stages of forest insects. While the effect of snowfall is like that of low temperature, hailstorm and heavy rainfall act mechanically. Lepidopterous larvae, beetles and their larvae, which are open feeders, feeding exposed on the foliage, are beaten and washed off by sudden heavy hailstorm and rain or killed by frost. They are either drowned or injured, thus preventing their ascendance to the crown of the trees. Rainfall also increases atmospheric humidity which influences the emergence period of insects, such as soil insects and wood borers. For example, winged form of termites and ants, wood borer adults of many species of Cerambycidae and Curculionidae, start emerging with the onset of monsoon rains.

WIND: Wind is yet another physical factor and its discernible effects are both mechanical and eco-physiological in nature. Strong wind shakes the crown of trees violently, blowing off and dislocating the insects, especially the more delicate forms (larvae) from their food plant and subsequently prevent them from gaining access to the food source and shelter.

NUTRITION

Nutrition (food) is the second important natural control factor, which, unlike physical factors of environment, is amenable and can be controlled with certain degree of accuracy. It constitutes the primary driving force for seeking not only the right type of food for its existence and development but also to see that it is available in sufficient quantities. Thus food most frequently act as a controlling factor on insects and influence the structure and activities of the insect communities (for that matter all animal commu-nities). Thus, if food conditions are controlled in a way that even if the weather conditions are favourable, it checks the pest epidemics. Nutritional control may be considered under the following aspects like, food abundance, competition and quality of food.

Food abundance plays an important role in all the living organisms which are capable of multiplying to the extent of food available in an area. Forest monocultures offer ideal conditions for population build up of insect species feeding on it. The competition may be intra-specific or inter-specific, the outbreak of secondary species under certain conditions competes with the major species for feeding space. In intra-specific competition, all the individuals of the same species compete for all factors in the habitat i.e., space and food. In case of outbreaks, food may become the limiting factor for their survival. The quality of food has a definite effect on the abundance of forest insects and their development. The principal factors affecting the quality of food are moisture content, alternate food plants and host resistance.

HOST RESISTANCE

All plant species possess certain physiological and genetic traits which impart some degree of pest resistance by making environmental inherent conditions of less favourable for the pest species. The resistance may be due to inherent conditions of unpalatability, such as hardness of leaf lamina, shoot, sap density and presence of certain chemical compounds, such as essential oils, acids, gums, resins, latex, etc. There are three mechan-ism of host resistance to pest recognized viz. non-preference, antibiosis and tolerance.

BIOTIC FACTORS

Biotic control comprises all such factors which tend to control the population of an organism (pest) through the activities of other organisms. The biotic factors, which play an important role in limiting the pest populations, are competition, parasitoids, predators and pathogens.

ARTIFICIAL OR APPLIED CONTROL OF TREE PESTS

It is defined as the introduction of external elements, such as introduction of biocontrol agents, chemical control, etc. into the forest environment and manipulation in sylvan ecosystem (silvicultural control), which though temporarily, is under siege of a pest. Most common control procedures envisaged are monitoring, silvicultural control, mechanical control, cultural control, legislative control, chemical (insecticidal) control and integrated pest management.

SILVICULTURAL CONTROL

The regulation of insect pest population by modifying forest stands or methods of growing and harvesting trees to suppress insect pests is called silvicultural control. It is a type of artificial control as it is achieved by augmenting natural controlling factors of the environment by deliberate human intervention. The silviculture control requires detailed knowledge about the prevailing ecological conditions of a forest that are favourable to the pests and then manipulate these conditions to the disadvantage of insect pests. Before taking-up any silvicultural pest management measures, it is necessary to have an idea of forest type and their relationship with insects. Various silvicultural practices have been recommended and practised for effective control of insect pests.

HOST SPECIES AND STAND MANIPULATION

Regulation of forest composition by mixed stand is known to foresters since long that more diverse type of forest has least insect problems. Increased ecological stability results from an increased interaction between insects and other biotic and abiotic factors in the forests. In general, outbreaks of insects are less common in a forest with a diverse species than in a forest with less diverse flora or consist of single species (monoculture). Monoculture should be avoided as far as practicable especially in case of species which are not gregarious. Monoculture not only increases the availabilities of quality and quantity of food for the pests but also discourages natural enemy complex. In general, clear felling system is more dangerous than any other silvicultural system like shelter wood or selection system. A large clear-filled area is more susceptible to insect attack than other areas. From the silvicultural point of view, it will be beneficial to distribute the area to be regenerated annually by means of small separate coupes in several felling series with variation in their age classes.

DENSITY AND TREE GROWTH MANIPULATION

REGULATION OF DENSITY BY THINNING: Stand density may influence the physical conditions of a forest. Degree of density is also important in determining the ability of forest tree to grow vigorously. Stand density is regulated by thinning which depends on a number of factors like pest and tree species, site conditions, tree age, area and cost-benefit ratio. Thinning must incorporate removal of dead and dying trees coupled with removal of refuse resulting from thinning of overstock stand. Dying dead trees and refuse are most susceptible to insect attack like defoliators, sap suckers and bark beetles.

Badly attacked plantations of *Toona ciliata* and *Bombax ceiba* have been recovered by regular thinning in young stands after the attack by shoot borer *Hypsipyla robusta* and *Tonica niviferana,* respectively (Beeson 1941). An early thinning of *Dalbergia sissoo* produces a more favourable ratio between the pest population and the quantity of foliage per tree. Attack of Cyrtotrachelus *dux* weevil on bamboo can be reduced by thinning the dense area (Singh and Bhandari 1988) because split clumps provide shelter to beetle in hot dry season. Thinning of average dense sal forest is essential to reduce the effect of Hoplocerambyx *spinicornis* beetles.

PURE STAND: It seems that some trees reach a certain "entomological rotation" age at which they are more a *pure stand.* The monoculture plantation is more susceptible to insect attack because of even aged structure, food abundance and lack of diverse enemy complex. All these congenial conditions will provide optimum environment for multiplication of the pest species. The annual epidemics of teak defoliator (Hyblaea puera) and teak leaf skeletonizer (Eutectona machaeralis) in pure stand of teak are the most common examples of adverse effect of monoculture plantations in India (Nair et al 1985). Poplar defoliators *Pygaera cupreata* and P. fulgurita and stem borer Apriona *cinerea* (Chatterjee and Singh 1968) provide other examples of insect epidemics in monoculture plantations. Many factors like water supply, soil, soil cover etc., affect the severity of economic damage in pure stands. Water supply affects the sprouting of *Dalbergia sissoo* and therefore, this knowledge can be manipulated to the disadvantage of defoliator *Plecoptera reflexa* by adjusting irrigation dates.

Monoculture of *Ailanthus* plantation suffers from two notorious pests namely *Atteva fabriciella* and *Eligma narcissus* (Varma 1986; Roonwal 1990). Termite species *Odontotermes obesus, O. distans* and *Microtermes unicolor* are major pests of *Eucalyptus* plantations in India (Thakur et al 1989). An even-aged monoculture plantation can be managed by developing a mosaic of age classes in small blocks. This also helps to reduce easy access of the pest to the food. A mixture of tree species can be achieved by cutting in groups or strips or by creating gap to enhance light penetration so that a light demander species could be established. Similarly, mixed age stands can be produced by cutting adjacent areas at intervals of about 15 meters.

MIXED STAND: In perspective of pest management, mixed stand is an association of several species of plants in which auxiliary flora assumes importance in proportion to their effect on the abundance of insect pests of timber trees. Generally, there are good and bad mixer. In a mixed stand, generally oligophagous or poliphagous insect species predominate over monophagous species and a polyphagous species rarely become a pest unless abnormal conditions in the environment exist. A mixed stand protects the main timber species from the attack of monophagous insects by reducing food supply and encouraging growth and survival of predators and parasites by providing them appropriate niches including food nectar of the adult parasites.

Different tree species of different age in a mixed forest often differ in their susceptibility to insect attack. When any insect species in a mixed stand exhibit a strong preference for a certain host species, selective elimination of that species and encouragement of a variety of non-host species may reduce the potential for injury to the primary host species.

ALTERNATE HOST AND GROUND COVER

Population of pests can be regulated to some extent by manipulation of weak links in their life cycle. One of those is the need for alternate host. A horizontal mixture of the main tree species and other over wood species of the same crown size has salubrious effect. Mixture of two different tree species at width of 15 to 30 meter can hardly be able to check the inter-exchange of winged insects and subsequently egg laying. An advantage may be obtained by such mixture to check the spread of larval stages that may spread from crown to crown. A more resistant mixture of block against insect attack can be created by choosing most suitable species to plant on a suitable soil in patches than the mixture arranged in lines or furrows. The abundance of certain insect species is often influenced by the occurrence and abundance of alternate hosts. Alternate hosts form an integrated part of the life cycle of various best species. The presence of alternate hosts in a plantation increases the chances of an infestation. For example, incidence of attack of collar weevil, *Pagiophleous longiclaris* in plantation of mahagony (*Swietenia macrophylla*), is primarily determined by the incidence of attack on *Toona ciliata* (Stebbing 1914). Teak canker grub *Dihammus cervinus* depends on the presence of *Clerodendron* and gamhar (*Gmelina arborea*) (Beeson 1941) and babool borer *Celosterna scabrator* breeds in *Acacia* species, *Eucalyptus* spp. and *Casuarina equistifolia* (Beeson 1941; Sen-Sarma and Thakur 1985; Chatterjee and Singh 1968).

FIRE AND LOGGING

There are number of other interesting examples of heavy infestation where forest composition has been changed by logging, fire or due to other causes. For example, heaviest defoliator attack was seen in Nilambur teak plantation when undergrowth in 2,600 acre was burnt (Beeson 1941). Ground fire was used to clear an area from refuse for regeneration purpose but it caused extensive damage to parasites and predators along with the pests. Soil inhabiting insects like termites, white-grubs and wood borer largely escape unhurt from ground fire. Ground fire is not advocated to control insect epidemic. The prevention of fire in a forest provides protection to natural enemies of forest pests.

INCREASING HOST VIGOUR

Host vigour can be improved by reducing the number of trees from the dense stands where competition has subsequently slowed down the growth of even the dominant trees. Reduction in basal area of an overstocked stand either by thinning or partial branch cutting will enhance the tree growth and vigour which in turn will lower the incidence of an insect outbreak. Vigorously growing trees are not liable to insect injury in comparison to weak trees. For example, vigorously growing trees of conifer forests are able to withstand bark beetle attack, while beetles and borers establish themselves in weakened trees. The successive insect attack occurs in forests as these developed from seedlings to pole and pole to mature trees. Some insects may thrive only on a certain stage of a tree. If a vigorous tree is attacked, it may soon recover from the insect attack. Tree vigour can be estimated by a variety of quantitative and qualitative methods. In general, when vigour is reduced, trees become stressed and supply low quantity of carbohydrate from their reserve for defensive functions. Thus, nutritional factors are greatly responsible for regulating insect activities.

SILVICULTURAL MANAGEMENT FACTORS

Over all, chief sound silvicultural management factors which ensure healthy and vigorous growth of plantation species are (i) site selection (ii) choice of forest tree species (iii) conditions in nursery (iv) transplanting of nursery stock (v) eradication of collateral hosts and (vi) encouragement of desirable and beneficial plant species. However, in practice, such ideal silvicultural conditions are seldom available to forest managers, who are often called upon to carry our plantation schemes on unsuitable sites. Most of these short comings can be partially mitigated through soil improvement practices and by using cover or nurse crops. Many of these practices are considered as cultural practices.

SITE SELECTION: Growing of a tree species at a suitable site is one of the most fundamental methods to reduce pest hazards through silvicultural practices. In fact, often this important factor is overlooked by foresters which results in poor development, reduced vigour and increased susceptibility of trees to insect attack. The most suitable example of this type in the history of Indian forest entomology may be the failure of *Acacia nilotica* plantation in Berar area of Madhya Pradesh (Beeson 1941). Unsuitable soil and deficient water supply were responsible for creating permanent breeding centers for the multiplication and spread of shoot and root borer *Celosterna scabrator.* The only practical remedy is to clear the infested areas and replant the area with a species which is not attacked by this pest. Therefore, choice of suitable site for planting is also an important consideration in silvicultural control. Under suitable site conditions, trees will be able to tolerate greater tree injury than growing under less favourable conditions.

CHOICE OF RIGHT SPECIES AT RIGHT PLACE: Sometimes, a site is poor because of acidic soil which inhibits tree growth. Such type of soil can be reclaimed by proper water drainage. Poor or problematic soils can also be amended with the help of fertilizer and gypsum to make suitable for tree planting with right species in a given area. Inadequate or poor forest site conditions have frequently been associated with increased injury and pest epidemics. For example, planting of *Dalbergia sissoo* should be restricted to suitable and well irrigated site and for planting of teak it should be insured that teak is not planted on the site previously occupied by *Clerodendron infortunatum* or adjacent to *Gmelina arborea* plantation as both the species act as collateral host for canker grub *Dihammus cervinus.*

BETTER SITE CHARACTERS: Establishment of new plantations, nurseries or reforesting the area after clear felling often cause precarious situation. In forests, the establishment of new plantings to create stand diversification often requires specific preparation of sites to ensure tree establishment and survival. Researches on bark beetles in European countries have shown clear correlations between site characteristics of poor quality soils and low moisture and outbreaks of pest insects.

MONITORING: When formulating any management strategies in forestry, monito-ring is an important step. Monitoring can take on many forms: scouting and sampling, pheromone trapping, tree baits and aerial surveillance. Sometimes, combinations of methods are used such as aerial surveillance to detect dead trees and ground checking to determine the cause of mortality. In some cases, more than one life stages of a pest is required before control measures are taken. For example, pheromone trapping may detect a large concentration of adult male moths in a given location. This is followed by a ground survey for the egg stage to verify that mating and reproduction has taken place. Regular observations, good record keeping and a thorough knowledge of the pest's life cycle are the foundation for an IPM program. Insect pheromones are chemicals that members of a species use to communicate with each other. Females of many species produce sex pheromones that attract males for mating. Traps baited with pheromones lures can be used to survey for invasive insects, or in some cases, to control limited infestations.

The chief advantage of silviculture control is cheaper as compared to direct control (chemical control) measures and this procedure is more in tune with the ecosystem, environmentally friendly, with inherent capacity to regulate the pest populations on long term basis. It aims at developing preventive control measures and encouraging forces of natural enemies of the pests. But the disadvantages of silvicultural control includes, a long term control measure and therefore, proves failure during emergencies, particularly when the pest has assumed epidemic status and its success depends on in-depth knowledge of the pest and its host. Unfortunately, precise information in most cases is lacking and unsound and ill planned silvicultural procedures may even prove counterproductive, with undesirable side effects.

MECHANICAL AND PHYSICAL CONTROL

Though non-conventional and highly specified, the mechanical and physical methods of control play an important role in pest management. These methods aim in reducing the pest population by employing certain mechanical devices or manual operations which ensure killing of insects by trapping or mechanical means. It act as barriers and prevent insects or its life stages from gaining access to host plants (mechanical control) or by manipulating certain physical conditions of environment more particularly the temperature and humidity, in a way that they prove destructive to the pest species (physical control) and thus preventing or minimizing the insect attack. These measures also involve the use of radiant energy in various ways, which could destroy the insect pests. In broader sense, the physical control measures comprise one or more of the following measures, viz. heat treatment, refrigeration, radiant treatment, trapping, sound percussion devices and self destruction or autocidal treatment.

Many insect pests of forest tree species are amenable to control by mechanical and physical methods, however, these methods rarely provide commercial control. These methods hardly differ from cultural methods except that they involve the use of special equipment or operation and are employed in situations where insects are to be eliminated from an area without any deleterious effect on the environment.

Hand picking of noxious insects or their life stages is the simple perhaps the most ancient effective method of insect control practised by man and which continues to be one of the most profitable methods for some insect pests (e.g. white-grubs and crickets in seed beds; beetles and larvae of some pests in forest nurseries). Use of hand nets or dragnets for catching small and noxious insects is an extension of hand-picking procedures (e.g. beetle defoliators, grasshoppers, plant hoppers, etc.). Similarly, wire probes or other flexible materials are also employed for poking tunnels of some wood borer species (e.g. *Batocera rufomaculata, Tonica niviferana, Zeuzera* spp., etc.). Jarring of branches of trees is also one of the mechanical methods employed in shaking off adult or larval population of species from the host plants. The purpose of control should be to maintain a balance between pests and hosts at such a level that economically important injury will not occur. Economic consideration is one of the most important factors to control a pest in a forest. In case of direct pests, even moderate levels of pest abundance cannot usually be tolerated. The appropriate level of control is determined by economic value of the forest stand.

Over-all, mechanical control in forestry generally means salvage or the removal of infested trees to prevent the spread of a pest. If the cut trees can be utilized, the control operations may be cost effective. In some instances, the infested trees may be too few or too remote for a cost effective operation. Cut-and-leave or cut-and-treat are options that are frequently employed for bark beetle control. The infested trees, along with a buffer, are simply cut down and left or treated by debarking, applying a pesticide, or burning. It is assumed that any surviving beetles will disperse and not create additional infestations. On high value trees, mechanical control could include simple hand picking, erecting insect or pathogen barriers or sticky traps.

SEMIOCHEMICALS (BEHAVIOUR-INDUCING CHEMICALS)

Chemical communication plays an important role in the life of an insect as the insect depends on it for host finding and a number of other interactions with the biotic environment, including the insect's own population. Semiochemicals are natural compou- nds produced and released by individuals of a species which elicit a behavioural response in members of the same or different species (Nordlund 1981). Semiochemicals which are used in intraspecific communication are referred to as pheromones or chemical substa-nces emitted by an individual to induce behavioural responses in other individuals of its own species are called pheromones. Behavioural responses to pheromones include sear-ching for mates by one sex (e.g., sex pheromones), aggregation of both sexes at a host plant (e.g., attractant or aggregation pheromones), and dispersal of both sexes away from a specific area (e.g., inhibitor or antiaggregation pheromones). Substances which cause a behavioural response in individuals of another species are called kairomones when they benefit the receiving individuals (e.g. attraction of parasitoids) or allomones when they benefit the emitting individuals (e.g. a repellent which keeps away a natural enemy).

There is considerable overlap with regards to the functions of a compound, i.e., the same compound may serve both intraspecific and interspecific functions. A wide variety of such chemicals is produced by insects for various purposes and attempts have been made to use them for management of pests. Thus sex pheromones have been used for trapping insects and for disruption of mating. Sex pheromones of several lepidopteran forest pests have been identified, synthesized and used for population management. Aggregation pheromones have also been used successfully for trapping bark beetles. Forest insects for which semiochemicals have been tested or used operationally for population suppression in developed countries of the West include the western pine-shoot borer (*Eucosma sonomana*), the gypsy moth (*Lymantria dispar*) and several species of bark beetles .

BIOLOGICAL CONTROL

Biological control is also referred as bioremediation and is defined as ingenious and resourceful employment of certain promising biotic elements (predators, parasitoids and pathogens) for suppressing pest species. Biological control refers to using natural insect enemies to control pest populations. Enemies can include vertebrate predators such as birds or mice, arthropod parasitoids such as the tachinid fly, arthropod predators such as spiders and mites. The phenomenon involves certain degree of deliberate human interference, which, unlike biotic natural control, is manipulated by man for regulating pest populations in an ecosystem. Simply, biological control is the reliance of natural enemies to keep a pest in check. Three principles or approaches are involved in biological control viz., importation, conservation and augmentation.

Normally importation is utilized against exotic pests where natural enemies from the pest's home land are introduced into the infested region. It is assumed that once the natural enemy is established it will be self-maintained. Conservation includes the identification and remediation of negative influences that suppress natural enemies already in the system, and the enhancement of habitat for natural enemies. Providing alternative food sources for natural enemies, such as pollen and nectar by encouraging selected flowering plants is the typical exmaple. Using more selective pesticides that have a minimum impact on natural enemies is another approach.

Augmentation is the release of enemy populations, which were artificially reared, for pest suppression. It is used when natural enemies are missing, too scarce, or slow to arrive. There are two strategies for augmentation releases. The first is "inoculative" where the released organisms and the progeny of the released organisms accomplish control of the pest. The second is "inundative", where the released organisms only accomplish control of the pest. Forestry has a great potential for utilizing biological control strategies because of the perennial nature of the system and the potential for alternative food sources. The fact that pesticides are generally not cost effective is another consideration.

PREDATORS AND PARASITOIDS

All insects have natural enemies. These may be vertebrate predators (birds, bats, reptiles etc.), insect predators, insect parasitoids, etc. They play an important role in the natural regulation of insect numbers and have been employed for artificial suppression of pest populations. The relative freedom of mixed tropical forests from pest outbreaks is generally attributed to the 'checks and balances' exerted by natural enemies in the complex natural community (Fig. 13.1). This inference rests essentially on circumstantial evidence and it is difficult to obtain direct proof. While the qualitative relationship between the insect pests and their various natural enemies has often been fully elucidated, the quantitative effects remain largely unknown. Based on theoretical considerations it is assumed, however, that in vegetation communities like forests and agroforests, natural enemies do play a significant role in preventing the population of pests from attaining damaging levels, that is, natural biological control. We may recognize its value only when we disrupt it, just as we seldom recognize the value of good health until we lose it.

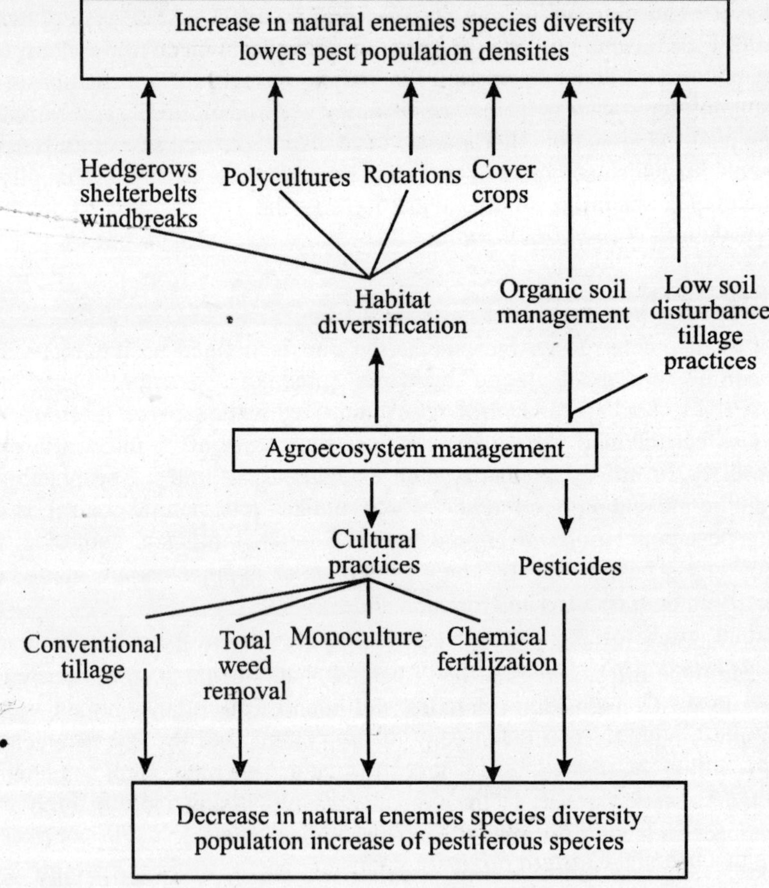

Fig. 13.1: Assortment of Agricultural Practices that enhance Beneficial Biodiversity in Agroforestry (Source: Altieri and Nicholls 2008)

The increased pest problems experienced in plantations is generally attributed to the disruption of natural enemy action. The attainment of pest status by some species when accidentally introduced into new geographical regions devoid of their natural enemies, and their suppression on introduction of the native natural enemies (classical biological control), is taken as proof of the effectiveness of natural enemies. When natural enemies are managed to control a pest, we call it biological control; in the strict sense it is applied or artificial biological control. The literal meaning of biological control can be extended to include any technique of human intervention employing biological means. The use of naturally occurring genetically resistant trees, transgenic trees or even spray application of commercially formulated bacterial or baculovirus preparations are all methods which make use of biological means of intervention. However, as commonly used, biological control means use of artificially introduced or augmented natural enemies, usually insect predators and parasitoids, for suppression of pest populations.

The effect of a resident population of parasitoids on millions of larvae that build up suddenly from immigrant moths will be insignificant. Some parasitoids do indeed exert some control over local populations, under certain conditions, but population outbreaks appear to be triggered by the plentiful food supply during the flushing period as well as the monsoon wind system which aids the immigration of moths. Natural enemies become unimportant under such circumstances. Unfortunately, the well-entrenched concept of 'balance of nature' and the successful examples of applied biological control have overemphasized the importance of parasitoids as regulators of pest populations. While they do regulate population outbreaks in some cases and under some circumstances, we must recognize that biological control is not a panacea. It also emphasizes the need to field-test the recommendations before advocating them, to safeguard the entomologists' credibility.

MICROBIAL CONTROL

Microbial control is another important component of biological control. Apart from parasitoids and predators, certain entomogenous microbial pathogens, comprising chiefly the bacteria, fungi and viruses are the well known biocontrol agents. These microorga-nisms, as the name suggests, constitute a specialized group of biocontrol agents, having extremely smaller size and eco physiological characteristics, which has imparted these organisms a distinct identity, quite different from insect parasitoids and predators. Nevertheless, the principles involved for profitable utilization (inoculum, augumentation and conservation) more or less remain the same.

BACTERIAL PATHOGENS

Among the bacterial pathogens, particularly the sporulating (sporeforming) bacterial diseases constitute an important group of microorganisms as they are highly resistant to drying and high temperatures. *Bacillus thuringiensis*, commercially named as Thuricide and abbreviated as *Bt*, is the best known bacterial pesticide with worldwide recognition. It has emerged as the most promising bacterium for control of lepidopteran and some coleopteran pests. Though broad spectrum, this biocontrol agent is selective and ecologically safe in most situations. In India, it has been successfully tested against over two dozen insect pests of forestry importance, achieving high mortality rate ranging from 30 per cent (*Lymantria mathura, Dioryctria abietella*) to 90 per cent (in *Ascotis selenaria,* Eutectona machaeralis, *Clostera fulgurita, Plecoptera reflexa* and *Selepa celtis*).

The related *B. sphaericus* is pathogenic to mosquito larvae and *B. popilliae* to scarabaeid beetles. Different strains of *B. thuringiensis* have been isolated with different levels of pathogenicity to various insects. *Bacillus thuringiensis* var. *kurstaki* has been found the most pathogenic to lepidopteran larvae. Bt formulations fall under the category of a bioinsecticide rather than a biological control agent. Commercial formulations of Bt have been marketed under a variety of trade names – Delfin, Dipel, Biolep, Bioasp, Biobit, Lepidocide, Thuricide, etc.

ENTOMOGENOUS FUNGI

The role of entomogenous fungi as effective biocontrol agent has been recognized throughout world. In India, a few fungal pathogens have been recorded infesting and killing larvae and pupae of some important forest insect pests. Two species of fungi have shown potential for applied biological control of tropical forest insects. These are *Beauveria bassiana* (white muscardine fungus) and *Metarhizium anisopliae* (green muscardine fungus) of the class Deuteromycetes ('imperfect fungi'). *Beauveria bassiana,* a well known insect pathogen has been found infecting and killing larvae of *Atteva fabriciella, Eligma narcissus, Hypsipyla robusta, Philosamia cynthia* and Scarabaeid beetles, causing large scale larval mortality.

Metarhizium anisopliae has a wide host range, covering species of Coleoptera, Lepidoptera, Diptera, Orthoptera, Hemiptera and Hymenoptera. In general, successful infestation by entomopathogenic fungi requires high atmospheric humidity, perhaps for spore germination, viability and sporulation after the host is dead. Because of this limitation, successful field control has been achieved only under some circumstances. Their potential needs to be further explored and conditions for successful use standardized. There is also scope for isolating more virulent strains. *Beauveria*, which can be mass-produced on artificial nutrient media, has shown potential for control of soil insects like whitegrubs in forest nurseries in China and India (Speight and Wylie 2001). *Metarhizium* has been found effective against the pine shoot-boring moth *Rhyacionia frustrana* in Cuba. It may also have potential against root-feeding termites.

VIRUSES

Of the numerous viral pathogens, which are known to cause diseases in insects, only those (viruses) belonging to families Baculoviridae and Reoviridae reportedly hold promise as biocontrol agents against insect pests. In India, nuclear polyhedrosis viruses (NPV), isolated from the diseased larvae of poplar defoliator *Clostera fulgurita,* have been successfully tested in the field and found very effective in controlling the pest population. More recently another NPV isolated from infected teak defoliator *Hyblaea puera*, has been tested in suppressing the defoliators population. Elsewhere, NPV of *Helicoverpa, Baculovirus heliothis, Pieris brassicae* and *Pieris rapae* are successful microbial pesticides.

CHEMICAL CONTROL

A key part of pest management is to use a pesticide only when it is needed to prevent an unacceptable amount of damage. Use of pesticides may not be justified if the cost of control or potential harm to the environment is greater than the estimated damage or loss. Pesticides are a very important tool, when large pest population threatens high value trees. Knowledge of the pest's life cycle, selection of an appropriate pesticide, proper timing of the application and use of the right application equipment will improve coverage and effectiveness. A pesticide application may be needed to control a pest outbreak or to eradicate limited infestations of an invasive species. Pesticides should be selected and used in a manner that will cause the least harm to non-target organisms in forests, seed orchards and nurseries, while still achieving the desired management goal.

Pesticides are labeled for specific pests, crops and land-use situations. Use of insecticides, fungicides and herbicides is common in managed seed orchards, forest nurseries, intensive short-rotation plantations and in Christmas tree production. In general, the most commonly used forest pesticides are herbicides used for site preparation, herbaceous weed control, etc. Insecticide applications are seldom used in general forest management because of high treatment costs and potential effects on non-target organisms. Situations justifying the widespread use of fungicides also are rare. In some cases, vertebrate animals must be controlled through trapping or hunting and similarly use of repellents and poison baits may be employed.

INSECTICIDES FOR FOREST PEST CONTROL

INSECTICIDES: Insecticide applications are rarely practical against forest insect pests. However, they may be useful in specific situations (For e.g., Chlorpyriphos 20EC against borers and termites in nursery, Methyl parathion 50EC against defoliators, Dimethoate 30EC, Imidachloprid and

Thiometaxm 25G against sucking pests) such as limited infestations of an invasive species. Insecticidal soaps, horticultural oils, Bt-insecticides, and systemics are common choices because they have very low potential to harm the environment, non-target species and applicators.

INSECTICIDAL SOAPS: They are made from salts of fatty acids. When sprayed directly on vulnerable stages of soft bodied insects, such as aphids and adelgids, they kill by damaging individual cells. However, they are not very effective against stages of the life cycle that are inactive or hidden, or against larger insects such as caterpillars and beetles. Insecticidal soap sprays must come into direct contact with the target pest and often results are best against specific life stages. Timing and thorough spray coverage is essential for best results. If the vulnerable stage is active over a long period of time, several applications are needed to control most insects. Insecticidal soaps should not be applied directly to water or near a water source. Hard water is not effective for mixing soap sprays, so softened or distilled water is used for best results. Tender young growth of evergreens and shrubs in the spring can be sensitive to insecticidal soaps.

Bt INSECTICIDES: Bt Insecticides (*Bacillus thuringiensis*) are protein toxins that are produced by a common soil bacterium. Many provide specific control of caterpillars without affecting other types of insects (beetles, sawflies, etc.). Bt insecticides disrupt the gut wall cells in the caterpillar digestive tract so a formulation of the insecticide must be sprayed on foliage that will be eaten by the caterpillars. Caterpillars stop feeding soon after eating the insecticide but usually do not die for several days. Bt insecticides work best against young caterpillars that are less than half-grown. These insecticides are relatively non-toxic to mammals and other animals.

SYSTEMIC INSECTICIDES: Imidacloprid is the common name of a systemic insecticide that is used to control sap-feeding insects, such as the hemlock woolly adelgid. It is a nerve poison but it is much more toxic to insects than to warm-blooded animals. Imidacloprid can be diluted in water and applied as a drench poured around the base of a tree, or injected into the soil. The insecticide is taken up by the roots and moved throughout the tree.

Before choosing a control method(s)/IPM, one should consider following points:

1. Correctly identify the organism/pest.
2. Assess the infestation and determine the potential economic damage.
3. Determine the available control methods.
4. Evaluate the benefits and risks of each method or combination of methods.
5. Are there threatened or endangered species or sensitive sites in the area to be treated?
6. Choose effective method(s) that will be least harm to humans and the environment.
7. Follow applicable local, state and federal regulations.
8. Correctly carry out the control practices and keep accurate records so results can be evaluated.

CONSTRAINTS TO FOREST PEST MANAGEMENT IN THE TROPICS

While the unique features described above are common to all forest pest management problems, whether in tropical or temperate countries, there are severe social, economic and policy constraints in the tropics (Nair 1986, 1991 and 2000). Most of these are the direct or indirect effect of the poor economic development of the tropical countries, whether in Asia, Africa or Latin America. The important constraints are the following:

- Small number of forest entomologists compared to the large number of pest problems.

- Lack of adequate training of entomologists in applied research and in the principles and techniques of pest management.
- General absence of demand from practising forest managers, largely due to lack of information on the economic impact of damage.
- Lack of adequate organizational and infrastructural facilities for entomological field work.
- Inapplicability of sophisticated pest management methods employed in developed countries, such as ultra-low volume aerial application of insecticides, aerial release of biocontrol agents, computer-based pest prediction and alert systems, due to technological and economic constraints.
- Centralization of research effort in a few government-controlled research centres which imposes physical limitations to field-oriented research and hinders the scope for diversification of research approaches.
- Poor research management and lack of incentives to researchers for carrying out applied research.
- In most tropical countries, forests are predominantly government owned, and forest pest management has low priority relative to the more pressing agricultural pest problems.

NEED FOR FURTHER RESEARCH

Knowledge of the complex interactions among trees, crops, and their associated fauna is necessary to determine the viability of a particular agroforestry practice. The effects of agroforestry designs and technologies on pests and diseases can be divided into biological (species-related) and physical effects of components (e.g., microclimate). The former is highly specific for certain plant–pest or plant–disease combinations and have to be studied on a case-by-case basis. The latter is easier to generalize, but even they depend on the regional climatic conditions. On the basis of results from intercropping studies, agroforesters expect that AFS may provide opportunities to noticeably increase arthropod diversity and lower pest populations compared with the polyculture of annual crops or trees by themselves (Schroth et al 2000). However, more work is needed in specific areas of research such as studies of the differences in arthropod populations between agroforestry and traditional agronomic systems, research into the specific mechanisms behind enhancement of pest management with agroforestry practices, and basic research into the life histories of target pests and potential natural enemies.

An understanding of what aspects of trees modify pest populations - shelter, food, or host resources for natural enemies, temporal continuity, microclimate alteration, etc. should help in determining future agroforestry design practices (Rao et al 2000). Well-designed agroforestry techniques can reduce crop stress by providing the right amount of shade, reducing temperature extremes, sheltering off strong winds, and improving soil fertility, thereby improving the tolerance of crops against pest and disease damage, while influencing the developmental conditions for pest and disease organisms and their natural enemies. Poorly designed systems, on the other hand, may increase the susceptibility of crops to pests. It is important to realize that the majority (75 per cent) of agroforests are located in developing countries managed by traditional farmers who cultivate a few hectares of land. They rely on low-energy, labor-intensive production methods and few agrochemicals (Altieri 1995). These resource-poor farmers have practised agroforestry for centuries: they used trees for fences and pest control, as well as for food, fodder, construction materials, and fuel (Altieri and Farrell 1984; Greathead 1988). These small farmers cannot afford high-input technologies or expensive agroforestry designs. The key challenge is to maintain a highly diverse farm with woodlands, forests and herbaceous edges, as in this way, allow traditional agroforesters to regulate pest populations by providing food and habitat for birds, spiders, parasites, and other natural enemies of pests. Although small farmers may lack the research tools used by scientists in industrial countries, traditional agroforesters do have valuable knowledge to contribute toward the design of sustainable AFS. They have developed practical systems for identifying damaging stages of pests, understanding their biologies, and applying management techniques to suppress their populations. This knowledge can be tapped through participatory research schemes whereby farmers and researchers engage in a true collaborative partnership.

■■■■

14 TREE CROP INTERACTIONS IN AGROFORESTRY

INTRODUCTION

Agroforestry is a collective name for land use systems and technologies where woody perennials (trees, shrubs, palms, bamboos, etc.) are deliberately used on the same land management units as agricultural crops and/or animals, in some form of spatial arrangement or temporal sequence (Lundgren and Raintree 1982). Tree-Crop Interface (TCI) is the space occupied by the crops and trees in an agroforestry system in which the interaction between the trees and crops takes place. The interactions refer to the influence of one component of a system on the performance of the other components as well as the system as a whole. When interaction between trees and animals occurs in an agroforesry system, it is referred as Tree-Animal Interaction or Tree-Animal Interface (TAI). In agroforestry, trees and agricultural crops are combined together and they compete with each other for growth resources such as light, water and nutrients. Tree-crop interaction is simply how the components interact and affect each other in a given environment. It occurs when more than one individual is present in a biological system and when resources are limited. The resources shared by the components may result in complementary or competitive effects depending upon the nature of the species involved in the systems, the manner in which they are grown and the climatic factors. Tree-crop interaction is an important aspect in agroforestry. The positive and negative effects of tree in agroforestry systems have been realized by field observations. Many of the quoted examples of soil fertility improvement are based on traditional systems that have been established for many years. Shelterbelts and windbreaks are well established examples of microclimatic improvement that helps to improve the crop yield. Several other benefits of trees have been realized for soil conservation, environmental protection, nutrient cycling, weed control, sustainability and these benefits may be direct or indirect.

The competition between trees and crops for the same limiting growth factors is more obvious when they are grown in close proximity. However, the extent of below ground competition is often not apparent. In agroforestry system, competition between tree and crop is usually a significant factor, which has a negative impact on extension of agroforestry. Quantification of competitive effect for soil moisture, light and nutrients in tree-crop system can play an important role in designing of an agroforestry system and can suggest the management practices to maximize the complementarity and minimize the competition.

KINDS OF TREE-CROP INTERACTIONS

Historically different groups of scientists have described the interactions differently. Agronomists and of late agroforestry researchers have used the terms below ground and above ground as adjectives to describe interactions between the components for growth factors observed through roots and those observed through leaves (Singh et al 1989; Ong et al 1991). ICRAF researchers have developed an equation for assessing the viability of agroforestry system. This equation is called tree crop interaction (I = F - C) where tree crop interaction (I) is determined by considering positive effects of tree and crop yield through soil fertility enrichment (F) and negative effects through competition (C) for growth resources between tree and crop.

In the equation, **I = F - C**

if F > C , The interaction is positive; The agroforestry system is productive and beneficial

if F < C , The interaction is negative; The agroforestry system is unproductive and harmful

if F = C , The interaction is neutral; The agroforestry system is neutral

The interaction takes place both above and below ground among tree, crops and animals. It is usually through light, water changes, and nutrient changes, shade effect and microclimate modifications. Component Interactions refers to the influence of one component of a system on the performance of other component as well as the system as a whole. Symbiosis is a relationship in which the two species live in close physiological contact with each other. In ecological view and based on the net effects of interactions, the following are the types of interactions:

- *Commensalistic*: Positive "+" effect on species one and no observable effect, "0" on species two (Symbolically + and 0)
- *Amensalistic*: Negative "-" effect on species one and no observable effect, "0" on species two (Symbolically - and 0)
- *Monoplistic/Predatory/Parasitic*: Positive "+" effect on species one and Negative "-" effect on species two (Symbolically + and -)
- *Neutralistic*: None of the species affects the other (Symbolically 0 and 0)
- *Inhibitory*: Negative effects on both species (Symbolically - and -)
- *Mutualistic/Synergetic*: Positive effects for both species (Symbolically + and +)

These concepts of observable net effects can also be expressed as complementary, supplementary and competitive which are used to describe economic interactions as well. Not all the tree-crop associations need be competitive or allelopathic. Under natural ecosystems, disparate groups coexist through spatial or temporal sharing of resource pools (utilizing different parts of it or utilizing it at different times) and through non-competitive adaptation. Using a bioeconomic perspective, three plant associations are described: complementary, supplementary and competitive.

- *Complementary*: When the interaction is positive, there is complementary between the components. Two components are complementary if increase in one results in a concomitant increase in the other.
- *Supplementary*: Complementary force is equal to competitive force. Two components are supplementary if increase in one does not decrease/increase the other.
- *Competitive*: There is a competition between components if interaction is negative. Two components are competitive if increase in one results in decrease in the other.

Based on the resources sharing by the components, tree-crop interaction may be positive or complementary or beneficial in one hand and negative or inhibitory or competitive or antagonistic on the other. Interaction may be among crop or tree components alone and between tree and crop components, the latter is more important in agroforestry. In a legume and non-legume based intercropping system, part of the nitrogen fixed in the root nodule of the legume may become available to the non-legume component which is an example of complementarity. Further, the complementarities in resource capture and sharing under mixed cropping/ intercropping system may be spatial or temporal. In temporal complementarity, the leaf canopies of component crops may occupy different vertical layers with taller component tolerant of strong light and high evaporative demand and shorter component favouring shade and high relative humidity. Multistoried cropping in coconut and planting shade trees in cacao and tea plantation are the best examples for complementarity.

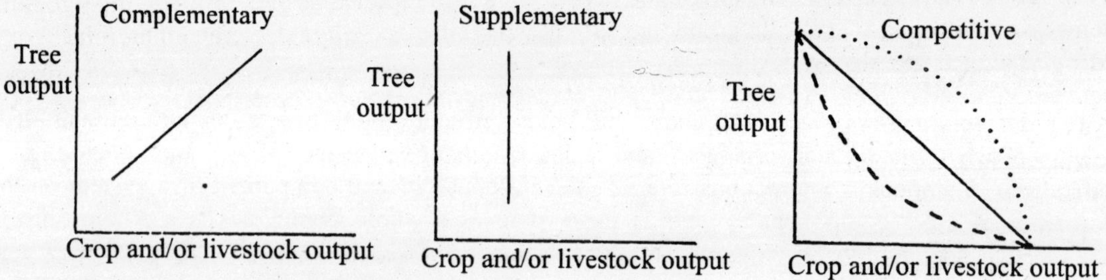

Fig. 14.1: Economic Interactions in Agroforestry Systems

Most complementary associations between tree and crop are attributable to either sparse canopy, deciduousness or induced canopy reduction. Partitioning the interactions into above ground and below ground groups can provide a sound basis for studying the process involved as well as suggesting improved management options for components and system. In similar manner under agroforestry system, exploitation of different layers of soil by the root systems of trees and crops may lead to better utilization of resources with much less competition. In temporal complementarity, the yield advantages provided by the mixture cannot always be explained by more effective use of growth resources at specific times. Indeed, there are substantial opportunities for temporal complementarity if species make their major demands on available resources at different times, thereby reducing the possibility of competition. Based on the effects of interaction, the tree-crop interactions can be classified into:

1. Positive Interactions
2. Negative Interactions

Positive Interactions

This deals with the beneficial effects of one component on another and their influence on the productivity of the overall system. The different positive effects in agroforestry are:

(a) Increased Production and Productivity

(b) Soil Improvement

(c) Nutrient Cycling

(d) Microclimate Amelioration

INCREASED PRODUCTION AND PRODUCTIVITY

There are ample evidences to show that the overall productivity of an agroforestry system is generally greater than that of an annual system although not necessarily greater than that of a forestry or grassland system. The basis for the potentially higher productivity could be due to the capture of more growth resources e.g. light or water or due to improved soil fertility. Some evidences on increase in productivity under various agroforestry systems are given below.

The agroforestry experiments conducted at Forest College and Research Institute, Mettupalayam, Tamil Nadu (India) reported that the productivity of the intercropping systems is always higher compared to monocropping or monoculturing. The increased productivity is evidenced when combining the best tree-crop combinations. The following tree-crop combinations viz., *Ceiba pentandra* + cowpea/ gymnema, Casuarina + turmeric, Eucalyptus + cotton/turmeric obtained higher productivity or returns (Divya et al 2010). The same results were also reported in the following intercropping systems viz., Jatropha + cowpea/groundnut, *Pongamia pinnata* + cowpea/groundnut/ blackgram, *Simarouba glauca* + cowpea/tulsi, Teak + cowpea/gymnema, Gmelina + cowpea, Silveroak + cowpea, *Acacia nilotica* + cowpea, *Acacia holosericea* + cowpea and Neem + cowpea (Divya et al 2006; Divya et al 2008a; Divya et al 2008b; Divya et al 2010). The maximum returns were obtained in the following industrial wood based agroforestry models viz., Casuarina + cowpea/groundnut, *Bambusa vulgaris* (green) + cowpea, *Melia dubia* + cowpea/groundnut (Divya et al 2011; Anon 2011).

Poplar (*Populus deltoides*) has become a popular species in agroforestry system. Deciduous nature of poplar allows agricultural crops to grow under it without adversely affecting the yield. Up to first two years, the returns obtained from sugarcane + poplar combination are higher than any other crop, as the preparatory and cultural operation for both of these are complementary. However, from the third year onwards, shade-loving crops like turmeric and ginger can be grown under it. During the winter (Rabi) season wheat can be grown till its harvest (Sharma and Singh 1992; Jagdish Chander 1998). Poplar tree matures for harvest after 7-8 years having a girth of 90-100 cm. In one acre, 160 trees fetch about Rs. 3,20,000. On an average, a farmer can earn about Rs. 40,000 $acre^{-1}$ yr^{-1}. The income from crops grown in association with poplar is additional. In fact, the growth of poplar significantly increased in intercropping compared to monoculture (Toky 1997).

In an agroforestry system integrating *Eucalyptus tereticornins* for pole wood on border rows of 40 m x 40 m plot, *Leucaena leucocephala* for organic mulch and manure in paired rows along the plots of shade loving turmeric generated cash return of Rs. 5,642 $ha^{-1}yr^{-1}$ compared with Rs. 2,997 from rainfed crops and thus proved economically more viable than normal system of growing field crops (Grewal et al 1995). The biomass production from trees adequately compensated the crop yield reductions. Land equivalent ratio of agroforestry land use was comparable or even better than monocropping systems (Pratap Narain et al 1998). Deb Roy (1991) reported that, in marginal soils, a silvipastoral system involving subabul (*Leucaena leucocephala*) and stylo (*Stylosanthes hamata*) as a pasture legume provides 8 to 9 tons of nutritious dry matter per hectare in a year and enriches the soil. At an interval of 4 or 5 years, stylo may be replaced with *Cenchrus ciliaris* grass.

Agroforestry gives more income to the farmer per unit area of land than pure agriculture or forestry. Several studies in different parts of the country suggested that agroforestry is more profitable to farmers than agriculture or forestry for a particular area of land (Mathur and Sharma 1983; Mathur et al 1984; Ralhan et al 1992; Singhal and Panwar 1992; Toky 1997; Jagadish Chander 1998; Divya et al 2010). In another study, a silvipastoral system involving *Eulaliopsis binata* in association with *Bauhinia purpurea* or *Grewia optiva* appear to be the most promising to sustain biomass production from the degraded bouldery riverbed lands of Doon valley (Samra et al 1999).

SOIL IMPROVEMENT

Agroforestry systems tend to protect soil from several adverse effects. Most agroforestry systems constitute sustainable land use and help to improve the soil in a number of ways. Some of these beneficial effects are evidenced in a number of experiments carried out in different parts of the world

(Nair 1984; Nair 1993; Young 1989, Divya et al 2010). Some of the beneficial effects of agroforestry systems for which enough experimental evidence is available include:

- Reduction of loss of soil as well as nutrients through reduction of runoff
- Addition of carbon and its transformation through leaf, twig, bark falls, etc.
- Nitrogen enrichment by fixation of nitrogen by nitrogen-fixing trees, shrubs, etc.
- Improvement of physical conditions of soil such as water-holding capacity, permeability, drainage, etc.
- Release and recycling of nutrients by affecting biochemical nutrient cycling
- Moderating effect on extreme conditions of soil acidity and alkalinity
- Creating more favourable microclimate by windbreak and shelterbelt effects

Under agriculture, repeated ploughing of topsoil leads to excessive leaching of nutrients during monsoon. Soil fertility is reduced by accelerated decomposition of organic matter and denitrifying processes. Loss of soil and nutrients due to erosion in monoculture is tremendous. Forestry systems maintain soil fertility through recycling of nutrients and prevention of soil erosion and loss of nutrients through leaching and runoff. Many leguminous tree species fix nitrogen from the atmosphere and return much more in leaf fall than they take from the soil. Leaves of the trees could be used as a green manure and help the farmers to increase the soil fertility. Agroforestry systems are therefore helpful in maintaining the soil productivity at optimum level, over a long period of time. Two important processes involved in soil improvement in agroforestry systems are:

(i) Soil and Water conservation

(ii) Soil Fertility Improvement

SOIL AND WATER CONSERVATION

The establishment of a vegetative cover is one of the most effective means of soil and water conservation. When the protective cover of vegetation on the soil is removed, the structurally unstable tropical soils are exposed to the beating action of rains. Losses due to erosion immediately after land clearing are normally alarmingly large. The potential role of trees in reducing run-off and erosion losses is well appreciated and understood. The different state of canopy progressively reduces the force of rain, thereby reducing the adverse effects of its impact on the soil. Furthermore the litter and the humus layers on the soil surface act as a cushion against erosion.

The soil conservation is particularly important in highlands with undulating topography and steep slopes, which are increasingly being brought under cultivation. In agroforestry, multistrata practices simulate the protective nature of natural forest. It has been estimated that soil erosion is accelerated to more than 900 times when tree plantations were clean weeded or the litter was burnt. In a semi-arid alfisol degraded land, established pasture reduced runoff to 26 per cent and soil loss to 8 per cent (Pathak 2000). Stubble mulching, or leaving on the surface plant residues in the field, is often used to reduce soil erosion because it not only binds the soil but increases the roughness of the soil surface.

Most of the beneficial effects of trees on soils are primarily due to their roots, crown, litter-fall and nitrogen-fixing ability; the effects depend on the density of the trees. Tree crowns protect the soil from the impact of raindrops and create specific microclimate (Pradhan 1973). The leaf litter from trees absorbs a large quantity of water, which means less quantity of water is available for surface runoff. Soil erosion from vegetated watersheds is always less in comparison to cultivated lands. The rate of infiltration of soil water is 3 to 5 times more in forested areas in comparison to agricultural lands (Ghosh et al 1974).

Contour hedgerows have consistently been demonstrated to be highly effective in controlling soil erosion even in as little as 18 months (Lal 1989; Maclean et al 1992). The woody hedgerows provide a semi-permeable barrier to surface movement of water, while mulch from the trees reduces the impact of raindrops on the soil and minimizes splash and sheet erosion (Young 1989). Mulching also provides an effective means of reducing soil evaporation and other improvement in microclimate (Kiepe and Rao 1994).

The most serious effect of soil erosion is the loss of organic matter and nutrients which results in reduction in crop yields. Trees reduce erosion through ground surface litter cover (Young 1989).When *Gliricidia* was grown as a hedgerow crop, maize and vegetables as the alley crops, the mean aggregate diameter and wet aggregate stability of soil were found to be increased with mean aggregate diameter of 1.3 mm in control and 3.1 mm in *Gliricidia* alley cropped plots. Thus, alley cropping is found to play an important role in soil conservation (Mapa and Gunasena 1995).

SOIL FERTILITY IMPROVEMENT

The primary objective of soil conservation is to improve or maintain soil fertility. To achieve this, control of erosion, maintenance of organic matter, maintenance of physical properties, organic matter addition and maintenance of nutrients are necessary. Tree crowns and roots protect the soil from the impact of raindrops and create a specific microclimate in the particular area (Pradhan 1973). Agroforestry practices play an important role in improving the fertility of the soils. Increase in amount of N, P and other macro and micro-nutrients are reported under the growth of 14 year old plantation of *Prosopis cineraria* as compared to a bare site of *Prosopis juliflora* of the same age (Shankarnarayan 1984). It is reported that forage species produce higher biomass under *Prosopis cineraria* tree canopy due to a high fertility status (Singh and Lal 1969). There are evidences that soil under the trees was richer in organic matter and nutrients than those in the surroundings and without tree cover. The soil fertility improvement was evidenced under intercropping systems. For example, the following tree-crop combinations viz., *Simarouba* + cowpea/blackgram, *Acaia nilotica* + cowpea/ horsegram, *Acacia holosericea* + cowpa/horsegram, *Azadirachta indica* + cowpea /horsegram, *Ceiba pentandra* +mint, *Jatropha curcas* + cowpea/groundnut, *Pongamia pinnata* + redgram/ cowpea, teak + cowpea, improved the soil fertility status (Divya et al 2010).

Samra et al (2000) observed an increase in organic carbon status of surface soil by 0.39 to 0.52 per cent under *Acacia nilotica* + *Saccharum munja* and 0.44 to 0.55 per cent under *Acacia nilotica* + *Eulaliopsis binata* after five years and suggested that *Acacia nilotica* + *Eulaliopsis binata* are conservative but more productive and less competitive with trees and suitable for eco-friendly conservation and rehabilitation of degraded lands of Shiwalik foot hills of subtropical northern India. Rehabilitation of degraded forests is possible through afforestation by adopting integrated land use planning with soil and water conservation measures on watershed basis (Samra et al 1996). An increase in the organic carbon content in the soil under intercropping as compared to sole cropping is observed. Due to intercropping after 5 years, the available soil nutrients viz., N, P, K and soil carbon content was increased. The reason for increase in N, P and K might be due to the fact that the addition of organic matter through leaf litter and efficient recycling of nutrients (Divya et al 2010).

NUTRIENT CYCLING

Nutrient cycling is the transfer from one component to another of nutrients already in the soil plant system. Transfer of nutrients from trees to crops in agroforestry systems is through prunings, leaf drop or root decomposition. Trees with long taproots recover nutrients lost to the subsoil through leaching and infiltration and recycle them to the surface in the form of litterfall. The litter after decomposition releases nutrients back to the surface soil. Nutrient cycling, on an annual basis, forms the major source of nutrients for plant use. In an agricultural ecosystem, the major sources of input include decomposition of agricultural waste and application of manures and fertilizers. The output of the nutrients involves harvest of grass and grains. When trees are grown with agricultural crops, nutrient cycling is affected due to nutrient uptake and recycling within the trees and release in the form of litter-fall. The accumulation of decayed leaves and branches increases the organic matter content and improve the physical, chemical and biological properties of the soil. Thus the soil is enriched for better crop production. Intensive cultivation of shallow rooted annual food crops on hilly land has resulted in rapid soil erosion which leads to other effects, such as siltation of rivers and reservoirs, uneven stream flow and pollution of water. These adverse effects have been controlled to a great extent by planting trees for upland stabilization (Vergara 1982).

Trees can provide nitrogen inputs in agroforestry systems by two processes:

(i) Biological Nitrogen Fixation

(ii) Deep Nutrient Capture

Biological nitrogen fixation can supply considerable nitrogen to crops through litterfall in soils. Deep nutrient capture is the uptake of nutrients by tree roots at depths where crop roots are not present (Mekonnen 1996). This is an additional nutrient input in agroforestry system because such nutrients are leached as far as the crop is concerned. They become an input on being transferred to the soil through tree litter decomposition (Hartermink et al 1996).

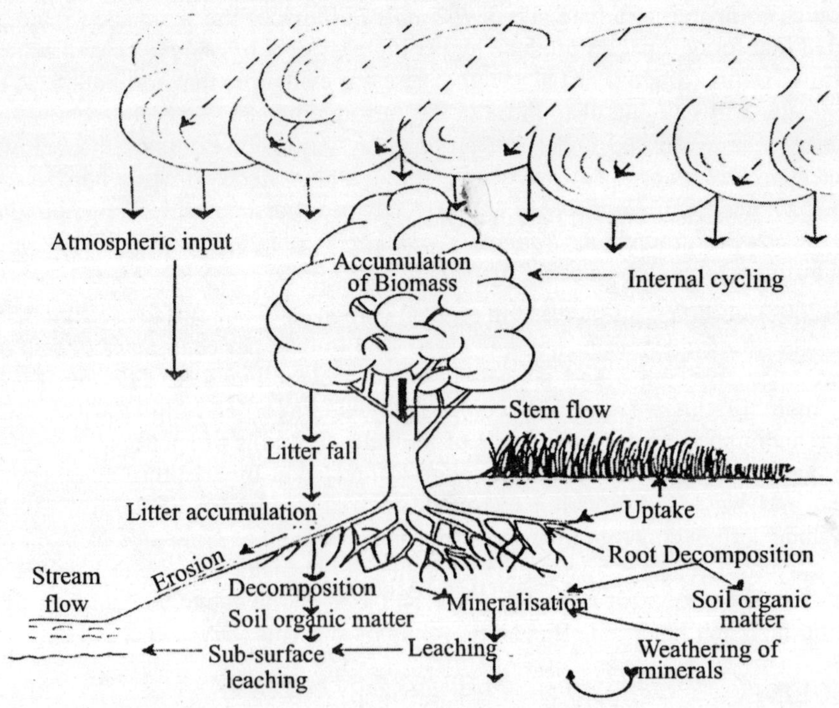

Fig. 14.2: Nutrient Cycling in Agroforestry

The conservation of soil and nutrients and release of nutrients through litterfall depends upon the agroforestry system, density of trees, species etc. Depending on the number of trees per ha and age, litterfall may vary from 0.5 to 6.5 ton $ha^{-1}year^{-1}$. Species of trees, especially nitrogen-fixing legumes, contribute substantially to sustaining an agroforestry system. Several leguminous trees, such as *Leucaena leucocephala*, *Acacia nilotica*, *Dalbergia sissoo*, *Gliricidia* spp., *Sesbania* spp., etc., and some non-legumes, e.g., *Casuarina equisetifolia*, *Alnus* spp., etc. are reported to fix about 50 to 500 kg of nitrogen per ha. Under some systems the leaves of trees could be used as mulch or green manure and improves the site further.

MICROCLIMATE AMELIORATION

The presence of tree changes the microclimate near the ground level by reducing the wind velocity, intercepting light and heat radiation and also moisture. Reduced wind velocity reduces evapo-transpiration from understorey plants. Interception of light radiation during daytime causes reduced photosynthesis and reduced evaporation. The use of trees as shelterbelts and windbreaks in areas that experience high wind or sand movement is well-established example of microclimate improvement that resulted in improved yields. Increased agricultural production due to windbreaks and shelterbelts in India has been widely reported (Rao and Sita Ram 1980). The influence of windbreaks on wind velocity, air temperature, relative humidity and solar radiation is also reported. It was found that the windbreaks reduced wind velocity (at a distance of 3 x H average height of windbreak) on leeward side. This reduction was about 36 per cent during sowing period and 51 per cent during harvesting period.

NEGATIVE INTERACTIONS

This deals with the non-beneficial effects of one component on another and their influence on the productivity of the overall system. The major negative effects in agroforestry are:

(a) Competition for Light, Water and Nutrients

(b) Allelopathy (+ve and -ve)

(c) Microclimatic Modifications for Pest and Diseases

COMPETITION

In an agroforestry system, trees being the dominant partners will compete with agricultural crops for nutrients, soil moisture and light, but the planting is so arranged in respect of location, spacing, and direction that perhaps and hopefully, tree cause the mini-mum adverse effect on crops. Harper (1977) most thoroughly examined the importance of understanding the mechanism of plant interaction. While describing the nature of interactions, he focuses that plant may influence its neighbouring species, not only adding or removing of some factor, but by affecting conditions such as temperature, light or wind movement and by altering the balance between beneficial and harmful insect. When the immediate supply of a single necessary factor falls below the combined demands of the plant, then the competition begins. The competition is also referred as "Ailelospoly".

COMPETITION FOR LIGHT

Many tree crops are inefficient in interception of radiant energy in their early years since full canopy formation in trees may take many years. Under such circumstances, solar radiation falling on bare soil is wasted and promotes only weed growth. Intercropping of ground cover crops with the trees in their early years will therefore help in better utilization of the resource. Reduction in crop yield from second year of tree growth under agroforestry systems has been reported. In considering the potential producti-vity of multiple cropping systems, especially intercropping systems involving trees, shrubs and agricultural crops, it is essential to be able to estimate the photosynthetically active radiation intercepted by each of the component crops at any given time and to integrate this over the time they occupy space. The productivity of a crop must ultimately depend upon its capacity to utilize solar energy.

Investigation on light interception and competition in agroforestry systems are generally scarce. However, some insights originate from the few available studies, including some on intercropping of herbaceous species. Shading was found to be more important than below ground competition in an intercropping study with pearl millet and groundnut in India (Willey and Reddy 1981). Similarly, Verinumbe and Okali (1985) showed that the competition for light was a more critical factor than root competition for intercropped maize between teak trees (*Tectona grandis*).

In an agrosilvicultural system, light availability was determined, in rows of upland rice and mung beans grown under trees of *Gliricidia sepium*, *Acacia auriculiformis* and *Acacia mangium*. The mean reduction of light incidence on rice and mung bean rows was low (3 - 10%) in the pruned alleys, but it was severe (53 - 59%) in the unpruned alleys. Rice and mung bean yields decreased linearly, with reduced light incidence. For every 1 per cent decrease in relative light incidence, rice yield decreased by 47 kg ha^{-1}, and mung bean yield decreased by 10 kg ha^{-1} (Miah et al 1995).

The studies on tree-crop interaction have shown that the yield of unirrigated wheat (grown in Rabi) and paddy (grown in Kharif) crops decreased consistently and progressively when *Grewia optiva*, *Morus alba* and *Eucalyptus* hybrid were grown along with them at 4 x 5 m spacing. Over a period of 13 years, there was an average loss of 29, 29 and 28 per cent in yield of wheat when grown in association with *Grewia optiva*, *Morus alba* and *Eucalyptus* hybrid respectively (Khybri et al 1992). In general, availability of light is the most important limiting factor in situations with relatively fertile soils and adequate water availability. The relative importance of light will decrease in moisture stress conditions as well as on sites with low fertility soils (Cannel 1983). The competition for light between trees and crops is also dependent on the tree and crop species, tree density, age of trees and spatial arrangement of trees.

In the irrigated fields of northern India, the shade of *Eucalyptus tereticornis* and mango grown on field bunds reduced the rice yield to the extent of 25 and 55 per cent respectively. Higher yield reduction under mango was attributed to dense canopy nature of mango (Tomar and Shrivastava 1986). In *Hardwickia binata* based agroforestry system under semi-arid conditions, sorghum was least affected compared to pearl millet, pigeon pea, soybean and cotton due to shade of trees (Khadse and Bharad 1996).

The studies on light interception in six year jack fruit, mango and tamarind based agroforestry systems showed that, the trees reduced the availability of light to finger millet crop due to interception of light by them. Jack intercepted more light compared to tamarind and mango due to its wide and denser canopy. Closer spacing of 8 x 8 m greatly reduced the availability of light due to higher light interception by trees compared to wider spacing of 12 x 8 m and 16 x 8 m. The yield of finger millet was reduced in agroforestry system compared to control. The reduction was more with jack and at closer spacing of 8m x 8m (Basavaraju 1997).

Cowpea was found to be a suitable intercrop for teak, silver oak and gmelina. Cowpea was found to be shade tolerant and performed well under all the three tree species (Divya et al 2010).The reduction in grain yield of the intercrops was noted as maximum reduction at closer spacing (3m x 3m) and lesser reduction at wider spacing (4m x 3m) when the trees were 4 years old. The maximum reduction in grain yield was observed in gingelly with the reduction of 46, 45, 35 and 33 per cent at 3m x 3m, 4m x 2m, 3m x 4m and 4m x 3m spacing respectively (Divya et al 2010).

COMPETITION FOR MOISTURE

Underground root competition for moisture, nutrients and space is relatively more important in agroforestry systems than above-ground crown competition. With the exception of areas with well distributed rainfall of azonal sites with a continuous supply of below ground water, water competition is likely to occur in most agroforestry systems at some period of time; this period may be as short as a dry spell of one or two weeks. The effect of these events depends on the severity of the drought and the drought tolerance of the plants. It also depends on the degree of competition for other resources, especially nutrients.

In alley cropping trials of *Leucaena* with cowpea, castor and sorghum under semi-arid conditions of India, competition for water appeared more important than shading effects (Singh et al 1989). The yields of cowpea, castor and sorghum were declined from 15 to 30 per cent of sole crop as the distance from the hedgerows of *Leucaena leucocephala* decreased from 5 to 0.3 m. The study suggested that alley cropping is not an appropriate technology for semi-arid regions as there is a greater competition for moisture between trees and crops.

Examining soil moisture effects of 3.5 year old *Eucalyptus tereticornis* on mustard and wheat yields next to the tree line in semi-arid India, Malik and Sharma (1990) reported reductions of over 30 per cent for the crops growing at a distance of less than 10m from the tree line. Thus despite the use of drought adapted plants, water competition is likely to play a major role in the productivity of agroforestry systems, especially in dry areas.

The study on the effect of boundary planting of *Prosopis juliflora* on sorghum under semi-arid conditions of India showed that the yield of sorghum was adversely affected up to a distance of 30 m from tree line due to competition offered by the roots of *Prosopis juliflora* for moisture and nutrients. From the above studies it can be inferred that the trees in agroforestry may compete with crops for moisture and nutrients. Shallow rooted and water demanding trees compete more for moisture and nutrients than deep rooted and less water demanding trees. The competition increases with age and density of trees. The crops grown in agroforestry system vary in their tolerance to competition for moisture and nutrients. The roots of most agricultural crops are limited to the upper 50 cm of the surface soil. Only a few species, such as *Cajanus cajan,* sugarcane, cotton, soyabean, sorghum, etc. might send their roots deeper than 50 cm.

It is reported from the arid zone of Africa, that *Faidherbia albida* has a negative impact on crops grown in association with it. It is relatively fast growing. The yield of cluster beans was adversely impacted during the third year by *F. albida.* This adverse impact was, attributed to the faster growth of *F. albida,* as well as its relatively shallower, horizontally spreading root system. Unlike the *Prosopis cineraria* roots which go very deep, the taproot of *Faidherbia albida* after penetrating 45 cm, allowed the emergence of about a dozen laterals at approximately 25 cm depth, and these grew horizontally up to 5.5 m in length. These shallow, lateral roots utilize much more soil moisture and nutrients and thus create an adverse edaphic climate for crops (Shankarnarayan et al 1987).

COMPETITION FOR NUTRIENTS

There are innumerable studies indicating how competition for nutrients can reduce crop yields. In most cases, the yield of the agricultural crop is the criterion by which the merit of an agroforestry system is assessed; yield depressions of this component therefore receive more attention than those of the associated tree species. Furthermore, since the crop is usually the smaller component (when compared individually), its root system will usually be confined to soil horizons that are also available to the roots of the trees; but the trees can exploit soil volume beyond reach of the crop. Therefore, the effects of nutrient competition will probably be more severe for the crop components. The theories and mechanisms of plant competition for nutrients have been reviewed by several workers, e.g., Tilman (1990). However, direct evidence as to where, and how severely, nutrient competition occurs is limited due to the difficulties of separating nutrient competition from competition for light, water and from allelochemical interactions (Young 1989). Additionally, soil and root studies are generally more difficult to conduct than above ground studies.

ALLELOPATHY

Inhibition of growth of one plant by chemical compounds that are released into the soil from neighbouring plants is called Allelopathy. The chemical substances are released by rain action (as leaf leachates) or through decomposition of litter. The released allelochemicals into the soil inhibits seed germination and establishment of certain crops. In most of the cases allelopathic effects are selective and vary with trees and crops.Generally the leaves are the most potent source of allelochemicals, however, the toxic metabolites are distributed in all other plant parts in various concentrations. The allelochemicals responsible for inhibitory effects are tannins in *Acacia leucophloea*, *Acacia nilotica, Adina cordifolia* (Swaminathan et al 1989; Bhatt and Todaria 1990), phytotoxins in Eucalyptus (Suresh and Rai 1987) and Casuarina (Story 1967) and other allelochemicals such as phenolics, terpenoids, organic cyanides.

Although allelochemicals are reported to be present in practically all plant tissues, the information on the nature active chemicals and their mode of action is lacking. While there is evidence that extracts of Eucalyptus can inhibit the germination and growth of crops when tested in laboratory conditions (de Moral and Muller 1969), evidence that enough of the chemicals are accumulated in the soil in field conditions to harm plant growth is rarely shown (Suresh and Rai 1987). For example, sunflower and cowpea are relatively unaffected by *Eucalyptus tereticornis* compared with sorghum (Suresh and Rai 1987). Bhabar grass has also continued to grow productively under *E. tereticornis* (Grewal 1992).

The allelopathic effect of leachates collected from different plant parts of *Simarouba glauca* was tested on germination and growth attributes of blackgram, cowpea, greengram and redgram. Among the leachates, root leachates registered maximum inhibition on germination, shoot length, root length and total dry weight of the above crops (Mohan Raj et al 2008). The leachates collected from different plant parts of *Ailanthus excelsa* was tested on germination and growth attributes of fodder sorghum, fodder cowpea, desmanthus and stylosanthes. Among the leachates, leaf litter leachates registered maximum inhibition on germination, shoot length, root length and total dry weight of the above crops than fresh leaf and root leachates. (Ravi et al 2008).

The germination, root length and dry matter production of the crops like Sorghum, Cowpea, Cotton and Sunflower were depressed due to allelopathic influence of the tree species such as *Eucalyptus tereticornis*, *Casuarina equisetifolia* and *Leucaena leucocephala*. Maximum reduction was obtained with top soil collected from Eucalyptus plantations. Among the intercrops, Sorghum proved most susceptible due to allelopathy than other crops (Divya et al 2010).

MICROCLIMATIC MODIFICATIONS FOR PESTS AND DISEASES

The effect of plant associations on pest and disease incidence is a potentially important but rather unexplored area. Bacterial and fungal diseases may increase in shaded, more humid environments (Huxley and Greenland 1989). For example, the incidence of *Phytophthora palmivora* on cacao increases greatly under conditions of heavy shading (Alvim 1977). The main reasons for this are probably greater relative humidity and decreased wind, both of which tend to favour fungal growth.

This situation is likely to apply to other crop plants susceptible to *Phytophthora.* However, reduced temperature and humidity fluctuations under shade can also have a suppressing effect on pests and diseases. For example, these conditions tend to reduce the spread of witches broom disease (*Crinipellis perniciosa*) on cacao (Lass 1985). It seems, then, that the balance between positive and negative effects will have to be assessed for each particular situation.

INTERACTION MANAGEMENT

The management of interaction effects between trees and other components of agroforestry systems depend on the characteristics of the species, their planting density, and spatial arrangements which is probably the most powerful method for capitalizing on beneficial effects of trees while reducing negative ones. However, in some cases, for example, when trees are used as supports for crop plants, the planting density of the trees is determined by the planting density of the crops. Therefore, in these cases, choosing wider plant spacing for trees with larger crowns may not be a valid option; under such conditions, knowledge of the light transmission characteristics of the tree crowns and of the options for tree management will become important.

Several characteristics could be identified as desirable attributes for trees in agroforestry systems; but often it is not possible to choose trees with all these characteristic, either because other plant are already established, or because production or protection goals favour the choice of other species. Whenever a tree species with all the desired characteristics is not available, tree crowns and roots can be manipulated through management operations, mainly by pruning and thinning. Other common management operations such as fertilization, application of mulch and manure and cut and carry fodder systems can also be employed. The different manipulations can be grouped as growth enhancing or growth reducing according to their effect on the targeted component. The balance of both positive and negative interactions determines the overall effect of interactions in the system. Selection of tree species having all the desirable characteristics for agroforestry is very important. Therefore, proper management of agroforestry systems is necessary to keep the negative effects of trees at minimum.

Yield benefits due to pruning increased gradually with increasing age of trees. Pruning of tree is essential to get regular crop yield from tree-crop system and there is a need to study the amount of canopy pruning with intensity so that tree growth as well as understorey crop yield should not be affected. Soil barrier is an option to reduce competitive effect between tree and crop but tree roots will again be active on upper surface of soil after three to four years. However, soil barrier is not a permanent solution to minimize belowground competition between tree and crop. The beneficial effect of soil barrier on yield increase in crops was comparatively less than pruning and irrigation. The important management practices to reduce negative effects in tree-crop interactions are:

(a) Manipulation of Densities and Arrangement of Trees

(b) Manipulation of Tree Crowns and Roots

(c) Choice of Agricultural Crops

MANIPULATION OF DENSITIES AND ARRANGEMENT OF TREES

This is the most powerful method for realizing beneficial effects of trees while reducing negative ones. The studies carried out in the semi-arid regions of India, showed that as the distance between the hedgerows of *Leucaena leucocephala* increased, the percent reduction in crop yield decreased. The reduction in the yield of sorghum, cowpea and castor was relatively less in wider row spacing (7.2 m) of *Leucaena leucocephala* compared to narrow spacing of 3.6 m (Singh et al 1989). In an alley cropping trial of *Leucaena leucocephala* with maize in the humid zone of Niger, the maize yield was lower with 2 m inter-hedgerow spacing than with 4 m spacing. This decline was associated with higher shading of maize at the narrow spacing (Lawson and Kang 1990). Tree orientation is also important to reduce the negative effect of trees on crops. Planting of trees in the East - West direction may reduce the shading effect of trees on the crops (Dhillon et al 1979).

The growth and grain yield of the crops such as groundnut, sunflower, gingelly, soyabean, black gram, green gram and cowpea were reduced under *Jatropha curcas* trees of different spacing when compared to poor crops. Among different spacing, maximum reduction in the intercrops was observed at closer spacing (2m x 2m) of Jatropha and lesser reduction was observed at wider spacing (4m x 3m) of Jatropha. The magnitude of reduction in the plant height and grain yield of intercrops was decreased with an increase in spacing (Divya et al 2006). Among the different spacing namely 2m x 1m, 2m x 1m (paired row, 2m x 2m and 1.5m x 1.5m tried for Casuarina, least reduction in the yield of intercrops was observed under wider spacing (2m x 2m) when compared to closer spacing (Divya et al 2011).

MANIPULATION OF TREE CROWNS AND ROOTS

The negative effects of trees can be further reduced by pruning and thinning of tree crowns and roots. In alley cropping systems, the pruning of trees and applying the biomass to the soil will reduce the trees' competitive ability and increase the growth and yield of the associated intercrops by providing green manure and by allowing more light to penetrate to the crop. In an alley cropping trial under semi-arid conditions of Karnataka, the yield of Rabi sorghum was higher at shorter cutting intervals (1-2 months) of *Leucaena leucocephala* hedge as compared to longer cutting intervals of 3-6 months (Korwar and Radder 1994). Although more frequent pruning and lower pruning height can minimize the shading effect of the hedgerows on the crop yield, they also reduce the effectiveness of the hedgerows for biomass production and nutrient recycling. Therefore, the optimum pruning regimes for the hedgerows and crop combinations has to be deter-mined, in order to obtain optimal pruning biomass yields, while minimizing the shading effect of the hedgerows and the labour requirement for prunings (Duguma et al 1988).

In situations of severe root competition for moisture and nutrients, especially in arid and semi-arid areas, root pruning operations or trenching may eliminate or at least reduce the negative effects of trees on the intercrops. In an alley cropping system with *Leucaena leucocephala* in semi-arid areas of India, the construction of root barrier of 0.5 m completely eliminated the yield reduction of cowpea, castor and sorghum grown in 10m wide alleys (Singh et al 1989).

In another trial, 33 per cent increase in the yield of Rabi sorghum was observed with root pruning of *Leucaena leucocephala* hedgerows by using country plough (Korwar and Radder 1994). The adverse effect of *Acacia tortilis* on clusterbean, greengram and fodder sorghum was reduced when trenches of 0.5 m deep at 0.5 m from tree line were opinioned (Chopra et al 1993). Root pruning plus lopping of *Hardwickia binata* trees reduced the competition for moisture and light respectively, thereby decreasing the adverse effect of trees on Rabi sorghum. The yield of sorghum was 65 per cent of the sole crop under pruned and lopped trees while the yield was only 36 per cent of sole crop under no managed trees (Korwar 1994). The beneficial effect of trenches on field crops may also be due to better water conservation in trenches, in additional to reduced root competition of trees for moisture and nutrients.

The reduction in agricultural crop yield caused by *Tamarindus indica* is reported to be severe due to the shading effect of the dense and large crowns of mature trees. It is reported that the taproot of a mature tree of *Eucalyptus globulus* penetrates as much as 6 m deep in the ground, and its superficial lateral roots spread out to about 8 m length. The superficial roots reduce the yield of crops to the tune of 30 to 40 per cent. It has been observed that about 75 per cent of the lateral roots can be prevented from invading the crop root zone by planting tall seedlings in pits 60 x 90 cm in case trees are to be grown on farm boundaries.

Among the various traditional agroforestry models in Chhattisgarh, maintaining *Acacia nilotica* (Babul) trees as scattered trees inside crop fields at high densities (100-125 trees/ha) on a 10-12 year rotation is very popular. The effect of Babul trees on associated rice crop growth, yield and soil physio-chemical properties was monitored in field plots located in Bilaspur district during 1993-94 period. The observations indicated that crop parameters excepting grain yield were not significantly affected by trees. Grain yield was reduced to the extent of 28-30 per cent. In case of paddy such influence was found up to 8 m from tree row. Timely root and crown pruning was well as thinning practices are suggested to ensure higher crop yields (Viswanathan et al 1988). Problems may be associated with the selection of wrong tree species for a particular situation. Some trees produce dense shade that decreases the productivity of adjacent agricultural crops. Examples are trees with spreading crowns such as cashews *(Anacardium occidentale)* or shrubs with low, spreading branches.

Field experiments were carried out in the Sahelian zone of Niger to compare trans-piration and soil water uptake among 3 windbreak tree species (*Acacia nilotica*, *Acacia holosericea* and *Azadirachta indica*). *Azadirachta indica* used least water, probably as a result of lower stomatal conductance, since windbreaks exhibited good physiological control over transpiration. The potential for competition for water was most severe with *A. nilotica* and *A. holosericea,* as they extracted large quantities of water through lateral roots, and at the location where trees could not access groundwater. At such sites, the effects of competition on crop productivity should be minimized by planting tree species with low water requirements and by pruning to limit tree transpiration.

CHOICE OF AGRICULTURAL CROPS

While selecting the trees and crops for agroforestry systems, the characteristics of tree and agricultural crops are important for achieving the maximum productivity.

Table 14.1: Characteristics of Individual Tree and Agricultural Crop in Agroforestry (Dwivedi 1992)

S. No.	Characteristics	Tree	Agricultural Crop
1.	Size	Large	Small
	Height	10 m to 40 m	Less than 1 m
	Root Depth	200 to 500 cm	20-50 cm
	Root Spread	50-150 m^2	0.25-1.0 m^2
	Crown Size	50-100 m^2	0.25-1.0 m^2
2.	Mass	500-5000 kg	0.25-2.5 kg
3.	Life	Long, usually	Short, usually
		25-150 years	Less than one year
4.	Density	250-750 trees/ha	10,000 to 150,000 plants/ha
5.	Utilization	Wood, sometimes leaves, fruits, flowers, etc	Grains, sometimes leaves, stem, etc.

There is a great need to identify the suitable agricultural and horticultural crops, which can grow well along with the tree plantations with limited solar energy available. Some of the crop plants which can grow well under shade include the following:

- Vegetables: Ginger, turmeric, potato, cucurbit, etc.
- Agricultural crops: Oats, maize, fodder crops and some varieties of soyabean, beans and ground nut
- Ornamental: Ferns, Orchids, Cactus, Hicotiana, Cineraria, Adiantum, Selaginella, Phyllocactus, Calceolaria, Cyclamen, Streptocarpus
- Grasses: Several species
- Shrubs: Tea, coffee, cotton, cassava, etc.

Trees are live and complex organisms, and agroforestry systems are also alive in the sense that they represent a complexity of interactions that is constantly changing. Agroforestry is rapidly being transformed from an empirical, largely anecdotal collection of beliefs and practices into an emerging science in the field of natural resource management. Though much work has been done on above ground interactions in an agroforestry systems, the below ground interactions has been given a least importance; because of complexities involved in quantifying it. Therefore, further thrust on research should be given to below ground interactions. Particular agroforestry system will always be highly site-specific and therefore the interaction studies should be of a site-specific one.

■■■■

15 DIAGNOSIS AND DESIGN IN AGROFORESTRY

INTRODUCTION

Agroforestry practice has been accepted by the farmers across globe as means to sustainability and environmental security. Changing scenarios in climate, economy and society driving all existing system to cope with and change accordingly. New concepts like organic farming, integrated farming, precision agriculture and conservation agriculture cropping up to be adopted by farmers under changing farm input subsidy, fluctuating prices and changing climatic regimes, requires farm analysis under the light of current and future vision with sound concepts of farming. Diagnosis and design concept is available for agroforestry systems under the light of emerging farming concepts for better designing of future farms.

Given all the different ways in which trees can be grown together with crops and/or animals in agroforestry systems, it is important to decide which agroforestry practices are best for a given user group or land use situation or which technologies exert a priority on scarce research resources. The challenge of identifying research and extension priorities in a field is complex and as initially underfunded as agroforestry was a central preoccupation of ICRAF's early programme of work, resulting in the Diagnosis and Design methodology (D&D).

The Diagnosis and Design (D&D) arose, in the words of J. B. Raintree, who directed its development at ICRAF, "Out of the demands of the agroforestry situation; it gives a special focus on agroforestry-related constraints and opportunities within existing land use systems, and highlights agroforestry potentials that might be overlooked by other methodologies. For example, for most Farming Systems Research/Extension (FSR/E) practitioners, the trees within the farming system tend to be invisible" (Raintree 1987a and 1987b). FSR/E (Shaner et al 1982; Hildebrand 1986), and the Land Evaluation Methodology (FAO 1976) are the earlier methodologies for the evaluation of land use systems. Diagnosis should be done before prescribing any treatment. There is no universal treatment to suit everyone. In the same way, the need of the different farm area and individual farmer is different. The climatic, edaphic, topographic, economic and socio-cultural factors of every region are unique in its characteristics. The success of a treatment is in its proper diagnosis.

DIAGNOSIS AND DESIGN (D&D)

Diagnosis and Design (D&D) is a family of procedures for the diagnosis of land management problems and potentials and the design of appropriate agroforestry interventions. D&D refers also to the body of heuristic knowledge and practice which is called upon in making the sequence of decisions taken in the course of a D&D application (Raintree 1993). D&D is a systematic and objective methodology developed by ICRAF to initiate, monitor, and evaluate agroforestry programmes. D&D methodo-logy is an adaptation of old or existing methodologies to the specific needs and condit-ions of agroforestry (Nair 1993). D&D is based on the philosophy that knowledge of the existing situation (diagnosis) is essential to plan and evaluate (design) meaningful and effective programs in agroforestry research for development. The methodology plays a strategic role in all the phases of the agroforestry research process. It borrows from other methodologies used by research or development agencies, such as baseline surveys and feasibility studies. However, D&D is unique in that it has been specially developed for the following purposes:

- To describe and analyze existing land use systems
- To design appropriate agroforestry technologies to alleviate those constraints
- To design appropriate research work, such as trials and funkier surveying

The basic unit of D&D analysis is the land use system (LUS). The LUS can be defined and analyzed at the level of a country, ecozone, farming system, crop system, or any other unit. The structure and function of any LUS are determined by climatic, physical, biological, technological, economic, social, and political factors. D&D focuses on the interactive effects these factors have on the LUS, and searches for opportunities for improved system development in the LUS.

ICRAF's D&D methodology can serve a variety of useful roles. It can, in the first place, provide a justification for agroforestry research by demonstrating land use constraints which the system can address (e.g., soil degradation, land scarcity, need for low-input technologies). At the same time, it reminds researchers that a particular agroforestry practice is just one land use system among many, and that other agroforestry or non-agroforestry interventions may be more appropriate in specific cases. Finally, D&D methods can guide researchers as they find ways to adapt agroforestry practices to local conditions. Macro D&D is a large-scale analysis of an ecozone within a country or a group of countries. Macro D&D is important for deciding on national agroforestry research-and extension agenda at-the national level. Micro D&D, in contrast, focuses on one land use system (LUS) within the larger ecozone that has special priority for agroforestry intervention. Micro D&D involves a detailed analysis of households and production systems in the LUS. It leads to guidelines for research that will address the constraints of the prioritized LUS.

AGROFORESTRY SYSTEMS RESEARCH PROCESS

The basic objective of ICRAF's agroforestry research is to develop technologies to solve farmers' problems in priority land use systems in specific ecozones. An agroforestry technology should be specified with reference to at least its principal components: MPT species, spatial arrangement, management regimes (i.e., management of the trees and associated components), and performance levels (i.e., technical and socioeconomic criteria). To this end, ICRAF has developed a research process that uses a systems perspective and an interdisciplinary approach. D&D exercises initiate the process, and the design of agroforestry technologies is the pivotal step. The process is called Agroforestry Systems Research. It is recommended for any D&D program in any areas of agroforestry. There are six main steps in the process:

- Macro D&D (national and ecozone level)
- Micro D&D (land use analysis at the production systems level)
- Technology design
- Component experimentation
- Technology testing

- Technology dissemination and adoption.

BASIC PROCEDURE OF D&D

The basic procedure of D&D contains five stages namely, (i) Prediagnostic stage, (ii) Diagnostic stage, (iii) Design and Evaluation stage, (iv) Planning stage, and (v) Implementation stage. The Table 15.1 precisely describes the basic procedure of D&D.

Table 15.1: Basic Procedure of D&D (Raintree 1987a)

D & D Stages	Basic Questions to Answer	Key Factors to Consider	Mode of Inquiry
Prediagnostic	Definition of the land use system and site selection (which system to focus on?)	Distinctive combina-tions of resources, technology and land use objectives.	Seeing and comparing the different land use systems.
	How does the system work? (how is it organised, how does it function to achieve its objectives?)	Production objectives and strategies, arrangement of components.	Analysing and describing the system.
Diagnostic	How well does the system work? (what are its problems, limiting constraints, problem generating syndromes & intervention points?)	Problems in meeting system objectives (production shortfalls, sustainability problems)	Diagnostic interviews and direct field observations.
		Causal factors, constraints.	Troubleshooting the problem subsystems.
Design & Evaluation	How to improve the system? (what is needed to improve system performance)	Specifications for problem solving or performance enhancing interventions.	Iterative design and evaluation of alternatives.
Planning	What to do to develop and disseminate the improved system?	Research and development needs, extension needs.	Research design, project planning.
Implementation	How to adjust to new information?	Feedback from on-station research, on farm trials and special studies.	Rediagnosis and redesign in the light of new information.

It stands to reason that any rapid appraisal methodology used to initiate research or development activities must be followed up by continued monitoring and evaluation of the situation as it develops over the course of the project. In the D&D framework, this internal monitoring and evaluation process takes the form of a reiteration of the basic diagnostic and design process to refine the original diagnosis based on more in-depth information resulting from continuous exposure to the site and to improve the technology design in the light of new information from on-farm trials with farmers, more complex and rigidly controlled on-station experiments and eventual extension trials in a wider range of potential sites (Raintree 1993).

The iterative D&D process provides a basis for close feedback and complementarity between the different components of an R&D project. By adjusting the plan of action to new information, the D&D learning process becomes continuous and self-corrective, ideally, it should not stop until the design is well-optimized and/or further refinements are judged not worth the additional effort. Until that point is reached, the D&D logic can be used to identify fresh opportunities for progressive refinement and adaptation of the experimental technologies to the needs of the client system. The farmers themselves are often the best source of relevant design improvements and, in any case, the ultimate fine tuning of the technology is beset left to the intended beneficiaries (Scherr 1988a and 1988b).

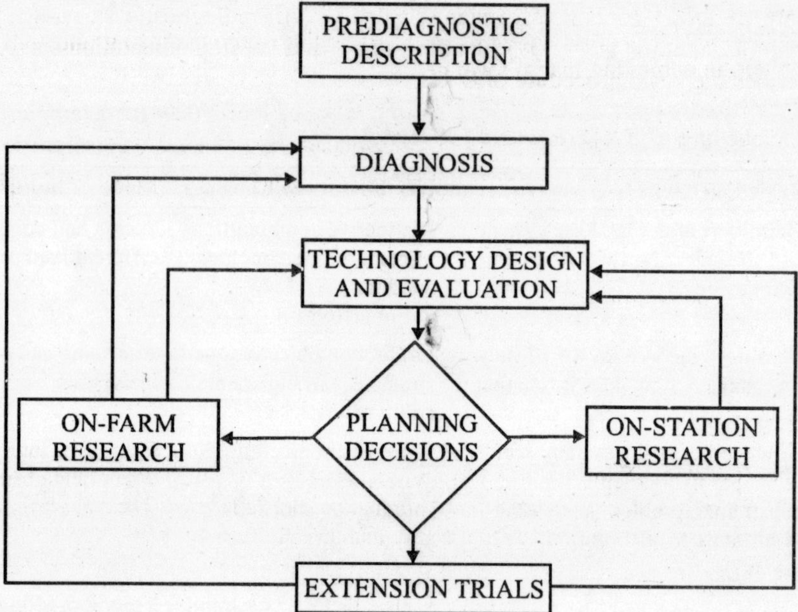

Fig. 15.1: Flowchart of Iterative Activities and Feedback in a D&D (Raintree 1987a and 1993)

KEY FEATURES OF D&D

D&D is a methodology that has been developed specifically for agroforestry applications, with emphasis on a comprehensive diagnosis of the problems, followed by design and implementation of appropriate interventions to solve the diagnosed problems. Its prominent features are:

- *Flexibility* - D&D is a discovery procedure which can be adapted to fit the needs and resources of a wide variety of land users.
- *Speed* - D&D has been designed to allow for a "rapid appraisal" application at the planning stage of a project with in-depth analysis occurring during project implementation.
- *Repetition* - D&D is an open-ended learning process. Since initial designs can almost always be improved, the D&D process need not end until further improvements are no longer necessary.

VARIABLE SCALE D&D PROCEDURES

D & D methodology is scale neutral i.e., it can be applied for different level of land use system. But the procedure needs some modification for applying at different levels. The D&D procedure can be divided into three types based on scale viz. macro level, meso level, micro level.

MACRO LEVEL

Usually Macro D&D is for ecoregions or state or some at country level. The macro-scale D&D involves the use of D&D procedures for a much larger application than the micro- and meso-levels

(Scherr 1989). It is a more complex exercise; apart from subject matter specialists here the role of policy makers is very important. A team of experts from various field including foresters, agriculture scientist, sociologist, economist and policy makers are vital for planning and decision making at this level. Consortium of national level institutions and guidelines from international organizations like ICRAF and FAO makes the effort more reliable. Application of remote sensing and GIS and other efficient technology help in achieving macro level goals.

Scientists/ subject matter specialists working in one level of D&D (say for example, Mesolevel D&D) should have clear idea and understanding about other levels of D&D (say Micro level limitations and Macro level information). Involvement of local people right from pre-diagnosis stage to implementation stage is necessary for proper D&D. The four main objectives of the Macro D&D are:

- To identify broad issues and problems constraining all the land use systems in a given ecozone
- To identify and prioritize areas for potential agroforestry interventions
- To identify research priorities and formulate research programs
- To identify needs, opportunities, and mechanisms for inter-institutional collaboration for technology development.

To meet these objectives, macro D&D uses rapid appraisal techniques. It relies heavily on secondary data which are verified and complemented by quick field surveys. The seven basic steps in a Macro D&D Exercise (Avila and Minae 1992) are:

(1) Identification of study ecozone
(2) Delineation of land use systems within the ecozone
(3) Description of land use systems
(4) Analysis of land use system constraints and potentials
(5) Analysis of potential agroforestry technologies
(6) Definition of agroforestry research needs
(7) Inter-institutional coordination

MESO LEVEL

The meso scale D&D is used to work with units larger than farms, such as watersheds and other landscape zones (Rocheleau 1985). Some problems needs solution above the household level approach, that is especially in the matter soil and water conservation. The matter of soil and water conservation or marketing of small timber or other farm produce etc., should be dealt under meso level. Design like setting up of Co-operatives at village level and tree growers association for procuring quality seedlings and their marketing can also be dealt under this level. Institutional linkage is usually found in this stage. In India, usually PRA (Participatory Rural Appraisal) exercise was carried out for diagnosis of problem and planning at village or watershed level. Management of Common Property Resources (CPR) by means of Energy plantation, Canal Plantation, Fodder Bank, Rangeland management, etc., for the common benefit of the villagers is also an important intervention at meso level.

MICRO LEVEL

The unit of management is household that is family farm. Even a small group of farmers' has land nearby with same set of problem or management was dealt with this level. Subsistence farming, addressing the basic needs of the farmer and his family is the prime objective. Diversity in terms of crop/ tree and/or cattle is usually present. Providing employment for the household throughout the year is necessary. Work and needs for women and men are different within the household. During the process of design knowledge about the locality, culture and societal norms should be kept in mind.

The objectives of micro D&D are similar to those of macro D&D. The major difference is that whereas macro D&D has a broad scope (i.e., an ecozone), micro D&D focuses on detailed analysis of one prioritized LUS. The three main objectives of micro D&D (Avila and Minae 1992) are:

- To describe and analyze an LUS in order to identify its constraints
- To design and evaluate agroforestry technologies to address the constraints
- To design and evaluate appropriate research programs aiming to develop these technologies

The basic principles and steps in a Micro D&D Exercise are same as in Macro D&D. Since resources are inevitably limited, a country, institution or project will have to be selective in deciding which LUSs should be subjected to micro D&D. The choice of the LUS for micro D&D depends on criteria such as:

- Political and economic importance of the system
- Technical potentials for improvement of the LUS
- Scientific expertise and other resources in the national collaborating institutions for carrying out research in the LUS

NEW FARMING CONCEPTS AND APPLICATION OF D&D TOOL

- Organic farming
- Integrated farming system
- Precision agriculture
- Conservation agriculture

ORGANIC FARMING

FAO suggested that "Organic agriculture is a unique production management system which promotes and enhances agro-ecosystem health, including biodiversity, biological cycles and soil biological activity, and this is accomplished by using on-farm agronomic, biological and mechanical methods in exclusion of all synthetic off-farm inputs".

INTEGRATED FARMING SYSTEM (IFS)

Integration of two or more appropriate combination of enterprises like crop, dairy, piggery, fishery, poultry, bee keeping etc., for each farm according to the availability of resources to sustain and satisfy the necessities of the farmer. IFS seems to be answer to the problems of increasing food production, increasing net farm income, improving nutritional status promoting natural resource management sustainable use of land, water and biota.

PRECISION AGRICULTURE

Across globe farming system has been blamed and praised paradoxically for increasing or decreasing impact of climatic risk to be posed by future climatic projections. Concept of precision farming had its genesis to address such issues with precise exploitation of natural resources. Precision agriculture in simple terms is the management of arable variability to improve economic returns and reduce environmental impacts. Precision Agriculture hinges on '5 + 1 Rs': Right input, Right time, Right manner, Right quantity, Right place and Right genotype/seed. When such system is going to take over farming system it is imperative to take D&D system of agro-forestry systems one step ahead to decide components in such manner which meet the future farm planning. This point is discussed here to throw some light on future policy issues guiding agriculture production and its management. Poor economies in tropical do not have the resources to atomize their farming system but certainly undergo resource restructuring in the crop field. The premier component of agroforestry system 'trees' has been viewed under different angle of precision light which aims to augment farm return with minimum environmental impact.

Table 15.2: Interface between emerging Farming Concepts and D&D Concepts

Farming Concepts	Details	D&D Analysis	
		Macro	Micro
Organic Farming	Production management system which promotes and enhances agro-ecosystem health, including biodiversity, biological cycles and soil biological activity, and this is accomplished by using on-farm agronomic, biological and mechanical methods in exclusion of all synthetic off-farm inputs	National level, agro-climatic and agro-ecological zone analysis and planning and compatibility analysis	Farm organic resource identification and utilization. Generation of farm organic sources
Integrated Farming System	Integrated farming system is a resource management strategy to achieve economic and sustained agricultural production to meet diverse requirement of the farm household while preserving the resource base and maintaining high environmental quality.	Policy to provide marketing and value addition facilities to varied products of IFS	Compatible enterprises identification to augment farm income
Precision Farming	Precision agriculture in simple terms is the management of arable variability to improve economic returns and reduce environmental impacts. Precision Agriculture hinges on '5 + 1 Rs': Right input, Right time, Right manner, Right quantity, Right place and Right genotype/seed.	Governments should have regional pilot planning and projects before making fully atomization	Precise selection of components with proper plan for utilization
Conservation Agriculture	Conservation Agriculture (CA) is an approach to managing agro-ecosystems for improved and sustained productivity, increased profits and food security while preserving and enhancing the resource base and the environment by adopting three principles minimal soil disturbance, permanent soil cover and crop rotations.	Conservation reserve source to plan during designing land use policy	Permanent soil cover through high ground trees and associated crops to be selected through proper diagnosis

CONSERVATION AGRICULTURE (CA)

It aims to achieve sustainable and profitable agriculture and subsequently aims at improved livelihoods of farmers through the application of the three CA principles: minimal soil disturbance, permanent soil cover and crop rotations. CA holds tremendous potential for all sizes of farms and agro-ecological systems, but its adoption is perhaps most urgently required by smallholder farmers, especially those facing acute labour shortages. It is a way to combine profitable agricultural production with environmental concerns and sustainability and it has been proven to work in a variety of agroecological zones and farming systems. It is been perceived by practitioners as a valid tool for Sustainable Land Management (SLM).

CONCLUSION

D&D is not a universal tool. Therefore keeping the understanding of this effective tool, one should device his own methodology with suitable modification. A logical diagnosis and an innovative design, tailor made for the particular set of problems, is the key for successful implementation of a project. D&D is a dynamic methodology and there is always a scope for improvement from the lessons learnt from the previous experience.

■■■■

16 ECONOMIC ANALYSIS OF AGROFORESTRY SYSTEMS

INTRODUCTION

Agroforestry trees play a key role in the economy of the rural populations and they add to diversification and contribute to nutrition and energy security of the people. In Asia, 25 per cent of rural people use agroforestry products (Leakey and Sanchez 1997). It comes out 566 million consumers, which is a formidable market by any means. The rising incomes in the developing economies are diversifying the food basket of the middle-income groups, which are creating demand for value added and nutrition foods including fruits and the processed foods. Owing to pressure for land development for various uses, intensification of land use by introducing agroforestry systems that integrate horticulture species like fruits is an ideal way. The main advantage of this system is that the farmer will have risk bearing capacity if one component say crop component performing below par. The agroforestry systems create huge employment potential.

Agroforestry systems have proven their financial viability and attractiveness as important land use alternatives in various settings throughout the world (Garrett 1997). Numerous economic studies of agroforestry systems have been conducted at global and national level. Generally, these studies examine the financial costs required to establish, manage, and produce various combinations of agricultural and timber crops, potential revenues from different agroforestry alternatives, and profitability of adopting agroforestry practices. Generally profitability of wide variety of agroforestry practices is analyzed and compared using NPV, Benefit/Cost ratio and ROR. The analyses are conducted from farmer's view point in an effort to bridge the information gap between agroforestry and landowners. It has been found in most cases that integration of agroforestry practices into traditional farming systems yields greater rates of return than monoculture practices alone.

Montambault and Alavalapati (2005) conducted extensive review and analysis of socioeconomic research in agroforestry literature between 1992 and 2002. Results showed a clear increasing trend in publications with more complex analyses, such as econometrics and optimization. The development of more sophisticated economic models creates applications that give more realistic and useful results for agroforestry practitioners. Indeed, the first World Agroforestry Congress organized in June 2004 at Orlando, Florida identified economics and policy as one of the key areas for enhancing the impacts of agroforestry. Investigations on the economic viability and technical feasibility of modern agroforestry practices (e.g., agrosilvicultural and silvopastoral) have revealed that these modern practices have the potential to increase the wood and agricultural yield, improve labour efficiency and reduce cash requirements during market lulls. Systematic and quantitative investigation of all benefits and costs associated with agroforestry production is necessary to convince economists and landowners that agroforestry offers positive monetary and non-monetary benefits.

Analyzing the economics of agroforestry practices is more complicated than that of annual crops for two main reasons. First, agroforestry practices are complex because they involve both trees and crops. Devising field trials to assess agroforestry practices and compare them with other practices is extremely difficult, requiring large plots and at times, large spaces between treatments. Second, there is usually a period of several years between the time the trees are established and the impact of agroforestry practices can be measured. Conducting trials and surveys with farmers over several years is expensive and problematic. For example, the greater the length of trial, the more likely that individual farmers will want to change trial parameters in response to changing circumstances or preferences. The more changes that each farmer makes, the less likely it is that treatments can be compared across farms (Coe 1998).

AGROFORESTRY INVESTMENT VALUATION

Agroforestry values are affected by many factors including site quality, topography, site preparation costs, land taxes, timber markets, proximity to the mill and markets, etc. Some agroforestry value indicators affected by these factors are land holding values, stumpage values, bid prices, market values, net present value (NPV), equivalent annual income (EAI), rate of return (ROR), land expectation value (LEV), etc. Evaluating potential capital investments in agroforestry-related practices begins with an estimation of the present value of all revenues and costs associated with the practice to yield a NPV for the investment. All costs and revenues are discounted to the present with costs then subtracted from revenues.

Agroforestry investments, due to their integration of forestry, agricultural, animal components and other activities must incorporate valuation techniques from all the components to estimate the economic attractiveness of an investment. This will also permit an analysis of the contribution of each component to overall system. Typically, economic analysis of agroforestry land use alternatives is conducted using NPV, EAI, ROR and NPV of perpetual rotations or LEV. The congruity of such values creates an easy mode of comparison for numerous agroforestry options (Grado and Husak 2005).

Many computer models have been designed to make financial comparisons of agroforestry options. They include MULBUD (Etherington and Matthews 1983), MIDAS (Kingwell and Pannell 1987), TREE$PLAN (Bulman 1991), FARMTREE (Loane 1991), FARMULA (Kubicki et al 1993) and USAEM (Pearson et al 1995). Of all computer models used to analyze agroforestry systems, spreadsheet-based models are the most commonly used because they allow a sensitivity analysis of changes to inputs on any operation in question (Thomas 1991). Although wildlife benefits created by agroforestry systems are considerable, many computer models are limited in their capacity to incorporate those benefits. Such deficiencies necessitate the development or adaptation of a suitable computer-based model that would include a valuation of wildlife and other non-market output benefits (e.g., savings from soil erosion reductions).

KEY FACTORS IN ECONOMIC ANALYSIS OF AGROFORESTRY

Economic analyses of agriculture and agroforestry developments are similar with a difference that agroforestry systems have more than one component with complementary and supplementary roles. Economic analysis of agroforestry in the past three decades focused on comparative analysis of agroforestry systems with that of farmer's traditional cropping/farming systems in financial terms with tools like net returns and benefit cost analysis. Since resources are limited, economic analysis of agricultural systems attains prominence. Some of the factors which affect the economics of a system are as given below.

LAND-MAN RATIO

Land resource is very scarce in India. India has only 2 per cent land resources to support about 15 per cent of the human and 16.5 per cent of the livestock populations of the world. Therefore, pressure on the land is very high. Per capita land area of India is one of the lowest in the world. In India, about 76 per cent of the population is rural and 148 million people are actively engaged in agriculture. The number of persons actively engaged in agriculture is one of the largest in the world. Therefore, land area in relation to the active person in agriculture is very low. Though land area under agriculture is 155 million ha, since a large number of persons are actively engaged in agriculture, the ratio is very small.

Table 16.1: Comparative Land/Man Ratio in India (Units in Ha) (Dwivedi 2001)

S. No.	Items	World	Asia	India
1	Per Capita Total Land Area	2.85	1.00	0.40
2	Total Land Area/Active Persons in Agriculture	15.74	4.34	2.20
3	Per Capita Agricultural land Area	1.01	0.41	0.23
4	Agricultural Land/Active Persons in Agriculture	5.58	1.79	1.00

Land holding is also very small. Holdings less than 1.0 ha constitute about 56.6 per cent of the total number of holdings. Small holdings (1-2 ha) are about 18.1 per cent, medium (4-10 ha) to large (more than 10 ha) holdings are only 25.5 per cent. About 40 million ha, i.e. 25 per cent of agricultural land is irrigated and remaining faces the vagaries of nature. Most people related to agriculture are poor and large number of them live below poverty line. Agriculture to them is subsistence.

POLICIES

The forest policy of 1988 aims to ensure the environmental stability and maintenance of ecological balance, including atmospheric equilibrium, which are vital for the sustenance of all life forms - human, animal and plant. The policy recommends increasing the productivity of the existing forests. Recognizing the importance of food production, the policy recommends that diversion of good and productive agricultural lands to forestry should be discouraged. The farmers especially small and marginal farmers should be encouraged to grow, on marginal degraded lands available with them, the wood species required for the industries. The present policy envisages growing of trees by the individual farmers not only to meet their own needs, but also the needs of the industries. However there is no mention about the pricing policy of the produce. Farmers are forced to sell their produce to industry at much cheaper rate in the event of high production and market saturation, as has been case with *Eucalyptus* in Haryana.

In some states, the trade in respect of certain tree species is under the control of government e.g. in MP, trade of teak and bamboo is under the control of government. Similarly in Tamil Nadu and Karnataka, trade of sandal wood is controlled by the state governments. Such restrictions adversely affect the price of the produce. In some states, transit of almost all wood, including forest produce is restricted. Such restrictions work as disincentives to better price of the produce.

MARKETING

In agricultural produce, government has price control mechanisms for controlling the price in the market. However such price control mechanism does not exist in wood, fodder and other produce Farmers are likely to be exploited by powerful industrial sectors. The experience of Haryana, Punj and western UP suggests that farmers did not get an appropriate price for *Eucalyptus* wood from the paper mills due to sufficient production from agrforestry. There is no storage facility of wood in India.

SHORTAGE OF WOOD AND STEEP PRICE RISE

During the last few decades, the human and the livestock populations have risen considerably with increased pressure on the land for more food, fodder, wood and other products. Agricultural land had increased during the last sixty years i.e. 118 million ha during 1951 increased to 142 million ha during 1981. Due to unrestricted grazing and fire, the good forest remains only about 38 million ha. The yields from the forest have reduced and the demands have increased considerably. The prices of agricultural commodities e.g., rice, wheat, millet, pulses, oilseeds, etc., have increased by 3 to 5 times over the last 25 years, while the prices of timber, firewood and charcoal have increased by 15 to 20 times (Dwivedi 1992). Growing of trees becomes a remunerative land use in comparisons to agriculture, particularly in areas where demands of poles and firewood are high. Since people want quick returns, they prefer fast growing species, such as *Eucalyptus* spp., *Populus deltoides*, *Casuarina equisetifolia* and *Leucaena leucocephala.*

WOOD CONSUMPTION PATTERN

Wood consumption has gradually increased in the country. With increase in income, there is almost an increase in wood consumption. With increase in standard of living, literacy and per capita income, there is proportionate increase in the consumption of industrial and commercial wood as consumption of paper, plywood, furniture increases considerably. The use of wood as fuel depends upon availability of alternative cheap fuel. There has been an increase in demand for wood in rural areas for non-traditional items such as furniture and industrial wood products. There is wide gap between present demand and production of various categories of wood in the country. The yields from the forest have reduced and demands have increased considerably.

ENVINRONMENTAL AWARENESS

The forest policy 1988 says that natural forests are to be managed primarily for conservation of the environment. More and more forest areas are being declared as national parks and sanctuaries. The total area under national parks and sanctuaries is about 14.1 million ha, which was only 2.6 million ha during 1970 (Anon. 1989). Some other areas are being declared as biosphere reserves or natural reserves. There is complete ban on green felling above 1,000 m elevation. Clear felling system is being discouraged since it doesn't provide environmental safeguards. All these factors led to reduced timber availability in the market. This awareness encourages afforestation of waste lands and planting of trees on farm lands and on sides of rails, roads and canals.

TIME SPAN

In tree based agroforestry system, output is obtained after several years of giving inputs. This is serious disadvantage for several reasons viz. (1) The available inputs could have been used for producing more immediate outputs, (2) The proposed production may not be available in desired quantum due to several risks during gestation period, (3) The produce finally obtained may no longer be in demand due to availability of various kinds of substitutes. Therefore, the procedure of discounting used in evaluating delayed returns does not really reflect the benefits in true measure.

RISKS

The risk in agriculture production system is greater than tree based production system as tree is less susceptible to drought and floods. At the same time, trees are long duration crops and returns and prices are subject to fluctuation. The factors affecting the prices of agricultural products are not the same as those which affect the prices of products from trees. The market infrastructure for agricultural commodities is well developed in comparison to timber and fuel market. Fodder markets are almost non-existent in several areas. All these factors are important in the economics of agroforestry.

VALUATION OF OUTPUTS

Agrofoerestry systems yield two types of main produce, i.e. material goods and services. Important services of agroforestry systems are sustainability attributed to biological nitrogen fixation, nutrient cycling, sheltering effect of wind breaks, contour planting, alley cropping, green manuring. There is no precise method for estimating the value of these services in spite of the fast-developing areas of environmental economics. The valuation of material output such as fuelwood, fodder, timber, fruits etc. will differ with subsistence and commercial objectives. When the produce is marketed, actual value can be realized. However in areas where significant marketing is not taken place, value of the produce cannot be realized.

ECONOMIC TOOLS

Analysis of Farm Management Efficiency: Three types farm management efficiency analyses farm income/comparative analysis, farm budgeting and gross margin analyses. Farm management analysis evaluates the production revenues and costs, utilization of labour, land and capital, constraints of production, etc.

Analysis Returns to Investment: To evaluate agroforestry activities involving high initial investment or projects for development funding, a returns on investment is critical criterion. This involves cost benefit analysis and payback analysis. Cost benefit analysis assesses the resource inputs, costs, and benefits over the lifetime of the proposed or existing agroforestry technology. Pay back analysis deals with the investment, rate and spread of returns in a project mode that would give a scope for comparing alternative projects.

ECONOMICS OF AGRISILVICULTURE SYSTEMS

ECONOMICS OF POPLAR AND EUCALYPTUS BASED AGROFORESTRY

In general, the farmers practice bund planting with poplar as well as eucalyptus in agroforestry. For the sake of arriving at the comparative economics of exclusive Poplar (*Populus deltoides*) and Eucalyptus based systems with average population maintained per ha, the B:C ratio was worked out. The cost benefit analysis of various systems is given in Table 16.2. Among the four systems, the farmers preferred the bund system due to its less detrimental impact on the grain/commercial crop pattern.

A study was conducted with the aim of evaluating the performance of poplar-based agroforestry in terms of income, employment and environmental impact from the farmers' perspective. Poplar based agroforestry was found to be economically viable and more profitable than many of the crop rotations followed in the study area. This land use system also found to be capable of providing employment opportunities on farms. Sensitivity analyses indicate that this system is not highly risky. However, costs charged by the promoting industry for technical advice substantially reduced the income from poplar plantation. With agroforestry experience, farmers can expect high dividends in subsequent rotations. Life-time matrices developed through group interviews were useful for thorough economics analysis of agroforestry projects, particularly in cases where data over a period of time are not available (Jain and Singh 2000).

Table 16.2: Cost-Benefit Analysis of Agroforestry System in Saharanpur (Kareemulla et al 2003)

System	Rotation (Years)	Tree Density Ha^{-1}	B:C Ratio (10% discounted)	Net Returns (Rupees Yr^{-1})
Poplar based bund system	8	146	2.84	17,282
Poplar based agrisilviculture	7	481	3.00	28,879
Eucalyptus based bund system	8	146	2.68	16,124
Field crops		Paddy-Wheat System		11,734

In lower Gangetic plain region, agricultural crops like paddy and wheat were grown with Eucalyptus hybrid under irrigated condition. 1,250 trees of Eucalyptus were planted per ha. Yield per ha worked out to be 1,600 kg/ha for paddy and 1,400 kg/ha for wheat. Sale price per kg is Rs. 4.50 for paddy and Rs. 4.00 for wheat. Expenses per ha is Rs. 5,040/- for paddy and Rs. 3,780/- for wheat. Total investment up to 10 years is Rs. 86,237 per ha and total benefits is Rs. 2,00,769 and net returns comes out to be Rs. 1,14,532 (Table 16.3). These observations depict that profits can be achieved by raising of paddy and wheat with Eucalyptus trees under irrigation condition. There is 15 per cent reduction in area under crops due to shade effect of trees from the fourth year onward.

Table 16.3: Input Output Analysis of Agrisilviculture System in Lower Gangetic Plain Region (Planning Commission 2003)

Year	Expenses per Ha (Rs.)				Benefits per Ha (Rs.)				Net Benefits Per Ha (In Rs.)
	Tree	Paddy	Wheat	Total	Tree	Paddy	Wheat	Total	
1	62500	5,040	3,780	71,320	0	7,200	5,600	12,800	-58,520
2	0	5,040	3,780	8,820	0	7,200	5,600	12,800	3,980
3	0	5,040	3,780	8,820	0	7,200	5,600	12,800	3,980
4	0	4,284	3,213	7,497	0	6,120	4,760	10,880	3,383
5	0	3,528	2,646	6,174	0	5,040	3,920	8,960	2,786
6	0	2,772	2,079	4,851	0	3,960	3,080	7,040	2,189
7	0	2,016	1,512	3,528	0	2,880	2,240	5,120	1,592
8	0	1,260	945	2,205	0	1,800	1,400	3,200	995
9	0	504	378	882	0	720	560	1,280	398
10	0	0	0	0	6,25,000	0	0	6,25,000	6,25,000
Net present value @15% DF				86,237				2,00,769	1,14,532

Assumptions

Tree: *Eucalyptus hybrid*; rotation 10 years

No. of trees/ha = 1,250

Income per tree = Rs.500 (in the 10th year)

Expenses per tree = Rs.50 (in the 1st year)

Analysis

Benefit cost ratio at 15% discount

B/C Factor: 2.33

Net present value in Rs. at 15% discount

NPV Factor: 1,14,532

IRR: 32%

Crops	**Paddy**	**Wheat**	
Yield per ha	1,600 kg	1,400 kg	**First Year (Initial) Investment**
Sale price per kg	Rs. 4.50	Rs. 4	Year 1 = Rs. 71,320
Expenses per ha	Rs.5,040	Rs. 3,780	

ECONOMICS OF KHEJRI BASED AGRISILVICULTURE SYSTEM

It is one of the chief indigenous tree of the Indian north-western plains and gently undulating ravine lands. The tree has monolayer canopy and deep taproot system and provides nutritious edible pods and fuelwood to the human beings, nutritious forage to the animals and fertility to the poor soils. Farmers usually grow jowar, pearl millet, moth bean, and cluster bean between the scattered Khejri (*Prosopis cineraria*) trees in arid region. Puri et al (1992 and 1994) observed 86 per cent higher grain yield of barley in association with Khejri. Besides, the yield of chickpea also improved and resulted in higher soil fertility and moisture conservation. Maximum net returns (Rs. 15,197/ha) and benefit cost ratio (3.73) were obtained when pearl millet in kharif was followed by toria in rabi between the khejri trees (Kaushik and Kumar 2003). Yield of fodder crops during both kharif and rabi seasons was higher in association with Khejri trees as compared to sole cropping.

ECONOMICS OF ALLEY CROPPING SYSTEM

The introduction of tree alleys is considered to be an addition to the traditional practice of monocropping agricultural crops. Before economic benefits of a tree alley cropping system can be defined, an accurate estimate of the market value for tree by-products must be derived. In India, no established agroforestry market exists for *Leucaena* and other alley cropping trees for the sale of wood or fodder, although there are well defined markets for similar products e.g. eucalyptus and casuarinas poles, *Acacia nilotica* firewood, poplar wood and other important tree products. It is likely that well established market structure will develop in future for other agroforestry tree species. Prices of these commodities can provide the basis for pegging the market value for tree products for their respective end uses primarily firewood and fodder.

Gross cash returns are higher in all cases of alley cropping systems. Alley-farmed sorghum yielded nearly twice the income of sole sorghum. The alley-farmed pigeon pea system yielded almost seven times the income of sole pigeon pea and this was due to the value of *Leucaena* by-products. Most of the *Leucaena*'s value is derived from its sale as a fodder, which appears to be of greater economic benefit than its application to the crops as a green mulch.

Table 16.4: Economic Analysis of Alley Cropping (TNAU 1984)

Parameters	**Yield/ Revenue**
Sole Crop Sorghum	
Grain Yield	1.55 tons/ha
Stover Yield	5.1 tons/ha
Total Cash Return	Rs. 3,250-7,200/ha
Sole Crop Pigeon Pea	
Grain	0.43 tons/ha
Stover	0.4 tons/ha
Total Cash Return	Rs. 1,800/ha

Parameters	Yield/ Revenue
Alley Cropped Sorghum-Leucaena	
Soghum Grain	1.09 tons/ha
Sorghum Stover	3.9 tons/ha
Leucaena Fodder (Crop Season)	7.2 tons/ha
Leucaena Fodder (Off Season)	3.1 tons/ha
Leucaena Fuel (Stems)	6.5 tons/ha
Leucaena Fuel (Stumps)	3.3 tons/ha
Leucaena Fuel (Seeds)	0.4 tons/ha
Revenue from Sorghum	Rs.2,340-5,660/ha
Revenue from Leucaena	Rs.7,480/ha
Total Cash Return	Rs. 9,820-13,140/ha
Alley Cropped Pigeon Pea-Leucaena	
Pigeon Pea Grain	0.13 tons/ha
Pigeon Pea Stover	0.3 tons/ha
Leucaena Fodder (Crop Season)	10.7 tons/ha
Leucaena Fodder (Off Season)	5.2 tons/ha
Leucaena Fuel (Stems)	10.7 tons/ha
Leucaena Fuel (Stumps)	0.3 tons/ha
Leucaena Fuel (Seeds)	0.9 tons/ha
Revenue from Sorghum	Rs. 1,720/ha
Revenue from Leucaena	Rs. 10,360/ha
Total Cash Return	Rs. 12,080/ha

ECONOMICS OF SILVIPASTURE SYSTEMS

Kareemulla et al (2002) had estimated the economics of introduction of silvipasture system in the degraded lands both for conservation and livelihoods. The results (Table 16.5) indicated that in an eight year rotation on one ha with *Albizia amara* + *Dichrostachys cinerea* + *Leucaena leucocephala* with *Chrysopogon fulvus* + *Stylosanthes hamata* + *S. scabra* the B:C ratio was 1.52 at a 12 per cent discount factor.

Table 16.5: Economics of Silvipasture System in One ha on Degraded Lands

Parameter	Indicator or Value
Gross returns (Rs.)	56,936
Cost of cultivation (Rs.)	36,835
Net returns (Rs.)	20,101
Annual average return (Rs.)	2,512
B:C ratio at current prices	1.55
B:C ratio at 12% discount prices	1.52

BAMBOO AND ANJAN GRASS BASED SILVIPASTURE SYSTEM

An indigenous technology for enhancing productivity of ravines has been evolved at Dehradun based Central Soil and Water Conservation Research and Training Institute (CSWCR&TI) Research Centre located at Vasad in Anand district, Gujarat, which has been adopted by various developmental organizations. The bamboo and anjan grass (*Cenchrus ciliaris*) based silvipasture system has been developed for enhancing productivity of ravines. Ravine lands can be successfully reclaimed by planting bamboo, a fast growing plant species, on the gully beds and anjan grass on the side slopes and the interspaces of gully bed for economic utilization of gullied land. This technology gives high returns and also checks water erosion, thereby preventing soil loss from ravines. The cost of planting bamboo and sowing grass slips in ravine works out to be about Rs. 22,000/- per hectare. During the initial years, green fodder of anjan grass can be harvested from the degraded ravines every year. Planting of grass also protects the ravine slopes and reduces soil loss and about 7.1 tons ha^{-1} $year^{-1}$ and green fodder can be obtained from the stabilized slopes. The grass yield from inter spaces of bamboo planted on bed portion is about 10 tons ha^{-1} $year^{-1}$ during initial 5 years. The grasses fetched an income of Rs. 3,000 to 6,000/- per ha over the period of 5 years at Vasad (Gujarat).

About 300 clumps of bamboo with 3,000-4,000 old and 1,000-1,500 new culms are available in one hectare bamboo plantation of over seven years old. About 30 per cent of old culms can be harvested easily per year. The bamboo planted on the ravine bed can fetch Rs. 6,000 to Rs. 27,000 ha^{-1} $year^{-1}$. The watershed with this silvipasture system absorbs more than 80 per cent of rainfall that is either utilized by the plant or percolated deep to recharge the ground water. Due to less runoff, soil loss is reduced to less than 1.0 tons ha^{-1} $year^{-1}$ only from about 20 tons ha^{-1} $year^{-1}$ from degraded ravines prior to plantation (Anon. 2009; ICAR 2013).

ECONOMICS OF AGRIHORTICULTURE SYSTEMS

Aonla (*Emblica officinalis*) is one of the important fruit crops of arid region. However despite its adaptability in arid climate, productivity and economic return of the crop are affected by low temperature and frost (Anon. 2006) which causes economic losses to the growers. In order to mitigate the risk of total crop failure, suitable multistorey crop combinations were evaluated at Central Institute for Arid Horticulture during 2004-2008 in the interspace of aonla orchard which could generate income, improve productivity, ameliorate and improve ecological situations (Awasthi et al 2009). Moth bean grown during kharif season was a common ground storey crop grown in rotation with rabi crops i.e. fenugreek, chick pea, mustard and cumin. Growth parameters of aonla were recorded significantly more with intercrops compared with sole plantation. Higher grain and straw yield were recorded in moth bean-chickpea (497 and 1,250 kg/ha) and moth bean-fenugreek (465 and 1,165 kg/ha) crop sequence.

PREFERRED AONLA BASED AGRIHORTICULTURE FOR ARID REGION

Aonla + Bael + Karonda + Moth bean + Gram

Aonla + Khejri + Cluster bean + Cumin

Aonla + Drumstick + Senna + Moth bean + Mustard

Aonla + Ber + Senna + Moth bean + Mustard

In Pratapgarh District of Uttar Pradesh aonla production spread in area of 12,000 ha (Pathak 2003). The present study was taken up to assess the present status of the nurseries and to analyze the costs and benefits of the enterprise. The average production capacity of the nursery is 30,000 saplings per annum. The annual cost of cultivation of an average aonla nursery worked out to be Rs. 1,24,399 costing Rs. 4/- per sapling (Table 16.6). Almost 92 per cent of the cost of cultivation was accounted for the operating (variable) costs and rest 8 per cent was on account of fixed costs. Among the variable costs, the highest was imputed salary for the nursery entrepreneur followed by labour and seed. The depreciation on tube well, pump set, pipeline and buildings were taken into account under the fixed costs.

Table 16.6: Cost of Cultivation for an Average Production of 30,000 Aonla Seedlings per Year (Pathak 2003)

Items/ Particular	Cost (Rs.)
Seed	11,900
Electricity	3,000
Labour	24,470
Salary of supervisor (Imputed wages for nurseryman)	36,000
FYM, fertilizer, irrigation, etc.	7,300
Plant protection	500
Interest on working capital @ (8%)	6,654
Opportunity cost of investment @ (8%)	24,600
Total variable costs	1,14,424 (92%)
Total fixed costs (installation of tube well, pipeline, etc.)	9,975 (8%)
Total cost per annum	1,24,399 (100%)

The financial analysis of nurseries was done at an investment of Rs.3.25 lakh per nursery being imputed value of land and capital investment on irrigation. The returns were estimated at an average sale price of Rs. 11.2 per sapling. The net annual returns worked out to be Rs. 1,27,186. The payback period was 3 years. The discounted B:C ratio was 1.14 and IRR was 23.69 percent. These financial ratios indicate the soundness and economic viability of the nurseries.

In agrihorticulture system of central Plateau & Hill regions, crops of groundnut and gram were grown with aonla (*Emblica officinalis*) under rainfed condition. Input/output analysis is given in Table 16.7. Two hundred trees of aonla were planted per ha with intercropping of groundnut and gram, and their yield were recorded of 800 kg and 1,000 kg per ha, with sale price of Rs. 8 and Rs. 9 respectively. Total cost of investment was Rs. 86,494 per ha whereas net benefit of Rs.7,410 per ha was obtained.

Table 16.7: Input Output Analysis of Agrihorticulture System in Central Plateau and Hill Region (Planning Commission 2003)

Year	Expenses per Ha (Rs.)				Benefits per Ha (Rs.)				Net Benefits/ha
	Aonla	Groundnut	Gram	Total	Aonla	Groundnut	Gram	Total	
1	12,000	5,000	5,500	22,500	0	6,400	7,000	13,400	-9,100
2	8,000	5,000	5,500	18,500	0	6,400	7,000	13,400	-5,100
3	8,000	5,000	5,500	18,500	0	6,400	7,000	13,400	-5,100
4	8,000	5,000	5,500	18,500	12,800	6,400	7,000	26,200	7,700
5	8,000	5,000	5,500	18,500	16,000	6,400	7,000	29,400	10,900

Year	Expenses per Ha (Rs.)				Benefits per Ha (Rs.)				Net Benefits/ha
	Aonla	Groundnut	Gram	Total	Aonla	Groundnut	Gram	Total	
6	8,000	5,000	5,500	18,500	19,200	6,400	7,000	32,600	14,100
7	8,000	5,000	5,500	18,500	16,000	6,400	7,000	29,400	10,900
8	8,000	5,000	5,500	18,500	12,800	6,400	7,000	26,200	7,700
Net present value @15% DF		86,494					93,903		7,410

Assumptions

Tree: *Emblica officinalis*; rotation 8 years

No. of trees/ha = 200

Yield per tree:

Yr 4=8 kg; Yr 5=10 kg; Yr 6=12 kg,

Yr 7=10 kg; Yr 8=8 kg

Sale price per Kg = Rs. 8

Expenses per ha: Yr 1 = Rs.12,000;

Yr 2 onwards = Rs. 8,000

Crops	**Groundnut**	**Gram**
Yield per ha	800 kg	1000 kg
Sale price per kg	Rs. 8	Rs. 7
Expenses per ha	Rs. 5,000	Rs. 5,500

Analysis

Benefit cost ratio at 15% discount

B/C Factor: 1.09

Net present value in Rs. at 15 % DF

NPV Factor: 7410

IRR: 27%

First Year (Initial) Investment

Year 1 = Rs. 22,500

TWO TIER CROPPING SYSTEMS BASED ON POPLAR PLANTATION

In north-western Indo gangetic plain, poplar (*Populus deltoides*) based agroforestry systems have been adopted by the farmers. Wheat, oats, sorghum, maize, sugarcane, berseem, turmeric, ginger and potato can be easily grown as intercrops in poplar plantations. The yield of intercrops decreases after two to three years of cultivation but this yield loss is often compensated by the sale of poplar wood at the end of rotation. It is apparent from the Table 16.8 that intercropping of poplar plantation with sugarcane and turmeric crops during six years of cropping period was found highly economical and resulted in 2-3 times more net returns as compared to poplar sole crop as well as intercropped with wheat, oilseeds/pulses on the rabi cropping seasons.

Table 16.8: Economics of Two Tier Cropping Systems based on Poplar Plantation (Chahal et al 2012)

S No	Treatment	Gross Returns from Poplar (Rs./ha)	Cost of Cultivation of Poplar (Rs./ha)	Net Returns from Poplar (Rs./ha)	Net Returns from Other Crops (Rs./ha)	Total Net Returns in Six Years (Rs./ha)	Average Net Returns (Rs./ha/yr)
1	Poplar + Sugarcane	3,76,464	24,600	3,51,864	34,264	3,86,128	64,355
2	Poplar + Turmeric	2,78,760	24,600	2,54,160	03,100	3,57,260	59,543

S No	Treatment	Gross Returns from Poplar (Rs./ha)	Cost of Cultivation of Poplar (Rs./ha)	Net Returns from Poplar (Rs./ha)	Net Returns from Other Crops (Rs./ha)	Total Net Returns in Six Years (Rs./ha)	Average Net Returns (Rs./ha/yr)
3	Poplar + Wheat/ Oilseed crops	2,03,184	24,600	2,00,784	16,781	2,17,565	36,261
4	Poplar + Rain-fed Wheat/ Lentil	1,18,800	24,600	94,200	18,116	1,12,316	18,719
5	Poplar	1,45,728	24,600	1,21,128	-	1,21,128	20,188

Poplar, sugarcane and turmeric based two-tier cropping system is highly remunerative and more economical than monocropping of poplar as well as the crops and good alternative to paddy-wheat rotation in Indo-Gangetic Plains.

CONCLUSION

Agroforestry has the potential to significantly improve the livelihoods, economic viability and agricultural production of small farmers in traditionally marginanalized areas. In addition, agroforestry can be used as a tool for changing attitudes towards natural resource management (NRM). Agroforestry produces multiple benefits, including increased income from selling surplus goods, improved environments and enhanced biodiversity, increased access to education through income generation, shelter from trees. Diverse impacts of agroforestry need to be evaluated for its economic and environmental contributions. It can be feasible by assessing the rates of returns on investment for various alternative agroforestry practices. There is need to assess the effect of agroforestry development on the society, region and environment.

■■■■

17 MARKETS AND MARKETING OF AGROFORESTRY PRODUCTS

INTRODUCTION

The agroforestry market is a place where buyers and sellers transact forest (timber and non-timber) products, agricultural and livestock products at a particular place and time. The marketing is considered as the study of entire gamut activities that direct the flow of goods and services from the primary producer to ultimate consumers. The importance of marketing includes optimization of resource use and output management in agroforestry; increase in farm income; widening of markets; growth of forest based industries; price signals to all stakeholders; adoption and spread of new technology; generating employment & income at a regional and national level; enhancing better living of the people; and creating a form, place, time and possession utilities. The classification of agroforestry markets is based on area, time, commodities, volume and competition dimensions such as local or village markets, regional markets, national markets & world market (based on area); short & long-period markets (based on time span); common markets, specialized markets (based on commodities); whole sale markets, retail markets (based on volume of transaction); perfect and imperfect markets (based on competition). Under imperfect markets based on supply side, oligopoly and monopolistic markets are found and based on demand side, monopsony, duopsony, oilgopsony and monopsonistic markets are found in different agroforestry products. The major participants in the markets are producers, traders, industry, government and consumers.

The major problems and issues of agroforestry products in view of production are long gestation period, natural based production, bulkiness of products, season oriented production, highly perishable nature of some products, variation in quality of produce and discrete supply, sporadic production points and lack of value addition at farm and intermediate levels. On the point of price, there are issues such as unstable prices or price fluctuation of agroforestry products, forced sale/ distress sale of tree growers, low collection charge/ low wage rate to workers in agroforestry activities, farmers' insufficient knowledge on market information such as price & quantity of arrival, monopoly of traders dictating the market prices of agroforestry products. Most of the farmers are not having computer knowledge and access. On view of promotion, there are problems such as lack of post harvest facilities at farm level; lack of capital to engage in promotional activities; assembling, grading, standardization are not fully possible at primary level which affects quality of the produce; lack of entrepreneurial skills among tree growers; lack of organization among owner/ producer/ promoters; and lack of infrastructure viz. go down & store house; on view of distribution channels such as existing cultural barriers, lack of transportation facilities, long way to reach the consumer, unorganized end users, more middle men and higher marketing cost.

Market facilitates the exchange of goods and services and its types are distinguished based on certain structural characteristics. Most of the markets for agroforestry products are imperfect markets with the added intervention of middlemen. The demand for agroforestry products are arising out of the demand for forest based products which are derived demand. The supply of sawlogs from private agroforestry areas are mainly based on cost of availability and cost of stumpage. The gap of demand and supply has widened over the years due to stringent forest policy, the supply of raw material is almost stopped from the conventional forest area and this paves the way to import forest products to feed our industries and consumers in our country. In timber and non-timber trade, producer is the sheer price followers. The government intervention is also noticed in marketing of forest products, but producers are presently not appropriately linked with end-users. This is the reason for low producers' share in consumers' rupees for agroforestry products and short supply. The industry based agroforestry models support the flow of products like pulp and paper wood, match wood and biomass in market to mitigate the short supply. The different marketing channels are also noticed in the markets, among the channels, industry linkage based marketing channel of contract farming is factitively supporting the tree growers in the markets. Mostly *Casuarina*, *Eucalyptus* for pulpwood and *Ailanthus* for match wood are intensively cultivated under contract farming and marketed by the buyback system with paper industries in Tamil Nadu viz., Tamil Nadu Newsprint and Papers Limited, Seshasayee Paper and Boards Limited. In order to develop well organized and efficient supply chain for agroforestry products, a proper linkage among all stakeholders with assured buyback system under perfect competition is warranted to eliminate the intervention of middlemen and fetches better net farm income. This will augment the supply of raw materials to agroforestry based industries on a sustainable basis to promote industry and households.

TIMBER MARKETS

The leading wood exporters are Scandinavian countries (Sweden, Norway, and Finland), Russia, Southeast Asian countries (Myanmar, Malaysia, Indonesia, Vietnam, Thailand, Cambodia, and Papua New Guinea), Australia, Democratic Republic of Congo, Ghana, Nigeria, Canada and Amazonian countries (Brazil, Peru, Chile, and Bolivia) where as leading importing Countries are Japan, China, India, UK, USA and some European Countries. Much of logging in tropical rainforests of developing countries is reported to be illegal. It is reported that nearly 65 per cent of the world's illegal supply of timber are from Southeast Asia. The illegal supply is chiefly directed to European Countries, USA, UK, China, Japan and Korea. The important timbers such as Kwila (*Afzelia australis*) from Papua New Guinea, Pyinkoda (*Intsia bijuga*) from Brazil, Pine wood (*Pinus radiata*) from New Zealand and (*Pinus sylvestris*) from Europe, Ebony (*Diospyros ebenum*) from Africa, Mora (*Mora excelsa*) from Guyana & South America, Tulip Tree or Yellow Poplar (*Liriodendron tulipifera*) and Purple Heart (*Peltogyne* spp.) from Brazil, Teak (*Tectona grandis*) from Ivory Coast, South America & Nigeria, Padauk (*Pterocarpus dalbergioides*) from Andaman and Myanmar, Sal (*Shorea* spp.) from Indonesia and Malaysia are traded throughout the World.

The population of India has grown more than 1 billion and per capita income approximately US$ 300, which by the year 2050 AD are likely to grow to nearly 2 billion and US$ 2000, respectively. The total wood requirement of the country in the year 1970 was 160 million cum and has ever since been growing and is expected to more than 400 million cum mark by 2015. The industrial wood demand will also rise from 13 million m^3 level of 1970 to 35million m^3 by 2015 AD and 110 million m^3 by the year 2050 AD (Chandra 2001). Most wood used in India is hardwood, particularly teak which accounts for approximately 50 per cent of domestic consumption. Imports of teak logs are mainly from Burma, Malaysia, Thailand and Africa. In south India, Thoothukudi port is one of six major ports importing timber particularly from Southeast Asian countries. Other ports are Kandala (Gujarat), Mumbai, Mangalore on the West Coast, Chennai and Kolkatta on East Coast. The leading traders are importing logs from abroad countries and on arrival keep them in timber warehouses. The traders have to pay a fee for storing their logs in warehouses in all major ports. The sale is not allowed at warehouses. The logs will be transported to wood depots or industries after sale. The destinations of logs depend on where there is demand. The imported Logs at Thoothokudi port may go to Northern districts of Tamil Nadu and Chennai port logs may find their way to the southern districts. Teak logs reaching west coast ports mainly from West African countries move all over the country. Movement of the logs is mostly by road for prompt delivery. Since import and export serve the demands of hinterland collectively, it is not possible to say the exact consumption of a particular region or state import and distribution of wood.

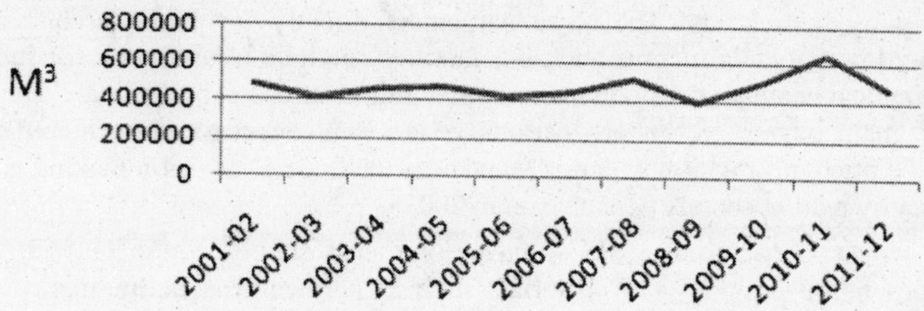

Fig. 17.1: Import of Logs at Thoothukudi Port (Tamil Nadu) and Warehouses

Most Indian wood imports, particularly logs are controlled by a small group of timber merchants who deal primarily in teak logs from Myanmar, Africa, Ghana, etc. and radiata pine logs from New Zealand. Architects and interior designers who source wood from these timber merchants have to go by merchant's business interest rather than insist on introducing new woods. Typically, a timber merchant will square imported or domestic logs into ten foot sections and sell these to carpenters and other end users. The total volume of logs into these Kandala, Mumbai and Mangalore ports is estimated to be 2,88,000 cubic meters per year. Most Indian timber merchants feel that supplies of African teak logs, primarily from Nigeria, may last for another 3 years and then they will have to source from elsewhere. Currently, the quality of teak from Burma and Africa is declining rapidly (Kondas 2012). Major imported wood species are tropical woods such as garjan, mahogany, and sapeli; plantation timber includes teak, eucalyptus, and poplar, as well as spruce, pine, and fir.

The top three export markets for India's wood products are the USA (23%), Germany (10%) and UK (9%). Serious fluctuations in the rupee and the US dollar exchange rate have affected imports and in recent weeks the 10 per cent rise in import costs of logs due to the depreciation of the rupee cannot easily be absorbed by industries say local observers. In principle, India imports roundwood logs in order to engage hundreds of sawmills and provide employment. In certain markets for *Eucalyptus* in Yamunanagar, it was found that supply curve was completely inelastic and buyers, normally commission agents, determined the price, depending upon their assessment of requirements of mills around. The recent ban on commercial forest cutting in the Northeast, Himachal Pradesh and Jammu & Kashmir, and huge imports of certain species of timber had also substantially influenced the timber market. The main species available in this region are teak, sal, shisham, eucalyptus, poplar, kail, fir, chirpine and deodar. The timber markets are of a mixed variety, combining both wholesale and retail markets. Most of the biggest traders own and operate sawmills as well. A number of traders are commission agents who act as middlemen organizing auction of the timber brought to the market by the farmers.

On the supply side, there are a number of State Forest Development Corporations who arrange the felling of trees, including annual one, as also grow their own forests mainly by the side of roads and railway lines and sell the timber by means of open auction at a number of auction centers. They send intimation of the auction to traders. The forest corporations have replaced private forest contractors who used to take lease of forests and undertake felling of trees and sale of timber. A number of states have seen fairly rapid growth of agroforestry mainly consisting of eucalyptus, poplar, shisham, etc. The bulk of timber demand is for the production of furniture, doors, windows, their frames, pulp and paper, packaging materials. The big builders, contractors, furniture units, ply units, paper mills, etc. are bulk consumers of timber. In India, most of the house construction activity is undertaken directly by households. They hire carpenters directly for making cupboards, doors, furniture, etc. Many lower middle income and poor families continue to use wood for fuel as well as for domestic cooking and heating. Thus one may say that industrial users, builders and contractors and household are the main buyers of timber. Various government agencies also enter the picture as buyers. Similarly the price information which is relevant to the traders' decision making is the price of the same and/or nearby market centers (Chaubey et al 2001).

Many State Forest Corporations (SFCs) have long-term supply contracts with large wood industries. The corporation provides raw materials such as pulpwood at below-market prices. The SFCs operate most of the timber auction throughout India. Observations suggest that these auctions obtain less than optimal prices. One option for gradual change in auctions might be introducing sale-based commissions for auction operators, so that they would have more incentive to maximize auction revenue. Another possibility would be to allow local communities to have the option of marketing their share through purchasing agents or other private channels (Jessica Mott 2001). The rapid spread of agro forestry in Haryana and Punjab has been possible because of wide relaxation permitted by timber transit rules. Fixing a support price for forest produce generated by the farmers could be assured and encouraging rural producers to intensify tree planting and protect the growers from exploitation by the dealers (Chandra 2001).

MAJOR MARKETS FOR FOREST, FARM AND AGROFORESTRY PRODUCTS

There are a number of well established large markets for forest, farm, agroforestry products_in different parts of the country from where wholesale trade is carried out and presented in Table17.1.

WOOD BASED PRODUCTS TRADE IN INDIA

The major import of wood products was from China and Malaysia (accounted 60 per cent) and the rest met by Italy, Germany, USA, Singapore, Sri Lanka, Hong Kong and Taiwan. The increase in imports resulted in a number of retail outlets for high-end imported furniture in major cities across India. With the introduction of economic reforms in the early 1990's and subsequently domestic ban on most logging, a unique market opportunity is slowly emerging for the import of lumber and wood products. In order to protect the local saw milling industry, the import of logs and squared logs free of import duty were allowed but still imposes high tariffs on most finished goods. India also produces 157 million sq.m of plywood and decorative veneers and 220,000 tons of particle board, and Medium Density Fibre (MDF) annually (Negi 1998). Most wood in India is used for the manufacture of doors, window frames, wall panels, moldings and furniture. Domestic manufacturing is highly fragmented and unorganized with much of the production, particularly for doors, windows and interiors still by individual carpenters on site. The lack of standard sizing is still a major impediment for Indian manufacturers but is expected to improve as more pre-fabricated products come into the market. The manufacture of pre-fabricated doors and windows is relatively new and the current market is estimated at US$ 20 million and growing at 10 per cent per annum. The total annual market for furniture in India is estimated to be 1.25 billion about 90 per cent of which is for wooden products. The branded (higher quality) wooden furniture industry is estimated to be 37 million and growing at 15 per cent annually. Wood peeling and slicing technology to make veneers for plywood is a universally accepted efficient way of wood utilization. Rubber wood once considered as firewood has moved into a world of niche products with export value. The export of paper wood products and newsprint were accounted as US$ 200,15,76,000 and US$ 67,45,000 respectively in 2011-2012. The import of pulp & waste paper and newsprint were accounted as US$ 52,70,51,000 and US$ 101,43,68,000 respectively in 2011-2012.

Table 17.1: Major Markets for Forest/Farm/Agroforestry Products in India (Negi 1998)

State	Market
Andhra Pradesh	Chittor, Hyderabad, Janaram, Karimnagar, Kothaqudem, Mancherial, Rajamundry, Vijayawada and Warangal
Assam	Dhoraj, Guwahati, Jetpur, Jorhat and Prilkia
Bihar	Daltonganj, Muzzafarpur and Ranchi
Gujarat	Amreli, Junagadh, Songarh and Umarpada
Haryana	Chika, Kurushetra, Kaithal, Ladwa, Peoha, Pundri, Radhaur, Shahabad, Thanesar and Yamunanagar
Madhya Pradesh	Akaltara, Balaghat, Bilaspur, Jabalpur, Jagdalpur, Mungeli and Ratnapur
Maharashtra	Ballarshah, Chandrapur, Nagpur, Pune and Thane
Odisha	Jeypore, Puri and Rayagada
Punjab	Ajnala, Amritsar, Hoshiarpur, Jalandhar, Kapurthala, Ludhiana and Pathankot
Tamil Nadu	Andhiyor, Denkanikottai, Dharmapuri, Chennai, Salem and Sengottai
Tripura	Agartala, Dhramnagar and Udaipur
Uttar Pradesh	Jhansi, Jwalapur, Kanpur, Ramnagar and Tanakpur
West Bengal	Calcutta

The British Colombia Trade and Investment Office (BCTIO) engaged in India to explore the lumber and wood products industry in 1998. Some private plantations, particularly of rubber wood also contribute to the domestic supply of lumber. Particleboard and Medium Density Fiber (MDF) boards have been finding more use in the Indian market. As an example, some Indian companies are now importing particleboard panels for use in kitchen cabinets from Italy, Australia and New Zealand. There has been a gradual decrease in import duties on lumber and wood products from 150 per cent in 1991 to 50 per cent in 1998. Even then, Indian tariffs are currently very high, particularly on wooden furniture. Until 1995, a 15 percent duty was imposed on log imports and this has been completely eliminated. This move was taken primarily to ensure that the Indian saw milling industry had steady supplies of timber. The duties on dimension lumber and finished products were at one time well over 100 per cent and have now been reduced to the 30-80 per cent ranges. Indian businessmen and manufacturers feel that over the next two to three years, these restrictive tariffs may continue to decline to more reasonable levels. The Indian import tariffs for products are 38, 43, 69, 63, 65, 69 and

81 per cent for dimension lumber, wood doors & windows, wood flooring , wooden railway sleepers, particle board & OSB , plywood veneered panels & veneer sheets, and wooden furniture respectively. Imports of fuelwood and wood in log form attract less import tariff whereas other wood products attract tariffs exceeding 30 per cent. Imports of fiber and particle board, veneer and sawn lumber have also increased to meet rising demand in the housing, construction, household, products and packaging sectors (Kondas et al 2012). Imports of processed wood products such as plywood, particle wood, veneer, etc., are exempted from phytosanitary control. The changes of wood and wood products over 12 years are presented in Table 17.2. It could be seen from the table that the import of wood logs and veneer had a negative change where as other products, wooden furniture, particle board and MDF had a positive change and also not much change was noticed on sawn lumber and plywood during 1996-2008.

Table 17.2: Changes of Wood and Wood Products in India during 1996-2008 (Kondas et al 2012)

Item	1996-97	2008-09	Percent Change
Wood Logs (HS 4403)	247	1,147	-24
Sawn Lumber (HS 4407)	5	31	0
Veneer (HS 4408)	6	21	-1
Plywood (HS 4412)	4	34	0
MDF/ HDF (HS 4411)	1	51	3
Particle Board (HS 4410)	4	71	3
Other Products	4	1,397	80
Wooden Furniture (HS 9403)	1	324	19
Total Wood Products	271	1,722	

WOOD PRODUCTION IN AGROFORESTRY AND ITS SALE AND MARKETING

WIMCO initiated and strive to establish poplar based agroforestry as a profitable enterprise with the hope of securing supplies of raw material for its match industries under a buy back guarantee on a fixed price at the time of harvest. Over the period of time the demand for poplar wood and its price escalated to levels, which made the material unaffordable to the match industry, and instead the produce has now become the main raw material for plywood manufacture in the Northwest region of India. However, fixing a support price for forest produce generated by the farmers could be assured and encouraging rural producers to intensify tree planting and protect the growers from exploitation by the dealers (Chandra 2001). Most of the wood based industries like pulp and paper, veneer and plywood, hardboard and safety match units are facing increasingly large deficits between requirements and supplies of wood based raw material from forests which are owned and managed by state governments. Agroforestry plantations could play a vital role in supplementing the wood production and conservation of natural forests. Simultaneously, the production of food grains, oil seeds, cash crops and fodder to be achieved (Rao 2001)

Joginder and Grover (2001) revealed that the market pattern for Eucalyptus plantation such as border plantation and block plantation was done at different stages of maturity. The farmers preferred pre-harvest sales due to high labour input involved in harvesting and malpractices prevailing in the wood markets. It was found that 72 per cent of the sample farmers under both categories resorted to pre-harvest sales. It was revealed that most of the producers while affecting pre-harvest sales did not have any estimate of the quantity of wood of their plantation. They struck the bargains in an arbitrary manner. Thus, they were badly exploited by the traders who had sufficient knowledge about the quantity and quality of the wood and the price to be fetched in the market. The important marketing channels for Eucalyptus are presented in Fig. 17.2.

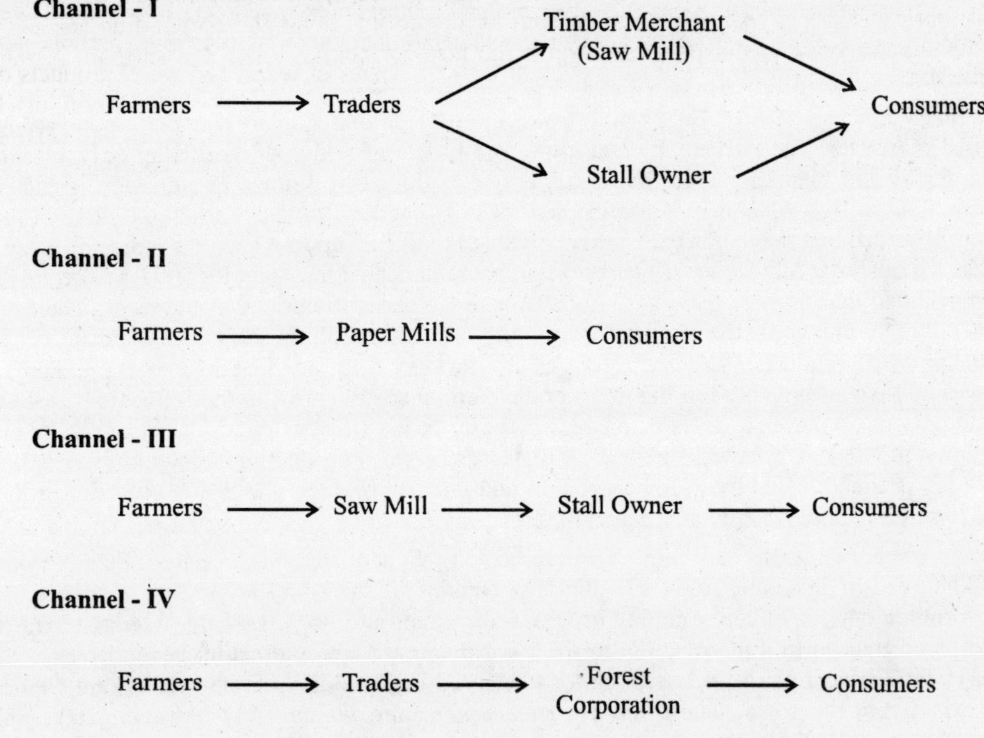

Fig. 17.2: Marketing Channels for Eucalyptus in Punjab

As a result of strong propaganda in favour of Eucalyptus and consequently it was heavy plantation during the seventies, the arrival of Eucalyptus created a glut in the wood markets during the eighties. The government intervention was, therefore, considered necessary to safeguard the interest of the farmers. The Punjab State Forest Department had a price support operation in case of Eucalyptus. The corporation purchased the Eucalyptus trees with a minimum girth of 20 cm from agroforestry plantations growers. The corporation further disposed its stock as fire wood, pulp and timber. The pulp was supplied to the paper mills. The corporation had also set up its own processing plants and converts the timber and other Eucalyptus wood into beams, frames of doors, windows and ventilators and furniture. Besides these, corporation entered into a contract with the industries manufacturing packing boxes, particle barrels and wood charcoal. Thus, the corporation was helping the Eucalyptus producers in disposing their produce, though the prices offered by it were not so remunerative.

The forest department corporation had a responsibility to purchase and felling of *Eucalyptus tereticornis* and *Populus deltoides* in farm lands at a predetermined price (Dhar and Ahuja 2001). Haryana Forest Development Corporation as the sole agency involved in the purchase, fell and dispose of *Acacia catechu* trees from The Shivalik belt of Haryana. The fuelwood was supplied to the Municipal Corporation, Delhi; Sal sleepers to the Indian Railways; mining props of *Eucalyptus* wood to Bharat Cooking Coal Ltd., Dhanbad and Mahanadi Coals Fields, Talcher; and Sal wood planks to Bhillai Steel Plants, Bhillai.

In U.P, Forest Corporation entered into the market to purchase standing Eucalyptus trees of the farmers with declared support price (diameter wise). In the beginning, the corporation did get a large quantity of Eucalyptus timber especially in western U.P. and the local contractors who were exploiting the farmers by way of offering ridiculously low prices of standing trees started feeling themselves to be outside the market. The farmers even started bargaining with the contractors with the result that they even got more prices than what was offered by the corporation. The corporation successively revised the support price of not only Eucalyptus but also for poplar, khair and shisham (Gaur and Sarkar 2001).

The steep fall in prices of Eucalyptus wood in Haryana and Punjab was due to monopoly of wood based industries. The farmers, by and large, had little knowledge of marketing or wood market grown in agrforestry plantations and therefore, easily exploited by village commission agents. In a study carried out in five villages in Haridwar district to know the awareness about the price of poplar tree, it was observed that the majority of farmers with a holding of up to 4 hectares indicated Rs.500 to Rs.1000 as the cost of a full grown poplar tree, and some indicated less than Rs.500.00 per tree. It can thus be concluded that farmers have very little knowledge about the prices of the trees. The farmers were generally ignorant about the market demand for various sizes of trees/logs and the different ways in which the wood product is utilized. Most of the farmers sold standing tree to a middleman or contractor who then arranged to fell the trees, grade them, made payment to the farmers and took it to the market. About 50 per cent cases of distress sales by farmers to contractors at throw away prices were common in villages. This was due to long distances between production point and market, timber being a bulky product, difficulty in transportation and poor knowledge about sale procedures, farmer preferred to sell it locally (Sharma and Suman 2001).

Village level contractors, saw-millers or middlemen purchase standing tree from the agroforestry farmers. The trees are generally sold on the basis of weight and the prices are settled after bargaining. They perform all activities like the felling of trees, conversion into logs, grading, making payment to the farmers and transporting the material to the local market. Such contractors generally exploit the agroforestry farmers and pay very low price for their produce. Mandis (private market) are wholesale markets existing in the town. These are permanent in nature where daily marketing takes place throughout the year. All sorts of facilities exist in these markets like sawing, storage, handling, transport and banking services. The tree growers generally harvest the trees and bring the timber to the mandis for sale. In these mandis, the wholesalers, commission agents and pucca arhatiya plays important roles in the sale of produce. The produce is generally purchased from farmers through brokers. In Haryana, Ambala and Yanmunanagar have developed into bigger markets for the purchase of farm grown timber. Poplar (*Populus deltoides*), *Eucalyptus* spp., *Dalbergia sissoo* and *Acacia nilotica* from adjoining states like Uttar Pradesh, Uttaranchal and Punjab are also brought to these markets for sale. Poplar trees are being marketed in the form of round logs as well as sawn timber. Due to non-availability from departments, industries like matchwood (WIMCO), and paper mills have established their depots or purchase it directly from the growers/farmers.

The marketing in Haryana and Western Uttar Pradesh involves various intermediaries functioning in between producer to consumers which takes away a maximum share of the value of the product. But if, farmers have direct access in consumer/industry, they may derive maximum profit for their produce. Mandis are the permanent market places where its marketing takes place round the year. The wholesalers and commission agents play role in the sale of produce. Poplar (*Populus deltoides*) is brought to these Mandis by the villagers themselves or at times procured by purchasing from the growers through brokers. The important poplar markets in Haryana are Yamunanagar, Ambala, Karnala, and Punipat. In Punjab, Amristar, Gurdaspur, Hoshiarpur, Jalandhar, Kurali, Ludhiana, Nangal, Patiala, Phillaur, Phagwara, Pathankot, and Roopnagar are the common markets.

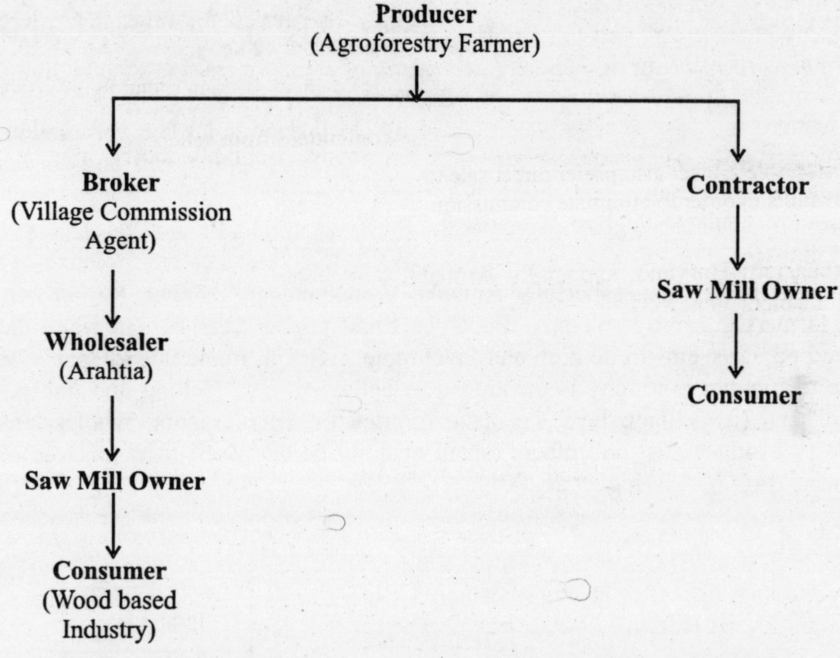

Fig. 17.3: Distribution Channels for Wood based Products in Haryana

In Punjab, wood markets are largely in the hands of private traders. The mature Eucalyptus and Poplar trees are being sold in the market. The tree growers normally sell their produce/wood to the private traders at the lump sum price or on the basis of weight. They also prefer to cut the trees themselves and bring the wood in log form as well as firewood on their own tractors to the nearby wood markets, where it is sold at auction by weight. The details of various practices of the wood markets (Mohindra 2001) of the State are summarized below:

WOOD MARKET IN PUNJAB

(a)	Method of sale	: Auction
(b)	Commission of Arhat charges collected from the seller	: 0 to 6.25%
(c)	Weighing charges in rupees per vehicle charged from the seller	(i) Truck Rs. 3 to Rs.10
		(ii) Trolley Rs.2 to Rs.8
		(iii) Cart Rs.2 to Rs.4
(d)	Unloading charges of the buyers	: Done by the sellers to premises
(e)	Deduction in weight due to dryness per quintal	: 5Kg to 10 Kg
(f)	Deduction in payment	: Payment same day about 2% cash discount. Quality discount at the time of unloading due to knotted portion is also charged.
(g)	General	(i) Produce is loaded, sorted or graded in lots, then transported to the premises of buyer.

(ii) Payment made for weight in multiples of 40 kg only. No payment for weight in kg exceeding multiples of 40 kg.

(iii) Payment made in round figures only.

(iv) Market fee @ 2% is also charged by market committees from seller/buyer.

Normally, tree grower also prefer direct sale to the sawmills in order to eliminate commission charge

In Northeastern Haryana, especially Ambala, Yamunanagar, Panipet, Karnal and Kurukshetra districts, the farmers generally sell their Eucalyptus and poplar trees as and when the demand for Eucalyptus and poplar seems to be high and favourable. Existing marketing channels being practised through three to five tier hierarchy is mentioned as below: (i) Self (felling and transportation to the market by owner), (ii) Village level agents or contractor, (iii) Traders (wholesalers, retailers or arahatia), and (iv) Industrialist and others (small quantity holders) sell their produce directly to any suitable agencies (Fig. 17.4) (Negi et al 2001).

Channel – I

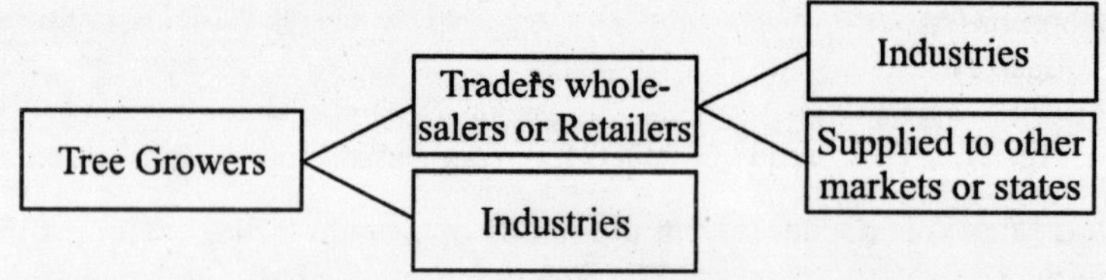

Channel – II

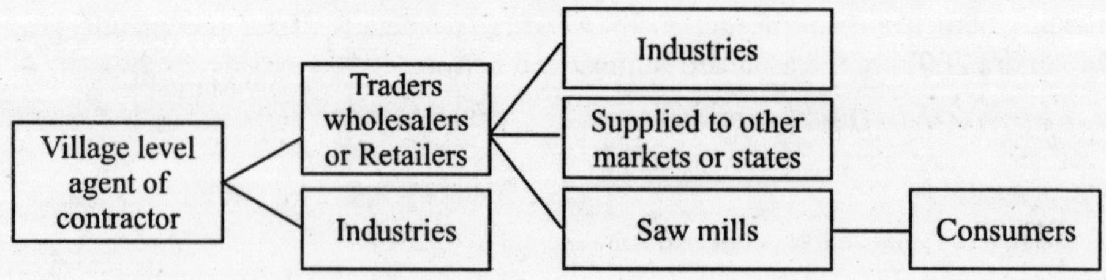

Channel - III

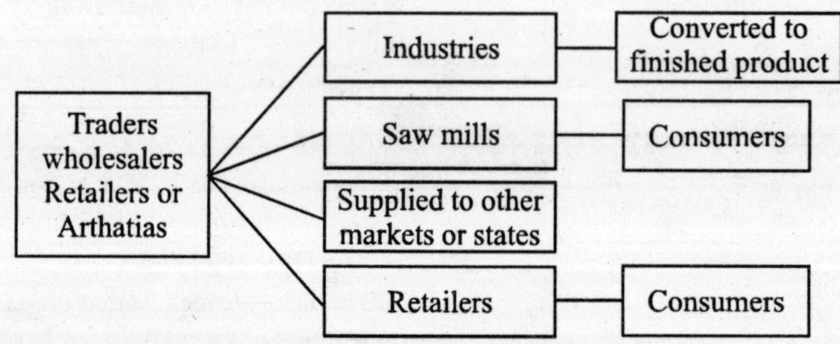

Channel - IV

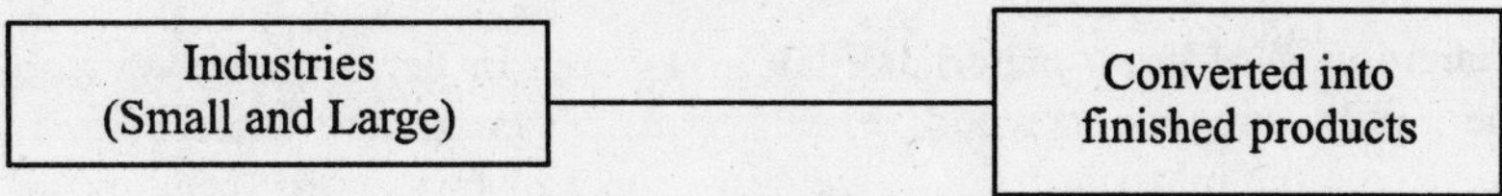

Channel - V

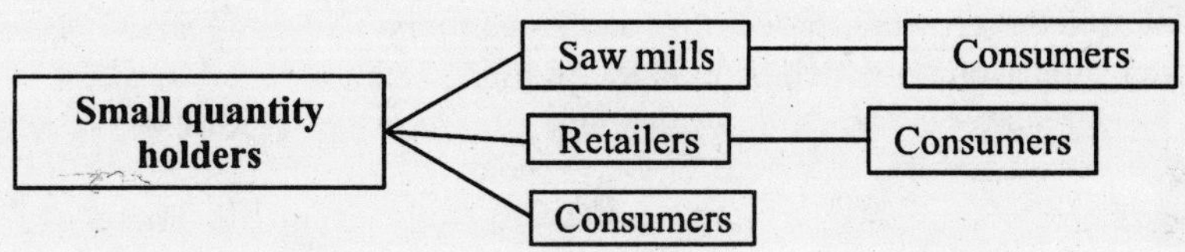

Fig. 17.4: Existing Marketing Channels of Eucalyptus and Poplar in Haryana

Monopolistic markets for NWFPs are controlled by trade by cartels in the cities and exploitation of collectors by the middlemen (Atulya and Mishra 2001). Bisen (2001) mentioned the marketing channel for medicinal plants based agroforestry model as in Fig. 17.5.

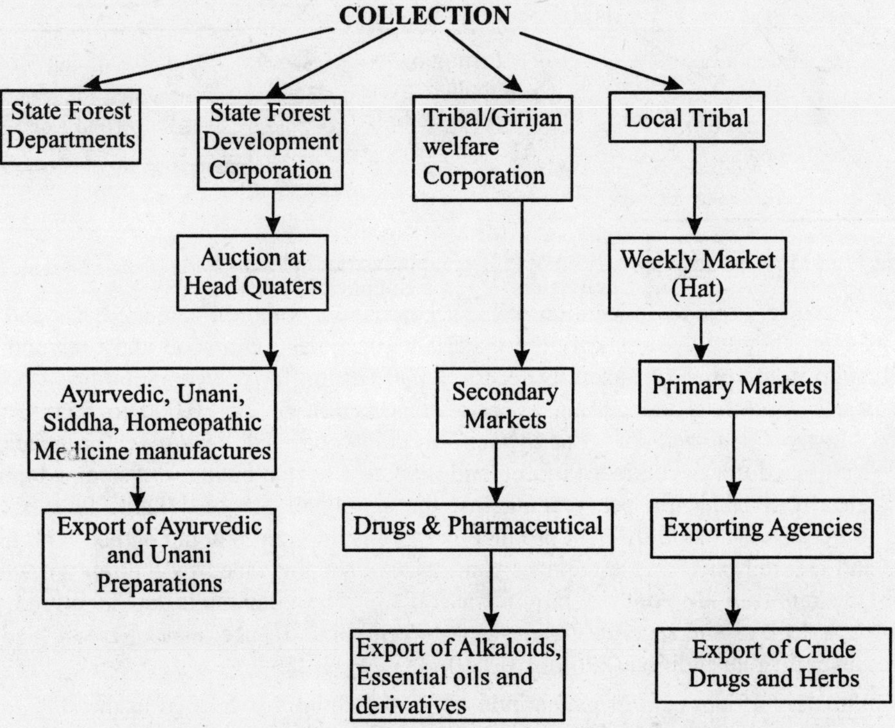

Fig. 17.5: Marketing Channel for Medicinal Plants based Agroforestry Model

INDUSTRIAL RAW MATERIAL BASED AGROFORESTRY PRODUCTS MARKETS AND MARKETING

Forest and plantation based raw materials play a key role in performance of wood based industries such as timber, pulp and paper wood, matchwood, plywood and veneer in Tamil Nadu. These industries directly get the material from forest department or farmers or from abroad countries. Before 2007-08, the majority of raw materials were delivered from Tamil Nadu Forest Plantation Corporation (TAFCORN), Karnataka and Kerala Forest Departments and rest from open market sources through traders. Contract farming is an effective and efficient system of success for production and consumption system of agricultural and allied sectors. It is an essential agreement between unequal parties, which include the growers, processors and consumers. Currently wood based industries have a multipartite model supply chain which is depicted in Fig. 17.6.

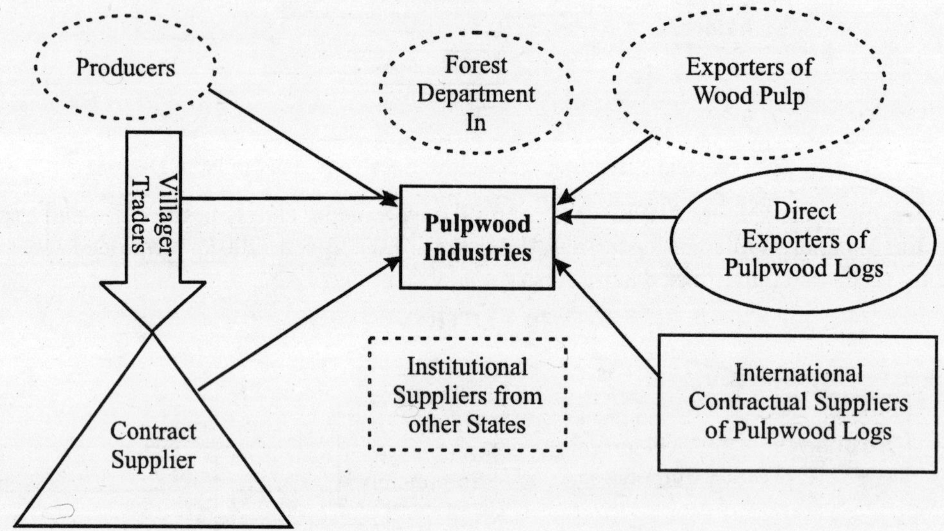

Fig 17.6: Multipartite Supply Chain of Pulpwood Industries in Tamil Nadu (Parthiban et al 2010)

The tree growers sell the plantation to local contractors some on acreage basis and some on a volume basis and they fell and classify the materials into poles, pulpwood and firewood. There is a wide fluctuation in the price of materials i.e. Rs. 1,200 per ton for firewood and Rs. 4,200-4,500 per ton for poles. The farmers are getting as an average return of Rs 60,000 to 85,000 per acre of plantation. Mostly *Casuarina* and *Eucalyptus* for pulpwood and *Ailanthus* for match wood are intensively cultivated under contract farming and marketed by the buyback system. Major pulpwood demand is nearby 8 lakhs ton per year against the supply of 3.5 to 4 lakhs tons per year from organized sources. Absents of efficient production to consumption system coupled with lack of price intensives and assured buy back agreement is a major constraint faced by the tree growing farmers. This could be removed by contract farming and also some constraints are identified in contract farming such as lack of site specific variety, lack of price intelligence, lack of credit facility, lack of mechanization and value addition (Parthiban 2010).

Approximately 65 per cent of match industry in the country is located in the state, contributing 95 per cent match box production. The demand of softwood for match industries is 18 lakhs tons per year. The general marketing channels for major agroforestry products in Tamil Nadu are presented in Fig 17.7 to Fig. 17.16.

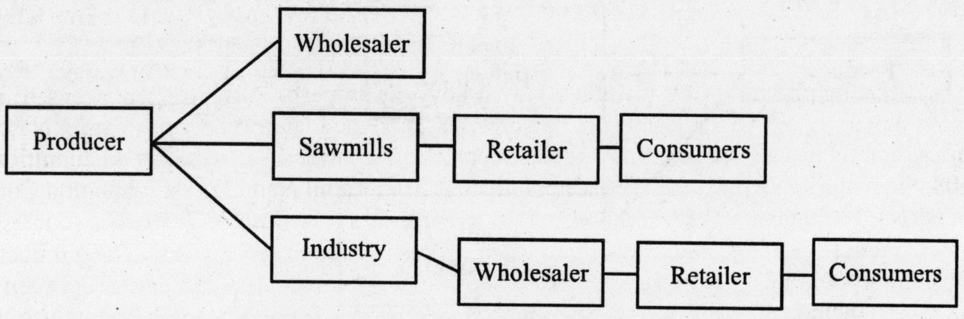

Fig. 17.7: Marketing Channels for Timber

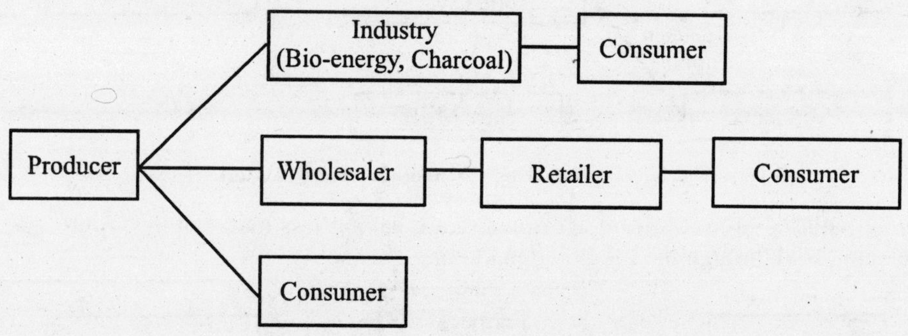

Fig. 17.8: Marketing Channels for Fire wood

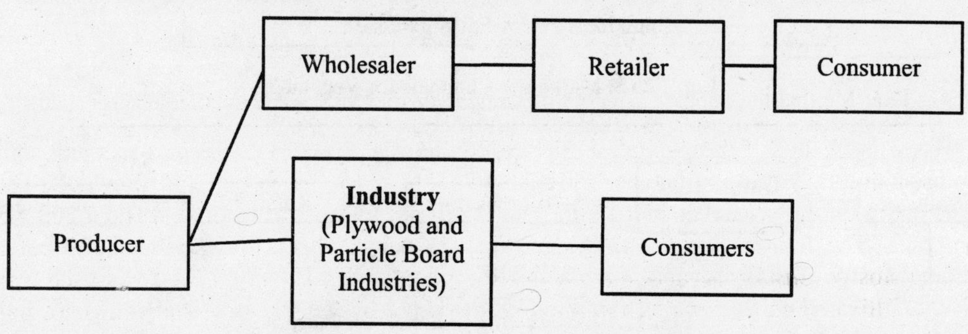

Fig. 17.9: Marketing Channels for Secondary Wood

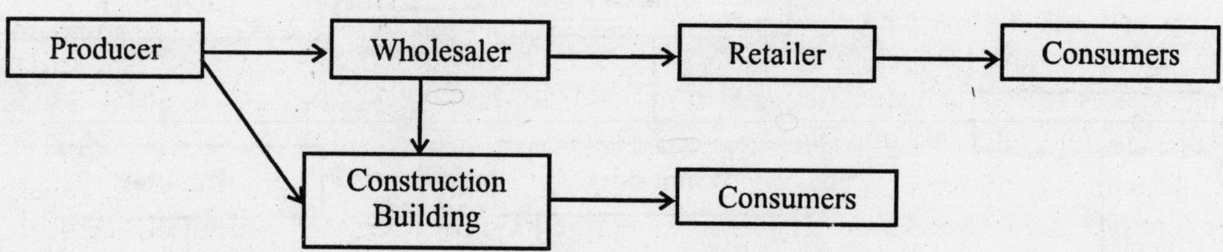

Fig. 17.10: Marketing Channels for Poles

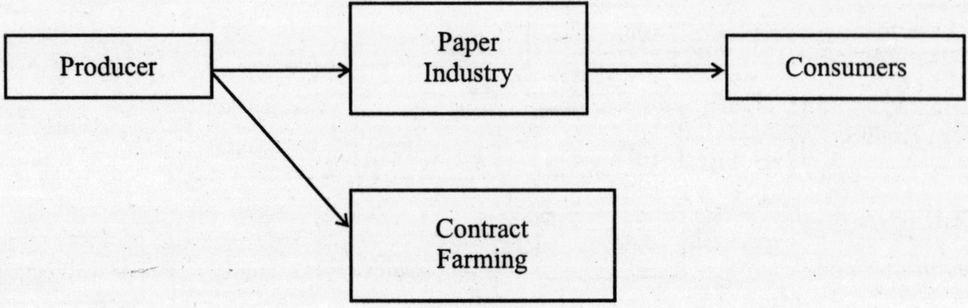

Fig. 17.11: Marketing Channels for Pulpwood

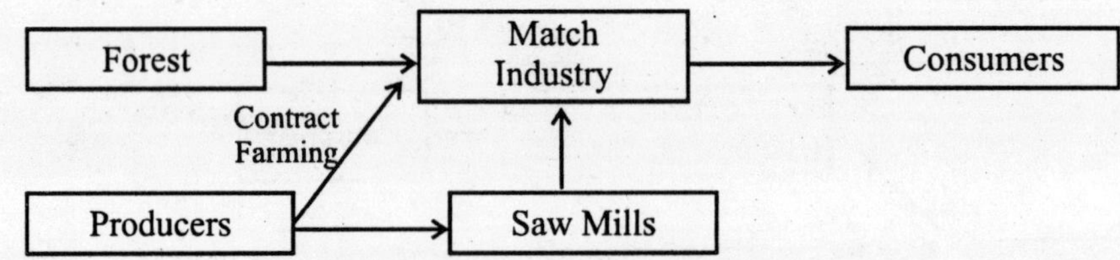

Fig. 17.12: Marketing Channels for Match Wood

Mostly agricultural products from agroforestry are having less marketable surplus. Even surplus products are being sold through the following marketing channels.

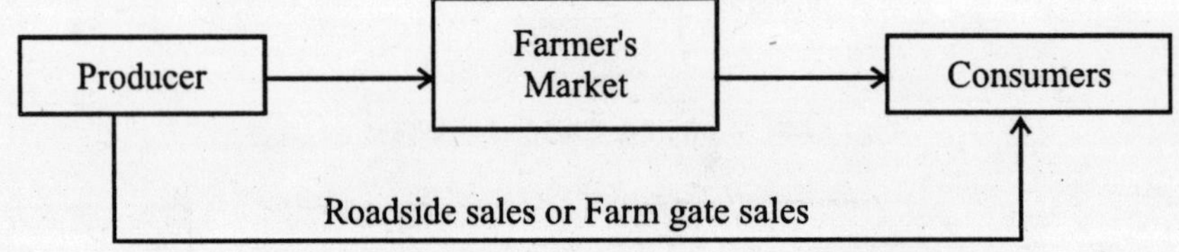

Fig. 17.13: Marketing Channels for Vegetables

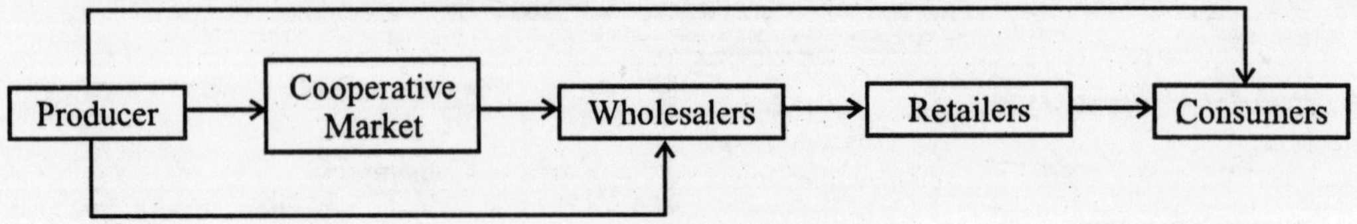

Fig. 17.14: Marketing Channels for Pulses

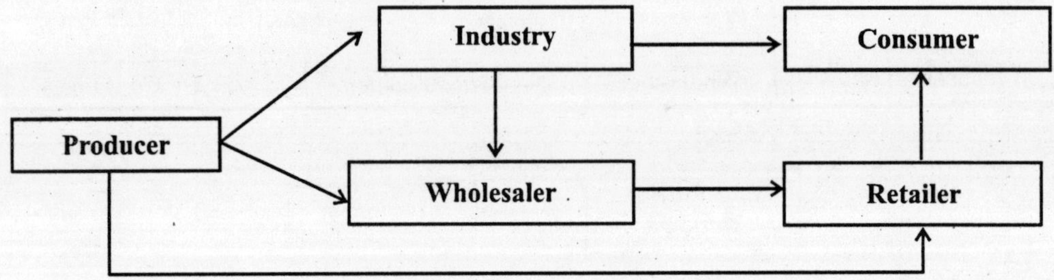

Fig. 17.15: Marketing Channels for Poultry

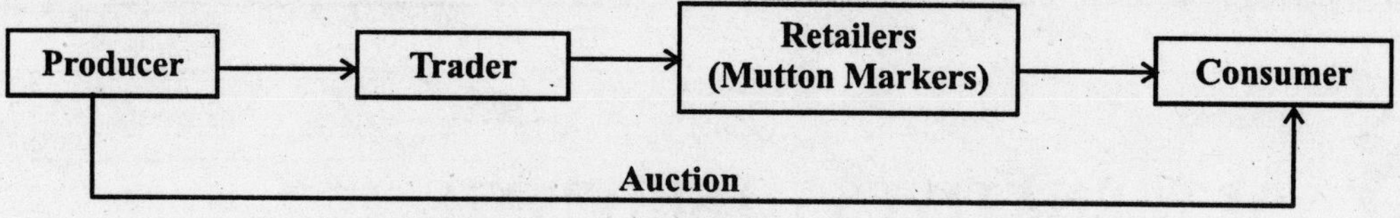

Fig. 17.16: Marketing Channels for Sheep and Goat

CONCLUSION

The pressure on traditional forest resources will be reduced by promoting tree cover in farm lands. It is necessary to adopt different production oriented agroforestry models at farm levels both extensively and intensively. The markets for agroforestry products are not effectively organized in the ground reality. Mostly traders and middlemen are the big player in the market, where as producers are simple price followers due to low producers' share in consumers' rupee. In order to develop well organized and efficient supply chain for agroforestry products, a proper linkage among all stakeholders with assured buyback system under perfect competition will eliminate the intervention of middlemen and fetches better net farm income. The minimum support price must be given for agroforestry products. This will augment the supply of raw materials to agroforestry based industries on a sustainable basis to promote better household and national economy.

■■■■

18 ALTERNATIVES TO SHIFTING CULTIVATION

WHAT IS SHIFTING CULTIVATION?

Shifting cultivation can be described as a land use system in which forested land is cleared and cropped for a few years, and then left for the regeneration of the bush (fallow period), the duration of the fallow period being much longer than that of the cropping period (Nair 1986). According to the definition produced at a seminar held in Nigeria in 1973 "the essential characteristics of shifting cultivation are that an area of forest is cleared, usually rather incompletely, the debris is burnt, and the land is cultivated for a few years – usually less than five – then allowed to revert to forest or other secondary vegetation before being cleared and used again" (FAO 1984). Although there are several variations in the practice in different parts of the world, in general all forms of shifting cultivation follow five stages: site selection, cleaning, burning, cropping and fallowing (Watters 1960). A common term frequently used is "slash and burn" cultivation (as the clearing is usually accomplished by the *slash-and-burn* method) and others terms are "swidden agriculture", "roving agriculture" and "jhumming".

The lengths of the cropping and fallow phases vary considerably; usually the fallow phase is several times longer than the cropping phase. The cultivation phase is usually short (2-3 years), but the regeneration phase, known as the fallow or bush fallow phase, is much longer (traditionally 10-20 years). The length of the fallow phase is considered critical to the success and sustainability of the practice. During this period the soil, having been depleted of its fertility during the cropping period, regains its fertility through the regenerative action of the woody vegetation (Nair 1993). Some of the more tangible criteria considered crucial for distinguishing shifting cultivation from other land use practices are as follows (FAO 1995):

- Cultivation is interrupted by a period of natural fallow; cultivation is neither permanent nor continuous.
- The duration of the fallow period and of the cultivation period may vary in length, but the fallow would be relatively long (usually more than five years).
- A wide variety of vegetation may grow on the fallow, but it would typically be some type of forest.
- The fallow period may or may not be sufficient to restore soil fertility since the minimum period required is extremely variable.
- The population density associated with sustainable shifting cultivation is relatively low, since there must be enough land per farmer to leave a portion of it to fallow for long periods.

- Due to increasing distance to reach cultivated plots during the fallow periods of the nearest fields, housing may be semi-permanent, or farmers may have permanent homes in villages and temporary shelter near their fields.

After fallow period which varies from 3-15 years, depending upon the place, population, land ratio and tribe, the farmer comes back for farming to the same piece of land which he left fallow a few years back. Thus, the cycle of cropping and fallow continues. The period between slash and coming back again to the same plot after completion of intervening fallow period constitute one *jhum* (a local term for shifting cultivation in North East India) cycle. With rising population, the *jhum* cycle in most areas which used to be 10-15 years earlier is now reducing to 2-3 years only (Tripathi and Barik 2003).

OVERVIEW OF SHIFTING CULTIVATION

Shifting cultivation is still the mainstay of traditional farming systems over vast areas of the tropics and subtropics. Estimates of area under shifting cultivation vary. One estimate still used repeatedly (FAO 1982) is that it extends over approximately 360 million hectares or 30 per cent of the exploitable soils of the world, and supports over 250 million people. Crutzen and Andreae (1990) estimated that shifting cultivation is practised by 200 million people over 300 million-500 million hectares in the tropics. Although the system is dominant mainly in sparsely populated and lesser developed areas, where technological inputs for advanced agriculture such as fertilizers and farm machinery are not available, it is found in most parts of the tropics, especially in the humid and sub-humid tropics of Africa and Latin America. In Africa, the main humid area where shifting cultivation remains the dominant form of farming is the middle belt of West Africa, between the coastal tree belt and the more permanently farmed northern plains. It is also still found in sparsely populated areas of Tanzania, Zambia and northern Mozambique and in the empty area of the Zaire basin. It is a major land use even in some parts densely populated Southeast Asia (Spencer 1966; Grandstaff 1980; Ruthenberg 1980; Kyuma and Pairinta 1983; Denevan et al 1984; Padoch et al 1985; Padoch and de Jong 1987; FAO 1984; Nair 1993).

Despite the remarkable similarity of the shifting cultivation practised in different parts of the world, minor differences exist, and are often dependent on the environmental and socio-cultural conditions of the locality and the historical features that have influenced the evolution of land use systems over the centuries. These variations are reflected, to some extent, in the various names by which the system is known in different parts of the world (Spencer 1966; Okigbo 1985)

Shifting cultivation is also said to have been widespread in Europe until a few centuries ago (Nye and Greenland 1960; Greenland 1974). Under resource-rich conditions, as in Europe, shifting cultivation has slowly been replaced by more technologically-oriented and profitable land use systems that bear no resemblance to the original system. In developing countries with low population densities, where the farmer had enough land at his disposal and freedom to cultivate anywhere he chose within a specified geopolitical unit or region, the ratio of the length of fallow period to cultivation phase reached 10 to 1. The system was stable and ecologically sound. However, under the strain of increasing population pressure, the fallow periods became drastically reduced and the system degenerated, resulting in serious soil erosion and a decline in the soil's fertility and productivity (Nair 1993).

The most remarkable differences in the practice of shifting cultivation are, perhaps, due to ecological conditions. In forest areas of the lowland humid tropics, the practice consists of clearing a patch of forest during the dry (or lowest rainfall) period, burning the debris *in situ* shortly before the first heavy rains, and planting crops, such as maize, rice, beans, cassava, yams, and plantain, in the burnt and decaying debris. The crops are occasionally weeded manually. After 2 or 3 years of cropping, the field is abandoned to allow rapid regrowth of the forest. The farmer returns to the same plot after 5 to 20 years, clears the land once again, and the cycle is repeated (Nair 1993).

Table 18.1: Local Terms for Shifting Cultivation in Different Parts of the Tropics (Okigbo 1985)

Region	Term	Country or Region
Asia	Ladang	Indonesia, Malaysia
	Jumar	Java
	Ray	Vietnam
	Tam-ray, rai	Thailand
	Hay	Laos
	Hanumo, caingin	Philippines
	Chena	Sri Lanka
	Karen	Japan, Korea
	Taungya	Myanmar
	Bewar, dhya, dippa, erka, *jhum*, kumri, penda, pothu, podu	India
Americas	Coamile	Mexico
	Milpa	Mexico, Central America
	Roca	Brazil
Africa	Masole	Zaire
	Tavy	Madagascar
	Chitimene, citimene	Zaire, Zambia, Zimbabwe, Tanzania
	Proka	Ghana

In India, the people of eastern and north-eastern region practice shifting cultivation on hill slopes. 85 per cent of the total cultivation in northeast India is by shifting cultivation (Singh and Singh 1992). Due to increasing requirement for cultivation of land, cycle of cultivation followed by leaving land fallow has reduced from 20-30 years to 2-3 years. Earlier the fallow cycle was of 20-30 year duration, thereby permitting the land to return to natural condition. Due to reduction of cycle to 2-3 years, the resilience of ecosystem has broken down and the land is increasingly deteriorating (Patra and Panda 1994).

SOIL DYNAMICS IN SHIFTING CULTIVATION

Before clearing the forest, a closed nutrient cycle exists in the soil-forest system. In this, most nutrients are stored in the biomass and topsoil and a constant cycle of transfer of nutrients from one compartment of the system to another operates through physical and biological processes of rainwash, litterfall, root decomposition and plant uptake. The amount of nutrients lost from such a system is negligible. Clearing and burning the vegetation leads to a disruption of this closed nutrient cycle. During burning the soil temperature increases, and afterwards more solar radiation is received on the bare soil surface resulting in higher soil and air temperatures (Ahn 1974; Lal et al 1975). This change in the temperature regime causes resultant changes in the biological activity in the soil. The addition of ash to the soil by burning causes important changes in soil chemical properties and organic matter content (Jha et al 1979). In general, exchangeable bases and available phosphorus increase slightly after burning; pH values also increase, but usually only temporary. Organic matter content is also expected to increase by burning, mainly because of the unburned vegetation left behind (Sanchez and Salinas 1981).

Table 18.2: Shifting Cultivation in North-Eastern India as reported by the Task Force on Shifting Cultivation (Ministry of Agriculture 1983)

State	Annual Area under Shifting Cultivation (sq.km)	Fallow Period (in years)	Minimum Area under Shifting Cultivation one time or other (sq.km)	No. of Families Practising Shifting Cultivation
Arunachal Pradesh	700	3-10	2,100	54,000
Assam	696	2-10	1,392	58,000
Manipur	900	4-7	3,600	70,000
Meghalaya	530	5-7	2,650	52,290
Mizoram	630	3-4	1,890	50,000
Nagaland	190	5-8	1,913	1,16,046
Tripura	223	5-9	1,115	43,000
Total	3,869	-	-	4,43,336

These changes in the soil after clearing and burning result in a sharp increase of available nutrients, so that the first crop that is planted, gets considerable benefit. Afterwards, soil becomes less productive and crop yields decline. The main reasons for the decline of yield are soil fertility depletion, increased weed infestation, deterioration of soil physical properties, and increased insect and disease attacks (Sanchez 1976). Finally, the farmers feel that further cultivation of the fields will be difficult and non-remunerative, they abandon the site and move on to others, knowing well that they can return to the site after the lapse of a few years, because the site so abandoned will be reinhabited by natural vegetation (forest fallow) and during the fallow period, the soil will regain its fertility and productivity (Nair 1986).

Thus, the cycle has been repeated in many regions and shifting cultivation has continued for centuries though at low productivity levels. However, over long periods of time when human population pressures have started to increase, the length of the fallow periods has diminished and farmers have returned to the fallow sites before these have had enough time for fertility to be sufficiently restored. The introduction of industrial crops and modern methods of crop production have also caused a shift in emphasis on the importance of fallow period in traditional shifting cultivation (Nair 1986).

POSITIVE AND NEGATIVE ASPECTS OF SHIFTING CULTIVATION

POSITIVE ASPECTS

The shifting cultivation helps to conserve the rich cultural diversity as it is interwoven into the culture and tradition of tribes. Shifting cultivation, being a labour intensive and low subsidy based farming system, provides an assured source of food security to the sustenance level farmers. The practice, in its traditional form may also contribute towards conservation of agrobiodiversity, particularly the germplasm of native crop varieties (Tripathi and Barik 2003).

NEGATIVE ASPECTS

Most shifting cultivation practices are subsistence level farming system having very low output/ input ratio compared to other farming systems/ methods (Tripathi and Barik 2003). Shifting cultivation pays back low remuneration when the production per unit area is converted into market value, and that this practice also involves extensive and systematic destruction of the forest and its produce, which constitutes a tremendous loss of valuable resources. Therefore, shifting cultivation is an economically inefficient land use practice (FAO 1995).

The practice causes total damage to the soil system. It is responsible for loss of soil nutrients and useful soil fauna and microbes. It also accelerates the soil erosion manifold. Burning of slash lowers soil acidity, organic matter and total nitrogen, and also contributes to air pollution. The clearing of forest areas at regular and frequent intervals for shifting cultivation results in the loss of primary forests and formation of secondary forests. This causes substantial loss of tree diversity and associated vegetation those are adapted to primary forests. Due to shortening of *jhum* cycle, quite often, the secondary forests also do not get adequate time to regenerate. The repeated use of land with short *jhum* cycle finally converts the *jhum* fallows into degraded wastelands. Therefore, shifting cultivation is considered to be the single most important factor causing deforestation and forest degradation (Tripathi and Barik 2003).

Shifting cultivation also accelerates the pace of environmental degradation and degradation of watershed conditions (FAO 1995). In India, there are 188 tribal districts in 26 states/ UTs. The total forest cover in the tribal districts is 4,11,881 km^2 which is 37.25 per cent of the geographical area of these districts. When compared to the 2009 forest cover assessment, the 2011 assessment shows a decrease of 679 km^2 of forest cover in all the tribal districts of the country put together (FSI 2011).

North eastern region of the country comprising eight states, namely; Arunachal Pradesh, Assam, Manipur, Meghalaya, Mizoram, Nagaland, Sikkim, and Tripura, is endowed with rich forest resources. The region, which constitutes only 7.98 per cent of the geographical area of the country, accounts for nearly one fourth of its forest cover. Because of its biodiversity richness, the region has been identified as one of the 18 biodiversity hot spots of the world. One distinct feature of land use is the prevalence of shifting cultivation in hilly parts of almost all the states of this region. Shifting cultivation has traditionally been the main source of livelihood of the tribal people and is intricately linked to their socio-cultural life (FSI 2011). The total forest cover in the region is 1,73,219 km^2, which is 66.07 per cent of its geographical area in comparison to the national forest cover of 21.05 per cent. Very dense, moderately dense and open forest constitute 14.73, 44.29 and 40.98 per cent respectively. When compared to the 2009 forest cover assessment, the 2011 assessment shows a decrease of 549 km^2 of forest cover in the north eastern region. The main reason for this decrease is attributed to the biotic pressure and shifting cultivation in the region (FSI 2011). Due to shifting cultivation practice on slopes, down-stream siltation of the water bodies takes place (Singh et al 1986).

CONTROLLING SHIFTING CULTIVATION: STRATEGIES AND INITIATIVES

STRATEGIES

Although shifting cultivation is a non-viable resource-utilization practice, tribals are still clinging to this primitive practice to sustain themselves and their families mainly due to non-availability of timely employment avenues. Hence the following strategies may help to wean away people from shifting cultivation (Rajiv Ranjan and Upadhyay 1999).

- Providing employment opportunities and income generation on a regular basis through proper utilization of the land ıesources, i.e. by equitable distribution of waste land among the tribals. But, the various schemes of the Government, under the tribal plan, will have to pump in sufficient resources for proper reclamation and development of the wasteland through agroforestry and silvipasture practices.

- By encouraging cooperative efforts for carrying out forest-based activities, i.e. basket making, rope making, cane furniture processing of minor forest produce, honey collection, etc. have to be made commercially viable by providing proper marketing facilities. This will not only discourage tribals from practising shifting cultivation but will also help them monetarily.
- By forming Village Forest Committees (VFC)/ Joint Forest Management Committees (JFMC) for the protection and development of the degraded forests. These committees by providing suitable incentives to the tribals, after the time of harvest can divert some of the tribals away from the shifting cultivation. Generating employment opportunities during the lean season of forestry operations will also prevent tribals from shifting to other areas.
- By ensuring implementation of total literacy campaign (TLC) in tribal areas on sustainable basis through services of various non-Governmental organizations and voluntary agencies, besides the regular Government machinery.

INITIATIVES

In India, several *jhum* and *jhumia* rehabilitation schemes have been implemented in the north-eastern region since independence by the state and central governments to control shifting cultivation. The examples of such schemes are Watershed Development Projects in Shifting Cultivation Areas (implemented by the Ministry of Agriculture, Government of India), Soil Conservation Schemes of Government of India, Tripura *Jhumia* Rehabilitation sceheme of Government of Tripura and New Landuse Policy Scheme of Government of Mizoram. However, most of these schemes could not achieve desired level of success due to the reasons furnished below (Tripathi and Barik 2003).

- Conceptualisation and formulation of schemes done unilaterally without involving the *jhumias* in the process of scheme development.
- Testing of scheme as pilot projects is necessary before its large scale implementation.
- Top-down approach does not work.
- Effective involvement of the target population to create a sense of belongingness.
- Poor research back-up.
- Integrated approach.
- Lack of co-ordination.
- Sustained and intensive training to both scheme operators and farmers.
- Popularisation of high value cash crops.
- Sustained and intensive monitoring and evaluation of schemes.

The concern of the Government of India for controlling the shifting cultivation is reflected in its agricultural and forestry policies from time to time. For instance, both the Forest policies notified during the post-independent period, viz., National Forest Policy 1952 (NFP 1952) and National Forest Policy 1988 (NFP 1988) have emphasized the need to control shifting cultivation and rehabilitate the affected areas. The NFP states that "tree lands in agricultural areas and their importance in rural economy" (section 17) and "concern for damage caused to forests by shifting cultivation" (section 23) and the NFP 1988 states that "concern for adverse effects of shifting cultivation on environment and land productivity" (section 4.7) and "to contain the areas already affected and rehabilitation through energy plantation and social forestry" (Tripathi and Barik 2003).

In order to deal with the problems of shifting cultivation, the Government of India has also constituted two task forces, one in the Ministry of Agriculture in the year 1983 and the other in the Ministry of Environment and Forests in 2001. Besides estimating the area under shifting cultivation in the country, the task forces recommended a sustainable approach to manage the shifting cultivation adopting a holistic and integrated approach (Tripathi and Barik 2003).

IMPROVEMENT AND/OR ALTERNATIVES TO SHIFTING CULTIVATION

Various approaches have been suggested as improvements and/or alternatives to shifting cultivation (FAO 1985), and most of them emphasize the importance of retaining or incorporating the woody vegetation into the fallow phase, and even in the cultivation phase, as the key to the maintenance of soil productivity. Depending on the ways in which the woody species are incorporated, the alternate land use system can either be a form of agroforestry (Nair and Fernandes 1985), or *taungya* system (FAO 1980) or even any other forms of improved, permanent production systems (Okigbo 1985). However, in general, any successful alternatives/improvements to shifting cultivation should:

- allow production of food and wood products simultaneously from the same piece of land in a sustainable manner; and
- be compatible with the socio-cultural aspirations and economic conditions of the people in order for such practices to be adaptable. The agroforestry approach to land use becomes quite relevant in this context (Nair 1986).

AGROFORESTRY

There is considerable agreement among scientists that agroforestry could, if practised more widely, provide a solution to some of the problems experienced with fallow cultivation in conditions of rising population density and increasing land shortage (Raintree 1980; Vergara 1981). Indeed, this is the assumption behind much of the work of the World Agroforestry Centre (ICRAF) in Nairobi. For example, both the International Institute of Tropical Agriculture and ICRAF are carrying out field trials with alley cropping of the fast-growing woody leguminous genus *Leucaena* (IITA: Hartmans 1981, Okigbo 1981; ICRAF: Raintree 1980).

The mountain eco-systems with shifting cultivation practice have to be made ecologically sustainable. Formulating an eco-development plan for the region for environmental sustainability, could consider completely replacing agricultural practice with agro forestry. Agricultural practices are at the cost of loss of biodiversity resources and estimates indicate that one unit of energy in agronomic production costs loss of greater energy from the forests. However, in the Central Himalayan eco-systems, where agriculture practice is more scientific compared to shifting cultivation, one unit of energy in agronomic production entails an expenditure of about 10-12 units of energy from the surrounding forests as firewood, fodder and leaf manure (Singh et al 1986). Loss of energy from the forests per unit of agricultural production may be far greater in shifting-cultivation areas. Agroforestry may be one of the solutions to redress this loss. The advantages of agroforestry area are (Singh et al 1986):

- The protective values of trees are far greater than those of annual crops
- Unlike annual crops which require frequent ploughing, tree plantations cause minimal soil disturbance
- Net above-ground primary productivity of forests is notably greater than that of agricultural crops and grassland

- By providing fodder and firewood, the farm forestry would create a favourable situation for the revival of the natural forests
- A revived forest cover could offer several other advantages

TAUNGYA SYSTEM

One of the alternatives to shifting cultivation is the *taungya* system of large-scale forest plantation establishment used by forest departments, in which food crops are interplanted with trees in the early years of the plantation. Two broad types of *taungya* can be distinguished, depending on the farmer's role (FAO 1980).

Individual *taungya*, or "own your own crop" (Olawoye 1975), where the farmer plants trees in return for use of the land for a limited period of time. Departmental *taungya*, or "farming for pay", where the farmer carries out jobs for the forestry department related to plantation establishment and agricultural crop production in return for a wage. While there are many accounts of the profitability of *taungya* from the point of view of forestry departments, which find it a low-cost system of establishing plantations, there are few accounts of how profitable or acceptable *taungya* is from the farmer's point of view (FAO 1980).

INDIVIDUAL TAUNGYA: Under a system of individual *taungya*, farmers are given access to plots of land on which they must plant trees which are the property of the forestry department. In return, they may cultivate the spaces between the trees with food crops, the produce of which is theirs. Depending on the crops and trees grown, and forestry department regulations, the farmer may cultivate the land with certain prescribed crops for between one and three years, up to the time the trees will have grown enough to shade out any food crops. At this point, the farmer has no other choice but to move to another area. If the forestry department is continuing to plant trees in the same way at this time, and if the department is pleased with the way the farmer has looked after the saplings, it may offer another plot for use. If not, the farmer will have to search for land elsewhere (FAO 1980).

DEPARTMENTAL TAUNGYA: Departmental *taungya* is distinguished from individual *taungya* by the fact that the produce from any agricultural crop belongs to the forestry department, which rewards the farmer with a wage. From the farmer's point of view, therefore, departmental *taungya* is like other types of paid agricultural labour and radically different from individual *taungya*. Under the latter system, the farmers depend on their food crops for survival and thus are highly motivated to take care in growing it, whereas under the former system the paid labourers are only motivated to the extent that they will continue to collect their wages (FAO 1980).

A further difference between the two systems is that whereas individual *taungya* attracts mainly local farmers, departmental *taungya* may also attract people from other areas, and even from towns, who are in search of a wage. In cases where these workers are inexperienced farmers, it is understandable if agricultural crop yields are low (FAO 1980).

OTHER ALTERNATIVES TO SHIFTING CULTIVATION

Certain other alternative programmes those can be practised in lieu of shifting cultivation are as follows (Tripathi and Barik 2003):

1. Ideal land use
2. Agricultural land use including terrace and wet terrace rive cultivation
3. Horticulture land use
4. Livestock farming

5. Fish farming
6. Agroindustries
7. Mushroom farming
8. Sericulture
9. Floriculture

However, while modifying the existing *jhum*, the following aspects must be taken into account (Tripathi and Barik 2003):

- Soil and water conservation
- Maintenance of soil fertility
- Crop diversity
- Food security
- High yielding varieties of crops
- Low volume and high value cash crops
- Market linkages
- Yield optimisation
- Conformity with ecological principles

And, the characteristics of an improved and modifies system should be as follows (Tripathi and Barik 2003):

- The modified system should improve the shifting cultivation to accommodate the needs of the growing population.
- The system should be suitable for farmers especially marginal and small scale farmers.
- The system should take into account the prevailing land tenure system.
- The system should be ecologically sound, economically sustainable and socially acceptable.

SUCCESSFUL EXPERIMENTS FOR POTENTIAL MANAGEMENT OF SHIFTING CULTIVATION

Several research and extension organisations in the North East India have been trying to develop alternatives to shifting cultivation since independence. Many of such efforts have resulted in developing ecologically and economically viable land use models having potential to work as alternatives to shifting cultivation. These successful experiments need to be adapted widely depending upon the local socio-economic and bio-physical conditions. Some of these successful experiments tried in NE India are as follows (Tripathi and Barik 2003).

1. NEPED (Nagaland Environment Protection and Economic Development) (experiment by Government of Nagaland)
2. ICAR 3-tier model (experiment by ICAR)
3. SALT models (sloping agriculture land technology) (experiment by GBPIHED)
4. SWEET (Sloping Watershed and Environmental Engineering Technology) (experiment by SFRI)
5. Intensive watershed based livestock production system (experiment by ICAR)

6. Modified shifting cultivation practices undertaken by *jhumias* with introduction of cash crops like large cardamom, medicinal plants, broom grass, beetle leaf and beetle nut, cinnamomum, fruit orchards and orchid cultivation (documented by RCNAEB and SFRI)

The Agroforestry project known as Nagaland Environment Protection for Economic Development (NEPED) funded by Canadian International Development Agency (CIDA) through India-Canada Environment Facility (ICEF) was initiated in 1995 to make Nagaland self-sufficient in agroforestry. The objectives of the project were: (a) identification of trees by local tribes and demonstration of method leading to more sustainable resource management; (b) addressing people's need, and evolving better management systems; and (c) promoting marketing initiatives. Under this project, experienced Government officials convince the villagers to set aside 6 ha of land, known as test plot, in *jhum* areas for over two-and-half years. Owner of this plot is required to plant 1,200 seedlings, along with usual crop, under supervision of village council and project team. These test plots become open school-cum-research station, as well as demonstration plots to teach new technology. Thus, NEPED project does not aim at eliminating *jhum* cultivation, but making it more stable and profitable. This novel project gives a more scientific way to tackle tribal-forest conflict (Rajiv Ranjan and Upadhyay 1999).

TECHNOLOGICAL GUIDELINES FOR IMPROVEMENT OF SHIFTING CULTIVATION

Depending upon the site conditions and suitability of the technologies, the following principles (Tripathi and Barik 2003) may be quite useful in developing ideal cropping systems as a modified form of *jhum* system.

1. Maximize both cropping and fallow period
2. Identify innovative technologies, institutions and policies that can address two fundamental challenges – poverty alleviation and biodiversity conservation
3. Adoption of wide spectrum of fallow management strategies
4. Slash and mulch (minimizing burning)
5. Green manure and cover crop
6. Accelerated fallows (fallow vegetation to be improved with nitrogen fixing trees, shrubs and interplanting of crops)
7. Enriched fallows (enriched with economic plants)
8. Interplanted fallows (nitrogen fixing trees/ shrubs and interplanting of crops)
9. Managed and enriched fodder fallows
10. Analog agroforests
11. Intercropping economic trees
12. Introduction of candidate species.

■■■■

19 TAUNGYA SYSTEMS

WHAT IS TAUNGYA SYSTEM?

Taungya system was originally the local term for shifting cultivation and was subsequently used to describe the afforestation method. The word is reported to have originated in Myanmar (Burma) and means hill (*Taung*) cultivation (*ya*) usually of a temporary character. It is a land management system which consists of the raising of forest crops in conjunction with temporary agricultural crops in which a short phase of food crop production is used as a silvicultural method to ensure the establishment of timber species (King 1968; Evans 1982). Generally, the practice of taungya system involves some form of government intervention in the establishment and control of forest plantations (Ruthenberg 1980). The original scheme was to give tribes, rural dwellers or farmers a piece of land that they could clear and cultivate for about 4-5 years while the trees are young. After the initial 4-5 years cropping period, the trees are managed as a monoculture forestry unit (Watson 1983), whereby the farmers are ordered, in accordance with the specifics of the taungya agreement, to discontinue cropping and maintenance of food crops in between the trees, most especially when there is closure of the tree canopy.

Wood production is the ultimate objective in the taungya system, the benefit in the short term of the farmer for practising it, as in shifting cultivation, is to produce food (Okigbo 1985). Farmers did not benefit from the output of the tree component cared for by the farmers in the early stages of establishment through the cultural practices carried out by the farmers. Today the taungya system is known by different names, some of which are also used to denote shifting cultivation, *Tumpangsari* in Indonesia; *Kaingining* in the Philippines; *Ladang* in Malaysia; *Chena* in Sri Lanka; *Kumri, Jhooming, Ponam, Taila,* and *Tuckle* in different parts of India; *Shamba* in East Africa; *Parcelero* in Puerto Rico; *Consorciarcao* in Brazil, etc. (King 1968). Most of the forest plantations that have been established in the tropical world, particularly in Asia and Africa, owe their origin to the taungya system (Von Hesmer 1966; King 1979a).

TYPES OF TAUNGYA SYSTEMS

DEPARTMENTAL TAUNGYA

Under this system, agricultural crops and plantation are raised by the forest department by employing daily paid labourers. The main aim of raising agricultural crops along with the plantation is to keep along with the land free of unwanted vegetation.

LEASED TAUNGYA

The forest land is given on lease to the person who offers the highest money for rais-ing agricultural crops for a specialized number of years and ensure care of tree plantation.

VILLAGE TAUNGYA

This is the most successful of all the three taungya systems. Under this system, the people who have settled down in a village inside the forest for this purpose raise crops. Usually each family has about 0.8 to 1.7 ha of land and has to raise trees and cultivate crops for 3 to 5 years.

TAUNGYA SYSTEM IN SOME TROPICAL COUNTRIES

Taungya is widely practised in some tropical countries with local variations in the system. The principle underlying the system of taungya is that in all cases, temporary fertility of the forestland is used by the farmers for cultivation of agricultural crops.

BURMA

Taungya system was started in 1856. However, from 1906 onward, taungya plantations were almost given up in most Forest Divisions and attention was turned more to the improvement of the natural stock by improvement felling. Due to the earlier taungyas, teak plantations had been scattered in such an indiscriminate way in small blocks that organized tending and thinning were almost impossible. Hence for sometime there was a great prejudice against any further planting operations.

INDONESIA

Taungya is called Tumpangsari and it is prevalent in central Java for establishment of teak plantations. The farmers are allowed to cultivate blocks of land for two years, growing dryland rice and maize and simultaneously maintaining teak seedlings.

NIGERIA

Forest regeneration is commonly achieved through taungya system. A tract for commercially logged land is allotted to a farmer who cuts the residual trees and undergrowth, burns them and raises his food crops along with tree seedlings planted by the Forest Department. The farmer ordinarily crops the land for two years, then turns it over to the Forest Department and he is assigned a new area for cultivation.

TAUNGYA SYSTEM IN INDIA

In 1856 when Diertrich Brandis was in Burma (then part of India), shifting cultivation was rampant and there were several court cases against the encroaching villagers. Brandis realized that shifting cultivation which was so detrimental to the management of timber resources could possibly be rendered useful to the development of forestry. Based on the well-known German system of *waldfeldbau* which involves cultivation of agricultural crops in forests, and on the success of teak taungya plantations in Taungoo and Tharrawaddy Divisions, Brandis encouraged the concept of 'regeneration of teak with the assistance of taungya' (Blanford 1925; King 1979b). Soon Forest Department distributed teak (*Tectona grandis*) and rice (*Oryza sativa*) seeds to farmers and two decades later, this system had proved so good that teak plantations could be established very much cheaper than otherwise. At the same time the taungya system was of educational value. The villagers no longer had to defend themselves in court for destroying the forests; instead they promoted afforestation of the cleared land by sowing teak seeds. After the clearing and before afforestation, the taungya fields were usually used for the intermediate cultivation of rice, tobacco (*Nicotiana tabacum*) and sesame (*Sesamum indicum*) in the valleys and sugarcane (*Saccharum officinarum*), cotton (*Gossypium* spp.) and maize (*Zea mays*) on the slopes. Thus a symbiotic agreement between the Forest Department and farmers developed.

The Forest Department later introduced taungya into British India too. In India, the first taungya plantation was initiated in 1910s. In the Siwaliks (Uttar Pradesh), twelve new settlements were created with the villagers' help. Each farmer was given 0.3 to 1 ha of forestland to clear and cultivate. During the first two years, the whole area was cultivated whereas during the next three years cultivation took place between the sal (*Shorea robusta*) seedlings. The farmers agreed to look after and weed the plantation in return for the use of the land for cultivation of food crops. The aim of taungya was to curb shifting cultivation, which was harmful to timber resources. Simultaneously, food was produced for the local people. The advantages of taungya were manifold, ranging from higher yields of agricultural as well as tree crops, protection from soil erosion if contour cultivation and terracing were practised on slopy lands and more regular employment for the local population. Taungya is extensively practised in several States in India. It is called Kumri in Tamil Nadu and Kumri or Poonam in Kerala. There is considerable variation in the practices followed in different regions. Some of the finest taungya plantations can be seen in Gorakhpur Forest Division of Uttar Pradesh.

TAUNGYA IN KERALA

Taungya is known by several names such as poonam, kumri and tuckle. It was introduced in 1922 in an area of 84 ha of teak plantation in Konni Division (George 1961). Since that time, taungya has been gradually extended to other areas. In 1932, teak taungya plantations were initiated in Begur area of Wynad Division. The taungya system practised during 1930s had the following procedure. The taungyadars were allotted their plots in May, usually about 0.4 ha to each family. They did a clean scraping of the area towards the end of May and ragi (*Eleusine coracana*) was sown at the beginning of the rains usually in the first or second week of June. The teak planted by the Forest Department was kept weeded by the taungyadars who also replaced casualties free of cost. Ragi was harvested either in November or December, by which time it was one metre high and very dense. With early stump planting, it was possible to get the teak seedlings upto over 45 cm before the ragi competed with them seriously.

Regarding teak taungya, one difficulty was that in two years, the teak saplings were so high that they interfered with the cultivation and taungyadars objected to being not given a new area for cultivation. The taungya system practised now is similar to the one just described. Instead of allotting smaller areas to farmers, larger blocks are now leased for taungya by tender and generally contractors are the taungyadars and not farmers. Also a variety of crops are grown such as rice, tapioca, ginger (*Zingiber officinale*), sesame and pulses. The taungya system is a development from shifting cultivation in that the agricultural crops are raised by destroying tree crops in shifting cultivation and in taungya, agricultural crops are grown in association with tree crops.

As the population increased considerably in the tropical regions, the use of forestland for agricultural purposes became a necessity and the commercialization of taungya led to a series of agrisilvicultural practices – all variations of taungya system – in most of the tropical countries. From these agrisilvicultural practices developed the concept of agroforestry involving integration of agriculture, animal husbandry and fisheries with forestry. In other words, agroforestry envisages the use of forest land for agriculture, animal husbandry and fisheries. However, the ways and means of achieving the desired ends by this integration will have to be developed. This evolutionary pattern of practices shows that taungya is possibly an intermediate step in the sequence shifting cultivation to agroforestry. In India, though taungya was intended to help the farmers (mostly tribals) directly or indirectly, they were alienated from taungya operations. This is particularly true in Kerala. Presently, contractors are mostly the taungyadars and the products mainly tapioca, paddy and ginger are moved in to from the production area. Though in Kerala taungya has indirectly encouraged encroachment of forest lands, in countries like Thailand and Indonesia, the taungya plantations act as buffer zones against encroachment since farmers (mostly tribals) are very much involved in taungya operations.

MANAGEMENT OF TAUNGYA SYSTEM

CROPS SUITABILITY IN TAUNGYA SYSTEM

There is some evidence that where tree crops are planted with agricultural crops, a wider espacement of the tree crops reduces mortality, increases the rate of growth and, at the same time, the presence of suitable agricultural species effectively reduces soil exposure. In addition, the farmer obtains higher yields per unit of plantation. Closer tree espacement frequently reduces early tree growth because of the increased competition, and necessitates earlier silvicultural tending. The agricultural crops which are grown in conjunction with the forest trees are generally chosen because of the agricultural and feeding habits of the cultivator.

Table 19.1: Trees and Crops of Taungya Systems in India (Tejwani 1987)

State	Tree Crops	Associated Agricultural Crops
Andaman	*Pterocarpus dalbergioides*	Sugarcane, maize
Andhra Pradesh	*Anacardium occidentale*, *Tectona grandis*, *Bombax ceiba*, Bamboo, *Eucalyptus* spp.	Hill paddy, ground nut, sweet potato
Assam	*Shorea robusta*	Paddy
Kerala	*Tectona grandis*, *Bombax ceiba*, *Eucalyptus* spp.	Paddy, *Manihot* spp., ginger, turmeric
Maharashtra	*Shorea robusta*	Sunhemp, jute, mesta, sunflower, castor
Tamil Nadu	*Tectona grandis*, Bamboo, *Santalum album*, *Tamarindus indica*, *Acacia nilotica*, *Acacia mearnsii*, *Ceiba pentandra*, Cashew, Rubber	Millets, pulses, groundnut, cotton, manihot, potato
Uttar Pradesh	*Shorea robusta*, *Tectona grandis*, *Acacia catechu*, *Dalbergia sissoo*, *Eucalyptus* spp., *Populus* spp.	Maize, paddy, sorghum, pigeon pea, soyabean, wheat, barley, chickpea, oilseed rape
West Bengal	*Tectona grandis*, *Shorea robusta*, *Schima wallichii*, *Cryptomeria japonica*, *Michelia* spp., *Quercus* spp.	Paddy, maize, millets, turmeric, ginger, ladies finger, pineapple, hemp

TREE SPACING IN TAUNGYA SYSTEM

Huxley (1983) has noted that optimal planting distance, and combinations of agricultural crops and trees are frequently site specific and therefore characterized by greater variability from place to place. Again, it markedly depends on the following factors:

- Local environmental conditions such as edaphic factors, slope, topography and climate
- Local economic or market demand
- Preferred agricultural tradition
- Land ownership regulations

One of the most critical decisions for silvicultural prescriptions in plantation establishment is the spacing. Tree spacing in taungya system is varied and largely depends on the objective of the plantation establishment, the type of tree species as well as prevailing environmental conditions. It is widely known that tree spacing varies from place to place even within the same climatic conditions. Many different tree spacing in taungya systems have been adopted for various reasons in the tropics. However, taungya farmers would prefer the widest, least dense spacing of trees possible, so as to give advantage to agricultural crops. In contrast, for successful forest regeneration and recovery of wasteland, closer and dense spacing is recommended (Huxley 1983). In the Forest Village taungya system teak was planted to 4 m x 4 m spacing but spacing was increased to 8 m x 8 m with the intention to incorporate a greater diversity of species and of structure that has a longer period of production of food crops for the socio-economic improvement of the farmers. Cultivators prefer the widest, least dense spacing of trees possible, to give advantage to crop growth and performance.

SOIL MANAGEMENT UNDER TAUNGYA SYSTEM

In most taungya systems, erosion hazards, rather than soil fertility, are likely to pose the greatest soil management problems. The long-term effect of the practice on soil fertility will, however, largely depend on the management practices adopted at the time of the initial clearing as well as subsequent re-establishment phases. In any case, soil fertility and the related soil management practices are, perhaps, only of secondary importance in determining the continuation of the traditional taungya system. In most cases, the biological problems of continuing cropping under an expanding overstorey tree canopy make it impossible to continue cropping after the initial two or three years.

TAUNGYA PLANTATIONS ESTABLISHMENT AND BIODIVERSITY CONSERVATION

Biodiversity is commonly understood to include the number, variety and variability of organisms living on earth. Biodiversity is a fundamental characteristic of life on earth and encompasses the "whole range of variation in living organisms". Environmental crisis presents complex economic and social problems of dimensions unlike those seen anywhere else in the world. The problems have many aspects, some interrelated, others caused or worsened by ill-conceived solutions, with a net result that there is now an accelerated depletion of natural resources with concomitant deterioration of living standards of the people especially in the rural communities (Shiva 1992). Human activity has always had an impact on biodiversity, but in recent centuries this impact has intensified to a position where we are in danger of undermining the primary functions of natural systems and to an extent that could ultimately threaten our own future (Larsen et al 2005). Losses of biodiversity have resulted from the destruction of natural habitats, over-exploitation of resources, pollution and changes in the composition of ecosystems due, for example, to the accidental or deliberate introduction of non-native species. The need to conserve biological diversity is perhaps of utmost importance in tropical countries where the lives of humans are intricately linked to the environment in a symbiotic relationship in which each species cannot survive without the essential contributions from the whole (Balick et al 1996). Biodiversity depletion triggers a chain reaction (Shiva 1992). The disappearance of a species is related to the extinction of many other species and in most cases threatens the life-support systems and livelihoods of people in the third world countries. The greatest threat to the protection of biodiversity is the biphasic issue of poverty and affluence which separately and through their interaction continue to cause rapid depletion of biological diversity. Poverty and the destruction of the environment are readily traceable to poor management of the natural resources and adoption to changed environment.

According to Shiva (1992), the primary causes of destruction of biological diversity include a combination of factors. Habitat destruction due to pressure on land to accommodate increased industrial demand for cheap tropical plants, and the cultivation of cash crops in place of subsistence farming. Invariably, the land preparation for taungya plantation establishment also involves clearing and burning of natural vegetation cover just as shifting cultivation which destroys biological diversity. In taungya system, the natural vegetation cover is replaced by a mixture of food crops planted between rows of planted tree seedlings. After the initial 4-5 year cropping period, the trees are managed as a monoculture forestry unit (Watson 1983). Monoculture taungya plantation unit is susceptible to insects and pest attack which may result in destruction of other species. The broad leaves of teak plants usually fall and cover the ground. The teak leaves which cover the ground have very low rate of decomposition, and therefore serve as mulch and prevent other plant species to grow under teak plantations. The slowly decomposing teak leaves dry up in the dry season and serve as fuel. This increases fire incidence in teak plantations, accelerated erosion and habitat destruction.

Pressure on land has increased to produce greater quantity to achieve previous level of incomes. Taungya farmers often demand for large track of land with the intention to increase food production and hence income for better standard of living. The dependence on the forest land for food production and commercial taungya plantation establishment will continue to increase in the future due to increase population, poverty, decrease farm sizes and lack of alternative employment in the rural communities (Emerton 1991). Commercial flora species in most forests in Africa had dwindled and will continue to dwindle. For example, in most forest in Ghana, the average density of many valuable indigenous tree species is now low and some are sometimes less than one commercial tree per ten hectares in even the "Primary" forest (Lamprecht 1989). The survival of tropical forest is therefore in a state of jeopardy due to exponential population growth in the tropical and sub-tropical regions. Coupling with growing mass poverty have created pressure on land which has engendered governments to commit land previously under natural ecosystems for human use. It is predicted that if current trends continue, most tropical forests will either be severely degraded or completely gone by the year 2020 (Kamugisha et al 1997).

Taungya plantation establishment and management will be therefore intensified to meet the demand for wood resources in Ghana, where the rate of forest cover loss remains high at 1.7 per cent per year (FAO 2000). The resultant effect will be loss of biodiversity conservation. Introduction of homogenous system of agriculture and replacement of diverse plant species and races with monoculture affect biodiversity. Taungya plantations, and particularly single species plantations, are at much greater risk of catastrophic insect and disease losses than natural forests. This is based on the observation that the introduction of exotics tends to alter natural balances that serve to keep pathogenic organisms in check in ecosystems. Taungya plantations are incapable of supporting biodiversity characteristics of native forests. The typical high stocking density and lack of structural diversity are the reasons for low species diversity within taungya plantations. Hence, less plant and animal diversity exist in taungya plantation forests than native forests within the same region. However, taungya plantations play positive role in the restoration of biodiversity to an impoverished landscape or marginal lands.

Pure teak stands have been associated with the deterioration of soil and erosion. This is usually the case when teak is planted on steep slopes where undergrowth has been systematically cleared or where excessive burning has taken place. Evidences of such environmental degradation under teak plantation in Ghana are common (Odoom 2002). Biodiversity loss has implications for the social and economic well-being of humanity. The natural forests contribute to the food needs of rural communities directly in the form of fruits, seeds, nuts and protein from bush-meat. If plant diversity is being reduced over time, due to taungya plantation expansion, then the consequences of this extend beyond the destruction of natural ecosystem alone. The establishment of taungya plantations on land that already carries natural forest will exacerbate deforestation rather than reducing it (ITTO 2000). Nevertheless, it is also true that plantations will play an increasingly important role in the global supply of cellulose fibre. Larsen et al (2005) had demonstrated that lower plant diversity leads to less primary productivity, less carbon storage and greater leaching of nitrates.

ADVANTAGES AND DISADVANTAGES OF TAUNGYA SYSTEM

ADVANTAGES

- *Artificially regeneration of the forest is done at cheaper rate*: Taungya system has been employed to establish plantations successfully at low establishment costs.
- *Problem of unemployment is solved to some extent*: Taungya is mainly considered as an effective means of reforestation, and its contribution to improving the welfare of the local people is subsidiary to the successful establishment of timber plantations. The taungya system provides a unique and attractive opportunity for absorption of unemployed rural labour. In particular, it represents the only alternative available to landless rural farmers.
- *Better utilization of land*: The available land is utilized in an efficient manner with tree cultivation and agricultural crops.
- *Remunerative to forest department*: Taungya system is considered to be one of the low-cost forms of government compensatory forest plantation establishment which enlists farmers who are willing to plant and care for forest crops in return for being allowed to grow food crops for 4-5 years while the trees are young.

DISADVANTAGES

- *Creation of certain legal problems*: Enforcing the policy of forest reserves becomes difficult and expensive where forestland is still plentiful. Shifting cultivators are still able to operate illegally and create legal problems.
- *Exploitation of human labour*: Although the taungya system is often cited as a popular and mostly successful agroforestry approach to establishing forest plantations, it has also been criticized as labour-exploitative. It capitalizes on the poor forest farmer's need for food and his willingness (often out of helplessness) to offer labour for plantation establishment free of cost in return for the right to raise.
- *Danger of epidemic*: As a result of combined effect of over-exploitation of forest resources, unsustainable farming practices, logging, wild-land fires and mining activities, the forest area has reduced and degraded. Forest loss affects the livelihoods and the environment of particularly the rural poor in different ways. These include shortages of fuelwood, non-timber forest products, accelerated soil erosion and reduction in agricultural productivity.
- *Loss of soil fertility*: Accelerated soil erosion in taungya system leads to loss of soil fertility.

IMPROVEMENTS TO TAUNGYA SYSTEM

FOREST VILLAGE SYSTEM

Several alternatives and improvements to taungya have been attempted in different places, most of them with the objective of providing better living and social conditions for the tenants. One of the most widely quoted examples is the Forest Village Scheme in Thailand, which has generated several reports (Boonkird et al 1984). The philosophy of the scheme was to encourage and support farmers to give up shifting cultivation in favour of a more settled agricultural system, while simultaneously obtaining their services for the establishment of forestry plantations. Each farm family who agreed to take part in the scheme was provided with a piece of land of at least 1.6 ha within the selected village unit for constructing a house and establishing a homegarden. The farmers were also permitted to grow crops between the young trees in the forest plantation unit that they helped to establish according to the plans of the Forest Industries Organization (FIO). The FIO then would appoint "development teams", of multidisciplinary experts for each forest village; the teams provided agricultural, educational, and medical services to the people of the village. The scheme has enabled the FIO to establish forest plantations at considerably reduced costs. In the early 1980s there was a total of about 4,000 ha of taungya forest plantations under cultivation in the FIO scheme. Economic returns from the scheme varied depending upon various local conditions in different regions of Thailand. Forest Village System is practised and combines the advantage of reforestation and subsistence agriculture. The original Burmese colonial taungya system considered improvement in social forestry in which there is intensive use of labour. This employment increases income and community stability (Gregersen 1982). There are many opportunities for workers to earn wages in the Forest Village system and since majority of the rural people have few assets other than their capacity to work, they seem to benefit from the Forest Village System.

The concept of the forest village has been tried, with varying degrees of success, in several other countries, e.g., Kenya, Gabon, Uganda, India, Nigeria, and Cambodia. Although it is more expensive (to the forestry departments) than the traditional practice of taungya, it is particularly suitable for countries with extensive natural forest resources and large numbers of shifting cultivators and landless farmers. Ideally, the system permits sustainable use of forest land for food production by landless people who would otherwise be engaged in forest encroachment.

MODIFIED TAUNGYA TYSTEM

Under the Modified Taungya System (MTS), being operated in Ghana, taungya farmers are fully involved in the establishment and maintenance of the plantations. The farmers tend food and tree crops that are planted in a mixture on the same plot until such a time that the canopy closure of the growing trees makes it impractical to continue with crop cultivation. After canopy closure, the taungya farmers continue to tend the trees until maturity. Other essential features of the Modified Taungya System are described below.

Farmers will essentially be owners of the products with the Forestry Commission, landowners and forest fringe communities as shareholders. The rights of taungya farmers are guaranteed under the Timber Resources Management Amendment Act, 2002 (ACT 617), which states that "no timber rights shall be granted in respect of land with private plantation; or land with any timber grown or owned by any individual or group of individuals". However, under the previous taungya system, the Forestry Commission was the owner with landowners as the only beneficiaries. Farmers benefited from the food crops but did not receive any benefits from the tree crops.

Farmers will carry out most of the functions, including pruning, maintenance and tending, while the Forestry Commission will be responsible for training the farmers to carry out the taungya functions efficiently. The Forestry Commission will also be responsible for supply of equipment and tools, stock inventory and auctioning or marketing of taungya products. Farmers are therefore eligible for a share of the benefits accruing from the plantation according to the benefit-sharing framework, which ensure greater benefit flows to participating farmers.

THE SHAMBA SYSTEM

The Shamba System is a highly developed form of taungya system practised on state forest land in Kenya where there is considerable integration of the cultivators into the Forest Department (Mburu 1981; Nair 1984). According to Oduol (1989), the Shamba system is a form of agroforestry system which has been popularly used in agricultural areas of Kenya since the early 1900s to establish forest plantations. Shamba system was meant to establish exotic forest plantations with minimum establishment cost, and it continued fairly successfully mainly because of the following factors:

- Availability of sufficient cultivable land
- Presence of a willing land-hungry farming population
- Availability of a market for surplus food produced
- Security or protection against wild animals

The Shamba system when properly practised could allow sustained, optimum production of food crops along with forestry species from the same piece of land and thus meet most of the social and economic needs of the farmers. The economic implications of Shamba system are derived from the sale of agricultural products. The system also provides more or less full-time employment to most Shamba farmers who have no other sources of income. For the Forestry Department, Shamba System is a cost-effective method of tree establishment. Apart from providing a weed-free area for tree planting, it also ensures weeding around young trees for 2-3 years so that tree survival rates increase. In addition to the weeding benefits, fire control is also an important reason for inter-planting agricultural crops with tree crops in taungya system (White 1985). Other social benefits provided by the Forestry Department in the specifics of the Shamba system agreement included free housing, medical facilities, schools, welfare and infrastructural development (Wanyeki 1981). The Shamba system seems to be the most appropriate land use system for rural development and poverty alleviation programmes. The combination of agriculture and forestry when properly implemented reduces environmental degradation and also provides means of reconciling the often-conflicting interest of tropical foresters and farmers. The system also satisfies most social and economic needs of subsistence farmers who participate in it, and save the forest from the destruction caused by shifting cultivation (White 1985).

TUMPANGSARI

This is a system of taungya employed by the Forestry Estate Enterprise in Indonesia for reforestation, especially teak (Siregar 1990) in areas with a pronounced dry season, and the expectation is for an 80-year rotation. In Tumpangsari, a contract is made between individual farmers and the Forestry Estate Enterprise. The contract specifies the area to be planted, the period of agricultural cultivation, and the kind of crops to be planted as well as the amount of money to be paid for tree planting and tending. In this system, each family is allotted 0.25-hectare plot to be intercropped with the planted tree seedlings for two years, and the agricultural crops include dry land rice, maize, pepper, peanut and soybean. Under certain conditions, cassava also is permitted. On sloping land, erosion control structures are built with stones or dead tees and branches. Other control measures include drainage ditches, and planting along contours (Wiersum 1982). Income from Tumpangsari is important in sustaining rural families and keeping them in the countryside. For example, in Indonesia, annual crops from modern taungya (Tumpangsari) contribute only 20 per cent of household revenue but farmers can sustain their families from related plantation activities such as logging operations and collection of forest product, and under-employment is less for those who participate in Tumpangsari (Kartasubrata 1986).

INTENSIFIED TAUNGYA CULTIVATION

Increasing use of high-yielding agricultural crop varieties and modern cropping practices were incorporated in the taungya system in Indonesian agricultural enterprise in the 1960. In 1972, the Indonesian State Forest Enterprise tested the feasibility of such agricultural improvement for the taungya system that proved successful with dry land rice production by increasing yield of rice from

700 kg to 3,740 kg per hectare (Wiersum 1981). The system was developed further based on research carried out by the Forest Faculty of the Gadjah Mada University and was developed with the following characteristic features:

- Utilization of selected high-yielding varieties of the main food crops
- Better practices for land preparation and soil tillage
- Minimum use of fertilizers
- Control of pest and diseases
- Correct timing of planting and fertilizing with respect to the rainfall

Intensified taungya system yielded appreciably good results and was gradually applied over many other areas. With the use of selected superior crop varieties and fertilization together with insecticides, yield of dry-land rice improved from about 700-1,500 kg ha^{-1} to 2,000-3,000 kg ha^{-1} and maize from between 1,000-1,500 kg ha^{-1} to over 2000 kg ha^{-1}.

■■■■

20 ALLEY CROPPING SYSTEM

INTRODUCTION

The concept of agricultural sustainability has received much attention recently. Agricultural sustainability and productivity have been a goal of many International and National Agricultural Organizations from its inception in the 1970s (CGIAR 1990). Since agricultural sustainability is a dynamic concept, CGIAR has defined sustainable agriculture as "The successful management of resources for agriculture to satisfy changing human needs, while maintaining or enhancing the quality of the environment and conserving natural resources". A comprehensive definition reached by the American Society of Agronomy describes sustainable agriculture as agriculture that "over the long-term enhances environmental quality and the resource base on which agricultural depends; provides for basic human food and fiber needs, is economically viable, and enhances the quality of life of farmers and society as a whole" (ASA 1989). In the simplest sense, sustainability is the ability of a production (cropping) system to produce a stable annual yield of the desired crop over a long period of time. The length of time acceptable as a measure of sustainability is undefined and could be set arbitrarily for each particular environment and set of socioeconomic circumstances.

One of the greatest challenges facing agriculture in the tropics is the need to develop viable farming systems for the rainfed uplands that are capable of ensuring increased and sustained crop production with minimum degradation of the non-renewable soil resource base. The incorporation of woody species into crop production systems is one option that has received significant attention in recent years (Kang et al 1990). Trees have commonly formed an important part of traditional farming systems. In the bush-fallow, slash-and-burn cultivation systems, farmers exploited the potential of trees and shrubs for soil fertility regeneration and weed suppression during the fallow phase and widely used them as source for staking material later during cultivation. Trees and shrubs are also widely grown in cropped areas particularly in compound farms. They are mainly used for soil fertility maintenance and as auxiliary sources of food, fodder, fuelwood and for other domestic needs.

Despite the importance of woody species in traditional farming systems, only limited information is available on the potential contribution, when used in agroforestry systems, of multipurpose woody species to the improvement and sustenance of crop production in the uplands of the humid and sub-humid tropics. To improve traditional systems, alternative farming methods with improved soil management technologies are being investigated by various research institutions worldwide. Emphasis is being given to technologies that promote sustainability and serve as intermediate steps to more permanent cultivation through the use of soil conservation, crop rotation, and intercropping techniques. There has also been increased awareness and research effort during the past four decades in the development of more sustainable and productive land use systems by intercropping selective woody (particularly nitrogen-fixing legume) species with food crops in agroforestry systems (Kang 1993).

WHAT IS ALLEY CROPPING?

B.T. Kang and co-workers defined alley cropping (also known as hedgerow intercropping or avenue cropping) as an agroforestry system in which food crops and pastures are grown in alleys formed by hedgerows of trees and shrubs, preferably legumes (Kang et al 1981 and 1984; Kang and Wilson 1987). Alley cropping is the planting of rows of trees and/or shrubs (single or multiple) at wide spacing, creating alleyways within which agricultural crops or horticultural crops are produced (USDA-NAC 1999). The hedgerows are cut back at planting and periodically pruned during cropping to prevent shading and to reduce competition with the associated food crops, but allowed to grow freely to cover the land when there are no crops. Alley cropping production system is classified by Nair (1985) as a zonal agroforestry system. The system has been described as an improved bush-fallow system, as it retains the basic features of the latter but is improved in the sense that the cropping and fallow phases take place at the same time and on the same land, thus allowing more intensive cropping for a longer period of time.

Alley cropping is a form of hedgerow intercropping in which nitrogen fixing trees (NFTs) or other fast-growing plants are grown with annual crops. Alley means space and thus, in alley cropping, trees are grown in wide rows in the form of hedges and the annual crop is planted in the inter-space or 'alley' between the tree rows. During the cropping phase, the tree hedgerows are pruned which are used as green manure or mulch to improve the organic matter status of the soil and to provide nutrients, particularly nitrogen, to the crop. The hedgerows are allowed to grow freely when there is no annual crop.

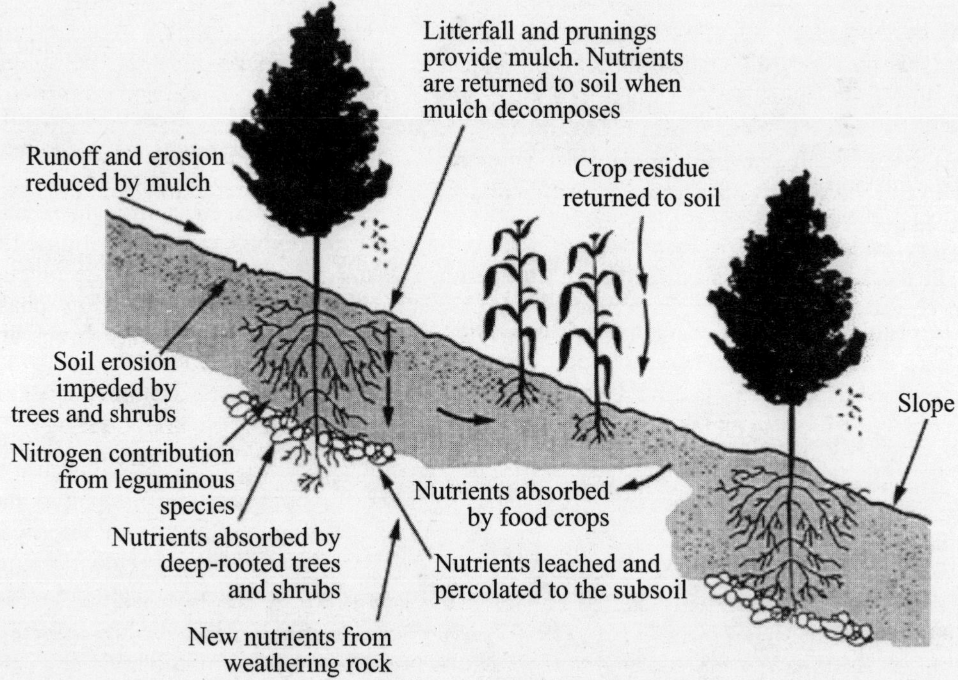

Fig. 20.1: Benefits of Nutrient Cycling and Erosion Control in an Alley Cropping System (Source: Kang et al 1986)

The underlying scientific principle of this technology is that, by continually retaining fast-growing nitrogen fixing trees and shrubs on crop-producing fields, the soil-improving attributes (such as recycling nutrients, suppressing weeds, and controlling erosion on sloping land) can be improved. Alley cropping thus retains the basic restorative attributes of the bush fallow through nutrient recycling, fertility regeneration and weed suppression and combines these with arable cropping so that all processes occur concurrently on the same land, allowing the farmer to crop the land for an extended period. Alley cropping is used to enhance or diversify farm products, reduce surface water runoff and erosion, improve utilization of nutrients, reduce wind erosion, modify the microclimate for improved crop production, improve wildlife habitat, and enhance the aesthetic of the area. Pioneering work on alley cropping was initiated at the International Institute of Tropical Agriculture (IITA), Ibadan, Nigeria by B.T. Kang and co-workers, in the early 1980s.

Table 20.1: Differences in Management between Traditional Bush Fallow and the Alley Cropping System (Kang et al 1986)

Bush Fallow	**Alley Cropping**
- Mixed native woody species	- Selected woody legumes
-Irregular planting pattern	- Planted in hedgerows
- Trees and shrubs cut back and burned before cropping to release nutrients	- Periodic pruning of trees and shrubs for use as mulch and green manure
- Growth controlled by fire	- Periodic hedgerow pruning
- Allows short-term cropping following fallow	- Allows continuous cropping

An important aspect of alley cropping is that there is a wide distribution of trees across a landscape. This increases the chance of capturing water which would have otherwise contributed to deep drainage. On the negative side, there is a large area of overlap between tree and crop root zones which would increase the competition between trees and annual species (Stirzaker and Lefroy 1997). In other words of Pedro Sanchez, the former Director General of ICRAF, "Alley cropping is like playing with fire, because it has strongly positive and strongly negative interactions and thus requires fine tuning in every situation".

The alley cropping system extended to include livestock by feeding a portion of the hedgerow foliage to animals was described as *alley farming* by Okali and Sumberg (1985). Both systems are similar, except that hedgerow management differs when the purpose includes animal feed production. Alley farming in its purest form essentially promotes crop-livestock integration and the production of mulch for improved crop production is still one of its objectives. In alley farming, most of the hedgerow foliage is used as mulch during the wet, crop growing season, particularly during the early growth phase, while a higher amount goes for animal feed during the dry fallow period. Thus foliage pruning must be strategically managed to satisfy both requirements.

ALLEY CROPPING: HISTORY AND MIXED RESPONSE

Formal research into alley cropping started as late as the late 1970s, at the International Institute of Tropical Agriculture in Nigeria. Kang et al (1990) record three factors which led to the development of this technology. First, multipurpose trees and shrubs already played a major role in traditional farming systems in tropics. Second, the traditional shifting cultivation was resulting in forests being cleared many times faster than they were being replanted. Third, there were large areas of infertile soils that erode and acidify under modern agricultural methods, making crop production unsustainable. In the wet tropics, cropping took place for only a few years after clearing before infertility and weed invasion made it impossible. The natural forest vegetation restored soil fertility, but as population pressure grew, it took an unacceptably long time to do so. In order to shorten the fallow period, farmers selectively retained trees during clearing and planted other trees after the cropping period, and so created improved or managed rotations. The development of alley cropping required a conceptual leap from the established rotation practice; tree and crop phases were now separated in space, rather than time, a situation which has many advantages as well as attendant difficulties.

On leached soils with rainfall exceeding 1,000 mm, field crops were limited by nutrients, rather than water, and the main role of the trees was to provide nutrients for the crop. Kang et al (1990) lists the desirable characteristics of a tree for alley cropping as ease of establishment; deep root systems; fast growth; tolerance to pruning and ability to coppice, ability to fix nitrogen; and nutritious to browsing stock. At least 5-10 t ha^{-1} yr^{-1} of prunings are needed to provide these nutrients, even for modest grain yields. To achieve this amount of prunings, tree rows were often spaced only 4-6 m apart and cut back several times per year to prevent shading of the crop. Research results from IITA showed that Leucaena/maize alley cropping could sustain maize yields of 2 t ha^{-1} and higher yields with the application of fertilizer (Kang et al 1981). Spurred on by encouraging results at IITA, alley cropping research spread to other parts of Africa (Arap-Sang 1986), India (Singh et al 1986; Rao et al 1991), South America (Szott et al 1991), the Philippines (Maclean et al 1992) and Indonesia (Parera 1986), supported by massive injection of funds from donor nations. Ten years later, experiences from around the world are mixed at best and frequently disappointing. This raised the questions – was alley cropping being tested in the wrong environments? were there problems with the way experiments had been carried out, or was the whole concept technically flawed? (Ong 1994; Coe 1994).

Today, it appears that alley cropping may have a role in the wet tropics with non-acid subsoils or on sloping land, but is marginal in comparison to a cereal/legume rotation when annual rainfall is less than 1,000 mm. Alley cropping is also not suitable in low base status soils, because the trees do not grow fast enough and there are not enough nutrients to recycle through the trees to the crops. In most situations, the trees are too competitive and cannot be pruned frequently enough; crops were shaded and a large proportion of nutrients released from prunings were recaptured by the trees before crops could take them up. There were certainly some design flaws in major alley cropping experiments (Stirzaker and Lefroy 1997). Ong (1994 and 1995) re-evaluated 7 major long term alley cropping trials from around the world. A recurring problem was that experiments assumed that there was little competition between trees and crops in a humid environment. Control plots were taken as plots which received no tree prunings, but closer examination showed that these “control plots” were riddled with roots from alley treatments 20 m or more away.

Alley cropping has largely failed in semi-arid India and Africa. These hard lessons have also opened the way for new ideas which are being followed up at World Agroforestry Centre. The first of these was to select trees based on their below-ground characteristics rather than their above-ground productivity. Researchers have found that some trees, e.g. *Grevillea robusta*, have a root system that could penetrate acid subsoils and capable of extracting phosphorus from low P soils by way of its proteoid roots (Harwood and Booth 1992). Another line of progress has been the investigation of quick measures of root distribution (van Noordwijk et al 1994). Finally there is a search for trees displaying unusual phenology. An example is *Paulownia tomentosa*, which is intercropped with wheat in more than 2 million hectares throughout China at a density of 40-70 trees ha-1. *Paulownia* is a fast growing tree which breaks into leaf very late in the season, when the wheat is already in head (Yunyin and Dalmacio 1991). These ideas have been encapsulated in the Alley cropping equation (Ong 1995), which provides a framework for quantifying competition and complementarity between trees and crops.

ESTABLISHMENT OF ALLEY CROPPING SYSTEM

CHOICE OF SPECIES

The characteristics of trees in alley cropping depend on the aim, objectives and priorities of the user. Climatic and soil factors should be given due consideration while making choice of species. The important characteristics of trees to be considered for alley cropping are as under:

- Trees should be relatively fast growing, multipurpose and easy to establish.
- Trees should be deep rooted so that it can take nutrients from deeper layers and can avoid competition with agriculture crops.
- Trees should not produce any growth-inhibitory substances (allelopathic effect) to minimize competition with crops.
- The growing season should complement the companion crop's growth period.
- Trees should withstand repeated coppicing so that they can be maintained in the form of hedges and should have high re-sprouting ability.
- Trees should have high leaf to stem ratio and small leaves or leaflets.
- Trees preferably should be nitrogen fixing with high foliar nitrogen concentration.
- Leaf should have rapid decomposition rate. However, if objective is for soil conservation and weed control, than high foliage productivity coupled with slow decomposition rate species are preferred. For example *Cassia siamea* or *Gliricidia sepium.*
- If objective is for livestock development, trees should have high fodder value and should be capable of producing good quality foliage throughout the year without/low toxins.
- Trees should have ability to withstand environmental stresses such as drought, water logging and extremes of pH.
- Trees should be free from pests and diseases.
- Trees should be easily controlled and not become invasive.
- Trees should produce a high-value products such as fuelwood, fodder, timber, fruits, etc.

As a hedge species, *Leucaena* has been the most widely used to date, but it is not ideal in erosion control as the leaves decay in one to two weeks, reducing soil cover. It has a demonstrated capacity to produce a dense hedge with high biomass production in climates ranging from humid to dry sub-humid. For the purpose of ground cover maintenance during the period of erosive rains, species with slower leaf decay such as *Cassia siamea* or *Gliricidia sepium* are to be preferred. Some of the promising species used worldwide for alley cropping are: *Calliandra calothyrsus*, *Cassia siamea*, *Erythrina poeppigiana*, *Flemingia macrophylla*, *Gliricidia sepium*, *Inga edulis*, *Leucaena leucocephala, Leucaena diversifolia*, *Paraserianthes falcataria, Sesbania sesban*, *Sesbania grandiflora.* All the above species are nitrogen fixing. *Alchornea cordifolia*, *Dactyladenia barteri*, *Gmelina arborea* and *Grevilea robusta* are some of the non-leguminous species used for alley cropping in many parts of the world.

All traditional agricultural or forage or horticultural crops can be grown with alley cropping. The primary factors determining which crops can be grown are the canopy density and sunlight requirement for the agronomic, horticultural, or forage crop.

HEDGEROW ESTABLISHMENT

For effective hedgerow planning, planting should be done on contours as it is effective in soil conservation. Contour maps are available from different sources for this purpose. However, if contour maps are not available, they can be prepared. The marking of contour lines is the first step for establishment of alley cropping. Contours are marked by using the simple 'A' Frame. It is in form of English alphabet "A". Contour frame is made of 5 to 6cm wide and 1 to 2 cm thick sawn timber pieces. Two such pieces are joined to make "A" of height 1.5-2 m with legs 1.2m apart (Legs spacement may be fixed as per the spacing from plant to plant in a row). To keep the legs in position they are strutted with a horizontal piece on which its middle point is marked. A plumb-bob is suspended from a nail fixed in the apex. When the thread of the plumb bob passes through the middle point mark of the horizontal piece, the two points on which the contouring frame stands, are on the same contour.

While working with "A" type wooden frame, one leg is kept at the place from where contour line has to be started. The other leg is moved up and down the hill till the plumb bob is in the centre. At this situation "both legs are on the same level. To proceed further, wooden frame is turned to next place. In similar way points falling on same line are marked and contour line is drawn. Contour marking and making bunds or trenches, always begins from top of the slope.

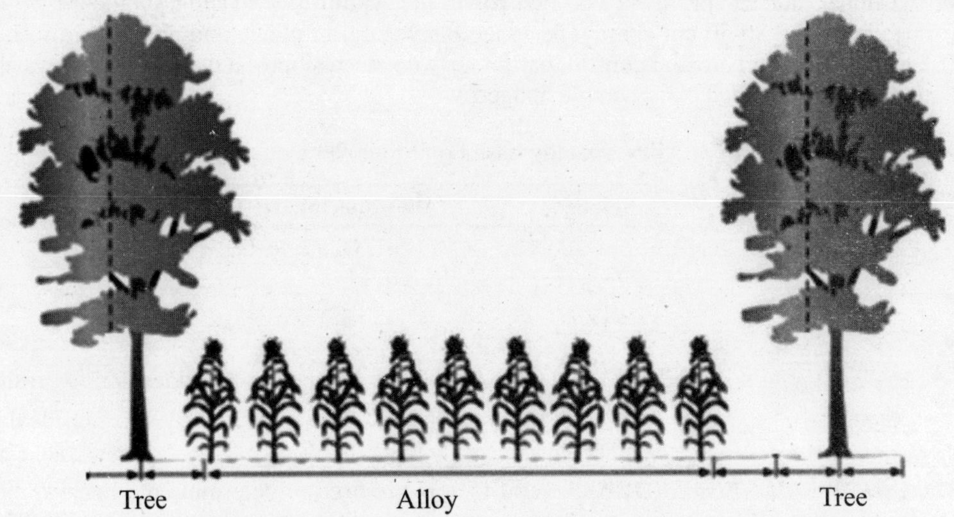

Fig. 20.2: Tree and Crop Planting Design in Alley Cropping

PLANTING

Hedgerows can be established from seeds or cuttings depending on the species being used. Seeds are less bulky and can be distributed more efficiently and at lower cost. Cuttings can be used where the seeds are less viable or are not available. However, owing to high tree population, establishment by cutting can be costly, inconvenient and impracticable. In dry sub-humid climate with less rainfall (<1,200 mm annually) establishment by seedlings is more preferred and ensure more success. Seedlings are raised in nursery and are transplanted at the onset of monsoon to ensure better survival. Seed can also be sown directly after giving seed treatment to break the dormancy. For example, hot water treatment is given to *Glircidia* seeds for optimizing the germination.

SPATIAL ARRANGEMENT

ROW ORIENTATION: Trees oriented in east-west direction maximize the sunlight received by an alley crop, provided the topography permits this arrangement. Trees may be planted in other orientations if prevailing winds have a negative influence on crop yields. On sloping lands, the trees should be planted along the contour.

SPACING WITHIN HEDGEROWS: The spacing between individual trees is critical in designing an alley cropping practice. Trees can be planted either as single or double rows. A single row minimizes loss of cropland, and can occupy only 0.5 m width if pruned low. Double rows, form a more substantial barrier with less chance of gaps occurring and are more productive than those of one row hedges. The spacing between trees is determined by the primary purpose of the alley cropping. Plants grown at wider spacing produce thicker stems while, close spacing encourage more leaf production and provides a more effective barrier to soil movement on sloping lands and creates a better microenvironment for crop growth. Closer spacing both within the row and between the rows also allows for improved distribution of nutrients to a greater proportion of the intercrop. The within-row plant spacing in hedgerows should be close. In humid and moist sub-humid climates, direct sowing of seed has been successful, following which, seedlings can be thinned to a spacing of the order of 10 cm. Where seedlings are planted, as is necessary in drier climates, 25 cm spacing appears to be sufficiently close and the gaps between stems are filled by crop residues and prunings coming to rest against the upper side of the rows. For two rows, the within-row spacing should be 50 cm. Trees within line should be 30-50 cm apart. The space between crop plants and any tree row should be at least 0.5 m. For optimum erosion control, particularly on steep slopes, a pair of hedgerows about 0.5-1 meter apart is recommended over a single hedgerow.

Table 20.2: Determination of Alley Spacing based on Slope Percentage (USDA-NAC)

Slope (%)	Alley Spacing (ft)
5-10	60
10-15	45
15-25	30
25-35	20
35-50	15
50-60	10

SPACING BETWEEN HEDGEROWS: Several factors are used to determine the interval between the row(s) of trees including slope length, field width, crop light requirements and equipment/machinery usage. Close spacing between the hedgerows reduces the amount of land available for the crop and can result in increased competition for the growth factors of light, moisture and nutrients between hedgerow and crop. Generally, spacing between hedgerows should be selected to strike a balance between reduction of competition between the hedgerow and crops, while maximizing production of organic matter and control of erosion. Spacing should be such that, hedgerows should not take more than 10-15% of the planting area. Hedgerow widths of over 10 m have been suggested for semi-arid climates. Guidance for the spacing of hedgerows comes at present only from the various formulae for vertical intervals between earth bunds or ditches found in national soil-conservation handbooks and textbooks. Inter-row distance along the ground surface can be calculated by the formula:

$$\mathbf{W = 100/S}$$

where, W = inter-row spacing in metres and S = slope angle in degrees

Thus, hedges would be 5 m apart on 20^0 (36%) slopes.

MANAGEMENT OF ALLEY CROPPING SYSTEM

GAP FILLING

To avoid gaps in the hedgerows, mortality should be replaced as early as possible as it becomes very difficult to establish new seedlings in the shade of the older hedgerow trees.

PRUNING

Prunings should be either laid against the upslope side of the hedgerows or should be distributed across the alleys. Laying prunings along the hedges immediately upon pruning serves to consolidate them as barriers and leave the alleys clear for tillage operations. Distributing prunings across the alleys helps in erosion control, providing ground cover and improving soil-fertility. For maximum cover, both woody stems and leaves should be distributed intact. Biomass yield is affected by intensity and height of pruning. More pruning adds more green manure to the soil. Less frequent pruning produce more fuelwood-diameter branches. The amount and frequency of pruning required depends on the shade tolerance of particular crop at particular stage of development. Frequent pruning reduces shading effect on annual crops and enhance crop productivity. Biomass and yield increases with decreasing pruning frequency. Pruning should be done when nutrient requirement of annual crop is high. The initial pruning should take place after the trees are well established. Depending on rainfall, it may take 6-12 months before the trees can first be cut back. During the first cutting, the woody portion is often laid at the base of the hedgerow trees to help with erosion control. Successive prunings should occur when the regrowing stems are still soft and leafy, usually every 3-6 months, depending on species used, rainfall and other factors. Biomass also increases with increasing pruning height up to 1m. A good rule is to maintain a hedge height equal to the distance between the hedge and nearest crop row (50 cm to 1 m).

WEED CONTROL

Weed control is essential during the establishment of the hedgerows. For the first three to five years, weeds need to be minimized usually near the tree rows. Weed should be removed manually during initial stages. Mulching can also be effective in controlling weed growth.

FERTILIZATION AND NUTRIENT MANAGEMENT

When using NFTs, nitrogen fertilizers are not necessary. However, during cropping phase, competition for nutrients is inevitable and should be minimized using external nutrients (FYM, fertilizers, etc) or by suitable management practices like root pruning.

PEST MANAGEMENT

Periodic inspection of the crops and trees is recommended to detect and identify possible pests. Insects and diseases can be significant factors in reducing the health and vigour of both the tree crop and the intercrop. The corrective actions should minimize the impacts on beneficial insects.

BENEFITS OF ALLEY CROPPING

CHECKING SOIL EROSION

In alley cropping, the tree rows act as a physical barrier to soil and water movement, resulting in significant reductions in erosion losses. The interception of rainfall by the tree canopy, network of roots and mulching by leaves also helps in checking soil erosion and improving water quality due to interception of sediment.

IMPROVED SOIL HEALTH

Addition of large amounts of organic materials from the prunings in alley cropping contributes organic matter to soil and releases nutrients as they decompose and have favourable effects on soil physical, chemical and microbial properties and hence on soil productivity. Fine roots add organic carbon to the soil and recycle the nutrients and improve their retention. The addition of mulch lowers soil temperature, reduces evaporation and improves soil fauna activity and soil structure resulting in better infiltration, reduced runoff and improved water use efficiency. The magnitude of these effects, however, varies with hedgerow species and management as this influences the quantity and quality of prunings. Tree root system also acts as a 'safety-net' of leached nutrients in deeper layer.

IMPROVED CROP HEALTH

Trees and shrubs improve crop production by slowing wind speed, reducing wind erosion, modifying the crop microclimate, addition of nutrients and organic matter to the soil/plant system. They also protect the inter-crops against wind damage and insect pests, moderate air and soil temperature extremes, and reduce moisture loss from the soil. Alley cropping can reduce crop evapotranspiration by 15-30 per cent and increase water content in the tillage layer by 5-15 per cent. Reducing wind effects also aids in pollination activities by beneficial insects thus increasing yields of some crops. It also helps in weed control during the fallow period by shading of the interspaces.

DIVERSIFYING FARM PRODUCTS AND SUPPLEMENTING INCOME

Alley cropping diversifies farm enterprises by providing short-term cash flow from annual crops while also providing medium to long-term products from the trees. Timber and non-timber products may contribute to income generation from the farm.

PROTECTING CROPS

Alley cropping protects crops from insect pests by reducing crop visibility, diluting pest hosts due to plant diversity, interfering with pest movement, and creating environments less favourable to pests and more favourable to beneficial insects.

DRAWBACKS OF ALLEY CROPPING

- Establishment investment for planning and installation
- Loss of cropping area to the hedgerows
- Risk of competition between tree and alley crops for light, water and nutrient
- Increased labour requirements and skills for maintenance of an alley cropping system and demands of different harvest seasons
- Requires a more intensive management system including specialized equipment for the tree management
- Difficulty in mechanizing agricultural operations
- Potential for the hedgerow species to become a weed and/or an alternate host for pests and pathogens, or harbour grain-eating birds
- Possibilities for increased termite activity, especially under dry conditions.

■■■■

21 TROPICAL HOMEGARDENS

WHAT IS HOMEGARDEN?

Homegardens are multi-species, multi-tier agroforestry production systems often in small parcels of land surrounding homesteads that integrate tree-crop-animal components and largely confined to humid tropics. By definition they are 'intimate, multi-story combinations of various trees and crops, sometimes in association with domestic animals, around homesteads' (Kumar and Nair 2004). They are space constrained subsistence farming systems running on traditional low input technologies while maintaining multiple outputs often meeting the livelihood and nutritional security of millions of people in the tropics. These homegardens maintain the biological diversity of native and exotic as well as managed or wild species, and play an important role in improving the quality of life and the economic and social welfare of people. Physiognomically homegardens resemble evergreen forests by virtue of the assemblage of diverse tree and crops arranged in multi-tiered-intimate-interlocked fashion (Ewel 1999). Hence, they are often described as most natural production systems which are ecologically and economically sound and socially acceptable. It evolved through generations of gradual intensification of cropping in response to increasing human pressure and the corresponding shortage of arable lands (Kumar and Nair 2004). These systems have probably evolved over centuries of cultural and biological transformations and through generations of innovation and experimentation (Wiersum 2006).

A typical homegarden is primarily a multiple output production system that spins around the homestead where selected trees, shrubs and herbs are grown for edible products and cash income, as well as for a variety of outputs that have both production and service values including aesthetic and ecological benefits. Homegarden products are directly linked with the livelihood security of the farmers who are primarily at subsistence level (Nair 2001). Traditionally, the homegardens mainly served to produce vegetables, fruits and other crops, which supplemented the staple food crops produced on open croplands. Apart from the numerous direct benefits that meets the basic human needs, homegardens deliver quiet a lot of ecosystem services that are assumed to be the key drivers of their sustainability. The biophysical advantages like efficient nutrient cycling, maintenance of biological diversity, and multiple products economic value, socio-cultural and socio-political advantages are the other traits that make homegardens important.

However, these fascinating systems while following sub-optimal management strategies still remain largely unexplored in terms of their productivity and sustainability and the scientific underpinnings of mystery of its sustenance over 1000s of generations. Today the studies on homegardens takes larger shifts on account of their perceived role in climate mitigation through their carbon sink functions and the low carbon footprints of the homegarden products.

DISTRIBUTION OF HOMEGARDENS

The warm humid tropics are assumed to be the seat of homegardens. Historically, their origin dates back to human settled agriculture, proceeding the era of shifting cultivation. From these pre-historic and probably scattered origins, homegardens has gradually spread to many humid regions in South Asia and Southeast Asia including Java (Indonesia), the Philippines, Thailand, Sri Lanka, India and Bangladesh. Randhawa (1980) reports that many travellers in the early 14th century describes the homegardens in Kerala, India with coconut (*Cocos nucifera*), black pepper (*Piper nigrum*), ginger (*Zingiber officinale*), sugarcane (*Saccharum officinarum*) and pulses (grain legumes). Nair and Kumar (2006) well described the distribution of homegardens. According to them the largest stretches of homegardens are found to be in the humid high rainfall regions 40° N and 30°S of equator. However, South Asia and Southeast Asia, the Pacific islands, East Africa and West Africa, and Mesoamerica are the regions where largest concentrations of homegardens can be found (High and Shackleton 2000; Nair and Kumar 2006; Montagnini 2006). Prominent among them are the Javanese homegardens in Indonesia and the Kerala homegardens of India.

Homegardens vary in their local vocabulary. For instance, the Javanese homegardens are known as 'Talun-Kebun' and 'Pekarangans', Kerala homegardens as 'Purayida Krishi' while that found in the Mt. Kilimanjaro in East Africa often described as 'Chagga' homegardens with coffee and banana integrated with multipurpose trees, Enset coffee homegardens in Ethiopia (Abebe et al 2006). Despite the presence, information on the area under homestead farming in different regions is by far limited. Nair and Kumar (2006) describe homegardening as the prominent land use in tropical countries such as Nepal, Sri Lanka, Bangladesh and Philippines. They provide conservative estimates that suggest 5.13 million ha of land under pekarangans in Indonesia, 0.54 million ha under homesteads in Bangladesh, 1.05 million ha in Sri Lanka, and 1.44 million ha in Kerala, India. Prominent regions in the Africa where homesteads practices are prevalent include Sudan, Ethiopia, Nigeria, Kenya, Tanzania, etc. while Peruvian Amazon homegardens, Brazilian, Mexican, Guatemala, Honduras, Nicaraguan and Costarican homegardens are the potential homegarden regions in the Latin America.

TREE-CROP DIVERSITY IN HOMEGARDENS

Homegardens are one among the many agroforestry approaches that are common to many ecological regions, but the nature of components that constitute the systems may vary depending upon site-specific factors (Nair 1991). Homegardens are intimate assemblage of tree-crop-livestock in variable spatial and temporal sequence. The key factor that distinguishes homegarden from other agricultural land use systems is the high species diversity (Swift and Anderson 1993; Kumar et al 1994; Abebe et al 2006). They are often christened as the epitome of biodiversity (Kumar and Nair 2004). Livelihood conditions are important factor influencing the structure and composition of homegardens. Homegardens scattered in different parts of the world are found to be agro-ecologically diverse in terms of composition and arrangement of various components. However, there exists remarkable similarity in the functional space occupied by various components. The species diversity is strongly influenced by functional priorities evolved through generations. Also, it is often dictated by the agroecological adaptability. Homegarden component diversity may be affected by geographic conditions as well. Compared to lowlands, homegardens in highland areas have lower plant diversity and simpler species composition (Karyono 1990). Similarly, diversity is more in wetter areas compared to arid and semi-arid regions. Various cultures also influence the structure and composition of homegardens. Abdoellah (1990) reported that vegetables and ornamentals were often more common in Sudanese homegardens. The diversity and density of woody species may decline with increasing age of crop fields, while diversity of woody species increase with increasing age and size of homegardens (Tolera et al 2008).

There are considerable differences in the species diversity with homegarden types. For instance, Javanese homegardens are rich in diversity as compared to those found in other regions. Studies on diversity of arboreal taxa in different homegardens clearly demonstrate their variation. For instance, Kumar et al (1994) reported 127 tree taxa in the homegardens of Kerala, India. Among them coconut and arecanut were the most prominent. *Ailanthus triphysa*, *Mangifera indica*, *Artocarpus heterophyllus*, *Tamarindus indica*, *Erythrina indica*, *Macaranga peltata*, *Thespesia populnea* and *Gliricidia sepium* were the other preferred tree species in Kerala homesteads. They also observed that the species diversity and size of homegardens strike an inverse trend. However, such patterns are not true for many homegarden types. For instance, the Brazilian homegardens show higher species diversity associated with large homesteads (Albuquerqu et al 2005). They reported a total of 54 woody species distributed among 46 genera and 23 families. *Annona squamosa*, *Prosopis juliflora* and *Psidium guajava* were the most abundant tree species. Reports also suggest species number and structural complexity is greater in humid areas than arid (Azudia and Leiva 2004). Much higher tree species diversity has been observed from elsewhere viz. 301 trees and shrubs from Mayan homegardens of Yucatan, Mexico (Rico-Gray et al 1991), 168 and 179 species from Peruvian Amazon (Padoch and de Jong 1991) and West Javan homegardens (Soemarwoto 1987) respectively. Studies in the south-central highlands of Ethiopia compared the tree species diversity among homesteads, cropped lands and natural forests and observed that the highest number of woody species (64) was recorded in homegardens, followed by crop fields (32) and the lowest number (31) in remnant natural forest (Tolera et al 2008).

It is observable that the functional priority often changes with regional preferences. Homegardens in Peruvian Amazon regions consists more of fruit trees such as *Mangifera indica*, *Eryngium foetidum*, *Syzygium* spp., *Cocos nucifera*, *Persea americana*, *Citrus sinensis*, *Anacardium occidentale*, *Artocarpus altilis*, *Annona squamosa*, *Spondias purpurea*, *Cedrela odorata*, *Theobroma cacao*, etc. (Perrault-Archambault and Coomes 2008). Gebauer (2005) from central Sudan also reported on the preference for fruit trees. As per their report, the five most common fruit trees were lime (*Citrus aurantifolia*), guava (*Psidium guajava*), mango (*Mangifera indica*), date palm (*Phoenix dactylifera*) and grapefruit (*Citrus paradisi*). In general, plant species composition within the homegardens is the result of continuous selection in which the family usually favours the planting of fruit trees with high productivity (Caballero 1992).

Agricultural crop diversity also varies considerably among homegardens. Prominent understorey crops in the Kerala homegardens include vegetables such as brinjal, ladies finger, cow pea, ash guard, bitter guard, snake guard, black pepper, tuber crops such as colocasia, elephant foot yam, diascoria. Banana is the common intercrop in homegardens throughout Kerala. Cassava, papaya, fodder grasses, pineapple, *Curcuma longa*, *C. aromatica*, *Zingiber officinale*, etc. are the other common intercrops (Kumar and Nair 2004).

Socioeconomic status and livelihood conditions may impact the species composition in the homegardens. For instance, the farmers at the subsistence level make use of the limited land for production of essential staple foods rather than supplementary crop production. Contrary to this well off people living in urbanized areas and having access to non-farm incomes may no longer maintain homegarden as part of their farming system, rather aesthetic considerations dominate their retention. Thus, not only the overall livelihood conditions, but also specific socioeconomic variables such as access to land or off-farm labour opportunities impact the homegarden structure and composition (Wiersum 2006).

Fig. 21.1: A Typical Kerala (India) Homegarden: Multistrata Composition of Various Components

STRUCTURAL ATTRIBUTES OF HOMEGARDEN

Despite the limitation in management area, the high diversity in the tree-crop and animal components in the homegardens is often coupled with their characteristic structural arrangement (Nair 1993). The layered canopy stratification and admixture of compatible species often mimicking evergreen forests, is the most conspicuous characteristics of all homegardens. They are carefully structured systems with every component having a specific place and function. The gradient of light and relative humidity creates different spatial niche enabling various species groups to exploit them. The stratification is often dictated by the life history of the different species. In general, all homegardens consist of an herbaceous layer near the ground, a tree layer at upper levels, and intermediate layers in between. The lower layer can usually be partitioned into two, with the lowermost dominated by different vegetable and medicinal plants and second layer being composed of food plants such as cassava, banana, papaya, yam, etc. However there can be changes in the vertical stratification and composition of components for different homegarden types.

Typical Kerala homegardens show 3-4 strata with ground layer consisting of herbaceous food crops, forage, medicinal and other crops while fruit trees and spice crops dominate the middle storey (Nair 1993; Kumar and Nair 2004). The upper canopies constitute mostly tall trees/palms such as coconut, teak, mahogany, and other fast growing multipurpose trees. Michon et al (1983) reported typical Javanese home garden occupy 4-5 strata. Lowest strata constituting less than 1m height contained 14 per cent of total canopy volume, 2nd layer 1-2 m (9%), 3rd layer 2-5 m (25%), 4th layer 5-10 m (36%) and 5th >10 m (16%). Herbaceous species form the layer near the ground. The upper layers are occupied by trees, while intermediate layer in between shared by medium sized trees, mostly fruit and medicinal trees. Quiet often, the lower layer may be partitioned to less than 1 m dominated by vegetables and medicinal plants, while second layer constitute food plant such as cassava, banana, papaya, yam and so on. The intermediate layers of 3-10 m height are dominated by various fruit trees and this layer is not structurally static but always dynamic. The upper tree layer divided to two viz. emergent fully grown timber and fruit trees occupying the uppermost layer (25 m height), medium sized trees of 10-20 m occupying the next lower layer.

The vertical structure of Brazilian homegardens reflects their degree of specialization and complexity. The lower stratum (1-3 m) consists mainly of medicinal and fruit trees, such as *Psidium guajava*, *Annona squamosa*, *Jatropha molissima*, as well as saplings of forage species (e.g. *Prosopis julifora* and *Capparis flexuosa*). The midstratum (3-7 m) is formed by combination of species with multiple uses. The upper stratum is generally formed by trees of approximately 7-12 m tall, such as *P. julifora*, *Schinopsis brasiliensis*, *Anacardium occidentale*, *Erythrina velutina* and *Anadenanthera colubrina* var. cebil (Albuquerque et al 2005).

Fig. 21.2: A Typical Large Kerala Homegarden: Coconut based Integrated Multistrata Farming

Homegardens in high humid, rainfall areas constitute more number of canopy stratification with more species diversity. For instance, the homegardens in the Andaman Islands of India maintain 5 characteristic strata (Pandey et al 2006). The top storey (15-20 m) was occupied by coconut whereas fourth storey (10-15 m) by arecanut and jackfruit. The third storey was dominated by trees like mango, cashew nut and tamarind (5-10 m). Second storey (2-5 m) however, was abundant with spice trees particularly nutmeg, cinnamon and clove. Fruit trees were closer to the house. Clove was found in both second as well as in third storey. Species that require specific niche like cinnamon, requiring high humidity, was grown mostly under the arecanut and occasionally under coconut. Simiiarly, banana usually grown in open receiving good water and sunlight usually where water of house drained. Short height annual crops like *Curcuma longa*, *Zingiber officinale*, *Manihot esculenta*, *Amorphophallus campanulatus*, grasses and pineapple (*Ananas comosus*) occupied the first storey (< 2 m).

The traditional Asian homegardens in general have in common a spatially separated farming system consisting of cultivated fields away from homes while maintaining homegarden surrounding residential houses. Homegardening in such farming systems is mostly subsidiary to staple food production and mainly focuses on vegetables, fruits, and condiments (Wiersum 2006; Soemarwoto 1987; Hoogerbrugge and Fresco 1993). But, instances of homegardens forming part of integrated farming system where staple food production is also an integral part is not uncommon. These 'complementary' homegardens typically consist of small (0.01 to 1 ha) plots around houses with a more or less randomly organized cropping pattern. Since, these type of homegardens are integral part of other components of the entire farming system, the component diversity and homegarden functioning are influenced by the nature and characteristics of the other components of the overall farming system (Karyono 1990; Abebe et al 2006).

FUNCTIONAL ATTRIBUTES OF HOMEGARDEN

The list of direct benefits accrued from homegardens are large that include food, fuel, fodder, fruits, timber, pulp, green manure, etc. Food production function is recognized as the most important one from homegardens especially at subsistence level. Various food products include edible fruits, nuts, grain, rhizomes and tubers, leaves, flowers to name a few. Homegardening being an agricultural practice at low or moderate input level, the produces are ecologically safe and hence contribute to the social health and hygiene.

Fruit and other food trees constitute predominant component of all homegardens (Nair 1993). Mango, guava, papaya, rambutan, mangosteen and other food producing trees such as *Moringa oleifera* and *Sesbania grandiflora*, spice crops such as cinnamon, clove, nutmeg dominate the Asian homegardens. Torquebiau (1992) in a review observed that dietary supplies from homegardens accounted for 3 to 44 per cent of the total calorie and 4 to 32 per cent of the protein intake. Homegardens function as a buffer source of food ensuring food security during the lean seasons. The homegardens are also significant sources of minerals and nutrients (Asfaw and Woldu 1997). Nutritional security is another area where homegardens considerably contribute. Year round availability of many high nutritional value crops takes care of the nutritive requirements (proteins, vitamins, and minerals) of the family. Studies revealed that families involved in year-round production and consumption of vitamin rich fruits and vegetables, alleviated deficiencies of iodine, vitamin A and iron (Molina et al 1993) and made children of garden owners less prone to xerophthalmia (Shankar et al 1998).

In addition homegardens provide enumerable products and services which are closely linked with the household needs of the farmer. For instance, they are potential source of wood and small timber for meeting the farmer requirements. Kumar et al (1994) observed that the average standing stock of commercial timber in the Kerala homegardens (India) has been estimated to range from 6.6 to 50.8 m^3 ha^{-1}. Homegardens are also good source of non-timber products such as bamboo. Studies on the standing stock of thorny bamboo (*Bambusa bambos*) in the homegardens in three districts of Kerala (India) suggest substantial stocking to the tune of 28,659 Mg (Thrissur); 1,24,389 Mg (Palakkad); and 86,267 Mg (Malappuram) on dry weight basis (Kumar et al 2005; Kumar 2008). Also, tropical homegardens function as predominant supplier of biofuels for rural households (Krishnankutty 1990; Wickramasinghe 1996; Levasseur and Olivier 2000; Shanavas and Kumar 2003; Kumar and Nair 2004). For example, reports suggest about 51 to 90 per cent of the domestic fuelwood needs in south and Southeast Asia is met from homegardens (Krishnankutty 1990; Torquebiau 1992).

ECOSYSTEM SERVICES OF HOMEGARDEN

Agroecosystem that remains relatively unexplored from an ecosystem service pers-pective is homegardens. They provide a large set of ecosystem services, cultural services being the category most valued. The quality and magnitude of ecosystem services provided by homegardens differ from those provided by other types of agroecosystems (Calvet-Mir et al 2012). Major ecological functions rendered by homegardens include maintenance of soil fertility, regulation of pests and pathogens, wildlife protection, clean water supply, carbon sequestration, maintenance of rural landscapes and rural lifestyles, and maintenance of recreational areas for ecotourism (Sandhu et al 2010; Swinton et al 2007; Zhang et al 2007).

Traditional agriculture practices such as homegardens are characterized by maintenance of high agro-biodiversity (Altieri 1999; Jackson et al 2007) which ensures coupling of agricultural productivity with the delivery of the other adaptable services that biodiversity provides. For instance, maintenance of agro-biodiversity in homesteads leads to enhancement in agroecosystems' resilience to ecosystem changes mooted by demogra-phic pressure (Jackson et al 2007; Pascual et al 2010). Homesteads essentially a low input agroecosystem rely primarily on site-specific biological, edaphic and climatic conditions there by reducing dependence on heavy inputs such as machinery, phytochemicals and power thereby reducing related hardships in terms of soil compaction, water pollution, greenhouse gas emissions, etc. (Altieri 1999).

Biophysical interactions triggered by trees provide substantial ecosystem services. Trees with their deep root system act as nutrient pumps and the fine roots dynamics contribute substantially to enrich the carbon content and nutrient status of the soil. The multilayered canopy stratification and the high litter production from the trees cover the soil and protect from insulation. The tree cover moderates the extremes of climate and biophysical extremes. Yet another environmental function rendered by homegardens is their carbon sink properties and in so doing take part in climate change mitigation. Homegardens are unique in that they address all the three mechanisms that qualify agroforestry as GHG reduction strategy viz. carbon sequestration, carbon substitution and carbon conservation (Montagnini and Nair 2004; Kumar 2006). The trees and other perennial components in the homegardens sequester substantial amount of CO_2. On a comparative scale homegardens sequester C much better than intensively managed crop lands (Nair et al 2009). For example, cropped land after slash and burn showed CS values 39 to 52 Mg C ha^{-1}(Sanchez 2000) while Indonesian homegardens, Sumatra sequester much higher to the tune of 55.8 to 162.7 Mg C ha^{-1} (Roshetko et al 2002). Aboveground carbon stocks of trees (>20cm girth at breast height) in the homegardens of selected 28 panchayaths of central Kerala, India were 24.32 Mg ha^{-1} (Kumar 2011). Homegardens also stands out as a promising example of climate change resilient land use practice.

MANAGEMENT CHARACTERISTICS OF HOMEGARDEN

As always, homegardens are less intensely managed systems. Scientific management of these traditional systems poses serious limits on account of intertwined nature of various components in time and space. The understorey space utilization for intercropping depends entirely on the tree density and light availability. A legitimate approach is to manipulate the tree environment rather than the tree itself. Hence practices such as branch pruning for improving understorey light conditions are a common management practice in homesteads. For instance, branch pruning of taller trees such as teak (*Tectona grandis*), *Terminalia paniculata*, *Swietenia macrophylla* is a common practice in homega-rdens of Kerala. Other conventional practices such as weeding, fertilization, crop density regulation are also in vogue in the various parts of the tropics. There has been renewed interest on tree husbandry in view of their economic stability and market demands. Especially the recent trends suggest tree based cash crops such as rubber and nutmeg rigorously managed in homegardens. Coconut is the most popular among the trees followed by arecanut and spices in high rainfall coastal regions like Kerala and the intensive management practices that include fertilization, use of plant protection chemicals, systematic weeding and management of interspaces in between the palms for intercropping, etc. are limited to such economically important crops only (Peyre et al 2006). Levels of management are again strongly influenced by the land size and socio-economic adaptability. A clear gradient in management intensity exist among homegardens in Kerala (Peyre et al 2006). For example, the small sized gardens are characterized by low to medium management with a concentration on internal inputs and with random arrangement of trees while the medium to big sized homegarden are subject to a more intensive management with use of both internal and external inputs such as chemical fertilizers, insecticides and purchased seedlings (Peyre et al 2006).

The type and intensity of management again may vary with objectives. The coffee-banana dominated Chagga homegardens in Tanzania follow time bound management practices for productivity enhancement. Various operations include opening up the canopy to ensure better fruiting of the coffee, spacing out the banana stools, and manuring the different crops (Fernandes et al 1985). As evident from the homegardens in Kerala, the management capabilities and preferences of the owner decide the number of banana clumps and coffee bushes in the Chagga gardens. Fernandes et al (1985) reported that the range of banana clumps per homegarden varies from 200 to 800 (330 to 1,200 per ha) and coffee trees from 300 to 1,000 (500 to 1,400 per ha). The management strategies followed in Javanese homegardens also vary with the system followed. The popular Javanese Pekarangan system is a mixture of annual crops, perennial crops, and animals on the land surrounding a house where as Kebun-Talun system is more production oriented and permits clearing the trees to cultivate field crops on a rotational basis. The Kebun phase confines to the crop cultivation with high economic value gradually merge with kebun-campuran phase where annuals are mixed with half grown perennials, period of low economic but high ecological value. The final phase (Talun) after the harvesting of the annuals is dominated by perennials of commercial value. The Kebun-talun will be converted to Pekarangan system when the house is developed upon it (Christanty et al 1986).

SHIFTING TRENDS IN HOMEGARDENING

Despite the manifold virtues as a promising land use system with potential to support livelihood security of millions of people in the tropics, homegardens are undergoing massive transformation that wear down their intrinsic characteristics. Fast changing agriculture scenario and the high market-orientation, due to government policies and demographic pressures do exert considerable pressures on homegardens (Kumar and Nair 2004). Urbanization and associated socioeconomic polarization has influenced the homegardens quiet a lot. Commercialization has resulted in drastic reduction in the structure and functions of homegardens world over. Indonesian 'pekarangan' (Abdoellah 1990; Abdoellah et al 2001), Kerala homesteads (Kumar and Nair 2004), are some of the examples for this. In Kerala, the multi-species structure of homegardens are shifting to cash crops based monocultures such as rubber and coconut, inflicting serious drain in plant diversity. Population pressure on the land and associated fragmentation, high cost of land, alternative options of land use are potential threats to retaining homegardens in the tropics. Homegardens are assumed to be the last refuge of many local varieties of crop/tree species. Urbanization and land fragmentation has exhausted many of these natural varieties. Report from West Java, Indonesia, suggests a loss of 27 varieties of mango during a 60-year period (Soemarwoto 1987) while Chagga homegardens of Tanzania experienced decline in diversity due to fragmentation (Rugalema et al 1994).

Socio-economic transformation triggered by globalization and market driven lifestyle has prompted people to do away with traditional agricultural practices which are 'labour intensive and time consuming'. The shift towards development oriented non-agricultural land use practices has started manifesting its far reaching consequences in all spectrum of human life such as escalated rates of landslides, soil and water conservation, poor drainage, massive soil erosion and associated loss of soil fertility. Need of the hour is to hold lessons from the advantages of homesteads as invaluable sustainable farming system that integrates diverse crop forms to meet the multitude of benefits to mankind.

CONCLUSION

Homegardens are unique traditional multistrata land use systems that integrate various life forms and practised within the small land parcels surrounding the home. They are ecologically sound, economically viable, socially acceptable and culturally pragmatic systems that touch upon the livelihood security of millions people in the tropics. The favourable biophysical environment promotes high level of association and comlementarity among the various components that guarantee sustainability of homegardens. The high floristic diversity attached to these systems has evolved through preferential selection followed over generations. Apart from the multitude of direct benefits, the array of ecosystems services rendered by the homegardens distinguishes them as robust land use system. The enormous potential of homegardens as carbon sinks qualify them as promising example of climate change resilient agriculture and mitigation strategy. Despite all these multifarious advantages, homegardens are yet to receive recognition that they deserve neither at the policy level nor at scientific front. Moreover, this time-tested agrarian tradition appears to fade out under the pressure of urbanization and globalization. Influx of monoculture, land fragmentation, alternative market avenue for homegarden products and fast changing socio-economic and cultural equations have put serious threats on the future of these wonderful systems.

■■■■

22 AGROFORESTRY SYSTEMS OF WORLD

OVERVIEW OF GLOBAL AGROFORESTRY

The world population has now surpassed 6.5 billion. Population growth and rapid economic expansion in Asia and Latin America have significantly accelerated growth in global food demand. However, over 1 billion people, most of them in East and South Asia and sub-Saharan Africa, continue to endure lives of extreme poverty. These trends have brought into sharp focus the critical roles of agriculture in poverty reduction, food security and maintaining environmental services. This is reflected in a number of recent global assessments including World Development Report (World Bank 2008), the Millennium Ecosystem Assessment (MEA 2005) and the International Assessment of Agricultural Science and Technology for Development (IAASTD 2008). Agriculture in developing countries must respond to a rapidly growing and diversifying demand for food, fibre, fodder and energy (ICRAF 2008).

Agroforestry provides the basis for producing such goods and services on farms and stabilizing agricultural landscapes and alleviating pressure on natural ecosystems. More than 1 billion rural smallholders grow trees on farms and in agricultural landscapes to access the goods and services that trees provide. The proportion of trees on farms and in forests varies considerably among countries, but two trends seem almost universal in the tropics: the number of trees in forests is declining and the number of trees on farms is increasing (FAO 2005). These trends are particularly reflected in the provision of firewood. Solid fuels are the primary energy sources for the world's poorest people. Over 3 billion people worldwide use solid fuels for cooking, boiling water, lighting and heating (Rehfuess et al 2006). These fuels account for more than 95 per cent domestic energy use in 25 of the world's low-income countries. Ensuring sustainable supplies of energy for smallholder farmers and their families is an ongoing challenge.

Agroforestry is uniquely suited to address both the need for increased food and biomass resources and the need to sustainably manage agricultural landscapes to provide critical ecosystem services. Moreover, it is suited to achieving these objectives in highly impoverished rural areas. Understanding the extent and distribution of trees on agricultural land, at the landscape level, including the numbers and characteristics of farmers and farming community within those landscapes, can help to assess the importance and role of agroforestry both to the livelihood of farming communities as well as to overall global agricultural production. Further, understanding the geographic, ecological and demographic distribution of agroforestry related land uses can also highlight those areas where increased tree densities could make a greater contribution to livelihoods or landscapes (Zomer et al 2009).

Several authors have produced estimates of the extent of particular agroforestry systems. Dixon (1995) estimated extent of silvopastoral and agrosilvopastoral systems in Africa and Asia as 585-1215 million ha (M ha). Nair and Nair (2003) estimated the extent of alley cropping, silvopasture, windbreaks and riparian buffers in the USA as 235.2 M ha. Kumar (2006) estimated the area of homegardens in South Indian and Southeast Asian homegardens as 8.0 M ha and Reisner et al (2007) estimated European silvorable systems to cover 65.2 M ha. International Assessment of Agricultural Knowledge, Science and Technology for Development (IAASTD) (2008) also listed the extent of agroforestry systems at different countries.

Table 22.1: Extent of Land Areas under Agroforestry (IAASTD 2008; Zomer et al 2009)

Country	Area (hectares)	Specific Information	Details
Indonesia	2.8 million	Jungle rubber agroforests	Comprise 80% of Indonesian rubber which are approximately equal to 24% of world production
Indonesia	3.5 million	All multistrata agroforests	Include jungle rubber, durian, benzoin, cinnamon, dammar and others
India	7.4 million	National estimate	-
Niger	5-6 million	Recently planted	-
Mali	5.1 million	90 % of agricultural land	-
Central America	9.2 million	Silvopastural system	Costa Rica, Nicaragua, Honduras, El Salvador and Guatemala
Central America	0.77 million	Coffee agroforests	Costa Rica, Nicaragua, Honduras, El Salvador and Guatemala
Spain/ Portugal	6 million	Dehesa agroforestry	-
Worldwide	7.8 million	Cacao agroforests	5.9 million ha in West and Central Africa, 1.2 million ha in Asia and 0.7 ha in South and Central America

Major agroforestry systems are distributed in different parts of the tropics and subtropics which are classified as humid lowlands, semi-arid lowlands and highlands. The common agroforestry systems in three major ecological regions of the tropics are presented in Table 22.2.

Table 22.2: Common Agroforestry Systems in the Tropics (Nair 1993)

Humid Lowlands	Semi-arid Lowlands	Highlands
Shifting cultivation	Silvopastoral systems	Soil conservation hedges
Taungya	Windbreaks and shelterbelts	Silvopastoral combinations
Plantation-crop combinations	Multipurpose trees for fuel and fodder	Plantation-crop combinations
Multilayer tree gardens	Multipurpose trees on farmlands	
Intercropping systems		

The most common agroforestry systems of tropical, sub-tropical and temperate areas of the world are described for the following regions:

- Agroforestry Systems in Africa
- Agroforestry Systems in Asia
- Agroforestry Sytems in Pacific Islands

- Agroforestry Systems in Latin American Tropics
- Temperate Agroforestry Systems in North America
- Temperate Agroforestry Systems in Europe
- Agroforestry Systems in Australia and New Zealand

AGROFORESTRY SYSTEMS IN AFRICA

Trees have commonly formed an important part of Africa's traditional farming systems. In the bush-fallow, slash-and-burn cultivation systems, farmers exploited the potential of trees and shrubs for soil fertility regeneration and weed suppression during the fallow phase and widely used them as source for staking material later during cultivation. Trees and shrubs are also widely grown in cropped areas particularly in compound farms. They are mainly used for soil fertility maintenance and as auxiliary sources of food, fodder, fuelwood and for other domestic needs (*Kang 1993*).

Table 22.3: Important Agroforestry Systems of Africa (Martin and Sherman 1992; Wilkinson and Elevitch 2000)

S.No.	System	Overstorey Species	Understorey Species	Location
1	Gully protection	*Tamarix*	Grasses	Tropical Africa
2	Understorey cropping	*Sesbania sesban*	Passion fruit (*Passiflora* sp.)	Tropical Africa
3	Homegarden	Fruit trees	Vegetables	West Africa
4	Dispersed trees	*Faidherbia albida*	Grain crops	Central Africa
5	Improved fallows	*Acacia mearnsii, Leucaena, Sesbania*	Grasses	Niger
6	Mixed perennials	Cacao, banana	Yam	Ivory Coast
7	Alley cropping	*Gliricidia sepium*	Root crops, grains	Nigeria
8	Vegetative strips	*Grevillea, Albizia, Leucaena*	-	Rwanda
9	Understorey cropping	*Erythrina* spp., *Inga* spp., *Cassia* spp.	Coffee	Uganda
10	Understorey cropping	*Cocos nucifera*	Cashew (*Anacardium occidentale*)	Kenya

PARKLAND SYSTEMS

Parklands are generally understood as landscapes in which mature trees occur scattered in cultivated or recently fallowed fields. In the ICRAF Agroforestry Systems Inventory, agroforestry parklands are included in the very general category of 'multipurpose trees on farmlands' (Nair 1985). Livestock production may be a significant or secondary component in these systems. Although no definitive estimate exists, it is probable that the parkland system, characterized by mature trees dispersed in cropped fields, is the largest single agricultural land use in sub-Saharan Africa. Across the entire Sudano-Sahelian zone of West Africa, one finds crops planted under varying densities of mature trees. It is quite possible that some of these older parklands are remnants of elaborate pre-colonial agricultural systems, with cropping intensities and input levels not found in the region today (Hervouet 1991).

The ability of these parklands, or two tiered systems, to enhance and stabilize crop production has been much studied over the past 25 years in West Africa. The systems that have received the most attention are the *Faidherbia albida*/grain systems predominating throughout the Sahelian zone and in some parts of East Africa. Many authors have shown an enhanced effect of these species, particularly *Faidherbia albida*, on grain crops growing underneath. Estimates of increased yields range as high as 100 percent compared with crops grown away from the trees (CTFT 1988). This effect is generally attributed to improved soil conditions under the tree because of litterfall. There is also new evidence that emerging crops benefit equally from lower soil and leaf temperatures under the light shade of the tree, which is defoliated at the start of the rainy season.

Parkland trees stand out as an important component of the spatial structure of the landscape. Parkland attributes include a regular distribution of relatively even-aged trees or shrubs and a low tree density so that tree cover is never continuous. Parklands are not, however, limited to the Sahel and Sudan zones of Africa. While they may not generally be called parklands, systems with scattered trees in fields with similar appearance and purpose are also widespread in Zimbabwe and Malawi and elsewhere in southern Africa. It has been described in the past about the frequent occurrence of *Vitellaria* parklands along the Niger River from Segou towards the east in Bambaraland and various tree species around houses and *Vitellaria* and *Parkia* trees in the surrounding agricultural landscape are reported. In Hausaland of Nigeria, well-maintained *Vitellaria*, *Parkia* and *Tamarindus* trees located in agricultural lands are most common (Boffa 1999).

Recognizing the widespread occurrence of *Faidherbia albida* in the vegetation of southwestern Senegal, it has been characterized as a climax association (or, more specifically, a stable plant community of anthropogenic origin, which replaces the climax vegetation, including various forms of open forest found in the Sudan zone, on silica and clay-rich soils), developed after the species was introduced by nomadic people and their livestock. It has also been reported the occurrence of *F. albida* parklands in Senegal, southern Burkina Faso and the inland delta of the Niger River, respectively.

Parklands are dominated by *Parkia*, *Vitellaria*, *Afzelia*, *Tamarindus* and *Acacia* spp. in Nigeria (Lely 1925). Parklands in northern Nigeria include *Parkia*, *Vitellaria*, *Ficus*, *Balanites*, *Adansonia*, *Hyphaene* and *Borassus* spp. and *F. albida*. Three types of parklands, dominated by *F. albida*, *Vitellaria-Parkia*, and *Parkia* spp., are described by FAO (1969) in northwestern Nigeria. In northeastern Nigeria, occurrence of *F. albida*, *Adansonia* and *Parkia* parklands (Leeuw and Tuley 1972), while *Ceiba* parklands around Zaria (Jackson 1970) are reported.

In northern Ghana, *Vitellaria-Parkia-Tamarindus* and *Adansonia-Tamarindus* parkland associations, with small parkland patches of *F. Albida* are reported (Taylor 1960). Similarly mature *F. albida, Parkia* and *Borassus* parklands, and *Vitellaria-Parkia* communities are described in Ghana (Ramsay and Innes 1963). Parkland communities dominated by *Vitellaria*, *Parkia*, *F. albida*, *Borassus* and *Hyphaene* are reported in the Chari-Logone lowlands of Chad and Cameroon (Pias 1955).

The Semi-Arid Lowlands of West Africa (SALWA) Programme of the World Agroforestry Centre (ICRAF) has contributed to gathering and generating information on parklands in its member states: Burkina Faso, Mali, Niger and Senegal. The Macro and Micro-Diagnostic and Design exercises conducted in the first half of the 1990s advanced the identification, description and general analysis of the various agroforestry land-use systems on a national basis (ICRAF 1990). National parkland reviews carried out within the African Research Network on Agroforestry (AFRENA) in SALWA countries represent an important step in synthesizing knowledge on these systems. Maps and parkland typologies have also been produced by ICRAF (1996) for the Dori area, Burkina Faso, and the middle Bani-Niger river basin and Gondo-Mondoro region of Mali.

Table 22.4: Parkland Species in Africa (Boffa 1999)

S.N.	Parkland Species	Major Occurrence	Countries
1	*Faidherbia albida*	Occurs in Sahel and Sudan zones of West Africa, as well as in Eastern and Southern Africa	Burkina Faso, Cameroon, Ethiopia, Mali, Niger, Senegal, Sudan, Ghana, Nigeria, Ivory Coast
2	*Vitellaria paradoxa*	Occurs in Semi-arid West Africa. There are two subspecies in the genus, the subspecies paradoxa occurring from Senegal to the Central African Republic and the subspecies nilotica in southern Sudan and Ethiopia, Uganda and northeast Zaire	Benin, Burkina Faso, Central African Republic, Chad, Ivory Coast, Ghana, Mali, Uganda
3	*Parkia biglobosa*	Naturally occurs in the dry forests of the Sudano-Guinean zones	Benin, Burkina Faso, Central African Republic, Ivory Coast, Mali, Nigeria, Senegal, Niger, Ghana, Togo, Cameroon
4	*Adansonia digitata*	Occurs in Sudano-Zambezian lowlands but it has extended into higher rainfall areas, possibly with human assistance	Senegal, Burkina Faso, Mali, Nigeria, Togo
5	*Borassus aethiopum*	Occurs along all the major rivers and other water bodies of the Sahel	Senegal, Mali, Burkina Faso, Niger, Ghana, Nigeria, Cameroon, Chad
6	*Acacia tortilis*	Sahelian (northern) parklands	Senegal, Mali, Burkina Faso
7	*Acacia senegal*	Rangelands in West Africa	Sudan
8	*Balanites aegyptiaca*	Important species on aeolian sands in the Sahel	Senegal, Mali, Burkina Faso, Niger, Nigeria
9	*Hyphaene thebaica*	Sahelian and Sudano-Sahelian zones	Senegal, Mali, Niger, Burkina Faso
10	*Azadirachta indica*	Sahelian countries	Burkina Faso
11	*Prosopis africana*	Chad-Cameroon border	Chad, Cameroon, Niger
12	*Ceiba pentandra*	Exotic speceis	Nigeria, Togo
13	*Cordyla pinnata*	Senegal and Mali	Senegal, Mali,
14	*Cordia africana*	Ethiopia	Ethiopia

Table 22.5: Dominant Species in West African Parklands (Pullan 1974)

Sahel	Northern Sudan	Southern Sudan	Northern Guinea
Acacia raddiana	*Faidherbia albida*	*Vitellaria paradoxa*	*Parkia biglobosa*
Balanites aegyptiaca	*Vitellaria paradoxa*	*Parkia biglobosa*	*Ficus* spp.
Hyphaene thebaica	*Parkia biglobosa*	*Faidherbia albida*	*Vitellaria paradoxa*
Acacia senegal	*Adansonia digitata*	*Borassus aethiopum*	*Daniellia oliveri*
Tamarindus indica	*Borassus aethiopum*	*Ficus gnaphalocarpa*	*Elaeis guineensis*
Piliostigma reticulatum	*Cordyla pinnata*	*Ceiba pentandra*	
Borassus aethiopum	*Tamarindus indica*	*Sterculia setigera*	
	Sclerocarya birrea		

LIVING FENCES

They are lines of trees or shrubs planted in close spacings on farm boundaries or on the borders of farmyards, pasture plots, animal enclosures or around agricultural fields. They have been used by farmers in areas where materials for fenceposts are scarce or expensive. They can be made of single or multiple, densely planted rows consisting of a mixture of plant species. Live fences can be divided into two basic categories: live fence posts and live barriers or hedges. Live fence posts are widely spaced, single lines of woody plants that are regularly pruned back and used instead of metal or wooden posts for supporting barbed wire, bamboo or other materials. Hedges are thicker, more densely spaced fences that generally include a number of different species and usually do not support other fencing materials.

They require less maintenance and are usually combined with other materials, such as barbed wire. Used all over the world, these structures are greatly beneficial to the people who use them. The primary purpose of live fences is to control the movement of animals and people, however, they have proven to be extremely diverse, low risk systems that provide farmers with numerous benefits.

Living fences are numerous in Africa, and highly diverse. A few of the more well known are:

- In Embu, Kenya, it was found that, when fed to cows, new foliage of *Calliandra calothyrsus* raised the butterfat in milk. In general, farmers often use the living fence as a source of wood and animal feed. In this particular region, it was found that pruning between 2 to 3 months created a higher leafy yield than pruning every 4 to 8 months (higher woody yield).
- Live fences were originally introduced in Burkina Faso in the early 1980s, and were made up of only one species: *Acacia nilotica.*
- *Jatropha* Fences in Mali: Farmers have been using this plant to make hedges for generations. The physic nut (*Jatropha curcas*) is a drought-resistant perennial that is not eaten by animals and can grow on marginal soils for up to 50 years. Among its uses are the medicinal properties of its oil, oil as biofuel for diesel engines and soap making. Malis development is affected in several ways from this plant.

Advantages of living fences include fuelwood, fodder, sustenance for people (such as fruit), windbreaks, fertilizer, ownership boundaries, mulch, erosion control and stabilizing the land, forage for small mammals, construction materials, medicines and shade.

Recommended Species for Live Fences in dry and semi-arid areas of West Africa (Louppe and Yossi 1999) are *Acacia mellifera*, *Acacia nilotica*, *Acacia senegal*, *Agave sisalana*, *Bauhinia rufescens*, *Citrus limon*, *Commiphora africana*, *Dichrostachys cinerea*, *Euphorbia balsamifera*, *Haematoxylon brasiletto*, *Jatropha curcas*, *Moringa oleifera*, *Prosopis juliflora*, *Ziziphus mauritiana* and *Ziziphus mucronata*.

WINDBREAKS

Tree rows as wind-breaks have long been used in North America, Europe and Asia for protecting crops and soil against wind damage and wind erosion. More recently, their effectiveness in increasing crop production has been demonstrated in other drier parts, particularly sub-Saharan Africa. In northern Nigeria, increase in millet yield has been reported behind a *Eucalyptus camaldulensis* windbreak. Yield increase in cotton, wheat, maize and rice has been reported in Egypt, where over 1,00,000 ha of croplands are protected by windbreaks. In the Majjia Valley of the central Niger, double-rowed *Azadirachta indica* (neem) wind-breaks spaced 100 m apart showed a 20 percent increase in crop yield (Vanderbeldt 1990). The successful establishment, protection and management of the Majjia Valley plantings have required nearly 15 years of sustained effort and funding from the Government of the Niger and donor agencies. The great interest generated by these data and experiences led to an increased emphasis on wind-breaks in West African development assistance projects during the late 1980s.

Throughout the African continent, farmers use windbreaks to protect crops, water sources, soils, and settlements on plains and gently rolling farmlands. Hedgerows of *Euphorbia tirucalli* protect maize fields and settlements in the dry savannas of Tanzania and Kenya. Tall rows of *Casuarina* line thousands of canals and irrigated fields in Egypt. In Chad and Niger, multi-species shelterbelts and windbreaks protect wide expanses of cropland from desertification. The practice is not new, although the design of multipurpose windbreaks for smallholders will require new input from agroforestry practitioners (Nair 1993).

ALLEY CROPPING

Alley cropping is an agroforestry system in which arable crops are grown between rows of woody shrub or tree fallows. The shrubs or trees are pruned periodically during the cropping season to prevent shading and to provide green manure for the companion food crop. Thus, the trees and shrubs grown in the hedgerows recycle nutrients and provide green manure, firewood and staking material. Alley cropping has the advantage over the traditional system of combining the cropping and fallow phases. Based on positive results from the humid tropics, particularly at the International Institute of Tropical Agriculture (IITA) in Ibadan, Nigeria (Kang et al 1984), alley cropping has generated much interest among donors. This is particularly so in West Africa where a collaborative network has been established by the IITA and the ICRAF and under various national programmes.

Alley cropping, which has been so successful on-station, has been widely extended to farmers in West Africa, particularly in and to the north of the Sudanian zone where a reduced incidence of trypanosomiasis now allows extensive animal husbandry. The fodder needs of these animals compete with the use of hedgerow loppings for green manure, a key component of the alley cropping technology. Furthermore, alley cropping has definite drawbacks in areas where water is a limiting factor. Narrow hedgerows, spaced at 4 m distances recommended for the humid and sub-humid tropics, are particularly competitive with crops in the semi-arid tropics (SAT). The future role of alley cropping in the SAT of Africa will depend on the development of management methods to control the sources of competition between hedgerows and crops, and a greater selection of woody or perennial grass species to use in the hedgerows. A better chance of success is likely if alley cropping is only considered for areas where the assured annual rainfall exceeds 750 mm on soils with good water holding capacity.

Trees in alley cropping of humid and sub-humid tropical Africa (Kang et al 1984) are *Albizia lebbeck*, *Cassia siamea*, *Leucaena leucocephala*, *Gliricidia sepium*, *Flemingia congesta*, *Inga edulis*, *Sesbania grandiflora*, *S. sesban*, *S. aculeata*, *S. rostrata*, *Calliandra* spp. and *Erythrina* spp.

SILVOPASTORAL SYSTEMS

In case of animal husbandry systems, the nomadic and transhumant systems are found in Sahel, where animals are moved from one site to another according to changing pasture conditions throughout the year, or herded to distant pastures during the rainy season to protect crops. Livestock play an important role in sustaining field fertility through manure production. In the Sahel, herders are paid to bed animals on arable fields. Livestock pressure must be balanced with rangeland vegetative productivity if it is to be maintained. However, most grazing lands in the SAT of Africa are public lands and are therefore freely accessible, a situation that may be leading to their overuse and irreversible decline.

Agroforestry interventions have demonstrated the usefulness of fitting the potentially excellent feeding value of trees and shrubs into silvopastoral systems. Fodder trees and other suitable perennial vegetation are considered as components of almost every agroforestry system for the SAT of Africa. Examples are the use of fodder trees in wind-breaks, or the management of parkland systems to produce a fodder component through pod production or lopping. Indigenous and exotic fodder trees for silvopastoral system in dry tropical Africa are *Faidherbia albida*, *Acacia nilotica*, *Acacia raddiana*, *Acacia senegal*, *Acacia tortilis*, other *Acacia* spp., *Azadirachta indica*, *Adansonia digitata*, *Atriplex nummularia*, *Balanites aegyptiaca*, *Dobera glabra*, *Salvadora persica*, *Tamarindus indica*, *Prosopis africana*, *Ziziphus mauritiana*, *Z. mucronata*, *Parkia biglobosa*, *Moringa oleifera*, *Afzelia africana*, *Combretum* spp., *Bauhinia* spp., *Cassia* spp., *Boscia* spp. and *Grewia* spp.

HOMEGARDENS

In many humid and sub-humid areas of Africa, homegarden (sometimes referred to as a backyard or kitchen garden) is part of the homestead and is one of several field systems operated by a rural household. In these cases, the homegarden forms the hub, with the homestead at its centre, from which paths lead to other field systems and production units devoted to annual crops for the market and for home consumption. In places where land is scarce, such as urban and peri-urban areas, the homegarden may be the only cultivation plot. And, in semi-arid regions, where water scarcity is a constraint, horticulture crops may be intercropped with staples on the family farm. In the wet lowlands, vegetable gardening usually becomes a dry-season activity. The extent of these activities is largely influenced by availability of a reliable water source. Population pressure can result in homegardens being located some distance from the village.

In Africa, as in other parts of the world, legumes, roots and tubers and even some oil crops are often cultivated as garden crops. In fact, some of them, particularly roots and tubers, play a major role as food crops in the homegarden. Even plantation plants, such as coffee, cacao, coconut, oil-palm, dates and many others, can be found in limited numbers in homegardens. A traditional African garden contains a mixture of perennial and annual crops, well adapted to the ecological conditions. The crops are often intercropped, with intercropping patterns varying from area to area. Crop mixtures found in a home garden are often the result of the deliberate selection of a wide variety of plants and tree crops that occupy different layers and play supportive roles. Often the home garden provides amenities for the household and the surrounding area. The garden provides the household with a mixture of food and cash crops. The trees in the garden provide shade where people can socialize or undertake household and income-generating activities. In addition, a complete garden can produce some fuelwood, spices and medicinal and ornamental plants (FAO 2001).

Mixed plant and tree cropping systems can greatly extend harvesting periods and thus ensure the continuous availability of some foods. Once established, tree species require only minimal labour and inputs for maintenance, and can provide a continuous food supply without needing to be replanted. Livestock, kept on a small scale (including sheep, goats, poultry and to some extent cattle and pigs), can provide food, income and manure. The biological diversity and complexity of home gardens decline with the transition from humid to semi-arid and arid areas of the Sahelian countries (where annual rainfall is below 500 mm). Insufficient water is a major constraint to successful gardening in dry areas, yet, even in these areas, crops can be kept growing through effective soil and water management. Plant species common in homegardens in sub-Saharan Africa are presented below (FAO 2001):

Grain Crops: maize (*Zea mays*), millet (*Pennisetum* spp.), rice (*Oryza sativa*), sorghum (*Sorghum bicolor*), wheat (*Triticum aestivum*), teff (*Eragrostis tef*)

Root Crops: sweet potato (*Ipomoea batatas*), cassava or tapioca (*Manihot esculenta*), taro or cocoyam (*Colocasia esculenta*), yam (*Dioscorea* spp.), carrot (*Daucus carota*), enset (*Ensete ventricosum*), hausa potato (*Plectranthus esculentus*), irish potato (*Solanum tuberosum*), oromo dinich (Amorphophallus abyssinicus)

Oilseed Crops: nug or niger seed (*Guizotia abyssinica*), oyster nut (*Telfairia pedata*), safflower (*Carthamus tinctorius*), sesame (*Sesamum indicum*), shea butter (*Butyrospermum paradoxum*), sunflower (*Helianthus annuus*)

Leafy Vegetables: african spinach (*Amaranthus* spp.), amaranth (*Amaranthus gangeticus, A. tricolor*), bitter leaf (*Vernonia amygdalina*), blackjack (*Bidens pilosa*), cabbage (*Brassica oleracea* var. *capitata*), cat's whiskers (*Cleome gynandra*), drumstick tree (*Moringa* spp.), jute (*Corchorus* spp.), kale (*Brassica carinata*), leek (*Allium porrum*), lettuce (*Lactuca sativa*), mustard (*Brassica campestris*), nettle (*Urtica* spp.), pumpkin (*Cucurbita mixta*), rape (*Brassica napus*), rosella (*Hibiscus sabdariffa*), runner bean (*Phaseolus coccineus*), spindlepod (*Gynandropsis gynandra*), water spinach (*Ipomoea aquatica*)

Fruit Vegetables: african eggplant (*Solanum aethiopicum, S. scabrum*), bitter cucumber (*Momordica charantia*), cucumber (*Cucumis sativus*), eggplant (*Solanum melongena*), melon (*Cucumis melo*), okra (*Hibiscus esculentus*), pumpkin (*Cucurbita pepo, C. moschata, C. maxima*), sweet pepper (*Capsicum* spp.), tomato (*Lycopersicon esculentum*)

Herbs, Spices and Medicinal and Other Crops: chilli (*Capsicum annuum*), coriander (*Coriandrum sativum*), false cardamom (*Aframomum melegueta*), fenugreek (*Trigonella foenum-graecum*), garlic (*Allium sativum*), marigold (*Calendula officinalis*), mustard (*Brassica nigra*), neem (*Azadirachta indica*), Onion/bunching onion (*Allium fistulosum*), sugarcane (*Saccharum officinarum*), sweet basil (*Ocimum basilicum*), tea (*Camellia sinensis*), tobacco (*Nicotiana tabacum*), turmeric (*Curcuma longa*)

Beans and Legumes: bambara groundnut (*Voandzeia subterranea*), broad bean (*Vicia faba*), chickpea (*Cicer arietinum*), cowpea (*Vigna unguiculata*), kidney bean or french bean (*Phaseolus vulgaris*), lima bean (*Phaseolus lunatus*), locust bean (*Parkia clappertoniana*), long bean (*Vigna sesquipedalis*), mung bean (*Vigna radiata*), pea (*Pisum sativum*), peanut (*Arachis hypogaea*), soybean (*Glycine max*), winged bean (*Psophocarpus tetragonolobus*)

Trees, Fruits, Nuts and Spices: avocado (*Persea americana*), banana/plantain (*Musa acuminata*), baobab (*Adansonia digitata*), breadfruit (*Artocarpus altilis*), cashew nut (*Anarcardium occidentale*), citrus (*Citrus* spp.), cacao (*Theobroma cacao*), coconut (*Cocos nucifera*), coffee (*Coffea arabica, C. robusta*), guava (*Psidium guajava*), jackfruit (*Artocarpus heterophyllus*), mango (*Mangifera indica*), oil-palm (*Elaeis guineensis*), papaya (*Carica papaya*), passionfruit (*Passiflora edulis*), pineapple (*Ananas comosus*), soursop apple (*Annona muricata*), tamarind (*Tamarindus indicus*)

Trees (Multipurpose and Soil Conservation): caliandra (*Calliandra calothyrsus*), drumstick (*Moringa oleifera*), *Flemingia congesta, F. macrophylla*, gliricidia (*Gliricidia sepium*), leucaena (*Leucaena* spp.), scarlet wisteria (*Sesbania grandiflora*), *Tephrosia tetraptera*

Fodder, Compost and Soil Conservation Crops: cowpea (*Vigna unguiculata*), guinea grass (*Panicum maximum*), velvet bean (*Mucuna pruriens*), setaria grass (*Setaria anceps*), vetiver grass (*Vetiveria* spp.)

TAUNGYA FARMING AND INTEGRATED TAUNGYA

Food crops are interplanted with trees in unit area of land for 2-3 years. Food crops cease to exist on the land when the tree crops close canopy. The system has proved effective in providing food for forestry workers and forage for cutting by cattle rearers. This is the main agroforestry model practice in the Forest Reserves for the past 60-70 years. Most of the State owned artificial plantations now being exploited were raised through Taungya system. The chief problem with this system is the need to plan a planting programme for long cycle trees with three or four years of crops (Igugu and Osemeobo 1995).

Inegrated Taungya is similar to Taungya farming but here when the tree canopy is closed, livestock grazing substitutes raising agricultural crops. The integrated approach aims at invoking the idea of land use practice whereby the activities on the land are stretched all the year round (Rander 1988).

AGROFORESTRY SYSTEMS IN ASIA

Agroforestry systems in Asia have evolved over centuries and are very diverse and complex. Today, as conditions change, farmers continue to innovate, experiment and improve these systems. To meet their needs of food, fodder and fuelwood, fibre and cash, farmers integrate agriculture crops, trees and livestock in their farming systems. This integration has resulted in a wide diversity of traditional agroforestry systems. Most of these systems are well-suited to the local agroecological conditions, the specific subsistence and cash needs of farmers, their social and cultural context and the environmental conservation needs. In areas where the people are still living in forest areas or still surrounded by abundant forests (e.g., parts of Laos and Indonesia), they practice shifting cultivation and harvest or collect many non-wood forest products for their own use or for markets. In settled agriculture, trees are extensively integrated with the farming systems. In densely populated areas, trees may be more valued and more valuable than in low-population areas where forests are still abundant. Agroforestry systems, particularly in high-population and fuelwood-deficit areas, can serve as buffer zones to mitigate the degradation of natural and plantation forests.

Trees in and around farms are planted judiciously, carefully and selectively. This is the most widely practised agroforestry system. Very specific tree and crop combinations are developed for all agroecological zones. If the farming system is based on a tree that takes long time to produce yields (e.g., coconut, areca nut, horticultural trees, rubber), farmers intercrop and/or integrate livestock to generate some early economic returns. Farmers living near the forests in densely populated areas have developed practices to integrate their farming systems with the adjoining forests to graze their cattle (or collect fodder) in the forest, or have developed forest gardens to meet their subsistence and cash needs. Agroforestry is most commonly practised in South Asia and Southeast Asia.

Southeast Asia, comprising vast region of Asia situated east of the Indian subcontinent and south of China. South Asia or Southern Asia or Indian Subcontinent is the southern region of the Asian continent, which comprises the sub-Himalayan countries. In Asia, Agroforestry is widely practised in India, Indonesia, Philippines, Thailand, Vietnam, Bangladesh, China and in some parts of other Asian countries. The detailed description of agroforestry practices in India is presented in Chapter-21 which gives good idea about the agroforestry in South Asia.

AGROFORESTRY SYSTEMS IN INDONESIA

Traditional agroforestry systems are found throughout Indonesia, including the kebun-talun and pekarangan systems in Java, and the multistoried agroforestry gardens in Sumatra. Introduced agroforestry systems also are common in many parts and are integrated into forest development programmes on forest lands, as well as being widely practised on private farm lands. Kebun-talun and homegardens in West Java generate relatively good income and are good sources of calcium, vitamin A and vitamin C. Farm-based agroforestry systems include pekarangan (homegarden) system, kebun-talun system, three-strata system and alley cropping. Forest-based agroforestry systems are shifting cultivation, improved fallow, multistoried agroforestry garden system and tumpangsari (taungya).

PEKARANGAN (*HOMEGARDEN*): The pekarangan (homegarden) is a mixture of annual crops, perennial crops and animals (including livestock) in the area surrounding a house. The homegarden system originated in Central Java and spread to East and West Java in the middle of the eighteenth century.

Top layer (>10 m)	: Coconut, *Albizia*, other trees
5-10 m layer	: Soursop, jackfruit, duku (*Lansium domesticum*), guava, mountain apple, cloves
2-5 m layer	: Bananas, papaya, other fruit trees
1-2 m layer	: Ganyong (*Canna edulis*), Xanthosoma, beans, spinach, cassava, gembili (*Dioscorea esculenta*)
Lowest layer (< 1 m)	: Taro, Xanthosoma, chili pepper, eggplant

KEBUN-TALUN SYSTEM: The kebun-talun system usually consists of three stages: kebun (garden), kebun campuran (mixed garden) and talun (mixed tree garden). The first stage, kebun, involves clearing the forest and cultivating annual crops. In the kebun stage, three vertical layers of annual crops predominate: the lowest layer consists of creeping plants that occupy the ground below a height of 30 cm. The layer from 50 cm to 1 m is occupied by vegetables, and the upper layer includes maize, tobacco, cassava or leguminous vines supported by bamboo sticks. After two years, tree seedlings start to grow, leaving increasingly less space for the annual crops. The kebun gradually evolves into the kebun campuran, in which the annuals are mixed among half-grown perennials. In the kebun campuran, shade-tolerant plants such as taro occupy the space below 1 m. Cassava forms the second layer from 1-2 m height and the third layer is occupied by bananas and trees. After harvesting the annual crops in the kebun campuran, the field may be abandoned for two to three years to become dominated by perennials. This stage is known as talun and is the climax stage of the kebun-talun system. The talun is dominated by a mixture of perennial trees and bamboos, forming three vertical layers.

THREE-STRATA SYSTEM: The three-strata system is a method of planting and harvesting grasses, legumes, shrubs and trees in such a way that animal fodder will be available throughout the year. The first layer, consisting of grasses and legumes is intended to supply fodder at the beginning of the wet season. The second layer, consisting of shrubs, is to supply fodder in the middle and the end of the wet season. The third layer, comprised of trees, is to supply fodder during the dry season. The three-strata system divides a piece of land into three parts: nucleus, blanket, boundary. The nucleus is maintained for food production. The blanket is divided into a number of compartments, with each compartment cultivated with various grasses and legumes. Fodder tree species such as *Gliricidia* or *Leucaena* are planted around the boundary.

ALLEY CROPPING: This technology has been developed in dryland farming programs in the semi-arid provinces of East Indonesia since the early 1980s. The technology consists of hedgerow planting on contour lines with legume species. Before 1986, the hedgerow mainly consisted of *Leucaena leucocephala*; but after the psyllid (*Heteropsylla cubana*) infestation, *Gliricidia sepium* and *Calliandra calothyrsus* were used. The hedgerows are intended for enhancing soil and water conservation. Between the hedgerows, annuals, perennial crops and grasses are planted.

SHIFTING CULTIVATION: Shifting cultivation (also called swidden or bush fallow agriculture) is widely practised in most Indonesian islands, except Java. Shifting cultivation includes a variety of practices occurring in a diversity of environments under many specific circumstances. In Apo Kayan

(East Kalimantan), almost all the forests cleared for agriculture are secondary forests and the fallow period is between 10 to 30 years. Occasionally, sites may be left unused for even longer periods (40 to 50 years) to prevent gradual declines in fertility and an increase in weedy species. The farmers recognize the merits of long-term swidden management. In Long Segar (also in East Kalimantan) shifting cultivators clear more primary forest than secondary forest for swiddens.

IMPROVED FALLOW: Improved fallow technology is one of the alternatives to control destructive shifting cultivation practices and to develop more sustainable dryland agricultural systems. Abandoned fields are planted to the fast-growing cover crop, *Pueraria javanica*, to rehabilitate soil fertility and as a perennial cash crop. Food crops can again be planted after 3-4 years, a much shorter fallow period than traditionally followed. This revised cropping pattern will produce rice and cassava within a short period; pineapple, ginger and peanut in the near term and *Gnetum gnemon* in the long-run.

MULTISTORIED AGROFORESTRY GARDEN SYSTEM: The system is practised in West Sumatra and characterized by an intensive integration of forest species and commercial crops, forming a multilayer, forest-like system. The intimate association of different species provides both subsistence and commercial products which supplement rice production.

Cultivated annual crops: Chili (*Capsicum annuum*), eggplant (*Solanum melongena*), maize (*Zea mays*), beans (*Vigna* spp., *Phaseolus* spp.), cucumber (*Cucumis sativus*)

Fruit Trees: Durian (*Durio zibethinus)*, jackfruit (*Artocarpus herterophyllus*), banana (*Musa* spp.), mango (*Mangifera indica*), guava (*Psidium guajava*), avocado (*Persea americana*), orange (*Citrus nobilis*), papaya (*Carica papaya*), *Nephelium lappaceum*, *Pithecellobium jiringa*, *Syzygium aqueum*, pineapple (*Ananas comosus*)

Timber Trees: Rosewood (*Dalbergia latifolia*), *Paraserianthes falcataria*

Fuelwood Trees: *Gliricidia sepium*, *Erythrina subumbrans*, *Leucaena leucocephala*

Spice: *Eugenia aromatica*, candle nut (*Aleurites moluccana*), *Myristica fragrans*, turmeric (*Curcuma longa*), pepper (*Piper nigrum*)

Other Trees: Coconut (*Cocos nucifera*), kapok (*Ceiba pentandra*), bamboo, sugar palm

TUMPANGSARI (*TAUNGYA*): Various tumpangsari methods exist. They are introduced agroforestry systems designed to meet the subsistence needs of households with limited access to cultivable lands. The implementation of tumpangsari in Java's teak forests is a good example. The spacing of the main crop (teak) is generally 1 m x 3 m; *Leucaena* is planted as a dense row in the middle between the teak rows. Additional crops can be planted in the 1.5 m space between the teak and the *Leucaena* rows. Similar activities can also be carried out in the establishment of plantations of other timber species, such as *Pinus*, *Agathis*, *Altingia* and *Swietenia*. Tumpangsari can also be found in rubber (*Hevea*) plantations, especially among smallholders. Other species used for interplanting are: *Acacia villosa*, *Calliandra calothyrsus* and *Gliricidia*.

AGROFORESTRY SYSTEMS IN PHILIPPINES

Agroforestry has been traditionally practised by different tribal groups in the Philippines for generations. This land use system is now recognized by the government as one of the alternatives to address the twin problems of meeting the needs of upland farmers and maintaining the integrity of the environment. It is one of the major components of the community-based forestry programmes of the government. Farm-based agroforestry systems include alley cropping, multistorey system, multistorey system+animals, trees along farm boundaries, protein or fodder bank and tree-crop grazing system. Forest-based agroforestry systems are taungya system and improved fallow system.

ALLEY CROPPING: Hedgerows of trees or shrubs (usually double hedgerows) are grown at intervals (usually 4-6 m) along the contours. The strips or alleys between the hedgerows are planted with agricultural crops (annuals and/or perennials). A good example of this system is the Sloping Agricultural Land Technology (SALT) in which one or two rows of woody perennials are grown as hedgerows, either from seeds or cuttings, along the contours. The recommended horizontal distance between contour hed-gerows is 46 m or about 1.5 m vertical distance. At an average of 5 m interval (horizontal distance) and 1 m width of hedgerows, about 20 per cent of the total area is occupied by the hedgerows. For the alleys, the recommended cropping pattern is to plant perennial crops (e.g., coffee, fruit trees, etc.) in every third alley while the other two alleys can be devoted to annual crops. The hedgerows are regularly pruned to a height of about 0. 5 m to minimize shading of agricultural crops in the alleys. Biomass from the prunings can be used as green manure or mulch to the alley crops or as fodder fed to livestock. Alley cropping with improved pasture grasses and/or fodder trees or shrubs is also practised.

Hedgerow Species: *Leucaena leucocephala*, *Gliricidia sepium*, *Desmodium rens-onii*, *Flemingia congesta*, *Cassia spectabilis*, *Calliandra calothyrsus*, *Desmanthus* spp.

Grasses: Napier (*Pennisetum purpureum*), vetiver (*Vetiveria zizanioides*), guinea grass (*Panicum maximum*) and *Setaria* spp.

MULTISTOREY SYSTEM: In this agroforestry system, mixed species occupy different canopy levels, with the upper layers occupied by trees or other woody perennials that provide partial shade to agricultural crops in the lower layers. This system is similar to the structure (multilayer) and composition (diverse species) of a tropical rainforest. Examples are coconut-coffee-pineapple-banana mix; Albizia-coffee/cacao mix; Gliricidia-coffee mix; and homegardens. This system can be adapted by interplanting shade-tolerant species under established tree and coconut plantations. Some common nursery trees for coffee and cacao plantations are *Gliricidia sepium*, *Alnus japonica*, *Leucaena leucocephala*, *Erythrina orientalis*, *Paraserianthes falcataria*, *Pterocarpus indicus* and *Albizia saman*.

MULTISTOREY SYSTEM + ANIMALS: This system is similar to the multistorey system except that free range grazing animals are added as a component. An example is the coconut (*Cocos nucifera*)- lanzones (*Lansium domesticum*) mixture, with horses or cattle, found in Laguna and Quezon provinces.

TREES ALONG FARM BOUNDARIES: Trees are planted along farm boundaries as boundary marker, live fence, live-fence post for tying barbed wire or bamboo slats, or as shelterbelts/ wind-breaks. When mature, some of these trees can be harvested and used as posts or as light construction materials. Pruning can also provide fuelwood, fodder or green manure. Live fences (or living fences) can be established by planting rows of trees or shrubs around a grassland area to enclose the grazing animals.

- Species used as live fence: *Leucaena leucocephala, Gliricidia sepium, Sesbania grandiflora*
- Species used as windbreak/shelterbelt: *Eucalyptus camaldulensis, Casuarina equisetifolia, Acacia auriculiformis*, Bamboo

PROTEIN OR FODDER BANK: Leguminous fodder trees or shrubs (e.g., *Leucaena*, *Gliricidia*, *Flemingia*) may be established in intensively planted small stands on the farm. These are usually fenced off and serve as a supplementary source of protein for livestock. The top and branch prunings are fed to animals through a "cut-and-carry" system. A good example is the intensive feed garden.

TREE-CROP GRAZING SYSTEM: Animals (e.g., cattle, carabao, goats, sheep) are allowed to graze freely underneath relatively mature tree plantations. The cattle are allowed to graze freely under *Aleurites moluccana* trees where improved forage grasses have been planted. This system has proven to be practical and economical because the land is fully utilized while being maintained and protected. The grazing animals keep the grass down, allowing for easier collection of the nuts.

TAUNGYA SYSTEM: Taungya is an agroforestry system in which newly established reforestation areas are interplanted with agricultural crops. As soon as the tree canopies close making the light intensity critically low fcr crop production, the farmers move to another reforestation area and the same process is repeated.

IMPROVED FALLOW SYSTEM: In the uplands, cultivated areas are planted with agricultural crops and then allowed to fallow for some time to allow the soil to rejuvenate. To shorten the fallow period, the area can be seeded with leguminous trees. Once the soil has been rejuvenated, these areas are again cleared for crops. This can be considered as an improved version of the traditional shifting cultivation practice. *Leucaena* is used as the species to shorten the fallow period; the trees are cut and the branches are piled along the contours to form a barrier structure, which traps the eroding soil. Through time, natural terraces are gradually formed, thus stabilizing the steep slopes.

AGROFORESTRY SYSTEMS IN THAILAND

Agroforestry has long been practised by farmers in Thailand. Shifting (rotational) cultivation, for example, has been used by some ethnic groups in the upland areas of northern Thailand for centuries. Similarly, homegardens and multistorey fruit orchards have been cultivated in many parts of the country for more than 100 years. Newly developed agroforestry systems, with a systematic combination of trees, crops and livestock and usually driven by market forces, emerged in areas where cultivated land is limited, beginning during the mid-1950s. Farm-based agroforestry systems include homegardens and live fences. Forest-based agroforestry systems are block planting on abandoned agricultural land, shifting cultivation, forest gardens and improved taungya.

HOMEGARDENS: Homegardens are multi-tiered systems found in family compounds surrounding the home. Permanently settled communities in upland areas throughout the country have cultivated homegardens for centuries. Most notable are the Karen and the Lua ethnic groups. In the north, homegardens are mostly located in communities in the foothills. Owing to small landholdings in the central region of the country, homegardens here tend to be small. Most homegardens have three to five layers. *Erythrina variegata* and banana are common tree species providing shade for crops and vegetables. *Moringa oleifera* and *Sesbania grandiflora* are commonly planted as multipurpose trees. In the south, big trees such as wild durian (*Durio* spp.), *Dipterocarpus* spp., *Parkia speciosa* and *Artocarpus integer* dominate the top storey of the gardens. *Eugenia caryophyllus* is commonly mixed with fruit trees to provide cash income.

LIVE FENCES: Live fences of bamboo and fast-growing trees are common in the central and eastern regions of Thailand where they serve as windbreaks and shelterbelts for crops and fruit orchards and as boundary markers. Commonly planted species are Thyrsostachys siamensis, *Bambusa nana*, *Bambusa flexuosa* and *Acacia* spp. In the south, *Azadiractha excelsa* is commonly planted in and around farm fields.

BLOCK PLANTING ON ABANDONED AGRICULTURAL LAND: Due to migration of rural people to cities, large marginal upland areas which used to be cultivated are being abandoned. This has presented an opportunity for small-scale commercial tree farming. People from urban centers are investing in land in rural areas and planting high-value trees such as teak (*Tectona grandis*), eucalyptus and *Azadirachta excelsa*. These species are long-term alternatives to large plantations for both economic and ecological reasons.

SHIFTING CULTIVATION: This system involves the cultivation of several areas of sloping land on a rotational basis (short cultivation, long fallow). It has long been sustainably practised in upland

and highland areas by several ethnic groups. Rice and various other crops are planted simultaneously to produce food and fodder.

FOREST GARDENS: Forest gardens include various indigenous forest and fruit trees, located away from but within walking distance of the family house. In forest gardens in central Thailand, Amomum xanthioides has been cultivated under moist evergreen forest for centuries. However, forest gardens are mostly found in the mountainous area of north and south. In the northern forest gardens, *Camellia sinensis* is cultivated under evergreen forests without having to cut trees in the forest. In the south, *Durio* spp. and *Dipterocarpus* spp. are the most dominant species of the top storey.

IMPROVED TAUNGYA: This system, combining trees and livestock, was first developed under the government reforestation programme, with an aim towards establishing teak plantations. It has since been expanded to include other species and to increase crop production. It is currently being practised in both government and private forest plantations. Forest tree species commonly planted include *Tectona grandis*, *Pinus kesiya* and *Pterocarpus marcrocarpus*. However, in some areas, fruit and rubber trees are also planted as the major trees. Normally, cash crops are intercropped with the trees in the early years of establishment. Livestock are usually allowed to graze in the plantation when the trees are more than three years old.

AGROFORESTRY SYSTEMS IN VIETNAM

In the uplands of Vietnam, agroforestry in various forms is widely used by farmers and forest enterprises. Of the total land area, 70 per cent is upland and almost 30 per cent of the population lives in upland areas. Almost all of the 54 ethnic minority communities are found in the upland areas. Farm-based agroforestry systems include homegardens, RVAC system, Forest/crop/irrigated rice system, woodlot, cash crops with MPTs, agricultural crops with commercial trees, hedgerow and contour planting. Forest-based agroforestry systems are shifting cultivation and taungya.

HOMEGARDENS: The homegarden is a traditional agroforestry system found throughout Vietnam from the lowlands to the highlands. On a relatively small piece of land around the house (usually only about 0.5 ha, but occasionally up to 5 ha), the land is used as efficiently as possible to produce a wide variety of products. Fruit, vegetables, root crops, fish, livestock, fodder, fibre, medicine, small timber fuelwood and various minor products are grown in a multilayered structure. One very common type of homegarden is the so-called RVAC system, combining forest trees, fruit trees, fishpond and livestock in an integrated farming system.

- *Fruit Trees*: jackfruit, banana, orange, lemon, pomelo, papaya, grapefruit, sapota, litchi, starfruit, coconut, cashew nut, durian, avocado, guava, water apple, custard apple, mangosteen, jujube
- *Other Trees*: coffee, cacao, bamboo, rattan, mulberry, tea, cinnamon,
- *Root Crops*: cassava, sweet potato, taro, canna, yam
- *Vegetables*: squash, soybean, pumpkin, cabbage, squash, spinach, black pepper, carrot, mungbean
- *Other Crops*: peanut, sugarcane, pineapple, ginger, medicinal plants
- *Animals*: fish, pig, goat, chicken, duck, cattle, silkworm, bee

FOREST/CROP/IRRIGATED RICE: Forest/crop/irrigated rice systems are often established in hilly and mountainous areas. A natural forest or plantation crowns the site and is usually managed by state forest enterprises or community groups. In some places, the upper part of the system includes a water reservoir which is to irrigate the lower areas and to generate electricity. Vegetables or cash crops are grown on terraces or along contour lines and the irrigation system allows farmers in the valley to grow paddy rice.

WOODLOT: Woodlots have been successfully established on underutilized or degraded land, using fast-growing trees, especially nitrogen-fixing multipurpose species. Along with the rehabilitation of wastelands, these woodlots supply fuelwood, small timber and other minor products for local consumption. Beekeeping and livestock grazing are often practised in the woodlot. Agricultural crops can be intercropped during the establishment of the woodlot tree species. Commonly used woodlot tree species are *Acacia auriculiformis*, *A. mangium*, *Cassia siamea*, *C. fistula*, *Erythrina* spp., *Eucalyptus tereticornis*, *E. camaldulensis*, *Gliricidia sepium* and *Melia azedarach*.

CASH CROPS UNDER MPTs: Cash crops, such as coffee, tea and black pepper, can be grown under trees which provide shade and other forms of support to the crops. Tree species commonly used in these systems are *Pinus kesya*, *P. merkusii*, *Cassia siamea*, *Leucaena leucocephala*, *Gliricidia sepium*, *Pterocarpus indicus*, *Moringa oleifera*, *Manglietia glauca* and *Aleurites fordii*.

AGRICULTURAL CROPS UNDER COMMERCIAL TREES: Several agroforestry systems use the space between commercial tree crops to intercrop agricultural crops (annuals). This intercropping is usually limited to the first few years of plantation establishment (e.g., rubber plantations, or fruit orchards of cashew nut, mango or jackfruit) until the tree canopy closes. This is usually done by a farm family who manages the plantation for another landowner (usually a state enterprise).

HEDGEROW AND CONTOUR PLANTING: Various combinations of techniques have evolved to allow cultivation on steep, sloping terrain without causing serious soil erosion. In Vietnam, farmers have adapted the SALT (Sloping Agricultural Land Technology) technique, originally developed in the Philippines. Locally suited species, like *Tephrosia* shrubs and *Indigofera* trees, have been incorporated. In Central Vietnam, trees are also planted on the contour lines, together with grasses and other crops. Natural terraces will gradually form as the tree rows form a barrier which prevents soil and plant material from washing down the slope.

SHIFTING CULTIVATION: Shifting cultivation, or slash and burn, is practised extensively by many people from more than 50 different ethnic groups in the hilly and mountainous areas of Vietnam. On a patch of upland forest, farmers slash shrubs and other low vegetation and then burn it. Hill rice is then grown for two to three years. The field is then abandoned and the farmer moves to another site to repeat the process. In the past, the farmer returned to a site only after 10 to 20 years, depending on the recovery rate of the fallowed fields. However, today this fallow period is often reduced to 5-7 years, resulting in much lower yields. In some places, cassava is cultivated for 1-2 years after the first two years of rice cultivation. But this practice leaves the soil without any cover for 3-4 months following each harvest, causing severe erosion.

TAUNGYA: The taungya system allows various intercrops to be grown (until the tree canopy closes) between newly established tree rows in forest plantations. This system has been used for many contract reforestation projects throughout Vietnam as a means to rehabilitate degraded natural forest lands. The forest trees may be grown for either timber or paper pulp production.

AGROFORESTRY SYSTEMS IN CHINA

One form or another of agroforestry has been practised in China since ancient times. China has a long history of agroforestry and systems often combined crops and trees to improve the form of the trees. Examples of this include the interplanting of Chinese scholar tree (*Sophora japonica*) with hemp to improve trees for road-side planting; chestnut (*Castanea* spp.) with soybean to make the trees grow upright or for shade or shelter benefits; the use of hemp with paper mulberry (*Broussonetia papyrifera*) to prevent frost damage to the mulberry; or the undersowing of mulberry (*Morus* spp.)

with tea (*Thea sinensis*). Shifting cultivation was also practised in ancient times (Wu and Zhu 1997). Now many successful agroforestry systems have been in practice, some of which have developed very fast throughout the country since new China was established.

INTERCROPPING AGRICULTURAL CROPS WITH TREES: Several systems of intercropping agricultural crops with trees have been practised in the plains. The main forms of intercrop are agricultural crops with *Paulownia*, chinese date (*Ziziphus jujuba*), fruit trees, willow, false indigo and white mulberry. The major examples include:

- *Paulownia* trees (four years old) and common wheat system (North China)
- *Paulownia* trees (twelve years old) and common wheat system (North China)
- Oilseed rape between *Taxodium decurrens* trees, central China.

FARMLAND SHELTER BELT: The farmland shelter belts are usually built together with roads and channels. Their sizes vary from 6-20 ha, most of which are narrow-spaced forest shelter belts. These systems have produced a large amount of timber and fodder. Such farmland shelter belts have been widely used in the North and Northeast of China as well as in parts of the southern subtropical areas.

FARMING UNDER TREES: Farming under trees is widespread practice with example of *Aleurites fordii, Sapium sebiferum, Juglans regia, Castanea bungeana, Camellia oleifera, Camellia sinensis, Hevea brasiliensis, Diospyros kaki* and *Fraxinus chinesis*. Successful trials have been conducted to plant traditional Chinese medicinal plants such as *Panax ginseng* in the Northeast, *Coptis chinensis* var. *brevisepala, Amomum villosum, Gastrodia elata* in the Southwest; and some edible fungi such as *Auricularia, Tremella, Dictyophora indusiata, Lentinus edodes* and *Pleurotus ostreatus* under various types of forests, such as bamboo forests that have shown a great production potential.

GARDEN TYPE AGROFORESTRY: Various models of garden agroforestry exist in the country according to the ecological, social and economic conditions of different districts. Timber trees such as Chinese toon, *Paulownia*, Chinese fir, plane, bamboo; economic trees such as *Philodendron*, palm, orange, papaya, loquat, as well as many undergrowth such as pistacho tree, *Atractylodes macrocephala* and beans are grown. Under *Paulownia* trees, many grapevines are cultivated, and under the grapevine-rack, multi-storey rabbit-sheds are established.

IMPORTANT SYSTEMS:

- Intercropping agricultural crops with *Paulownia*: This promising agroforestry system has become important in the flatland areas of north China. *Paulownia* is a fast-growing tree with a deep root system (minimizing root competition with annual crops). It seems to enhance the suitability of the microclimate for agricultural crops and also produces timber within a short period (about 10 years).
- Intercropping agricultural crops with *Ziziphus jujuba*: In an intercropping system, the mixed planting of the Chinese date with an annual crop such as wheat can increase the yield of both crops. The vitamin-rich Chinese date is very popular in China and is found from temperate to subtropical areas.
- Multiple-layer artificial population: Several variations of multiple-layer agroforestry systems have been found to perform well in the tropical regions of China. Rubber-tea and rubber-camphor-tea combinations have been the most productive. Traditional Chinese medicine species (e.g., *Cinchona ledgeriana*) have also been used successfully.

- Tea-based agroforestry, Rubber-based agroforestry, Timber tree-based agroforestry, Timber and medicinal plants agroforestry are also gaining popular in China. Masson pine, slash pine, loblolly pine, Chinese fir, Chinese tallowtree, tung oil tree, poplar and sassafras trees are widely planted under these systems. Jujube, peach, pear, persimmon, plum and other deciduous fruit trees are also grown.

AGROFORESTRY SYSTEMS IN PACIFIC ISLANDS

The Pacific Islands comprise 20,000 to 30,000 islands in the Pacific Ocean. The Pacific Islands lying south of the tropic of Cancer are traditionally grouped into the three divisions of Melanesia, Micronesia and Polynesia.

- Melanesia means *black islands*. These include New Guinea (the largest Pacific island and second largest island in the world after Greenland, which is divided into the sovereign nation of Papua New Guinea and the Indonesian provinces of Maluku, Papua and West Papua), New Caledonia, Zenadh Kes (Torres Strait Islands), Vanuatu, Fiji, and the Solomon Islands.
- Micronesia means *small islands*. These include the Marianas, Guam, Wake Island, Palau, the Marshall Islands, Kiribati, Nauru, and the Federated States of Micronesia. Most of these lie north of the equator.
- Polynesia means *many islands*. These include New Zealand, the Hawaiian Islands, Rotuma, the Midway Islands, Samoa, American Samoa, Tonga, Tuvalu, the Cook Islands, Wallis and Futuna, Tokelau, Niue, French Polynesia, and Easter Island. It is the largest of the three zones.

The region's islands are classified into two groups, high islands and low islands. Volcanoes form high islands, which generally can support more people and have a more fertile soil. Low islands are reefs or atolls, and are relatively small and infertile. Melanesia, the most populous of the three regions, contains mainly high islands, while most of Micronesia and Polynesia are low islands. The developing countries of the Pacific Region consist of 27 countries and territories with a total land area of 5,42,000 km^2. Though highly variable in physical characteristics and land use pattern, these countries, in general, give a higher priority for production of export crops over other agricultural crops, so that they are net importers of food. Natural forests of the region are luxuriant and floristically rich, but the level of their commercial exploitation is low.

The destruction of trees and the failure to plant and protect trees in the process of modern agricultural development has led to a loss of trees in agricultural systems throughout the Pacific Islands. This process is referred to as "agrodeforestation". Agrodeforestation is the removal of trees or the de-emphasis on the planting and/or protection of trees in agricultural ecosystems. This process opposes agroforestry or agroforestation, the protection and planting of trees as an integral part of agricultural systems. As a result of the deterioration of traditional agroforestry-based food systems, urbanized Pacific Island populations now have some of the highest rates of nutritional disorders and nutrition-related non-communicable disease in the world (Randy Thaman et al 2013).

Swidden cultivation is the mainstay of subsistence livelihood, but with the increase in population, the system is now causing severe environmental and land use problems. The major agroforestry systems and practices in the region include various forms of combination of tree crops such as coffee, coconut and cacao with nitrogen fixing trees such as *Casuarina*, *Gliricidia* and *Leucaena*, and food crops (mostly tubers) such as cassava, taro, sweet potato and yams. Additionally, some improvements to swidden (shifting) cultivation are also being tried in several places, the most common being the use of Casuarina oligodon before abandoning the swidden as in PNG highlands, and intercropping food crops in tree crop stands. The immediate opportunity in agroforestry in the region lies in making better use of the unexplored potential of a large number of locally available trees and agricultural crops. Several governments of the region are now promoting the agroforestry approach. Intercropping in tree crop stands and the taungya system seem to promising methods for the Pacific region (Vergara and Nair 1985). The major agroforestry practices in Pacific Islands are silvopastoral systems (trees and livestock), windbreaks, live fences or border plantings, hedgerow intercropping, improved fallow and land rehabilitation, woodlots, sequential cropping systems and dispersed trees/understorey cropping (Elevitch and Wilkinson 2000).

SILVOPASTORAL SYSTEMS

In the Pacific Islands, grazing of cattle under commercial timber species has been practised in many settings. In Papua New Guinea, cattle have been used to control weeds and reduce fire danger under plantations of Caribbean pine (*Pinus caribaea*), hoop pine (*Araucaria cunninghamii*) and *Eucalyptus* spp. In Vanuatu, cattle have been grazed under laurel (*Cordia alliodora*) as well as under *Pinus caribaea* plantations. In Solomon Islands, cattle have been grazed under *Eucalyptus deglupta* plantations. The most widespread silvopasture system found in Pacific Islands is the grazing of cattle under coconut (*Cocos nucifera*) plantations, which has been practised since colonial times throughout the Islands including the Solomon Islands, Tonga, Niue, Vanuatu and Fiji (Clarke and Thaman 1993).

WINDBREAKS

Windbreaks are common on many Pacific Islands, and numerous species and planting patterns are used. While there is tremendous variation, three examples are: A three-row windbreak of dwarf Brazilian banana (*Musa balbisiana*) on staggered 2.5 m × 2.5 m (8 ft x 8 ft) spacing (Guam); a two-row windbreak of timber trees narra (*Pterocarpus indicus*) and sissoo rosewood (*Dalbergia sissoo*) on staggered 2.5 m x 5 m (8 ft × 15 ft) spacing (Guam)(Lawrence 1999); and a two-row windbreak of Dunn's white gum (*Eucalyptus dunnii*) and *Podocarpus* spp. on staggered spacing (Hawaii). A single-row windbreak is also effective if composed of a very wind-strong species with uniform branches to the ground. In single-species windbreaks, one of the most commonly used species is *Casuarina equisetifolia*.

In Pacific Islands, many kinds of trees have been used as windbreaks. Some native and Polynesian species make good windbreak components; for example in Hawaii milo (*Thespesia populnea*), false kamani (*Terminalia catappa*), hau (*Hibiscus tiliaceus*) and kamani (*Calophyllum inophyllum*) are used (Shigeura and McCall 1979). A few examples of food-producing windbreak trees include dwarf Brazilian banana (*Musa balbisiana*), tamarind (*Tamarindus indica*) and jackfruit (*Artocarpus heterophyllus*). Timber-producing windbreak species include *Eucalyptus* spp., mahogany (*Swietenia macrophylla*), narra (*Pterocarpus indicus*) and neem (*Azadirachta indica*).

LIVE FENCES/ BORDER PLANTINGS

Live fence posts and live barriers/hedges are two common types of border plantings. Live fence posts are commonly found in conventional barbed wire fences. Farmers may plant stakes of easy-to-root species such as *Gliricidia sepium*, *Erythrina* spp., *Spondias* spp. and *Bursera simaruba*. *Gliricidia sepium* is the most common live fence post species because of the ease with which large stem cuttings root and its multiple uses such as forage, green manure, etc. Some farmers use a number of different tree and shrub species to establish dense and often thorny live barriers/ hedges to protect their crops. If well established and maintained, these natural barriers can deter both animal and human trespassers from entering into the farm.

Species used to form live barriers and hedges include thorny species such as natal palm (*Carissa grandiflora*), kiawe (*Prosopis pallida*) and Manila tamarind (*Pithecellobium dulce*). Other good live fence species are sisal (*Agave americana*) and various bamboos. There are also a number of tropical fruit trees that have been incorporated into live fencing systems either as fence posts or within live fence hedgerows including guava, citrus, bush plum (*Canarium* spp.), *Inga edulis*, *Spondias mombin*, *Moringa oleifera* and a variety of palm species. Certain plants with medicinal properties are also used for live fences such as *Prunus africana*, *Colubrina* spp., *Commiphora* spp. and neem (*Azadirachta indica*) (Elevitch and Wilkinson 2000).

HEDGEROW INTERCROPPING AND CONTOUR HEDGEROWS

Alley cropping has been used and researched for several decades, and has received recognition for its potential as a sustainable technique for producing annual food crops. An on-farm study of alley cropping in a tropical fruit orchard in Holualoa, Island of Hawaii showed soil fertility improvement in two years time with no fertilizer inputs except for the mulch from the hedgerows. It also showed a significant increase in total nitrogen and potassium in the soil as well as improvement in soil pH (more neutral) (Elevitch and Wilkinson 1998). Species used for hedgerow intercropping must be fast growing, able to regenerate easily after pruning and nitrogen-fixing. Common hedgerow species include *Calliandra calothyrsus*, *Gliricidia sepium*, *Flemingia macrophylla*, *Erythrina subumbrans*, *Albizia saman*, *Leucaena leucocephala*, *Pithecellobium dulce*, *Leucaena diversifolia*, *Cajanus cajan* and *Paraserianthes falcataria*.

IMPROVED FALLOW/ LAND REHABILITATION

Improved fallow has its origin in shifting cultivation in the tropics where farmers scatter the seeds of fast growing trees in order to shorten the length of the fallow period (Nair 1993). Species used in improved fallow systems should be nitrogen-fixing and/ or able to produce large amounts of litter with high nutrient content, and they should have deep root system. Ideally, improved fallow species should also be able to produce useful by-products such as firewood, poles, edible seeds, etc. Species used for this purpose include *Acacia auriculiformis*, *Albizia lebbeck*, *Calliandra calothyrsus*, *Casuarina equisetifolia*, *Gliricidia sepium*, *Pithecellobium dulce*, *Sesbania grandiflora*, *Sesbania sesban*, etc.

WOODLOTS (TIMBER, FIREWOOD, FODDER BANKS)

Woodlots are usually solid stands of trees and consist of trees planted close together, and uniformly spaced. Desirable species for woodlots are fast growing, nitrogen-fixing and provide useful products. Fodder banks should be dense, nearly pure stands. Recommended spacing varies from 1 m x 1 m to 5 m x 5 m. Choice of spacing depends on management objectives. Total biomass yields per area increase at higher densities. Wider spacing is generally used when both fodder and small diameter wood, for fuel or poles, are desired. Closer spacing maximizes fodder production, but may make access for harvest or grazing difficult. Spacing of 1m x 1m for woodlot is common for many species. The important woodlot species for fodder and firewood available throughout the Pacific region are (Winrock International 2013) *Acacia auriculiformis*, *Adenanthera pavonina*, *Albizia chinensis*, *Alphitonia zizyphoides*, *Calliandra calothyrsus*, *Casuarina equisetifolia*, *Eucalyptus camaldulensis*, *Eucalyptus urophylla*, *Gliricidia sepium*, *Securinega flexuosa* and *Leucaena leucocephala*.

The hardy timber-producing species such as sissoo rosewood (*Dalbergia sissoo*), brown salwood (*Acacia mangium*), Tibet tree (*Albizia lebbeck*), neem (*Azadirachta indica*), pheasantwood (*Cassia siamea*) and *Eucalyptus* spp. are also used in woodlots (Otsyina et al 1996).

SEQUENTIAL CROPPING SYSTEMS

Sequential cropping is a practice wherein short-term crops are planted with and eventually replaced by long-term trees. The long-term tree crops can include either fruit/nut or timber species. Sequential cropping may be used during the establishment phase of a forestry or orchard project, or in other cases where annual or short-term crops are being replaced by long-term tree crops. Sequential planting in a forestry system of Pentecost, Vanuatu is a good example. In Vanuatu, cash crops are grown between the line plantings of laurel (*Cordia alliodora*). Crops have included sweet potato, cassava, yam, kava (*Piper methysticum*), coffee and cardamom. Subsistence gardens have also been established in *Cordia alliodora* plantings by local land growers and forest workers (Clarke and Thaman 1993).

DISPERSED TREES/ UNDERSTOREY CROPS

Understorey cropping combines trees and shade-tolerant crops in a permanent arrangement. The practice can be integrated during the design of forestry, orchard or other tree-crop system, or may be used to increase economic returns from a native forest or conservation area. Some farmers may add overstorey trees to provide shade for existing crops. In pacific Islands, it is common to find taro, kava (*Piper methysticum*), or pasture under coconut plantations. In Hawaii, a variety of culturally valuable plants such as kava, herbaceous medicinals, and vines for garlands are cultivated in the understorey of native forests.

Table 22.6: Understorey Cropping System in Pacific Islands

S.No.	Overstorey Species	Understorey Species	Location & Reference
1	Macadamia nut (*Macadamia integrifolia*)	Coffee (*Coffea* spp.)	Hawaii
2	Laurel (*Cordia alliodora*)	Sweet potato (*Ipomea batatas*), cassava (*Manihot esculenta*), yam (*Dioscorea* spp.), kava (*Piper methysticum*) and cardamom (*Elettaria cardamomum*)	Vanuatu (Clarke and Thaman 1993)
3	Coconut (*Cocos nucifera*)	Kava (*Piper methysticum*) and taro (*Colocasia esculenta*)	Fiji (Clarke and Thaman 1993)
4	Native forest	Garland/ lei plants (e.g., maile (*Alyxia* spp.) or medicinal plants	Hawaii
5	Coconut (*Cocos nucifera*)	Pasture	Polynesia (Dalla Rosa 1993)
6	Leucaena (*Leucaena leucocephala*), coconut (*Cocos nucifera*) and banana (*Musa* spp.)	Cacao (*Theobroma cacao*)	Papua New Guinea (Clarke and Thaman 1993)
7	Coconut (*Cocos nucifera*) and physic nut (*Jatropha curcas*)	Vanilla (Vanilla fragrans)	Tonga (Clarke and Thaman 1993)
8	Kamarare (*Eucalyptus deglupta*)	Cacao (*Theobroma cacao*) and coffee (*Coffea* spp.)	Papua New Guinea (Clarke and Thaman 1993)

AGROFORESTRY SYSTEMS IN LATIN AMERICAN TROPICS

Amerindians have practised their traditional forms of agroforestry for millennia without apparent detriment to the rainforest. These systems are generally referred to as swidden agriculture, or sometimes as slash and burn or bush fallow farming (Warner 1991). A parcel of forest (swidden plot)

is cut and the area burned to release nutrients. A mixture of short-term and annual crops, often followed by perennials, are grown until soil fertility becomes inadequate to maintain production and/or competition from successional plant species becomes too great. The farmer then prepares a new site while the old field returns to fallow. During the fallow period large quantities of nutrients are stored in the plant biomass and are released during the burning of the fallow vegetation or secondary forest when the land is cleared for the new cropping cycle (Barton 1994).

Tabel 22.7: Agroforestry Systems in Latin American Tropics (Martin and Sherman 1992; Wilkinson and Elevitch 2000)

S.No.	System	Overstorey Species	Understorey Species	Location
1	Dispersed trees	laurel (*Cordia alliodora*), *Gliricidia sepium* and *Erythrina* spp.	Coffee and cacao	Central America
2	Understorey cropping	Leguminous trees, *Faidherbia albida*	Grains, pasture	Central America
3	Understorey cropping	laurel (*Cordia alliodora*), earpod (*Enterolobium cyclocarpum*) or monkeypod (*Albizia saman*)	Coffee	Costa Rica
4	Understorey cropping	*Erythrina* spp.	Cacao (*Theobroma cacao*)	Costa Rica
5	Dispersed trees	*Cordia almifolia*	Coffee	Costa Rica
6	Dispersed trees	*Erythrina* spp.	Coffee	Costa Rica
7	Understorey cropping	*Inga* spp.	Coffee, banana, root crops	Puerto Rico
8	Dispersed trees	*Inga* spp.	Coffee, cacao	El Salvador
9	Dispersed trees	*Brosimim*	Variety of crops, pasture	Mexico
10	Mixed perennials	Orange, avaocado, banana	Coffee, root crops	Puerto Rico
11	Understorey cropping	Peach palm (*Bactris gasipaes*)	Black pepper (*Piper nigrum*) and cacao	Brazil
12	Dispersed trees	Silver oak (*Grevillea robusta*)	Coffee	Northeast Brazil

Seven distinguishing characteristics of indigenous Latin American agroforestry systems are described: they integrate indigenous species; they make use of natural environmental variation; they utilize natural successional processes as management tools; they include a large number of species; plots are integrated into a diversified farmstead; there is variation among farms using the same system; and each farm is designed to meet all the family needs for independent survival within the context of the local community land use. Sequential agrosilvicultural systems create 5 basic structures:

- Fields of staples
- Homegardens
- Orchards
- Managed fallow thickets
- Artificial forests

However, it is useful to classify these systems on the basis of the dominant crop component since this has a strong effect on the structure of the system, and also tends to follow cultural and geographic divisions. On this basis, the four agrosilvicultural systems are described:

- Maize (*Zea mays*) initiated systems such as the milpa system of Middle America (e.g. the Mexican Huastec Maya, the Totanac of northern Veracruz, Mexico, and the Mexican Yucatecan farmers)
- Manioc or Cassava (*Manihot esculenta*) initiated systems in Central and South America and the Caribbean (e.g. chacras in Costa Rica)
- Wheat (*Triticum aestivum*) initiated system of Totomicapan, Guatemala, which has pine (*Pinus* spp.) trees in wheat fields
- Sugarcane (*Saccharum officinarum*) system (e.g. those maintained by the Haustec of Mexico and the Kuna of Panama)

INDIGENOUS SUBSISTENCE AGROFORESTRY SYSTEMS

BORA AGROFORESTRY: The Bora, Amerindians of the Amazon in eastern Peru and southern Colombia subsist from swidden agriculture, fishing, collecting and hunting. They use both primary and secondary forest for their swidden plots. Fallows are felled during the period of lowest rainfall and burned after drying. Useful palms and trees, as well as certain valuable timber species, e.g. Spanish cedar (*Cedrela odorata*) are spared during clearing. Coppice regrowth of certain species is also encouraged. The main staple of the Bora is cassava (*Manihot esculenta*). It is intercropped with pineapple (*Ananas comosus*), fruit trees and many other annual crops (Denevan et al 1984). Indeed, the crop combination of a Bora swidden varies enormously with the needs of a family and labour availability. Generally, families will have six or more fields in various stages of succession to fallow. Those fields with less diversity tend to be abandoned earlier than those in which perennial tree crops have been introduced. Several tree species planted in Bora fields are well adapted to growing in dense secondary forest. Umari (*Poragueiba sericia*) and mocambo (*Theobroma bicolor*) are often evident in old fallows. Spanish cedars are, however, transplanted into swiddens and fallows, and when mature, floated down river to market. This timber can produce a substantial cash return 30 years from planting (Barton 1994).

KAYAPO AGROFORESTRY: The Kayapo Amerindians of northern Mato Grosso and southern Para States, Brazil have a sophisticated understanding of different ecological zones and use it in farming and settlement decisions (Posey 1985). Their territory includes grasslands, mountains and forest, and their intimate knowledge of these ecological zones and their subtle variations enables them to choose settlement sites in transitional zones. Kayapo thus reside in areas of maximum species diversity. Each zone will provide natural products and attract different game species at different times of the year. Their swidden plots are set out in the form of concentric rings; the major crops being sweet potatoes, yams and plantains. The centre of the swidden is dominated by sweet potatoes (*Ipomoea batatas*), a secondary ring begins with maize followed by cassava and sweet potatoes. The outer ring includes yams, cupa (*Cissus gongylodes*), bananas, pineapples, urucu (*Bixa orellana*) and fruit trees (Hecht and Posey 1989). Kayapo swiddens continue to produce harvests of yams and taro (*Colocasia esculenta*) for 5-6 years, bananas for 12-15 years and cupa for more than 30 years. Old fields are also important for their concentrations of medicinal plants.

KA'APOR AGROFORESTRY: The Ka'apor Amerindians inhabit a tract of dry land which is not subject to annual floods in the dense forest in northern Maranhao State, Brazil. Here the clearance of new swiddens is often accompanied by a change in settlement site (Balee and Gely 1989). Once a site has been selected a communal swidden is cleared and cultivars from the old site are transported to the new. The first food crops planted are fast growing cultivars of cassava, banana and sweet potato. The inhabitants of the old site move to the new between 7 and 21 months after the first planting. At this point, an area is cleared with a radius of 5 m around each house (houses are usually situated within the swidden), i.e. this part of the field is harvested first and a house garden is established (trees and herbs). Thereafter individual families clear smaller individual swiddens in the surrounding forest.

Ka'apor recognize six major vegetational zones (or types of forest or manipulated forest): (a) house gardens; (b) young swiddens up to two years from burning; (c) old swiddens 2-40 years from initial burn; (d) old fallows 40-100 years from first burning; (e) mature forest and (f) swamp forest. The house garden is part of the original community swidden. It consists of herbs and trees, many of which are light demanding and will only grow in house gardens. Young swiddens (between 0.5 and 1.5 ha) are characterized by cassava, sweet potato, papaya, cashew and banana cultivation. They are also an important source of firewood. Older swiddens (over four years) returning to secondary forest have numerous palms, vines, mango and citrus trees, and other fruits. Old swiddens may return to cultivation after a period of 15 years, although established fruit trees from the previous cycle will be protected from the burn. Old fallows (40-100 years) display signs of earlier disturbance, with enhanced densities of fruit trees and palms (Barton 1994).

AMUESHA AGROFORESTRY: The Amuesha inhabit the tropical rainforests of the Palcazu valley of east/ central Peru and live from subsistence agriculture, supplemented by fishing, gathering and occasional hunting (Salick 1989). They farm small pockets of fertile soil found on the slopes of the valley (Salick and Lundberg 1990), as much of the fertile alluvial soils of the valley floor have been taken by European immigrants. The Amuesha practice swidden agroforestry, but contact with Europeans and the rest of Peru after the construction of a sealed road has led to some modifications and participation in the wider economy. They classify and choose land according to the soil's ability to support different forms of vegetation, and fallow times reflect differential fertility status. Following a burn (often unsuccessful in this high rainfall region) they grow:

- beans and groundnuts on seasonally flooded beaches
- maize, cassava, plantain and fruit trees on dry lowlands
- rice and cassava on upland fields
- taro on less fertile uplands
- variety of useful plants in homestead gardens

All fields are intercropped with a great diversity of minor crops. In the dry lowlands, cassava is planted and harvested for up to three years. Plantains are also introduced and often underplanted with fruit and fuelwood species. Plantains are harvested until the fallow species begin to dominate. For a number of years, weeds are kept from growing round the useful trees which are protected during subsequent burns so that they provide a useful harvest. Lowland fallows are short (3-5 years) reflecting the fertile nature of these sites.

MARKET-ORIENTATED AGROFORESTRY SYSTEMS

HUASTEC AGROFORESTRY: The Huastec Maya farm the communally owned and densely populated foothills of the Sierra Madre Oriental mountains of Mexico (Alcorn 1989a). Their agroforestry management strategy is a highly productive indigenous practice, given that landholdings vary in size from 1 ha to 15 ha with population densities of around 100/km^2. They have a mixed economy, producing agricultural goods for subsistence and the market, and are wage labourers outside the community. They plant maize on fallows cleared from secondary regrowth of three to seven years of age. During the burn, selected useful species are protected and firewood trees are pollarded to provide support for climbing beans and a continuous supply of firewood. Most fields produce only one crop of maize and then return to fallow, although older fallows will have sufficient fertility for two crops. Maize is intercropped with beans, sesame, squash, yam bean, sweet potatoes and cassava (Barton 1994).

Weeding is carried out only once, when care is taken to leave any useful seedlings that have emerged. After harvesting maize, part of the plot is selected as a forest garden (te'lom), whilst the greater part of the original clearing is left to regenerate naturally. The garden is made up of native species that were protected during the initial burn as well as transplanted native species, coffee, garden crops and fruit trees. A garden, from 0.25 to 3.0 ha or more, can be in continuous production for at least 80 years. As new ones are created, however, other older ones are cycled back into swidden (maize) production. Gardens are created and managed flexibly, depending upon the time spent working off-farm. Around the edges of the gardens where there is sufficient light, cacti, bananas and cassava are grown. The typical garden may contain up to 300 species, 90 per cent of which produce useful products (construction materials, poles, furniture, firewood, food, medicine and livestock feed). Coffee and sugarcane are the most important commercial products although fruits such as avocado, mango, orange and others may be marketed in small quantities (as much as half of total fruit production may go unharvested). Hand-sawn cedar is sold along with potted parlour palms as house plants (Alcorn 1989b).

RIBERENO AGROFORESTRY: The riberenos, inhabiting Tamshiyacu, 30 km southwest of Iquitos in eastern Peru are descendants of both immigrants who settled during the rubber boom of the last century, and Amerindians of the region. Their agriculture, in this lowland forested region, displays similar patterns to that of indigenous people of the surrounding area, such as the Bora, but has been adapted to produce goods for the market (Padoch and de Jong 1987). Their swidden agroforestry cycle begins with the clearing of fields from primary or secondary forest, although valuable species preserved during earlier cropping cycles are spared. Cassava and plantains are grown both for subsistence and sale. Other crops are interplanted with cassava, the most important of which is pineapple. In the second year perennial plants such as umari are introduced as cassava is harvested, and pineapples become the predominant crop. By the fourth year, pineapple yields decline and are replaced by early maturing fruit trees. These include short-lived species such as cashew, uvilla (*Pourouma* cecropiifolia), and caimito (*Pouteria caimito*). Umari (*Poragueiba sericia*) and Brazil nuts (*Bertholletia excelsa*) are the only products of the swidden in the later stages of the cycle.

Orchards are weeded for up to 25 years, and the secondary forest is not allowed to regenerate, with the exception of a few economically useful species which are protected when they spontaneously establish. If fields are well maintained, they may continue in fruit production for up to 50 years. As soon as production begins to fall significantly, the umari and other trees present are cut, dried and converted to charcoal for sale (Brazil nut trees are often spared); the plot then being left fallow for five to ten years. The plot is invaded with secondary regrowth which is cleared and the swidden cycle recommences (Barton 1994).

SANTA ROSA AGROFORESTRY: Inhabitants of the old plantation village of Santa Rosa on the Ucayali river 150 km upstream from Iquitos, Peru are also of mixed origin. Villagers inhabit dry land and practise their swidden agriculture at these sites, but also cultivate the floodplains, in contrast to other local farmers (Padoch and de Jong 1991). After farm sites are cleared and burnt, staple annuals and semi-perennials for subsistence use are introduced, as poor communications deny the villagers access to markets. Cropping and maintenance cease after two or three years. Fruit trees are introduced during the first year of cropping, and early fruiting species produce by the second year. By the fifth or sixth year, the plot has become an orchard. A great variety of fruit and timber trees develop in the fallow. However, the major source of cash comes from rice and plantains which are cultivated on the river plains and the seasonally inundated mudflats fertilized by annual deposits of silt.

RIBERENO FLOODPLAIN AGROFORESTRY: The riberenos of San Jorge, 45 km southwest of Iquitos in Peru, have a very diverse economy. They farm, fish, gather forest products, hunt and work for wages. Traditionally they cultivated the floodplains of the Amazon and its tributaries which are seasonally fertilized by the rich silts deposited in the short annual floods (Hiroaka 1989); swidden agroforestry is practised on rarely flooded parts of the lowlands. Swidden management begins with the clearance of high or secondary forest. Cassava is intercropped with plantains, maize, beans, a number of minor crops and fruit trees. Weed and pest control are carried out up until the third year of cultivation by which time yields of annual crops begin to decline and weeding becomes increasingly time consuming. The field is left to the fallow natural succession, although the bases of fruit trees are occasionally weeded. The orchard swidden continues to produce for at least ten years, and provides an ideal hunting ground for small game; this is important in the flood season when fish become dispersed and catches small. A family clears and plants a new swidden plot every one or two years, ensuring a constant supply of the staples, cassava and plantain. The swiddens are generally used for subsistence, but increasingly staple foods and crops are marketed. The annually flooded areas produce rice (the principal cash crop), maize, beans, cucumber, tomato and vegetables (Barton 1994).

ASSAI PALM FOREST SYSTEMS: The riverside dwellers (riberenos) of the Amazon estuary in Brazil inhabit the floodplain and use the forest cover in a variety of ways. They have, to a great extent, maintained native species to assure sustained production of the assai palm (*Euterpe oleracea*). In contrast to the extensive upland areas of the Amazon, the floodplains do not display a high biotic diversity and are dominated by the assai palm (Anderson 1990). This plant is an important source of palm heart and its fruit is processed into a beverage that has a ready market throughout Para state, although this market has been recently disrupted by a cholera epidemic. The inhabitants of the estuary, while collecting fruit and hearts from wild palms, introduce assai seeds or seedlings into their shifting cultivation plots promoting high density stands which utilize the fallow period. Their agriculture is characterized by three types of land use. House gardens are made near the dwelling and are used for the production of a wide variety of native and exotic shrubs and domesticated animals. Swidden plots used for the production of staple crops such as rice, beans and maize are found some distance from the dwelling to minimize disturbance by domestic animals. Between these and the natural forests (used for hunting, gathering of fruit, latex, palm heart, wood, fertilizer, honey, fibres and medicines) there exist partially managed forests where desirable trees are favoured, such as the assai palm, rubber, fruit trees and firewood species. These forests are thinned involving the removal of vines and understorey vegetation which improves access. Mango, cacao (*Theobroma cacao*), coconut and other trees may be introduced as seeds or seedlings.

CABOCLO AGROFORESTRY: The term caboclo refers to a Brazilian of Indian-European-African parentage (Moran 1974). Caboclos have been generally ignored by development planners, having been considered of low social status. However, they make their livelihood from marketing extractive forest products whilst maintaining subsistence agriculture to survive the boom or bust nature of the market. The caboclo have adapted to the Amazon environment over three centuries and have an intimate knowledge of the natural resources of the region in terms of its agricultural potential and as a source of naturally occurring products for extraction. Although they inhabit areas accessible by rivers and good for fishing, they practice swidden agriculture on the dry lands. Because of their cultural links with both the indigenous populations of the region and the new colonists, the caboclo provide a very good example of the ways that indigenous subsistence and new marketing livelihoods can be combined in a sustainable way (Barton 1994).

JAPANESE AGROFORESTRY SYSTEMS: Japanese farmers have been cultivating remote forest regions of Tome Acu in Para State, Brazil since the 1920s. Over time their agricultural systems have developed from subsistence based farming to commercial agriculture, fully integrated into the national and world economies (Subler and Uhl 1990). They replace forest vegetation with a mixture of both native and exotic species of economic value, producing a variety of high value cash crops through a mixture of annual and perennial plants. The systems are intensive, relying on large inputs of labour, capital and materials in the form of inorganic and organic fertilizers. The system is a multilayered mixture of perennials and annuals in succession. A typical example would be the planting of rubber seedlings following forest clearance. This area is then intercropped with rice and maize. After harvesting the grain, the rubber is interplanted with cotton. After the cotton harvest black pepper and papaya are planted in rows between the rubber followed by a crop of pumpkins. After two years the papaya has completed its productive phase while the black pepper begins production. The pepper continues production for 3-4 years by which time the rubber begins to produce latex.

OTHER AGROFORESTRY SYSTEMS

In addition, the other prevalent land use systems are artificial forests (which range from managed forests of indigenous species to orchards with a few indigenous species and single tree crop plantations mixed with a few native species); dooryard gardens (homegardens); silvopastoral and agrosilvopastoral systems (under plantation crops or forest trees); wetland systems (e.g. mangrove swamps); and live fences.

TEMPERATE AGROFORESTRY SYSTEMS IN NORTH AMERICA

ALLEY CROPPING

Alley cropping is the cultivation of food, forage or specialty crops between rows of trees. It is a larger version of intercropping or companion planting conducted over a longer time scale. Alley cropping can provide profitable opportunities for row crop farmers, hardwood timber growers, nut growers and Christmas tree growers. The benefits of alley cropping are (AFTA 2013):

- *Income Diversification*: Crop production during the years before nut trees come into bearing or hardwood timber is harvested creates cash flow and diversifies farm income, thereby improving the return on long-term investments in trees.
- *Marginal Land Improvement*: By planting rows of nut or timber trees on land where annual crop production is low due to erosion or other limitations, marginal croplands may be converted to higher value woodlands.
- *Shelter*: Rows of trees reduce wind speed, thereby controlling wind erosion. They also create sheltered microclimates that improve the yield and quality of crops growing in the alleys.
- *Wildlife*: Alley cropping increases the biodiversity of cropland which creates new habitat for wildlife.

A good example for alley cropping model is hardwood timber or nut trees with crops. Hardwoods such as black walnut (*Juglans nigra*) or pecan (*Carya illinoinensis*) may be combined with annual crops such as wheat, corn or soybeans. Forage crops can be cultivated between tree rows for harvest or livestock grazing, or shrubs palatable to livestock can be grown for fodder, e.g. tagasaste (*Chamaecytisus palmensis*), orchard grass and alfalfa. Possible alley crops also include specialty crops, e.g. herbal or medicinal, perennials such as conifers for Christmas trees, and even nursery stock, e.g. using the Pot-in-Pot system. Combining specialty crops with conventional field crops is another example. Another alternative is to plant rows of perennial specialty crops within a field of tall field crops such as corn to diversify income while maintaining annual crop production (UMCA 2013).

FOREST FARMING

Forest farming is the intentional cultivation of edible, medicinal or decorative specialty crops beneath native or planted woodlands that are managed for both wood and understorey crop production. It does not include the gathering of naturally-occurring plants from native forests, also known as wildcrafting. Forest farming can provide profitable opportunities for forest and woodland owners, nut growers, sugar maple (*Acer saccharum*) growers, and herb growers. The benefits of forest farming are:

- *Extra Income and Diversification*: Cultivation of specialty crops in a forest setting provides new sources of annual or periodic income before, or instead of, timber harvests.
- *High Value Products*: A wide variety of plants can be grown in a forest to produce natural or processed (value-added) edible, herbal, medicinal, decorative and craft products.
- *Ecologically Responsible*: The intensively-managed cultivation of forest plants that are endangered in the wild due to over-harvesting is the responsible alternative to wildcrafting.

A successful forest farming practice begins with sustainable forest management and timber stand improvement strategies, which can help create an environment for shade crops like mushrooms and medicinal herbs to flourish. These practices may also offer income potential for value added forest products, such as wood blanks and furniture-grade lumber. Shade-loving plants that are naturally adapted to grow under forest conditions are candidates for forest farming cultivation. Medicinal plants under forest farming are getting more attention. Market prices for ginseng (*Panax ginseng*), black cohosh (*Cimicifuga racemosa*) and other medicinal herbs are rising. These shade-loving plants thrive in the canopy created by a forest farming practice. Cultivating specialty mushrooms through forest farming is another profitable process (UMCA 2013).

RIPARIAN FOREST BUFFERS

Riparian forest buffers are strips of trees, shrubs and grass planted between cropland or pasture and surface water courses. Buffers protect water quality, reduce erosion and flooding. Riparian forest buffers can provide beneficial opportunities for row crop farmers, ranchers, horticulturists, and dairy and livestock producers. The benefits of riparian forest buffers are:

- *Water Quality*: Buffers have been shown to reduce non-point source pollution by absorbing and filtering animal wastes, sediments, nutrients and pesticides from crop and rangelands.
- *Erosion and Flood Control*: Riparian buffers combined with bioengineering practices help stabilize streambanks, slow out-of-bank flood flows, and protect cropland from flooding.
- *Extra Income*: Multirow riparian buffers can be managed to allow the periodic harvest of timber or non-timber products while still maintaining buffering capacity. This helps offset the loss of productive crop or pasture land to buffers.
- *Wildlife Habitat*: Riparian forest buffers provide shade, shelter and food for fish and aquatic organisms, as well as terrestrial wildlife.

Red osier dogwood (*Cornus sericea*) and curly willow (*Salix matsudana*) are examples of florist materials that can be grown for additional income in a riparian buffer practice. Woody florals can be sold to retail or wholesale florists by the stem or the bundle at competitive prices.

SILVOPASTURE

Silvopasture is the intentional combination of trees, forage and livestock managed as a single integrated practice. In a typical silvopasture practice, perennial grasses and/or grass-legume mixes are planted between rows of trees for livestock pasture. The trees not only provide a long-term investment for nut crops or a timber harvest, but also provide the animals shade in the summer and a windbreak in the winter. In turn, the forage base provides feed for beef cattle which ultimately provides livestock sales for short-term income. A silvopasture practice diversifies farm income; can minimize the need for chemical or mechanical vegetation control; and can reduce hay and feeding costs for livestock. The benefits of silvopasture are:

- *Return on Investment*: By adding a livestock production component, silvopasture creates a stable source of cash flow prior to nut or timber harvest (e.g. pine, poplar or pecan) and diversifies the forestry enterprise. For livestock graziers, add a long-term timber while maintaining forage production provides more income than grazing alone. In some areas, pine straw harvest can also provide supplemental income.
- *Silvicultural Benefits*: Nitrogen-fixing forage species, pasture fertilization and animal manure all help improve the soil and tree nutrition. Grazing controls competing brushy species and reduces fire hazard. Competition between trees is less at the wider spacing employed with silvopasture. The result is greater timber yield.
- *Livestock Benefits*: Trees create a sheltered microclimate to protect animals from heat and cold. Shelter also improves forage quality and lengthens its growing season. The prunings of some trees can also be used as fodder, e.g. poplar. The result is better livestock growth.
- *Wildlife*: By creating a more biologically diverse system, wildlife habitat is also created .

Softwood trees are planted on existing pastures or range in widely-spaced single or double rows. This arrangement reduces shading and root competition while maintaining forage production over a longer portion of the timber rotation. Legumes grown in the alleys between nut trees such as black walnut (*Juglans nigra*) and pecan (*Carya illinoinensis*) can provide high quality forage for livestock. Grazing can be used instead or in addition to mowing to manage the orchard floor to reduce erosion and improve wet-weather trafficability. Trees should be at least 10-12 feet tall before grazing starts. Some pesticides used in nut orchards may restrict grazing following application. In native or planted forests, forage species can be sown under an existing tree plantation that has been thinned and pruned, and cattle or sheep introduced under a managed, rotational grazing system.

WINDBREAKS

Windbreaks are planned and managed as part of a crop and/or livestock operation to enhance production, protect livestock, and control soil erosion. Field windbreaks protect a variety of wind-sensitive row, cereal, vegetable, orchard and vine crops, control wind erosion, and increase bee pollination and pesticide effectiveness. The benefits of windbreaks are:

- *Crop Yield and Quality*: By microclimate modification, field and orchard windbreaks have been shown to increase the yield of many different crops. Windbreak protection can be especially valuable in orchards and vineyards of high-value horticultural crops.
- *IPM* (Integrated Pest Management): By incorporating plants that attract beneficial insects, windbreaks can also serve to increase biological control of crop insect pests.
- *Non-Timber Products*: By using trees and shrubs that produce specialty food or decorative products, e.g. chokecherry or corkscrew willow, windbreaks can provide extra income.

- *Animal Survival and Livestock Performance*: Tree-sheltered havens, or living barns, within pastures can make the difference between death and survival for livestock subject to harsh weather conditions, e.g. newborn and newly-shorn sheep. Windbreaks around feedlots have been shown to improve the health and weight gain of cattle and sheep in cold climates.
- *Timberbelts*: Multirow windbreaks of fast-growing species such as poplar can be sequentially thinned for timber products while maintaining continuity of shelter.
- *Soil and Water Conservation*: Windbreaks are proven effective in reducing wind erosion of light-textured soils. By dispersing snow evenly across fields, windbreaks help make more moisture available for crops.
- *Snow Control*: Properly placed windbreaks prevent snow buildup around buildings and on roadways.
- Reduce energy costs and improve comfort
- *Energy Conservation*: Rural homes protected by windbreaks need 10-20% less energy for heating and cooling compared to unsheltered homes.
- *Human Comfort*: Windbreaks around farm dwellings and rural communities improve quality of life for the inhabitants by reducing wind speed along with noise and dust. Established windbreaks increase property values and enhance aesthetic benefits.
- *Wildlife Habitat*: Trees and shrubs provide much-needed food and habitat for game birds and other wildlife.
- *Carbon Sinks for Carbon Credits*: It has been estimated that for each acre planted in field windbreaks, over 21 metric tons of carbon dioxide will be stored in the trees by age 20 (AFTA 2013).

TEMPERATE AGROFORESTRY SYSTEMS IN EUROPE

Agroforestry is the integration of trees, crops and/or livestock on the same area of land. Trees can be inside parcels or on the boundaries (hedges). Agroforestry can be applied to all agricultural systems, in all parts of Europe. Agroforestry systems are obtained by planting trees on agricultural land or introducing agriculture in existing woodland (e.g. silvopasture). Europe has a unique heritage of traditional agroforestry systems with a high environmental and cultural value. The earliest stages of agricultural history were dominated by shifting cultivation, with alternating periods of agriculture and forestry. This evolved into more settled systems involving woodland grazing and silvopasture with transfer of fertility from woodlands to cultivated crops via manure (Von Maydell 1995; Eichhorn et al 2006).

In Europe, tree fodder from species such as ash, elm and poplar was collected and stored to feed livestock. Trees of high value, such as fruit trees, oaks and beech trees for acorns and beech mast, and ash trees for fodder, were retained during forest clearance and formed the basis of early agroforestry systems (Dupraz and Newman 1997). These early systems integrating trees and agricultural production began to decline during the Middle Ages, when crop rotation was developed as a method to maintain soil fertility, replacing reliance on the transfer of nutrients from woodlands to croplands. Agroforestry systems are still important in several regions of the Mediterranean. For example, in the Segura river basin in south-eastern Spain, an area of marginal agricultural land, agroforestry systems cover 60 per cent of the basin territory and support 62 per cent of the livestock population (Correal et al 2008). The greatest diversity of systems is in Greece, with agroforestry systems covering an estimated 3 million hectares or 23 per cent of the whole country (Papanastasis et al 2008).

Europe has a high potential for innovative modern agroforestry systems developed by research centers across Europe during the last two decades. While the majority of Europe's agroforestry practrces are currently focused in the Mediterranean, and it is practised in Greece, Spain, the UK, Hungary, Germany, Italy, the Netherlands, Switzerland, France and Slovenia. Wide range of agroforestry types exist, including silvopastoralism – Europe's predominant form of agroforestry – as well as alley cropping, forest farming, silvoarable agroforestry systems and the use of trees for shelter. The six basic types of agroforestry existing in Europe today are:

- *Silvopasture:* Combining trees with forage and animal production. It comprises forest or wood land grazing and open forest trees. This is the major agroforestry system of Europe especially in North of Europe.
- *Silvorable Agroforestry*: Widely spaced trees inter-cropped with annual or perennial crops. It comprises alley cropping, scattered and line belts. This is the major agroforestry system in central Europe.
- *Forest Farming*: Forest areas used for production or harvest of natural standing speciality crops for medicinal, ornamental or culinary uses.
- *Mutipurpose Trees Systems*: Fruit and other trees randomly or systematically planted in cropland or pasture for the purpose of providing fruit, fuelwood, fodder and timber, among other services, on farms and rangelands. Orchards with multiple fruit trees are best example.
- *Improved Fallow*: Fast growing, preferably leguminous woody species planted during the fallow phase of shifting cultivation. The woody species improve soil fertility and may yield economic products.
- *Riparian Buffer Strips*: Strips of perennial vegetation (tree/shrub/grass) natural or planted between croplands/pastures and water sources such as streams, lakes and wetlands, and ponds to protect water quality.

Traditional systems include fruit tree systems, olives, shelterwoods and woodland grazing, pollards, pannage, Dehesas of Spain and Montados of Portugal, hunting forests and parkland, erosion control and shelterbelts, hedgerows and timber tree systems (Smith 2010). But all of these systems are in rapid decline.

FRUIT TREE SYSTEMS: Fruit and nut silvorable systems covered large areas in central Europe. Typical examples include almond trees with cereals or fodder in Sicily; fig trees with cereals in Crete and Aegean islands; low density fruit tree plantations known as *pre-vergers* in Northeast France provide grazing land and may be intercropped in first five to 15 years of a 30-year cycle (Eichhorn et al 2006); high-growing fruit trees (*boguards*) with an understorey of grass for cattle grazing in the Netherlands (Oosterbaan and Kuiters 2008); *pomeradas* of humid areas of northern Spain consisting of apple tree planted in lines or scattered throughout meadows and croplands; intercropping soft fruits or vegetables in the early years of developing full-stature orchards in the Kent region of UK; mixed vineyards incorporated trees as mechanical support for grapevines in southern Europe; *Streuobst* system of central Europe in which tall fruit trees of different types and varieties dispersed on croplands, meadows and pastures in irregular association (Eichhorn et al 2006);

OLIVES: In pre-Roman times, rows of olive trees were intercropped with wheat in alternate years in Greece and Italy. Olive trees, grown in rows or as scattered, are intercropped with cereals, vegetables and fodder crops and systems combining olives with grape vines are still found in Spain and Greece (Eichhorn et al 2006).

SHELTERWOODS AND WOODLAND GRAZING: Woodland pasture system in northern Europe, Mediterranean regions and the New Forest in southern England are the examples of pasturing in woodland (Smith 2010).

POLLARDS: It is the practice of cutting branches from trees two to three meters above ground level, to obtain leaf fodder for feeding livestock (fodder pollards) and/or wood for fuel and other uses (wood pollards) on a two to six year rotation. It was common in northern Europe and mountainous areas such as Pyrenees, Alps, high pasture areas of the Basque country, western Norway, Greece, Crete and Sicily, and in pars of Scandinavia (Read 2006). Important fodder pollard trees include scot elm (*Ulmus glabra*), ash (*Fraxinus excelsior*), silver birch (*Betula pendula*) downy birch (*Betula pubescens*) and goat willow (*Salix caprea*) (Austad and Hauge 2006). Wood pollard trees are oak (*Quercus robur*), beech (*Fagus sylvatica*) and hornbeam (*Carpinus betulus*) in Epping Forest of England for fuelwood (Dagley 2006). Elm (*Ulums campestris*), osier willow (*Salix viminalis*) and mulberry (*Morus alba*) were source of firewood in northern plains of Italy (Ferrini 2006).

PANNAGE: It is the practice wherein pigs are released into beech and oak woodlands to feed on the acorn and beech mast, and into fruit orchards to eat fallen fruit. It was prevalent in central Europe and Denmark, and practice continues today in the New Forest of southern England (Shaw 1940; NFNP 2010).

DEHESAS OF SPAIN AND MONTADOS OF PORTUGAL: Dehesas, or montados are a Mediterranean land use system based on widely spaced oaks (*Quercus rotundifolia, Q. ilex, Q. pyrenaica, Q. suber*) which produce acorns, wood, charcoal and cork, with sheep, goats, pigs and cattle grazing and browsing beneath the trees. Similar systems are also found in Greece, marginal lands of Italy and Sardinia (called *seminativo*), and also in South America (called *espinal* and uses *Acacia caven*) (Moreno and Pulido 2008; Castro 2008)

TIMBER TREE SYSTEMS: Poplar intercropped with cereals in France and poplar integrated with livestock in the Netherlands for the production of matchstick veneer are the examples (Eichhorn et al 2006).

EXAMPLE FOR TRADITIONAL MODELS

- Fruit trees in lavender (Alpes de haute-Provence)
- Holm oak providing forage in pasture land (Spain)
- Acorns of cork oak providing forage in plots of land cultivated for fodder production (Spain)
- Dual purpose walnut trees intercropped with maize (Isère, France)
- Walnut trees and intercropped sunflower (Drôme, France)
- Olive and vine system (Hérault, France)
- Market gardening under peach orchards (Pyrénées-Orientales, France)

EXAMPLE FOR MODERN MODELS

- Wheat intercropping in poplar agroforestry (Vézenobres - France)
- Wheat intercropping in walnut agroforestry (Restinclières - France)
- Walnut trees in lavender field (Provence - France)
- Hazel and Willow alley cropping (Silvorable system) at Wakelyns, Eastern England

- Combined food and energy system with integration of bioenergy production from belts of alder, willow and hazel with crop and pasture production (Taastrup, Denmark)

AGROFORESTRY SYSTEMS IN AUSTRALIA AND NEW ZEALAND

As a further extension of plantation forestry, agroforestry is now an established practice in Australia and New Zealand. In the last two decades, more farm lands have been planted with trees, mostly on pasture and mostly by small-scale growers. The objective of agroforestry is to increase the total productivity and profit from the land, and to meet many non-economic goals such as erosion control, weed suppression, livestock welfare, and aesthetic enhancement. There are three main kinds of agroforestry practised in Australia and New Zealand:

- Farm-based: Trees planted on existing farms
- Forest-based: Animals grazing in existing forests
- Timberbelts: Shelterbelts and windbreaks on farms managed to produce high quality timber

AGROFORESTRY SYSTEMS IN AUSTRALIA

Agroforestry can confer many commercial and environmental benefits to Australia. To date development has been impeded by lack of clear identification of regions and particular agroforestry systems with the greatest opportunities for commercial investment and environmental outcomes. As global wood shortages increase and new markets emerge for products such as forests for carbon off-sets, large scale expansion of agroforestry will become a viable possibility for Australia. In temperate regions of Australia, agroforestry systems have been developed over the last 30 years to address the problems of land degradation including salinization and soil erosion. Agroforestry in Australia include scattered trees in pastures, tree belts and woodlots.

AGROFORESTRY WITH WIDELY SPACED TREES AND GRAZING: The objectives of this system developed in Western Australia are to produce high quality timber and pasture and to provide shelter for stock. In this agroforestry system, pines and eucalypts with wide spacing and pastures are grown together. The competition is less between trees and pastures in agroforestry compared with high competition in dense monocultures of pines, eucalypts or other tree species. The nitrogen fixed by pasture legumes is available for trees. Research of widely-spaced tree systems has broadened to include eucalypts. There are three main reasons for this:

- Farmers have shown a preference for eucalypts compared with pine agroforestry
- There is a need to extend wide spaced agroforestry to sites unsuitable for pine
- The demand for high quality hardwood sawlogs is likely to increase strongly.

The management of eucalypt agroforestry is essential the same as for pine agroforestry. The pruning costs are likely to be significantly less for eucalypts than for pine. In Western Australia, the studies have shown that agroforestry to be much more profitable in the long term than grazing practise alone in the high and medium rainfall regions of the State. Similarly agroforestry is a profitable practice compared with conventional agriculture. Agroforestry is a practical and economic option for controlling salinity.

TREE BELTS: Tree belts are systems in which trees provide shelter for crops, pastures or livestock. The main purpose is agriculture and trees fulfil a service role. The trees in tree belts can also be a source of timber, fuelwood, fodder etc.

WOODLOTS: Farmers grow trees in separate blocks as woodlots along with agricultural fields. Now the practice is expanding fast due to shortage of fuelwood and demand for wood in commercial industry. Farm woodlots essentially involve the growing of trees as a field cash crop. Woodlots are small plantations of less than 10 ha, often much less, that are established by the individual farmer for the production of wood, fuel, fodder and other products.

In Australia, agroforestry is practised in many parts by planting the trees in farms [Southern Wet (>550 mm); Southern Dry (275-550 mm); East Coast Tropics and Sub-Tropics (>800 mm); Northern Savanna and Semi-Arid (>275 mm)]. The trees are planted in pastures, tree belts and woodlots for the following reasons:

- Hardwood sawlogs
- Softwood sawlogs
- Pulpwood
- Bioenergy
- *In situ* fodder
- Hardwood carbon plantings
- Softwood carbon plantings

LIST OF TREES FOR AGROFORESTRY PLANTING IN AUSTRALIA (Source: RIRDC 2008)

- Hardwood Sawlogs: *Eucalyptus globulus, E. nitens, E. saligna, E. cladocalyx, E. camaldulensis, E. tereticornis, E. grandis, E. dunnii, E. pellita, E. pilularis, Corymbia citriodora, C. maculata*
- Softwood Sawlogs: *Pinus radiata, P. pinaster, P. brutia, P. canariensis*, PEExPCH (*P. elliottii* var. *elliottii* x *P. caribaea* var. *hondurensis*)
- Pulpwood: *Eucalyptus globulus, E. nitens, E. grandis, E. dunnii, E. camaldulensis, Acacia mangium, Pinus pinaster*
- Bioenergy: *Eucalyptus globulus, E. cladocalyx, E. tricarpa, E. grandis, E. camaldulensis, Acacia mangium*
- *In situ* Fodder: *Acacia saligna, Artiplex amnicola, Artiplex nummularia, Chamaecytisus palmensis*
- Hardwood Carbon Plantings: *Eucalyptus globulus, E. cladocalyx, E. grandis, E. camaldulensis*
- Softwood Carbon Plantings: *Pinus radiata, P. pinaster, P. brutia*, PEExPCH hybrid

AGROFORESTRY SYSTEMS IN NEW ZEALAND

Exotic pines, willows, poplars and cypresses are now common in rural New Zealand. Trees are more in evidence in the broad valleys and plains, with willows lining the banks of waterways, and pines, poplars and cypress forming tall shelterbelts that criss-cross the landscape. The first farmers to arrive on the treeless plains of Canterbury and Wairarapa quickly planted exotic trees for shelter around their homesteads and stockyards. Later they planted shelterbelts to protect their crops and fields.

RADIATA PINE: Radiata pine (*Pinus radiata*) is the most common tree planted on farmland. It gained popularity as a windbreak tree because it grows fast and can survive on a range of soil types. Its useful life as a shelterbelt tree is around 50 years. Radiata pine has been the preferred species planted on steep eroding hill country.

- *Radiata Pine Shelterbelt:* Radiata pines grown close together and trimmed into hedges are excellent for sheltering stock from cold winds that can cause high losses, especially at lambing time. Radiata pines grow quickly and require little maintenance apart from pruning.
- *Radiata Pine Woodlot*: Radiata pine is the most common species grown in small woodlots. The trees can be harvested and sold or used on the farm for fencing, construction or as firewood.
- *Radiata Pine Silvopasture*: By growing trees with a wide enough distance between them for good pasture to develop, it is possible to raise a crop of trees on the same land that stock graze. However, the wood from trees grown in these situations is of a lower quality than that of trees grown in a pine plantation. The typical examples include: Rows of radiata pine in pasture land; Lines of widely spaced radiata pine plantations on a farm raising milk cows; Low density plantation of radiata pine on prairie (100 trees per hectare); Hedge of radiata pines pruned to 6 m.

MACROCARPA AND OTHER CONIFERS: Large spreading macrocarpas (*Cupressus macrocarpa*) are a common sight in rural New Zealand, usually growing alongside homesteads and farm buildings. Macrocarpa (also known as Monterey cypress) was brought to New Zealand in the 1860s, and planted for shelter. It grows well throughout lowland New Zealand on fertile and moderately fertile sites. The cultivation of trees in farms is limited by canker disease.

- *Macrocarpa Shelter* and *Woodlot*: Macrocarpa is more tolerant of wind and salt spray than radiata pine, and is the preferred shelter tree for coastal farms. It has spreading branches with dense foliage, so stock can shelter and remain dry under a macrocarpa hedge. Macrocarpa is highly sought after as a decorative and building timber. It is durable outdoors and can be used without any preservative treatment. Most macrocarpa timber in New Zealand has come from old, untended woodlots and shelterbelts. Some plantations were established by farm foresters during the 1970s and 1980s, but, as trees succumbed to canker, the species has fallen out of favour with growers.
- *Other Cypress Species*: There are about 18 species of cypress in New Zealand. Mexican or lusitanica cypress (*Cupressus lusitanica*) is fairly resistant to cypress canker and is now the preferred cypress for farms. Hybrid cypresses show some resistance too, and Leyland cypress (*Cupressocyparis leylandii*) - a hybrid between macrocarpa and *Chamaecyparis nootkatensis* - is also grown in farms.
- *Douglas Fir*: Douglas fir (*Pseudotsuga menziesii*) is the second-most common plantation tree in New Zealand. Its timber is highly regarded for building and construction work. A very hardy tree, it is the preferred species for farm forestry in the foothills of the Southern Alps and upland Southland - regions that are too cold for radiata pine.
- *Japanese Cedar*: Japanese cedar (*Cryptomeria japonica*) is a fast-growing, slender conifer, which is sometimes used for shelter, especially in northern New Zealand.

POPLARS AND WILLOWS: The poplars that now grace erosion-prone hillsides for soil conservation are most likely to be hybrid clones of various parentages or the Chinese poplar (*Populus yunnanensis*) resistant to poplar rust. Willows have been extensively used for river control, stabilizing river and stream banks. Willow hybrids and hybrid poplars are used for river control work, as well as for soil conservation on hill slopes. Selected poplars and willows are also used for fast-growing shelter belts, especially around orchards and market gardens. The main species used are *Salix matsudana* and its *alba* hybrid, and black poplar (*P. nigra*) hybrids are also common. Willows and poplar foliage can provide nutritional fodder for stock during droughts.

EUCALYPTS: Sydney blue gum (*Eucalyptus saligna*) was widely planted in the upper North Island during the 1970s. Mountain ash (*E. regnans*) - one of the tallest trees in the world - was planted in cooler areas around the central North Island in the 1970s and 1980s. Alpine ash (*E. delegatensis*) is also grown in cold districts. Since the late 1990s brown barrel (*Eucalyptus fastigata*) has been the preferred gum species and healthiest of the eucalypts, as it has been least attacked by pests and is tolerant of cold conditions. Shining gum (*Eucalyptus nitens*) has been planted in steep gullies. Shining gum can be used for firewood or for hardwood timber. Apart from providing an extra source of revenue to the farm, the woodlot helps to prevent soil erosion, which protects the quality of the water.

WATTLES: Various tree wattles (*Acacia* species) are grown in New Zealand. On farms, most were originally planted for firewood or quick-growing shelter. Black wattle (*A. mearnsii*) was originally introduced for the tannin its bark produces. However, it was attacked by a fungus that causes brown galls to develop on its branches and it became uneconomic to grow for the tanning industry. Blackwood (*A. melanoxylon*) produces one of the finest woods for cabinet and furniture making, and is increasingly planted in farm woodlots. It needs shelter from strong winds and year-round moisture. It is suitable for planting in gullies and beside streams, and can be planted in light scrub.

OAK: English oaks (*Quercus robur*) were first planted in the 1820s and have been grown throughout the country. They thrive in fertile soils with year-round moisture, but do not grow well in windy environments. Although occasionally grown as shade trees in paddocks, they are more commonly planted along farm driveways.

■■■■

23 AGROFORESTRY SYSTEMS OF INDIA

INTRODUCTION

In India, agroforestry has been a way of life and social, religious and cultural ethics have been closely linked with the planting and protection of various trees. Tree worship is still continued in many societies of India. Tree biodiversity has been preserved in many areas because of its role in farming and other land use systems. However, with increase in demographic pressure resulting in over exploitation of resource base leading to degradation and vast gap between demand and supply of woody produces, agroforestry has once again been identified as the most appropriate land use option. India has a long tradition of agroforestry and many different types of indigenous agroforestry systems can be found in different parts of the country. Most of these systems are, however, very location specific and information on them is mostly anecdotal. Therefore, their benefits have remained vastly underexploited. Trees play an important role in ecosystem in all terrestrials and provide a range of products and services to rural and urban people. As natural vegetation is cut for agriculture and other types of development, the benefits that trees provide are best sustained by integrating trees into agricultural system - a practice known as agroforestry. Farmers have practised agroforestry since ancient times. Agroforestry focuses on the wide range of trees grown on farms and other rural areas. Among these are fertilizer trees for land regeneration, soil health and food security; fruit trees for nutrition; fodder trees for livestock; timber and energy trees for shelter and fuel wood; medicinal trees to cure diseases and trees for minor products viz. gums, resins or latex products. Many of these trees are multipurpose, providing a range of benefits.

According to the 2011 report of the Forest Survey of India, the forest and tree cover of the country is 78.29 million ha, which is 23.81 per cent of the geographical area of the country. The forest cover in the hilly districts is only 39.74 per cent compared with the desired 66 per cent area. The National Agriculture Policy of 2000 emphasized the role of agroforestry for efficient nutrient cycling, nitrogen fixation, organic matter addition and for improving drainage and underlining the need for diversification by promoting integrated and holistic development of rainfed areas on watershed basis through involvement of community to augment biomass production through agroforestry and farm forestry. The Task Force on Greening India for Livelihood Security and Sustainable Development of Planning Commission in 2001 has also recommended that for sustainable agriculture, 10 million ha of irrigated and 18 million ha of rainfed lands to be put under agroforestry. This all will, besides ensuring ecological and economic development, provide livelihood support to about 350 million people. The practice of agroforestry can help in achieving these targets. Gap of demand and supply of forest produce in India is widening and forests are unable to fulfill the demand. Agroforestry can play an important role in filling this gap and conservation of natural resources.

The origin of agroforestry practices, i.e. growing trees with food crops and grasses, is believed to have been during Vedic era (Ancient period, 1000 BC) the agroforestry as a science is introduced only recently. The systematic research in agroforestry geared up after the establishment of the International Council for Research in Agroforestry (ICRAF) in 1977 which is now known as "World Agroforestry Centre". In India, organized research in agroforestry was initiated in 1983 by the establishment of All India Coordinated Research Project (AICRP) on Agroforestry by ICAR at 20 centres and later establishment of the National Research Centre for Agroforestry at Jhansi in 1988. At present, 39 centres of agroforestry are working in the country.

Table 23.1: Main Characteristics and Agroforestry Emphasis in Five Agro-Ecological Zones of India (Nair and Dagar 1991)

Characteristics	Himalayan Region	Indo-Gangetic Alluvial Plains	Arid and Semi-Arid Region	Humid and Sub-Humid Region	Tropical Coastal Island Region
Climate	Varies from temperate cold alpine to sub-tropical; rainfall 8 to 350 cm in west, 200-400 cm in east	Very hot summer and cold winter; rainfall 30 to 200 cm; ground frost common in winter season	Dry cold to very hot sub-humid; rainfall in arid 10-65 cm; in semi-arid 70-125 cm	Humid to sub-humid; rainfall 100 to 400 cm, Monsoonic	Tropical humid lowlands; rainfall 60-310 cm
Geographical Spread	Extreme north-eastern to north western regions covering parts of Sikkim, Arunachal Pradesh, Assam, Nagaland, Manipur, Megha-laya, Mizoram, U.P., Himachal Pradesh, Jammu and Kashmir; elevation 1000 to 3000m	Below foot hills of Himalaya covering about 47 million ha of Punjab, Haryana, Delhi, U.P., Bihar, parts of W. Bengal; elevation 150 to 600m.	Spreads over Rajasthan, Gujarat, Punjab, Haryana; parts of U.P., M.P., A.P., Maharashtra, Karnataka and Tamil Nadu	Spreads over parts Assam, Meghalaya, Mizoram, Tripura, Bengal, Odisha, M.P., and the southern states; elevation 150 to 1500m	Coastal region of West Bengal, Odisha, A.P., Tamil Nadu, Kerala, Karnataka, Maharashtra and the Islands
Soil and Vegetation	Light to heavy textured, immature, mostly inceptisols; Alpine scrub, Sub-alpine coniferous, Tropical evergreen forests, Alpine grasslands	Mostly alluvial with patches of saline and alkali soils; Planted woodlots of *Eucalyptus*, *Populus*, *Acacia*, etc., Xerophytic trees and shrubs occur naturally in degraded soils	Sandy to clay loam; Xerophytic thorny scrub forests and woodlands and Psammophytic scrub vegetation, grazing lands	Alluvial to clay loams, Oxisols, Alfisols; Evergreen, semi-evergreen and deciduous forests	Entisols, Oxisols, Inceptisols, Alfisols; Evergreen, semi-evergreen, moist deciduous, littoral and mangrove forests
Main Land Use Systems	Grazing lands, forestry, horticul-tural and forestry species with agricultural crops, shifting cultivation in North East	Agricultural crops (wheat, pulse, sugarcane, oil seeds), comm-ercial woodlots, MPTs with crops and	Trees and shrubs in crop fields, along boundaries, permanent grazing lands having trees and shrubs with stunt growth	Cash crops (tea and large cardamom), agricultural crops, shifting cultivation, forests	Plantation crops and multistoried cropping, field crops, fish culture, forestry

Characteristics	Himalayan Region	Indo-Gangetic Alluvial Plains	Arid and Semi-Arid Region	Humid and Sub-Humid Region	Tropical Coastal Island Region
	States	along boundaries, degraded grazing lands			
Main Land Use and Ecological Problems	Excessive deforestation, degraded lands, over-grazing, land sliding and shortage of fuel and fodder	Soil salinity, waterlogging and decline in soil fertility due to over-cultivation, fuel and fodder shortage	Drought, overgrazing, salinity, extension of sand dunes, low water table, soil erosion, degraded lands	Deforestation, soil erosion, soil acidity and consequent problems, decline in soil fertility, shortage of fuel and fodder, shortening of fallows	Deforestation and degradation of environment, coastal erosion, acid soils and related soil problems, soil erosion along slopes and shortage of fodder
Major Agroforestry Emphasis	Planting woodlots on hills, contour farming, blended with MPTs, management of pastures, raising orchards	Soil reclamation using MPTs, NFTs, suitable grasses, growing NFTs/ shrubs with crops along field boundary, raising woodlots on degraded lands	Raising shelterbelts, windbreaks, growing NFTs in field, raising woodlots (NFTs) on degraded lands	Intercropping with nitrogen-fixing trees, alley cropping, improved fallows	Afforestation of coastal regions with littoral and mangrove species, aquaculture with mangroves, multistoried cropping system, alley cropping

HISTORY OF AGROFORESTRY PRACTICES IN INDIA

In about 700 BC, man changed from a system of hunting and food gathering to food production. Shifting cultivation in India is prehistoric and partly a response to agro-ecological conditions in the region. Horticulture as co-existent with agriculture is found to have been prevalent in India from early historic period (500 BC to 1st century AD) when a certain amount of share in garden crops started to have been enjoyed by the king for providing irrigation. Some stray references occur in different texts of the Vedic literature. The cultivation of date-palm, banana, pomegranate, coconut, jujube, aonla, bael, lemon and many varieties of other fruits and requirement of livestock in agriculture and mixed economy of agriculture and cattle-breeding may be traced in proto-history chalcolithic periods of civilization. But in India, the plant husbandry (intentional sowing or planting for production of desirable plants or plant domestication) happened to start under progressively arid climatic zone from about pre-Neolithic period. The role of many common trees such as khejri (*Prosopis cineraria*), aswattha (*Ficus religiosa*), palasa (*Butea monosperma*) and varana (*Crataeva roxburghii*) in Indian folk-life has been mentioned in ancient literature of Rig Veda, Atharva Veda and other ancient scriptures. Traditional agroforestry systems manifest rural people's knowledge and methods to benefit from complimentary uses of annuals and woody perennials on the sustained basis. It also indicates that farmers have a closer association with trees than any other social group and promoters of forests.

The farmers and land owners in different parts of the country integrate a variety of woody perennials in their crop and livestock production fields depending upon the agro-climatic conditions and local needs. Most of these practices are, however, very location specific and information on these are mostly anecdotal. Therefore, their benefits have remained vastly under exploited to other potential sites. It has now been well recognized that agroforestry can address some of the major land use problems of rainfed and irrigated farming systems in India, and that a great deal can be accomplished by improving indigenous systems. With the current interests in agroforestry worldwide, attempts are being made in India to introduce agroforestry techniques using indigenous and exotic multipurpose and nitrogen-fixing woody perennials.

TRADITIONAL AGROFORESTRY SYSTEMS OF INDIA

Agroforestry is widespread in all ecological and geographical regions of India. The systems and practices of agroforestry vary enormously in their structural complexity and species diversity, their productive and protective attributes, and their socio-economic dimensions. They range from apparently "simple" forms of shifting cultivation (especially in the northeastern parts of the country) to "sophisticated" hedgerow intercropping systems; from systems involving sparse stands of trees on farm lands (e.g., "Khejri" tree (*Prosopis cineraria*) in the arid regions of western India) to high-density, complex multistoried homegarden of the humid lowland (as in Kerala); from systems in which trees play a predominantly "service" role (e.g., shelterbelts) to those in which they provide the main "commercial" production [e.g., intercropping with plantation crops). Reviews and descriptions are available for major agroforestry systems in the country (e.g., ICAR (1981); Singh (1987); Tejwani (1987)] as well as of individual systems and practices [e.g., Nair (1983, 1989 & 1993); Kumar and Abrol (1984); Nair and Sreedharan (1986); Shankarnarayan et al (1987); Mittal and Singh (1989); Singh et al (1988, 1989a & 1990); Rao et al (1990); Sharma et al (1990)]. An assessment of these voluminous descriptions of the many indigenous agroforestry systems, which are mostly qualitative, shows that the systems are extremely site specific with very few examples of their extrapolatability. Table 23.2 summarizes the major types of agroforestry systems and the nature of their main benefits and role in different agro-ecological zones.

Table 23.2: Major Types of Agroforestry Systems and Benefits in different Agro-Ecological Zones of India

Major Types of Existing AF Systems	Benefits of Agroforestry Practice	References
Himalayan Region		
1. Fruit trees in combination with agricultural crops and fodder trees	Production of fruits and fodder	Dadhwal et al (1989); Toky et al (1989); Singh (1987); Singh et al (1989b)
2. Fodder trees with pastures	Fodder	
3. Trees and grasses for soil conservation	Soil conservation, fuel, timber, fodder	
4. Cash crops in forests/ shade of trees	Cash/ Shade	
Indo-Gangetic Plains		
1. Trees for soil reclamation (sodic/ saline/degraded soils)	Reclamation of degraded soil, production of fuel wood & fodder	Grewal and Abrol (1986); Singh et al (1988, 1989a & 1990); Tomar and Gupta (1985); ICAR (1981); Ahmed (1989)
2. Fodder trees in degraded grazing lands	Fuel, fodder	
3. Trees on boundaries of agricultural fields	Fuel, timber, fodder, cash, shade, minor products	
4. Fodder banks	Soil conservation, fodder, fuel	
5. Block plantation	Cash, lumber	

Major Types of Existing AF Systems	Benefits of Agroforestry Practice	References
Arid and Semi-Arid Region		
1.MPTs in agricultural fields	Fodder and fuel wood, cash, MFP	Tejwani et al (1996); Bhimaya et al (1961); Ahuja et al (1978); Muthana et al (1985); Shankaranarayan et al (1987); Dagar (1978 & 1987); Grewal and Abrol (1986); Yadav (1980); Singh et al (1988)
2.Trees for soil conservation	Reclamation of soil fuel, fodder, MFP	
3.Windbreaks	Sand dune stabilization	
4.Fodder banks	Soil conservation, fodder, fuel	
5.Trees in pasture lands	Fodder, fuel	
Humid and Sub-Humid Region		
1.Homegardens	Production of multiple outputs	Nair (1979, 1983, 1989, 1990 & 1993); Nair and Sreedharan (1986)
2.Plantation-crop combinations	Cash economy	
3.MPTs in agricultural fields	Minor products	
4.Trees and grasses for soil conservation	Soil conservation, fodder, fuel	
Coastal and Island Region		
1. Plantation crop combinations	Production of multiple outputs, cash	Nair (1979, 1983, 1989, 1990); Nair & Sreedharan (1986); ASA (1987); Dagar (1982); Sharma et al (1990)
2. Multistoried cropping	-do-	
3. Homegardens	-do-	
4. Trees with aquaculture	Fish, fuel, fodder	
5. Trees on pasture lands	Fodder, soil conservation	
6. Mangrove plantation	Shore/Env. protection, fuel, fodder	
7. Sheterbelts and Windbreaks	Shore/Beach stabilization	
8. Boundary trees on agricultural fields	Fodder, fuel, minor products	

The common denominator and scientific foundation of all these various agroforestry systems is the multipurpose tree (MPT). The success of agroforestry will depend upon the extent to which the productive, protective and service potential of MPTs is understood and exploited (through research) and realized (through development and extension efforts). As can be expected, the number, type, distribution and uses of MPTs in agroforestry systems in India are extremely variable. Even a listing, let alone characterization, of the various MPTs used in the different agroforestry systems in various agro-ecological regions will entail several hundred species.

SHIFTING CULTIVATION (SLASH AND BURN SYSTEM)

It refers to farming system in north-eastern high rainfall areas in which land under natural vegetation (usually forests) is cleared by slash and burn method, cropped with common arable crops for a few years and then left unattended when natural vegetation regenerates. Traditionally the fallow period is 10-20 years but in recent times it is reduced to 2-5 years in many areas. Due to the increasing trends of population pressure, the fallow period is drastically reduced and system has degenerated causing serious soil erosion depleting soil fertility resulting to low productivity. In northeastern India, many annual and perennial crops with diverse growth habits are being grown.

At times annual crops such as potato, rice, maize and ginger are grown in monoculture or mixed culture along with *Pinus kesiya*. Another important attribute of the system is secondary succession of vegetation during fallow period. The tree species which may be considered suitable for afforestation of fallow areas or to intercrop with arable crops must be fast growing preferably nitrogen-fixing and must efficiently recycle available nutrients within the system shortening time required to restore fertility. These may include species of *Acacia*, *Albizia*, *Alnus*, *Casuarina*, *Erythrina*, *Faidherbia*, *Gliricidia*, *Inga*, *Leucaena*, *Parkinsonia*, *Pilhecellobiuin*, *Prosopis*, *Robinia* and *Sesbania*. Thus intercropping under or between fast-growing trees in a fallow phase is one of the approaches while finding alternative to shifting cultivation. A farming-system approach based on the watershed management has been advocated as an alternative to shifting cultivation.

TAUNGYA SYSTEM

The Taungya system is like an organized and systematically managed shifting cultivation. The word is reported to have originated in Myanmar (Burma) in 1850s. "Tauang" means hill and "Ya" means cultivation, i.e. hill cultivation. It involves cultivation of crops in forests or forest trees in crop-fields and was introduced to Chittagong and Bengal areas in colonial India in 1890. Later it had spread throughout Asia, Africa and Latin. The taungya system was introduced into India by Brandis in 1890 and the first taungya plantations were raised in 1896 in North Bengal and the north-eastern hill region. In southern India, the system is called 'kumari'. It is practised in areas with an assured annual rainfall of over 1200-1500 mm.

HOMEGARDENS/HOMESTEADS CULTIVATION

A homestead is an operational farm unit in which a number of crops including tree crops are grown with livestock, poultry and/or fish production, mainly for the purpose of meeting the routine basic needs of the farmer. It is an age old practice in coastal states particularly in Kerala, Tripura, Assam, North-eastern states and parts of West Bengal and Andaman and Nicobar Islands. Although the homegardens appear to be a mixture of trees, shrubs and herbs, a certain general pattern seems to exist. There is wide variation in the intensity of trees, species and crops based on the size of holding, needs of the people residing in homesteads and micro-climate. Coconut, arecanut, guava, mango, citrus, tamarind, jack fruit, papaya, banana, *Moringa*, *Sesbania*, custard apple and many multipurpose trees are the major trees found grown in homegardens.

Domestic animals and poultry are the main components of homesteads. Therefore, sometimes forages like stylo (*Stylosanthes guianensis*/ *S. hamata*), guinea grass, Guatemala grass, napier grass and others are also grown frequently. The components are so intimately mixed in horizontal and vertical strata as well as in time that a complex interaction exists among soil, plants, other aboveground and below ground components, nutrition and environmental factors. There is a critical competition both for light and nutrition. The holder chooses his crops and crop combinations based on his wisdom, needs and perceptions acquired over generations of experience.

Many multipurpose trees having productive and protective functions are integrated into the system in different spatial and temporal arrangements. Most of these trees closely interact with agricultural crops when grown. These trees include teak (*Tectona grandis*), jack fruit (*Artocarpus* spp.), *Casuarina equisetifolia*, *Mangifera india*, *Ceiba pentandra*, *Leucaena leucocephala*, *Grevillea robusta*, *Bambusa arundinacea*, *Erythrina variegata* and *Gliricidia sepium* (both good support to black pepper). *Thespesia populnea* is common in low lying homesteads and the wood is commonly used for agricultural implements. Mangroves form an essential part of homesteads of backwater areas in lowlands. These are commonly used as fuelwood. Coconut and pandanus (*Pandanus tectorius*) can be commonly seen near canals and backwaters. Palmyrah tree (*Borassus flabellifer*) is common multipurpose palm in coastal Andhra Pradesh. Fish and shrimp culture in backwater channels and in association with mangroves is the main activity in homestead. Thus, multitude of crop species in the homesteads helps to satisfy primary needs of the farmer such as food, fuel, fodder, timber and cash. This, in spite of high intensity of cropping, also helps to conserve fertility by nutrient cycling or organic manuring or mulching and increased microbial activity in the rhizosphere of crops. Among social benefits are high family labour utilization and risk minimization. This system also helps in checking soil erosion (due to high intensity of vegetation cover), environmental health and conserving biodiversity.

PLANTATION BASED AGROFORESTRY SYSTEMS

Modern commercial plantation crops like rubber, coffee, poplar, eucalypts and oilpalm represent a well managed and profitable stable land use activity in tropics. The scope for integrative practices involving plant associations in these commercial plantations is limited, except during the early phases of plantation when some intercropping is feasible. The commercial production of these plantations is aimed at having a single commodity. Some of the plantation crops like coconut, palms, etc., have been cultivated since very early times but their economic yield remained low. However, the research attention and commercial yields of these crops have increased substantially in recent times.

Contrary to common belief, a substantial proportion of tropical plantation crops is grown by small farmers. Most of the coconut production in India and other countries come from small holdings, in which coconut tree is integrated with a large number of annual and perennial crops like clove, cardamom, coffee, cacao, cassava, yams, fodder grasses and legumes. Grazing under coconut and cashew nut is also common. In India, small holders grow cashew-trees (in wider space) with other crops.

In the last two decades, most of the data have come from coconut-based systems in India related to intercropping under coconuts, integrated mixed farming in small holdings, grazing under coconuts, factors favouring intensification of land use with coconuts and multi-storey tree-gardens. Important cereals grown with coconut include rice, finger millet and maize; pulses such as pigeon pea, green gram, black gram, cowpea and soybean; oilseed crop such as groundnut; root crops such as sweet-potato, yams, elephant foot-yam and taro; spices and condiments such as ginger, turmeric, cinnamon, clove, chilli and black pepper; fruits like pineapple, mango, banana and papaya; cash crops such as cotton and sugarcane; plantation crops such as arecanut, cacao and coffee; many trees such as species of *Erythrina*, *Ceiba*, and *Cordia* find a place in the system.

SCATTERED TREES ON FARM LANDS

The practice of growing agricultural crops under scattered trees on farm lands is old and does not seem to have changed for centuries. Though world-wide list of such trees is long, some of them have received more attention than others e.g. *Prosopis cineraria* in north-western India and Poplars in north India. The species diversity in these systems is very much related to ecology. With increase in rainfall, the species diversity and system complexity increases. Thus, there is a proliferation of more diverse multistoried homegardens in humid areas and less diverse, two tiered canopy of configurations (trees + crop) in drier areas.

Trees are grown scattered in agricultural fields for many uses such as shade, fodder, fuel wood, fruit, vegetables and medicinal uses. Some of the practices are very extensive and highly developed in India. For example, growing *Prosopis cineraria* and *Ziziphus* spp. in arid areas; *Acacia nilotica* in Indo-Gangetic plains; *Grewia optiva* and other tree species in hills of Uttarakhand and HP; *Eucalyptus globulus* in southern hills of Tamil Nadu and *Borassus flabellifer* in peninsular coastal regions. There are strong convictions for the acceptance of these trees on agricultural fields since time immemorial.

Farmers retain trees of *Acacia nilotica*, *Acacia catechu*, *Dalbergia sissoo*, *Mangifera indica*, *Ziziphus mauritiana* and *Gmelina arborea* are preferred in Gujarat with crops. Farmers in sub-humid terai region of Uttarakhand and Uttar Pradesh prefer *Dalbergia sissoo*, *Syzygium cumnii* and *Trewia nudiflora*. In Bihar, *Dalbergia sissoo*, *Litchi chinensis* and *Mangifera indica* are frequently grown on fields. Other most common trees found on farmer's fields are *Azadirachta indica*, *Moringa oleifera*, *Tamarindus indica*, *Ceiba pentandra*, *Anacardium occidentale*, *Cocos nucifera* and fruits like banana, custard apple, guava and pomegranate.

TREES ON FARM BOUNDARIES

Trees grown in agricultural fields are also often and usually grown on farm boundaries. In northern parts of India, particularly in Haryana and Punjab eucalypts and poplars are commonly grown along the field boundaries or bunds of paddy fields. Other trees which are found grown as boundary plantations or live hedge include *Dalbergia sissoo* and *Prosopis juliflora*. Farmers of Sikkim grow bamboo (*Dendrocalamus*, *Bambusa*, etc.), all along the irrigation channels. In coastal areas of Andhra Pradesh, *Borassus* is the most frequent palm. In Andamans, farmers grow *Gliricidia sepium*, *Jatropha* spp., *Ficus*, *Ceiba pentandra*, *Vitex trifolia* and *Erythrina variegata* as livehedges. At many places, succulents like *Agave* and many cactoids are grown as common live-fence. Many of the boundary plantations also help as shelterbelts and windbreaks, particularly in fruit orchards. In Bihar, *Dalbergia sissoo* and *Wendlandia exserta* are most common boundary plantations.

WOODLOTS

In many parts, farmers grow trees in separate blocks as woodlots along with agricultural fields. Now the practice is expanding fast due to shortage of fuelwood and demand of poles or pulpwood in industry. For example, bamboo poles are in great demand for orange orchards in Nagpur area and *Eucalyptus* and *Populus* for match industries. These days woodlots are being raised mostly on large farms due to increase of labour costs and labour management, lack of irrigation and risk of crop investments. Woodlots of *Casuarina*, bamboo, poplar, eucalypts, *Leucaena leucocephala*, red sandal (*Pterocarpus santalinus*) and *Dalbergia sissoo* have become popular in many parts of the country. In recent times, due to increase in population pressure these woodlots have shrunk at a fast rate.

SYSTEMS FOR SOIL CONSERVATION OR AMELIORATION

About 150 million ha of land in India is subjected to serious wind and water erosion, of which 69 million ha are critically affected. About 4 million ha is suffering from degradation due to ravines and gullies; 11.3 million ha as riverine land, coastal sandy areas and steeply sloping lands and more than 9 million ha is salt affected. The deep and narrow gullies are best controlled by putting them to permanent vegetation after closure to grazing. Afforestation with suitable tree species like *Acacia nilotica*, *Azadirachta indica*, *Butea monosperma*, *Prosopis juliflora*, *Dalbergia sissoo*, *Tectona grandis*, *Bambusa* spp., *Dendrocalamus* and other adaptable species such as grasses like *Dichanthium annulatum*, *Bothriochloa pertusa*, *Cynodon dactylon* and *Sehima nervosum* will help in stabilizing ravines and gullies and checking their spread.

SHELTERBELT

Arid regions witness very high wind velocity throughout the year. Farmers build shelterbelt (kana bundi) by either small dead wood or local vegetation to check wind velocity within safer limits. *Crotalaria burhia*, *Leptadenia pyrotechnica* and *Aerva pseudotomentosa* bushes are planted in 20-25 m apart in rows across the wind direction. Between the lines of these shrubs grasses such as *Cenchrus ciliaris*, *C. setigerus*, *Lasiurus sindicus*, etc., are planted on leeward side of each break. This permanent vegetation helps to accumulate the sand near them which is again spread in the field. This also helps to increase the crop yields along the lines.

TREES ON RANGELANDS

Salvadora oleoides, *Capparis decidua*, *Acacia nilotica*, *A. leucophloea*, *P. cineraria* and *Prosopis juliflora* also are most frequent trees on common community grazing lands. In tropics, coconut is most common tree on pasture lands. Cattle raising usually involves grazing on these pastures. In some cases special, fodder plants including legumes are also cultivated.

AGROFORESTRY SYSTEMS IN DIFFERENT AGRO-ECOLOGICAL ZONES OF INDIA

Crop growth and yield are the function of many factors like weather, soil type and its nutrient status, management practices and other inputs available. Of these, weather plays an important role, probably more so in India where aberrant weather such as drought, flood, etc., is a rule rather than an exception. Efficient crop planning requires proper understanding of agro-climatic conditions. This calls for collection, collation, analysis and interpretation of long-term weather parameters available for each region to identify the length of the possible cropping period taking into consideration the availability of water. The agro-climatic zone planning aims at scientific management of regional resources to meet the food, fiber, fodder and fuel wood without adversely affecting the status of natural resources and environment. While assessing the resource base required is the holistic perspective, development will have to be achieved through an appropriate mix of crop production and allied activities including horticulture, forestry, animal husbandry and agro-processing etc. Improved farming systems and technologies will replace traditional crop production approach to minimize regional variations in rural incomes.

One of the essential characteristics of agroforestry systems is their site specificity. Although the distinct agroforestry practice that constitutes the innumerable agroforestry systems in different places are only a few, the same practice takes various forms in different places depending on site-specific biophysical and socio-economic conditions (Nair 1989 & 1990). The ecology and climate of India are extremely varied so that almost all agro-ecological zones in the world can be found in one part or the other in the country. Obviously, this enormous diversity makes it difficult to classify the country into a few distinct agro-ecological zones. For reviewing the status of agroforestry in India, Singh (1987) identified major agro-ecological regions.

The government of India in 1989 has classified the country into agro-ecological regions on the basis of climate, soil type, topography, water resources, and irrigation facilities. The All India Coordinated Research Project on Agroforestry of the Indian Council of Agricultural Research (ICAR) identified major agro-ecological zones in India (ICAR 1990). Thus, various agro-ecological classification schemes have been adopted for different purposes. Understandably, any classification will have limitations and drawbacks; there cannot be a universally acceptable demarcation of agro-ecological zones in a vast and diverse country like India. With the 328.2 million hectares of the geographical area, the country presents a large number of complex agro-climatic situations. The five major agro-ecological zones (AEZ) of India are:

1. Himalayan Region
2. Indo-Gangetic Plains Region
3. Arid and Semi-Arid Region
4. Humid and Sub-Humid Region
5. Coastal Plains and Island Region

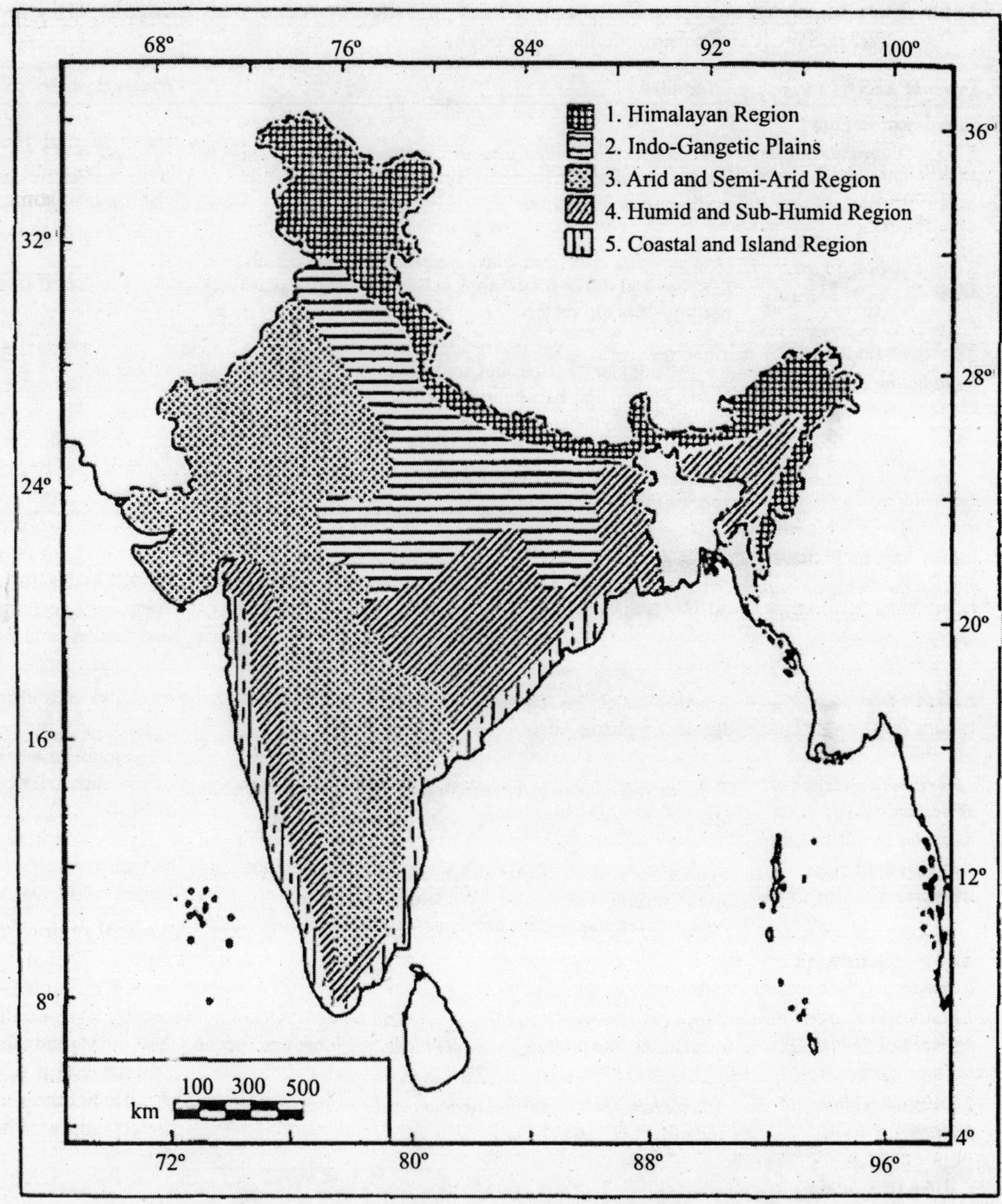

Fig. 23.1: Agro-Ecological Zones of India (Source: Nair and Dagar 1991)

Table 23.3: Main Agroforestry Systems practised in different AEZs with Examples (Dhyani et al 2007; CSWCRTI-Training Manual, Dehradun)

Types of AFS	Examples	Major Benefits
Himalayan Region		
1.Trees in agricultural fields or on field bunds	*Grewia optiva* and other trees in Jammu & Kashmir, Himachal Pradesh, Garhwal & Kumaon Himalaya, Sikkim and other north-eastern hill region.	Production of food, fruits, fodder etc., and stabilization of bunds
2.Intercropping in fruit orchards	Orange, other citrus and guava intercropped with cereals, tuberous and rhizomatous crops in Sikkim, Meghalaya and other north-eastern hill region.	Production of fruits, food etc.
3.Plantation crops under shade of trees	Large cardamom, coffee, etc., under alder in Sikkim and Nagaland; tea under leguminous trees in Assam and West Bengal; betel vine and black pepper on arecanut and other trees in Meghalaya and other north-eastern states.	Food, spices, fuel wood, timber etc.
4.Silvipastoral practices (fodder trees with pastures)	Tree leaf fodder from *Grewia* spp., *Celtis australis* and other species in the Himalayas.	Fodder, fuelwood etc.
Indo-Gangetic Plains		
1.Trees for soil reclamation (sodic /saline/degraded soils)	*Prosopis chilensis, Acacia nilotica, Parkinsonia aculeata* and other tree species for problem soils.	Reclamation of degraded lands, production of fuel and fodder
2.Fodder trees in degraded grazing lands	*Albizia lebbeck, Bauhinia purpurea, Dalbergia sissoo* and other species in grazing lands.	Fuel, fodder, timber
3Trees on boundaries of agricultural fields	*Populus* spp., *Tamarindus indica, Ceiba pentandra, Eucalyptus* spp., *Dalbergia sissoo*, etc.,	Fuel, fodder, timber, shade, cash, minor products
4.Fodder banks and Woodlots	Fodder trees - *Azadirachta indica, Melia azaderach, Albizia* spp., *Syzygium cuminii*, etc.; Woodlots - *Casuarina equisetifolia, Eucalyptus, Bamboos, Dalbergia sissoo*, etc.,	Fodder, soil conservation, fuel, etc.,
Arid and Semi-Arid Region		
1.Multipurpose trees in agricultural fields	*Prosopis cineraria, Ziziphus* spp., *Acacia senegal, Ailanthus excelsa, Acacia nilotica, A. leucophloea* and other tree species.	Fodder, fuelwood, fruits, cash, shade and minor tree products
2.Trees for reclamation of degraded soils	*Albizia lebbeck, Casuarina, Acacia tortilis, Azadirachta indica* and other tree species.	Reclamation of soil, fuel. fodder, minor products.
3.Windbreaks and shelterbelts	*Acacia nilotica, A. senegal*, other *Acacias, Azadirachta indica, Cassia* spp., *Prosopis chilensis, Tamarix* spp. etc.,	Sand dune stabilization
4.Fodder banks	*Prosopis cineraria, Ziziphus nummularia, Salvadora persica, S.oleoides, Acacia nilotica*, etc.,	Fodder, fuel and soil conservation etc.,
5.Trees on rangelands (Silvopastoral system)	*Prosopis chilensis, Acacia senegal, A. tortilis, P. cineraria, Ziziphus* spp., *Azadirachta indica* etc.; trees in pasture lands.	Fodder, fuel, shade & timber

Types of AFS	Examples	Major Benefits
Humid and Sub-Humid Region		
1.Homegardens/ homesteads	Homegardens in Kerala, Assam, W. Bengal, etc., have a mixture of trees, shrubs herbs, etc.; Coconut, arecanut, *Erythrina*, *Gliricidia* with black pepper, cacao, coffee, cassava and other cash crops.	Production of multiple outputs
2.Multi-tier system or plantation crops combination	Coconut, arecanut, *Erythrina* and other trees with coffee/ banana, pineapple/ papaya, cacao/ coffee and black pepper/ betel vine /large cardamom.	Cash and multiple outputs
3.Multipurpose trees in agricultural fields	*Acacia auriculiformis*, *Aegle marmelos*, *Albizia* spp., *Anogeissus latifolia*, *Anthocephalus chinensis*, *Artocarpus* spp., *Casuarina equisetifolia*, *Diospyros melanoxylon*, etc.	Fodder, minor products
Coastal and Island Region		
1.Plantation crop combination/ multistoried	Coffee under *Erythrina lithosperma*; cacao with coconut; black pepper on *Gliricidia* spp., *Grevillea robusta*; cardamom under *Toona ciliata*, *Artocarpus*, etc.,	Production of multiple outputs, cash
2.Trees with aquaculture	Composite fish culture in ponds and multipurpose trees in homesteads.	Fish, fuel, fodder, timber
3.Mangroves plantations, homesteads	*Acanthus ilicifolius*, *Avicennia officinalis*, *Cerbera odollam*, *Rhizophora conjugata*, etc.,	Shore protection, fuel, fodder, environmental protection
4.Shelterbelts and windbreaks	*Casuarina equisetifolia*, *Azadirachta indica*, *Prosopis chilensis*, *Acacia Senegal*, etc.,	Shore/beach stabilization Fodder, fuel, shade, minor products
5.Trees on boundaries of agricultural fields	*Aegle marmelos*, *Albizia* spp., *Azadirachta indica*, *Bamboos*, *Bombax malabaricum*, *Calliandra calothyrsus*, *Cassia* spp., etc.	

The research on conservation agroforestry based production systems envisage to conserve and improve the land and also optimize the combined productivity of the trees and the agricultural crops. Following are some of the most common and important agroforestry based production systems practised with emphasized crop combination which are enlisted for five agro-ecological zones of India:

AGRI-SILVICULTURE SYSTEM

Under this system, the land is used for the concurrent production of agricultural crops including fodder crops and fuel/ fodder trees and shrubs in combination. The major agroforestry practices of this system are alley cropping, shelterbelts and windbreaks, fuel woodlots, multipurpose trees on crop lands, etc.

SILVI-PASTORAL SYSTEM

In this system, the land is used for a combination of forest tree species (fuel/fodder trees) with forage grasses/ fodder crops for the management of domestic animals. The major practices are trees on pasture and range lands, plantation crops with pastures and livestock, live fences, fodder banks, etc.

AGRI-SILVI-POSTORAL SYSTEM

In this system, the land is managed for the concurrent production of agricultural crops with fuel/ fodder trees and grass/ forage species for rearing of domestic animals. Major practices are homegardens with animals, alley farming, woody hedge rows around homesteads, etc.

AGRO-HORTI-PASTORAL SYSTEM

A combination of horticultural crops (fruits and vegetables), fields crops and tree species (fuel/ fodder) are integrated in this system. The system may also include apiculture with trees, aqua-forestry and multipurpose tree lots.

Table 23.4: A list of Trees, Crops and Grasses in various Agroforestry Systems of different AEZs of India (Dhyani et al 2007, CSWCRTI-Training Manual, Dehradun)

Trees (Timber/ Fodder/ Fuel)	Fruit Trees	Grasses	Agricultural Crops
Himalayan Region			
Bauhinia variegata	Mandarin Orange, Guava,	*Chrysopogon fulvus*	Wheat
Grewia optiva	Peach, Apple, Mango,	*Eulaliopsis binata*	Maize
Celtis australis	Apricot, Almond, Lemon,	*Dichanthium annulatum*	Pepper
Albizia chinensis	Sweet Orange, Banana,	Hybrid Napier	Potato
Morus serrata	Papaya, Walnut, Cherry	*Panicum maximum*	Peas
Ficus roxburghii		*Cenchrus ciliaris*	Tomato
Toona ciliata		*Digitaria decumbense*	Cauliflower
Pistacia integerrima		*Heteropogon contortus*	Beans
Ficus palmata			Mustard
Melia azedarach			Sesamum
Bombax ceiba			Taramira
Prunus puddum			
Morus alba			
Indo-Gangetic Plains			
Azadirachta indica	Mango, Litchi, Guava, Bael,	*Diplachne fusca*	Sesamum
Dalbergia sissoo	Tamarind, Mulberry, Amla,	*Dichanthium annulatum*	Wheat
Acacia robusta	Ber, Lemon, Pomegranate,	*Cenchrus ciliaris*	Taramira
Eucalyptus hybrid	Orange, Papaya, Banana	*Mentha arvensis*	Pearlmillet
Populus deltoides			Mustard
Sesbania spp			Paddy
Tectona grandis			Potato
Prosopis juliflora			
Casuarina equisetifolia			
Terminalia arjuna			
Arid and Semi-Arid Region			
Prosopis cineraria			
Acacia nilotica	Mango, Lemon,	*Chenchrus ciliaris*	Sorghum
Acacia tortilis	Sapota, Custard Apple	*Dichanthium annulatum*	Pigeon pea
Albizia lebbeck	Coconut, Ber, Amla, Jamun,	*Panicum antidotale*	Black gram
Prosopis juliflora	Guava, Datepalm,	*Cenchrus setigerus*	Taramira
Holoptelea integrifolia	Phalsa, Pomegranate,	*Brachiaria ramosa*	Safflower

Trees (Timber/ Fodder/ Fuel)	Fruit Trees	Grasses	Agricultural Crops
Acacia senegal *Acacia albida* *Eucalyptus tereticornis* *Hardwickia binata*	Papaya, Sweet Orange	*Digitaria adscendens* *Dactyloctenium scindicum*	Pearl millet Castor Cluster bean Tobacco Cowpea
Humid and Sub-Humid Region			
Eucalyptus tereticornis *Acacia nilotica* *Acacia leucophloea* *Acacia planifrons* *Ailanthus triphysa* *Casuarina equisetifolia* *Sesbania grandiflora* *Pterocarpus marsupium* *Acacia auriculiformis* *Albizia falcataria* *Artocarpus heterophyllus* *Sesbania aegyptiaca* *Bauhinia purpurea* *Gmelina arborea* *Trema amboinensis* *Tamarindus indica* *Mussaendra macrophylla* *Gliricidium sepium*	Guava, Mango, Banana, Coconut, Arecanut, Amla, Cacao, Lemon, Sweet Orange, Moringa, Tamarind, Grape Vine	*Cenchrus ciliaris* Hybrid Napier Guatemala grass Guinea grass *Setaria* grass *Stylosanthes guianensis*	Maize Mustard Pigeon pea Safflower Cassava Elephant foot yam Turmeric Phaseolus Paddy Black gram Horsegram Cowpea Niger Groundnut
Coastal and Island Region			
Azadirachta indica *Cassia siamea* *Morus alba* *Pongamia pinnata* *Tamarindus indica* *Eucalyptus globulus* *Albizia lebbeck* *Sesbania sesban* *Acacia senegal* *Acacia albida* *Hardwickia binata* *Albizia procera* *Prosopis juliflora* *Dalbergia sissoo* *Ceiba pentandra*	Mango, Guava, Ber, Custard Apple, Papaya, Sapota, Mandarin, Sweet Orange, Lemon, Banana	Hybrid Napier *Chloris burnii* *Cenchrus ciliaris* *Dichanthium annulatum* *Stylosanthes hamata* *Cenchrus setigerus* *Setaria* grass Guinea grass	Sorghum Safflower Bengal gram Maize Cowpea Pearlmillet Pigeonpea Castor Clusterbean Horsegram

AGRI-HORTICULTURE SYSTEM

Under this production system, fruit trees and crops are grown together in various ways. Depending upon the pattern and configuration, these companion crops are known as intercrops, under planting, hedge row planting or alley cropping. Some of the common horticulture based production systems which are highly remunerative are: Agri-horticulture, Agri-silvi-horticulture, Silvi-horticulture, Horti-pastoral system.

In these systems, agricultural/ pastoral crops are normally grown between rows of fruit trees planted at a spacing of 5 to 7 m as intercrops. The agricultural crops provide seasonal revenue while fruit trees are managed for 30-35 years giving regular returns from fruits and in some cases fuelwood from pruned wood and fodder. Several kinds of shade loving crops (turmeric, ginger, etc.), are also planted under the canopy of fruit trees. However, shade is not always desirable for agricultural crops such as maize, sorghum, ground nut, etc.

MULTIPURPOSE FOREST TREE PRODUCTION SYSTEM

In this system, the forest tree species are raised and managed on farm lands for their ability to produce not only wood but also leaves and/ or fruits that are suitable for fodder and food purpose. Indian council of Agricultural research (ICAR) through its All India Coordinated Research Project (AICRP) on agroforestry at 39 centres all over the country have identified and enlisted the following most commonly cultivated species as a prioritized species for different agro-ecological regions of India.

Himalayan Region

Gerwia optiva, *Morus alba*, *Celtis australis*, *Eucalyptus* spp., *Alnus nepalensis*, *Acacia* spp., Pinus spp., *Paraserianthes falcataria*, *Albizia* spp., *Terminalia* spp., *Populus ciliata*, *Toona ciliata*, *Acacia catechu*, *Robinia pseudoacacia*, *Pinus kesiya*, *Michelia* spp., *Morus* spp.

Indo-Gangetic Plains

Populus ciliata, *Acacia* spp., *Dalbergia sissoo*, *Anthocephalus cadamba*, *Casuarina equisetifolia*, *Prosopis chilensis*, *Eucalyptus* spp., *Acacia nilotica*, *Populus deltoides*, Bamboos, *Gmelina arbórea*

Arid and Semi-Arid Regions

Prosopis chilensis, *Azadirachta indica*, *Dalbergia sissoo*, *Leucaena leucocephala*, *Acacia* spp., Bamboos, *Hardwickia binata*, *Ailanthus excela*, *Eucalyptus* hybrid

Humid and Sub-Humid Regions

Gmelina arborea, *Acacia* spp., *Dalbergia sissoo*, Bamboos, *Tectona grandis*, *Azadirachta indica*, *Acacia nilotica*, *Leucaena leucocephala*, *Hardwickia binata*

Tropical Coastal and Island Region

Tamarindus indica, *Acacia mangium*, *Artocarpus hetrophyllus*, Bamboos, *Terminalia* spp., *Casuarina equisetifolia*, *Azadirachta indica*, *Ceiba pentandra*, *Tectona grandis*, *Gliricidia* spp.

AGRO-CLIMATIC ZONES OF INDIA

The important rational planning for effective land use to promote efficiency is well recognized. The ever increasing need for food to support growing population in the country demands a systematic appraisal of our soil and climatic resources to recast effective land use plan. Soils and climatic conditions of a region largely determine the cropping pattern and crop yields. Reliable information on agro-ecological regions and homogeneity in soil site conditions is the basic to maximize agricultural production on sustainable basis. This kind of systematic approach may help the country in planning and optimizing land use, preserving soils, and environment. India exhibits a variety of land scopes and climatic conditions which are reflected in the evolution of different soils and vegetation. There also exists a significant relationship among the soils, land form, climate and vegetation. The object of agro-climatic zones is to delineate such regions as uniform as possible in respect of physiography, climate, length of growing period (LPG) and soils for macro level and land use planning and effective transfer of agrotechnology.

Table 23.5: Fifteen Agro-Climatic Regions/ Zones in India (Source: Planning Commission)

S.No.	Agro-Climatic Regions/ Zones	State-wise Distribution
I	Western Himalayan Region	Himachal Pradesh, Jammu & Kashmir, Uttarakhand
II	Eastern Himalayan Region	Arunachal Pradesh, Assam, Manipur, Meghalaya, Mizoram, Nagaland, Sikkim, Tripura, west Bengal.
III	Lower Gangetic Plain Region	West Bengal
IV	Middle Gangetic Plain Region	Uttar Pradesh, Bihar
V	Upper Gangetic Plain Region	Uttar Pradesh
VI	Trans Gangetic Plain Region	Chandigarh, Delhi, Haryana, Punjab, Rajasthan
VII	Eastern Plateau and Hills Region	Chhattisgarh, Jharkhand, Madhya Pradesh, Maharashtra, Odisha, West Bengal
VIII	Central Plateau and Hills Region	Madhya Pradesh, Rajasthan, Uttar Pradesh
IX	Western Plateau and Hills Region	Madhya Pradesh, Maharashtra
X	Southern Plateau and Hills Region	Andhra Pradesh, Karnataka, Tamil Nadu
XI	East Coast Plains and Hills Region	Andhra Pradesh, Odisha, Puducherry, Tamil Nadu
XII	West Coast Plains and Ghat Region	Goa, Karnataka, Kerala, Maharashtra, Tamil Nadu
XIII	Gujarat Plains and Hills Region	Gujarat, Dadra & Nagar Haveli, Daman & Diu
XIV	Western Dry Region	Rajasthan
XV	Island Region	Andaman & Nicobar Islands, Lakshadweep

An ecological region is characterized by distinct ecological responses to macro-climatic conditions as expressed in vegetation, fauna and aquatic systems. An agro-ecological region is the land unit on the earth surface covered out of agro-climatic region, which is based on land form, kinds of soils and soil conditions, climate and LGP (length of growing period). Within a broad agro-climatic region, local conditions may result in several agroecosystems, each with its own environmental conditions. However, similar agroecosystems may develop on comparable soil and landscape positions. Thus a small variation in climate may not result in different ecosystems, but a pronounced difference is seen when expressed in vegetation and reflected in soils. India has been divided into 24 agro-climatic zones by Krishnan and Mukhtar Sing in 1972 by using "Thornthwait indices". The planning commission, as a result of mid-term appraisal of planning targets of VII plan (1985-90), divided the country into 15 broad agro-climatic zones based on physiography and climate. The emphasis was given on the development of resources and their optimum utilization in a suitable manner within the frame work of resource constraints and potentials of each region.

IMPROVED AGROFORESTRY SYSTEMS IN AGRO-CLIMATIC ZONES OF INDIA

IMPROVED AGROFORESTRY SYSTEMS

Various research institutions in India have developed the improved agroforestry systems in different areas. The following are the most common systems, based on the nature of their components:

1. Agrisilviculture (trees + crops)
2. Boundary plantation (trees on boundary + crops)
3. Block plantation (trees + crops)
4. Energy plantation (trees + crops during initial years)
5. Alley cropping (shrubs + crops)
6. Agrihorticulture (fruit trees + crops)
7. Agrihortisilviculture (trees + fruit trees + crops)
8. Agrisilvipasture (trees + crops + pasture/animals)
9. Hortipasture (fruit trees + pasture/animals)
10. Silvipasture (trees + pasture/ animals)
11. Forage forestry (forage trees + pasture)
12. Shelterbelts (trees +/- crops)
13. Windbreaks (trees +/- crops)
14. Live fence (shrubs/under trees on boundary)
15. Homestead (multiple combinations of trees, fruit trees, etc.)
16. Entomoforestry (trees + sericulture)
17. Aquaforestry (trees + fish)

Especially during the last 3-5 centuries spectacular improvements have been made in tree plantation. The most important agroforestry practice is known from the Kangeyam tract of Tamil Nadu, where *Acacia leucophloea* + *Cenchrus setigerus* in silvi-pasture system is perfected. Similarly, in ravines of Yamuna and Chambal, trees, shrubs and bamboos with grasses are planted for rearing milk-producing Jamunapari breed of goats and sheep. Scattered trees with khejri or mehndi in association with bajra, jowar and chillies are grown in the semi-arid area of Tamil Nadu. Tree plantation is continued as demarcation and control unit against wind erosion throughout the country. Plantation of khejri trees for various uses on farm lands is a common practice in Rajasthan.

In coastal areas, *Casuarina equisetifolia* and other trees are grown in association with crops on farm lands for cash and to generate small timber. Live hedges are common as an agroforestry practice in which *Lawsonia inermis* (mehndi), *Agave sisalana* and *Euphorbia* species are common. In paddy-growing areas, *Pongamia glabra* and *Sesbania grandiflora* are grown, lopped annually and their leaves applied to fields as green manure. Application of green manure to paddy field is common in Madhya Pradesh and Uttar Pradesh. In Western Ghats, *Terminalia* leaves are harvested, spread on land, burnt and then paddy, ragi and millets are sown. Multistorey homesteads or homegardens are in existence in Kerala, Karnataka, Tamil Nadu, Tripura, Assam and other North Eastern states as an important agroforestry practice.

Table 23.6: Improved Agroforestry Systems for various Agro-Climatic Zones of India (Source: Data compiled from AICRP on Agroforestry Centres, NRCAF 2007)

Agro-Climatic Zone	Agroforestry System	Tree Component	Crop/ Grasses
1.Western Himalaya Region	Silvipasture (RF)	*Grewia robusta*	*Setaria* spp.
		Morus alba	*Setaria* spp.
	Agrihorticulture	*Malus pumila*	Millets, Wheat
	Agrihorticulture	*Prunus persica*	Maize, Soybean
2. Eastern Himalaya Region	Agrisilviculture	*Anthocephalus cadamba*	Paddy
	Agrihorticulture	*Alnus nepalensis*	Large cardamom/ Coffee
	Silviculture	Bamboos, *Parkia roxburghii*, *Morus alba*	-
	Silvipasture	*Bauhinia variegata*, *Ficus* spp., *Morus alba*	Napier
3. Lower Gangetic Plains	Agrisilviculture (Irri)	*Eucalyptus*, *Albizia lebbeck*	Paddy
	Agrihorticulture (Irri)	Mango/ Banana/Litchi	Wheat, Paddy, Maize
	Silvipasture	*Morus alba*, *Albizia lebbeck*	*Dichanthium*, *Pennisetum*
4. Middle Gangetic Plains	Agrisilviculture (Irri)	*Populus deltoides*	Sugarcane-Wheat
	Agrisilviculture (Irri)	*Eucalyptus* spp.	Rice-Wheat
	Agrisilviculture	*Dalbergia sissoo*	Sesamum
	Agrihorticulture (Irri)	Mango/ *Citrus* spp.	Rice-Wheat
	Silvipasture	*Albizia lebbeck*	*Chrysopogon*, *Dichanthium*
5. Trans Gangetic Plains	Agrihorticulture (Irri)	*Emblica officinalis*	Black gram/ Green gram
	Agrisilviculture	*Azadirachta indica*	Black gram-Wheat/Mustard
	Silvipasture	*Bauhinia variegata*, *Albizia lebbeck*	*Cenchrus*, *Pennisetum*
6. Upper Gangetic Plains	Agrisilviculture (Irri)	*Populus deltoides*	Wheat, Bajra fodder
	Agrisilviculture (Irri)	*Eucalyptus* spp.	Rice-Wheat
	Silvipasture	*Bauhinia variegata*, *Albizia lebbeck*	*Chrysopogon*, *Poa*
7. Eastern	Agrisilviculture	*Gmelina arborea*	Paddy, Linseed

Agro-Climatic Zone	Agroforestry System	Tree Component	Crop/ Grasses
Plateau & Hills	Agrisilviculture	*Acacia nilotica*	Paddy
	Silviculture	*Acacia mangium, A.nilotica*, Bamboos	-
	Silvipasture	*Leucaena leucocephala*	*Chrysopogon, Pennisetum, Dichanthium*
8. Central Plateau & Hills	Agrihorticulture (Irri)	*Psidium gujava*	Bengal gram/ Ground nut
	Agrihorticulture (RF)	*Emblica officinalis*	Black gram / Green gram
	Agrisilviculture	*Acacia nilotica/ Leucaena leucocephala/ Azadirachta indica/ Albizia lebbeck*	Soyabean, Black gram-Mustard/Wheat
	Silvipasture (RF and degraded lands)	*Albizia amara, Leucaena leucocephala, Dichrostachys cinerea*	*Chrysopogon* spp., *Stylosanthes hamata, S. scabra*
	TBOs (RF)	*Jatropha curcus*	-
9. Western Plateau & Hills	Agrihortisilviculture (Irri)	*Tectona grandis, Achrus sapota*	Paddy, Maize
	Agrihorticulture	*Areca catechu*	Black Pepper, Cardamom
	Silviculture	*Prosopis juliflora, Ailanthus* spp.	-
	Silvipasture	*Acacia mangium, Albizia amara*	*Cenchrus*
10. Southern Plateau & Hills	Agrisilviculture (RF)	*Eucalyptus* spp.	Cotton
	Agrisilviculture (Irri)	*Eucalyptus* spp.	Chilli
	Silviculture (RF)	*Leucaena leucocephala, Eucalyptus* spp.	-
	Agrihorticulture	*Tamarindus indica*	Chilli
	TBOs	*Pongamia pinnata*	-
11. East Coast Plains & Hills	Agrisilviculture (RF)	*Ailanthus excelsa, Acacia leucophloea*	Cowpea
	Silviculture	*Casuarina equisetifolia, Leucaena leucocephala*	-
	TBOs	*Pongamia pinnata*	-
	Silvipasture	*Artocarpus* spp.	*Chrysopogon*, Napier, *Cenchrus*
12. West Coast Plains & Hills	Agrisilviculture (RF)	*Acacia auriculiformis*	Black pepper
	Agrihorticulture (RF)	*Artocarpus heterophyllus*	Black pepper
	Agrisilviculture (RF)	*Acacia auriculiformis*	Paddy

Agro-Climatic Zone	Agroforestry System	Tree Component	Crop/ Grasses
	Agrihorticulture	*Cocos nucifera/ Areca catechu*	Paddy
	Agrisilviculture	*Casuarina equisetifolia*	Paddy
	Silvipasture	*Hardwickia binata, Albizia lebbeck*	*Cenchrus*
13. Gujarat Coast Plains & Hills	Agrisilviculture	*Azadirachta indica*	Cowpea
	Silviculture	*Prosopis juliflora, Acacia nilotica*	-
	Agrisilviculture	*Ailanthus excelsa*	Green gram
	Silvipasture	*Leucaena leucocephala*	*Cenchrus, Setaria*
14. Western Dry Region	Agrisilviculture	*Prosopis cineraria, Azadirachta indica*	Pearl millet
	TBOs (RF)	*Jatropha curcus*	-
	Silvipasture	*Albizia lebbeck*	*Cenchrus*
15. Island Region	Agrihorticulture	*Cocos nucifera*	Paddy
	Silvipasture	*Bauhinia, Erythrina, Leucaena*	*Cenchrus, Pennisetum*

* Note: Irri: irrigated; RF: Rainfed; TBO: Tree Borne Oilseeds

RECENT SCENARIO OF AGROFORESTRY DEVELOPMENT IN INDIA

The diverse challenges and constraints as growing population, increasing food, feed and fodder needs, natural resource degradation, climate change, new parasites, slow growth in farm income and new global trade regulations demand a paradigm shift in formulating and implementing the agricultural technologies. Agroforestry has traditionally been a way of life and livelihood in India for centuries. A major role for agroforestry is now emerging in the domain of environmental services such as biodiversity conservation, watershed protection, carbon sequestration and mitigating climate change effects besides help in poverty alleviation. The role of agroforestry in soil conservation, bio-amelioration and climate moderation is most widely acclaimed and one of the compelling reasons for including trees on farm lands. Thus, the present challenges of food, nutrition, energy and environment security can be met through different agroforestry systems developed for various agro-climatic zones. Agroforestry has a great potential to provide employment to rural and urban population through industrial application and value addition.

Agroforestry plays a vital role in the Indian economy by way of tangible and intangible benefits. Agroforestry meets half of the demand of fuel wood, 2/3rd of the small timber and also the subsistence need of households. It is playing the greatest role in maintaining the resource base and increasing overall productivity in arid and semi-arid regions. Changing priorities in avenues like bio-fuels, employment generation, carbon sequestration and value addition are now being focused through agroforestry. Industries have taken up poplar, eucalypts, neem, bamboos, *Acacia*, *Casuarina*, *Ailanthus* and teak for commercial agroforestry due to their great market potential. Agroforestry has helped in rehabilitation of degraded lands, bioremediation of problem soils and reclamation of waterlogged soils through bio-drainage. At the same time, it has been realized that agroforestry is the only alternative to meet the target of increasing forest cover to 33 per cent from the present level of 23.81 per cent.

Table 23.7: Area under various types of Tree Plantations in India (Dhyani et al 2007)

S.No.	Plantation Types	Area (million Ha)
1.	Production and social forestry	23.00
2.	Externally aided social forestry	
	(a)Farm Forestry/ Agroforestry	1.24
	(b)Village Woodlots	0.50
	(c)Strip Plantations	0.07
	(d)Rehabilitation of Degraded Lands	0.24
	Sub-total	2.05
3.	By NGOs/Vas	0.04
4.	Tree Growers Co-operatives	0.04
5.	Conservative Forestry	0.19
	Total	**25.32**

India is bestowed with bounty of natural resources. Innumerable combinations of edapho-climatic conditions, number of ethnic groups and their vivid life styles lead to development of highly specialized land uses. Ever increasing human population pressure exerting tremendous pressure on land, water, vegetation and domestic animals is leading to accelerated exploitation of resources and degradation. Since land resource is limited, integrated land use combining woody perennials and herbaceous vegetation/ shrubs on the same unit of land is need of the hour. As such, agroforestry land use has attained prime importance. Traditional agroforestry comprised of retaining trees on field bunds and/or inside fields or leaving a chunk of holding under natural vegetation. After advent of fast growing tree species, industrial agroforestry came into being with plantation of poplar, eucalypts, subabul, *Acacia* hybrids and *Casuarina*. These species allowed cropping unabated and simultaneously yielded rich dividends to the farmers in terms of hard cash. Agroforestry land use is known to conserve land and water; efficiently use water, fertilizer, space (below ground as well as aboveground); ameliorate micro-climate and soil.

Agrisilviculture and agrihorticulture systems can be adopted in arable lands while silvipasture, energy plantation and other forms of agroforestry can go well on non-arable lands. By 2030 when population pressure goes beyond 135 million, agroforestry will have to be an important option. So far suitable tree-crop combinations have been identified in different agro-climatic zones of country and their package of cultivation practices are standardized but the same has not reached to farmer's field for want of awareness and adequate infrastructure. Knowledge gained in agroforestry has to be translated on farmer's field for averting degradation and enhancing productivity of not only herbaceous vegetation but woody perennials as well. Agroforestry products particularly wood based are either consumed locally or traded conventionally. The market is highly un-organized. There are restrictions on harvesting of trees as well. Commercial agroforestry in tandem with wood based industries have shown immense potential as they are exempted from the felling restrictions. Poplar, eucalypts, *Casuarina* and *Leucaena* based commercial agroforestry systems could flourish in different regions of the country when promoted by industries. Similarly, fruit based industries coupled with food supply chain and creation of infrastructural facility like refrigerated storage, value addition and transport facilities have changed marketing of fruits, vegetables, medicinal and aromatic plants.

Number of industries (wood based or crop based) can join hands together to process variety of products from agroforestry systems. There is a vast potential for investments and creation of employment in agroforestry sector through involvement of corporate sector. Bioenergy is another area where industrial groups can play an important role to meet the requirements for energy as well as for employment generation.

Agroforestry practices for conservation of resources, enhancing productivity and reversing the trends of land degradation in diverse agro-climatic situations can help substantially in meeting various demands. It is noteworthy that the system being the same, species choice changes with the climate and edaphic variations. Tree species priority in different agro-climatic zones shows that *Populus deltoides* and *Eucalyptus* species have a higher priority due to fast growth, short cycle, adaptability and market demand. In India, agroforestry system clearly shows that six prominent tree species are commonly adopted in most agroclimatic regions: *Populus deltoides*, *Eucalyptus* spp., *Acacia nilotica*, Bamboos, *Prosopis cineraria* and *Casuarina equisetifolia*. The species are important as they are highly adapted in the farming systems; have short cycle of harvest; multipurpose in nature; and have industrial value and market potential (NRCAF 2007).

■■■■

24 AGROFORESTRY SYSTEMS OF HIMALAYAN REGION

INTRODUCTION

The Indian Himalaya lies between 27° to 36°N latitude and 74° to 97°E longitude, i.e. from Jammu and Kashmir in the west to Arunachal Pradesh in the east. A tremendous diversity in ecology, terrain, altitude, climate, resource availability, ethnicity, agricultural activities, flora and fauna is found in the Indian Himalaya. This huge geographical entity is one of the most important resource bases for the Indian subcontinent. Indian monsoon, the lifeline of Indian agriculture, is shaped by this huge geographical barrier located in north of the country. Because of its enormous size, location and climate, a huge human population along with their livestock lives here. Steep topography, prolonged and severe cold winter, shallow soil, lack of irrigation, etc., have limited the choice of agricultural activities. These constraints have limited the arable agriculture to an average altitude of 3,000 m above the mean sea level.

Beyond 3,000-3,500 m elevation, where arable agriculture is not remunerative, livestock rearing is the only agricultural activity. Grazing lands mostly comprise sub-alpine and alpine pastures, forest openings and the demarcated forests. If these forests are not maintained in good health, agriculture production along hillsides and bottomlands would suffer considerably. But these upland forests can not just be protected by laws and decrees. No application of the whip will stop rural people to raid forests for obtaining their day- to-day needs at any risk, if there is no alternative source. In such situation, cultivation of trees on the farms in form of agroforestry is one of the alternatives to concurrently produce food and tree products. Although traditional agroforestry systems exist in the Himalayas since time immemorial, yet, scientific agroforestry approaches to sustainable land use in this region are of recent origin (Sood et al 2010). Well developed agroforestry systems have been integral to traditional hill agro-ecosystem for the innumerable techno-socio-economic benefits that they provide (Puri and Nair 2004). Farmers in the rural areas of Indian Himalayan Region have integrated and practised agroforestry based agriculture, reaping rich benefits, individually and for the community as well. Therefore, agroforestry has been a part of peasants' subsistence strategies.

AGROFORESTRY: A SUSTAINABLE LAND USE IN HIMALAYAN REGION

The mutually reinforcing relationship between Agriculture Production System and Forest Production System is more understood now than ever before. What was stated in 1950s literally holds true today. The foresight of late K.M. Munshi, the then Agriculture Minister can not be overlooked. His reference to forest trees as a source of food deserves a special mention. It was he who, for the first time in the country, sought to recognize the inter-relationship between agriculture and forestry. This approach to land use which we call agroforestry i.e. tree + crops together, as one of the farm management options has tremendous potential. This approach will not jeopardize the food production. But, in many cases, can provide continuous vegetative cover to help conserve soil, water resources and increase in crop yields as well as fuel and fodder biomass. Agroforestry is thus, a collective name for land use systems involving trees combined with crops and/or animals on the same unit of land. The common features of agroforestry are:

- It combines production of multiple outputs with protection of the resource base.
- Places emphasis on the use of multiple indigenous trees and shrubs.
- It involves the interplay of socio-cultural values more than in most other land-use systems.
- It is structurally and functionally more complex than monoculture.

Woody perennials form the essential component of any agroforestry system (Nair 1989b). Herbaceous plants are found in most agroforestry systems with the possible exception of the systems where trees and/or shrubs are primary components or in other systems like apiculture with trees (Nair 1989b). Animals are present only in some agroforestry systems e.g. silvipasture or agrisilvipasture (Carne 1993). The following patterns of tree growing in agroforestry systems exist in the Himalayan region of India:

(1) Trees on cropland
   - Trees in homegardens (homestead)
   - Trees on fields (e.g. intercropping system)
   - Trees on bunds
   - Trees on boundaries of agricultural holdings

(2) Trees on pasture/grassland

(3) Trees in fruit orchards

(4) Trees on wasteland

(5) Scattered trees retained in and around on jhum lands (shifting cultivation area)

These agroforestry practices/systems complement hill farming in terms of manure, fodder and fuel needs of the farmer. It forms the backbone for practising integrated farming systems which is necessary for self dependence and sustainable agriculture. The purpose of tree growing on cropland bunds have multiple socio-ecological benefits, e.g. to meet consumption of family or cattle, means for building soil fertility, as wind barriers, to generate cash through sale of leaves, fruits, etc. in the market.

COMPOSITION OF TRADITIONAL AGROFORESTRY CROPS AND TREES IN HIMALAYAN REGION

AT ALTITUDE <1800 M

Agriculture Crops: Wheat, rice, maize, barley, millets, gram, lentil, mustard, linseed, potato, sugarcane, black gram, soyabean, oil seeds, colocasia, turmeric, ginger, brinjal, tomato, chilli, beans, cauliflower, cabbage, lady's finger, pumpkin, bottle gourd, etc.

Fruit Crops: Mango, guava, ber, citrus, pear, walnut, litchi, pomegranate, peach and plum

Forest Trees: *Acacia catechu*, *Albizia lebbeck*, *Aegle marmelos*, *Artocapus lakoocha*, *Bauhinia variegata*, Bamboo, *Bombax ceiba*, *Butea monosperma*, *Cassia fistula*, *Celtis australis*, *Cinnamomum tamala*, *Dalbergia sissoo*, *Ehretia acuminata*, *Emblica officinalis*, *Ficus roxburghii*, *Grewia optiva*,

Terminalia bellerica, Terminalia chebula, Terminalia tomentosa, Leucaena leucocephala, Litsea polyantha, Machilus spp., *Mallotus philippinensis, Morus* spp., *Myrica nagi, Populus deltoides, Prunus cerasoides, Pyrus pashia, Syzygium cumini, Robinia pseudoacacia, Ulmus laevigata, Toona ciliata*

AT ALTITUDE BETWEEN 1800 TO 2200 M

Agriculture Crops: Wheat, maize, barley, coarse cereals, potato

Fruit Crops: Apple, cherry, walnut, pear, apricot, almond, plum

Forest Trees: *Morus* spp., *Myrica nagi, Pinus wallichiana, Quercus luecotricho-phora, Quercus himalayana, Populus ciliata, Prunus armeniaca, Prunus cerasoides, Pyrus pashia, Rhodendron arboretum, Ulmus wallichiana*

AT ALTITUDE > 2200 M

Agriculture crops: Potato, chicory, hops, peas, kuth, black cumin

Fruit crops: Almond, apple, apricot, walnut, raisin grapes

Forest Trees: *Pinus wallichiana, Pinus gerardiana, Populus ciliata, Alnus nepalensis, Aesculus indica, Betula alnoides, Cedrus deodara, Populus alba, Quercus dilatata, Rubus idaeus, Salix* spp., *Rhododendron* spp.

TRADITIONAL AGROFORESTRY COMBINATIONS IN HIMALAYAN REGION

SUB-TROPICAL AND INTER-MEDIATE

Trees: *Acacia nilotica, Ailanthus excelsa, Albizia lebbeck, Grewia optiva, Bauhinia variegata, Zyziphus mauritiana, Dalbergia sissoo, Dandrocalamus strictus, Melia azedarach, Morus alba, Toona ciliata, Populus deltoides, Sesbania aegyptiaca, Leucaena leucocephala*

Crops: Wheat, maize, pulses, millets and medicinal & aromatic crops like ashwagandha, sarpgandha, brahmi, solanum, mulhathi, damask rose (variety-Jwala), mints, babchi, safed musli, etc.

TEMPERATE ZONE

Trees: *Robinia pseudoacacia, Salix* spp., *Celtis australis, Ulmus wallichiana, Populus ciliata, Quercus leucotrichophora, Alnus nepalensis*

Crops: Wheat, oats, maize, pulses, vegetables and medicinal & aromatic crops like kuth, damask rose (variety- Himrose), atees, patees, valerian, *Dioscorea* spp., chirayta, karroo, etc.

COLD ARID ZONE

Trees: *Salix* spp., *Populus* spp., *Juniperus communis, J. wallichiana*

Crops: Vegetables, sorghum and medicinal & aromatic crops like sea-buck thorn (*Hippophae* spp.), *Artemisia* etc.

AGROFORESTRY FOR LIVELIHOOD SECURITY AND SUSTAINABLE DEVELOPMENT IN HIMALAYAN REGION

Himalayan mountains characterized by fragility, poverty, inaccessibility, marginality (resource scarcities and uncertainties) and harsh climatic conditions which in turn leads to limited livelihood options for the people. To add to the problem, increasing population and market influx has degraded large tracts of land which can not support agriculture. A majority of small farmers who are poor and who do not want to migrate still depend on such unproductive lands for their livelihood. In such a situation, integrating more multipurpose trees, in already prevailing agroforestry practice as a component in the traditional farming systems, can make a difference. Trees grown along with crops in the form of agroforestry provide a biomass reserve upon which people can fall back for subsistence and income in times of crop failure, unemployment and other kinds of hardships, contingencies or to meet exceptional needs. Agroforestry has also a capacity to divert the population pressure on existing forests for tree-based needs of the people as these tree products are obtained from the system itself.

Many tree species grown in agroforestry systems in the Indian Himalayan region act as source of the income for the people. Tree products from agroforestry tree species like *Quercus* spp., *Syzygium cumini*, *Ulmus laevigata*, *Myrica nagi*, *Toona ciliata*, *Albizia lebbeck* in Western Himalayas and *Elaeocapus floribundus*, *Artocarpus heterophyllus*, *Dillenia indica*, *Terminalia chebula*, *Dendrocalamus* species in many parts of Eastern Himalayas provide considerable amounts of household income. Trees also provide income to buy food items for their domestic use.

Trees in agroforestry systems provide subsistence goods and supplement inputs for farming activities. For example, small timber and branches from trees (e.g. *Albizia lebbeck, Celtis australis, Bauhinia variegata, Ulmus laevigata, Litsea* spp., *Quercus* spp., *Elaeocarpus floribundus, Dendrocalamus* spp.) grown in agroforestry are used in making ploughs and implement handles almost throughout the Indian Himalayas. Trees grown in agroforestry systems play an important role in household food security. Tree based foods are widely consumed by the inhabitants of the Indian Himalayas as regular food supplement in diet. These tree foods add variety to diets, improve palatability and provide essential vitamins, minerals, proteins and calories. For examples fruits/fruit products of *Myrica nagi*, *Ficus roxburghii*, *Syzygium cumini*, *Juglans regia*, *Pyrus pashia*, *Morus* spp., *Prunus amygdalus* and *Aesculus indica* are eaten by people of Western Himalayas which provide additional minerals and calories. Tree products like tree food or fodder are extensively used to supplement seasonal shortage of these items. For example, young fruits and buds of trees grown in agroforestry systems like *Ficus palmata* and *Bauhinia variegata*, provide vegetables during the dry season when there is a shortage of other vegetables in Western Himalayan region. In Eastern Himalaya young tender bamboo culms and fruits of *Mimusops elengi*, *Artocarpus heterophyllus* are relished as vegetable. The bamboo is used for making residential huts by the tribals of Eastern Himalayas. Similarly *Livistona jenkinsiana* is grown in agroforestry systems in Eastern Himalayas and its leaves are used in making the roofs of the residential huts.

Trees in agroforestry systems provide fuelwood which reduces the pressure on women to carry head loads of fuelwood from far away forests. Trees help to meet contingencies by directly providing recurrent, seasonal or occasional flow of food, fodder, timber, and other products or earn cash thereby reducing household vulnerability to risk. Trees existing in agroforestry systems in the Himalayas provide green fodder during the period of shortage of the fodder to the livestock e.g. *Grewia optiva*, *Terminalia chebula*, *T. bellerica*, *Celtis australis*, *Quercus* spp., *Ficus roxburghii* etc. Leaves are lopped in winter in Western Himalayas for livestock when other green fodder is scarce.

Trees protect crops from direct or indirect effect of erosion, help in reduction in rate of dam siltation and conserve soil and can contribute to maintaining site productivity. Tree cover on the farms helps to restore nutrient in the upper layer of the soil and enhances the site productivity. The tree cultivation would not only help in increasing tree cover at local, regional and global level but also provide concomitant environmental benefits. Thus it can be inferred that trees play an important role in achieving household livelihood objectives of the people in the Indian Himalayas and also ameliorate the environment.

IMPORTANT AGROFORESTRY PRACTICES OF HIMALAYAN REGION

Important agroforestry practices in Himalayan region include agrisilviculture, agrihorticulture, silvopastoral system and livestock production, alley farming, and medicinal and aromatic plants based agroforestry. The various systems and their benefits are discussed below.

AGRISILVICULTURE

The intercropping arrangement of annual food crops with woody perennials is a land use system that enhances productivity and ensures sustainability. At the same time, it helps to sustain a favourable and equitable microclimate, offer protection against wind and water erosion, and ensure that plant nutrients, leached out of reach of the crops, are returned to the surface. In the case of species that produce cattle feed, the trees will further make it possible to fatten more livestock on the farmland and to obtain soil enriching manure in the bargain. They will also yield firewood, small timber, poles, thus enabling the farmer to stop burning dung and obtain some cash income from the sale of wood. It may in fact be profitable to cover as much as 5-10 per cent of the existing farmland with trees. The improvement in agriculture yields that could be expected from the beneficial influence of the trees would undoubtedly in many cases exceed the percentage of land occupied by the trees, and in addition the wood produced by the trees would help to overcome the firewood crisis. In foothills of Himalayan region, farmers have adopted the Poplar, *Eucalyptus*, *Melia* based agroforestry in a big way having economic return per hectare 3-5 times more than that from sole crop agriculture (Joshi 1996).

AGRIHORTICULTURE

Agrihorticultural system where fruit trees and crops are grown together in various ways and depending on planting pattern and configuration, its companion crops are known as intercrops, under planting or alley cropping. In intercropping, agriculture crops are generally grown between rows of fruit trees. The agriculture crops meet home requirement and seasonal revenue, whereas fruit trees managed for 30-40 years provide regular returns of fruit. In some cases, fuelwood and fodder is also obtained from prunings. The locals dwelling in higher Himalayan valleys prefer horticulture trees like apple, plum, peach, hippophae for their agroforestry systems along with their medicinal crops to maximize their benefits. In Kashmir, cultivation of pulses and vegetable crops in established orchards of apple, peach, almond, plum, etc. is a traditional agrihorticulture system. In lower zone, tall and perennial medicinal trees that need to be planted at a wider spacing such as *Olea cuspidata*, *Sapindus mukorossi*, *Terminalia chebula*, etc. can be intercropped with annual crops in early years until the tree canopy covers the ground. Shade tolerant and rhizomatic crops such as ginger, turmeric, garlic, etc. can be grown on a longer term basis in widely spaced plantations. The intercrops give some income to farmers during the period when the main trees have not started production.

SILVOPASTORAL SYSTEM AND LIVESTOCK PRODUCTION

Tree foliage is an important source of feed for livestock because trees provide a reliable source of feed during conditions of floods and drought (David and Aaron 1998). The potential of integrating trees and livestock in hills is pressing due to increasing paucity of grazing lands because of desertification. Silvopastoral system which is a mixed land use system for forage, livestock and wood production is gaining importance in hills where certain areas cannot be dedicated to conventional agriculture without producing severe, and sometimes irreversible soil degradation. The mixture of trees, grasses and legumes conserve soil and moisture. The legumes benefit soil by nitrogen fixation and in the mixture they help growth of grasses and trees besides improving forage quality. Trees and shrubs besides providing fuelwood and timber, also supply fodder during lean period. The system works well with improvement in land productivity by 2.5 times as compared to traditional system. After 5-7 years, grazing could be allowed thus reducing the cost of grass harvesting. Trees have to be chosen for both its suitability for the grass grown and the livestock to be grazed. Trees and grass species should be selected in such a way that they are compatible to each other. Ideal tree species are: (i) Fast growing and compatible with grass species, (ii) Highly palatable and nutritious with no side effects, and (iii) Multipurpose and vigorous coppicing ability. The grasses should be highly productive, good quality and longer duration with high palatability. The legume component in the pasture needs to be very hardy, easily colonizing, palatable, nutritious and with strong regeneration properties through roots or self grown seeds.

ALLEY FARMING FOR CONSERVING WATER AND SOIL RESOURCES

Kang and Wilson (1987) defined alley farming as the growing of arable crops between hedgerows of planted shrubs and trees, preferably leguminous species, which are periodically pruned to prevent shading the companion crops. This farming system has a potential in hills where cultivation is on slopes and when hedges are planted on contours, it can be a successful soil erosion control measure. Studies conducted (Kidd and Pimentel 1992) by alternating three rows of *Leuceana leucocephala* with three rows of maize perpendicular to the slope of 10 to 15 per cent revealed that the system can control erosion if all the maize residues and the pruning weighing 2,000 - 2,500 kg, are left on the surface of the land. Erosion rate under these conditions with 1,000 mm of rainfall would be about one ton/ha/year. Earlier Banda et al (1994) studied Maize-Leucaena intercrop on steep slope of 44 per cent reported similar results. Plots protected by Leucaena hedgerows lost two ton/ha/year topsoil compared with 80 ton/ha/year for an unprotected plot.

MEDICINAL AND AROMATIC PLANTS IN HIMALAYAN AGROFORESTRY SYSTEMS

In India, the supply of 70 per cent of medicinal and aromatic plants is from tropical areas with dry and deciduous vegetation, viz. Western and Eastern Ghats, Aravalis and Sub-tropical regions of the Himalayas, whereas the 30 per cent of medicinal plants which include species having high medicinal values are found in the temperate and evergreen forests. Medicinal and aromatic plants growing in forest areas require partial shade, moist soil with high humus content, high humidity and mild temperature. For example, *Taxus baccata*, *Valeriana jatamansi*, *Bergenia*, *Viola*, *Picrorhiza*, *Dioscorea*, *Saussurea*, etc. grow as understorey in the forests of deodar (*Cedrus deodara*), fir (*Abies pindrow*) and spruce (*Picea smithiana*). Other species like sarpgandha (*Rauvolfia serpentina*), ashwagandha (*Withania somnifera*), kalihari (*Gloriosa superba*), cowhedge (*Mucuna pruriens*), aloe (*Aloe barbadensis*), etc. are common in mixed forests of sub-tropical regions. Cultivation of such plants in their natural zone of occurrence can be taken up in thinned forests and as intercrops in new forest plantations.

Many plants which already exist in our traditional agroforestry systems have medicinal and aromatic values. There are number of medicinal and aromatic plants species which are present in such systems. Some medicinal and aromatic plants viz. *Ocimum sanctum* (tulsi), *Murraya koenigii* (karipata), *Azadirachta indica* (neem), *Moringa oleifera* (drum stick tree), *Mentha* spp. (mints), *Catharanthus roseus* (periwinkle), *Aloe barbadensis* (aloe), *Coriandrum sativum* (coriander) etc. are common in Indian households. Himalayan people use different parts of many trees to cure many ailments reducing health risks of the household e.g. *Pistacia integerrima*, *Terminalia bellerica*, *Terminalia chebula*, *Embilca officinalis* and *Syzygium cumini*. Ratanjot (*Jatropha curcas*), an important source of bio-diesel having medicinal value (purgative), is commonly used as live fence in farmlands in foothills. *Pinus roxburghii* (chir pine) is also an important, remunerative aromatic tree species found commonly in grasslands of J&K, Himachal Pradesh and Uttrakhand in addition to forests and plantations (Gupta et al 2007). *Acacia catechu* (khair) having high medicinal value in addition to its value for fodder and timber, is another one such species found commonly in grasslands and farmlands in north-west Himalayas. There are number of such species which exists in different agroforestry systems but sometimes their medicinal/aromatic uses become secondary when they are used for timber, small wood, fodder, etc. So the tree species like *Azadirachta indica* (neem), *Moringa oleifera* (drum stick tree), *Sapindus mukorossi* (retha), *Terminalia chebula* (harad), *Terminalia bellerica* (bahera), *Emblica officinalis* (amla), *Aegle marmelos* (bael), *Syzygium cumini* (jamun) etc. can be intercropped with annual crops (food/medicinal/aromatic) in the early years until the tree canopy covers the ground. The tree species can be managed at low height or may be planted at larger spacing so to obtain the annuals throughout.

Medicinal plants that have been successfully intercropped with trees (*Acacia auriculiformis*, *Albizia lebbeck*, *Eucalyptus tereticornis*, *Gmelina arborea* and *Leucaena leucocephala*) are *Chlorophytum borivilianum*, *Rauvolfia serpentina*, *Curcuma longa*, *Curcuma aromatica*, *Curculigo orchioides* and *Zingiber officinale* (Chadhar and Sharma 1996; Mishra and Pandey 1998; Prajapati et al 2003). In general shade tolerant species can be grown successfully under the canopy of trees. The intercropping of aromatic crops such as *Cymbopogon* spp., *Mentha* spp., *Pelargonium* spp. gave two to four times greater economic returns than sole crops which indicate that aromatic crops can be grown in Agroforestry systems. There were records of significantly higher net returns from intercropping of medicinal and aromatic plants with commercial timber species (Chauhan 2000; Dutt and Thakur 2004).

CONCLUSION

Agroforestry is uniquely suited to improve food and fuel security besides ecosystem services, but risk is the greatest obstacle to the farmer adopting any agroforestry system. Traditional agroforestry techniques can help lower perceived risks and real risks. It is better to build upon existing techniques than to introduce new ones. Farmers may be more willing to adopt agroforestry if they are familiar with the use of trees in agriculture and know that integrating tree and crop production has been successful for other people. The profitability should be one of the main factor for developing agroforestry systems followed by other minor factors such as demand for produce, access to market, availability of planting material and specific local uses, which can influence the farmers to select species for planting on their lands. It is suggested that the tree species to be selected for hills, should be based on the altitude, quality of land, suitability of climate, growth rate and profitability. With integration of medicinal trees and herbs, agroforestry now has become a sustainable land use system for effective increase of the tree cover, thereby reducing the pressure on existing forests. There is a need to take various measures to design suitable, productive and socially acceptable agroforestry systems as a source of sustainable livelihood in the Himalayas.

■■■■

25 URBAN AGROFORESTRY SYSTEMS

INTRODUCTION

Contrary to the popular belief, more than half of the global human population resides in the cities worldwide. The world's urban population today is around 3.2 billion people which are more than the world's total population in 1960. The world urban population exceeded the rural population in 2008 for the very first time in history. The urban human population worldwide is projected to rise from 3.3 billion to 6.4 billion between 2007 and 2050 (United Nations 2008) and by 2030 nearly 61 per cent of the people will be dwelling in the urban and sub urban areas. Meeting the food and nutritional requirements for such a huge urban population will be one of the most challenging issues for the urban managers. UNDP has projected that rising population in the urban areas will result in the rapid expansion in urban areas, which already account for 4 per cent of global land cover. In 1900, there were 6.7 rural dwellers to each urban dweller; now there is less than one and projections suggest three urban dwellers to two rural dwellers by 2025 (Cohen and Garrett 2009).

In the last 50 years, many countries in the world have experienced unprecedented levels of urban development. During this time, the general failure of traditional industrialized planning models to be successfully adapted in these countries has resulted in the need to explore alternative models for urban settlement. In this way, the incorporation of tree based agricultural systems (agroforestry) into urban settlements has considerable potential to address many of the problems associated with rapid urbanization such as food security, waste management, environmental degradation and unemployment. Research has already shown how urban agroforestry systems can improve food security, increase access to nutritional foods, recycle organic waste, create employment and protect fragile ecological systems. However in towns and cities, urban agroforestry systems are rarely developed beyond a homegarden setting (East 2008).

With infertile soils, severely restricted land and water resources, many developing nations face some of its greatest challenges. Due to rapidly expanding urban populations, it is currently being investigated for the development of future planned urban settlements. In such a scenario, potential exists to extend urban agroforestry systems beyond a homegarden setting and explore alternative models for sustainable urbanization. The promotion and development of urban agroforestry systems in future planned urban settlements will give production benefits as well as ecosystem services. Although urban agroforestry is a highly sustainable land use, it faces two main challenges viz. people's perception that urban agroforestry systems are a relatively low value land use and the general inability of the Government to effectively regulate urban land uses. In many countries, urban agroforestry systems are deliberately included in urban settlement plan considering nutritional contribution of urban agroforestry species in future planned urban settlements. While considerable constraints must be overcome to ensure the long term viability of planned urban agroforestry systems at a whole of settlement scale, it is argued that such an approach is one of the most cost effective, culturally acceptable and environmentally responsible methods for addressing a range of urban issues in the developing countries and is therefore an essential component to the design of future planned urban settlements.

URBANIZATION

Urbanization is one of the major features and issues of the developing countries. Urbanization can be understood as the increase in proportion of the urban dwellers when compared to the rural population though the word is also used to refer the expansion in urban areas and the growth in urban population. One of the major causes of urbanization is the migration of the people to cities in search of better employment opportunities, better wages and better quality of life. Migration of people is a serious issue as it results in the cramming of people resulting in a range of economic, political, social, cultural and environmental impacts. Yet urbanization is unavoidable. Economic growth is strongly related to urbanization and fact remains that most of the prosperous and developed nations are predominantly urban and the developing nations less urbanized. Most cities in Asia and Africa and many in Latin America and the Caribbean are facing severe consequences due to urbanization. The population living in very poor and overcrowded conditions and in informal settlements like slums lack adequate provision for water, sanitation, drainage, primary healthcare and schools and these are the results of the rapid and non uniform economic growth. This happens partly due to lack of resources and partly due to improper governance and political issues.

Urbanization causes drastic changes in socio economic set up and the environment. It also influences agriculture and biodiversity as it fragments the natural vegetation, modify climate, hydrology of the area, alters the richness and abundance of species. Another major issue would be the changes in the demand and supply of agricultural produce as there will be major changes in the dietary patterns of the urban dwellers. As a city grows, the cost of housing and infrastructure also increases resulting in scarcity of resources like water, land and food materials. In addition to this, there will be changes in the employment patterns also as there will be an increasing aptitude towards 'white collar' jobs. In 1980, the population employed in industry and services exceeded that employed in the primary sector viz. agriculture, forestry, mining and fishing. Land use pattern also faces transformations and conversion of the agricultural lands to non agricultural purposes becomes rampant as they are seldom prohibited by rules and laws. Often cities and townships develop over the localities' most productive agricultural lands. Even when there is a reduction in agricultural production and the supply of the agricultural produce from the local areas, there will be huge market demand for a range of high-value food crops from the high income urban dwellers. Meeting the demands of the local public by the imports from different parts of the country may not be a viable option on a long run due to the very high costs involved. It is in this context that the urban agriculture and urban agroforestry gain importance.

The environmental challenges posed by urbanization are also many. Issues related to green house gas emissions and climate change will be the top ranked among them along with the impacts on hydrology and biodiversity. Urban heat islands are now common features of the cities around the world. The immediate solution for managing the ill effects of urbanization is green vegetation or urban forestry. The role of vegetated land or green spaces within developed areas in amelioration of climate is undisputed. Tree planting and green spaces like parks or biological gardens in the urban areas help to reduce environmental impacts, by buffering rain water drainage, influencing local climate and enhancing air quality (Bolund and Hunhammar 1999). Green vegetation supports biodiversity (Gilbert 1989) and the access to green spaces in the urban areas have positive effects on human health and psychological well-being in society and quality of life (Fuller et al 2007). Setting apart land separately for urban agriculture and urban forestry is a constraint due to scarcity of land in the urban areas. The best option would be to combine the two in the same piece of land as urban agroforestry systems which help to meet the food and nutritional requirements of the urban population and at the same time serve as green spaces amidst the urban infrastructures.

URBAN GREEN DEVLOPMENT AND URBAN FORESTRY

Although trees have been an important part of human settlements throughout history, only recently their full value to urban dwellers has been recognized. Trees and green spaces play an important role in improving city living conditions. In the past, urban trees in developed countries were considered almost exclusively on the basis of its aesthetic merits. Now, a closer look is being given to the environmental services and quantifiable economic benefits they provide (Kuchelmeister and Braatz 1993). Trees, parks, woodland and other green spaces in or near urban areas improve the quality of life and environment in urban societies. Urban greening facilitates a sustainable relationship between urban dwellers and their environment. Sustainability is achieved through short and long term economic, social, and ecological well-being. Urban greening is important while considering the rising energy costs, increased concern about global warming, and increased consumption of non-renewable resources. It provides more livable environ-ment. Cities with full tree canopies and extensive park systems and trails have higher property values, are safer, and cooler.

Generally in urban areas, collective masses of trees are found within the boundaries of cities, towns or neighborhoods. Such publicly or privately owned trees are growing in parks, near schools, within residential areas, on the grounds of institutions and along streets. These tree collections may have very different appearances. They may be small tracts of trees preserved during development that become a naturalistic looking greenbelt. Some sites have a designed landscape, made up of carefully placed trees and shrubs surrounded by turf, as in many urban or community parks. Some trees are on "leftover" land, an untended collection of plant volunteers and scruffy vegetation. Each of these urban tree areas not only looks different but the ecological functions of each vary. Scientists have proven that trees in all urban settings provide extensive environmental benefits such as carbon dioxide capture and sequestration, air pollution reduction, reduced energy use, noise abatement, glare reduction wildlife habitat, water quality improvements, recreation and aesthetic value. Beyond ecological and aesthetic benefits, urban forestry has a role in helping resource-poor populations to meet basic needs, particularly but not exclusively in developing countries.

Urban forestry is the care and management of urban forests i.e., tree populations in urban settings for the purpose of improving the urban environment. Urban forestry advocates the role of trees as a critical part of the urban infrastructure. Urban foresters plant and maintain trees, support appropriate tree and forest preservation, conduct research and promote many benefits trees provide. Urban forestry is practised by municipal and commercial arborists, municipal and utility foresters, environmental policymakers, city planners, consultants, educators, researchers, community activists and urban residents. Urban forestry is the management of trees for their contribution to the physiological, sociological and economic well-being of urban society. Urban forestry deals with woodlands, groups of trees, and individual trees, where people live - it is multifaceted, for urban areas include a great variety of habitats (streets, parks, derelict corners, etc) where trees bestow a great variety of benefits and problems (Grey and Deneke 1986). Miller (1988) defined urban forestry as an integrated, city wide approach to the planting, care and management of trees in the city to secure multiple environmental and social benefits for urban dwellers.

URBAN AGRICULTURE

The industrial revolution ushered in new perspectives on the separation between urban and rural land uses. In particular, it was increasingly assumed that the urban residents should buy food, rather than grow it themselves. It was a common perception of most countries that city is a sterile and sanitized environment which should exist separately to rural food production. The notion of producing food in an urban environment (urban agriculture) has much to contribute to the debate on sustainable urban development. Urban agriculture is the generic term used in reference to a variety of food production practices in and around city boundaries including hydroponics, aquaponics, urban forestry and homegardening (Dresher et al 2006). Research into urban agriculture has gained increasing support following the global shift towards urban living, particularly in developing countries.

The concept of urban agriculture in cities is not new but rather has existed in the form of household gardens since Neolithic times (Cleveland and Soleri 1987). Empirical investigations into urban food production show how urban agriculture can be used to significantly increase the self-sufficiency and decrease the environmental impact of cities. Findings from various Asian, Latin American and Pacific Islands case studies are often used to support the claims of promoting urban agriculture as one of the most efficient tools for transforming urban wastes into food, jobs and a healthier, cleaner urban environment. However, concept of urban agriculture has its own problems and some of the major criticisms associated with promoting urban food production include potential pollution of the urban environment from agricultural activities, theft of urban produce, increased toxicity of crops resulting from absorption of chemically harmful urban pollutants, increasing pressure on already stretched urban infrastructure, the high cost of land in urban areas and unnecessary interference with urban land markets. While such criticisms are biased towards preserving the traditional "urban/rural divide" as a product of free market economics, they largely ignore the waste reduction potential of urban agriculture or the potential for positive environmental benefits (East 2008). As a natural system, agriculture is perhaps the most sustainable of all human activities using raw inputs such as waste, sunlight, soil, water and labour to create produce with a higher economic value (Overton et al 1999).

URBAN AGROFORESTRY

Urban and rural agroforestry have not been separately defined in any literature. Needless to say, most of the agroforestry systems are more relevant in the rural than in the urban areas. Due to the increasing awareness on the ill effects of urbanization and the need to provide quality life for all the income groups in the urban areas, urban agroforestry systems are being advocated as a panacea and though it is being practised only in a very limited scale. Most of the examples cited in the literature are from the Pacific Islands where urban agroforestry systems are a well-established practice and a part of the sustainable urban development. In the tropical regions, trees and agroforestry have always been, and will continue to be, central to social, economic, nutritional, spiritual and ecological well-being. Even with increasing urbanization, trees still dominate the landscape, and "urban agroforestry" is an emerging practice though it is well-established in some areas such as Pacific island towns. Trees are perhaps as important to urban people as they are to rural people, and the preservation, promotion, and improvement of urban agroforestry could be one of the most direct and economically, socially, ecologically and nutritionally appropriate means of bringing about sustainable development.

Urban agroforestry is a term that is used to denote tree based systems in an urban context or within urban areas. Where practised, it is defined as the deliberate planting and protection of trees and forestry in and around agricultural systems within city boundaries in order to improve or maintain the short term or long term economic productivity, cultural unity and ecological stability of agricultural ecosystems (East 2008). Thaman (1987) defined urban agroforestry as the planting, protection or preservation of trees for their economic, social and ecological value as part of agricultural and horticultural systems in urban areas, not only adjacent to houses and other buildings but also on undeveloped land within urban areas. Urban agroforestry includes two forms:

- Homegarden urban agroforestry
- Urban agroforestry on undeveloped and underutilized lands

All the traditional land use systems are influenced to a great extent by the biophysical and socio-cultural characteristics of the locales where they are practised (Kumar 2006). The scope of agroforestry in the urban areas are limited to either the land adjacent to the residence known as dooryard urban agroforestry or the practice of agroforestry on idle or underdeveloped land within the urban areas (Thaman 1983). Agroforestry in undeveloped and underutilized lands includes large scale commercial intercropping of tree plantations such as coconut plantations; animal husbandry and urban agrosilvipastoralism.

BENEFITS OF URBAN AGROFORESTRY

PRODUCTION BENEFITS

Beyond their aesthetic and ecological value, trees can contribute to the satisfaction of energy requirements as well as the daily food requirements of urban dwellers, particularly in the case of the poorest elements of society. Food production from agriculture component and fruits from tree component from urban agroforestry systems are common in many cities in Asia, Latin America and Africa (Yeung 1987; Streiffeler 1987). Who and how many people practice it as well as what form it takes differ greatly from place to place. It is most often practised in the urban fringe area by low-income families but, in places such as Africa and the Pacific Islands, urban agroforestry is widespread within cities.

Although in most places the emphasis is not on the production of staple foods, through the production of vegetables, fruits and condiments, urban agroforestry can contribute to the improvement of the nutritional value and variety of city dwellers' diets through its diversified food products. Fruit trees are often an important component of urban home gardens. In some places, trees are planted to help supplement fuelwood and fodder needs and even to provide raw materials for handicrafts. The role of agroforestry in improving productivity and diversifying production will become much more important in the future. In many developing countries, the low and moderate income households benefit through urban agroforestry systems. Urban agroforestry will not only be important to household nutrition but may also offer a source of income.

Fuelwood from woody components of urban agroforestry systems is the important energy source for urban low-income populations. Wood fuel provides between 25 and 90 per cent of urban household energy supplies and it is particularly important as a source of energy in smaller urban centres in developing countries, especially in dry zones (Kuchelmeister 1998). Poor urban households spend a significant proportion of their cash income in obtaining wood energy. If the urban poor population continues to grow, an increase in the consumption of traded wood fuel is likely to be a consequence. Under favourable circumstances, fuelwood from non-rural forests and agroforestry systems can contribute significantly to fuelwood supply.

Timber availability and adequate timber supply are the problems for a growing number of households in developing countries. Principle sources of timber in urban areas are plantations, street trees, shelterbelts or windbreaks and greenbelts, parks and gardens. In many cities, timber harvesting is combined with intensive outdoor recreation activities. Systematic planting of street trees for timber production is widely practised in China and Malaysia. Some cities in industrialized countries offset the costs of tree care through harvesting of trees.

ECOSYSTEM SERVICES

One of the major problems in urban areas is poor air quality. Plants are effective sinks for airborne pollutants and plants help to remove pollutants from the air in three ways: absorption by the leaves or the soil surface; deposition of particulates and aerosols on leaf surfaces; and fallout of particulates on the leeward (downwind) side of the vegetation because of the slowing of air movement. Trees intercept dust: a belt of trees measuring 30 meters in width has been found to intercept almost all dust in the air. Trees also often mask fumes and disagreeable odours by replacing them with more pleasing scents or by actually absorbing them. Trees also help to increase the relative humidity of urban air through evapotranspiration.

Modifying temperature extremes is another ecosystem service of urban agroforestry. Trees, shrubs and other vegetation help to control temperature extremes in urban environments by modifying solar radiation. The shade of one large tree may reduce the temperature of a given building to the same extent as would 15 air conditioners do in unshaded building. Energy saving through tree-planting around houses ranges from 10 to 50 percent for cooling and from 4 to 22 percent for heating (NAA/ISA 1991). Excessive noise levels in most major cities contribute to both physical and psychological damage. Trees can help both by absorbing and refracting or dissipating noise such as that produced by the heavy vehicular traffic which characterizes urban areas.

Urban vegetation can help in the protection of urban water supply, wastewater treatment systems and storm water management. Most poor cities face significant wastewater treatment challenges and could integrate stabilization ponds into park systems and reuse wastewater for urban vegetation. Reusing city wastewater not only recharges aquifers but also reduces the demand exerted on scarce water reserves. The greatest potential of wastewater reuse is in arid zones in developing countries (Braatz 1994; Kuchelmeister 1998).

Trees and forests are a means of soil conservation, preventing landslides in fragile ecosystems with steep terrain, little vegetation and harsh seasonal rains, and thus protecting people's lives and homes. Green areas have a vital role in urban biodiversity. Suburban wetlands can be some of the most productive natural ecosystems and can provide important habitats for fauna. Incorporating green areas in networks will improve biological conservation and biodiversity; greenbelts and greenways (linear parks) can serve as biological corridors (IUCN 1994).

URBAN HOMEGARDENS/ DOORYARD URBAN AGROFORESTRY

Urban agroforestry is most commonly practised in urban homegardens. The concept of urban homegarden is same as that of homegardens in the rural areas itself and the only difference being that in the former the trees are integrated along with the other components in an urban setup. Homegardens are one of the oldest and traditional food producing systems (Nair 1993). The term homgarden itself is used to denote an array of practices involving food producing systems at the household level in both rural and urban areas. They are the most popular agroforestry systems as they are low cost systems affordable to all which are mainly practised mainly in the tropical and subtropical regions of the world (Nair 1993). The key characteristics of homegardens include multi-tired, multipurpose, annual and perennials mainly cultivated by family labour within the boundaries of the residence (Fernandez and Nair 1986). The practice of homegardening is a very old one and the cultivation techniques and principles have evolved over ages. Though it originated in the rural areas of India and Indonesia, it is being widely practised by the urban dwellers and will continue to gain popularity in the years to come

(Kumar and Nair 2004). Traditional tropical gardens typically exhibit a wide diversity of perennial and semi-perennial crops, trees and shrubs, well adapted to local microclimates and maintained with a minimum of purchased inputs. The same principle of integrating diversity of perennial, semi-perennial crops, shrubs and trees, well adapted to the local climatic conditions can be imitated in the urban areas also with minimum purchased inputs.

While the Urban home gardens provide food security and economic security to the urban poor, they also help to intensify the food production in the urban areas and provide commodities of own choice and quality. Urban agroforestry systems are sources of fresh, nutritious, and pesticide-free food and fodder, many other products like fuel and medicinal plants. Such dooryard systems also eliminate the need for transportation, processing, storage and refrigeration which will ultimately bring down the cost of the produce. Providing common spaces for cultivation of vegetables in the dooryards of apartment and living spaces have already started gaining popularity especially in the developed countries. The scope of urban homegardens varies from a few household plants in the poor homes to large agribusinesses. Urban homegardens can be practised in the adjacent areas of the residences or around houses (front, back and side yards, walls and fences) and wherever unoccupied land is available and in community spaces; on abandoned public and private lands; along roads and railways, at airports and on otherwise uninhabitable floodplains; wetlands and steep slopes; and areas adjoining schools, community buildings and common spaces in the urban areas. Commercial gardening contributes to the large percentage of total non-grain urban food supply in many developing countries adding significantly to urban food self-sufficiency. More intensive urban and peri-urban gardens also create jobs for women and unemployed youth. Urban homegardening also provides opportunity for waste recycling and productive uses of city waste and garbage (East and Dawes 2009).

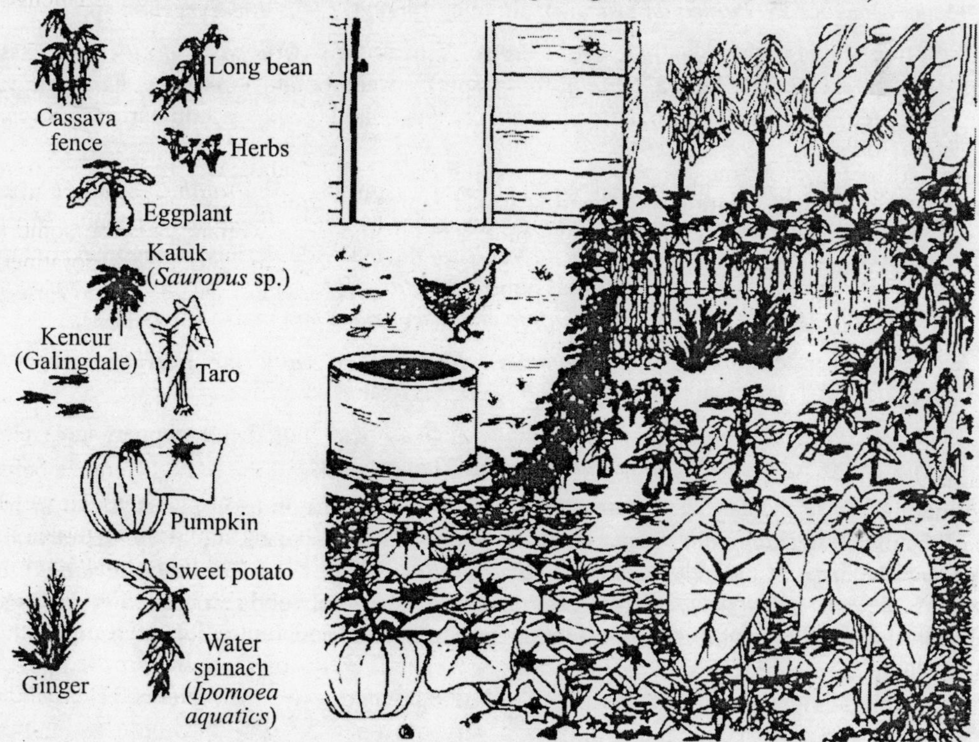

Fig. 25.1: General Structure of a Traditional Homegarden (FAO 1995)

Few examples of successful urban home gardening quoted from literatures (Marsh 1998):

- In Lima, Peru, during the past two decades the Ministry of Agriculture, FAO and the (UNICEF) have promoted household and community kitchen gardens to avert widespread hunger.
- The Centre for Education and Technology in Santiago, Chile promotes 20 m^2 gardens, plant containers stacked in pyramids and the use of walls for vines to double available space for horticulture.
- In Havana, Cuba an estimated 26,000 home and vacant lot gardens are cultivated in addition to State and private urban farms, communal gardens run by organizations and workplaces, and organic non-soil medium gardens.
- In China, 18 major cities satisfy 80 per cent of their vegetable demand and a significant percentage of total chicken and fish demand through urban agriculture.
- Urban and peri-urban gardens are the principal source of vegetables for major urban markets of Hanoi and Ho Chi Minh City, Vietnam.
- Urban and peri-urban gardens are the most reliable suppliers of urban markets and a vital source of income for many households in Moscow.

In a random survey conducted in the urban agroforestry in Pacafic Islands, it was observed that the major components of the urban homegardens were (Thaman 1985):

- Food trees/ plants like *Musa* clones, coconut (*Cocos nucifera*), breadfruit (*Artocarpus altilis*), pandanus, citrus trees (*Citrus* spp.), papaya (*Carica papaya*), guava (*Psidium guajava*), avocado (*Persea americana*), almond, *Annona*, *Ficus* and *Syzygium* spp.
- Non-tree staples which includes root crops such as taro (*Colocasia esculenta*), cassava (*Manihot esculenta*), tannia (*Xanthosoma* spp.), sweet potato (*Ipomoea batatas*), yams (*Dioscorea* spp.), giant taro (*Alocasia macrorhiza*) and giant swamp taro (*Cytosperma chamissonis*).
- Non-tree food plants like onion (*Alliums* spp.), spinach (*Amaranthus* spp.), amaranth (*Amaranthus gangeticus*, *A. tricolor*), pineapple (*Ananas comosus*), peanut (*Arachis hypogaea*), cabbage (*Brassica oleracea* var. *capitata*), a wide variety of legumes, pumpkin (*Cucurbita mixta*) okra, tomato, passion-fruit (*Passiflora* spp.), sugarcane (*Saccharum officinarum*), eggplant (*Solanum aethiopicum*, *S. scabrum*) and maize (*Zea mays*).
- Spices such as chilli (*Capsicum annuum*), coriander (*Coriandrum sativum*), ginger and mint.
- Beverage, stimulant and depressant plants such as betel nut, betel pepper, kava (*Piper methysticum*), tobacco (*Nicotiana tabacum*) and lemon grass.

In addition to these plants, there were also a variety of useful non-food plants found in home gardens. These include *Pandanus* spp., used in plaited ware; paper mulberry (*Broussonetia papyrifera*) used for bark or tape cloth; annatto (*Bixa orellana*), and koka (*Bischofia javanica*), both sources of dyes; *Leucaena leucocephala*, an important renewable fuelwood resource; a great range of medicinal plants and other plants of considerable technological, economic, social, ecological and ornamental value.

URBAN AGROFORESTRY SYSTEMS ON UNDEVELOPED AND UNDERUTILIZED LANDS

Lands left unutilized and underdeveloped are very common sights in our cities. Road frontages, river banks, canal banks, unoccupied allotments, rights-of-way for proposed or existing paths and roads, urban savannah land, open land in general, hillsides, swampland/ marshy areas, etc. are some examples of such underdeveloped areas. Substituting such areas with agroforestry systems will not only add to the tree cover in the city but also beautify our urban and sub urban areas. Cultivation and/or protection of trees and non-tree plants on idle or undeveloped urban and pert-urban land is widespread and constitutes an important source of food and tree products such as timber, fence posts, fuelwood, medicines, food, leaves, flowers, fruits and nuts (Thaman 1976, Thaman 1977, 1984 and 1985).

The most common species in "undeveloped areas" are again the staple root crops, particularly cassava, sweet potato, taro, yam and tannia. Important trees include *Musa* spp., papaya, mango, guava, bread-fruit, coconuts, *Citrus* and *Syzygium* spp., *Pometia pinnata*, *Terminalia catappa*, and a range of pioneer non-food species such as *Hibiscus tiliaceus*, *Macaranga* spp., *Morinda citrifolia*, *Bischofia javanica* and *Leucaena leucocephala*, plus the recently introduced fiddlewood species (*Citharexylum spinosum*) and jambolan (*Syzygium cumini*) and native *Eucalyptus* spp.

PROBLEMS OF PRACTISING URBAN AGROFORESTRY SYSTEMS

Even to a layman's knowledge the concept and idea of urban agroforestry and their role as food systems is highly appealing but the practice of urban agroforestry systems is still not very popular. Research works on the integration of agroforestry systems in the urban areas are also scanty. The probable reasons include:

- High cost of land in the urban areas and unavailability of land
- High cost of labour
- Lack of quantitative data on the nature, extent, ecological, cultural and economic significance of urban agroforestry systems
- Non-availability of the tested, local affordable technologies, plants, planting materials, cultural practices, etc.
- Unless emphasis and acknowledged at the policy level, it is very unlikely that such models succeed as they come under a larger preview of sustainable urban development which involves decision making and policy formulation at the higher levels.

■■■■

26 AGROFORESTRY PRACTICES FOR ARID AND SEMI-ARID REGIONS

INTRODUCTION

Agroforestry systems in India have occupied more area due to its wide spread in farm lands and variety of forest areas under local forest management and ethno-forestry practices (Pandey 2007). As per current assessment, total forest cover of India is 692,027 Km^2 (21.05% of the geographical area) and total tree cover (trees outside forest) is estimated to be 90,844 Km^2 (2.76 % of the geographical area) (FSI 2011). Even though Indian forests are poorly stocked and their mean annual increments are poor, the trees outside forests supply considerable share of timber, fuelwood and non-timber forest produce needs. For example, with merely 3.5 per cent of geographical area of Haryana's state under forests, the state has become self sufficient in small wood, fuelwood and industrial timber by establishing large scale plantations on farmlands. Trees in agroecosystems have increased the extent of area under forest and tree cover to 6.63 percent. Similarly, the case of Kerala suggests that the state has a surplus of 0.027 million m^3 of wood in terms of consumption. While the total wood production in the state is 11.714 million m^3, the forests provide only about 10 percent and trees in homegardens and mixed cropping multi-tier agroforestry system contribute to the remaining 90 percent (Krishnankutty et al 2008). Referring to the environmental condition and ecological suitability of systems, certain types of agroforestry systems can be more appropriate for certain ecological conditions i.e., there can be separate sets of agroforestry systems for arid and semiarid lands, tropical highlands, lowland humid tropics, etc (Nair 1993). This chapter explains the agroforestry practices for arid and semi arid region. The following table describes the major characteristics of dry region.

Table 26.1: Major Characteristics of Dry Regions (Arid and Semi-Arid) (Nair 1989)

S.N.	Factors	Description
1	Climate	Hot/ cold; one or two wet seasons and at least one long dry period; rainfall 1,000 mm
2	Vegetation and soils	Savannas with low or medium-high trees and bushes; thorn scrub and steppe grasslands; Vertisols, alfisols and entisols
3	Geographical spread	Semi arid and arid parts of Indian subcontinent
4	Main land use systems	Arable farming, extensive ranching or nomadic pastoralism, perennial crop husbandry towards the more humid areas, forestry
5	Main land use and ecological problems	Drought (in areas with less rainfall), soil fertility decline caused by over cultivation, over grazing, degradation of deciduous woodland, fuelwood/ fodder shortage, population increase
6	Major agroforestry emphasis	Fuelwood/ fodder production, soil fertility improvement, windbreaks and shelterbelts, food production, sustainable natural resource utilization, value chain

MAJOR CROPPING SYSTEM AND TREE CHOICE FOR ARID AND SEMI-ARID ZONE FARMERS

The following table lists the major cropping system and tree choice of farmers of varying economic strata in arid and semi-arid agro-climatic regions.

Table 26.2: Major Cropping System and Tree Choice of Varying Economic Strata Farmers (Pathak et al 2000)

Agro-Climatic region	Choice of Trees	Cropping Systems
Small Farmer		
Western Himalayan Region	*Populus, Robinia, Salix, Grwia optiva, Toona ciliata, Dalbergia sissoo, Morus alba*	Paddy-mustard, Paddy-berseem, Paddy-fallow, Paddy-oats, Wheat-maize, Wheat-soybean, Wheat, Pulses, Pea-tomato, Berseem, Pulses Maize-wheat, Rice
Eastern Himalayan Region	*Albizia* spp., *Alnus nepalensis, Artocarpus heterophyllus,* Bamboo, *Parkia roxburghii, Mangifera indica, Areca catechu*	Agri-horticulture: Paddy, Maize, Colocasia, Gourds, Pumpkin, Potato, Mustard, Pea, Oilseeds with Banana, Peech, Plum, Cabbage, Orange, Lemon
Southern Plateau and Hill Region	*Acacia nilotica, Azadirchta indica, Prosopis cineraria, Tamarindus indica, Annona squamosa, Mangifera indica, Anacardium occidentale, Musa paradisiaca, Cocos nucifera, Citrus aurantifolia*	Ground nut + red gram, sunflower, Sole cropping of Sorghum, Pearl millet, Paddy, Groundnut; Single crop (mono): Paddy-paddy, Paddy-fallow, Groundnut-fallow, Sorghum-fallow, Groundnut-chilli, Paddy-vegetables, Green gram-paddy, Groundnut-pulses, Tobacco-pulse crop, Groundnut -ragi, Pearl millet-ragi
Gujarat Plains and Hill Regions	*Azadirchta indica, P. cineraria, Mangifera indica, Acacia nilotica, Citrus* spp.	Pearl millet, Castor, Pulses-castor
Medium Farmer		
Western Himalayan Region	*Populus, Robinia, Salix, Grewia optiva, Toona ciliata, Dalbergia sissoo, Morus alba,* Other fruit, fodder, fuel, timber spp.	Paddy-mustard, Paddy-berseem, Paddy-fallow, Paddy-oats, Wheat-maize, Pea-tomato, Wheat-soybean, Wheat, Pulses, Berseem, Maize-wheat, Rice, Maize, Soybean, Rice-Maize, Vegetables, Cereals, Pulses
Eastern Himalayan Region	*Areca catechu*, Bamboo, Fruit trees	Rice, Jute, Oilseeds, Wheat, Pulses, Vegetables
Southern Plateau and Hills Region	*Acacia nilotica, Azadirachta indica, P. cineraria, Tamarindus indica, Annona squamosa, Mangifera indica, Anacardium occidentale, Musa paradisiaca, Cocos nucifera, Citrus aurantifolia*	Double cropping: Potato Sole cropping: Sorghum, Sugarcane, Tobacco, Sunflower Mixed cropping: Sorghum + red gram, Sunflower, Soyabean Green gram-groundnut, Sesame-Maize, Green gram-cotton, Groundnut-chilli, Paddy-vegetables, Green gram-paddy, Groundnut- pulses, Tobacco- pulses, Paddy-paddy, Groundnut-ragi, Pearl millet-ragi, Paddy-fallow
Gujarat plains and Hills region	*Azadirachta indica, Mangifera indica, Acacia nilotica, Emblica officinalis, Eucalyptus, Citrus* spp.	Sole cropping: Pearl millet and cluster bean, Pearl millet + pulses, Wheat, Mustard, Gram and Vegetables

Agro-Climatic region	Choice of Trees	Cropping Systems
Large Farmer		
Western Himalayan Region	Fruit trees, fuel, timber, energy plantation	Wheat, Maize, Rice-Maize, Cash crops, Floriculture
Eastern Himalayan Region	*Shorea robusta*, *Tectona grandis*, *Anthocephalus cadamba*, bamboo, *Michelia* spp., *Areca catechu*, *Artocarpus heterophyllus*	Rice, Jute, Oilseeds, Pulses, Vegetables Rice-rice, Mustard/potato/tomato, Radish, Maize-soyabean/mustard/radish
Southern Plateau and Hills Region	*Mangifera indica*, *Acacia nilotica*, *Azadirachta indica*, *Leucaena leucocephala*, *Tamarindus indica*, *Eucalyptus* spp., *Acacia auriculiformis*, *Annona squamosa*, *Anacardium occidentale*,*Musa paradisiaca*, *Cocos nucifera*, *Citrus aurantifolia*	Intercropping: Chilli + cotton + onion Sole crop: Cotton, Sorghum, Sugarcane, Sunflower Double cropping: Maize-wheat, Mixed cropping: (Sorghum + Red gram), Sun flower, Soybean Potato-wheat, Green gram-sunflower, Green gram-cotton, Groundnut-chilli, Paddy-vegetables, Green gram-paddy, Groundnut-pulses, Tobacco-pulses, Paddy-paddy, Groundnut-ragi, Pearl millet-ragi, Paddy-fallow
Gujarat Plains & Hills Region	*Mangifera indica*, *Acacia nilotica*, *Ailanthus excelsa*, *Tectona grandis*, *Emblica officinalis*, *Eucalyptus* spp.	Sole crop: Pearl millet, Cluster bean and Pulses, Wheat, Gram, Mustard and Vegetables

IMPORTANCE OF MAJOR AGROFORESTRY PRACTICES IN ARID AND SEMI-ARID REGIONS

The trees/shrubs on agricultural lands have been referred as a vegetation type (Champion and Seth 1968), however they are presently recognized as agroforestry system under trees outside forest/ tree cover category by Forest Survey of India. Fuelwood shortage is a major problem in most parts of the semi-arid and arid tropics and agroforestry potentials in fuelwood production are well documented. Similarly, desertification and fodder shortage, which are the other major land use problems in this zone, could be addressed to some extent through the agroforestry approach (Rocheleau et al 1988). Rural folk in Thar Desert have been practising arable cropping in association with these scattered trees on crop fields since time immemorial. This combined protective, productive system of integrating trees into farming system rooted in the principles of ecology, productivity, economics and sustainability. In fragile ecosystems like Thar Desert, agriculture alone cannot be a dependable enterprise, hence desert dwellers with their traditional wisdom are integrating trees into farming since ages in order to confer stability and generate assured income (Rathore 2004). As most of the trees are drought resistant, these are able to provide fuel, fodder and other products, when arable crops fail due to drought and other common feature of the hot arid region.

In case of cold arid region like Ladakh and Nubra valley, traditional agroforestry system in the form of agrisilviculture with the boundary plantation of Willow (*Salix* spp.) and Poplar (*Populus* spp.) are common. The local inhabitants of this cold region have their traditional agroforestry systems in meeting diverse subsistence needs (medicinal, fuel and fodder) particularly in xeric condition occur during prolonged winter season (Butola et al 2012). Hence, agroforestry in cold arid region originates as life support system.

In case of semi-arid region, the practice of agroforestry originates to increase the system productivity and sustainability. *Leucaena* based silvopastoral system, *Emblica officinalis* based horti-agriculture system and *Casuarina/ Eucalyptus* based silviagriculture system are some examples for them. The main agroforestry systems in semiarid zone are also influenced by population pressure; home gardens and multilayer tree gardens are found in the wetter areas with high population pressure. But generally speaking, the predominant agroforestry systems in this zone are (Mann and Saxena 1980; Shankarnarayan et al 1987):

- Various forms of silvopastoral systems
- Windbreaks and shelterbelts
- Multipurpose trees on crop lands, notably *Acacia* spp. based in semi-arid and *Prosopis cineraria* based in the dry areas of India

AGROFORESTRY PRACTICES IN HOT ARID REGIONS

Arid environments are diverse by land forms, soils, fauna, flora, water balance and human activities, while water is a common scare commodity. Much of rainfall is lost by evapotranspiration and as a result, ground water is recharged only by seepage through soil profile. However, it is a common phenomenon in arid zones of the world that ground water is frequently used at the rate that exceeds recharge. The situation in arid tropics of India, which is spread over an area of 31.7 million ha, is also experiencing the same situation. Tropical agriculture including agriculture in arid tropics, particularly subsistence agriculture is vulnerable, as farmers do not have adequate resources to adapt to climate change. Land use options that increase livelihood security and reduce vulnerability to climate and environmental change are necessary. Traditional resource management adaptations, such as agroforestry systems, may potentially provide options for improvement in livelihoods through simultaneous production of food, fodder and firewood as well as mitigation of the impact of climate change (Mann and Muthana 1984). Agroforestry may play a significant role in mitigating the atmospheric accumulation of greenhouse gases (DeFries and Rosenzweig 2010). It also has a role to play in helping farmers adapt to climate change.

The hot arid regions of India lie between 24° and 29° N latitude, and 70° and 76° E longitude, covering an area of 31.70 million hectares, and involving seven states: Rajasthan, Gujarat, Punjab, Haryana, Andhra Pradesh, Karnataka, and Maharashtra. In total, 11.8 per cent of the country is under a hot arid environment. The arid regions of Rajasthan, Gujarat, Punjab, and Haryana together constitute the 'Great Indian Desert', better known as the 'Thar Desert'. As arid western Rajasthan accounts for 61 per cent of hot arid region of the country, it is considered principal hot arid region. The hot arid western Rajasthan extends for 640 km from northwest to southeast, with an average width of 300 km from east to west. In the east of Jaisalmer, the desert forms a lowly elevated plateau, 150-325 m above sea level, with numerous rock masses jutting out of the desert sand. The only river of some significance is the Luni. This river rises in the Aravalli hill range near Ajmer, and after an initial west-southwesterly course towards Barmer, it flows southwest until it drains into the northeastern corner of the western Rann of Kutchh. The principal tributaries of the Luni river are the Lilri, Raipur Luni, Guhiya, Bandi, Sukari, Jwai and Jojri. Neither the Luni River nor its tributaries are perennial. Because of lack of adequate river and drainage systems in the Thar, there are no well defined channels for the discharge of water in the entire region.

EDAPHO-CLIMATIC CONDITIONS

Low annual precipitation (100-300 mm); very high temperature during the summer season (mean maximum temperature 4°C) touching a maximum of 48° to 50°C; short (December to mid-February) cool and dry winters (the mean winter season temperature varies from 10° to 14°C); high wind speed

(30-40 km/hr); high evapotranspiration; and general low humidity (an aridity index of 0.045-0.19) are important climatic limitations of hot arid region (Sharma and Tewari 2005). Sand dunes are a dominant land formation of the region. More than 58 per cent of the area is sandy and intensities of dunes vary from place to place. In general, soils contain 1.8-4.5 per cent clay, 0.4-1.3 per cent silt, 63.7-87.3 per cent fine sand, and 11.3-30.3 per cent coarse sand. They are poor in organic matter (0.04-0.12%), and low to medium in phosphorus content (0.05-0.10%). The nitrogen content is mostly low, ranging between 0.20 and 0.07 per cent and infiltration rate is very high, at 7–15 cm/hr (Dhir 1997). Because of the complete absence of any aggregation, the soils are highly erodible by wind and water.

NATURAL RESOURCES

Since the vast arid expanse of Ghaggar is blank on Chalcolithic and Iron Age maps of the Indian subcontinent, it seems that human occupation of the region was thin, and depended upon hunting and limited pastoralism (Dhir 1982). The population growth rate for the decades 1971-1981 and 1981-1991 were 36.7 and 30.7 per cent, respectively. The population in the Thar during 1971 was 48 persons/km^2 which reached 127 persons/km^2 in 2011. Moreover the population density is quite high compared with the global average of 6-8 persons/km^2 for arid zones. The sources of rural livelihood are caste occupations such as carpentry, black smithy, oil pressing, pottery making, tanning, leather work, dyeing, gold smithy, etc.

In spite of erratic and unevenly distributed rainfall, agriculture is the mainstay of rural populations, and mixed crop-livestock, mixed livestock-crop farming, and livestock farming, form the spectrum of economic activities. The distribution of agricultural land holdings by size is fairly large at 4-6 ha. Next to land, livestock constitutes the most important asset of the cultivators. On an average, a household has 1.32 buffaloes, 2.28 cows, 1.48 cow young stock, 0.62 buffaloes, 0.38 buffalo young stock, 6.48 sheep, 6.50 goats and 0.49 camels (Sharma and Tewari 2005). There are sparsely distributed trees mainly of *Prosopis* and *Acacia,* and underneath growth of arable crops and/or grasses, and other herbaceous flora (especially during the *Kharif* season, as agriculture is predominantly rainfed) in long stretches interspersed with distantly distributed village settlements and *dhanis* (a unique settlement pattern of the arid western Rajasthan intended for the life of agriculture families during the active cropping period away from the village, but nearby their fields) are features of the region. In fact, desert of arid western Rajasthan is a vegetated desert having a very rich floral diversity of about 682 plant species, out of which 131 are known for their economic uses.

TRADITIONAL AGROFORESTRY PRACTICES IN HOT ARID REGION

In general rainfall appears to be governing factor for evolution of traditional agroforestry systems. On the basis of rainfall, four types of major traditional agroforestry systems can be classified as presented in Table 26.3 (Narain and Tewari 2005). The tree/shrub density of the system forming species is also clearly governed by rainfall regime. Data indicated that with decrease in rainfall, the density of woody component of the system decreased substantially and as well as the productivity of arable crops and grasses were declined.

Generally rainfall plays determinant role in productivity. *P. cineraria* - *A. nilotica* based agroforestry system was found to be highly productive and *Ziziphus* spp. - *P. cineraria* - *Salvadora* spp. system was least productive (Table 26.4). Pearl millet, main cereal crop of the region, is a most compatible crop under any tree and interestingly yields of the crop under the canopy of *Ziziphus* spp. - *P. cineraria* and *Ziziphus* spp. - *P. cineraria* - *Salvadora* spp. systems substantially increased in comparison to sole cropping. However, legumes such as moth bean, cluster bean and mung bean exhibited slight yield reduction below the tree canopies, 5.0 per cent (*Ziziphus* spp. - *P. cineraria* system) to 19.5 per cent (*P. cineraria* system); 12.5 per cent (*Ziziphus* spp. - *P. cineraria* - *Salvadora* spp. system) to 18.2 per cent (*Ziziphus* spp. - *P. cineraria* system); and 10.0 per cent (*P. cineraria* system) to 33.7 per cent (*P. cineraria* - *A. nilotica* system), respectively. Yield reduction under tree canopies for sesamum ranged from 26.3 per cent (P. *cineraria* - *A. nilotica* system) to 46.1 per cent (*P. cineraria* system). *Prosopis cineraria* based agroforestry systems yielded more forage and herbage production.

Table 26.3: Traditional Agroforestry Practices in Arid Region of Western Rajasthan

Rainfall (mm)	Agroforestry Practice	Trees/Shrubs (No./ha)	Density of Prominent Tree Species (%)
<200	*Ziziphus* spp. - *P. cineraria* - *Salvadora* spp. based	17.2	87.2 (65% *Ziziphus* spp.)
200-300	*Ziziphus* spp. - *P. cineraria* based	91.7	100.0 (91.7% *Ziziphus* spp.)
300-400	*P. cineraria* based	14.2	80.0
>400	*P. cineraria* - *A. nilotica* based	31.4	80.5

Table 26.4: Yield (t/ha) of Major Crops under Traditional Agroforestry Practice in Western Rajasthan

Crops Systems	Agroforesry Practices			
	***P. cineraria* based**	***P. cineraria* - *A. nilotica* based**	***Ziziphus* spp. - *P. cineraria* based**	***Ziziphus* & other spp. based**
Pearl Millet				
Below the canopy	7.9	10.5	2.0	2.0
Away from canopy	7.9	10.8	1.8	1.2
Moth Bean				
Below the canopy	2.1	2.8	1.9	1.4
Away from canopy	2.6	3.0	2.0	1.6
Cluster Bean				
Below the canopy	1.8	2.5	0.9	0.8
Away from canopy	2.1	2.9	1.1	1.0
Mung Bean				
Below the canopy	0.9	1.4	-	-
Away from canopy	0.1	2.1	-	-
Sesamum				
Below the canopy	2.1	1.4	1.9	-
Away from canopy	3.9	1.9	2.6	-

Traditional *P. cineraria* - *A. nilotica* based fuelwood and leaf fodder production is maximum, while it was minimum for *Ziziphus* spp. - *P. cineraria* - *Salvadora* spp. system (Table 26.5). The wood production under these systems also follows the rainfall gradient.

Table 26.5: Fuelwood and Tree Fodder Production of Traditional AF Practices in Western Rajasthan

Agroforestry Practices	Plant Density (trees/ha)	Fuelwood (t/ha/yr)	Leaf Fodder (t/ha/yr)
P. cineraria - A. nilotica based	31.4	1.23	0.29
P. cineraria based	14.2	0.60	0.20
Ziziphus spp. - *P. cineraria* based	91.7	0.62	0.90
Ziziphus spp. - *P. cineraria - Salvadora* spp.	17.2	0.18	0.14

IMPROVED AGROFORESTRY PRACTICES IN HOT ARID REGION

Since 1970s, several improved agroforestry practices to enhance overall productivity and economic returns of the system have been developed and standardized. Following are some improved AF practices found promising and remunerative for arid India.

AGRIHORTICULTURE SYSTEM: Improved Ber (*Ziziphus mauritiana*) varieties and legume as understorey crop is the well accepted agrihorticulture system in arid Rajasthan. Table 26.6 explains the productivity of the agroforestry system over sole cropping in agrihorticultural system with ber (@ 400 trees/ha) + Mung bean (Gupta 1997). Similarly Aonla (NA - 7 var., 10 x 8 m spacing) and pomegranate (Jalore seedless var., 5 x 5 m spacing) with intercrops henna, horse gram and moong beans are recommended AF practice for Pali region of arid Rajasthan (Rohilla et al 2009).

Table 26.6: Productivity of Agrihorticulture System

Treatment	Annual Rainfall (mm)	Fruit Yield (kg/ha)	Grain Yield (kg/ha)	Net Profit (Rs/ha)
Sole crop	200	-	520	4800
Intercropped with Ber	200	800	200	8000

HORTIPASTORAL/ SILVOPASTORAL SYSTEM: Ber based hortipastoral systems have proved highly remunerative in Thar desert on farmers' field. The ideal spacing of ber in this system is 6 m x 6 m and suitable grass species is *Cenchrus ciliaris*. Arid region of Rajasthan is reported to have 89 species of grasses. The species like *Lasiurus sindicus* is efficient builder of biomass with energy use efficiency of 1.4-2.0 per cent (Harsh et al 1992). An improved silvopastoral system by *Hardwickia binata* at 3 m x 3 m spacing with *Cenchrus ciliaris* grass showed the average carrying capacity of 4.1 sheep/ha/year against 3.7 for sole *C. ciliaris* pasture and 1.6 for sole *H. binata* plantation. In this type of improved silvopastoral practice, in addition to grass + top feed production to the tune of 3.06 t/ha/year, 0.26 t/ha/year of fuelwood is also obtained.

Grasses when intercropped between trees spaced at 5 m x 5 m gave a dry forage yield of 12.5 to 20.0 q/ha depending upon rainfall distribution and had no adverse effect on tree. *Cenchrus* species have minimum effect on retardation of tree species. *Grewia tenax* and *Ziziphus nummularia* planted at 10 m x 4 m and 10 m x 2.5 m respectively with *Cenchrus ciliaris* found optimal for enhancing the productivity of silvopastoral system. Meanwhile the grass production was found maximum under *Prosopis cineraria* and *Eucalyptus camaldulensis* in Marwar region (Rohilla et al 2009). The above proven AF practices ensure the productivity augmentation in arid region.

IMPROVEMENT OF WATER USE EFFICIENCY BY AGROFORESTRY

The tree canopies in agroforestry systems intercept the rain and reduce runoff (Khan et al 1995). In a study at CAZRI, it was found that in *Acacia tortilis* silvopastoral stand canopy interception was 21.4

per cent, whereas in *Colophospermum mopane* silvopastoral it was 13.1 per cent. Rainfall interception was positively related with canopy cover and negatively related with throughfall. Average surface run off in *A. tortilis* based silvopastoral stand was 53 per cent higher than in *C. mopane* based silvopastoral. This indicated that hydraulic response to rain is dominated by plant species character, however the per cent annual runoff and soil erosion was very low in situations with trees on agricultural fields in comparison to bare soil condition. Thus, the enhanced use efficiency of rain water by woody species in agroforestry systems improves agricultural productivity. The initial observation on stem flow and through flow of arid trees varied greatly, which recorded maximum stem flow in *Eucalyptus* followed by *Azadirachta indica*, *Acacia senegal* and minimum in *Prosopis cineraria*.

MODERATION OF MICROCLIMATE BY AGROFORESTRY

Genetic potential of many crops and varieties can only be realized when environmental conditions are close to optimum. Trees on farm bring about favourable changes in the microclimatic conditions by influencing radiation flux, air temperature, wind speed, saturation deficit of understorey crops, all of which will have a significant impact on modifying the rate and duration of photosynthesis and subsequent plant growth, transpiration, and soil water use (Monteith et al 1991).

Following examples present microclimatic moderation effect in harsh climatic conditions of arid regions of western Rajasthan. Air temperature exhibited appreciable variation under the canopy of *A. tortilis* during monsoon period (Tewari et al 1989). A decline of 0.1° to 0.7° C was evident beneath the canopy cover of *A. tortilis* than that of surrounding open areas at 07.00 hrs. Around 14.00 hrs, this decline was of the order of 0.6° to 2.0° C. Similarly, it was found during monsoon season that the soil temperature just beneath the tree cover was lower by as much as 10° to 16° C in top soil zone and 4° to 5° C at 30 cm depth when compared to open field conditions, thereby indicating better soil-thermal regime. Seven year old *A. tortilis* trees provided more humidity on cluster bean grown with the trees. This finding proves the efficiency of agroforestry practices for maintaining favourable moisture status on arable lands (Ramakrishna and Shastri 1977).

AGROFORESTRY SYSTEMS ENHANCING LIVELIHOOD OPTIONS

Arid regions are particularly notable for ethano-forestry practices and indigenous knowledge system on growing trees on farm lands. Agroforestry contributes to livelihood improvement in arid regions of India, where people have a long history of accumulated local knowledge (Pandey 1998). The economics of previously discussed traditional agroforestry systems of arid western Rajasthan indicated that net B:C ratio of such systems is generally on positive sides (Table 26.7).

Table 26.7: Economics of Traditional Agroforestry Practices of Arid Western Rajasthan

AF Practices	Expenditure (Rs./ha)	Returns (Rs./ha)			Gross Returns (Rs./ha)	Net Returns (Rs./ha)	Net B:C Ratio
		Crops	Fuel Wood	Leaf Fodder			
P. cineraria - *A. nilotica* based	1,850	4,103	1,230	870	6,203	4,353	2.3
P. cineraria based	1,550	3,670	600	420	4,690	3,140	2.0
Ziziphus spp. - *P. cineraria* based	1,550	1,506	620	600	2,726	1,176	0.7
Ziziphus spp. - *P. cineraria* - *Salvadora* spp.	1,500	1,400	500	500	2,400	900	0.6

In fact, multipurpose trees in agroforestry systems provide a rational formula for drylands. In hot arid regions, trees on farmlands are the organisms with an ecological investment strategy (Oldeman 1983). The biomass and energy are immobilized in long lived trees and this majority of accumulated production is harvested at longer interval than in case of arable crops which accounts for non-tangible benefit to the society at large.

Potential of agroforestry in arid regions are breaking the poverty and food insecurity. In addition to providing food and feed, there is numerous other non-timber tree products (excluding gums and resins) collected from trees on farm lands. The most spectacular evidence in context of livelihood improvement potential of agroforestry in arid region came through *Acacia senegal* based localized traditional agroforestry system in parts of Barmer and Jodhpur districts. *A. senegal* is source of gum Arabic which has very high commercial value. In addition to harvesting crop grain for food and crop straw for fodder, the farmers harvested substantial quantity of gum arabic through CAZRI gum inducer. The participatory mode of proven technologies like the gum exudation from *Acacia senegal* assures the possibility of win-win situation to the technology provider and the technology practitioner (Ratha Krishnan et al 2012). Meanwhile seeds of *Acacia senegal* are also used in preparing vegetable. The lifeline tree of desert, *Prosopis cineraria* providing fuel and fodder, nutritious vegetable in the form of pod, locally known as 'sangri' while, pod floor of *Prosopis juliflora* is as good as wheat floor. During drought years their value and demand is increased tremendously. In fact, scattered trees on agricultural fields in arid regions of western Rajasthan are multipurpose in nature and they have ability to satisfy the expectations and aspiration of rural folk regarding production of their basic needs while provides unique basis of life support in hot arid regions. However processing of traditionally important arid foods into more useful and convenient product can improve livelihood security of the people residing in the desert region (Madhu Goyal and Sharma 2009).

AGROFORESTRY PRACTICES IN COLD ARID REGIONS

Agroforestry land use of cold condition is not having great variety of systems and practices as like tropics, only a few agroforestry systems are practised in these regions. The two most common systems have been the agrisilvicultural use of windbreaks and shelterbelts to prevent soil erosion in the plains, and silvopastoral practices with livestock in many different woodland and range ecosystems. Agrisilvicultural combinations of nut or fruit trees and herbaceous crops are an increasingly common third system. Socioeconomic conditions of this cold region have also strongly influenced land use practices. Although small farms were historically dominant in the cold regions mainly for their home needs, still in many regions, there has been a significant trend towards large, family, corporate, or communal farms where production is largely concentrated on a few crops for local and distant markets. Agroforestry applications on such farms have often focused on one or two high-value crops and include high levels of mechanization. Combinations of trees and agriculture are viewed opportunistically, i.e., as a means to improve economic profitability. Thus, these cold arid agroforestry characteristics are in contrast to those of tropical agroforestry practices, which are most frequently found on small individual farms, or sharecropped on community lands.

DISTRIBUTION AND CHARACTERISTIC FEATURES OF INDIAN COLD DESERT

Indian cold deserts spread in Himachal Pradesh (Lahaul & Spiti and Kinnaur districts), Jammu & Kashmir (Leh and Kargil districts), Uttarakhand, Sikkim and Arunachal Pradesh. Among that the Ladakh region is one of the most elevated (2,900 m to 5,900 m asl) and coldest region (lowest temperature -30°C). It covers more than 70,000 square km geographical area of Jammu and Kashmir, lies between sparsely populated along the river banks of different valleys namely Indus, Nubra,

Changthang, Zanskar and Suru valley. The mean annual precipitation is less than 50 mm, received mostly in the form of snowfall. The region faces fast blowing winds 40-60 km/hr mainly in the afternoon hours. The soil moisture remains frozen during winters and low relative humidity during the summer months. The region has barren topography. The soils of the region are gravelly and sandy loams on the alluvial fans to sandy and slit clay loams on the Indus plains. Loose sandy loam texture, high percentage of stones and granules, low water holding capacity, high bulk density and low soil fertility may be due to result of uneven distribution of plantation or sparse vegetation (Butola et al 2012).

The region has very short growing season as the land remains landlocked for more than 6-7 months every year during extreme winter. Due to small land holdings, local population subsist on limited crops and largely depends on natural resources for meeting diverse subsistence needs which has led to over exploitation. However, main source of income are rearing of Pashmina goats or Changthangi Goat (*Capra aegagrus hircus*), which provide cashmere wool for making well-known Pashmina shawl, and eco-tourism activities. Natural wealth of the region is under various biotic (over-harvesting, grazing, trampling, invasion of alien flora, etc.) and abiotic (natural calamities, habitat fragmentation and degradation through increasing human settlements, climate change, tourism activities, etc.) pressures causing high ecological imbalance in the region.

AGRICULTURAL CROPPING PATTERN IN COLD ARID REGION

Mixed crop-livestock farming and livestock farming forms the spectrum of economic activities in rural setting of the cold arid region. Farming systems in Ladakh region mainly consider with internal resource recycling, self-sufficiency, value complementarities, diversity and efficient utilization of resources (Pandey and Singh 1984). In spite of harsh climatic conditions, Ladakhis have managed to develop a remarkably productive agricultural system, with yields often comparable to and even out competing those of European intensive regimes (Mankelow 2003). Yet, the availability of cultivable soil and, more critically, of water tend to limit both productivity of the land and areas that can be put under cultivation, and hence to some extent the size of villages. Droughts occur, forcing farmers to drastically reduce the area under cultivation (Norber-Hodge 2000). Since the growing season is short (in most regions less than ninety days), fast-growing species of barley, wheat and peas are planted in association of MPTs mainly on farm bunds. At lower altitudes, one finds buckwheat, apricot and apple orchards. Livestock, mainly sheep, goats, yaks, cows, dzo, donkeys and horses, plays a central role, providing fuel, transport, labour, wool, milk, meat, hides and heat in winter. But most of all, animals allow the recycling of nutrients derived from agricultural wastes and pastures. With the exception of donkeys and horses, domesticated animals usually spend the summer grazing in high-elevation pastures. Dung is collected from the pastures, fields and stables to be used as fuel and as fertilizer. Agricultural work is done with the help of animals, wind and water.

The land utilization pattern of Leh, Nubra and Nyoma showed about 2.64 per cent area as arable and 9.66 per cent area under vegetation cover. This kind of lower percentage of arable land is restricted only to flat valleys and in lower slopes where water availability is ensured. In high ridges, the wild plant species depend upon the good amount of snowfall because of an acute scarcity of water on the barren mountains. The traditional crops of the region include barley, grim, wheat, buck-wheat system, millets, mustard, oat, vegetables (pea, potato, onion), spice and medicinal plants. Kitchen gardens are more common to meet their daily family requirements.

As important cropping pattern, local farmers practising diversified farming by including vegetable crops including *Brassica rapa*, *B. oleracea* (both cauliflower and cabbage), *B. nigra*, *B. caulorapa*, *Chenopodium album*, *Cucurbita maxima*, *Cucumis melo*, *Solanum lycopersicum*, *Solanum melongena* and *S. tuberosum* under cultivation. Besides cultivated crops, some wild plants in the region are used as vegetable which include *Amaranthus spinosus, Capsella thomsonii, Allium thomsonii, Lactuca dolichophylla, Chenopodium foliosum, Lepidium latifolium, Orobanche hansii* and *Polygonum aviculare*. Additionally, many legumes associated with agricultural crops act as valuable sources of fodder and soil enrichment (Butola et al 2012).

TRADITIONAL AGROFORESTRY PRACTICES OF COLD ARID REGION

Agroforestry systems in various parts of the temperate region especially for Himalayan India are already reported by Dadhwal et al (1989) and Toky et al (1989). Traditional agroforestry system of this region exists in the form of agrisilviculture system that is the combination of agricultural crops with boundary plantations of willow and poplar tree species. They are main source of fuelwood and fodder. Otherwise the requirement of fuelwood during winters when temperature goes very low is met through cutting of dry as well as green plants or collection of fallen twigs in nearby forests. This unsustainable practice of harvesting has posed huge pressure on the wild stock. Almost all the woody species are used as a source of fuelwood in the valley. According to an estimation, every year these species are contributing 400 tons of leaf litter to ground and thus, being great source of organic carbon and responsible for sequestration of more than 75,000 tons of carbon (Kumar et al 2009).

A wild plant, Seabuckthorn (*Hippophae rhamnoides*), a multipurpose thorny shrub used for food, fuel, fodder, medicine and for fencing the fields is an important species. Being a nitrogen fixing species, it is planted in Igoo - PHE canal as rehabilitation measures to the soil condition of the area and to support the agricultural crops. The valley portion are well vegetated with thickets of seabuckthorn compared to mountain slopes and remaining part of Ladakh region. In lower slopes, the integration of agricultural crops are rarely seen with fruit trees as apple (*Malus pumila* or *M. sylvestris*), apricot (*Prunus armeniaca*), peach (*Prunus persica* Batsch), mulberry (*Morus alba*) and walnut (*Juglans regia*). Wild strawberry (*Fragaria vesca*) has been introduced in combination with agricultural crops. The raising of some fruit tree species in kitchen garden is also well established tradition of the region.

IMPROVED AGROFORESTRY PRACTICES FOR COLD ARID REGION

The commercial fruit orchards are the best example for improved AF practice in cold arid regions. In that, fruit trees are usually widely spaced and orchard practices in the last 100 years often excluded agricultural activities other than forage production for limited grazing within the orchards. In most fruit orchards, cultivation of vegetable and other crops during the establishment phase reduces the need for vegetation management such as mowing and herbicide application. Fertilizers applied to the orchard trees or vegetables are available to the other crops. Produce from the orchards may be used for home consumption or market sale. As the trees develop and shade the orchard floor, annual crops can be replaced by forage species. At that time, the orchard can be opened to grazing as the trees would be large enough to escape damage by animals. Combinations of fruit trees and other species are traditional practices in regions such as the mid-elevation Himalaya mountains in the Indian subcontinent. For example, citrus is intercropped with winter vegetables and gram (*Cicer arietinum*) for 2-3 years, and beans and peas are often grown in dwarf-apple orchards for 5-6 years, or beneath apricots, peach, plum, and nectarine for 2-3 years (Tejwani 1987). Vegetable production is eventually reduced as the fruit trees mature and shade the orchard floor. The main tree species used in these Himalayan agroforestry systems are listed in Table 26.8.

Table 26.8: Common Tree Species in Agroforestry Practices of Western Himalayas (Dadhwal et al 1989; Toky et al 1989)

Tree Types	Scientific Name	Local / Common Name
Fodder Trees	*Albizia chinensis*	ohi
	Bauhinia variegata	kachnar
	Celtis australis	khirik
	Ficus roxburghii	timla
	Grewia optiva	bhimal
	Morus serrata	kimu
Fruit Trees	*Citrus* spp.	orange, lemon
	Litchi chinensis	litchi
	Mangifera indica	mango
	Prunus armeniaca	apricot
	Prunus domestica	plum
	Prunus dulcis	almond
	Prunus persica	peach
	Psidium guajava	guava
	Pyrus communis	pear
Timber and Fuelwood Trees	*Bombax ceiba*	simal
	Dalbergia sissoo	shisham
	Eucalyptus spp.	safeda
	Ficus palmata	fig
	Melia azedarach	darek
	Pinus roxburghii	chirpine
	Pistacia integerrima	kakkar
	Prunus puddum	pazza
	Quercus spp.	oak
	Shorea robusta	sal
	Toona ciliata	toon

AGROFORESTRY IN EASTERN HIMALAYAN REGION

The indigenous Himalayan people, Lepcha and Limbu, used to collect large cardamom from natural forests, which were later on domesticated. Among several shade tree species in large plantation area, alder (*Alnus nepalensis*) is found to be the most abundant and preferred tree, which is non-leguminous and nitrogen fixing tree. Besides large cardamom, many food crops like maize, millet, potato, barley, chilli and colocasia are grown with alder. The tree not only provides shade to arable crops, timber and fuel wood but also ameliorates soil and protects it from erosion on hilly slopes. A yield of 400-500 kg/ha of cardamom has been reported in a year under alder trees. If a village with 100 families could set aside about 120 ha of land to grow alder trees, all families would be able to get sufficient fuelwood every year and at the same time raise crops under alder in about 30 ha sole crop area of a year. Cultivation of coffee, ginger, cardamom, turmeric and medicinal plants under the shade of naturally growing trees in Meghalaya is one of the best examples of successful combination of trees with annual crops.

Besides the above system, many trees are lopped for their green fodder which is rich in crude protein and calcium. These are found grown on terraces which are widely spaced and helps in better yield reduction of intercrops. Different trees provide fodder at different times of year sustaining supply of fodder throughout the year. Among the introduced tree species, *Leucaena leucocephala* (var. S-11) does well at low to mid hills and it could produce 9 tons/ha dry foliage annually under the stocking density of 2,500 trees/ha. In low to mid hills, mandarin-orange is a commercial fruit crop which is intensively intercropped with annual food crops mainly maize. In Meghalaya, pear, plum and peach while in Sikkim apple trees are intercropped with potato maize, vegetables, peas and mustard. At some places pineapple is also grown as intercrop. Among monocrops, tea plantation is common in pockets but now cherry (*Prunus cerasus*) is also planted as live-fence. Shade trees like *Albizia chinensis* may be grown for desired quality of tea which has been reported to add 2.5 to 5.0 ton/ha organic matter to soil annually containing 63-126 kg N, 18-36 kg P_2O_5, 22-44 kg K_2O, 32-64 kg CaO and 16-32 kg MgO per ha. Department of Agriculture in Mizoram has developed its own contour trench-farming for Jhum areas on hills where top-portion is of undisturbed forest, middle-portion is with horticultural crops and down the hill, terraced rice is cultivated with pineapple and grasses on contours. Many species of bamboo, palms (coconut, arecanut and species of *Licuala*, *Phoenix*, *Wallichia*, *Corypha*, *Caryota*) and rattans (*Calamus* spp.) are cultivated widely in north-eastern areas as mixed or boundary plantations. Pastures in forests are also a common practice in this region. In Sikkim hilly area, Nevaro (*Ficus auriculata*) based silvopastoral systems have showed production potential of 15.6-51.6 tons/ha annually and goats could be reared successfully. It has been found that sericulture based system is highly relevant for these areas. Mulberry with French bean-groundnut followed by mustard is a profitable cropping system with mulberry, guava/lemon/pear and pineapple in paired rows and grasses on the bunds is an ideal system for silk production and additional income from fruits and cattle-rearing. Paddy-cum-sericulture is said to be more viable as the cash returns are more frequent.

AGROFORESTRY IN WESTERN HIMALAYAN REGION

In western Himalayas 60 to 70 per cent requirement of the firewood is met from arboreal components and several multipurpose trees along the bunds of agricultural lands or scattered trees on the pasture lands. *Grewia optiva*, *Celtis australis*, *Bauhinia variegata*, *Albizia chinensis*, *Bombax ceiba*, *Melia azedarach* and *Toona ciliata* are common MPTs while plum (*Prunus domestica*), apricot (*P. armeniaca*), peach (*P. persica*), almond (*P. dulcis*) and pear (*Pyrus communis*) are common fruit trees of agri-horti-silvicultural and agrihorticulture systems. Hedge-row intercropping is feasible and important on sloping hilly lands when pruned biomass during cropping season can be used for fodder and fuelwood. The success of kinnow based horti-silvi-agricultural system at Dhaulakuan in foothills of Western Himalaya has been practised successfully. Kinnow was planted at a spacing of 5 m x 5 m. The interspaces between rows were utilized for sowing *Leucaena leucocephala* in hedge rows or scattered trees and maize, soybean, vegetables and wheat were grown as annual crops. To accommodate wood for packing of horticultural produce in region, a horti-silvopastoral system was also developed growing trees of Santa Rosa plum at a spacing of 8 m x 8 m and interspaces and boundary of field were utilized for growing *Populus deltoides*, and the results have been reported as encouraging. Since fruits and animal products are back bone of Western Himalayan economy, strategies to develop agroforestry systems based on fruit trees supplemented with cattle, keeping in view the soil conservation aspect, are must. Hence, fruit trees must be grown with forages, crops, vegetables and MPTs on small watershed basis. The timber woodlots for supporting cottage industries may help to improve socio-economic status of rural people.

MAJOR SETBACK IN TRADITIONAL AGROFORESTRY PRACTICES OF COLD ARID

Following are the major setbacks to practice improved AF practices in cold arid India.

- Limited choice of trees (only Willow and Poplar) to include as perennial component in AF system.

- Non-availability of improved tree species cultivars.
- Very poor linkage of research, extension and marketing, etc.
- Non-awareness of propagation, plantation techniques of introduced tree, agriculture crops.
- Lack of integrated sustainable multidisciplinary approach.
- Research strategies are either ineffective or insufficient for the region
- Shifting of local people from agriculture to seek opportunities in other enterprises and ecotourism activities.
- Moisture (very low precipitation) and temperature limitations
- Climatic uncertainties

AGROFORESTRY PRACTICES IN SEMI-ARID REGIONS

In semi-arid regions, agroforestry provides a means by which the imbalances brought about by deforestation may be redressed, environmental services restored, and productivity increased relative to that offered by traditional cropping systems (Jeremy 2000). In tropical semi-arid regions, there is substantial potential for increasing water use by implementing agroforestry as existing cropping systems often use less than half of the available rainfall due to substantial losses caused by evaporation from the soil surface, runoff and drainage (Wallace 1986). Low rainfall use efficiencies occur primarily because the fraction of rainfall used for transpiration is low and the ratio of evaporation to transpiration is high when the crop canopy is sparse (Unger et al 1988). Ong et al (1992) reported that a sorghum/pigeonpea intercrop on the Deccan plateau in India used 41 per cent of the annual rainfall, while the rest was lost as runoff (26%) and deep drainage (33%). In agroforestry systems, runoff may be reduced by the barrier effect of closely planted tree rows, as outlined above, while deep drainage may be limited by water extraction by tree roots located beneath the crop rooting zone (van Noordwijk et al 1996).

The choice of tree species for use in agroforestry is likely to be based on a combination of product value, interactions with crops, availability and price of seedlings, management inputs, and the environmental services provided. Attention has recently turned to indigenous tree species, partly because of the perception that fast-growing exotic species are extravagant in their use of water (Florence 1986), but also due to a realization of their value to local communities (Nair 1997) and importance in providing environmental services, whilst moving the source of tree products away from natural forests (Leakey and Simons 1998). Indigenous species may be less susceptible to attack by pests and have better survival rates, but often exhibit lower growth rates than exotic species providing wood of similar quality (Haggar et al 1998). However, there is great potential for the improvement of the quality and growth rates of wild indigenous trees through domestication, genetic improvement, and development of suitable management strategies (Leakey and Simons 1998). A study conducted at RRS, CAZRI, Bhuj, Gujarat with three tree species (neem, subabul and Israeli babool) and two *Cenchrus* species showed that neem and subabul with *Cenchrus ciliaris* based AF system increased the wood biomass and fodder yield (321 and 303 kg/ha) (Devi Dayal et al 2009).

The use of nitrogen-fixing trees to improve soil fertility has received much attention in systems containing widely scattered trees such as *Faidherbia albida* in the arid and semi-arid tropics, and in hedgerow intercropping in the sub-humid tropics, where *Leucaena leucocephala* and *Gliricidia sepium* may rapidly increase soil fertility (Ong 1996). However, benefits have only been found where the soil is fertile and there is sufficient rainfall and labour (Sanchez 1995). In the semi-arid tropics, hedgerow intercropping has not met with the success initially anticipated as the fertility improvements obtained have failed to offset competition for water and nutrients (Rao et al 1997). *G. sepium*, contributes substantial quantities of nitrogen to the soil-plant system in the humid and sub-humid tropics (Heineman et al 1997). Similarly, Avery and Rhodes (1990) reported that less than one per cent of the nitrogen fixed by *L. leucocephala* trees was taken up by sorghum in agroforestry systems under nitrogen-limiting soil conditions.

In Central India, *Dalbergia sissoo, Acacia nilotica, A. leucophloea, A. catechu, Albizia lebbeck, Azadirachta indica, Butea monosperma, Pongamia pinnata, Holoptelea integrifolia, Balanites roxburghii* and *Dichrostachys cinerea* form important constituent of silvopastoral systems. On black cotton soil region, very tall grasses supporting this system include *Heteropogon contortus, Chrysopogon fulvus, Themeda triandra, Iseilema laxum, Dichanthium annulatum*, and the industrial grass *Cymbopogon martinii*. These can make a sustainable silvopastoral system. In an assessment of soil conservation values of some grasses and herbs in two phase, it was found that *Dichanthium annulatum* showed maximum conservation value (up to 89), followed by *Cynodon dactylon* and *Bothriochloa pertusa*. In second phase *Dichanthium caricosum* showed maximum value (94.4), followed by *Sehima nervosum, Cymbopogon martinii, Chrysopogon fulvus, Iseilema laxum* and *Heteropogon contortus* (all having conservation value of more than 83).

In semi-arid regions of Peninsular India, the systems are more complex as the problem of frost does not exist, therefore, a vast number of trees (both fruit yielding and MPTs) exist on agricultural fields. *Borassus flabellifer, Tamarindus indica, Acacia leucophloea, A. catechu, Casuarina equisetifolia, Senna siamea, Eucalyptus tereticornis, Albizia lebbeck*, and many others are frequent trees on farm. *Leucaena leuocephala* has also been adopted as the common hedgerow crop in many areas.

COMPONENTS INTERACTION IN SEMI-ARID AGROFORESTRY

In semiarid regions, agroforestry practice exists to enhance the system productivity. When several components utilize the same reserves of growth resources such as light, nutrients, water, and CO_2, negative interactions, often through competition, are likely to occur in every plant association (Newman 1983). This competition can be separated into that caused by direct interference (real competition), and that caused by exploitation of shared resources which is mediated by other plants or shared predators (apparent competition) (Cannell 1990). Shading was found to be more important than below ground competition in an intercropping study with pearl millet and groundnut in India (Willey and Reddy 1981). Similarly, Verinumbe and Okali (1985) showed that competition for light was a more critical factor than root competition for intercropped maize between teak trees (*Tectona grandis*) in Nigeria. While the availability of light may be the most limiting factor in many situations, particularly those with relatively fertile soils and adequate water availability, the relative importance of light will decrease in semi-arid conditions as well as on sites with low fertility soils. Since crops differ in their responses to poor nutrition, competition for light or water may either be reduced or amplified by a shortage of nutrients (Cannell 1983).

There are innumerable studies indicating how competition for nutrients can reduce crop yields. In most cases, the yield of the agricultural crop is the criterion by which the merit of an agroforestry system is assessed; yield depressions of this component therefore receive more attention than those of the associated tree species. Furthermore, since the crop is usually the smaller component (when compared individually), its root system will usually be confined to soil horizons that are also available to the roots of the trees; but the trees can exploit soil volume beyond reach of the crop. Therefore, the effects of nutrient competition will probably be more severe for the crop components.

With the exception of areas with well-distributed rainfall, water competition is likely to occur in most agroforestry systems at some period of time; this period may be as short as a dry spell of one or two weeks. The effect of these events depends on the severity of the drought and the drought tolerance of the plants. It also depends on the degree of competition for other resources, especially nutrients. In alley cropping trials of *Leucaena* with cowpea, castor, and sorghum under semi-arid conditions in India, competition for water appeared more important than shading effects (Singh et al 1989). Corlett et al (1989), in a semi-arid study from India, reported similar results for an alley cropping mixture of *Leucaena* and millet. Establishment of a root barrier (0.5 m depth) next to the *Leucaena* hedges practically eliminated the yield reduction of the adjacent millet. Examining soil moisture effects of 3.5 years old *Eucalyptus tereticornis* on mustard and wheat yields next to the tree line in semi-arid India, Malik and Sharma (1990) reported reductions of over 30 per cent for the crops growing at a distance of less than 10 m from the tree line. Thus, despite the use of drought adapted plants, water competition is likely to play a major role in the productivity of agroforestry systems, especially in dry areas.

One way of tree crop interaction in semi-arid is through allelopathy, which refers to the inhibition of growth of one plant by chemical compounds that are released into the soil from nearby plants. The effects of these chemicals on other plants are known to be dependent principally upon the concentration as well as the combination in which one or more of these substances is released into the environment (Putnam and Tang 1986).

Like the plant-plant interaction, other most important negative interactions between animals and plants can be classified as direct effects. Low quality of, or toxic components within, tree fodder can adversely affect livestock production. Conversely, mechanical damage of trees or deterioration of soil properties, e.g., through compaction, can have a negative impact on the woody perennial component. While tree fodder holds great promise, particularly as a dry season supplement in semi-arid areas, its value should not be overestimated. Many species contain secondary compounds that reduce the feed value. The presence of high levels of phenolic compounds (tannins) or strong odours found in the leaves of species such as *Senna siamea* and *Gliricidia sepium* may reduce palatability or acceptability of the fodder. In addition, digestibility can be low and the leaves may contain toxins or toxic concentrations of certain micronutrients. The most widely discussed problem is probably that of the toxic compound mimosine found in *Leucaena* fodder. Other particularly harmful compounds include cyanogenic glucosides in *Acacia* species, or robitin in *Robinia* (Ivory 1990).

AGROFORESTRY POLICY RESEARCH PRIORITIES FOR ARID AND SEMI-ARID REGIONS

The science in agroforestry is well known while the component selection and management varies with place, socioeconomic status, market and climatic factors. The ultimate feasibility of any agroforestry practice will depend on the actual impact that has on farmer's economic and physical well being. No matter how convincingly we argue in favour of agroforestry either in terms of long term organic matter maintenance and nutrient recycling, such attributes will remain largely invisible to farmer. Meanwhile all agroforestry innovations demand changes in the labour practices of the farming system. Furthermore, labour requirements are the main factor for rural people before they decide whether or not to adopt a new agroforestry practice (Hoskins 1987). The change in policy may play major role for increased adaptability of AF in arid and semi-arid region. Following are some priorities for policy research in arid and semi-arid region to favour agroforestry practice.

- Optimal land use strategies and policies necessary to encourage their adoption
- Policies affecting planting, harvesting, transport and sale of trees outside forest
- Tenure and customary rights to trees
- Markets and value chain for tree products

■■■■

27 AGROFORESTRY PRACTICES FOR SALT AFFECTED LANDS (BIOSALINE AGROFORESTRY)

INTRODUCTION

All soils contain some amount of soluble salts, which is really essential for the healthy growth of plants. If the quantity of soluble salts and/or Na on the exchange complex in a soil exceeds a certain value, the growth and/or the yield of most crops is adversely affected. Such soils, which contain excess soluble salts, adversely affect the plant growth are called the salt affected soils. They occur where potential evapotranspiration greatly exceeds precipitation, which is in arid and semi-arid regions. They are grouped under Aridisols order as well as Intrazonal order or Halomorphic sub-order (Thorp and Smith 1949). In irrigated areas, drainage, leaching and soil amendments are used to reclaim these soils. This traditional approach to ameliorate salt affected soils is difficult and expensive. The alternative way is the biological approach, which involves selection and planting of salt tolerant plants and trees in plantation or agroforestry practices for fodder, fuelwood and/or timber production. Salt tolerant plants and trees have the potential to ameliorate salt lands and can be grown using poor quality water.

Planting trees in salt affected soils requires a specialized approach with reference to site development, choice of species and level of management. Though many tree species are astonishingly tolerant to adverse conditions, yet to raise viable plant cover on naturally inhospitable sites requires special attention. The stress faced by the trees depends upon the nature and extent of the problem in soil, which itself is governed by the soil reaction, nature and amount of soluble salts, physical barriers like hard pan, quality of underground water and depth of water table, etc. So understanding the soil problem is a pre-requisite for successful plantation/ agroforestry practices on salt affected soils. Since the constraints faced by the trees differ, depending upon the type of the soil problem i.e. sodicity or salinity, the planting techniques and choice of species also differ in these two situations.

Finding ways to make marginal land and saline wasteland productive will permit better quality land to be farmed in more rational ways and lessen the risk of degradation. The standard approach is to increase the salt tolerance of trees and plants with concerted research, but the gains come slowly. The alternative approach is to make use of those plants and trees (halophytes) that have high tolerance for salinity. Substantial progress has been made in evaluating halophytes for their potential to rehabilitate salt affected lands. In recent years, it has been demonstrated that revegetation of saline habitats with halophytic species is profitable and provides many additional benefits. There are about 6,000 species of terrestrial and tidal halophytes in the world which offer promising options for reclaiming salt affected lands (Le Houerou 1991). In general, these species are neglected and usually considered impediments rather than opportunities for agricultural development. There are major problems to be overcome to determine and select the best species and ecotypes to be used in the vast areas of degraded salt affected lands around the world. Increased research on the selection of halophytic trees and plants of economic uses, with appropriate management, could result in the rehabilitation and revegetation of salt affected land.

DISTRIBUTION OF SALT AFFECTED LANDS

About seven per cent or approximately 1,000 million hectares (Mha) of the world's land area is affected by salinity and sodicity (Dudal and Purnell 1986). The global extent of salt-affected land, as calculated from the FAO Harmonized World Soil Database, amounts to 1.1 Gha. As per FAO/UNESCO Soil Map of the World, the total area of saline soils is 397 Mha and sodic soil area is 434 Mha. Although the actual rate of salinization is not well studied, global irrigation-induced salinization is estimated at a rate of 0.25 to 0.5 Mha per year. It is also estimated that 20 Mha of land is rendered to zero or negative productivity each year. This situation is not only being made worse by further land deterioration but also by population increase. Estimates given by Dregne et al (1991) and Oldeman et al (1991) suggests that more than 76 Mha of land worldwide has become salt affected due to human causes (secondary salinity); of this 43 Mha is irrigated land (world's semi-arid and arid regions) and 31 Mha is non-irrigated land. Secondary salinization has resulted through inappropriate water management in a variety of landscapes, such as by over-irrigation in semi-arid climates and by clearing of deep-rooted native vegetation in dry land and irrigated area.

Salt-affected soils are located all over the world but they mainly occur in semi-arid and arid regions and along the coastal belt (as the result of sea water intrusion) of also humid regions. Important to note is also that there is continuous salinization of land (especially irrigated agricultural land in semi-arid and arid regions). Salt affected soils are often found in countries which have regions encompassing, (1) semi-arid and arid climates (e.g. India, Pakistan, Iran, USA, Australia, etc) and (2) seasonally dry tropical climates (e.g. Thailand, Sri Lanka, etc.).

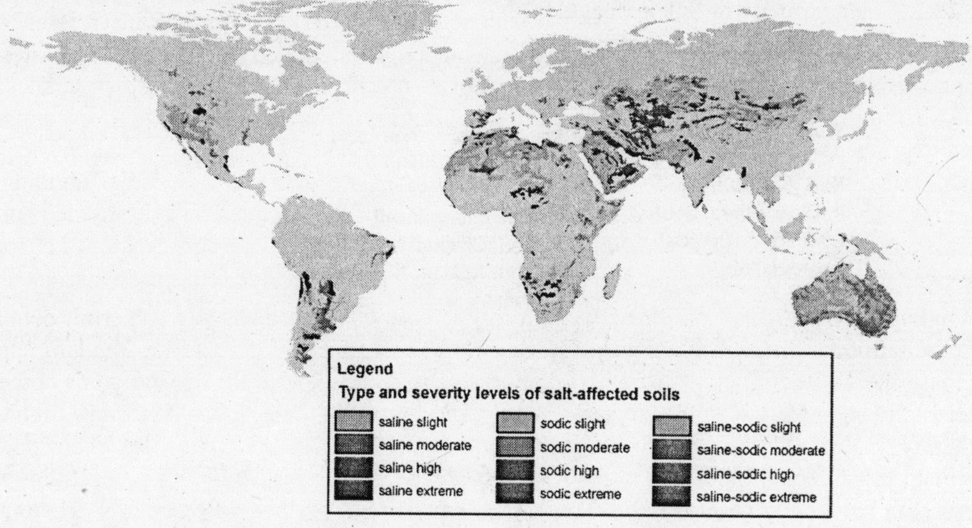

Fig. 27.1: Distribution of Salt-Affected Soils in World (Source: FAO)

In India, the available literatures report the area of salt affected land in the range of 7 to13 Mha. To the estimates of Central Soil Salinity Research Institute (CSSRI), Karnal (Haryana), such soils occupy about 10 Mha, of which a major fraction (say two-third) is sodic or alkali in nature and occurs in the Indo-Gangetic Plain, followed by the Deccan plateau dominantly supporting black soils; the rest (about one-third), largely occurring in the semi-arid and arid coastal regions, are saline soils. The saline-sodic soils (recently termed as sodic) of the Indo-Gangetic Plain, occur dominantly in the Agro Ecological Region's (AER's) of 4 and 9 that receive an annual rainfall ranging from 500 to 800 mm. The saline soils, on the other hand, predominate in the AER's of 2, 4, 6, 8, 16 and 19 that receive an annual rainfall of < 500 mm, except in the coastal areas. The worst affected states are Gujarat, Rajasthan, Haryana, Uttar Pradesh, Bihar and the coastal regions.

There are three principal mechanisms of salinization: salt accumulation, seepage and wind deposition. Salinization by salt accumulation occurs when leaching is reduced and salt accumulates at the surface or at some depth in the soil profile, where following erosion, it may become exposed. Salinization also can occur when salt is leached into a perched water table and then seeps to a lower point in the landscape. Salinization by wind deposition relies on a suitable source of salt deposits. All three of these salinization processes are likely to be affected by climate change. Coastal salinity depends primarily on sea level and its tidal, seasonal, and long term fluctuations; the geology, geomorphology, and relief of the coastal area including river deltas and estuaries; and the climate (temperature, precipitation, rate of evaporation, and spatial and temporal variability in them).

CLASSIFICATION OF SALT AFFECTED SOILS

Soils of arid and semi-arid regions contain high concentration of either soluble salts or exchangeable Na or both due to poor leaching of base forming cations. The soil pH is generally >7 and soil colloidal complex dominated by Ca^{++}, Mg^{++}, K^{+} and Na^{+} ions. On the basis of soil pH, SAR, ESP and EC, salt affected soils are classified into four groups: saline soil, non-saline alkali soil, saline-alkali soil, degraded alkali soil

Table 27.1: Characteristics of Salt Affected Soils

Particulars	Saline Soil	Non-Saline Alkali Soil	Saline-Alkali Soil	Degraded Alkali Soil
Other Names	White Alkali, Solanchak, Szek and in India, their local names are Reh, Kari, Kallar, Khar, Pokkati, Shora, Soulu, Loni, Thur (Punjab), Brown alkali soil	Black alkali, Solonetz, Szik and in India, their local names are Black alkali, Typical usar, Alkali soil, Sodic soil, Chaupansoil (Bombay), Chaundusoil, Kshar soil, Rakkar, Bara-bari soil (Punjab)	Usar soil	Solod soil, Soloth soil
Soil pH	< 8.5	8.5-10	> 8.5	6 (surface soil), 8.5 (sub surface)
SAR	< 13	> 13	> 13	<13 (surface), >13(sub surface)
ESP	< 15	> 15	> 15	> 15
EC of Saturated Extract (dSm^{-1})	> 4	< 4	> 4	< 4
Soluble Salts (%)	> 1%	< 1%	> 1%	< 1%

Particulars	Saline Soil	Non-Saline Alkali Soil	Saline-Alkali Soil	Degraded Alkali Soil
Dominant Salts	SO_4^-, Cl^-, NO_3^- of Na, Ca, Mg and K CO3⁻and HCO_3^- are also found in small amount	Na_2CO_3		Na_2CO_3 in lower layer
Content in Soil	High soluble salts of Na	High exchangeable Na on the soil colloidal complex	Na clay and soluble salts	H^+ in upper layer and Na^+ in lower layer
Soil Colour	White due to SO_4^- and Cl^- of Na, brown due to $NaNO_3$	Black colour due to Na_2CO_3 content	-	Black colour in lower layer due to Na_2CO_3
OM Content	Slightly less than normal soil	Very less amount due to dispersion of OM	Low	Low
Climate	Arid and semi-arid regions, Rainfall < 550 mm annually	Semi-arid & Sub humid region, Rainfall > 550 mm annually	Semi-arid & Sub humid region	High rainfall areas
Formation Process	Salinization	Alkalinization	Salinization, Alkalinization	Solodification or solodization
Physical Condition of Soil	Flocculated or granulated condition Good permeable to air and water Soil structure is optimum	Deflocculated or un-granulated condition Poor permeable to air and water Soil structure is very poor	Flocculated and Deflocculated condition	Low rate of infiltration and permeability It develops columnar or prismatic structure Compacted soil
Recommended Fertilizer	$(NH_4)_2$ SO_4, CAN, SSP, TSP	$(NH_4)_2SO_4$, NH_4NO_3, DAP	NH_4NO_3, DAP, TSP	Basic fertilizer i.e. $Ca(NO_3)_2$, $NaNO_3$
Harmful Effect	Available moisture is less Poor seed germination due to poor aeration Soils are generally barren Wilting coefficient is very high Soil mass is structureless and dusty Conc. of K is low Conc. of B, F, NO_3 are high which affect the plant growth Toxic effect of CO_3, Cl, SO_4 of Na Root injury Leaf tip burnt / chlorosis	Dispersion of soil particles Poor soil physical condition Na generated Ca and Mg deficiency N, Zn, Fe, Cu are low P, K, B are high Toxicity due to high pH, CO_3 or HCO_3 of Na, and B Crust formation during dry Sticky during wet	Poor internal drainage due to hard clay pan formed by dispersed Na-clay Poor plant growth	Very low P and due to leaching of Phosphorous

Bhumbla and Abrol (1979), however, proposed that only two categories of salt affected soils be recognised depending upon the nature of soluble salts, plant response, physico-chemical properties and management practices required for their reclamation, these are: (1) Saline soils, and (2) Sodic soils or alkali soils. The saline-sodic soils, according to these authors, behave like sodic soils and as such should be classified as sodic soils, and that a saturated soil paste pH of 8.5 should be used to distinguish sodic (alkali) soils from saline soils.

SALINE SOILS

- These soils have EC of the saturation extract exceeding 4 dSm^{-1} and have a variable ESP usually in the low range with pH < 8.5.
- Saline soil contains a concentration of neutral soluble salts sufficient to interfere seriously with the growth of most plants. The salts are usually chlorides and sulphates of Na, and chloride of Ca and Mg. Although Na^+ concentration may be somewhat higher than those of Ca^{++} and Mg^{++}.
- The SAR is less than 13 in saline soils.
- These soils support patchy growth of crops and sometimes, visible signs of salt injury such as tip burn or chlorosis (pale yellow colour) of leaves can be seen.
- Soluble salts, which are invariably present in large quantities, form a white efflorescence on the soil surface.
- Since, the neutral salts present in this type of soil can absorb moisture from the soil surface. These soils look wet, although, this water would not be available to the plants.
- Saline soils are generally deficient in available N, medium to high in available K.
- Poor seed germination due to poor aeration and high concentration of the boron, fluoride and nitrate.

ALKALI OR SODIC SOILS

- Sodic soils do not contain any great amount of soluble salts. The detrimental effects of these soils on plants is due to the toxicity of Na^+, HCO_3^- and OH^- ions; reduced water infiltration and aeration. The high pH of sodic soils is largely due to the hydrolysis of Na_2CO_3 or adsorbed Na. The pH is always above 8.5 often rising to 10.
- These soils have dispersed soil structure.
- Following an irrigation or rain event, water stagnates for longer period particularly in the lower spots.
- Upon drying, these soils develop 1-2 cm wide cracks,
- OM present in the soil solution of highly alkali soils gets dissolved and is deposited on the soil surface through capillary action causing dark black surface.
- The cropped fields have spotty growth.
- When the alkali status is high, crops show scorching and leaf burn typical of Na toxicity or the yellowing of leaves resulting from poor soil physical properties. In extreme cases, the soils remain barren with particularly very little or no vegetation.
- Less air and water enters the soil resulting in poor environment for root growth.
- Internal drainage is restricted resulting in temporary waterlogging.
- Formation of surface crust affects seedling emergence.
- Compacted top soil and sub-soil creates problems for root penetration.
- OC content of most alkali soils is quite less compared to normal soils of the region.
- Alkali soils are deficient in soluble Ca and poor in exchangeable Ca. Plants grown in these soils die more often due to lack of Ca rather than the toxic effect of Na.

- Alkali soils are deficient in available N because poor in OM, slow transformation of ammonical N to NO_3-N, high loss by volatilization and denitrification and less symbiotic fixation of atmospheric N.
- Alkali soils are rich in total and exchangeable P.
- The clay fraction of alkali soils is dominated by Illite mineral, which is a rich source of K.

Table 27.2: Properties of Saline and Sodic Soils

Particulars	Saline Soil	Sodic Soil
Surface soil under dry condition	Friable soil surface	Hard surface
Sub surface hard pan	Sometimes thin hard pan	Generally thick hard pan
Collected water in surface	Clean	Muddy
Collected water in sub-surface	Bad	Good
Absorption of rain water or irrigation water	Soon	Waterlogged on surface soil due to presence of hard pan

RECLAMATION OF SALT AFFECTED LANDS

PHYSICAL MEASURES OR MECHANICAL MEASURES FOR RECLAMATION

- *Land Grading*: Land grading helps in applying shallow depth of water per irrigation and uniform application over the whole land surface.
- *Construction of Bunds*: Strong bunds (45-60 cm high) are needed to conserve the rainfall within the reclaimed lands.
- *Scrapping*: It includes scrapping of surface soil to 15-30 cm depth suitable for patchy field.
- *Leaching*: Leaching of soluble salts below the plant root zone, Leaching process is more useful in salt affected soil where water table is down. Summer season is more suitable for leaching.
- *Flushing*: Removal of soluble salts from soil surface by runoff from upper portion to lower portion is called flushing. This process is more useful for salt affected areas where water table is high and soils having gentle slope.
- *Trenching*: Using 50 cm x 50 cm x 50 cm trenches with 2m distance found to be effective. This is suitable in places where sub-soil is good.
- *Washing*: Washing out of soluble salts from surface layer. This is more suitable where water is cheap and available in excess amount.
- *Drainage*: Facilitating proper water drainage in the salt affected areas.

AGRONOMICAL PRACTICES OR CULTURAL PRACTICES FOR RECLAMATION

It includes deep ploughing with the help of soil turning plough i.e. victory plough to open hard layer, growing of salt tolerant crops, suitable crop rotation, increased plant population / unit area, sowing of seeds in furrow, use of acidic nature fertilizers, salt free water for irrigation purpose, using light irrigation than heavy irrigation, sprinkler irrigation and application of organic residue mulch @ 4 tons per ha on surface soil to reduce the evaporation rate.

BIOLOGICAL MEASURES FOR RECLAMATION

It includes application of FYM @ 50 tons per ha (0.5% N + 0.25% P_2O_5 + 0.5% K_2O), growing of green manuring crops i.e. Dhaincha (*Sesbania aculeata*), use of molasses i.e. sira, pressmud, use of 2.5 tons per ha of dry powder of Satyanasi weed (*Argemone mexicana*), use of Usartor Masala which is an organic amendment.

CHEMICAL RECLAMATION

GYPSUM: Main aim of reclamation of sodic soil is to replace exchangeable Na with Ca. Of all Ca compounds, gypsum is considered best and cheapest for reclamation of sodic soil because it has high soluble Ca salt and easily available in the market. The Chemical formula for Gypsum is $CaSO_4.2H_2O$. It has the composition of 23.2 per cent Ca, 18.6-22.5 per cent S and 20.9 per cent water. The amount of gypsum required to be added in a sodic soil to lower the ESP to a desired value is known as gypsum requirement (GR). It is expressed as milliequivalents (me) of Ca /100 g soil. GR is governed by degree of sodicity, crops to be grown, soil texture, drainage condition and quality of irrigation water.

$$GR = \frac{CEC \times [ESP\ (initial) - ESP\ (final)]}{100}$$

where,

ESP (initial) = ESP value before application of gypsum

ESP (final) = It is kept at 10

CEC = Cation exchange capacity as Cmol(p^+)/kg soil

GYPSUM APPLICATION METHOD: Agricultural grade gypsum (70% pure) should be powdered and sieved (sieving with 2 mm size sieve). It is applied and mixed thoroughly in 10-12cm soil depth with shallow tillage followed by planking. About 25 tons per ha of FYM should be applied. The field should be irrigated and kept ponded for 15 days with 6-8 cm standing water.

For planting tree seedlings, the pit should be dug upto 1 m depth and excavated soil should be discarded. Fresh soil mixed with gypsum should be filled into the pit and FYM should be added. Finally salt resistant tree seedlings should be planted in the pit.

PYRITE: The chemical formula for pyrite is FeS_2 (Iron sulfide). It has the composition of 20-22 per cent Fe, 22-24 per cent S, 35-40 per cent Si, 6-8 per cent Al, 2-3 per cent C and other elements such as MgO, CaO, Zn, Cu and Mn. Pyrite requirement (PR) for reclamation of sodic soil is calculated using the formula, PR=GR x 0.727 (GR is the Gypsum Requirement).

Other methods include application of sulfur, application of H_2SO_4 (suitable where alkalinity is very high but costly and risky method) and application of ferrous sulfate.

BIOSALINE AGROFORESTRY

Biosaline agroforestry is the cultivation of trees and/or crops on salt-affected soils, a type of degraded land. It's an alternative land use option for these soils because some tree species are less susceptible to soil salinity and sodicity than agricultural crops and because the cultivation of trees can help regenerate these soils. The main problem related to salt affected soils is that most agricultural crops are very sensitive to soil salinity and sodicity and that crop yields are significantly reduced. Because agricultural crop yields are low and physical remediation of the salts is often prohibitively expensive for most small scale farmers, particularly in developing countries, conventional agriculture on salt-affected sites is generally not economically viable.

Various products from biosaline agroforestry can be used for fodder or forage including the pods and leaves of trees and shrubs, and grasses. Pods that are high in protein and sugars are produced by several tree species, e.g., *Prosopis juliflora* and *Acacia nilotica*, and are suitable for animal consumption. In addition to salt tolerance, the nutritive value and palatability are also important for livestock production on/from salt-affected soils. However, the crude protein and digestible fibre content of salt-tolerant fodder/forage plants is variable and the mineral composition of the plants may in some cases be disadvantageous.

Wood from salt-affected soils can be used for nearly any application as like wood from non-salt-affected soils without modifications. Examples are fuelwood, charcoal and timber. Application for pulp and paper production is also thought to be unaffected, but co-firing biomass from salt-affected soils with coal to produce electricity or gasifying it for liquid fuel production is likely to be limited. This is mainly the result of higher salt content in the wood leading to corrosion of the equipment. However, only few data points on the composition of biomass from salt-affected soils are available in the literature. More research on the wood composition from salt-affected soils is needed to better assess the possible applications of it.

Table 27.3: Organizations Performing Research on Biosaline Agroforestry/Forestry/ Agriculture (Kuiper and Olenburger 2005)

Organization		Country	Website
CSSRI	Central Soil Salinity Research Institute	India	www.icar.org.in/cssri/cssri.html
CSIRO	Commonwealth Scientific and Industrial Research Organisation Forestry and Forestry Products	Australia	www.ffb.csiro.au/
ICBA	International Centre for Biosaline Agriculture	UAE	www.biosaline.org
DDC	Desert development centre of the American University of Cairo	Egypt	www.aucegypt.edu/academic/ddc/
IAR	The Institutes for Applied Research Ben-Gurion University of the Negev	Israel	www.bgu.ac.il/IAR/index.php
KIS	Kuwait Institute for Scientific Research	Kuwait	www.kisr.edu.kw
USSL	United States Salinity Laboratory	United States	www.ussl.ars.usda.gov
PARC	Pakistan Agriculture Research Council	Pakistan	www.parc.gov.pk
LDD	Land Development Department	Thailand	www.ldd.go.th/index_lddeng.htm
TERI	The Energy resources Institute	India	www.teriin.org
ODE	Ocean Desert Enterprises	Netherlands	www.oceandesertenterprises.com

POSITIVE IMPACTS OF BIOSALINE AGROFORESTRY

Biosaline agroforestry's most important opportunity lies in controlling soil salinity and sodicity. In addition, by increasing biomass growth and reducing other degradation processes (e.g. soil erosion, dispersion, and leaching), the cultivation of trees on salt-affected land can improve soil conditions. An important aspect is the potential soil carbon sequestration of biosaline agroforestry. Scientific literature has found a large variation in sequestration rates, which is primarily explained by differences in the soil type, initial soil conditions, climate, and the tree species (as a result of different litter production rates). But even at the lower level of the range found in literature, soil carbon sequestration is an important benefit of biosaline agroforestry. Other improvements in soil conditions include reducing water erosion (and thereby nutrient losses) through improving water infiltration, reducing impacts by water droplets, intercepting rain and snow, and physically stabilizing soil through roots and leaf litter.

Not only can trees reduce the loss of nutrients, but they can actually increase the supply of nutrients within the rooting zone of crops through biological fixing of atmospheric nitrogen (N_2) and retrieving nutrients from below the rooting zone of crops, this is important because many salt affected soils are also characterized by low soil fertility. Biosaline agroforestry systems may potentially have the positive effect of improving water infiltration and soil moisture retention and may provide an opportunity for improving yields of agricultural crops intolerant to waterlogging. The latter is the result of including trees in the agricultural production system that can help remove excess water and thereby reduce waterlogging. Various studies have confirmed that, when properly implemented, these so-called biodrainage systems can lower groundwater tables. However, biodrainage can also have risks for the water and soil conditions.

An important aspect to also consider is that even severely degraded land such as salt-affected land is still often used, for example, for fuelwood collection and livestock grazing and may still provide ecosystem services comparable to managed landscapes. These uses and functions must be accounted for prior to establishing agroforestry or forestry production in order to avoid or at least minimize the direct and indirect land use change and its potentially negative environmental and social consequences. However, it should also be noted that biomass production on salt-affected soils can be combined with food and feed/forage production by, for example, intercropping, rotational woodlots, and hedge rows.

POTENTIAL RISKS OF BIOSALINE AGROFORESTRY

There are also environmental risks associated with biosaline agroforestry, particul-arly the potential invasiveness of salt-tolerant tree species and the impact on groundwater levels and quality. Salt-tolerant tree species may become invasive and weedy and spread into neighbouring, non-salt-affected areas. This is especially (but not exclusively) the case for the introduction of exotic species that are potentially more salt-tolerant than local species and which have no natural predators or competitors. In addition, salt-tolerant species are often less susceptible to other stresses (e.g. droughts, fires and waterlogging). In order to reduce the probability of invasion and weediness, native salt-tolerant plants should be used whenever possible. If exotic species are planted, they must be screened with great care for their invasiveness and weediness potential before introducing them into a new environment. Agroforestry plantations with introduced species must also apply appropriate management techniques in order to minimize the risk of invasion. Management must be specific to the tree species because different species have different stress tolerances and seed dispersal mechanisms.

Regarding the risk for groundwater levels and quality, biodrainage in hydrological discharge areas can result in salt accumulation underneath the plantation. Salinization of groundwater and soils as a result of afforestation can occur because of increased evapo-transpiration and groundwater consumption by trees. However, evidence is limited, and more research is needed on quantifying this effect (particularly in the long term) and on determining the conditions under which this effect occurs. Related to the removal of excess water is also the risk of an exacerbation of water shortages in water-scarce regions due to the additional competition for water. In addition to the direct uptake of groundwater by trees, irrigation of biosaline agroforestry will increase the pressure on water resources. Competition for high quality water can be reduced by irrigating with brackish water not useful for other purposes. However, applying brackish irrigation water can cause even more severe soil salinity problems when the amount of irrigation water does not allow for sufficient leaching of the salts. Therefore, prior to establishing biosaline agroforestry plantations, local water resources and potential impacts by the agroforestry system must be assessed.

SALT TOLERANT CROPS, TREES AND GRASSES FOR BIOSALINE AGROFORESTRY

GROWING OF SALT TOLERANT CROPS

Highly Salt Tolerant Crops: Barley, cotton, dhaincha, lucern, paddy (transplanted), sugarbeet, senji, safflower, toria, tobacco, turnip, taramira

Medium Salt Tolerant Crops: Arhar, berseem, bajra, carrot, cauliflower, cabbage, pea, wheat, oat, sorghum, maize, mustard, tomato, paddy (direct sown)

Low Salt Tolerant Crops: Moong, urd, cowpea, lentil, gram, groundnut, beans, radish, onion

SOWING/PLANTING PRACTICE: About 25 per cent higher seed rate/planting rate over the recommended seed/planting rate for normal soils which would ensure a good crop stand. In case of transplanted crops, the number of seedlings/hill should be increased. Furrow planting might help in obtaining better crop stand and yield under saline condition.

NUTRIENT MANAGEMENT: About 25 per cent higher compared to dose of N fertilizer in normal soils should be applied in split doses to reduce N loss through volatilization and denitrification. Application of P along with FYM has been found to be beneficial, as this combination markedly alleviates the salt injury symptoms in crops. K and micronutrients (Fe, Mn, Zn, Cu) should be applied on the basis of soil test values. Use of acidic nature fertilizer in suitable amount is recommended.

IRRIGATION: The retardation of evaporation is an important feature in the control of soil salinity. Slowing evaporation will not only save moisture but also will retard the translocation upward of soluble salts into the root zone. Where irrigation is practised, frequently light irrigations with salt free water are often necessary.

GROWING OF SALT TOLERANT TREES

The tree species which can thrive better on salt affected soils are *Acacia auriculiformis, Acacia nilotica, Albizia procera, Casuarina equisetifolia, Emblica officinalis, Pongamia pinnata, Prosopis cineraria, Zizyphus nummularia, Albizia lebbeck, Cassia siamea, Prosopis juliflora, Butea monosperma, Melia azedarach, Salvadora oleoides, Acacia farnesiana, Azadirachta indica, Eucalyptus hybrid, Parkinsonia aculeata, Terminalia arjuna*, etc.

The relative tolerance of different species to soil alkalinity is reported as:

- Alkalinity Sensitive Trees (pH<9): *Ficus virens, Populus deltoides, Bombax ceiba, Tectona grandis, Bauhinia variegata, Hardwickia binata, Lagerstroemia speciosa, Thespesia populnea.*
- Alkalinity Moderately Tolerant Trees (pH 9-9.5): *Acacia auriculiformis, Albizia procera, Morus alba, Bambusa spp., Moringa olifera, Leucaena leucocephala, Melia azedarach, Sesbania sesban, Grevillea robusta, Azadirachta indica, Kigelia pinnata, Dalbergia sissoo, Cassia siamea, Pithecellobium dulce, Salvadora persica, Capparis decidua, Punica granatum.*
- Alkalinity Tolerant Trees (pH 9.5-10): *Albizia lebbeck, Butea monosperma, Acacia catechu, Pongamia pinnata, Tamarindus indica, Terminalia arjuna, Eucalyptus tereticornis.*
- Alkalinity Highly Tolerant Trees (pH>10): *Casuarina equisetifolia, Tamarix articulata, Acacia nilotica, Prosopis juliflora, Prosopis cineraria, Tamrix articulata, Achras sapota.*

SUITABLE TREE SPECIES: Examples of tropical tree species tolerant to soil salinity, soil sodicity, or both are *Acacia nilotica*, *Eucalyptus camaldulensis*, *Eucalyptus tereticornis*, and *Prosopis juliflora*. Some tree species have adapted to waterlogging conditions by developing root air channels (aerenchyma) and adventitious (nodal) roots. Examples are *Casuarina*, *Tamarix aphylla* and *Eucalyptus camaldulensis*. There are also temperate tree species including, for example, some poplar and willow species suitable for salt affected lands. In areas where salinity is not associated with shallow water table condition, species like Australian kikar (*Acacia auriculiformis*) and subabul (*Leucaena leucocephala*) can be grown. Where high salinity and shallow water table exist, species like casuarina (*Casuarina equisetifolia*), pharanas (*Tamarix articulata*) and pahari kikar (*Prosopis juliflora*) perform better.

PLANTING TECHNIQUES: The three methods generally adopted are ridge-trench method, planting in furrow-cum-irrigation channel and sub-surface planting in auger holes.

(a) *Ridge-Trench Method*: In coastal saline soils, where soil becomes saline due to ingress of sea water or the salinity of the surface soil increases due to shallow groundwater, ridge plantation of trees is useful. Since, rainfall is generally high near the sea coast, it helps in leaching of salts from the ridges, providing stress free environment in the root zone. It also helps in good aeration. Trees are planted on ridges (50-100 cm high). Tranches between the ridges are used for draining/storing excess water.

(b) *Planting in Furrow-cum-Irrigation Channels*: Adoption of this practice in situations where saline water was used for irrigation has also given very encouraging results. High moisture content in the furrows pushes the salts laterally to relatively drier ridges. Young saplings are planted in the furrows (about 30 cm deep). These furrows are also used to irrigate the trees. Sub-surface planting in furrows aided by leaching during irrigation helps in creating a low salinity zone around the plants. The water in the furrow transports the salts away from the active root zone of the trees.

(c) *Sub-Surface Planting in Auger Holes*: Planting of trees saplings in auger holes, 30-50 cm below the soil surface helps in avoiding the zone of excessive salt accumulation since roots are exposed to intermediate salinity levels, which corresponds to the transmission zone salinity of the soil. This technique is best suitable where top soil layer is characterised by maximum salt and sub-surface soil contains relatively very less amounts of salts. The best time to make auger holes is during the months of May and June. Auger holes of 20-25 cm diameter and 100-150 cm deep are made using tractor operated augers. These holes are filled back with a mixture of original alkali soil + 4-5 kg gypsum + 10-12 kg FYM + 8 kg river coarse sand before planting.

TREE SPACING: Species with a spreading crown need more space than the species with erect type crown.

Close planting would partly cover the risk of high mortality. If the primary purpose of planting trees is firewood production or energy plantation then 2m x 1m or 2m x 2m spacing should be preferred as in the case of woodlot with agroforestry trees. In agroforestry systems with understorey cropping, the minimum spacing between rows and between trees should be kept at 4m and 2m, respectively. If the tree is spreading type like *Prosopis* spp., *Acacia* spp., etc., the spacing should not be less than 5m between rows and 3m between trees.

USE OF AMENDMENTS: If irrigation water has high RSC, frequent use of gypsum or pyrite to counteract the harmful effect of CO_3^{--} and HCO_3^- ions in the irrigation water is beneficial. In fine textured soils, addition of coarse sand and rice husk improves the physical environment resulting helps in more efficient leaching of salts.

APPLICATION OF NUTRIENTS: Application of N to the non-N fixing trees significantly improves the growth. For non-N fixing trees, a basal dose of N (50 g urea/plant) should be mixed in the soil mixture used to fill the auger holes. $ZnSO_4$ @ 10 g/auger hole should also be applied at the time of planting. Later, 50g urea should be applied in two equal splits every year, one during the monsoon season and the second in February-March. N application in N–fixing tree species could be avoided. Application of P and K could be avoided, as these are present in sufficient amount in alkali soils.

IRRIGATION: To establish the trees, 6-8 irrigation should be applied in the first year and 3-5 in the second year of plantation with salt free water. Even at latter stages wherever available, protective irrigation should be provided during the winter season to save the plants from frost damage.

GROWING OF GRASSES

Grasses are more tolerant to salt affected soil environment compared to field crops. Karnal grass (*Diplachne fusca*), rhodes grass (*Chloris gayana*), bermuda grass (*Cynodon dactylon*) and para grass (*Brachiaria mutica*) are highly tolerant to alkali conditions and can withstand ESP up to 80-90. Cultivation of grasses also improves the physical and chemical characteristics of the soils such that after a few years of pastoral use, other crops could be grown. For example, *Cynodon dactylon* can be grown in soils having ECe up to 20 dSm^{-1} where, the problems of high soil salinity, high water table and saline water (EC 20 dSm^{-1}) existed. *Cynodon dactylon* grass can be grown successfully even irrigated with saline water. Other promising grasses for rehabilitation of saline soil are *Sporobolus helvolus*, *Dichanthium annulatum* and *Chloris barbata.*

AGROFORESTRY PRACTICES IN SALT AFFECTED LANDS

Many of the species that may be suitable for the establishment of tree plantations on salt affected soils are used in agroforestry systems, especially in arid regions. In agroforestry systems, trees are used as sources of fodder during the dry season when there is no grass, serve as windbreaks and hedges and provide shade. If the trees are used as fodder or as shade trees, they can be planted scattered over the fields that are used as pastures, but can also be planted in blocks or woodlot or timber belts. The trees can be directly browsed by cattle or the leaves and pods can be harvested to feed cattle. A well known species that is favoured in agroforestry systems is Prosopis cineraria and it fixes large amounts of nitrogen and does not affect the growth of plants under the canopy (Qureshi and Barrett-Lennard 1998).

Salt tolerant trees can also be used in agroforestry systems with the purpose of managing dryland salinity. Dryland salinity generally results from the build-up of salts in surface soil, usually as a result of a rising water table and subsequent groundwater seepage. Water tables can rise due to the removal of deep rooting native vegetation. Due to the removal of vegetation large volume of rainfall are leaking through upper soil and recharge the ground water. This results in rising water tables especially in discharge sites bringing salts to the surface layer of the soil. Trees can be used to lower these risen water tables. They are relatively deep rooting and can therefore reduce water leakage. They may use water all year around whenever it is available, and intercept a significant fraction of rainfall before it reaches the soil. Strategic tree planting in biosaline agroforestry can contribute to lowering risen salt water tables (Kuiper and Olenburger 2005).

SEQUENTIAL AGROFORESTRY SYSTEM

Under this agroforestry practice, trees and agricultural crops are grown in sequence instead of growing them simultaneously. This practice helps to raise the fertility status of the soil which has gone down either by continuous cropping or due to soil salinization. Short duration, fast growing and nitrogen fixing trees like *Leucaena leucocephala*, *Sesbania* spp., etc. can be grown for few years and then cut for fodder and fuelwood. After harvesting the trees, the land can be used for cultivation of agricultural crops.

SILVOPASTORAL SYSTEM FOR SALT AFFECTED LANDS

Salt tolerant tree species and grasses are planted together on barren salt affeced soils. Plantation of *Prosopis juliflora* along with karnal grass for 6 years significantly improved the surface soil layer of an alkali soil. The soil improved to an extent that after 6 years less tolerant but more palatable fodders species i.e. shaftal (*Trifolium resupinatum*), berseem (*Trifolium alexandrinum*) and senji (*Melilotus parviflora*) could be grown successfully. When tree and grass are grown together for 8-10 years, the soil can improve to an extent that after the removal of trees or even along with the trees, field crops can be grown. Raising a crop along with the trees is relatively better option in view of soil amelioration and crop nutrition.

Trees such as *Acacia nilotica*, *Eucalyptus tereticornis*, *Salvadora* spp., *Prosopis juliflora*, *Pithecellobium dulce* and *Parkinsonia aculeata* are planted on ridge and karnal grass is established in the trenches between ridges. This system conserved rainwater in trenches during the monsoon season, which in turn increased the biomass of trees and intercrop of karnal grass. Since, alkali soils are displaced in nature, this practice helped in minimising soil loss, which used to go with the runoff. Highly salt tolerant and high biomass producing grass species such as Aeluropus lagopoides, *Sporobolus helvolus*, *Cynodon dactylon*, *Brachiaria ramosa*, *Dactyloctenium aegyptium*, *Dichanthium annulatum* and *Digitaria ciliaris* can be grown in silvopastoral system.

AGRI-HORTICULTURE SYSTEM FOR SALT AFFECTED LANDS

Recent experiments at N.D. University of Agri. & Tech., Faizabad and CSSRI, Karnal with some salt tolerant species of horticultural plants have shown that there are about half a dozen fruit plants, which can be grown in salt affected soils (pH>9) by adopting proper site preparation techniques and by using appropriate doses of organic and inorganic amendments. The fruit species, which can be established successfully in salt affected lands are ber (*Ziziphus mauritiana*), guava (*Psidium guajava*), pomegranate (*Punica granatum*), amla (*Emblica officinalis*), karaunda (*Carissa carandas*), imli (*Tamarindus indica*), jamun (*Syzygium cumini*), bael (*Aegle marmelos*), etc.

To plant horticultural trees, pits of 60 cm × 60 cm × 60cm are prepared manually. In the centre of each pit an auger hole of 20-25 cm diameter and 140-160cm deep is made using a tractor-mounted auger. These pits cum auger holes are filled back with a mixture of original soil along with 8-10 kg gypsum, 10 kg FYM and 20 kg river sand. To improve upon survival and quality of fruits, saplings are planted followed by *in situ* grafting for desired characters. Regular application of irrigation water is essential besides additional irrigation required to keep the orchard soil moist to offset frost damage.

ALLEY CROPPING SYSTEM

Alley cropping system combines trees and shrubs with pastures and agricultural crops to maximize productivity and water use. Planting layout and density is flexible; 100 to 450 trees/ha in multiple rows up to 50 m apart will provide a satisfactory ratio of trees to pasture/crops. Tree belts in alley cropping systems have the potential to use saline ground water. The key issue generally is to maximize the distance between tree belts in order to reduce production losses due to competition between trees and pasture or crops and to balance the local recharge of ground water with ground water use by trees in belts. Qureshi and Barrett-Lennard (1998) indicate that it is better to use a few wide belts of trees than using a lot of narrow belts of trees. Using a few wide belts of trees reduces the area at which an edge effect is experienced by the pasture or the crop. In this way, the area at which loss of crop growth caused by competition with invading tree roots and by shading takes place is minimized. Nitrogen fixing trees with high salt tolerance, which leaves can be used as green manure and fodder, are preferred for alley cropping system in salt affected lands. The examples include *Leucaena leucocephala*, *Sesbania* spp., etc.

■■■■

28 AGROFORESTRY PRACTICES FOR WETLANDS AND WATERLOGGED AREAS

WHAT ARE WETLANDS?

Wetlands are defined as lands transitional between terrestrial and aquatic ecosystems where the water table is usually at or near the surface or the land is covered by shallow water (Mitsch and Gosselink 1986). Wetlands are among the most productive ecosystems besides being a rich repository of biodiversity, and are known to play a significant role in carbon sequestration. Wetlands usually occur in depressions or along rivers, lakes, and coastal waters where they are subjected to periodic flooding. Some wetlands also occur on slopes associated with the ground water seeps. Conceptually, wetlands lie between well-drained upland and permanently flooded deep waters of lakes, floodplains of rivers and coastal environs. The first scientific national inventory of wetlands in the India was carried out at 1:250,000 scale by Space Applications Centre (ISRO) at the behest of the Ministry of Environment and Forests (MoEF), Govt. of India, using IRS satellite data (1992-93 timeframe). The major wetland types found in India include river/stream, inter-tidal mudflat, reservoir, tank, and lake/pond. India has also some of the unique wetlands like mangrove and coral reef.

Wetlands being dynamic and influenced by both natural and man-made activities, need frequent monitoring. Regular updating of the status of the wetlands is all the more significant in view of accelerating pressure on the very existence of these resources due to developmental activities and population pressure being witnessed currently. The value of the world's wetlands are increasingly receiving due attention as they contribute to a healthy environment in many ways. They retain water during dry periods, thus keeping the water table high and relatively stable. During periods of flooding, they mitigate flood and to trap suspended solids and attached nutrients. Thus, streams flowing into lakes by way of wetland areas will transport fewer suspended solids and nutrients to the lakes than if they flow directly into the lakes. The removal of such wetland systems because of urbanization or other factors typically causes lake water quality to worsen. In addition, wetlands are important feeding and breeding areas for wildlife and provide a stopping place and refuge for waterfowl. As with any natural habitat, wetlands are important in supporting species diversity and have a complex of wetland values.

CLASSIFICATION OF WETLANDS

Entire country including the main land and islands territories has been considered for inventory and assessment of wetlands. Total wetland area estimated is 15.26 Mha, which is around 4.63 per cent of the geographic area of the country. Total 2,01,503 wetlands have been mapped at 1:50,000 scale. In addition, 5,55,557 ha area of small wetlands (<2.25 ha) have also been identified. Excluding rivers/streams, the total wetland area estimated to be 10 Mha.

Table 28.1: Wetland Classification System (Panigrahy et al 2011)

Level-I	Level-II	Level-III
Inland	Natural	Lake/Pond
		Oxbow lake /Cut-off meander
		High altitude wetland
		Riverine wetland
		Waterlogged
		River/Stream
	Man-made	Reservoir/Barrage
		Tank/pond
		Waterlogged
		Salt Pan
Coastal	Natural	Lagoon
		Creek
		Sand/beach
		Intertidal mud-flat
		Salt marsh
		Mangrove
		Coral
	Man-made	Salt pan
		Aquaculture pond

As per the definition of wetland, diverse types of classes get included like lagoon, beach, mangrove, coral reef, salt pan, aquaculture pond, waterlogged, ox-bow lake, reservoir, lake, tank inter-tidal mudflat etc. For the current inventory, a classification system based on IUCN/RAMSAR definition and amenable from remotely sensed data is used to categorize the wetlands. Total 19 types of wetlands were delineated using a hierarchical system that excludes rice fields. This classification categorises inland and coastal wetlands at level-I followed by natural and man-made wetlands as level-II, which were further categorized into 19 types of wetlands.

Table 28.2: Category-Wise Wetland Distribution in the Country (Panigrahy et al 2011)

S. No.	Wetland Category	Total Wetland Area (ha)	% of Wetland Area
1.	Inland Wetlands-Natural	66,23,067	43.404
2.	Inland Wetlands-Man-made	39,41,832	25.83
	Total-Inland	1,05,64,899	69.22
3.	Coastal Wetlands-Natural	37,03,971	24.27
4.	Coastal Wetlands-Man-made	4,36,145	2.86
	Total-Coastal	41,40,116	27.13
	Sub-Total	1,47,05,015	96.36
5.	Wetlands (<2.25 ha)	5,55,557	3.64
	Total	1,52,60,572	100

Table 28.3: Type-Wise Wetland Areas of India (Panigrahy et al 2011)

S. No.	Wetland Category	Total Wetland Area (ha)	% of Wetland Area
1	Lake/Pond	7,29,532	4.78
2	Ox-bow lake/Cut-off meander	1,04,124	0.68
3	High altitude wetland	1,24,253	0.81
4	Riverine wetland	91,682	0.60
5	Waterlogged (Natural)	3,15,091	2.06
6	River/stream	52,58,385	34.46
7	Reservoir/barrage	24,81,987	16.26
8	Tank/Pond	13,10,443	8.59
9	Waterlogged (Man-made)	1,35,704	0.89
10	Salt pan (Inland)	13,698	0.09
11	Lagoon	2,46,044	1.61
12	Creek	2,06,698	1.35
13	Sand /Beach	63,033	0.41
14	Intertidal mud flat	24,13,642	15.82
15	Salt marsh	1,61,144	1.06
16	Mangrove	4,71,407	3.09
17	Coral reef	1,42,003	0.93
18	Salt pan (Coastal)	1,48,913	0.98
19	Aquaculture pond	2,87,232	1.88
	Sub-total	1,47,05,015	96.36
	Wetland (<2.25 ha)	5,55,557	3.64
	Total	1,52,60,572	100.00

DISTRIBUTION OF WETLANDS IN INDIA

India, with its annual rainfall of over 130 cm, varied topography and climatic regimes, supports and sustains diverse and unique wetland habitats. Natural wetlands in India consists of the high altitude Himalayan lakes, followed by wetlands situated in the flood plains of the major river systems, saline and temporary wetlands of the arid and semi-arid regions, coastal wetlands such as lagoons, backwaters and estuaries; mangrove swamps; coral reefs and marine wetlands, and so on. In fact with the exception of bogs, fens and typical salt marshes, Indian wetlands cover the whole range of the ecosystem types found. In addition to the various types of natural wetlands, a large number of man-made wetlands also contribute to the faunal and floral diversity. These man-made wetlands, which have resulted from the needs of irrigation, water supply, electricity, fisheries and flood control, are substantial in number. The various reservoirs, shallow ponds and numerous tanks support wetland biodiversity and add to the countries wetland wealth. It is estimated that freshwater wetlands alone support 20 per cent of the known range of biodiversity in India (Deepa and Ramachandra 1999). Wetlands in India occupy about 15.26 million hectares, excluding areas under wet paddy cultivation (Directory of Indian Wetlands). Majority of the inland wetlands are directly or indirectly dependent on the major rivers like, Ganga, Bhramaputra, Narmada, Godavari, Krishna, Kaveri, Tapti. They occur in the hot arid regions of Gujarat and Rajasthan, the deltaic regions of the east and west coasts, highlands of central India, wet humid zones of south peninsular India and the Andaman and Nicobar & Lakshwadeep islands.

STATE-WISE WETLAND DISTRIBUTION IN INDIA

The country has thirty five States/Union Territories (UTs). State-wise distribution of wetlands showed that Lakshadweep has 96.12 per cent of geographic area under wetlands followed by Andaman & Nicobar Islands (18.52%), Daman & Diu (18.46%) and Gujarat (17.56%) which have the highest extent of wetlands. Puducherry (12.88%), West Bengal (12.48%), Assam (9.74%), Tamil Nadu (6.92%), Goa (5.76%), Andhra Pradesh (5.26%), and Uttar Pradesh (5.16%) are wetland rich states. The least extents(less than 1.5 % of the state geographic area) have been observed in Mizoram (0.66%) followed by Haryana (0.86%), Delhi (0.93%), Sikkim (1.05%), Nagaland (1.30%), and Meghalaya (1.34%).

WETLAND REHABILITATION MEASURES

The wetland areas occur under a great variation in annual rainfall. The operations which help in achieving successful afforestation are: (1) improving drainage to remove excess water, (2) reducing salinity, and (3) planting with suitable species. In stagnant shallow waters (saline, brackish or fresh), species such as *Typha elephantina*, *Phragmites australis*, *Eichhornia crassipes* and various species of *Cyperus*, *Fimbristylis*, *Ipomoea, Paspalum*, *Juncus*, *Vetiveria*, *Echinochloa* and *Scirpus* are common. *Sacciolepis interrupta* is common tall grass in Odisha, Bihar and Tamil Nadu and *Urochondra setulosa* forms clumps in swamps of Gujarat. *Saccharum spontaneum* is common throughout in swamp areas. Free-floating *Azolla*, *Lemna*, *Spirodela* and other aquatic plants are quite frequent. Many unproductive waterlogged areas have supported good forest stands after improving drainage in countries such as Finland, Sweden, the Netherlands, Germany and the United Kingdom.

Lines and Neustein (1966) reported that improving the drainage facilities by deepening the drains increased the growth of Sitka Spruce by about 20 per cent in Sourthern Scotland. Excess saline patches may be removed by proper soil working. The ridge and furrow method and open trenches of soil working have given good results in India (Shrivastava and Qureshi 1966). The most common practice of soil working in Britain is to plough for temporary drainage and provision of planting turfs. The furrows are then traversed by a deep drainage plough (pulled by crawler tractors). The depth of cross-drains varies from 45-90 cm depending on the peat depth. The spacing varies from 20 to 50 m depending on the site and alignment varies with the catchment of the area and soil type. Tall planting has given good results. Jack (1966) reported that 3-year old seedlings of Sitka Spruce have given better results than 2 year old seedlings when planted in the blanket-bog area. Addition of chemical fertilizers also increase the survival per cent.

SELECTION OF SUITABLE TREES AND CROPS FOR WETLAND AGROFORESTRY

Wetlands, including swamps and waterlogged soils, amount to several hectares in India. Long under-utilized, they could be made productive by adopting various AF practices, such as aquaforestry, etc.

MEASURES OF SOIL IMPROVEMENT

In wetlands too much water in the soil is the major obstacle to land use for either forestry or agricultural purposes. Several measures are generally practised to improve the site conditions.

DITCHING: In moist land, ditching is commonly used to lower the water table by draining the excess water. In some countries, tractors are used to turn over the subsoil for raising dykes on one or both sides of the ditches, which are then planted. The depth and width of the ditches mainly depend on the water table and its fluctuation. The higher the water table with greater fluctuation, the deeper and wider the ditches. The water table and its fluctuations are generally controlled below 50 cm from the ground surface.

TERRACING: In waterlogged lands, the water table is closed to or even above the ground surface. Just ditching is not enough to drain the excess water. The soil removed from the water logged and immerged lands is used to raise spaces (terraces) on one side and to create ponds on the other. The size of the terrace and pond varies greatly in accordance with the water situation and agroforestry practice. In general, water and land are such proportions as 0.5:1.5, 1.0:1.0 and 1.5:1.5. In immerged land, terraces should be higher and ponds should be bigger and deeper. For planting trees with a deep root system, the terraces should be much larger and higher. If fishery is considered as a part of the agroforestry system, bigger ponds with deeper depths are needed.

SELECTION OF TREE SPECIES

Ecological adaptation and economic benefits are two important factors governing the selection of tree species for a wetland agroforestry system. Despite terracing, excess water is still a major problem for tree planting and crop cultivation in wetlands. Tree species for wetlands should have the following characteristics:

- Moisture tolerant with a relatively shallow root system and in particular capable of survival in water logged soil for a certain length of time.
- In view of growth and production of interplanted or underplanted crops, tree species should be deciduous with a narrow and thin crown.
- For agroforestry purposes, fast-growing multipurpose trees should be selected.

TREE SPECIES FOR WETLANDS

According to the ecological requirements and economic conditions, the following tree species are commonly used for wetland agroforestry practices: *Salix babylonica* (Water willow), *Salix xuchonensis* (Basket willow), *Alnus cremastogyne* (Long peduncled alder), *Alnus trabeculosa* (Trobeculate alder), *Paulownia tomentosa* (Hairy Paulownia), *Morus alba* (Mulberry), *Taxodium distichum* (Swamp Cypress) and *Taxodium scandens* (Pond Cypress). Of these, *Taxodium scandens* is the best. It is a deciduous conifer with a small and thin crown. Its root system is shallow and water tolerant and is able to survive in immerged soils for several months or year-round. Its stem is straight, fast growing and good in wood quality. It sheds its needles in late October and renews its foliage in April, providing full light for underplanted winter crops. The next in water tolerance, growth rate and timber quality are: water fir, swamp cypress and peduncled alder. These four species are very popular for wetland agroforestry. Similarly Jackfruit, Monkey Jack and Crambola are fruit species which can tolerate waterlogged conditions to some extent.

SELECTION OF CROPS

To obtain full utilization of environmental resources and to increase economic benefits, an appropriate crop should be interplanted with or underplanted after the establishment of a tree plantation in accordance with site and climatic conditions. Wheat, barley, oat, rape or broad-bean are commonly planted in late autumn and harvested by the end of the next spring. Soyabean, Cotton and Watermelon are seeded in late spring or early summer and harvested in autumn or in late summer. Vegetables such as radish carrot, cabbage, etc. can be cultivated between the major crops. In some areas, strawberry and mint are also planted. As the trees grow and their crown enlarges, causing considerable shade on the floor, especially during the growing season, some shade tolerant crops or medicinal plants such as ginger, edible mushrooms, etc. are selected for under planting. In shallow ponds and streams, aquatic plants such as rice, lotus, water-chestnut can be planted as crops. For deeper ponds, water-lily, water-lettuce, water-caltrop are cultivated as floating crops and ponds with a water depths of more 1 m are good for breeding fishes.

MAJOR AGROFORESTRY PRACTICES FOR WETLANDS

The following major types of agroforestry systems with various modifications are generally practised in wetlands:

- Agrisilviculture system (Forestry-Agronomy)
- Agrosilvopastoral system (Forestry-Agronomy-Animal Husbandry)
- Agrosilvoacquaculture system (Forestry-Agronomy-Fishery)
- Agrosilvoaquapastoral system (Forestry-Agronomy-Animal Husbandry-Fishery)
- Silvipisciculture system (Forestry-Fishery)

AGRISILVICULTURE SYSTEM (FORESTRY-AGRONOMY)

This is a very simple system of agroforestry commonly practised both in lowlands and uplands. Trees are planted in spaces of 2 m x 2 m either before or after cropping. Crops are planted alternatively according to seasons, market situations and owner's plan. Such practices could last 4 to 5 years until the canopy becomes fully closed.

SALIX TETRASPERMA BASED AGROFORESTRY IN LOW LANDS: *Salix tetrasperma* is commonly called as Indian willow. It is medium sized tree of wet and swampy places in Terai zone of West Bengal. In terai zone of West Bengal, *Salix tetrasperma* is seen growing in middle of the pond of many households. During winters when the ponds are dry, the light textured soil doesn't retain water for long time; the tree bears lush green foliage. On contrary during rainy season when the ponds are filled with water, it sheds its leaves. The phenomenon is just inverse to the other species which generally sheds leaf during winters and bear leaves during summer and rainy season. This may be attributed due to soil aeration. During rainy season, due to stagnant water anaerobic situation is created which leads to leaf shedding in order to combat poor soil aeration where as during winters, on account of good soil aeration, plant sends forth new flush and turns lush green. This is definitely a tree adaptation to combat anaerobic condition for survival.

Agroforestry practices in terai zone of West Bengal commences with the onset of rainy season when farmers start cultivating their land for raising paddy. Since rainy season is prolonged in this zone starting from May to October farmers can take two crops of paddy. Heavy rainfall of about 3,300 mm and above in a period of six months creates flood like condition and almost all the areas are submerged in water. In such situation, paddy is the only crop which can thrive better. It is because of this reason that farmers of this region cultivate only paddy and is also part of their staple diet. During winters land lie vacant due to scarcity of water. Vegetable and irrigated rice are grown near rivers or where irrigation facilities are available. Continuous monocropping of rice had resulted in depletion of nutrients of soil in already nutrient starved soil. The need is to develop a land use system which is sustainable in terms of production and does not pose competition to the agricultural crop and also improve the soil nutrient status. In such cases, agroforestry is the best viable option for achieving this target. Taking advantage of typical *Salix tetrasperma* based agricultural system in terai zone in experimental field of Department of Forestry, Uttar Banga Krishi Vishwa Vidyalaya, Cooch behar (West Bangal) was established. The trees were spaced at 6 m x 3 m and after three years of plantation, mustard was sown in winter within the rows of *Salix tetrasperma*. It was seen that the trees were adorned with lush green leaves during this season. Whereas, paddy was growing within the rows of *Salix tetrasperma* during June-July. It was clearly visible that the trees were almost without leaves, further, observations reveal that more the depth of stagnant water more quickly and completely the tree sheds its leaves. The advantage of *Salix tetrasperma* inclusion in agroforestry system in low lands of terai zone of West Bengal is demonstrated. It can be a boon to develop local specific agroforestry land use system (Panwar et al 2011).

AGROSILVOPASTORAL SYSTEM (FORESTRY-AGRONOMY-ANIMAL HUSBANDRY)

In addition to forestry-agronomy, activities of animal husbandry are also carried out directly or indirectly within the system. Some under planted crops can be wholly or in parts used as forage or feeds to livestock and poultry or domestic animals or poultry simply allowed to feed directly on the harvested fields either between crops or on fallow land. Their excrements are good manure for trees and crops.

AGROSILVOACQUACULTURE SYSTEM (FORESTRY-AGRONOMY-FISHERY)

Forestry and terrestrial agronomy are practices on the terrace lands while the surrounding ponds and streams are good for fishery and aquatic crops according to the depth of water. In shallow ponds and streams, lotus, wild rice stem, water-chestnut and arrowhead can root to the bottom soil, while the floating plants such as water-caltrop, water-lily, water-lettuce can be cultivated for forages. The big, deep ponds and streams can be used for breeding fishes.

AGROSILVOAQUAPASTORAL SYSTEM (FORESTRY-AGRONOMY-ANIMAL HUSBANDRY-FISHERY)

This is very complicated system of agroforestry which integrates all the above three systems. To illustrate, a mulberry tree is planted and leaves are used to feed silkworms whose excrement is good feed for fish. Their larvae are rich in nutrition and can be used to feed poultry. Silk is the end product. At the same time different crops can be underplanted for desirable purposes in the mulberry stands. Similarly, if timber tree species are planted in a relatively wide space, a number of crops can be raised underneath for several years for different purposes-some for food, some for forage and some for vegetables. Thus agroforestry provides an excellent opportunity for putting wetlands under production to sustain life and the environment. However, an appropriate choice is necessary.

SILVIPISCICULTURE SYSTEM (FORESTRY-FISHERY)

Silvipisciculture technique has been tried in the sunderban deltaic areas of West Bengal. The region is criss-crossed by creeks and channels of varying width and depth. The settled area consists of more than 50 islands which are protected by earthen embankments. People in this area largely depend on cultivation and fishing. The mangrove area has been retained as a national park and the remaining area is under systematic forest management where selective felling and enrichment planting is usually carried out. Several non-mangrove tree species, e.g., *Acacia nilotica*, *Prosopis chilensis*, *Leucaena leucocephala*, *Casuarina equisetifolia* and mangrove species, e.g., *Avicennia* spp., *Bruguiera gymnorrhiza*, *Rhizophora mucronata* etc., are planted and have given good success. For cultivation of fish, the area is enclosed with earthen embankments. Inside the embankment, a system of ridges and canals has been created. Several varieties of fish are cultivated in these areas. It has been observed in the initial trials that the fish which preyed on shrimps and prawns but subsequently mixed cultures of prawns and mullets are established with the fish yield varying from 150-200 kg/ha (Lahiri 1987).

The land receives intensive treatment. The rain-water is collected by making bunds which helps in developing non-mangrove species. In the lower part of the bund, the mangrove species are grown. Trenches are provided for Pisciculture. This land use creates more job opportunities and helps the local population to improve their earnings. Another example of silvipisciculture is seen in Central and Southern India where tank-shore afforestation has been taken up on a large scale under a social forestry programme. The tank fills up during the monsoon but the water level recedes during the summer. *Acacia nilotica* is widely planted in these tanks, utilizing about 1/3 to 1/2 of the submerged area. Some of these tanks have been taken up recently for pisciculture also. The tanks are therefore providing area for water storage, forestry and fishery. The trees help to meet the timber and firewood needs and fishery improves the nutritive value of the diet.

OTHER AGROFORESTRY OPTIONS FOR WETLANDS

INTEGRATED RIPARIAN MANAGEMENT THROUGH VEGETATIVE BUFFERS (TREES-GRASSES-SEDGES)

Riparian buffers are managed forest and shrubs in areas bordering lakes, streams, rivers, and wetlands. Integrated riparian management systems are used to enhance and protect aquatic and riparian resources as well as generating income from timber and non-timber forest products. Similar to shelter and timber belts, integrated riparian management systems can employ a wide variety of tree and shrub species, with specific plantings tailored to suit the specific growing conditions and production opportunities.

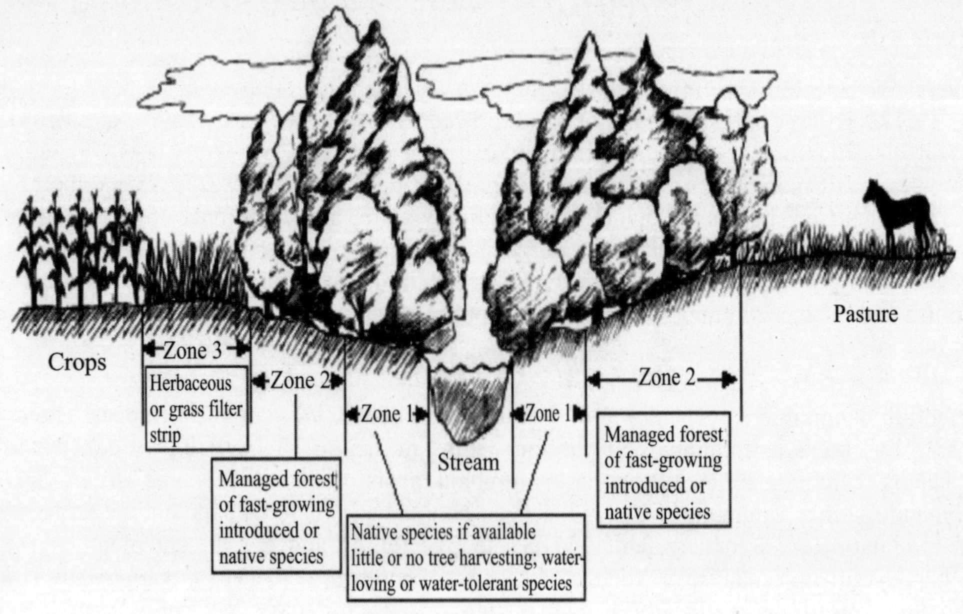

Fig. 28.1: Illustration of Model Multi-species Riparian Buffer Strip

SALIX - A VIABLE OPTION FOR VEGETATIVE BUFFER: Willow belongs to the genus *Salix* and family Salicaceae. The word Salix is derived from celtic '*Sal*' meaning *near* and '*lis*' meaning *water*.

Willows range from prostrate shrubs to large tree over 30 meters high, but most are shrubs or small trees. They are usually found on damp ground or along river and stream margins. In India, there are around 33 species and most of them are categorized as shrubs except *Salix tetrasperma*, *S. alba*, *S. daphnoides*, *S. fragilis* and *S. babylonica*. Species of genus *Salix* commonly growing along streams and swamp areas have been proposed as essential elements for vegetation filters. Willow stands support a high density for breeding bird communities and wetland willow thicket provides stopover sites for bird species in both riparian and wetland habitats.

Physiological characteristics of *Salix* such as high rates of evapotranspiration, efficient nutrient uptake, tolerance of flooded conditions and high biomass productivity are crucial for the use of willows for vegetation filters/buffers. High accumulation of organic pollutants in their roots compared with other macrophytes and algae, the ability to transport oxygen down to the root zone through *aerenchyma* formation (helps for anaerobic respiration) contributing to better conditions for bacterial growth and harvestable biomass providing necessary lignin for composting operations add to the value of willows included into constructed wetlands. Willows as fast growing tree species, planted as vegetation filters will facilitate excess water and nutrient uptake, reduce soil erosion, produce wood for fuel and cricket bat industry, provide habitat for numerous organisms above and below the water level and enhance a site's aesthetic value and greening visual appearance.

FORAGE BUFFER STRIPS

Natural wetlands contain a variety of perennial plant species in the riparian zone. If the surrounding upland is in annual crop, it is often desirable to establish an additional band of perennial forage outside the wetland area. Part of the reason relates to the fact that the outer edge of the riparian zone may vary over time between wet and dry cycles. Therefore, adding a perennial forage buffer helps to ensure that the entire riparian zone remains in perennial vegetation and provides good ecosystem functions. Of the entire wetland area, this outermost zone will be the most feasibly used for forage production, because it will be least susceptible to flooding. The width of this outermost band may vary considerably depending on many factors. In the prairie region, forage buffer widths of 10 to 30 meters have been recommended (Huel 1998). Wider buffers could be rationalized for the following situations or functions:

- Gently sloping wetlands that have greater variability in land area susceptible to having excess moisture
- Grassed waterways or channelized water widens or fans out before entering a riparian area
- Increasing the buffers ability to enhance wildlife habitat and biodiversity
- Increasing the buffers ability to filter sediment, nutrients, and pesticides

The use of this forage buffer could be hay or pasture, although pasture will be limited to early spring or late fall if the surrounding upland is in annual crop. If high nutrient levels are threatening the wetland area, harvested hay or silage that is transported and fed offsite also provides an extra advantage in helping remove some of these nutrients. Vegetative buffer strips is highly beneficial for marginal wetlands since it is easier to establish, easy to maintain and provides direct benefits like fodder and thatching material for economic value. In India, *Vetiveria zizanioides*, *Stylosanthes scabra*, *Brachi-aria mutica*, *B. brizantha*, *Panicum maximum*, *Pennisetum purpureum*, *Pennisetum americanum* are used and found effective in erosion control and dry matter production.

GRASSED WATERWAYS

Often much of the surface runoff entering wetlands are through channelized flow nearby from agricultural field. This type of flow also occurs when one wetland overflows resulting in runoff to another wetland downstream. These areas should have a flat bottom with gentle side slopes, stabilized with a sod forming perennial grass. This will ensure that the flow remains in the waterway and does not cause gully erosion and deposition of sediments in wetlands. In India, *Panicum repens* is the best suited grass for waterways followed by *Brachiaria mutica*, *Cynodon plectostachyus* and *Paspalum notatum*. The suitability of the grass is based on the density of the cover, ease of establishment and forage yield obtained in wetlands.

TREES AND SHRUBS FOR FORAGE ON PRAIRIE WETLANDS

Trees and shrubs are a natural part of prairie wetlands. Species like willow and aspen are often found in the riparian zone. These species have considerable flood tolerance. This tolerance is exhibited not only in the ability of individual trees to survive a certain amount of flooding, but also the ability of these species to generate new trees through root suckering. For example, while prolonged flooding lasting for one year or more can kill a stand of aspen within a wetland area, new tree shoots will often emerge upslope of the flooded area. Trees and shrubs also provide excellent nesting and foraging habitat for many resident and migrating birds. In addition, other forms of wildlife use these areas as cover and shelter habitat. Beneficial insects such as bees and other pollinators are attracted to these sites. The use of trees for commercial biomass production has recently gained interest on agricultural land (i.e. agroforestry). Planting trees for agroforestry around wetlands may help to extract and remove excess nutrients. Trees may be even more effective than some forage, by using more extensive and deeper root systems to recapture deep leached nutrients. A major concern with growing trees in this part of the landscape is the risk of flood damage. This depends on the risk of a flood event, the duration of flooding and soil saturation, the flood tolerance of the tree species, and the expected harvest interval of the tree crop. For example, species such as willow that are harvested on a frequent time interval would be considerably less risky than hybrid poplar harvested only after reaching a mature age.

COCONUT, RICE AND PINEAPPLE CULTIVATION ON WETLANDS/ LOWLANDS OF KERALA

Pineapple is highly sensitive to water stagnation and high moisture regimes. Hence it is important to provide good drainage, if grown in wetlands. In paddy lands, pineapple is planted in paired rows at 45 x 30 cm spacing on ridges taken at 60-90 cm height, depending on the water table and drainage requirement. The ridges are separated by drainage channels having 60 cm width. The width of the ridges varies from 120-150 cm. Wherever water stagnation and poor drainage are expected, a wider and deeper channel is given in between ridges. Coconut trees are planted along bunds of wetlands.

FIREWOOD PLANTATION IN WETLANDS OF ASSAM

In Assam, addition to fish resources, wetlands are direct sourcing of firewood and traditionally households grow bamboo, bael (*Aegle marmelos*), otenga (*Dillenia indica*), kadam (*Anthocephalus cadamba*), *Pitha*, *Koroua*, bhelu (false hemp tree), moss, *Bombax malabarica*, within the homestead areas which are least susceptible to waterlogging. They also undertake cultivation of dhaincha (*Sesbania aculeata*) a green manuring crop as a firewood resource in the farm fields which together reduces fuelwood extraction by 32.7 per cent from the wetlands. Growing dhaincha in the wetland farm field simultane-ously help farmers in maintaining soil productivity by providing low-cost nitrogen inputs to their crop production system and as a good companion crop for wetland rice cultivation.

MANGROVES AS A MULTIPURPOSE ECOSYSTEM AND NATURAL AQUAFORESTRY SYSTEM

Mangroves may be considered as peculiar agroforestry systems or, more precisely, aquaforestry systems, associating animal components (fish, prawns, crabs, shells, etc.) with a specific flora of trees and shrubs developed in the intertidal zone of coastal areas and sheltering larvae and immature individuals of this fauna. Mangroves are considered to constitute one of the most productive ecosystems and, as such, are multipurpose but also complex and fragile ecosystems. Because of their faunal and floristic richness, recent years mangroves have been over-exploited, notably for their wood by the local population of fishermen. This is the case for the mangrove of Kakinada, Andhra Pradesh, where trees (*Avicennia* spp., *Excoecaria agallocha* and *Lumnitzera racemosa*, in particular) constitute the main source of fuelwood and are still an important source of materials in the building of traditional houses, shelters and fences, in spite of the interdiction to cut trees imposed by the Forest Department (Denis 2003).

Fig. 28.2: Wetland Taro System with Bund Plantation of Banana and Other Trees

WETLAND TARO SYSTEM AND BUND PLANTATION

The starch food crop, wetland taro (*Colocasia esculenta*, var. *aquatilis*) has many wetland cultivars. It is grown in and around the Pacific as well as in India and Africa in similar locations to wetland rice. Larger plantings require field levelling, puddling by ploughing, disking, harrowing and grading. It is also necessary to have constant water circulation to avoid foot rot. Since many tropical peat swamps, particularly in their centres, have deoxygenized water, favourable conditions only occur near their edges where peat merges into mineral deposits. Nutritionally taro is a better choice of crop than rice, as the amino acid content is as high as in potatoes and the crop should be grown more widely in wetlands. Swamp taro is also originated from Assam or north-eastern part of India as the region has a natural rich diversity of Colocasia germplasm (Sand and Barua 2000). In eastern part of India, swamp taro/ *Colocasia esculenta* locally known as Pani Kachu/Lati/Kachu Lati grows naturally in swampy environment. Hence it is known as swamp taro. Depending on varieties , entire plant parts from leaves, petioles, rhizomes are consumed as green vegetable especially in states like Assam, Bihar, Eastern U.P and West Bengal (Goswami and Sen 2001; Sand and Barua 2000). There is potential for cultivation of this crop in about 1.2 million hectare waterlogged /marshy land in these states. A medium harvested swamp taro crop can give up to Rs. 43,000 per ha.

SWAMP TARO CULTIVATED IN PACIFIC ISLANDS AS TRADITIONAL AGROFORESTRY SYSTEM: Intensive taro cultivation systems require a heavier application of labour and time relative to output. In the high volcanic islands of Hawaii, *Colocasia* - taro is intensively cultivated in the wetlands or taro patch. The taro patches is created in valley bottoms where streams are diverted to maintain a constant flow of fresh water (Harley 2006).

Fig. 28.3: Swamp Taro Cultivation in Pacific Islands as Traditional Agroforestry System

BENEFITS OF WETLAND AGROFORESTRY PRACTICES

Benefits of agroforestry practices may also be employed to realize a number of economic as well as other associated environmental services from wetlands ecosystems, including:

- Carbon sequestration
- Pollution reduction belts through bioremediation methods – sewage, odour, dust, and noise reduction
- Waste water or manure management (e.g. utilizing urban waste water on intensive, short rotation forests for wood fibre production)
- Green space and visual aesthetics
- Enhancement or maintenance of wildlife habitat

WATERLOGGED LANDS

Even though availability of water is essential for better productivity, but the excess water situation in agricultural field is a serious challenge to agricultural productivity in addition to negative ecological implications like green house gas emission from waterlogged areas and marshy lands. The seepage losses from the conveyance and along distribution system, excess water application in the crop field and recharge of ground water table, inadequate drainage, etc. are among reasons for waterlogged condition in agricultural fields. The diversion of flood waters through existing canal systems to low lying areas causes large scale inundation of agricultural fields rendering them unproductive for agriculture.

National Commission on Agriculture (1976) has defined an area to be waterlogged when the water table rises to an extent that the soil pores in the root zone of crop become saturated resulting in restriction of the normal circulation of air, decline in the level of oxygen and increase in the level of carbon dioxide. As a consequence, yield of crops commonly grown on the land is reduced well below the normal for the land, or, if the land is not cultivated, it cannot be put to its normal use because of high subsoil water table. The depth of water table at which such conditions arise depends on the height of capillary fringe and the type of crops grown, type of soil, quality of water, and the period for which the water table remains high. Waterlogging has become more pronounced in India with the increase in the irrigated areas and shift in emphasis from protective irrigation to intensive irrigation as a means of attaining greater agricultural production. The problem exists in varying degrees in most parts of the country but it has taken a serious turn in the irrigated areas of Indo-Gangetic plains which have very flat countryside and inadequate outfalls. An area of about 3 million ha is affected by waterlogging and another 5 million ha gets frequently flooded. Waterlogged lands may be defined as lands where soil is badly or imperfectly drained. These lands are generally found in high to low rainfall areas. These waterlogged lands may be classified into the following categories.

CLASSIFICATION OF WATERLOGGED LANDS

- Waterlogging conditions caused by stagnant water
- Waterlogging conditions caused by impeded but mobile water
- Swampy areas
- Marshy areas and saline muddy areas

ADVERSE EFFECTS OF WATERLOGGING

The waterlogged lands are rendered infertile because of ill effects of waterlogging to the soil, to the crops and to the environment. Waterlogging conditions are also caused by excessive and faulty canal irrigation in areas adjoining agricultural fields. Such conditions are found in Punjab and Haryana. Waterlogging conditions are also in the depressions or borrow pits, along canals, roads, railway tracks, etc., during rainy seasons. Littoral salty marshes and swamps are confined to the coastal areas mainly in the estuaries of river in creeks and lagoons and along the lower reaches of tidal swamps. Ghosh (1997) reports special type of waterlogging conditions in Malda forest division of West Bengal. The lands are fertile with gangetic alluvial deposit, but most of the areas are low lying, a good part of which remain dry for about eight months, whereas during the monsoon these areas become totally submerged by the accumulation of runoff from the surrounding catchment, where the water depth is upto 5 m. Owing to refractory characteristics, these areas become difficult to afforest mainly due to physiological drought, defective aeration, excessive salt accumulation and poor nutrient availability. Under natural conditions, nothing grows except tall grasses such as *Saccharum spontaneum*, *Vetiveria zizanioides*, *Themeda* spp., *Phragmites karka* and *Avistida* spp. Among the tree species *Salix alba*, *Barringtonia acutangula*, *Syzygium cumini*, *Bruguiera* spp., *Rhizophora*, *Ceriops* spp., etc., are found. The yield of crops is adversely affected if the water table is within 30 cm (sugarcane), 60 cm (rice), 90 cm (gram and barely), 90-125 cm (wheat), 120 cm (fodder), 125 cm (maize, bajra and cotton) and 210-240 cm (lucerne). Usually, the land with subsoil water table within 1.5 m of the natural surface is taken as waterlogged. In fine grained soils, capillary rise may be appreciable causing saturation of the root zone even with relatively low water table. From revenue assessment angle, a land is said to be waterlogged when the yield is 25 per cent or less of the normal yield.

ADVERSE EFFECTS ON SOIL

- *Lack of Aeration*: Due to presence of excess water, soil pores within the root zone of plants are saturated to cut off the normal circulation of air. The capacity of soil bacteria is affected resulting in retardation in the formation of organic acids to dissolve plant food material.
- *Reduced Soil Temperature*: Waterlogged soil is slow to warm up. Low soil temperature restricts root development, depresses biotic activity in the soil resulting in lowered rate of production of available nitrogen.
- *Creation of Salinization*: High salinization and deposits of sodium salts in the soil at or near the ground surface is created which may be toxic or lead to the formation of alkaline condition. Salt accumulation in the root zone of crops result in corrosion of the roots, reduced yield and ultimate destruction of the plant.

- *Inhibiting Activity of Soil Bacteria*: Excess water presence in the soil reduces the supply of oxygen and in turn cessation of nitrogen process.
- *Denitrification*: Denitrification occurs because of the competition for nitrogen by the soil microorganisms due to lack of aeration.
- *Retarding Cultivation*: Difficulty in carrying out normal cultivation operations in waterlogged soil.

ADVERSE EFFECTS ON CROPS

- *Delayed Cultivation Operation*: Normal cultivation operations of tilling and ploughing are adversely affected due to presence of excess water in the soil.
- *Growth of Weeds*: Water-loving wild plants grow profusely and have competition with the crops, thereby affecting the growth of useful crops. Weed removal also entails extra investment on the part of the cultivator. In extreme waterlogged condition, only wild growth is available.
- *Diseased Crops*: Waterlogging conditions cause physiological disease to crops. Decay of roots, external symptoms on the foliage, fruits, etc., are common.
- *Loss of Cash Crops*: Cash crops desired to be grown in the land cannot be cultivated.
- *Low Yield*: Maturity period of crops is reduced resulting in low yield. Waterlogged soil warms up slowly, absorbs much less heat compared to dry soil there by hampering the seed germination and seedling growth.

CAUSES OF WATERLOGGING

The waterlogging conditions are produced because of many factors. Large areas in the Punjab and Uttar Pradesh have become waterlogged because of canal irrigation. Littoral salty marshes and swamps are confined to the coastal areas mainly in the estuaries of rivers, and in creeks. In high rainfall zones where drainage is poor, land become waterlogged generally in depressions and along river and stream beds. The factors tending to raise the water table termed as inflow are: (1) Infiltration from rainfall in saturated phase; (2) Seepage from canals in saturated phase; (3) Infiltration from canal irrigation in saturated phase; (4) Percolation from well or floor irrigation; and (5) Subsoil flow from upper regions. The factors tending to lower the water table, termed as outflow are: (1) Infiltration into rivers; (2) Subsoil flow to the lower regions; (3) Increased soil evaporation (4) Transpiration by plants; and (5) Seepage drains. Various causes of waterlogging are:

- *Over Irrigation*: Over irrigation and intensive irrigation result in waterlogging due to heavy percolation and rise of water table.
- *Seepage from Canals*: Excessive seepage from unlined canal system and water courses result in the rise of water table leading to waterlogging conditions.
- *Inadequate Drainage*: In irrigated areas, water losses from canal system and water courses continuously contribute to water. If drainage is inadequate, waterlogging conditions are developed.
- *Impervious Barrier*: There may be an impervious stratum below the top layer which obstructs the seepage from going down resulting in creation of false water table or perched water table leading to waterlogging.

- *Poor Irrigation Management*: Poor irrigation management by cultivators results in development of waterlogging.
- *Obstruction of Natural Drainage*: Interception of natural drainage by the construction of canals, roads, railways, water courses, etc., results in development of waterlogging.
- *Excessive Rainfall*: Excessive rainfall, in the absence of adequate drainage, causes waterlogging. Rain apart from irrigation, is the major source of waterlogging.
- *Flat Topography*: Flat terrain with depressions lead to waterlogging.
- *Occasional Spills by Floods*: Occasional flooding of the countryside and flood water not quickly drained off give rise to water table resulting in waterlogging.
- *Abandonment of Percolation Wells*: Advent of canal irrigation has resulted in abandonment of percolation wells giving rise to water table.
- *Land Locked Patches without Outlets*: In land locked patches, waterlogging develops due to absence of outlets to drain excess rain water.
- *Unsuitable Cropping Pattern*: Unsuitable cropping also leads to waterlogging.
- *Physiography of Watershed*: Physiography of watershed, i.e., shape, topography and drainage pattern has an important bearing on the drainage problem of the watershed. In areas with high value of shape index compactness coefficient and drainage density, the problem of waterlogging is not as serious as in the catchments where the values of these factors are low.

ANTI-WATERLOGGING MEASURES

PREVENTIVE MEASURES

Preventive measures envisage reduction of inflow into the subsoil water table. Various measures in this context are:

1. *Arresting Percolation from Rainfall*: Quick disposal of storm water in excess of crop requirement is a must for canal irrigated lands.
2. *Arresting Percolation from Canal*: (a) Lining of canals to control excessive seepage from unlined canals is a very important preventive measure; (b) Lowering of water level in canals. It is desirable to lower designed full supply level in the canal, where possible, with a view to cut down seepage losses; (c) Provision of pushta on canal blank; and (d) Interception drains along the canal 15 to 30 m away from the canal bank, are effective to arrest seepage from the canals.
3. *Arresting Percolation from Irrigation in the Fields*: (a) Restriction of irrigation. High water table areas may be restricted to receive kharif irrigation only and during rabi season, the cultivators may irrigate from open wells and tube wells. In high water table areas, controlling intensity of irrigation or permitting only a fraction of the total culturable commanded area to receive irrigation water; (b) Lining of water courses; and (c) Economical use of water by adopting khaljiari system of fields, leveling the fields and refraining from applying more water than the soil can hold.

CURATIVE MEASURES

Curative measures envisage extraction of excess water from the root zone of crops to restore the condition conducive for normal growth of crops. Various measures in this context are:

- *Pumping from Subsoil*: Pumping from subsoil through shallow tube wells, deep tube wells, seepage-cum-pumping stations, open wells to lower water table and use the pumped water for irrigation as it is or by mixing with canal water.
- *Land Grading*: Land grading is useful for betterment of waterlogged lands.
- *Cropping Pattern*: Vegetation and crops with high rate of evapotransipiration may be patronized to serve as natural drainage system. Eucalyptus plantation is helpful in lowering water table.
- *Optimum Water Use*: It is through conjunctive use of surface and ground water.
- *Better Irrigation Methods*: Sprinkler irrigation and drip irrigation methods be introduced to reduce percolation losses. Improvement of irrigation methods by giving only the optimum depth of irrigation water.
- *Sub-surface Drainage*: Sub-surface drainage is the most efficient and permanent type of anti-waterlogging measure.
- *Puncturing Underground Barriers*: Geological formation such as buried ridges as may be interfering with the subsoil flow leading to waterlogging conditions may be punctured to lower water table.
- *Efficient Drainage System*: An efficient drainage system is essential for quick disposal of the storm water and excess irrigation water.
- *Removing Obstruction in Natural Drainage*: The natural drainages should be improved by removing debris etc., from the waterway as also providing adequate waterway under the road bridges, drainage crossings, etc.
- *Reclamation of Soil*: Reclamation of waterlogged soil by various reclamation methods is an important curative measures.

In ending, it can be said that the method of pumping is a very effective anti-waterlogging measure. A judicious combination of canal and tube well irrigation system offers an ideal solution of the waterlogging problem in most areas. Lining of main canals and branches is also an important anti-waterlogging measure. Surface drains, as a general work of reclamation need to be provided at the earliest, to minimize inflow of surface water in the sub-soil strata by deep percolation. More problematic waterlogged areas shall be more intensively drained by providing underground drains and comprehensive programmes of providing tube wells in the command area. Improvement of drainage will also go a long way to solve problem of controlling the rise of water table in order to prevent waterlogging. Conjunctive use of surface and ground water also retards rise in water table.

REMEDIAL MEASURES FOR WATERLOGGED AND MARSHY LANDS

Remedial measures are (a) drainage either surface or sub-surface or both to remove surplus water; (b) lining of canals to prevent seepage and rise; (c) sinking tube wells and utilizing water for irrigation thereby lowering water table. Waterlogged lands contiguous to the seacoast may be ameliorated for crop cultivation or used for fish culture (NCA 1976). Once the land has been drained, it is advisable to provide a cover with suitable tree species tolerant to high water table. Tree species found suitable for water logged areas are: *Acacia auriculiformis*, *Acacia nilotica*, *Dalbergia sissoo*, *Eucalyptus tereticornis*, *Eucalyptus robusta*, *Pongammia pinnata*, *Syzygium cumini*, *Albizia procera.* In waterlogged and salinized gully beds of Chambal (Kota), Para grass (*Brachiaria mutica*), a forage grass has been found to be successful. The waterlogged lands could be reclaimed by adopting following reclamation measures (Khan 1990).

In Uganda and East Africa, swamps have been afforested with *Eucalyptus robusta* in the wettest parts (where it is able to withstand submersion for as long as a fortnight during the rains) and with *Eucalyptus saligna* in the drained swamps in replacement of *Eucalyptus robusta* after the first fuel rotation of 8 to 10 years. In waterlogged areas of India, some species which have yielded good results are *Eucalyptus robusta*, *Syzygium cumini*, *Salix* spp., *Populus nigra*, *Terminalia arjuna* and *Acacia nilotica*. In marshy areas species such as *Barringtonia* spp., *Bischofia javanica*, *Eucalyptus robusta*, *Eucalyptus rudis*, *Lagerstroemia speciosa*, *Casuarina equisetifolia* have been successful. *Diospyros embryopteris*, *Pterospermum acerifolium*, *Bischofia javanica* and canes are generally grown in freshwater swamps. *Arundo donax*, *Sapium sebiferum* etc., have been successful in swampy areas in many parts of the country.

AFFORESTATION TECHNIQUES AND AGROFORESTRY PRACTICES FOR WATERLOGGED AREAS

Afforestation is another potent measure to reclaim the waterlogged land. The afforestation with selected species of trees i.e., *Eucalyptus* helps in lowering the water table on account of increased transpiration from new vegetation. Some crops like rice, jute, sugarcane dhaincha, etc., and selective grasses like dalis, napier, guinea, kari and murta which can withstand waterlogging conditions, should be grown on such area (Singh and Singh 1995). Trees which have high transpiration rates are preferred for planting in waterlogged areas. Some of the species which have given good results are *Eucalyptus robusta*, *Anthocephalus chinensis*, *Syzygium cumini*, *Salix* spp., *Terminalia arjuna*, *Acacia nilotica* spp., etc. The marshy areas species such as *Barringtonia acutangula*, *Lagerstroemia speciosa*, *Pongamia pinnata*, *Casuarina equisetifolia* have been found to be successful. *Diospyros embryopteris*, *Pterospermum acerifolium*, *Bischofia javanica* and canes are generally grown in fresh water swamps. In *terai* areas of west Bengal *Lagerstroemia speciosa* and *Bischofia javanica* have been successfully grown on waterlogged sites where no other species could establish. The soil profiles carry in better growth of Eucalyptus hybrid possess medium internal drainage while profile with inferior growth are either associated with rapid to excessive drainage or with impeded drainage leading to pronounced mettling (Yadav and Jai 1964).

In muddy and flat mangrove areas of Gujarat, only *Avicennia officinalis* is planted. Natural regeneration of *Avicennia officinalis* is supplemented by artificial regeneration. The following grasses and legumes were most successfully adopted in waterlogged areas such as *Brachiaria mutica*, *Iseilema laxum*, *Dichanthium* spp. and *Sesbania aegyptiaca*. The suitable species of trees /shrubs and grasses/legumes for rehabilitation of special problem areas such as waterlogged areas along with their economic use are given in the Table 28.4.

Table 28.4: Suitable Trees/Shrubs and Grass/ Legume Species for Waterlogged Areas (Sharda et al 2006)

Trees/Shrubs	Uses	Grasses/ Legumes	Uses
		Frequently Waterlogged	
Barringtonia acutangula	2,3,5	*Arundo donax*	4,8,15
Bischofia javanica Lagerstroemia	2,3,5	*Brachiaria mutica*	1,4,8
flosreginae	2,3,5,7,8	*Vetiveria zizanioides*	4,8,16
Terminalia arjuna	2,5,7	*Iseilema laxum*	1,4,8
Psidium guajava	2,10	*Dichanthium caricosum*	1,4,8
Syzygium cumini	5,7,3, 10	*Saccharum munja*	4,8,16

Trees/Shrubs	Uses	Grasses/ Legumes	Uses
Salix tetrasperma	2,8	*Saccharum spontaneum*	4,8,16
Eucalyptus globulus	2,3,7,9,13	*Cynodon dactylon*	1,4,8
E. robusta	2,3,7,9,13	*Leptochloa fusca*	1,4,8
E. rudis	2,3,7,9,13	*Tripsacum dactyloides*	4,8
Ficus glomerata	1,2,3,5		
Ipomea carnea	2,6,8		
Occasionally Waterlogged			
Eucalyptus tereticornis	2,3,9,13	*Dichanthium annulatum*	1,4,8
Pongamia pinnata	1,2,3,8	*Bothriochloa intermedia*	1,4,8
Acacia nilotica	1,2,3,6,11,14	*Eulaliopsis binata*	1,4,8,12,13
Acacia auriculiformis	2,3,9,8	*Heteropogon contortus*	1,4,8
Senna siamea	2,3,4,6,8,9	*Paspalum notatum*	1,4,8
Dalbergia sissoo	1,2,3,5,8,14	*Paspalum dilatatum*	1,4,8
Sesbania grandiflora	1,8,10,14	*Chloris gayana*	1,4,8
Sesbania sesban	1,2,7,8,14	*Ischaemum rugosum*	1,4,8
Bauhinia purpurea	1,2,5,11,12	*Panicum* spp.	1,4,8
Bauhinia variegata	1,2,7,10,12		
Bauhinia racemosa	1,2		
Grewia optiva	1,2,12		
Terminalia bellerica	2,3,5,7		

*1. Fodder; 2. Fuel/charcoal; 3. Timber/pole; 4. Mulch; 5. Shade; 6. Fence; 7. Medicine; 8. Soil conservation; 9. Wind break; 10. Food/fruit; 11. Gum; 12. Fibre; 13. Pulp; 14. N-fixation; 15. Toys; 16. Thatching.

DEVELOPMENT OF BIO-DRAINAGE SYSTEM THROUGH AGROFORESTRY

Management of excess water in agricultural field is a major challenge globally as well as in India. The extent of water logging and associated soil salinity has already thrown a serious challenge for sustainability of irrigated agriculture. Flat and saucer shaped topography of land, high rainfall, and inadequate drainage often lead to waterlogging. To combat drainage problem of such an extent, the technology needed to be less expensive and environment friendly for its sustenance. Reliance on capability of vegetation to reduce water table has been reported promising both in India as well as in other countries. As physical drainage measures, several engineering approaches like tube well, surface drainage and sub-surface drainage have been adopted to deal with the problem of waterlogging. But they have limitations like high cost of construction and subsequent maintenance cost, etc. Moreover, large scale engineering intervention altering topography could be a source of environmental concern.

In this context, use of vegetation for drainage appears promising. It is reported that trees could be used to manage rising water table and salinity problem. In naturally existing ecology, the various components of hydrological systems like rainfall, evapotranspiration by above-ground vegetation, storage of water in soil profile as well as drainage, etc. exist in equilibrium. High rainfall scenarios results in short-term increase in surface run-off or drainage flow with rise in ground water table and increased storage of water in soil profile. However above-ground vegetation has an important role in controlling moisture storage in soil profile due to evapotranspiration. The root systems of trees could intercept saturated zone or unsaturated capillary fringe above water table to control shallow water table. These plants are known as phreatophytes.

The clear impact of lowering of water table is reported to be evident even within 10 years of establishment of vegetation by several countries. The approach is relatively cheaper, sustainable and ecologically compatible relying on natural capability of vegetation to transpire water. Introduction of suitable intercrops in bio-drained area is also an issue for strategic application of bio-drainage as a tool for reclamation of waterlogged area for agricultural production. It is perceived that combination of bio-drainage with conventional drainage might make drainage planning in irrigation commands socioeconomically more relevant and environmentally sustainable.

Several plants species are used for this purpose from salt bush (*Atriplex*) to tall trees like *Eucalyptus*, *Casuarina*, *Tamarix*, etc. The main physiological feature of such vegetation is profuse transpiration whenever the root system comes in contact with ground water. However compared to shorter vegetations, tall tree species have greater advantages like high aerodynamic roughness resulting into increased water loss through transpiration. This effect is more pronounced when trees are planted in a row to give clothesline effect (Chorley and Barry 1971). Moreover, deeper root systems of trees could reach out for soil water more extensively at a greater depth for draining out waterlogged soils. Reliance on capability of vegetation to reduce water table has been reported promising both in India (Chhabra and Thakur 1998; Naik and Manjunath 2000) as well as in other countries (Bhutta and Choudhry 2000; Zang et al 1999; Stirzaker et al 1999).

AGROFORESTRY MODEL FOR BIO-DRAINAGE OF WATERLOGGED SOILS IN HARYANA

Rise in groundwater table followed by waterlogging and secondary soil salinization is a serious problem in canal-irrigated areas of arid and semi-arid regions. To combat the problem, an agroforestry model for biodrainage was tested in waterlogged fields of Haryana (northwest India), where 10 per cent area (0.44 m ha) is waterlogged resulting in reduced crop yields and abandonment of agricultural lands. In this model, four parallel strip-plantations of clonal *Eucalyptus tereticornis* (Mysure gum) were raised in December 2002 on four ridges constructed in the north–south direction in 4.8 ha canal-irrigated waterlogged fields of farmers. The strip-plantations were spaced at 66 m and each strip-plantation contained two rows of trees at a spacing of 1 m x 1 m, resulting in a density of 300 plants per ha. Levels of groundwater table were measured over 3 years (April 2005-April 2008) in 22 observation wells installed in two transects across the strip-plantations. The groundwater table underneath the strip-plantations remained lower than that in the adjacent fields and the drawdown in groundwater table was 0.85 m in 3 years. The annual rate of transpiration by 240 surviving trees per ha was 268 mm per annum against the mean annual rainfall of 212 mm. The strip-plantations sequestered 15.5 tons carbon per ha during the first rotation of 5 years and 4 months. Benefit-cost ratio of the first rotation of strip plantations was 3.5 : 1 and it would be many folds for next 3 to 4 rotations due to negligible cost of coppiced *Eucalyptus*. Wheat yield (April 2008) in the interspace of strip-plantations was 3.4 times that in adjacent waterlogged areas without plantation. It was mainly because of lowering of the water table and improvement in soil properties resulted in reclamation of waterlogged areas (Ram et al 2011).

BIODRAINAGE SYSTEM WITH TREE SPECIES IN ODISHA

Sub-Himalayan plateau, Indo-Gangetic plains, deltaic Odisha are among major areas plagued with such type of problems due to waterlogging and sub-optimal agricultural productivity due to waterlogged situation. In India out of 8.4 m ha waterlogged area, about 2.46 m ha is waterlogged due to over irrigation and rest is due to run-off congestion. In Odisha about 85,000 ha area is affected by waterlogged condition. Nearly 58 per cent of Mahanadi command is under influence of waterlogging (MOWR 1991) mainly in deltaic region. The deltaic region of Odisha suffers from waterlogging mainly due to following three reasons: (1) High and unusual rainfall within short span coupled with insufficient drainage leading to flood like situation in wider area; (2) Localized brief or prolonged water accumulation in the topographic depressions; and (3) Sea water intrusion in deltaic areas. Under these three given scenarios prevailing in deltaic region of Odisha, use of bio-drainage vegetation to control excess water situation in this region appears feasible only in 'type-2' scenario where accumulation of water or waterlogged situation occurs in topographical depressions which lacks adequate drainage facilities and 'type-3' scenario where the natural drainage is incapacitated by ingression of sea water.

The rate of drainage might be slower compared to discharge rate following engineering drainage measures. Unlike engineering method, slow but continuous drainage is another feature of bio-drainage and is influenced by the moisture absorption ability from soil profile and its transpiration efficacy. Moreover resultant biomass due to growth of vegetation is additional benefit over reclaimed field. Therefore vegetation has potential to complement engineering methods for removal of excess water from soil profile and can be utilized to reduce the extent of waterlogging. Scope for integration of agronomic, silvicultural practice as well as aquaculture was also tried to enhance overall productivity of the system. Present study also suggests that aforesaid strategy could be useful in keeping shallow ground water beyond the root zone of intercrops. Use of bio-drainage tree species like *Casuarina* in coastal areas and *Eucalyptus* in inland topographically depressed areas could keep the water table beyond the root zone of several cultivable crops like rice, groundnut, pulses, watermelon, etc. The beneficial effect of reducing the water table beyond root zone is further reflected through advancing the rabi cultivation activity by about 2-3 weeks earlier than the field without bio-drainage vegetation. This accelerated drainage has helped the farmer to advance their rabi cultivation by a period of 15-20 days.

The use of 4 m x 4 m spacing for growing plantation has facilitated intercultural operations like tillage with conventional bullock operated ploughing or power tillers and was found useful in accommodating intercrops. The elasticity of wood of *Casuarina* plantation in coastal areas provided resilience to plantation against constant strong wind. But in *Eucalyptus*, wider spacing and shallow root depth in soft waterlogged soil made it vulnerable to strong wind and causes lodging particularly during post-monsoon period when cyclone and storm is very common in deltaic Odisha during October-November period every year. However, narrower spacing of Eucalyptus might facilitate faster fall in water level providing better stability to vegetation preventing lodging, but would reduce convenience of intercrop cultivation. A clear cut draw down of ground water up to 109 cm underneath *Casuarina* vegetation and up to 115 cm beneath *Eucalyptus* vegetation, in comparison to non-vegetated area without biodrainage vegetation, was evident at the end of the season underneath four year old vegetation. This fact suggested that such ability can be used successfully for reducing water table in waterlogged areas. The ability of *Eucalyptus* and *Casuarina* to survive under prolonged period of 'above ground level water' condition also corroborated sustainability of the system in high rainfall low lying areas. The co-existence of intercrops like pulses, groundnut, watermelon, etc. could make farming under bio-drainage more remunerative. With average drainage rate of 16.5-22.19 lit/day (measured with sap flow system in the experiment), it is about 1.03-1.38 mm per day accelerated drainage from soil profile which could be achieved from the system. The aggregation of effective functional absorbing root system of bio-drainage plantations within first 60 cm of the soil profile would provide favourable regime for intercrop cultivation. Maintenance of desirable soil water balance within root zone would facilitate cultivation of many common rabi crops. The integration of pisciculture is also a profitable intervention which significantly increased the productivity of water and economic return of the whole system. This approach shall play an important role in enhancing the productivity of waterlogged areas in eastern India (Chowdhury et al 2011).

POND BASED FARMING SYSTEM FOR DEEP WATERLOGGED AREAS

Deep waterlogged areas due to poor drainage, saucer shaped topography and high monsoon rainfall in some parts of east coast of India remain waterlogged (>1m surface waterlogging) and unproductive. To stabilize and obtain net income from such waterlogged ecosystem, pond based farming technology (deep water rice in Kharif + salt tolerant vegetables like watermelon, lady's finger, spinach, chilli in winter + on-dyke vegetables and fruit trees of banana, papaya on bunds + fish inside pond) was conceptualized and implemented in representative deep waterlogged areas (12.5 m water depth) of Coastal Odisha, India (Mohanty et al 2008).

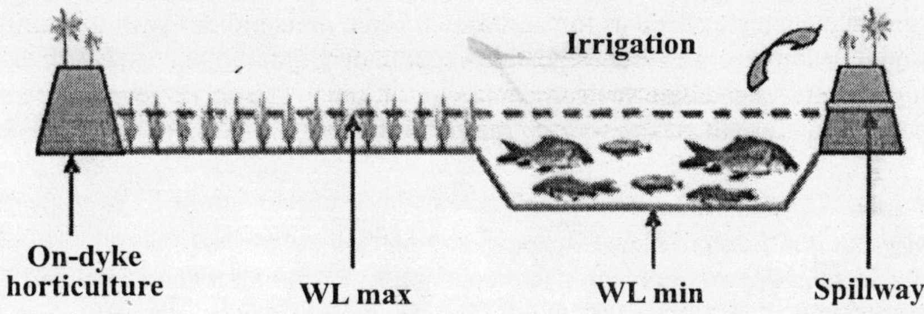

Fig. 28.4: Bio-drainage System in Waterlogged Areas

SUSTAINABLE INTEGRATED FARMING SYSTEM FOR WATERLOGGED AREAS

Eastern India received high amount of rainfall, inflow from highlands, canal irrigation and saucer shaped physiography near the coast in deltaic alluvial region which are most important reasons for waterlogging in 3.28 million ha land. Waterlogging, high rainfall and excessive ground water resources can be converted into poverty alleviating opportunities in many parts of eastern India. Converting farming of rice monoculture into integrated farming system will be environmentally sustainable, economically viable and risk avoiding strategy. Diversified broad based farming system has a tremendous potential to provide a range of nutritional sources for ensuring good health and generate employment all around the year. In view of above, an integrated farming system of fish and prawn culture, cultivation of paddy, vegetables, fruits and poultry rearing was successfully evaluated on farmer's field for 15 years. Integrated farming system can be classified into three major systems with twelve major models. The most popular model preferred in India and suitable for waterlogged situations are mainly pond-dyke integration, fish-rice-duck/poultry-vegetable.

In these systems, marginal lands/waterlogged lands/wetlands are generally brought into productive use where pond serves as a focal point for direct/indirect links between other components. It is reported that 50 kg of fish biomass can produce enough pond humus to fertilize nearly 6,500 m^2 of crop land. Further, under standard aquaculture practice, 10-25 cm silt per year is deposited at the pond bottom which is enriched with organic matter (2.4%), total nitrogen (2.0-2.1%), phosphorous (0.16-0.2%) and potassium (1.0-2.39%). Approximately 250 kg of pond sediment can produce enough fodder which can utilize to produce 2-3 kg of grass carp. In this system, animal/poultry wastes are used to fertilize fish pond and crop lands. The land in return produces crops, which serve as food for animal, fish and human beings (Samra et al 2003).

Suitable Fish and Prawn Species: Three major Indian carps, i.e., *Catla catla*, *Labeo rohita* and *Cirrhinus mrigala* and prawn (*Macrobrachium rosenbergii* and *M. malcolmsonii*) and exotic carps like silver carp and common carp may be stocked for culture in the integrated farming system.

Suitable Rice Varieties: Improved high yielding, tall, long-duration, submerge and pest resistant variety of rice with in-built characteristics of photoperiod sensitivity, strong seedling vigour such as Durga, Gayatri, Tulasi (for Odisha), Sabita (for Assam and West Bengal), Sudha (for Bihar) and Jalapriya (for Eastern U.P) can be suitable for this system.

ON-DYKE HORTICULTURE

The dykes to be constructed for preventing escape of fish from the integrated system, may be used for growing vegetables and other fruit trees like papaya and banana to make the system more economically viable. Vegetables such as gourd, radish, brinjal, leaf vegetables during pre-Kharif and Kharif season and vegetables such as tomato, french bean, radish, bitter gourd, cucumber, cauliflower, cabbage, brinjal, pumpkin and leaf vegetables (coriander, amaranthus and Indian spinach) can be grown during winter. Vegetables such as snake gourd, bitter gourd, ridge gourd, bottle gourd or ash gourd can be ground throughout the year on raised platforms.

BUND PLANTATION

In the initial years, the farmers planted papaya and banana on the land along with mango (*Mangifera indica*), teak (*Tectona grandis*), arecanut and coconut. In the later years, papaya and banana were tried which did not grow well because of shade of the trees and were also vulnerable to high wind velocity. Different kinds of fruit trees in limited number like guava, pomegranate, sapota, litchi, jamun (hybrid) were also grown on the bunds. Different species such as bay leaf (tejpatta), ginger and turmeric were taken as intercrop. Also floriculture was tried in the initial years. It thus has served as an example of successful agricultural enterprise through integrated farming system approach with right kind of scientific approach driven by market demand.

OTHER BIOMASS PRODUCTION TECHNOLOGY FOR SEASONAL DEEP WATERLOGGED AREAS

DEEP WATER RICE PRODUCTION TECHNOLOGY

In some parts of the coastal areas, prolonged waterlogging (>1m) during rainy season reduces tillering and growth of normal rice varieties. Sometimes flash flood inundates the standing crop at any stage of growth for 8-10 days at a stretch, resulting in heavy mortality. The crop is damaged completely if this situation occurs at early vegetative stage. Proper establishment of crop before onset of flooding and adoption of deep waterlogging tolerant rice varieties are of paramount importance to realize net return from crop. Hence, deep water rice production technology for seasonal deep waterlogged areas was developed with seeds of deep water rice varieties (Hangeswari, Saraswathi, Ambika, Sabita). Improved deep water rice cultivar Hangeswari produced 2.4-2.5 t/ha yield in kharif season in deep water situation. Higher and stable incomes from deep waterlogged areas in eastern India are reported. Based on survey in kharif 2009 at Odisha, it was revealed that about 180 farmers in 105 ha of land adopted deep water rice.

WATER CHESTNUT PRODUCTION TECHNOLOGY

Water chestnut (*Trapa bispinosa*) or Singhara Phal or Pani Phal is one of the few neglected but economically important aquatic crops grown in difficult parts of India and even in several other South East Asian countries like Bangladesh, Thailand, Myanmar, etc. The origin of water chestnut roots in India as there is mention about the crop in ancient literature. In India, water chestnut is cultivated in north from Jammu and Kashmir, Uttar Pradesh up to Madhya Pradesh, on east from Bihar, West Bengal, Odisha up to Gujarat and Maharashtra in western part of the country. The crop is marshy grown in roadside depressions in addition to localized topographical depressions where water remains stagnant at least up to month of January (Hazra et al 1996; Ahmed and Singh 1999). Otherwise the low-lying areas of shallow lowland ecosystem in the country mainly in eastern India itself provides ideal condition for cultivation of this crop mainly during kharif season. Net profit from water chestnut can reach up to Rs. 20,000-25,000 per ha. Water chestnut cultivation already been adopted in 900 ha area in Odisha covering six districts. It has a further potential for adoption in 1.10 million ha of non-arable waterlogged areas of Odisha. Fruits are generally consumed as raw or after boiling as a vegetable. After sun drying, nut flour is also used a source of non-cereal carbohydrate diet. A significant portion of the nut is processed for using the flour for food or for textile sizing (Alam et al 2001).

PHRAGMITES-SACCHARUM GRASS AND OTHER AQUATIC BIOMASS PRODUCTION

Waterlogged and marshy localities having high clay to loamy soils are common in Terai Uttaranchal, Punjab, Rajasthan, North Bihar, N. Bengal, Assam, Manipur, Tripura, Andaman, Odisha and Deltaic Region. Vegetation is characterized by tall, coarse grass species like *Phragmites karka*, *Saccharum* spp., *Imperata cylindrica*, *Desmostachya bipinnata* and *Vetiveria zizanioides*. These grasses are highly used for thatch purposes and generate employment through handicrafts making and cottage industries.

■■■■

29 AGROFORESTRY PRACTICES FOR NON-WOOD FOREST PRODUCTS AND RURAL DEVELOPMENT

INTRODUCTION

Agroforestry is an age old practice followed by every rural household for supporting their local needs which also maintains the microclimate and aesthetic of the surrounds. Agroforestry focuses on the wide range of trees grown on farms and other rural areas. Among these are fertilizer trees for land regeneration, soil health and food security; fruit trees for nutrition; fodder trees for livestock; timber and energy trees for shelter and fuel wood; medicinal trees to cure diseases and trees for minor products viz. gums, resins, or latex products. Many of these trees are multipurpose, providing a range of benefits. Scientific agrofoestry emphasized for increase in the productivity from unit area by incorporation of tree components in intensive agricultural practices like agrisilviculture, agrihorticulture, silvopastoral, agrisilvihorticulture, etc. These practices by and large applied to large farm holdings but small land holders are interested in homegarden type model which is most primitive and followed in all around the tropics (Leakey 1996). These homegardens provide many non-wood forestry components like nuts, fruits, leaves, fibres, gums, dye, flavours, medicinal plants, etc. to meet out the needs.

Cultivating trees in combination with crops and livestock is an ancient practice. Agroforestry practices differ considerably from country to country as farmers adapt to needs and circumstances. In India, most of the NWFPs are collected from the forest and collection & marketing is done through state forest corporations. With the shrinakage in forest cover and ban on harvesting and extraction of majority of produce except tendu leaves, some of the TBO's like sal, mahua, karanj, neem, etc., and gums, resin, and grasses are collected by FDCs or TRIFED mainly engaging tribals. This practice is nowadays not lucrative and collectors didn't get a good share and limited access to these produce. Intrestingly every state of India (more than 25) established FDCs but in these 25-30 years of existence very few were able to deliver services for development of Non-Wood Forest Proucts (NWFPs) and rural development (Tewari 2008). Non-Wood Tree Products (NWTPs) refer to products arising from individual tree species and excludes the products of the wider forest that are not tree derived. Non-wood is used in preference to non-timber since it excludes not only timber but also fuelwood and poles. In essence, the term NWTPs is restricted to products that are tree derived but for which it is not necessary to fell the tree. They are currently receiving focused attention from ethnobotanists, conservationists, policy-makers and breeders, as they embody the goals of conservation, development and production activities. With forested areas declining to accommodate agricultural expansion, and with human populations rising, the logical endpoint must be a choice between more people moving into the forest or moving more trees out of it. Many of the tree species that would move out of the forest onto farms would be those that provide NWTPs.

First World Agroforestry Congress (Florida, United States) in 2004 advocated for agroforestry to increase household income; to promote gender equity; to empower women; to improve the health and welfare of people, and; to promote environmental sustainability. It also urged all governments to highlight the role of agroforestry in poverty reduction strategies, to provide financial support and to develop policies that promote the adoption of associated practices. Recently in Indian Science Congress on 4th January, 2013 at Kolkata, Prof.M.S.Swaminathan quoted that the farming is a life giving profession and farmers are life givers to us. He has suggested three major steps to overcome the crisis such as: (i) Diversification of source of income, through Rural Cropping System Research (crop - crop/ food crop - fodder) and Farming System Research (crop - livestock); (ii) A mechanism of social support in the form of a fund, which he referred as 'Life Saving Fund'; (iii) Errors minimized agronomy to avoid or reduce degradation, which should be profit maximizing and risk minimizing. It has been suggested for ensuring high input support, crop diversification, special technological packages, adoption of credit policy and financial assistance, which will improve farming practices and help the farmers. Creation of value addition and small scale construction of storage structures at village level, consolidation of land and stopping fragmentation, increasing water productivity, reducing post-harvest loss, and concept of group farming & cooperative system will collectively strengthen farming systems and effectively control farmers' suicide.

GREATER ROLE OF AGROFORESTRY

The Indian National Agriculture Policy of 2000 emphasized the role of agroforestry for efficient nutrient cycling, nitrogen fixation, organic matter addition and improving drainage. It also underlined the need for diversification by promoting integrated and holistic development of rainfed areas on watershed basis through involvement of community to augment biomass production through agroforestry and farm forestry. The Task Force on Greening India for Livelihood Security and Sustainable Development of Planning Commission (2001) has also recommended that for sustainable agriculture, agroforestry may be introduced over an area of 14 million ha (M ha) out of 46 M ha irrigated areas that are degrading due to soil erosion, waterlogging and salinization. For integrated and holistic development of rainfed areas, agroforestry is to be practised over an area of 14 million ha out of 96 M ha rainfed areas. This all will, besides ensuring ecological and economic development, provide livelihood support to about 350 million people. The practice of agroforestry can help in achieving these targets. Therefore in the quest of optimizing productivity, the multi-tier system approach came into existence. Gap of demand and supply of forest produce in India is widening and forests are unable to fulfill the demand (Chopra 1993). Agroforestry can play an important role in filling this gap and conservation of natural resources. Agroforestry is a "social forestry" - its purpose is sustainable development. Practices are focused on meeting the economic, environmental and social needs of people on their private lands. At the farm level, agroforestry is a set of practices that provide strong economic and conservation incentives for landowner adoption. Incorporated into watersheds and landscapes, agroforestry practices help to attain community/society goals for more diverse, healthy and sustainable land use systems.

Available information indicates that agroforestry can provide a greater range of environmental benefits thân conventional types of annual crop cultivation. In present era of diversification of agricultural practices due to ill effect of monoculture and intensive farming, environment friendly agroforestry practices for non-wood forest products and rural development are increasing for broadly food and social security. In many parts of globe, this practice is already followed but so far systematic incorporation of agroforestry species and components have been not followed. NWFP has been presently emerged as a promising alternative to wood products in natural forest management. The extraction of NWFP from natural forest is also now restricted to check over exploitation and genetic erosion. The domestication and the commercialization of these NWFPs tend to emerge as an alternative strategy for their extraction from agricultural land.

NON-WOOD FOREST PRODUCTS FROM AGROFORESTRY

The International Conference on Domestication and Commercialization of Non-Timber Forest Products in Agroforestry Systems, hosted by ICRAF, was held in Nairobi, Kenya, from 19 to 23 February 1996. This was the first world-level meeting to be held exclusively to draw attention to issues dealing with domestication and commercialization of non-timber forest products in agroforestry systems.The forestry mission of the Food and Agriculture Organization of the United Nations is to enhance the contribution of trees and forests to global human well-being. Since its founding in 1945, FAO has grown to include 174 member countries. This broad membership allows FAO to address issues facing the entire world's forests - boreal, temperate, subtropical and tropical; forests in developed and developing countries; dry forests and humid ones; high altitude forests and mangroves; even trees on farms and in cities. The programme "Promotion and Development of Non-Wood Forest Products (NWFP)" at the Forest Products Division of FAO's Forestry Department, aims to enhance the value of non-wood forest products and services through improved harvesting, utilization, trade and marketing. The sustainable utilization of NWFP, combined with an equitable distribution of the benefits obtained by closely involving local people, contributes significantly to the wise management of the world's forests, to the conservation of their biodiversity and to a precise appraisal of their socio-economic values. Income generation and the contribution of NWFP to poverty alleviation and food security are important elements. The programme further includes data collection, information dissemination, technology transfer, networking and strengthening of partnerships on NWFP development, training and policy advice.

Incorporating NWFPs in production systems is not a new practice. Various forms of agroforestry associations have developed around NWFPs and form the very basis of a suitable indigenous agriculture. It is only in recent years that agroforestry research is considering the prospects of these indigenous systems for forest species. In Southeast Asia, a complex agroforestry systems for the management of forest resources have been developed for centuries by local people ranging from the production of locally consumed fruits to highly valuable industrial products, such as resins and latexes. These agroforestry systems are transformations of both selected forest resources and a true forest structure from the sphere of 'nature' to that of 'agriculture'. This process can thus be analyzed as a particular domestication strategy, which could integrate conventional species domestication techniques: selection, reproduction and plantation practices; to an original form of ecosystem 'domestication'. Prospects for further developing this agroforest strategy for the domestication of forest species, particularly NWFPs, are very important. The social, economic and institutional implications of such an integration of NWFP resources to agricultural development are to be analysed, based on various examples of agroforest development and focusing on the efficiency of this 'appropriation' strategy by smallholder farmers.

DOMESTICATION OF NON-WOOD FOREST PRODUCTS IN AGROFORESTRY

Domestication of forest species for purposes of commercial cultivation seems to have bright prospects. Forestry definitely needs to find an answer to the exhaustion of wild resources while in the same time rationalizing the production of marketable NWFPs. There is great demand for new crops and new markets. Sustainable development has to mitigate the effects of deforestation by increasing the planting of trees on cleared lands, and farmers have to find substitutes for the natural resources lost through deforestation. However, domestication is neither a fast nor an easy process. It took many years of farmers' work of trials and tests to develop a range of food crop varieties and decades of agroscientific research to create productive clones of industrial tree crops. Modern plant domestication relies on relatively sophisticated techniques. Modern agricultural and forestry science evolved in Europe, in a pastoral and cereal-growing civilization, which influenced most of the present agricultural paradigms and agroecosystem models. Our perceptions of domestication and agricultural development are deeply influenced by a tacit preference for cereal-based models, which may not be the most useful for the domestication and development of tropical forest resources. There are many definitions of 'domestication' and it is clear that the process has evolved as man has become more intimately associated with plants and animals. The steps in this evolution can, in general, be identified as: extractivism for own use; extractivism for barter with others; extractivism for sale locally, regionally or internationally; retention of desirable species in farm land; planting around settlements; creation of plantations and ranches; genetic selection, cloning and breeding; and biotechnology (FAO 2011a).

This pattern is not universal, relatively few tree species have progressed beyond the fourth or fifth step. However associated with the later steps, we find the development of extensive commercial activity, the start of processing and the expansion of trade. The domestication of natural resources involves a move from gathering in the wild on communally owned land to the deliberate cultivation of NWFPs on tenured farm land. The domestication of a chosen species then involves genetic selection and the management of varieties or cultivars. Through selection, yield and quality are improved so that the price paid for the product increases. Profits from these improved cultivars growing under personal control are the incentive to plant and manage the resource sustainability. There are however several constraints to domestication and the formal development of markets for NWFPs which include:

- lack of infrastructure in the rural areas, making access to markets difficult
- low volume of products
- poor or variable quality of products
- no continuity of supply
- poor handling and storage qualities
- limited knowledge of the product among consumers

One group of NWFPs with commercial potential is the essential oils. Here again, the potential for the exploration of essential oils from wild plants is severely restricted, as there are several problems that make the essential oil industry wary of getting involved. For example, the quality of some oils can vary from region to region. For example, when Israel tried to break into the geranium oil industry, it obtained its germplasm from France, which is a top producer. However, the Israeli oils were found not to be acceptable to the market until a special niche was created for it. As with other NWFPs, essential oils also suffer from supply problems. One success story, however, comes from South Africa, where it was found that *Tagetes minuta* oil had characteristics that the market needed, although continuity of supply was difficult to guarantee without its cultivation and also farmers preference for crop substitution (Wickens 1991).

The design of the agroecosystem in which domesticated species are grown is as essential as the choice for particular plant selection, breeding and reproduction techniques. This is particularly important when switching from annual crops to trees and from fields to forests. Most 'modern' agroecosystems in the world relate to a single agri-cultural model: intensive, highly specialized stands of homogeneous crops. Even in agro-forestry models are either fuel-based, fodder-based or industrial wood based intensive e.g. eucalyptus and poplar tree models with wheat or rice in northern India. However, other available models exist, which should be examined under the new perspective of domestication of NWFPs of forest species, particularly in agroforestry. It is essential to understand that the importance of an ecosystem approach to domestication goes far beyond biological or technical considerations. It also has obvious sociopolitical and institutional dimensions. Scientific agroforestry models not only change the face of forest landscapes but also national economies. It is essential to analyze in the context of NWFPs domestication, as it deeply linked to forest communities and their socioeconomic life. Domestication is part of a resource appropriation process, and resource appropriation by a powerful fraction of the population might lead to dispossession of the weaker fraction.

Agroforestry is recognised as a promising land-use technology and an interface between agriculture and forestry, especially in the developing countries of the tropics and subtropics. With recent research, the principles underlying age-old agroforestry practices are gradually being understood, and improved practices are introduced. The scientific foundation of agroforestry is the multipurpose tree (MPT). Two important attributes of MPTs are their ability to produce multiple products, from one or more plant parts, and their ability to provide benefits and services that may be intangible but nonetheless environmentally significant. Among the many recognised products and benefits of MPTs, most agroforestry projects are oriented towards the production of fuelwood and fodder. Very little has been done to exploit the non-wood forest products (NWFPs) in agroforestry systems. Considering that agroforestry places a strong emphasis on smallholder, low-input, multiple-output systems and many of the trees and other species that produce NWFPs are amenable to integrated management, agroforestry as a land-use approach and NWFPs as an output are closely integrated. Thus it is somewhat paradoxical that the role of agroforestry in the promotion of NWFPs has not been examined in any detail so far. It can be argued that because MPTs have a central role in agroforestry systems, and because the emphasis of MPTs is on non-wood products and services, most agroforestry systems deal inescapably with non-wood forest tree products. But the concept of NWFPs is such that the "dominant" tree products such as fuelwood, fodder, and charcoal that have received the major share of attention in agroforestry and social forestry, are excluded from this category (although, botanically, some of them are "non-woody" in nature). Common NWFPs include food and food additives; fibre and flosses; phytochemicals and aromatics; oils; resins, latex and other exudates; decorative articles; and products such as fodder, etc.

Agroforestry and NWFP extraction are predominantly subsistence practices; they make minimal use of costly external inputs, but heavy use of human labour; and each has location-specific characteristics. However, there are also subtle differences between NWFP production and the practice of agroforestry. First, agroforestry systems are, generally, more intensively managed, with a higher degree of species domestication, than most NWFP species. Second, agroforestry systems place emphasis on not only the productive role of their woody components, but also of their protective roles such as soil amelioration and environmental protection. This need not necessarily be so in the case of NWFP species. A third difference is that although agroforestry is often described as an interface between agriculture and forestry, most agroforestry practices are on the so-called agricultural and marginal lands and some in the buffer zones around forests, with very little or no agroforestry in the forests. NWFPs, with some notable exceptions are extracted mostly from forest and/or marginal land, with relatively lesser quantities from agricultural lands. Identification of such similarities and contrasts between common agroforestry practices and NWFP extraction can help in examining opportunities for integrating NWFPs and agroforestry systems. By placing NWFPs in a production system context, this may also help to explore new benefits of NWFP species (Nair and Merry 1995).

NWFPs YIELDING PLANTS FOR POTENTIAL AGROFORESTRY USE

The farmers and land owners integrate a variety of woody perennials in their crop and livestock production fields depending upon the agroclimates and local needs for diversified products. With the current interests in agroforestry worldwide, attempts are being made to introduce agroforestry techniques using indigenous and exotic multipurpose and nitrogen-fixing woody perennials. Agroforestry is widespread in all ecological and geographical regions of tropical regions. The systems vary enormously in their structural complexity and species diversity, their productive and protective attributes and their socio-economic dimensions. They range from apparently simple forms of shifting cultivation to complex homegardens; from systems involving sparse stands of trees on farm lands (e.g. *Prosopis cineraria* called khejri tree in arid regions of western India) to high-density complex multistoried homesteads of humid lowlands; from systems in which trees play a predominantly 'service' role (e.g. shelterbelts) to those in which they provide main salable products (e.g. intercropping with plantation crops).

A homestead is an operational farm unit in which a number of crops including tree crops are grown with livestock, poultry and/or fish production, mainly for the purpose of meeting the routine basic needs of the farmer. It is an old-age practice of domestication of NWFPs in coastal states particularly in Kerala, Tripura, Assam, North-eastern states and parts of West Bengal and Andaman and Nicobar Islands. Although the homegardens appear to be a mixture of trees, shrubs and herbs, a certain general pattern seems to exist. There is wide variation in the intensity of trees, species and crops based on the size of holding, needs of the people residing in homesteads and microclimate. Coconut, arecanut, guava, mango, citrus, tamarind, jackfruit, papaya, banana, drumstick, sesbania, custard apple and many multipurpose trees are the major trees found grown in homegardens. Many multipurpose trees having productive and protective functions are integrated into the system in different spatial and temporal arrangements which provide multiple products.

Ethnobotanical inventories of indigenous knowledge of plants and their environment have great value in identifying species for domestication and commercialization. Indigenous people themselves are enterprising and have been selling and starting to domesticate these inventoried products without any intervention from outside commercial interests or from governmental or non-governmental organizations. These accomplish-ments already achieved by indigenous peoples and other small-scale farmers should be further supported. To promote and improve markets to benefit local gatherers and vendors, it is important to consider resource tenure, distance to markets, access to transport, seasonal variations in supply of products and tree management. Tree management could include their integration into agroforestry systems. Numerous literatures are available regarding different tree species yielding NWFPs in India and world. Important NWFP yielding plants (flavours and fragrances or essential oils, gums, resins, latex, fibres and flosses, tannins and dyes, edible plant products and important bamboos, rattans and palms) which are/can be used in agroforestry sytems for adoption in different ecological conditions are given below (Mehta 1981; ICFRE 1990; GFU 2008; MPSMFPCF 2013; NTFP 2013; MPSG 2013; IUFRO 2013; FAO 2013).

FLAVOURS AND FRAGRANCES (ESSENTIAL OILS)

Grass Oils: Lemon grass oil (*Cymbopogon flexuosus*), West Indian lemon grass oil (*Cymbopogon citratus*), Palmarosa oil (*Cymbopogon martinii* var. *motia*), Ginger grass oil (*Cymbopogon martinii* var. *sofia*), Citronella oil (*Cymbopogon nardus* var. *winterianus*), Vetiver oil (*Vetiveria zizanioides*)

Wood Oils: Sandalwood oil (*Santalum album*), Agar oil (*Aquillaria agalocha*), Deodar oil (*Cedrus deodara*), Pine oil/ Turpentine oil (*Pinus roxburghii*), Linaloe oil (*Bursera delpechiana*), Cedar oil (*Juniperus macropoda*)

Leaf Oils: Eucalyptus oil (*Eucalyptus globulus*), Lemon scented gum oil (*Eucaylyptus citriodora)*, Camphor oil (*Cinnamomum camphora*), Pine needle oil (*Pinus roxburghii*), Mint oil (*Mentha* spp.), Patchouli oil (*Pogostemon patchouli*), Geranium oil (*Pelargonium graveolens*), Wintergreen oil (*Gualtheria fragrantissima*)

Root/Rhizome Oils: Costus oil (*Saussurea lappa*), Indian valerian oil (*Valeriana wallichii*), Curcuma oil (*Curcuma aromatica*), Cyperus oil (*Cyperus scariosus*), Sweet flag oil (*Acorus calamus*)

Flower Oils: Jasmine oil (*Jasminum officinale*), Rose oil (*Rosa damascena*), Keora oil (*Pandanus odoratissimus*), Tagtes oil (*Tagetes indica*), Champa oil (*Michelia champaca*), Lavander oil (*Lavandula officinalis*), Cassie perfume (*Acacia fernesiana*), Clove oil (*Syzygium aromaticum*),

Other Essential Oils: Citrus oil/Lime oil (*Citrus aurantifolia*), Petgrain oil (*Citrus aurantium* and *Citrus limettioides*), Orange oil (*Citrus sinensis*), Mandarin oil (*Citrus reticulata*), Cinnamoum oil (*Cinnamomum zeylanicum*), Nutmeg oil (*Myristica fragrans*), Salai gum oil (*Boswellia serrata*)

GUMS, RESINS AND LATEX

Acacia Gums: Babul (*Acacia nilotica*), True gum arabic (*A. senegal*), Khair (*A. catechu*), *A. modesta*, *A. fernesiana*, *A. leucophloea*

Other Gums: Karaya gum (*Sterculia urens*), *Cochlospermum religiosum*, Semal (*Bombax ceiba*), *Astragalus heratensis,* Kapok (*Ceiba pentandra*)*,* Hog tragacanth (*Prunus amygdalus*), Cherry gum (*Prunus armeniaca*), Bengal kino (*Butea monosperma*), Gum kino (*Pterocarpus marsupium*), Jhingan gum (*Lannea coromandelica*), Salai gum (*Boswellia serrata*), Moringa gum (*Moringa pterygosperma*), Dhaura/ Ghatti gum (*Anogeissus latifolia*), Semla gum (*Bauhinia retusa*), Madiata gum (*Adenanthera pavonina*), Gutta percha gum (*Dichopsis polyantha*), Eucalyptus kino (*Eucalyptus globulus*), Lemon scented gum (*Eucalyptus citriodora*).

Minor Gums: *Albizia chinensis, A. odoratissima, A. procera, Anacardium occidentale, Azadirachta indica, Bauhinia purpurea, B. racemosa, B. variegata, Chloro-xylon swietenia, Feronia limonia, Mangifera indica, Terminalia alata, T. bellerica, Agele marmelos, Ailanthus excelsa, Pithecellobium dulce, Spondias pinnata, Borassus flabellifer, Buchanania latifolia, Calophyllum apetalum, C. elatum, C. inophyllum, Chukrasia tabularis, Delonix regia, Elaeodendron glaucum, Gardenia turgida, Gaurga pinnata, Macaranga peltata, Prosopis juliflora, P. cineraria, Zizyphus sativa.*

True Resins: Balck dammar (*Canarium strictum*), Rock dammar (*Hopea odorata*), Green dammar (*Shorea tumbuggaia*), White dammar (*Vateria indica*), Amber (*Pinus succinfera*), Lacquer (*Rhus vernicifera*)*,* Shell lac (*Laccifer lacca*), Mastic (*Pistacia lentiscus*), Sand arac (*Callitris quadrivalvis*), Storax (*Atingia excelsa*)

Other True Resins: *Abies excela, A. webbiana, Ailanthus malabarica, Cannabis sativa, Carica papaya, Daemonorops kurzianus, Juniperus communis, Liquidambar orientalis, Styrax benzoin*

Oleo Resins: Salai gum (*Boswellia serrata*), Gurjan oil (*Dipterocarpus turbinatus*), *Kingiodendron pinnatum, Cedrus deodara, Erythroxylon monogynum, Melanorrhoea usitata, Pinus insularis, Pinus roxburghii, Pinus wallichiana, Shorea robusta*

Gum Resins: Gamboge (*Garcinia morella*), Myrrah/Gugal (*Commiphora mukul*), Galbanum (*Ferula galbaniflua*), Asafoetida (*Ferula asafoetida*), *Diospyros peregrina*, Dikamali/ Cumbi gum (*Gardenia gummifera*)

Latex: Chilte (*Cnidoscolus* spp.), Chicle (Manilkara zapota), Rubber latex (*Hevea brasiliensis*)

TREE BORNE OIL SEEDS (TBOs)

Jatropha curcas, Pongamia pinnata, Madhuca indica, Simmondsia chinensis, Azadirachta indica, Simarouba glauca, Shorea robusta, Mangifera indica, Garcinia indica, Salvadora oleoides, S. persica, Ziziphus mauritiana, Prunus armeniaca, Aleurites montana, Aleurites fordii, Thespesia populnea, Moringa oleifera, Calophyllum inophyllum, Schleichera oleosa, S. trijuga and *Actinidaphne hookeri*

FIBRES AND FLOSSES

Fibres: *Agave sisalana, Streculia villosa, Abroma augusta, Abutilon* spp., *Ananas cosmosus, Antiaris toxicaria, Boehmeria nivea, Borassus flabellifer, Cannabis sativa, Cordia dichotoma, C. rothii, Giradinia heterophylla, Grewia glabra, G. elastica, G. optiva, Hibiscus* spp., *Malachra capitata, Marsdenia tenacissima, M. volubilis, Phormium tenax, Sansevieria roxburghiana, Sesbania bispinosa, Sida rhombifolia, Sterculia foetida, S. urens, Themeda arundinacea, Trema orientalis, Typha elephantina, Urena lobata, Oreocnide integrifolia*

Flosses: Indian Kapok (*Bombax ceiba*), Kapok or Silkcotton (*Ceiba pentandra*)

DYES/ COLOURANTS

Wood Dyes: Kutch dye (*Acacia catechu*) and other dyes from *Artocarpus heterophyllus, A. lakoocha, Pterocarpus santalinus, Caesalpinia sappan*

Bark Dyes: *Terminalia tomentosa, Acacia concinna, A. farnesiana, A. leucophloea, Alnus* spp., *Casuarina equisetifolia, Manilkara littoralis, Myrica esculenta, Ventilago madraspatana*

Flower and Fruit Dyes: *Mallotus philippensis, Woodfordia floribunda, Bixa orellana, Butea monosperma, Toona ciliata, Nyctanthes arbortristis, Mammea longifolia, Wrightia tinctoria, Crocus sativus*

Root Dyes: *Berberis aristata, Datisca cannabina, Morinda tinctoria, Punica granatum, Rubia cordifolia*

Leaf Dyes: *Indigofera tinctoria, Lawsonia inermis*

TANNINS

Bark Tans: *Acacia mearnsii, A. dealbata, A. nilotica, A. auriculiformis, Cassia fistula, Cassia auriculata, Rhizophora mucronata, Ceriops roxburghiana, Terminalia arjuna, Shorea robusta*

Fruit Tans: *Terminalia chebula, T. bellerica, Acacia nilotica, Emblica officinalis, Zizyphus xylocarpa, Caesalpinia coriaria*

Leaf Tans: *Anogeissus latifolia, Emblica officinalis, Carissa spinarum*

Gall Tans: *Tamarix* spp.

MEDICINAL PLANTS

Root Drugs: *Podophyllum hexandrum* (Bankakru), *Saussurea lappa* (Kuth), *Aconitum heterophyllum* (Atis), *Acorus calamus* (Bach), *Rheum emodi* (Rhubarb), *Picrorhiza kurroa* (Kuru), *Valeriana wallichii* (Banahfsa), *Asparagus adscendens* (Satavar), *Glycyrrhiza glabra* (Mulathi), *Berberis* spp., *Dioscorea deltoidea* (Medicinal yam), *Rauvolfia serpentina* (Sarpgandha), *Nardostachys jatamansi, Colchicum luteum, Abroma angusta, Hemidesmus indicus, Sarsaparilla* spp., *Urginea indica*

Bark Drugs: *Cinchona* spp., Holarrhena antidysenterica, *Soymida febrifuga, Alstonia scholaris, Terminalia arjuna*

Flower, Fruit and Seed Drugs: *Artemisia* spp., *Strychnos nux vomica, Aegle marmelos, Ricinus communis, Cassia fistula, Chenopodium ambrosioides, Pyrethrum* spp., *Casealpinia crista, Tamaridus indica, Terminalia bellerica, T. chebula, Emblica officinalis, Plantago ovata*

Leaf Drugs: *Ephedra gerardiana, Vitex nugendo, Gultheria fragrantissima, Mentha arvensis, Hyoscymus niger, Ocimum kilimandscharium, Azadirachta indica, Cannabis sativa, Atropa acuminata, Datura* spp., *Swertia chirata*

EDIBLE PLANT PRODUCTS

Edible Fruits: Chironji (*Buchanania lanzan*), Aonla (*Emblica officinalis*), Tamarind (*Tamarindus indica*), Bael (*Aegle marmelos*), Wood apple (*Feronia elephantum*), Jackfruit/Kathal (*Artocarpus heterophyllus*), Barhal (*Artocarpus lakoocha*), Kala lakuch (*Artocarpus gomezianus*), Jamun (*Syzygium cuminii*), Custard apple/Sarifa (*Annona squamosa*), Ramphal (*Annona reticulata*), *Carissa opaca*, Karaunda (*Carisa carandas*), Timla (*Ficus auriculata*), Pakar (*Ficus infectoria*), Parphuta (*Ficus nerifolia*), Bedu (*Ficus palmata*), Gular (*F. glomerata*), Ber (*Zizyphus jujuba*), Jangali jalabi/ Madras thorn/ Manila tamrind (*Pithecellobium dulce*)

Edible Nuts: Cashew nut (*Anacardium occidentale*), Chilgoza (*Pinus gerardiana*), Walnut (*Juglans regia*), Almond (*Prunus dulcis*)

Edible Flowers: Mahua (*Madhuca indica*), *M. longifolia, Musa* spp, *Bombax ceiba, Bauhinia purpurea, Alangium salvifolium, Ficus glomerata, Sesbania grandiflora, Rhododenron arboreum*

Roots and Tubers: *Amorphophallus campanulatus, Dioscorea belophylla, D. oppositifolia, Ipomoea aquatica, Bamboos*

Leaves: *Agave americana, Aloe vera, Moringa oleifera, Antidesma diandrum, Fern species*

Edible Bamboo Shoots: *Bambusa balcooa, B. bambos* (Kanta bans/Spiny bamboo), *B. polymorpha* (Betwa), *B. tulda, B. nutans, B. pallida, Dendrocalamus asper, D. brandisii, D. giganteus* (Giant bamboo), *D. hamiltonii* (Kagzi bans), *D. longispathus, D. strictus* (Lathi bans/ Male bamboo), *D. latiflorus, D. oldhami, D. membranaceus, Phyllostachys pubescens, P. praecox, P. iridescens, P. bambusoides, P. edulis, P. dulcis, Schizostachyum, Thyrsostachys*, etc.

Fungi: *Agaricus campestris, Morchella esculenta, Volvaria terastius, Collybia albuminosa*

CANES/ RATTAN

They are stems of climbing palms of genus *Calamus* consisting about 30species. Climbing often attain a length of even 100 m. They are usually cylindrical and have uniform thickness, solid and pale white to yellow in colour. Outer surface is hard, smooth and shinning. Cane is used in variety of ways due to its pliability, strength and long length. It is used in walking sticks, polo sticks, umbrella handles, baskets and cane furniture. The important species of cane are *Calamus acanthospathus, C. guruba, C. tenuis, C. viminalis, C. rotang, C. andamanicus, C. brandisii, C. flagellum, C. gambeli, C. gracilis, C. latifolius, C. pulstris, C. rheedei, Daemonorops kurzanius, Plectocomia himalyana.* Among these, *Calamus tenuis* is the common cane of north India.

PALMS

Palms are monocots, included in the section of Angiosperms characterized by bearing a single seed leaf. Scientifically, palms are classified under the family Arecaceae (or Palmae) and are perennial, distinguished by having woody stems. Palms are important NWFP with numerous uses such as food, beverages, oil, fuel, feed, fertilizer, building material, furniture, games and toys, ornamental uses, medicinal uses etc. The important palm genus are *Areca, Arenga, Bentinckia, Corypha, Hyphaene, Licuala, Livistona, Loxococcus, Oncosperma, Phoenix, Pinanga, Wallichia, Caryota, Nypa* and *Borassus.*

The above mention list is indicative and includes the commonly used Indian plants but there are many more plants in the world used for food and survival. The proper screening of prioritized plants for their domestication/ introduction and utilization is necessary for future food security of the nation. Tropical forests are full of tree species that people use every day of their lives for their own needs (food, fodder, medicines, building materials, resins, dyes, flavours); the so-called non-wood tree products (NWTPs). These species, however, have generally been less studied than their industrial counterparts (*Eucalyptus, Pinus* spp.). Leakey and Newton (1994) referred collectively to these neglected taxa as the 'Cinderella' species since their potential is yet to be unveiled. They also regarded as LKTS (Lesser known tree species).Yet, if carefully promoted and marketed they raise hopes of alleviating rural poverty and increasing the sustainability of agroecosystems.

NWFPs BASED AGROFORESTRY

The possible impacts of NWFPs based agroforestry are:

1. Poverty alleviation through increased income by diversifying the production of agroforestry products for home consumption and market.
2. Nutritional security by incorporation of wild fruits, vegetables, nuts, spices, herbs etc NWFPs with traditional agricultural cereal and millet crops.
3. Empowerment to women farmers and small holdings rural residents in processing and value addition of NWFPs through development of rural units.
4. Checking irrational harvest and pressure on forest by providing NWFPs on farms.
5. Increasing buffering capacity of farmers against the climate change and participation in carbon trading.
6. Development of rural small scale and cottage units based on diverse NWFPs through cooperatives.
7. Augmenting accessibility to medicinal plants for primary health care and revitalization of local health traditions.
8. *Ex situ* conservation / domestication of lesser known plants in farm for the future improvement of these crops.
9. Improving the agrobiodiversity *vis-a-vis* ecological sustenance of the agroecosystem and control of insect-pest.
10. Blending of ethnoagriculture with modern agrotechniques for commercialization of NWFPs from agroforests.

Like any other agroforestry practices NWFPs based agroforestry is also intentional systematic combinations of trees, shrubs, herbs with crops and/or livestock that involve intensive management of the interactions between the components as an integrated agroecosystem. It also satisfies the criteria of Intentional, Intensive, Interactive and Integrated approach for wider adoption and production. The hair line demarcation only that it emphasizes more on adoption for poverty alleviation and rural development of small and marginal farmers rather intensive production by large farmers for selected NWFPs. It provides a different land use option, compared with traditional arable and forestry systems.

It makes use of the complimentarily relationship between trees and crops in vertical strata, so that the available resources can be effectively utilized. It is a practice that supports the environment and has an obvious landscape benefit. It allows for the diversification of farm activities and makes better use of environmental resources.

Due to increase in population of human and livestock, there is increasing demand for food, fodder, fuel rather diversified food and produce for cloth, paper, dyes, tans, herbs, wild fruits, gums, etc. particularly in developing countries. A management system therefore, needs to be devised that is capable of producing food from marginal agricultural land and is also capable of maintaining and improving quality of producing environment. NWFPs based agroforestry have different perspectives in ecological, economic and social spheres. Diversification of the activities of arable farmers, with the building-up of an inheritance of multipurpose trees in ecological sequence adds with continuous revenue from farm. Tree species that are little used for industrial woods like timber, furniture, paper, plywood, etc., but are of high value NWFPs and provide ecological services, could be grown in agroforestry systems. NWFPs based agroforestry systems on croplands/farmlands bring significant economic return even in incidence of total crop failure, common to single-cropping also incomes due to improved and sustained productivity (Mercer 1992). Besides the economic benefits, social benefits occur from increase in products for nutrition and health due to increased quality and diversity of food. It provides rural and poor farm communities an improved and sustained livelihood.

The NWFP based agroforestry systems can be adopted in any type of traditional and modern agrofoestry systems by integration of components and modification in spatial and temporal arrangement.

- Scattered Trees: Trees are grown scattered in agricultural fields for many uses such as shade, fodder, fuel wood, fruit, vegetables and medicinal uses.
- Trees on Farm-boundaries: Multipurpose trees mainly for fruits and other products can be grown on farm boundaries. In northern parts of India, Jamun (*Syzygium cumini*), Ber (*Ziziphus* spp.*)*, Lasora (*Cordia myxa*), *Morus alba*, *Azadirachta indica, Acacia senegal*, *Borassus*, bamboo (*Dendrocalamus*, *Bambusa*), etc. all along the irrigation channels. In Andamans, farmers grow *Gliricidia sepium*, *Jatropha* spp., *Ficus* spp., *Ceiba pentandra*, *Vitex trifolia* and *Erythina variegata* as live hedges. At many places succulents like *Agave* and many cactoids are grown as common live-fence.
- Forest Farming: The use of farm woodlots and forests by farmers for obtaining products such as wood, fuel, building materials, sugar, honey and food has been historical. The more popular forest farming practices that are presently researched and promoted by agroforestry institutions consist in the production of specialty products in wooded areas. These products include ginseng (*Panax* spp.), maple (*Acer* spp.), mushrooms and pine straw which have become multi-million dollar business enterprises. In addition, there are economic potentials from forest farming from the production of walnuts, honey, and fruit and social benefits for rural development.
- Parkland System: It is the most widespread system of agricultural production. Trees are present everywhere on the farmers' fields. Trees of useful species are selected in fallow land for retention, before bringing the land into cultivation. Thus tree regeneration is the result of the practice of fallowing. The main reason for maintaining woody species in the fields is the demand for food and traditional medicines, for both humans and animals. These products, which are currently undervalued, contribute to the enhancement of the nutrition of rural populations, a better family income (particularly for women) and the economy of the country.

Domestication has recently been accepted as a phrase applicable to agroforestry, as a corollary to the Green Revolution of crop plants, the "*Woody Plant Revolution*" is now being heralded. Almost without exception, the tree germplasm that is used or envisaged for immediate use in agroforestry is wild and unimproved. Domestication of agroforestry trees for non-wood tree products (NWTPs) requires information on yield data, selection criteria, botanic knowledge and consumer acceptance. The domesticates of tree species that provide NWTPs have the primary purpose of enhancing the productivity and sustainability of agroforestry systems (Leakey and Maghembe 1994). In turn, it is envisaged that the domesticates may contribute to the conservation of genetic resources of such species and also the preservation of natural habitats (e.g., rainforests, woodlands, savannah, etc.) in which their progenitors reside. It is further desired that the provision of improved germplasm will accelerate adoption and expansion of agroforestry technologies. FAO (1995) lists advantages and disadvantages of the domestication of NWTPs. Advantages include reliable production, relieving pressure on forests, income generation, ease of harvesting, improved growth rates, increased value of the crop. The disadvantages focus on increased susceptibility to pests, loss of ecological function, reliance on new sources of wild seed, added-value benefit to large corporate entities. The development of new tree crops on a sufficient scale in sustainable agroforestry systems is a challenge to link genetic improvement research with marketing to ensure income generation and food security. Domestication is both about product development and product creation. Agroforestry systems where NWTPs trees could play a role are numerous and include homegardens, scattered trees in fields, multistrata systems and boundary plantings.

Before embarking on the domestication of a tree, the agroforester needs to ascertain whether it is likely that there will be a market for its products, while the industry that develops the market wants to know that there is a minimum reliability of supply for a uniform product of a given quality, before committing capital to developing that market. Nevertheless, some progress has been made in thinking about the impact on small-scale, resource-poor farmers of domesticating and commercializing NTFPs. All the available literature on NWFP production deals with one or more of the following topics in decreasing order of abundance:

- An inventory (listing) of the major species, giving such details as their local/vernacular, as well as scientific names, and the specific part(s) of each plant that is economically used;
- Socio-cultural values, conservation strategies, and policy issues;
- Species description of a botanical/ethnobotanical nature;
- Economic assessment of some practices and products; and chemical and biochemical descriptions of certain products and/or processes.

CONSTRAINTS FOR DEVELOPMENT OF NWFPs

Thus, information on the quantity and value of NWFPs are either non-existent or unreliable and most are estimates. Considering that NWFPs represent a hitherto ignored area of scientific and academic interest, this state of affairs is not unexpected. Because of the descriptive nature of most NWFPs and agroforestry literature, it is difficult to make a quantitative assessment of the extent of NWFPs use in agroforestry systems. ICRAF's global inventory of agroforestry systems during the 1980s, which provides the most comprehensive account of agroforestry systems in developing countries, lists 380 perennial woody species reported as components of existing agroforestry systems and their main uses in different regions. Most of the available reports describe systems where NWFP-yielding plants are grown in association with commercial tree crops. Avilable literature identifies a number of localized, under-exploited tree species that produce NWFPs in a variety of ecosystems and outline the potential for their improvement through agroforestry. NWFPs have a major role in many indigenous agroforestry systems; and agroforestry offers a viable approach to realizing the potential of several under-exploited NWFPs (World Bank 2011; FAO 2011b).

Chandrasekharan (1993) has identified a large number of issues related to the development of NWFPs and suggested that nothing goes right or proper for NWFPs at the present time. Inventory of NWFPs - A serious hindrance to NWFP development is the lack of authentic and reliable statistics on the classification, production, and value of NWFPs. Product grouping followed in statistical reports and the aggregation of products in International Standard Trade Classification (ISTC) makes it impossible to separate out NWFPs by specific products and sources. Many NWFPs fall under the category of "vegetable materials and vegetable products". For example, all edible nuts are grouped together such that this category is dominated by such important items as cashew, groundnuts, and almond. Similar situations exist for other groups of other products as well (e.g. oils, fibres, fruits, and so on). Because of this, even if statistical data on NWFPs are available, they cannot be reported adequately under existing data reporting systems. This is somewhat similar to the data reporting systems for plantation crops such as black pepper, cacao, cashewnut, coffee, oil palm, and rubber in agroforestry systems. A "solution" to this problem was found by reporting data on mono-crop and mixed-crop stands of these species separately. While sole stands account for most commercial production of these crops, mixed-crops are mostly in smallholder agroforestry systems (Nair 1983 & 1993).

Table 29.1: Constraints and Areas of Action for the Development of NWFPs (Chandrasekharan 1993)

S. N.	Constraints	Areas of Action
1	Neglect of scientific management and conservation strategies	Assessment of NWFP resources by appropriate categories
2	Inappropriate and unsustainable harvest methods	Enhance planning for integrated forest management
3	Lack of land-use planning	Domestication for forest conservation and improved production
4	Lack of integration in forest management	Improve pre and post-harvest techniques
5	Lack of processing and storage facility	Increase value-added processing at local level
6	Potential competition from domesticated or synthetic sources	Stabilize supply for local and international markets
7	Lack of clear policy support and bias against NWFPs	Improve database on NWFP species and products
8	Lack of research and technology development for NWFPs	Strengthen research and enhance technology transfer
9	Poor data and statistics on NWFPs	Augment supporting institutions
10	Insufficient incentive for sustainable management	Access international assistance

Agroforestry inventory methods can be applied to NWFPs, and there is an urgent need to compile the basic statistics of NWFPs. In order to make systematic efforts in the improvement of NWFPs, it is essential that there should be a logical classification scheme. As mentioned above, NWFPs are currently being identified in long lists of product categories. However, the plant parts that yield these diverse products, the management attention, harvesting regime, valuation criteria, and several other factors are all different for different products. A classification scheme should be devised to group the products according to common characteristics or action-oriented parameters. This will then enable to devise appropriate development strategies for each group of products. For example, the development strategy for the fruit and food-producing plants could be different from that of raw material-producing plants. There is an advantage in choosing biological characteristics rather than agroecological factors as the first set of criteria for such a classification, because of the application of these characteristics over wide regions. Ideally, the inventory data should lead to the selection of broad-based criteria. In the absence of such data bases, available information suggests that characteristics such as plant morphology, nature of products; frequency and method of harvesting, and type of processing needed, could be taken as some such criteria.

Since tree domestication by breeding is a long and slow process, vegetative propagation and clonal selection that have been developed for tropical trees are particularly promising techniques in this context. An important point here is that the NWFP domestication process and sustainable use have to be strongly linked with social and cultural characteristics of the people. All plant domestication efforts are linked to economic needs for the products of such plants. This is especially true in regard to NWFPs. Additionally, cultural and social values, and indigenous knowledge of the local use of species are key issues in NWFP domestication. Realistic assessment may be made of these values, and the information will need to be integrated into the strategy for domestication. Domestication and conservation of genetic resources are two issues that need to be considered jointly. Characteristically, domestication leads to substantial decline, and in some cases, total extinction of the wild populations. Both domesticated and wild populations may coexist for some time, but eventually the wild sources will be lost. This conjecture is based on the fact that the wild forms that may have given rise to the cultivated forms of such commercially important tree crops as breadfruit, citrus, coconut, and mango are either non-existent or unknown (Simmonds 1985). In such a situation, it should not be allowed to happen in the case of today's NWFPs. NWFPs have to be developed for situations that are non-conventional for forestry and this approach is implicit in the concept of agroforestry.

There is growing concern that the harvest methods for NWFPs are unsustainable. Increased marketing and attention to NWFPs has in some cases led to their destruction. For example, in Central America evidence of unsustainable harvest is shown by the over-tapping of chicle latex from the *Manilkara zapota* in Belize under the high level of demand during the years 1930-1950 (Alcorn 1994). Many harvest methods are not ecologically sustainable giving an example of a 78 per cent decrease in bamboo yields in the Tamil Nadu State of India, and several other authors are giving examples of unsustainable harvest of NWFPs. Thus, in addition to the economic issue related to the harvest of NWFPs the biological sustainability must be evaluated. Extractive sustainability at harvest that has no long-term detrimental impact on the resource is compared to non-harvest natural population. This may be described as an ethnoecological approach to understanding the potential products from the rain forest which attempts to consolidate the ecological requirements for resource maintenance and human needs. Sustainability is a corner stone of the foundations of agroforestry this means that all agroforestry practices should ensure long-term production without degradation of the resource base on which the production is dependant. Ecological harvesting of agroforestry components is therefore implicit in the concept of agroforestry, and destructive harvesting of agroforestry components has not surfaced as a serious issue (UNEP 2011).

Markets for NWFPs can be differentiated into two categories: (1) the local markets and cottage industries; and (2) the industrial and/or export markets. In many instances, the dynamics of local markets are not well understood. Other problems in local markets include periodic flooding of products which drive prices down especially significant for products that have low elasticities of demand, and high levels of substitute products. Local markets present an important source of rural income and cottage industries will improve marketability of the products. However, in the case of a strong product demand, the problem may arise when cultivators or extractors specialize to fulfil the market needs, leading to situations of risk due to dependence on one product as a primary source of income. International/export markets are often fickle and yet demanding in products availability and quality. When producing or collecting for the export market, the products must meet the requirements of (1) reliability of supply; (2) consistency of quality, including regulated size and appearance; and (3) transportation costs and sanitation conditions, that are far more stringent than for local markets (Raintree and Francisco 1994). Clay (1992) lists the following general principles for the development of NWFPs in the markets:

- Diversify product and reduce dependence on a few products
- Diversify the markets for raw and processed forest products
- Add value locally
- Capture value that is added further from the source
- Proposed solutions must equal the scope of the problems
- No single forest group can provide enough commodities for even a small company and most profitable method to sell the product and the ability to supply the market must be established
- Controlling a large market share of the commodity allows considerable influence over the entire market;
- Make a decent profit in the market place,
- Certification of environmental sustainability

DEVELOPMENT OF NWFPs THROUGH AGROFORESTRY

The implication is that if a NWFP is to compete on the international market it is subject to the unpredictable price movements that accompany this market. This is also true for smaller domestic markets and thus careful consideration must be placed on the markets and market stability if investment is to be encouraged in industries based on NWFPs. Guidelines can only be developed on the basis of accumulated experience and data, but development experience and quantitative data on production and value of NWFPs are lacking. This appears to be a vicious circle. In order to overcome this dilemma and move forward, we need to make some strategic planning and embark on a realistic and relevant action programme. It is essential to explore the possibilities of how and to what extent the experience from agroforestry can be used to initiate the efforts in NWFP development (Nair and Merry 1995). The major role of NWFPs in many indigenous agroforestry systems and the potential role of agroforestry as an approach to NWFP development have already been identified. The accumulated experience in agroforestry has been analyzed to establish some strategies for NWFP development. Systematic and scientific efforts need to be undertaken in a number of critical areas of NWFP development. These include:

- Undertaking a comprehensive inventory of the production and economic value of NWFPs
- Development of a systematic, logical, and broad-based classification scheme to serve as a basis for formulating organised research and development plans
- Identification of promising NWFP species and domestication of their wild forms, with due consideration of the cultural and social context of the species and their products

- Conservation of genetic resources to avoid potential danger of eventual extinction of wild forms of preferentially domesticated species
- Standardization of product valuation methods for NWFPs
- Development of sustainable harvest methods and local as well as international markets for NWFPs
- Strengthening of policy and institutional support for NWFP development

Table 29.2: A Tentative basis for preparing Guidelines for Development of NWFPs through Agroforestry (Source: Nair and Merry 1995)

S.No.	Issue	Suggested Action	Agroforestry Options/Experience
1	Lack of quantitative data on production and values of NWFPs	Inventory of NWFPs	Experience from agroforestry can be used to develop a comprehensive plan for a systematic inventory; can be organised on the basis of pilot studies from selected regions.
2	Diversity of issues, species, and social conditions necessitate a scheme for coordination and prioriti-zation of development programmes	Classification of NWFPs according to validated criteria	Agroforestry experience is quite relevant. Once the issues, products, and conditions are grouped, efforts can be focused on representative examples of each. A multiple-matrix type of classification scheme will need to be considered.
3	Multiplicity of species and products	Identification of most promising NWFP species in specific ecological and socio-economic strata	Again, the procedure that has been adopted in agroforestry would be relevant. Special attention will need to be focused on the most promising species: e.g. rattan, fruit trees, and palms in humid lowlands; *Acacia* spp., *Adansonia digitata*, *Borassus aethiopum*, and *Vitellaria paradoxa* in Sub-Saharan Africa; and *Acacia mearnsii, Alianthus* spp., *Alnus* spp., *Ficus* spp., and *Robinia pseudoacacia* in tropical highlands.
4	Most NWFP species exist only in "wild" forms	Species domestication and improvement	This involves a long drawn-out process with slow, time-consuming procedures. Some of the procedures adopted from agroforestry/tropical fruit tree development can be made use of: e.g. germplasm collection and evaluation, vegetative propagation, and clonal selection. The ideotype concept could be applied and "model" species for specific categories of trees that yield certain types of products could be developed; social concerns need to be taken into account.
5	Poor and erratic yields from NWFPs	Development of appropriate management techniques	This is an area in which agroforestry techniques are most relevant. Experience in agroforestry and NWFPs shows that NWFP species are components of agroforestry systems in most places. The low-input, integrated, multiple-output nature of agroforestry offers excellent opportunities for incorporation of NWFPs as components of new and improved agroforestry systems. A whole host of plant management and manipulation techniques need to be developed for integrating little-exploited and promising NWFPs into agroforestry systems.
6	Available genetic resources need to be conserved	Genetic conservation	Combined systems such as agroforestry (as opposed to monocultural systems such as plantations) offer the

S.No.	Issue	Suggested Action	Agroforestry Options/Experience
			best scope for the conservation of genetic resources. *In situ* conservation will be a preferred strategy, and this can best be accomplished through agroforestry.
7	Harvest methods of most NWFPs are destructive	Development of sustainable harvest methods	Agroforestry is mostly a managed system involving domesticated plants with the harvest of most agroforestry products conducted in a sustainable manner. While developing harvest methods of NWFPs, example of sustainable harvests of agroforestry products could be taken as relevant models.
8	There is considerable loss of products due to poor pre- and post-harvest techniques	Development of efficient pre-and post-harvest techniques for NWFPs	The examples from managed agroforestry systems are relevant in this case also. An important step in improving harvest and processing techniques would be to modify the current method of extractive "harvesting" to more managed types of harvest techniques. The effect of post-harvest techniques on product quality and storage will also need to be investigated.
9	Product valuation methods are weak or non-existent	Establish norms of value-added transformation, so that more of the final product value is captured in the region/nation concerned	This is a concern common to agroforestry as well; less work has been done in this aspect of agroforestry development. As NWFPs become components of agroforestry systems the established norms of agroforestry-product-valuation might be applied to the newly developed NWFPs as well.
10	Market structure is weak, and there is no assured supply of products	Build reliable market mechanism for assured supply of NWFPs	Agroforestry experience does not have much to offer in this context; one reason is that agroforestry products are either "basic needs" that are consumed locally or well-established commercial products such as coffee and cacao, for which efficient market-supply mechanisms exist. However, this is also an area of concern for some agroforestry products.
11	There is no coherent policy for NWFP development in most countries	Policy reform to recognize the value, importance, and need for developing NWFPs on a sustainable basis	The agroforestry experience is quite relevant in this context. Compared to the situation that existed about two decades ago when agroforestry was a non-entity, today most countries have recognized, at least in principle, that agroforestry is a viable approach to land management. Concerted efforts by technical experts, policy planners, and non-governmental and donor agencies are needed for such a development.
12	Institutional infrastructure for NWFP development is non-existent	Build up institutional capability within existing forestry and agri-cultural set-up to deal with NWFP development	Again, drawing from experience of agroforestry development, it will be prudent to organise NWFP development units within existing bureaucracy rather than trying to start new structures for NWFPs.
13	There is no research support for NWFP development	Develop an efficient and competent	Understanding the principles of agroforestry and employing these principles for improvement of such

S.No.	Issue	Suggested Action	Agroforestry Options/Experience
		research support system for NWFP development	systems or developing new ones was the strategy for agroforestry development. Experience to date shows that this strategy has paid off well. Efforts to develop NWFPs are unlikely to succeed unless they are based on sound principles. As is the case with agroforestry, the effort should be on applied rather than fundamental research. Scientific evidence that is available in the relevant areas of biological and social sciences needs to be synthesized for their application to solving or addressing NWFP issues. A sound research strategy should include procedures to address the 12 issues outlined earlier (and other potentially important ones). Research protocols and methods that are appropriate for each issue/topic could then be developed by groups of disciplinary experts. A central, coordinating entity (such as ICRAF for agroforestry) could be the nerve-centre for these research efforts.

While addressing these issues of NWFP development, agroforestry experiences can help in three ways:

1. The procedures, methodologies, and criteria that have been developed for agroforestry can be applied with necessary modifications and adjustments. For example, research methods and procedures encompassing biophysical and socioeconomic characteristics are now available in agroforestry for systems inventory, component characterization, field experimentation and data analysis, biological manipulation of components and systems, economic evaluation, and policy evaluation.
2. Operational procedures for species domestication, product valuation, and holistic evaluation of systems are also available.
3. Because of the high emphasis that is implied on the system's rather than the component's performance in agroforestry systems, these systems could provide excellent opportunities for species domestication and genetic conservation of NWFPs.

Exploiting these procedures, methods, opportunities, and experiences of agroforestry will greatly facilitate systematic and rapid development of NWFPs.

CONCLUSION

Agroforestry has made tremendous strides in recent years, but many challenges remain in terms of its wider application. It is necessary to identify and measure the range of benefits, given that they are not well documented. Moreover, additional research is required to quantify the benefits to various stakeholders, to deal with the variability in benefits, to assess the effects and trade-offs of different policies and to examine the impact of agroforestry practices on forest protection, particularly in the tropics. Determining which practices are most suited to particular groups, such as women and poor people, is another area that warrants attention. Many success stories appear to be confined to small areas. Thus, emphasis needs to be placed on ways to replicate these on a larger scale to reach more households. Other issues involve identifying policies, institutional innovations and extension strategies that facilitate the spread of agroforestry and increase economic benefits. With research and extension services declining throughout the tropics, ways to promote farmer experimentation and enhance farmer to farmer communication need to be found as well. Measures are required to overcome lack of planting materials (seeds, seedlings or cuttings) and lack of information. Improving marketing and adding value to raw products are critical for enhancing the livelihoods of agroforestry farmers. In this regard, private sector contracting mechanisms should be extended to countries and commodities where they do not exist. More market analysis is also needed to assess how consumer preferences can be satisfied without simply increasing production. Community based institutional mechanisms are needed to help farmers acquire information and business skills, market produce and promote quality.

Poor forest labourers, who were previously unorganized, have been brought under the umbrella of various organizations like the Forest Labour Cooperative Societies, the Large-Area Multipurpose Societies (LAMPS), Tribal Development Corporations (TDC), Minor Forest Produce Federations (MFPF), and other organizations. ICRAF has a very diverse client group. The resource-poor farmers of the tropics plant trees that provide NWTPs for a variety of reasons (food, income generation, risk aversion through diversification). Fitness of purpose of agroforestry trees is the prime objective of domestication, and this is best ensured through provision of a choice of priority species to farmers. Recognizing that through domestication many NTFPs can be improved qualitatively and quantitatively to be more attractive to farmers, more marketable and so contribute to the alleviation of malnutrition and poverty.

■■■■

30 BAMBOO BASED AGROFORESTRY SYSTEMS

INTRODUCTION

Bamboo represents one of the world's best natural and renewable resources with a large number of uses and applications which serve as an eco-friendly alternative to rapidly depleting wood resources. Bamboo is a versatile multipurpose forest produce which has over 1,500 uses and tremendous versatility. Bamboo is used in nearly every aspect of daily life and classified under non-wood forest products (NWFP). Its importance is better felt and understood in areas where it abounds or where timber and other traditional construction materials are not readily available or are extremely expensive. Bamboo plays a vital role in the domestic economy of many South Asian and SE Asian countries and has been in use since times immemorial. It is perhaps the best known plant in the orient. The total number of bamboo species is represented by 75 genera and 1,250 species which are reported to occur in the world (Soderstrom and Ellis 1987).

Bamboo is used for house construction, agricultural tools and implements, as food material, fodder and even for medicinal purposes. Besides being a convenient source of cellulose for paper manufacture and rayon, it supports a number of traditional cottage industries in the world. Bamboo is also emerging as a major source of raw material for several industries, viz. bamboo mat boards, bamboo mat veneer composites, bamboo mat corrugated sheets, etc. They are of economic and high cultural significance in East Asia and South-East Asia where they are used extensively, as a building material as well as a food source. It's plethora of essential uses has led to the use of terms such as "green gold", "Poor man's timber", "bamboo, friend of the people" and "the cradle to coffin timber". As a reforestation species, it helps to control erosion and stabilize the river banks. It also helps for preserving the ecological balance of an area. Bamboo has also been used for ornamental purposes and as a landscape material. Bamboo is also becoming increasingly popular in the horticultural industry as house plant (Rao et al 1990).

Several hundreds of million people worldwide depend on bamboo and rattan for their livelihoods. According to International Network for Bamboo and Rattan (INBAR), annual world trade in bamboo and rattan is currently estimated at US$ 14 billion (Swarup and Gambhir 2008). India's share is only US$ 100 million. These developments have created a new interest in bamboo across the world.

BAMBOO MORPHOLOGY AND GROWTH

Bamboo is a perennial, giant, woody grass belonging to the family Poaceae (Gramineae) under the group angiosperms and order monocotyledon. Bamboos are classified under the subfamily

Bambusoideae. About half of the total world bamboo species grow in Asia, most of them within the Indo-Burmese region, which is also considered to be their area of origin. Bamboos are found in diverse climates, from cold mountains to hot tropical regions. Most of the bamboos need a warm climate, abundant moisture and productive soil, though some do grow reasonably well in cold weather. According to Grosser and Liese (1971), bamboos grow particularly well in the tropics and subtropics, but some taxa also thrive in the temperate climate of Japan, China, Chile and the USA. Smaller bamboo species are mostly found in high elevations or temperate latitudes and the larger ones are abundant in the tropic and subtropic areas. Bamboo can grow from sea level to as high as 3,000 masl.

The majority of bamboos thrive at temperature ranges of 48^0 F to 97^0 F (8.8^0 C to 36^0 C). The most common rainfall requirement range is 50 to 160 inches (1,270 to 4,050 millimeters) per year. Most bamboos are found on sandy loam to loamy clay soils, derived from river alluvium or frequently from the underlying rock. Usually bamboo prefers well-drained soils but it is also found on swampy or wet stream beds. No bamboo is reported on saline soils (Huberman 1959). Optimum pH range for bamboo is 5.0-6.5. Bamboo is quite adaptable and some bamboo species from one country have been introduced to other countries.

The general structure of bamboo is divided into two major portions viz. rhizomes and culms. The rhizome is the underground part of the stem and is mostly sympodial (clumping bamboo) or, to a much lesser degree, monopodial (running bamboo). In monopodial type, main rhizome continues its horizontal growth through its apical part, sending off aerial culms from the nodal buds. In sympodial type, main rhizome turns into aerial shoot, developing lateral buds which eventually become aerial. The upper ground portion of the single erect stem is called culm. It is the portion of the bamboo that contains most of the woody material. Most of bamboo culms are cylindrical and hollow, with diameters ranging from 0.25 inch to 12 inches, and height ranging from 1 foot to 120 feet (Lee et al 1994). It is without any bark and has a hard smooth outer skin due to presence of silica. A group of culms in single bamboo plant is called clump. The sympodial bamboos are *Arundinaria*, *Bambusa*, *Dendrocalamus*, etc. Monopodial bamboos include *Phyllostachys*, *Pleioblastus*, *Sasa*, *Pseudosasa*, *Semiarundinaria*, *Sinobambusa*, etc.

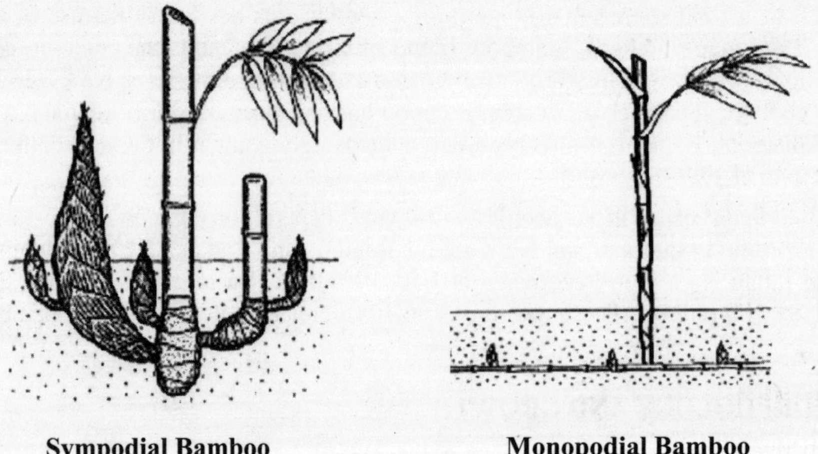

Sympodial Bamboo **Monopodial Bamboo**

Fig. 30.1: Types of Bamboo

Bamboo is a fast growing species and a high yield renewable resource. Bamboo growth depends on species, but generally all bamboo matures quickly. Aminuddin and Latif (1991) stated that bamboo might have 40-50 culms in one clump, which adds 10-20 culms yearly. Bamboo can reach its maximum height in 4-6 months with a daily increment of 15-18 cm (5-7 inches). Culms take 2-6 years to mature, which depends on the species. It is suggested that with good management of the bamboo resource, the cutting cycle is normally 3 years. Bamboo's growth is rapid than any other plant on the planet.

BAMBOO OCCURRENCE AND FLOWERING

Bamboos are an important component of wet evergreen, moist deciduous and dry deciduous forests in the tropical parts of southeastern Asia, principally in India, Burma, Cambodia, Sri Lanka, Indonesia, Laos, Malaysia, New Guinea, Pakistan, Philippines, Thailand and Vietnam. Genera found in these countries include *Arundinaria*, *Bambusa*, *Thyrsostachys*, *Gigantochloa*, *Oxytenanthera*, *Dendrocalamus*, *Cephalostachyum*, *Melocanna*, etc. At higher altitudes and in temperate climates in Asia, as in Bhutan, China, Japan and Nepal, many of these genera are represented by different species, and other genera such as *Phyllostachys* and *Sasa* are common in China and Japan. In this connection, it is interesting to note that pure bamboo forests are common on slopes where shifting cultivation has been carried on in Burma, East Pakistan and other parts of Asia. Two indigenous bamboos occur in tropical parts of Africa viz. *Oxytenanthera abyssinica* in dry deciduous formations, and *Arundinaria alpina* in the humid highlands of the equatorial zone. Average productivity of bamboo from forest is around 2 tons ha^{-1} $year^{-1}$. The bamboo flowering is a unique and very rare occurrence in the plant kingdom. Three types of flowering are reported in bamboo viz. gregarious, sporadic and annual.

Gregarious flowering is when all populations of a particular species of bamboo flowers all at once. For most species of bamboo, this can happen at intervals anywhere between 60-130 years. These long flowering intervals are largely a mystery and still astound many botanists with no definitive explanation. When a bamboo flowers gregariously, it expends a tremendous amount of energy for producing flowers and seeds. This mass flowering stresses the bamboo to such an extent that it will die. In the gregariously flowering species the interval from germination of the seed to the next general flowering establishes the life cycle, which seems to be fairly constant. Life cycles have been determined for certain species as follows: *Arundinaria falcata* (28-30 years), *Bambusa arundinacea* (32 years), *Dendrocalamus strictus* (32 years), *Dendrocalamus hamiltonii* (30-40 years), *Bambusa tulda* (35-40), *Bambusa polymorpha* (60 years), *Phyllostchys nigra* (60 years) and *Melocanna bambusoides* (45-50 years).

When bamboo flowers sporadically or intermittently, it is called sporadic flowering. It is typically brought on by environmental factors rather than genetics. Unlike gregarious flowering, sporadic flowering does not happen on a mass scale and the bamboo seldom dies after flowering. Sporadic flowering bamboos are *Dendrocalamus strictus, D. hamiltonii*, *D. longispathus*, *D. giganteus*, *Bambusa tulda*, *Oxytenanthera nigrociliata*, *O. albociliata*, *Arundinaria*, *Dendrocalamus falcata* and *Cephalostachyum pergracile.* There are few species of bamboo that flower annually. The annual flowering bamboos are *Arundinaria wightiana*, *Bambusa liniata*, *Ochlandra rheedi* and in the species of the Western Hemisphere.

BAMBOO RESOURCES OF INDIA

As per the FAO report on world forest resources, India is the second richest country of the world after China in terms of bamboo genetic resources. Harvest of 13.5 million tons of bamboo every year has been reported in India. North East region of India accounts for majority of the India's total bamboo resources. Sharma (1987) has reported about 130 species belonging to 23 genera of bamboos from India. Forest Survey of India (2011) has reported that 125 indigenous and 11 exotic species of bamboos belonging to 23 genera are available in India. Rai and Chauhan (1998) reported that the forest area containing bamboo is 8.96 million ha which constitutes 11.71 per cent of the recorded forest area. But recent estimates of FSI (2011) state that India's area under bamboo is around 14 million ha (exactly 139,577 Km^2). They are found in almost all parts of the country form the tropical to the temperate regions and the alluvial plains to the high mountains, the only exception is the Kashmir region where bamboo do not occur naturally.

The principal bamboo genera occurring in India are *Arundinaria*, *Bambusa*, *Chimonobambusa*, *Dendrocalamus*, *Dinochola*, *Gigantochloa*, etc. More than 50 per cent of the bamboo species occur in Eastern India viz. Arunachal Pradesh, Assam, Manipur, Meghalaya, Mizoram, Nagaland, Sikkim, Tripura and West Bengal. Other areas rich in bamboos are the A&N Islands, Bastar region of Chhattisgarh and the Western Ghats. Arunachal Pradesh state has the largest area of bamboo bearing forest (16,083 Km^2). *Bambusa* and *Dendrocalamus* are found under tropical conditions, whereas *Arundinaria* and alliees are found in the temperate region. *Dendrocalamus strictus* is predominantly found in dry deciduous forests, while *Bambusa* species flourishes best in moist deciduous forests. The most important bamboo species of the semi evergreen forests of Andamans is *Gigantochloa rostrata*. The important bamboos in Eastern and North Eastern region are *Bambusa tulda*, *Dendrocalamus hamiltonii* and *Melocanna bacciferra* (FSI 2011). Among the occurring bamboo species in India, 67 per cent belongs to clump forming (sympodial) and 33 per cent belongs to non-clump forming (monopodial).

BAMBOO PROPAGATION

Bamboo can be propagated through sexual or asexual methods. Sexual propagation involes the production of new bamboo plants through seeds. However, this methods is unreliable because of unavailability of seeds due to long flowering intervals in bamboo. Sometimes viability of seed is unknown and poor. Asexual propagation is with the use of vegetative planting materials such as rhizome, offset, whole culm, culm cutting, branch cutting, marcotted culm and micropropagation. Propgation through asexual or vegetative means can be done anytime of the year provided that there is sufficient supply of water and the planting materials are protected against desiccation and excess heat. The propagules are considered to be successfully established when they form new rhizomes and start to develop new shoots.

SEXUAL METHODS

Direct Sowing of Seeds: This type of propagation happens in nature. But bamboos do not produce seeds every year. They flower only once in 30 to 50 years or sometimes more depending on species. Hence this method of propagation is not possible or popular. Nevertheless, it is the cheapest method of reproducing forests. Most of bamboo forests are regenerated by this method.

Raising Seedlings in Nursery: This method is possible only when seeds become available. Because of the poor viability, seeds need to be collected and sown in nusery beds without dealy. The seeds are sown in germination bed/nursery bed and when the seedlings are about three months old, they are picked out into polythene bags or root trainers containing suitable growing media. They are kept under partial shade and irrigated daily. Finally they are planted in the field.

Wildlings: If wildlings or naturally growing bamboo seedlings in the forests are available, these can be collected and used as planting material.

Macroproliferation: It is sometimes termed as seedling multiplication. In order to increase the planting stock before transfer to the field, macroproliferation technique is employed. A bamboo seedling of 5-6 months old produces 5-6 small tillers/culms. Plantlets with single tiller/single stem and small piece of rhizome and roots are seperated from bamboo seedling for producing more planting materials.

ASEXUAL METHODS OR VEGETATIVE PROPAGATION

Rhizome Seperation and Planting: This method will ensure maximum survival. It is preferred when there is a need for rapid production of planting materials or when the planting material is intended to be planted immediately in the field without undergoing the tedious and time-consuming process of producing potted, rooted seedlings. The procedure involves seperating rhizome portion from the mother clump using digging bar and sharp edge cutting tools. This method will allow only a limited number of planting materials to be extracted from the mother clump. It will also affect the productive capacity of the remaining clump. Rhizome planting is mainly limited to non-clump forming (monopodial) bamboos.

Offset Planting: The most effective and very common way of propagting bamboo is by means of offsetting a culm from a clump. The term "offset" denotes the bamboo propagule each composed of the lower part of a single culm with the rhizome axis basal to it. Generally one to 2 year old offsets give better results in the field. Two nodes of the culm with rhizome are sufficient for survival of offset and most commonly practised. Offsets are normally obtained and transplanted just before the rainy season. Although this method has high success rate, it is not the most practical method for large scale propagation, nor is it the best way of obtaining planting material that will be directly planted in the field. This is due to the labour intensiveness of the method and the time and logical costs involved in obtaining and moving a large number of offsets to the field. The method of offset propagation is however very useful when one desires to raise few bamboo plants in a homestead or farm.

Culm or Stem Cutting: Propagation of bamboo through culm or stem segments is known as culm cutting or stem cutting technique. Generally single nodal or 2-3 nodal culm cuttings bearing healthy buds obtained from one to two year old culms are used for propagation. Culm cuttings are buried under the soil with upright or at an angle or horizontally. If the species is hollow bamboo, then hole is made on the culm cutting, water is filled inside the hollow space and covered with polythene film before burying the culm cutting under the soil. The new roots and shoots will emerge in the node. Treatment of culm cuttings with growth regulators will induce root formation. Large-scale multiplication is possible by this method and plants raised from culm cuttings will develop to clumps very fast. Sometimes multinodal cutting or whole culm is buried under the soil for producing multiple plants.

Branch Cutting: This method is ideal for thick walled bamboo species. Branches can be easily seperated from the culm with the use of sharp edge tools such as knife or crosscut saw. 2-3 nodal branch cuttings can be made using hand secateurs, treated with growth regulators and planted in the potting medium for inducing rooting. Comparatively, this method is more advantageous than culm cutting because the culm remains useful. For rapid and maximum production of rooted cuttings, only those branch cuttings with pronounced root initials at the bases should be used.

Layering and Marcotting: Layering is a method of bringing a culm or branch in contact with soil so that propagation occurs. For obtaining success in layering, suitable media is required. *Marcotting or air layering* is the propagation technique where a stem is induced to form roots while still on the parent plant.

Micropropagation: Tissue culture is playing a major role in producing *true-to-type* bamboo plants. It offers enormous potential in producing large quantities of desired planting material in a short time. However, it is essential that enough care is taken in selection of initial material, production of plants, nursery development and field plantation. The *in vitro* methods offer an attractive alternative to conventional methods for the mass propagation of bamboos.

DEVELOPMENT OF BAMBOO BASED AGROFORESTRY SYSTEMS

Recently, in most cases, tree plantations are emphasized for reclamation of degraded lands and of course, some success has been achieved. However, the most important task in reclamation is to improve the ecosystem as a whole. Agroforestry has been advocated as an alternative for development of degraded lands and increasing productivity per unit area. Bamboo is considered to be the most suitable species for agroforestry models on degraded lands in Indian context. Bamboos have been propagated by the villager of the state through rhizome in their farm bund and in homestead gardens. Bamboos are grown under agroforestry systems for the three main reasons: ecological security, sustainable food security and livelihood improvement.

- *Ecological Security*: Conservation of forests is achieved through timber substitution by bamboo. Bamboos are the alternate materials to non-biodegradable & high energy consuming materials like metals and plastics.
- *Sustainable Food Security*: Bamboo based agroforestry system ensures maintenance of soil fertility of adjoining agricultural lands and bamboo shoots are used as food.
- *Livelihood Improvement*: Generation of employment in planting and primary processing for manufacturing mat based composites and other market driven bamboo products.

AGRISILVICULTURAL SYSTEM

Farmers have been usually planting the bamboo at 5m x 5m to 7m x 7m spacing in agroforestry systems. Sometimes wider spacing is adopted for bamboo plants between rows. In between the rows of bamboo clumps, farmers usually grow tuber crops like sweet potato (*Ipomoea batatas*), cassava or tapioca (*Manihot esculenta*), taro or cocoyam (*Colocasia esculenta*), yam (*Dioscorea* spp.), turmeric, ginger, vegetable crops, food crops etc. in the agrisilvicultural models. For bamboo based agrisilvicultural systems, *Bambusa bambos*, *Bambusa nutans*, *Bambusa balcooa*, *Bambusa tulda*, *Bambusa vulgaris*, *Dendrocalamus strictus*, *Dendrocalamus hamiltonii*, etc. are the most suitable species (Malik 2004). It is to be noted that spacing of bamboo plants between rows is greater than the spacing of plants within row for better growth and higher production of intercrops. In China, intercrops are grown under bamboo plantation. The bamboo forest produce the highest income and profit with intensive management measures, such as organic and inorganic fertilization, pest and disease control, annual weeding and silvicultural management. In bamboo (*Phyllostachys* spp.) + *Paulownia* + crop model, trees grow better and give good economic income; and the common agricultural crops intercropped are soybean, sweet potato, radish, potato, pea, watermelon, etc.

Seshadri (1985) has reported that cultivation of soybean as an intercrop of bamboo (*Dendrocalamus strictus*) during the first six years is technically feasible and economically viable and the period of intercropping can be extended further in wider spacing of the bamboo clumps and judicious manipulation of the bamboo canopy. The scope of bamboo in agroforestry is wide because of the uncertain weather condition and increasing cost of labour involved in agriculture these days (Balaji 1991). Dwivedi (1992) has reported the successful bamboo based agroforestry models in three regions of India: bamboos of different species with wheat, rice, maize, jowar, bajra, pulses, oilseeds, etc. (Cenral Region); *Dendrocalamus hamiltonii* with rice, tobacco, chillies and sugarcane (Southern Region); *Dendrocalamus hamiltonii* with paddy (North Eastern Region). Behari (2001) have developed successful seven agroforestry models with three bamboos (*Bambusa bambos*, *B. nutans* and

Dendrocalamus strictus) and intercrops soybean (*Glycine max*), niger (*Guizotia abyssinica*), moong (*Phaseolus aureus*), wheat (*Triticum asetivum*), urad (*Phaseolus mungo*), pigeon pea (*Cajanas cajan*) and mustard (*Brassica campestris*).

Fig. 30.2: Bamboo based Agroforestry

Shanmughavel and Francis (2001 and 2002) have recommended intercropping of pigeon pea, soybean and turmeric in bamboo (*Bambusa bambos*) plantations based on comparative growth and yield. Jha and Lalnunmawia (2004) have reported intercropping of ginger under three fertilized edible sympodial bamboos was beneficial for both the components under degraded soil conditions of NE India. Viswanath et al (2007) have reported the highest NPV and LEV in intercropping of ginger under 6m x 6m spacing of bamboo (*Dendrocalamus brandisii*) based agroforestry system. National Bamboo Mission of India has also conducted research on development of bamboo based agroforestry systems in Jharkhand and observed the better growth of bamboo (*Bambusa nutans*) with intercrop of turmeric, colocasia and oal; number of culms per clump is also higher in the agroforestry system (Malik 2011).

SILVOPASTORAL SYSTEM

Researchers have also observed better growth of fodder grasses like sudan grass, guinea grass (*Panicum maximum*), deenanath grass (*Pennisetum pedicellatum*) in the bamboo based silvopastoral system. These agroforestry models have been proved to be beneficial for livestock rearing in the farming system. The most suitable bamboo species used under silvopastoral system include *Bambusa bambos*, *Dendrocalamus strictus*, *Bambusa nutans*, *Bambusa balcooa*, *Bambusa vulgaris*, etc.

Fig. 30.3: Bamboo based Agrisilvicultural Model

SILVIHORICULTURE SYSTEM

Bamboos are commonly grown with horticultural crops especially fruit trees to get multiple benefits and diversified products. The systematically planted bamboo plants intercropped with that of traditional horticultural trees like mango (*Mangifera indica*), cashewnut (*Anacardium occidentale*), jackfruit (*Artocarpus heterophllus*), kokum (*Garcinia indica*) and rubber (*Hevea brasiliensis*) in Konkan region of Karnataka have given encouraging results with better performance for adopting bamboo based silvihorticulture models and bamboo is reported to be the most profitable among the crops studied and cashew and mango are ranked next to bamboo (Wagh and Rajput 1991). Similarly many vegetable crops are grown as intercrops in between rows of bamboos in most places under silvihorticulture system.

BAMBOO HEDGES AND BOUNDARY PLANTING

Bamboos are planted on farm boundaries regardless of land holding size. Cultivation of bamboo in the borders of agricultural farms is a strategy to minimize competition between bamboo and other crops and to facilitate freedom of movement for farmers within the garden. Bamboo hedges or bamboo planting along the borders of agricultural fields are practised to delineate the boundaries between land owners and to protect the field against damage by free grazing domestic livestock and wild animals. The important bamboo species used for bamboo hedges and boundary planting are *Bambusa bambos*, *B. nutans*, *B. balcooa*, *B. vulgaris*, *Dendrocalamus strictus*, *Dendrocalamus hamiltonii*, *Dendrocalamus brandisii*, etc.

HOMEGARDEN SYSTEM

A homegarden represents an operational farm unit which integrates trees, bamboo with field crops and sometime domestic animals, around the homestead. This farming system has attained international popularity due to the fact that the basic objective of the system is to ensure the sustained availability of multiple products besides generating employment and cash income. Homegardens exemplify many agroforestry characteristics like the intimate mix of diversified agricultural crops, bamboo and multipurpose trees. They fulfill most of the basic needs of the social populations. The multi-storeyed configuration and high species diversity of the homegardens help to reduce the environmental deterioration commonly associated with monocultural production systems. Multi-tiered and multi-species structure is essentially the major feature identified for homegardens.

The tropical homegardens constitute a principal source of natural resources such as timber, biofuels and bamboo. Occurrence of bamboo as one of the plant components in homegardens is documented in Asia-Pacific. In ancient days in India, the Assam state had houses which were usually concealed in dense groves of bamboo, plantain and jackfruit trees where as the villages of Odisha were characterized by the linear cluster of houses with clumps of jackfruits, mangoes, bananas, bamboos and plantains. In Kerala state of India, the homestead has commonly *Bambusa bambos, B. vulgaris* and *Dendrocalamus strictus* species. Number of bamboo clumps in the homesteads depends primarily on the size of the homesteads. For instance, in small homegardens of less than 0.5 hectare, one bamboo clump is more common; whereas 0.5 to 1.5 hectare, two bamboo clumps and large homestead of more than 1.5 hectare, 5 to 6 clumps are common. However number of clumps in the homegardens is also decided by the farmer's crop preference with less number of clumps in homesteads dominated by the commercial crops than in mixed species homesteads.

UTILITY OF BAMBOO PLANTATIONS

TIMBER: *Bambusa bambos*, *B. balcooa*, *B. textilis*, *Dendrocalamus strictus*, *D. giganteus*, *Gigantochloa* spp., *Schizostachyum* spp., *Phyllostachys pubescens,* etc.

EDIBLE SHOOTS: *Bambusa bambos*, *B. tulda*, *B. balcooa*, *B. nutans*, *B. pallida*, *B. polymorpha*, *Dendrocalamus asper*, *D. brandisii*, *D. hamiltonii*, *D. longispathus*, *D. strictus*, *D. latiflorus*, *D. oldhami*, *D. membranaceus*, *Phyllostachys pubescens*, *P. praecox*, *P. iridescens*, *P. bambusoides*, *P. edulis*, *P. dulcis*, *Schizostachyum*, *Thyrsostachys*, etc.

PAPER, PULP AND RAYON: *Bambusa bambos*, *B. tulda*, *B. vulgaris*, *B. textilis*, *Dendrocalamus longispatus*, *D. hamiltonii*, *D. strictus*, *Phyllostachys pubescens*, *Melocanna bambusoides*, *Gigantochloa* spp., etc.

THATCHING, WALLING, MATS AND HANDICRAFTS: *Bambusa bambos*, *B. tulda*, *B. textilis*, *B. polymorpha*, *B. blumeana*, *B. vulgaris*, *Dendrocalamus longispathus*, *D. giganteus*, *Ochlandra stridula*, *Melocanna bambusoides*, *M. baccifera*, *Gigantochloa* spp., *Cephalostachyum* spp., *Schizostachyum* spp., etc.

ORNAMENTAL: *Phyllostachys heteroclada* (water bamboo), *Phyllostachys nigra* (black bamboo), *P. viridis*, *P. vivax*, *Bambusa ventricosa* (Buddha's belly bamboo), *Bambusa vulgaris* (yellow or golden bamboo), some *Sasa* spp., etc.

SOIL AND WATFR CONSERVATION: *Bambusa*, *Dendrocalamus*, *Sinarundinaria*, *Yushania*, *Chimonobambusa*

VALUE ADDED PRODUCTS FROM BAMBOO: Following value added products are developed from bamboo grown under agroforestry:

- Bamboo mat board
- Bamboo mat veneer composite
- Bamboo mat moulded products

- Bamboo mat corrugated sheets
- Bamboo mat moulded skin board
- Bamboo mat ridge cap
- Plywood
- Bamboo flooring
- Particle board
- Bamboo wood

INTERNATIONAL NETWORK FOR BAMBOO AND RATTAN

The International Network for Bamboo and Rattan (INBAR) is an intergovernmental organisation dedicated to improving the livelihoods of the poor producers and users of bamboo and rattan, within the context of a sustainable natural environment. INBAR is established by treaty deposited with the United Nations. INBAR evolved out of an informal network of bamboo and rattan researchers, set up in 1984 by Canada's International Development Research Centre (IDRC). INBAR connects a global network of partners from the government, private and non-profit sectors in over 50 countries. INBAR promotes sustainable development with bamboo and rattan by consolidating, coordinating and supporting strategic and adaptive research and development. INBAR's mission is to improve the well-being of producers and users of bamboo and rattan, while maintaining a sustainable resource base by supporting innovative research and development.

INBAR has regional offices that are extending network across Asia, Africa and Latin America. INBAR has regional offices that are extending network across Asia, Africa and Latin America. On 6th November 1997, nine founding member states and observers from 6 other countries witnessed the signing of the INBAR Establishment Agreement at the Great Hall of the People in Beijing. Founding Countries are Bangladesh, Canada, China, Indonesia, Myanmar, Nepal, the Philippines, Peru and Tanzania; and Observer Countries are Italy, Japan, Korea, the Netherlands, Pakistan and Thailand. At present, 38 countries are holding membership in INBAR and had acceded to INBAR's Establishment Agreement: Argentina, Bangladesh, Benin, Bhutan, Burundi, Cameroon, Canada, Chile, China, Colombia, Cuba, Ecuador, Ethiopia, Ghana, India, Indonesia, Jamaica, Kenya, Madagascar, Malaysia, Mozambique, Myanmar, Nepal, Nigeria, Panama, Peru, Rwanda, The Philippines, Senegal, Sierra Leone, Sri Lanka, Surinam, Tanzania, Togo, Tonga, Uganda, Venezuela and Vietnam. With the exception of Canada, all of these countries have major bamboo and/or rattan resource bases, and a considerable population of disadvantaged people living in conjunction with them, and who could use them to improve their lives and livelihoods.

NATIONAL BAMBOO MISSION

Government of India has launched National Bamboo Mission (NBM) which is a Centrally Sponsored Scheme, in which the contribution of the Central Government is 100 per cent. The Scheme is implemented by the Division of Horticulture under the Department of Agriculture and Cooperation in the Ministry of Agriculture, New Delhi. It covers majority of the North Eastern region and the eastern states of West Bengal, Jharkhand, Odisha and Bihar, which is the main region of bamboo of India. The NBM was thus envisaged to make India a global bamboo power besides removing multiple constraints the bamboo sector has been facing. These constraints include lack of scientific methods for propagation and cultivation, lack of post harvest treatment and technology for product development and value addition, inadequate trained manpower, inadequate infrastructure for large scale harvesting in the event of gregarious flowering, and poor market linkages. The objectives of National Bamboo Mission (NBM) are:

- To promote the growth of the bamboo sector through as an area based regionally differentiated strategy
- To increase the coverage of area under bamboo in potential areas, with improved varieties to enhance yields
- To promote marketing of bamboo and bamboo based handicrafts
- To establish convergence and synergy among stake-holders for the development of bamboo
- To promote, develop and disseminate technologies through a seamless blend of traditional wisdom and modern scientific knowledge
- To generate employment opportunities for skilled and unskilled persons, especially unemployed youths.

■■■■

31 FRUIT TREES BASED AGROFORESTRY SYSTEMS

INTRODUCTION

Growing of fruit trees started in prehistoric times. It was accidental or deliberate dispersal of seeds of highly valued fruit trees in the vicinity of hunter and gatherers to open up bonanza of nutritious food. The very evidence of integration of fruit trees in croplands is homegardens are practised since long time. The modernization of agriculture and allied sciences has open up integration of fruit trees in cropland. Agroforestry is a land use that involves deliberate retention, introduction, or mixture of trees or other woody perennials in crop and animal production fields to benefit from the resultant ecological and economic interactions (Nair 1992). About 50 per cent of the total agricultural land in the country is estimated to be marginal land and could be more productively used by revegetating with suitable species of trees and grasses in association with crops. Dwivedi (2002) stated that status of agroforestry extension perspectives in agroforestry research in India is determined by the very character of the problems to be solved by the land use technologies. In this context, Dwivedi and Shukla (1998) stated that Government regulations on felling of trees from agroforestry lands and transport of agroforestry produce has to be carefully deregulated to benefit various strata of farming community. However, fruit based agroforestry systems are the best option for the farmers.

Numerous benefits of agroforestry have been reported and substantial effort has gone into quantifying the underlying processes. Species level biodiversity and ecosystem processes in the subtropics are better maintained with partial canopy cover of a limited number of species in the overstorey (Perfecto et al 1996; Michon and de Foresta 1999) compared to monocultures (Gallina et al 1996). It has also been well documented that fruit-tree-based agroforestry systems in the north temperate zone are able to provide the public environmental services better than monocultured annuals or intensively managed orchards (Herzog 1998). Additionally, agroforestry technologies may improve nutrient cycling (Glover and Beer 1986; Nair et al 1999; Schroth et al 2001), buffer understorey temperature extremes (Barradas and Fanjul 1986), or enhance soil water balance while reducing erosion (Rao et al 1998). One central agroforestry hypothesis is that trees provide benefits to farmers and to ecosystems by capturing and using water, light, and/or nutrient resources that would remain unexploited in tree-less systems (Cannel et al 1996).

Homegardens or homestead gardens or sometimes called Kandy garden are age old horticultural based agroforestry system. In the densely populated flood plains, rural communities depend on the groves developed around a cluster of houses or homestead for supply of fruits, fodder, fuel wood, medicinal plants, etc. Such home gardens help in revealing the pressure on forest lands and slow down the degrading on natural resources. The homegardens, whether complex with numerous strata and diverse species of crops that provide a legion of products for use or sale (Fernandes and Nair 1986; Kumar and Nair 2004) or simple temperate gardens composed of an understorey of annuals under dispersed overstorey perennials, have a common thread linking them. In all of them, the tree components overwhelmingly produce consumable or saleable products on a regular or seasonal basis. Thus, within fruit-tree-based agroforestry systems, the importance of competitive resource losses to crops is likely to be diminished because the tree component produces a valuable good in exchange for its competitivity with the crop over a similar time scale.

FRUIT TREE BASED AGROFORESTRY

Fruit tree based agroforestry involves intentional and simultaneous association of annual or perennial crops with perennial fruit-producing trees on the same farm unit. Because of the relatively short juvenile (pre-production) phase of fruit trees, high market value of their products, and the contribution of fruits to household dietary needs, fruit tree based agroforestry enjoys high popularity among resource limited producers worldwide (Bellow 2004). Most examples of fruit tree based agroforestry have developed over long periods of time in response to interactions between agroecological conditions, plant diversity and farmer resources and needs. Because of this, the system performance at any given location will depend to a great extent on several site specific features. Nevertheless, the system performance also follows some general characteristics such as their potential benefits and limitations that are applicable over wider regions. An understanding of such general characteristics of these systems is helpful for adaptation and extension of the system to other highland areas with similar production environments.

The fruit based agroforestry system can be defined as a planting system comprising combinations of plants with various morpho-phenological features to maximize the natural resource use efficiency and enhanced total factor productivity. The system comprise of a combination of perennial and annual plant species as different components in the same piece of land arranged in a geometry that facilitates maximum utilization of space in four dimensions (length, width, height and depth) leading to maximum economic productivity of the system.

WHY FRUIT TREES BASED AGROFORESTRY SYSTEM?

Fruit tree based agroforestry systems are well preferred by farmers over other agroforestry systems for the following reason:

- Fruit trees are efficient enough in providing higher economic return even under stressed growing conditions prevailing under the upland situations than the other annual crops.
- The approach aims at improving productivity by effective utilization of air space which is not utilized in single tier system. The multitier system aims at sustainable management of natural resources like soil, water, space and environment.
- The summation of productivity of the component crops can increase the total productivity of the land where poor soil fertility coupled with low water holding capacity of the soil contributes towards low productivity of any crop in this region.
- Higher labour requirement per unit area of multitier system contribute towards creation of job opportunities at site.
- High return per unit area under upland conditions is the ultimate result of fruit based agroforestry system.

CONCEPT OF FRUIT BASED AGROFORESTRY SYSTEM

The fruit based agroforestry system is a self sustainable system where solar energy can be harvested at different heights, soil resources can be efficiently used and cropping intensity is increased. The system consists of three main components, viz., main crops, filler crops and intercrops which occupy three different tiers in space of the production system.

MAIN CROPS

The main crops are the fruit species having a larger canopy size and prolonged juvenile as well as productive phase. They utilize the upper most layer of the multitier system from which the economic productivity is obtained. Generally the crops utilize the entire land after 20-25 years whereas only 25-30 per cent of land is effectively used up by the main crop up to 10 years. Under the multitier system, these plants are planted at wider spacing.

FILLER CROPS

The filler crops are the fruit species which are precocious in nature, prolific bearers having short stature. They utilize the middle layer of the multitier system from which economic productivity is obtained. These plants are planted with the purpose to generate additional income from the land during the initial 10 years of the orchard (during the juvenile and initial bearing stage of the main crops) by utilizing the unused land and space. The plants are generally hardy in nature and have shorter economic life than the main crops.

INTERCROPS

The intercrops occupy the lower most layer of the multitier system and are grown in the remaining unused land of the multitier system. Generally the intercrops are the location specific annual crops, selected as per the climatic and socio-economic suitability. The intercrops also include the dependant crops like creepers which are grown with the support of main or filler crops. During initial years of the multitier system any crops can be taken whereas during the later years shade tolerant crops can be grown as inter crops.

Table 31.1: Crop used as Intercrop in Fruit based Agroforestry System

Types	Crops
Cash crops	Clove, pepper
Spices and condiments	Turmeric, ginger, green chillies, betel and coriander
Fruit crops	Papaya, banana, pineapple, passion fruit, phalsa as a filler crops
Tuber crops	Sweet potato, colocasia, dioscorea, elephant foot yam
Cereals	Maize, oat, wheat, barley
Pulses	Cowpea, green gram, winged bean
Oil crops	Groundnut, soybean, mustard
Pasture	Various species of palatable grasses and legumes
Vegetables	Cabbage, cauliflower, capsicum, pea, tomato, brinjal, bhendi, cucurbits
Flower crops	Marigold, calendula, china aster, godetia, sweet william, gompherna, annual chrysanthemum, carnation

Fruit trees and crops are grown together in various ways. Depending upon the pattern and configuration, these companion crops are known as intercrops, under planting, hedge row planting or alley cropping. In intercropping, agricultural crops are normally grown between rows of fruit trees planted at a spacing of 5 to 7 m apart. The agricultural crops provide seasonal revenue, while fruit trees are managed for 30-35 years giving regular returns of fruit and in some cases fuel wood from pruned wood and fodder. Several kinds of crops are also under planted with the canopy of fruit trees providing shade. However, shade is not always desirable for agricultural crops such as maize, sorghum, groundnut, etc. Following are the crops, generally under planted in the orchards.

ADVANTAGES OF FRUIT BASED AGROFORESTRY SYSTEMS

- This system mainly focuses on higher income per unit area. Well-maintained and established orchards bring better returns than field crops from the same piece of land.
- The farmer can practice intercropping during the early stages of the fruit trees.
- The system is helpful in generating more employment especially during this off-season when the crops are not cultivated.
- Fruit growing also provides scope for ancillary industries like fruit processing, canning, preservation, dehydration, essential oils, package transport and refrigeration.
- Fruits are protective foods, necessary for the maintenance of human health. Many fruits are a rich source for energy giving carbohydrates. Besides, fruits are valuable source of minerals, vitamins and enzymes, which are required for better functioning of human body and for fighting diseases.
- Many fruits are also known to possess specific medicinal values. Pectin and cellulose found in several fruits stimulates intestinal activity in human body.
- Fruit trees are deep rooted and hardy. These provide minimum guaranteed returns even in drought years (atleast fodder and wood) when arable crops fail.
- Trees utilize off-season rainfall in addition to the rainfall during the cropping season.

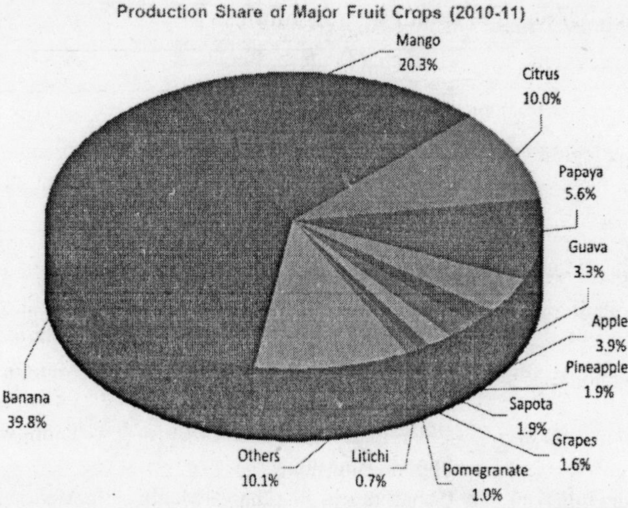

Fig. 31.1: Production Share of Major Fruit Crops (2010-11)

FRUIT BASED AGROFORESTRY SYSTEMS – INDIAN SCENARIO

India is the second largest producer of fruits (71.52 million MT) obtained from 6.33 million ha area and contributing about 12 per cent share in global fruit production arising from 11.5 per cent area under horticultural crops. The economic importance of horticultural produce has been increasing over the years due to increasing domestic and international demand. Area, production, productivity, availability and export have increased manifolds. This has provided ample opportunities for utilization of waste/marginal lands, employment generation and effective land use strategies.

Table 31.2: Area and Production Estimates of Fruits in India (2010-11) (NHB 2011)

Fruits	Area in 1000 ha	Production in 1000 MT	Productivity (MT/ha)
Banana	830	29,780	35.9
Mango	2,297	15,188	6.6
Citrus	846	7,464	8.8
Papaya	106	4,196	39.6
Guava	205	2,462	12.0
Apple	289	2,891	10.0
Pineapple	89	1,415	15.9
Sapota	160	1,424	8.9
Grapes	111	1,235	11.1
Pomegranate	107	743	6.9
Litchi	78	497	6.4
Others	1,265	7,583	6.0
Total	6,383	74,878	11.7

Table 31.3: Fruit Trees in Different Agroclimatic Regions/ Zones in India (Anon. 2012)

S.No.	Agroclimatic Regions/ Zones	States Represented	Fruit Trees Preferred
1	Western Himalayan region	Himachal Pradesh, Jammu & Kashmir, Uttarakhand	Apple, cherry, strawberry, peach, plum, kiwi, apricot, almond & walnut.
2	Eastern Himalayan region	Arunachal Pradesh, Assam, Manipur, Meghalaya, Mizoram, Nagaland, Sikkim, Tripura, West Bengal	Orange, lemon, banana, Kagzi lime, pine apple, papaya & cherry
3	Lower Gangetic plain region	West Bengal	Litchi, mango & guava
4	Middle Gangetic plain region	Uttar Pradesh, Bihar	Litchi, papaya, mango, guava and Jamun
5	Upper Gangetic plain region	Uttar Pradesh	Mango, papaya, guava, peach, jamun & aonla
6	Trans Gangetic plain region	Chandigarh, Delhi, Haryana, Punjab, Rajasthan	Kinnow, guava, aonla & mango
7	Eastern plateau and hills region	Chhattisgarh, Jharkhand, Madhya Pradesh, Maharashtra, Odisha, West Bengal	Mandarin, mango, guava, aonla, pomegranate, lemon, custard, apple, papaya & Jatropha.

8	Central plateau and hills region	Madhya Pradesh, Rajasthan, Uttar Pradesh	Aonla, ber, mango, mandarin
9	Western plateau and hills region	Madhya Pradesh, Maharashtra	Mandarin, lemon, Malta, pomegranate, papaya, banana, grapes, mango,
10	Southern plateau and hills Region	Andhra Pradesh, Karnataka, Tamil Nadu	Mango, banana, grapes, guava, sapota, & citrus
11	East coast plains and hills region	Andhra Pradesh, Odisha, Pondicherry, Tamil Nadu	Mango, sapota, banana, custard apple and pine- apple
12	West coast plains and ghat region	Goa, Karnataka, Kerala, Maharashtra, Tamil Nadu	Mango & citrus
13	Gujarat plains and hills region	Gujarat, Dadra & Nagar Haveli, Daman & Diu	Sapota, banana, guava, dates, mango, grapes
14	Western dry region	Rajasthan	Ber, pomegranate, kinnow & Mausambi
15	Island region	Andaman & Nicobar Islands, Lakshadweep	Papaya, mango & sapota

Farming system consisting fruit trees have been practised all over in the country in different forms. It may be sole main crop or integrating one or more different functional units along with it in the same land unit. Thus, as per the main component, the system may known as mango, citrus, aonla etc. based agroforestry system. Some fruit plants are grown as field crops for example banana, papaya, pineapple, strawberries etc. Here are some of the fruit based agroforestry system are listed and discussed below considering their economic returns, yield of the components. Major fruit based agroforestry system practised are as follows:

- Mango based agroforestry system
- Guava based agroforestry system
- Citrus based agroforestry system
- Aonla based agroforestry system
- Apple based agroforestry system
- Peach based agroforestry system
- Litchi based agroforestry system
- Sapota based agroforestry system

MANGO BASED AGROFORESTRY SYSTEM

Mango is the most important fruit crop cultivated in the Indian subcontinent for well over 4,000 years and it is a major fruit crop in all fruit growing states of India, as it occupies maximum area, i.e., 2.31 million ha accounting for about 37 per cent of the total area under fruits in the country. Since mango takes several years to grow to its full size, intercropping to utilize the interspaces is desirable. Legumes like sunhemp, cowpea and vegetables are suggested. Mango fruit crop provides opportunities of utilizing the land spaces to its maximum, particularly during its initial years of establishment (Rajput et al 1986 and 1989). However, wheat, oat, sorghum and pearl millet proved harmful to mango due to their common time of grain formation with flower bud initiation and the competition for moisture and nutrients. Among different intercrops (groundnut, green gram and cowpea) tried under four-year-old mango orchard at Hyderabad, groundnut was a successful intercrop

(CRIDA 1995). There was considerable reduction (53 to 56%) in pod yield of groundnut (506-539 kg/ha) in agrihorticultural system over sole groundnut (1,148 kg/ha). The leguminous intercrops can be taken in mango orchard up to at least 8 years after planting with minimum reduction in yields of crops. In Karnataka, under young mango plantations ragi and groundnut intercropping is very common.

To standardize the suitable mango based cropping system for uplands of sub-humid regions of Eastern Indian Plateau experimental study was conducted by Das et al (2009) for about ten years in a row. The study revealed that the maximum yield of mango (7.56 kg/plant) in the ninth year was recorded with intercropping of French bean. The maximum content of soil organic carbon was recorded in case of intercropping of paddy in 0-30 cm layer (0.82%). However, in tenth year, the maximum paddy equivalent yield (10.6 t/ha) was recorded in Mango + Guava + Stylosanthes combination. Further to standardize suitable intercrop under grown up mango orchards, a study was carried out. Different intercrops like black gram, cowpea, French bean, paddy and turmeric were grown with recommended doses of nutrients in prevailing shade conditions under sixteen year old mango trees. There were no adverse effects of the intercrops which could be obs-erved on the mango trees even after eight years of experimentation. Turmeric was found to be the most profitable intercrop under the adult bearing mango trees (Table 31.4).

Table 31.4: Profitability of Intercropping under adult bearing Mango Trees (Average of Two years) (Das et al 2009)

Intercrop	Yield (q/ha)		Net Income (Rs. Per ha)			Benefit: Cost (Rs.)
	Intercrop	Mango	Intercrop	Mango	Total	
Black gram	1.64	40.0	1,230	35,250	36,480	6.36:1
Cow pea	12.77	46.0	2,860	41,250	44,110	6.34:1
French bean	10.76	51.0	1,956	46,250	48,206	6.21:1
Paddy	1.14	45.0	-1,430	40,250	38,820	6.75:1
Turmeric	74.72	47.6	18,160	42,850	61,010	3.50:1*

*Due to high cost of production of turmeric, low benefit:cost ratio was observed

Lal et al (2004) have also suggested that okra-toria or cowpea-toria cropping system may be adopted in interspaces between mango trees for better economic returns. The best intercrop combination in terms of total economic returns was mango + brinjal that yielded 19.38 MT ha^{-1} of brinjal along with 2.88 MT ha^{-1} of mango. Besides brinjal, bottle gourd (13.54 MT ha^{-1}), cauliflower (9.23 MT ha^{-1}) and cabbage (8.50 MT ha^{-1}) were effective in sustaining additional income especially during pre-production phase and off-years in mango orchards (CISH 2011). In another study, the returns were 2.45 – 3.41 times in mixed cropping as compared to monocrops, where, ginger, bay leaf and pepper were intercropped with mango (Asati et al 2009).

In another study, wheat was intercropped under four different varieties of mango viz. Amrapalli, Mallika, Dasheri and Langra. The grain and straw yield of wheat was recorded on the basis of first, third and fifth row spaced at 25 cm apart (R × R) from the mango trees. There was significant difference in grain and straw yield in between rows and also under different varieties of mango (Table 31.5). It may due to the plant canopy of the trees which cause shading effect. The plant canopy and collar diameter of the Amrapalli was least among all the four mango varieties. The increase in grain yield was seen as the distance increases from the trees (Gill and Ajit 2006).

Table 31.5: Average Grain Yield and Straw Yield (q ha^{-1}) of Wheat under different rows from the interspaces of Mango Varieties (Gill and Ajit 2006)

Row	Mango Varieties				Mean
	Amrapalli	**Mallika**	**Dasheri**	**Langra**	
I	25.2 (37.8)	20.1 (32.1)	23.3 (37.0)	22.9 (34.2)	22.9 (35.2)
III	26.9 (44.2)	22.9 (39.5)	25.1 (40.0)	24.7 (36.9)	24.9 (40.1)
V	28.0 (45.8)	25.9 (39.1)	25.8 (40.4)	26.0 (40.0)	26.4 (41.3)
Mean	26.7 (42.6)	22.9 (36.9)	24.7 (39.1)	24.7 (37.0)	24.7 (38.8)

*Figures in bracket are straw yield of wheat (q ha^{-1})

Mango (*Mangifera indica* var. Amrapali) trees were planted in the year 1996 with distance of 6 m × 6 m and 8 m × 8 m in east-west direction. After nine years of planting, the average crown spread of trees in east-west direction was 5.013 m and 5.443 m and in north-south direction it was 5.067 m and 5.707 m in 6 × 6 m and 8 m × 8 m spaced plots, respectively. Elephant Foot Yam (EFY) was planted as intercrop in both the spacing's along with control (without trees) in the month of March at two depths (10 cm and 15 cm). It was found that the height was maximum in closer spacing (66.07 cm) at 10 cm depth and least in open (46.73 cm) sown at 15cm depth. Plant girth was maximum (13.37 cm) of plants, which were sown at 10 cm depth in 6 m × 6 m spacing and least (7.43 cm) when the plants were grown in open and at a depth of 15 cm. The maximum yield of 8.60 kg per plot was obtained, when, intercrop was sown at a depth of 10 cm in 8 m × 8 m spaced mango trees. The least yield of EFY was 6.37 kg per plot, when, it was grown at a depth of 15 cm in 6 m × 6 m spaced mango tree orchard (Panwar et al 2011).

In Uttar Pradesh, five to seven year old mango fields were selected for two tier cropping system. The alleys of mango trees were intercropped with important vegetable crops like potato, brinjal, pumpkin, vegetable pea, garlic and onion. The Dasheri mango trees were at spacing of 8 m × 8 m in rows. The associated crop of potato planted at the distance of 50 cm apart on 12 ridges between two rows of mango. Onion planted in 40 rows at 15 cm apart in the interspaces of two rows of mango. Three rows of pumpkin were planted between two rows of mango at 200 cm × 50 cm × 75 cm apart. Vegetable pea planted in 20 rows at 30 cm apart between two rows of mango and brinjal planted at 60 cm apart in 10 rows between two mango rows. Garlic was adjusted with plantation of 40 rows at 15 cm apart between two mango rows. The intercrop under mango based cropping system yielded 225, 218, 278, 92, 73 and 255 q ha^{-1} of potato, brinjal, green fruits of pumpkin, vegetable pea, garlic and onion respectively. Thus, this system has maintained cash flow and improved the economic status of farming community in the region (Singh et al 2010).

Ravitchandirane and Haripriya (2011) conducted an experiment to find out the effect of intercropping and organic manures on the yield of mango cv. Alphonso. Mango trees were intercropped with aloe and periwinkle. Highest yield of 15.71 and 10.91 kg tree^{-1} respectively were found with the supply of vermicompost @ 5t ha^{-1} + groundnut oil cake @ 500 kg ha^{-1} + panchagavya 3% spray. All the organic manure treatments for intercrops exerted a positive influence on the yield of main crop. The results of this experiment also showed that the yield of medicinal plants aloe and periwinkle were not affected when grown as intercrops in young mango orchard. This was due to the fact that at the initial stage of fruit trees there are no adverse effect on the yield of intercrops. Evaluation of mango based agrihorticulture systems indicated that cowpea-toria sequence was quite remunerative with gross income of Rs.16,850 ha^{-1} (Rathore et al 2012).

GUAVA BASED AGROFORESTRY SYSTEM

Guava is a popular fruit tree of tropical and sub tropical climate. Guava is considered as one of the exquisite and nutritionally valuable and remunerative crop. Guava needs about 4-5 years to develop full canopy under medium rainfall and marginal fertility conditions and thus allows intercrops in the initial years. Cowpea, cluster bean and green gram are suitable intercrops during kharif and cucurbits in summer (CRIDA 1999). Bheemaiah et al (1998) observed that sunflower intercropped with guava + curry leaf showed higher gross returns, net returns and benefit cost ratio than sole cropped sunflower. Horti-pasture system, a combination of fruit trees and pasture, is identified as one of the potential alternative land use options in undulating shallow rainfed soils. At CRIDA, Hyderabad yield reduction of stylo (*Stylosanthes*) was found to be lesser under wide spacing of trees (8 m × 5 m) as compared to close spacing trees (5 m × 5 m). Hence, stylo should be grown under wider spaced fruit tress (Osman and Rao 1999) (Table 31.6). *Cenchrus ciliaris* out yielded stylo and took less time for establishment compared to stylo (Rao 2009).

The field study conducted during 2002 in 5 years old guava & curry leaf plantations revealed that the seed yield of soybean when grown as intercrop either in guava or curry leaf plantation was comparable with that of sole cropping of soybean (Subrahmanyam 2009). Whereas, the total net returns obtained from the system (tree + crop) was found maximum under intercropping of soybean in curry leaf which increased the total net returns from the system in comparison with soybean grown as sole crop without trees (Table 31.7). Hence integration of fruit or economically important trees with arable crops is intended to maximize the land use efficiency to generate supplemental income to medium level farmers in semi arid tropics.

Table 31.6: Fresh Yield of Forage and Fruit in Guava based Horti-pasture system at CRIDA, Hyderabad (Rao 2009)

Spacing (m)	Fodder Yield (t/ha)		Fruit Yield (kg/plant)
	Stylo legume	Cenchrus	
5 × 5	5.22 (40.4)	2.43 (3.9)	95.4
8 × 5	6.56 (25.1)	2.14 (16.1)	99.7
Control	8.76	2.55	-
Mean	6.84	2.38	97.5
CD (0.05)	0.88	NS	NS

Table 31.7: Seed Yield of Soybean and Total Net Monetary Returns from Guava and Curry Leaf based Agrihorticulture System (Subrahmanyam 2009)

Cropping System	Seed Yield (Kg/ha)	Total Net Returns (Rs/ha)
Intercropping of Soybean in guava	739	8,021
Intercropping of Soybean in curry leaf	740	4,845
Sole cropping of Soybean	800	3,269

In Guava and *Acacia mangium* based silvihortipastoral system, *Acacia mangium* is planted at 6 m × 6 m interval and horticultural component guava in the middle at 6 m × 6 m interval. Fodder grasses like guinea and stylo are found to be more remunerative under this silvihortipastoral practices. Research results under AICRP Agroforestry at Bhubaneswar centre reveals that on an average guinea grass produced 124.6 q/ha of green fodder as compared to stylo which could yield only 98.0 q/ha under same management conditions from two cuttings. This practice is well prevalent in coastal districts of Odisha (Anon. 2003). The results of an investigation conducted in eastern parts of Himalayas (Anon. 1988) indicated that the high net return was obtained when guava (*Psidium guajava*) is grown with ginger. No adverse effects of intercrops on the growth of guava were observed.

Table 31.8: Crop Yield and Net Return of Guava based Agroforestry System (Anon. 1988)

Treatment	Crop Yield (q/ha)	Net Return (Rs./ha)
Guava + groundnut	14.4	Medium
Guava + soybean	14.5	Low
Guava + rice	19.5	Low
Guava + ginger	92.8	High
Guava + turmeric	106.9	Medium

CITRUS BASED AGROFORESTRY SYSTEM

Borthakur et al (1979) have demonstrated the model of an ideal agroforestry system as an alternative to 'Jhumming' or shifting cultivation in north eastern hill region. In the alternative model, lower 1/3rd area of hilly slope is terraced and utilized for raising field crops keeping farmers requirements in view. Crops like rice, maize, ragi, mustard, sesamum, black gram and root crops like ginger, turmeric, etc., may be grown on such terraces. Usually, these terraces are stabilized by few mechanical measures and planting with fodder grasses or legume on risers. Fodder grasses and legumes suitable for terraces risers are stylosanthes (4.73 t/ha dry matter yield). The middle 1/3rd of the hill slopes are used for forest including fuel, fodder, plantation. Citrus fruits such as lemon could be intercropped with cowpea, garden pea, French bean and sweet potato in agrihorticulture system. Even lemon and pineapple were found to grow together. The average net income realized from one hectare farming was the highest in three-tier system followed by in complete terracing, in puertorican terraces and in Jhum cultivation.

In Meghalaya (Anon. 1987), growing ginger in mandarin orchards has been highly profitable than in mandarin grown with rice, soybean, groundnut and turmeric. According to another study by Dhyani and Chauhan (1989), pineapple with black pepper and arecanut resulted in highest net return after six years of plantation of tress than arecanut with black pepper. In the temperate and sub alpine zone of Himalayas, plum with potato/cabbage gave a high net return. In subtropical hills, the combination of schima with large cardamom could give a high net return than large cardamom alone. Mandarin (*Citrus reticulata*) is an important commercial fruit crop in Sikkim. Mandarin orchards are intensively intercropped with many species of cereals, pulses and vegetable crops. Ginger is the most remunerative intercrop in Sikkim. Singh and Pradhan (1989) observed that various intercrops with mandarin can yield 609 kg/ha (ginger), 149 kg/ha (black gram and rice bean).

Dayal et al (1996) have developed a horticulture-based AFS for small farmers by integrating plants yielding a number of products and having synergistic effects on each other. Subabul (*Leucaena leucocephala* "K8") was raised for fuel, fodder and live hedge on the border row, followed by lemon (*Citrus aurantifolia*) "Baramasi" on the peripheral row, papaya (*Carica papaya*) and okra (*Hibiscus esculentus*) as understorey crops in the main plot. During first five years (1985-89) the system provided well spread out, handsome net returns (Rs.15,000-21,000/ha/year) with limited irrigation and Rs.10,000-18,000/ha/year for three years (1990-92) under rainfed conditions, compared with Rs.4,900-9,800/ha/year with limited irrigation and Rs.1,740-4,600/ha/year under rainfed conditions, from agricultural crops (maize-wheat).

AONLA BASED AGROFORESTRY SYSTEM

Aonla or Indian gooseberry (*Emblica officinalis*) has hardy nature, suitable to various waste lands, high productivity per unit area (15-20 t ha^{-1}), nutritive and therapeutic value. Hence, it has become an important fruit. Aonla has sparse foliage which allows 87.5 per cent area for intercropping during the initial 10 years. The rainfed aonla based multitier cropping system with guava as filler plant was investigated to standardize suitable intercrop for maximization of profitability. During the 9th year of establishment of the system, the maximum paddy equivalent yield was recorded (6.05 t ha^{-1}) in case of intercropping of groundnut. With respect to plant growth parameters, significant effect of different cropping systems could be recorded on plant volume of guava which was the maximum (44.48 m^3) in case of intercropping of paddy. The soil properties differed significantly in different cropping systems with respect available potassium in both 0-15 cm and 15-30 cm soil layer. However, with respect to available phosphorous, it differed significantly in 0-15 cm soil layer. Intercropping of deenanath grass and ragi resulted in significant increase in the content of potassium (317.18 and 312.36 kg ha^{-1} K, respectively) in 0-15 cm soil layer, whereas, the content of phosphorous was maximum in case of intercropping of ground nut (12 kg ha^{-1} P) in 0-15 cm soil layer. Thus, fruit based multitier cropping systems were found to be effective alternatives to the traditional rice based monocropping system for increasing the profitability under rainfed as well as irrigated upland conditions of the eastern plateau and hill region (Das et al 2009).

The effect of irrigation on the performance of aonla under agroforestry system was investigated (NRCAF 2011). In kharif and rabi 2010-11, fodder cowpea var. B.L.-1 was grown in interspaces and wheat variety WH-147 had been sown respectively. The observations on plant growth indicated that plant height ranged between 3.01 to 3.56 m, collar diameter between 9.28 to 10.29 cm while, canopy diameter ranged between 3.24 to 4.17 m across the treatments. The differences in growth parameters under different treatments were negligible. Similar was the case in fruit yield which ranged from 3.09 to 4.28 kg per plant. Plant growth and fruit yield of aonla showed no adverse effect of intercropping (Table 31.9). Yield of intercrop in kharif 2010 showed fresh biomass yield of about 33 t ha^{-1} in both the treatments which was lower than sole crop (52.5 t ha^{-1}). Reduction in crop biomass yield in association with aonla trees may be due to shade caused by tree and competition for nutrient and moisture. Growth parameters of crop showed due reduction in plant population, number of branches and average length.

Table 31.9: Mean Growth and Yield of Aonla under different Treatments (NRCAF 2011)

Treatments	Plant Height (m)	Collar Diameter (cm)	Canopy Diameter (m)	Fruit Yield (kg $tree^{-1}$)
(T_1) Aonla rainfed	3.56	9.95	4.17	3.57
(T_2) Aonla summer irrigation	3.56	10.29	3.94	3.93
(T_3) Aonla irrigation year round	3.30	9.63	3.90	4.28
(T_4) Aonla rainfed + Crop	3.41	9.54	3.55	3.09
(T_5) Aonla summer irrigation + Crop	3.01	9.28	3.24	3.68
(T_6) Crop Sole	-	-	-	-

Leaf litter production per tree ranged from 3.37 to 5.58 kg depending upon tree canopy. On an average 52.7 g N, 4.48 g P and 30.1 g K per tree was added in soil across the treatments under aonla based agroforestry system (Table 31.10). This is sizable amount of nutrient that is being replenished by tree.

Table 31.10: N, P and K Content in the Leaf Litter of Aonla (NRCAF 2011)

Treatment	Leaf Litter (kg tree^{-1})	N (g tree^{-1})	P (g tree^{-1})	K (g tree^{-1})
(T_1) Aonla rainfed	5.58	64.3	5.47	36.7
(T_2) Aonla summer irrigation	4.98	57.4	4.88	32.8
(T_3) Aonla irrigation year round	4.88	56.2	4.79	32.1
(T_4) Aonla rainfed + Crop	4.05	46.6	3.97	26.6
(T_5) Aonla summer irrigation + Crop	3.37	38.8	3.30	22.2
Mean	**4.57**	**52.7**	**4.48**	**30.1**

Agroforestry can play a major role in bringing the desired level of diversification along with sustainability. The farm industry linkages have also helped the systems to be more sustainable than the traditional cropping systems (Kareemulla et al 2002). Resource degradation leading to an unsustainable production system has demanded our attention for sustainable practices to assure continued production. In this context, aonla or Indian gooseberry (*Emblica officinalis*) based agri-horticultural system has immense potential to utilize and conserve rainfed area for betterment of poor farmers. Aonla being a deep-rooted deciduous tree species has a wide range of adaptability to grow in any type of soil. It is considered a highly tolerant and potential fruit species suitable for growing under salt-affected and wasteland/ravine lands. Aonla provides higher economic returns with little investments in plantation establishment and its management. Cultivation of agricultural crops with aonla provides an opportunity for maximum use of land surface and helps to distribute the risk due to adverse climate. The tree canopy of aonla with sparse foliage allows filtered light and permits intercropping even after the trees are fully grown. Intercropping not only generates an extra income, but also helps check soil erosion through ground coverage and improves the soil physio-chemical condition (Ghosh and Pal 2010).

Table 31.11: Yield of Aonla and Intercrops under Agrihorticultural Systems (Das et al 2011)

	Aonla Fruit (t/ha)			**Average**	**Intercrop (q/ha)**			**Average**
Treatment	**2007**	**2008**	**2009**		**2007**	**2008**	**2009**	
Aonla (Sole)	5.75 (20.8)*	8.75 (31.6)	7.04 (25.4)	7.18 (25.9)	-	-	-	-
Aonla + turmeric	11.63 (42.0)	15.71 (56.7)	12.57 (45.4)	13.30 (48.0)	260.4 [7.5]	248.2 [9.6]	227.1 [12.0]	245.2 [9.7]
Aonla + ginger	9.16 (33.1)	12.96 (46.8)	9.70 (35.0)	10.61 (38.3)	110.5 [12.2]	100.9 [16.5]	92.3 [19.3]	101.2 [16.0]
Aonla + Arbi	10.25 (37.0)	14.02 (50.6)	10.86 (39.2)	11.71 (42.3)	107.6 [15.7]	102.4 [18.2]	88.7 [25.3]	99.6 [19.7]
Mean	9.20 (33.2)	12.86 (46.4)	10.04 (36.3)	-	-	-	-	-

* Figures in small brackets indicate the yield of aonla fruit in 'kg/tree'.

**Figures in square bracket indicate the reduction in the yield of intercrop in percentage compared to sole-crop

Das et al (2011) investigated yield, soil health and economics of aonla based agri-horticultural systems in eastern India. The fruit yield of aonla was 9.20 t ha^{-1} and 12.86 t ha^{-1} during the sixth and seventh year of plantation with 2,367.9 mm rainfall with 71 rainy days and 1,535 mm rainfall with 66 rainy days respectively. In the eighth year, there was decline in the fruit yield up to 77 per cent (10.04 t ha^{-1}) due to poor rain and regular drought receiving 753.7 mm rainfall with 47 rainy days. The magnitude of crop yield losses in agri-horticultural systems increased with the age of the trees (Table 31.11). This indicated that yield was inversely proportional to crown diameter. The reductions in yield of intercrops between sixth and eighth years of cultivation were 7.5-12.0 per cent for turmeric, 12.2-19.3 per cent for ginger and 15.7-25.3 per cent for arbi (*Colocasia*) compared to the yield in open field. Among the different intercrops, turmeric performed better with the least reduction (9.7%) in yield compared to the yield in open without aonla. Increased competition with age was due to the increased size of the trees and their ability to mop up greater resources at the expense of crops (Prasad et al 2010). The yield reduction in the crop under plantation could be due to the lower photosynthetic rates of crop in shade under plantation (Chaturvedi and Jha 1998).

APPLE BASED AGROFORESTRY SYSTEM

Apple is grown in hill slopes and seasonal intercropping accompanied by soil working is not encouraged in order to prevent soil erosion. Where slope of the land is more than 10 per cent, various grasses and legumes are grown as permanent covers. The tree basins are kept clean and sod culture in the orchards is practised in the hills of Himachal Pradesh. The grasses are regularly cut when 7-10 cm high and the residue is left in the orchards for decomposition. The cultivation of legumes and grasses as intercrop is becoming popular in orchards. The important legumes are red clover (*Trifolium pratense*), white clover (*T. repens*) and lucern (*Medicago sativa*). Among the grasses, orchard grass (*Dactylis glomerata*), fescues (*Festuca arundinacea*) and timothy (*Phleum pratense*) are common. Orchard grass is grown where moisture is high, while fescues is suitable for dry areas or southern aspects. In apple, sometimes filler trees of peaches or dwarf apple plants are also grown. At the young age of orchard plantations, vegetables like cole crops, potato and tomato are grown to supplement the income. However, this practice is abandoned when the orchard starts bearing after 5-7 years to avoid competition. In Shimla hills and Kullu Valley areas of H.P., cultivation of gladioli for flower and bulb production as intercrop is carried out in the apple orchards since it can tolerate shading.

PEACH BASED AGROFORESTRY SYSTEM

In the Doon valley of Uttarakhand, Singh et al (1990) reported that both rabi and kharif crops can be grown with peach. On the basis of five-year data, they reported that turmeric and ginger do particularly well. Lal et al (2004) in an experiment, cultivated black gram and toria under peach based agroforestry systems, accrued net returns of Rs. 10,000/ha/year and B:C ratio was recorded as 1.65. In the sub-montane regions at Dehradun, peach can be grown on terraces cut out on hill slopes along with vegetables as understorey crops in purely rainfed conditions. Vegetables like tomato and brinjal in the summer (kharif) and cabbage, radish and cauliflower in winter (rabi) are grown extensively in many parts of the Himalayas along with peach. Peach based system introduced for the first time in the district Sharanpur, UP, India as an alternative to sugarcane dominated cropping system. Three peach + vegetables based systems are practised in 4 ha area. They are peach (Saharanpur prabhat) + Brinjal (Navkiran) + chilly (AK 47); Peach (Saharanpur prabhat) + spinach (rajdhani) and Peach (Saharanpur prabhat) + Cucumber (Alamveer) - black gram (PU 35). Peach based system viz. peach + cucumber-cucumber-blackgram-spinach provided highest net profit of Rs 3.43 lakh/ha, followed by peach+brinjal+chilly of Rs 3.39 lakh per ha. Intercropping of guar, cowpea or pearl millet with peach; turmeric with papaya; *Chrysopogon fulvus* or Napier grass with aonla or ber has been identified as an economically-viable and eco-friendly system for rehabilitation of these lands. Aonla gave the highest yield of 86 kg fruits per tree in association with pigeon pea but the yield was reduced by 17 and 23 per cent respectively in association with *Chrysopogon* and napier grass.

LITCHI BASED AGROFORESTRY SYSTEM

Litchi is a slow growing plant and takes about 15-16 years to develop canopy and cover the area. The planting of guava, custard apple, lime/lemon in the centre, between and within the rows of litchi have been found to give additional income in the initial stage of planting without competing with the main crop. Papaya is also planted as filler plant on same unit of land at the spacing of 2.5 m × 2.5 m. Initially cowpea, French bean, okra, brinjal or other suitable crops of the regions are grown as intercrops in between plants. Under mature litchi orchards, shade loving plants/crops viz. ginger, turmeric, elephant foot yam grown successfully and provides additional income per unit area. In Bihar, different litchi based agroforestry system are practised. Litchi with banana combination found to be successful for quality production in successive years. Intercrops in combination such as okra and gladiolus, cowpea-potato-onion and cowpea-French bean-okra rotation followed under litchi during initial years of establishment. Okra-gladiolus found to be the best in terms of net return and plant growth.

In multitier system comprising the main crop, filler crop and intercrop in interspaces was litchi, guava and cowpea, French bean, niger and horse gram respectively, were investigated for their suitability. Intercropping did not show any marked adverse effect on mean monthly increment in the height of plant, girth of trunk and area of canopy spread of main crop as well as filler crop. Fruit yield of filler crop (guava) was also not influenced by intercropping. Among the intercrops, cowpea was found to be the most profitable (Rs.24,000 ha^{-1}). Sixth year onwards drastic yield reduction in all the intercrops was observed. The reason could be the shading effect of main and filler crop and more could be due to main crop as the litchi is evergreen fruit tree. At the end of 8th year, the cowpea continued to record its supremacy as an intercrop over other intercrops and yielded 3.03 t ha^{-1} green pods. Cowpea was recommended for intercropping under the litchi based multitier cropping system upto the initial 8 years of the orchard. Afterwards, intercropping of shade tolerant crops like ginger, turmeric and elephant foot yam can be followed (Das et al 2009).

Chaturvedi and Jha (1998) studied yields of intercrops under two rotations, viz. paddy-wheat-green gram-ginger and maize-toria-green gram-pointed gourd. These crops were intercropped under an age series of litchi plantations during years 1-9 following planting. Average height and diameter of litchi trees under 1 to 9 year-old plantation varied between 1.0 ± 0.20 to 5.47 ± 0.33 m and 5.10 ± 1.30 to 36.3 ± 3.82 cm, respectively. Relative light intensity under tree plantations decreased with increasing canopy depth and tree height. They found there was a marked reduction in yield of all intercrops with increase in age of the plantations. The benefit:cost ratio for the rice-wheat-mung bean-ginger plus litchi system was 2.17, and for maize-toria-mung bean-pointed gourd plus litchi was 2.73, indicating that the latter cropping system was more profitable. Thus, intercropping during different growth stages of litchi trees substantially increased the overall productivity of land from the first year of planting.

SAPOTA BASED AGROFORESTRY SYSTEM

A multi-component agroforestry system with sapota as the base crop, teak in the sapota line and agricultural crop in the interspace was developed for high rainfall areas having irrigation facilities. Broad spacing provided to this crop provided an opportunity to cultivate an intercrop in the initial years. Sapota was planted at a recommended spacing of 10 m × 10 m in rows across the slope. Three rows of teak plants were planted at distance of 3 m × 2 m and 2 m × 3 m in between two rows of sapota trees. Field crops, viz. horsegram, jowar and bajra, were grown in the interspaces of sapota + teak alleys. Sapota crop served as an insurance against failure of field crops. The same technology was adopted by another farmer with a modification, i.e. sapota being replaced by mango. Fruit bearing in sapota and mango started from the seventh and eighth year respectively. Presently, the sapota is yielding 30 to 40 kg/plant which fetches for Rs. 22,000 to 25,000/ha. The fruit yield from mango is 30-50 kg/plant and that fetches Rs.36,000 to 60,000/ha. The income generated from field crops in both the cases is about 2,500 to 3,500/ha. The value estimation of teak reveals that teak pole gives considerable economic return. With age, the crown size of perennial component (sapota/mango and teak) increased and consequently of field crops was not in cultivation from 2007 onwards. The system generated employment to an extent of 180 man days per year. The socio-economic status of the farmers improved as farmer is earning on an average of 23,500/ha/year with sapota and 48,000/ha/year with mango based system as against 3,000/ha/year only during initial period from the same land (Anon. 2012).

MULTITIER SYSTEMS CONSISTING OF TWO OR MORE FRUIT CROPS

Tomar and Bhatt (2004) conducted a field experiment in a six year old plantation of guava (*Psidum guajava* cv. Allahabad safeda), Assam lemon (*Citrus limon* cv. local) and peach (*Prunus persica* cv. TA 170) during 2002 and 2003, on acid alfisols under rainfed conditions in Umiam, Meghalaya, India, to study the performance of upland rice cultivars as intercrops in the existing fruit plantation and to study the overall productivity of agri-horticulture systems. Their report showed that on an average, the maximum net monetary benefit per hectare was recorded from peach intercropped with rice (Rs.48,044), followed by guava (Rs. 27,887) and Assam lemon (Rs. 20,991), irrespective of rice cultivars. Peach, guava and Assam lemon exhibited 5.09, 2.95 and 2.22 fold higher net returns, respectively, compared to control. Peach based AFS was found promising for rainfed agricultural conditions in the northeastern region of India in general and Meghalaya in particular.

Three years' study based on 12 farmers of district Kurukshetra infers that the farmers can adopt favourable cropping patterns in newly planted orchards which are not harmful to the growing canopy and provide better returns. The most favourable cropping sequence was bottle gourd-pea-cucumber which provided net return of Rs.57,150 ha^{-1} followed by bottle gourd-tomato (Rs.53,775 ha^{-1}). The income from crops decreased considerably during third year after orchard plantation (Table 31.12). The expenditure incurred on the establishment of 180, 140, 180, 275, 225 and 180 plants of mango, chiku, aonla, guava, pear and ber along with filler plants peach and papaya was Rs.27,740, Rs.28,785, Rs.31,820, Rs.26,050, Rs.32,700 and Rs.32,820 ha^{-1} respectively. The total expenditure including gap filling, hoeing, ploughing and insect management for three years were Rs.46,190, Rs.45,435, Rs.49,545, Rs.43,375, Rs.49,500 and Rs.48,095 ha^{-1} respectively (Hari Om et al 2008).

FLOWER CROPS AS INTERCROP IN ORCHARDS

A field study to work out the best intercropping option with two marigold varieties viz. African and Jaffri in newly planted orchards of chiku, guava and aonla was carried out during 2004-05 and 2005-06 in district Kurukshetra, Haryana, India. The number of plants of chiku, guava and aonla transplanted by the farmers were 140, 275 and 180 ha^{-1}, respectively. Peach and papaya plants were used as filler plants in chiku and aonla orchards respectively. The study revealed that the total expenditure on establishment of 140, 275 and 180 plants of chiku, guava and aonla was Rs. 28,785, Rs.26,050 and Rs.31,820 respectively. It was observed that Jaffri variety of marigold was more remunerative and gave an advantage of Rs. 9,900 ha^{-1} and Rs. 14,700 ha^{-1} in first and second year over

African variety, when grown as intercrop in newly planted orchards of chiku, guava and aonla (Hari Om et al 2008).

Table 31.12: Net Returns from various Cropping Patterns in Intercropping with Orchard (Hari Om et al 2008)

Kharif (June-Sept)		Rabi (Oct-Jan)		Zaid (Feb-May)		Total Net Return (Rs./ha)
Crop	**Return (Rs.)**	**Crop**	**Return (Rs.)**	**Crop**	**Retun**	
Bottlegourd	22,950	Tomato	30,825	-	-	53,775
Spongegourd	19,400	Tomato	30,825	-	-	50,225
Bottlegourd	22,950	Pea	23,075	Cucumber	11,125	57,150
Spongegourd	19,400	Pea	23,075	Cucumber	11,125	53,600
Sorghum fodder	4,075	Pea	23,075	Cucumber	11,125	38,275
Bottlegourd	22,950	Palak	10,450	Cucumber	11,125	44,525
Bottlegourd	22,950	Methi	9,888	Cucumber	11,125	43,963
Spongegourd	19,400	Palak	10,450	Cucumber	11,125	40,975
Spongegourd	19,400	Methi	9,888	Cucumber	11,125	40,413
Sorghum fodder	4,075	Palak	10,450	-	-	14,525
Sorghum fodder	4,075	Methi	9,888	-	-	13,963
Sorghum fodder	4,075	Berseem	5,975	-	-	10,050

AGRIHORTISILVICULTURE

In this system fruit trees are grown along with crops and multipurpose trees (MPTs). Under rainfed situation hardy fruit trees like ber, aonla, pomegranate, guava could be grown along with dryland crops like pigeonpea, til, mothbean, mustard etc. Grafted ber (Var., gola, seb, mundiya, banarasi kasak) may be planted at 6 m × 6 m with 2 plants of subabul in between. Under partial irrigation, guava, pomegranate, lemon, kinnow have been successfully grown at 6 m x 5 m along with crop like wheat, groundnut and subabul (200 plants/ha) for quick leaf fodder and fuel wood production. For further protecting fruit crops from desiccating hot summer and cold, winter planting of subabul/ sesbania at every 2 m apart as wind breaks are carried out. Alternate plants of subabul/sesbania could be harvested for quick fodder and fuel wood production every 3rd year. Relative grain yield was 70-85 per cent even in 3rd and 4th year.

In agroforestry systems from land management point of view, the combination of the broad-leaved trees and pineapple, oranges with large cardamom, arecanut, black peeper and pineapple on slopes wherever the soil, climate is suitable would not only supplement the income from orchard but also help in the conservation of sol and runoff. In Manipuri valley, agrihortisilviculture system is highly remunerative and preferred by the farmers. Among the tribal farmers, crops like paddy, mustard, sugarcane, soybean, beans, vegetables (mainly pumpkin or potato), walnut, cherry, banana, guava, etc., are grown. In this region, forest tree species are grown to produce wood and leaves for fodder in combination with high income horticultural crops like pineapple. In a study at Langoi Hill, Lamphelpat, pineapple was grown with *Eucalyptus tereticornis*, *E. citriodora*, *Leucaena leucocephala* and *Parkia roxburghii.* This study shows that the tree do not have any shade effect on the yield o pineapple. Pineapple with *Parkia roxburghii* agroforestry system gave the highest yield of pineapple (Anon. 1990).

AGRIHORTICULTURE

Integration of fruit crops in croplands is referred to as agrihorticultural land use system of agroforestry. Aonla, ber, guava, citrus etc. are promising fruit crop in the dry land region, fruit crops are first preference of farmers under agroforestry system on account of short gestation period, regular income, risk cover and aesthetic value. Agrihorticulture land use is an important component of agroforestry. In Custard apple, pomegranate and Aonla based agroforestry at Hyderabad, it was observed that the yields of rainfed intercrops of sorghum, groundnut and mung bean in association with fruit crops (pomegranate and custard apple) were reduced by 23-26 per cent compared to the sole crops (CRIDA 1999). The yield reduction was higher with custard apple compared to pomegranate. Among the different systems, groundnut grown with pomegranate or custard apple gave highest gross income (Rs.19,540-19,770/ha). Custard apple + mungbean recorded highest yield advantage (54%) over respective sole crops (Rao 2009).

Most prevalent tree based agroforestry system in District Kinnuar, HP was agrihorticulture (Kumari et al 2008). Average area under the model was 8.44 bigha and crops grown were potato/peas/ maize/jau/ogla/fafra/rajmash /urd/shimla-mirch/green chillies+apple/almond/apricot. Pastoral horticultural area averaged about 2.2 bigha and crops cultivated were grasses + apple/almond/pears/ plum/ apricot. The benefits of different agroforestry models were in maximum the form of fruits (52.2) followed by fuel (39.2) and fodder (38.2). Agrihorticulture systems can be used to improve economic condition of the farmers. In the North East region, peach based agrihorticulture systems were evaluated on class II lands. Both annual and companion crops were grown. The net returns doubled when turmeric was grown with peach. Mixed cropping is preferred over pure crops in rainfed condition as risk of failure of crop is minimized. If annual crop fails, tree crop can provide some income.

The modern agroforestry practices which evolved are not only commercially oriented but aimed to meet specific needs. Modern agroforestry systems could be adapted for multiple uses such as shade, fodder, fuelwood, fruits, timber, industrial raw materials like oils and drugs, gums, medicinal plants and most importantly their importance as biofuels. Agroforestry is an age-old practice in the states of north eastern region. Out of the various agroforestry practices, tribes of the region adopt a combination of pineapple and black pepper with arecanut, plum with potato or cole crops, indigofera with large cardamom and schima with ginger or turmeric. These are the most productive horticulture based agroforestry systems. The above system may be coupled with runoff harvesting and recycling and other biological components may be added. There is great scope of pisciculture, dairying, bee keeping and cut flowers industries, which will further increases the income of the marginal farmers through agro-based industries.

FRUIT TREES BASED AGROFORESTRY RESEARCH IN Dr.YSPUHF, HP

The dominant agroforestry models in low hill subtropical zone of Himachal Pradesh, are either kinnow based or mango based. Poplar and eucalyptus are also grown by the farmers on the bunds of the field in order to generate the additional farm income. Sometime block plantation of poplar and eucalyptus are also carried out by the farmer to fulfill the need of the wood based industry. It has been found that in sole cropping system, the farmers are getting net return of Rs.10,603.50 and Rs.14,845.50 per hectare by growing wheat (*Triticum aestivum*) var. PBW 343 and gobhi sarson (*Brassica campestris* var. Neelam (HPN-3)), whereas, highest net return of Rs.47,169.69 was obtained from 12 year old kinnow plantation. But on the other hand, when wheat and gobhi sarson were grown along with kinnow in agrihorticulture system, the net return found to be increased over sole crop and the increase was noticed in the tune of 22.6 per cent (Rs.53,069.62) in case of wheat + kinnow, whereas, in case of gobhi sarson it was recorded 20 per cent (Rs.56,407.55). As regarding the cost, an increase of 42.52 per cent in case of wheat and 20.29 per cent in case of gobhi sarson with kinnow was observed. Hence, analysis revealed that cultivation of gobhi sarson is more profitable as compared to wheat with kinnow. However, the net profits in both the cases were higher as compared to the sole crop of kinnow. Economics of agrihorticulture system revealed that both gross as well as net returns increased per hectare basis. Agroforestry models are thus important to enhance the land use efficiency as well as productivity (Sharma et al 2008).

From the study it was emerged out that for the development of kinnow based agrihorticulture system, kinnow should be spaced at 7-8 m R × R and 4-5 m P × P instead of regular square planting of 6 m × 6 m. This rectangular planting geometry will ensure wider row to row space for growing agricultural crops and will minimize the shade effect on agricultural crops and at same time it will accommodate almost equal number of kinnow plants per hectare and thus not affect the kinnow yield per hectare.

Agri-horti-silviculture system was developed which consisted of kinnow-mandarin (*Citrus nobilis* × *C. deliciosa*) 400 plants per hectare, subabul (*Leucaena leucocephala*) in quincunx method at varying plant density (0, 100, 167 and 277 plants ha^{-1}) having equal P × P and R × R in each plot as woody perennials and agricultural crops viz. wheat and mash under irrigated conditions. It is second success story for Poanta valley, District Sirmour (HP). The system was evaluated on the farmers' fields with the application of only standard inputs as per the recommendation of State Horticulture and Agriculture departments for kinnow-mandarin and agricultural crops, respectively. The yield data of different agroforestry system components i.e. kinnow-mandarin, wheat, mash as sole as well as integrated with each other and with leucaena for fuel and fodder is presented in Table 31.13. The system was developed in 1982 and evaluated after 10 years for two consecutive years i.e. 1992-93 and 1993-94. The fruit yield (number of fruits per hectare), average fuelwood and fodder yield of leucaena and grain yield of wheat and mash were recorded.

Table 31.13: Fruit, Fodder, Fuelwood and Grain Yield under Agri-horti-silviculture System (Chauhan et al 1997)

AF System	Number of Fruits ha^{-1}	Fodder Yield (kg ha^{-1})	Fuelwood Yield (q ha^{-1})	Wheat Yield (q ha^{-1})	Mash Yield (q ha^{-1})
K (400) + A	1,42,914	-	-	8.90	4.50
K (400) + S (100) + A	1,59,056 (+11.30)	14.76	8.98	8.55 (-3.93)	4.26 (-5.56)
K (400) + S (166) + A	1,74,084 (+21.81)	25.04	15.93	7.95 (-10.67)	3.63 (-19.33)
K (400) + S (277) + A	1,32,014 (-7.63)	37.52	23.39	6.95 (-21.91)	3.21 (-28.67)

1. Parentheses represent the percentage increase (+) or decrease (-)
2. Approximately number of kinnow per kg = 8-10
3. K = Kinnow-mandarin; S= Leucaena; and A= Agriculture crops i.e. wheat in rabi and mash in kharif
4. Figure in brackets in first column shows number of trees

The maximum number of fruits (1,74,084 ha^{-1}) which is the only economical part of the system for sale, fodder yield (25.04 qha^{-1}) and fuelwood yield (15.93 qha^{-1}) were obtained when kinnow was spaced at 5 m x 5 m and leucaena trees were planted at a distance of 10 m × 5 m in between the rows of kinnow. The percentage increase in the agri-horti-silviculture system was 21.81 per cent over the horti-agriculture i.e. kinnow with agriculture system. It is evident from the table that there is relatively decrease in agriculture crop yield in the system when density of leucaena trees is increased; on the other hand there is a significant increase of kinnow fruits by incorporating leucaena plants. The leucaena plants fixes the atmospheric nitrogen for improvement of the soil fertility which is available to the kinnow trees and also provide additional benefits in terms of fuelwood and fodder for livelihood subsistence of the farmers. It is therefore recommended that horticultural crops planted at standard distance should be incorporated with leguminous fodder trees for enhancing the income as well as livelihood subsistence in the region.

Tree-Crop interaction studied by Kumar (1995) in agroforestry system for the economic and biophysical criteria of scoring for designing economically viable, sustainable, more productive and diversified land use system, the different tree-crop combinations comprising specific functional units are found suitable/ promising in the following order of preference:

- **If net returns dominated over diversification**

  Plum - soybean > Plum - Mulberry - Soybean > Mulberry - Soybean > Plum - Mulberry - Pomegranate - Soybean > Pomegranate - Mulberry - Soybean > Soybean > Pomegranate - Soybean

- **If product diversification dominates over net returns**

  Plum - Mulberry - Pomegranate - Soybean > Plum - Mulberry - Soybean > Pomegranate - Mulberry - Soybean > Plum - soybean > Mulberry - Soybean > Pomegranate - Soybean > Soybean

- **Overall preference (Net returns, total costs, total biomass, product diversification and soil nutrients)**

  Plum - soybean = Plum - Mulberry - Soybean > Plum - Mulberry - Pomegranate - Soybean > Mulberry - Soybean > Pomegranate - Mulberry - Soybean > Pomegranate - Soybean > Soybean

On the basis of study carried in Dr.YSPUHF, Nauni-Solan, HP, India (Lal 2003) concluded that agroforestry system not only assures diversification but also provides higher economic returns in comparison to monocropping. Agri-horti-silviculture system comprising of Peach (*Prunus persica*) and *Morus alba* with wheat-soybean and lentil-blackgram cropping sequences can be better option for higher monetary benefits than agrisliviculture system integrating wheat-soybean with *Grewia optiva*. Out of two cropping pattern followed under agri-horti-silviculture system, the cropping pattern with lentil-blackgram sequence gives higher returns as compared to wheat-soybean on per hectare basis. Despite negative effect on growth, yield and some physiological attributes, particularly at closer distance to the tree rows, agroforestry is still beneficial as it results in higher available soil nutrients and improves soil physical properties. Moreover it furnishes higher economic returns because of diversified produces comprising of 1 Peach + 1 Morus + field crops (Rs.47,719.76/ha/yr) and 2 Morus + 1 peach + field crops (Rs.56,578.96/ha/yr) for wheat-soybean and lentil-blackgram cropping patterns than that of open control (without trees) wheat-soybean (Rs. 23,669.90/ha/yr) and lentil-blackgram (Rs.33,826.70/ha/yr). Net returns (monetary gains) were comparatively more under agroforestry system than that of open control. Lentil-blackgram cropping pattern provided higher returns as compared to wheat-soybean under agri-horti-silviculture system.

Study carried out in Dr.YSPUHF, Nauni-Solan, HP, Inida (Kumar Silpa 2011) indicates that the two medicinal plants *Andrographis paniculata* and *Stevia rebaudiana* thrive well under plum based agroforestry system. Growth, yield & physiological attributes of medicinal plants were positively affected by application of different doses of organic manures. The use of organic manure + chemical fertilizers has been found to be more efficient than organic manures alone. Also, the production potential and economic returns were higher inside the agroforestry system than monocropping. Among two herbs, *Stevia rebaudiana*, however, provided better total economic returns (Rs. 99,952.50 ha^{-1}) as compared to *Andrographis paniculata* (Rs.89,939.30 ha^{-1}) and, therefore, it can be suggested that growing of *Stevia rebaudiana* under plum based agroforestry system is economically more profitable.

The study was conducted on apple based agroforestry in Kullu district of Himachal Pradesh. It revealed that the average cost of cultivation of apple was Rs.3,88,850.70 ha^{-1} and the average net benefit from the orchard by selling fruit was Rs.10,45,523 ha^{-1} (Table 31.14). The integration of high value crops i.e. tomato, pea and French bean in the system not only offered diversification in different growing seasons but also generated surplus income without affecting the fruit yield of the orchard (Table 31.15). However, there was significant loss (Rs.57,493.92 ha^{-1}) in the old traditional system when wheat was integrated as a cereal crop under apple. The average data of two consecutive years revealed that growing of pea as a substitute crop for wheat in rabi season benefited the farmers to the tune of Rs.2,84,676 ha^{-1} and Rs.5,31,966 ha^{-1} in the kharif from the cultivation of tomato (Anjulo 2009). Hence, it is recommended that high value vegetable crops, such as beans, peas, tomato, cauliflower, cabbage, broccoli as off-season vegetable can be integrated in the temperate orchards to increase the farm income.

Table 31.14: Apple Fruit Yield, Expenses and Return over two years under Agrihorticulture System (Anjulo 2009)

Year	No. of Trees/ ha	Average Fruit Yield (kg t^{-1})	Fruit Yield (q ha^{-1})	Selling Price (Rs.kg^{-1})	Gross Return (Rs. ha^{-1})	Total Expenses (Rs. ha^{-1})	Net Benefit/Loss (Rs. ha^{-1})	Benefit Cost Ratio
1st	278	194.06	539.48	22	11,86,870	3,88,850.7	7,98,019.3	3.05
2nd	278	189.06	525.58	32	16,81,877	3,88,850.7	12,93,026.0	4.32
Average	-	-	-	-	14,34,374	3,88,850.7	10,45,523.0	3.69

1 US $ = 46.80 INR as per rate in December 2009

Table 31.15: Crop Average Yield, Expenses and Return from Annual crops in Apple based Agri-Horticulture System (Anjulo 2009)

Crop	Average Yield (q ha^{-1})	Selling Price (Rs. kg^{-1})	Gross Return (Rs. ha^{-1})	Total Expenses (Rs. ha^{-1})	Net Benefit (Rs. ha^{-1})	Benefit Cost Ratio
Wheat	21.00 (39.58)	10	21,000.00 (39,580.00)	97,073.92	-26,073.92 (-57,493.92)	0.22 (0.41)
Pea	88.00 (117.88)	50	4,40,000.00 (5,89,400.00)	1,55,324.00	+2,84,676.00 (4,34,076.00)	2.83 (3.79)
Tomato	736.00 (624.00)	10	7,36,000.00 (6,24,000.00)	2,04,033.92	+5,31,966.08 (4,19,966.08)	3.61 (3.06)
Fresh beans	45.00 (51.50)	50	2,25,000.00 (2,57,500.00)	99,940.00	+1,25,060.00 (1,57,560.00)	2.25 (2.58)

1 US $ = 46.80 INR

Value in parentheses represents the control i.e. sole crops with orchards

HORTI-MEDICINAL AGROFORESTRY SYSTEM

The yield of fruits increased when aromatic and medicinal plants i.e. Tulsi (*Ocimum sanctum*), Ashwagandha (*Withania somnifera*) and Kalmegh (*Andrographis paniculata*) were grown between the rows of high density 5 year old peach plantations in the mid-hill Himalayas. There was a significant increase in peach yield i.e. 7.77, 11.78 and 9.02 per cent on sole peach yield when tulsi, ashwagandha and kalmegh integrated in the peach plantation, respectively. The fruit and biomass yield of aromatic and medicinal plants in tons/ha and percentage increase in yield under different agroforestry components is presented in Table 31.16 and it has been observed that under peach plantation, there was a significant increase of biomass at the rate of 21.06, 46.38 and 26.23 per cent on the sole crop of tulsi, ashwagandha and kalmegh. The biomass of aromatic and medicinal plants below ground was also increased at the rate of 12.12, 21.54 and 13.13 per cent in tulsi, ashwagandha and kalmegh, respectively. The biomass removal and retention of different aromatic and medicinal plants is higher than the sole-crops and it is highest i.e. 48.72 per cent including underground (which is the only material used for medicine) in ashwagandha, 20.99 per cent of tulsi, and 30.14 per cent in kalmegh. Highest peach fruit yield was observed with ashwagandha i.e. 4.46 t ha^{-1} against 3.99 t ha^{-1} in sole crop. The growing of aromatic and medicinal plants under agrihorticulture systems is highly profitable. Verma and Thakur (2010) also recommended the cultivation of *Withania somnifera* (ashwagandha) under peach, *Morus* and *Setaria* based horti-silvi-pastoral based agroforestry systems for high income and productivity.

Table 31.16: Peach Yield, Biomass Production, Removal and Retention levels of different Agroforestry Systems (Tripathi 2012)

AFS Components	Fruit Yield (t ha^{-1})	Above Ground Biomass (t ha^{-1})	Below Ground Biomass (t ha^{-1})	Biomass Removed (t ha^{-1})	Biomass Retention (t ha^{-1})
Peach (Sole)	3.99	13.40	2.98	3.71**	12.67
Tulsi with peach	4.30 (7.77)	2.30 (21.06)	0.37 (12.12)	1.96 (20.99)	0.71 (16.39)
Ashwagandha with peach*	4.46 (11.78)	1.01 (46.38)	0.79 (21.54)	1.16 (48.72)	0.64 (14.29)
Kalmegh with peach	4.35 (9.02)	2.31 (26.23)	0.17 (13.13)	1.77 (30.14)	0.71 (14.57)
Tulsi (Sole)	-	1.90	0.33	1.632	1.61
Ashwagandha (Sole)	-	0.69	0.65	0.78	0.56
Kalmegh (Sole)	-	1.83	0.15	1.36	0.62

Parenthesis represents the percentage increase

*Only the biomass of medicinal and aromatic plants was taken

** Biomass removed by pruning

■■■■

32 MEDICINAL AND AROMATIC PLANTS BASED AGROFORESTRY SYSTEMS

MEDICINAL AND AROMATIC PLANTS: A VALUABLE RESOURCE

All ancient civilization depended on medicinal plants for well being for their human population, livestock and plants. Medicinal plants were used in healthcare either for prevention or cure, or for some rituals and belief. Medicine man of the tribes or village was most revered during old time. These plants collectively called as medicinal and aromatic plants (MAPs). Even some MAP have been historically infused in cuisine of some cultures such as neem (*Azadirachta indica*), osimum and turmeric in Indian, zinger and garlic in Oriental, herbs such as basil, oregano, rosemary, coriander in Mediterr-anean. MAP species is still playing an important role in healthcare, wellness as well as lifestyle of people around the world. World Health Organization (WHO) estimates, about 80 per cent of the world population rely on traditional systems of medicines for primary health care. China, Cuba, India, Sri Lanka, Thailand, and a few other countries have endorsed official use of traditional systems of medicine in their healthcare programmes. These traditional systems mostly incorporate MAPs as a main ingredient.

MAPs listed for ailments in various indigenous systems are as follows: folklore (4,720), Siddha (1,122), Ayurveda (1,773), Amchi (600), Unani (753) and Allopathy (30) (Rabe and Staden 1997; Prajapati et al 2003; Kalaichelvi and Swaminathan 2009). There is an estimate that 12.5 per cent of known plants species documented worldwide are reported to have medicinal value which is evident by use of plants and synthetic analogues built on prototype compounds isolated from plants in modern pharmacopoeia with a global share of 25 per cent (Schippmann et al 2002). About 42 per cent of the sales of the top 25 selling drugs world-wide are either biological, natural products, or entities derived from natural products (Kalaichelvi and Swaminathan 2009). Even Up to 60 per cent of the drugs prescribed in Eastern Europe consist of unmodified or slightly altered higher plant products. Indian subcontinent alone is a vast repository of medicinal plants and traditional communities are using 7,000-7,500 plants for curing different diseases (Dev 1997; Samy and Ignacimuthu 1998; Samy and Ignacimuthu 2000; Kamboj 2000; Ballabh and Chaurasia 2007).

SIGNIFICANCE AND UTILIZATION OF MEDICINAL AND AROMATIC PLANTS FROM FOREST

From ancient time to present, forest has been a primary source of medicinal plant for traditional medicine as well as pharmaceutical industry. Forest provides highly prized medicinal plants in alpine and sub-alpine, temperate, humid and sub-tropical areas throughout the world. In most developing part of the world, medicinal plants are harvested regularly in large quantity and exported legally as well as illegally. Though limited study made in this regards, there are some examples. In Nepal, over 15,000 tons of medicinal herbs representing some 100 species are harvested every year from wild for commercial and industrial purposes (Bhattarai 1997). In China, out of 1,000 commonly used medicinal plants, 80 per cent in terms of number of species and 60 per cent in terms of total quantity come from wild sources (He and Sheng 1997). In sub-Saharan Africa, rural families are increasingly turning to forests for medicinal plants and other NTFPs to cope with HIV/AIDS (Barany et al 2001). In Amazonia, at least 1,300 plant species from forests are being used as medicines, poison or narcotics (Schultes and Hoffman 1979). In India 600-700 species are being used by Indian industries and according to WHO more than 90 per cent of these herbs are being harvested from wild. A survey carried out among companies involved in trade and production of herbal remedies and other botanical products revealed that although companies reported 60 to 90 per cent of material was from cultivation, the number of species cultivated relative to the volume is small (Laird and Pierce 2002). In Germany, out of 1,543 species traded, only 50 to 100 species are exclusively obtained from cultivation (Schippmann et al 2002).

There is a huge demand of MAPs in several industries apart from the local consumption which reflects in export and import database of UNCTAD COMTRADE (Table 32.1). World Bank reports trade in medicinal plants, botanical drug products and raw material is growing at an annual growth rate between 5 to 15 per cent. The herbal industry shares about US$ 62 billion with good growth potential in global pharmaceutical market. In India the value of botanicals related trade is about US$ 10 billion per annum with annual export of US $1.1 billion while China's annual herbal drugs production is worth US$ 48 billion with export of US$ 3.6 billion. The USA is the largest market for Indian botanical products whereas Japan, Hong Kong, Korea and Singapore are the major importer of the herbal drugs from China. Increasing demand for medicinal plants due to population increase in developing countries has been met by indiscriminate harvesting of medicinal plants from wild. As a result many species have become extinct and various other species are endangered. This is a concern for conservation status of MAPs as most of them come from forest. MAP species have been reported to be disappearing rapidly due to habitat destruction, forest degradation and heavy exploitation for pharmaceutical industry coupled with destructive harvesting form wild.

Table 32.1: World's Top Ten Leading Countries of Import and Export of Pharmaceutical Plants (Source: UNCTAD COMTRADE Database, United Nations Statistics Division, New York)

S.No.	Import Details of MAPs			Export Details of MAPs		
	Country of Import	**Quantity (tons)**	**Value (USD)**	**Country of Export**	**Quantity (tons)**	**Value (USD)**
1	Hong Kong*	67,000	291,200,000	China	147000	281,800,000
2	Japan	51,350	136,000,000	Hong Kong*	63150	228,800,000
3	The USA*	49,600	135,500,000	India	33900	566,50,000
4	Germany*	45,350	110,200,000	Germany*	15100	700,50,000
5	Rep. Korea	32,250	52,300,000	The USA*	13500	115,500,000
6	France	21,350	52,000,000	Mexico	13000	11,250,000
7	China	13,650	41,600,000	Egypt	11750	13,850,000
8	Italy	11,700	42,850,000	Chile	11600	28,200,000
9	Pakistan	11,050	11,150,000	Bulgaria	10050	14,500,000
10	Spain	9,100	27,650,000	Singapore	9600	56,600,000

Note: * Main trade centers *Export figures include re-export*

Many species, which are now economically valuable, are becoming endangered due to over-exploitation from the wild. Some examples include '*yohimbe*' (*Pausinystalia yohimbe*, bark is used to treat male impotence) in Central Africa (Sunderland et al 2000), golden seal (*Hydrastis canadensis*) collected from hardwood forests in eastern North America (Hill and Buck 2000) and African cherry (*Prunus africana*, bark used to treat prostatitis) in Cameroon and Madagascar. A survey in the Shinyanga region in northwestern Tanzania indicated that the top ten priority medicinal species that people traditionally have been harvesting from natural woodlands have become scarce and are near extinction (Dery et al 1999). Over-exploitation of these species is leading to unsustainable depletion of natural resources and narrowing of their genetic base. In India, 16 medicinal plants including the highly valued *Atropa acuminata*, *Dioscorea deltoidea* and *Rauvolfia serpentina* are listed as endangered species in the northwestern Himalayas. The situation is grim with regard to high in demand medicinal herbs especially *Picrorhiza kurroa*, *Podophyllum hexandrum*, *Aconitum heterophyllum*, *Valeriana jatamansi*, *Gentiana kurroo*, *Swertia chirayita*, *Hypericum perforatum*, *Nardostachys jatamansi*, *Ephedra gerardiana*, etc.

CONSERVATION OF MAPs IN FARM CULTIVATION

Agroforestry, though being practised from ancient time, now getting focus as it produces agricultural crops as well as other crops or natural products using efficient temporal and spatial adjustment at farm land. Medicinal plants are adapted to different forest ecosystems and these can thrive in agroforestry system either as a tree crop or under the canopy of taller trees. Also, MAP can be cultivated as a pure crop on farm land and wasteland. By including MAP in agroforestry system conservation concerns of the MAP in the forest can be addressed and its production can meet the growing demand of pharmaceutical industry. Although many species are employed in traditional medicine, the number of MAP species currently in formal cultivation for commercial production does not exceed a few hundred worldwide. For example, the species cultivated are only 100 to 250 in China (He and Sheng 1997), just over 100 in India (Prajapati et al 2003; Kalaichelvi and Swaminathan 2009) and about 40 in Hungary (Bernáth and Nemeth 1999). Some cultivated varieties of MAPs in India are given in Table 32.2.

Table 32.2: Improved Varieties of MAPs for Cultivation in India (Hegde et al 2002)

MAP Name	Variety Name	Organization where Developed
Isagbol (*Plantago ovata*)	Gujarat Isagbol-1 Gujarat Isagbol-2	Medicinal and Aromatic Plant Centre, Gujarat
Senna (*Cassia angustifolia*)	Tirunelveli Senna ALFT-2	Tamilnadu Agricultural University, Coimbatore
Opium Poppy (*Papaver somnife[illegible]m*)	Jawahar Apheem-16	JNKV, Bahaduri Research Farm NPBGR, New Delhi
	Trishna (IC-2)	Medicinal and Aromatic Plant Centre, Dept of Agronomy, RAU, Udaipur
	NOP-4 (Kirtiman)	Medicinal and Aromatic Plant Centre, Dept of Horticulture, Faizabad
	NBRI-3	Plant Breeding Lab, Nat. Bot. Res. Inst., Lucknow
	Sweta (GS2)	CIMAP, Lucknow
Khasi Katri (*Solanum viarum*)	Glaxo variety, Arka Sanjeevani	IIHR, Bangalore
Steroidal Yam (*Dioscorea* spp.)	Arka Upkar	IIHR, Bangalore
Sarpagandha (*Rauvolfia serpentina*)	RS 1 J	NKVV, Indore (Madhya Pradesh)

Given the demand for a continuous supply of medicinal plants and accelerated depletion of forest resources, increasing the cultivation of MAP species appears to be an important strategy for meeting the growing demand and relieving harvest pressure on wild populations (Lambert et al 1997). The high price paid for some of the species make them potential new crops (e.g., *Warburgia salutaris, Garcinia kola, G. afzelii* and *G. apunctata* in Africa) that can be grown in sole as well as agroforestry systems (Cunningham 1997). Domestication and cultivation of Medicinal plants will reduce the pressure on depleting natural resources. Cultivation of MAP offers advantages from marketing prospective such as: sustained supply of raw material; availability of reliable and authentic material; better marketing of products by rapport building between growers and wholesalers/companies; controlled post-harvest handling and therefore rigorous quality control and product regulation based on consumer preference. Although cultivation of traditionally used medicinal plants has been suggested for over a decade, there has been little response in many countries. Though some countries like India started works in this regard by developing agronomic practices and improved varieties (Tables 32.2, 32.3 and 32.4). A farmers' cooperative in the northern lowlands of Costa Rica has successfully demonstrated cultivation of the medicinal herb 'raicilla' (*Cephaelis ipecacuanha*) in natural forests for export to the Netherlands and Germany (Hager and Otterstedt 1996). The lack of understanding with respect to cultivation, economics of cultivation and market opportunities are considered the main limiting factors in commercialization.

Table 32.3: Cultivation Practices of some Medicinal Plants (Hegde et al 2002)

Name	Period	Seeds/ha (Kg)	Spacing (cm)	Duration	Productivity (ha)
Sadaphuli (*Catharanthus roseus*)	June - July	2.5	45 x 30	160 days	10-15 q roots 20-25 q herbage
Sarpagandha (Rauvolfia *serpentina*)	June - July	5.5	45 x 30	18 months	6-10 kg seeds 15-20 q roots
Ashwagandha (*Withania somnifera*)	June - July	1.0	75 x 60	18 months	12-15 q roots
Isagbol (*Plantago ovata*)	Nov - Dec	4.0	Broadcast	120 days	10-12 q seeds 10-12 q roots
Senna (*Cassia angustifolia*)	June - July	10.0	60 x 45	150 days	2-3 q seeds

Table 32.4: Cultivation Practices of some Medicinal Trees (Hegde et al 2002)

Name	Spacing	Flowering	Productivity/ Tree (kg)	Life of Trees (years)
Amla	7 x 7	June-July	50-150	-
Hirda	7 x 7	Dec-March	40-50	25-30
Bael	6 x 6, 8 x 8	June-Dec	20-40	25-30
Shikakai	2 x 1, 2 x 2	-	2-4	25-30
Ritha	7 x 7, 10 x 10	Oct	30-40	25-30
Shevga	3 x 3, 4 x 4	-	1,000-3,000 fruits	-

MEDICINAL AND AROMATIC PLANTS BASED AGROFORESTRY SYSTEMS

Medicinal Plants based agroforestry systems also known as silvo-medicinal agroforestry system can provide a wide array of products ranging from food, fodder, fruit, fibre, pulp and medicinal plants for consumption and trade on one hand and conserving the biodiversity by reducing pressure on the natural resources on the other hand. This can provide an opportunity to diversify the existing wheat-paddy rotation. Majority of medicinal plants are found in the forest and are shade tolerant and therefore agroforestry systems offers a convenient strategy for promoting their cultivation and conservation. In, silvo-medicinal system, medicinal trees are grown as overstorey crop in combination with other crops or shade tolerant medicinal plants are integrated as lower strata species in a short cycle in existing forest/plantations/farm trees. Different prevailing medicinal and aromatic plant based agroforestry systems can be grouped into following types:

1. Medicinal and aromatic plants in pure stands/ orchard
2. Medicinal and aromatic plants in intercropping
3. Homestead gardens
4. Traditional system of agroforestry

MEDICINAL AND AROMATIC PLANTS IN PURE STANDS/ORCHARD

Many plants in traditional agricultural systems in the tropics have medicinal value. These can be found (either planted or carefully tended natural regenerations) as scattered trees in croplands and grazing lands, and on field bunds (Table 32.5). Many *Acacia* species found in Africa such as *Acacia nilotica*, *A. seyal*, *A. senegal* and *A. polyacantha*, as well as several species in African croplands (e.g. *Faidherbia albida*, *Vitellaria paradoxa*, *Adansonia digitata*, *Markhamia lutea* and *Melia volkensii*) have medicinal value (ICRAF 1992). Similarly, 'arjun' (*Terminalia arjuna*) in India, chinaberry (*Melia azedarach*) in Asia and *Erythrina* species in Latin and Central America combine many uses including medicinal. A number of plants used as live fences around home compounds such as henna (*Lawsonia inermis*) in India (Singh et al 1996), *Ipomoea carnea* ssp. *fistulosa* in Bolivia and Asian countries (Frey 1995) and *Euphorbia tirucalli* around crop fields in Africa (ICRAF 1992) have medicinal values. Many trees yielding aromatic products and essential oils can be established as pure stands in similar way to medicinal trees.

MEDICINAL AND AROMATIC PLANTS AS OVERSTOREY TREES

Medicinal tree species that grow tall and develop open crowns can also be used for this purpose. However, research needs to identify the medicinal trees that can be grown in association with the plantation crops and develop management practices for them. In Ivory Coast, 19 of the 41 tree species planted as shade trees in coffee and cacao provide pharmaceutical products for traditional medicine. Tall and perennial medicinal trees that need wider spacing such as *Prunus africana*, *Eucalyptus globulus* (for oil), sandalwood (*Santalum album),* ashok (*Saraca indica*), bael (*Aegle marmelos*), custard apple (*Annona squamosa*), amla (*Emblica officinalis*), drumstick or moringa (*Moringa oleifera*), *Azadirachta indica*, *Terminalia chebula*, *Terminalia arjuna* and soapnut tree (*Sapindus mukorossi*) can be intercropped with annual crops in the early years until the tree canopy closes. Some of the medicinal trees may allow intercropping for many years or on a permanent basis, depending on the spacing and nature of the trees. The intercrops generate income for farmers when the main trees have not started production.

Table 32.5: Medicinal Trees in Agroforestry Systems of the Tropics

Tree Species	Common Name	Agroforestry System	Medicinal Use of Tree species
Acacia nilotica	Babul	Field bunds, scattered trees in crop-lands woodlots and grazing lands in Africa and India	Gum for treating diarrhea dysentery, diabetes, sore throat, bark to stop external bleeding
Azadirachta indica	Neem	Woodlots, scattered trees, shelterbelts in Africa and India	Digestive disorders, malaria, fever, hemorrhoids, hepatitis, measles, syphilis, boils, burns, snakebite, rheumatism
Erythrina variegata	Coral tree	Shade tree in coffee, live fence	Bark useful to treat stomach ache and applied to swellings; juice to stimulate lactation; warm poultice of leaves to relieve rheumatic joints
Parkia biglobosa	Tree bean, locust bean	Parklands in West Africa	Piles, malaria, stomach disorders, jaundice
Prosopis cineraria	Khejri	Scattered trees in croplands in arid to semi-arid areas mostly in northwestern India	Flowers for blood purification and curing skin diseases, bark against summer boils, seprosy, dysentery, bronchitis, asthma, leucoderma and piles
Tamarindus indica	Tamarind	Field bunds and scattered trees in crop-lands in semi-arid India and Africa	Fruit pulp as refrigerant, carminative and laxative. It is also recommended in febrile diseases and bilious disorders.

MEDICINAL AND AROMATIC PLANTS AS INTERCROPS

MEDICINAL INTERCROPS

Medicinal plants growing in forests often require or tolerate partial shade, moist soils (high in organic matter), high relative humidity and mild temperatures. The cultivation of MAPs can be undertaken in thinned forests and cleared forest patches, or as intercrops in new forest plantations, timber and fuelwood plantations, fruit trees and plantation crops. The trees may benefit from the inputs and management given to the intercrops. Short stature and short cycle medicinal plants and culinary herbs are particularly suited for short-term intercropping during the juvenile phase of trees. Some well known medicinal plants such as safed musli (*Chlorophytum borivilianum*), rauvolfia (*Rauvolfia serpentina*), turmeric (*Curcuma longa*), wild turmeric (*C. aromatica*), *Curculigo orchioides*, and ginger (*Zingiber officinale*), have been successfully intercropped with fuelwood trees (e.g., *Acacia auriculiformis*, *Albizia lebbeck*, *Eucalyptus tereticornis*, *Gmelina arborea*, and *Leucaena leucocephala*) (Chadhar and Sharma 1996; Mishra and Pandey 1998; Prajapati et al 2003). Only 10 out of 64 herbaceous medicinal plants tried in intercropping with two-year old poplar (*Populus deltoides*) spaced 5 m apart gave poor performance (Kumar and Gupta 2004), indicating that many medicinal plants can be grown in agroforestry systems. Wherever markets are established, medicinal plants are remunerative alternative intercrops to the traditional agricultural crops. Intercropping of medicinal plants in coconut (*Cocos nucifera*) and arecanut (*Areca catechu*) stands is an age-old practice in India and other parts of south Asia and south-east Asia. These palms allow 30 to 50 per cent of incident light to the underneath, which is ideal for some medicinal plants, including cardamom (*Elettaria cardamomum*).

Kacholam or galang (*Kaempferia galanga*) – a medicinal herb – is traditionally intercropped in mature coconut gardens in Kerala, India. Kacholam intercropped in a 30 year-old coconut plantation produced 6.1 Mg ha^{-1} of rhizomes compared with 4.8 Mg ha^{-1} as a sole crop (Maheswarappa et al 1998). Twelve year old coconut trees did not adver-sely affect the growth and yields of a number of medicinal species grown as intercrops compared to the yields in the open (Nair et al 1989). In coconut based multistoried agroforestry stystems, coconut is grown as overstorey crop and in lower strata different fruit trees and medicinal plants are grown as understorey crops. Coconut plants are planted at a spacing of 8m × 8m, fruit trees like guava and lemon are grown at a row spacing of 3m × 3m. Medicinal plants viz. *Aloe vera*, *Asparagus racemosus* and *Kaemp-feria angustifolia* are grown as intercrops which provide additional income. In Karnataka and Kerala states, India, *Areca catechu* (arecanut palm) is commonly intercropped with ginger, turmeric, black pepper (*Piper nigrum*) and cardamom (Korikanthimath and Hegde 1994). Some of these intercrops may cause small reduction in arecanut yields but the combined returns from both the components are greater than from arecanut alone. Another plantation crop intercropped with medicinal plants is rubber (*Hevea brasili-ensis*), for example with *Dioscorea floribunda* in the state of Assam in India (Singh et al 1998) and with *Amomum villosum* in Yunnan province of China (Zhou 1993).

In India, cultivation of medicinal plants in association with forest trees species have been evaluated and commercial cultivation in agroforestry systems have been recommended by several workers. In *Pinus roxburghii* plantations, *Andrographis paniculata*, *Spilanthes acmella*, *Mucuna pruriens* and *Solanum khasianum* can be grown to provide additional income. Similarly in teak stands, *Aloe vera*, Lemon grass, *Coleus*, Citronella, Palmrosa, Patachuli, Stevia can be grown and these crops provide yield of 85-100 per cent relative yield to pure crop. Some of the medicinal plants which can be included in agroforesty system in Northern India are presented in Table 32.6.

Table 32.6: Medicinal Plants suitable for Agroforestry Systems in Northern India

Medicinal plant	Common Name	Part Used	Yield (Kg/ha)	Market price (Rs/kg)
Aloe vera	Ghee Kaur	Leaves	390-410	2 (leaves)
Anethum graveolens	Sova	Leaves	9.8-12	1500 (oil)
Cassia angustifolia	Senna	Leaves	15-20	15-20 (leaves)
Catharanthus roseus	Sadabahar	Root, Leaves	20-22	18-20 (leaves)
Ocimum sanctum	Ban Tulsi	Whole plant	25-30	180 (dried leaves)

Bari and Rahim (2012) reported that aloe vera (*Aloe indica*), asparagus (*Asparagus racemosus*), misridana (*Kaempferia angustifolia*) in coconut based multistoried agrofore-stry system (MAF) was raised in Bangladesh. Though yield of aloe vera was low, the gross returns were highest under coconut based MAF systems compared to sole cropping. It was due to the substantial additional contribution by coconut, guava and lemon fruits in MAF, while extra benefit was absent under sole cropping condition. In China, the cultivation of medicinal plants is an old practice under the name of "silvo-medicinal" systems. In northeast China, ginseng (*Panax ginseng*) and other medicinal plants are grown under pine (*Pinus* spp.) and spruce (*Picea* spp.) forests. In central China, many medicinal plants are planted with *Paulownia tomentosa*. In southern China, medicinal herbs are often planted under bamboo and *Cunninghamia lanceolata* (Chinese fir) forests.

American ginseng (*Panax quinquefolius*), a medicinal herb exported to China from the United States and Canada is grown as an understorey in red maple (*Acer rubrum*) forests (Nadeau et al 1999) or deciduous hardwoods such as black walnut (*Juglans nigra*) and sugar maple (*Acer saccharum*), instead of growing under artificial shade with considerable expense (Hill and Buck 2000). Indeed, cultivation of ginseng and several other medicinal plants in the forests is a common and growing form of forest-farming practice of agroforestry in North America (Table 32.7). Light demanding understorey species (e.g., *Echinacea* spp.) may be intercropped initially to provide early returns from plantations and after canopy closure, shade-tolerant species such as ginseng and goldenseal can be intercropped (Teel and Buck 2002). Studies in New Zealand have indicated that the American ginseng can be successfully grown under *Pinus radiata* with best growth under a tree stand of 130 stems ha^{-1} (Follett 1997). In addition to providing shade, the trees may also benefit the understorey component from hydraulically lifted water. Gupta (1986) listed a number of indigenous understorey herbs and shrubs that can be produced as part of forest farming or in new forest plantations to improve economic returns from forests in India (Table 32.7).

Table 32.7: Commercially Valuable Medicinal Plants as Understorey Component in Forest and Tree Plantations

Species Name	Common Name	Plant Type	Part Used	Medicinal Use	Location
Aconitum heterophyllum	Atis	Tall herb	Rhizome	Hysteria, throat disease, astringent	Alpine and sub-alpine Himalayas
Amomum subulatum	Large c ardamom	Perennial herb	Seeds	Stimulant, indigestion, vomiting, rectal disease	Sub-Himalayan rage, Nepal, Bhutan
Amomum villosum	Saren	Perennial herb	Seeds	Gastric and digestive disorders	China
Caulophyllum thalictroides	Blue cohosh	Perennial herb	Roots	Gynecological problems, bronchitis	North America
Cimicifuga racemosa	Black cohosh	Perennial herb	Roots	Menses related problems	North America
Chlorophytum borivilianum	Safed musli	Annual herb	Tubers	Male impotency, general weakness	India
Costus speciosus	Crepe ginger	Cane	Leave, stem rhizome	Purgative, depurative and as a tonic	India
Dioscorea deltoidea	Himalayan yam	Vine	Tubers	Source of saponins and steroids	India, Pakistan
Echinacea purpurea	Coneflower	Perennial herb	Roots, Rhizome	Enhancing immune system	North America
Hydrastis canadensis	Goldenseal	Perennial herb	Rhizome	Tonic	North America
Panax ginseng	Ginseng	Herb	Roots	Tonic	China, Korea, Japan
Panax quinquefolius	American ginseng	Perennial herb	Roots	Tonic	North America
Cephaelis ipecacuanha	Raicilla, Ipecae	Shrub	Roots	Whooping cough, bron-chial asthma, amoebic dysentery	Brazil, India, Bangladesh, Indonesia
Rauvolfia serpentina	Rauvolfia	Shrub	Roots	Hypertension, and certain form of insanity	Sub-montane zone India
Serenoa repens	Saw palmetto	Shruby palm	Fruits	Swelling of prostrate gland	Southeastern USA

AROMATIC INTERCROPS

Studies have been conducted in India since mid-1980s on the feasibility and economic aspects of intercropping aromatic plants with timber trees. Experiments at different sites in India demonstrated that aromatic species such as *Mentha* spp. (*M. arvensis*, *M. piperita*, *M. citrata* var. *citrata*, *M. spicata*, *M. cardiaca* and *M. gracilis*) and *Cymbopogon* spp. [lemon grass (*C. flexuosus*); Java citronella (*C. winterianus*); and palmarosa (*C. martinii*)] can be grown intercropped with *Populus deltoides* or *Eucalyptus* spp. for three to five years after planting the trees (Table 32.8).

Some of the above aromatic species and *Pelargonium* spp. (geranium) can also be intercropped with another essential-oil-yielding tree *Eucalyptus citriodora* (lemon- or citron-scented gum) (Singh et al 1998). Tree growth was generally little affected or improved probably due to the inputs given to the intercrops. Herb growth and oil yields of these aromatic plants were not affected during the first two years of intercropping, but they decreased slightly from the third year compared with sole crops (Singh et al 1990). Essential oil profiles of citronella and palmarosa grown in open and partial shade were not significantly different, but the oil of menthol mint (*M. arvensis*) produced in the open was richer in menthone while that produced under shade was richer in menthol (Singh et al 2002). This implies the need to consider the effect of intercropping on oil quality of some aromatic plants. These intercropping systems gave two to four times greater economic returns than sole cropping of the components. These studies indicate that aromatic plants can be profitably grown in association with linearly growing fuel and timber trees such as *Leucaena leucocephala*, *Casuarina* spp., and *Grevillea robusta*.

Table 32.8: Intercropping of Aromatic Plants with Trees

Location	Trees species	Spacing	Aromatic plant	Intercrop Yield as per cent of Sole Crop		Comments
				Herbage Yield	Oil Yield	
Pantnagar, India	*Populus deltoides*	5m × 4m	*Cymbopogon flexuosus*	82	82	Results averaged over 5 years
				96	86	
		2.5 m×2.5 m	*C. martini*	96	90	
	Eucalyptus hybrid		*C. flexuosus*	100	86	
			C. martinii	90	90	
Pantnagar India	*Populus deltoides*	5m × 4m	*Mentha arvensis*	85	90	Results averaged over 3 years
			M. piperita	85	88	
			M. citrata	90	90	
			M. spicata	87	87	
			C. flexuosus	99	97	
			C. martini	99	98	
			C. winterianus	98	99	
Hyderabad India	*Eucalyptus citriodora*	3m × 2m	*C. flexuosus*	87	87	Results averaged over 2 years
			C. martini	87	87	
			C. winterianus	112	91	
			Pelargonium spp.	91	90	
Nepal	*Dalbergia sissoo*	10m × 2-8m	*C. winterianus*	-	100	Results averaged over 2 years

MAPs IN HOMESTEAD GARDENS

Homestead or homegardens are complex agroforestry systems involving many plant species, characterized by different morphology, stature, biological function and utility. They are mostly practised in the humid and sub-humid tropics. Homegardens of individual holdings generally cover small parcels of land and are established around homesteads. Medicinal plants are an invariable component of homegardens, whether they are in the Peruvian Amazon (Lamont et al 1999), on the slopes of the Mt. Kilimanjaro in Tanzania (O'Kting'ati et al 1984), or in the humid and semi-arid Cuba (Wezel and Bender 2003). Holy Basil or '*tulsi*' (*Ocimum sanctum*), drum stick (*Moringa oleifera*) and curry leaf (*Murraya koenigii*) are backyard plants in many Indian households and they are routinely used for common ailments or in food preparations. In the Soqotra Island, Yemen, endemic medicinal plants such as *Aloe perryi*, *Jatropha unicostata* and *Commiphora ornifolia* are cultivated in homegardens (Ceccolini 2002). In the past, these systems were mostly used to meet the household needs of small-scale farmers in the forest margins. Although in recent times, increased urbanization, transport and market opportunities have helped stimulate the production of cash-crops from homegardens. Food, fruit and timber species often dominate homegardens and occupy the middle and upper strata, while medicinal plants, spices and vegetables occupy the lower strata. Following three medicinal plant categories are recognized in homegardens:

- Species used exclusively for medicine
- Horticultural or timber species with complementary medicinal value
- Weedy medicinal species

While the first two categories are deliberately planted, the latter group grows spontaneously. The species composition, plant density and level of management vary considerably depending on the soil, climate, market opportunities and cultural background of the people. Multistrata systems involve fewer species (three to 10 species) than in home gardens in definite planting arrangement and can be designed for home consumption as well as commercial production. Agelet et al (2000) made a detailed analysis of medicinal plants found in 155 homegardens in the mountain zones of Catalonia (north-eastern Iberian Peninsula, Spain). The gardens contained nine distinct categories of species: plants exclusively cultivated for medicinal purpose (23) mostly close to the house; the medicinal wild plants favoured by homegarden structure and care (105); and seven kinds of horticultural plants with complementary medicinal values (117). There was, however, loss of about 56 taxa or 23 per cent of the total over the years. The species composition differs depending on cultural background, distance from markets and influence of tourism. Medicinal plants accounted for about 27 per cent of total plant species in the homegardens in Amazon (Padoch and de Jong 1991), 56 per cent in northern Catalonia (Iberian Peninsula) (Agelet et al 2000) and 45 per cent in the floodplains of the river Jamuna in Bangladesh (Yoshino and Ando 1999).

MAPs IN TRADITIONAL AGROFORESTRY PRACTICES

Tribal populations have tremendous traditional knowledge regarding use of natural resources. Nyishis tribes of Arunachal Pradesh, India traditionally follow multistoried agroforestry system which includes about 80 species of ethno-medicinal importance belonging to 69 genera and 45 species. Some of the important plants of medicinal use are *Areca catechu*, *Ageratum conyzoides*, *Alstonia scholaris*, *Artemisia nilagirica*, *Bauhinia variegata*, *Cassia alata*, *Centella asiatica*, *Clerodendrum viscosum*, *Erythrina indica*, *Moringa oleifera*, *Piper betel*, *Solanum khasianum*, *Spilanthes acmella*, *Terminalia myriocarpa*. Similarly in Mizoram state of India, a total number of 54 medicinal plants species belonging to 52 general and 35 families have been recorded from different agroforestry system. Nine plants from homegardens and agroforestry systems were identified as new from the region. Out of 54 plants recorded, 14 (7 as Nearly Threatened; 3 as Vulnerable; 3 as Critically Endangered and 1 as Endangered) were of threatened status. *Aeschynanthus sikkimensis*, *Melia azedarach* and *Artemisia indica* are used in fever while *Andrographis paniculata*, *Asparagus racemosus*, *Aquilaria malaccensis*, *Curcuma caesia*, *Emblica officinalis*, *Garcinia cowa*, *Jasminum nervosum*, *Mesua ferrea*, *Oroxylum indicum* and *Terminalia chebula* in diarrhoea and digestive system problems. *Blumea lanceolaria* is used as anticancerous while *Catharanthus roseus* was used in high blood pressure. *Clerodendrum colebrookianum* is used in hypertension while *Elaeagnus caudata* and *Saraca asoca* have got gynecological applications. *Catharanthus roseus* and *Cinnamomum tamala* are used against the diabetes by tribal people. In Yunnan province, China, traditional 'Dai and Jinuo' agroforestry systems involve the medicinal crop *Amomum villosum* in the forest areas cleared of undergrowth and forest is thinned to permit 30 to 40 per cent shade. Seedlings or cuttings are planted, which produce an average dried fruit yield of 375 kg ha^{-1} per annum (Saint-Pierre 1991; Zhou 1993).

In the Taungya system (slash and burn cultivation), newly established forest plantations can be intercropped with medicinal and aromatic plants similar to food crops until the trees cover the ground. The participation of the local people with right to share benefits of the plantations, especially ownership to crops, has helped governments to establish and protect large-scale tree plantations without conflict with the local people in many Asian countries. The same approach can be employed for the cultivation of medicinal plants in the new plantations. In the rehabilitation of degraded forest lands, participatory planning and implementation with local communities and economic benefits from an early stage onwards will ensure commitment of the people (Rao et al 2004).

CONCLUSION

MAPs indeed have an important role in people's healthcare and wellness in developing as well as developed world. The use of plant products as nutrition supplements and in the cosmetic and perfume industry has increased the value of medicinal and aromatic plants in recent years. People around world are now insisting on natural way of life and consequently demand of MAP is increasing. This is posing some problem of existence of MAP in wild. Its cultivation may be some answer for sustainable supply and also help to preserve inter and intra specific diversity. To support MAP species conservation in the wild and there is need of a certification system for cultivated MAP species so that its cultivation should be promoted. Apart from growing as a sole crop on agricultural farm, alternative land use system such as agroforestry can be an alternative to cultivate it. Several threatened medicinal plants are recorded from homegarden and other traditional agroforestry systems. MAPs are well adapted in forest ecosystems for their requirement for light, temperature, moisture, soil characteristics, etc. so they can be accommodated in many tier of agroforestry system. MAP intercropping with new plantations of coffee, tea, banana, coconut, palm, rubber, apple, mango and cacao offers scope for cultivation of forest medicinal trees that are in demand. It is imperative that this effort should be supported by research on agronomic practices of MAP species and their improvement in traditional, existing as well as new agroforestry system.

■■■■

33 FODDER PRODUCTION FROM AGROFORESTRY TREES

INTRODUCTION

Agroforestry is the only alternative land use method for the production of fodder and pasture. It involves the use of fodder trees that are permanent, and which will replace annual grasses; trees which will provide their own nitrogenous fertilizers; trees which will extend their roots deep beneath the ground to the water table. Agroforestry is the land use system with multipurpose trees that give high feed yields all year round, every year, even where grasses cannot grow. Many rural farmers in the world keep livestock, particularly cattle, buffalo, goats and sheep for various animal products. Livestock provide milk, meat, leather or wool and manure to fertilize agriculture crops. Through selling some of the animals or their products farmers also earn cash money. Farmers expect quality produce from their animals. Therefore the animals need quality feed and water in order to live, grow, work or produce milk. When animal are fed better be in good condition and the farmers benefit best from their animals.

Feeds, like leaves and grass, contain water and dry matter (DM). The dry matter contains energy (carbohydrates), protein and other substances like minerals. Even when a cow, goat or sheep is not producing milk, it needs energy to breathe, walk and maintain all kinds of body processes, and it needs protein to grow. This basic need of feed which is necessary just to maintain a stable condition of the body is called maintenance requirements. For young animals to grow fast or animals to produce high milk yields, extra energy and protein is needed above the daily maintenance requirements. However, due to dry periods and cash problems, feed shortage is often experienced and feeds of low quality are used. The farm animals are not able to get enough energy and protein to grow (fast) or to produce (extra) milk. Sometimes animals are even fed less then the daily maintenance requirements and as a result will be in a bad condition. Fodder trees and shrubs become then important as a source of energy and protein to keep the animal's body healthy, improve growth rates and even increase milk and wool production.

SIGNIFICANCE OF TREE FODDER

The reasons for growing trees as a fodder source are (Andevan 2004):

- *Source of Protein*: Protein is required for growth and milk production in the daily feeding ration. But livestock is often fed on low-quality roughage or grasses, such as napier grass, which contain low amounts of protein. Therefore it is recommended to supplement the basic diet of dairy cows and cattle wiih concentrates (dairy meal), which contain enough protein (about 160 gram in 1 kg). Farmers lack cash to buy concentrates, and they are not satisfied with the nutritive value, and it is difficult or costly to transport it, to their homes. Tree fodders are rich in proteins and are easily collected from trees.

- *Dry Season Supplement*: Because of the deep rooting system, trees and shrubs remain green for a longer period, because they can tap water beyond the reach of grass roots. Fodder trees remain green and can provide the required energy and protein when other sources such as grasses are dry.
- *Other Uses*: Tree species enhance soil fertility by providing mulch or by fixing nitrogen or they also provide timber, firewood, fruit, fibre and shelter.

Table 33.1: Chemical Composition of Important Fodder Tree Species (% of Dry Weight basis)

Tree Species	Crude Protein	Ether Extract	Crude Fibre	N-free Extract	Total Ash	Phos-phorus	Calc-ium	NDF	References
Acacia auriculiformis	13.10	3.29	25.23	51.95	6.43	-	-	-	Datt et al 2008
Acacia catechu	18.2	-	34.32	-	-	-	-	53.62	Negi et al 2003
Acacia leucophloea	14.81	1.92	24.49	47.97	-	0.15	2.28	-	Rai and Mishra 2003
Acacia nilotica	16.9	5.5	23.9	-	6.6	-	-	33.1	Dey et al 2006
Acacia tortilis	18.1	-	-	-	-	0.2	-	-	Debroy et al 1980
Ailanthus excelsa	19.87	3.53	12.82	51.81	11.97	-	2.11	-	Singh and Patnayak 1977
Albizia lebbeck	16.81	2.85-4.68	26.47-37.52	35.99-42.76	7.11-11.54	0.14-0.25.	1.10-2.71	-	Pal et al 1979
Albizia procera	19.0	3.58	29.41	42.29	5.75	-	-	-	Datt et al 2008
Anogeissus latifolia	8.30	2.9	24.2	-	8.40	2.0	22.0	49.0	Bakshi et al 2011
Anogeissus pendula	10.01	-	19.29	-	5.42	-	-	29.13	Rai et al 2006c
Artocarpus heterophyllus	15.9	3.9	32.5	-	11.1	-	-	36.3	Dey et al 2006
Azadirachta indica	17.04	2.74	29.41	46.27	8.44	-	-	-	Datt et al 2008
Bauhinia purpurea	15.7	-	-	-	-	0.18	-	-	Debroy et al 1980
Bauhinia variegata	15.7	-	-	-	-	0.18	-	-	Debroy et al 1980
Butea monosperma	15.02	-	35.6	-	11.08	-	-	50.61	Ramana et al 2000
Capparis horrida	16.87	-	22.39	-	14.14	-	-	38.94	Palian et al 2004
Celtis australis	10.7	-	-	-	4.90	-	-	44.00	Sultan et al 2008
Dalbergia.sissoo	13.09	3.35	20.67	53.81	9.08	-	2.24	-	Datt et al 2008
Dichrostachys cinerea	13.1	-	28.2	-	-	-	-	32.7	Rai et al 2006
Emblica officinalis	17.3	19.94	50.67	41.39	3.28	0.24	3.28	-	Gupta 1993
Ficus benghalensis	10.9	3.9	36.9	-	14.8	-	-	42.5	Dey et al 2006
Ficus glomerata	15.2	4.6	30.9	-	11.7	-	-	39.3	Dey et al 2006
Flacourtia indica	8.9	-	34.28	-	-	-	-	50.26	Negi et al 2003
Gliricidia sepium	22.3	4.2	19.70		7.10	2.3	11.9	49.10	Markel et al 1998
Grewia flavescens	16.1	-	36.2	-	-	-	-	50.68	Negi et al 2003

Tree Species	Crude Protein	Ether Extract	Crude Fibre	N-free Extract	Total Ash	Phos-phorus	Calc-ium	NDF	References
Grewia oppositifolia	22.5	4.4	34.6	-	9.6	-	-	57.9	Dey et al 2006
Leucaena leucocephala	23.33	1.92	12.83	55.77	6.15	-	-	-	Datt et al 2008
Madhuca indica	8.8	1.3	30.4	52.43	7.79	0.16	0.22	-	Kehar et al 1959
Maytenus emarginata	16.6	-	35.42	-	11.07	-	-	41.95	Palian et al 2004
Melia azedarach	29.25	-	15.47	-	8.78	-	-	21.81	Ramana et al 2000
Mimosa hamata	13.69	-	15.23	-	11.07	-	-	34.92	Palian et al 2004
Moringa oleifera	25.0	10.6	7.9	-	8.40	-	-	-	Richter et al 2003
Morus alba	17.49	2.61	12.41	56.04	11.45	-	-	24.4	Dey et al 2006
Morus rubra	25.89	-	15.67	-	15.74	-	-	22.8	Ramana et al 2000
Pongamia pinnata	19.36	-	30.85	-	9.19	-	-	52.59	Datt et al 2008
Populus spp.	10-11.3	-	-	-	8.5-10.5	-	-	30.5-32.5	Azim et al 2001
Prosopis cineraria	12.31	1.60	17.71	52.33	16.00	0.36	2.78	58.70	Khirwar et al 2003
Prosopis chilensis	18.30	1.82	21.79	39.94	18.15	1.20	4.86	64.18	Khirwar et al 2003
Quereus incana	9.5	4.9	44	-	4	-		55.1	Dey et al 2006
Robinia pseudoacacia	14.5-23.5	-	-	-	6.5-9.8		-	37.7-44.2	Azim et al 2001
Salix babylonica	9.8-12.2	-	-	-	7.70-10.9	-	-	33.9-34.5	Azim et al 2001
Sesbania grandiflora	33.4	-	-	-	-	0.34	-	-	Deb Roy et al 1980
Sesbania sesban	25.6	6.2	21.7	45.2	10.8	-	-	32.2	Muthu kumar et al 2005
Shorea robusta	10.5	-	-	-	-	1.42	0.19	-	Osti et al 2006
Tamarindus indica	13.87	4.27	21.99	52.9	6.97	0.24	1.78	-	Momin and Ray 1943
Tectona grandis	10.33	2.92	24.49	53.00	8.26	-	-	-	Datt et al 2008
Terminalia bellerica	11.15	-	-	-	-	3.35	0.41	-	Osti et al 2006
Ziziphus nummularia	13.1	-	40.27	-	-	-	-	52.42	Negi et al 2003
Ziziphus oenoplia	13.52	-	18.92	-	16.15	-	-	39.33	Palian et al 2004
Ziziphus xylopyrus	13.2	-	36.07	-	-	-	-	53.15	Negi et al 2003
Quercus leucotrichophora	10.2	4.8	31.3	-	5.2	1.5	9.9	57.0	Sen 1938
Quercus glauca	9.6	4.1	29.0	-	7.6	2.3	18.7	-	Sen 1938
Quercus floribunda	96	4.5	29.1	-	5.1	3.0	16.1	-	Sen 1938

SELECTION OF TREES FOR FODDER PURPOSE

Not all types of trees can be used for fodder production. The following characteristics should be used to select trees for fodder production (Abel et al 1997):

- Leaves should have a high nutritive value, which means that they should contain a more of protein
- Trees should produce huge quantity of leaves
- Tree should withstand lopping, pruning, pollarding, and coppicing
- Tree should be fast growing and regrow easily after frequent pruning
- Tree leaves should not contain toxins (beyond the permissible limit)
- Tree leaves need to have a high palatability and digestibility
- Trees must preferably be tolerant to drought, resistant to pests and diseases
- Trees should not compete too much with agriculture crops. For example, they should form deep roots in order to avoid competition with shallow rooted crops for water

FODDER TREES PLANTING

The following points should be taken into consideration before choosing number of trees to be planted for fodder purpose:

- Type and number of animals to be kept at the farm and amount of feed do they need
- Type of feed that is available and quantity of feed that is in short during different season
- Quantity of fodder tree leaves required to reduce the feed shortage
- The expected leaf production of a fodder tree

Once it is known about the amount of fodder tree leaves required, and the yield of different types of trees, it is possible to calculate the total number of trees to be planted on a farm. The number of trees to be planted is now calculated in the following steps:

- First, the quantity of fresh leaves needed for one day are multiplied by the total number of days (365) in one year to get the quantity of leaves, which is required for a whole year.
- The annual fresh leaf requirements are then divided by the minimum 'annual fresh leaf production' of the tree species, which the farmer likes to plant. It is also divided by the maximum production per tree. The outcome is the minimum and maximum number of trees to be planted at the farm to feed the number of animals.
- At a 'double zig-zag spacing' of 0.15 m, the length of the hedge is also calculated by multiplying the spacing with the minimum and maximum number of trees.

In the above-mentioned method for calculating the required number of trees to feed the animals all year round, it is assumed that the trees produce leaves continuously. In fact leaf production will fluctuate during the year. In dry periods the production will be less than in the rainy season. The animals also do not produce to their full capacity year round and thus do not always need to be supplemented. In areas where soils are poorer and rainfall is less than normal, the maximum number of trees needs to be planted. Addition of manure will increase the yields of the field crops. Also it should be kept in mind to plant different fodder tree species. Few trees should be grown for seed production to meet requirement in future. These trees should be left unpruned and collect the mature pods before the seeds drop. The seeds can also be sold to earn some cash (Andrevan 2004).

ROLE OF NITROGEN FIXING TREES

Nitrogen fixing trees grown in the agroforestry systems are capable of fixing about 50-100 Kg N/ha/year. Experience in Punjab, Haryana, Uttar Pradesh, Gujarat and some parts of the southern states indicate that tree combining with agriculture crop production system is more productive. Total production and value of fuel, fodder and small timber in degraded lands are reported to be many times more than the coarse grains usually produced on them. Appropriate agroforestry systems improve soils physical properties, maintain soil organic matter and promote nutrient cycling. Nitrogen fixing trees are mentioned as one of the most promising component of agroforestry system. The leaf litter after decomposition forms humus, releases nutrients and improves various soil properties and it also reduces the fertilizer needs (Roshetko and Gutteridge 1996).

TREE BASED AGROFORESTRY SYSTEMS FOR FODDER PRODUCTION

The different tree based agroforestry systems for fodder production are as follows (Rai and Ajit Sharma 2007):

AGRISILVICULTURE

Agrisilviculture is a production technique which combines the growing of agricultural crops with simultaneously raised and protected forest crops. In this system, trees are integrated with crops (food grains, oil seeds, fibres, pulses, fodder crops, etc.). This system can be adopted in irrigated and rainfed conditions. This practice of agroforestry has been in existence in various primitive forms since man learned to clear forests and cultivate land, and has different names in different parts of the world.

BOUNDARY PLANTATION

In this system, trees are planted on the boundaries of the field to get green leaf fodder during lean period without much affecting the yield of grain crops.

ALLEY CROPPING

Alley cropping is the planting of trees in two or more sets of single or multiple rows with agronomic, horticultural, or forage crops cultivated in the alleys between the rows of woody plants. It is used synonymous with the term hedgerow intercropping. In this system, fast growing and coppicing trees are planted in a single or multiple rows making a dense hedge rows effective in reducing runoff, increasing infiltration and reducing soil loss through barrier effect. This system is mostly used in high rainfall (humid and sub-humid) areas especially slope lands to protect soil from erosion and also leaf fodder and fuelwood are obtained through lopping frequently.

AGRISILVIHORTICULTURAL SYSTEM

In this system, between two plants of fruit trees one or two fast growing forest tree component (fodder trees) is introduced to utilize the wider spaces more efficiently without affecting the growth and yield of fruit trees.

AGRISILVIPASTORAL SYSTEM

Under this system, trees are integrated with grain crops, pasture and animals. This system provides opportunity in arable high land, humid tropics for mixed tree crop livestock husbandry. The crops, trees and grasses are grown on the same unit of land simultaneously which provide food and fuelwood to human being and feed and fodder to the livestock.

HORTIPASTORAL SYSTEM

In this system fruit tree species are integrated with livestock as they are usually planted on a wide spacing, which offers good light penetration especially in young crops. The shade level is regulated by lopping of trees, which are fed to livestock. The major objective of this system is to yield fruits. However, tree leaves are obtained as by-product to feed the animals.

SILVIPASTORAL SYSTEM

The term silvipasture denotes "silvi" means trees and "pasture" means grasses or grass + legume mixture. But in real sense it is an ideal combination of grasses, legumes and trees for optimizing land productivity, conserving plants, soils and nutrients and producing forage, fuelwood and timber on sustainable basis. Thus, this involves replantation, substitution or intervention in existing vegetation by desirable species. Actually silvipastoral systems are land use systems in which trees are combined with livestock and pasture for forage and fuelwood on the same unit of land.

HARVESTING OF TREE FODDER IN AGROFORESTRY SYSTEMS

HARVESTING TECHNIQUES OF TREES

The following techniques can be used for harvesting leaves from the fodder trees (James 1994):

- *Pruning and Lopping*: These are methods in which the side branches of a tree are cut. Pruning is different from lopping in that the branches are cut from the base. Lopping is not always done starting from the lower part, but can be done more haphazardly.
- *Pollarding*: If all branches and the crown of a tree are cut off but the whole stem is left (about 2 meter from ground), this is known as pollarding.
- *Coppicing*: Many species of trees and shrubs have the ability to resprout after the whole tree has been cut leaving 20-30 cm stem (stool). If this ability is used for regeneration of the tree, the practice is known as coppicing.

FIRST PRUNING OF TREES

Fodder trees can only be pruned, when the thickness (diameter) of the stem at 50 cm above the ground level is 8-10 cm. The altitude, soil fertility, rainfall and the type of tree determine how fast the tree grows and thus the period in which the stem reaches the required thickness. For fast growing trees, the required diameter will in general be reached within one year.

FOLLOWING PRUNINGS

After the first pruning, the following prunings or loppings should not be done too regular, otherwise the yields per pruning and the life span of the tree will be reduced. The farmer should experiment to experience how a tree copes with a certain pruning frequency. The intervals may be longer or shorter depending on the climate, rainfall, soil fertility and altitude. In general, avoid cutting trees too often or at a low height at the beginning, during and immediately after the wet season. Too much pruning of trees during these periods exposes them to a high chance of disease attack. It is wise to let the trees grow and save the leaves for the dry season.

HARVESTING FODDER TREES IN DRY AREAS

In dry areas, it is useful to allow the stem to reach more than 10 cm in diameter before pruning or allowing browsing by livestock. The main root is then supposed to have grown deep enough to reach the lower water table, thus allowing the tree to get enough water during the dry season. Thereafter, pruning should not exceed 2-3 cuts per year to be sure that the strength of the tree is not lost. In areas with long dry seasons, water shortage can lead to the loss of fodder material, unless cutting times are chosen wisely. If the aim is to maximize the yield of fodder at the height of the dry season, experiments have shown that the final wet season cut should be made six months earlier.

PRUNING TOOLS

Secateurs or pruning clippers is the recommended tool for pruning or lopping the trees. Much damage can be done because of bad pruning techniques such as pruning with a machete. When the branches are cut at the base with a machete, a large gash is left. This causes a 'dieback' and has a negative effect on the regrowth of the tree. If the farmer has only a machete for pruning, the cut should be upward (from the underside up) not downward. It should be made sure that the machete is very sharp.

Table 33.2: Fodder Tree Species, Region, Fodder Availability Period and Preference to Livestock

Tree Species	Region	Fodder Availability Period		Preference to Livestock
		Leaf Fodder	Pod Fodder	
Faidherbia albida	SAR	May-Feb	Mar-June	Goat, Sheep, Camel
Acacia nilotica	AR,SAR, IGP, CZ,COZ	May-Feb	Apr-June	Cattle, Goat, Sheep, Camel
Acacia senegal	AR, SAR	Apr-Dec	May-June	Goat, Sheep, Camel
Acacia tortilis	AR, SAR	May-Feb	May-June	Goat, Sheep, Camel
Ailanthus excelsa	AR, SAR	May-Mar	-	Goat, Sheep
Albizia lebbeck	AR, SAR, IGP, CZ, NER	Apr-Nov	Aug-Nov	Goat, Sheep, Cattle
Albizia procera	SAR, CZ, NER	Jul-Mar	Jan-Mar	Cattle, Goat, Sheep
Anogeissus latifolia	SAR, AR	Apr-Oct	Jan-Mar	Cattle, Sheep, Goat and Camel
Anogeissus pendula	SAR, AR	Apr-Oct.	Jan-Mar	Cattle, Goats, Sheep and Camel
Artocarpus lakoocha	IGP, NER, STZ	Apr-Nov	-	Goat, Sheep, Cattle
Azadirachta indica	AR, SAR, IGP, CZ, NER	Throughout the year	July-Aug	Goat, Sheep, Cattle
Bauhinia purpurea	STZ, CZ, NER, IGP	Apr-Nov	-	Goat, Sheep, Cattle
Bauhinia variegata	IGP, NER, CZ, STZ	Apr-Nov	-	Goat, Sheep, Cattle
Dalbergia sissoo	SAR, IGP	Feb-Dec	Apr-Sep	Goat, Sheep, Cattle
Ficus religiosa	SAR, CZ, IGP	July-Nov	-	Cattle, Sheep
Gliricidia sepium	SAR	Mar-Nov	-	Goat and Sheep
Gmelina arborea	SAR, CZ, IGP	July-Nov	-	Cattle, Sheep, Goat
Grewia optiva	STZ	May-Nov	-	Cattle
Hardwickia binata	SAR, CZ	Apr-June	Feb-Mar	Cattle, Sheep, Goat
Leucaena leucocephala	SAR, IGP, CZ	Throughout the year	Dec-May	Cattle, Goat, Sheep

Tree Species	Region	Fodder Availability Period		Preference to Livestock
		Leaf Fodder	Pod Fodder	
Melia azedarach	IGP, SAR, STZ	July–Nov		Goat, Sheep, Cattle
Morus alba	IGP, CZ, STZ	Throughout the year	-	Cattle, Goat, Sheep
Prosopis cineraria	AR, SAR	Throughout the year	May-June	Goat, Sheep, Camel, Cattle
Quercus incana	NER	Mar-Nov	-	Cattle, Sheep, Goat
Robinia pseudoacacia	STZ	Mar-Nov	-	Cattle, Sheep, Goat
Sesbania grandiflora	IGP, CZ, NER	July-Nov	-	Cattle, Sheep, Goat
Sesbania sesban	SAR, IGP, CZ, NER	Throughout the year	Dec-Mar	Cattle, Goat, Sheep
Tecomella undulata	AR, SAR	Mar-Nov	Apr-June	Cattle, Sheep, Goat, Camel
Terminalia bellerica	COZ, STZ	July-Nov	-	Cattle, Sheep
Terminalia tomentosa	COZ, STZ	July-Nov	-	Cattle, Sheep
Ziziphus nummularia	AR, SAR	Throughout the year	-	Cattle, Goat, Sheep

(AR= Arid region, SAR= Semi arid region, IGP= Indogangetic plains, CZ= Central zone, COZ= Coastal zone, NER=North eastern region, STZ= Sub-temperate zone)

FEEDING TREE FODDER TO FARM ANIMALS

The part of the trees fed to livestock is usually leaves, but pods, seeds and small twigs may also be fed. In general, the animals like fodder tree leaves, though they may have to get used to some species.

Browsing or 'Cut-and-Carry': Fodder tree leaves can be browsed or 'cut-and-carried' to stall-fed (zero-grazed) animals. Browsing is possible but it should be done in a rotational manner. This means that the trees are browsed for a couple of days and left for six or more weeks to recover. Overgrazing should be avoided otherwise fodder trees may die. The cut-and-carry system is recommended even though it involves a lot of work. For most tree species, the fodder leaves should be fed to the animals within an hour of harvesting. Pruning, coppicing and pollarding is common in the sub-humid zones. In the medium and dry zone, browsing is much more common.

Mixture of Grass and Fodder Trees: Grass can be fed as much as the animal can eat. In addition to grass, fresh fodder tree leaves can be fed in limited quantities

When to feed fodder trees?: If there is plenty of fodder trees available at the farm, the leaves can be fed to the animal throughout the year in limited quantities. If there are a limited number of trees, it is recommended to feed a dairy cow, dairy goat and sheep the month before and the month after giving birth. Oxen should be fed one or two months before and during the period of land preparation.

Tree Fodder Diet for Dairy Cows: To the napier grass diet of a crossbred dairy cow, 8 kg of fresh fodder tree leaves can be added. This mixture provides for body maintenance and production of 10 liter of milk or more, depending on the type (genetic potential) of the cow and the quality of napier grass. If 2 per cent kg maize germ is added to this mixture, on a daily basis, it will maximize the benefit from the use of fodder tree leaves. When this ration is fed to the crossbred cow in its early lactation, it might produce about 15 litre of milk per day (depending on genotype and management conditions).

HARMFUL SIDE-EFFECTS OF TREE FODDER

Trees and shrubs have many ways to protect themselves from being eaten by herbivores (plant eating animals), for example through extremely hard thorns or spines and toxic compounds. The toxic compounds can have a negative health effect and even can cause death. Many plants also contain tannin, a certain substance, which is responsible for reduced digestibility of feed. It fixes protein, preventing it from being degraded by rumen microorganisms, or from being absorbed in the intestinal duct of the farm animal. This means that the protein in the feed can not be used by the animal for growth or milk production. Farmers are therefore cautioned on the amount of fodder tree leaves, which they include in the daily ration of cows, goats and sheep. This is because it could be harmful to the animals when fodder tree leaves are consumed in excess. Screening tree and shrub species for harmful tannins and toxins is very difficult. Many of these harmful ingredients are very difficult to identify. Therefore it is wise to get information from extension workers or farmers who have experience in feeding tree leaves to their animals. They know if certain types of trees are harmful to livestock and what quantities can be used in the daily ration of animals. Some fodder tree leaves are toxic to horses, chicken, rabbits and pigs. It is advised not to feed fodder tree leaves to these animals before more information is obtained. The advance research works are required in more detail from the Research Institutions for screening of tree fodders for the presence of toxins and their adverse effects on farm animals (Roothaert 2000).

■■■■

34 AGROFORESTRY PRACTICES FOR FUELWOOD PRODUCTION

INTRODUCTION

In developing countries, people derive at least 90 per cent of their energy requirements from wood and charcoal. Due to its huge demand, this natural essential resource is seriously depleted and threatened and developing world is facing critical crises of firewood shortage seriously. In developing countries, crop residues account for most of the biomass energy e.g., China (55%), Egypt (77%) and Bangladesh (90%). In response to fuelwood scarcity, smallholder farmers usually use crop residues and livestock manure as supplementary energy sources. Consequently, little or no organic matter is returned to the soil, which escalates the vicious cycle of soil nutrient depletion, declining crop production, and forest degradation. This practice accelerated nutrient depletion mainly through crop harvesting and soil erosion. Deforestation of natural forests for more land, which was cleared for crop production, also leads to the land degradation. Moreover, the processes of urbanization, modernization, agriculture expansion and uncontrolled wood harvesting from natural forests also contribute to declining fuelwood (i.e., firewood and charcoal) supply. The fuelwood collection in the rural areas for cooking and energy requires a considerable amount of time and effort. It has been estimated that in India, Indonesia, Ghana, Peru, families spend on an average 1.5 to 5.0 hr of time everyday for collecting fuelwood to use as fuel (Peskin et al 1992). It is also estimated that the poor spend 15 to 25 per cent of their income for fuel (FAO 1978).

To meet the increasing needs of burgeoning populations of human as well as livestock both, it is needed to have an alternate farming system that supplies wood, food and fodder. Agroforestry is an answer and an alternative land use system to solve the crises of fuelwood in every tropical countries and in the world as a whole. It is "a dynamic, ecologically based, natural resources management system that, through the integration of trees on farms and in the agricultural landscape, diversifies and sustains production for increased social, economic, and environmental benefits for land users at all levels" (Nair 1993). The integration of trees and crops in agroforestry leads to a more diverse system with multiple ecosystem functions including supply of food, fuelwood, fodder, and medicines; carbon storage; and conservation of soil and biodiversity (Sileshi et al 2007; Nair et al 2008). It is broadly categorized into simultaneous systems in which trees and crops are spatially integrated on the same piece of land, and sequential systems, in which trees are rotated with crops as in improved fallows (Young 1997; Rao et al 1998). On-farm fuelwood production using agroforestry, is a promising approach to reduce pressure on natural forests, to reclaim degraded farmlands, and to increase food production among small and marginal farmers (Ramadhani et al 2002; Kwesiga et al 2003; Jama et al 2008) and to ensure fuelwood supply from the farm.

OVERVIEW OF FOREST AND FUELWOOD STATUS

The wood harvesting and deforestation to provide the basic human necessities in the country and in global world are increasing with an alarming rate. In the present scenario, the forests cover is less than one - third of the total geographical area of the country and as per the FSI estimates (FSI 2011), the forest and tree cover of the country is 78.29 million ha (Mha) which is 23.81 per cent of the total geographical area of the country. This includes 2.76 per cent of tree cover outside the forest. The area covered by very dense forests (VDF) is 8.3471 Mha (2.54%), that with moderately dense forests (MDF) is 32.0736 Mha (9.76%) and open forests (OF) is 28.7820 Mha (8.75%). The distribution of forest in the country is not appropriate according to the National Forest Policy's (1988) aim of 33 per cent. The demand of our burgeoning population is estimated 225 million tons of food grains, 201 million tons of fuelwood, 64 million cubic meter of timber and 745 and 584 million tons of green and dry fodder respectively (NRCAF 2000). However, availability of food grain is sufficient but country is producing 43 million m^3 of timber wood and 115 million tons of fuelwood only (FSI 2001). The gap of demand and supply of forest produce in India is widening and forests are unable to fulfill the demand (Table 34.1). Agroforestry can play an important role in filling this gap and conservation of natural resources.

Table 34.1: Demand and Supply of Forest Produce in India (FSI 2001)

Parameters		Timber (million m^3)	Fuelwood (million ton)	Industrial Wood (million ton)	Green Fodder (million ton)	Dry Fodder (million ton)
Demand		64	201	28	593	178
	Forest	12	17			
Supply	Farm Forest	31	98	12	145	482
	Total	43	115			
Deficit		21	86	16	448	304

Moreover, more than 200 million people, mostly poor, are highly dependent on forests for meeting their subsistence and livelihood needs (SFR 2009). They lived in a symbiotic relationship with forest through ages and forests play a very significant role in their economy and their socio-cultural activities are woven around forests. Firewood problem is enormous and is reflected in the rising price of charcoal, even more rapidly than the price of imported fuel. To respond to scarcity of firewood, gathering wood is now an entire day's task in many parts of the country, a burden to women and children. In the absence of firewood, anything that burns is used (crop residues, twigs, barks, etc.) in the tribal and rural sector. Hence, the fuelwood trees as natural resources, if it is effec-tively managed and cultivated, could help in the alleviation of firewood scarcity problem.

AGROFORESTRY FOR MEETING FUELWOOD DEMAND

There is no single solution to firewood shortage, but some mingle of fuel conservation, tree planting and new technologies could certainly enhance its production in the country. The concept of 'agroforestry' is to fulfilling the demand of wood, fodder, fuelwood, small timber, fruit, fibre, fertilizer, etc. through integration of forestry with agriculture for increasing productivity per unit area per unit time. It is a systematic land use system that involves trees, arable crops and/or livestock seeking positive synergism on the same land unit in space or time. This practice in arable lands can help substantially in meeting such demands and solve some of our pressing problems of wood and food besides achieving national target of 33 per cent area under tree or forest cover. Agroforestry is "a sustainable land management system which increases the overall yield of the land, combines the production of crops (including tree crops) and forest plants and/or animals simultaneously or sequentially on the same unit of land and applies management practices that are compatible with the cultural practices of the local population" (Bene et al 1977; King and Chandler 1978).

Agroforestry is a system of optimizing the benefit from tree-crop-livestock interactions and it provides an opportunity towards the solution to the fuelwood crisis. It involves elements of agriculture and forestry, wherein woody perennials are deliberately mixed with agricultural crops and/or animal production units. The current world agroforestry overview indicates that there are several tree-crop-animal combinations of agroforestry systems and practices in different ecological and geographic regions of the world. The immediate logical response to the firewood shortage, one that will have many incidental ecological and socio-economic benefits, is to plant more trees on farms, along roads, in shelterbelts and on used lands throughout the rural areas. For most agroecological regions, fast growing tree/shrub species that can be managed in agroforestry systems for fuelwood production have been identified. Most of the tree species managed in agroforestry systems are either trees or shrubs that can meet many requirements of fuelwood for farmers, growers and smallholders.

The forestry and agroforestry practices like farm forestry, social forestry, commercial forestry, plantation, energy plantation, tree planting, etc. could certainly help in fuelwood production and in solving the problem of fuelwood crises in the country and in world. The potential indigenous fuelwood trees in India are *Acacia nilotica*, *Azadirachta indica*, *Casuarina equisetifolia*, *Calliandra calothyrsus*, *Dalbergia sissoo*, *Prosopis cineraria*, *Prosopis juliflora*, *Leucaena leucocephala* and *Ziziphus mauritiana* and exotic species like *Acacia auriculiformis*, *Acacia tortilis*, *Eucalyptus camaldulensis* and *Eucalyptus tereticornis* were studied and found potential for providing wood fuel at the highest efficiency, shorter rotation and also their high adaptability to diverse habitats and climates (Pathak 2002). Under the agroforestry, social forestry and farm forestry activities, it is recorded that 25.72 Mha (7.82%) area are planted with the various need based location specific trees in the country for the supply of wood and fuelwood (Dhyani et al 2009). Some of the important multipurpose trees that can be managed in the farming systems as firewood biomass production are described, looking to their adaptability to different areas, their calorific value, and agroforestry potential for their cultivation.

SCOPE AND APPROACHES FOR FUELWOOD PRODUCTION IN LAND USE SYSTEMS

AGROFORESTRY

In this context, the concept of 'agroforestry' comes in handy in fulfilling these deficiencies through integration of forestry with agriculture for increasing productivity per unit area per unit time. Agroforestry has great potential to rehabilitate and reclaim degraded lands for sustainable production. It is one of the important alternatives farming system for improvement in degraded sites for the following reasons:

- It can play a very vital role in utilization of all the natural resources in a most effective manner for sustainable crop production by maintaining the long term production potential of land
- It is an important tool for arresting land degradation and amelioration of lands
- It is a land management system that optimizes the overall land productivity and reduces resources depletion through the utilization of positive interaction between crop-tree-animals in its temporal and spatial dimensions

- Agroforestry can be seen as phases in the development of productive agroecos-ystem and an alternate land use system, wherein, farmers can manipulate and manage their land by growing trees for the service or products

Some fuelwood based agroforestry system are well developed and established like Babul based agroforestry system, Albizia based agroforestry system, Leucaena under alley cropping, Neem based agroforestry system and some fuelwood trees on farm bunds like Arjun, Butea, Kapok, Drumstick, etc.

FARM FORESTRY

It is the practice of forestry on farm in the form of raising rows of trees in blocks or on bund/ boundaries of farm/fields. The concept of farm forestry can be utilized for the production of fuelwood trees and consequently the protection of other farm vegetation from hot wind and desiccation by planting trees of fuel importance on all side of farm field. The suitable fuelwood tree species are *Acacia nilotica*, *Albizia lebbeck*, *Albizia procera*, *Butea monosperma*, *Leucaena*, *Casuarina*, etc.

ALLEY CROPPING

It is an agroforestry practice in which rows of closely planted woody plants with annual crops planted in alleys in between hedges. The main object is to maintain or increase crop yields by improvement of the soil status and microclimate. The best example for growing fuelwood species under alley cropping as hedge row is subabul (*Leucaena leucocephala*).

SOCIAL FORESTRY

It is a system of sustained production of forest resources, which is executed by the people, of the people and for the people. It is the practice of forestry on lands outside the conventional forest areas for the benefit of rural and urban people and has the objective of improving the quality of life of the rural and urban population by supplying the fuel for the villagers, fodder for their cattle, small timber for agricultural implements and other tree based produce for the rural poor. The species suitable for social forestry are *Acacia nilotica* (babul), *Azadirachta indica* (neem), *Albizia lebbeck* (siris), *Acacia auriculi-formis*, *Cassia siamea*, *Casuarina equisetifolia*, *Dalbergia sissoo* (shisham), *Anogeissus latifolia*, *Prosopis juliflora*, *Terminalia arjuna*, *Eucalyptus* spp. etc.

ENERGY PLANTATION

Planting of fuelwood species in high density is called as energy plantation. High density short rotation (HDSR) forestry with close planting of 40,000 to 1,00,000 stems per hectare to be worked on a rotation of one to two years is suggested. It provides firewood from tree species and commonly used fuelwood trees are *Acacia nilotica* (babul), *Albizia lebbeck* (siris), *Cassia siamea*, *Casuarina equisetifolia*, *Dalbergia sissoo* (shisham), *Prosopis juliflora*, *Eucalyptus* spp. etc.

SHORT ROTATION FORESTRY (SRF)

The short rotation forestry (SRF) is another form of tree farming which is the prac-tice of cultivating fast growing trees that reach their economically optimum size between eight and 20 years old. This technique of tree cultivation could rapidly increase the production of energy plants in the country, increase wood fuel supply in sustained basis and make better use of wasteland and degraded lands. Short rotation forestry plays a vital role in socio-economic development as important source of subsistence, employment, revenue earning and raw material for various industries. The short rotation tree species like Eucalyptus, babul, subabul, etc. have the potentials to provide greater volumes in terms of biomass from the same unit of land area along with the other farming practices.

CHOICE OF FUELWOOD SPECIES IN AGROFORESTRY

In India, the majority of the population lives in the rural areas where fuelwood, charcoal, crop residue and animal wastes provide most of the energy requirement. Many people, especially in rural areas of India, use wood for energy purposes rather than for other uses. It is estimated that wood still accounts for up to 90 per cent of the total energy consumption. It is important to select the most suitable trees/shrubs. The species found growing in nearby natural forest, or growing well in neighbouring plantations are good choices for selection of fuelwood trees. It is important to know that the indigenous species grow better than the exotic trees in the area. Moreover, the introduction of exotic species is required to understand the suitability to the soil, rainfall, and other requirements of the species and matching with the conditions of the planting area. The following factors should be kept in mind while selecting trees for the fuelwood purpose:

- Fast growing (adapt quickly and yield a maximum volume of wood in a short time)
- Nitrogen fixing ability
- Establishes easily
- Low nutrient requirement
- Soil stabilizing and maintaining soil fertility
- No toxicity to soil
- Require less care and management
- Suitable to short rotation fuelwood production systems
- Resistance to diseases and pests
- Tolerance to drought and other climatic stresses
- Good coppicing ability
- Ability to produce wood with high calorific value
- Multipurpose uses that contribute to the farm income
- Wide adoptability and adaptability (to grow successfully in a wide range of environments, including different soil types, rainfall regimes, amounts of sunlight and terrains)
- Produce trees without thorns
- Tree have small stems and branches with less diameter
- Produce wood that splits easily
- Low wood moisture content or dry relatively quickly
- Produce minimal and non-toxic smoke (low ash and sulphur content) when burned
- Not spit or spark while burning
- Good combustion values
- Produce dense wood and burn long time

Some species often used for fuelwood are *Acacia auriculiformis*, *Casuarina equisetifolia*, *Gliricidia sepium*, *Calliandra calothyrsus* and *Leucaena leucocephala*. The growers and farmers can grow fuelwood trees in various combinations with other crops as in agroforestry and in block plantations for the production of fuelwood. The firewood species which are of common and important from the fuelwood production point of view in various agro-climatic region in the country are elaborated in the Table 34.2 along with their cultivation and propagation techniques and calorific values.

Table 34.2: Promising Trees/Shrubs for Fuelwood Production

S.No.	Plant Species	Family	Propagation	Calorific Value (Kcal/kg)
	Acacia auriculiformis (Wattle)	Leguminosae	NR: Seed AR: DS, TP	4,800-4,900
	Acacia catechu (Cutch tree)	Leguminosae	NR: Seed AR: DS, TP ST:24 hr water soaking	5,142-5,244
	Acacia mangium (Mangium)	Leguminosae	NR: Seed AR: DS, TP ST: None	4,800-4,900
	Acacia mearnsii (Black wattle)	Leguminosae	-	3,500-4,000
	Acacia nilotica (Babul)	Leguminosae	NR: Seed AR: DS, TP ST : 40 hr water soaking	4,870-4,950
	Acacia senegal (Gum Arabic)	Leguminosae	NR: Seed AR: DS ST: hot water	-
	Acacia tortilis (Israeli babul)	Leguminosae	NR: Seed scanty AR: DS, TP ST-30 min. hot water	4,400
	Albizia lebbeck (Kala siris)	Leguminosae	NR: Seed scanty AR: DS, TP ST-30 min. hot water	5,163-5,166
	Albizia procera (White siris)	Leguminosae	NR: Seed AR: DS, TP	-
	Alnus nepalensis (Indian alder)	Betulaceae	NR: Seed AR: DS	-
	Anogeissus latifolia	Combretaceae	NR: Seed AR: TP ST: None	4,948
	Azadirachta indica (Neem)	Meliaceae	NR: Seed, coppice & root suckers- AR: DS, TP, stump planting	
	Butea monosperma	Leguminosae	NR: Seed, root sucker AR: DS, stump planting	4,909
	Cajanus cajan	Leguminosae	AR: DS ST: None	-

S.No.	Plant Species	Family	Propagation	Calorific Value (Kcal/kg)
	Calliandra calothyrsus (Calliandra)	Leguminosae	NR: Seed AR: DS, SP ST-hot water than cold water soaking	-
	Cassia siamea (Kassod tree)	Leguminosae	NR: Seed AR: DS, SP ST-10-30 min acid scarification	-
	Casuarina equisetifolia (Sea oak)	Casuarinaceae	NR: Seed but poor AR: T, air layering, rooting cladodes ST- water soaking	4,950
	Dalbergia sissoo (Shisham)	Leguminosae	NR: Seed AR: DS, TP, coppices, suckers ST-48 hr water soaking	4,908-5,181
	Eucalyptus camaldulensis (Red gum)	Myrtaceae	NR: Seed AR: TP ST: None	3,172-5,680
	Eucalyptus tereticornis (Mysore gum)	Myrtaceae	NR: Seed AR: TP ST: None	3,172-5,680
	Gliricidia sepium (Gliricida)	Leguminosae	NR: Seed AR: cuttings ST: Hot water soaking	4,900
	Gmelina arborea (Gamhar)	Verbenaceae	NR: Seed AR: DS, TP, suckers ST-Alternate wetting & drying	4,763-4,800
	Grevillea robusta (Silver oak)	Proteaceae	NR: Seed AR: DS	4,800-4,950
	Leucaena leucocephala (Subabul)	Leguminosae	NR: Seed plenty AR: DS, TP, suckers coppice ST-None	4,200-4,600
	Mangifera indica (Mango)	Anacardiaceae	AR: TP	-
	Melia azedarach (China berry)	Meliaceae	NR: Seed AR: DS, TP ST-Water soaking	5,043-5,176
	Morus alba (Mulberry)	Moraceae	NR: Scanty AR: Cutting	4,371-4,773
	Pithecellobium dulce (Jungal jalebi)	Leguminosae	NR: Seed AR: Cuttings ST: None	

S.No.	Plant Species	Family	Propagation	Calorific Value (Kcal/kg)
	Prosopis cineraria (Khejri)	Leguminosae	NR: Seed AR: root suckers ST: 24 hr water soaking	5,000
	Prosopis juliflora (Mesquite)	Leguminosae	NR: Seed AR: Seed, root suckers ST: 24 hr water soaking	5,000-5,500
	Sesbania grandiflora (Sesbania)	Leguminosae	NR: Seed AR: DS, Cutting ST: None	4,350
	Syzygium cumini (Jamun)	Myrtaceae	NR: Seed AR: DS, Cutting , TP	5,834
	Tamarindus indica (Tamarind)	Leguminosae	DS, EP ST-water soaking	4,909-4,969
	Terminalia tomentosa	Combretaceae	NR: Seed AR: DS, SP	4,923
	Ziziphus mauritiana (Indian jujube)	Rhamnaceae	DS,BC,EP ST-Water soaking	4,878

Note: NR- Natural Regeneration; AR- Artificial Regeneration; DS- Direct Seeding; TP- Transplanting; EP- Entire Planting; ST- Seed Treatment

MANAGEMENT OF FUELWOOD TREES

The cultivation of fuelwood tree species for wood energy does not require continuous management, only good silvicultural practices for increased firewood production and optimizing the benefits from the tree-crop interface are needed. Firewood production in integrated agroforestry systems is appropriate to this fuelwood supply point of view. Using appropriate species on appropriate sites and climate, wood energy can be produced in commercial way. In farming systems, trees for firewood production can be established in farm boundaries, on contours, scattered on farms or as rotational woodlots. Correct spacing of planted trees is vital to integrated production. Normally 2m x 2m spacing is a standard espacement for fuelwood production. If properly managed, fuelwood species can be self-renewing.

Most tree species have high coppicing abilities and can be managed on short rotations. The ability to coppice or grow rapidly from stumps is important for firewood species which allows repeated harvests without the cost and effort for replanting seedlings each time. Moreover, because of the multiple roles of the species, other products can also be obtained (fodder, green manure, minor produce, etc.). Most woods burn, but there are properties that differentiate their relative value for fuel. Density is the most general measure of wood burning quality. The heavier the wood (when dry), the greater its calorific value. Hence proper choice of species is critical for firewood crop establishment. On the basis of tree growth and management the development of a firewood production system is divided into different stages:

1. Pre-planting phase: It includes site preparation, improvement of soil fertility and weed control

2. Establishment phase: From planting until the time of the first pruning or thinning
3. Silvicultural management phase: During which all the thinning and pruning are done
4. Maturing phase: During which the trees are left virtually unattended until they reach maturity
5. Harvesting phase: Felling and conversion

CONCLUSION

Thus, agroforestry has been widely advisable in the disadvantageous area as a sustainable low input system for replenishing soil fertility and increasing production. In India, fuelwood markets are dominated by wood of *Acacia nilotica*, *Tamarindus indica*, *Prosopis* spp. and other local species, in spite of the large-scale tree planting efforts for fuelwood production by state agencies using exotics such as *Leucaena*, *Casuarina* and *Eucalyptus* (Vandenbeldt 1990). The above facts lead us to suggest that the agroforestry, in the form of homegardens and farm forestry and other forms are the important source of woodfuel and fuelwood for domestic consumption in many areas in the country. Besides above, it is also requiring joint effort of Research Institutions, Government, Communities and NGOs to make efforts and activities to alleviate the growing shortage of fuelwood such as:

- Encouragement of the use of alternative energy sources such as biogas and solar heating.
- Efficient utilization of fuelwood.
- Intensive firewood production on farms/farmers' field, with the farming system incorporating, fast-growing tree species of acceptable calorific value identified by research, are required to plant trees on contour strips, on farm boundaries, etc. to produce firewood.
- Management of firewood crops on farms could help slow deforestation caused by firewood harvesting.

35 AGROFORESTRY AND BIODIVERSITY CONSERVATION

INTRODUCTION

Biological diversity includes all plants, animals, microorganisms, the ecosystems of which they are part, and the diversity within species, between species, and of ecosystems. Biodiversity is determined by the interaction of many factors that differ spatially and temporally. Functional diversity describes the variety of ecological functions of species or groups of species in an ecosystem. The provision of goods and services by ecosystems is underpinned by various aspects of biodiversity, although the relationship is a complex one. Goods and services provided by biodiversity have significant economic value, even if some of these goods and most of the services are not traded by the market. The value of biodiversity-dependent goods and services is difficult to quantify. Biodiversity-dependent goods and services may be worth trillions of dollars annually, but most of these services carry no price tags to alert the society. The current levels of human impact on biodiversity are unprecedented, affecting the planet as a whole, and causing large-scale loss of biodiversity. Agroforestry is one of the potential land use systems for increasing the biodiversity in agricultural landscape.

The world at present is facing innumerable problems such as increasing population, ecosystem degradation, declining agricultural productivity, and changing environment conditions. For its sustaining, we need to find some solutions, to ensuring food security particularly to our population. Agroforestry is a traditional practice of combining trees with agricultural crops/pasture can contribute substantially through their multiple outputs. Thus properly designed agroforestry can be helpful in poverty alleviation, food security, maintaining ecosystem health, managing pest and weeds, conserving biodiversity, and mitigating greenhouse effects by carbon sequestration. In the last few decades, there has been a growing interest in agroforestry because of its biodiversity and ecosystem services it provide. Thus the plantation trees are increasing in farming systems. A recent global assessment of tree cover found that 48 per cent of the world's agricultural land had at least 10 per cent of tree cover (Zomer et al 2009). In 2009, Ministry of Agriculture of the Government of Kenya enacted new farm forestry rules which require 10 per cent of all farms to be covered with trees. This was in response to recognition of deforestation, the increase in agricultural land area, and the high motivation of farmers to plant trees. A high percentage of tree cover is found in nearly all continents of the world, highest being in Central America and Southeast Asia. Although Africa shows a smaller percentage of tree cover at continental level, the most widespread farming system in Africa is the so-called agroforestry parkland (scattered trees in cropland), making Africa a typically "treed continent" in agricultural areas (Boffa 1999).

AGROBIODIVERSITY

Biodiversity loss in agricultural landscapes affects not just the production of food, fuel, and fibre, but also a range of ecological services supporting such as clean water supplies, habitats for wild species, and human health. The world's population of 6.3 billion people is projected to grow to 9 billion by 2050. To fulfill demand of food, more land will be converted to agriculture, which will increase pressure on biodiversity. The United Nations Convention on Biological Diversity (UNCBD) defines biodiversity as "the variability among living organisms from all sources, including, inter alia, terrestrial, marine and other aquatic ecosystems, and the ecological complexes of which they are part. This includes diversity within species, between species, and of ecosystems". The UNCBD further defines agrobiodiversity as biodiversity, which is important for agricultural production, including crop and livestock genetic diversity, wild biodiversity closely associated with domesticated species, and other wild biodiversity sharing the resources (Garrity et al 2006). The goal of the agrobiodiversity science plan and implementation strategy is to establish the scientific basis needed to address the trade-offs between food production, biodiversity conservation, ecosystem services, and human well being in agricultural landscapes. Four key research foci of the agrobiodiversity science plan integrate the biological and social sciences (Jackson et al 2005):

(1) To assess biodiversity in agricultural landscapes and the anthropogenic drivers of biodiversity change

(2) To determine the factors that increase biodiversity in agricultural landscapes and anticipating the impacts of social and environmental change

(3) To identify the goods and services provided by agrobiodiversity at various levels of biological organization, e.g., genes, species, communities, ecosystems, and landscapes

(4) To evaluate the socioeconomic options for the sustainable use of biodiversity in agricultural landscapes

Implementation of the agrobiodiversity science plan will involve collaboration between geneticists, ecologists, anthropologists, and economists, to cross ecosystem boundaries to understand the environmental and social drivers of biodiversity change, ecosystem services provided by biodiversity in agricultural landscapes, and how to use this information for policy-relevant strategies to meet human needs. In 2001, a second phase of the programme was launched with the aim of further refining its three main objectives (Jackson et al 2005):

(1) Discovering biodiversity and predicting its changes

(2) Assessing impacts of biodiversity changes on ecosystem functioning and services

(3) Developing the science of the conservation and sustainable use of biodiversity

Agricultural biological diversity or agrobiodiversity includes all components of biological diversity of relevance to food and agriculture and all components that contribute to sustaining the key functions of agroecosystems. Thus, agrobiodiversity encompasses the variety and variability of animals, plants and microorganisms that are necessary for sustaining key functions of the agroecosystem, including its structure and processes for, and in support of, food production and food security (FAO 1999a). The Convention on Biological Diversity (CBD) recognized the importance and significant contributions of agrobiodiversity in the functions of agroecosystems. In the international policy arena, agricultural biodiversity was addressed for the first time in a comprehensive manner by the conference of parties of the Convention on Biological Diversity in 1996. Since 1992, Agenda 21 and the Convention on Biological Diversity (CBD) have set the social and legal framework for the sustainable use and conservation of biological diversity. Within the CBD process, the topic of agrobiodiversity was addressed for the first time at the 3rd session of the Conference of the Parties to the Convention (COP3) in 1996 and specified at COP5, where ecological services of agrobiodiversity was focused.

If we have a closer look at the relationship between biodiversity and agricultural productivity, four relative positions of agricultural systems can be identified as below (van Noordwijk et al 2006):

- High biodiversity, low productivity, as in the 'agroforest landscapes' of the humid tropics
- Low biodiversity, low productivity, as in many fragile ecosystems under the impact of 'land degradation'
- High biodiversity, high productivity – an impossible dream scenario, but approachable in biodiversity-friendly agriculture
- Low biodiversity, high productivity, as found in intensively managed agricultural landscapes that have avoided thousands of critical environmental degradation where the relatively low levels of agrobiodiversity may be highly appreciated by society

SIGNIFICANCE OF AGROBIODIVERSITY

The case studies demonstrate a wide range of agrobiodiversity: rare and locally adapted breeds of livestock, diverse landraces like rice and potatoes, and threatened varieties of fruit and vegetables. These all areas are associated with several years back with examples such as 7,000 years in the case of Peru, 2,500 years in the Canyon de Chelly and 2,000 years in the Philippines. In Nepal, Europe and Ethiopia there is evidence of many hundreds of years of adaptation. A case of the Peruvian Potato Park, here economy is largely dependent on the potato. The people of the Philippines rice terraces are dependent on rice cultivation. In Quijos and European case study, many local varieties of crops, fruit and livestock have been lost or are in decline but there is a consensus that they could be an important element in reviving the rural economy. In the Philippines, complex rituals that bind the community together are associated with the cultivation and harvesting of rice, and its preparation in the form of food and drink. In Nepal, there is a conscious effort to revive and maintain the richness of the agrobiodiversity heritage by holding 'diversity fairs' and several events are held in Garrotxa in Spain. In England, Rhön, Garrotxa and Gaspé are local varieties of livestock and fruits which have been less important now in economic terms but they retain much of their social and cultural significance (Amend et al 2008).

Bioversity international suggested agrobiodiversity is the result of natural selection and human intervention over millennia and play important role in sustainable development for the followings:

- It provides food, fodder, fibre, fuel, medicines and other products for subsistence/ sale
- It provides sustained ecosystem services
- It provides genetic material for breeding new varieties of plants and animals
- It provides social, cultural, aesthetic and amenity values to the people

Literature, experience and research have shown that agrobiodiversity can have following roles (Thrupp 1997):

- Increases productivity, food security, and economic returns
- Reduces the pressure of agriculture on fragile areas, forests and endangered species
- Makes farming systems more stable and sustainable
- Contributes to sound pest and disease management

- Conserves soil and increases soil fertility and health
- Contributes to sustainable intensification
- Diversifies products and income opportunities
- Helps to maximize effective use of resources and the environment
- Reduces dependency on external inputs
- Improves human nutrition and provides sources of medicines and vitamins
- Conserves ecosystem structure and stability of species diversity

CHANGES IN AGROBIODIVERSITY

United Nations Food and Agriculture Organization (FAO) estimates show that more than 90 per cent of crop varieties have disappeared from the farmers' fields in the past 100 years. Agricultural plants are continuing to disappear at two per cent a year. Livestock breeds are being lost at five per cent annually. The current extinction rate of species range from approximately 1,000 to 10,000 times higher than natural extinction rates. As a result of these rates of decline, over 50 pollinator species are listed as threatened or endangered and wild honeybee populations have dropped 25 per cent since 1990. Locally varied food production systems are under threat, including local knowledge and the culture and skills of women and men farmers. With this decline agrobiodiversity is disappearing, the scale of the loss is extensive. With the disappearance of harvested traditional species of crops, varieties and breeds, a number of other growing species not directly in use have also been disappeared. FAO (1999b) has informed the following agricultural changes during past 100 years:

- Since the 1900s, 75 per cent of plant genetic diversity has been lost as farmers worldwide have left their multiple local varieties and landraces for genetically uniform, high-yielding varieties
- Thirty (30) per cent of livestock breeds are at risk of extinction (six breeds are lost each month)
- Seventy five (75) per cent of the world's food is generated only from 12 plants and five animal species
- Of the 4 per cent of the 2,50,000 to 3,00,000 known edible plant species, only 150 to 200 are used by humans
- Only three crops i.e., rice, maize and wheat – contribute nearly 60 percent of calories and proteins obtained by humans from plants
- Animals provide 30 per cent of human requirements for food and agriculture and 12 per cent of the world's population live almost entirely on products from ruminants

Some reasons for which *in situ* conservation has also been neglected for several years are:

- Primary reason for neglecting *in situ* conservation was concerned over genetic erosion in traditional farming systems (Harlan and Martini 1936) and the belief that replacement of landraces by modern cultivars is inevitable
- If genetic erosion is novel, inevitable and inexorable, then only means of preserving crop germplasm would be in gene bank
- Farmers cannot be trusted to maintain such valuable resources
- There is rapid and uncontrolled loss of germplasm from traditional agriculture due to replacement of traditional varieties by the modern varieties

- Farmer's conservation methods are rejected because of the assumption that they would condemn certain areas to perpetual poverty for the benefit of others
- *In situ* conservation may prove to be expensive as it requires subsides to make farmer do something that otherwise he wouldn't have done
- Finally, as long as the short term and immediate benefits are the focus of the scientist, *in situ* conservation approaches will be rejected

The reasons for lacks of popularity or not adopting on site conservation are several but going by the experience of decades of off-site conservation the advantages from *in situ* conservation of landraces cannot be ignored. Recently, greater attention and advocacy for onsite conservation may be due to the realization of the facts that,

- Collection of germplasm is a continuous process for evaluation trial. There is loss of collections due to genetic drift
- Research in centers of crop diversity has shown that the adoption of improved varieties does not necessarily lead to the abandonment of local, farmer varieties
- Diffusion of modern varieties is not uniform, and many areas that are rich in crop genetic resources are by passed by crop improvement programs
- Participation by farmers in conservation is part of achieving such equity
- Crop scientists have begun to recognize that the conservation of knowledge systems and evolutionary processes for crop require in situ conservation
- There is a continuous need to collect the germplasm for off-farm conservation and collections inevitably experience loss due to genetic drift and other causes
- Commonness and rarity are not very well understood for any crop population and so there are all chances of negligence of certain valuable resources
- Evolutionary processes are halted as a result of off farm conservation
- *In situ* method may be less expensive if methods other than direct subsidies are devised
- On site conservation may compliment for offsite conservation especially, for wide range of characters that are outside the breeder's current interest

THREATS TO AGROBIODIVERSITY

Since long back from prehistoric times, humans had a considerable impact upon biodiversity. Many large animals and forest systems have been exploited for their needs to extinction. According to Global Biodiversity Assessment, biodiversity is currently decreasing at an unprecedented high rate. FAO (2009) has suggested many plant and animal genetic resources (PAGR) are currently under increasing threat and many more have already disappeared from farming systems around the world. Furthermore, as a direct result of growing commercialization and industrialization of farming systems (e.g., via the 'Green Revolution') agroecosystems are increasingly characterized by high levels of intensification with low levels of agrobiodiversity (Thrupp 2000; Drucker et al 2005; Jackson et al 2007). Since the beginning of agriculture 12,000 years ago, agrobiodiversity has been evolving through a process of human and natural selection. It is estimated that humans have made use of about 7,000 plant species (FAO 1997). According to FAO assessment of world's animal and plant genetic resources published in 2007 and 2009, out of 7,600 livestock breeds, 20 per cent are classified as being at risk and 9 per cent are already extinct. Most of these are located in developing countries where the risk of loss is highest, and between 2000 and 2006 alone 62 breeds became extinct (FAO 2007).

Deforestation, changes in land use patterns, natural disasters are contributing to abundant habitat fragmentation/destruction, of the crops and their wild relatives (Saxena et al 2002). Social disruptions or war also pose a constant threat of genetic wipe-out of such promising diversity (OECD 1996). Over exploitation also introduced invasive/alien species. Global warming and pollution have also been recognized as one of the causes for loss of biodiversity (Myers 1994). There are several factors responsible for decline in agrobiodiversity:

- Developmental activities
- Deforestation
- Land use patterns
- Natural disasters
- Increase demand for food, land, and other agro-based resources
- Urbanization lifestyle of farmers
- Conversion of agricultural land
- Natural calamities
- Land degradation
- Habitat fragmentation.
- Climate change considered to speed agrobiodiversity loss.
- Shifted demand of urban populations from traditional to modern foods
- Supermarkets demand
- Increasing technological and financial support to introduced high yielding varieties
- Large scale modification of farming condition for high yielding varieties,
- The market preferences of consumers for uniform grains or vegetables or foods
- Possibilities of insects, pests, disease spreading

Singh et al (1997) reported that the diversity of traditional multipurpose tree communities has been decimated for the promotion of apple plantations in the western Himalaya. Rai (1995) reported plantation of *Alnus nepalensis* and *Albizia stipulata* in the eastern Himalaya during the last 50 years responsible for erosion of biodiversity of traditional trees. The reasons for erosion in agrobiodiversity in the hills are:

- Degradation of natural forests which sustaining hilly farming system
- Changing attitudes towards plant species
- Cultivation of medicinal plants in high hills
- Migration of hilly people to plain for employment
- Subsidized cost for seed for high yielding varieties by Government
- Attraction for cash crop
- Lack of incentives for traditional crops
- Lack of market of traditional crops
- Fruit cultivation has been adopted in a big way in the temperate belt
- The destruction or conversion of habitats

- Changing social and religious norms, and cultivation methods
- Changes in food habits

CONSERVATION STRATEGIES OF AGROBIODIVERSITY

The Convention on Biological Diversity (CBD) within its broader framework defines two conservation strategies for conserving genetic resources: *ex situ* conservation in gene banks and *in situ* conservation of ecosystems in protected areas and reserves. The choice of conservation strategy depends mainly on the nature of the material to be conserved i.e., the life cycle, mode of reproduction, size and the ecological status (OCED 1999). Two major approaches for crop diversity conservation are:

IN SITU CONSERVATION OF AGROBIODIVERSITY

UNEP (1992) defined *in situ* conservation as "the conservation of ecosystems and natural habitats and the maintenance and recovery of viable populations of species in their natural surroundings and in the case of domesticated and cultivated species, in the surroundings where they have developed their distinctive properties". *In situ* (on-farm conservation) is a highly dynamic form of plant genetic resources (PGR) management. The potential threat that loss of genetic diversity poses to the world's food fuelled so called PGR conservation movement which started in the early 1970s with the establish-ment of IBPGR which now called Biodiversity International has creation of 1740 *ex situ* storage collections of gene banks around the world (Bioversity 2009; FAO 2010c).

The forest areas where *in situ* conservation of biological diversity is the prime objective have increased by more than 95 million hectares since 1990. These forests now account for 12 per cent of the total global forest area, or more than 460 million hectares. Most, but not all of them, are located inside protected areas (FAO 2010a). The Convention on Biological Diversity has given highest priority to this approach of conservation which includes species protected in the wild as well as landraces (Primack 1993). *In situ* conservation serves as a continuous source of germplasm for *ex situ* conservation. FAO (1996) has suggested four kinds of multidisciplinary research to successfully run the *in situ* conservation.

(1) Ethnobotanical and socio-economic research to understand and analyze farmers' knowledge, selection/breeding and utilization and management of plant genetic resources with the approval of the involved farmers with applicable requirements for protection of their knowledge and technologies

(2) Population and conservation biology to understand the dynamics of the local landraces and farmer's varieties (population differences, gene flow, degree of inbreeding and selection pressure *etc.*)

(3) Crop improvement research in mass selection and simple breeding without significant losses in local biodiversity

(4) Extension studies for lesser-known crops including their seed production, marketing and distribution

However, the germplasm maintained under *in situ* conservation are highly vulnerable to the threat posed by genetic drift, inbreeding, habitat loss, competition from exotic species and pest infestation. The criteria for site selection for *in situ* conservation within the study areas are (Tan and Tan 1998):

- Wide range of diversity of a single or few crop species within a given site
- Ecological heterogeneity

- Possibility to control or monitor the site
- Easy access for monitoring and management

EX SITU CONSERVATION OF AGROBIODIVERSITY

Ex situ conservation refers to the conservation of germplasm away from its natural habitat. This strategy is particularly important for crop gene pools which can be achieved by propagating/ maintaining the plants in genetic resource centre, botanical gardens, tissue culture repositories or in seed gene banks (OCED 1999). The *ex situ* conservation concept was developed first time in 1958, long time before its official adoption under the Convention on biological diversity signed in 1992 in Rio de Janeiro, Brazil. *Ex situ* conservation methods for plants animals and microorganisms proved to act as valuable tools in studying and preserving biological resources for different purposes. The need for conserving agrobiodiversity on farms has gained attention more recently. Some 1750 gene banks have been established globally to conserve genetic resources *ex situ*. With some 7.4 million accessions (FAO 2010b). But there are gaps in the collections because all species seeds cannot be stored long-term and alternative methods, such as field gene banks and *in vitro* storage, are costly.

There are increasing efforts to sample and store local crop varieties and wild relatives *ex-situ* in seed collections (FAO 2009). Such gene banks can function as a low-cost conservation instruments in order to safeguard genetic resources for many years, thereby avoiding loss through changing environmental or economic conditions (Wood and Lenne 1997). Including the advantages of *ex situ* conservation, there are limitations of relying only on this approach:

- Many important species are under-represented because of the recalcitrant nature of the seeds
- Genetic shifts or alterations cannot be ruled out due to inappropriate storage conditions
- Since the crops are grown with external application of fertilizers and pesticides, and use of heavy machinery, the plants slowly get accustomed to more congenial conditions, the roots architecture and assimilatory properties get modified since nutrients are easily available and availability of porous well ploughed soil
- *Ex situ* conservation does not maintain evolutionary processes that created the crop germplasm. The genetic resources are not exposed to natural or artificial pressure and therefore no chance exist for further evolution or adaptations

Some of the genetic material grown by farmers may no longer be available in the field, but has been collected and stored in gene banks and breeding stations. The NBPGR and ICAR have several hundred thousand accessions of gene bank. Such *ex-situ* collections of genetic material can be used as base material for genetic up gradation of agriculture (Kothari 1999). But such collections also suffer from severe limitations: they are very expensive, lack adequate space to store the complete genetic diversity found in agriculture, and suffer loss of viability of stored germplasm.

To support agrobiodiversity, conducive policies and institutions are needed, such as legal frameworks and regulations that promote the use of agrobiodiversity, including appropriate seed legislation, national development strategies and agricultural policies. Extensive networking needs to be established between the different stakeholders, from farmers, NGOs, scientists and researchers, to political institutions, government agencies, education institutions and national and international networks, and the networking strengthened at all stages. System-wide policies and related legal instruments on biodiversity conservation should be harmonized, and global action plans for conservation and use of agrobiodiversity developed. Strong, complementary and synergistic collaboration of the major institutions and organizations in this field is needed. Some examples are: Bioversity International; Food and Agriculture Organization of the United Nations (FAO); the International Treaty on Plant Genetic Resources for Food and Agriculture (ITPGRFA); the FAO

Commission on Genetic Resources for Food and Agriculture (CGRFA); the Convention on Biological Diversity (CBD); the World Intellectual Property Organization (WIPO); the System-wide Genetic Resources Programme (SGRP) of the Consultative Group on International Agricultural Research (CGIAR); the CGIAR Genetic Resources Policy Committee (GRPC); the Central Advisory Service on Intellectual Property (CAS-IP); the World Health Organization (WHO). Responsibilities and duties should be clearly defined in order to avoid unnecessary duplication of work, while some 'coordinated overlapping', although with different foci, may be useful.

AGROFORESTRY AND BIODIVERSITY CONSERVATION

Agroforestry is identified as integrated land use system which enhances plant diversity and reduces habitat loss and fragmentation (Noble and Dirzo 1997). Agroforestry is a collective name for land use systems and technologies where woody perennials are deliberately used on the same land management units as agricultural crops and/or animals, in some form of spatial arrangement or temporal sequence (Lundgren and Raintree 1982). Agroforestry applies to private agricultural, forest lands and communities land. These are highly disturbed, human-dominated land use systems. These lands can be targets for agroforestry which also include highly-erodible, flood-prone, economically marginal and environmentally sensitive lands. Agroforestry provides strong incentives for adoption of conservation practices and alternative land uses, and supports a collaborative watershed analysis approach to management of landscapes containing mixed ownerships, vegetation types and land uses.

Farmers show their appreciation of diversity by retaining and managing trees on their farmlands. It has been reported an increase in demand in the adoption of agroforestry by smallholder farmers in Africa, Latin America, South and South East Asia and Pacific Island countries. The impact of agroforestry adoption on livelihoods of farmers in these countries includes increase in crop yields, increase in income, increased savings resulting in change of wealth and soil improvement (Tscharntke et al 2010). A literature revealed that agroforestry contributes to the conservation and protection of biodiversity, including that of both wild species and species more directly related to agricultural production (Boffa 1999; Buck et al 2004). Some of the studies suggested following relationships between agroforestry and biodiversity (Martin 2012):

Agroforestry promotes plant diversity: Agroforestry provides way for several combinations of trees and crops of valuable species for human population need which enriches biodiversity of the area.

Agroforestry protects conservation areas: Agroforestry reduces pressure on forests and protected conservation areas by providing forest products to local dependant people. Ramadhani et al (2002) found that 5-year old woodlots of *Acacia crassicarpa* in the Tabora district of Tanzania produced five times as much wood as mature 'miombo' woodlands. Garrity et al (2002) found that agroecosystem around the Mount Kitanglad Range National Park in Mindañao, the Philippines, provides support for a link between agroforestry and reduced pressure on protected areas.

Agroforestry provides habitat for wild species: Agroforestry creates some time good habitat for wild species. The integration of trees with agriculture crops help in nesting, protective cover against predators, access to breeding territory, access to food sources in all seasons, and encourage beneficial species such as pollinators. Buck et al (2004) reviewed 12 studies in agroforestry systems and found it provided habitat for diverse populations of birds.

Agroforestry reduces invasive alien species: Agroforestry reduces the invasion of alien invasive species. Evidence across the world indicates that agroforestry projects have contributed to the ecological problems associated with alien invasive species. ICRAF has adopted a policy that focuses on reducing the risk of introducing invasive alien species as part of new agroforestry research and development programmes.

Use of species in agroforestry: Usually trees selected for agroforestry are multipur-pose in use. These trees are generally used for fodder, fuel wood fruit trees etc.

CONSERVATION OF TREE DIVERSITY IN AGROFORESTRY

The selection of tree species in agroforestry is important decision. Usually, species useful to the growers are preferred. People plant trees for fruit, fodder, fuel, shade and various other purposes. Since a large variety of trees are available which could grow in different edaphic and climatic conditions. Nowadays growing demand of market opportu-nities for tree products results in greater attention for the tree species in agroecosystems. Dwivedi (1992) has suggested following factors the choice of tree species in agroforestry:

SITE FACTOR: Climate, soil, physiography, biotic etc. are the important site factors. These factors affect the growth and performance of the species. Therefore, proper selection of species according to site factor is needed.

EFFECT OF ANNUAL CROPS: The selection of tree species is important, because of their effect on the agriculture crops. Selection of trees should be made based on least and no damages to agriculture crops. Thus ideal trees for agroforestry should be:

- The root system of trees should be deep, thus will be less competition for resources with food crops
- Smaller, conical and cylindrical crown of trees should be preferred because smaller crown also help in accommodating a large number of trees. Oval and umbrella shaped crown will affect the annual crop
- Trees bole should be straight and long i.e., *Eucalyptus*, *Casuarina equisetifolia*, etc.
- Tree should be deciduous, therefore no competition for light, nutrient and moisture during leafless period

PURPOSE OF PLANTATION: Trees are normally selected for several purposes, such as fuelwood, fodder, timber pole, fruits, fiber etc. Trees can also be used for industrial need and environmental protection (water and wind erosion, stabilizing landslides, landslips, disturbed soil and reclamation of wastelands).

MARKETABILITY OF PRODUCE: In earlier times, species are grown for multipurpose and not for marketability, because the tree products' price were very low. In last few decades, the prices of firewood, fodder, minor forest produce, etc. increased with well established market. In this regard, only those tree species which are having fast growth and short term economic return values are used for agroforestry purposes. Some trees have market values e.g. *Eucalyptus* spp., *Populus* spp., *Leucaena leucocephala*, *Casuarina epuisetifolia*, because usually these trees are felled in a period of 8-10 years.

EFFECT ON SITE: The selected trees should have the ability to protect and improve the site conditions. Tree should be selected, where leaves are easily decomposed.

EASE OF ESTABLISHMENT: In agroforestry, several trees are grown and some of them are easy to establish. Some of the good examples of trees are *Eucalyptus*, *Populus*, *Dalbergia sissoo*, *Salix*, etc.

RESISTANCE TO INSECT PEST AND DISEASE: Usually fast growing tree species are liable to attack by insect pest and fungi than slow growing species.

Fuelwood Trees

Grewia oppositifolia, *Celtis australis*, *Ficus roxburghii*, *Quercus leucotrichophora*, *Alnus nepalensis*, *Morus serrata*, *Bauhinia variegata*, *Acacia nilotica*, *Acacia auriculiformis*, *Acacia tortilis*, *Prosopis juliflora, Azadirachta indica, Casuarina equisetifolia, Calliandra calothyrsus, Dalbergia sissoo, Prosopis cineraria, Leucaena leucocephala, Ziziphus mauritiana, Eucalyptus camaldulensis, Eucalyptus tereticornis, Albizia lebbeck, Albizia procera, Butea monosperma,* Senna siamea, *Gliricidia sepium*

Timber Trees

Populus deltoides, Tectona grandis, Dalbergia sissoo, Azadirachta indica, Melia azedarach, Eucalyptus spp., *Pistacea integrrima, Prunus puddum, Quercus* spp., *Shorea robusta, Toona ciliata, Pinus roxburghii, Pyrus pashia, Ficus palmata, Rhododendron arboreum, Myrica esculenta, Cedrus deodara, Lyonia ovalifolia, Prunus cerasoides, Toona ciliata*

Fodder Trees

Faidherbia albida, Acacia senegal, Acacia auriculiformis, Acacia catechu, Acacia leucophloea, Acacia nilotica, Acacia tortilis, Ailanthus excels, Albizia lebbeck, Albizia procera, Albizia chinensis, Bauhinia variegata, Bauhinia purpurea, Hardwickia binata, Celtis australis, Ficus roxburghii, Grewia oppositifolia, Morus alba, Morus serrata, Moringa oleifera, Prosopis cineraria, Prosopis chilensis, Prosopis juliflora, Sesbania grandiflora, Sesbania sesban, Anogeissus latifolia, Anogeissus pendula, Leucaena leucocephala, Ziziphus nummularia.

Fruit Trees

Citrus spp., *Artocarpus heterophyllus, Artocarpus altilis, Litchi chinensis, Mangifera indica, Prunus armeniaca, Prunus domestica, Prunus dulcis, Prunus persica, Psidium guajava, Pyrus communis, Persea americana, Carica papaya, Tamarindus indicus, Passiflora edulis, Musa paradisiaca, Syzygium cumini, Durio zibethinus, Annona squamosa, Annona reticulata, Punica granatum, Aegle marmelos, Emblica officinalis, Ziziphus maurtiana, Manilkara zapota*

Fertilizer Trees

Faidherbia albida, Leucaena leucocephala, Gliricidia sepium, Calliandra calothyrsus, Flemingia spp., *Erythrina* spp., *Casuarina equisetifolia, Sesbania sesban, Sesbania grandiflora, Albizia lebbeck, Acacia tortilis, Acacia senegal, Acacia auriculiformis, Pithecellobium dulce, Prosopis juliflora, Prosopis cineraria, Prosopis chilensis, Dalbergia sissoo, Alnus nepalensis*

Bamboos

Bambusa bambos, B. tulda, B. balcooa, B. nutans, B. vulgaris, B. pallida, B. polymorpha, B. textilis, Dendrocalamus asper, D. giganteus, D. brandisii, D. hamiltonii, D. longispathus, D. strictus, D. latiflorus, D. oldhami, D. membranaceus, Phyllostachys pubescens, P. praecox, P. iridescens, P. bambusoides, P. edulis, P. dulcis, Melocanna bambusoides, M. baccifera, Schizostachyum spp., *Thyrsostachys* spp., *Gigantochloa* spp., *Cephalostachyum* spp., *Sinarundinaria* spp.

■■■■

36 SOIL FERTILITY IMPROVEMENT AND NUTRIENT CYCLING IN AGROFORESTRY

INTRODUCTION

Maintaining the soil fertility is a key prerequisite to sustain crop productivity in the tropics. Several studies perceive that soil fertility decline is widely spread in tropical regions and that it is caused by inadequate nutrient replenishment and high losses as compared to natural ecosystems. Soil quality is of fundamental importance for agricultural production, and soil fertility management is increasingly becoming a central issue in the decisions on food security, poverty reduction and environment management. For the purpose of safe ecological stewardship and achieving global food security, emphasizing soil fertility management is becoming more and more important. Even so, the crucial role of soil fertility management for sustainable resource management and food security has been recognized only quite recently. Agricultural systems with insufficient nutrient input on land with poor to moderate potential are the root cause of human-induced soil degradation. Declining soil fertility indicated by declining soil organic matter levels means crop yields are increasingly at risk of being limited by nutrient shortages. Farming practices which maintain or increase soil organic matter involve growing high yielding crops and pastures and retaining residues to maximize organic matter returns. It also helps to maintain soil structure for optimal water entry, root growth and contributing to the reserve of nutrients in the soil.

Poorer soil fertility often limits production in good seasons and a cycle of decline can develop where a lesser return of organic residues leads to poorer soil structure, higher risk of erosion, depletion of soil organic matter reserves and lower yields. While there has been much soil degradation in past decades, it is simplistic to blame poor fertilizer management alone. Intensity of cropping, cultivation practices and greater productivity (and profitability) all have a role to play. In fact, many of the fertilizer management practices have had demonstrable, positive effects on soil fertility. Declining soil fertility has long been recognized as a major obstacle to sustaining current levels of agricultural production in tropics. Indeed, as fertility declines in the face of rising population and disabling policies, there is the near certainty of exponential deterioration in fertility and decline in yields. Unfortunately, soil fertility research indicates difficulties in finding permanent solutions to the crisis and its impact, initially on rural households but ultimately on national economies.

According to a study by the World Agroforestry Centre (ICRAF) 43 per cent of the planet's agricultural lands (more than a billion hectares) has more than 10 per cent tree cover (Zomer et al 2009). A lesser but still significant area of agricultural land, 160 million hectares, has more than 50 per cent tree cover. The potential of trees to bring improvements in nutrition, income, housing, health, energy needs, and environmental sustainability in the agricultural landscape has guided ICRAF's mission, with the presence of trees being the principal component of an "evergreen agriculture" (ICRAF 2008). Within the array of benefits brought by trees, an important element is the positive effect of trees on soil properties and consequently benefits for crops. Relation between trees and soil, based on agroforestry systems research, as well as studies in non-agricultural or natural environments, demonstrates significant effects of trees on soil. In this scenario, agroforestry systems have been indicated as one of the more promising alternatives to achieve a more sustainable agriculture, in greater equilibrium with the environment (Porrow 2009).

SOIL FERTILITY AND PRODUCTIVITY

Soil fertility is a measure of a land's capacity to sufficiently provide the nutrients that are needed to grow crops. *Soil* fertility is the characteristic of *soil* that supports abundant plant life. When soil fertility is considered in terms of the highest practical level of productivity, the focus is mostly on physical and chemical aspects of soil. The fertility of soil can be considered in different ways, depending on land use. In intensively managed agricultural and horticultural systems, and even in forestry, soil fertility can be defined in terms of the value of products produced relevant to inputs used (including economic aspects of nutrient budgeting). Alternatively, the emphasis may be on quality or productivity. There are three main components of soil fertility - physical, chemical and biological. The level of soil fertility results from the inherent characteristics of the soil and the interactions that occur between these three components.

- The fertility of soil is related to its capacity to produce a product.
- The fertility of soil is related to its capacity to support a particular natural community of plants.
- The fertility of a soil is related to its capacity to maintain consistent output with minimal input.

Soil fertility deals with the nutrient status or ability of soil to supply nutrients for plant growth under favourable environmental conditions such as light, temperature and physical conditions of soil. Soil productivity is defined as the capability of the soil for producing a specified quantity of plant produce per unit area and the ability to produce sequence of crops under a specified system of management. Integrated Nutrient Management (INM) is a way of combating nutrient depletion. It is an approach to soil fertility management that combines organic and mineral methods of soil fertilization with physical and biological measures for soil and water conservation. FAO uses the term Integrated Plant Nutrition Systems (IPNS) which has very much the same meaning. An agricultural ecosystem differs from a natural ecosystem in that plant nutrients are constantly being removed and exported. It is a holistic view of plant nutrient management, since it takes into consideration the totality of the farm resources that can be used as plant nutrients. It is based on the following principles:

1. Maximize use of organic materials
2. Judicious use of inorganic fertilizer
3. Minimize losses of plant nutrients

The objective of any soil and crop management program is sustained profitable production. The strength and longevity of any civilization depend on the ability to sustain and/or increase the productive capacity of its agriculture. From soil productivity and soil fertility standpoint, soil conservation is essential for long-term sustainability. Soil management practices that contribute to or encourage soil degradation will reduce soil productivity and impair progress towards sustainability. Soil erosion represents the greatest threat to sustain soil productivity. Physical removal of nutrient-rich, high organic matter from top soil and oxidation of SOM with tillage reduce the productive capacity of the soil. Exposed sub-soil is often less productive. Soil erosion is a symptom of poor soil management, whether it is inadequate plant nutrients or improper cropping systems.

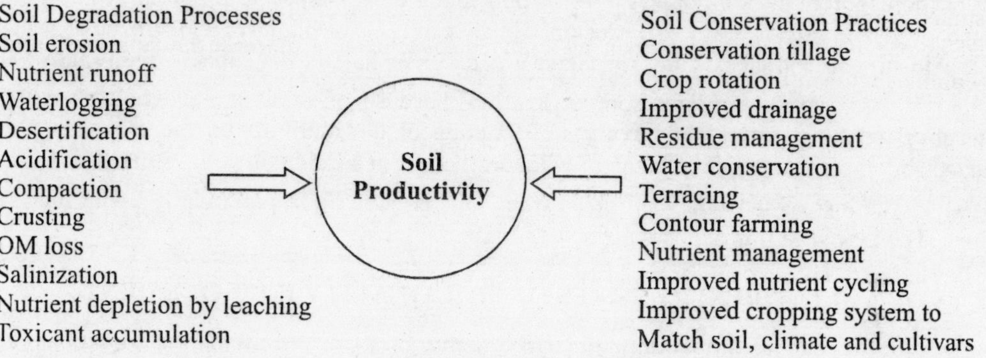

Fig. 36.1: Soil Productivity Linkage with Soil Degradation Processes and Soil Conservation Practices

FACTORS RESPONSIBLE FOR POOR SOIL FERTILITY AND PRODUCTIVITY

Following factors are responsible for poor soil fertility and productivity:

SOIL EROSION: Soil erosion results in huge loss of plant nutrients in suspension or solution, which are removed away from one place to another, thus, causing depletion or enrichment of nutrients. It is of primary interest to agricultural scientists because of its combined effects on crop production.

SALTS AND ALKALINITY: Every year, good agricultural lands are reported to have gone out of cultivation due to salinity and alkalinity. Ultimately, thousands of hectares of land are becoming unproductive because of this problem.

INFECTIOUS ORGANISMS: Man's desire to maximise production by controlling one or other disease, pest or micro-organism sets in train a new series of changes and disturbs the equilibrium. Measures against seed borne diseases are often used as seed coating with disinfectants and fungicides. But the mercury in fungicides and arsenic in pesticides is the major cause of poor soil fertility.

INDUSTRIAL WASTES: Fluorides and sulphur dioxide causing leaf injury to plants and sulphur dioxide is known to reduce the pH of the soil.

USE OF PESTICIDES: The changes which pesticides produce in soils have not been fully investigated. Their movement in soil leaching water and absorption by plants and changes in microbiological activity in soils are the problem areas for investigation. However, SOM is very effective in detoxification.

FERTILIZERS: *Nitrogen*: Major loss of N is through soil erosion and through denitrification. But losses through leaching can be considerably more particularly under conditions such as rice cultivation. In neutral and acidic soils, NH_3 volatilization occurs from urea. Surface application of fertilizers increases NH_3 volatilization. NH_3 volatilization from bare soil, coarse textured soil, calcareous or alkaline soil is large. Toxicity of N due to application of high dose of urea and $(NH_4)_2SO_4$ fertilizers in band placement, NO_3 accumulation in soil at pH value of >7.8 and high concentration of free NH_3 and NO_3. Application of high dose of $(NH_4)_2SO_4$, anhydrous NH_3 or ammonium sulphate nitrate fertilizer up to long period resulting loss of Ca from surface soil and finally produce acidity in soil.

Phosphorus: The loss of P from the fertilizer through leaching are very small because of phosphatic fixation phenomenon in the soil, unless a soil is pure stand or is subject to flooding. The major source of phosphate loss in soil is from soil erosion which results in mechanical movement of potentially available P to streams and lakes. Clay soils have greater phosphate fixing capacity than sandy soils.

Potassium: K losses from fertilizer sources are low because of rapid movement from soluble to exchangeable and fixed forms and luxury consumption of the nutrients by the crops. In poor and unfertilized soil, there is a greater chance of K losses than from a rich and well fertilized soil.

Ca, Mg & S: The loss of secondary nutrients from soils is quite high, particularly under high rainfall conditions. The loss of SO_4 in sandy soils is very high.

EFFECTS OF AGROFORESTRY TREES ON SOIL QUALITY IMPROVEMENT

In farming systems, soil fertility is defined as the "ability of a soil to provide the conditions required for plant growth" (Stockdale et al 2002). The soil is not only seen as a substrate for crops but also as a key-component of the ecosystem services supply chain. Thus, Larson and Pierce (1994) suggest the concept of soil quality, "represented by a range of physical, chemical and biological properties of the soil within its particular environment that together provide a medium for plant growth and biological activity, regulate and partition water flow and storage in the environment, and, serve as a buffer in the formation and destruction of environmental hazardous compounds". The trees have three main types of effects on the soil quality. They influence soil structure, soil fauna and flora, and the nutrients content of the soil.

EFFECTS ON SOIL STRUCTURE

The presence of trees in a crop field influences soil structure via two main phenomena. Firstly, the trees bring organic matter to the soil (litter fall, twigs, roots turnover, pruning by-products) and significant increases in OM have been observed in many agroforestry practices especially alley cropping systems planted with nitrogen fixing leguminous trees. Secondly, the trees develop important root systems that explore and improve notably deep and compacted soil layers that crop roots do not reach (Schroth 1999). These two phenomena have major effects on soil structure. By increasing the cover of the soil, the presence of the trees limits the effects of wind and rain erosion. The increase of organic matter and the development of root system limits soil compaction, improves water infiltration & the porosity of the soil, and increases basal soil respiration (Chander et al 1998). Moreover, they increase the weight and diameter of soil aggregates and so, increase soil stability and decrease erosion risks in response to rainfall. Finally, the trees improve the soil holding capacity for water and nutrients. These effects can be beneficial for the crops because the roots ability to permeate the soil and the plant capacity to uptake and export water and nutrients from the soil are improved (Jose et al 2008). Moreover, they can create better conditions for soil biology.

EFFECTS ON SOIL MICRO AND MACRO BIOLOGY

Different observations of microbial community dynamics in ecosystems characterized by the presence of trees showed a correlation between the proximity to trees and an increase of both fungal and bacterial active biomass in the top fifteen centimeters of the soil. Moreover, the fungi/bacteria ratio and the number of earthworms are higher close to the trees because of the increased C-rich organic matter input. A likely conclusion from these observations is that the trees provide physical habitat and nutrient-rich substrate for fungi and bacteria. Actually, the supply of organic matter in the topsoil creates a better environment for microorganisms in the rooting zone (Jose et al 2008) but the latter are essential actors of the biochemical degradation of organic matter. As part of the recycling process, the soil microorganisms convert pruning by-products, litter fall and crop residues into crops-available and mineralized-form of nutrients (Lejon et al 2005). Seiter et al (1999a) and Lejon et al (2005) focused on the composition of microbial biomass and the types of mycorrhizal associations. The authors say that although the trees are influenced notably by the acidity, the quality and the quantity of organic matter input and the root exudates that differ between conifer and hardwood species, they themselves influence the speed of organic matter decomposition and the type of nutrients that are released. Among the microbial, rhizospheric and mycorrhizal organisms that interact with tree root systems, two distinct groups are particularly relevant in relation to their influence on the nutrient availability: the nitrogen-fixing and phosphorus-solubilizing organisms and/or associations (Schroth 1999).

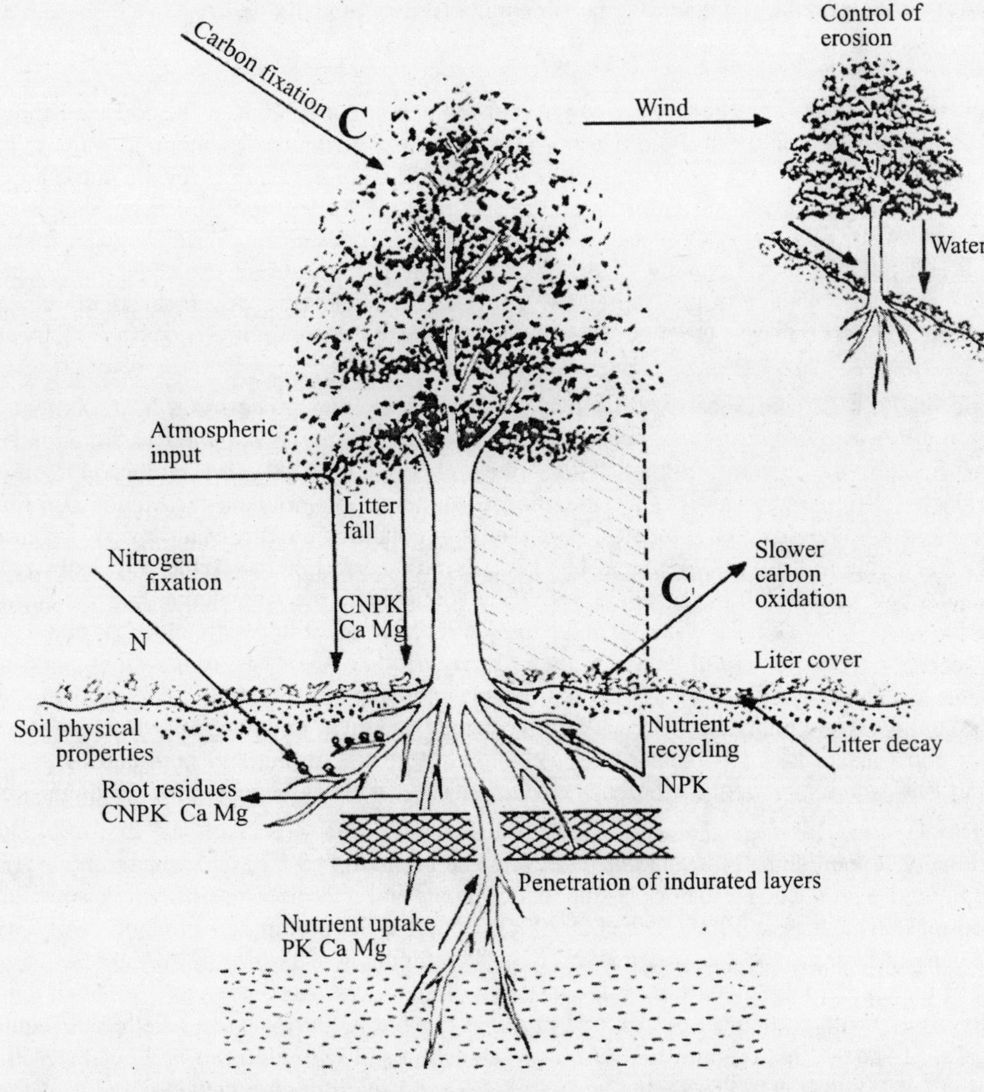

Fig. 36.2: Processes by which Trees improve Soil (Source: Young 1989)

The diversity and density of macroinvertebrates in the soils of natural forests are considered high, and the conversion of these ecosystems to other forms of land use can cause severe changes in their populations (Lavelle et al 1995). Barros et al (2008) cite soil macroinvertebrates as sensitive indicators of natural systems, land use and management, with agroforestry systems presenting a greater abundance and diversity when compared to other forms of agricultural land use. Tapia-Coral et al (1999) found greater densities and biomass of Diplopoda and Isopoda in the litter layer of agroforestry systems than in adjacent secondary forests. Secondary forests, however, presented greater values for Isoptera, principally during the rainy season. This is an indication that such systems can support sufficiently abundant and diverse communities of these organisms that play an important role in ecological processes such as breakdown and cycling of organic matter. However, agroforestry systems encompass a range of practices, such that there is considerable variation in the quality of organic matter produced and the effects of vegetation on soil humidity and temperature regime, with significant effects on the soil macrofauna community (Barros et al 2008).

EFFECTS ON SOIL NUTRIENTS CONTENT

The presence of trees in the ecosystem can influence the nutrient content of the soil via its impacts on the soil structure and on the soil biology but also due to the particular functioning of trees. There are three levels of interactions. First of all, trees can be providers of new inputs for the nutrient cycle. For instance, tree canopies collect atmospheric depositions that are incorporated into the soil when the leaves fall. Also, deep roots of the trees can capture nutrients and minerals from the rock and the subsoil and pump it to the canopy so that it is an input in the nutrient cycle. As well, certain tree species associate with nitrogen-fixing or phosphorus-solubilizing mycorrhizal or rhizospheric organisms and make these nutrients available for the plants of the entire ecosystem (Chander et al 1998).

Secondly, trees can be active drivers in the nutrient cycling processes. Kho (2008) studied the nitrogen cycle in agroforestry. In such agroecosystems, nitrogen is supplied to the topsoil by the fertilizer and/or the manure applied in the alleys (Jose et al 2000), the biological fixation from atmosphere (N-fixing tree species) and the mineralization of organic matter (which is also influenced by the trees) (notably by root exudates, improved soil conditions and root turnover) (Lehmann et al 1999). According to Kho (2008), the problem is that nitrates are highly soluble and easily transported by runoffs and leaching through soil layers. They contaminate the water table and the rivers which make the water unsuitable for drinking and cause eutrophication of downstream ecosystems. To avoid such seepages, trees are helpful thanks to their "safety net", i.e. their vast root system that can uptake nutrients in the deeper soil layers where crop roots cannot take (Jose et al 2008). By actively capturing leached nutrients and pumping them back to the canopy and then, to the topsoil, the trees decrease the erosion and nutrient losses (Chander et al 1998), help to maintain nutrient pool and the soil fertility (Young 1997), and improve the efficiency of nutrients use in the agroecosystems (Rigueiro-Rodriguez et al 2009).

Finally, some tree roles in the nutrient cycle can be improved by human intervention. Litter fall and pruning by-products are major organic matter inputs and their decomposition releases nutrients in the topsoil (Chander et al 1998; Jose et al 2008). But what the pruning by-products really matters to avoid nutrient losses at the ecosystem level (Kho 2008; Seiter et al 1999b). If they are used as chipped wood or mulch, the nutrients are recycled in the agroecosystem. According to Kho (2008), in certain conditions and with particular species, tree prunings content in nutrients can meet crops requirements in N, Ca, Mg and K. In addition, root turnover can enrich soil with nutrients and notably with carbon (Schroth 1999; Gupta et al 2009) such as trees influence C/N ratio (Augusto et al 2004). But the root management practices such as root prunings accelerate the root turnover and enhance the redistribution of nutrients through lateral roots (Chander et al 1998). In fact, the roots contain a lot of nutrients that travel easily along the whole root system. By pruning the roots, agroforesters induce root decomposition and the release of nutrients in the topsoil throughout the entire crop field.

TREES AND SOIL CARBON

Trees add organic matter to the soil system in various manners, whether in the form of roots or litterfall or as root exudates in the rhizosphere (Bertin et al 2003). These additions are the chief substrate for a vast range of organisms involved in soil biological activity and interactions, with important effects on soil nutrients and fertility. In participating in these complex processes, trees contribute to carbon accumulation in soils, a topic that is increasingly present in global discussions on the mitigation of greenhouse gases associated with global warming and climate change. Although carbon (C) constitutes almost 50 per cent of the dry weight of branches and 30 per cent of foliage, the greater part of C sequestration (around 2/3) occurs belowground, involving living biomass such as roots and other belowground plant parts, soil organisms, and C stored in various soil horizons (Nair et al 2010).

In a study that gathered information from sites around the world, Nair et al (2009) found values for soil organic C stocks ranging from 6.9 to 302 Mg ha^{-1}. Despite the great amplitude of these values, attributed to the variation between systems, ecological regions, and soil types, the study revealed a general trend of increasing soil C sequestration in agroforestry when compared to other land-use practices, with the exception of forests. Although the ability of soils to accumulate C is generally related to characteristics that are little influenced by management, such as texture (clay soils typically accumulate more C than sandy soils), some management practices can influence soil C sequestration, particularly the insertion of trees in agricultural systems (Pinho et al 2012). Soils in various sites studied by Takimoto et al (2008) in the African Sahel were not markedly different among each other in terms of their characteristics such as pH, bulk density, and particle size, such that variations in their C contents seemed to be related to the influence of trees. In eight years old alley-cropping systems with five different species, for example, the authors found that greater C content is nearer to the trees. However, the greater part of this C was found in the form of particles of size between 250-2000 μm, fractions that are considered to be large and less stable. In systems where trees where present for more than 30 years (parklands), there was a predominance of soil C in smaller fractions (<53 μm), which are more stable and thus represent a more "protected" form of C.

In homegardens in India, Saha et al (2010) found levels of soil organic C to be 30 per cent and 114 per cent greater than in coconut plantations and rice paddies, respectively, when homegardens where small (less than 0.4 ha) and levels 18 per cent and 94 per cent greater in larger homegardens (larger than 0.4 hectares). Although the authors do not elaborate on the reasons for decrease in soil C in the larger homegardens, this may be related to a drop off in the deposition of organic residues, which generally occurs in proximity to dwellings (Pinho et al 2012). In the Northeast of Spain, Howlett et al (2011) studied the levels of soil C in silvopastoral systems composed of different species, and found that systems with birch (*Betula pendula*) presented greater levels of soil C than systems with pine (*Pinus radiata*). This difference was attributed to the fact that *Pinus radiata* creates an understorey environment that is less conductive to plant growth by the formation of a slowly decomposing duff layer, which inhibits germination and growth of other species and, over time, reduces C inputs to the soil.

CONCEPT OF NUTRIENT CYCLING

One of the main principles of soil management in agroforestry is to make the best use of its resource-conserving and resource sharing potentials. Therefore, it is extremely useful to have a proper nutrient budget for the whole system based on nutrient requirements of individual components and nutrient dynamics within the system. Nutrient cycling processes, which take place to varying degrees in all land use systems, become particularly relevant in the agroforestry context because of the likely effects of trees on such processes. 'Closed' nutrient cycles are known to operate in mixed evergreen natural forests. A simplified model of the nutrient cycle in the tree-based ecosystem is presented in Fig. 36.3.

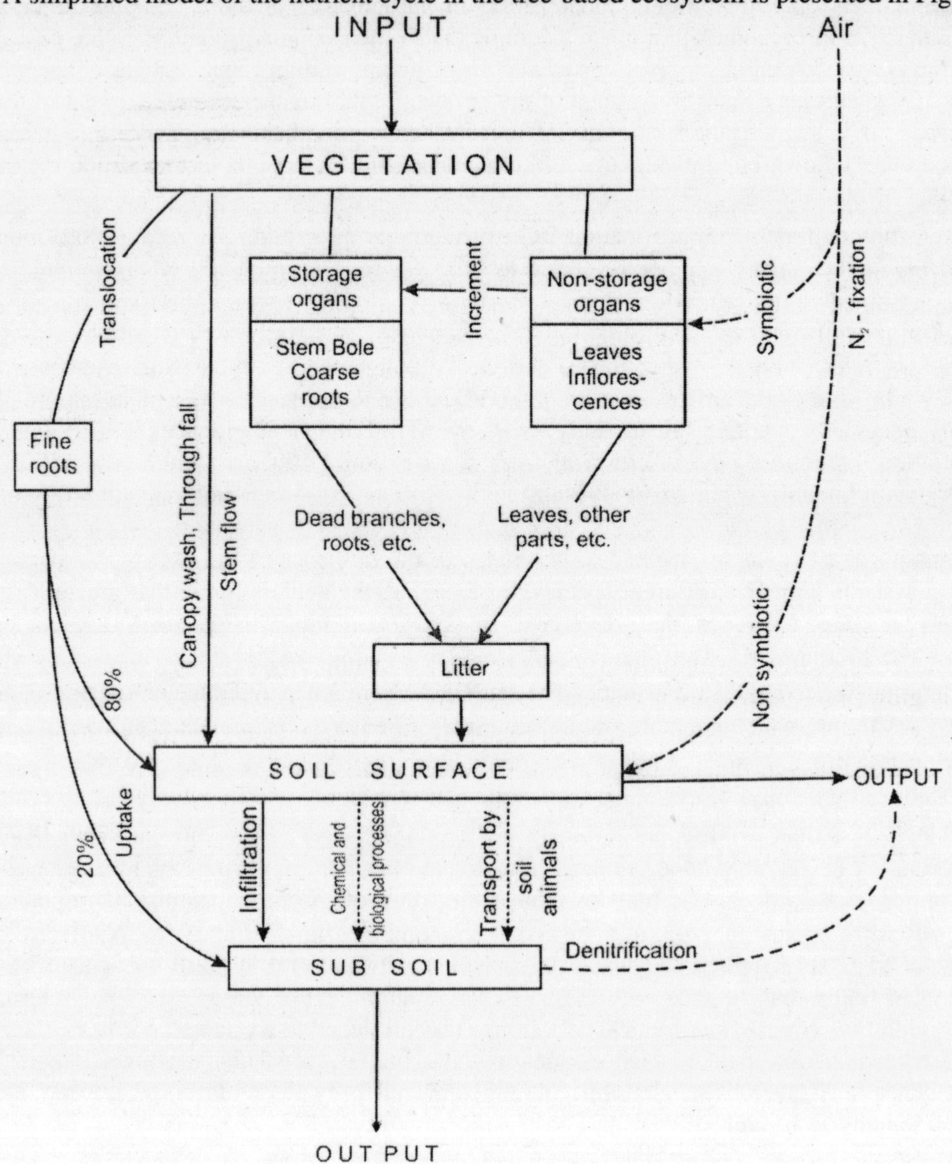

Fig. 36.3: A Simplified Model of Nutrient Cycling in Tree-based Ecosystem (Source: Nair 1984)
(Dotted lines indicate the biological process taking place with the involvements of other organisms)

The model consists of the soil-plant system that is partitioned into several compartments. The crown surface forms the boundary of the system where input of bio-elements occurs through precipitation. The soil surface is the entry point for inputs into the soil compartment, occurring through fertilizers, rainfall and stream flow. The surface layer may be considered the zone of intensive root activity, with the sub-soil constituting the extensive root activity zone. The lower end of the extensive root layer is the boundary of the ecosystem to the hydrosphere and lithosphere. Bio-elements transported beyond this layer are lost from the ecosystem and appear as output from the system.

Nutrients taken up by the plant are either stored in an increment (storage) compartment or are used for the production of non-storage organs. It is well known that fertilizer application results in an input of nutrients and the accompanying ions into the solid phase of the uppermost layer of the soil. Depending on the water content of the soil and the solubility of the fertilizers, they pass into the solution phase of the same soil layer, and then spread simultaneously, depending on the mobility of the ions in different layers of the soil, into the plant stand via uptake. Based on the flow rate of percolating water and the soil properties, part of the nutrients in the soil solution is washed out of the nutrients- absorbing zone and represents a loss (output) from the system.

Plant nutrients are therefore involved in a constant and somewhat closed cycling within the soil and plant compartments of tree-based ecosystems, with minimal output (loss) from the system. The extent to which the system is 'closed' or 'open' depends on various factors, the decisive one being the tree/crop proportion. Even in agroforestry systems with a high proportion of woody perennials, the frequency and method of harvesting their produces, the structure and density of the root system, the type of magnitude of disturbance (including management) to the system, and the spatial and temporal arrangement of species are among the important factors that contribute to the effectiveness of the nutrient cycling processes in soil productivity. Some tree species have also been used in reclamation of problematic soils, such as saline and alkaline soils, for example, *Prosopis juliflora*, *Acacia tortilis*, etc., through uptake of salts and addition of organic matter.

NUTRIENT CYCLING IN NATURAL FOREST ECOSYSTEMS (IN EQUILIBRIUM)

The existence of nearly closed nutrient cycles between a mature, tropical moist forest and the soil on which it grows has been recognized and studied for about 60 years (Hardy 1936; Vitousek and Sanford 1986). Nutrient inputs from atmospheric deposition, biological nitrogen fixation and weathering of primary soil minerals are in balance with nutrient losses owing to leaching, denitrification, runoff and erosion. Nitrogen (N), phosphorus (P), potassium (K), magnesium (Mg), calcium (Ca) as well as sulphur (S) and micro-nutrients are absorbed by forest roots and returned to the soil via the decomposition of litter and roots as well by throughfall and stemflow.

Tropical humid forests accumulate huge quantities of nutrients in their vegetation as the forest grows, with a mature forest reaching steady state values of 700 to 2,000 kg N, 30 to 150 kg P and 400 to 3,000 kg K, Mg or Ca per hectare. The soil also contains large quantities of nutrients. The efficient cycling of nutrients from the soil to the biomass and back to the soil makes possible the lush tropical forest growing on acid, relatively infertile soils of the humid tropics, as long as there are no major biomass removals from the system. Improper harvesting of forests results in major disruptions of this process because large quantities of nutrients are removed from the system and nutrient cycles are disturbed (Sanchez and Palm 1996).

NUTRIENT CYCLING IN AGRICULTURAL SYSTEMS (IN DISEQUILIBRIUM)

Agricultural systems differ from natural systems in one fundamental aspect: there is a net output of nutrients from the site via crop harvest removals. This nutrient removal can result in net negative balances if nutrients are not replaced. Nutrient depletion can be offset by fertilizers, manures from outside the field and other nutrient inputs. This is generally the case in commercial farms of the developed world where such additions, coupled with the recycling of crop residues, have resulted in major nutrient accumulations. Such nutrient buildups, however, sometimes result in the pollution of groundwaters and in algal blooms in streams.

The magnitude of nutrient mining as a result of crop harvests in tropics is huge. Net losses of about 700 kg N, 100 kg P and 450 kg K per hectare during the past 30 years have been estimated for 100 million ha of cultivated land in Africa. Smaling's (1993) studies on nutrient balances throughout Africa lead the authors to conclude that soil fertility depletion is the fundamental biophysical reason for the declining per ha food production in smallholder farms in Africa (Sanchez et al 1995). Nutrient mining in tropical regions, therefore, contrasts sharply with nutrient capital buildups in temperate regions.

NUTRIENT CYCLING IN AGROFORESTRY SYSTEMS

Two key principles set agroforestry systems apart from agricultural or forestry systems: competition and complexity. They in turn determine two desirable properties: profitability and sustainability. All four combine biophysical and socioeconomic issues. The biophysical bottom line of agroforestry is how to manage the competition between the tree component and the crop and/or livestock components for light, water and nutrients to the benefit of the farmer. Although agroforestry systems have been classified in a myriad different ways, there are two functionally different types, simultaneous and sequential systems. Simultaneous agroforestry is where the tree and the crop components grow at the same time and sufficiently close to each other to allow competition for light, water or nutrients. Examples of this type are alley cropping (hedgerow intercropping), contour hedges, parklands, boundary plantings, home gardens and several silvipastoral systems. Simultaneous systems can vary greatly in the relative proportions of trees and crops and in their spatial arrangement. Spatially mixed systems such as coffee and cacao plantations include upper-storey trees that provide nutrients, from litter fall and periodic prunings, and shade to the lower storey tree crops. Spatially zoned systems include hedgerow intercropping where annual crops are grown in alleys between rows of trees that are pruned periodically to provide nutrients to the crops in the alleys.

Competition for nutrients is maximized in simultaneous agroforestry systems, particularly in short-duration systems such as alley cropping. The available literature, comprising more than 100 alley cropping experimental data sets ranging from semiarid to humid climates in widely different soils, has recently been analyzed. Many factors affect alley cropping performance: the choice of tree and crop species, the alley width, the pruning regime, biomass production, the number of crop cycles, the time and frequency of prunings, tillage, fertilization, weed dynamics, etc. The overall conclusion appears to be clear: the chances that alley cropping will work are limited and very site-specific because, in most cases, the tree's competition for water and nutrients is likely to exceed the fertility benefits from the leguminous mulch additions. Sequential agroforestry systems are those where the maximum growth rates of the crop and the tree components occur at different times, even though both components may have been planted at the same time and are in close proximity. Examples of this type are shifting cultivation, improved fallows, taungya and some multistrata systems. Competition for growth resources is minimized in sequential agroforestry because the peak demands for light, water and nutrients occur at different times for each component.

One of the main tenets of agroforestry is that trees enhance soil fertility, the capacity of the soil to provide essential nutrients for plant growth. There is often confusion between key terms such as nutrient inputs, nutrient outputs, nutrient balances, nutrient cycling and nutrient capital. They all refer to a soil-plant system, usually at the scale of the farmer's field. Nutrient inputs are additions originating from outside the system, such as nitrogen fixed from the air by legumes or the use of chemical fertilizers. Animal manures are inputs if the manure was produced outside the system Nutrient outputs are those that leave the system through crop harvest removals, soil erosion, leaching, gas volatilization and other processes. The nutrient balance is the difference between nutrient inputs and outputs. Nutrient cycling refers to the transfer of nutrients already in the soil plant system from one component to another, for example the release of nitrogen from soil organic matter as ammonium or nitrate and its subsequent uptake by plants (Sanchez and Palm 1996).

Forest ecosystems represent closed and efficient nutrient cycling systems, meaning that they have high rates of turnover and low rates of out puts or losses from the system. In other words, they are self-sustaining. On the other hand, common agricultural systems are often open, meaning that the turnover within the system is relatively low and losses as well as inputs are comparatively high. The Sources of nutrient addition in agroforestry system that should be taken into account are:*Trees:* Litter (leaf fall/pods/fruits + small twig etc.)

- *Crop:* Manures and fertilizers, mineralization, N- fixation, weathering, crop litter, etc.
- *Rainfall* also accounts for nutrient additions via through fall, stem flow and atmospheric dust

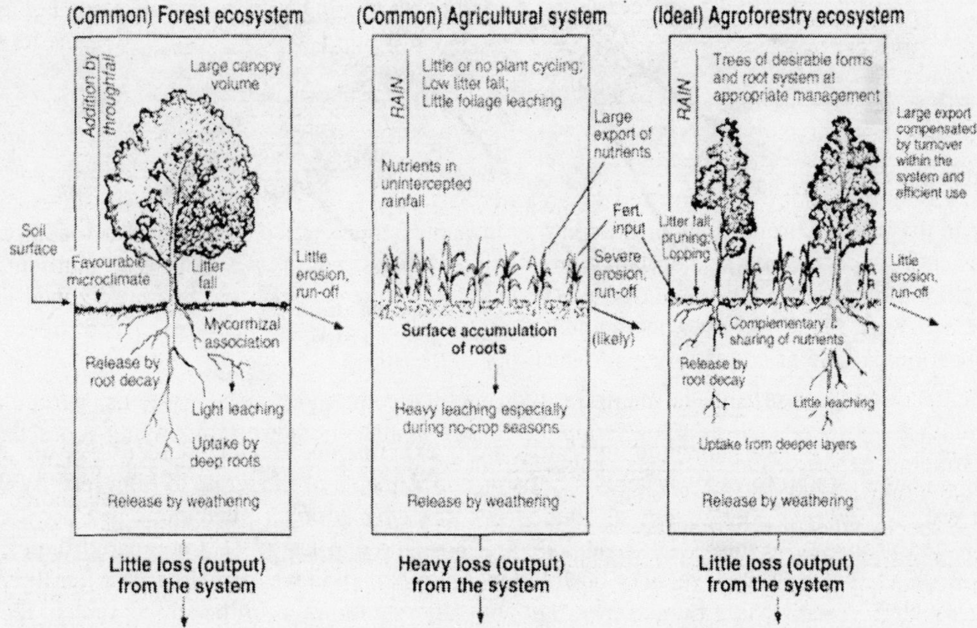

Fig. 36.4: Comparison of Pattern of Nutrient Cycling in Forest, Agriculture and Agroforestry (Source: Nair 1984 and 1993)

Other processes involved in nutrient cycling are the return of crop residues such as stover back to the soil; manure and urine deposited by cattle in the system; the incorporation of leguminous green manures into the soil and the transfer of nutrients from trees to crops in agroforestry systems through prunings, leaf drop or root decomposition. The fewer the nutrient losses from the system, the fewer the inputs needed from outside the system to balance the budget. The nutrient capital refers to the soil reserves of nutrients that will be released gradually over a time scale of years or decades. One fundamental principle of sustainability is to return to the soil the nutrients removed through harvests, runoff, erosion, leaching, denitrification and other loss pathways (Palm 1995; Sanchez 1995).

Agroforestry, the growing of trees with crops and/or livestock on the same piece of land, is believed to promote a more efficient cycling of nutrients than agriculture. This hypothesis is based partially on studies of the efficient cycling of nutrients from litter to trees in natural ecosystems, and the assumption that trees in agroforestry systems will likewise transfer nutrients to intercropped plants. This is supported by observations of higher crop yields near trees of *Faidherbia albida* in the African Sahel and where trees have been recently removed, as in the case of bush and tree fallows. Trees, therefore, can have an effect on soil fertility; however, one must also consider the relative importance of other factors such as soil structure, soil organic matter and the competition for light, water and nutrients before expecting a positive effect on soil fertility in a particular system.

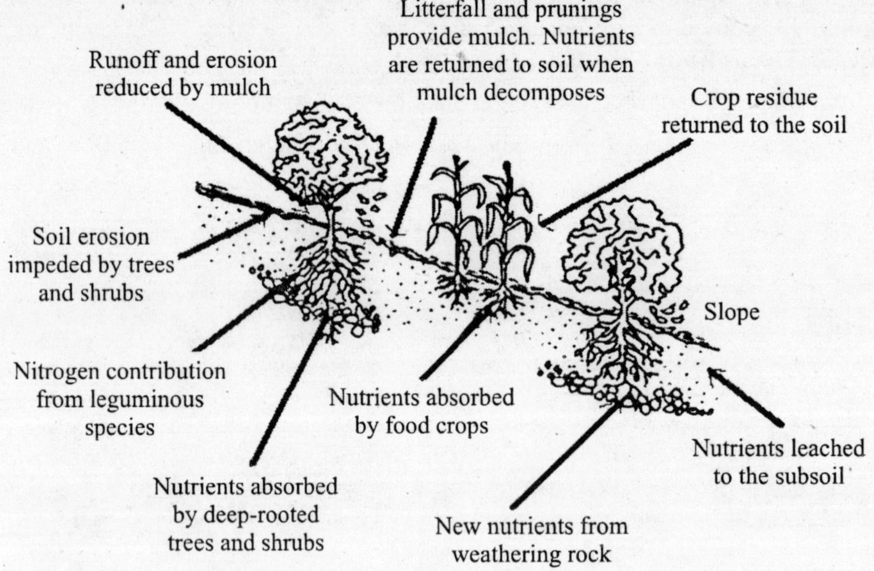

Fig. 36.5: Pattern of Nutrient Cycling under Agroforestry (Source: Nair 1993)

Sanchez (1987) cited encouraging results from experiments conducted to assess the nutrient cycling potential of agroforestry systems in alfisols and Andepts of moderate to high fertility. Studies on the use of *Erythrina poeppigiana* as shade trees in *Coffea arabica* plantations in Cost Rica have also yielded promising results (Glover and Beer 1986). Juo and Lal (1977) compared the effects of *Leucaena leucocephala* fallow versus a bush fallow of an Alfisol in western Nigeria. After three years, during which *L. leucocephala* biomass was cut annually and returned to the soil as mulch, the cation exchange capacity and levels of exchangeable calcium and potassium were significantly higher in *L. leucocephala* fallow than in the bush fallow. The results of investigations in the Amazon basin of Peru showed that managed leguminous fallows significantly increased soil nutrient (N and P) levels (Szott et al 1991). Agamuthu and Broughton (1985) showed that nutrient cycling in oil palm plantations where leguminous cover crops (*Centrosema pubescens* and *Pueraria phaseoloides*) were used was more efficient than in plantations where there was no cover crop. Besides addition of about 150 kg N ha^{-1} yr^{-1}, the loss of nitrate nitrogen through leaching was significantly lower in the plantations with cover crops.

Young (1989) reviewed different studies on nitrogen content of litterfall and prunings in various agroforestry systems of humid and moist sub-humid climates and compared them from natural vegetation communities. In alley cropping systems, some species are capable of supplying 100-200 Kg N ha^{-1} yr^{-1} (even more in some cases) if all the prunings are left on the soil which is approximately the same as the amount of nitrogen that is removed during harvest in cereal/legume intercropping system. In coffee and cacao plantations with shade trees, the return from litter and prunings is 100-300 kg N ha^{-1} yr^{-1}, which is much higher than the amount removed during harvest or derived from nitrogen fixation.

■■■■

37 AGROFORESTRY PRACTICES FOR SOIL AND WATER CONSERVATION

INTRODUCTION

Soil and water are most essential for the growth and sustenance of plant life. Soil is important as it provides, foothold for plants and majority of nutrients needed by them. Water is essential as it forms larger part of the living matter and acts as a nutrient carrier. Though, both soil and water resources are available in plenty, they are not distributed equally in quality and quantity in every part of the world and are not inexhaustible. Their abuse would mean a great loss resulting in poverty. It takes centuries to form one inch layer of soil, but it does not take long to lose it by erosion. Research has shown that loss of soil from unprotected land is as much as 125 tons per hectare every year and may be as high as 300 tons in a single year. The weight of one hectare of soil 2.5 cm deep is about 325 tons. Similarly, rain water which can sustain a good crop, if not conserved properly will not only cause scarcity and famine, but also wash away the soil which is a valuable national asset. There are many examples which show how once fertile plains and valleys have become deserts or barren lands due to neglect by mankind. It is the prime responsibility of each generation to conserve soil, which is the main capital of the farmer as well as the nation, at all costs and pass it on in good condition from one generation to another.

The fertility status and the productivity of soil are largely dependent on the top soil which, besides being a producer of biomass, is important for many other well-known important functions. The loss of top soil due to erosion is largely responsible for land degradation and produce considerable negative impacts both on-site as well as off-site. Water erosion accounts for 53 per cent of the degradation of the 107.4 million ha degraded land in India (MOA 1994). It is estimated to erode 5,334 million tonnes of soil annually (16.35 Mg $ha^{-1}yr^{-1}$), 29 per cent of which is lost to the sea (Dhruvanarayana and Ram Babu 1983). Erosion is especially serious in mountainous areas such as the Himalayas, where soil loss is as high as 80 Mg $ha^{-1}yr^{-1}$ on steep slopes (Singh et al 1992). Soil loss of 20 Mg $ha^{-1}yr^{-1}$ is common on mild slopes in the foot hills and valley regions (Narain et al 1992). Thus conservation of natural resources like soil and water are major priority areas of concern which needs immediate attention. Conservation aspect needs to be addressed at top priority, but it may not be adoptable unless reasonable production is also ensured. The need is therefore to devise/refine land use systems and management practices which can enhance the productivity besides conserving the resources on sustainable basis. In this context, agroforestry offers affordable alternative for providing significant tangible (viz., improved production of food, fuelwood, fodder, timber and other products) and intangible (viz., carbon sequestration, hydrological balance, soil fertility recovery and slope stability) benefits (Young 1989).

SOIL EROSION

The United Nations Convention to Combat Desertification (UNCCD) has defined soil erosion as one of the main causes of land degradation and desertification. Soil erosion means the processes by which soil is removed from one place by forces such as wind, water, waves and human activities and eventually deposited at some new place (Choudhury and Jansen 1999). Soil aggregates can gradually break down into smaller particles. If this happens on sloping land, these particles may be transported down the slope. The water which runs from a slope during a heavy rain is called run off. Important factors influencing soil erosion are the rainfall pattern, slope steepness (gradient), slope length, soil type, existing erosion control structures, cropping practices and time. The stronger the rain and the steeper the slope, the more and the faster water erosion occurs. In contrast to natural erosion which happens slowly, human-induced erosion can happen fast with large amounts of soil being removed. If this happens, it can be a serious threat to agricultural production and the environment. Erosion always has on-site effects, i.e. consequences at the place from where the soil is removed and off-site effects, i.e. consequences at places which are affected by the transport of eroded soil or where the removed soil is deposited.

GEOLOGICAL EROSION

Normal or geologic erosion is a normal feature of any landscape. It occurs where soil is in its natural environment surrounded by its natural vegetation without human disturbance. Geologic erosion has been taking place naturally for millions of years and it helps to create balance in uncultivated soil that enables plant growth. There is always equilibrium between the removal and formation of soil, so that unless some outside agents disturbs the equilibrium, the mature soil preserves more or less, a constant depth and character indefinitely. Geologic erosion takes place gradually but so slowly that ages are required for it to make any noticeable alteration in the major features of the earth`s surface.

ACCELERATED EROSION

It is an erosion process occurring at an increased rate under conditions of ecological disequilibrium. Accelerated erosion is the most dangerous type and it needs concerted efforts through careful planning and implementation of appropriate control measures. Accelerated soil erosion is the removal of the surface soil from areas denuded of their natural protective cover as a result of human and animal intrusion and takes place at a much faster rate than that it is built up by the soil-forming process. This accelerated detachment rapidly ravages the land and this type of soil erosion is considered serious. Nature requires, on an average, about 1,000 years building 2.5 cm of top soil but wrong farming methods may take only a few years to erode it from lands of average slope.

WIND EROSION

The wind can remove the valuable fine soil on the land surface. Wind erosion takes place normally in arid and semi-arid areas that is devoid of vegetation, where the wind velocity is high. The soil particles on the land surface are lifted and blown off as dust storms. When the velocity of the dust-bearing winds is retarded, rough soil particles are deposited in the form of dunes. Thus, fertile lands are rendered unhealthy for cultivation. In other places, fertile soil is blown away by winds and the subsoil is exposed, as a result the productive capacity of the soil is considerably reduced.

WATER EROSION

It is a two-part process involving the detachment and transport of soil particles. The water erosion process consists of discrete stages from rain drop impact to the formation of gully erosion. Each stage has its own processes and characteristics. Controlling or preventing water erosion requires an understanding of each step in the erosion process. Water erosion causes severe soil erosion and this category of soil erosion can be distinguished in four forms, namely splash erosion, sheet erosion, rill erosion, and gully erosion.

Soil erosion = f (Erosivity of rain) x (Erodibility of soil)

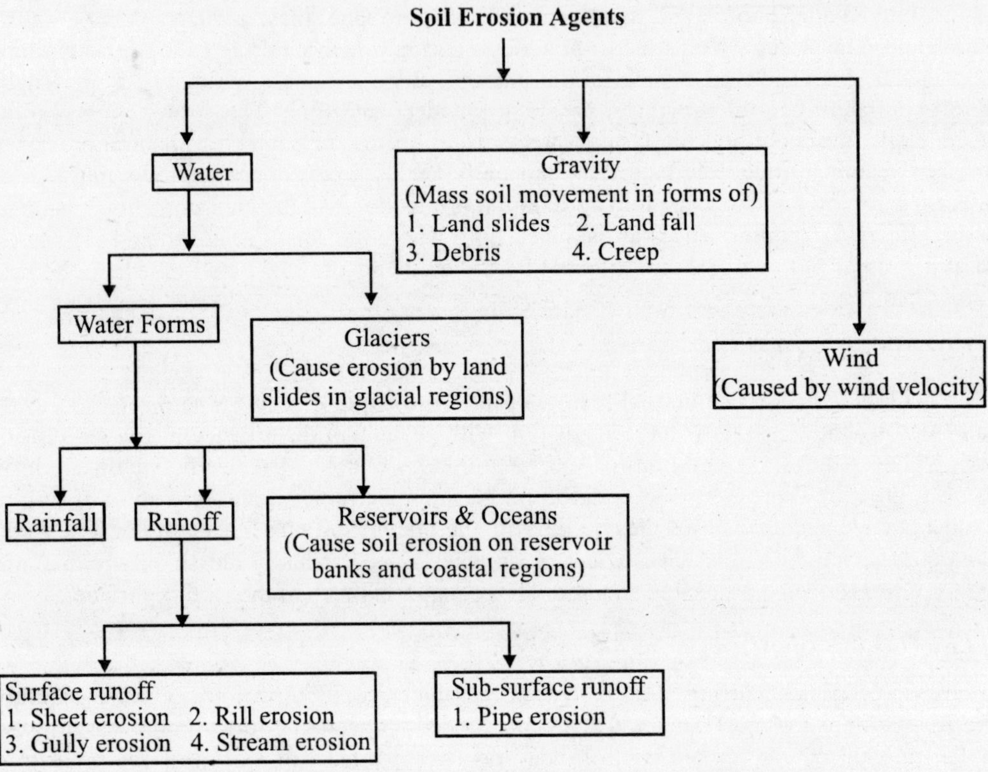

Fig. 37.1: Soil Erosion Agents (Source: Munish Kumar and Uttam 2012)

EROSIVITY OF RAIN (R-Factor): It is defined as the power of overland flow runoff to erode soil material. This is partly a property of the rainfall, and partly of the soil surface. Rainfall erosivity is a climatic factor which can be determined from local rainfall data. Rainfall erosivity is highly related to soil loss. Increased rain erosivity indicates greater erosive capacity of the overland water flow. Soil erosion by running water occurs where the intensity and duration of rainstorms exceeds the capacity of the soil to infiltrate the rain. Where rainfalls are intense, it is particularly urgent to adopt conservation and management techniques to protect the soil during the rainy season. Rainfall erosivity depends primarily on rainfall intensity and amount. Rainfall erosivity is an important parameter for soil erosion risk assessment under future land use and climate change. For calculating erosivity, the modified version of the Fournier index (FI) has been used.

$$\text{Fourier Index} = \sum_{i=1}^{12} Pi^2 / p$$

Where 'Pi' is the precipitation total in month i, and 'p' is the mean annual precipitation total.

Table 37.1: Classification based on Fournier Index

Class Description	Fourier Index (FI) Range
Very low	< 60
Low	60-90
Moderate	91-120
High	121-160
Very high	>160

ERODIBILITY OF SOIL (K-Factor): The soil erodibility factor is a quantitative description of the inherent erodibility of a particular soil. It is a measure of the susceptibility of soil particles to detachment and transport by rainfall and runoff. For a particular soil, the soil erodibility factor is the rate of erosion per unit erosion index from a standard plot. The factor reflects the fact that different soils erode at different rates when the other factors that affect erosion (e.g., infiltration rate, permeability, total water capacity, dispersion, rain splash, and abrasion) are the same. Texture is the principal factor affecting K-factor but structure, organic matter, and permeability also contribute. The soil erodibility factor ranges in value from 0.02 to 0.69.

TYPES OF WATER EROSION

SPLASH EROSION: Splash erosion is the first stage of the erosion process. It occurs when raindrops hit bare soil. The explosive impact breaks up soil aggregates so that individual soil particles are 'splashed' onto the soil surface. The splashed particles can rise as high 60cm above the ground and move up to 1.5 meters from the point of impact. The particles block the spaces between soil aggregates, so that the soil forms a crust that reduces infiltration and increases runoff.

SHEET EROSION: Sheet erosion refers to the uniform movement of a thin layer of soil across an expanse of land devoid of vegetative cover. Raindrops detach soil particles, which go into solution as runoff occurs and are transported downstream to a point of deposition. Deposition occurs when runoff slows to the point where soil particles can no longer remain in suspension. Tilled agricultural fields and construction sites are subject to sheet erosion.

Table 37.2: Classification of Gullies based on the Size

	Very Small Gullies (g_1)	Small Gullies (g_2)	Medium Gullies (g_3)	Deep and Narrow Gullies (g_4)
Deep	upto 3 m	upto 3 m	3-9 m	>9 m
Bed Width	< 18 m	> 18	> 18 m	Varies
Side Slope	vary	8-15 %	8-15%	Mostly steep

RILL EROSION: When sheet flows begin to concentrate on the land surface, rill erosion occurs. While sheet erosion is generally invisible, rill erosion leaves visible scouring on the landscape. This type of erosion occurs when the duration or intensity of rain increases and runoff volumes accelerate. Rills may become stable through soil consolidation; however, they are still the major sediment transport route for soil detached on the inter-rill areas. Improved understanding of the ability of rain-impacted flows in rills to transport sediment is needed to improve the estimates of sediment transport and delivery.

Table 37.3: Classification of Gullies based on the Shape (Munish Kumar and Uttam 2012)

Particulars	U-shaped Gullies	V-shaped Gullies
Location	Alluvial plains where surface and sub-surface soils are easily erodible	Hilly regions where land surface is tough to resist the cutting
Cross section	U shaped	V shaped
Slope	Slopy field	High slopy field
Velocity of flow	Less	High
Catchment's area	Large	Small
Runoff volume/ unit time	More	Less
Lateral spacing of gullies	Large	Small
Control measures	Permanent drop structure	Check dams

GULLY EROSION: Rill erosion evolves into gully erosion as duration or intensity of rain continues to increase and runoff volumes continue to accelerate. *Gullies* are steep-sided trenches formed by the coalescence of many rills. A gully is generally defined as a scoured out area that is not crossable with tillage or grading equipment.

OTHER FORMS OF EROSION

TUNNEL OR PIPE EROSION: Tunnel or pipe erosion occurs when surface water moves into and through dispersive sub-soils. Dispersive soils are poorly structured so they erode easily when wet. The tunnel starts when surface water moves into the soil along cracks or channels or through rabbit burrows and old tree root cavities. Dispersive clays are the first to be removed by the water flow. As the space enlarges, more water can pour in and further erode the soil. As the tunnel expands, parts of the tunnel roof collapse leading to potholes and gullies. Indications of tunnel erosion include water seepage at the foot of a slope and fine sediment fans downhill of a tunnel outlet. Remediation actions include breaking open existing tunnels, revegetation, and increasing soil organic matter. Extensive earthworks may be required.

TILLAGE EROSION: Tillage erosion moves soil from the top of the field downward, exposing subsoil at the crest while burying soil at the bottom. After many years of tillage, topsoil accumulates at the bottom of the slope. No soil leaves the field due to tillage erosion, but the effects for productivity and increased yield variability can be huge.

STREAM CHANNEL EROSION: Stream channel erosion consists of both stream bed and stream bank erosion. Stream bed erosion occurs as flows cut into the bottom of the channel, making it deeper. This erosion process will continue until the channel reaches a stable slope. The resulting slope is dependent on the channel materials, and flow properties. As the stream bed erodes, and the channel deepens, the sides of the channel become unstable and slough off resulting in stream bank erosion. Stream bank erosion can also occur as soft materials are eroded from the stream bank or at bends in the channel. This type of stream bank erosion results in meandering waterways. One significant cause of both stream bed and stream bank erosion is due to the increased frequency and duration of runoff events that are a result of urban development.

RIVER BANK EROSION: This is a common phenomenon in the rivers of the eastern and western Himalayan regions. During floods, the rivers undermine their banks and big chunks of land mass are engulfed by flood water. This leads to broadening of river beds and changes in their courses.

LANDSLIDES OR SLIP EROSION: During monsoon season or heavy down-pour, the hills get saturated with water resulting in instability of the land mass. The complete surface portion of soil may slip down from its actual position, causing landslides. Landslides are very common in hills and huge quantities of soil are washed away in streams and rivers. Hills devoid of vegetation are more prone to slip erosion, causing damage to agricultural lands, orchards, communication systems and human lives.

COASTAL EROSION: The strong waves of sea strike against the seashore causing large scale soil erosion. In this case, erosion is a combined effect of wind and water.

PREDICTING SOIL EROSION

UNIVERSAL SOIL LOSS EQUATION (*USLE*): By taking all factors into account, a prediction equation was developed for calculating the soil loss, called as Universal Soil Loss Equation (USLE). It accounts all parameters affecting the soils loss and predicts the annual soil loss. It computes sheet erosion. It is not used when slope are steeper than 20 per cent.

A = RKLSCP (Wischmeier and Smith 1965)

Where,

A = Estimated gross erosion (t/ha/year)

R = Rainfall erosivity factor

K = Soil erodibility factor

L = Slope length factor

S = Slope gradient factor

C = Crop cover or vegetation management factor

P = Support conservation practice factor

Demerits of USLE: This equation does not compute the sediment yield from the watershed directly. Annual sediment yield can be determined which is less important for design of water storage structures. This equation is used in small watershed to prediction of soil loss.

MODIFIED UNIVERSAL SOIL LOSS EQUATION (*MUSLE*): This equation computes sediment yield either monthly or seasonal. This equation is used in large watershed for prediction of soil loss.

$Y = 95\ (Qq)^{0.56}$ KLSCP (Williams 1975)

Where,

Y = Sediment yield for an individual storm (t/ha)

Q = Volume of runoff (ha-m)

q = Peak flow rate (m^3)

E = f (IKCLV) (Chepil and Woodruff 1965)

Where,

E = Average annual soil loss (t/ha/year)

I = Soil erodibility

K = Roughness

C = Control of soil conservation measure

L = Slope length

V = Vegetation

WHAT IS SOIL CONSERVATION?

Soil conservation basically means a way of keeping everything in place, literally as well as in a more abstract sense of maintaining the functions of the soil in sustaining plant growth. Soil conservation practices involve managing soil erosion and its counterpart process of sedimentation, reducing its negative impacts and exploiting the new opportunities it creates. Young (1989) defined soil conservation as a combination of controlling erosion and maintaining soil fertility. Soil conservation is the combination of different integrated approaches with four major disciplines such as forestry, agronomy, engineering and soil science. In the past, the focus has been on trying to keep the soil at its place by plot-level activities only. Currently, the attention has switched to landscape level approaches where sedimentation is studied along with erosion, and the role of channels (footpath, roads, streams) is included as well as the filters that restrict the overland flow of water and/or suspended sediment.

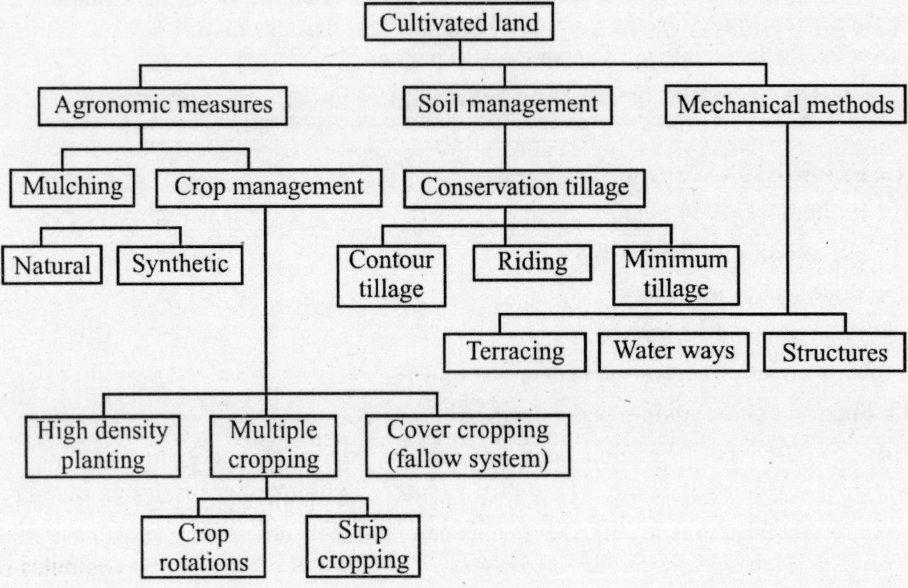

Fig. 37.2: Soil Conservation Strategies for Agroecosystem (Source: Morgan 1986)

Conservation, development and management of the land resources ensure the physical, chemical and biological health of soil profile and water resources which are of prime importance. It also helps for soil moisture conservation and flood control in the case of water resources. In a predominantly agricultural system, the objective of improving the productivity, profitability and prosperity of the farmers and achieving agricultural development on an ecologically sustainable basis can be attained only when conservation, development and management of soil and water resources are assured.

Types of soil conservation measures:

- *Agronomic*: such as plant/soil cover, conservation farming methods, contour farming
- *Vegetative*: such as agroforestry, planting barriers (vegetative strips), live fences, contour hedges, windbreaks
- *Structural*: such as terraces, Fanya Juus, Fanya Chini, banks , bunds, cut off drains, barriers, check dams, gully plugs, contour trenches, ditches, broadbeds and furrows
- *Overall Management*: such as area closures, selective clearing

ROLE OF AGROFORESTRY IN SOIL AND WATER CONSERVATION

Agroforestry plays a significant role in soil and water conservation, especially in hilly areas, where severe soil erosion is major area of concern due to faulty cultivation practices. Trees in the agroforestry systems provide good vegetative cover to the soil, reduce runoff and soil loss, and enrich soil physical, chemical and biological properties and also provide fodder, fuelwood, fruit and other products. Agroforestry can be used to diversify and intensify farming systems. Agroforestry, through integration of indigenous trees-crops and the domestication of various tree/shrub species has been proposed as one of the ways that is beneficial to the soil and water conservation and environment amelioration. Agroforestry possibly encompasses some of these properties of perennial vegetation in controlling erosion from agricultural fields (Young 1989)

IMPORTANCE OF AGROFORESTRY IN SOIL AND WATER CONSERVATION

Agroforestry helps in optimizing the use of resources through the principles of recycling, internalizing the input production, reducing risk and conserving natural resources. It can reduce erosivity of rainfall and erodibility of soil through dissipation of energy of raindrops by canopy at low heights, surface litter, obstructing runoff, root binding and improves soil organic matter, physico-chemical and biological properties. Trees in agroforestry slow down water movement, reduce surface flow, and facilitate water infiltration into the ground. Roots help to hold the soil in place, create macro pores to increase infiltration and enhance soil moisture holding capacity. Canopy of the trees in agroforestry provide shade which prevents the soil to become too dry and help in maintaining micro-climate. Canopy also helps in lessening the impact of rain drops on the soil, thereby reducing soil erosion and increasing water absorption into the ground. Litter from agroforestry prevents runoff and allows the water to percolate into the soil thereby helping ground water recharge. Dead plants decompose to form humus, organic matter that holds the water and provides nutrients to the soil. The hydrological cycle and microclimate is also affected by agroforestry. Trees in agroforestry help filtering the water thereby, keeping it as sediment free as possible and maintaining high water quality. The surface cover, debris and tree roots in agroforestry trap sediments and stop their down slope movement and helps in stabilizing slopes and preventing shallow landslides. The effect of perennial vegetation in controlling erosion however, depends on a number of factors such as canopy cover, ground vegetation, litter effects, root effects and changes in the physio-chemical properties of the soil (Bhatt et al 2001).

AGROFORESTRY PRACTICES FOR SOIL AND WATER CONSERVATION

Agroforestry has shown encouraging results in the Himalayan foothills and valley regions for enhancing productivity and arresting the process of land degradation. It offers affordable alternatives to poor farmers in place of expensive conventional conservation measures. Some of the agroforestry practices for soil and water conservation are discussed as under:

PLANTATION CROPS

A wide range of agroforestry systems fall under the practice of plantation crop combinations (crop with MPTs). In India, large areas of the humid tropics are characterized by moderate to steep slopes and agricultural plantation crops such as tea, coffee, cacao, oil palm, rubber and pineapple are frequently grown on these areas. Tea (*Camellia sinensis*) is grown under shade of *Albizia chinensis*, *A. odoratissima*, *A. lebbeck*, *A. procera*, *Acacia lenticularis*, *Derris robusta*, *Grevillea robusta*, *Acacia* spp., *Erythrina lithosperma* and *Indigofera teysmannii*. coffee (*Coffea arabica* and *Coffea canephora*) plants are grown under the shade of *Erythrina lithosperma* as temporary shade while, permanent shade trees include *Ficus glomerata*, *F. nervosa*, *Albizia chinensis*, *A. lebbeck*, *A. moluccana*, *A. sumatrana*,

Dalbergia latifolia, *Artocarpus integrifolia*, *Bischofia javanica* and *Grevillea robusta*. Cacao (*Theobroma cacao*) is grown under the shade of coconut and arecanut, and *Dipterocarpus macrocarpus* (in forest). Black pepper *(Piper nigrum)* is grown with support from *Erithrina indica*, *Garuga pinnata*, *Spondias* spp., *Mangifera* spp., *Gliricidia maculata* and *Grevillea robusta*. Small cardamom (*Elettaria cardamomum*) and large cardamom (*Amomum subulatum*; *A. aromaticum*) grow in forests under temporary shade tree of *Maesopsis eminii*. Large cardamom is grown under the shade of natural forest as well under planted shade trees viz. *Alnus nepalensis*, *Schima wallichii*, *Cinchona* spp., *Lagerstroemia* spp., *Albizia lebbeck*, *A. julibrissin*, *Castanopsis tribuloides*, *C. hystrix*, *C. indica*, *Terminalia myriocarpa* and *Bischofia javanica* (Tejwani 1994).

In Sikkim, among the tree species grown with large cardamom, *Alnus napalensis* is the predominant. *Alnus nepalensis,* a fast growing tree not only provides fuel and timber but also fixes atmospheric nitrogen and enriches the soil by adding leaf litter which in turn benefits the large cardamom (Singh et al 1989). The trees in this practice are either spaced randomly or regularly. Plantation crop combinations are helpful in augmenting the income of farmers besides conserving the soil and water resources efficiently. Erosion control in these types of system depends on the spacing/density and management practices of the trees. Where the shade trees are widely spaced, as common in tea plantations and some coffee systems, their effect is not substantial, and erosion control depends on good management of the plantation crop itself. Tree canopy with close spacing is effective, but the capacity of these dense, mixed agroforestry systems to control soil erosion depends on maintaining surface litter and ground cover. The management practice of keeping the ground bare through chemical weed control, for ease of maintenance, is highly undesirable from an erosion control point of view. In hilly region especially in northeast, land management point of view, the combination of the broad-leaved trees and pineapple with large cardamom; arecanut, black peeper and pineapple on slopes, wherever the soil and climate are suitable, would not only supplement the income from orchard but also help in the conservation of soil and water (Tomar et al 2007).

MULTISTOREY HOMEGARDENS

Homegarden is an old-age practice in many parts of the country particularly in Kerala, northeastern states, parts of West Bengal, and Andaman and Nicobar Islands. In this practice, number of trees, crops, livestock, poultry and/or fish production are grown together for the purpose of meeting the basic needs of the farmer. Different components occupy different strata in homegardens which are helpful in conserving the soil and water. Top most strata is occupied by forest trees, the middle by fruit trees and the ground crops are generally agricultural crops, spices etc. Coconut, arecanut, guava, mango, citrus, tamarind, jackfruit, papaya, banana, *Moringa*, *Sesbania* and custard apple are the major trees grown in home-gardens. They provide fruits and vegetables, and also support climbers. Coffee and cacao are also found in these gardens. Among food crops, cassava, yam, sweet-potato, taro, elephant-yam, and vegetables are grown. Spices includes cardamom, cinnamon, clove, ginger, nutmeg, pepper and turmeric. Pineapple is common fruit grown as ground storey crop. Thus, home gardens are also considered as excellent tools for biodiversity conservation (Tejwani 1994). This is highly productive, fully sustainable and very practicable system. The practice is helpful in reducing erosion owing to the cover of herbaceous plants and litter.

HEDGEROW INTERCROPPING AND BARRIER HEDGE

Hedgerow intercropping (also called alley cropping) has multiple objectives including fertility maintenance, and may be practised on flat or sloping land. In this practice, perennial, leguminous trees or shrubs are grown simultaneously with an arable crop. The trees are managed as hedgerows and annual crops are grown in wide rows between the trees. During the cropping phase, the trees are pruned and the prunings are used as green manure or mulch on the crop to improve the organic matter status of the soil and to provide nutrients, particularly nitrogen, to the crop. The hedgerows are allowed to grow freely to shade the inter-rows when there are no crops. Alley cropping retains the basic restorative attributes of the bush fallow through nutrient recycling, fertility regeneration and weed suppression. It combines these attributes with arable cropping so that all processes occur concurrently on the same land, allowing the farmer to crop the land for an extended period.

Gliricidia sepium, *Calliandra calothyrsus*, *Leucaena leucocephala*, *Flemingia macrophylla* and *Erythrina subumbrans* are some of the important species used in hedgerow intercropping. The functions of the hedges in soil conservation are to check soil loss through the cover effect and by laying prunings on the ground surface in the cropped alleys. It can reduce runoff and soil loss through the barrier effect, improve infiltration and soil fertility through the decay of prunings and root residues, and develop terraces progressively. During summer, the trees protect the crops from intense solar radiation and high wind velocity, and reduce the rate of evapotranspiration.

BOUNDARY PLANTING AND LIVE FENCES

Field boundaries, when aligned along the contour, are an effective means of erosion control. It will be good if this can be combined with productive and service functions through boundary planting or live fences. Multipurpose tree species like *Acacia nilotica*, *Azadirachta indica*, *Holoptelea integrifolia*, *Moringa oleifera*, *Mangifera indica*, *Pithecellobium dulce* and *Annona squamosa*, having multiple uses, viz. fuel wood, fodder, poles and small timber and minor forest produces, are grown on agriculture field boundaries in close spacing under rainfed as well as irrigated conditions. The locally available tree species are generally selected for planting on the field boundaries or as per the priority need of the farmers. Farmers of northeastern region grow bamboo all along the irrigation channels. In different parts of India, *Eucalyptus* and *Populus* are commonly grown along the field boundaries or bunds. Other trees which are found grown as boundary plantations or live hedge include *Acacia nilotica* and *Dalbergia sissoo*. Many of the boundary plantations also act as shelterbelts and windbreaks and help in checking wind erosicn, ultimately conserving the soil. In Bihar, *Dalbergia sissoo* and *Wendlandia exserta* are most common boundary plantations. *Casuarina equisetifolia* is extensively planted on field bunds and along sandy coastal areas in Odisha.

Trees are also retained as live fence in different parts of the country. For example, *Jatropha*, *Ficus*, *Ceiba pentandra*, *Vitex trifolia*, *Erythrina*, *Carissa*, etc., are used as live-hedges at close (2 m) spacing. At many places, *Agave* and many cactoids are grown as common live-fence. Pruned material of live fence is used as fodder, green manure and fuelwood. Live fences through their dense stem help in checking sediment losses and helps in preventing the soil erosion. Deep roots of the live fences helps in uptaking the water from deeper soil layer. Further, they are also helpful in enhancing soil health. Besides, live posts are used as fence posts to support barbed wire for protection from biotic interference and providing fodder, fuelwood and green manure. Tree barriers are also effective in conserving soil and moisture. A study conducted at Dehradun indicated that sediment deposition along hedge rows during a period of three to nine years of tree age varied from 184 to 256 t ha^{-1} which is equivalent to 15 to 20 mm of soil depth. *Eucalyptus tereticornis* and *Eulaliopsis binata* raised in Shiwalik foothills (@ 2500 trees ha^{-1}) in paired rows with under storey grass planted at 50 cm x 50 cm spacing allowed no soil loss with an annual return of about Rs. 4000 ha^{-1} yr^{-1} from commercial grass alone besides additional returns from Eucalyptus and proved to be more remunerative than traditional rain-fed crop.

PLANTATION OF SOIL BINDING GRASSES

Slopes having loose material and heavy soil erosion needs to be put under plantation of soil binding perennial grasses for resource conservation and biomass production. Grasses, when grown as vegetative barriers, serve as good filter strips to check erosion and enhancing productivity of agriculture crops. Grass species such as *Panicum maximum*, *Vetiveria zizanioides*, *Eulaliopsis binata*, *Thysanolaena maxima* (broom grass) and *Saccharum* spp. are found to be effective. *Panicum maximum* and *Pennisetum purpureum* (napier grass) have been found suitable in the Shiwalik and lower hills. Maize and wheat yield were found to increase by 23-40 per cent and 10-20 per cent respectively when planted with vegetative grass barriers. Vegetative barriers of grasses can reduce the runoff and soil loss by 18-21 per cent and 23-68 cent respectively on slopes varying from 2-8 per cent.

WINDBREAKS AND SHELTERBELTS

The role of windbreaks and shelterbelts in controlling wind erosion in semi-arid regions is well established. Windbreak or shelterbelt is a densely planted strip of tall vegetation usually between 50-200 feet wide oriented at right angle to the directions of prevailing winds in order to reduce their velocity near the ground. This can reduce the wind speed by 20-25 per cent on an average and slow the evaporation rate by about 10 per cent during day time and about 4 per cent during night. Shelterbelts also favourably influence temperature. Farmers build shelterbelt by either small dead wood or local vegetation to check wind velocity within safer limits. These are integral part of arid land use systems. *Acacia nilotica*, *Albizia lebbeck*, *Azadirachta indica*, *Tamarix articulata*, other *Acacia* spp., *Casuarina equisetifolia* and *Parkinsonia aculeata* are the common tree species used as shelterbelts. *Calligonum polygonoides*, *Crotalaria burhia*, *Leptadenia pyrotechnica* and *Aerva pseudotomentosa* bushes, and grasses such as *Cenchrus ciliaris*, *C. setigerus*, *Lasiurus sindicus* are used in windbreaks. Dead wood and shrubs are also grown in checker board system to check the movement of sand dunes. This permanent vegetation helps to accumulate sand and check the soil erosion. This also helps to increase crop yields along the lines. Shelterbelts and windbreaks are also grown in orchards in agri-horti system for protecting fruit trees and to enhance aesthetic and environmental values.

SILVOPASTORAL PRACTICES

Soil erosion on pastures is often more severe than on croplands. Severe sheet erosion and gullying are both common. The initial cause is degradation of the vegetation through overgrazing, which leads to a sparse sometimes almost zero ground cover, leaving the soil open to erosion. It is not uncommon for 10 cm or more of topsoil to be removed. Such erosion occurs in semi-arid regions depending primarily on grazing and on land used for pasture in areas of mixed farming. Silvopastoral practices include scattered trees on pastures (e.g. systems with *Faidherbia albida* or other *Acacia* species), combinations of plantation crops with pastures (e.g. cattle under coconuts, sheep under rubber), live fences, fodder banks, windbreaks and shelterbelts, and hedgerow intercropping on pastures. The potential of windbreaks to control wind erosion is well established. It would be of great value if means were found for applying silvopastoral practices to the control of water erosion. If this is attempted simply by planting trees, without other changes in the management of degraded pastures, it will not be successful. The basic principles of pasture management, such as restriction of livestock numbers and rotational grazing, are a prerequisite to erosion control as to any other aspect of silvopastoral systems.

In a sound pasture management, trees may contribute to erosion control in a number of ways. It may be possible to use live fences to control livestock movement and assisting rotational grazing. The direct effect of the tree canopy in reducing raindrop impact is unlikely to be substantial. However, the greatest potential is through indirect means. A known function of trees in silvopastoral systems is to supply protein-rich fodder at times of year when grass is absent or indigestible. This can be through direct browse, as by sheep, goats and game animals, or through cut-and-carry fodder. By reducing grazing pressure, such methods can lead to a better vegetation cover and thus less erosion at the critical period, the start of the rains. As with silvopastoral practices in general, these considerations apply to semi-arid and sub-humid grazing land, and to areas of the humid tropics where sloping land is used for grazing, as is common in Latin America. A wide perspective on the potential of silvopastoralism in the semi-arid zone is given by Baumer (1987).

The importance of silvopastoral system has now been accepted throughout the world for increasing the productivity of degraded/problematic lands and rehabilitation of degraded lands. It helps in checking soil erosion, conserving more in-situ moisture, maintaining higher survival per cent, yielding more fodder and fuelwood biomass. Silvopastoral systems can be raised on marginal lands with suitable fodder grasses like *Dichanthium annulatum*, *Leptochloa fusca*, *Brachiaria mutica*, *Panicum antidotale*, *Sehima nervosum*, *Chloris gayana*, *Cenchrus ciliaris*. A number of MPTs viz. *Grewia optiva*, *Morus serrata*, *Celtis australis*, *Robinia pseudoacacia*, *Ulmus wallichiana*, *Quercus* spp., *Bauhinia variegata*, *Melia azedarach*, *Parkinsonia aculeata*, *Eucalyptus tereticornis*, *Cassia siamea*, *Acacia auriculiformis*, *Dalbergia sissoo*, *Albizia procera*, *Acacia nilotica*, *Thespesia populnea*, *Ailanthus excelsa*, *Derris indica*, etc. have been found suitable for establishment of silvopastoral systems for rehabilitation of degraded land. To improve grass yields and fodder quality of pastures, sowing/planting of legumes like *Stylosanthes hamata*, *Medicago* and *Trifolium alexandrinum* can be done (Singh and Chaturvedi 2011).

In the Shiwalik foothills, silvipasture systems involving *Eulaliopsis binata*, *Saccharum munja* or Vetiveria zizanioides with *Acacia nilotica* were found to be promising. Further, *Eucalyptus tereticornis* + *Eulaliopsis binata* has been identified as an economically-viable and eco-friendly system for rehabilitation of these lands. The study conducted for 14 years with four multi-purpose tree species viz., *Albizia lebbeck*, *Grewia optiva*, *Bauhinia purpurea* and *Leucaena leucocephala* and two grass species viz., *Chrysopogon fulvus* and *Eulaliopsis binata* at CSWCRTI, Dehardun revealed that gravelly and bouldery lands could be effectively utilized by putting them under perennial vegetation for improving soil fertility and biomass production (Vishwanatham et al 1999; Samra et al 1999; Raizada and Singh 2010). These species can be grown on a wide variety of sites and lopping can be carried out when the trees have attained sufficient growth.

AGRI-HORTICULTURE SYSTEMS

Suitable agri-horticultural systems are being adopted for rehabilitation of degraded lands in different agro-ecological regions. These systems are most important in terms of economic returns to the farmers and their preferences. Fruit trees are grown on terrace bunds along with agriculture crops at linear spacing of 5-8 m. Wide variety of fruits (citrus, banana, mango, apple, pineapple, walnut, plum, peach, cherry, etc.), vegetables (potato, pea, capsicum, cabbage, cauliflower, etc.), spices (ginger, chillies, cardamom, saffron, etc.) and flowers (orchids, gladiolus, marigold, chrysanthemum, etc.) can be intercropped. These trees commence the production five years after plantation. In some cases, fuel/ fodder/timber trees are also retained on the field bunds and managed under agri-horti-silvicultural system. In northeastern region, agri-horticultural systems involving cultivation of ginger with fruit plants, such as mandarin and guava on mild slopes have been found profitable, besides protecting and conserving the hill soils. Pineapple can be easily associated with multi-purpose trees in paired rows on hill slopes, where it checks soil erosion and also gives considerable cash returns within a short period. Intercropping of guar, cowpea or pearlmillet with peach; turmeric with papaya; *Chrysopogon fulvus* or napier grass with aonla or ber have been identified as economically viable for soil conservation (Anon. 2010). Intercropping of turmeric with kinnow or peach; and cowpea, blackgram and toria with mango and litchi have also been found to be promising (Dhyani 2011).

RECLAMATION FORESTRY

Land capability class IV and above soils are termed as non-arable land. The carrying capacity of these lands is very low and most of them are unfit for crop production. These lands can be suitably reclaimed for resource conservation and productivity enhancement using different agroforestry measures. There are opportunities to combine reclamation with production. The first step is to establish a full forest cover, including at least some nitrogen-fixing species, initially with protection from grazing and allowing all plant residues to reach the soil. As soon as a check of erosion and satisfactory build-up of soil organic matter have been achieved, agroforestry provides ways of combining continued erosion control with productive use. The techniques and products can vary widely.

DEGRADED RIVERBEDS: Riverbeds generally comprise 65-80 per cent boulders, stones and pebbles and 20-35 per cent coarse and fine sand material. In foot hills of Uttarakhand, large area of 2.94 lakh ha are categorized as bouldery wasteland of which about 30 per cent (0.3 M ha) falls in this category in Doon valley. Silvopastoral system of *Dalbergia sissoo* and *Chrysopogon fulvus* was found to give 3.37 t ha^{-1} yr^{-1} of wood and 5.4 t ha^{-1} yr^{-1} of air dry grass. *Grewia optiva* + hybrid napier system was also found to be economically viable with BC ratio of 1.6 and payback period of 8 years when calculated for a 15 year cycle at 12 per cent discount rate (Raizada and Singh 2010; Singh and Dadhwal 2012). Silvipasture system involving *Eulaliopsis binata*, *Saccharum munja* or Vetiveria zizanioides with *Acacia nilotica* was found to be promising in Shiwalik foothills. Plantation of *Albizia lebbeck* at 4 m x 4 m with *Chrysopogon fulvus* at 1 m x 0.5 m produced oven dry lopped fodder of 75.7 q ha^{-1} yr^{-1}, air dry fuelwood of 9.7 q ha^{-1} yr^{-1} and oven dry grass of more than 40 q ha^{-1} yr^{-1}. Combinations of *Eulaliopsis binata* with *Grewia* and *Bauhinia* fodder trees were also found promising. Pollarding of *G. optiva* was found to give 2.0 t ha^{-1} green leaf and 4.6 t ha^{-1} of wood while combination of pollarded *G. optiva* with *Panicum maximum* yielded 20.5 t ha^{-1} of total biomass. In another experiment, highest green fodder yield of 11.0 t ha^{-1} was obtained in Eucalyptus + *Chrysopogon fulvus* combination. Dense plantations at 2 m x 2 m gave maximum biomass yield in *G. optiva* and *Dalbergia sissoo* (Anon. 2010). Plantation of *Jatropha curcas* was also found to be successful in bouldery lands. Aromatic grasses involving lemon grass and java grass were also found to enhance the productivity of bouldery lands. After two years of experimentation, it was observed that survival percentage of lemon grass and java grass was 95 and 80 per cent respectively. The yield of lemon and java grass ranged from 440-550 q ha^{-1} and 260-340 q ha^{-1} respectively. The average net benefit from lemon and java grass was Rs. 42,250 and Rs. 32,750 ha^{-1} yr^{-1} respectively (Tomar et al 2012).

TORRENT CONTROL: Torrents are common in the foothills of the Himalayan region. They cause extensive damage to life and property in the valley reaches as a result of frequent changes in their course due to meandering and lateral migration. They are a constant threat to natural resource base as well as human settlements. Torrents are characterized by the narrow width and high gradient at the upper reaches of its catchment area and wide spreading beds with ill-defined wide bank in the downstream area. It usually remains dry during major part of the year, but during monsoon they are prone to flash floods and swift currents with high velocity which brings lot of silt and debris which deposits in the bed of the torrents in hazard manner thereby making the stream shallower every year and change the course of flow and damage nearby agriculture field and land all along the banks. Every year, torrents engulf more and more productive area. The torrent banks could be effectively protected by planting *Ipomea carnea*, *Vitex negundo*, *Arundo donax*, *Salix*, *Populus*, hybrid napier and a mixture of grasses. Vegetative spurs of *Erythrina suberosa*, *Bombax ceiba* and *Lannea grandis* were also found to be effective (Sharda et al 2012).

LANDSLIDE PRONE AREAS: Landslide is the downward movement of mass rock or soil as a result of slope failure. It affects dams, reservoirs, human settlements, agriculture lands, forest, orchards and other developmental activities. Areas affected by landslide in the Himalayan region can be reclaimed and utilized for biomass production by adopting suitable bio-engineering technologies. A landslide project was undertaken by CSWCRTI at Nalotanala in 4 ha landslide area on Dehradun - Mussoorie road. The area was rehabilitated by using bio-engineering measures. Bare erodible slopes were protected by planting *Ipomoea carnea*, *Vitex negundo* and napier with *Erythrina suberosa*, *Dalbergia sissoo* and *Acacia catechu* which proved to be very effective. The area was completely revegetated and rehabilitated within a period of 10 years. The sediment load was brought down from 320 t ha^{-1} yr^{-1} to 5.5 t ha^{-1} yr^{-1} and runoff from 54 per cent to 32 per cent (Sharda et al 2012).

MINESPOIL REHABILITATION: CSWCRTI, Dehradun undertook a research project in a mined watershed of 64 ha near Shahastradhara, Dehradun. Bio-engineering measures and geo-jute at a cost of Rs. 15,000 per ha were undertaken for rehabilitation of the area. *Eulaliopsis binata*, Thysanolaena maxima, *Saccharum munja*, *Chrysopogon fulvus*, *Pennisetum purpureum* (grasses), *Ipomoea carnea*, *Vitex negundo* (bushes), *Leucaena leucocephala*, *Acacia catechu*, *Salix alba*, *Lannea grandis* (trees) were found to be suitable. Planting of MPTs, shrubs and grasses helped in improvement of soil characteristics over a period of time (Dhyani et al 1988). After 14 years, the pH came down by 0.6 units (8.1 to 7.5), organic carbon increased from 0.13 to 0.42 per cent, whereas $CaCO_3$ content decreased from 54.6 to 31.0 per cent and bulk density from 1.63 to 1.47 Mg m^{-3}. The debris flow was reduced from 550 to 6 t ha^{-1} $year^{-1}$ and runoff from 57 to 37 per cent with reduced flood peaks and increased base flow. The vegetal cover increased from 10 to 90 per cent due to reforestation over a period of 20 years. In addition, PWD could save Rs. 1.0 lakh yr^{-1} which they used to spend for removing debris from the road leading to the tourist place. Because of the various treatment measures, the main drainage channels which used to dry up by the end of October, have now become perennial. The dry weather flow recorded in the main drainage channel was to the tune of 265 and 100 cum day^{-1} in the months of November and February, respectively. The augmented flow is now being used for irrigation purposes by the farmers. Runoff water from treated mine site was found within the permissible water quality limit.

SHIFTING CULTIVATION AREA: In northeastern region of India, large-scale deforestation associated with shifting cultivation has led to widespread land degradation, which is adversely affecting the agricultural production. Farming activities on hill slopes are the major areas of concern due to severe soil erosion and loss of flora and fauna. In hill and mountain ecosystem, agri-horticultural systems involving cultivation of ginger with fruit plants, such as mandarin and guava on mild slopes have been found profitable, besides protecting and conserving the hill soils in the NEH region. Khasi mandarin based agroforestry has also great potential for increasing crop productivity and sustainability through restoration of soil fertility by checking erosion and improving its physico-chemical and biological properties (Tomar 2006). Pineapple is a good crop associated with MPTs in paired rows on hill slopes, where it checks soil erosion and also gives considerable cash returns within a short period. Under silvi-horti-system with *Acacia auriculiformis* and *Leucaena leucocephala*, pineapple produced an yield of 11.7 t ha^{-1} and 13.4 t ha^{-1} respectively (Dhyani et al 1995). Tree-based land use systems proved suitable where soil depth ranged between 1.0-2.5 m, whereas dairy farming was found profitable under shallow and highly-leached soil conditions (soil depth <0.75 m). These land use systems showed tremendous fertility restoration capacity by protecting soil, improving soil organic matter and consistent replenishing of nutrients. The vegetative barriers with agroforestry species were able to retain 10 to 61 t ha^{-1} yr^{-1} soil, which otherwise could have washed down and resulted in siltation of streams.

CONTOUR FARMING

Contour farming is one of the methods used for soil and water conservation. The contour planting conforms to the principles of soil and water conservation to conserve rainwater and to check severe erosion. Planting should be done on contours as it is effective in soil conservation. Contour maps are available from different sources. However, if contour maps are not available, they can be prepared. Contour lines are marked by using the simple 'A' Frame. Planting can directly be done on lines or using contour trenches. Contour trenches are excavated along the contours to break the slope length for reducing the velocity of surface runoff and the water retained in the trenches help in conserving the moisture. The trenches are dug by cutting a trench of 30 cm deep and 45 cm wide along the contour at an interval varying from 20-50 m according to percentage of slope. The trench interval is longer on the lower slope and smaller on the higher slope. The trench is dug manually and excavated soil from the trench is dumped on the down stream along the trench to form a bund. Various grasses are generally cultivated on the bunds to stabilize the loose soil. To reinforce the trench, different tree species, pineapple, etc can be planted in few rows just below the bund (Fig. 37.3).This method could be applied even in lower edges of *jhum* fields, whereas, upper slopes could be left for restoration purposes, followed by horticulture (Thansanga1997; Satapathy 2003).

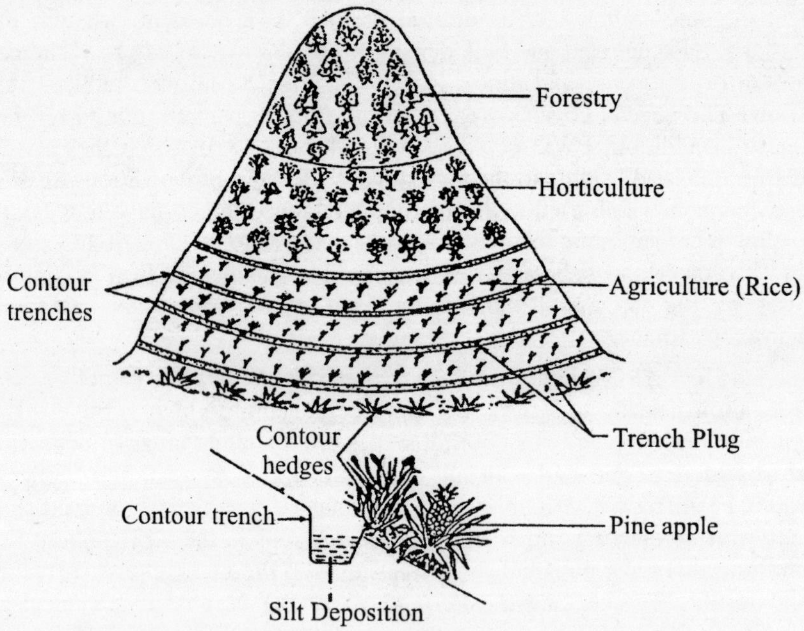

Fig. 37.3: Contour Farming in Mizoram (Source: Thansanga 1997)

WATERSHED BASED SYSTEMS

Watershed is a hydrological unit of an area which drains runoff to a single outlet. Watershed based system includes all kind of interventions for reducing soil loss, improve productivity, sustainability and livelihood security. It is a holistic approach aimed at overall development of its natural, human and animal resources. The Government of India has accorded highest priority to the holistic and sustainable development of rainfed areas and wastelands through watershed development programmes. Watershed program- mes mainly aims at raising the productivity of rainfed agriculture and non-arable lands by encouraging the sustainable management and optimal use of surface and ground water resources and using land according to the land capability classes consistent with soil and water conservation. Conceptually watershed has three main parts: Catchment area, submergence area and command area (Tomar et al 2002; Joshi et al 2004). These parts have specific role to play in watershed and hence these parts should be managed accordingly.

Three-tier model of land use seems most suitable on watershed basis, where restoration forestry/pure silviculture is maintained on hilltop, followed by silvipasture, agrihorticulture and pure agriculture down to the slope, respectively. But this system becomes impractical, when the farmers based on hilltop, used to cultivate the area for his/her subsistence in and around the hilltop itself. Due to this fact, the protective trees cover of hill tops have been removed almost in all the states of Northeastern hill region. A study was conducted by ICAR Research complex, Barapani by dividing watershed into three tier system viz., upper 1/3 area under pasture and silviculture for rearing livestock (goats and pigs), middle 1/3 area under horticulture including orange (*Citrus* spp.), guava (*Psidium guajava*) and pineapple (*Ananas comosus*) and remaining 1/3 lower area under agriculture for cultivation of cereals, pulses, vegetables, spices, fodder, etc. The middle portion of micro watershed of which 50 per cent area was put under orange, 25 per cent under guava and 25 per cent area was put under Assam lemon. Agri-horti-silvi-pastoral system with livestock, dairy farming and agro-pastoral system recorded an input output ratio of 1:2.14, 1:2.08 and 1:2.05 respectively and was recommended as a viable alternative to shifting cultivation in northeast states (Singh 1987). There is great scope of pisciculture, dairying, bee keeping and cut flowers industries which will further increase the income of the marginal farmers through agro-based industries.

38 AGROFORESTRY PRACTICES FOR WASTELAND DEVELOPMENT

INTRODUCTION

Land degradation and expansion of wastelands are recognized as the main challenges of people living in dryland area and they also receive global attention. Degradation caused by overcultivation, overgrazing, deforestation and inefficient irrigation affects most of the world's drylands. Land degradation refers to a temporary or permanent decline in the productive capacity of the land or its potential for environmental management. Net Productivity change reflects both natural and human induced processes of degradation and improvement. Some types of land degradation are, for all practical purposes, irreversible. Examples are severe gullying and advanced salinization. In these cases, the long-term biological and environment potential of the land has been compro-mised. Displacement of soil material (erosion) is also irreversible, although its long-term effects on productive capacity depend on the depth and quality of soil remaining.

Most types of land degradation, however, can be prevented or reversed viz. adding nutrients to nutrient-depleted soil, rebuilding topsoil through soil amendments, re-establishing vegetation (reforestation), buffering soil acidity. The practicality of rehabilitating degraded landscapes depends on the costs relative to the value of output or environmental benefits expected. Where farmers wish to intensify agricultural production on a sustainable basis, it may be necessary to undertake such land-enhancing or land-protecting measures even in non-degraded landscapes. Degradation is a process of change over time. From policy perspective, it is critical to distinguish those lands that are currently undergoing degradation, in order to assess the need for action to stabilize or reverse the process. Other lands may have reached a degraded state relative to their natural condition many decades or centuries ago, but are currently in a stable or improving condition. For practical purposes, the land degradation problem refers to the former areas.

The international community has long recognized that desertification and land degradation are major economic, social and environmental problems of concern to many countries in all regions of the world. In 1991, the United Nations Environment Programme (UNEP) concluded that the problem of land degradation in arid, semi-arid and dry sub-humid areas had intensified. As a result, the United Nations Conference on Environment and Development (UNCED) supported a new integrated approach to the problem emphasizing action to promote sustainable development at the community level. As a result, the United Nations Convention to Combat Desertification (UNCCD) was launched in 1996 to tackle desertification and drought. The three Rio conventions, which include UNCCD, Convention on Biological Diversity (CBD) and the United Nations Framework Convention on Climate Change (UNFCCC) are all relevant to the prevention and control of land degradation. Joint work programmes for the three Rio conventions are now starting to achieve multiple global benefits, such as poverty alleviation, sustainable ecosystems and minimising greenhouse gas emissions. Since all of the Rio conventions are closely linked to each other, consolidating the policy and planning work required to carry out the responsibilities of the conventions allows for maximum efficiency and synergy in planning. Addressing land degradation also contributes significantly to the Millennium Development Goal (MDG) of reducing by half the proportion of people in poverty by 2015 and ensuring environmental sustainability.

LAND DEGRADATION

The Global Environment Facility (GEF) defines land degradation as any form of deterioration of the natural potential of land that affects ecosystem integrity either in terms of reducing its sustainable ecological productivity or in terms of its native biological richness and maintenance of resilience. UNCCD defines land degradation as the reduction or loss, in arid, semi-arid and dry sub-arid areas, of the biological or economic productivity and complexity of rain fed cropland, irrigated cropland, or range, pasture, forest and woodlands resulting from land use or from a process or combination of processes, including processes arising from human activities and habitation patterns, such as:

(i) Soil erosion caused by wind and/or water

(ii) Deterioration of the physical, chemical, and biological or economic properties of soil and

(iii) Long-term loss of natural vegetation

Various types of soil and land degradation have been explained by many authors including FAO/UNDP/UNEP. As only to be expected, authors differ in their approach to describing and classifying land degradation. It is generally recognized that land degradation has five main components. (i) soil degradation, (ii) vegetation degradation, (iii) water degradation, (iv) climate deterioration (v) loss due to urban/industrial develop-ment. Each of these major components could be subdivided into more specific types of degradation. The land degradation is classified according to specific phenomenon prevalent in the land with respect to physical, chemical and biological condition.

- Physical: compaction, crusting, sealing, loss of structure, accelerated erosion
- Chemical: salinization, toxicities, nutrient depletion and imbalances, acidification
- Biological: reduction in biodiversity, loss of ecological component

The causes of land degradation has been grouped into three categories (Lal and Antony Raj 2012): natural degradation hazards, direct causes of land degradation, underlying causes of land degradation.

(i) Natural degradation hazards: cyclones, drought, volcanic activities, sea level rise, etc.

(ii) Direct causes of land degradation: deforestation, overcutting of vegetation, shifting cultivation, overgrazing, non-adoption of soil conservation practices, extension of cultivation in the lands of lower potential and /or high natural hazards, improper crop rotation, unbalanced fertilizer and pesticide use

(iii) Underlying causes of land degradation: population increase, attitude, economic pressure, land tenure, land shortage, poverty

The most important on-farm effects of land degradation are declining potential yields. The threat of degradation may also be reflected in the need to use a higher level of inputs in order to maintain yields. Serious degradation sometimes lead to temporary or permanent abandonment of land. In other cases, degradation induces the farmers to convert land to lower-value uses. FAO/UNDP/UNEP has grouped the effects of land degradation as effects upon production and consequences for the people.

EFFECTS ON LAND DEGRADATION UPON PRODUCTION

- Reduced vegetation cover to the soil and biodiversity loss
- Top soil loss
- Watershed destruction with subsequent water shortages
- Reduced return of organic matter and less biological activity in the soil
- Increased pollution from increased use of agrochemicals
- Land is abandoned (where degradation is severe)
- Crop yields are reduced
- Inputs and costs of prodution are increased (where farmers attempt to combat reduced yields by increased inputs)
- Responses to inputs are decreased
- Flexibility of land management is decreased
- Risk is increased
- Labour, technical and financial resources are diverted to reclamation

CONSEQUENCES OF LAND DEGRADATION FOR THE PEOPLE

- Landlessness is increased
- Food supplies are reduced or less reliable
- Labour requirements are increased
- Incomes are decreased
- Migration of young people in search for employement opportunities

WHAT IS WASTELAND?

The term wasteland designates various types of unutilized land. It also indicates small pieces of land, usually on less fertile tract, which fail to yield a return to the farmers. It is also defined as the land which is lying uninhabited, uncultivated, land left after some use or land which is no longer serving any purpose. Wasteland are those lands which for one reason or another do not fulfill their life sustaining potential through the impact population pressure both human and livestock and short sighted development policies and have progressively lost their ecological and economic functions; intensifying poverty and human misery. Excessive felling of forests, especially in ecologically sensitive areas, over grazing by pastures indiscriminate use of forest and fertile land for industrial and other development in appropriate agricultural practices have all contributed to the creation and spread of wasteland.

- **Dictionary** definition (non-technical) of wasteland is "An empty area of land, especially in or near a city, which is not used to grow crops or built on, or used in any way and/or a place, time or situation containing nothing positive or productive, or completely without a particular quality or activity".
- **The National Wasteland Development Board (NWDB)** defined wasteland as "degraded land which can be brought under vegetative cover with reasonable effort and which is currently under utilized and land which is deteriorating for lack of appropriate water and

soil management or on account of natural causes". This is the most widely accepted definition of wasteland.

- **ICAR** proposed that "wastelands are lands which due to neglect or due to degradation are not being utilized to their full potential. These can result from inherent or imposed disabilities or both, such as location, environment, chemical and physical properties, and even suffer from management conditions".
- **Technical Task Group Report** of the National Wastelands Development Board defined wasteland as "a land which is presently lying unutilized due to different constraints".
- **Society for Promotion Wasteland Development (SPWD)** defined wastelands as "wasted areas presently in the possession of public or private sector agencies/Institutions/Government/Panchayat and even under the ownership of individuals and which are being utilized for less than their optimum potential".
- **National Remote Sensing Agency (NRSA)** defined wasteland as "those lands which are presently not being used to their optimum potential due to some constraints".
- **Bhumbla and Khare (1984)** defined wastelands as "those lands which are ecologically unstable, whose top soil has been nearly completely lost, and which have developed toxicity in root zones for growth of most plants, both annual crop and trees".

Wastelands are the degraded and unutilized lands except current fallows due to different constraints. Poor land practices have led to malnutrition and decline in production capacity of the soil. It is estimated that in wastelands the biomass production is less than 20 per cent of its overall potential. It includes areas affected by water logging, sheet and gully erosion, ravine, riverine lands, shifting cultivation, salinity and alkalinity, wind erosion, shifting sand dunes, extreme moisture deficiency, coastal sand dunes etc. These degraded lands are ecologically unstable with almost complete loss of top soil and are unsuitable for cultivation due to decline in their quality and productivity. In general, wastelands may be put under two categories: cultivable wasteland and uncultivable wasteland.

CULTIVABLE WASTELAND: The land which is capable or has potential for the development of vegetative cover and not being used due to different constraints of varying degrees such as erosion, water-logging, salt effects etc.

UNCULTIVABLE WASTELAND: The land which cannot be developed for vegetative cover e.g. rocky areas, deserts and snow-covered glacial areas.

WASTELAND ASSESSMENT

As early as mid sixties scientists working on wasteland management realized the need to categorize the wasteland in accordance with their intrinsic characteristics and causative factors. Among various characteristics, data on soil texture, soil depth, pH values, slope, erosion status and other inhibiting factors like salts, water-logging, flooding, rockiness, stoniness etc. are essential. Complete inventory of reasons, responsible for lands lying waste, were assessed by collecting information from various sources. They serve as points for assessment of any wasteland.

- Poor fertility of the soil due to rocky, gravely, sandy, saline, alkali, water-logging, shallowness and eroded nature of the soil
- Steep and undulated slopes
- Shifting cultivation
- Frequent droughts

- Lack of irrigation facilities
- Frequent submergence and flooding hazards
- Lack of resources
- Poor economic conditions
- Lack of labour during peak periods
- Uneconomical return under cultivation
- Certain domestic and legal difficulties
- Absentee landlords holdings.

WASTELAND MAPPING TECHNIQUES

Through mapping, the findings of these questions are recorded in a pictorial form under a specified scale. These maps will serve anyone who wants to study about a particular area of wasteland or to utilize it for some purpose or other. The following important mapping systems are available in India: conventional, aerial photo-interpretation and remote sensing. The first system can be stated as a real time system where as the latter two are the tools to improve the time and cost efficiency.

CONVENTIONAL SYSTEM: It is based primarily on ground transect using suitable base maps for plotting boundaries. For detailed investigations, the surveyor transverses from field to field with a base map to collect the soil and land-use data after closely spaced field observations to identify and map wasteland.

This approach is very effective but very time consuming. One field surveyor can collect information from only 16,000 ha per annum. For mapping the wastelands on regional basis, the reconnaissance type of survey could be adopted with Survey of India topographical maps at 1:50,000 scale as base material. Detailed data on texture, depth, profile, morphology, physical and chemical characteristics are gathered. The ownership details and other socio-economic factors have, however, to be collected by actual visits to the village and local enquiries.

AERIAL PHOTO-INTERPRETATION (API): Black and white panchromatic aerial photographs of 1:25,000 scale and smaller are very suitable for wastelands mapping. False colour APs may improve the level of mapping as far as wastelands are concerned since the cropped lands, orchards and forested areas can be segregated more efficiently with their varied colour manifestation. The derived map could be at a semi-detailed level. However, the data on socio-economic factors, ownership details and other relevant information have to be collected by field visits personal enquiries and verification of village records only.

REMOTE SENSING: The satellite era ushers new scopes in land use mapping. The periodicity and larger perspective view have their own advantages in mapping wastelands and monitoring the changes using satellite data in a big way involving several organizations. Mapping with satellite data is quicker and cost-effective as compared to any other methods. Ability to effectively monitor the changes is the special advantage in this system because of the periodicity of the data. It is expected to gain the accuracy of 74.1 to 95.3 per cent. Ownership details cannot be included in the remote sensing data unless resorted to other methods such as grid survey, revenue records etc.

WASTELANDS IN INDIA

India has a total geographical area of 328.2 million hectare with a cultivated area of 141 million hectare. The country is situated between 8º04' to 37º06' N latitudes and $68^0$07' to 97º25' E longitudes.

Over the years, the country's landmass has suffered from different types of land degradations. Land degradation is caused by biotic and abiotic pressures. An ever increasing population places enormous demands on the land resources. This is particularly true in India, which has only 2.4 per cent of the world's geographical area but supports over 16 per cent of the world's population. It has 0.5 per cent of the world's grazing area but has over 18 per cent of world's cattle population. These pressures have led to drastic changes in the proportion of land utilized for agricultural activities, urbanization and industrial development.

Table 38.1: Wasteland Category/ Classes in India 2008-2009 (DoLR and NRSC 2011)

S No	Wasteland Class	Area (Sq. Km)	Percentage to TGA
1	Gullied and/ or ravinous land- medium	6,145.96	0.19
2	Gullied and/ or ravinous land- deep/very deep	1,266.06	0.04
3	Land with dense scrub	86,979.91	2.75
4	Land with open scrub	93,033.00	2.94
5	Waterlogged and marshy land- permanent	1,757.07	0.06
6	Waterlogged and marshy land- seasonal	6,946.31	0.22
7	Land affected by salinity/ alkalinity- moderate	5,414.53	0.17
8	Land affected by salinity/ alkalinity- strong	1,391.09	0.04
9	Shifting cultivation area- current jhum	4,814.68	0.15
10	Shifting cultivation area- abandoned jhum	4,210.46	0.13
11	Under utilised/degraded forest- scrub dominated	83,699.71	2.64
12	Agricultural land inside notified forest land	15,680.26	0.50
13	Degraded pastures/ grazing land	6,832.17	0.22
14	Degraded land under plantation crops	278.53	0.01
15	Sands- riverine	2,111.96	0.07
16	Sands- coastal sand	654.47	0.02
17	Sands- desert sand	3,934.80	0.12
18	Sands- semi-stabilized to stabilized (>40m) dune	9,279.75	0.29
19	Sands- semi-stabilized to stabilized moderately high (15-40m) dune	14,273.03	0.45
20	Mining wastelands	593.65	0.02
21	Industrial wastelands	58.00	0.00
22	Barren rocky area	59,482.29	1.88
23	Snow covered and/or glacial area	58,183.44	1.84
	Total	**4,67,021.16**	**14.75**

Intensive agricultural practices that rely heavily on water, chemical fertilizers and pesticides have caused waterlogging and salinity in many parts of the country. The expansion of the irrigation system without adequate steps for treatment of the catchment areas has exacerbated this. The quest for increased agricultural productivity has led to intensive cultivation of marginal lands causing their degradation. These pressures on land are compounded by the fact that over 69 per cent of our geological area falls within dry zone as per the Thornwaite classification. Land degradation has a direct bearing on the productivity of soil, its vulnerability to rainfall variations, scarcity of drinking water, fodder and fuel wood. Given the inter-linkages of crop production, livestock economy and environment, land degradation has a major impact on the livelihoods of the people, especially in rural areas.

Some of the most degraded lands in the country are the common property resources and these resources include community pastures, community forests, wastelands and co-mmon dumping and threshing grounds. In spite of concerted efforts to check defores-tation and the various afforestation schemes taken up during successive Plan periods, large tracks of forest continue to be classified as degraded. Of the total forest area, 31 million hectare suffers from some form of degradation and 14.06 million hectare suffers from extreme degradation and are part of 63.85 million hectare of wastelands reported by the NRSA. Desertification and Land Degradation Atlas of India has reported that total area under land degradation in India is 105.48 million ha (32.07 per cent) and total area under desertication is 81.45 million ha (24.78 per cent) (*Space Application Centre 2007*)

In addition to the wastelands identified by the NRSA, other areas such as deserts, drought-prone, flood-prone and tribal areas have been subjected to severe forms of degradation. The capacity of these lands is limited due to environmental factors. Pressures of human and livestock population have further compromised them.

National Commission on Agriculture made the first step for assessment of area affected by the land degradation and reported 148 million ha land area are affected in India. Other organizations have also taken up the assessment of land degradation in India viz. Ministry of Agriculture, ICAR's National Bureau of Soil Survey and Land Use Planning (NBSS & LUP), National Wasteland Board, National Remote Sensing Agency-Hyderabad (NRSA), *Space Application Centre of ISRO*. NBSS & LUP followed process based degradation mapping methodology in agriculture and non-agriculture areas, derived from 1:250,000 soil map prepared from the satellite data supported with soil profile studies. The NRSA followed remote sensing techniques, identifying land use and physical condition of the surface features for mapping non-agriculture areas (non-arable) on 1:50,000 scale using satellite image with adequate field checks.

Table 38.2: Wasteland Assessment by Different Organizations (Gautam and Narayan 1988; Space Application Centre 2007; DoLR and NRSC 2011)

Agency	Estimated Extent (M ha)	Criteria for Delineation
National Commission on Agriculture (NCA 1976)	148.09	Based on secondary data
Ministry of Agriculture (1978-Soil &Water Conservation Division)	175.00	Based on NCA's estimates. No systematic survey undertaken
Society for the Promotion of Wasteland Development (SPWD) – (Bhumbla and Khare 1984)	129.58	Based on secondary data
NRSA (1985)	53.28	Mapping on 1:1 million scale based upon remote sensing techniques
Ministry of Agriculture (MOA 1985)	173.64	Land degradation statistics for the states
Ministry of Agriculture (MOA 1994)	107.43	Elimination of duplication of area. Area reclaimed counted

Agency	Estimated Extent (M ha)	Criteria for Delineation
NBSS and LUP (1994)	187.70	Mapping based upon Global Assessment of Soil Degradation (GLASOD) guidelines
NBSS and LUP (2004) revised	146.82	1:1 million scale soil map
Dept. of Environment (Vohra 1980)	95.00	-
National Wasteland Development Board (1985)	123.00	-
NRSA (2000)	63.85	Remote sensing techniques
Space Application Centre (2007)	105.48	Remote sensing techniques
DoLR and NRSC (2010)	47.23	Remote sensing techniques
DoLR and NRSC (2011)	46.70	Remote sensing techniques

PROCESSES OF WASTELAND FORMATION IN INDIA

(Space Application Centre 2007)

VEGETAL DEGRADATION: It is observed as deforestation, forest blanks, shifting cultivation area and also degradation in grazing area/ grassland as well as in scrubland. At places, agriculture is observed within forest land, this has been classified under vegetal degradation within forest area. Vegetal degradation processes are found in both cold and hot regions.

WATER EROSION: It is observed in both hot and cold areas, in various land covers and with varying severity levels. The sheet erosion, rill erosion and shallow gullies mostly within agricultural lands are categorized as low category of severity. While deep & wide gullies and ravines are categorized as high category.

WIND EROSION: It pertains to the Aeolian activities. It denotes the spread of sand by virtue of lift and drift effect of wind, even upto lofty altitudes of Himalayas. Various categories of sand cover and their severity are classified based on the depth and spread of sand sheet, sand dunes and barchans.

SALINIZATION AND ALKALINIZATION: It is fundamentally the chemical property of soils. It occurs mostly in cultivated lands, especially in the irrigated areas. At places salinity is clearly observed on satellite images, while the alkalinization is not seen and mostly inferred based on ground truth data and published maps.

WATERLOGGING: The undrained land areas tend to accumulate standing water for longer duration of time on the surface, this condition is called waterlogging. The severity of waterlogging is determined based on the period for which the water remains stagnant.

MASS MOVEMENT OR MASS WASTING: It is predominantly observed in mountai-nous areas, especially in cold dryland regions and it is defined as the process of deserti-fication which leads to the down slope movement of rock, regolith and debris through the action of gravity viz. landslides, scree slopes

FROST HEAVING: It is defined as the process of intense frost and freezing of water operating in glacial and periglacial environment and evolves peculiar forms of rock, regolith and soil, for example, a typical irregular pattern ground are seen.

FROST SHATTERING: It is defined as the freeze and thaw action operating mostly in periglacial environment. When water, passing through the crevices and pores within the rock freezes, it expands by almost ten times. This puts enormous pressure on the surrounding rocks which shatters them.

MAN MADE: All those land degradation processes which is included directly or indirectly by human intervention and are not natural, are categorized as man made desertification processes. It includes mining/quarrying, brick kiln, industrial effluents, city waste, excess use of fertilizers and pesticides. It occurs across various land use/ land cover classes.

EXAMPLES OF LAND DEGRADATION IN INDIA

- Mining sites are abandoned after excavation work is complete leaving deep scars and traces of over-burdening.
- In states like Jharkhand, Chhattisgarh, Madhya Pradesh and Orissa, deforestation due to mining have caused severe land degradation.
- In states like Gujarat, Rajasthan, Madhya Pradesh and Maharashtra, overgrazing is one of the main reasons for land degradation.
- In the states of Punjab, Haryana, Western Uttar Pradesh, over irrigation is responsible for land degradation due to water logging leading to increase in salinity and alkalinity in the soil.
- The mineral processing like grinding of limestone for cement industry and calcite and soapstone for ceramic industry generate huge quantity of dust in the atmosphere. It retards the process of infiltration of water into the soil after it settles down on the land
- In recent years, industrial effluents as waste have become a major source of land and water pollution in many parts of the country.

NATIONAL ACTION PROGRAMME TO COMBAT DESERTIFICATION

India became a signatory to the UNCCD on 14 October 1994 and it came into effect on 17 March 1997. One of the obligations of all developing country parties to the convention, including India, is to prepare the National Action Programme to Combat Desertification and to mitigate the effects of drought. The Ministry of Environment and Forests, as the National Focal Point for the implementation of the Convention, has prepared the National Action Programme to Combat Desertification (NAPCD) through the setting up of a High-level Inter-Sectoral National Steering Committee (NSC) in July 1999. The NSC constituted four Working Groups (WG) on various relevant to desertification viz. Desertification Monitoring and Assessment, Sustainable Land Use Practices for Combating Desertification, Local Area Development Programme, Policy and Institutional Issues.

Similarly Desertification Cell has been established under Ministry of Environment and Forests comprising multi-disciplinary team including agriculture, forestry, environment science, geo-hydrology, rural development and social sciences. It would serve on all matters like preparation of NAP, monitoring its implementation, organizing reviews, feasibility studies and evaluation and impact studies.

FEATURES OF NAPCD

- Water is a diacritical element for building people's confidence and satisfaction level, reclamation of degraded lands for sustainable biomass production ultimately leading to a better quality of life and enabling conditions through empowerment of the local communities.
- Bottom up approach and project planning, evaluation and monitoring by *Panchayat Raj* Institutions (Grass root level elected local self Governments in which women have at least 30 per cent representation) on all aspects of land development including rehabilitation of degraded areas

- Convergence of resources and services. All resources available under different schemes will be channelled through *Panchayats* through single window service to the communities.
- Gaps in all the on-going schemes of different departments will be identified and resources will be provided to fill them up for generating good impact at the ground level.
- The first five-year period will be experimental and include pilot projects and activities.
- Concurrent and continuous monitoring will be done and mid-course correction will be effected as the experiences are gained
- The problem faced by farmers will be reported to R&D institutions for finding solutions. Thus both lab-to-land and land-to-lab flow of information will be ensured

WASTELAND DEVELOPMENT

The first and foremost step in wasteland development is checking land degradation through soil and water conservation measures such as bunding, trenching, drainage line treatment and vegetative barriers (engineering and vegetative structures). The next step is to increase the biomass availability by measures such as pasture development, planting and sowing of legumes, planting hardy plants, promotion of agroforestry and horticulture, planting bamboo, planting fast growing multipurpose trees (MPTs) and establishing woodlots for meeting fodder, fuelwood and small timber demand.

Wasteland management includes measures for land conservation, land rehabilitation and land improvement.

- *Land Conservation*: It is the measures taken to maintain the functions of the land. The land is allowed for productive use, but within limits. Land is maintained to stay within system tolerance and to allow system to reach an equilibrium. If it is necessary, land is removed from use, or even positive measures are taken to rehabilitate.
- *Land Rehabilitation*: It is the restoration of the previous land functions of the degraded land
- *Land Improvement*: It is an increase in the environmental function or productive potential of land.

WHAT IS ECOLOGICAL RESTORATION?

Ecological restoration can be defined as the process of assisting the recovery and management of ecological integrity, including a critical range of variability in biodiversity, ecological processes and structures, regional and historical context and sustainable cultural practices. Restoration can be passive, in which the degrading agent(s) is identified and removed or active, in which management techniques such as planting, weeding, burning and thinning are undertaken with a particular image of desired structure, composition, or pattern in mind.

The need for restoration assumes some level of impairment in an ecosystem because of low ecological integrity. Although natural disturbances can cause impaired conditions, often human activities are responsible and thus a change in the management can theoretically bring about improved conditions. There are number of land restoration options that, if applied wisely, could boost agricultural production and slow down land and environmental degradation.

MULTI-FACETED ACTIONS TO COMBAT LAND DEGRADATION

The integrated approach is necessary to combat land degradation. The following approaches are considered necessary for the land improvement.

- Integrated management of natural resources
- Soil conservation and restoration
- Improving agricultural production and stock-farming systems
- Agricultural water management
- Sustainable forest management
- Agroforestry
- Watershed management
- Environmental monitoring, drought early warning and forecasting/response system
- Promotion of sustainable land management techniques geared to climate change adaptation
- Biodiversity preservation
- Adding value to agricultural, forest, farmed products
- Diversifying activities so as to relieve pressure on resources
- Increasing land and labour productivity with available technology
- Developing technology to overcome agricultural production constraints

Table 38.3: Types of Land Degradation and Improvement (Sara J. Scherr and Satya Yadav 1996)

Component	Degradation	Improvement
Physical soil management	Crusting, compaction, sealing, wind erosion, water erosion, devegetation, overtillage	Soil conservation barriers (live, inert), terracing, revegetation of denuded lands, tree protection, soil decompaction, breaking up of pans, covercrops, windbreaks, soil deposition, improved tillage methods
Soil water management	Impeded drainage, waterlogging, reduced waterholding capacity, reduced infiltration, salinization	Irrigation, water harvesting, field drainage, draining of waterlogged areas, filter strips
Soil nutrient and organic matter management	Alkalinization, acidification, nutrient leaching, removal of organic matter, burning of vegetative residues, nutrient depletion	Fertilization, composting, green manuring, animal manuring, Flushing of saline and alkaline soils, liming of acid soils
Soil biology management	Overapplication of agrichemicals, industrial contamination	Introduction of biotic organisms, nitrogen fixing microorganisms
Vegetation Management	Decline in vegetative cover, decline in biodiversity, decline in species composition, decline in availability of valuable species	Increased vegetative cover, increased species diversity, increased species composition, Improved availability of valuable species

WASTELAND DEVELOPMENT THROUGH AGROFORESTRY

Agroforestry refers to land use systems in which trees or shrubs are grown in association with agricultural crops, pastures or livestock and in which there are both ecological and economic interactions between trees and other components. An agroforestry practice is a distinctive arrangement of components in space and time, characterized by environment, plant species and arrangement, management and social and economic functioning. Agroforestry ensures maximum productivity per unit of land on sustainable basis. It improves the soil fertility and productivity, reduces the soil run off and provides more income to land owners than monocropping system. Adoption of agroforestry practices such as agrisilviculture, silvopastoral and agrisilvipastoral systems with an appropriate combination of trees, grasses and agricultural crops develop the wastelands very fast. It ensures moisture improvement, soil improvement and biomass improvement in the wastelands.

TREES, SHRUBS AND GRASSES FOR WASTELAND DEVELOPMENT

It is well recognized that a large area in the country is degraded due to soil erosion by water. Soil erosion is caused mainly by lack of vegetative cover, slope, nature of soil, rainfall and its intensity, cropping system and land management. It is also caused by deforestation, shifting cultivation, overgrazing and improper cropping of undulated lands. If the intensity and degree of soil erosion are high, it may lead to the formation of deep gullies and ravines. Appropriate measures are required to check runoff and soil loss which include planting vegetation (trees, grasses, shrubs), agroforestry, mulches, bunding, terracing, contour cultivation, graded trenching, crop rotation practice with different crops, no tillage strategy etc.

Similarly drylands need to be reclaimed using drought tolerant species. Sand areas and sand dunes should be stabilized using appropriate tree and grass species. Area affect- ed by wind erosion need to be protected by establishing wind breaks and shelter belt. Salt tolerant grasses and tree species are planted in lands affected by salinity and alkalinity.

Species selection is the important criteria for the success of any wasteland development programme which is based on compatibility to local site conditions, survival and productivity. The genetic quality of native species plays a key role for local adaptability and to withstand adverse environmental conditions. It is also important for suitability to soils with different depth and water retention capacities. The plant establishment mainly depends on the development of deep and good root system. The inherent characteristics of a species, to propagate or regenerate itself vegetatively viz. root suckers, when damaged, are also important for survival.

Tree Species: *Faidherbia albida, Acacia tortilis, Acacia senegal, Acacia catechu, Acacia auriculiformis, Acacia ampliceps, Acacia bivenosa, Acacia nilotica, Prosopis cineraria, Prosopis juliflora, Parkinsonia articulata, Tamarix articulata, Ziziphus mauritiana, Zizypus nummularia, Calligonum polygonoides, Cassia siamea, Albizia lebbeck, Leucaena leucocephala, Gliricidia sepium, Sesbania sesban, Erythrina* spp., *Casuarina equisetifolia, Inga dulcis, Grewia tenax, Anogeisus rotundifolia, Euphorbia caducifolia, Tamarix aphylla, Cercidium floridum, Eucalyptus tereticornis, Eucalyptus camaldulensis, Hardwickia binata, Pongamia pinnata, Tecomella undulata, Azadirachta indica, Dendrocalamus strictus, Bambusa bambos, Agave sisalana, Agave americana, Calliandra* spp., *Gmelina arborea, Grevillea robusta, Dalbergia sissoo, Paulownia* spp., *Terminalia* spp., *Anthocephalus* spp., *Madhuca* spp., *Pterocarpus* spp., *Emblica officinalis* and *Aegle marmelos*

Grasses and Forage Crops: *Vetiveria zizanioides, Cynadon dactylon, Dichanthium annulatum, Panicum repens, Panicum antidotale, Panicum maximum, Cenchrus setigerus, Cenchrus ciliaris, Lasiurus sindicus, Sehima nervosum, Bracharia brizantha, Bracharia mutica, Sporobolus marginatus, S. airoides, S. helvolus, Fagonia cretica, Eragrostis* spp., *Taverniera cuneifolia, Barleria acanthoides, Eremopogon foveolatus, Heteropogon contortus, Lepidagathis* spp., *Aristida* spp., *Oblong folia, Tribulus terrestris, Brachiaria ramosa* and *Cymbopogon jwarancusa.*

Medicinal Plants: *Aloe vera, Balanites aegyptiaca, Commiphora wightii, Euphorbia antisyphilitica, Haloxylon* spp. and *Cassia angustifolia*

IMPROVED FALLOW SYSTEM

Natural processes will normally restore productivity of nutrient depleted degraded lands when sufficient time is given. Traditionally, farmers have used the practice of "fallow" to allow crop land to rest without crops and be rejuvenated naturally. When the fallow is enriched with fast-growing trees, shrubs or vines, the practice is called "improved fallow". Improved fallow is an agroforestry practice that has its origins in slash-and-burn agriculture. Farmers use improved fallow to accelerate the process of rehabilitation and thereby shorten the length of their fallow periods. The technology can be applied to any agricultural land that is not under cultivation in order to accelerate recovery, increase nutrient reserves, and improve the potential for future productivity on the site.

To create an improved fallow system, farmers scatter seeds or plant seedlings of fast-growing plants after harvest of the crops from the site. Normally, nitrogen-fixing plants are used, because they are vigorous, deep-rooted, tolerant of drought, and have the ability to accumulate atmospheric nitrogen. The trees and shrubs are left to occupy the site for several months or years. During the fallow period, the plants accumulate nitrogen from the air and from deep layers of the soil, and drop their leaf litter to enrich the soil and conserve moisture. When the trees are removed at the end of the fallow period, their roots remain in the soil to decompose gradually, releasing additional nutrients to the subsequent crops. The important plant species used in the improved fallow systems are *Inga edulis*, *Cajanus cajan*, *Crotalaria* spp., *Sesbania sesban*, *Gliricidia sepium*, *Erythrina* spp., *Cassia siamea*, etc. (Wilkinson and Elevitch 2013).

VEGETATIVE BARRIERS OR BARRIER HEDGES

Runoff velocity can be reduced drastically by planting vegetative hedges, using, trees, shrubs and grasses on the contour at regular intervals. These hedges can increase the time for water to infiltrate into the soil, and facilitate sedimentation and deposition of eroded material by reducing the carrying capacity of the overland flow. Vegetative hedges or narrow vegetation strips serve as porous filters. These hedges may reduce runoff amount and can drastically decrease soil loss. If leguminous plants are used, the land can be rehabilitated in a shorter period of time with soil improvement, biomass improvement and moisture improvement. The commonly used species are *Gliricidia sepium*, *Calliandra calothyrsus*, *Leucaena leucocephala*, *Acacia angustissima*, *Desmodium rensonii*, *Flemingia macrophylla*, *Cassia spectabilis*, *Tithonia diversifolia*, etc. The grass species widely used are *Axonopus micay*, *Brachiaria brizantha*, *Brachiaria decumbens*, *Brachiaria mutica*, *Cenchrus ciliaris*, *Eragrostis curvula*, Molasses grass, *Panicum antidotale*, *Panicum coloratum*, *Panicum maximum*, *Paspalum conjugatum*, *Paspalum decumbens*, *Paspalum notatum*, *Pennisetum purpureum*, *Setaria* spp. and *Vetiveria zizanioides*.

HEDGEROW INTERCROPPING/ ALLEY CROPPING

To deal with the unique management problems of the soils, scientists at the International Institute of Tropical Agriculture (IITA) in the 1970s incorporated woody species in crop production systems. This ultimately led to the development of the alley cropping system (Kang et al 1981). In alley cropping, food crops and woody species are intercropped, food crops are grown in the alleys formed by hedgerows of planted trees and shrubs, preferably legumes. The hedgerows are cut back at planting and periodically pruned during cropping to prevent shading and to reduce competition with the food crops. The prunings are used as green manure or mulch. The hedgerows are allowed to grow freely to cover the land when there are no crops.

One major advantage of alley cropping is that the cropping and fallow phases take place concurrently on the same land, thus allowing the farmer to crop the land for an extended period without a break. Though alley farming was originally designed for use by small farmers, it is believed that it is sufficiently flexible to be adapted for mechanized farming using appropriate machinery. The major tree species used in alley farming are *Acacia auriculiformis*, *Gliricidia sepium*, *Leucaena leucocephala*, *Acioa barteri*, *Calliandra calothyrsus*, *Flemingia macrophylla*, *Paraserianthes falcataria*, *Alchornea cordifolia*, etc. (Kang 1993).

SILVOPASTORAL SYSTEM

Soil erosion on pastures is often more severe than on croplands. Severe sheet erosion and gullying are both common. The initial cause is degradation of the vegetation through overgrazing, which leads to a sparse, sometimes almost zero, ground cover, leaving the soil open to erosion. It is not uncommon for 10 cm or more of topsoil to be removed. Such erosion occurs both in semi-arid regions dependent primarily on grazing, and on land used for pasture in areas of mixed farming. Silvopastoral practices include scattered trees on pastures (e.g. systems with *Faidherbia albida* or other *Acacia* species), combination of plantation crops with pastures, live fences, fodder banks, windbreaks and shelterbelts and hedgerow intercropping on pastures. The potential of windbreaks using fodder trees to control wind erosion is well established. If silvopastoral system is attempted simply by planting trees in pastures, without other changes in the management of degraded pastures, it will not be successful. The basic tenets of pasture management, such as restriction of livestock numbers and rotational grazing, are prerequisite to erosion control as to any other aspect of silvopastoral systems.

The major tree Species used in silvopastoral systems are *Faidherbia albida*, *Acacia tortilis*, *Acacia leucophloea*, *Acacia senegal*, *Acacia catechu*, *Acacia auriculiformis*, *Acacia nilotica*, *Prosopis cineraria*, *Prosopis juliflora*, *Parkinsonia articulata*, *Ziziphus mauritiana*, *Cassia siamea*, *Albizia lebbeck*, *Leucaena leucocephala*, *Gliricidia sepium*, *Sesbania sesban*, *Erythrina* spp., *Casuarina equisetifolia*, *Inga* spp., *Eucalyptus tereticornis*, *Eucalyptus camaldulensis*, *Hardwickia binata*, *Pongamia pinnata*, *Azadirachta indica*, *Calliandra* spp.,etc.

BOUNDARY PLANTING AND LIVE FENCES

Live fencing is one form of agroforestry that can provide a range of products and services on degraded farm lands. Although live fencing systems are traditional, the extent of the many potential benefits and the number of different systems are currently not well understood or even not well documented. Planting of multipurpose trees and shrubs (MPTS) around the farm is a very common practice. They provide protection, privacy, and valuable products to the farmers. Trees are planted within property line as fence, or as demarcation of farm lots. Tree plantations on plot boundaries have several advantages: creation of a fence limiting the movement of animals, protection against the wind and harshness of the sun, occasional forage resource.

The most common species are *Acacia mellifera*, *Acacia nilotica*, *Prosopis juliflora*, *Acacia senegal*, *Agave sisalana*, *Bauhinia* spp., *Dichrostachys cinerea*, *Euphorbia balsamifera*, *Haematoxylon brasiletto*, *Jatropha curcas*, *Ziziphus mauritiana*, *Ziziphus mucronata*, *Vitex* spp., Bamboo, etc.

TREES, CROPS AND GRASSES SUITABLE FOR AGROFORESTRY PRACTICES IN DEGRADED LANDS

GULLIED AND RAVINE LANDS

Trees and shrubs suitable for silvopastoral system in gullied lands are *Acacia nilotica*, *Acacia catechu*, *Azadirachta indica*, *Albizia lebbeck*, *Dendrocalamus strictus*, *Dalbergia sissoo*, *Prosopis chilensis*, *Prosopis juliflora*, *Eucalyptus* spp., etc. The important barrier hedge species are *Impomoea carnea*, *Agave americana*, *Arundo donax*, *Vitex negundo*, etc. The grass species suitable for silvopastoral system are *Cenchrus ciliaris*, *Pennisetum pedicellatum*, *Saccharum spontaneum*, *Dichanthium annulatum*, *Cynodon dactylon*, *Heteropogon contortus* and *Stylosanthes* spp.

WATTERLOGGED AND MARSHY LANDS

Proper drainage is important to rehabilitate these lands. Suitable tree species for waterlogged and marshy areas are *Eucalyptus robusta, Eucalyptus rudis, Syzygium cumini, Salix* spp., *Populus nigra, Terminalia arjuna* and *Acacia nilotica, Barringtonia* spp., *Bischofia javanica, Lagerstroemia flos-reginae* and *Casuarina equisetifolia.*

SALINE AND ALKALINE LANDS

The tree species suitable for agroforestry practices in saline and alkaline lands are *Prosopis juliflora, Prospopis chilensis, Acacia nilotica, Acacia arabica, Acacia catechu, Acacia auriculiformis, Acacia tortilis, Tamarix articulata, Eucalyptus* spp., *Ziziphus* spp., *Albizia lebbeck, Terminalia arjuna, Azadirachta indica, Morus* spp., *Ailanthus excelsa, Casuarina equisetifolia, Inga dulcis, Syzygium cumini, Dalbergia sissoo, Salvadora persica, Leucaena leucocephala,* etc. Guava, pomegranate, custard apple, ber and sapota are the suitable fruit trees for salt affected lands. The most suited grasses are *Brachiaria mutica, Panicum antidotale, Diplachne fusca, Chloris gayana, Cynodon dactylon, Panicum* spp., *Pennisetum purpureum, Leptochloa* spp., and *Sporobolus* spp.

DEGRADED PASTURES

Commonly planted trees are *Acacia nilotica, Acacia leucophloea, Acacia catechu, Prospopis juliflora, Prosopis cineraria, Albizia lebbeck, Leucaena leucocephala, Gliricidium sepium, Sesbania sesban* and *Erythrina* spp. The recommended grass species are *Dichanthium annulatum, Panicum antidotale, Cenchrus setigerus, Cenchrus ciliaris* and *Chrysopogon fulvus.* The common planted legumes are *Stylosanthes* spp., *Phaseolus* spp., etc.

BARREN ROCKY AND STONY WASTELANDS

The recommended trees are *Eucalyptus tereticornis, Dendrocalamus strictus, Melia Azedarach, Albizia lebbeck, Acacial catechu, Ailanthus excelsa, Hardwickia binata, Cassia siamea, Prosopis chilensis, Inga dulcis* and *Dalbergia sissoo.*

MINING AND INDUSTRIAL WASTELANDS

The recommended tree species are *Prosopis juliflora, Acacia auriculiformis,* other Acacias, *Azadirachta indica, Bombax ceiba, Casuarina equisetifolia, Eucalyptus tereticornis, Eucalyptus camaldulensis, Tamarindus indica, Agave* spp., *Ailanthus excelsa, Inga dulcis,* etc.

■■■■

39 WINDBREAK AND SHELTERBELT AGROFORESTRY SYSTEMS

WINDBREAK AGROFORESTRY SYSTEM

A windbreak is a narrow row of trees planted in fields bordering a farm plot (IDF 1981). Webster on-line dictionary defines windbreak as 'hedge or fence of trees designed to lessen the force of the wind and reduce erosion'. It is reported that the term "windbreak" was first used in popular English literature sometime before 1886. It is mcre popular and more widely used in developed countries like USA, UK, Russia and Germany than in developing countries like India. On the contrary, windbreak as an agroforestry system has greater and more vital role to play in developing countries, more particularly with reference to increasing land area under agroforestry with minimal sink in cultivation area under agriculture as well as to make agroecosystems as more climate change resilient systems through enhanced productivity; reduced evapotranspiration and in turn increased water use efficiency of the agroecosystem; reduced crop damage particularly in plantain cultivation in Tamil Nadu, India; and increasing carbon sequestration in biomass and in soil. However, 'windbreak' has been used as a broad term to include both shelterbelts as well as any other vegetative environmental buffers (Tyndall and Wallace 2011). 'Windbreaks' being broader term and is commonly used in the context of agroforestry system for both windbreaks and shelterbelts.

COMMON BENEFITS OF WINDBREAK

Windbreak can reduce soil erosion, increase crop yields and protect livestock from heat and cold. Windbre ks can shield buildings and roads from drifting snow. They beautify the landscape and provide travel routes and habitat for wildlife. Windbreaks can also be sources of wood and food. The benefits of windbreaks are described as below:

MODERATION OF MICROCLIMATE: By reducing wind speed, windbreaks modify the climate in the areas they shelter. The effects of reduced wind speed are: (i) Moderated soil and air temperatures, (ii) Increased relative humidity, (iii) Reduced evaporation and increased soil moisture, and (iv) Changes in the distribution of snow. Though the effectiveness of an individual windbreak in reducing the wind speed may vary considerably, it is often recognized that under average condition (shelter density 50 to 60 per cent or wind permeability about 40 to 50 per cent), the sheltered area may go up to 15 to 25 times the height of the windbreaks on the leeward side and 2 to 3 times on the windward side when the wind is 40 km per hour (Dwivedi 1992).

REDUCING SOIL EROSION: Windbreaks can reduce soil erosion by: (i) Reducing the occurrence of winds that are strong enough to carry soil away, and (ii) Reducing the loss of soil moisture, which binds soil particles together and makes them less likely to be blown by wind.

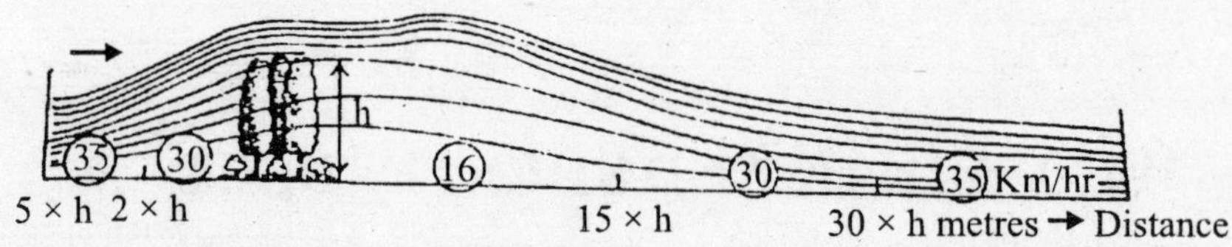

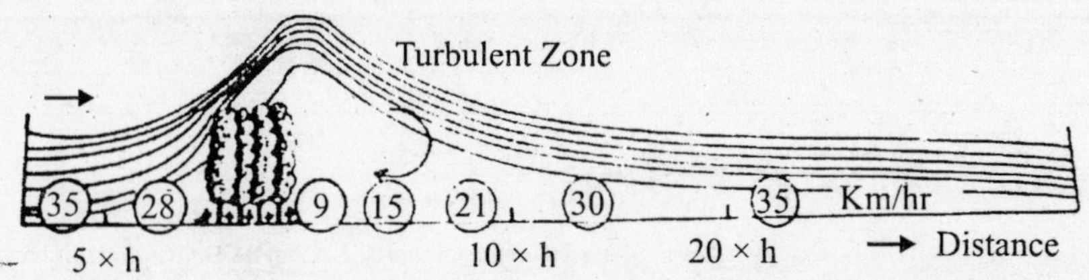

Fig. 39.1: Effect of Moderately Penetrable and Dense Windbreaks (Source: Dwivedi 1992)

ENHANCING CROP PRODUCTIVITY: Windbreaks increase crop quality and yield in sheltered areas by: (i) Providing lower temperatures in the day and warmer temperatures in the night, (ii) Increasing relative humidity and helping to retain soil moisture, and (iii) Reducing physical damage caused by high wind.

BENEFITING HEALTH OF LIVESTOCK: Windbreaks increase the health of livestock and the survival rates of young animals by protecting livestock from heat in summer and cold wind in winter. Protection from extreme cold also increases productivity by allowing food energy to be used for growth and milk production, rather than for maintaining body heat. In this way, windbreaks can reduce the amount of food animals require to keep warm in winter. Because animals graze less when exposed to heat, the shade provided by windbreaks also helps to increase productivity in summer.

PROTECTION TO BUILDINGS: Windbreaks help to save energy and to reduce the cost of heating and cooling by protecting buildings from winter wind and summer sun. Windbreaks can reduce winter heating costs up to 25 per cent. The reduction in summer air conditioning costs from windbreaks can be quite dramatic. The cooling effect of one mature deciduous tree is equal to 10 room-sized air conditioners. Windbreaks can also protect greenhouses from heat loss in winter. Studies suggest that heat loss from greenhouses doubles as wind speed increases from zero to 24 kilometres an hour. Windbreak protection can decrease heat loss by 10 to 15 per cent.

PROTECTION FROM HARSH EFFECTS OF SNOW: Dense windbreaks are useful for protecting roads and farm yards from drifting snow. They retain snow and shape it into deep, narrow drifts. Windbreaks of open structure, which distribute snow evenly over fields, are useful for protecting crops, reducing soil erosion and increasing soil moisture. Studies suggest that a layer of snow 20 cm deep completely protects the soil from freezing. A single row of trees is most effective because it allows air to flow through the gaps between trees. This type of windbreak can distribute snow evenly over a distance 25 times the height of the trees.

ENHANCED FOOD AND WOOD PRODUCTION: In addition to their role in providing shelter, windbreaks can be designed to provide both food and wood. They can produce fruit, nuts, firewood, posts, poles, veneer and sawlogs.

BENEFITS OF WINDBREAK REPORTED IN RESEARCH STUDIES

Windbreaks on field boundaries effectively control injuries to the tender crops from sand blasting and hot wind. It reduces wind velocity by 20-46 per cent and soil loss by 76 per cent (Gupta et al 1997). Windbreaks are an important tool for farming in semi-arid areas. In the western Rajasthan, 3-row windbreak (*Senna siamea* and *Albizia lebbeck*) is highly effective in reducing wind speed and loss of nutrients. It has been found that wind protection provided by shelterbelts (tree windbreaks) can increase pod yields of *Phaseolus vulgaris* both early and late in the seasons. In Egypt, it has been estimated that sheltered field of cotton, wheat, summer maize and rice increased yield by 36, 38, 47 and 10 per cent respectively, over the unprotected fields. Besides crop protection and enhanced productivity, the windbreaks contribute to the organic matter content of the soil through leaf fall. There are several outstanding examples of successful windbreaks around the world. Since 1949, a 3000 km long windbreak of *Casuarina equisetifolia* has been established in China along the coast bordering the South China Sea. This has provided shelter for crops growing on the leeward side and has also stabilized drifting sand in the area. In Egypt, Casuarina shelterbelts have also been grown to protect agricultural land mainly from wind erosion.

Campi et al (2009) studied the effects of tree windbreak on microclimate and wheat productivity in a Mediterranean environment. The 3-year field study was carried out by integrating agronomic and microclimatic approaches. Study concerned durum wheat growing in open field, in a typical Mediterranean environment. A *Cupressus arizonica* windbreak (3 m in height) bordered at North the experimental field. The analysis of the microclimatic observations shows that, when wind blew from the North direction, the windbreak presence influenced the wind speed until the distance 12.7H (H is the windbreak height), and temperature increased in a distance of 4.7H from the barrier. On the basis of the soil water content, continuously measured by TDR technique, evapotranspiration (ET) was daily determined and season ET calculated. Windbreaks mitigated ET for a distance of 12.7 times the windbreak height. Out of this area, the ET was 16 cent higher than the ET measured near the windbreak belt (<4.7H). Yield performances changed accordingly the distance from the windbreak. Within the distance of 18 times the windbreak height, wheat productions were higher than those obtained in the zone not influenced by the windbreaks. Within the protected area, wheat water use efficiency (WUE, calculated as the ratio between yield and seasonal evapotranspiration) attained the maximum value of 1.15; out of the windbreak protection, WUE was 0.70 kg m^{-3}. Since windbreaks reduce ET, farms of the Mediterranean environments should be re-designed in order to consider the windbreaks as possible issue of sustainability. By considering the above facts, windbreak agroforestry system can be promoted to make the agroecosystems as climate change resilient system.

PLANNING AND DESIGNING OF WINDBREAKS

Windbreaks are designed to perform specific jobs. The best tree or shrub species to use, the spacing between trees, the size of the windbreak and its location are determined by the characteristics of the land and the job the windbreak is expected to do. When planning a windbreak, it is needed to consider the shape and orientation of the property, wind speed and direction, and the way snow accumulates. The positions of buildings, roads, power lines, property lines, ditches, trees and wooded areas are other important factors. The growing period and the amount of care required by different tree species should also be considered.

A windbreak of optimum porosity (up to 25%), would consist of a single row of even-height trees with touching canopies and an understorey of bushy species that do not lose their leaves even in bouts of high winds. Windbreak spacing should be roughly 15 times the average height of the mature trees for greatest effectiveness. The area of protection in any single windbreak system depends upon several factors:

- Orientation – angle of the windbreaks to the direction of the wind
- Interval – distance between two successive windbreaks
- Continuity – number and size of gaps or passages in the windbreak
- Height – average height of the trees in the windbreak rows
- Density – relating to width and porosity of the windbreak
- Form of species – referring to height and shape of crown

The general guidelines for planning a windbreak are described below (Natural Resources Conservation Services, USDA):

1. Position the windbreak perpendicular or as nearly perpendicular to the most troublesome winds as possible.
2. For wind protection only, position the row containing the tallest tree species approximately 2 to 5 H from all primary areas or objects in need of protection. ('H' is the tree height.) Where the objects to be protected are located uphill from the windbreak, the row may need to be located closer to provide adequate wind protection. All areas in need of wind protection should be located within 10 H of the tallest tree row. Since 'H' will change until the tallest tree row reaches maturity or becomes ineffective, it is suggested that the estimated height of the tallest tree species at 20 years of age be used for planning purposes. When a primary area in need of protection is located more than 10 H leeward of a primary windbreak, supplementary windbreaks of one to three rows should be made at 8 to 10 H intervals leeward of the primary windbreak.
3. For wind and snow protection, there are different guidelines for windbreaks. It is still desirable to locate primary areas or objects in need of protection 2 to 5 H from the tallest tree row. However, in some areas, this could result in the burying of primary areas under deep drifts. To minimize problems associated with deep drifts, it is important to properly position the most windward row.

   The windward row of a windbreak designed to provide protection from wind and drifting snow should be located 100 to 200 feet from primary areas or objects in need of protection. The distance will vary from region to region and should be based on the average length of snow drifts adjacent to typical windbreaks after severe storms. If blowing snow can come from wide, open fields, the distance recommended is approximately 200 feet to the windward row of primary windbreaks. In most other areas, 100 to 150 feet is adequate.
4. The number of legs (i.e. windbreak belts) in a windbreak planting will depend upon the number of directions from which troublesome or problem winds occur. Windbreaks with one leg can be effective for controlling troublesome winds and/or drifting snow coming from one direction

Fig. 39.2: Area Protected by One Primary Windbreak (Source: Clifton et al 1986)

The Fig. 39.2 illustrates the amount of area protected by a primary windbreak planted on the north side of on area in need of protection. The area to be protected would be vulnerable to winds from all directions except from the north.

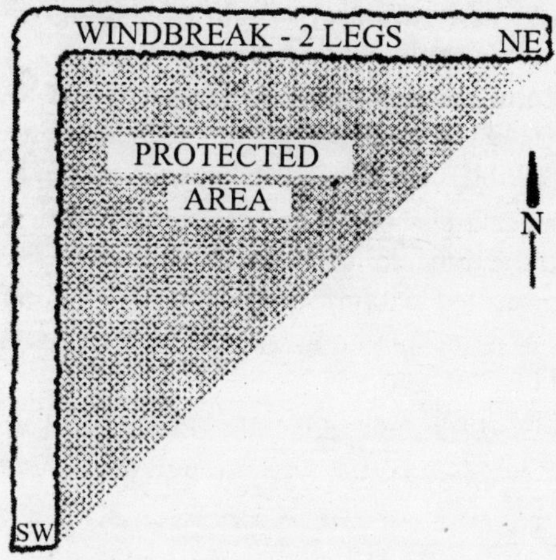

Fig. 39.3: Area Protected by Two Primary Windbreaks

The Fig. 39.3 illustrates the amount of area protected by primary windbreaks planted on the north and west sides of area in need of protection. It also illustrates why the standard (L) shaped windbreak on the north and west rides of farmsteads and feedlots is necessary.

1. Access lanes or roads which cut through a windbreak designed only for wind protection should be at an angle to prevailing or troublesome winds. Access lanes or roads which are parallel to the prevailing wind direction will cause the winds to funnel through gaps. Lanes or roads located at an angle to prevailing or troublesome winds will not allow wind to funnel into areas in need of protection.

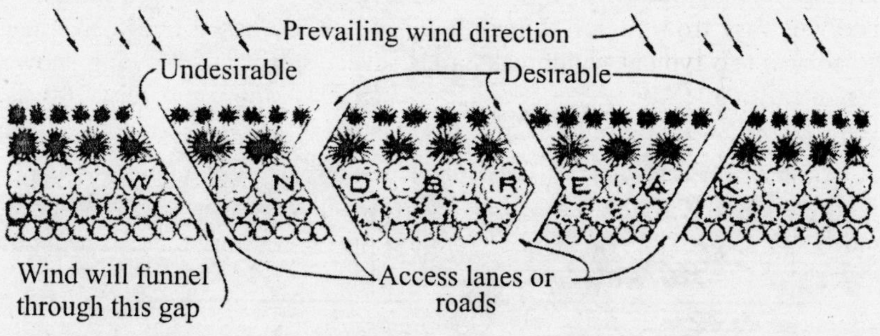

Fig. 39.4: Alignment of Access Lanes and Roads in Windbreak (Source: Clifton et al 1986)

Access lanes or roads which must be used during winter months in areas subject to severe snow blowing, should not cut through a windbreak. Lanes and roads located in such a manner are prone to deep snowdrifts. It can also be difficult to remove snow from such lanes or roads. It is recommended that access roads be located at the ends of windbreaks in areas beyond where snowdrifts form. The lanes or roads should be a minimum of 100 to 500 feet from the ends of the windbreaks to minimize snowdrift problems.

2. The soils of the potential planting site must be known. If there is more than one soil type, there may be a need to plant different tree and shrub species on each soil or to plant around areas of problem soils.

3. The location of property lines, subsurface drain fields, and septic fields must be known. Property lines can severely restrict the location of plantings. In some cases where severe snow drifting can be a problem and sufficient space is not available to properly locate a planting, it might be advisable not to plant all or a portion of a windbreak. Subsurface drain lines or septic fields should be relocated or avoided.

4. Any existing windbreaks, clumps or groves of trees or shrubs, and, in some cases, large individual trees, should be outside the boundaries of new wind-break plantings. New plantings should not be located within 50 feet of any existing trees if space is available. Where space is limited existing plantings should be incorporated into the design of new plantings or they should be removed.

5. The standard L-shape is used for the windbreak. Only species adapted to wet soils should be planted in the area of the windbreak that runs through the wet soils.

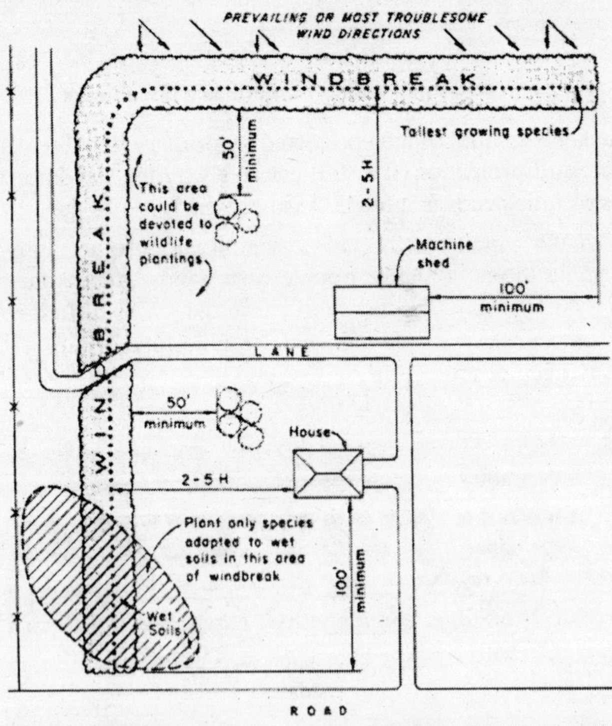

Fig. 39.5: Windbreak Design for Wind Protection (Source: Clifton et al 1986)

OPTIMUM NUMBER OF ROWS AND SPACING FOR WINDBREAKS

It is recommended that minimum number of rows will vary from one row in areas where only wind protection is of concern, to 8 to 10 rows where high winds and severe snowdrift problems are common. However, research and experience have shown that dense windbreaks (those that exceed 65 per cent density) are needed to provide adequate protection from wind and drifting snow around buildings, feedlots, and so forth. Therefore, the density of the species planted should be more important than the number of rows. A deciduous tree in full leaf has a density of 65 per cent or greater. A deciduous tree without leaves has a density from 40 to 65 per cent. These densities vary depending upon the thickness of the tree crown (number of branches and branching pattern).

To minimize the potential impact of any future insect or disease problem, it is recommended that the rows be of different species. Combinations of evergreen species, evergreen species and deciduous trees or shrubs or deciduous trees and shrubs can all give desired densities. Where adequate space is available, it is highly recommended that thought be given to using a variety of species in more than the minimum number of rows. The added variety and number of rows will significantly enhance wildlife, beautification and general landscape values. Generally, two to three rows of trees of the desired density can provide adequate protection. If different plant species are to be planted in different rows of windbreaks, the arrangement of species in plantings should be such that shrubs, short trees and slower growing trees should be located in the outer rows of plantings. They should not be positioned between two taller growing species. By placing these slow growing species in the outer rows, it assures them of adequate light and their development.

Another important aspect of planting trees in more than one row is optimum spacing required for planting. In windbreak plantings where competing vegetation is to be controlled by mowing, mulching, or herbicides, the minimum between-the-row spacing should be approximately 8 feet. On the other hand, if weed control is aimed by ground shading by the tree rows itself, then the spacing between-the-row can be 3 to 5 feet. Within-the-row spacing is also quite variable. The spacing for most shrub species vary from 3 to 6 feet, deciduous tree species 6 to 15 feet and evergreens 6 to 20 feet in temperate regions. With-the-row spacing for tree species in tropical region can be 6 feet to 12 feet in multiple rows planting.

SELECTION OF SPECIES FOR WINDBREAKS

The primary consideration for the selection of the species must be matching of the tree species to the site conditions particularly with reference to the adaptation of the species to the planting location. Secondly, the selected species need to have comparatively better tolerance/resistance to pest and disease problems. Importantly, the species to be promoted for windbreaks should have desired crown density with good number of branches and sufficient density of foliage. The species should have fast growth rate so that it will be able to provide needed protection from wind within 3 to 5 years in tropical regions and within 10 years in temperate regions.

Soils play a very important role in the selection of species for the site. For tree and shrub species to become well established and to grow in a satisfactory manner, they must be adapted to the soils of the windbreak site. Some species of trees and shrubs are adapted to a wide range of soil conditions. Others do well only on wet or moist sites. Some can tolerate high lime conditions; others cannot. When dealing with sites with more than one soil, it is best to plant species that will grow satisfactorily on all the soils. However, this is not always possible. Where soil properties are drastically different, it is best to plant only those species which are adapted to each soil. Failure to do so can result in large gaps or weak spots in windbreaks. These gaps or weak spots can have a serious effect on the way windbreaks function. Some of the tree species used for windbreaks are *Eucalyptus*, *Senna*, *Prosopis*, *Leucaena*, *Casuarina*, *Acacia*, *Grevillea*, *Syzygium*, *Dalbergia*, etc.

IDEOTYPE BREEDING FOR WINDBREAKS

Agroforestry systems aim to integrate agriculture and forestry crops to generate economically productive, ecologically diverse and socially acceptable sustainable land-use systems. In such integration of agriculture and forestry crop, the most important desirable traits are narrow canopy, less branches, acute branch angle, self-pruning ability and deep root production rather than lateral root production. These traits help to minimize tree-crop competitive interactions for light, water and nutrients.

Contrary to the selection of trees for wide-row alley cropping system, suitable ideotype for "windbreak and shelterbelt" agroforestry systems need to have more branches and foliage biomass and should not have self pruning ability. The branch angle also needs to be near right angle. Nicodemus et al (2001) reported that certain phenotypes of *Casuarina equisetifolia* and *C. junghuhniana* have shown certain branching characteri-stics which are ideally sought for windbreaks. Institute of Forest Genetics and Tree Breeding (IFGTB), Coimbatore, India initiated efforts to select superior phenotypes of *Casuarina equisetifolia* which are suitable for developing a windbreak agroforestry system. The selected phenotypes will be useful to address the recurrent problem of crop damages particularly in plantain growing belt of the Coimbatore district, which is caused by strong gusty winds during monsoon period. Twenty one (21) phenotypes were selected by adopting "point grading method", while giving greater score for higher productivity, trees having more branches were given additional weighted scores. These phenotypes along with one productive clone released by IFGTB and seedlings were planted in clone bank of IFGTB. After a period of six months, all these 24 phenotypes were evaluated for branch characteristics and height growth parameter. The results are presented in Table 39.1. Interestingly, the top three clones in terms of height growth *viz.* IFGTB-WBC 8, IFGTB-WBC 17 and IFGTB-WBC 18 recorded greater number of branches, branch length as well as wider branch angle. Further, these phenotypes were narrowed down to 10 superior phenotypes on the basis of initial height growth and rooting potential, as rooting potential among different clones varied from 40 to 94 per cent with a overall mean value of 83 per cent. These 10 superior phenotypes were multiplied clonally and planted in four farmlands under windbreak agroforestry system (Fig.39.7). The between-the-row spacing used in this clonal trial was 1 m and within-the-row spacing was 2 m. The planting pattern adopted was zig-zag configuration also known as 'Quincunx' planting as in Fig.39.6.

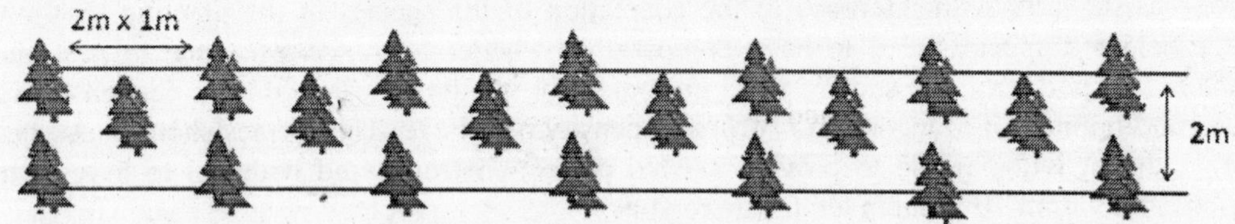

Fig. 39.6: Quincunx Planting Configuration in Windbreak Agroforestry System

This clonal trial was evaluated one year after planting for height and diametrical growth as well as for branching pattern and consistency of branching traits across different site conditions. In this effort, IFGTB has now developed productive clones of *Casuarina* exclusively suitable for windbreak agroforestry system which will have more number of branches, greater branch thickness, wider branch angle along with greater growth rate and biomass productivity and those clones are proving as suitable clones for windbreak.

Ho KuenYih and Tsai WeiJe (2011) investigated the adaptive behavior of *Casuarina* windbreak stands along the southwestern coast of Taiwan and concluded that vitality of the tree crowns, soil salinity, soil pH, tree diseases, tree flowering, and root damage were the 6 major factors affecting differences in the adaptive performance of *Casuarina* windbreak stands. Further, they identified 3 factors of tree status, crown vitality, and changes in the soil environment as influential indicators.

Fig. 39.7: Superior Clones of *Casuarina equisetifolia* in Windbreak Agroforestry System

Table 39.1: Evaluation of Different Phenotypes of *Casuarina equisetifolia* selected for Windbreak Agroforestry Systems (Krishnakumar and Buvaneswaran 2012)

Clone	Number of Branches up to 2 m Height	Mean Branch Length (in cm)	Mean Branch Angles (in Degree)	Mean Height Growth (cm)
IFGTB-WBC 1	27	31.95	61.20	181.3
IFGTB-WBC 2	20	36.79	57.98	174.0
IFGTB-WBC 3	9	16.80	51.63	61.7
IFGTB-WBC 4	33	27.3	58.08	129.0
IFGTB-WBC 5	40	39.17	64.31	239.0
IFGTB-WBC 6	39	52.69	69.81	223.3
IFGTB-WBC 7	48	42.75	65.00	254.3
IFGTB-WBC 8	**54**	**61.66**	**61.72**	**293.7**
IFGTB-WBC 9	47	38.63	59.37	242.7
IFGTB-WBC 10	49	35.91	59.61	204.7
IFGTB-WBC 11	36	41.99	59.28	175.0
IFGTB-WBC 12	39	37.79	55.18	163.7
IFGTB-WBC 13	37	31.37	48.78	151.3
IFGTB-WBC 14	43	40.15	50.71	188.0
IFGTB-WBC 15	37	38.37	65.72	178.7
IFGTB-WBC 16	27	35.86	56.06	172.7
IFGTB-WBC 17	**42**	**57.00**	**78.13**	**314.3**
IFGTB-WBC 18	**43**	**64.87**	**67.12**	**269.7**
IFGTB-WBC 19	11	24.80	45.96	88.3

Clone	Number of Branches up to 2 m Height	Mean Branch Length (in cm)	Mean Branch Angles (in Degree)	Mean Height Growth (cm)
IFGTB-WBC 20	23	37.51	47.26	195.3
IFGTB-WBC 21	28	44.50	70.96	206.0
CE-Seedlings	30	37.63	66.11	170.7
CJ-Seedlings	24	29.87	59.82	156.7
IFGTB-CE 3	16	31.99	60.81	126.7

SHELTERBELT AGROFORESTRY SYSTEM

A shelterbelt is defined as belt of trees and/or shrubs maintained for the purpose of shelter from wind, sun, snowdrift, etc. Shelterbelts are generally more extensive than windbreaks and cover areas larger than a single farm and sometimes a whole region on a planned pattern (Dwivedi 1992). In the context of definition of agroforestry, in strict sense, shelterbelt may not fall under agroforestry system. However, there are reports to show the interactive effect of shelterbelt on the enhancement of productivity of agricultural crops in areas leeward side of shelterbelts. The following are the main characteristics of shelterbelts (Negi 1986):

- *Shape and Composition*: Shelterbelts will have a typical triangular shape. This can be usually be brought about by raising tall trees in the centre.
- *Density and Width*: A certain degree of penetration by wind is planned, as by raising a solid wall, the protection decreases very fast on the leeward side. Shelterbelts of upto 50 m width are considered ideal under tropical conditions.
- *Height and Spacing*: The ratio of height and width should roughly be 1:10. However, this figure may vary with local conditions.
- *Orientation*: Orientation of shelterbelt depends upon the direction and velocity of the prevailing winds. Shelterbelts may be raised in quadrangles if wind directions tend to change very often.
- *Length*: Length is an important consideration so far as shelterbelts are concerned; because at the fringes of these zones eddies are formed. This results in an increase in the wind velocities. The minimum length of a shelterbelt should be about 25 times its height.

The tree species recommended for shelterbelts are *Acacia arabica*, *Acacia leucopholea*, *Dalbergia sissoo*, *Lannea coromandelica*, *Eucalyptus* spp., *Tamarix articulata*, *Parkinsonia aculeata*, *Prosopis juliflora*, *Prosopis spicigera*, *Casuarina equisetifolia*, etc. The shrubs recommended include *Calotropis procera*, *Clerodendron phlomoides*, *Cassia auriculata, Dodonaea viscosa*, etc. The grass species include *Saccharum spontaneum*, *Saccharum munja*, *Panicum antidotale*, *Cenchrus* spp., etc.

The Great Plains Shelterbelt was a project to create windbreaks in the Great Plains states of the United States, that began in 1934. President Franklin D. Roosevelt initiated the project in response to the severe dust storms of the Dust Bowl, which resulted in significant soil erosion and drought. The United States Forest Service believed that planting trees on the perimeters of farms would reduce wind velocity and lessen evaporation of moisture from the soil. By 1942, 220 million trees had been planted, stretching out 29,900 km in a 100-mile-wide zone from Canada to the Brazos River. Even as of 2007, the federal response to the Dust Bowl, including the PSFP (Prairie States Forestry Program which planted the Great Plains Shelterbelt) and creation of the Soil Erosion Service, represents the largest and most-focused effort of the U.S. government to address an environmental problem.

In Germany, shelterbelts were established in 10 m wide and 6.1 m high and spaced 199 m apart for protection as well as for enhancing yield of agricultural crops. In Italy, 4.8 to 7.0 m wide shelterbelts were established for protecting wheat crops. In Guangdong province of China, *Casuarina* based shelterbelts were established over 4034 ha in 1970s for reducing wind damage, halting shifting sand dunes and reclamation of agricultural lands. As a step taken by India to avoid large-scale destruction in the event of tsunami, several rows of *Casuarina equisetifolia* trees have been planted on the sandy beaches under the shelterbelt project that began July 2005. Using $6 million funding from the World Bank, a total of 4,700 hectares of the shelterbelt and 1,400 hectares of mangroves are being created. When finished, the shelterbelt will stretch from Kanyakumari to Chennai (India) in more than 1,000 km. Professor Narasimhan of Madras Christian College reported in his communication to FAO that though *Casuarina* is an attractive option for planting in the coastal landscapes, the mixed species approach, especially using native species, should be encouraged taking community needs into account. The east coast is characterized by unique vegetation called tropical dry evergreen forest. Species of this forest such as *Manilkara hexandra*, *Morinda pubescens*, *Pongamia pinnata*, *Gmelina asaitica* and *Streblus asper* can form important components for bio-shield and can be linked to the livelihood of the neighbourhood communities.

INCORRECT

Slower growing and shorter growing species planted between faster growing species

CORRECT

Slower growing and shorter growing species are planted in the outer rows

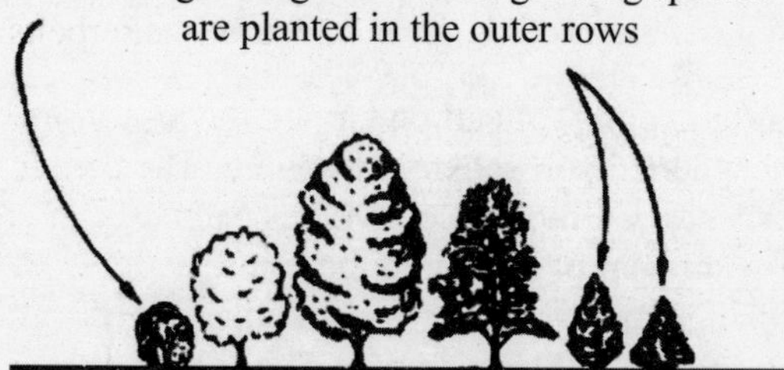

Fig. 39.8: Shelterbelt Design

It is important to note that windbreak and shelterbelt agroforestry systems will not replace the agriculture crops while increasing the tree cover. But, there exists greater scope for enhancing windbreak and shelterbelt agroforestry system in many areas of tropical regions considering their huge benefits. The potential area available for windbreak agroforestry system needs to be quantified.

40 AGROFORESTRY FOR CARBON SEQUESTRATION, CLIMATE CHANGE MITIGATION AND ADAPTATION

GLOBAL CARBON CYCLE

CARBON

Carbon is an element found in all living substances as well as in many inorganic materials. Carbon is a key element which makes up 18 per cent of our bodies. All living things are made of elements, the most abundant of which are, oxygen, carbon, hydrogen, nitrogen, calcium and phosphorous. Of these, carbon is the best at joining with other elements to form compounds necessary for life, such as sugars, starches, fats and proteins. Together, all these forms of carbon account for approximately half of the total dry mass of earth's plants and trees. It also exists in the environment in non-living things like rocks, oil, natural gas, coal and air. It is the basic building block of all life and the environment we live in.

Carbon is a ubiquitous element on Earth. Most of the Earth's carbon is stored in rocks, but this carbon is essentially inert on the 100's to 1,000's year timescales of interest to humans. The rest of the carbon is stored as CO_2 (carbon dioxide) in the atmosphere (2%), as biomass in land plants and soils (5%), as fossil fuels in a variety of geologic reservoirs (8%) and as a collection of ions in the ocean (85%). These are the "active" reservoirs of carbon of interest.

CARBON CYCLE

Carbon cycle is the movement (or cycling) of carbon among the reservoirs or storage places within the earth system: atmosphere, lithosphere, hydrosphere, biosphere. The carbon cycle is a bio-geochemical cycle since it involves the biosphere. Carbon cycle involves exchange of carbon between the land, the oceans and the atmosphere. The amount of carbon in these reservoirs is so large that it is expressed in gigaton (Gt) also known as pentagram (Pg) which is equivalent to 10^{15} grams or 10^{9} metric tons or a billion metric ton (1 metric ton equals 1,000 kilograms). The carbon cycle is one of the most important biogeochemical cycles and is based primarily on CO_2. It involves the conversion of inorganic CO_2 into organic compounds, and its subsequent re-mineralization to CO_2.

Carbon cycle involves the cycling of a chemical element - carbon - between organic (CHO) and inorganic (CO_2 and carbonate/bicarbonate compounds) states. There are three important carbon cycles in the earth system:

- Short-term organic carbon cycle, with emphasis on the interactions between the atmosphere and the biosphere: it has terrestrial (land) and marine (ocean) comp-onents. It includes processes that transfer carbon from one reservoir to another in a matter of years, including photosynthesis, plant and animal respiration, and the movement of CO_2 across the air-sea interface.

- Long-term organic carbon cycle, with emphasis on the formation and destruction of fossil fuels and other sediments containing organic carbon.
- Long-term inorganic carbon cycle with emphasis on calcium carbonate ($CaCO_3$, limestone), by far the largest of the carbon reservoirs. This cycle is linked to the carbonate-silicate cycle, supplying the calcium ions necessary for the formation of limestone.

The residence time of carbon varies widely among different reservoirs. On average, a carbon atom spends about 5 years in the atmosphere, 10 years in terrestrial vegetation, and 380 years in intermediate and deep ocean waters. Carbon can remain locked up in ocean sediments or fossil fuel deposits for millions of years. Fast cycling processes that take place in months or a few years have rapid effects but only influence small CO_2 reservoirs, so they do not change long-term CO_2 levels significantly. Slow processes that take place over centuries, millennia, or geologic epochs have greater influence on CO_2 concentrations over the long term.

Carbon cycle research integrates ecology and the earth sciences. Although the importance of CO_2 as a greenhouse gas was described by Svante Arrhenius in 1896, the importance of the carbon cycle was fully realized five decades ago. Accurate and direct measurements of the concentration of CO_2 in the atmosphere began in 1957 at the South Pole and in 1958 at Mauna Loa, Hawaii. Starting in 1957, the direct measurements of atmospheric CO_2 levels have revealed a rapid increase. In 1988, the severe consequences of this increase were acknowledged at an international level. Today carbon cycle research is a very active field, with many issues yet to be solved. It is unclear where all of the huge carbon emitted continuously by human activities ends up. New methodologies are being developed and explored such as the use of satellite imagery in conjunction with model calculations and precise measurements of chemical reactions involving carbon.

CARBON SOURCES AND SINKS

The carbon sources emit the CO_2 into the atmosphere. The increase in atmospheric CO_2 concentration results mainly from human activities which include primarily burning of fossil fuels and deforestation, but also cement production and other changes in land use and management such as biomass burning, crop production and conversion of grasslands to croplands. Vegetation fires can be a significant source of CO_2 to the atmosphere on annual time scales.

Two processes remove CO_2 from the atmosphere: photosynthesis by land plants and marine organisms, and dissolution in the oceans. There is an important distinction between these processes in terms of permanence. CO_2 taken up through photosynthesis is converted into organic plant material, whereas CO_2 dissolved in the oceans is transferred to a new carbon reservoir but remains in inorganic form. Organic carbon in plant tissues can remain sequestered for thousands or millions of years if it is buried in soils or deep ocean sediments, but it returns to the atmosphere quickly from material such as leaf litter. Similarly, CO_2 dissolved in the oceans will stay a long time if sequestered in deep water, but will escape more readily back into the atmosphere if ocean mixing brings it to the surface. Carbon dioxide is continuously exchanged between the atmosphere and the ocean. Carbon dioxide entering the surface ocean immediately reacts with water to form bicarbonate (HCO_3^-) and carbonate (CO_3^{2-}) ions. Carbon dioxide, HCO_3^- and CO_3^{2-} are collectively known as dissolved inorganic carbon (DIC).

Oceans and land ecosystems thus serve as both sources and sinks for carbon. Until recently these processes were in rough equilibrium, but the balance is being disrupted today as human activities add more carbon to the atmosphere and a large fraction of that anthropogenic carbon is transferred to the oceans. Of the total emissions from human activities during the period 2002-2011, 46 per cent accumulated in the atmosphere, 26 per cent in the ocean and 28 per cent on land. During this period, the size of the natural sinks has grown almost at the same pace as the growth in emissions, although year-to-year variability is large.

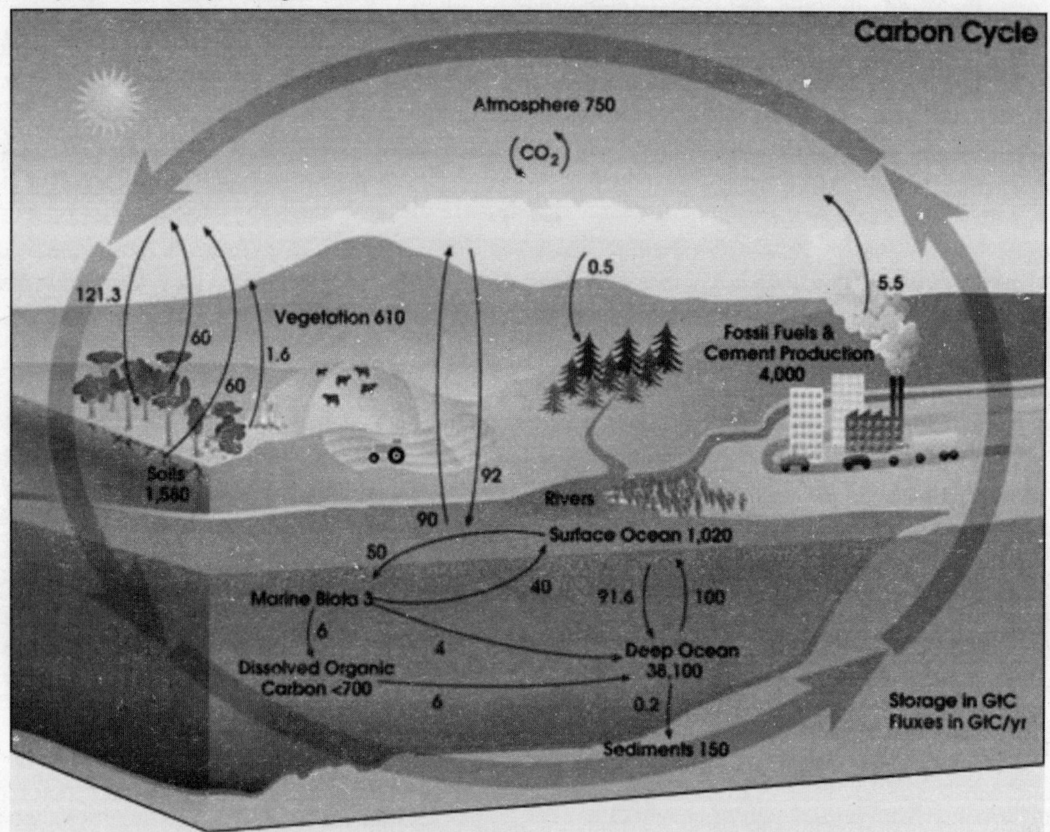

Fig. 40.1: Global Carbon Cycle (Source: NASA Earth Observatory)

CARBON POOLS

Carbon is present in the earth's atmosphere, soils, plants, fossil fuels, oceans and crust. When viewing the Earth as a system, these components can be referred to as *carbon pools* or *stocks* or *reservoirs* because they act as storage houses for large amounts of carbon. Carbon is stored on our planet in the following major pools:

- as organic molecules in living and dead organisms found in the biosphere
- as the gas carbon dioxide in the atmosphere
- as organic matter in soils
- in the lithosphere as fossil fuels and sedimentary rock deposits such as limestone, dolomite and chalk
- in the oceans as dissolved atmospheric carbon dioxide and as calcium carbonate shells in marine organisms

Major Carbon Pools

- The Earth's Crust : 10,00,00,000 Pg
- Fossil Fuels in the Earth Crust : 4,000 Pg
- Oceans : 38,000 Pg
- Atmosphere : 750 Pg
- Plants : 560 - 610 Pg
- Soil : 1,580 Pg

CARBON FLUXES

The movement of any material from one place to another is called a flux and carbon flux is the transfer of carbon from one pool to another. Carbon fluxes are usually expressed as a rate in which an amount of carbon being transferred over a certain period of time. In the integrated system, fluxes connect reservoirs together to create cycles and feedbacks.

Major Carbon Fluxes

- Photosynthesis (Influx) : 120 Pg
- Plant Respiration (Outflux) : 60 Pg
- Soil Respiration (Outflux) : 60 Pg
- Litter Fall (Outflux) : 60 Pg
- Ocean Atmosphere Exchange (Influx) : 92 Pg
- Ocean Atmosphere Exchange (Outflux) : 90 Pg
- Fossil Fuel Combustion (Outflux) : 5.5 - 6 Pg
- Land Cover Change (Outflux) : 0.9 Pg

CARBON BUDGET

The Earth's carbon cycle is in a constant state of motion. Through processes that take place over seconds, days, years and millennia, carbon is constantly being transferred between all the various pools discussed above. If all sources are equal to all sinks, the carbon cycle can be said to be in equilibrium or in a state of balance and there is no change in the size of pools over time. The global carbon budget is the balance of the exchanges (incomes and losses) of carbon between the carbon reservoirs.

Just as in the constructing of a financial budget, a carbon budget is simply a list of all pools along with estimates of their size and of all the fluxes that constitute inputs and outputs. Presently, a budget of the Earth's carbon cycle shows that it is far from being in a state of balance. While random variation in natural processes such as climate and forest fires often results in some degree of imbalance on an annual basis, the very large imbalance in today's carbon cycle is due to the processes of fossil fuel combustion and land cover change.

GLOBAL CO_2 EMISSIONS

Human activities, especially the burning of fossil fuels such as coal, oil and gas, have caused a substantial increase in the concentration of carbon dioxide in the atmosphere. This increase in atmospheric CO_2, from about 280 to more than 395 ppm over the last 250 years, is causing measurable global warming. There is a consensus that under existing emission scenarios, global mean temperature will increase by one degree Celsius by the year 2025 and three degree Celsius by the year 2100. It is expected that atmospheric CO_2 levels will continue to rise and may exceed 500 ppm by 2050. Not all gases are equal in their green house potential. CO_2 is most common and most important per unit of emissions than other GHGs. We need to distinguish between atmospheric carbon (C), atmospheric CO_2 and atmospheric carbon dioxide equivalent (CO_2e) warming potential.

The concentration of CO_2 in the atmosphere is now about 395 ppm and the concentration of CO_2e is about 455 ppm. Each ppm of atmospheric CO_2 corresponds to about 2.12 Pg carbon and 7.78 Pg of CO_2. Over the past 50 years, about 55 per cent of each emitted Pg of carbon has stayed in the atmosphere and the rest has been absorbed by land or oceans. The rate at which atmospheric CO_2 is increasing has risen from about one ppm per year in 1960s to two ppm for 2000s and it is still continuing to increase every year. At present, atmospheric carbon increases by about 4.5 Pg per year on average. 1 kg carbon (C) equals to 3.67 kg carbon dioxide (CO_2) and one Pg C equals to 3.67 billion tons CO_2 or 3.67 $GtCO_2$.

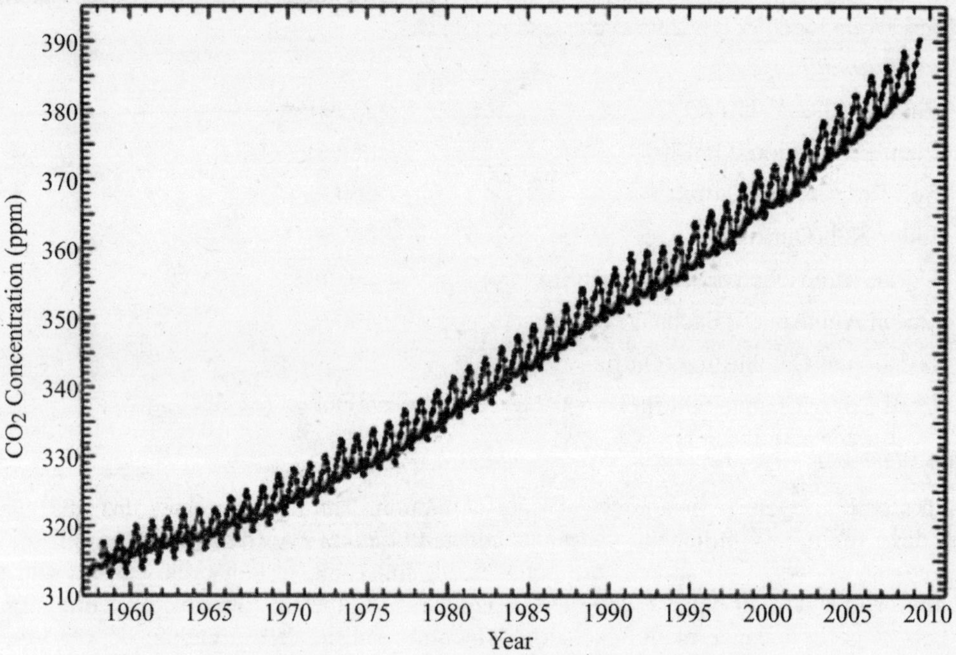

Fig. 40.2: Trends in CO_2 Concentration in Atmosphere
(Source: Mauna Loa Observatory, Hawaii & South Pole Antarctica)

The accumulation of atmospheric CO_2 in 2011 was 3.6±0.2 Pg C, with a total cumulative of 161.3 Pg C since the beginning of atmospheric high precision measurements in 1959 and 240 Pg C since 1750. The rates of atmospheric CO_2 accumulation are influenced by both the anthropogenic emissions and the net uptake by natural sinks (ocean and land), and their inter-annual variability is large. Global fossil and cement emissions have contributed 9.5±0.5 Pg C in 2011. Emissions growth in 2011 was dominated by China which was responsible for 80 per cent of the global emissions growth in 2011. Top four emitters in 2011 covered 62 per cent of global emissions with the contribution of China (28%), United States (16%), EU27 (11%) and India (7%). World average per capita emissions in 2011 were 1.4 t C/p. The countries and the per capita emission: China (1.8 t C/p), United States (4.7 t C/p), EU27 (2.0 t C/p), India (0.5 t C/p). Share of global emissions in 2011 is contributed by coal

(43%), oil (34%), gas (18%) and cement (5%). By extracting fossil fuels (oil, gas and coal) from deep in the Earth, we are overloading the atmosphere with carbon, and changing our climate in irreversible ways.

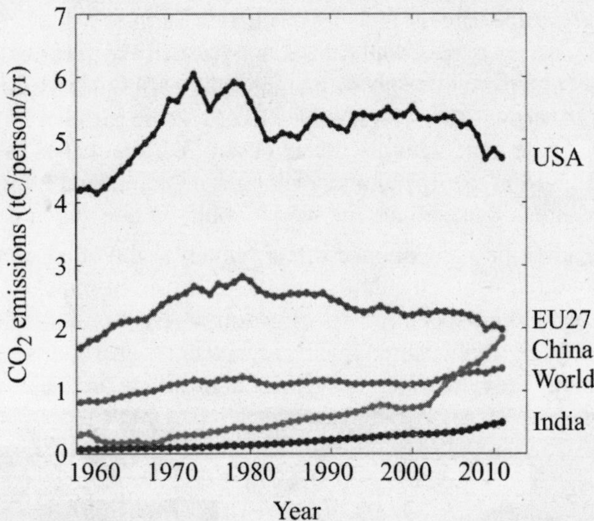

Fig. 40.3: Top Fossil Fuel Emitters - Per Capita
(Source: CDIAC, US Oak Ridge National Laboratory)

Table 40.1: Global Carbon Emissions Statistics (Global Carbon Project 2012)

Region/ Country	**Per Capita Emissions in 2011 (tons C/ person)**	**Total Emissions in 2011 (Pg C, % of global)**
Global (with bunkers)	1.4	9.5
Global (no bunkers)	1.3	9.1
	Developed Countries (Annex B)	
Annex B	3.0	3.6 (40%)
United States of America	4.7	1.5 (16%)
EU27	2.0	1.0 (11%)
Russian Federation	3.2	0.46 (5.1%)
Japan	2.6	0.32 (3.6%)
Germany	2.5	0.20 (2.2%)
	Developing Countries (Non-Annex B)	
Non –Annex B	0.9	5.4 (60%)
China	1.8	2.5 (28%)
India	0.5	0.62 (6.8%)
Iran	2.3	0.17 (1.9%)
South Korea	3.3	0.16 (1.7%)
South Africa	2.8	0.14 (1.6%)

GREENHOUSE EFFECT AND GLOBAL WARMING

Certain gases – mainly carbon dioxide (CO_2), but also methane (CH_4), nitrous oxide (N_2O), and volatile organic compounds (VOCs) – play an important role in regulating the temperature of the earth's atmosphere. These are known collectively as the greenhouse gases (GHGs). Most visible and ultraviolet light coming from the sun is able to reach the Earth's surface by passing through the atmosphere. Some of this energy is re-radiated back to space in the form of long-wave or infra-red radiation. Natural amounts of greenhouse gases maintain the earth's temperature at a habitable level by absorbing some of this energy – the 'greenhouse effect'. There are two common meanings of the term "greenhouse effect". There is a "natural" greenhouse effect that keeps the Earth's climate warm and habitable at 33°C (from -18°C to +15°C) thus making it habitable and allowed the present form of life to evolve. Without this greenhouse effect, the world would be freezing at about -17°C.

There is also the "man-made" greenhouse effect, which is the enhancement of Earth's natural greenhouse effect by the addition of greenhouse gases from the burning of fossil fuels. However, human activities are increasing the concentrations of greenhouse gases in the atmosphere, leading to a gradual warming of the earth's atmosphere. It is this enhanced greenhouse effect that is generally referred to as global warming. Gases that trap heat in the atmosphere are called greenhouse gases.

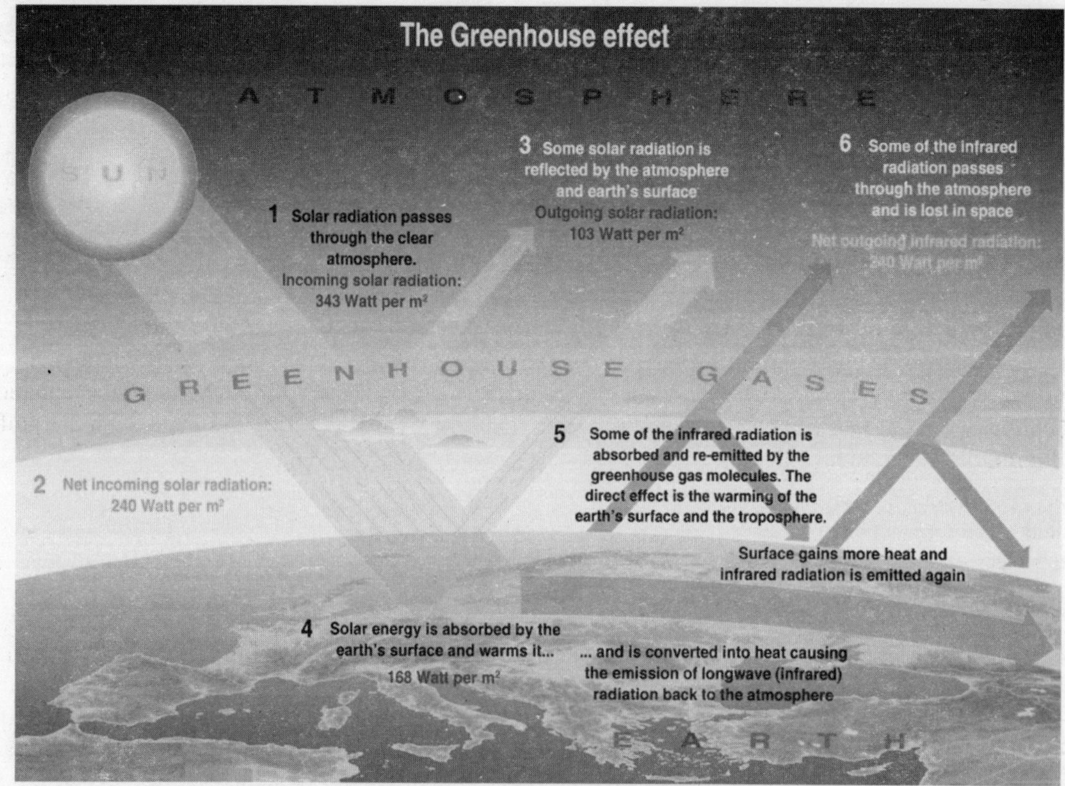

Fig. 40.4: Greenhouse Effect and Global Warming (Source: UNEP)

MAJOR GREEN HOUSE GASES

CARBON DIOXIDE (CO_2): Carbon dioxide enters the atmosphere through burning fossil fuels (coal, natural gas and oil), solid waste, trees and wood products, and also as a result of certain chemical reactions (e.g., manufacture of cement). Carbon dioxide is removed from the atmosphere (or "sequestered") when it is absorbed by plants as part of the biological carbon cycle.

METHANE (CH_4): Methane is emitted during the production and transport of coal, natural gas, and oil. Methane emissions also result from livestock and other agricultural practices and by the decay of organic waste in municipal solid waste landfills.

NITROUS OXIDE (N_2O): Nitrous oxide is emitted during agricultural and industrial activities, as well as during combustion of fossil fuels and solid waste.

FLUORINATED GASES: Hydrofluorocarbons (HFCs), perfluorocarbons (PFCs), and sulfur hexafluoride (SF_6) are synthetic, powerful greenhouse gases that are emitted from a variety of industrial processes. Fluorinated gases are sometimes used as substi-tutes for stratospheric ozone-depleting substances (e.g., chlorofluorocarbons (CFCs), hydrochlorofluorocarbons (HCFCs), carbon tetrachloride (CCl_4) and halons). These gases are typically emitted in smaller quantities, but because they are potent greenhouse gases, they are sometimes referred to as High Global Warming Potential (GWP) gases.

CLIMATE CHANGE

When we use the term 'climate' what we actually mean is the average weather experienced in a region over a long period of time. Meteorological variables other than temperature, such as wind and rainfall patterns are constituents of climate. The climate on earth has undergone many changes in the past and this is entirely natural. However, the rate at which the climate has been changing over the past 50 years has led to a consensus amongst scientists that this recent change is likely to be as a result of man's activities. This is what we refer to as 'Climate Change'.

Thermometer measurements show that Earth's average surface temperature has risen substantially over the past century, and especially over the last three decades. These data are corroborated by a variety of independent observations showing warming in other parts of the Earth system, including the oceans, the lower atmosphere, and ice-covered regions. The enhancement of the 'greenhouse effect' by rising concentrations of these greenhouse gases is largely responsible for the issue of global warming. However, rising temperatures are only one consequence of the enhanced greenhouse effect, with changes to weather patterns and many aspects of the climate such as rainfall and humidity predicted. Collectively, these are referred to as 'climate change'.

CLIMATE CHANGE DEFINITION

Intergovernmental Panel on Climate Change (IPCC) defines climate change as "a change in the state of the climate that can be identified (e.g. using statistical tests) by changes in the mean and/or the variability of its properties, and that persists for an extended period, typically decades or longer. Climate change may be due to natural internal processes or external forcings, or to persistent anthropogenic changes in the composition of the atmosphere or in land use".

United Nations Framework Convention on Climate Change (UNFCCC), in its Article 1, defines "climate change" as "a change of climate which is attributed directly or indirectly to human activity that alters the composition of the global atmosphere and which is in addition to natural climate variability observed over comparable time periods". The UNFCCC thus makes a distinction between "climate change" attributable to human activities altering the atmospheric composition, and "climate variability" attributable to natural causes.

IMPACTS OF CLIMATE CHANGE

Science has made enormous progress toward understanding climate change. As a result, there is a strong, credible body of evidence, based on multiplelines of research, documenting that Earth is warming. Strong evidence also indicates that recent warming is largely caused by human activities, especially the release of greenhouse gases through the burning of fossil fuels. Global warming is closely associated with other climate changes and impacts, including rising sea levels, increases in intense rainfall events, decreases in snow cover and sea ice, more frequent and intense heat waves, increases in wildfires, longer growing seasons, and ocean acidification. Individually and collectively, these changes pose risks for a wide range of human and environmental systems. While much remains to be learned, the core phenomenon, scientific questions, and hypotheses have been examined thoroughly and have stood firm in the face of serious scientific debate and careful evaluation of alternative explanations. Projections of future climate change anticipate an additional warming of 2.0 to 11.5ºF (1.1 to 6.4ºC) over the 21st century, on top of the 1.4ºF already observed over the past 100 years. Warming will be greatest over land areas and higher latitudes. Projected impacts of future climate change include:

- Sea level is rising. During the 20th century, sea level rose about 15 cm (6 inches) due to melting glaciers and polar ice sheets, and expansion of warmer seawater. Models predict that sea level may rise as much as 59 cm (23 inches) during the 21st Century, threatening coastal communities.
- Water availability will decrease in many areas that are already drought-prone and higher temperatures will cause a higher rate of evaporation and more drought in some areas of the world.
- A higher fraction of rainfall will fall in the form of heavy precipitation, increasing the risk of flooding and, in some regions, the spread of water-borne illness.
- People and ecosystems in coastal zones will be exposed to higher storm surges, intrusion of saltwater into freshwater aquifers, and other risks as sea levels rise.
- Coral reefs will experience widespread bleaching as a result of increasing temperatures, rising sea levels, and ocean acidification.
- Climate change will affect agricultural yield directly because of alterations in temperature and rainfall, and indirectly through changes in soil quality, pests and diseases.
- Frequent heat waves will become more common in more areas of the world and warmer temperatures will directly affect human health by increasing cases of heat stress.
- Climate change will affect the structure and function of forest ecosystems. It may worsen many of the threats to forests such as pest outbreaks, fires, drought, etc. Climate changes directly and indirectly affect the growth and productivity of forests and it also affects the frequency and severity of many forest disturbances.

There is less certainty in other projections, such as how the combination of greenhouse gas increases, temperature increases, precipitation changes, and other climate and climate-related changes will affect agricultural crops and natural ecosystems in different regions. Furthermore, different sectors, populations, and regions will vary in their exposure and sensitivity to the impacts of these changes. In general, research suggests that the impacts of climate change will more harshly affect poorer nations and communities.

Table 40.2: Impacts of Climate Change on Vegetation (Trees and Crops)

Variable	Beneficial Effect	Detrimental Effect
Atmospheric Carbon Dioxide	• Increase in growth rate • Reduction in stomatal conductance • Increased water use efficiency	• Increase in leaf area • Possible effects on ground vegetation • Reduction in timber quality • Possible nutrient imbalances
Ozone Pollution	• None	• Reduction in growth rate • Impaired stomatal regulation - increased susceptibility to drought
Temperature	• Longer growing season • Increased potential productivity • Lower risk of winter cold damage • Less snow damage • Potential use of species that are not hardy at present	• Less growth and productivity of plants • Delayed hardening • Risk of frost damage possibly increased • Longer growing season reducing soil water recharge period • More rapid development and increased fecundity of insect and mammal pest • Potential for exotic/alien pests to spread
Rainfall	• Reduced intensity of some foliar pathogens	• Waterlogging limiting access for forest operations • Reduced tree stability • Root death increasing susceptibility to drought and soil-borne pathogens • Drought induced mortality.
Wind	• None	• Increased risk of wind damage
Cloud Cover	• Increased potential productivity	• Increased diurnal temperature range - increased risk of frost damage

FOREST CARBON CYCLE

Most of today's carbon uptake is due to historical land use changes. Carbon exchange is sensitive to climate and especially to growing season length changes. Much of the global carbon uptake is in forest ecosystem and which is linked to fire suppression and recovery of forests from historical harvest. Carbon management is also linked to watershed management. Land management activities can play a critical role in limiting the build-up of carbon dioxide in the atmosphere, especially in the near-term. To stabilize the atmospheric concentration of carbon dioxide, it will require significant emissions reductions globally, which can only be achieved by either reducing energy emissions or by capture and storage (carbon sequestration).

Forests exchange large amount of CO_2 and other gases with the atmosphere and store carbon, in various forms, in trees and soils. Carbon stored in plants or soils is called "sequestered carbon". Carbon returned to the atmosphere when it has been used by trees or other organisms as energy for life is called "respired carbon". Much of the CO_2 in the air above a forest is taken in by trees through the process of photosynthesis, where it becomes one of the building blocks for tree growth or energy for life. Some carbon goes right back into the atmosphere as the tree respires (breathes out) but, if it stays, then it may remain sequestered in the tree throughout its life – whether that is 10 or 500 years. When a tree dies or loses a leaf or branch containing carbon, it generally falls to the forest floor where it will be decomposed by bacteria and fungi, and either be respired back to the atmosphere or made into soil carbon.

Carbon is returned daily to the atmosphere when it is decomposed and respired by soil organisms. But, much of it remains in complex chemical forms that resist decomposition and persist for hundreds to thousands of years. Soil carbon is an important carbon storehouse. It accounts for as much carbon as is presently found in plants and the atmosphere combined. Fire plays an important role in the forest carbon cycle. When a fire occurs, a portion of the trees, plants, grasses and other biomass are consumed and converted to CO_2 and other gases, and another portion is converted to charcoal, an essentially permanent form of storage. Only 10 to 30 per cent of the biomass in a forest is actually consumed by a fire and the majority remains on-site. Live trees will continue their role in the carbon cycle. Dead trees will slowly decompose and release carbon to the atmosphere or make new soil carbon. Regrowth after a fire will recapture carbon from the atmosphere, reversing the fire's emissions. About one to 10 per cent biomass killed in a fire is converted to charcoal, a uniquely stable form of carbon that will persist for thousands of years. The different forms of carbon are: carbon absorbed, carbon stored and carbon released.

- *Carbon Absorbed*: Young, healthy forests absorb carbon more rapidly than older, dense forests. Older forests release carbon at the same rate that they absorb it, neutralizing their effect on global warming. Sustainably managing forests is an effective way to store carbon. Trees also produce oxygen that we all need.
- *Carbon Stored*: As a tree grows, it stores carbon in its trunk, branches and roots. Sustainably managed forests continuously store and absorb carbon. Trees store cabon for a long time. When trees are harvested, the carbon continues to be stored in wood products. Harvested forests are replanted and the cycle begins again.
- *Carbon Released*: Catastrophic fires release carbon that has been stored in trees into the atmosphere. Manufacturing and automobiles also contribute carbon to the atmosphere by burning fossil fuels. Natural processes like volcanoes and decompositon also release carbon to the atmosphere.

Until humans began burning fossil fuels, the carbon cycle was closed to new inputs of carbon and carbon was continually recycled. Earth's plants and animals evolved over thousands of years under this level of CO_2 and a slowly changing climate, creating the forest ecosystem. Now billions of tons of fossil fuels are being extracted each year to meet the energy demands of a growing global population, adding new carbon to the atmosphere and changing the climate. Prior to fossil fuel use, this carbon was locked underground for millions of years and was not part of the carbon cycle. Current levels of CO_2 are 25 per cent higher than before the industrial revolution. As a result of this elevated levels of carbon, our forest ecosystems are changing. They are changing the way they grow in response to elevated CO_2, and they are changing in response to new climate patterns, including warmer temperatures and different levels of precipitation. These changes also affect the way that they store and release carbon, sometimes reducing the amount that goes into tree carbon or soil carbon.

Elevated levels of CO_2 in the atmosphere are changing the carbon cycle and the climate. Because forests have evolved under a climate that has changed at slow rates over the past several thousand years, they cannot be expected to function in the same way or provide the same habitat under elevated CO_2 levels and rapidly changing climatic conditions. As such, our forest management goals should include preserving healthy forest ecosystems and the natural role of forest fires and restoring forests toward resilient conditions when they have been depleted.

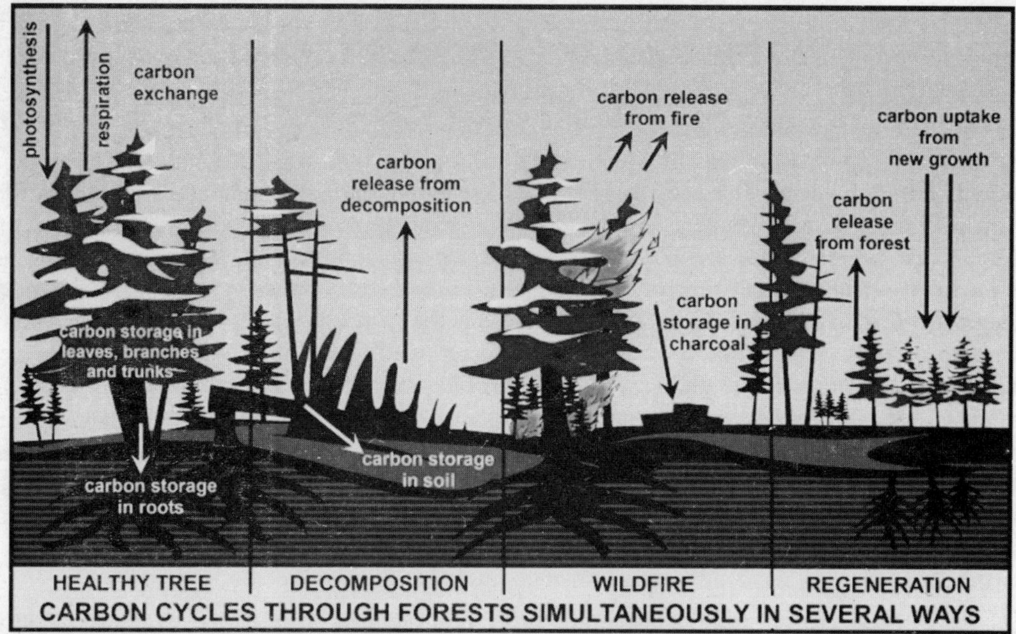

Fig. 40.5: Forest Carbon Cycle

CARBON SEQUESTRATION

TERRESTRIAL CARBON STOCK

'Terrestrial carbon stock' is the term used for the carbon stored in terrestrial ecosystems, as living and dead plant biomass (aboveground and belowground) and in the soil, along with usually negligible quantities as animal biomass. Aboveground plant biomass comprises all woody stems, branches and leaves of living trees, creepers, climbers and epiphytes as well as understory plants and herbaceous growth. For agricultural lands, this includes trees (if available), crops and weed biomass. The dead organic matter pool (necromass) includes dead fallen trees and stumps, other coarse woody debris, the litter layer and charcoal (or partially charred organic matter) above the soil surface. The belowground biomass comprises living and dead roots, soil fauna and the microbial community. There also is a large pool of organic carbon in various forms of humus and other soil organic carbon pools. Other forms of soil carbon are charcoal from fires and consolidated carbon in the form of iron-humus pans and concretions. For peatland, the largest carbon pool is found in the soil. Peat soils can store 10-100 times more carbon per unit area than other areas and are thus of special interests for the global carbon cycle.

CARBON BALANCE OF PLANTS

The carbon balance of plants is controlled by the quantity of CO_2 fixed through photosynthesis and the rate at which fixed C is returned to the atmosphere by respiring plant tissues. Light, temperature and the availability of soil water and nutrient constrain these physiological processes and thus influence productivity of plants and the ecosystems in which they occur. Only a small fraction of the radiant energy reaching the Earth's surface (2%) is used by green plants to assimilate atmospheric CO_2 into organic compounds. These compounds are used to construct new plant tissue, maintain existing tissue, create storage reserves, invest in mechanical structure or provide defense against insects and pathogens. In simple terms, trees grow (or gain C) when the amount of C fixed through photosynthesis exceeds the amount of C lost from respiring leaves, branches, stems and roots.

Understanding the factors influencing the balance between photosynthesis and respiration is of ecological importance, because plants with greatest net C gain under a specific set of environmental conditions are often the best competitors. It follows that fast-growing tree species require rapid rates of photosynthesis, but the converse is not always true. Factors affecting leaf carbon gain are photosynthesis and dark respiration, light, temperature, water and soil nutrient availability.

PHOTOSYNTHESIS AND DARK RESPIRATION: Processes of photosynthesis and respiration occur in the leaves of all plants, and the balance of these more-or-less opposing physiological processes controls the net C gain of leaves. Gross photosynthesis is the total amount of C, plants assimilate from the atmosphere. However, a portion of that fixed C is returned to the atmosphere from the leaves as CO_2 during dark respiration. Net photosynthesis is the balance between gross photosynthesis and leaf dark respiration, and it represents the amount of C available for growth and tissue maintenance. The maximum photosynthetic capacity of forest trees ranges from 2 to 25 μmol m^{-2} sec^{-1}, generally lower than that of most agronomic crops. Shade-intolerant deciduous species have some of the highest photosynthetic capacities of all forest trees (upto 25 μmol m^{-2} sec^{-1}) whereas those of shade-tolerant deciduous species are much lower (3 to 6 μmol m^{-2} sec^{-1}). Although some conifers have high photosynthetic capacities, most are modest when compared to those of deciduous trees.

LIGHT: Net photosynthesis varies substantially with light intensity. Sun leaves of plants attain high rates of net photosynthesis than shade leaves. Although shade leaves photosynthesize at a lower rate than the sun leaves, they can contribute as much as 40 per cent of the C assimilated by tree canopies. Similarly leaves of different species show different rate of net photosynthesis. Plants and trees (in-tolerant species) attaining high rates of net photosynthesis, typically contain large amounts of chlorophyll and the enzymes used for C-fixation which represent the biochemical machinery plants use to convert light into chemical energy. Leaves with high net photosynthetic rates also respire rapidly. Because rates of net photosynthesis vary with light intensity while dark respiration remains constant, the ability of a species to persist in low light environments is related to the balance of these physiological processes.

TEMPERATURE: Air temperature directly controls leaf C gain by influencing rates of gross photosynthesis and dark respiration. The difference between these two physiological processes (i.e., net photosynthesis) reaches a maximum between 15°C and 20°C for most temperate trees. Tropical trees show higher optimum temperatures, between 30°C and 35°C. Rates above or below this range rapidly decline because low and high temperatures limit the process of C assimilation. Leaf carbon gain approaches zero as temperature rises between 35°C and 43°C. At temperatures above 43°C, rapid rates of dark respiration combined with slow rates of gross photosynthesis result in a negative C balance and the net loss of C from the leaf.

WATER: Water is important for plant growth and its availability directly influences photosynthesis and leaf C gain. Plants are faced with a fundamental trade-off because C fixation and water loss occur simultaneously in leaves. Trees open their stomata allowing CO_2 to diffuse form the atmosphere to sites of C fixation in the leaf mesophyll. At the same time, water vapour diffuses out of stomata in response to differences in water potential between the leaf and atmosphere. Declines in leaf water content from 5 to 10 per cent often have little influence on photosynthesis, but a drop in leaf water potential can sharply reduce C fixation. At leaf water potential of -1.0 to -2.5 MPa, stomatal closure begins to limit the diffusion of CO_2, thus slowing rates of C fixation.

SOIL NUTRIENT AVAILABILITY: Photosynthesis as well as other biochemical processes in the leaf are influenced by the supply of soil nutrients. Nitrogen is a constituent of chlorophyll, other light-harvesting pigments and the enzymes involved in C fixation. Their presence makes leaves rich in compared to most other plant tissues. Thus, the supply of N can have a large influence in leaf's ability to fix C from the atmosphere.

Nowadays there is much attention on factors that influence the fixation and release of carbon by trees and forest ecosystem. This is mainly because the rising CO_2 concentrations of the Earth's atmosphere and projected changes in global climate have the potential to alter the present day geographic distribution of forests and the rate at which they sequester C from the atmosphere. Such a change is of both ecological and economic importance, as the production of plant matter in forest ecosystems not only influences the global cycling of C but also represents an indispensable source of food, fuel, fodder and fibre for human populations around Earth.

TERRESTRIAL CARBON PROCESSES

GROSS PRIMARY PRODUCTION (GPP): The total amount of carbon that is fixed in the process of photosynthesis from the atmosphere by plants in an ecosystem is called GPP. The plants convert CO_2 into carbohydrates during photosynthesis, which is carbon assimilation by photosynthesis ignoring photorespiration. GPP is measured on photosynthetic tissues, primarily leaves, on an hourly timescale and integrated to an annual account. Terrestrial GPP has been estimated to be 120 Gt C/yr.

NET PRIMARY PRODUCTION (NPP): Annual plant growth or net production of organic matter by plants is the difference between photosynthesis and autotrophic respiration (Ra), and is referred to as net primary production (NPP). NPP is the fraction of GPP resulting in plant growth, and can be measured through sequential harvesting or by measuring plant biomass on a daily or weekly timescale, provided turnover of all components (e.g., fine roots) is included. Global terrestrial NPP has been estimated to be 60 Gt C/yr, i.e. about half of GPP is incorporated in new plant tissue. The other half is returned to the atmosphere as CO_2 by autotrophic respiration, i.e. respiration by plant tissues.

NET ECOSYSTEM PRODUCTION (NEP): It denotes the net accumulation of organic matter or carbon by an ecosystem. NEP is the difference between the rate of production of living organic matter (NPP) and the decomposition rate of dead organic matter (heterotrophic respiration - Rh), which determines the amount of carbon lost or gained by the ecosystem without disturbances, such as harvests and fire. Global NEP is estimated at about 10 Gt C/yr. NEP can be measured in two ways: one is to measure changes in carbon stocks in vegetation and soil over time, using an annual timescale; the other is to integrate hourly/daily fluxes of CO_2 into and out of vegetation and integrate up to the yearly timescale. NEP should be integrated up to a decadal (10 year) timescale.

NET BIOME PRODUCTION (NBP): It denotes the net production of organic matter in a region containing a range of ecosystems (a biome) and includes, in addition to heterotrophic respiration, other processes leading to loss of living and dead organic matter. It is the carbon accumulated by the terrestrial biosphere when carbon losses from non-respiratory processes are taken into account, including fires, harvests/removals, erosion and export of dissolved organic carbon by rivers to the oceans. NBP is a small fraction of the initial uptake of CO_2 from the atmosphere and can be positive or negative; at equilibrium it would be zero. Compared to the total fluxes between the atmosphere and biosphere, global NBP is comparatively small at 0.7 to 1.0 Gt C/yr. NBP is the critical parameter to consider for long-term (decadal) carbon storage. NBP is estimated to have averaged 0.2 +/- 0.7 Gt C/yr during the 1980s and 1.4 +/- 0.7 Gt C/yr during the 1990s.

AGROFORESTRY AND CARBON SEQUESTRATION

Agroforestry is the intentional integration of woody plants into crop and livestock systems to improve soil, water, air quality and biodiversity while supporting sustainable production of food, feed, fibre and energy. It represents a significant opportunity for sequestering C on agricultural lands in that a substantial proportion of the C is sequestered in woody biomass, thus creating a system that sequesters a large amount of C per unit area and for longer duration than many other practices (Montagnini and Nair 2004). Since agroforestry is not explicitly inventoried within the natural resource inventories, its potential contributions to C sequestration have been estimated based on assumptions of where these practices would suitably occur for services other than C sequestration.

Globally, it has been estimated that agricultural lands contain approximately 12 per cent of the world's terrestrial Carbon (Dixon 1995). The amount of C sequestered largely depends on the agroforestry system put in place, the structure and function of which are, to a great extent, determined by environmental and socio-economic factors. Several other factors are also influencing carbon storage in agroforestry systems such as tree species and its management practices. Agroforests are very complex systems, and have higher plant diversity. These complex systems of trees and agricultural crops are widely practised in several other parts of the world i.e., Latin America, Southeast Asia and Equatorial Africa and are among the most sustainable cropping systems in the tropics. The trees play various functions, including shading crops to reduce evapotranspiration, erosion control and nutrient cycling. The regular harvesting of crops, wood and other products, decline in soil fertility can be considered as minimal in complex agroforestry systems. Abundant litter and/or pruning biomass returned to the soil combined with the decay of roots, contribute to the improvement of soil physical and chemical properties.

CARBON SEQUESTRATION POTENTIAL OF AGROFORESTRY

Agroforestry is an appealing option for sequestering carbon on agricultural lands because it can sequester significant amounts of carbon while leaving the bulk of the land in agricultural production. These practices represent a category of conservation/ produ-ction activities that can sequester large amounts of carbon while providing a multitude of additional benefits (Montagnini and Nair 2004). Simultaneously, it can help the land owners and society to address many other issues such as economic diversification, biodiversity and water quality facing these lands. Despite this fact, agroforestry remains underrecognized as a greenhouse gas mitigation option for agriculture in many regions of the world. Reasons for this include the limited information base and tools agroforestry can currently offer compared to the decade's worth of investment in agriculture and forestry; and agroforestry's cross-cutting nature that puts it at the interface of agriculture and forestry which is not strongly promoted by either discipline. Agroforestry research is beginning to establish the scientific foundation required for building carbon accounting and modeling tools but more progress is needed before it is readily accepted within agricultural greenhouse gas mitigation programmes and further, incorporated into the broader scope of sustainable agricultural management. Enhanced communication of agroforestry's carbon cobenefit, as well as the other benefits afforded by these plantings, will help elevate agroforestry awareness (Schoeneberger 2005).

The amount of carbon sequestered per unit area by agroforestry is substantial due to the large amount of carbon sequestered in the woody biomass. However, unlike afforestation, agroforestry plantings do not result in a change in land use to forest. Indeed, the appeal of agroforestry as a carbon sequestering activity on agricultural lands rests in large part on its ability to sequester significant amounts of carbon on a relatively small land base (Ruark et al 2003). Agroforestry can give the land owner the biggest net gain of carbon per unit area generally without compromising agricultural activity. Although the carbon fixed within a single agroforestry planting is small, taken within a whole-farm context the amount can become significant.

CARBON SEQUESTRATION CAPACITY ESTIMATES

- Carbon-sequestration potentials vary from 0.29 metric tons of carbon per hectare per year in fodder banks (trees and shrubs grown as food for livestock) of the West African Sahel to 15.21 metric tons of carbon per hectare per year in mixed species stands of Puerto Rico (Nair et al 2009).
- One model estimated a global potential to sequester 1.1-2.2 Gt of carbon per year over 50 years (Dixon 1995). However, using the median carbon sequestration potential used by Dixon in that model (1.88 metric tons per hectare per year) and the estimate of total land area under agroforestry (1,023 million hectare) made by Nair *et al* (2009) results in a carbon sequestration potential of 1.9 Gt of carbon per year over 50 years.
- In order to increase the amount of carbon sequestration, new agroforestry projects must be implemented in the remaining 3,953 million hectares of croplands and pastures in the world. The IPCC estimates a global amount of 630 million hectares of unproductive croplands and grasslands that could be converted to agroforestry and could potentially sequester 1.43 megatons of carbon dioxide per year by 2010 and 2.15 megatons of carbon dioxide per year by 2040 (IPCC 2000). One megaton is equivalent to one million metric tons.

AGROFORESTRY CARBON POOLS

The UNFCCC defines carbon sequestration as the process of removing C from the atmosphere and depositing it in a reservoir. It entails the transfer of atmospheric CO_2 and its secure storage in long-lived pools (UNFCCC 2007). From the agroforestry point of view, C sequestration primarily involves the uptake of atmospheric CO_2 during photosynthesis and the transfer of fixed C into vegetation, detritus and soil pools for secure or long-term storage (Nair et al 2010). The "intensive, intentional, integrated and interactive" nature of agroforestry plantings (Gold et al 2000) results in species combinations, use of plant materials, stocking levels, management and therefore, carbon flows that are quite different than in the forest stands. Growth and carbon allocation patterns in agroforestry trees reflect the more "open-grown" or "edge forest" conditions created by agroforestry arrangements. The majority of woody biomass equations available for developing carbon estimates were derived from forest stands and, not unexpectedly, found to underestimate woody biomass in agroforestry plantings (Zhou 1999). The carbon sequestration in agroforestry occurs in two major segments: aboveground and belowground. The major agroforestry carbon pools are: aboveground woody biomass, belowground root biomass, understory vegetation, litter/deadwood, soil carbon.

ABOVEGROUND TREE CARBON POOL: This component represents the most easily and reliably reported pool in agroforestry plantings and captures majority of carbon sequestered by this system. Overstorey trees are prime importance and virtually, all the C sequestration projects require aboveground tree biomass estimation. In fact, most registries only accept aboveground live tree C. This forest carbon pool has been measured frequently since this is the "merchantable timber" of the forester and the timber and paper industry. In-field inventory methods are standardized and well developed. There are also a number of available mathematical models as well as look-up tables.

BELOWGROUND ROOT CARBON POOL: Belowground biomass is a major carbon pool. However, belowground biomass is difficult to measure. Data generated under the aboveground accounting schemes can be utilized to include belowground coarse root biomass. Actually, direct measurements of belowground coarse root biomass are extremely difficult and costly, have high uncertainty, and lack a standard protocol. There is a constant relationship between aboveground biomass and belowground biomass and root-to-shoot ratio is therefore commonly used to estimate belowground living biomass.

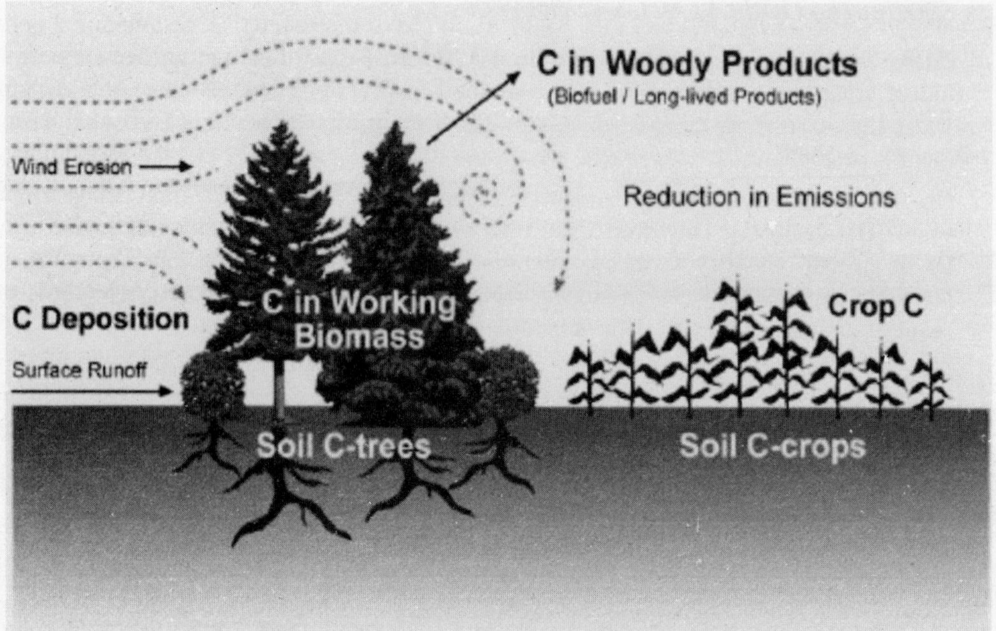

Fig. 40.6: Carbon Storage in Agroforestry (Source: Schoeneberger 2005)

UNDERSTORY/HERBACEOUS VEGETATION: Understory vegetation and herbac-eous vegetation also sequester carbon. This biomass consists of standing underground vegetation, shrubs, herbs and cover crops. This pool is measured by destructive method in sample plots.

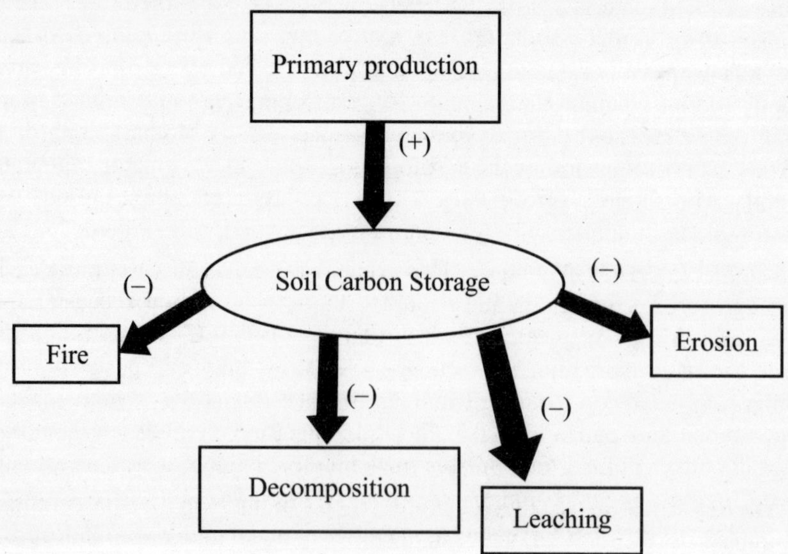

Fig. 40.7: Major Processes Influencing Soil Carbon Pool in Agroforestry

SOIL CARBON POOL: This pool is required by a number of the projects and may account for a fairly substantial amount of carbon. This pool typically is one that can grow rapidly during early years of afforestation and thus has some interest. This pool is not difficult to measure in the field and there are relatively good models and tables available for estimation.

CARBON STORAGE IN AGROFORESTRY PRACTICES

According to recent projections, the area of the world under agroforestry will increase substantially in the near future. Undoubtedly, this will have a great impact on the flux and long-term storage of C in the terrestrial biosphere (Dixon 1995). Agroecosystems play a major role in the global carbon cycle and contain approximately 12 per cent of the world terrestrial carbon (Smith et al 1993; Dixon et al 1994). Several agroforestry studies have shown that the inclusion of trees in agricultural landscapes often improves the productivity of systems while providing opportunities to create carbon sinks (Winjum et al 1992; Krankina and Dixon 1994). The amount of carbon sequestered largely depends on the agroforestry system put in place, the structure and function of which are, to a great extent, determined by environmental and socio-economic factors. Other factors influencing carbon storage in agroforestry systems include tree species and system management.

Table 40.3: Carbon Storage Potential of Different Biomes (DOE 1999)

Biomes	**Primary Method to increase Carbon Sequestration***	**Potential Carbon Sequestration (Pg C per year)**
Agricultural Lands	Management (H)	0.85 – 0.90
Biomass Croplands	Management (H)	0.50 – 0.80
Grasslands	Management (M)	0.50
Rangelands	Management (M)	1.20
Forests	Management (M)	1.0 – 3.0
Urban Forests and Grasslands	Creation and Maintenance (M)	-
Deserts and Degraded Lands	Manipulation (H)	0.80 – 1.30
Terrestrial Sediments	Protection (L)	0.70 – 1.70
Boreal Peatlands and Other Wetlands	Protection (L)	0.10 – 0.70
Total		5.65 – 10.10

* The primary method of carbon sequestration is rated as high (H), medium (M) and low (L) levels of sustained management intensity required over a long term. Global potential sequestration rates were estimated that might be sustained over a period of 25-50 years.

Agroforestry systems have the potential to sequester more carbon per unit area. With adequate management of trees in cultivated lands and pastures, a significant fraction of the atmospheric carbon could be captured and stored in plant biomass and in soils. However, increasing carbon stocks in a given period of time is just one step and the fate of those stocks is what ultimately determines sequestration. In agroforestry systems, carbon sequestration is a dynamic process and can be divided into phases. At establishment, many systems are likely to be sources of GHGs (loss of C and N from vegetation and soil). Then follow a quick accumulation phase and a maturation period where tons of carbon are stored in the boles, stems, roots of trees and in the soil. At the end of the rotation period, when the trees are harvested and the land returned to cropping (sequential systems), part of the carbon will be released back to the atmosphere (Dixon 1995). Therefore, effective carbon sequestration in agroforestry can only be considered if there is a positive balance from an initial stock after few decades (Feller et al 2001).

Table 40.4: Carbon Sequestration Potential of Agroforestry Systems in Different Ecoregions of the World (Dixon et al 1993; Krankina and Dixon 1994)

Continent	Ecoregion	System	Potential Carbon Sequestration (Mg C ha^{-1})
Africa	Humid tropical high	Agrosilvicultural	29 - 53
South America	Humid tropical low	Agrosilvicultural	39 – 102*
South America	Dry lowlands	Agrosilvicultural	39 - 195
South East Asia	Humid tropical	Agrosilvicultural	12 - 228
South East Asia	Dry lowlands	Agrosilvicultural	68 - 81
Australia	Humid tropical low	Silvopastoral	28 - 51
North America	Humid tropical high	Silvopastoral	133 - 154
North America	Humid tropical low	Silvopastoral	104 - 198
North America	Dry lowlands	Silvopastoral	90 - 175
Northern Asia	Humid tropical low	Silvopastoral	15 - 18

* Carbon storage values were standardized to 50-year rotation

Realistically, carbon storage in plant biomass is only feasible in the perennial agroforestry systems, which allow full tree growth and where the woody component represents an important part of the total biomass. One comparative advantage of these systems is that sequestration does not have to end at wood harvest. Carbon storage can continue way beyond if boles, stems or branches are processed in any form of long-lasting products. Similarly producing firewood from agroforestry systems may still present interesting opportunities in CO_2 mitigation through:

- The protection of existing forests and other natural landscapes
- The conservation of soil productivity
- The reduction of fossil energy consumption by using wood as energy sources

Carbon allocation in tree components is of fundamental importance in understanding the adaptive strategies of trees in agroforestry system. Consequently, the selection of tree species and the design or management of agroforestry systems has to take carbon sequestration potential of different components of trees into account besides other benefits. On an average, tree stem is the largest carbon store but combining the litter including branches and the roots, the carbon proportion becomes more than 50 per cent. As the above litter part is added to the soil, thus depicting huge potential of trees in soil organic carbon build up in agroforestry systems. The mean carbon storage values for trees are valid till the plantations exist and replaced after each harvest. Therefore, the average tree carbon storage over many rotations is suitable for long term storage. The young and short rotation trees store most of the carbon as soil organic carbon (Sanjeev et al 2009). The continuity in the agroforestry practices supports the vast potential of fast growing tree species due to higher productivity in comparison with natural forests.

Table 40.5: Carbon Storage Potential of different Agroforestry Trees (Sanjeev et al 2009)

Agroforestry Tree Species	Stem Carbon (t ha^{-1})	Aboveground Litter Carbon (t ha^{-1})	Root Carbon (t ha^{-1})	Carbon added to Soil (Roots+Litter) (t ha^{-1})	Total Carbon Storage (t ha^{-1})
Acacia catechu	1.96	2.06	1.49	3.55	5.51
Acacia nilotica	1.61	1.83	1.16	2.99	4.60
Artocarpus fraxinifolius	6.05	2.58	2.64	5.21	11.26
Anthocephalus cadamba	3.39	2.56	2.25	4.80	8.20
Bombax ceiba	1.39	1.35	1.11	2.46	3.85
Dalbergia sissoo	2.10	3.03	1.39	4.42	6.52
Eucalyptus tereticornis	4.20	2.74	2.42	5.16	9.36
Gmelina arborea	3.39	1.34	1.51	2.84	6.24
Melia azedarach	1.50	1.30	1.16	2.46	3.96
Populus deltoides	4.80	3.67	2.27	5.95	10.74
Syzygium cumini	0.73	1.01	0.42	1.43	2.16
Terminalia arjuna	2.39	2.27	1.00	3.28	5.67
Toona ciliata	1.53	1.27	1.36	2.63	4.17

Agroforestry practices can be better climate change mitigation option than ocean and other terrrestrial options because of vast production and protective benefits. By including trees in agricultural production systems, agroforestry can increase the amount of carbon stored in lands devoted to agriculture, while still allowing for growing of food crops (Kursten and Bruschel 1993). The tree components in agroforestry systems can be significant sinks of atmospheric carbon due to their fast growth and high productivity. Thus, promoting agroforestry can be one of the options to deal with problems related to land use and global warming. The amount of carbon sequestered will largely depend on the type of agroforestry system and also on its structure and function which to a great extent, are determined by environmental and socio-economic factors. Also tree species and system management can influence carbon storage in agroforestry systems (Albrecht and Kandji 2003). Average carbon sequestration potential in agroforestry has been estimated to be 25 ton C ha^{-1} over 96 million ha of land in India and 6-15 ton C ha^{-1} over 75.9 Mha in China (Sathaye and Ravindranath 1998). Watson et al (2000) estimated carbon gain of 0.72 Mg C ha^{-1} yr^{-1} on 4000 million ha land under agroforestry, with potential for sequestration 26 Tg C yr^{-1} by 2010 and 45 Tg C yr^{-1} by 2040.

Table 40.6: Biomass Carbon Stock (t ha^{-1}) under different Agroforestry Systems (Anil Kumar 2010)

Agroforestry Systems	Above Ground (AG) Carbon Stock in Trees (t ha^{-1})				Total AG Carbon in Trees	Crop/ Grass Carbon	Roots Carbon	Total Carbon (t ha^{-1})
	Stem	Branch	Leaf	Litter				
Populus deltoides + Wheat	21.38	0.92	0.73	0.20	23.23	2.80	5.83	31.86
Eucalyptus hybrid + Wheat Boundary Plantation	9.59	0.16	0.07	0.02	9.84	2.71	1.87	14.42
Populus deltoides + Wheat Boundary Plantation	5.49	0.17	0.13	0.05	5.84	2.92	1.53	10.29
Populus deltoides + Lemon grass	20.63	0.68	0.53	0.20	22.04	3.45	5.48	30.97

ADVANTAGES OF AGROFORESTRY AS A CARBON SEQUESTRATION STRATEGY

If the entire earth's area that is suitable for agroforestry (585-1215 x 10^6 ha) were converted to accommodate some form of tree cultivation, 1.1-2.2 Pg C could be sequestered annually over 50 years, which could offset about 10-15 per cent of the current annual C emissions. There is tremendous merit in promoting agroforestry projects for C sequestration especially if the ancillary benefits of cultivating trees on farm are considered:

- Given the improbability of obtaining any single mitigation method, adding modest contributions together appears to be a more realistic way of achieving CO_2 reduction targets.
- The implementation of agroforestry projects will have other environmental and socio-economic benefits way beyond the mere climatic concerns. Increasing soil C greatly benefits agricultural productivity and sustainability, and the presence of trees in agricultural landscapes enhances biodiversity, improves microclimate and beauty and provides watershed services.
- The financial cost of C sequestration through agroforestry appears to be much lower (approximately $1-69/Mg C, median $13/Mg C) than through other CO_2 mitigating options. Economic analysis showed that these costs could be easily offset by the monetary benefits from agricultural and tree products. Trading in carbon credits is another option, which could make C sequestration through agroforestry more attractive to land users.

Agroforestry options that seek to restore degraded croplands and pastures are effective in storing carbon in the soil. Pre-cultivation carbon stocks on the present area of cultivated land are estimated at 222 Pg, which may be considered as an upper limit for soil C sequestration. However, instances of increased C stocks relative to those of native soils have been achieved after intensive management of specific biomes. There is enough scientific evidence to suggest that soil C levels can be considerably increased at a global scale if management options that improve land productivity are pursued. Agroforestry trees improve land cover in agricultural fields in addition to providing C inputs (root biomass, litter and prunings) to the soil. The reduction of soil erosion and the improvement of soil structure is a crucial process in the soil C dynamics. Soil C sequestration through agroforestry can significantly contribute to climate change mitigation especially if leakages can be prevented with complementary conservation practices such as minimum tillage, to slow down decomposition and leaching.

With adequate management of trees in cultivated lands and pastures, a significant fraction of the atmospheric C could be captured and stored in plant biomass and in soils. The perennial agroforestry systems (tree crop combinations, agroforests, boundary planting), which allow full tree growth and have an important woody component, store considerable quantities of C in the living biomass. If the necessary arrangements are made to promote small scale industries, the use of high quality wood trees in agroforestry projects can contribute significantly to climate change mitigation while providing small holders with additional cash income. At national level, this can help create jobs and generate foreign exchange through export of wood products.

C sequestration in soil is also realistic with most agroforestry practices including short-rotation systems such as improved fallows and hedgerow intercropping. Degraded soils and wastelands occupy a large proportion of the earth's area and there is general belief that converting them into agroforestry would be a major global opportunity to absorb a significant portion of the atmospheric CO_2. However, cultivating trees or crops in sub-standard soils still remains a major challenge. Improved fallow is a promising technology for increasing C stocks in degraded soils. Perennial systems like homegardens can store and conserve considerable amounts of C in living biomass and also in processed wood products.

Table 40.7: Total Carbon Storage under Agroforestry Systems in Different Regions of India (Fanish and Priya 2012)

Region	Agroforestry System and Components	Total C Storage (t C ha^{-1})	References
Semi-arid Region	Silvopastoral System (age 5 years)		Rai et al (2001)
	Acacia nilotica + natural pasture	9.5-17.0	
	Acacia nilotica + established pasture	19.7	
	Dalbergia sissoo + natural pasture	12.4	
	Dalbergia sissoo + established pasture	17.2	
	Hardwickia binata + natural pasture	16.2	
	Hardwickia binata + established pasture	17.0	
North-western India	Silvopastoral System (age 6 years)		Karur et al (2002)
	Acacia/ Dalbergia/ Prosopis + Desmostacya	6.8-18.5	
	Acacia/ Dalbergia/ Prosopis + Sporobolus	1.5-12.3	
Central India	Block Plantation (age 6 years)		Swamy et al (2003)
	Gmelina arborea	24.1-31.1	
Arid Region (Rajasthan)	Agrisilvicultural System (age 8 years)		Singh (2005)
	Emblica officinalis + Vigna radiata	12.7-13.0	
	Hardwickia binata + Vigna radiata	8.6-8.8	
	Colophospermum mopane + Vigna radiata	4.7-5.3	
Semi-arid Region	Agrisilvicultural System (age 11 years)		NRCAF (2005)
	Dalbergia sissoo + crop	26.0	
North-western Himalayas	Different Pastoral Systems		AICRPAF (2006)
	Silvipastoral System	2.17	
	Agrihortipastoral System	1.15	
	Hortipastoral System	1.08	

ASSESSMENT OF CARBON SEQUESTRATION IN AGROFORESTRY

With the increasing importance of carbon stock assessments in policy and the possible consequences for economic incentives (C markets), it is relevant that local stakeholders are aware of data collection and processing. The carbon stock assessment in agroforestry intends to provide essential baseline information for the negotiation of carbon credits with potential carbon buyers. Rapid Carbon Stock Appraisal (RaCSA) is one such carbon accounting tool developed for measuring C stocks. The basic data for RaCSA must come from efforts at a more local level to measure the carbon stocks in the landscape. Such a more localized assessment can be undertaken by following the RaCSA protocol. The basic steps of data collection and measurement of trees are not particularly difficult and do not require expensive or complex equipment but consistency and attention to detail are necessary. The cost of carbon measurement depends on the design of the system. The carbon stock assessment can be substantially more cost effective if local expertise can be developed and standards of reporting and verification can be maintained. The assessment of carbon sequestration is normally carried out for the following components in agroforestry: (1) aboveground vegetation, (2) understory/ herbaceous vegetation, (3) belowground living biomass (roots), and (4) belowground soil.

ABOVEGROUND VEGETATION

The aboveground biomass comprises all woody stems, branches, leaves of living trees, creepers, climbers and epiphytes as well as herbaceous undergrowth. For agricultural lands, this includes crop and weed biomass. The total standing aboveground biomass of woody vegetation elements is often one of the largest carbon pools. There are two ways of measuring the biomass of aboveground vegetation viz. destructive sampling and allometric equations. Aboveground measurements of carbon sequestration are direct derivatives of aboveground biomass (AGB) measurements, assuming that 50 per cent of the biomass is made up by carbon. The AGB is often derived by summing up the amount of harvested and standing biomass.

Table 40.8: Parameters and General Methods used in Agroforestry Carbon Stock Measurement

S.No.	Parameter	Methods
1	Large living trees with a stem DBH >30 cm and small trees with a stem DBH of 5 - 30 cm	Non-destructive measurement of stem diameter (and height) and apply allometric biomass equation
2	Understory/ herbaceous vegetation comprising plants with a stem diameter < 5 cm	Destructive
3	Belowground living biomass (roots)	Use a constant value for relation of above to belowground biomass ratio
4	Belowground soil	Destructive and soil analysis

DESTRUCTIVE METHOD OF BIOMASS ESTIMATION (*WHOLE-TREE HARVESTING*): Estimation of tree biomass by whole-tree harvesting is an old approach. It consists of cutting down sample trees, separating various parts (stem, leaves, branches, etc.), digging out and washing the roots, determining their dry weights from sample of each part, and adding them up to get the total biomass. After dividing up the harvested representative trees into their various components (stem, branches, dead branches, branchlets, leaves, roots and fine roots), and determining their dry weight, the carbon content in each is measured. Using the data, allometric equations are developed as regression models with the measured variables such as diameter at breast height (DBH), total tree height or commercial bole height, and sometimes wood density, as the independent variables and total dry weight as the dependent variable. The destructive method of determining tree biomass, though comparatively accurate, is extremely time and labour intensive, especially for large trees. It is often used to validate other methods such as the estimation of C stock using non-destructive in-situ measurements and remote sensing (Nair 2011).

NON-DESTRUCTIVE METHOD OF TREE BIOMASS ESTIMATION: Dixon et al (1993) made estimations by measuring the volume of stem wood and multiplying it with species-specific wood density; that number was then multiplied by 1.6 to get an estimation of whole-tree biomass; Carbon content was assumed as 50 per cent of the estimated whole-tree biomass, and root biomass was excluded. This rough estimation is used for more extensive estimations of tree biomass globally. More recently, databases for tree characteristics such as wood density for agroforestry species developed at the World Agroforestry Centre are being used in such calculations.

TREE BIOMASS ESTIMATION USING ALLOMETRIC EQUATIONS: This is the most widely used method for measuring carbon in the tree biomass (woody vegetation > 5 cm dbh) by non-destructive sampling. Allometric biomass equations are established for many tree species for various site conditions using DBH, height and wood density as independent variables. If you measure the DBH and height of the all tree species using sample plots, the tree biomass can be calculated for each tree species using an appropriate allometric equation. Carbon content is measured as 50 per cent of the calculated tree biomass. Palms, bamboos and lianas need a separately established equation, as their stem diameter does not increase by secondary thickening and thus does not reflect actual canopy size. If possible, the equations used for estimating biomass should be developed for each location, species or group of species, and trees of similar sizes and ages. If no site specific or tree specific equations are available, then the following equations may be used. Once an algometric equation has been established for different classes of trees in a vegetation, it is only needed to measure DBH (or other parameters used as a basis for the equation) for estimating the biomass of individual trees. The sum of biomass estimates for all trees within the measurement transect can be converted to a biomass in Mg ha^{-1}.

Table 40.9: Allometric Equations for Estimating Biomass (Plants with DBH >5 cm)

Life Zone (rainfall in mm/yr)	Equation	Range (cm)	Number of Trees	R^2
Dry (<1500)	$W = 0.139\ D^{2.32}$ (Brown 1997)	5-40	28	0.89
Moist (1500-4000)	$W = 0.118\ D^{2.53}$ (Brown 1997)	5-148	170	0.90
	$W = 0.049\ \rho\ D^2\ H$ (Brown et al 1995)			
	$W = 0.11\ \rho\ D^{2+c}$ with c (default 0.62) based on			
	$H = a\ D^c$ (Ketterings et al 2001)			
Wet (>4000)	$W = 0.037\ D^{1.89}\ H$ (Brown 1997)	4-112	160	0.90

* Note: W = tree biomass (Kg/tree); D = dbh (cm); H = height (m); ρ = wood density (g cm^{-3})

UNDERSTORY/ HERBACEOUS VEGETATION

In destructive sampling, all vegetation less than 5 cm dbh is harvested within 1m x 1m quadrat. The vegetation in this given area is cut and weighed (fresh weight) and subsamples of parts of vegetation (understory biomass, coarse litter, branches, flowers and fruits are taken, weighed fresh in the field and weighed again after oven-drying. The samples are subjected to C_{org} or C fraction analysis. The total dry weight and carbon content are calculated using the following formulas:

$$\text{Total Dry Weight (kg m}^{-2}\text{)} = \frac{\text{Total Fresh Weight (kg)} \times \text{Subsample Dry Weight (g)}}{\text{Subsample Fresh Weight (g)} \times \text{Sample Area (m}^{2}\text{)}}$$

$$\text{Carbon Conent of Understory} = \text{Dry Weight Biomass} \times \text{Carbon Conc.}$$

(Where, Carbon Conc. (%) = g C/ 100 g biomass; Dry Weight Biomass = g/ m^{-2})

BELOWGROUND LIVING BIOMASS (ROOTS)

Studies of tree growth physiology across a broad range of tree species on many sites indicate that there is a relatively constant relationship between above and belowground living tree biomass. The root-to-shoot ratio is therefore commonly used to estimate belowground living biomass. The ratios differ considerably among species (e.g., higher in palms than in dicot trees) and across ecological regions (e.g., higher in cold than in warm climate). In the absence of measured values, man researchers assume that the belowground biomass constitutes a defined portion of the aboveground biomass and the values so assumed ranged from 25 to 40 per cent (Nair 2011). Generally, this ratio of the above to belowground biomass is taken approximately 4:1 or, in other words, ~20 per cent of total tree biomass is belowground. As such, estimation of the carbon in belowground coarse root biomass (including tap root) is easily achieved by utilizing aboveground biomass estimates and multiplying by 0.25. This highly variable root biomass is also best estimated using available empirical equations. Carbon content in root is measured as 50 per cent of the calculated root biomass.

$$\text{Belowground Root Biomass} = \text{Aboveground Tree Biomass} \times 0.25$$

BELOWGROUND SOIL

The estimation of belowground soil organic carbon (SOC) dynamics in agroforestry is crucial for understanding the impact of the system on C sequestration. The organic carbon occurs in soils in a number of different forms. Difficulties of separating these different forms and their complex interactions make measurement, estimation and prediction of soil C sequestration a difficult task. The most common method of estimating the amount of C sequestrated in soils is based on soil analysis, whereby the carbon content in a sample of soil is determined (mass per unit mass of soil, such as g C per 100 g soil) and expressed in megagrams (Mg = 10^6 g or tons) per hectare. Soil organic carbon is often measured on a whole soil basis (Nair 2011). The Walkley-Black procedure is used extensively in the past and is perhaps used even now, is no longer recommended because of concerns about the accuracy of determination as well as environmental concerns due to the impact of the use of potassium dichromate. Currently, the SOC is measured by quantifying the amount of CO_2 produced through heating in a furnace or the change in weight of the sample is measured after heating (Kimble et al 2001).

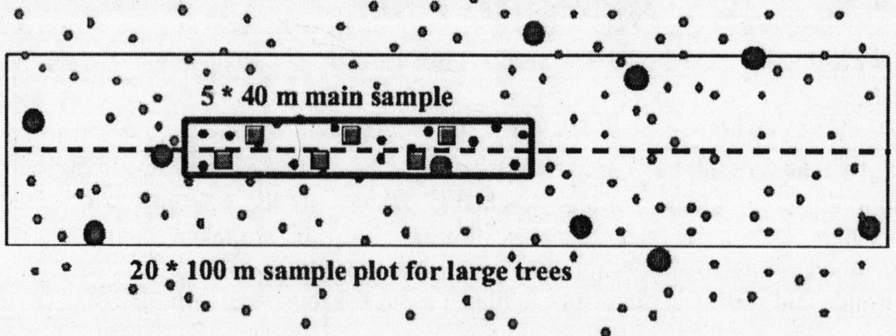

Trees > 30 cm diameter at 1.3 m, inside or outside plot

Trees 5< ..< 30 cm diameter at 1.3 m, inside or outside plot

Understorey & litter layer sample plot

Fig. 40.8: Nested Plot Design for Sampling various Carbon Stocks

SAMPLING PROCEDURE FOR CARBON ASSESSMENT

A nested sampling approach is followed, assessing large diameter trees (with a stem diameter > 30 cm) in rectangular plots of 20 m x 100 m = 2000 m^2, smaller trees (stem diameter 5 – 30 cm) in subplots of 5 m x 40 m = 200 m^2 within these, and understory vegetation in smaller sub-subplots of 1 m x 1 m = 1 m^2. Sample plots are selected within a plot of at least 1 hectare, avoiding borders of the plot, unless specifically indicated in the sample design. Plot location is randomized if there are marked discontinuities in the vegetation. It should be ensured the sample plots do not all fall in the area with densest or least vegetation. The rectangular plots are chosen as they tend to include more of the within-plot heterogeneity, and thus be more representative than square or circular plots of the same area. The larger the total area sampled the more accurately the estimate reflects the population of interest. Instead of sampling a large, contiguous area, it is better to divide the sampling into several, smaller areas within the field of study (randomly chosen or based on some prior stratification)

AGROFORESTRY AND CLIMATE CHANGE MITIGATION

The Intergovernment Panel on Climate Change (IPCC) was established in 1988 for the purpose of controlling GHGs in the atmosphere and to provide the world with a clear scientific view on the current state of climate change. IPCC is the leading body for the assessment of climate change, established by the United Nations Environment Programme (UNEP) and the World Meteorological Organization (WMO). United Nations Organization also created a global forum called "United Nations Framework Convention on Climate Change (UNFCCC)" during Rio Earth Summit in 1992 to tackle the problems of climate change. UNFCCC entered into force on 21 March 1994 and the ultimate objective of the Convention is to stabilize GHGs concentrations in the atmosphere at a level that will prevent dangerous human interference with the climate system. As a part of solution to this problem of increasing levels of CO_2 and other GHGs in the atmosphere, Kyoto Protocol was evolved in the COP 3 meeting held in Japan during 1997 by UNFCCC. It was adopted on 11 December 1997 and entered into force on 16 February 2005. The detailed rules for the implementation of the Protocol were adopted at COP 7 in Marrakesh in 2001 and are called the "Marrakesh Accords."

Climate Change Mitigation: It may be defined as any attempt to reduce the rate at which greenhouse gases are accumulating in the atmosphere. The two options that can be considered are:

- Reducing emissions
- Locking up the main source, carbon dioxide, through carbon sequestration.

Due to natural inertia within the system, a delay or lag time exists between the accumulation of greenhouse gases in the atmosphere and increasing temperatures (global warming). Thus, the changes in climate that we are experiencing now are as a result of past emissions in addition the carbon dioxide that is already in the atmosphere dictates that it is inevitable that temperatures will continue to rise for the next 50-100 years. The successful stabilization of greenhouse gas emissions is therefore vital if we are to reverse this trend for future generations.

In addition to woodlands and forests which contain substantial carbon (C) in the soil, trees and other vegetation, agroforestry are globally important to the carbon, water and energy cycles. Removing trees from forests releases CO_2 into the atmosphere, while growing trees in agroforestry systems absorb CO_2 from the air. Other greenhouse gases (GHG) such as methane and nitrous oxide are also exchanged between trees and the atmosphere, so forests and agroforestry are key components of the planet's GHG balance. Therefore, the functioning and management of agroforestry trees are critical to efforts to reduce climate change ('climate change mitigation') and reduce the net GHG emissions into the atmosphere ('emissions abatement').

Agroforestry attracted special attention as a C sequestration strategy following its recognition as carbon sequestration activity under the afforestation and reforestation (A&R) activities of the Kyoto Protocol (Nair 2011). Agroforestry systems spread over one billion ha in diverse ecoregions around the world and these woody perennial based land use systems have relatively high capacities for capturing and storing atmospheric CO_2 in vegetation, soil and biomass products. According to the Intergovernmental Panel on Climate Change, agroforestry offers important opportunities of creating synergies between both adaptation and mitigation actions with a technical mitigation potential of 1.1-2.2 Pg C in terrestrial ecosystems over the next 50 years. Additionally, 630 million ha of unproductive croplands. and grasslands could be converted to agroforestry representing a C sequestration potential of 0.586 Tg C/yr by 2040 (1 Tg = 1 million tons). This total C storage in the aboveground and belowground biomass in an agroforestry system is generally much higher than the land use without trees under comparable conditions (Mohan Kumar and Nair 2011).

AGROFORESTRY AND CLIMATE CHANGE ADAPTATION

Climate Change Adaptation: UNFCCC defines climate change adaptation as *the actions taken to help communities and ecosystems cope with changing climate condition.* The IPCC describes it as *adjustment in natural or human systems in response to actual or expected climatic stimuli or their effects, which moderates harm or exploits beneficial opportunities.* The UNDP calls it *a process by which strategies to moderate, cope with and take advantage of the consequences of climatic events are enhanced, developed, and implemented.*

IPCC's prediction of temperature increase between 1.1°C and 6.2°C by the end of the century will most likely create extreme changes in temperature and precipitation. On the other hand, human impact of ecological change may also be severe, leaving people more exposed and vulnerable. World Bank (2008) has estimated that over one billion people live on less than two dollars a day and 75 per cent of these poor people reside in rural areas. Three-quarters of rural dwellers in developing countries are directly involved in agriculture and ecosystem services. Climate change is expected to decrease agricultural productivity in the developing world by 10-20 per cent over the next 40 years. Farmers in developing countries who depend on rain-fed agriculture and natural resources for food production and income generation may particularly experience greater levels of poverty and hunger as their sources of livelihood become increasingly exposed to climate-linked risks. On the other hand, farmers have always had to cope with climate variability and extreme weather events to survive. Adaptation strategies in agriculture have mostly revolved around changes in natural resource management, such as planting climate-resilient crops and/or managing water resources more efficiently (Chaudhury et al 2011).

IPCC defines climate change vulnerability as "the degree to which a system is susceptible to, or unable to cope with adverse effects of climate change, including climate variability and extremes". Subsistence farmers in the developing world are particularly unable to cope with such climate variability, as they do not have the capital to invest in new adaptive practices with which to protect their homes and families. Adaptation to climate change is an adjustment in ecological, social or economic systems in response to observed or expected climatic changes. Adaptive capacity is the ability of individuals and groups to adapt or adjust to climate variability and change; and accommodate shock and stress to systems. Adaptive capacity also includes community's capacity to take advantage of the benefits. Social protection will become more and more important as an adaptation strategy to prevent further increase in poverty and hunger because frequency of climate-related events keeps on increasing.

Agroforestry has been proposed as a potential strategy for helping subsistence farmers reduce their vulnerability to climate change (World Bank 2008). It is the intentional use of trees in the cropping system to increase farm productivity, diversify income sources and provide environmental services (Garrity et al 2006). For example, nitrogen fixing trees can be intercropped among rows of food crops to improve farm productivity by providing limited nutrients to crops. Agroforestry techniques also include the use and sale of tree crops such as fuel wood, fruit and timber. Agroforestry trees can also reduce soil erosion in agricultural lands by providing long-term vegetation cover. Finally, trees are considered to be less sensitive to climate-related shocks such as floods and droughts due to their deep root systems. Many research studies have confirmed the specific role of agroforestry techniques which can play a vital role in helping farmers to reduce their vulnerability to climate change. In a nutshell, agroforestry improves farmer well-being by improving farm productivity and incomes.

Agroforestry extension services will play an important role in helping farmers understand the value of tree planting. Case study reports of many agroforestry practices have highlighted the role of agroforestry in boosting adaptive capacity. When farmers choose to plant trees on farms, it will help to improve the soil and water quality, leading to increased ecological resilience against drought. Most literatures on adaptation in agricultural landscapes mention the value of agroforestry systems as an adaptation strategy. Although many case studies demonstrate agroforestry as the adaptation strategy and adaptive capacity of farmers, the way in which adaptive capacity relates to social protection, which includes enhancement of income and capability, is very different. Farmer's income is protected or enhanced because they are able to protect their crops that they sell to earn income by planting fertilizer trees. This also makes economic sense as they are cheaper option compared to chemical fertilizers.

Social protection measures that have allowed agroforestry systems to thrive have increased adaptive capacity. The role of agroforestry practice is critical in supporting a system that not only institutionalizes income diversification, but also allows the trees to be planted in agricultural landscape for improving soil quality. The deliberate planting of trees in agricultural lands minimize the impact of drought. The protection of trees on farms provides additional income, protects soil moisture and allows farmers to reduce their levels of poverty.

CHALLENGES AND MAIN THRUST AREAS IN CARBON AGROFORESTRY

CHALLENGES AND CONSTRAINTS IN CARBON AGROFORESTRY

Despite some of the potential benefits associated with agroforestry systems and carbon sales, there are also challenges and trade-offs. The challenges and constraints associated with carbon agroforestry are:

- Agroforestry systems already incorporate tree cover by nature. Therefore, such systems presently occur even without additional incentives associated with selling carbon credits.
- The amount of carbon stored in agroforestry systems is minimal and hence may only provide small improvements in income if sold. In order for carbon markets to provide an incentive for shifting to agroforestry systems increases in income from carbon credit sale must at least offset the declines in agricultural production due to the increased tree cover.
- Carbon storage in trees may only be temporary if the trees are converted in the future.
- The initial costs associated with entering the carbon market can be high.
- Legal, social and institutional challenges such as land and tree tenure and government policies. Very few countries have regulatory frameworks in place for governing carbon rights.

- Limited access to carbon markets
- Logistical issues associated with defining and demarcating individual/family plots. The costs associated with doing this will likely be high; however, sampling approaches used for group certification may be able to reduce costs if multiple smallholders are grouped together to achieve economies of scale. In the cases of coffee and cacao this may work through existing organizations.
- The costs associated with measuring and monitoring carbon stocks are high
- There is still global uncertainty on REDD policy/mechanisms which is limiting investment.

Another challenge is the fact that currently the exact global area covered by different agroforestry systems is unknown making it difficult to predict the extent to which agroforestry practices may be able to counter carbon emissions. This is further exasperated by the fact that carbon sequestration in agroforestry systems also depends on the system and the environment in which it exists (Montagnini and Nair 2004). Overall, there is a need for standardized methodologies for estimating above and below-ground carbon, greater understanding of emissions of other trace gases and more robust methods to analyze the costs/benefits of agroforestry based GHG mitigation (Albrecht and Kandji 2003). The World Agroforestry Centre (ICRAF) has also acknowledged that there are some technical concerns associated with assigning carbon credits to agroforestry systems, but feels that these challenges are being resolved. In particular, ICRAF has been working on developing a system using satellite technology to calculate carbon stock in agricultural and forest lands.

RESEARCH NEEDS IN CARBON AGROFORESTRY

The current land-based methods of C assessment may have led to an overestimation of C sinks in many areas of the globe. These inaccuracies will be compounded further if we consider that changes in C fluxes are likely to occur in the next 50 years as a result of shift in global climate, land use and land cover. The magnitude and direction these changes will take, reamain largely unknown. If carbon fluxes in agroforestry systems are well documented, this is not the case for other trace gases such as nitrous oxide (N_2O) and methane (CH_4).

Legumes play a prominent role in agroforestry and are effective in improving the nutrient status of nitrogen-depleted soils. Recent studies have shown that nitrogen inputs derived from agroforestry pratices such as improved fallows can entirely substitue mineral N fertilizers. However, if the tree-derived N exceeds the agronomic requirements of subsequent crops or is not used efficiently, there is a risk of volatilizaton in the form of N_2O. N_2O is one of the most important trace gases and has a global warming potential (GWP) of 200-300 times higher than that of CO_2. There is a growing concern that the widescale use of woody legumes might result in massive release of N_2O gas into the atmosphere. Similarly, ungulate production and cultivation of rice paddy in agroforestry systems can produce significant quantities of CH_4 on a global scale. More research will be needed to clearly understand the implications of agroforestry vis-à-vis the emission of trace gases.

It is apparent that carbon storage in agroforestry systems and its eventual consequences on the global climate are fraught with a number of uncertainties. Consequently, there is still some work to be done to improve our understanding of C sequestration and GHG mitigation. Critical data and research needs in agroforestry include:

- Quantifying C dynamics in agroforestry systems
- Developing effective strategies for measuring and monitoring C sequestration in soil and woody components

- Consideration of other C stocks often left out in estimates which include deep soil carbon (especially when trees are involved) and carbon in durable wood products.
- Predictive models to accommodate future climate and land-use changes and their implications for CO_2 mitigation through agroforestry systems.
- Adequate understanding on issues such as pests and diseases and the emission of other GHGs, especially N_2O and CH_4 in agroforestry systems
- More powerful methods to implement cost/benefit analyses of agroforestry-based GHG mitigation and to define incentives for wide scale adoption of agroforestry.
- Developing/ implementing a national inventory of agroforestry.

■■■■

41 TREE BORNE OILSEEDS (TBOs) BASED AGROFORESTRY SYSTEMS

INTRODUCTION

Agroforestry is a land use system in which woody perennials (trees, palms and bamboos) are deliberately used on the same land management unit as agricultural crops, animals or both, either in some form of spatial arrangement or temporal sequence. In agroforestry systems, there are both ecological and economic interactions between the different components. 'Agroforestry' is a generic term for different types of systems, for example, agrisilvicultural system, silvopastoral system, etc. Agroforestry is a collective name for land management systems that optimize the economic and ecological benefits created when trees and/or shrubs are integrated with crops and/or livestock. Agroforestry systems increase species diversity within farming systems, providing for human needs while supporting wildlife, soil microorganisms, rural communities, farmers, economic interests, watersheds, clean air concerns, biodiversity, and more. Trees are valuable elements in agricultural systems because by their nature, they lend themselves to making connections to other plants, animals, people, soil, etc.

The major benefits of agroforestry to people includes (i) greater long-term economic stability through diversified products, (ii) reduced need for purchasing off-farm inputs, (iii) broader opportunities for rural enterprises, (iv) reduced risk to the farmer, (v) increased overall yields, (vi) year-round production, and (vii) local creation of resources like firewood, fodder, construction materials, etc. More importantly, the trees in farmland act as 'bio-insurance' at times of failure of agricultural crops. Besides above mentioned benefits, agroforestry systems have major role to play in the context of combating global warming by way producing 'biofuels' from tree borne oil seeds. The scope of such agroforestry tree species is increasing nowadays. Tree borne minor oilseeds have been accorded very high priority as a source material for biodiesel production in the country. India is endowed with a vast potential for oilseeds of tree origin, the important of them being sal (*Shorea robusta*), mahua (*Madhuca indica*), neem (*Azadirachta indica*), rubber (*Hevea brasiliensis*), karanja (*Pongamia pinnata*), kusum (*Schleichera oleosa*), khakan or pilu (*Salvadora oleoides*), undi (*Calophyllum inophyllum*), dhupa (*Vateria indica*), ratanjyot (*Jatropha curcas*) etc. These oilseed bearing trees are found widely and distributed throughout the country. The present availability of oilseeds from them is estimated to be about 5 million tons annually. However, only 20 per cent of the total availability is utilized for commercial applications (Kumar 2003).

TREE BORNE OILSEEDS (TBOs)

There are hundreds of species of plants/trees/shrubs found wild or cultivated inside as well as outside the forest areas containing sizeable amount of vegetable oil having domestic and industrial utility. Out of these species, the oilseeds of tree origin like Jatropha (*Jatropha curcas*), Karanja (*Pongamia pinnata*), Jojoba (*Simmondsia chinensis*), Neem (*Azadirachta indica*), Mahua (*Madhuca indica*) etc. have uses like illumination, lubricants and alternative source of biodiesel. They have the capacity to grow and establish in varied agro-climatic conditions and in the wastelands of the country. There is need to exploit the existing potential and also to augment the current potential to achieve the self reliance in biofuel requirement. The TBOs possess 20-60 per cent oil content in seeds/kernel, which are edible fat and non-edible for industrial uses. Some of them can be utilized as source of bio-diesel production (Paramathma and Pandey 2012).

Table 41.1: Potential of Tree-borne Oilseeds in India (Paramathma and Pandey 2012; Sivaprakash et al 2013)

S.No.	Common Name	Botanical Name	Oil Content (%)
1	Sal	*Shorea robusta*	12
2	Jatropha or Ratanjyot	*Jatropha curcas*	30-40
3	Mahua	*Madhuca indica*	35
4	South Indian Mahua	*Madhuca longifolia*	35
5	Neem	*Azadirachta indica*	20
6	Rubber	*Hevea brasiliensis*	45
7	Karanja	*Pongamia pinnata*	27
8	Kusum	*Schleichera oleosa*	33
9	Khakan or Pilu	*Salvadora oleoides*	33
10	Undi	*Calophyllum inophyllum*	60-65
11	Dhupa	*Vateria indica*	19
12	Jojoba	*Simmondsia chinensis*	50
13	Wild Apricot	*Prunus armeniaca*	45-50
14	Tumba	*Citrullus colocynthis*	21
15	Indian Zuzube or Ber	*Ziziphus mauritiana*	33
16	Wild Plum	*Ximenia americana*	49-60
17	Paras Pipal	*Thespesia populnea*	20
18	Drumstick	*Moringa oleifera*	35
19	Nahar	*Mesua ferrea*	45
20	Tung	*Aleurites fordii*	50-60
21	Amoora	*Aphanamixis polystachya*	35
22	Silkcotton	*Bombax ceiba*	20
23	Siltimur	*Litsea cubeba*	22
24	False Nutmeg	*Myristica malabarica*	40
25	Kokum	*Garcinia indica*	25
26	Paradise Tree	*Simarouba glauca*	60-70
27	Kapok	*Ceiba pentandra*	20-25
28	Aisandra	*Diploknema butyracea*	60
29	Oil Palm	*Elaeis guineensis*	56

The availability of TBOs can be enhanced considerably without any extra land and inputs if proper network for procurement from seed collectors is established. There is a considerable scope to enhance the collection of seeds from the existing trees by developing infrastructure facilities such as seed procurement centers equipped with facilities for drying, decorticating, cleaning/grading, depulping, storing and oil extraction near the areas of collection of TBOs. Establishment of biodiesel processing units near the procurement centers will further help in reducing the cost of transportation of the raw oil to the biodiesel processing plant. This should result in reasonable remuneration to the primary seed collector and also help in getting a quality product by reducing losses caused due to delayed and improper handling of the material at different stages in the existing trade of TBOs in India (Rajvanshi et al 2007). In addition to the existing potential of TBOs, there is about 60 million hectares of wasteland of which 30 million hectares can be suitably utilized for growing plantations of biofuel plants. TBOs from the wasteland can make a significant and important contribution to the energy requirement of the country in the days ahead.

TBOs BASED AGROFORESTRY SYSTEMS

Traditionally, farmers grow/retain trees like *Azadirachta indica*, *Pongamia pinnata*, *Madhuca indica*, etc. as scattered in their fields to have green manure, fuelwood, small timber for agricultural purposes. *Jatropha curcas* has been cultivated in bunds as live hedge. Among the TBOs, *Azadirachta indica* and *Pongamia pinnata* are widely adopted species for agroforestry systems. TBOs have not been a key component of any traditional agroforestry system and often they may not have been established deliberately by farmers. The exception is Jatropha, which is established by farmers as a live fence. This requires planting them in a row at close spacing. Other species like *Pongamia pinnata*, *Madhuca* spp. and *Calophyllum inophyllum* probably came up as wildlings and were retained by farmers. Their presence is usually limited to a few trees in the middle of crop fields or along the borders. Oilseed production is not the intention of retaining these trees and they are mostly there for service functions like providing shade and protection as fence (Daniel and Hegde 2007). These tree species, which can be considered as potential tree borne oil yielding species based on the oil content in their seeds, can be integrated into the agricultural lands through agroforestry practices for additional benefits. But it needs careful agroforestry research investigation such as compatibility with agricultural crops and other components, interactions studies, impact on soil fertility, economic feasibility, environmental services, etc. before recommending them for large scale planting in agricultural lands. Nowadays, farmers are encouraged to cultivate tree borne oilseed trees in their lands.

There are enormous wasteland areas available in India as well as in global level and developing these wastelands by undertaking agroforestry plantations of tree borne oilseeds has great scope in future. Promotion of TBOs plantation can generate tremendous job opportunity among the rural masses, simultaneously augment the availability of edible and non-edible oils. The oils of *Jatropha curcas* and *Pongamia pinnata* are potential substitute of diesel and the indigenous production of bio-diesel will resist the outflow of valuable foreign exchange caused by the import of crude petroleum. It may lead to a revolutionary transformation of the gloomy economic and energy scenario with an era of economic bloom and prosperity of all sections of society (Paramathma and Pandey 2012).

SCATTERED TREES ON FARM LANDS

The practice of growing agricultural crops under scattered trees on farm lands is old and does not seem to have changed for centuries. Among TBOs, some trees have received more attention than others. The important TBOs, which are retained in agricultural lands as scattered trees, are *Azadirachta indica*, *Pongamia pinnata*, *Madhuca indica*, etc. The species diversity in these systems is very much related to ecology. The tree diversity is more in humid areas and less diverse in drier areas.

Trees are grown scattered in agricultural fields for many uses such as shade, fodder, fuelwood, fruit, vegetables and medicinal use. There are strong convictions for the acceptance of these trees on agricultural fields since time immemorial.

BOUNDARY PLANTATION/ LIVE FENCES/ LIVE HEDGES

Trees grown in agricultural fields are also often and usually grown on farm boundaries. Many of the boundary plantations also help as shelterbelts and windbreaks. At many places, *Jatropha curcas* is grown as common live-fence.

AGRISILVICULTURAL SYSTEM AND SILVOPASTORAL SYSTEMS

Research trails are being carried out to see the compatibility of TBOs with agricultural crops such as cereal crops, black gram, green gram, cowpea, fodder sorghum, groundnut, shade tolerant vegetables and tuber crops. Many research findings has recommended for the cultivation of agricultural crops during the initial stages of tree growth before tree canopy becomes dense. Similarly the intercropping studies are being taken to study the impacts of growing TBOs and fodder grasses together in the same land which have shown positive results.

NEEM (*AZADIRACHTA INDICA*)

DISTRIBUTION

Neem is a versatile multipurpose tree native of dry forest areas of India, Pakistan, Sri Lanka, Malaysia, Indonesia, Thailand, Myanmar and Africa. In India, it occurs throughout the drier parts including the states of Uttar Pradesh, Bihar, Odisha, Maharashtra, Gujarat, Rajasthan, Karnataka, Andhra Pradesh and Tamil Nadu.

ENVIRONMENTAL REQUIREMENTS

Neem is found in areas with annual mean maximum temperature up to 32-42°C and minimum temperature of 4-21°C. Temperatures below 4°C and frost are unfavourable. Neem grows well in the plains and in lower foothills. But, found even up to 1,830 meters elevation. Neem comes up well in areas with an average annual rainfall of 450-1200 mm. It grows well in loamy, clayey and black cotton soil. It also thrives better than other species on dry, stony, clayey and shallow soils. It can also come up on soil, where there is hard calcareous or clay pan just below the soil surface. It does not grow well on sodic soil and grows moderately on alkaline soil. Water logging and poorly drained soils are unsuitable for neem. Neem is a light demander. It is frost tender species and cannot withstand excessive cold, especially in seedling stages. Well extensive root system makes neem drought hardy. It is a fire tender species.

ESTABLISHMENT OF PLANTATION

Artificial regeneration is done through direct sowing or out planting nursery raised seedling. Direct sowing is adopted for raising plantation for fuelwood and for the reclamation of wastelands. This could be done through dibbling in bushes, broadcasting, line sowing, sowing on mounds or ridges, and sowing in trenches. In dry localities pits of 45 cm^3 and in moist locality pits of 30 cm^3 are dug and kept ready planting. Seedlings of 6-12 months old are out planted into the pits during the rainy season. Pits can be filled with native soil along with 5 kg FYM and 25-50 g of DAP at planting for better establishment and growth. The spacing normally applied for neem is 5 m x 5 m or 10 m x 10 m.

MANAGEMENT PRACTICES

Although neem requires less care during the initial stages especially in dry localities, watering and weeding are very essential during the first two years for proper establishment and growth. Watering once in 10 days will help the tree to tide over the drought period. In saline soils, watering during summer and periods of drought is necessary. Mulching around the tree basins with pebbles or available mulch in that area will help in conserving moisture. Young seedlings should be free from weeds for early establishment and growth. Drawing of tractor drawn tillers in between trees will help to control weeds and conserve moisture. In the naturally raised plantation, thinning has to be done at the end of first year to space the seedling apart for initial growth and establishment. At the end of 2-3 year the trees are spaced at 3 m x 3 m or 5 m x 5 m depending on the locality. In the artificially raised plantation at the end of 5th year, selective thinning can be taken up to cull out inferior trees and space the good trees at 10 m x 10 m spacing.

GROWTH AND YIELD

Neem is grown for its seeds and timber. The tree starts fruits from the 4th or 5th year and gives steady yield from the 10th year. Each tree at this age can yield 20-25 kg of fruits per tree. The ratio of seed to pulp is about 1:2. The rotation age for timber is 35-40 years and for fuelwood is about 8 years. The timber yield is about 108-137 m^3 per ha.

NEEM BASED AGROFORESTRY SYSTEM

As reported, neem tree as woody component in agroforestry has both positive and negative aspects. The positive aspects may include its ability to withstand pollarding, coppicing ability, being an abode for pollinators, production of good timber and usage in biopesticides. Some of the negative aspects include less rate of growth, deep and extensive root systems and dense rounded canopy. Neem in traditional agroforestry systems are found as scattered trees on croplands with stand density of 80-90 trees per ha and having rotation period of 10-15 years. The major agricultural crops grown are cotton, millet, sorghum, black gram, green gram, coriander, sunflower. However, there exist other opportunities of growing neem tree as wind breaks and shelter belts, silvipasture and in wide row interplanting. Though these traditional practices existed for very long period, very less or no research & development interventions has been made to sustain or to improve the practices for higher economic productivity.

RESEARCH ON NEEM BASED AGROFORESTRY PRACTICES

Experiment on Screening of MPTs for Agroforestry under rainfed conditions has been conducted in Dindigul, Anna District, Tamil Nadu (India). The species tried were Neem, Eucalypts, Glyricidia, *Albizia amara*. The results revealed that no significant reduction in crop yield up to third year for the crops tried - black gram, fodder sorghum and groundnut. Minimum yield reduction for groundnut in 5th year under Neem (12%) and maximum under eucalyptus (41%) were found.

Trials with Neem under Improved Agroforestry Practices: Trials have been set up with neem in row planting/wide row interplanting system in three on farm locations in Coimbatore District, Tamil Nadu (India) under the NOVOD project. Various agricultural crop combinations like fodder grass, pulses, green manure have been interplanted with neem planted in 4 m x 8 m espacement. The two year growth data on tree growth have showed that there is considerable potential for raising neem in rainfed conditions in farmlands for providing economic stability to the farmer.

Crown Management Studies under Neem based Agroforestry System: Further, to study the effect of pruning under the neem based agroforestry systems, five year old established neem plot was taken for this study. The neem plot was established in the espacement of 8 m x 4 m and to study the effect of shade, 100 per cent pruning was carried in alternative rows. With usual agricultural practices, intercropping with black gram, green gram, cowpea and sorghum was carried out. The yield was increased over 35 per cent under the pruned area compared to the un-pruned area.

Table 41.2: Effect of Crown Management on Yield of Agricultural crops under Neem based Agroforestry

S.No.	System	Category and Yield (kg/ha)	
		Without Pruning	**With Pruning**
1	Neem + Green gram	303	465 (35% more yield)
2	Neem + Black gram	552	849 (34% more yield)
3	Neem + Cowpea	159	256 (38% more yield)
4	Neem + Sorghum	1,104	1,604 (31% more yield)

PUNGAM (*PONGAMIA PINNATA*)

DISTRIBUTION

An Indo-Malaysian species, a medium-sized semi-evergreen tree, common on alluvial and coastal situations from India to Fiji.

ENVIRONMENTAL REQUIREMENTS

Pungam withstands temperatures ranging from 0°C to 50°C and annual rainfall of 500-2,500 mm, the tree grows wild on sandy and rocky soils, including oolitic limestone, but will grow in most soil types, even with its roots in salt water. It is regarded as both a saline and drought tolerant species. It comes up from sea level to 1,200 m altitude

ESTABLISHMENT OF PLANTATION

Seedlings can be easily transplanted from the nursery after about a year. Root suckers are rather plentiful as well. It is a rapid-growing coppice species that can be cloned. The recommended spacing is 5 m x 5 m. Trees are grown in 30-year rotations for fuel in West Bengal.

GROWTH AND YIELD

Trees reach adult height in 4 or 5 years, bearing at the age of 4-7 years. A single tree is said to yield 9-90 kg seeds per tree, indicating a yield potential of 3,600 to 36,000 kg seeds per ha. The average oil content in the seed is 27-32 per cent. With upgraded germplasm, it may be possible to obtain 2.0 MT oil and 5.0 MT firewood per hectare per year on a renewable basis. Since, this tree is widely grown over Tamil Nadu and in many parts of India, as a shade or ornamental tree, this tree could be well utilized for both wasteland afforestation combined with biodiesel production.

PUNGAM BASED AGROFORESTRY SYSTEM

In Forest College and Research Institute, Mettupalaym (India), an experimental plot with pungam was established with tree spacing of 5 m x 5 m and several intercrops were evaluated. When the trees were one year old, the reduction in yield of intercrops was highest for soyabean (23%) and lowest for cowpea (5%). This yield increased as the age of the tree increased to 3 years. It is concluded that among the agricultural crops evaluated, cowpea, groundnut and black gram could be the suitable intercrops for pungam based agroforestry systems.

Table 41.3: Yield of Intercrops (Kg/ha) under Pungam based Agroforestry System (Divya et al 2013)

Intercrops	Age of Trees (Years)					
	First Year		Second Year		Third Year	
	Inter Cropping	Pure Cropping	Inter Cropping	Pure Cropping	Inter Cropping	Pure Cropping
Black gram	710 (11%)	800	489 (16%)	480	341 (20%)	425
Green gram	505 (8%)	550	356 (21%)	450	285 (29%)	400
Red gram	1,120 (7%)	1200	465 (21%)	590	264 (31%)	384
Cowpea	570 (5%)	600	440 (10%)	490	450 (12%)	510
Groundnut	900 (7%)	970	860 (12%)	980	800 (16%)	950
Soybean	310 (23%)	400	233 (33%)	350	513 (35%)	790
Sunflower	380 (15%)	450	310 (23%)	400	385 (34%)	580
Grain Amaranth	680 (16%)	800	620 (23%)	800	250 (36%)	390

Note: Values inside the bracket show yield reduction

JATROPHA (*JATROPHA CURCAS*)

DISTRIBUTION

Jatropha is a multipurpose and drought resistant large shrub/small tree belonging to the family Euphorbiaceae. Despite being a native species to tropical America, it is now being cultivated extensively throughout Africa and Asia. In the recent past, there has been substantial social and political importance for growing Jatropha in India, as a means of economic upliftment and poverty alleviation through use of oil extracted from seeds as a source of bio-fuel. Further, Jatropha can help to increase the farm income and alleviate poverty in rural areas.

ENVIRONMENTAL REQUIREMENTS

Being a drought hardy species, it can be grown in the arid/ semi-arid areas with an average rainfall between 300 and 1,000 mm. However, it grows well in areas with higher amount of rainfall. It occurs mainly at lower altitudes (0 to 500 m).

ESTABLISHMENT OF PLANTATION

Jatropha is normally propagated through seeds. The seeds are sown in polybags and maintained in nursery for a period of 3 months. It can also be propagated vegetatively using branch cuttings (2-3 cm thickness and 15-30 cm length). However, trees propagated by cuttings show a lower longevity and possess a lower drought and disease resistance than those propagated by seeds. It is planted during rainy season with a normal spacing varying from 2 m x 2 m, 3 m x 2 m and 3 m x 3 m in block plantations. However, the spacing of 3 m x 3 m can be ideal for block plantations maintained for longer periods. For hedgerow or boundary plantations, the spacing can be 1 m between plants. The plant can be maintained for 30 years economically.

WATER REQUIREMENT

In general, the water requirement of the species is low. The crop requires a life irrigation on the third day after field planting. After this, irrigation at an interval of 20 to 25 days may be done based upon need. In general, around 6-8 irrigations are required per year. Though Jatropha survives under rainfed and extreme dry conditions, it is observed to be a poor performer and poor yielder under these conditions. Hence irrigation is recommended at least at fortnightly intervals to ensure year round production of seeds.

FERTILIZER APPLICATION

Manure is applied only in the first year. From the second year onwards, fertilizers are applied @ 20:120:60 kg per acre of NPK respectively in two equal doses in June-July and Oct-Nov after flowering. Gibberellic acid 50 ppm is sprayed at the time of flowering to induce and increase flowering and better pod development.

PRUNING

Pruning has to be taken up for the first three years to increase number of fruiting branches. Early pruning at 6 to 8 months after planting or just before flowering is advisable. First pruning must be at 45 cm from ground level which will enhance more number of auxiliary branches. Second pruning must be made in the newly developed shoots by cutting two thirds of the new flush leaving one third in the plant. Similar procedure has to be followed on the new flush upto the completion of the third year.

GROWTH AND YIELD

Jatropha starts bearing from first year onwards. However, stabilized yield is realized only from 3rd year. A cumulative yield of 500 and 2,000 kg per ha can be realized in the first two years under rainfed and irrigated conditions respectively. From 3rd year onwards, the expected yield from one hectare plantation is 2,500 kg under rainfed and 7,500 kg under irrigated conditions. Obtaining optimal yields is dependent mainly upon use of improved planting material and matching the germplasm to sites with fertile soils and adequate moisture that will allow it to express its genetic potential. Although Jatropha may look promising as a tree/shrub for marginal lands, without added nutrients, moisture and improved germplasm, marginal yields only can be expected.

JATROPHA BASED AGROFORESTRY SYSTEM

Studies conducted at BAIF to examine the suitability of Jatropha for agroforestry purposes showed that the yield reduction on horse gram and finger millet was negligible during the first four years. In fact, the annual crops seemed to benefit from the sheltering effect of Jatropha when there were strong dry winds. Agroforestry systems can be in the form of (a) scattered trees, (b) rows or strips of trees and (c) trees in border rows or strips. The ratio of trees to crops in the system can be determined by adjusting the spacing between the trees as well as spacing between rows or strips. The present practice of a single row of Jatropha live-fence can be extended to a strip of 3-5 m. This strip can accommodate 2-3 rows of Jatropha and this would effectively mean devoting up to 20 per cent of land to the tree component. This arrangement will take up relatively less land for the tree species while the protection function of the live-fence will be strengthened. The outcome of this arrangement would be a loss of about 25 per cent in the yield of the crops in the middle of the field while the fence produces oilseed. Farmers may not opt for this system if the land is fertile because the loss in crop production is unlikely to be compensated by the TBOs in the fence. On the other hand, it can be a favourable alternative for lands where crop yields are low and less reliable.

The other two systems, scattered trees and strips within the cropping area, are also unsuitable for fertile lands because the trees can reduce the crop yields. On such lands, fruit trees may be acceptable as the produce is of high value. But the returns from TBOs are unattractive for farmers to forego existing crops that are proven and more remunerative. Depending upon the fertility, a suitable tree:crop ratio can be decided upon and the tree spacing can be arrived at. If the tree proportion in the system is assumed to be 50 per cent, the Jatropha yield, as discussed earlier for average production conditions, will be about 1.0 ton per ha. This translates to an income of Rs.10,000 per ha (2,000 kg per ha at the rate of Rs.5 per kg of seed), which is not appealing enough to attract growers in large numbers. But it is also a fact that present returns from vast areas of wastelands in the country are lower than this amount. Most of these lands belong to small marginal farmers in dry areas who routinely take up cultivation of some crop during the rainy season. Jatropha is a suitable option for them. There are others who raise crops on a part of their land while the less fertile part usually remains barren. Such land can be brought under TBOs as it requires very little management after establishment and will not be a constraint on farmer's time and resources (Daniel and Hegde 2007).

In an experiment conducted in 4-year-old Jatropha based agroforestry system at Forest College and Research Institute, Mettupalayam (India), maximum reduction in yield was observed for gingelly and the yiled reduction was minimum for groundnut. The study concludes that cowpea and groundnut are the suitable intercrops and the optimum spacing for Jatropha could be 4 m x 2 m or 4 m x 3 m.

Table 41.4: Yield of Intercrops (Kg/ha) under 4-year-old Jatropha based Agroforestry System (Divya et al 2013)

Tree Spacing	Black gram	Green gram	Cowpea	Groundnut	Gingelly	Sunflower
3 m x 3 m	393 (32%)	306(32%)	458(16%)	728(15%)	304(46%)	391(39%)
4 m x 2 m	457(21%)	392(29%)	465(15%)	789(12%)	317(45%)	441(32%)
3 m x 4 m	419(28%)	401(27%)	479(12%)	795(7%)	378(35%)	412(36%)
4 m x 3 m	472(19%)	430(22%)	483(12%)	798(7%)	386(33%)	467(27%)
Pure crops	580	550	546	858	580	644

Note: Values inside the bracket show yield reduction

PARADISE TREE (*SIMAROUBA GLAUCA*)

DISTRIBUTION

It is an exotic species introduced from El-Salvador of Central America. It is a versatile multipurpose tree, which can grow well even in the degraded soils. This tree is regarded as highly suitable for growing on both arable and non-arable wastelands. It needs no special care and requires minimum protection as it is generally not browsed by cattle, goats and sheep.

ENVIRONMENTAL REQUIREMENTS

The temperature preference of the species is 17°-35°C with an annual rainfall of 500-2,200 mm. It grows well up to 1,000 above MSL in all types of well-drained soil with pH 5.5-8.0. However, a minimum of 1.0 m deep soil is preferred for its growth. Soils of shallow depth with canker underneath are relatively unfavourable for its growth. It is a light demander and capable of withstanding the drought condition. It is however, very frost tender especially in the seedling and sapling stages. It is a medium coppicer and fire tender. It can withstand pollarding.

ESTABLISHMENT OF PLANTATION

Six months old seedlings are suitable for planting in the main field. Seedlings are planted at a espacement of 6 m x 6 m (277 trees/ha) in pits of 45 x 45 x 45 cm size half filled with top soil and 2.0 kg compost + 20 g phosphorus + 20 g nitrogen.

MANAGEMENT PRACTICES

The plants grow well with protective watering. Every year, ploughing and crescent basins formation to be taken up prior to the onset of the monsoon to harvest rain water and in turn to induce flowering and fruiting. Timely weeding in the first two years helps the better establishment of saplings. Softwood cleft grafting and air layering gives 80 per cent success, which helps in growing male and female trees in the desired ratio (1:6).

GROWTH AND YIELD

The tree starts flowering and fruiting at about three years of age. Flowering is annual, beginning in December and continuing up to February. The tree starts bearing when they are 4-6 years old and reach stability in production of another 4-5 years. The droplets (blackish purple in pink genotypes and brownish yellow in green genotypes) are ready for harvest by March/April. Individual fruits have a development and ripening period of 1-2 months. Even though bearing commences in the 3rd or 4th year, the tree gives the economic yield of 20 kg per tree from 10th year onwards.

SIMARUBA BASED AGROFORESTRY SYSTEM

In an experiment conducted in 3-year-old Simaruba based agroforestry system at Forest College and Research Institute, Mettupalayam (India) maximum yield reduction was observed for black gram and red gram and the minimum yield reduction was observed for cowpea.

Table 41.5: Yield of Intercrops under 3-year-old Simaruba based Agroforestry System (Divya et al 2013)

Intercrops	Grain Yield (Kg/ha)	
	Pure Cropping	Intercropping
Black gram	1,134	1,054 (7%)
Green gram	1,172	1,107 (5%)
Cowpea	1,256	1,226 (2%)
Red gram	970	902 (7%)
Mean	1,133	1,072

Note: Values inside the bracket show yield reduction

ALEXANDRIAN LAUREL (*CALOPHYLLUM INOPHYLLUM*)

DISTRIBUTION

The tree is native from East Africa, through India and Southeast Asia to the Philippines, Taiwan, and the Marianas. Southward its range extends through Melanesia to Australia and through southern and eastern Polynesia. The habitat is primarily coastal and adjacent to lowland forests. It is essentially a littoral tree of the tropics, occurring above the high-tide mark along sea coasts of northern Australia and extending throughout Southeast Asia and southern India.

ENVIRONMENTAL REQUIREMENTS

It grows in warm temperatures in wet or moderate conditions. It is not suited to high elevations, cool areas, or very dry conditions. Elevation ranges from 0-200 m in Hawaii and up to 800 m at the equator. Mean annual rainfall required is 1,000-5,000 mm with summer, winter, or uniform rainfall patterns. Dry season duration (consecutive months with <40 mm rainfall) is 4-5 months. Mean annual temperature of 18°-33°C is preferred with mean maximum temperature of hottest month of 22°-37°C and mean minimum temperature of coldest month of 12°-17°C. It tolerates a wide range of soils. It grows best in sandy well drained soils in coastal areas but will tolerate clays, calcareous soils, and rocky soils. The soil is generally dry at the surface, but the water table is usually only a few decimeters down, although the water it taps is often brackish. It is also found higher up the rivers along river margins. The tree demands light and prefers full sun, and only less shade is tolerated. It is sensitive to frost and fire. It tolerates occasional waterlogging in coastal areas.

ESTABLISHMENT OF PLANTATION

In plantation trials in Indonesia, the spacing of seedlings is usually 2 m x 3 m. The species coppices moderately. The tree is said to be brittle and liable to damage by wind. Frequent weeding is necessary until the crop is established. The tree is planted for shade and for reforestation and afforestation. It also acts as an efficient shore protector in most places. Although slow growing, it is a popular tree for roadside and avenue planting in India.

GROWTH AND YIELD

It is a high potential bio-diesel plant. Seed yields 65 per cent oil from its dry weight and 3,744 kg/ha from 400 trees. It has the highest per tree oil yield from 75 plant species tested so far. It can be used in conventional diesel engines (without any alterations) in its pure form or as a blend with mineral oil. It is also a durable multi-purpose timber (density 560-900 kg/m^3). With recently discovered plant properties like anti-HIV and anti-cancer active compounds, *Calophyllum inophyllum* can be placed amongst the most important multipurpose trees.

KUSUM (*SCHLEICHERA OLEOSA*)

DISTRIBUTION

It occurs naturally from the foothills of the Himalayas and the western Deccan to Sri Lanka and Indo-China. It was probably introduced to Malaysia and has naturalized in Indonesia. It is occasionally cultivated throughout the tropics, especially in India.

ENVIRONMENTAL REQUIREMENTS

It occurs spontaneously in dry, mixed deciduous forest and savanna with only scattered trees, sometimes gregariously. It grows on rather dry to occasionally swampy locations on various, often rocky, gravelly or loamy, well drained, preferable slightly acid soil. It is a fire-resistant species. Seedlings are frost sensitive and light-demanding. It occurs usually at low altitudes, but can be found up to 900 to 1,200 m. It requires annual rainfall of 750-2,500 mm. It come up with the absolute maximum temperatures of 35°-4°C and absolute minimum temperatures of 2.5°C.

ESTABLISHMENT OF PLANTATION

Natural regeneration is by seed and root suckers. Propagation is by direct sowing in thoroughly prepared soil or by stump planting. In nurseries in West Bengal (India), seed is sown 7.5 cm apart immediately after collection. Stumps are prepared after one year, when the seedling stem is about 1 cm in diameter. The stem is cut back to about 4 cm, the roots to 25 cm. Plant holes should be about 30 cm deep and wide. Regular weeding and protection from grazing is required.

MANAGEMENT PRACTICES

In India, it is used as host for the lac insect (*Laccifer lacca*). The product is called kusum lac and is the best in quality and in yield. In Central India, it is much planted as a wayside tree.

GROWTH AND YIELD

S. oleosa produces root suckers freely and pollards well. In Bihar (India), trees grow to a height of about 7 m and a stem diameter of 10 cm in 16 years; in Uttar Pradesh (India) coppice shoots reach a height of 2 m in one year, in South Kanara (India) 5 m in 3 years. In India, a mature tree yields 21-28 kg depulped seed per year. After crushing the depulped seed, the oil is extracted by boiling or pressing. The oil yield obtained by boiling is 32-35 per cent of the kernel weight, by pressing 25-27 per cent. Where wild *S. oleosa* occurs abundantly, it remains important as a fuelwood, but its growth is too slow to be planted for fuel. Where seed is available in large amounts, pressing and refining of oil combined with the manufacturing of seed cake as cattle feed may be viable, although the quantity currently processed is well below its potential.

MAHUA (*MADHUCA LONGIFOLIA*)

DISTRIBUTION

The tree grows throughout the greater part of India except the temperate and arid areas and southern parts of the Indian peninsula. It is a common tree of the deciduous forests of Madhya Pradesh, Maharashtra, Gujarat, Central India, Indian peninsula, Chota Nagpur, Odisha and is characteristic of dry plains. It is extensively cultivated near villages because of its multi-utility. Two varieties are recognized viz., (i) *Madhuca longifolia* var. *longifolia* which grows in South India, and (ii) *Madhuca longifolia* var. *latiflolia* which grows in North India.

ENVIRONMENTAL REQUIREMENTS

It is a tree of dry tropical and subtropical climate. In its natural habitat, the absolute maximum temperature varies from about 41°C to 48°C, the absolute minimum temperature from about 1°C to 8°C and the normal annual rainfall from about 750 to 1,875 mm. The man relative humidity in its natural range varies from about 40 to 80 per cent in January and from about 60 to 90 per cent in July. The tree grows on a wide variety of soils, but prefers sandy soils. It grows well in alluvial soils of the Indo-Gangetic plains. In sal forests, it is found growing on high clay and even calcareous soils. It is a strong light demander and gets suppressed under shade. Seedlings are frost tender, but mature trees are frost hardy. It is a drought resistant crop.

ESTABLISHMENT OF PLANTATION

One year old seedlings are used for planting in main field. Seedlings can also be containerized. In that case, one month old seedlings are pricked into containers. Planting is done in 30 cm^3 pits at a spacing of 4 m x 4 m. Planting of stumps is done in crow bar holes or 30 cm^3 pits. Spacing can be up to 9 m x 9 m to 10 m x 10 m. One year old stumps establishes more successfully than seedlings.

MANAGEMENT PRACTICES

Mahua is worked on 25 to 30 year coppice cycle. It coppices well, if felled in hot season. Protection from fire and fencing of plantations are essential at early stages along with clean weeding and soil working.

GROWTH AND YIELD

The seed and oil potential of this species in India is 500,000 tons and 180,000 tons respectively. The tree matures from 8 to 15 years and fruits upto 60 years. It produces a mean annual increment of 3 to 5 m^3 per ha. Seeds are plentiful from 2nd or 3rd year onwards.

JOJOBA (*SIMMONDSIA CHINENSIS*)

Jojoba is an exotic hardy desert shrub which can withstand extreme arid conditions of temperature and very low annual rainfall. However, irrigation and manuring are needed for good yield and economic return. Further, shallow soils and temperature factor have major impact on flowering and fruit yields in A.P. and other Southern States. The possibilities of promoting jojoba in such areas may be examined more critically. Rajasthan, Gujarat, Haryana and A.P. are potential states wherein, it can be propagated through seeds and cutting during October and February. The jojoba plants start yielding consistently in the fourth year and it can yield upto 200 years. The jojoba seeds contain about 50 per cent of oil. Jojoba oil and its derivatives have diversified uses viz. cosmetics, pharmaceuticals, lubricants, food, electrical insulators, foam control agents, plasticizers, fire retardants, transformer oil. It is also used for obtaining waxes which is used for polishing floors, furniture, automobiles and in candle and textile industries. The cake can be used as cattle feed.

WILD APRICOT (*PRUNUS ARMENIACA*)

Wild apricot, locally called chullu, is found in the dry temperate region of J&K, H.P. and Uttaranchal. It is distributed in the North Western Himalayas particularly in the valley of Kashmir, Chenab, Kullu, Shimla and Garhwal hills. It is propagated through seeds in the month of October and through cutting in February-March. The fruit of wild apricot mixed with doze of cultivated types are utilized in production of apricot jam, apricot nectar and apricot papad. The fruit is also used by the tribals for the preparation of distilled alcoholic liquor. The kernel which constitutes about 20-40 per cent of fruit contains about 45-50 per cent oil. The oil is used as a substitute for almond oil. The oil is also used for food purposes, cosmetics and pharmaceutical preparations and in medicines for ear ache and other ailments. The cake is used as a feed for livestock.

TUNG (*ALEURITES* SPECIES)

Tung is an exotic deciduous rapid growing tree found in tropical and sub-tropical regions of Eastern Asia and Malaysia. Tung trees are grown for its valuable quick drying oil (A better drying oil than linseed oil). China and United States are the major tung oil producing countries. An elevation of 2,500-3,500 meter and rainfall of about 50-70 inches are best suited to its growth. Out of six species of tung, two species viz. *Aleurites montana* and *Aleurites fordii* have been well acclimatized in Mizoram. More than 5 million grown up tung trees are reported available in Mizoram state.

The plants can be raised from seeds, either by direct sowing or by transplanting. Propagation by budding and grafting has also been suggested. To increase the yield, it is advisable to select seeds from dominantly female flowering trees, but vegetative propagation of high yielding trees will give better yields. The seedlings are transplanted after six months at a spacing of 3.8 m in rows 9 m apart.

Ordinarily 200-250 trees could be maintained in a hectare. The tree is comparatively fast growing and can attain a height of 2.4 to 3 m in the third year, when some of them begin to flower and bear fruit. The full bearing capacity is attained at the 10th year and the tree continues to bear fruits for another 20 years. A ten year old tree generally yields 20-25 Kg of fruits per season. The fruits yield 50-60 per cent of nuts or seeds. In India, the seeds yield 55-60 per cent of kernels which in turn yield 49-65 per cent of oil. The wide range in oil content indicates adequate scope for selection and improvement in oil yield for biodiesel production.

INDIAN JUJUBE OR BER (*ZIZIPHUS MAURITIANA*)

The Indian jujube is native from the Province of Yunnan in southern China to Afghanistan, Malaysia and Queensland, Australia. It is cultivated to some extent throughout its natural range but mostly in India where it is grown commercially and has received much horticultural attention. *Z. mauritiana* is a medium sized tree that grows vigorously and has a rapidly developing taproot, a necessary adaptation to drought conditions. The species varies widely in height, from a bushy shrub of 1.5-2.0 m tall, to a tree of 10-12 m tall with a trunk diameter of about 30 cm. It is hardy tree that copes with extreme temperatures and thrives under dry conditions with an annual rainfall of 15–225 cm. It also grows well on laterite and medium black soils with good drainage or sandy, gravelly, alluvial soil of dry river-beds where it is vigorously spontaneous. In India, there are 90 or more cultivars available depending on the habit of the tree, leaf shape, fruit form, size, colour, flavour, quality and fruiting season.

It is widely grown from seeds which may remain viable for two and half years. The hard stone restricts germination and cracking the shell or extraction of seeds hastens germination. Without pretreatment, the seeds normally germinate within six weeks whereas extracted seeds only need one week to germinate. The germination can be improved by soaking seeds in sulfuric acid. Ber seedlings do not tolerate transplanting, therefore the best alternatives are to sow the seeds directly in the field or to use polythene tubes placed in the nursery bed. Seedlings are ready for budding in 3-4 months. In addition, seedlings from the wild cultivars can be converted into improved cultivars by top-working and grafting. Nurseries are used for large scale seedling multiplication and graft production. The seedlings should also be given full sunlight. The seedlings may need as long as 15 months in the nursery before planting in the field. Scientists in India have standardized propagation techniques for Ber establishment. Budding is the easiest method of vegetative propagation used for improved cultivars. Different types of budding techniques have been utilized with ring-budding and shield-budding being the most successful.

The major production regions for Indian jujube are the arid and semi arid regions of India. Trees in northern India yield 80-200 kg of fresh fruit per tree per year when the trees are in their prime bearing age of 10-20 years. The fruit is eaten raw or pickled or used in beverages. It is quite nutritious and rich in vitamin C and it is second only to guava and much higher than citrus or apples. The fruits are applied on cuts and ulcers and employed in pulmonary ailments and fevers. The fruits are mixed with salt, chilli and peppers and given for indigestion and biliousness. The dried ripe fruit is a mild laxative. The seeds are sedative and are taken, sometimes with buttermilk, to halt nausea, vomiting and abdominal pains in pregnancy. They check diarrhoea and used as poultice on wounds. Mixed with oil, they are rubbed on rheumatic areas. The fatty-acid methyl ester of *Z. mauritiana* seed oil meets all of the major biodiesel requirements. The average oil yield is 4.95 kg oil per tree or 13700 kg oil per hectare.

WILD PLUM (*XIMENIA AMERICANA*)

False sandalwood is synonymous for wild plum and it is also known as wild olive, sour plum and tallow nut in English and kadaranji and siruyilandai in Tamil. *Ximenia americana* is drought resistant tree and mostly available solitary. It is available in open country, savannah, gallery forest, along coastal areas, in the understorey of dry forests, in dry woodlands, or on riverbanks. It can be grown in the area of 0-2000 meter altitude, having mean annual temperature of 14-30°C and mean annual rainfall of 300-1,250 mm. It can be grown in often poor and dry soils including clays, clay loam, loamy sands, sandy clay loam and sandy soils. It is a bush-forming shrub or small tree of 2-7 m high. Trunk diameter seldom greater than 10 cm and bark colour dark brown to pale grey and smooth to scaly. It is easily propagated through seed and cuttings. The tree regenerates naturally from seed and coppice. Fresh seeds should be sown for good germination.

The fruits being pleasant to eat raw and can be used to make juice, jams and jellies or an intoxicating drink. The pulp of seed and fruit contains hydrocyanic acid and it is advisable not to chew the seed. Seed oil contains oleic, linoleic, linolenic, arachidonic, eicosatrienoic, erucic and nervonic acids. Kernel oil is used as a vegetable butter and as a ghee substitute. Firewood and charcoal are the chief uses of the wood, because the trunk is usually too small to make it useful for timber. Volatile oil from leaf yields benzaldehyde, hydroxy benzyl cyanide and isophorone. Kernels are purgative and the oil is suitable for bio-diesel production (Paramathma and Pandey 2012).

■■■■

42 TREE GROWTH AND YIELD MODELLING IN AGROFORESTRY

INTRODUCTION

Modelling has been used in natural sciences since centuries in one way or the other. However, system modelling as a discipline was established in the current century only. Models have been built in the physical, biological and social sciences. The building blocks for these models have been taken from calculus, algebra, statistics, geometry, probability and nearly every other field within mathematics. Various researchers have defined modelling in their own way. Mclone and Andersons (1976) described modelling as the representation of our so called 'real-world' in mathematical terms so that we may gain a more precise understanding of its significant properties, and which may hopefully allow some form of prediction of future events. Hall (1963) has said that the goal of modelling is to understand reality mathematically. Meyer (1985) defined model to be an object or concept that is used to represent something else. Its reality is scaled down and converted to a form we can comprehend. He further adds that a mathematical model is a model whose parts are mathematical concepts, such as constants, variables, functions, equations, inequalities, etc.

Agroforestry modelling is a complex process that essentially involves at least two elements viz. tree and crop, where tree is the long term and crop is the short term compo-nent. Agroforestry models, usually, predict tree growth and yield and the reduction/ enhancement in crop yield as influenced by the tree component in temporal and spatial sequence. Models providing the most accurate prediction of height included basal area, trees per ha, dominant stand height, and diameter at breast height (DBH) (Sharma and Parton 2007).

A model is a way to represent, in mathematical terms, the relationship that exists in the real world. You can use models to describe what might happen in future based on certain conditions and assumptions. You intuitively use models every day although you probably do not think about it consciously. For example you know how much time you must allow to get home from your work place. If you choose a mini bus it usually requires less time than a regular bus. If you wanted you could have described that relationship with a mathematical model. Models are handy tools because they parallel the way we human beings think.

Once you describe relationships that exists in the real world you can then extend the model you developed to predict what will happen if those relationships does not remain the same. Thus, modelling is done to aid the conceptualization and measurement of complex systems and some times to predict the consequences of an action that would be expensive, difficult or destructive to do with the real system. Thus, you can easily do "what-if" modelling. Just change your model assumptions and conditions, repeat the calculations, and observe the predicted outcomes.

ARTISTRY OF MODELLING

PRINCIPLE OF MODELLING

- Situations are modelled for various purposes. Foremost among these is the need to predict new results or new features, which may be of the form of extension of existing results, or of a more radical nature. The predictions are often of conditions likely to exist at some future date. There may, on the other hand, be predictions of events for which direct experimental evidence is unobtainable.
- Not all situations are modelled for this purpose. In some, it is sufficient to be able to describe the working of the system by mathematical means in order to obtain a greater understanding - many of great physical theories do this, although they make predictions as well. What is not usually involved in such descriptions is an element of control.
- Indeed, many models in forestry/agroforestry studies are designed to assist management in the making of decisions.

DEVELOPMENT OF MODEL

Development of a model essentially involves two stages:

- Model Estimation
- Model Validation

Construction of a precise model, not only includes sound mathematical/statistical formulation based on theoretical considerations of the process being modelled but equally important are practical observations and from this view point, it is desirable that one should have two independent data sets; one to be used for model estimation and the other for model validation. However, if the practical situations do not permit to have two independent data sets, then the single data set should be judiciously utilized for model estimation and validation. Geisser (1975) suggested that out of a single data set, a random sample (with out replacement) of about 80 per cent data points should be selected and it should be utilized for model estimation and the remaining 20 per cent data points may be kept as the second independent data set to be used for model validation.

MODEL ESTIMATION

At this stage the model is formally developed on the basis of theoretical considerations and experimental observations. Model estimation involves three steps:

- Simplification or idealization
- Construction or formulation
- Parameter estimation

SIMPLIFICATION OR IDEALIZATION

It includes translation of a real life situation into a problem amenable to mathematical treatment. This is a crucial stage since the real situation is usually complex, involving many processes. Some features will appear significant, many others irrelevant. This stage essentially involves variable identification. A major decision often taken at the very onset of modelling concerns the nature of mathematical variables involved. Basi-cally variables are of two types. One class represents known features that are quantities capable, at least in theory, of precise measurement and control; they are called deterministic variables. The other representing the unknown features that are quantities the measu-rement of which can never be known precisely, but which illustrate a random feature; these are called stochastic variables. A model which contains stochastic variables must, by definition, be concerned with mathematical techniques appropriate to probability theory; deterministic variables often, but by no means exclusively, call for the use of calculus, differential equations and stochastic techniques. Some situations do not reveal their nature at the onset; others have variables of both types in them. It is fundamental to the construction of the model that the nature of variables be properly diagnosed.

CONSTRUCTION OR FORMULATION

Once the significant features have been identified, the next stage is to translate these into mathematical entities, and to postulate relations between these identities. This is generally the most difficult stage and one in which it is virtually impossible to define a formal course of action. At this stage, the equations and other mathematical relations are set up in the model.

PARAMETER ESTIMATION

After the biological problem has been translated into mathematical one i.e. once the model has been constructed/formulated in terms of equations or functions, the next step is to estimate the parameters of the model.

MODEL VALIDATION

Once a model is constructed it needs to be validated. Indeed, some form of valida-tion is usually carried out through out the formulation; the equation or other mathematical relations set up in the model are continually checked with the initial situation. Validation takes a number of forms. In the first place, the mathematics itself which constitutes the model must be self consistent and obey all usual laws of mathematical logic. In the second place, model validity rests in its ability to represent the situation initially described; however, whether a model passes this test is to a large extent subjective. A model may have to represent reality, but it is not 'itself' reality. Model validation also involves two steps:

- Optimization/refinement
- Model diagnostics or Statistical validation techniques

OPTIMIZATION/ REFINEMENT

Once the model has been constructed, its parameters are to be estimated from the experimental observations. In agroforestry/forestry modelling studies, the growth equations usually involves three or four parameters. The estimates of the parameter along with confidence interval and asymptotic standard errors (ASE) are computed. Smaller the confidence interval, more accurate the parameter estimates. For assessing the certainty of the parameter estimates, Wald confidence region and C-W confidence curves (Cook and Weisberg 1990) should be plotted. The agreement between the two methods indicates the certainty of estimates. However, if the parameters in the model are highly inter-correlated, this might indicate that the model has too many parameters or the model is over parametized. The effect of over parametization in the full model may be solved by fixing one parameter (usually the one for which Wald and C-W curves coincides) and estimating the others for which C-W curve shows asymmetry in distribution. Thus in the partial model (with one parameter fixed) the estimates of the parameters gets refined leading to optimized values and minimum possible confidence intervals.

MODEL DIAGNOSTICS OR STATISTICAL VALIDATION TECHNIQUES

Once the model has been optimized, the last phase of model development is the statistical testing procedures that can ensure, to a certain extent, the safe use of model for predictive (interpolative/extrapolative) purposes. Standard statistical texts (Gomez and Gomez 1984; Mead et al 1993), in context of statistical inference and hypothesis testing, recommends BLUE (Best Linear Unbiased Estimate) i.e an estimator that is unbiased and has the minimum variance in the class of unbiased estimators. On the similar lines we seek for unbiased model with minimum value of MAR (Mean Absolute Residual) and RSS (Residual Sums of Square). Leaving aside usual residual diagnostics, numerous other statistical procedures have been devised for validating a model. For example, residual diagnostics, to be performed as a compulsory measure with any proposed model, in addition to the usual coefficient of determination R^2 (observed vs predicted) and MSE (Mean Square Error), it should include the following statistics and graphical plot:

(1) To verify that the residuals are normally distributed, the residuals should be plotted against their expected values for normal distribution

(2) To check that the residuals are independent, the autocorrelation plot of the residual should be prepared

(3) To validate that the residuals have constant variance, the residual should be plotted against the estimate

(4) To confirm that the residuals are not continuously being over/under estimated, the residuals should be plotted against the independent or explanatory variate.

In continuance to the above mentioned classical procedures, some of the most recent and worth mentioning statistical model validation procedures are:

(1) Neter et al (1996) proposed that to test the unbiasedness of a model 't' test should be used.

(2) Chen et al (1998); Nigh and Sit (1996) proposed polynomial regression procedure on the deviation of the predicted value from the test value. The procedure applies a fifth order polynomial regression analysis, followed by the back word elimination procedure to examine whether the mean of the residuals derived from fitting is associated with a linear, or second, third, fourth or fifth order polynomial pattern. This procedure tests whether residuals differ significantly from zero.

(3) Amaro (1998) proposed that a linear regression between observed value (y-obs) and estimated or predicted (y-pred) value should be fitted i.e. (observed) = **a+b***(predicted), and for a perfect model, we expect that, '**a**' should be zero and '**b**' should be one. $R^2_{adjusted}$ should also be one.

(4) Kayahara et al (.1998) used the statistic root-MSPR to measure the precision of the model. The term root-MSPR is defined as the square root of the mean squa-red prediction error and lesser the value of root-MSPR, better is the precision.

TREE MODELLING TECHNIQUES

The basic techniques start with the sampling of trees that represents the tree popul-ation in a system, then the harvesting of trees and finally the recording of observations on the harvested trees. Some of the basics of tree data recording are discussed below.

SAMPLING OF TREES

The most accurate method of assessment of biomass can be the complete harvest, however cutting and weighing every tree is not possible. Sampling of trees is necessary to avoid such difficulties. In laying out an agroforestry experiment, it should be ascertain that at the end of experimental period the number of trees per treatment would be 9-12 for single site. This is the minimum number; we may select more when trees are clearly variable due to soil variability, etc. Tree can be selected with stratification by size to ensure the whole size range is covered. Although random sampling is easier to do than using recommended stratified sampling. The random method is potentially dangerous, because the selected trees may not cover the whole range. When using random sampling, we may take minimum 16 trees as sample per treatment. For tree selection, we rank all the trees according to diameter in descending order and delete the smallest tree(s) from list until remaining number of trees are divisible by 3 i.e. if there are 17 trees in the plot, we should take larger 15 and divide them into tree groups. Within each group, randomly select 1, 2, 3 trees as per the minimum number of tree required for each treatment. All the trees should be cut 20 cm above the ground.

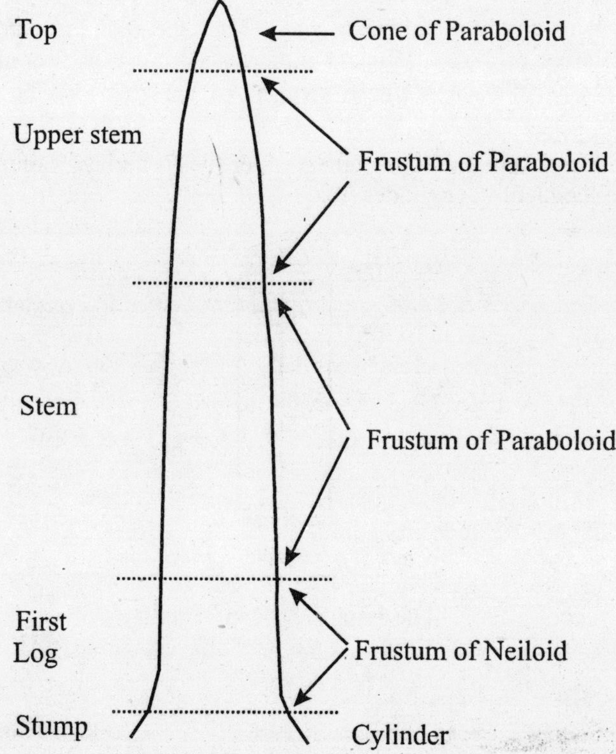

Fig. 42.1: Volume Formulas suggested by Portion of the Stem

MEASUREMENT OF HEIGHT

In forestry concept, surveyor can measure the approximate height of a standing tree with a transit, or accurate height using hypsometers, altimeters, abney's level, etc. but in case of destructive sampling when a tree is already felled and lying on the ground, a tree height measuring instruments is of course unnecessary, as its height can be measured directly with a measuring tape or a meter scale. This height is from base to maximum top in the canopy of the tree.

MEASUREMENT OF DIAMETER

Word diameter implies that trees are circular in cross section, but this is only approximation. In many cases, the section is much wider in one direction than in another. In such case, diameter at the long and short axes, average gives the exact diameter. When the cross section is irregular in shape, the convenient procedure is to find and measure the maximum axis and axis just right angle to the previous one. A tree caliper is required for the purpose. Another way is to measure the circumferences of the tree with a flexible tape. The circumference divided by (3.142) gives the diameter.

DIAMETER AT BREAST HEIGHT (DBH)

In case of standing tree, a standard position above than base diameter is desirable and is arbitrarily defined as diameter at 1.37 m above the ground. When there are more than one stem at 1.37 m point, the dbh is calculated using the equation: dbh = SQRT $(d_1^2 + d_2^2 + d_3^2 \ldots)$ when d_1, d_2, d_3, ... are the diameters of different stems at 1.37 m height.

TREE VOLUME ESTIMATION

Tree volumes can be estimated in a number of ways. Most methods estimate different portions of the tree volume using different formulas. Fig.42.1 indicates the most commonly used formula for each portion of a tree. Volume of each of these sections of the stem can be estimated using specific equations.

CYLINDER: The cylinder formula is the simplest of all the formulas. It requires only the area of one end ($A_b = \pi r^2 h$) and the length of the cylinder (h).

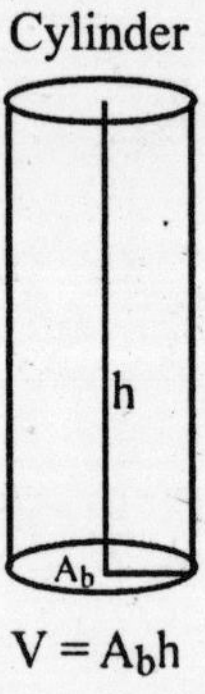

Fig. 42.2: Volume of Cylinder

CONE

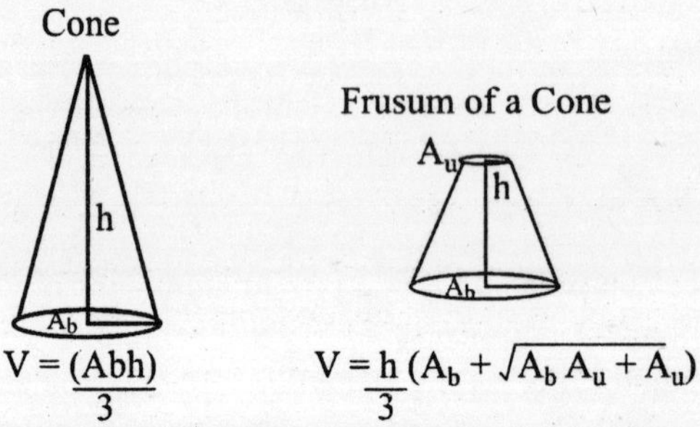

Fig. 42.3: Volume of Cone and Frustum of a Cone

PARABOLOID

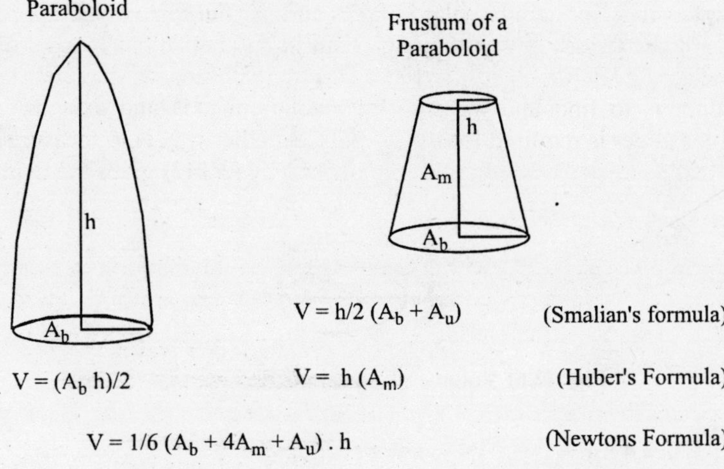

Fig. 42.4: Volume of Paraboloid and Frustum of a Paraboloid

NEILOID

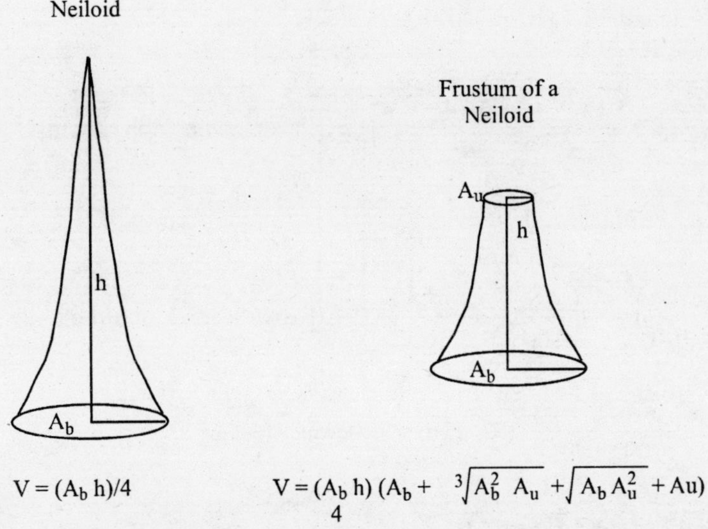

Fig. 42.5: Volume of Neiloid and Frustum of a Neiloid

VOLUME CALCULATION OF SECTIONAL PIECE OF MAIN BOLE

For estimating the volume, observation of lower (d_l), middle (d_m) and upper (d_u) diameter is recorded for every two-meter piece of bole of tree. In general, Newton's Formula (Husch et al 1982) is used for estimating volume.

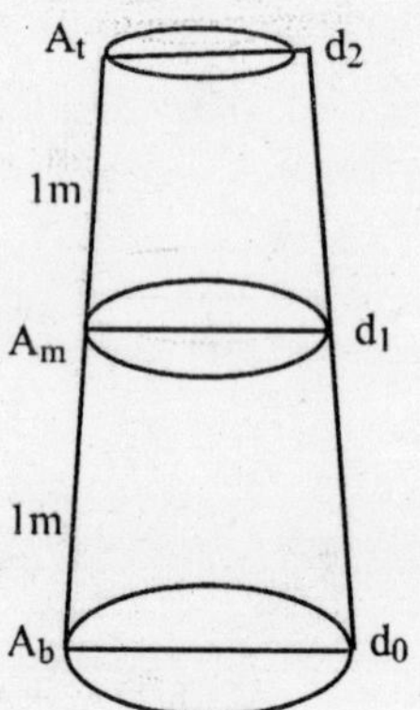

Fig. 42.6: Volume of Sectional Piece of the Main Bole

$$Volume = \frac{1}{6}(A_l + 4A_m + A_u) * h \text{ [Newton's Formula]}$$

$$or \quad V = \frac{1}{6}(\pi r_l^2 + 4\pi r_m^2 + \pi r_u^2) * h$$

$$= \frac{1}{6}\left(3.14 * \left(\frac{d_l}{2}\right)^2 + 4 * 3.14 * \left(\frac{d_m}{2}\right)^2 + 3.14 * \left(\frac{d_u}{2}\right)^2\right) * h$$

$$= \frac{1}{6}\left(\frac{3.14}{4} * d_l^2 + 4 * \frac{3.14}{4} * d_m^2 + \frac{3.14}{4} * d_u^2\right) * h \text{ ['h' in meter, 'dbh' in cm]}$$

$$= \frac{1}{6} * 0.784 * (d_l^2 + 4d_m^2 + d_u^2) * h$$

$$= \frac{1}{6} * 0.784 * \left(\left(\frac{d_l}{100}\right)^2 + 4\left(\frac{d_m}{100}\right)^2 + \left(\frac{d_u}{100}\right)^2\right) * h \text{ [Now 'h' and 'dbh' both are in mts]}$$

$$= \frac{1}{6} * 0.784 * \left(\frac{d_l^2 + 4d_m^2 + d_u^2}{100 * 100}\right) * h \text{ [1m = 100cm]}$$

$$= \frac{h}{6} * 0.0000784 * (d_l^2 + 4d_m^2 + d_u^2)$$

$$= \frac{2}{6} * 0.0000784 * (d_l^2 + 4d_m^2 + d_u^2) \text{ [For our case, h = 2m, i.e. log of 2mts]}$$

$$= \frac{1}{3} * 0.0000784 * (d_l^2 + 4d_m^2 + d_u^2)$$

$$Volume = 0.0000261 * (d_l^2 + 4d_m^2 + d_u^2) \text{ [where } d_m = \frac{d_l + d_u}{2}\text{]}$$

VOLUME CALCULATION OF UPPER STEM OF TREE

For calculating the volume of upper end of tree, the formula for paraboloid is used which is *Volume* = (0.00003927) * $(d^2 l)$

$$\text{Volume of Paraboloid} = \frac{1}{2} * A * l$$

$$= \frac{1}{2}(\frac{\pi}{4} * d^2) * l \qquad [\text{Area} = \pi r^2]$$

$$= \frac{1}{2}(\frac{3.14}{4} * d^2 * l)$$

$$= \frac{1}{2}(0.784 * d^2 * l) \qquad [l \text{ in mts, d in cms}]$$

$$= \frac{1}{2}(0.784 * (\frac{d}{100})^2 * l) \quad [1\text{m} = 100\text{cm}]$$

$$= \frac{1}{2}(0.0000784 * d^2 * l)$$

$$= 0.00003927 * d^2 * l$$

Where 'd' is the lower diameter and 'l' is the length of upper end of tree

INDEX/PRODUCTIVITY CALCULATION

The formula for index calculation

Index = (volume of one tree) * (number of tree in one hectare)

Since 1 hectare = 10,000 m^2

Number of trees in one hectare = 10,000/(spacing m^2)

$$\text{Volume Index} = \frac{\text{Volume } (m^3/\text{tree}) \times 10000}{\text{Spacing (m x m)} \times \text{Age (yrs)}} \quad (\text{unit: } m^3/\text{ha/yr})$$

$$\text{Biomass Productivity} = \frac{\text{Biomass (kg/tree)} \times 10000}{\text{Spacing (m x m)} \times \text{Age (yrs)} \times 1000 \text{ (to convert into tons)}}$$

$$\text{Biomass Productivity} = \frac{\text{Biomass (kg/tree)} \times 10}{\text{Spacing (m x m)} \times \text{Age (yrs)}} \quad (\text{unit: tons/ha/yr})$$

BIOMASS COMPONENTS

To evaluate useful biomass yields in an experiment, separate harvested biomass components of interest may be classified into well-defined size classes and can be converted into dry weight. Sampling of components within each size class has to be done for using it to estimate biomass production of each component. In deciding, to partition biomass component, one can partition the trees according to the end use and market requirements for different products and economic feasibility of the trees species being assessed. Relevant practices use branch size categorization as:

- < 1cm diameter (small branches, twigs, foliage etc.)
- 1 cm and < 5cm (small branches for fuel)
- 5 cm and < 10 cm (large branches for fuel and stems for timber)

MOISTURE CONTENT AND OVEN DRY WEIGHT

Moisture content is an important variable to define and determine carefully. Over dry weight is strongly recommended. The main advantage of oven dry weight is that, it permits the reliable comparison of materials with greatly differing moisture contents and is a widely used means of reporting wood weight. To convert wet weight to total dry weight, we calculate,

$$\text{Total Dry Weight} = \frac{\text{Sample Dry Weight}}{\text{Sample Fresh Weight}} \times \text{Total Fresh Weight}$$

A drying temperature of 50-103°C is recommended for most of the samples, depending upon the objectives. If the object is to determine the oven dry weight of woody biomass, 103°C (or just below the boiling point of water) is recommended. Leafy biomass, litter and twigs with less than 1 cm diameter normally reach their constant weight after 24 hours and dried at a temperature above 80°C. Woody material with a diameter more than 1 cm and less than 5 cm requires about 48 hours. Larger material will require even more time. Sample should be approximate 1 per cent of the total weight. 100 g sample for moisture content in bush or small tree is minimum and may be more in case of larger trees. To obtain adequate representation of variation with in tree, wood samples from different parts of tree within each size category may be mixed to give over all moisture content value.

TREE GROWTH AND YIELD MODELLING STRATEGIES

The above model classification of agroforestry models depicts a clear picture of available models, however an agroforestry model basically involve two components namely tree and crop. The tree component being long duration and requiring prediction of tree growth/yield at an early stage, more emphasis has to be accorded to the tree component and also to tree growth modelling. Tree growth models can be classified on the basis of growth function being used (linear or non-linear), approach of modelling (classical or simulation) and finally model optimization (simple R^2 criterion or thorough statistical validation and refinement).

The existing tree growth modelling research in our country may be broadly classified into four broader categories viz. empirical; linear; sigmoid; and quadratic/ parabolic. Empirical models, the simplest of all kinds involves maximizing the R^2 values of the fitted model by merely increasing the variants of independent variable(s). The linear models may suffer from the problem of negative estimation of size for lower range of explanatory variate (Ajit et al 2000) that can be settled by imposing lower bound on explanatory variate. The sigmoid models lead to constant estimation of size particularly for the higher range of explanatory variate (Ajit et al 1999) that may be tackled by using either a very wide range of explanatory variate or simulation modelling approach. The quadratic/parabolic models forbear with the illogical estimation of size particularly in the middle range of explanatory variate that may be resolved by defining the acceptable range of the explanatory variate. By "Negative Estimation" refers to the predictive values of size/dependent variable (viz. biomass or volume of tree) using the developed model that comes out to be negative for a specific range of explanatory/independent variate. "Constant Estimation" refers to case where the value of the explanatory variable (usually dbh, height, etc.) increases but the predictive values of the size remain almost constant. "Illogical Estimation" means that the predictive values of size decrease with the increase in the values of explanatory variate.

LINEAR FUNCTIONS

Linear tree growth models employ linear regression to describe the change in the size/response variate with respect to explanatory variate. The response variate may represent tree volume, biomass, etc. and the explanatory variate includes age, total height (H), diameter at breast height (DBH), or a combination of these variables e.g. D^2H, to describe the change in response variable (Y) with respect to explanatory variable (X).

The use of linear model is very common in forestry/agroforestry (Kumar et al 1996; Jain et al 1998; Chaturvedi and Sood 1995). Basically, these models provide the mathematical building blocks of simple/complex biological growth.

PROBLEM IDENTIFICATION AND SOLUTION

The statistical model employed in linear regression is of the form Y = a + b*X + e, where Y is the response variate, X the explanatory variate, 'a' and 'b' are regression coefficients and 'e' the error term. Statistical procedure to estimate 'a' and 'b' assume that the error term is normally distributed with mean zero, constant variance and independent between cases (Drapper and Smith 1982). The corresponding mathematical equivalent of simple linear model is the straight line Y=a+b*X, where 'a' denotes the intercept on Y-axis, and 'b' the slope of the line (b=tan θ), where θ is the angle which this line makes with the +ve direction of X-axis. The method of least squares is used to fit this straight line and to estimate the parameters 'a' and 'b' using the given data set. There are numerous graphical and statistical methods to support the fitting of the equation and consequently model validation, viz. the plot of residuals against their expected values; the autocorrelation plot of residuals; the plot of residuals against the estimate; the plot of residuals against the independent variate (Mayer and Butler 1993; Janssen and Titus 1995; Soares et al 1995). All of these are based on the comparison of observed and estimated values and the diagnosis of residuals. Once these statistical assumptions are fulfilled, the linear model may be used for predictive purposes.

However, in true sense the biological growth process is not linear, which is the basic assumption in linear models. Due to this reason, simple linear growth models fail to precisely predict the change in the response variable, particularly for the lower range of explanatory variable. For this range, these models sometimes estimate misleading values and even vague, viz. negative values of response variable. This problem associated with linear tree growth models has been addressed as 'negative estimation' of tree size. Thus, the linear model reflects a slightly unrealistic picture of tree growth, which is sigmoid in nature. Growth of the tree results from two opposing factors: firstly the positive component associated with biotic potential, photosynthesis activity, absorption of nutrients, constructive metabolism, etc. and secondly the negative component representing the restraints imposed by competition, limited resources, stress, respiration and ageing factors. These two components when combined together results in the sigmoid shape of tree growth.

The apparent reason for 'negative estimation of size' with linear models (Gupta et al 1996; Jain et al 1996; Jain et al 1998), even with satisfactory R^2 values is that the data set utilized to fit the regression equation represents that range of explanatory variable, belonging to the second phase of growth and which is linear in nature. However, when we use this linear equation for interpolation/extrapolation particularly for lower range the results of prediction are not encouraging and sometimes vague, viz. negative estimation of size. In the last phase (third) they lead to overestimation of size.

The apparent reason for high R^2 values associated with linear models is that the data set usually belongs to the second phase of tree growth and which is linear in nature. However, reckoning into the prediction capabilities of these models, the results are not encouraging. For the initial (first) phase of tree growth they result in underestimation and even negative estimation and for the last phase (third) they lead to overestimation of size. It is therefore recommended that linear models, in context of tree growth, should be used with caution and in case the fitted simple linear equation results in negative value of 'a' and positive value of 'b', then the lower bound of explanatory variable should be computed as proposed here, e.g. $X > a/b$ to avoid the negative estimation of tree size. Thus when the phenomenon is inherently undesirable, as the negative estimation of size in simple linear models for lower range of explanatory variable, to put implied lower bound on the raw values is most appropriate. To avoid the negative estimation problem in published tree models in India, a lower bound proposed on the values of the explanatory variate, i.e. D or D^2H as the case may be. It is suggested that the equations must not be used, to the prediction purposes, for the values of the explanatory variate below the lower bounds since in that case it will predict negative values of dependent variate i.e. volume/ biomass.

NON-LINEAR FUNCTIONS

Functions other than linear (linear means x to the power one only) may be broadly classified as non-linear. Non-linear functions are in general more powerful than linear ones. The published non-linear tree models can be broadly categorized into three main groups, namely allometric, parabolic and sigmoid. Accordingly nonlinear equations has been classified and presented below into three categories.

ALLOMETRIC/ POWER FUNCTIONS

Allometry is used to describe the morphological evaluation of species, and is based on the relation between a tree biomass, volume and of any part of the tree component. The general form of the allometric equation is $\mathbf{y = ax^b}$ where **y** is a measure/process in question viz. volume/biomass, **x** is size (usually DBH), **b** is the allometric exponent (which tells about the relationship between x & y), and **a** = constant (the allometric coefficient). Allometric equation, that compensates for nonlinear functions is: $\log(y) = \log(a) + b\,[\log(x)]$. Allometric modelling may be theoretically, physiologically, and statistically superior to alternative methods of modelling used in tree biomass estimation. These equations behave reasonably well within the observed range. However they lead to steep overestimation just outside the observed range of independent variable.

SIGMOID FUNCTIONS

Growth functions, whose curves are S-shaped, are called sigmoid functions. These curves can be divided into three regions; In the first one, the growth pace is very fast i.e exponential and it slows down in the second phase to follow linear and in the last phase it again considerably goes down and follow monomolecular. Altogether these three phases takes the shape of S-distribution. The prominent among them are the equations of exponential decay (Gompertz 1825) for the age distribution in human population, logistic equation (Verhulst 1838), Richards function (Richards 1959; Richards 1969), etc. The sigmoid growth functions have a lot of potentialities that reflect their suitability in tree growth models. To mention a few: most of the parameters of these functions have biological interpretation, viz. the parameter 'a' represents the asymptotic value of the response variable, 'b' narrates the rate of approach to the asymptote, 'c' is the shape parameter. However, the parameters 'p' and 'q' have as such no direct biological interpretation. Generally, these functions lead to statistically unbiased and efficient models of tree growth. They mimic, the complex tree growth process, more precisely and more accurately and parameters of these models have mathematical description and biological interpretation. However these models are not transportable to other data sets for extrapolative predictions since the asymptotic value of predicted size is approximately equal to experimentally observed maximum value of size. For these models theoretically $y \rightarrow C$, the asymptotic value as $x \rightarrow \infty$, but practically y→C even slightly outside the actual observed range of 'x' and much before the infinite value of 'x'.

While fitting the sigmoid growth models to the data, we may sometimes encounter the problem of 'constant estimation of tree size' depending upon the observed range of explanatory variate. Smaller the observed range of explanatory variate, greater is the severity of problem i.e. more is the error of prediction in extrapolated zone. The term 'constant estimation of size' means that when we use the fitted sigmoid models for extrapolation, particularly for higher range of explanatory variable, the predicted values of the response variable comes out to be almost uniform. In other words, as the value of explanatory variable increases, the value of the predicted response increases, but the pace of increment is almost negligible and it ultimately leads to nearly constant estimation of the predicted response. The apparent reason for 'constant estimation' problem is the small observed range of explanatory variate. When the observed range of explanatory variable is large enough, there is no such problem of constant estimation of size with sigmoid functions and due to this reason, for tree growth models with height or DBH as response variate and time as explanatory variate, we may avoid this problem since it is possible to record such data for a long duration. However, for tree yield models which necessitate destructive sampling with timber volume or biomass as response variate and DBH or height as explanatory variate, one would have to wait for years together to harvest trees at different intervals for obtaining a large observed range of explanatory variable in systematically planned experiments for such studies. In usual plantation forestry/agroforestry experiments, the trees are in general more or less of the same age group, and that is why in a single harvesting, there is practically no variability in the observed values of the explanatory variate and which ultimately leads to 'constant estimation' problem while fitting sigmoid functions with such data.

It is worth mentioning here that linear/curvilinear/allometric models some times may lead to impractical predictions (even negative estimation) for lower range of explanatory variable and sigmoid models result in unrealistic predictions (constant estimation) for higher range. However, one of the possible solutions to overcome the 'constant estimation' intricacy is to use the concept of simulation to generate the data outside the observed range and then fitting the sigmoid function to the generated data. The proposed approach, a combination of curvilinear and sigmoid functions, overcomes the drawback of both the models to a certain extent and results in more realistic predictions for lower and higher ranges. The reason being that the generated data set now includes the feasible 'asymptotic value' of size, which a tree may attain.

PARABOLIC FUNCTIONS

A power function with degree two is known as parabolic function. The problem observed with parabolic functions of the type $Y = bX^2$ is again illogical estimation intricacies. For the initial range, as the DBH increases the biomass/volume first decreases from a certain medium values and then it again increases with increasing DBH. In fact, in the function $Y=bX^2+a+cX$ the positive constant term (+a) and the positive linear term (+cX) result in minimizing/decreasing the estimation error term where as the negative constant term (-a) and the negative linear term (–cX) leads to maximizing/increasing the estimation/prediction error.

TREE HEIGHT AND VOLUME EQUATIONS

Because tree height measurement is costly, it is common to measure all trees for diameter but to measure only a sample of trees for height and then to use a model based on these data to predict tree height for all trees. Tree height is necessary as tree volume (V) is commonly derived from a function of diameter (DBH) or (D) and tree height (H), i.e. (V = f[D,H]). A wide range of height equations are available in literature and Antony Joseph Raj (2012a) has suggested following tree height equation forms to estimate the height of trees using diameter values ('H' denotes height and 'D' denotes DBH):

$H = b_0 + b_1D$ (Linear)

$H = 1.3 + b_1D + b_2D^2$ (Quadratic)

$H = b_0 + b_1\ln(D)$ (Log linear)

$H = b_0 + b_1/D^{b_2}$ (Inverse linear)

$H = b_0D^{b_1}$ (Power)

$H = b_0 \exp(b_1/D)$ (*Exponential*)

$\ln(H) = b_0 + b_1\ln(D)$ (Log - log)

$\ln(H) = b_0 + b_1/D$ (Log inverse linear)

$1/\{(H-1.3)^{0.4}\} = b_0 + b_1/D$ (Pettersen Curve)

$D/\{(H-1.3)^{0.4}\} = b_0 + b_1D$ (*Alternative Pettersen Curve*)

Similarly, there is need to predict tree volume or stand volume and one way is to develop tree volume equations that just require tree diameter (D) and either tree height (H) or upper stand height. Antony Joseph Raj (2012a) has suggested following tree volume equation forms based on DBH alone and based on DBH and height:

Commonly used typical local volume equation forms (based on DBH)

$$Y = b_0 + b_1D^2$$

$$Y = b_0 D^{b_1}$$

$$Y = b_0D + b_1D^2$$

$$\ln(Y) = \ln(b_0) + b_1 \ln(D)$$

$$\frac{Y}{D} = b_0 + b_1D$$

The parameters b_0 and b_1 may be estimated by simple linear regression.

Commonly used general volume equation forms (based on DBH and height)

$$Y = b_0 + b_1D^2H$$

$$Y = b_0 + b_1D^2H + b_2D^2 + b_3H$$

$$Y = b_0 + b_1D^2 + b_2H + b_3D^2H$$

$$Y = b_0 + b_1 (D - b_2)^2 (H - b_3)$$

$$Y = b_0 + b_1 (D^2 - b_2)(H - b_3)$$

$$Y = b_1 D^{b_2} H^{b_3}$$

$$\ln(Y) = b_1 + b_2 \ln(D) + b_3\ln(H)$$

TREE GROWTH MODELLING WEBSITE IN INDIAN PERSPECTIVE

Tree growth modelling website has been designed with the basic objective of providing a platform where the user can have the basic concepts used in development of tree growth models at one end and the comprehensive collection of already developed model for a specific tree species at other end so as to use them directly for prediction purposes. At the initial phase of development of this site, Eucalyptus species modelling have been concentrated. This website has been hosted at the server of IASRI (Indian Agricultural Statistics Research Institute), New Delhi and can be accessed at the URL: http://mirror.iasri.res.in/net/tgm/index.htm. This site is a unique collection of published biomass and volume equations, on Eucalyptus off course in Indian context only. The general portion of this site describes definition of modelling by various authors and principles of modelling; types of modelling describing about various types of models like descriptive models, prescriptive models, predictive models and simulation models; model development process depicting various steps of model development process like model estimation and model validation; tree growth modelling techniques describes essential steps of modelling tree growth like sampling of trees, measurement of height, diameter measurement, volume estimation, height measurement, estimation of biomass components, etc.; models relevant to agroforestry provides the working outlines of the models relevant to agroforestry worldwide and finally have added tree growth modelling publication displaying a compilation of publication on tree growth modeling along with full length papers available in pdf format for downloading. This website has been designed to help for organizing, managing and sharing research information on Eucalyptus modelling in India. HTML, DHTML have been used for creating the dynamic pages of the database at the front end and the source codes written in HTML for designing the home page of Tree Growth Modelling Web site. The unified database has been equipped with interactive web pages to browse information on Eucalyptus introduction, modelling introduction, types of modelling, model development process, detailed information on biomass and volume models in India related to Eucalyptus species (Ajit and Nighat 2008).

CONCLUSION

This chapter on tree growth modelling discusses at length the perspectives (*which model?* i.e. the scope of agroforestry models), needs (*why modelling?* i.e. is it necessary to go for modelling?) and issues (*how modelling?* i.e. the artistry and intricacies of modelling along with validations approaches). Modelling strategies (i.e. simplification or idealization of problem, construction/ formulation of the biological problem into mathematical equations/functions, estimation of parameters, optimization/ refinement of the model and finally the validation of the model) has been elaborated in detail. The pros and cons of tree growth modelling have been elaborated and the limitations of models (linear models - negative estimation; sigmoid models - constant estimation; and parabolic models - illogical estimation) along with the possible solutions have been discussed (simple linear models - lower bound on explanatory variate; sigmoid model - simulation modelling approach; and parabolic model - permissible prediction range). There is no doubt in accepting that even with the investment of huge money by many institutes in both research and extension, it is still difficult to provide predictive insight into which type of tree/crop combination is appropriate for a specific location and a particular farmers objective. However, it is endorsed that agroforestry models can provide a deep sagacity in the tree growth pattern and the effect of tree component on crop yield (reduction/promotion) in a temporal framework.

■■■■

43 REMOTE SENSING AND GIS APPLICATION IN AGROFORESTRY

INTRODUCTION

Human being was and is directly or indirectly dependent on natural resources specifically soil, water and vegetation (plants) for survival as well as luxurious life. Forest as part of the natural resource covers a large of land globally and contributes a wide range of products and services from fuelwood up to stabilizing the whole ecosystem through carbon sequestration. This very important resource has been serving human being from the beginning of human survival on the earth as food, shelter, raw material of construction, etc. However, due to its free availability and abundance at the beginning, it was considered as infinite resource and hence no protective measure was taken against its diminishing with time. Later after the shortage of the resource became critical, natural resource scientists and researchers started to device different approaches and mechanisms by which to combat the ever increasing deforestation and shortage of forest products and one among the different approaches, currently used in countries with agricultural community such as most of the developing countries, is agroforestry.

Agroforestry is the production of trees and non-tree crops or animals on the same piece of land by arranging the different components either spatially or temporally. It is the integration of trees, plants, and animals in conservative, long-term, productive systems. The main reason why agroforestry system is very important at this time is that agroforestry systems make maximum use of the land and every part of the land is considered suitable for useful plants. Emphasis is placed on perennial, multiple purpose crops that are planted once and give benefits over a long period of time. Such benefits include construction materials, food for humans and animals, fuels, fibres, and shade. Trees in agroforestry systems also have important uses such as holding the soil against erosion and improving soil fertility. With increase in extent and variety, management becomes difficult, time taking and expensive unless improved technologies are applied to facilitate the situation. Thanks to the advance in technology, nowadays, resources that are spatially extended on a wide area of land like the forests are being properly managed with the help of remote sensing and GIS.

REMOTE SENSING TECHNIQUES AND COMPONENTS

Remote sensing is the name for a group of techniques of collecting information about an object and its surroundings from a distance without physically contacting the objects. In general, this result in some form of image which is further processed and interpreted to extract useful information for some purpose. It can also be defined as the science (and to some extent, art) of acquiring information about the Earth's surface without actually being in contact with it. This is done by sensing and recording reflected or emitted energy and processing, analyzing, and applying that information. Remote sensing comprises aerial photography, as well as microwave sensing and multi-spectral scanning. It usually relies upon measurement of electromagnetic radiation reflected or emitted from the features of interest. We can make this definition a bit narrower by including that the object be investigated is located on or near the earth's surface, and the sensing instrument is more or less above the object, and at a 'substantial' distance from it. And the information is carried by electromagnetic radiation, some property of which is affected by the remotely sensed object.

The definition of remote sensing shows that there are two main subdivisions: data acquisition and information extraction. The theoretical basis for data acquisition lies primarily within the domain of the physical and mathematical sciences whereas information extraction is based primarily on the mathematical, statistical, biological and psychological sciences. However, the nature of the data used in that analysis depends upon the scientific basis and technological characteristics of the sensor and acquisition conditions. In a simple term every human being are actively involved in "Remote Sensing" all the time. Familiar activities like watching a football game or observing birds flying on the sky are actually remote sensing activities. Remote sensing deals with the detection and measurement of phenomena with devices sensitive to electromagnetic energy such as:

- Light (cameras and scanners)
- Heat (thermal scanners)
- Radio Waves (radar)

COMPONENTS OF REMOTE SENSING

- *Energy Source or Illumination* (A) - The first requirement for remote sensing is to have an energy source which illuminates or provides electromagnetic energy to the target of interest.
- *Radiation and the Atmosphere* (B) - As the energy travels from its source to the target, it will come in contact with and interact with the atmosphere it passes through. This interaction may take place a second time as the energy travels from the target to the sensor.
- *Interaction with the Target* (C) - Once the energy makes its way to the target through the atmosphere; it interacts with the target depending on the properties of both the target and the radiation.
- *Recording of Energy by the Sensor* (D) - After the energy is reflected by, or emitted from the target, we require a sensor (remote - not in contact with the target) to collect and record the electromagnetic radiation.
- *Transmission, Reception and Processing* (E) - The energy recorded by the sensor has to be transmitted, often in electronic form, to a receiving and processing station where the data are processed into an image (hardcopy and/or digital).
- *Interpretation and Analysis* (F) - The processed image is interpreted, visually and/or digitally or electronically, to extract information about the target, which was illuminated.
- *Application* (G) - The final element of the remote sensing process is achieved when we apply the information we have been able to extract from the imagery about the target in order to better understand it, reveal some new information, or assist in solving a particular problem (Kumar 2005).

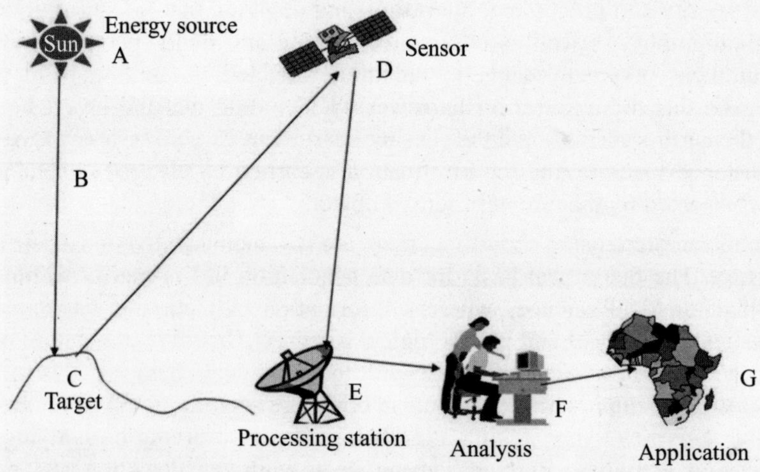

Fig. 43.1: Remote Sensing Process and Components

REMOTE SENSING PROCESSES

Two basic processes involved in electromagnetic remote sensing of earth resources are: data acquisition and data analysis. The elements of the data acquisition process are energy sources, propagation of energy through the atmosphere, interactions of energy with earth surface features, retransmission of energy through the atmosphere, airborne and/or space borne sensors, the generation of sensor data in pictorial and/or digital form. This in short means that we use sensors to record variations in the way earth surface features reflect & emit electromagnetic energy.

The data analysis involves examining the data using various viewing and interpretation devices to analyze pictorial data and/or a computer to analyze digital sensor data. Reference data about the resources being studied are used where available to assist in the data analysis which could help the analyst to extract information about the type, extent, location, and condition of the various resources over which the sensor data were collected. This information is then compiled in the form of hard copy maps and tables or as computer files that can be merged with other "layers" of information in Geographic Information System (GIS). At the end, the result is presented to planners or decision maker who will apply it to their decision making processes.

TYPES OF REMOTE SENSING

Remote sensing can be categorized as a system in different ways. Among these the most common are:

BASED ON SOURCE OF ENERGY

In any of its form, electromagnetic energy is very important for remote sensing and without it remote sensing cannot succeed. Based on source of energy, remote sensing is classified into active systems and passive systems.

ACTIVE SYSTEMS: These systems supply their own source of energy (radiation) to illuminate features of interest. They generate their own energy and then record the reflection of that energy from the earth's surface. They are "active" in the sense that they provide their own source of energy and are independent of solar and terrestrial radiation. Because they sense energy provided by the sensor itself, such instruments have the capability to operate at night and during cloudy weather. Imaging radars, which transmit a microwave signal towards the earth's surface from an aircraft or satellite and then use the reflected energy to form an image, are examples of active systems. Another very good example of an active system is a camera utilizing flash during photography.

PASSIVE SYSTEMS: These are systems that sense naturally occurring (emitted thermal or reflected solar) radiation. These systems don't supply their own illumination but rather sense the naturally available energy. Sensors of this system depend upon radiation derived from the sun, rather than the sensor itself.

The choice between active and passive remote sensing systems is influenced by a number of factors. A passive system will be inappropriate at certain wavelengths at which insignificant amounts of radiation occur naturally. An active system may be technically infeasible if the amount of power which has to be radiated in order to obtain a measurable reflected signal is too high.

BASED ON WAVELENGTH REGIONS

Based on the wavelength region they operate, remote sensing systems can be categorized into three as:

- Visible and Reflective Infrared Remote sensing
- Thermal Infrared remote sensing
- Microwave Remote sensing

The energy used in visible and reflective infrared remote sensing is the sun. The sun radiates electro-magnetic energy with peak wave length of 0.5μm. Remote sensing data obtained in the visible and reflective infrared mainly depends on the reflectance of objects on the ground surface. Therefore, information about objects can be obtained from the spectral reflectance. The source of radiant energy used in thermal infrared remote sensing is the object itself, because any object with a normal temperature emits electromagnetic radiation with a peak at 10μm.

BASED ON THE PLATFORM USED

In remote sensing, we use sensors to detect and record electromagnetic energy of the target. Depending on the visibility of the target; the obstructions expected and the coverage required, the sensors need to be placed at different range with respect to the target object. Based on this, the commonly used remote sensing systems can be categorized into three as:

GROUND BASED: Sometimes sensors (Cameras) are placed on fixed poles, towers and vehicles to record data for some time of a certain area of interest.

AERIAL (*AIRCRAFT*): Aircraft of various types provide an exceptionally convenient platform for remote sensing. With a suitable choice of vehicle, a range of heights can be covered from a few dozen meters to many kilometres above most of the earth's surface. The range of speeds available is more or less continuous from zero (hovering helicopter) to several hundred meters per second. The choice of operational conditions for an airborne sensor may be determined by intentional operational decisions, and the effects of errors and perturbations in the motion.

SPACE (*SATELLITES*): Placing a satellite in orbit above the earth is clearly more expensive than mounting an airborne remote sensing campaign. However, they are advantageous in terms of increased platform speed and swath width, as well as continuity of mission. In general, the data coverage obtained from a satellite mission is better than that obtainable from an airborne mission, and the fact that a satellite may continue to function for five years means that the data are also homogeneous. Satellite system with detectors which produce digital data may also be preferred because:

- The digital data can be transmitted over radio or microwave communications and stored on magnetic tapes, so that it is easily processed and analyzed by a computer.
- The satellite is in orbit around the earth so that the same area can be covered on regular basis for change detection.
- Once the satellite is launched, the cost for data acquisition is less than that for an aircraft data.
- Satellites have relatively very stable geometry, meaning that there is less chance for distortion or skew in the final image.

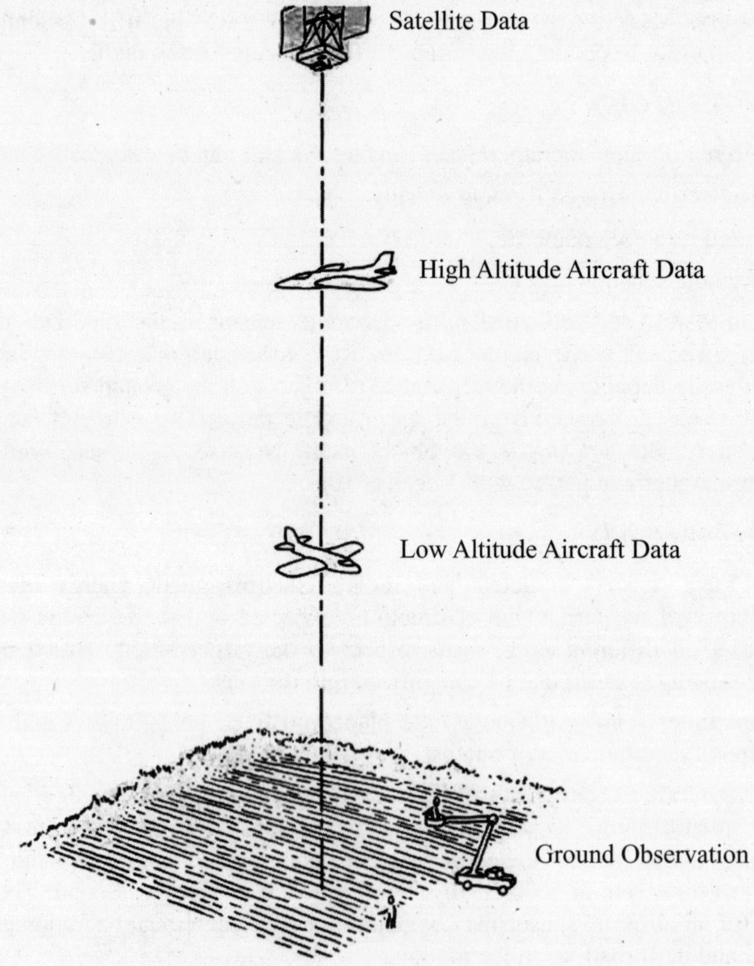

Fig. 43.2: Multistage Remote Sensing Concept

PHOTOGRAMMETRY AND AERIAL PHOTOGRAPH

Photogrammetry is defined as the art, science, and technology of obtaining reliable information about physical objects and the environment through processes of recording, measuring, and interpreting photographic images and patterns of recorded radiant electromagnetic energy and other phenomena. Two basic classifications of photography used in the science of photogrammetry are terrestrial and aerial. Terrestrial photographs are taken with ground based cameras, the position and orientation of which are often measured at the time of exposure.

Aerial photography is commonly classified as either vertical or oblique. Vertical photos are taken with the camera axis directed as nearly vertically as possible. In practice, there are no truly vertical photographs because the camera axis is rarely held perfectly vertical due to unavoidable aircraft tilts. Most aerial photographs are classified as either oblique or vertical, depending on the orientation of the camera relative to the ground during acquisition. Oblique aerial photographs are taken with the camera pointed to the side of the aircraft. High oblique photographs usually include the horizon while low oblique photographs do not. Oblique photographs can be useful for covering very large areas in a single image and for depicting terrain relief and scale. However, they are not widely used for mapping as distortions in scale from the foreground to the background preclude easy measurements of distance, area, and elevation. Vertical photographs taken with a single-lens frame camera is the most common use of aerial photography for remote sensing and mapping purposes.

The amount of detail shown in an aerial photograph is dependent, among other things, on the scale of photograph. "Map scale" is ordinarily interpreted as the ratio of a map distance to the corresponding distance on the ground. In a similar manner, a "photograph scale" is an expression that states that one unit /any unit of distance on a photograph represents a specific number of units of actual ground distance. Scales may be expressed as unit equivalents, representative fractions, or ratios. For a vertical photograph taken over flat terrain, scale is a function of the focal length of the camera (f) used to acquire the image and the flying height above the ground (H′) from which the image was taken.

$$S = \frac{f}{H'}$$

Where,

S = scale

f = focal length of the camera

H′ = altitude of the aircraft above the terrain

Another way of expressing the above equation for scale is:

$$S = \frac{f}{H - h}$$

Where ,

S = scale

f = focal length of the camera

H = flight height above sea level

h = elevation of the terrain above sea level

AIR PHOTO INTERPRETATION

Photographic interpretation is the act of examining photographic images for the purpose of identifying objects and judging their significance. Aerial photographs contain a detailed record of features on the ground at the time of exposure. A photo interpreter systematically examines the photos and, frequently, other supporting materials such as maps and reports of field observations. Based on this study, an interpretation is made as to the physical nature of objects and phenomena appearing in the photographs. The process is not restricted to making decisions concerning what objects appear in photo-graphs; it also usually includes a determination of their relative locations and extents. This requires the application of at least some of the elementary photogrammetric measurement techniques discussed earlier. Success in photo interpretation varies with the training and experience of the interpreter, the nature of the objects or phenomena being interpreted and the quality of the photographs being utilized. Commonly, users or interpreters follow the following steps to properly interpret aerial photograph:

- *Detection*: Detection of a photographic image is possible when it possesses elements that differ from those of surrounding images. Tonal contrast plays a very important role in image detection.
- *Recognition*: The importance of the object is judged with respect to the objectives of the photo interpretation. No further attention is paid to images that are recognized as not directly interesting to the interpreter.
- *Delineation* of the selected images separates them from images of different nature. This is useful for purposes of mapping and measurement.
- *Measurements* on the images are essential for the identification of the objects. Also measurements allow classification of the examined features.
- *Deduction and Comparison* is the step preceding the identification. The information gathered from the images is compared with the knowledge of the interpreter.
- *Identification* of the objects through the analysis of the photo images usually is the exclusive purpose of the photo interpretation.
- *Classification* of the identified objects is necessary for reasons of economy and simplicity. The objects are grouped in a limited number of classes or strata.
- *Codification* of the delineated classes is necessary for the recording on the aerial photographs. This facilitates mapping and additional measurements.

A systematic study of aerial photographs usually involves several basic character-istics of features shown on a photograph. The exact characteristics useful for any specific task and the manner in which they are considered depend on the field of application. However, most applications consider the following basic characteristics:

Shape: It refers to the general form, configuration, or outline of individual objects. In the case of stereoscopic photographs, the object's height also defines its shape. Shape is very useful in recognizing objects from their photographic images. For example, a railroad is usually readily distinguished from a highway because its shape consists of long straight tangents and gentle curves as opposed to the curvy shape of a highway.

Size: Objects' size on photograph must be considered in the context of the photo scale. Objects can be misinterpreted if their sizes are not evaluated properly. A tukule, for example, might be misinterpreted as a heap of hay or crop if size is not considered.

Pattern: It relates to the spatial arrangement of objects. The repetition of certain general forms or relationships is characteristic of many objects, both manmade & natural, and gives objects a pattern that aids the photo interpreter in recognizing them. An outdoor drive in theatre, for example, has a particular layout and pattern of parking spaces that aid in its identification. Likewise, the ordered spatial arrangement of trees in an orchard is in distinct contrast to that of forest tree stands.

Tone (or hue): It refers to colour or relative brightness of objects on photographs. It is related to reflectance of light from objects. Water, which absorbs nearly all incident light, photographs black, where as a Portland cement concrete highway reflects a high percentage of light and consequently produces very light tones. Tonal patterns also vary according to the drainage condition of soils (higher tone areas will be drier, darker tone areas will be wetter). Without tonal differences, the shapes, patterns, & textures of objects could not be discerned.

Texture: It is the frequency of tonal change on the photographic image. Texture is produced by an aggregation of unit features that may be too small to be discerned individually on the photograph, such as tree leaves and leaf shadow. It is a product of their individual shape, size, pattern, shadow, and tone. It determines the overall visual "smoothness" or "coarseness" of image features. As the scale of the photograph is redu-ced, the texture of any given object or area becomes progressively finer and ultimately disappears. An interpreter can often distinguish between features with similar reflectance based on their texture differences. An example would be the smooth texture of green grass as contrasted with the rough texture of green tree crowns on medium scale air photos.

Shadows: They are important to interpreters in two opposing respects: (1) the shape or outline of a shadow affords an impression of the profile view of objects (which aids interpretation), and (2) objects within shadows reflect little light and are difficult to discern on photographs (which hinders interpretation). For example, the shadows cast by various tree species or cultural features (bridges, towers, etc) can definitely aid in their identification on apricots. Also, the shadows resulting from even subtle variations in terrain elevations, especially in the case of low sun angle photographs, can aid in asses-sing natural topographic variations that may be diagnostic of various geologic landforms.

Site: It refers to topographic or geographic location and is particularly important aid in the identification of vegetation types. For example, certain tree species would be expected to occur on well-drained upland sites, whereas other tree species would be expected to occur on poorly drained lowland sites. Also, various tree species occur only in certain geographic areas.

Association: It refers to the occurrence of certain features in relation to others. Naturally some features are found in association with others. For example, a church with its compound might be difficult if it is located in a forest, but would be easy to identify if it is sited in an area of low forest cover.

SATELLITE REMOTE SENSING

The Earth is constantly under observation by many satellites orbiting the planet and collecting data. They are engaged in an activity called "remote sensing", the act of obtaining information about something without being in direct contact with it. The satellite images, as well as the actual predictions, are obtained through remote sensing of the Earth. The satellites don't gather the information themselves; they simply orbit the Earth and provide platforms from which the sensors can observe large areas of the surface. Airplanes also provide platforms for remote sensing and some sensors operate from land. Remote sensing satellites are launched by government agencies of many countries and are usually equipped with sensors that serve a particular purpose. Satellites have been providing multispectral images of the Earth continuously since the early 1970's. A unique 40 years data record of the Earth's land surface now exists. This unique retrospective portrait of the Earth's surface has been used across disciplines to achieve improved understanding of the Earth's land surface and the impacts of humans on the environment. Satellite data have been utilized in a variety of government, public, private and national security applications. Examples include land and water management, global change research, oil and mineral exploration, agricultural yield forecasting, pollution monitoring, land surface change detection, cartographic mapping, etc.

Satellite based remotely sensed digital data have been used to map forest and tree resources since the inception of the LANDSAT satellite programme in 1972. Satellites provide synoptic images and homogeneous data, which can be geographically registered over time and which can be therefore an efficient tool for providing high quality forest management information. Since the 1980s, with the availability of scanners, improve-ments in computer software and development of image processing algorithms, there have been many studies of remote sensing applications in forestry and agroforestry (Satheesh and Mohanraj 2012).

WHAT IS GEOGRAPHIC INFORMATION SYSTEM (GIS)?

There is no clear-cut definition for Geographic Information System (GIS). Different people defined GIS according to capability and purposes for which it is applied. GIS is "a computer assisted system for capture, storage, retrieval, analysis and display of spatial data, within a particular organization" (Clarke 1986). GIS is "a powerful set of tools for collecting, storing, retrieving at will, transforming and displaying spatial data from the real world" (Burrough 1987). It is "an information technology which stores, analyzes and displays both spatial and non-spatial data" (Parker 1988). However, GIS can be defined more commonly as "a computerized information storage, processing and retrieval systems that have hardware and software specially designed to cope with geographically referenced data and corresponding attribute information (tables, charts and statistics)". A more comprehensive and easy way to define GIS is the one that looks at the disposition, in layers (Figure 43.3), of its data sets. It is a "group of maps of the same portion of the territory, where a given location has the same coordinates in all the maps included in the system". This way, it is possible to analyze its thematic and spatial characteristics to obtain a better knowledge of this zone.

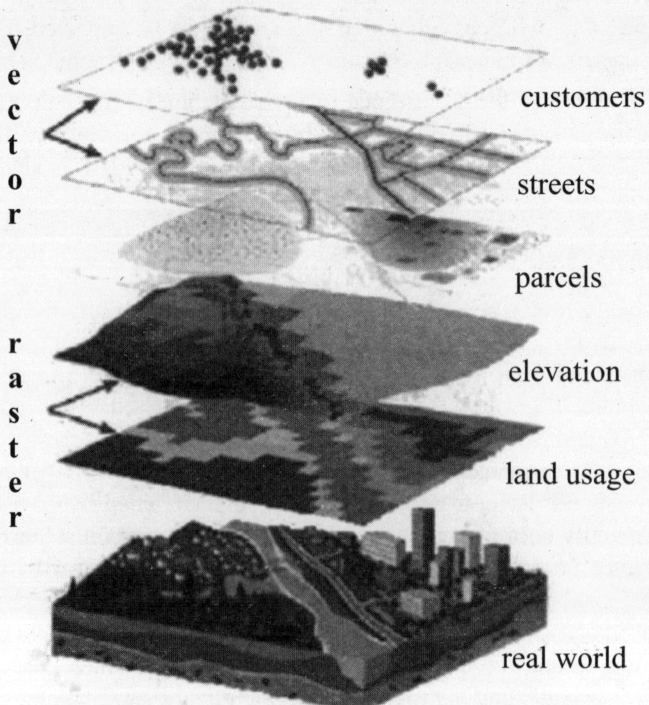

Fig. 43.3: The Concept of Layers in GIS

COMPONENTS OF GIS

GIS comprises of four components:

(1) Hardware: used to store, process and display.

(2) Software: used to control and perform operations.

(3) Data: on which GIS operations are performed (Spatial, non-spatial).

(4) Expertise (Live ware): the human element required to drive the system to meet needs.

HARDWARE

The general hardware components of Geographic Information System are the main computer system or the Central Processing Unit (CPU), the terminal, keyboard and the visual display unit (VDU), digitizer, disk drive, plotter, printer, etc. The computer or the CPU is linked to a disk drive storage unit, which provides the space for storing data and programs. The digitizer or other device is used to convert data from maps and documents into digital form and send them to the computer. A plotter or other kind of output device is used to present the results of data processing, a tape drive is used for storing / retrieving data or programs on magnetic tapes or with other systems. The user interacts with the computer and the peripherals (a general term for digitizer, printers, plotters and other apparatus linked to the computer) via terminals.

SOFTWARE

The GIS software is required for driving the hardware. Common interfaces in GIS are menus, graphical icons and commands. Most Common GIS Softwares available in the market are: ESRI products like Arc/Info, Arcview, ArcGIS, Map Info from Map Informatics Inc, Intergraph, ILWIS, etc.

DATA

Data includes both spatial and non-spatial data on which GIS operations are performed to derive new information. Spatial data from various sources such as Remote sensing images, Aerial Photographs or Map Data can be integrated with corresponding non-spatial data in GIS.

PERSONNEL/EXPERTISE

Equally important as the computer hardware and software, the brain ware refers to the purpose and objectives, and provides the reason and justification, for using GIS (Huisman et al 2009).

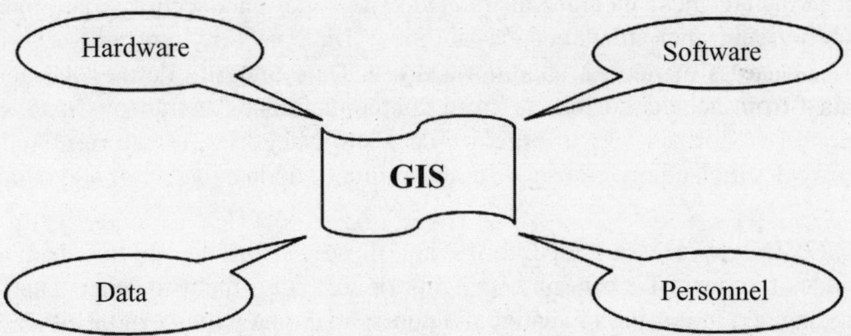

Fig. 43.4: GIS Components

CAPABILITIES OF GIS

A GIS is a computer-based system that provides four sets of capabilities to handle geo-referenced data.

- Data inputs
- Data management (data storage and retrieval)
- Manipulation and analysis
- Output (visualization)

DATA CAPTURE AND ENTRY

The function of data input to convert data from their existing form into one that can be used by GIS. Geo-referenced data are commonly provided as proper maps, table of attributes and associated attribute data, aerial photos and satellite imagery.

DIGITIZATION: The process of converting the analog information into digital information is called digitization. Information on a map can be captured by digitizing each feature. Thus the process of digitization is which converts the spatial features on a map into digital format. Therefore, point, line and area features that compose a map are converted into X & Y co-ordinates, where a single co-ordinate represents a point and a string of co-ordinates represent a line.

Digitization can be done either using digitizing tablet (Head up digitization) or on the screen digitization using mouse through scanning and can be stored either as scanned image or vector coverage. Digitizing tablet is a device in graphic peripherals, which allows the user to select point from a map, and automatically enters its X, Y co- ordinates in the computer. Digitizing tablet consists of tiny wires both horizontally and vertically inside the board forming a grid. When the tablet is placed on the board, the cross hair on the tablet allows the user to locate visually a point on the map. Once the point is identified electrical impulse is triggered. Since both tablet and digitizing table is connected to the computer, the exact location of the impulse in relation to the nearest intersection of the electronic grid is known. This allows the computer to translate the location of the point on the table into the location represented on the map, because the map has been registered on table previously by the user. Since most systems are interactive the location of the tablet on the table is simultaneously displayed on graphic display screen.

DATA STORAGE AND MAINTENANCE

The data storage and management are the functions of database management system of GIS. They are concerned with the way the data is structured, handled, accessed and perceived by the user of the system. Effective data management includes all related aspect like data security, data integrity, data filling and accessibility and data maintenance abilities. Data security ensures security against modification of GIS or access of data to unauthorized use. Data integrity defines the ability of the system to protect data from accidental loss or from contamination by extraneous data. Filling and accessibility provides an authorized user to organize data into categories, directories, study area etc. Data maintenance provides the authorized user with the ability to update, delete or add data to the GIS database.

DATA PROCESSING: Data processing operations are those performed on the data to produce information. It includes removal of errors and updating or matching them to other data. Errors can arise during an encoding and in-putting of spatial and non-spatial data which can be either incomplete or double, in the wrong place at the wrong scale, distorted or linked to the wrong non-spatial data. Besides, data may be over defined and may need to be reduced in volume. Data editing is interactively performed to ensure that all the errors are corrected, updated and properly verified to achieve the required accuracy which is vital to analysis.

MANIPULATION AND ANALYSIS (SPATIAL QUERY AND ANALYSIS)

The most distinguishing parts of a GIS are its functions for spatial analysis, i.e. operators that use spatial data to derive new geo-information. Data conversion is only part of the input phase of GIS. What next is the ability to interpret and to analyze qualitatively and quantitatively the information that has been collected. Spatial analysis tools are used to model, make predictions and reach conclusion about the problem of interest. Such analysis involves combining data from multiple spatial data categories and performing analysis/statistical operations on the GIS datasets to transform the data into information suitable for a given application. Typical operation includes overlaying different thematic maps, computing areas, performing proximity searches, buffer zone creation, performing logical operations, scale changing, etc. Other techniques are creation of 3D perspective view using elevation data and generation of slope maps, network analysis, costing, etc.

GIS DATA TYPES AND MODELS

GIS DATA TYPES

GIS data have two forms namely spatial data and attribute data. Spatial data explains the location of the target and answers the question where, while the attribute data describes the target and answers the question what. GIS operates under the assumption that the relevant spatial phenomena occurs in a two or three dimensional Euclidean space. Euclidean space can be informally defined as a model of space in which locations are represented by coordinates (x,y) in 2D and (x,y,z) in 3D, while distance and direction can be defined with geometric formula. We might define a geographic phenomenon as a manifestation of an entity or process of interest that:

- can be named or described
- can be geo-referenced
- can be assigned a time (interval) at which it is/was present

GIS DATA MODELS

In GIS environment, we usually use models to represent the real world and the two data models common in GIS are vector data model and raster data model.

VECTOR BASED MODEL: In the vector based model, geospatial data is represented in the form of co-ordinates. In vector data, the basic units of spatial information are points, lines (arcs) and polygons. Each of these units is composed simply as a series of one or more co-ordinate points, for example, a line is a collection of related points, and a polygon is a collection of related lines.

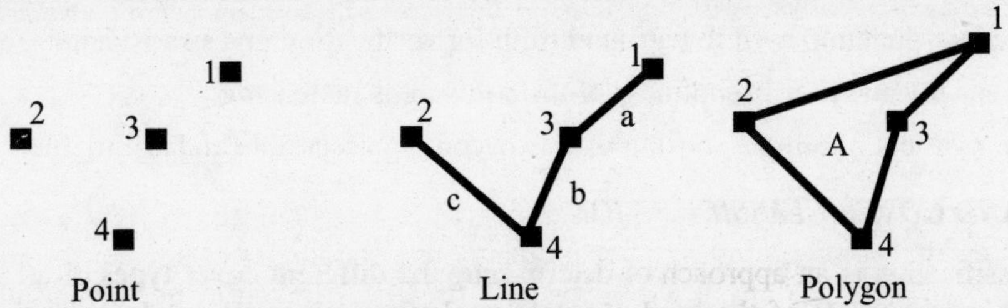

Fig. 43.5: Vector Representation of GIS Data

RASTER BASED MODEL: Raster based spatial models regard space as a tessellation (resembled mosaic) of cells, each of which is associated with a record of classification or identity of the phenomena that occupies it. The raster model represents the 2D location of phenomena as a matrix of grid cell and each cell is known as pixel (Short form of Picture Element). Since the cells are of fixed size and location, raster tend to represent natural and human made objects in a blocky fashion. The information content in one cell depends upon the size of the cell. As a result, if the cells are sufficiently small, the information present in one cell will be more. This is called resolution of the image. The raster model or grid cell is relatively simple approach to data representation both conceptually and operationally, and hence has been popular since the earliest days of GIS development.

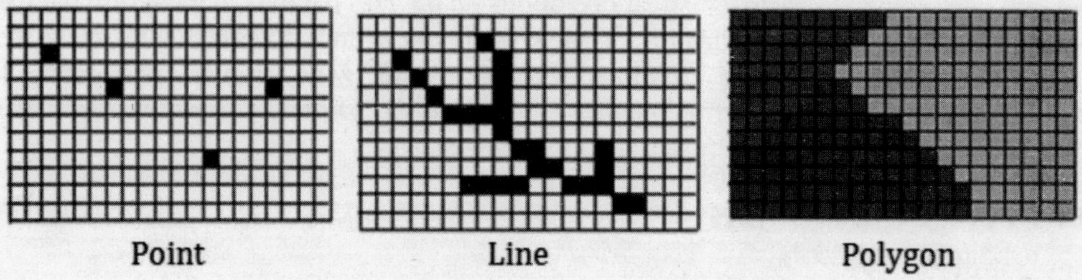

Fig. 43.6: Raster Representation of Discrete Features

ROLE OF REMOTE SENSING AND GIS IN AGROFORESTRY

In Agriculture and Forestry, data collection and analysis are important and commonly practised activities for proper resource management. For long time, these activities were executed through field survey using different instruments such as chain, compass, theodollite, etc. For small areas of land, these methods can work accurately reasonably regarding the time, budget and labour, but when we think of a large and extended area covering larger regions, it becomes difficult and sometimes impossible. Moreover, the data collected in such a way gives information about that particular period while natural and manmade processes taking place on the surface of the earth is dynamic. As a result, changes that may take place immediately after data collection may not be considered during analysis. Land use land cover classification using the above mentioned method becomes more complex and the accuracy decreases.

Remote sensing as a science and art that includes both aerial photograph and satellite imagery together with GIS gives a multitude of capabilities to handle and manage such complicated spatial data with better accuracy and at a faster rate for a large area of land. The importance of both GIS and remote sensing is increasing from time to time due to the following main reasons:

- Remote sensing images especially satellite images cover large area of land
- Inaccessible areas for ground survey can easily be viewed and analyzed
- Image interpretation with few ground truth for verification gives satisfactory results
- The capacity and user friendliness of the software is increasing
- Greater variety of remote sensing data is becoming accessible and affordable

LAND USE/ LAND COVER CLASSIFICATION

Land cover classification is an approach of determining the different cover types of an area for better and appropriate management of the land resource ecologically, socially and economically. Starting from the human history of owning private land resources and his interest to know the extent of the different types of such resources, different techniques have been under use. Determination of the resources and estimating their extent was done using the available technology. With advance in technology, remote sensing and GIS are now the preferred technologies to facilitate the process of land cover analysis. Remotely sensed data is acquired by using aircrafts of different altitude or from satellites on space. The difference in reflectance value of the different earth surface features gives the interpreters or users of the remote sensing data the possibility to interpret the image and identify the different features. Interpretation of images and aerial photographs can be performed either visually by the interpreter or automatically using different softwares designed for that purpose. Visual image interpretation is done with the help of photo interpretation elements such as the tone, texture and pattern mentioned under air photo interpretation. On the other hand, the different softwares use various spectral properties of the features and classify the image. Image classification can be either unsupervised or supervised and produce a thematic map of the different land cover types.

IDENTIFICATION OF AREA CURRENTLY UNDER AGROFORESTRY

The choice of the above mentioned two types of image classification depends on the background knowledge of the user about the area under consideration. When there is a gap of knowledge of the area covered by the image, applying unsupervised image classification can give clue on the number of possible land cover classes covered by the image. This step is usually followed by selecting samples of all different classes and identify what land cover type they are. Finally supervised classification can be done. For proper planning of development activities in addition to location of the area of interest, the extent should be determined with a reasonable accuracy. At a faster rate and with a better accuracy than the traditional survey method, determination of the extent of intervention area can be managed in GIS.

IDENTIFICATION OF POTENTIAL AGROFORESTRY AREAS

Apart from identifying the current extent agroforestry intervention area and its type, there is also possibility of identifying possible areas of intervention for future planning of new agroforestry expansion. The type of agroforestry intervention required for an area depends on various site conditions and is affected by different parameters of the intervention area. Some among the major parameter to select the best suiting agroforestry system are precipitation, temperature, soil condition and topography. Under field conditions, measuring these parameters is time consuming and needs high budget and human labour. GIS is a system with the main capability of solving such limitations of the field survey and the process of determining the best suitable combination of the above mentioned parameters can be achieved using GIS. With the advance in GIS and remote sensing technology, some of the necessary inputs for the process such as the Digital Elevation Model (DEM) which is important for topographic analysis, is available free and can be downloaded online. With some meteorological data from the available meteorological stations around the area of interest, determination of the climatic conditions of the rest of the area is also possible in GIS environment. Well developed research institutes such as the World Agroforestry Centre (ICRAF) have developed a good database regarding the agroforestry species for various local conditions based on the environmental and edaphic factors which may have impact on the success of Agroforestry practice. GIS as a tool has the capacity of automatically identifying the suitable areas provided that the parameters are clearly stated.

CHANGE DETECTION AND TREND ANALYSIS IN AGROFORESTRY

In remote sensing, aerial photograph has a long history and as a result historical photographs are available for many countries and regions. Due to the main reason that aircraft has been invented long time before satellite, it was the only way of monitoring the environment and used as a base for map making for a long time. Since introducing new technologies such as new agroforestry system or technology to a certain area needs background knowledge of the local community in related fields and historical aerial photographs can give important information of an area regarding land cover and its pattern in the past. From the land cover and pattern of the farming system, one can deduce necessary information about the past history of agroforestry which can give a clue for future planning and selection of the alternative techniques.

Land use/ land cover is most of the time under dynamic process especially in developing countries where most people is dependent on land resources for survival and additional income. Since most of the actions performed by human being have their own reason, understanding the change from the past trend is important to properly plan future development activities in a sustainable manner. Comparison of historical photos/images with the recent one is one way of detecting change in remote sensing and GIS. Remote sensing have the capability of performing land cover change automatically provided that images of an area of at least times are available.

SUITABILITY ASSESSMENT OF AGROFORESTRY TREES AND CROPS

Resource planners need to be able to determine the best locations for growing agroforestry speciality crops and multipurpose trees which will help land owners and farmers to adopt agroforestry. Suitability assessment matches potential species with ideal growing conditions. A suitability assessment makes a match by overlaying data maps with information like soil type and climate factors generated from geographic information systems (GIS). Each map or "layer" is ranked based on suitability of a particular crop or tree. Suitability assessment has the power to identify the range of agroforestry speciality products for any given tract of land, giving landowners/farmers the power to integrate the best species in agricultural lands. Resource planners have used suitability assessments for decades to site landfills, wildlife reserves and mining areas. Agroforesters, however, have yet to use these technologies extensively. For agroforestry to achieve its full potential, a shift in policy needs to provide adequate financial, institutional and technical support for its development. Depending on the scale of analysis, suitability assessments can serve in the promotion, adoption, planning and design of agroforestry systems. Suitability assessments conducted at a state-level and national-level can help influence policy makers to incorporate agroforestry into nation's or state's priority areas (Bentrup and Leininger 2002).

■■■■

44 AGROFORESTRY EDUCATION, RESEARCH AND EXTENSION

INTRODUCTION

Agroforestry education, research and extension are important not only for sustainable development but also to develop technologies, skills and expertise, which translate into tailor-made solutions to problems related to the environment, forestry and natural resource management. Indian Council of Agricultural Research (ICAR) and Indian Council of Forestry Research and Education (ICFRE) are providing support for forestry education, research and extension and for developing technical capabilities and strengthening infrastructure for agroforestry as a major programme. Most of the agricultural universities are imparting forestry and agroforestry education at graduation, post-graduation and doctoral level. An All India Coordinated Research Project on Agroforestry was launched by ICAR in 1983. This is the first major national initiative of this magnitude anywhere in the world. National Research Centre for Agroforestry was established by ICAR in 1988 at Jhansi in central India, and agroforestry research and education has been taken up in a substantial manner in most of the state agricultural universities. The Indian Council of Forestry Research and Education (ICFRE) that was set up along the lines of ICAR to coordinate and promote forestry research at the national level also has agroforestry as a major programme area in most of its eight national research institutes and three advance centers. In addition to these public-sector undertakings, some private and non-governmental institutions are involved in research that is of relevance to agroforestry. Notable among them is the Tata Energy Research Institute, Delhi, which has a major research interest in biomass-productivity enhancement and large-scale multiplication of elite trees. Transfer of technology from laboratory to the fields is being done through Krishi Vigyan Kendras (KVKs) in ICAR and Van Vigyan Kendras (VVKs) in ICFRE system at the national level. State forest departments are also disseminating agroforestry in their respective states.

At international level, African Network for Agriculture, Agroforestry and Natural Resources Education (ANAFE) and Southeast Asian Network for Agroforestry Education (SEANAFE) focus on improvement of agroforestry education in a sustainable manner for the contribution of social and economic development of the people. They also focus to develop human resources for agroforestry through collaboration among educational institutions. World Agroforestry Centre (ICRAF) is working closely with government extension systems, NGOs, and development projects to catalyze greater adoption and impact. The major contributions of World Agroforestry Centre are in providing research support and technical advice, studying approaches to dissemination, and helping organizations to share their experiences. The Center for Agroforestry at the University of Missouri (UMCA), established in 1998, is one of the premier centers of its kind in the world dedicated to agroforestry research, teaching and technology transfer.

AGROFORESTRY EDUCATION IN INDIA

Natural tree resources in India are diminishing rapidly, as they are over exploited to meet the demand of firewood, fiber, fodder, timber, etc. of rising population. In areas where the villages are located away from forests areas, human resources are mainly spent collecting fodder and firewood for livelihood rather than on farming or exploitation of forest products for economic benefit. The increasing population (human and animal) will cause heavy shortages of firewood, fodder and food. This has forced planners, scientists and educators to think seriously on various aspects of education, research and extension programmes in agroforestry for the benefit of the rural masses. Agroforestry is especially relevant because there is no scope of increasing the land area for forests. In India, agroforestry research on scientific lines started in 1979, when a seminar was organized by the Indian Council of Agricultural Research (ICAR) at Imphal, Manipur (ICAR 1981). The seminar, suggested the research direction and organizational pattern for strengthe-ning agroforestry research and education. Establishment of an All-India Coordinated Research Project with centers in states and union territories for a comprehensive study on the various aspects of agroforestry was recommended. Consequently, the Indian Council of Agricultural Research sanctioned an All-India Coordinated Research Project on agroforestry with 20 centers in different agro-ecological regions under the Sixth and Seventh Five Year Development Plans in the 1980's. Later it was realized that without education in agroforestry as a land use system, it would be very difficult to solve the problem of fuel and fodder shortages and to carry out satisfactory research and extension activities. In the search of scientific management of forests, the formal education in forestry started in 1867 with the training of foresters. The forestry education in pre-independence India may be classified into four categories, viz., Training of Officers of the Superior Services, Range Officers, Foresters and Forest Guards.

The first theoretical course was started in Forestry in 1881. The Government of India took over The Central Forest School, Dehradun in 1884 and renamed it as "Imperial Forest School". Two years training for Rangers was started in 1891 at the College of Science, Poona, thereafter; Rangers' College was started at Dharwar. With the establishment of Imperial Forest Research Institute in 1906, the status of the school was raised and it was renamed as Imperial Forest College, Dehradun. The training of the officers of the Indian Forest Service continued at the British Universities till 1925. The training of IFS started in India at Imperial Forest Research Institute (IFRI) in 1926 and continued till 1932 when the lack of demand for trained forest officers compelled its closure. The demand for trained forest officers to manage the forests in the provinces resulted in the creation of the Indian Forest College (IFC) in 1938, where Provincial Forest Officers were trained. The level of training was upgraded in 1940 and was given an all India outlook in order to manage the vast national assets. The course was termed as the Superior Forest Service Course which offered "Associate of Indian Forest College (AIFC)" Diploma, replacing the Provincial Service course. This course was continued till 1975.

Following the creation of Indian Forest Service again in 1966, the training of IFS officers started in 1968 in IFC along with the State Forest Service (SFS) trainees till 1975. After 1975, the IFC was exclusively set aside for training of Indian Forest Service probationers with a separate course of its own and the increased demand for training State Forest Service Officers necessitated establishment of three separate State Forest Service colleges at Burnihat in 1976, at Coimbatore in 1980 and at Dehradun in 1981. Government of India subsequently upgraded the IFC to the status of a national academy named after the late Prime Minister Mrs.Indira Gandhi as Indira Gandhi National Forest Academy (IGNFA) in 1987. The training of foresters and forest guards has always been the responsibility of the State Governments except for a brief period between 1952 to 1960 when a Regional Foresters' School was run by Government of India at Southern Forest Rangers College

(SFRC), Coimbatore which catered to the needs of the States of Madras, Andhra Pradesh, Mysore, Kerala and Andaman & Nicobar Islands. Currently, there are about 48 training schools in the States for imparting training to the Foresters and Forest Guards. Various indigenous agroforestry systems were included in curricula from the beginning of agricultural education in India. Many institutions still teach it to undergraduate and postgraduate students of agronomy, soil conservation and horticul-ture. Until 1979, the importance of this aspect was not fully realized because shortage of fuel and fodder from natural forests and pasture resources was still not obvious.

Agroforestry education was started formally with the establishment of forestry departments in different agricultural universities. ICAR started forestry education at the University level in the country from the State Agricultural Universities (SAUs) in the year 1985 with the inception of B.Sc. Forestry as a four year degree programme. ICAR selected 13 State Agricultural Universities of India for starting B.Sc. Forestry degree programme during the academic year 1985-86. Agroforestry is only part of the forestry teaching programme. Indian Council of Forestry Research and Education (ICFRE), is the apex organization of forestry in the country undertaking the holistic development of forestry education through need-based planning, promoting, conducting and coordinating research, education and extension covering all aspects of forestry including agroforestry. Master and Doctoral programmes in various disciplines of forestry including agroforestry are run at Forest Research Institute University at Dehradun. In India, full dedicated M.Sc.Agroforestry and Ph.D.Agroforestry programme are offered in very few Institutions such as Sam Higginbottom Institute of Agriculture Technology and Sciences (Formerly Allahabad Agricultural Institute) Deemed University-Allahabad, Banaras Hindu University-Varanasi, Bundelkhand University-Jhansi, etc. The List of Educational Institutions (Institutes, University, College and Department) who offer Graduate and Post Graduate Level courses in forestry along with agroforestry as a major programme area are presented below:

LIST OF UNIVERSITIES, COLLEGES AND DEPARTMENTS OFFERING FORESTRY AND AGROFORESTRY EDUCATION (Source: Antony Joseph Raj 2012b)

- ASPEE College of Horticulture and Forestry, Navsari Agricultural University, Navsari, Gujarat
- Forest Research Institute-Deemed University, Dehradun, Uttarakhand
- Forest College and Research Institute (FC&RI), Mettupalayam of Tamil Nadu Agricultural University, T.N.
- College of Forestry, Dr.Y.S. Parmar University of Horticulture and Forestry, Nauni, Solan, H.P.
- College of Forestry, Ponnampet of University of Agricultural Sciences-Bangalore, Karnataka
- College of Forestry, Sirsi of Unviersity of Agricultural Sciences-Dharwad, Karnataka
- College of Forestry, Kerala Agricultural University, Thrissur, Kerala
- College of Forestry and Environment, Sam Higginbottom Institute of Agriculture Technology and Sciences (Formerly Allahabad Agricultural Institute)-Deemed University, Allahabad, U.P.
- College of Forestry and Hill Agriculture, Ranichauri of G.B.Pant University of Agriculture and Technology, Uttarakhand
- College of Forestry, Dr.Panjabrao Deshmukh Krishi Vidyapeeth, Akola, Maharashtra
- College of Forestry, Dr.Balasaheb Sawant Konkan Krishi Vidyapeeth, Dapoli, Maharashtra

- College of Horticulture and Forestry, Jhalawar of Maharana Pratap University of Agriculture and Technology, Rajasthan
- College of Horticulture and Forestry (CH&F), Central Agricultural University, Pasighat, Arunachal Pradesh
- College of Forestry & Hill Agriculture, Tehri Garhwal of Uttarakhand University of Horticulture and Forestry, Uttarakhand
- Faculty of Forestry, Birsa Agricultural University, Kanke, Ranchi, Jharkhand
- Faculty of Forestry, Sher-e-Kashmir University of Agricultural Sciences and Technology, Shalimar, Srinagar, Kashmir
- Division of Agroforestry, Sher-e-Kashmir University of Agricultural Sciences and Technology, Jammu
- School of Studies (Forestry & Wildlife), Bastar University, Bastar, Chhattisgarh
- Department of Forestry, *North Eastern* Regional *Institute* of Science and Technology (NERIST) Deemed University, Nirjuli, Arunachal Pradesh
- Department of Forestry and Natural Resources, Punjab Agricultural University, Ludhiana, Punjab
- Department of Forestry, Kumaun University, Nainital, Uttarakhand
- Department of Forestry, Orissa University of Agriculture and Technology, Bhubaneswar, Odisha
- Department of Forestry, CCS Haryana Agricultural University, Hisar, Haryana
- Department of Forestry, C.S.Azad University of Agriculture and Technology, Kanpur, U.P.
- Department of Forestry, Indira Gandhi Agricultural University, Raipur, Chhattisgarh
- Department of Forestry, H.N.B. Garhwal Universiy, Srinagar (Garhwal), Uttarakhand
- Department of Forestry, Wildlife and Environment Science, Guru Ghasidas Vishwavidyalaya University, Bilaspur, Chhattisgarh
- Department of Forestry, Uttar Banga Krishi Vishwavidyalaya, Cooch Behar, West Bengal
- Department of Wildlife Science, Aligarh Muslim University, Aligarh, U.P.
- Department of Agroforestry, Bundelkhand University, Jhansi, U.P.
- Jawahar Lal Nehru Krishi Vishva Vidhalaya (JLNKVV), Jabalpur, M.P.
- Ch. Sarwan Kumar Himachal Pradesh Krishi Vishvidhalaya (CSK HPKV), Palampur, H.P.
- Mizoram University, Tanhril, Aizawl, Mizoram
- University of Horticultural Sciences, Venkataramnagudem, West Godavari, A.P.
- Banaras Hindu University, Varanasi, U.P.
- Doon (PG) College of Agriculture Science and Technology (DCAST), Dehradun, Uttarakhand
- Dolphin (PG) Institute of Biomedical & Natural Sciences, Dehradun, Uttarakhand

LIST OF FORESTRY ORGANIZATIONS PROVIDING SPECIALIZED EDUCATION AND TRAINING

- Indian Institute of Forestry Management (IIFM), Bhopal on forest management
- Wildlife Institute of India (WII), Dehradun on wildlife research and management
- Indian Plywood Industries Research and Training Institute (IPIRTI), Bangalore on plywood manufacturing and mechanical wood processing
- Institute of Paper Technology (IPT), Saharanpur on pulp and paper technology

GLOBAL EXPERIENCES IN AGROFORESTRY EDUCATION

Traditional forestry and agriculture teaching failed to notice that, in fact, a number of farmers were rearing trees, crops and livestock together in number of countries as they practised agroforestry. That omission began to be corrected in the 1970s, albeit slowly. The pace fastened as deforestation and land degradation marked a decline in natural capital worldwide, and agroforestry came as an option by combining production with resource conservation. Governments, researchers, educators and NGOs joined hands to understand, develop, and promote various agroforestry systems. It was realized that agroforestry could contribute directly to food security, health and nutrition through improvement of land productivity and services. As a result, the number of educational institutions that provided agroforestry education and training increased very rapidly during the 1980s and 1990s. This trend continued up to 2004 and several international conferences tackled issues of agroforestry education.

An opposite trend can also be seen at the technical (certificate and diploma) level. Traditionally, many countries established technical colleges under their ministry of forestry or agriculture, primarily to provide training to their field staff. That link between education and employment is no longer as strong, partly a consequence of structural reform programmes that have reduced public sector employment. This is a main reason for the decline of technical education, especially in Africa. In consequence, fewer graduates with agroforestry competence are available at that level. Growing international concern for the environment articulated in Agenda 21 (United Nations 2004) and numerous other conventions and agreements on biodiversity, climate change, marine protection and desertification spotlighted the potential contribution of agroforestry in environmental conservation.

Agroforestry is being rapidly incorporated into the curricula of several universities and colleges worldwide. Agroforestry can easily be linked with many aspects of land use such as agriculture, horticulture, forestry and environmental management. There is much debate about how best to teach agroforestry. Current experiences show that it ranges from a topic within a subject area to a fully-fledged programme resulting in agroforestry qualifications. This is to be expected because agroforestry, though an old practice, is a young and rapidly growing area of science and technology.

Twenty five-thirty years ago, very few educational institutions taught agroforestry. The number of universities and colleges now teaching agroforestry has risen to the hundreds. Consequently, agroforestry has become a subject or even a full programme at universities and colleges throughout the tropics, as well as in some subtropical and temperate countries. Not only has agroforestry education increased in terms of the sheer number of courses and programmes, it has also become a component of programmes such as environmental conservation and rural development. Many universities and colleges in South Asia, Europe, Australia, Latin America and North America have developed agroforestry education at three different levels as a stand-alone degree or diploma programme, as a core or elective subject within various programmes, or as a single topic in other subjects.

The African Network for Agriculture, Agroforestry and Natural Resources Education (ANAFE) comprises of African colleges and universities teaching agriculture and natural resource sciences. Supported by the World Agroforestry Centre (ICRAF) in Nairobi, Kenya, the network was established in 1993 and is currently made up of 134 member institutions (universities and colleges) in 35 African countries. The objective of the Organization is to improve in a sustainable manner, the contribution of agricultural & natural resource management education to social and economic development of the African people. The Southeast Asian Network for Agroforestry Education (SEANAFE) was established in April 1999 with its headquarters in Bogor, Indonesia. SEANAFE's mission is to develop human resources for agroforestry and integrated natural resource management through collaboration among educational institutions. Presently, the network has 85 member institutions in five country networks; in Indonesia, Laos, Philippines, Thailand and Vietnam.

Table 44.1: Agroforestry Education at Different Levels (Rudebjer et al 2005)

Level of Education	Description
Agroforestry at the Technical Level	
Certificate-level	Institutions in some countries include agroforestry content in certificate programmes, primarily in forestry and agriculture. Their durations vary from a few months to two years. Usually, agroforestry is offered as specific subjects, or as topics within other subjects. However, certificate-level training in many countries in Africa and Southeast Asia is now in decline (Temu et al 2005). On graduation, certificate holders prove eligible for employment as frontline extension or development workers, technical assistants or, in some countries, teachers.
	Given the important role played by technicians and extension workers in natural resource planning and management, curricula at certificate level should be both broad and applied. In addition to principles and practices of agroforestry, curricula should include topics such as; the role of agroforestry in rural development; seed systems; socioeconomics, including gender in agroforestry; ecological aspects of agroforestry; agroforestry intervention approaches; extension methods marketing; and business management.
Diploma-level Training	Training at diploma level can vary in length from one year for in-service programmes to three years for regular diploma programmes, though with marked differences between countries. As with certificates, diplomas that include agroforestry are awarded by both agricultural and forestry institutions. Typically, agroforestry is taught as a separate subject within a diploma programme, such as the diploma in forestry taught in the Kenya or Zimbabwe Forestry Colleges, or the diploma in agriculture from Botswana Agricultural College. In some institutions, agroforestry is a full diploma programme, as with the post-graduate diploma in agroforestry from the University of the Philippines Los Baños, or the diploma in agroforestry from Nyabyeya Forestry College, Uganda. Mbalmayo Forestry College in Cameroon offers both certificate and diploma programmes in agroforestry. Though agroforestry curricula from different places and institutions will never be identical, some core topics should be standard for all diploma courses. Diploma graduates often play important roles as field technicians, extension workers or teachers, and their training should reflect this. For example, the agroforestry diploma at the University of the Philippines Los Baños includes appropriate on-farm technologies, cottage industry development in agroforestry, and community organization and development for agroforestry. These important topics are unfortunately often missing in diploma curricula. Another topic that should be considered is agroforestry monitoring and evaluation (Kung'u and Temu 2003).

Level of Education	Description
	Diploma graduates play a key role in sharing agroforestry knowledge with the ultimate users – the farmers. Graduates should thus be able to facilitate a process of shared learning, in participation with local communities, institutions and government units. This is a quite different approach to the 'technology transfer' paradigm that is still widely taught. Hastening the shift to greater participation is one motivation for encouraging review and development of curricula.
Agroforestry at the Professional Level	
Bachelor's Degrees in Agroforestry	Agroforestry graduates with a bachelor's degree frequently work in research and in mid-level positions in the extension service. This type of training is offered under a bewildering array of programmes throughout the world. Commonly, as with the University of Florida, USA, agroforestry is offered as a subject in a forestry bachelor's. In Uganda, agroforestry is offered as a subject in several different B.Sc. programmes, including forestry, community forestry and agriculture. Agroforestry may also be found as a subject within bachelor's degrees in animal science, environmental science and soil science. Degrees as varied as landscape architecture and home economics can sometimes feature agroforestry. Yet more programmes include agroforestry as a topic within other subjects. This is probably more usual in institutions focused primarily on agriculture. More rarely, agroforestry is offered as a full undergraduate programme in its own right. The Philippines is a leading example. There, more than 30 universities and colleges, such as the Don Mariano Marcos State University, offer a bachelor's degree in agroforestry or a major in agroforestry. These programmes take three to four years. Given this huge variation in undergraduate programmes, it comes as no surprise that agroforestry content differs widely. Topics like 'an introduction to agroforestry' and 'classification of agroforestry systems' are normally included in most curricula. Unfortunately, important topics – like the socioeconomic aspects of agroforestry, multipurpose trees and shrubs, agroforestry practices and technologies, the role of agroforestry in soil and water conservation, and research design and methodologies – are found in very few curricula (Kung'u and Temu 2003). Other weak spots exist. For one, curricula should include topics that will enable the graduates to implement agroforestry projects, since it has been shown that many B.Sc. graduates will eventually become project managers or research scientists (Nair et al 1990). Topics such as marketing and entrepreneurship are also often omitted. Curricula are frequently oriented towards the supply side of agroforestry, such as the production of trees and shrubs. They thereby ignore the changing demand for products and services from those trees and shrubs. More generally, given the fields of work adopted by many agroforestry graduates, there is a real need for transferable professional skills like proposal writing and communication. Where these are not covered in other course components, there is a strong argument for their inclusion in agroforestry, since good communication with agroforestry researchers, donor agencies, farmers and students will be essential in an agroforester's professional life.
Master's Degrees in agroforestry	Agroforestry at master's level is relatively new, but already offered under all manner of different programmes. Most common is where agroforestry forms a component of another programme. For example, most universities offering master's programmes in forestry teach agroforestry as a separate subject. In contrast, institutions in Africa offering master's programmes in agriculture or animal health normally include agroforestry as a topic within another subject (Kung'u and Temu 2003). One exception is Sokoine University of Agriculture, Tanzania, which offers agroforestry as a subject in agriculture, forestry and animal science M.Sc. programmes. Combined

Level of Education	Description
	courses also exist, such as at Kenya's Kenyatta University, which offers a combined master's programme in agroforestry and rural development. Agroforestry and sustainable development is offered in Universidad Autonoma Chapingo, Mexico. Dedicated agroforestry programmes in M.Sc. are available in some countries. Maejo University in Thailand, Kenyatta University in Kenya, Makerere University in Uganda and the University of Florida in the USA all offer postgraduate programmes in agroforestry. So do University of Malawi, Ibadan University in Nigeria, Sokoine University in Tanzania, University of Kwame Nkrumah in Ghana, Wageningen Agricultural University in the Netherlands, University College of North Wales in Bangor, UK, and Oregon State University in the USA. There are many others. In South Asia and Southeast Asia, several universities are currently developing new master's programmes in agroforestry.

AGROFORESTRY RESEARCH IN INDIA

Agroforestry as a land use system is as old as agriculture. In a country like India with varied agro-climatic zones, various combinations of trees with arable crops, fruits and animal husbandry exist. Agroforestry range from apparently 'simple' forms of shifting cultivation observed especially in the northeast to the complex multistoried homegardens of the humid coastal areas in the south. Agroforestry has been a way of life in India from ancient time. The two great Indian epics Ramayana and Mahabharata (based on events that have supposed to have happened in 7000 B.C. and 4000 B.C. respectively) have the mention of Ashok Vatika, a type of what are called homegardens today. In spite of this long-standing social, ethnic, and religious significance of trees in India and the practice of various traditional forms of agroforestry in different parts of the country (Tejwani 1994), recognition of trees as components of farming systems has been rather limited (with the notable exception of interest in farm woodlots during the past two to three decade). This is partly because of the traditional divide between agriculture and forestry in government business, just as it is elsewhere in the world. The several traditional land use practices that have been in existence in India for a long time were ignored while organized research endeavors in agriculture and forestry were developed along strict disciplinary lines – again, a situation not unique to India.

India is among a few countries where organized agroforestry research started almost at the same time when such initiatives started on the global scene in the 1970s. An All India Coordinated Research Project on Agroforestry was launched by ICAR in 1983 in 20 centers all over the country and the number of centers now has risen to 37. Further, a National Research Centre for Agroforestry was established by ICAR in 1988 at Jhansi in central India, and agroforestry research and education has been taken up in a substantial manner in most of the state agricultural universities. In addition to research, providing extension workers, community-development personnel, farmers, and scientists with training in various aspects of agroforestry is a major activity in all research centers. There are six major research objectives under AICRPAF. Each center conducts research on one or more of these broad objectives, with appropriate local modifications. The six objectives are:

(1) Diagnostic survey and appraisal of existing farming systems and agroforestry practices

(2) Collection and evaluation of promising multipurpose tree and shrub species for fuel, fodder, and small timber

(3) Management aspects of various agroforestry technologies

(4) Social, anthropological, and economic aspects of agroforestry

(5) Agroforestry for environmental protection, wasteland reclamation, and community development

(6) Post-harvest technology and non-timber forest products in relation to agroforestry

Today, about 2,000 scientific and technical personnel are involved in agroforestry research and education in India. Given the geographic, ecological, demographic, social, cultural, political, and economic complexities of a country like India, its rather unparalleled experience in agroforestry research and development will be of considerable interest to the international community. With India's much acclaimed track record of green revolution and the success story of creating a well-organized infrastructure in agricultural research, India's experiences in agroforestry research for development present a unique case study. Despite the considerable efforts for promoting and implementing agroforestry practices in India, improved agroforestry technologies have not been widely adopted in the country, except perhaps for the poplar (*Populus* spp.) based system in the northwestern parts. Some controversies were also generated about agroforestry. Critics argue that such activities are taking over land that previously provided them with the means to feed themselves and earn a living. They claim that planting large areas with eucalypts (*Eucalyptus* spp.), acacias (*Acacia* spp.), pines (*Pinus* spp.) and poplars (*Populus* spp.) have adversely affected the environment and ruined smallholder farmers (FAO 1995; Nambiar et al 1998). Others say the practices will help protect natural forests and provide economic growth. Part of the problem can, arguably, be related to the location-specificity of agroforestry systems. But agricultural systems are site specific too; yet modern agricultural technologies have gained widespread adoption in India.

Table 44.2: Major Agroforestry Practices in India (Puri and Nair 2004)

S.No.	Agroforestry Practice	Agro-ecological Region/ States
1	Boundary planting and live hedges	In all regions
2	Farm woodlots	Throughout the country
3	Homegardens	Mainly tropical west coast region, especially Kerala, southern Karnataka, Andaman & Nicobar Islands
4	Industrial plantations with crops	Intensively cropped areas in northern and northwestern regions: Haryana, Himachal Pradesh, Punjab, Uttar Pradesh; also in southern states (Andhra Pradesh, Karnataka, Kerala, and Tamil Nadu)
5	Scattered trees on farms, parklands	All regions, especially semiarid and arid regions
6	Shaded perennial systems with plantation crops	Mainly humid tropical region in the southern region; also Assam, W. Bengal
7	Shelterbelts and windbreaks	In wind-prone areas, esp. coastal, arid, and alpine regions
8	Silvopasture: Fodder trees, Intercropping/ grasses with fruit trees Seasonal forest grazing	Throughout the country Subtropical and tropical; orchards in grazing hilly regions Semiarid and mountainous ecosystem
9	Taungya System	Eastern region
10	Tree planting for reclamation of saline soils	Waste Lands of semiarid and canal-irrigated regions, mostly in the northern and northwestern regions
11	Woodlots for soil conservation	In hilly areas, along sea coast and ravine lands

AGROFORESTRY RESEARCH IN WORLD

The practice of raising trees and crops together has been prevalent for many centuries in different parts of the world, especially under subsistence farming conditions. Homegarden, a major agroforestry practice today and one of the oldest forms of agriculture in Southeast Asia (Kumar and Nair 2006), is found to have been associated in the moist tropical region about 13,000–9,000 BC (Wiersum 2004). It is said that in Europe agroforestry was started around 4,000 BC, when domestic animals were allowed in forests for grazing. The dehesa (animal grazing under trees) system of Spain is said to be 4,500 years old. Only during the past three decades, the indigenous forms of growing trees and crops/ animals together have been brought under the modern scientific land use scenarios (Nair et al 2008). There were several reasons for taking a new look at the old traditional practices. The Green Revolution of the 1970s (Evenson and Gollin 2003) largely did not reach the poor farmers in the tropics. The faulty land management practices resulted in increased deforestation, firewood shortage, degradation of land, and loss of biodiversity. The search for strategies to solution of these problems focused the attention on the age-old practice of combining trees, crops, and livestock on the same land unit, and an appreciation of their inherent advantages led to the establishment of an international center, ICRAF, in 1977 (Steppler and Nair 1987). The center, now known as the World Agroforestry Centre, is one of the CGIAR (Consultative Group of International Agricultural Research) institutions. Agroforestry thus began to be recognized and incorporated into national agricultural and forestry research agendas in many developing countries during the 1980s and 1990s (Nair 1993). In temperate regions, ''modern'' agroforestry had a slower evolution than in the tropics. It started with an increased perception on the part of the general public about the environmental consequences of high-input agriculture and forestry. The single-species emphasis of food and wood production in commercial systems has caused considerable decline of biological diversity, e.g. compare the diversity of corn (*Zea mays*), soybean (*Glycine max*) or pine (*Pinus* spp.) production systems with the approximate 100 species found in an oak (*Quercus* spp.).

Agroforestry has had a slower evolution as a science in the temperate regions, than in the tropics (Garrett et al 2000). The environmental conscience of general public in developed nations started demanding greater environmental accountability of land use practices and the application of ecologically and socially compatible management approaches. As a result, the concept of agroforestry gained acceptance in the industrialized nations during the 1990s. Agroforestry systems and their application are also prevalent in countries and regions such as the United States, China, Australia and New Zealand, and southern Europe. Agroforestry systems in different parts of the world vary in nature, complexity, and objectives (Nair 1993). The economic advantage of diversified income is a major motivation for practising such systems in both tropical and temperate regions. In general, subsistence farming and emphasis on the role of trees in improving soil quality of agricultural lands are characteristic of tropical agroforestry systems. Environmental sustainability is a major driving force for the development and adoption of agroforestry in the industrialized nations, where monocultural production of agriculture and forestry commodities has led to reduced biodiversity and loss of wildlife habitat, increased non-point source pollution of ground and surface water, and deterioration of family farms. Such problems are the legacy of maximizing production of agricultural products without sufficient knowledge of, or regard for, impacts on future productivity, the environment, and society in general. Today, the idea of incorporating the structure and functions of natural ecosystems into the design of managed ecosystems is gaining wide acceptance. The contributions of MPTs and agroforestry systems in general are usually grouped under two broad categories i.e. production of commodities and ecosystem services. The production refers to increase of outputs such as food, fodder, firewood, timber, and non-timber products, whereas the ecosystem services refer to the services such as carbon sequestration, biodiversity conservation, and water-quality enhancement.

A large number of indigenous MPTs have been identified and their multiple roles in providing food and nutritional security, medicines, cash income, and a whole host of other products and benefits have been recognized (Elevitch 2006). A vast majority of them, however, have not been domesticated let alone exploited commercially (Akinnifesi et al 2008). A major opportunity as well as challenge in agroforestry lies in domesticating the indigenous MPTs. Agroforestry has emerged as an integrated applied science that has demonstrated potential for addressing some of the land management and environmental problems the world over. The 1st World Congress of Agroforestry held in Orlando, Florida, USA, June-July 2004 signifies this coming of age of agroforestry (Nair 2007). A Congress Declaration affirmed by more than 500 delegates from 82 countries states that the adoption of agroforestry over the next decade will "greatly enhance the achievement of the United Nations Millennium Development Goals" by increasing household income, promoting gender equity, improving health and welfare of people, and enhancing environmental sustainability (Nair et al 2005). The 2nd World Congress of Agroforestry was organized in Nairobi, Kenya in August 2009.

Ever since the Kyoto Protocol, agroforestry has gained increased attention as a strategy to sequester carbon from both developed and developing nations. The available estimates of carbon stored in agroforestry range from 0.29 to 15.21 Mg C/ha/year above ground, and 30-300 Mg C/ha up to 1 m depth in the soil (Nair et al 2010). Since the industrial revolution, atmospheric CO_2 has increased by more than 40 per cent, from 280 ppm in 1750 to about 392 ppm in 2012 and at the current rate it is expected to surpass 400 ppm by 2015 (Tans 2012). Agroforestry has been recognized as having the greatest potential for carbon sequestration of all the land uses analyzed in the Land-Use, Land-Use Change and Forestry (LULUCF) report of the IPCC (2000). Agroforestry is well suited to produce biomass for bio-power and bio-fuels (Jose et al 2012). Heavy reliance on foreign based fossil fuels has sparked an interest in domestic renewable energy sources in many countries.

Advances in agroforestry can contribute significantly to the achievement of virtually all of the Millennium Development Goals (MDGs) and the WEHAB (Water, Energy, Health, Agriculture and Biodiversity) initiative. Agroforestry focuses on the role of trees on farms and in agricultural landscapes to meet the triple bottom line of economic, social and ecological needs in today's world. Recognition of this role in overcoming key problems from local to global levels is growing. The World Agroforestry Centre has identified seven key challenges related to the MDGs and WEHAB that agroforestry science and practice can materially address:

(1) Help eradicate hunger through basic, pro-poor food production systems in disadvantaged areas based on agroforestry methods of soil fertility and land regeneration

(2) Lift more rural poor from poverty through market driven, locally led tree cultivation systems that generate income and build assets

(3) Advance the health and nutrition of the rural poor through agroforestry systems

(4) Conserve biodiversity through integrated conservation-development solutions based on agroforestry technologies, innovative institutions, and better policies

(5) Protect watershed services through agroforestry based solutions that enable the poor to be rewarded for their provision of these services

(6) Assist the rural poor to better adapt to climate change and to benefit from emerging carbon markets, through tree cultivation

(7) Build human and institutional capacity in agroforestry research and development

The research of the World Agroforestry Centre works towards more productive, diversified, integrated and intensified trees and agroforestry systems that provide livelihood and environmental benefits, including:

- Enriching the asset base of poor households through farm-grown trees
- Maximizing the productivity of agroforestry systems and the complementary benefits of trees to the productivity of crops and livestock
- Improving the income of poor households by better linking them to markets
- Expanding the multi-functionality of agricultural landscapes by balancing increased productivity with the sustainable management of the natural resource base
- Maintaining or enhancing the supply of ecosystem services in agricultural landscapes, particularly regarding water, soil health, carbon sequestration and biodiversity

World Agroforestry Centre conducts research in six eco-regions across sub-Saharan Africa (eastern, southern, western and central) in South and Southeast Asia, and Latin America. The programmes are driven by the specific needs and opportunities in each region. There are many examples of successful agroforestry practices – that is adoption at significant scale. This is perhaps most visible in more humid areas, where many agroforestry systems have proliferated, often with a tree crop foundation. These can be found in Indonesian rubber agroforests, Ethiopian and Central American coffee agroforestry systems, Chagga homegardens, and cocoa agroforestry in West Africa. Also widespread is the integration of fruits and timber into farms, sometimes in mixed systems sometimes in segregated orchards or woodlots. This is common in India, East African Highlands, Cameroon humid zone, and Philippine uplands to name a few. Indigenous parkland systems appear to be regaining vibrancy through protected regeneration in the Sahel, Tanzania and Ethiopia. Other technologies such as fodder shrubs and fertilizer trees have had recent successes in East Africa/India and Southern Africa respectively. These represent a blend of traditional practices and new innovations. Since agroforestry in its entirety is common, it can be said to be a successful land use system which is relevant to many areas of the world.

AGROFORESTRY EXTENSION

Agroforestry extension is a people-oriented activity combining forestry and agriculture for food and wood. Agroforestry extension is a process in which the efforts of Government or other organizations and the people are combined. It can be described as an informal process of learning and teaching method based on the need or problems of the people in a particular locality or site. Agroforestry extension is an out-of-classroom effort to inform and educate the masses on the value and importance of trees and multiple cropping and motivate them to take active part in various farming systems or activities. In agroforestry, the extension laboratory is the farm where practitioners can demonstrate to the farmers the proper spacing and patterns of the different crop combinations. In fact, agroforestry extension is a multidisciplinary work. It is not the monopoly of the foresters, agriculturists, agronomists, sociologists, or others. All extension personnel, however, must have common knowledge and understanding of the values and importance of forest resources.

In agroforestry, there are various production technologies, such as agrisilviculture, silvipasture, agrisilvipasture, homegardens, alley cropping, boundary plantation, etc. which should be transferred and brought to the farmers for the improvement of their socio-economic status. Agroforestry is a new concept at the professional levels but it is an age-old practice in different forms among the rural people in different places. The tribal people have been practising shifting cultivation and the Forest Department used taungya system for reforestation. The taungya system of plantation is developed as an agroforestry practice but it has not been able to reach those people who are practising shifting cultivation. If the taungya system of agroforestry practice could be extended to the tribal people, their socio-economic conditions would have been much improved. The production of crops and trees could also increase. Thus agroforestry extension can play a vital role to develop an alternative land use practice for the improvement of social and economic conditions of the people.

In India, almost every agricultural university has the directorate of extension to transfer technological packages including appropriate models to the intended target groups especially the farmers through their Krishi Vigyan Kendras (KVKs). ICAR coordinates various extension activities of ICAR institutes, centers and KVKs of agricultural universities at zonal and national levels through its Zonal Project Directors. The KVKs has the mandate of Extension of all subjects including agroforestry through publications like bulletins, brochures, pamphlets, newsletters and annual reports. KVKs are also doing their off-farm programmes by regularly organizing Kisan Melas, Field Days, Field Demonstrations and Training of Stakeholders including line department officials. In future, emphasis on transfer of technology through print and TV media will be more crucial. Recently on the lines of KVKs (ICAR), ICFRE has established Van Vigyan Kendra (VVKs) in each state for dissemination of various technologies developed by the council and its institutes, and state forest departments (SFDs) to the user groups including farmers and forest based industries.

■■■■

45 ON-FARM RESEARCH IN AGROFORESTRY

INTRODUCTION

In agroforestry research projects, it is highly desirable to initiate on-farm research activities as early as possible. Naturally, before agroforestry experiments can move on-farm, a certain amount of time will be devoted to purely on-station research. This is necessary for screening and evaluation of multipurpose trees, and for experimentation with management practices (Atta-Krah and Francis 1987). During 1980s, ICRAF researchers developed a "farming systems" approach to on-farm agroforestry research which was based on the Diagnostic and Design (D&D) methodology (Raintree 1987a; Raintree 1987b; Scherr 1990). This kind of consultative approach has resulted in a significantly increased participation of farmers in research with professional researchers. Nevertheless, ICRAF researchers discovered that the consultative paradigm embodies considerable limitations as an approach to developing truly adoptable agroforestry technologies. One of the key requirements for on-farm research is an ability to identify successful farmer innovations when they occur and to facilitate feedback to the community about what it, collectively, can learn from the experiments. This is important in moving the group learning curve beyond what a single farmer could achieve in isolation. The ICRAF researchers and cooperating extension workers keep an eye out for farmer innovations in the on-farm trials and try to understand how and why farmers modify the experimental technologies.

WHAT IS ON-FARM RESEARCH?

On-farm research is an indispensable tool for developing and validating agroforestry technology. On-farm research (OFR) can be defined in its simplest terms as research carried out on farmer's fields and in a farmer's environment. From this simple definition, one can identify four key elements in OFR. These are:

- Farmer
- Farmer's land
- Farmer's involvement
- Farmer's environment

FARMER

In OFR, it is essential to specify the type of farmer for which a particular intervention is aimed, whether for development or for testing. Thus if one is developing a technology for low-resource-base smallholder farmers, it will be incorrect to sample commercial, large-scale farms for the OFR. Equally, for a technology which requires use of costly inputs - seeds, fertilizer, herbicide, insecticide, etc., - one would have to deal with medium and large-scale farmers who could afford the essential inputs for the technology, rather than deal with smallholder, low-resource farmers. The "type of farmer" issue is not linked only to resource base, but may also be linked to the production system. For example, in developing a fodder intervention package, it would be necessary to look for a farmer community where both livestock and crop production are important, so that the technology will be relevant. The central issue, therefore, is to define and describe the type of farmer for which the technology to be developed or tested is appropriate and relevant.

FARMER'S LAND

For research to be classified as on-farm, it should be carried out on a plot of land belonging to the farmer and within the farm environment of the farmer. Any experiment carried out on a plot of land outside the experimental station could be described as off-station research.

FARMER'S INVOLVEMENT

The nature of farmers' involvement in any OFR activity is very important as it influences the interpretation of output and results obtained. OFR scientists seek varying degrees of farmer's involvement in OFR. The exact nature and degree of farmer involv-ement is determined by the objective of the OFR and the nature of the research in terms of components, systems, or technologies being assessed. The degree of farmer involve-ment also has an effect on the design of the experiment and the interpretation of results obtained. The four possible ways in which farmers are usually involved in OFR are:

- Land lord /tenant relationship
- Passive on-looker involvement
- Active involvement - researcher controlled
- Active involvement - farmer controlled

FARMER'S ENVIRONMENT

The farmer does not live as an independent entity. He lives within a family structure, which in itself is embedded within a community structure. Thus, the farmer's input, assessment, and eventual adoption of a system will have to be viewed and assessed within the framework of the community in which the farmer operates. The sociocultural, anthropological and economic environment within that community will have to be taken into account in the design of the technology, in its testing, and eventually in the assessment of its acceptability to farmers. The second aspect of the farmer's environment has to do with the cropping and farming system in which the farmer operates and the bio-physical base within which the farming activity goes on. For example, farmers' fields may have many more problems associated with soil fertility and drainage than the research station fields where on-station experiments have been conducted. Farmers may also be practising a much more complex cropping system than is used in on-station trials. All these factors underline the need for OFR, and the need to take into account the farmer's environment in the design and testing of technologies.

WHY ON-FARM RESEARCH?

While some amount of time devoted to purely on-station research will be necessary, (in order to screen and evaluate multipurpose trees, and to experiment with management practices), quick initiation of on-farm research will help the research and development process. Research should be aimed at solving farmers' problems and at involving farmers in the research. On-farm research plays essential roles in the following areas:

- Testing and validation of agroforestry systems under local farmer's conditions
- Development and adaptation of agroforestry systems for local farmers' conditions
- Demonstration and extension of agroforestry systems in local farming communities

TESTING AND VALIDATION

On-farm research is commonly used as a means to ensure that technologies developed on-station will be relevant to the problems and priorities of the targeted client adopters. In the case of agroforestry systems, the target adopters are typically resource-poor small farmers. To validate on-station results, OFR is carried out to assess the performance of particular systems on-farm, with or without the farmer's involvement. Such research will likely lead to the observation of yield gaps or shortfalls (Fig. 45.1). Research is then aimed at identifying constraints causing the gaps and eliminating or narrowing the gaps. On-farm trials allow assessment of the system according to a broad range of criteria. Analysis of such trials should be based not only on productivity and profitability, but on all other factors that are likely to influence the acceptability of the system to the farmer. These may include farmers' resources, the community's economic and social infrastructure, etc.

DEVELOPMENT AND ADAPTATION

On-farm research is often used to generate new or modified systems. Moving to farmers' fields and interacting with farmers allows the researcher to have an appreciation of the farmers' conditions and problems. It also provides a great opportunity for the identification of problem areas and researchable issues that may arise following farmer's use of developed systems. This leads to a continuous process of refining, improving, and re-testing the system. The length of time required for standardization and adaptation of the systems to various specific farmer situations may be shortened through contributions from the participating farmers. The farmers' own adjustment mechanisms and experiences will be important inputs in the adaptive process. The farmers have an important role to play at this stage of the research, and when possible they should be encouraged to experiment with the system and to suggest improvements.

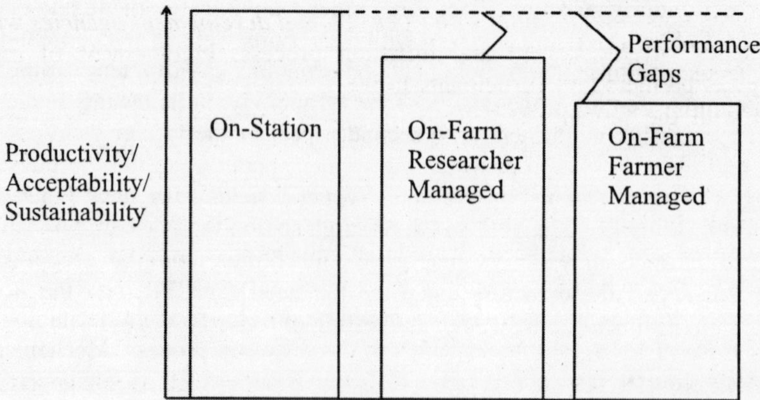

Fig. 45.1: Comparison of Technology Performance On-station and On-farm with Different Levels of Farmer Involvement

DEMONSTRATION AND EXTENSION

For any new system or technology to be accepted by farmers, it has to be shown to be superior to the existing system. The most reliable means of proving this is through OFR, in which the farmer is involved and the trial is run within the farm environment. Such OFR trials provide an excellent opportunity to compare the performance of the proposed system and farmer's traditional practice in a reliable way (Fig. 45.2).

When researchers and farmers are testing agroforestry system in the field, they are also demonstrating the system to the local community. On-farm research thus creates links with extension in at least three ways. Firstly, if on-farm research clearly *demonstrates the viability of the system*, it may create a "neighbourhood effect", whereby innovation waves spread outward from the research sites. Since on-farm research is likely to be conducted in many locations across the region, the innovation waves will spread from many centres and thus speed up both generation and diffusion of the systems. Moreover, horizontal (farmer to farmer) diffusion is likely to take place, due to lateral learning within each research location.

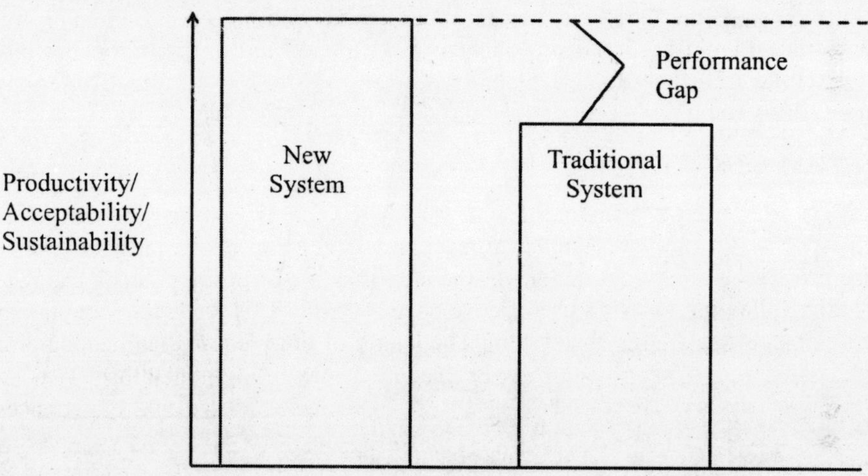

Fig. 45.2: Demonstration of the New System's Improved Performance

Secondly, OFR promotes *collaboration with extension and development agencies* which helps in improving the efficiency of the system and diffusion process. Involvement of extension and development agencies as partners and participants in the system generation process will bring them directly into contact with the farmers. It will also acquaint them with the salient features of the technology while it is being generated.

Thirdly, the OFR stage may test the *suitability of the existing institutional framework* for proper delivery of the systems/technology to the users. For example, in most countries, crop, livestock, and forestry extension services are independent with little collaborative activity. Moreover, the crop extension service tends to be much better organized than the other two. Since the scope of agroforestry cuts across all three fields, appropriate mechanisms could be developed at the technology generation stage to integrate the roles of these various agencies in the diffusion process. Mechanisms for using non-governmental and traditional institutions in the diffusion process such as village associations and local leadership structures, could also be studied at the OFR stage.

TYPES OF ON-FARM RESEARCH

There are basically two types of on-farm research

(1) Experimental OFR

(2) Developmental OFR

The range of objectives for which OFR may be carried out is very wide. Thus, it is inconceivable that a single OFR activity will embrace all the objectives. It is essential in carrying out OFR to define the objectives clearly. It is important to make a clear distinction between the assessment of a technology for its biological, technical, and economic potential, on the one hand, and assessment for its workability, acceptability, and potential adaptability by farmers on the other. Often, different types of OFR activity are required for achieving different objectives (Fig. 45.3). The interpretation of results from different types of OFR should take into account the limitations imposed on the system by the specific objectives and methodologies followed.

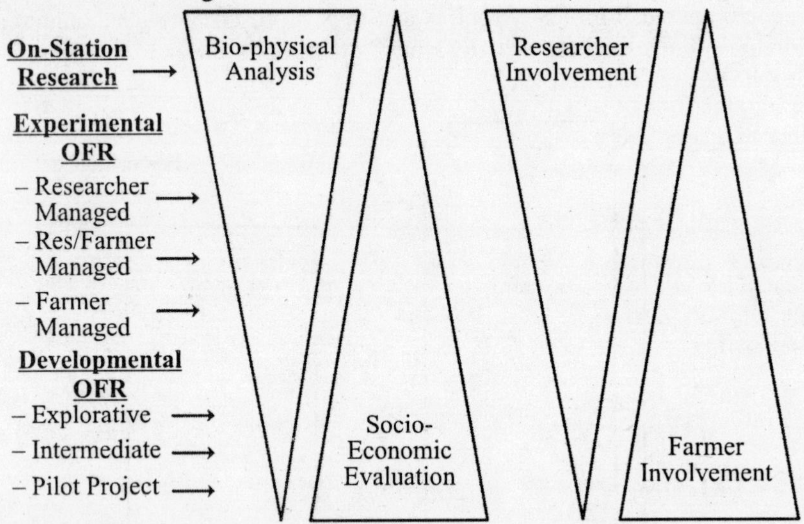

Fig. 45.3: Research Objectives and Levels of Farmer Involvement in Different Types of OFR

EXPERIMENTAL ON-FARM RESEARCH

This is the more commonly known and practised of the two types of OFR. It is performed for bio-physical, technical, and economic assessment of alternative systems or treatments within the framework of standard experimental designs. Bio-physical assessment aims at determining the system's biological and physical yield and productivity, while economic assessment enquires into the availability of labour, cash, and other resources for meeting the projected needs of the alternative system, and looks into the level and dependability of profit. Experimental OFR trials emanate directly from on-station research. Their structure and design are very similar to those used on-station. Generally, however, on-farm experimentation is kept as simple as possible to ensure effective farmer understanding of issues and meaningful involvement and contribution. Depending on the nature of farmer/researcher involvement in the trials, experimental OFR may be further classified into three different types, namely:

(1) Researcher-managed trials

(2) Joint Researcher and farmer-managed trials

(3) Farmer-managed trials

RESEARCHER-MANAGED TRIALS

Researcher-managed trials are very similar in structure to on-station trials. The researcher is responsible for directing and implementing the treatments in accordance with the chosen design and methodology of the trial. A single farmer's field could be used for such a trial, though this may be repeated on another farmer's plot (if required). The farmer and researcher may have a *landlord/tenant relationship*, which represents the lowest degree of farmer involvement. It applies to the situation where a researcher obtains a plot of land from a farmer's holding to carry out an OFR activity in which the farmer has no part to play. The farmer may also have no direct interest in what is going on, and may consider his or her involvement only as having given land (on lease, on loan, or as gift) for research activity.

JOINT RESEARCHER AND FARMER-MANAGED TRIALS

These are trials in which management and operation are the joint responsibility of farmer and researcher. Such trials need to be made simpler than the researcher-managed trials, since an increased level of farmer's involvement is required. Simplicity insures a better understanding of the trial by the farmer. The farmer's role may be termed *active involvement* (*researcher-controlled*), as the farmer is directly involved in carrying out some or all of the management operations in the trial. However, the farmers' contribution is very clearly defined and controlled by the researcher. Farmer is therefore unable to use his/her initiative, and does what the researcher has programmed for him/her to do in terms of treatment applications and management requirements.

FARMER-MANAGED TRIALS

In farmer-managed OFR, the farmer is responsible for carrying out almost all management operations for the trial. An even higher level of simplicity is thus required, and the number of unit plots within a single farmer's field is kept at a minimum to avoid complications for the farmer.

DEVELOPMENTAL ON-FARM RESEARCH

This type of OFR activity has received less attention than the experimental type. It involves: (1) the introduction of particular systems within the farmer environment, and (2) the assessment of the workability of the system and its acceptability by farmers. Developmental OFR operates within a framework of research-extension collaboration. Its main purpose is the extrapolation of the tested results to the target area. An attempt is made to fine-tune the system and to determine the required support structures prior to wide-scale extension of the systems. Through the developmental OFR process, farmers of the targeted area are gradually exposed to a new system, and their management of the system is monitored in order to identify problem areas and researchable issues. Developmental OFR makes use of extension techniques and methodologies for the introduction of the concept or system and development of farmer's awareness. For this reason, developmental OFR requires the joint involvement of researchers, farmers and extension agents. The farmer's involvement evolves gradually to the point where he or she considers the experiment to be his/her own, and is free to make modifications and adjustments in accordance with his/her own circumstances.

PHASES IN DEVELOPMENTAL ON-FARM RESEARCH

There are three phases in the developmental OFR process, namely:

(1) The exploratory phase

(2) The intermediate phase

(3) The pilot project phase

EXPLORATORY PHASE

The exploratory phase is the stage where a new system or concept introduced into a community. This allows the farmers to gain an accurate image and a practical understanding of the system. This exploratory phase thus has a demonstration objective. It begins with the identification of individual farmers within the community with whom the researchers work closely to put the system on the ground. During this phase, researchers' involvement is very high as the farmer's perception of the system is almost nil. Only a few farmers (1-5) are selected for these trials; they do not necessarily have to be in the same village.

INTERMEDIATE PHASE

The intermediate phase begins after the exploratory trials are established and management of the system has commenced. This phase, like the exploratory phase, is also targeted at individual farmers, but requires a greater involvement of the farmer in the establishment and management of the experiment. Farmers participating in the intermediate trials will have a clearer perception of the system because of the existence of the exploratory (demonstration) units which provide a visual dimension for discussions on the system's structure and potential. The number of farmers used in this stage could be 3-5 times the number in the exploratory trials. The exact number is usually determined by resource availability.

PILOT PROJECT PHASE

The pilot project phase takes off after the intermediate trials have been conducted and farmers' understanding and capability in management have been sufficiently established. At this point, direct involvement of researchers in management and other farm operations is withdrawn, and the farmer involvement is greatly increased. The main objective of the pilot project is to place the system within a community framework and to enable assessment of its relevance, workability, and acceptability by the farmers. More specifically, a pilot project aims at:

- Evaluating benefits to farmers and community from the adoption of the new system
- Assessing institutional and social requirements for the accelerated adoption of the system
- Identifying constraints and researchable problems in adoption of the system by individual farmers and the community
- Redesigning the production programme as necessary for wide-scale implementation

During the pilot project stage, the focus is on the community rather than on individual farmers. Participating farmers are responsible for all farm activities and for management of the experimental plots. The involvement of extension agents, which is required to a lesser degree in the earlier stages, also reaches its peak during the pilot project phase. The extension officer becomes the key link between the farmer and the researcher.

EXPERIMENTAL DESIGNS IN ON-FARM RESEARCH

Experimental design is a way to arrange treatments so that error and bias are reduced and the data may be accurately analysed using statistics. Design and analysis fit together to make a meaningful whole. *Replication*, *randomization*, and *use of a control* are essential in designing an experiment because they help to separate out treatment effects from natural levels of background variation. Without these three factors, any data gathered will be just about worthless. In research, *error* refers to anything besides treatment effects that is measured. Setting up the experiment in a structured way helps to reduce the amount of error in the experiment; replication, randomization, and use of a control allow to use statistics to actually separate out error from treatment effects in the measurements.

Yield varies from year to year due to different rainfall, temperature, and other factors. Yield also often varies from location to location within a field because of high and low spots, inherent fertility, variation in soil types, etc. This natural variation is why replication is essential in experiments. The more times a treatment are duplicated, the more likely it is that measurements will reflect the effect of the treatment rather than the effect of natural variation in the field. However, there is a point of diminishing returns after which increasing replication doesn't give increased accuracy. There are statistical methods to calculate the exact number of replications necessary for any given experiment, but in most cases between four and six replications are adequate for on-farm experimentation. There are different ways to replicate an experiment. One way is to have multiple plots at one location. Another way is to replicate the experiment on many farms. Another way is to replicate across time, performing the experiment in multiple years.

To prevent unanticipated sources of bias from entering the data measurements, treatments must be randomized. This means that the order of treatments cannot be the same in every replication. Standard experimental formats or designs are usually used in on-farm research. The criteria used to select which design fits which experiment depends on the number of treatments under investigation. If one wants to compare two levels of a treatment, he can use a design called a *paired comparison*. An easy statistical analysis, the t-test, can be performed on the data to detect any significant differences. If one wants to add more levels of treatments, he can use a *randomized complete block* design. A *split plot* design allows seeing how different treatments interact. These designs can provide more information than the paired comparison but also require more sophisticated statistical analyses and more space in the field. Because each treatment must be replicated at least 4 times, each treatment increases the research area required. The following designs are mostly used in on-farm research:

- Paired comparison
- Randomized complete block
- Split plot

PAIRED COMPARISON

This classical on-farm design is characterized by having long strips side-by-side in the field, replicated up to six times. Each pair of strips should be located in an area that is fairly homogeneous or similar. This makes it easy to apply treatments along the entire strip without having to start or stop in the middle of the field. A general principle is to avoid "edge effects" by taking measurements from the centre of the plot.

RANDOMIZED COMPLETE BLOCK

The randomized complete block (RCB) design is used for experiments looking at three or more levels of a treatment in areas with topographic variation on a gradient, such as on a slope. It is similar to the paired comparison in that all treatments are grouped together in blocks that are replicated across the field. The purpose of "blocking" the treatments is to maintain as much uniformity as possible in each block and keep environmental variation outside of the blocks. Blocking doesn't help when variation in the field is random, but can reduce error when variation runs along a gradient such as a slope, irrigated field, changing soil texture, or other factor. Block borders ought to run perpendicular to the gradient. The treatments are randomized within each block to avoid bias. In this design, blocks are synonymous with replications because a complete set of treatments is replicated in each block.

SPLIT PLOT

The split-plot design looks at how different levels of a treatment interact with another set of treatments by applying sub-treatments over main treatments. Statistically speaking, you sacrifice precise information on the main treatment for more precise measurements of the sub-treatments simply because the sub-treatments are replicated more than the main treatments. Though fairly easy to set up in the field, analysing the data can be somewhat complex. Because of the greater number of treatments, adequately replicating a split-plot experiment can take much more space in the field.

ON-FARM ADAPTATIONS

As mentioned earlier, researchers and extension workers should be alert to adaptations made by farmers that could lead to improvements in existing agroforestry practices viz. alley farming practices. Clearly, this requires a common sense approach. When a particular farmer does something unusual, the researcher/ extension worker should first discover the reason. If the farmer was simply ignorant of recommended practices, and was mismanaging the system, the researcher should educate him or her so as to improve the chances of success. However, if the farmer had a rationale for the modification, the researcher should look at the new system carefully, and discuss with the farmer the possible harms and benefits. The farmer may be modifying the system to better achieve his/her own objectives, while not harming crop productivity or long-term sustainability. Through this sort of interaction, researchable issues can emerge and suggest new directions for on-station and/or on-farm research. On-farm research is a powerful decision-making tool for agroforestry farmers. Economic data can be included in the results and useful cost benefit analyses of different agroforestry farming practices may be generated. Groups of farmers can join together in research clubs and assist each other in investigating new systems, practices and inputs.

■■■■

46 EXPERIMENTAL DESIGNS IN AGROFORESTRY RESEARCH

INTRODUCTION

Agroforestry is "a dynamic, ecologically based, natural resources management system that, through the integration of trees in farms and in the landscape, diversifies and sustains production for increased social, economic and environmental benefits for land users at all levels" (Leakey 1996). Agroforestry is a summary term for practices that involve the integration of trees and other large woody perennials into farming systems through the conservation of existing trees, their active planting and tending, or the tolerance of spontaneous tree regrowth (Schroth et al 2004). It is a kind of agricultural practice where fast growing trees are grown along with a crop. It is an integrated approach of using the interactive benefits from combining trees and shrubs with crops and/or livestock. It combines agricultural and forestry technologies to create more diverse, productive, profitable, healthy and sustainable land use systems.

Agroforestry is an integrated and intensive agricultural production system that includes trees and shrubs as an essential component to achieve environmental, economic and social goals. This means that trees are not incidental to the farm operation but rather contribute to improved productivity, yield, profitability and sustainability (AAFC 2012). Agroforestry is the management of trees for forest products. It can be an integral component of productive agriculture. It has application to multiple use agricultural land, or as a single land use. It may include existing native forests and forests established by landholders. It is a flexible concept, involving both small and large sized land holdings. Agroforestry has the potential to combine four elements of land management: Agriculture, Land Protection, Forestry and Conservation. In a very similar way, agroforestry is defined as the combination of *agriculture* and *forestry*; rather than treating these as separate options for land use, it brings both together, recognizing and promoting tree use on farms (WAC 2013). Any agroforestry system must therefore have the following:

(1) Woody perennials (trees, shrubs, hedges, bamboo, etc.)

(2) Herbaceous plants (local crops), and/or livestock; Agrisilviculture, Silvipasture (both agroforestry)

(3) Spatial or temporal (rotational) arrangement

(4) The woody and non-woody components are deliberately arranged together by the farmer and it is not accidental

(5) The woody components are interacting with non-woody components ecologically and economically.

In the definition of agroforestry we recognize three characteristics and attributes:

- Three common characteristics
  - Deliberate growing of woody perennials and crops or animals
  - Significant interaction (ecological and/or economic) between the components
  - Desirable benefits
- Three attributes
  - Maintain or increase productivity of land (productivity)
    - o Increased output of tree products
    - o Increased yield of associated crops
    - o Decrease in cropping system inputs
    - o Increased labour efficiency
  - Conserve the productive potential of land (sustainability)
  - Improved or new AF technologies to conform local farming practices (adoptability)

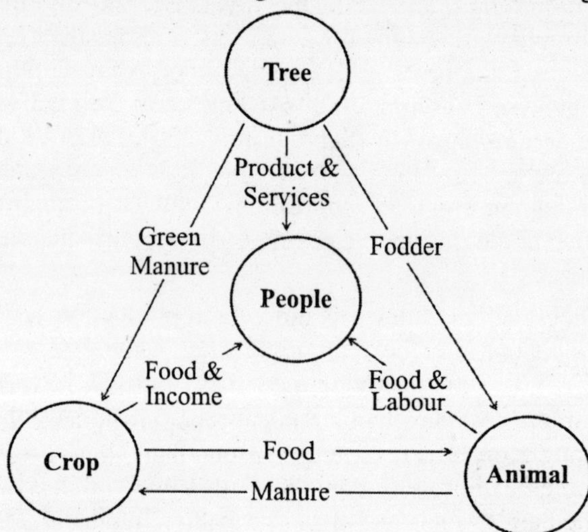

Fig. 46.1: Interaction of Major Components in an Agroforestry Land Use

WHAT IS EXPERIMENTAL DESIGN?

Experimental design is a planned interference in the natural order of events by the researcher. He does something more than what he carefully observes the events. This emphasis on experiment reflects the higher regard generally given to information so derived. There is good rationale for this. Much of the substantial gain in knowledge in all sciences has come from actively manipulating or interfering with the stream of events. There is more than just observation or measurement of a natural event. A selected condition or a change (treatment) is introduced. Observations or measurements are planned to illuminate the effect of any change in conditions. Researchers want to make inferences about what produced, contributed to, or caused events. To gain such information without ambiguity, some form of experimental design is ordinarily required. As a consequence, the need for using rather elaborate designs ensues from the possibility of alternative relationships, consequences or causes. The purpose of the design is to rule out these alternative causes, leaving only the actual factor that is the real cause.

The plan that we choose is a design which forms an essential part of research strategies. The design itself entails:

- selecting or assigning subjects to experimental units
- selecting or assigning units for specific treatments or conditions of the experiment (experimental manipulation)
- specifying the order or arrangement of the treatment or treatments
- specifying the sequence of observations or measurements to be taken

An *Experimental Design* is the laying out of a detailed experimental plan in advance of doing the experiment. Well-chosen experimental designs maximize the amount of "information" that can be obtained for a given amount of experimental effort. The choice of an experimental design depends on the objectives of the experiment and the number of factors to be investigated. In experimentation, we attempt to monitor the effects of certain inputs or material on the subject matter, of interest. The inputs could be different plants/trees planted under identical conditions, while the effects to be monitored could be the changes in soil fertility status, the yield of agricultural crops planted between the tree rows, the productivity of the animal being fed with the foliage from the tree crops, or the tree crop performance. The allocation of treatments (inputs) to the experimental units (plots) may be loosely referred to as the design.

The experimenter decides which individual unit is to receive which particular treatment according to a laid down procedure. The choice of procedure will often determine the basic design. What is important, however, is that during an experiment, the researcher has a choice as to how and when to apply the treatments. (In a survey situation, in contrast, there is no such choice). The choice of design is influenced by several considerations, notably the objectives, the amount of resources, and the time available. In all cases, however, the emphasis is on the reduction of unknown error and the elimination of systematic bias.

AGROFORESTRY EXPERIMENT DESIGN REQUIREMENTS

Agroforestry systems are more complex than the mono-cropping system. Several characteristics of the trees like slow growth, long term effects on their surroundings, long life, age of trees, the area over which the influence of trees extend, etc. complicate the issue of designing experiments for these systems. In addition to the basic principles i.e. randomization, replication and local control (Jaggi**), there are several other factors that need to be taken care of while planning agroforestry experiments. For example, within one experimental plot there may be crop rows, individual trees or hedges with different treatments applied to each and each component may respond to its own treatment as well as to the treatments applied to other components. Generally there are certain experimental design requirements peculiar to agroforestry. For the selection of appropriate experimental designs, the experimenter will need to respond to the following twelve issues (Nokoe 1992):

(1) What are the specific study objectives?

(2) What are the variables to be observed (i.e. the dependent variables)?

(3) Are these dependent variables quantifiable and/or measurable? If these are not measurable, what criteria will you use for later comparisons among treatments?

(4) What are the independent variables that is to say, the treatments to be applied?

(5) Are these treatments fixed or random? In other words, do you have several treatments to choose from or do you have a fixed number of treatments among which specific comparisons are desired?

(6) Are the levels of treatments qualitative or quantitative?

(7) How many replicates of each treatment can be available?

(8) Will all the replicates be available at the same time?

(9) How much land or materials (to which the treatments are to be applied) are available?

(10) Are the subject materials or available land uniform enough to receive all treatments at a time?

(11) What will be the sampling unit? That is, indicate how small or large the area or material to be observed (In response, one may simply define the area)

(12) How often will data (from dependent variables) be collected?

A good research design starts by asking a good research question. A good research or experimental design encompasses different steps (Fig. 46.2).

Thinking about research questions helps us to ensure that we are learning something needed and new and the activities will generate valid evidence efficiently. A good research question must be:

1. *Clear and focused*: The question will drive the design of the approach use to answer it. Hence it must be unambiguous, specific and bounded. Poor research question: Do people value fruit trees? Better: What levels of investment in fruit trees at farm level would be appropriate to meet varying farm household objectives? How does this vary with farm resource endowment? How would such a level of investment affect land and labour allocation to other farm enterprises?

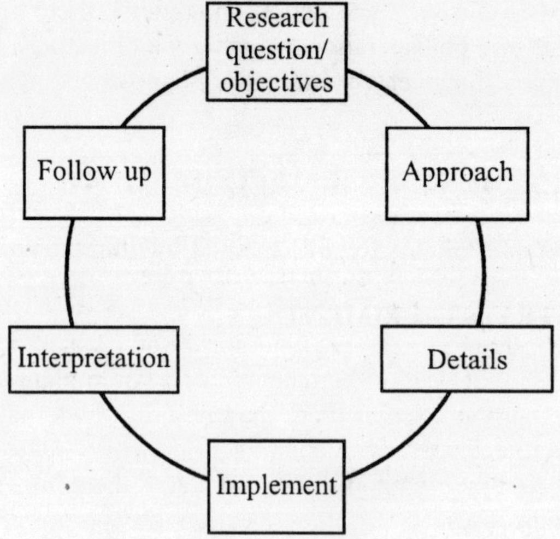

Fig. 46.2: Elements of Good Research Design

2. *Non-trivial*: Question and answer should be complex and not obvious. Poor: Could planting fruit trees increase availability of fruit? Better: Do women in farm families in the project area value fruit trees as contributors to: (a) nutrition, (b) income (c) other? How does this vary with: (1) farm resource endowment (2) farming objectives?

3. *Answered to make progress*: You envisage a stream of activities to meet project activities but cannot move ahead because there are unknowns. Link to the project conceptual

framework or theory of change. Poor: Do people use indigenous fruit trees? Better: What are cost-effective ways of protecting young trees from livestock, fire and drought?

4. *Linked to a hypothesis*: There should be a real, open debate about the question – people informed by global and local knowledge should be arguing about the way the world works. We collect the data and answer.
5. *Of interest beyond the project*: Answers to a good question will be useful beyond the project and will be published. Poor: What percentages of people in the project area eat home-grown fruit? Better: How is level of consumption of home-grown fruit determined by farm resource endowment, gender and other farm enterprises?
6. *Answerable within the scope of the project*: Remember the limitations of time and resources; the limitations of observational data (surveys), particularly cross-sectional (one-off) surveys; and studies might generate answers from secondary data, models, literature and experts as well as primary field data.
7. Answered through a trial or experiment
8. Contributes to project objectives (tree species and management effects on crops and water)

The main issues need to be addressed in an experimental design includes:

- Identification of the objectives of the experiment in the form of specific questions.
- The selection of treatments to provide answers to the questions. Statistical concepts of treatment structure are used to provide maximum information relating to these questions.
- Choice of experimental units and amount of replication.
- Control of the variability between sets of units through systems of blocking and/or by using ancillary information (i.e. covariate information) collected on the units.
- Allocation of treatments to particular units within the overall structure of units, involving, where possible, an element of randomization.
- Collecting data that are appropriate for the objectives of the research. These may be on the whole experimental unit, or may involve sampling within the unit.

TYPES OF AGROFORESTRY EXPERIMENTS

There are different types of agroforestry experiments. The several types of experimental trials in agroforestry are:

- variety trials
- provenance trails
- field germplasm trials
- multipurpose tree screening and evaluation trials
- agroforestry management trials (e.g., mulching effects, crop productivity)
- livestock integration trials (e.g., feed supplementation, animal productivity)
- fertilizer trials
- cultural/agronomic trials
- chemical (other than fertilizer) trials
- socio-economic assessment (e.g., economic returns) trials

Before discussing the types of experiments and their designs, it is vital to know the components of an agroforestry system: land, trees and non-trees.

LAND

Agroforestry is not a system of pots on a balcony or in a greenhouse. It is a system by which land is managed for the benefit of the landowner, environment and long-term welfare of society. While appropriate for all landholdings, this is especially important in the case of hillside farming where agriculture may lead to rapid loss of soil. If the farmer owns the land, he/she has a vested interest in thinking conservatively, how the land can be maintained over long periods of time. Unfortunately, farmers who rent land may have less interest in the long-term benefits of agroforestry and may even fear that making improvements will raise the rent or result in the lease being terminated.

TREES

In agroforestry, particular attention is placed on multipurpose trees or perennial shrubs. The most important of these trees are the legumes because of their ability to fix nitrogen and thus make it available to other plants. The role of trees on the small farm may be sources of fruits, nuts, edible leaves, and other food; sources of construction material, posts, lumber and thatching; sources of non-edible materials including sap, resins, tannins, insecticides, and medicinal compounds; sources of fuel; beautification; shade; soil conservation, especially on hillsides and improvement of soil fertility. In order to plan for the use of trees in agroforestry systems, considerable knowledge of their properties is necessary. Desirable information for each species includes its benefits, adaptability to local conditions (climate, soil, and stresses), the size and form of the canopy and root system, and suitability for various agroforestry practices.

NON-TREES

Any crop plant can be used in agroforestry systems. The choice of crop plants in designing such systems should be based on those crops already produced in a particular region either for marketing, feeding animals, or for home consumption, or that have great promise for production in the region. In keeping with the philosophy of agroforestry, however, other values to be considered in crop selection include proper nutrition, self-sufficiency and soil protection. Thus, selection of crops requires a judgment based on knowledge of the crops, adaptations, production uses, as well as family needs, opportunities for barter, and markets. Similarly any farm animal can be used in agroforestry systems. The choice of animal will be based on the value the farmer places on animal-derived benefits including income, food, labour, non-food products, use of crop residues, and manure.

Agroforestry experiments may be broadly classified under following four heads (Jaggi et al 2001):

(1) Multipurpose tree (MPT) screening and selection trials: Usually experiments are designed to screen several promising germplasms (often of several species, but sometimes of varieties of one or two species), the objective being to identify the most promising among them based on their early performance and also on specific outputs and benefits. There are many factors which will contribute to the performance of the tree or "system" under investigation. Some of these will be related to tree management factors such as configuration, spacing, and pruning and harvesting. Since ultimately, these factors will be controlled by the farmers, it is necessary to find out to what degree these can be modified by them without affecting the technology. One of the factors that the farmer will not be able to modify is the choice of tree species. Initially, a series of species may be proposed but once a choice is made, the farmer will be "stuck" with it. The following questions need to be addressed in choosing one or more tree species:

- Are the trees adapted to environmental conditions?
- What function(s) are the trees expected to provide?
- What product(s) are needed from the trees?
- Which adoption constraints can be overcome by species choice?

Some important points to consider when establishing a species screening trial are:

- Using well documented seed sources for species, varieties and provenances.
- Designing the screening trial and the measurements that will be made in the context of the intervention or technology under consideration.
- Carefully choosing plot size and duration of the trial because above and below ground interactions between species will rapidly develop.
- It should be aware that the number of entries being evaluated will affect statistical design.

(2) Component and system management trials: The objective of these trials is to improve specific agroforestry technologies like in alley cropping, selection of hedgerow species, hedgerow spacing, hedge row to crop row distance, etc.

(3) Component interaction studies: These include understanding and quantifying the interrelationships between components of agroforestry systems (mostly tree-crop interactions) to investigate the sharing of resources below or above ground.

(4) Prototype evaluation trials: These trials are undertaken with the objective of evaluating specific packages of agroforestry technologies under realistic field conditions. They represent a transition between research and extension and are undertaken in farmers' fields.

The arrangement of tree and crop components in relation to one another within plots deserves important consideration, especially in interaction studies (Jaggi**). Therefore, use of an appropriate design is very important aspect of agroforestry experimentation. A general recommendation or solution cannot be given for all the situations since a specific problem requires a particular design. The plot size is also an important consideration in agroforestry experiments. The size of the plot depends on many factors like the objectives of the experiment, types of measurement to be made, the expected duration of the experiment, likely ultimate size of the trees, the requirement for extra space to avoid interference between plots, etc. For example, MPT selection trials involve large number of species or provenances and if the focus is on tree survival and growth, each species can be allocated a small plot (20-30 m^2). Larger plots (50-200 m^2) are needed for experiments designed to test species for particular agroforestry technologies or to study the effects of management practices (Roger and Rao 1990). Choosing suitable sample from which to measure the response is another important aspect of these experiments.

AGROFORESTRY EXPERIMENTAL DESIGNS

To obtain valid results from agroforestry experiments, it is important to follow scientific principles. The research methods should take into account the unexplainable variability in the experimental material. In view of the large number of factors to be studied in the system, the number of combinations may become too unwieldy to be evaluated in conventional designs. However conventional designs are being used in some situations. Nokoe (1992), Jaggi et al (2001) and Jaggi** have given an overview of the designs used in agroforestry experimentation.

COMPLETELY RANDOMIZED DESIGN (CRD)

It is the simplest design. In CRD, each experimental unit has an equal chance of receiving a certain treatment. The completely randomized design for 'p' treatments with 'r' replications will have 'rp' plots. Each of the 'p' treatment is assigned at random to a fraction of the plots (r/rp), without any restriction. A Useful assumption for the application of this design is homogeneity of the land or among the experimental materials. This design is rarely used in most trials involving woody vegetation, but could be used under laboratory and possibly green house conditions. The total source of variation (error) is made up of differences between treatments and within treatments.

RANDOMIZED COMPLETE BLOCK DESIGN (RCBD)

The most widely used design in agroforestry experiments because of its simplicity is randomized complete block design (RCBD) or randomized block design (RBD). This design is characterized by the presence of equally sized blocks, each containing all of the treatments. The randomized block design for 'P' treatments with 'r' replications has 'rp' plots arranged into 'r' blocks with 'p' plots in each block. Each of the 'p' treatments is assigned at random to one plot in each block. The allocation of a treatment in a block is done independently of other blocks. For example in an experiment planned to study the type of crops that could be grown under different multipurpose trees, the objective is to identify suitable crop under different tree species. The design recommended is RCB Design with five crops viz. Maize, Sorghum, Cowpea, Wheat and Barley each with two varieties viz. local and improved taken under each of the three trees viz. *Acacia nilotica*, *Acacia tortilis* and *Dalbergia sissoo*, the replications in the design are two. There are many situations in agroforestry where a randomized complete block (RCB) design may not be appropriate. For example, in multipurpose tree (MPT) evaluation studies or in germplasm evaluation trials, the number of treatments may be too large so that it is difficult to locate a block with sufficient uniformity to accommodate a complete set of all treatments to be tested. In such cases, Incomplete Block designs such as lattice designs, alpha designs, and confounded designs may be adopted.

LATIN SQUARE DESIGN (LSD)

The randomized complete block design is useful for eliminating the contribution of one source of variation only. In contrast, the Latin Square Design (LSD) can handle two sources of variations among experimental units. In LSD, every treatment occurs only once in each row and each column. For example, cropping history is the first source of variation in four large blocks and in addition to this if the land has fertility gradient at right angle to the "cropping history". One may tackle this problem by using a Latin Square Design. Each treatment (in this case, the *Gliricidia* accessions) is applied in "each" cropping history as well as in "each" fertility gradient In our example, restriction on space allows us to have a maximum of only 16 plots, when, say, 64 might have been ideal. The randomization process has to be performed in such a way that each accession appears once, and only once, in each row (cropping history) and in each column (fertility gradient).

The four blocks correspond to the four different cropping histories. The Latin Square (LS) design thus minimizes the effect of differences in fertility status within each block. The total sources of variation are made up of row, column, treatment differences, and experimental error. For field trials, the plot layout must be a square. This condition imposes a severe restriction on the site as well as on the number of treatments that can be handled at any one time. However, the principle can be extended to animal experime-ntation where a physically square arrangement does not necessarily exist. For instance, if the intention is to assess the nutritional effects of the accessions when fed to animals, the latter could be divided into four age and four size classes. The LS arrangement will thus be used to ensure that each age class and size class receives one and only one of each accession type. The LS design can be replicated leading to what is commonly referred to as "Replicated Latin Squares".

INCOMPLETE BLOCK DESIGNS

One precondition for both the RCB and LS designs is that all treatments must appear in all blocks and all rows (For RCB) or columns (For LS). Sometimes with large number of treatments (say 20 accessions), each requiring relatively large plot sizes, this condition may not be practicable. Latin Square and RCB then fail to reduce the effect of heterogeneity(s). The designs in which the block phenomenon is followed but the condition of having all the treatments in all blocks is not met, are called Incomplete Block designs. In Incomplete Block situations, the use of several small blocks with fewer treatments results in gains in precision but at the expense of a loss of information on comparisons within blocks. The analysis of data for incomplete block designs is more complex than RCB and LS. Thus where computation facilities are limited, incomplete block designs should be considered a last resort. Among incomplete block designs, lattice designs are commonly used in species and variety testing.

FACTORIAL DESIGN

Only one factor (i.e., one accession or other treatment) is involved in CRD and RCBD. However, more than one factor will often need to be studied simultaneously. Such experiments are known as factorial experiments. The treatments in factorial experiments consist of two or more levels of the two or more factors of production. Suppose we are interested in studying the yield of an agricultural crop in an alley farm where four different leguminous tree species and three cultural methods are of interest. The leguminous tree species could be *Acacia* spp., *Cassia* spp., *Leucaena* spp. and *Gliricidia* spp. The cultural treatment could include two weedings, one weeding and no weeding; the agricultural crop is maize planted between hedgerows of the same tree species. For a complete factorial set of treatments, each level of each factor - must occur together with each level of every other factor. Thus in the present case we ensure that each cultural method is applied to each tree species. Since there are 4 species and 3 cultural methods, the total number of treatments will equal 12. In reality, what we have here is 12 treatments, with one treatment being made up of 2 factors having 4 and 3 levels, respectively. One might say, in this case, the factors are crossed.

This is not an "experimental design" but rather a "treatment design," because the 12 treatment combinations could be applied to any of the designs discussed previously. If we take the simplest design, the unrestricted randomized design, and four replications, then the conduct of an experiment with 4 leguminous species and 3 cultural methods will imply the randomization of "12 treatments" in 48 plots. If it is a Block design, we will have to ensure that each of the 12 treatments appears in all the blocks. The advantages of the factorial arrangement are many. One major advantage is the reduction in the number of experiments, and a second the possibility of studying the interactions among the various factors. A significant interaction implies that changes in one factor may be dependent on the level of the other factor. If this happens, interpretation of the results has to be done cautiously to avoid inaccurate general statements on the individual factors.

SPLIT-PLOT DESIGN

Another commonly used design is the split-plot design. Split-plot experiments are factorial experiments in which the levels of one factor, for example tree species, are assigned at random to large plots. The large plots are then divided into small plots known as "sub-plots" or "split plots", and the levels of the second factor are assigned at random to small plots within the large plots. This arrangement is often useful when we wish to combine certain treatments (as in factorial and nested), some of which require larger plots than others for practical and administrative convenience. Examples are situations requiring spraying insecticides, irrigation, tillage trials, etc. Usually, the treatment on which maximum information is desired is placed in the split-plot or in the smallest plot. It is important to emphasize that the split-plot is not a design as such but rather refers to the manner in which treatments are allocated to the plots.

For example, in an alley-cropping experiment under the component and system management trials, for testing hedgerow species and hedgerow pruning management, species could be allocated to main plots and pruning management (frequency, height of cutting, etc.) to subplots of a split-plot design. Split-plot experiments are particularly useful if the treatments are susceptible to 'neighbour effects' as in irrigation experiments or in experiments involving tree species of varying growth habits, especially height. But these experiments should be avoided as far as possible because the treatments in the main-plots are compared with less precision than those in the subplots. The split-plot design could be further extended to Split-split plot design. The strong recommendation is that split plot designs should only be used when practical considerations of plot size require it. Related to the split-plot design is the Strip-plot design that is specifically suited for two factor experiments when both factors require large plot sizes and when the interaction between the two factors is desired to be measured more precisely than the effects of either of the two factors.

SYSTEMATIC DESIGNS

In some circumstances, randomization is impracticable or unwise. For example in spacing trials, the extreme treatments must not adjoin. The designs for such situations are called "systematic designs" and these can be used in agroforestry experiments (Huxley 1985b). Many factors of interest in agroforestry research do not allow randomization, such as slope, orientation, rainfall or distance from tree or hedge. These designs despite their limitations have been proposed for agroforestry research because they minimize the requirement for field and experimental resources. The concept of systematic design is that over a series of plots the level of the quantitative factor varies systematically but slowly (e.g.10% change) through the sequence of plots. Each particular treatment level is then surrounded by treatment levels differing only slightly, and therefore guard areas are unnecessary.

The harvested area on the systematic design is much more than in any other randomized design conducted on the same piece of land. For example, in alley cropping trials, the spacing between hedgerows (alley width) can increase gradually across a site rather than having plots of varying alley widths located at random. Another example is an alley cropping trial where distance of crop rows from hedgerows is the treatment variable so that the crop row nearest the hedgerow will be compared against other rows that are progressively further from the hedgerow. The direction of the systematic change of the treatment level within each plot can and should be randomly selected.

Fan designs (Nelder 1962) and Parallel Row layouts (Bleasdale 1967) are the systematic designs normally used to investigate sole crop spacing problems with a wide variety of plant material. For agroforestry mixed cropping experiments, including tree-crop interface studies, spacing of both components is an important variable and systematic design can be used to explore the effects. For example to select a suitable tree/bush species and agricultural crop associates, a simple Parallel Row design for the tree species, with replicated strips of the various agricultural crops sown across this at right angles could be a suitable layout. Fan designs are a kind of systematic design that involves planting on rays emanating from the centre of an arc of a circle, the angle between successive rays usually being about 50. In tree spacing trials, the distance between trees could be increased systematically. If the object of the experiment is to compare inter-row spacing, the plant density along a row can be kept constant.

The statistical analysis of the data from systematic design is little complex. Whatever kind of systematic design or fan is used, the data should not be regarded as coming from a randomized experiment. The analysis of variance would not be the most appropriate initial step in the analysis. What it expresses is a relationship between, say, growth and the character i.e. area, inter-row spacing or rectangularity, etc. Hence the appropriate statistical tool is fitting of a response function to represent the effect of the systematic factor for each set of data. The subsequent analysis is based on the comparisons of parameters of the fitted response functions. Such comparisons may be between replicates of the systematic plot or between different treatments applied to the different plots within which the systematic treatment level variation occurs. Mead (1967) has examined the general problem of spacing and has given a mathematical model for the estimation of inter-plant competition. Randomization in the systematic design can be introduced at a higher level (Nelder 1962). This is done by repeating the sets of systematically arranged treatments at different locations within the experimental site. Some more systematic designs are described below.

Y - DESIGN

The kind of field layout used for tree-crop interface (TCI) investigations, i.e. component interaction studies, will depend on the precise research objectives and the resources available, including the extent and nature of the site. Orientation may have a significant role at TCI because of the shading effect of tall trees on the under storey crop component. Therefore, orientation as a factor may also be considered. Huxley (1985a) suggested that TCI and orientation effects can be studied in simple geometrical designs. Huxley described a Y design (or 120° design) with three strips (or arms) each containing the tree and crop components juxtaposed to each other. This design enables a study of the TCI in a relatively small space and reduces interfaces between the different arms, especially with tall trees. This design takes care of the orientation effects.

STAR DESIGN

Rao et al (1991) have described a TCI experiment designed to investigate the effect of row orientation and to evaluate the potential of hedgerow intercropping. The layout was designed to allow a test of a wider range of orientation than the Y design and the experimental design used is a Star Design. In Star design, from the centre of the layout there are eight arms (or plots) at 45° angle with rows oriented in all directions giving the shape of a Star.

TWO-WAY SYSTEMATIC DESIGN

Rao et al (1990) have used a two-way systematic design to study the potential of hedgerow intercropping in semi-arid India. This design was intended to give a range of spacing between hedges and concomitant changes in the proportion of annual crops.

AUGMENTED DESIGN

Pinney (1991) has made use of augmented design for on-farm trials or prototype evaluation trial. On-farm trials are an essential means of technology transfer and can be viewed as a means of providing feed-back from the farm to research station. It is a powerful and appropriate strategy for agroforestry research, especially for the applied type. The technology tested at research station is tested under farmer conditions to obtain information about its performance and acceptability by the farmers.

DESIGN TO STUDY THE MULTIPURPOSE TREE EFFECT

For soil sampling, the area under the canopy of each tree is divided into four radial transects (Takimoto et al 2008). The factors studied can be depth and distance. For each selected trees, it is designed for soil sampling each at two depths: 0 – 20 cm and 20 – 50 cm; and at three horizontal distances from the tree trunk: near 1 m (under canopy), 3 m (at the boarder of canopy), and 10 m from the trunk (control) at four different directions East, West, North and South around the tree. The centre of the circle in (Fig. 46.3) represents a single tree, whereas the circle represents the area covered by the canopy of the tree and the square shaped in four different directions are sample plots having size of 0.5 m x 0.5 m each in which composite soil samples could be collected.

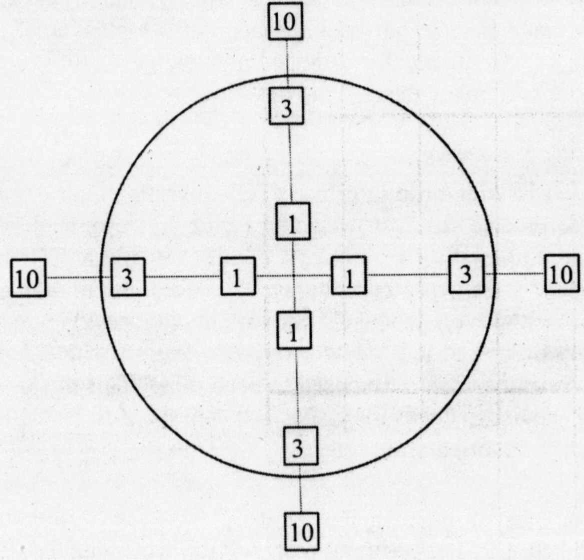

Fig. 46.3: Experimental Design for Soil Sampling under a Tree

47 INTERNATIONAL AND NATIONAL ORGANIZATIONS OF AGROFORESTRY

INTERNATIONAL ORGANIZATIONS

WORLD AGROFORESTRY CENTRE

The World Agroforestry Centre (formerly the International Centre for Research in Agroforestry-ICRAF) is part of the alliance of the Consultative Group on International Agricultural Research (CGIAR) centres dedicated to generating and applying the best available knowledge to stimulate agricultural growth, raise farmer's incomes and protect the environment. It was established in 1977 with its headquarters in Nairobi, Kenya. It has five regional offices located in India, Indonesia, Kenya, Malawi and Mali, and conduct research in eighteen other countries around the developing world. The World Agroforestry Centre is dedicated to generating and applying the best available knowledge to stimulate agricultural growth, raise farmers' incomes, and protect the environment.

The Centre's vision is a rural transformation in the developing world as smallholder households strategically increase their use of trees in agricultural landscapes to improve their food security, nutrition, income, health, shelter, energy resources and environmental sustainability. The Centre's mission is to generate science-based knowledge about the diverse roles that trees play in agricultural landscapes, and use its research to advance policies and practices that benefit the poor and the environment. It conducts research in agroforestry, in partnership with national agricultural research systems with a view to developing more sustainable and productive land use. The focus of its research is countries/regions in developing world, in particular the tropics. The World Agroforestry Centre is guided by the broad development challenges: poverty alleviation that entails enhanced food security and health, improved productivity with lower environmental and social costs and resilience in the face of climate change and other external shocks. Website: http://www.worldagroforestry.org/

CONSULTATIVE GROUP ON INTERNATIONAL AGRICULTURAL RESEARCH

Consultative Group on International Agricultural Research (CGIAR) is a global partnership established in 1971 that unites organizations engaged in research for a food secure future. CGIAR research is dedicated to reducing rural poverty, increasing food security, improving human health and nutrition, and ensuring more sustainable manage-ment of natural resources. It is carried out by 15 Centers that are members of the CGIAR Consortium, in close collaboration with hundreds of partner organizations, including national and regional research institutes, civil society organizations, academia, and the private sector.

The 15 Research Centers generate and disseminate knowledge, technologies, and policies for agricultural development through the CGIAR Research Programs. The CGIAR fund provides reliable and predictable multi-year funding to enable research planning over the long term, resource allocation based on agreed priorities, and the timely and predictable disbursement of funds. The multi-donor trust fund finances research carried out by the Centers through the CGIAR Research Programs. CGIAR have over 8,000 scientists and staff, unparalleled research infrastructure and dynamic networks across the globe. CGIAR collections of genetic resources are the most comprehensive in the world. CGIAR collaborate with research and development partners to solve development problems. Website: http://www.cgiar.org/

The mission and objectives are:

- To identify significant global development problems that science can help solve
- To collect and organize knowledge related to these development problems
- To develop research programs to fill the knowledge gaps to solve these development problems
- To catalyze and lead putting research into practice, and policies and institutions into place, to solve these development problems
- To lead monitoring and evaluation, share the lessons we learn and best practices we discover
- To conserve, evaluate and share genetic diversity
- To strengthen skills and knowledge in agricultural research for development around the world

CENTRE FOR INTERNATIONAL FORESTRY RESEARCH

Centre for International Forestry Research (CIFOR) is a nonprofit, global facility dedicated to advancing human wellbeing, environmental conservation and equity. It conducts research that enables more informed and equitable decision making about the use and management of forests in less-developed countries. It was established in May 1993 and has its headquarters in Bogor, Indonesia. CIFOR is one of 15 Centers within the Consultative Group on International Agricultural Research (CGIAR).

CIFOR's research and expert analysis help policy makers and practitioners shape effective policy, improve the management of tropical forests and address the needs and perspectives of people who depend on forests for their livelihoods. In implementing its research, CIFOR is collaborating with other Centers within the CGIAR - the World Agroforestry Centre, the International Center for Tropical Agriculture (CIAT) and Bioversity International - to conduct the CGIAR Consortium Research Programme *Forests, Trees and Agroforestry* (also known as CRP6). CRP6 aims to achieve a clear objective: enhancing the management and use of forests, agroforestry and tree genetic resources across the landscape, from forests to farms.

Website: http://www.cifor.org/

INTERNATIONAL INSTITUTE OF TROPICAL AGRICULTURE

The International Institute of Tropical Agriculture (IITA) is one of the world's leading research partners in finding solutions for hunger, malnutrition, and poverty. IITA is a non-profit international organization founded in 1967 and has headquarters at Ibadan, Nigeria. Its award-winning agricultural research for development (R4D) addresses the needs of the poor and vulnerable in the tropics. It works with public and private sector partners to enhance crop quality and productivity, reduce risk to producers and consumers, and generate wealth from agriculture. The Institute's R4D covers the following themes: biotechnology and genetic improvement, natural resource management, plant production and plant health, and social science and agribusiness.

For the last 45 years, IITA's agricultural R4D has delivered over 70 per cent of CGIAR impact in sub-Saharan Africa. It has achieved this impact by focusing on key tropical food crops, such as banana and plantain, maize, cassava, soybean, cowpea, tree crops, and yam. With the mission to enhance food security and improve livelihoods through R4D, IITA tackles these challenges by pursuing these interrelated objectives: improving food security, increasing the profitability of foods and other agricultural products, reducing risks to producers and consumers, and helping national entities to expand agricultural growth. Website: http://www.iita.org/

INTERNATIONAL LIVESTOCK RESEARCH INSTITUTE

The International Livestock Research Institute (ILRI) works with partners worldwide to enhance livestock pathways out of poverty, principally in sub-Saharan Africa and Asia. The products of these research partnerships help people in developing countries keep their farm animals alive and productive, increase and sustain their livestock and farm productivity, find profitable markets for their animal products, and reduce their risk to livestock-related diseases. ILRI is a non-profit institution founded in the year 1994 and has its headquarters in Nairobi, Kenya, a principal campus in Addis Ababa, Ethiopia, and offices in West and southern Africa and in South, Southeast and East Asia.

ILRI is organized in three major research themes viz. Biotechnology Research Theme; Markets, Gender and Livelihoods Theme; People, Livestock and the Environment Theme and a regional Biosciences eastern and central Africa Hub. Cross-cutting gender and systems research as well as communications, knowledge management, capacity development and partnership programs support these research themes. ILRI leads the CGIAR Research Program on Livestock and Fish, leads a component of the CGIAR Research Program on Agriculture for Nutrition and Health on the prevention and control of agriculture-associated diseases, and contributes to five other CGIAR Research Programs. Website: http://www.ilri.org/

INTERNATIONAL CROPS RESEARCH INSTITUTE FOR THE SEMI-ARID TROPICS

The International Crops Research Institute for the Semi-Arid Tropics (ICRISAT) is a non-profit, non-political organization that conducts agricultural research for development in the drylands of Asia and sub-Saharan Africa. ICRISAT and its partners help empower the poor people to overcome poverty, hunger and a degraded environment through better agriculture. ICRISAT was established in 1972 and has its headquarters in Hyderabad, Andhra Pradesh, in India, with two regional hubs (Nairobi, Kenya and Bamako, Mali) and country offices in Niger, Zimbabwe, Malawi, and Mozambique. ICRISAT envisions a prosperous, food-secure and resilient dryland tropics. To achieve this, its mission is to reduce poverty, hunger, malnutrition and environmental degradation in the dryland tropics. It approaches this through partnership-based international agricultural research for development that embodies *Science with a Human Face*.

ICRISAT's research-for-development agenda is achieved through the CGIAR Research Program on Dryland Systems, the CGIAR Research Program on Policies, Institutions and Markets, the CGIAR Research Program on Grain Legumes and the CGIAR Research Program on Dryland Cereals. These are implemented in ways that benefit smallholder farmers, enabling them and their families to go beyond subsistence farming to produce surpluses that can be stored and sold to markets, paving the way for prosperity in the drylands. A socio-economic process called inclusive market-oriented development (IMOD) on which ICRISAT's strategy is anchored. ICRISAT has defined six developmental outcomes that it believes will help the poor to move along the IMOD path: food sufficiency, intensification, diversification, resilience and health & nutrition, and women empowerment. Website: http://www.icrisat.org/

INTERNATIONAL CENTER FOR AGRICULTURAL RESEARCH IN THE DRY AREA

Getting agricultural research innovations into use in the dry areas is critically important, especially since these regions cover 40 per cent of the earth's surface and are home to 2.5 billion people – a significant percentage of the world's population. This is the core work of the International Center for Agricultural Research in the Dry Area (ICARDA), which works with partners worldwide, developing innovations to improve food security and the livelihoods of the rural poor. ICARDA was established in 1977 and has its headquarters in Aleppo, Syria. ICARDA's work targets non-tropical dry areas in developing countries, and also produces international public goods with potential for global application. Research covers crops, the management of natural resources (water, land, biodiversity), small ruminant production, farming systems (intensification, diversification, integration between farming system components), and socio-economics and policy research.

The innovations include crop improvement research and production of new varieties; technology and policy packages that countries can apply to increase the productivity and income of smallholder farmers; water harvesting; conservation agriculture methods; diversification of production systems; integrated crop/rangeland/ livestock production systems; poverty mapping; and plant Genetic resources. ICARDA's genebank holds over 135,000 accessions from over 110 countries: traditional landraces, improved germplasm, and a unique set of wild crop relatives. Research and capacity development is coordinated by a network of regional programs and country offices in: West Asia, South Asia and China, North Africa, Nile Valley and sub-Saharan Africa, Central Asia and the Caucasus, Arabian Peninsula, and Highlands Network (parts of Afghanistan, Iran, Pakistan and Turkey). Website: http://www.icarda.org/

INTERNATIONAL FOOD POLICY RESEARCH INSTITUTE

Established in 1975 with headquarters at Washington DC, the International Food Policy Research Institute (IFPRI) identifies and analyzes national and international policies and strategies for meeting the food needs of the developing world on a sustainable basis, with particular emphasis on low-income countries and on the poorer groups in those countries. While the research effort is geared to the precise objective of contributing to the reduction of hunger and malnutrition, the factors involved are many and wide-ranging, requiring analysis of underlying processes and extending beyond a narrowly defined food sector. The Institute's research programme reflects worldwide collaboration with governments, as well as private and public institutions, interested in increasing food production and improving the equity of its distribution. Research results are disseminated to policymakers, opinion formers, administrators, policy analysts, researchers, and others concerned with national and international food and agricultural policy. Website: http://www.ifpri.org/

BIOVERSITY INTERNATIONAL

Bioversity International is the only global non-profit organization that places the use and conservation of agricultural biodiversity in smallholder farming systems at the center of its work. Bioversity International has its headquarters in Rome, Italy and works in areas of significant levels of rural poverty to help farmers throughout Asia, in Latin America and the Caribbean. Bioversity focuses on rain-fed farming systems, primarily managed by smallholder farmers, in areas where large-scale agriculture is not a viable option. Its research influences policy decisions and investment in agricultural research, from the local level to the global level. From local traditions to global treaties, Bioversity's work creates new opportunities for economic development, health and sustainability. Website: http://www.bioversityinternational.org/

INTERNATIONAL CENTER FOR TROPICAL AGRICULTURE

The International Center for Tropical Agriculture (CIAT), working in collaboration with hundreds of partners across the developing world, is dedicated to developing technologies, innovative methods, and new knowledge that better enable farmers, mainly smallholders, to improve their crop production, incomes, and management of natural resources. Established in 1970 and with headquarters at Cali, Colombia, as one of the four original research centers in the Consultative Group of International Agricultural Research (CGIAR), CIAT now works in more than 50 countries worldwide. Progress in CIAT's crop improvement research also depends on unique collections of genetic resources (65,000 crop samples in all) which it holds in trust for humanity. CIAT conducts research on two major issues that cut across tropical crops and production niches: (1) sustainable management of tropical soils, and (2) decisions and policies that are critical for coping with challenges such as climate change, environmental degradation, gender inequities, and weak links between farmers and markets.

Center's research in agrobiodiversity, soils, and policy is carried out by about 200 scientists working in Latin America and the Caribbean as well as in 28 countries Africa and 5 in Southeast Asia. Guided by a vision of eco-efficiency, CIAT's work amply demonstrates how economically and ecologically efficient use of resources can raise agricultural productivity to improve livelihoods while minimizing negative environmental impacts. Reflecting farmers' centuries-old concern with prudent resource use, the Center's eco-efficiency approach is especially necessary now in the face of multiple crises centering on the global climate, environment, and economy. Website: http://ciat.cgiar.org/

INTERNATIONAL DEVELOPMENT RESEARCH CENTRE

The International Development Research Centre (IDRC) (French: *Centre de recherches pour le développement international*-CRDI) is a Canadian Crown Corporation created by the Parliament of Canada in 1970 that supports research in developing countries to promote growth and development. The aim of IDRC is to create innovative, lasting local solutions that build "healthier, more equitable, and more prosperous societies". IDRC also builds the capacity of people and institutions in developing countries to undertake the research that they identify as most urgent. IDRC works with researchers as they confront the challenges of the 21st century within their own countries and contribute to global advances in their fields. In doing so, IDRC supports networking and knowledge sharing between scientific, academic, and development communities in Canada and developing countries. IDRC head office is in Ottawa and regional offices are located in Montevideo, Uruguay; Cairo, Egypt; Nairobi, Kenya; and New Delhi, India. IDRC funds research under four broad themes: agriculture and the environment; health and health systems; social and economic policy; and science, technology, and innovation.

Website: http://www.idrc.ca

INTERNATIONAL UNION OF FOREST RESEARCH ORGANIZATIONS

The International Union of Forest Research Organizations (IUFRO) is a non-profit, non-governmental international network of forest scientists, which promotes global cooperation in forest-related research and enhances the understanding of the ecological, economic and social aspects of forests and trees. It has its headquarters in Austria. IUFRO was founded as the International Union of Forest Experiment Stations in 1892. IUFRO focuses on science-based management, conservation and sustainable development of the world's forest resources for the benefit of present and future generations.

IUFRO has identified three strategic goals and related objectives for the period 2011-2014. IUFRO addresses Research and Institutional Goals at the same time. The Research Goals strongly focus on the following six thematic areas: Forests for People, Forests and Climate Change, Forest

Bioenergy, Forest Biodiversity Conservation, Forest and Water Interactions and Resources for the Future. Website: http://www.iufro.org/

IUFRO's three strategic goals are

- To strengthen research and expand IUFRO's capacity for interdisciplinary cooperation
- To strengthen coordination within the scientific community and increase visibility of science-based research findings
- To further strengthen IUFRO's work at the science-policy interface

INTERNATIONAL INSTITUTE FOR ENVIRONMENT AND DEVELOPMENT

International Institute for Environment and Development (IIED) is one of the world's most influential international development and environment policy research organizations. It was founded in 1971 by merging the concept and cause of sustainable development and has the headquarters in London, UK. IIED is one of the world's most influential policy research organizations working at the interface between development and environment. IIED contributes to many international policy processes and frameworks, including the Intergovernmental Panel on Climate Change, the Millennium Ecosystem Assessment and the UN conventions on climate change and biological diversity. International Institute for Environment and Development (IIED) promotes sustainable development through collaborative research, policy studies, consensus building and public information, and has extensive information available on its Website: http://www.iied.org/

INTERNATIONAL INSTITUTE OF RURAL RECONSTRUCTION

The International Institute of Rural Reconstruction (IIRR) was established in 1960. IIRR has over 80 years of history in participatory, integrated and people-centered development. The Institute has enhanced the capacity and confidence of over 100,000 development managers, practitioners and community leaders across Asia, Africa and Latin America and has a long history of documenting and disseminating field-based experience through its publications. Website: http://www.iirr.org/

INTERNATIONAL INSTITUTE FOR SUSTAINABLE DEVELOPMENT

The International Institute for Sustainable Development (IISD) is a Canadian-based, public policy research institute that has a long history of conducting cutting-edge research into sustainable development. It was established in 1990 and has its head office in Winnipeg, Manitoba, Canada. The institute champions global sustainable development through innovation, research and relationships that span the entire world. It contributes to the international negotiations on the sustainable management of forests with targeted communications aimed at improving the transparency and clarity of the discussions in a variety of forum viz. UNFF, CSD, WCFSD. Website: http://www.iisd.org/

EUROPEAN FOREST INSTITUTE

European Forest Institute (EFI) is an international organization established by European States and so far 23 European States have ratified the Convention on EFI. EFI has over 120 Associate and Affiliate Member organizations in more than 30 countries, five Regional Offices and one Project Centre. Its headquarters is in Joensuu, Finland and Policy Advice office is located in Barcelona, Spain. EFI's mission is to promote, conduct and cooperate in research of forestry and forest products at the pan-European level and to make the results of the research known to all interested parties, notably in the areas of policy formulation and implementation, in order to promote conservation and sustainable management of forests in Europe. Website: http://www.efi.int/portal/

The five Regional Offices of EFI are:

- **Mediterranean Regional Office (EFIMED)**, located in Barcelona, Spain established in 2007
- **Central European Regional Office and the Observatory for European Forests (EFICENT-OEF)** established in 2009 and it has distributed locations in Germany, France and Switzerland
- **Atlantic European Regional Office (EFIATLANTIC)**, located in Bordeaux, France established in 2009
- **Central-East European Regional Office (EFICEEC)**, located in Vienna, Austria established in 2010
- **North European Regional Office (EFINORD)**, located in Copenhagen, Denmark established in 2010

FOOD AND AGRICULTURE ORGANIZATION OF THE UNITED NATIONS

The Food and Agriculture Organization of the United Nations (FAO) is a specialized UN agency leading international efforts to defeat hunger and poverty. Serving both developed and developing countries, FAO acts as a neutral forum where all nations meet as equals to negotiate agreements and debate policy. FAO helps developing countries and countries in transition to modernize and improve agriculture, forestry and fisheries practices, ensure good nutrition, achieve food security for all, better the lives of rural populations and contribute to the growth of the world economy.

In 1943, forty four governments gathered in Hot Springs, Virginia, United States for United Nations Conference on Food and Agriculture committed themselves to find a permanent organization for food and agriculture. First session of FAO Conference in Quebec City, Canada on 16 October to 1 November 1945 established FAO as a specialized United Nations agency. FAO is composed of seven departments including Forestry Department. It has its headquarters in Rome, Italy and present in over 130 countries. Website: http://www.fao.org/

UNITED NATIONS DEVELOPMENT PROGRAMME

The United Nations Development Programme (UNDP) is the United Nation's global development network, an organization advocating for change and connecting countries to knowledge, experience and resources to help people build a better life. It is based on the merging of the United Nations Expanded Programme of Technical Assistance, created in 1949, and the United Nations Special Fund, established in 1958. UNDP was established in 1965 by the General Assembly of the United Nations and has its headquarters in New York.

UNDP's mandate is to fight poverty. As the development organization of the United Nations, UNDP works to achieve this through supporting development projects in a sustainable way. The overarching goal of reducing poverty has been more concretely defined through the Millennium Development Goals. UNDP's network links and coordinates global and national efforts to reach these goals and helps developing countries attract and use aid effectively. Website: http://www.undp.org

UNITED NATIONS ENVIRONMENT PROGRAMME

The United Nations Environment Programme (UNEP) is an international organization established in 1972 to coordinate United Nations environmental activities, assisting developing countries in implementing environmentally sound policies and practices. It was founded as a result of the United Nations Conference on the Human Environment in June 1972 and has its headquarters in the Gigiri neighborhood of Nairobi, Kenya. UNEP also has six regional offices and various country offices. The UNEP's mission is to provide leadership and encourage partnership in caring for the environment by inspiring, informing and enabling nations and peoples to improve their quality of life without compromising that of the future generations. Website: http://www.unep.org/

WORLD BANK GROUP – FORESTS AND FORESTRY

The World Bank Group's (WBG) mission is to fight poverty and improve the living standards of people in the developing world and has its headquarters in Washington D.C. It is a development bank established in 1944 and now with a closely associated group of five development institutions. It provides loans, policy advice, technical assistance and knowledge sharing services to low and middle income countries to reduce poverty.

Website: http://www.worldbank.org/

THE WORLD BANK FORESTS TEAM: It takes a multi-sectoral approach to forest management that takes into account the impact on forests and forest people of activities, policies and practices outside the forest sector. The Forests Team is comprised of professionals housed in both Agriculture and Rural Development Department and Environment Department, as well as expertise from regional staff and external partners. Furthermore, the Forests Team is part of the Bank's Natural Resources Management Thematic Group which encompasses Land Resources Management, Natural Resource Management Institutions and Irrigation and Drainage.

THE FOREST POLICY AND STRATEGY OF THE WORLD BANK: It was approved in October 2002 and has been implemented to help attain the goal of poverty reduction while safeguarding the environmental values intrinsic to sustainability. The strategy is built on three pillars: harnessing the potential of forests to reduce poverty; integrating forests into sustainable economic development; protecting global forest values

PROGRAMME ON FORESTS: Programme on Forests (PROFOR) is an independent programme hosted at the World Bank within the Environmentally and Socially Sustainable Development Network. It is a multi-donor partnership formed to pursue a shared goal of enhancing forest's contribution to poverty reduction, sustainable development and protection of environmental services. Initially established in 1997 to implement the outcomes of the Intergovernmental Panel on Forests (IPF), PROFOR was relocated from the United Nations Development Programme to the World Bank in 2002.

GLOBAL ENVIRONMENT FACILITY

The Global Environment Facility (GEF) unites 182 countries in partnership with international institutions, civil society organizations (CSOs) and the private sector to address global environmental issues while supporting national sustainable development initiatives. Today the GEF is the largest public founder of projects to improve the global environment. An independently operating financial organization, the GEF provides grants for projects related to biodiversity, climate change, international waters, land degradation, the ozone layer and persistent organic pollutants. The Global Environment Facility Secretariat and the Evaluation Office are located in Washington, DC, USA.

The Global Environment Facility was established in October 1991 as a $1 billion pilot program in the World Bank to assist in the protection of the global environment and to promote environmental sustainable development. The United Nations Development Programme, the United Nations Environment Programme and the World Bank were the three initial partners implementing GEF projects. In 1994, GEF was restructured and moved out of the World Bank system to become a permanent, separate institution. Since 1994, however, the World Bank has served as the Trustee of the

GEF Trust Fund and provided administrative services. The GEF was entrusted to become the financial mechanism for the UN Convention on Biological Diversity, UN Framework Convention on Climate Change, Montreal Protocol of the Vienna Convention on Ozone Layer Depleting Substances, Stockholm Convention on Persistent Organic Pollutants and the United Nations Convention to Combat Desertification. Website: http://www.thegef.org/ gef/

COMMONWEALTH FORESTRY ASSOCIATION

Commonwealth Forestry Association (CFA), UK is the world's longest established international forestry organization, tracing its history back to the founding of the Empire Forestry Association in 1921. Today it unites 1,200 members in 78 Commonwealth and non-Commonwealth countries. The objective of the CFA is to promote the conservation and sustainable management of the world's forests and the contributions they make to peoples' livelihoods through improved communication amongst forestry professionals and all those with an interest in the world's forests. A particularly important aspect of the CFA is the links that are being forged between forestry professionals and civil society. Website: http://www.cfa-international.org/

ASSOCIATION FOR TEMPERATE AGROFORESTRY

The Association for Temperate Agroforestry focuses on temperate agroforestry with an emphasis on North America. Formed in 1991, the Association for Temperate Agroforestry Inc. (AFTA) is a private, non-profit organization based at the University of Missouri Center for Agroforestry at Columbia. With members in the US, Canada and overseas, AFTA pursues its mission through activities such as networking, information exchange, public education, and policy development. The intended audience includes university researchers, educators and extensionists, private farm and forestry groups, public policy makers, and private landowners. AFTA is primarily a voluntary organization. Many individuals from universities, public agencies, private organizations and businesses donate their time and agroforestry-related expertise to help AFTA conduct its educational programs.

AFTA's educational programmes are aimed at facilitating the exchange of scientific and practical information about temperate agroforestry, primarily in North America. Since 1989, the biennial North American Agroforestry Conference is being organized which is a major scientific forum for those involved in agroforestry research, extension and application in the US, Canada and overseas. Website: http://www.aftaweb.org/

EUROPEAN AGROFORESTRY FEDERATION

The European Agroforestry Federation (EURAF) aims at promoting the use of trees on farms as well as any kind of silvopastoralism throughout the different environmental regions of Europe. EURAF has about 250 members from 18 different European countries. The European Agroforestry Federation will be comprised of national agroforestry organizations and It aims to promote agroforestry, including lobbying for agroforestry adapted policies at the European scale. Website: http://www.agroforestry.eu/

ASIA-PACIFIC AGROFORESTRY NETWORK

The genesis of the Asia-Pacific Agroforestry Network (APAN) can be traced to the Nineteenth FAO Regional Conference for Asia and the Pacific in 1988. The Ministers of Agriculture and Forestry present at the conference endorsed the development of a regional initiative in agroforestry to address three sets of pressing problems: marginal lands and marginal people, shifting cultivation and sustainable development. FAO was asked to formulate a Project Document to support the establishment of a regional agroforestry network. The original title of the project was Agroforestry Systems

Research and Development in the Asia and Pacific Region and a scheduled duration of 20 months. The project began in April 1991 with the appointment of the Regional Coordinator. On 2 May 1991, the project secretariat was formally inaugurated at the Forestry Research and Development Agency (FORDA) in Bogor, Indonesia. The ten original member countries were: Bangladesh, India, Indonesia, Laos, Nepal, Pakistan, the Philippines, Sri Lanka, Thailand and Viet Nam. All countries designated national focal points and coordinators to collaborate on the project. At the first Project Advisory Committee meeting in November 1991, the name of Asia-Pacific Agroforestry Network was adopted.

The long-term development objective was to assist member countries to achieve sustainable production within traditional farming systems and to raise the socio-economic status of smallholder farm households through the improvement and adoption of environmentally sound and economically viable agroforestry practices, activities and programmes. Since 1991, the annual APAN Advisory Committee (APANAC) meetings have been hosted on a rotating basis by Indonesia, Nepal, Viet Nam, Sri Lanka, Laos and China. These meetings have proved to be important mechanisms for the participatory planning of regional and national network activities. They have strengthened the commitment of APAN national coordinators and other partners and increased the participation of national coordinators in evaluating the progress and problems of network activities.

FORUM FOR AGRICULTURAL RESEARCH IN AFRICA

Forum for Agricultural Research in Africa (FARA) is an umbrella organization bringing together and forming coalitions of major stakeholders in agricultural research and development in Africa. FARA complements the innovative activities of national, international and sub-regional research institutions to deliver more responsive and effective services to its stakeholders. It plays advocacy and coordination roles for agricultural research for development, while the national agricultural research systems (NARS), advanced research institutions (ARIs) and international agricultural research centres (IARCs) develop improved technologies along the research-to-development continuum in their respective countries and coverage areas.

FARA was established in 2001 with its headquarters in Accra, Ghana. FARA in its *Vision* and *Mission* statements emphasize its commitments to the Millennium Development Goal of eradicating extreme poverty and hunger, and agriculture-led development. FARA's *Super Objective* is to sustainably reduce African food insecurity and poverty and enhance environmental conditions. This is to be done by sustainably establishing high broad-based agricultural growth in Africa (*General Objective*). FARA will make this contribution by achieving its *Specific Objective* of sustainable improvements to broad-based agricultural productivity, competitiveness and markets. Website: http://www.fara-africa.org/

AFRICAN NETWORK FOR AGRICULTURE, AGROFORESTRY AND NATURAL RESOURCES EDUCATION

The African Network for Agriculture, Agroforestry and Natural Resources Education (ANAFE) comprises of African colleges and universities teaching agriculture and natural resource sciences. Supported by the World Agroforestry Centre (ICRAF) in Nairobi, Kenya, the network was established in 1993 and is currently made up of 134 member institutions (universities and colleges) in 35 African countries. The network consists of four regional chapters known as RAFTs (Regional Agricultural Fora for Training - one each in Eastern and Central Africa (ECA), Southern Africa (SA), the Sahelian countries (Sahel), and the Africa Humid Tropics countries (AHT)) - and 21 National Chapters known as NAFTs (National Agricultural Fora for Training). Activities in each RAFT are spearheaded through a Focal Institution. The ANAFE Secretariat is based in Nairobi, Kenya and supports the networking functions. The objective of the Organization is to improve in a sustainable manner, the contribution of agricultural & natural resource management education to social and economic development of the African people. Website: http://anafe-africa.org

SOUTHEAST ASIAN NETWORK FOR AGROFORESTRY EDUCATION

The Southeast Asian Network for Agroforestry Education (SEANAFE) is a network of 85 member institutions in five country networks; in Indonesia, Laos, Philippines, Thailand and Vietnam. It was established in 1999 and has its headquarters in Bogor, Indonesia. SEANAFE has been working towards improving agroforestry education, training, research and extension, and contributing to socioeconomic development, empowerment of farming communities and sustainable natural resource and environ-mental management in the Southeast Asian region. SEANAFE works closely with established networks, government agencies, and regional and international development organizations in building capacities in agroforestry and natural resources education. It also welcomes collaboration with civil society and other non-government organizations involved in promoting the practice and science of agroforestry through either formal or non-formal education. SEANAFE is focusing on three major projects: markets for agroforestry tree products; agroforestry landscape analyses; and accelerating forest and environmental policy implementation. Website: http://www.seanafe.org

SWEDISH INTERNATIONAL DEVELOPMENT AGENCY

Swedish International Development Agency (SIDA) is a government funding organization under the Swedish Foreign Ministry. It has its headquarters in Stockholm with many offices in other countries with ongoing activities. Swedish development aid follows three thematic priorities; democracy and human rights, environment and climate change and gender equality and womens' role. SIDA carry out enhanced development cooperation with a total of 33 countries in Africa, Asia, Europe and Latin America. The selection of cooperation countries are based on political decisions made by the Swedish government. SIDS's mission is to allocate aid and other funding. Website: http://www.sida.se/English/

CANADIAN INTERNATIONAL DEVELOPMENT AGENCY

The Government of Canada is committed to making its international assistance more efficient, focused and accountable. Canadian International Development Agency (CIDA) is Canada's lead funding agency for development assistance. CIDA's aim is to manage Canada's support and resources effectively and accountably to achieve meaningful, sustainable results and to engage in policy development in Canada and internationally, enabling Canada's effort to realize its development objectives. CIDA's mission is to lead Canada's international effort to help people living in poverty. Three priority themes that agenda and guide CIDA's work are increasing food security; securing the future of children and youth; and stimulating sustainable economic growth.

Website: http://www.acdi-cida.gc.ca/home

AUSTRALIAN CENTRE FOR INTERNATIONAL AGRICULTURAL RESEARCH

Australian Centre for International Agricultural Research (ACIAR) works to improve the well-being of people in developing countries and Australia through international collaboration in research and related activities that develop sustainable agricultural systems and appropriate strategies for natural resource management.

Website: http://aciar.gov.au/

INTERNATIONAL INSTITUTE OF TROPICAL FORESTRY

The International Institute of Tropical Forestry (IITF) is part of the United States Department of Agriculture (USDA) Forest Service that was founded in 1939. The IITF is located in Rio Piedras, Puerto Rico on the grounds of the University of Puerto Rico, Agricultural Experimental Station. The Institute is dedicated to tropical forestry on an international level. Within the Forest Service's mission of caring for the land and serving people, the IITF's mission is to develop and exchange knowledge critical to sustaining tropical ecosystem benefits for human kind. Website: http://www.fs.fed.us/global/iitf/

REGIONAL ORGANIZATIONS

COMMONWEALTH SCIENTIFIC AND INDUSTRIAL RESEARCH ORGANIZATION

The Commonwealth Scientific and Industrial Research Organization (CSIRO) is Australia's national science agency and one of the largest and most diverse research agencies in the world. It was founded in 1916 originally as the Advisory Council of Science and Industry. CSIRO's corporate headquarters are located at Limestone Avenue in the Australian Capital Territory in Campbell, Australia. CSIRO's research is performed by the 13 Divisions. CSIRO is involved in more than 740 research activities, working with leading scientific organizations in over 80 countries with partners and customers ranging from foreign governments, small companies to large multi-nationals and international foundations. Website: http://www.csiro.au/

USDA NATIONAL AGROFORESTRY CENTER

The USDA National Agroforestry Center (NAC) had its origins in the 1990. It began as a Forest Service Research and State & Private Forestry effort in 1992 and expanded into a partnership with the Natural Resources Conservation Service (NRCS) in 1995. It is administered by the Forest Service's, Washington, DC, Office of Research and Development. NAC offices are located in Lincoln, Nebraska and Blacksburg, Virginia. NAC accelerates the application of agroforestry through a national network of partners. It conducts research, develops technologies and tools, coordinates demonstrations and training, and provides useful information to natural resource professionals. The US Department of Agriculture's National Agroforestry Center supports agroforestry practices which integrate trees and agriculture, primarily in temperate areas. Website: http://nac.unl.edu/

UNIVERSITY OF MISSOURI CENTER FOR AGROFORESTRY

University of Missouri Center for Agroforestry (UMCA) in Columbia, USA is among the global centers in agroforestry research, education and outreach. Its website contains a wealth of up-to-date information on temperate agroforestry research along with a number of practical "hands on" guides. The Center for Agroforestry at the University of Missouri, was established in 1998, and is one the world's leading centers contributing to the science underlying agroforestry, the science and practice of intensive land-use management combining trees and/or shrubs with crops and/or livestock. Website: http://www.centerforagroforestry.org/

CENTER FOR SUBTROPICAL AGROFORESTRY

The Center for Subtropical Agroforestry (CSTAF) is a multidisciplinary, multi-institutional initiative established within the School of Forest Resources and Conservation (SFRC), University of Florida, USA to undertake activities in research, extension, and education and training related to agroforestry. The Center focuses initially on the southern and southeastern parts of the United States, but it is envisioned to be an eventual world leader in agroforestry. The Center for Subtropical Agroforestry is committed to developing agroforestry systems that are economically and environmentally interesting to landowners in the Southeast United States.

Website: http://sfrc.ifas.ufl.edu/cstaf/

AGRICULTURE & AGRI-FOOD CANADA'S AGROFORESTRY DEVELOPMENT CENTRE

The Agriculture & Agri-Food Canada's Agroforestry Development Centre (AAFC-ADC) focuses on research, development, promotion and adoption of agroforestry science and practices across Canada and in temperate regions. Website: http://www4.agr.gc.ca/AAFC-AAC/display-afficher.do?id=1186517615847&lang=eng

NITROGEN FIXING TREE ASSOCIATION

Nitrogen Fixing Tree Association (NFTA) is a non-profit organization based in USA that encourages more and better use of nitrogen fixing trees (NFTs) to benefit people, especially small farmers in the developing tropics. People, who are interested in receiving publications about trees in sustainable land-use systems, material to establish trials, or small grants for development projects, can consider becoming an associate of the Nitrogen Fixing Tree Association (NFTA).

Besides conducting its own research, NFTA assists the research of others through the Cooperative Planting Program (CPP) and other activities. The CPP helps people plant trials to determine which NFT species are appropriate for their area's environment and people's needs. They can provide help in selecting species, seeds, and planting, management, and analysis guidelines. A database provides quick access to information on the range, ecology and uses of NFTs. The goal of Development program is to help people design and manage better agroforestry projects. The program includes training, grants of $300 to support small scale projects such as nurseries and demonstration areas, and on-site technical assistance by staff members or NFTA associates.

Website: http://www.fadr.msu.ru/rodale/agsieve/txt/vol2/1/art7.html

ALTERNATIVES TO SLASH-AND-BURN

Alternatives to Slash-and-Burn (ASB) is a global partnership of over 50 institutions around the world with a shared interest in conserving forests and reducing poverty in the humid tropics supported by ICRAF, Nairobi, Kenya. Website: http://www.asb.cgiar.org

THE AGROFORESTRY RESEARCH TRUST

The Agroforestry Research Trust based in UK researches temperate agroforestry and all aspects of plant cropping and uses, with a focus on tree, shrub and perennial crops. Website: http://www.agroforestry.co.uk/

APPROPRIATE TECHNOLOGY TRANSFER FOR RURAL AREAS

Appropriate Technology Transfer for Rural Areas (ATTRA) provides technical assistance to farmers, extension agents, market gardeners, agricultural researchers, and other agro professionals in the US. Website: http://www.attra.org

CAB INTERNATIONAL

CAB International (CABI), UK has its mission to help improve human welfare worldwide through the dissemination, application and generation of scientific knowledge in support of sustainable development, with emphasis on agriculture, forestry, human health and the management of natural resources. Website: http://www.cabi.org/

CENTRO AGRONOMICO TROPICAL DE INVESTIGACION Y ENSENANZA

Centro Agronomico Tropical de Investigacion y Ensenanza (CATIE) works to improve the well-being of humanity through scientific research and postgraduate education applied to the development, conservation and sustainable use of natural resources. It has its headquarters at Costa Rica. Website: http://www.catie.ac.cr/

THE INTERNATIONAL ANALOG FORESTRY NETWORK

The International Analog Forestry Network was established in 1996 as a response to the need to maintain an exchange of knowledge, experience and up-to-date information between groups interested in learning, promoting and applying an Analog Forestry system in their own area. The primary objective of the Network is to restore ecosystems environmental stability and biodiversity, through research, design and application of the Analog Forestry system. It has its headquarters at Costa Rica.

Website: http://www.analogforestrynetwork.org/

PERMANENT AGRICULTURE RESOURCES

Permanent Agriculture Resources (PAR) carries out agroforestry education and research in the Pacific, provides workshops and publications including The Overstory, a free e-mail journal.

Website: http://www.agroforestry.net

EDUCATIONAL CONCERNS FOR HUNGER ORGANIZATION

Educational Concerns for Hunger Organization (ECHO) based in USA has many online publications related to agroforestry. It offers very useful information and resources to practitioners around the world, with an emphasis on self-sufficiency systems. ECHO is dedicated to solve the problems giving rise to food shortages by empowering people to make wise use of their resources. Website: http://www.echonet.org/

FOREST & LANDSCAPE DENMARK

Forest & Landscape Denmark (FLD) works to improve the livelihood of people in developing countries by supporting them in growing trees. Forest & Landscape Denmark, It is based in The Royal Veterinary and Agricultural University, Hoersholm, Denmark.

THE EUROPEAN TROPICAL FOREST RESEARCH NETWORK

The European Tropical Forest Research Network (ETFRN) in Wageningen, The Netherlands facilitates access to the European capability on tropical forest research. Website: http://www.etfrn.org/

AGROFORESTRY NET

Agroforestry Net, Inc. is a nonprofit organization dedicated to providing educational resources about agroforestry, trees, and sustainable stewardship of land and water. The site contains agroforestry project information, publications, and links to other resources on the web. Website: http://www.agroforestry.net

SUSTAINABLE HARVEST INTERNATIONAL

Sustainable Harvest International (SHI), USA provides farmers and communities in the tropics with long-term assistance implementing environmentally and economically sustainable technologies.

Website: http://www.sustainableharvest.org/

EVERGREEN TRUST

Evergreen Trust in London, UK promotes the use of trees in Africa.

Website: http://www.evergreentrust.org.uk/trust/agroindex.html

FOREST, FARM, AND COMMUNITY TREE NETWORK

Forest, Farm and Community Tree Network (FACT Net) is the new incarnation of the Nitrogen Fixing Tree Association (NFTA) which has provided invaluable information to thousands of practitioners and researchers in the use of nitrogen fixing and multipurpose trees for more than 15 years.

Web: http://www.winrock.org/fnrm/factnet/factnet.htm

NATIONAL ORGANIZATIONS

INDIAN COUNCIL OF AGRICULTURAL RESEARCH

The Indian Council of Agricultural Research (ICAR) is an autonomous organisation under the Department of Agricultural Research and Education (DARE), Ministry of Agriculture, Government of India. Formerly known as Imperial Council of Agricultural Research, it was established on 16 July 1929 in pursuance of the report of the Royal Commission on Agriculture. The ICAR has its headquarters at New Delhi. The Council is the apex body for co-ordinating, guiding and managing research and education in agriculture including horticulture, fisheries and animal sciences in the entire country. With 99 ICAR institutes and 53 agricultural universities spread across the country this is one of the largest national agricultural systems in the world.

The ICAR has played a pioneering role in ushering Green Revolution and subsequent developments in agriculture in India through its research and technology development that has enabled the country to increase the production of foodgrains by 4 times, horticultural crops by 6 times, fish by 9 times (marine 5 times and inland 17 times), milk 6 times and eggs 27 times since 1950-51, thus making a visible impact on the national food and nutritional security. It has played a major role in promoting excellence in higher education in agriculture. It is engaged in cutting edge areas of science and technology development and its scientists are internationally acknowledged in their fields. Website: http://www.icar.org.in/

NATIONAL RESERCH CENTRE FOR AGROFORESTRY

The National Research Centre for Agroforestry (NRCAF), Jhansi as unit of ICAR was established in 1988. It undertakes basic & applied research for developing and delivering technologies based on sustainable agroforestry practice on farms, marginal & wastelands for different agroclimatic zone in India

Website: http://www.nrcaf.ernet.in/

The mandate of the Centre is:

- To undertake basic and applied research for developing and delivering technologies based on sustainable agroforestry practices on farms, marginal and wastelands for different agroclimatic zones in India.

- To co-ordinate network research with the SAUs/ ICAR Institutes/ other related research institutes for identifying technologies which can be transferred from one region to another.
- To provide training in research methodologies, and use and application of technologies developed at various levels.
- To develop technological packages of different agroforestry practices for various agroclimatological zones for transfer to farm field and wastelands.
- To act as repository of information on the subject.
- To collaborate with relevant national and international agencies for achieving the mandate.
- To provide consultancy.

OTHER IMPORTANT ICAR INSTITUTES

- Central Arid Zone Research Institute (CAZRI), Jodhpur
- Central Research Institute of Dryland Agriculture (CRIDA), Hyderabad
- Central Soil and Water Conservation Research & Training Institute (CSWCRTI), Dehradun
- Central Soil Salinity Research Institute (CSSRI), Karnal
- Indian Grassland and Fodder Research Institute (IGFRI), Jhansi
- Central Institute of Arid Horticulture (CIAH), Bikaner
- Central Institute of Sub Tropical Horticulture (CISH), Lucknow
- Central Institute of Temperate Horticulture (CITH), Srinagar
- Central Plantation Crops Research Institute (CPCRI), Kasargod
- Central Agricultural Research Institute (CARI), Port Blair
- ICAR Research Complex for NEH Region, Barapani
- Indian Agricultural Statistical Research Institute (IASRI), New Delhi
- Indian Institute of Horticultural Research (IIHR), Bangalore
- Indian Institute of Natural Resins and Gums (IINRG), Ranchi

INDIAN COUNCIL OF FORESTRY RESEARCH AND EDUCATION

Indian Council of Forestry Research and Education (ICFRE), Dehradun is an autonomous body of Ministry of Environment and Forests (MoEF), Government of India established in the year 1988. ICFRE is an apex body in the national forestry research system and has been undertaking the holistic development of forestry research through need based planning, promoting, conducting and coordinating research, education and extension covering all aspects of forestry. The Council deals with the solution based forestry research in tune with the emerging issues in the sector, including global concerns such as climate change, conservation of biological diversity, combating desertification and sustainable management and development of resources. Topical research by the Council enhances public confidence in the ability of forest managers and researchers to successfully handle challenges related to natural resource management. ICFRE has eight Regional Research Institutes and four Research Centres located in different bio-geographical regions of the country to cater the forestry research needs of the nation. Website: http://www.icfre.org/

REGIONAL RESEARCH INSTITUTES UNDER ICFRE

- Forest Research Institute (FRI), Dehradun was established in 1906
- Institute of Forest Genetics and Tree Breeding (IFGTB), Coimbatore was established in 1988
- Institute of Wood Science and Technology (IWST), Bangalore was established in 1988 but its origin goes back to 1938 when, the then Government of Mysore had set up a Forest Research Laboratory (FRL) at Bangalore.
- Tropical Forest Research Institute (TFRI), Jabalpur came into existence in 1988, although its origin goes back to 1973 when a Regional Centre of Forest Research Institute, Dehradun was established at Jabalpur
- Rain Forest Research Institute (RFRI), Jorhat was established in 1988
- Arid Forest Research Institute (AFRI), Jodhpur was established in 1988
- Himalayan Forest Research Institute (HFRI), Shimla was established as High Level Conifer Regeneration Research Centre during 1977. The institute made its humble beginning from this Centre at the time of re-organization of forestry research in ICFRE during 1998
- Institute of Forest Productivity (IFP), Ranchi was established in 1993

ADVANCED RESEARCH CENTRES UNDER ICFRE

- Centre for Social Forestry and Eco-Rehabilitation (CSFER), Allahabad was established in 1992 as an advanced centre under the umbrella of ICFRE and presently, it is a branch of Forest Research Institute (FRI), Dehradun.
- Centre for Forestry Research and Human Resource Development (CFRHRD), Chhindwara came into existence in 1995. It was declared in 1996 as a satellite Centre of Tropical Forest Research Institute, Jabalpur
- Forest Research Centre (FRC), Hyderabad was initially established in 1997 as Advanced Centre for Biotechnology and Mangrove Forests functioning under the administrative control of Institute of Wood Science and Technology Bangalore. It was formally inaugurated as FRC in 2000
- Advanced Research Centre for Bamboo and Rattans (ARCBR), Aizawl was established in 2004

STATE FOREST RESEARCH INSTITUTE

State Forest Research Institute (SFRI), Jabalpur the first state level forest research institute in the country, came into existence on 27 June 1963, as per the recommendations of the 10th Silvicultural Conference held at Dehradun in 1961. The mandate of the institute is to provide scientific technical support to the M.P. forest department and various forestry related institutions, as well as other stakeholders and forestry sector as a whole, in the state. Website: http://mpsfri.nic.in/

KERALA FOREST RESEARCH INSTITUTE

Established in 1975, the Kerala Forest Research Institute (KFRI) is an institution of the Government of Kerala under its Science and Technology Department. The Institute is dedicated to research in tropical forestry, environment and biodiversity conservation. In 2003, when the State owned R & D institutions were re-organized to constitute the Kerala State Council for Science and Technology (KSCSTE), KFRI also became part of the latter. The Main Campus of KFRI is situated at Peechi, Thrissur District. Website: http://www.kfri.res.in/

OTHER STATE FOREST RESEARCH INSTITUTES

- State Forest Research Institute, Kolapakkam, Chennai
- State Forest Research Institute, Kanpur, Uttar Pradesh
- State Forest Research and Training Institute (SFRTI), Raipur, Chhattisgarh
- State Forest Research Institute, Itanagar, Arunachal Pradesh
- State Forest Research Institute, Janipura, Jammu (J&K)

INSTITUTIONS AND UNIVERSITIES OFFERING AGROFORESTRY EDUCATION & RESEARCH

In India, Agroforestry education and research are conducted in the Central Universities, State Agricultural Universities, Traditional Universities, Deemed Universities (approved by UGC/MHRD/ ICAR/ ICFRE) and Private Academic Institutions (approved by UGC). Agroforestry education and training is offered to B.Sc. Forestry, B.Sc. Agriculture, B.Sc. Horticulture and other Undergraduate degree programmes by including the agroforestry subject in the UG curriculum. At Post graduate level, the specialized degrees are offered in selective Institutions viz. M.Sc. Agroforestry or M.Sc.Forestry (specialization in Agroforestry) and Ph.D. Agroforesry or Ph.D. Forestry (specialization in Agroforestry).

AGROFORESTRY/FORESTRY JOURNALS

AGROFORESTRY SYSTEMS

Agroforestry Systems is a peer-reviewed international scientific journal published by Springer Netherlands presenting original research results, critical reviews and short communications on all aspects of agroforestry, including both biophysical and socioeconomic aspects. Short communications typically report on incomplete or initial results. The journal also periodically publishes book reviews and other professional information. Coverage includes investigations of a fundamental or applied nature, research methodologies and techniques, and analytical descriptions of little-studied but potentially promising agroforestry and other integrated systems involving trees and crops and/or livestock. Till date 87 Volumes 242 Issues 2,265 Articles are available from 1982-2013. Website: http://link.springer.com/journal/10457

PLANT AND SOIL

Plant and Soil publishes original papers and review articles exploring the interface of plant biology and soil sciences, and offering a clear mechanistic component. This includes both fundamental and applied aspects of mineral nutrition, plant-water relations, symbiotic and pathogenic plant-microbe interactions, root anatomy and morphology, soil biology, ecology, agrochemistry and agrophysics. Articles discussing a major molecular or mathematical component also fall within the scope of the journal. All contributions appear in the English language. Plant and Soil is published under the auspices of the Royal Netherlands Society of Agricultural Science. Till date 364 Volumes 678 Issues 13,668 Articles are available from 1949- 2013.

Website: http://link.springer.com/journal/11104

AGROFORESTRY TODAY

Agroforestry Today is a quarterly publication of World Agroforestry Centre (ICRAF) focusing on agroforestry research. This is a journal aiming at field practitioners, and as such is very relevant to extension workers.

Website: http://www.ciesin.org/IC/icraf/agrotod.html

ALLELOPATHY JOURNAL

Allelopathy is new science, which indicates inhibitory or stimulatory biochemical interactions between the two plant species. World over, the Allelopathy Journal is only International Journal in this new field of science. The Journal invites original Research Papers, Reviews and Short Communications on all aspects of allelopathy and related areas in both aquatic and terrestrial ecosystems. The scope of Journal is multidisciplinary and accepts papers related to allelopathy from following disciplines: Agroforestry and Forestry, Agronomy, Biochemistry, Biotechnology, Botany, Chemistry, Ecology, Entom-ology, Fresh water Biology, Genetics and Plant Breeding, Horticulture, Limnology, Microbiology, Nematology, Plant Pathology, Soil Science, Vegetable crops, Zoology.

Website: www.allelopathyjournal.org

AGRICULTURE, ECOSYSTEMS & ENVIRONMENT

Agriculture, Ecosystems & Environment is an international journal for scientific research on the interaction between agroecosystems and the environment. It publishes scientific articles dealing with the interface between agroecosystems and the natural environment, specifically how agriculture influences the environment and how changes in that environment impact agroecosystems. Preference is given to papers from experimental and observational research at the field, system or landscape level, complemented as appropriate by dynamic and statistical modelling, that bridge scientific disciplines, integrate knowledge, and are placed in an international or wide comparative context. Website: http://www.journals.elsevier.com/agriculture-ecosystems-and-environment/

INTERNATIONAL TREE CROPS JOURNAL

International Tree Crops Journal publishes scientific research on tree species in agroecosystem and forestry ecosystem. Website: http://www.tandfonline.com/toc/tftl19/9/4

ECOSYSTEMS

Ecosystems is an international journal published by Springer-Verlag and it publishes research papers related to ecosystem science, fundamental ecology, environmental eco-logy and environmental problem-solving. The scope of ecosystems extends from bounded systems such as watersheds to spatially complex landscapes. It focuses on landscape ecology, global ecology, biogeochemistry, aquatic ecology, soil science, hydrology, ecological economics and conservation biology.

Website: http://link.springer.com/journal/10021

ECOSYSTEM SERVICES (SCIENCE, POLICY AND PRACTICE)

Ecosystem Services, associated with the Ecosystem Services Partnership (ESP), is an international, interdisciplinary journal that deals with the science, policy and practice of *Ecosystem Services* in the following disciplines: ecology and economics, institutions, planning and decision making, economic sectors such as agriculture, forestry and outdoor recreation, and all types of ecosystems.

Website: http://www.journals.elsevier.com/ecosystem-services/

THE INTERNATIONAL FORESTRY REVIEW

The International Forestry Review (IFR) is a peer-reviewed scientific journal that publishes papers, research notes and book reviews on all aspects of forestry and forest research, with an emphasis on issues of transnational significance. It is published four times per year by Commonwealth Forestry Association, UK. The website: http://www.cfa-international.org/international_forestry_review.php

UNASYLVA

Unasylva is an international journal of forestry and forest industries, produced in separate English, French and Spanish editions and it is published by Food and Agriculture Organization of the United Nations (FAO). Unasylva covers all aspects of forestry: policy and planning; conservation and management of forest-based plants and animals; rural socio-economic development; species improvement; industrial development; international trade; and environmental considerations, including the role of forests and trees in maintaining a sustainable base for agricultural production at the micro and macro levels as well as the effects of environmental change on forestry. All issues of Unasylva are available online free of charge at www.fao.org/forestry/unasylva

NFT HIGHLIGHTS

NFT Highlights are published by Nitrogen Fixing Tree Association (NFTA) annually which are 2-page reviews of a NFT species' ecology, botany, establishment, management and uses. In addition, NFTA also publishes Leucaena Research Reports and Nitrogen Fixing Tree Research Reports, each containing scientific papers, publication reviews and other information.

THE OVERSTORY

The Overstory is a free e-mail agroforestry journal for researchers, professionals, practitioners, extension agents, students, and enthusiasts. A free e-mail journal, which is concise and informative, focusing on trees and their roles in agriculture, natural ecosystems, and human culture and economy. One issue is sent each month focusing on a concept related to designing, developing, and learning more about trees and agroforestry systems. The website: http://agroforestry.net/overstory/index.html

ASIA-PACIFIC AGROFORESTRY NEWSLETTER (APANews)

APANews is a newsletter produced in support of the Asia-Pacific Agroforestry Network (APAN) that provides information on agroforestry research, promotion and development, education and training to researchers, practitioners and policy-makers in the region.

Website: http://www.fao.org/asiapacific/rap/nre/links/apanews/en/

ANNALS OF PLANT SCIENCES

Annals of Plant Sciences (APS) is an International, Peer-reviewed journal, which aims to include high quality research articles dealing with all aspects of plant sciences.

Website: http://ebioscholar.com/ojs/index.php/ap/index

THE TEMPERATE AGROFORESTER

The Association for Temperate Agroforestry's (AFTA) quarterly newsletter, The Temperate Agroforester, contains reports on research and demonstration trials, landowner success stories, conference reports, book reviews, links to information sources in print and online, and a calendar of agroforestry-related events.

Website: http://www.aftaweb.org/temperate_agroforester.php

JOURNAL OF TROPICAL FOREST SCIENCE

The Journal of Tropical Forest Science (JTFS) is an international reviewed journal published by Forest Research Institute-Malaysia (FRIM) concerning the science, technology and development of tropical forests and forest products. The journal welcomes articles reporting original fundamental or applied research on tropical forest biology, ecology, chemistry, management, silviculture, conservation, utilization and product development. The journal is published four times a year.

Website: http://www.frim.gov.my/?page_id=1826

INDIAN JOURNAL OF AGROFORESTRY

Indian Journal of Agroforestry is published by Indian Society of Agroforestry, National Research Centre for Agroforestry, Jhansi. It publishes research papers containing original research results, critical reviews and short communications on all aspects of agroforestry.

RANGE MANAGEMENT & AGROFORESTRY

Range management Society of India (RMSI) publishes the Journal "Range management & Agroforestry" in two issues every year. It publishes papers related to management, improvement, conservation and utilization of rangelands, pastures, forage crops and forage production systems. It also publishes papers on management of degraded / wastelands through alternate land use systems like agroforestry/ dryland agriculture.

Website: http://www.indianjournals.com/ijor.aspx?target=ijor:rma&type=home

INDIAN JOURNAL OF FORESTRY

Indian Journal of Forestry is a quarterly Journal from Dehra Dun, India which publishes the research papers from all disciplines of forestry.

THE INDIAN FORESTER

The Indian Forester, a pioneer journal in the field of forestry research, established in 1875, has been serving the cause of scientific forestry research since its inception. It is closely associated with the continuous advancement of forestry, and is a storehouse of information for those who want to indulge deep into the progress of forestry in the country. The Indian Forester started its career as the outcome of a resolution of the forestry conference held at Allahabad in January 1874. The credit for the success of the journal goes to stalwarts like Sir Dr. Brandis, Schlich, Pearson, Ribbentrop, Gamble, Wroughten, Lace, Wilmot, and other dedicated foresters. It publishes research papers from all disciplines of forestry and published from Dehra Dun, India.

Website: http://www.indianforester.co.in/

ANNALS OF FORESTRY

Annals of Forestry is published from Dehra Dun, India and it publishes research papers, short communications, research notes, thoughts and reflections, and book review related to various branches of forestry.

JOURNAL OF TROPICAL FORESTRY

Journal of Tropical Forestry is published by Society of Tropical Forestry Scientists (Jabalpur, India) from 1985. It publishes the research papers from all disciplines of forestry.

JOURNAL OF TREE SCIENCES

Journal of Tree Sciences is published by Indian Society of Tree Scientists (Solan, HP, India) since 1982. It publishes the research papers from all disciplines of forestry.

■■■■

BIBLIOGRAPHY

GENERAL REFERENCES

Publications, Technical Reports and Website of World Agroforestry Centre (ICRAF), Nairobi, Kenya

Publications, Technical Reports and Website of FAO (Food and Agricultural Organization of the United Nations)

Publications, Technical Reports and Website of IPCC (Intergovernmental Panel on Climate Change)

Publications, Technical Reports and Website of ICAR (Indian Council of Agricultural Research), New Delhi

Publications, Technical Reports and Website of NRCAF (National Research Centre for Agroforestry), Jhansi

Publications, Technical Reports and Website of Ministry of Environment and Forests, GOI, New Delhi

Publications, Technical Reports and Website of ICFRE (Indian Council of Forestry Research and Education), Dehradun

Publications, Technical Reports and Website of DoLR (Department of Land Resources), Ministry of Rural Development, GOI, India

Publications, Technical Reports and Website of Planning Commission, GOI, New Delhi

Publications, Technical Reports and Website of Forest Survey of India, Dehradun

Publications, Technical Reports and Website of National Remote Sensing Centre, Hyderabad

Publications, Technical Reports and Website of Indian Institute of Forest Management, Bhopal

Albrecht A, Kandji S and Verchot L 2004 Carbon Sequestration in Tropical Agroforestry Systems, World Agroforestry Centre (ICRAF), Nairobi, Kenya

Anonymous 2004 Reclamation and Management of Salt-Affected Soils, CSSRI, Karnal, 160 page

Antony Joseph Raj and SB Lal 2012 Forestry: Principles and Applications, Scientific Publishers, Jodhpur, Inida, 805 pages

Arora CL 2002 Analysis of Soil, Plant and Fertilizer, In Fundamentals of Soil Science

Barton D 1994 Indigenous Agroforestry in Latin America: A Blueprint for Sustainable Agriculture? NRI Socio-economic Series 6, Chatham, UK: Natural Resources Institute

Bhargava GP 1989 Salt Affected Soils of India, Oxford and IBH Publishing Company, New Delhi, 261 pages

Coulson RN and Witter JA 1984 Forest Entomology, Ecology and Management, Chap-5 Concepts of integrated Pest management, John Wiley & sons, New York, 97-121 pages

Dagar JC 2009 Opportunities for alternative land uses in salty and water scarcity areas, International Journal of Ecology and Environmental Sciences **35**(1) 53-66

Dagar JC and Tomar OS 2002 Utilization of salt affected soils and poor quality waters for sustainable biosaline agriculture in arid and semiarid regions of India, 12th ISCO Conference, Beijing

Daniel JN and Hegde NG 2007, Tree-Borne Oilseeds in Agroforestry, In *Proceedings of the National Seminar on Changing Global Vegetable Oils Scenario: Issues and Challenges before India* (ed) DM Hegde, Indian Society of Oilseeds Research, Hyderabad, India, 263-276 pages

Dhyani SK, Karemulla K, Ajit and Handa AK 2009 Agroforestry potential and scope for development across agroclimatic zones in India, Indian Journal of Forestry **32**(2): 181-190

Dwivedi AP 1992 Agroforestry - Principles and Practices, Oxford and Publishing Company Pvt. Ltd., New Delhi, India, 365 pages

Elevitch C and Wilkinson K 1999 A Guide to Orchard Alley Cropping for Fertility, Mulch and Soil Conservation, www.agroforester.com

Elevitch CR and Wilkinson KM 2000 Agroforestry Guides for Pacific Islands, Permanent Agricultural Resources, Holualoa, Hawaii, USA, webpage: http://www.agroforestry.net

FACT Net 1992 Hedgrow intercropping with upland root crops, A publication of Forest, farm, and community Tree Network (FACT Net), Winrock International, www.winrock.org/ forestry/FACTPUB

FAO 1995 Resource management for upland areas in Southeast Asia - An information kit, FAO Regional Office for Asia and the Pacific and Asia-Pacific Agroforestry Network

FITEC (New Zealand Forestry Industries Training Organizations) 2013 Forestry Insights, webpage: http://www.insights.co.nz/sustainable_plantations_wrp.aspx

Foort JV, Delobel V and Bongers F 2012 The Potential of Agroforestry for Soil Quality Improve-ment in European Temperate Agroecosystems, Technical Notes, Wageningen University

Franklin W Martin and Scott Sherman 1992 Agroforestry Principles Technical Note, ECHO, USA

Ghose GK 2008 Bamboo the wonderful grass, A.P.H. Publishing Corporation, New Delhi, India, 4-16 pages

Gray B 1972 Economic tropical forest entomology, *Annual Review of Entomology* **17** 313-354

Gupta RK, Tomar OS and Minhas PS 1995 Managing Salty Soils and Waters for Afforestation, Bulletin No. 7/95, CSSRI, Karnal, 23 page

Hairiah K, Dewi S, Agus F, Velarde S, Ekadinata A, Rahayu S and Van Noordwijk M 2011 Measuring Carbon Stocks Across Land Use Systems: A Manual, Bogor, Indonesia, World Agroforestry Centre (ICRAF), SEA Regional Office

Hairiah K, Sitompul SM, Van Noordwijk M and Palm C 2001 Methods for Sampling Carbon Stocks Above and Below Ground, Bogor, Indonesia, World Agroforestry Centre (ICRAF), SEA Regional Office

Hodge S, Garrett H and Bratton J 1999 Alley Cropping: An Agroforestry Practice, Agroforestry Note-12, USDA Forest Service, website: www.unl.edu/nac/afnotes/ac-1

INBAR 2001 *INBAR News Magazine* **8**(1) 6-11

Jha LK and Sen-Sarma PK 1994 Forest Entomolgy, Chapter 20: Control of Forest insect pests, New Delhi: Ashish Publishing House

Kanwar, JS 1978 Soil Fertility: Theory and Practice, ICAR, New Delhi.

MacDicken KG 1997 A Guide to Monitoring Carbon Storage in Forestry and Agroforestry Projects, Arlington, United States: Winrock International Institute for Agricultural Development

Mahat TBS (ed) 1993 Agroforestry Training Course Module for Bangladesh, Proceedings of the Workshop held at the Bangladesh Agricultural Research Council, Dhaka (27 February - 4 March), Training Support Series-2, Dhaka, Bangladesh

Mehta T 1981 A Hand Book of Forest Utilization, Dehradun: International Book Distributors

Mohan V Pathological Problems of Economically Important Forest Tree Species in India – An Overview, Division of Forest Protection, Institute of Forest Genetics and Tree Breeding (IFGTB), Coimbatore

Nair KSS 2007 Tropical forest insect pests: ecology, impact and management, Cambridge University Press, New York, 154-182 pages

Nair PKR (ed) 1989 Agroforestry Systems in the Tropics, Kluwer Academic Publishers

Nair PKR 1983 Agroforestry with coconuts and other tropical plantation crops, In *Plant Research and Agroforestry* (eds) Huxley PA, ICRAF, Nairobi, Kenya, 79-102 pages

Nair PKR 1984 Soil Productivity Aspects of Agroforestry, International Centre for Research in Agroforestry, Nairobi, Kenya

Nair PKR 1993 An Introduction to Agroforestry, Kluwer Academic Publishers, The Netherlands

Nair PKR, Gholz HL, Duryea MI 1990 Recommendations of the International Workshop on Professional Education and Training in Agroforestry, University of Florida, December 1988, In *Agroforestry Systems 12*, Amsterdam: Kluwer Academic Publications, 133-139 pages

Negi SS 1996 An Introduction to Forest Pathology, Dehradun: IBD

Pinho RC, Miller RP and Alfaia SS 2012 Agroforestry and the Improvement of Soil Fertility: A View from Amazonia, Applied and Environmental Soil Science, Volume 2012, Article ID 616383, 1-11 pages (doi:10.1155/2012/616383)

Puri S and Panwar P 2006 Agroforestry Systems and Practices, NIPA, New Delhi

Nath S, Das R, Chandra R and Sinha A - Bamboo based agroforestry for Marginal lands with special reference to productivity, market trend and economy, Technical Report, Institute of Forest Productivity, Ranchi, Jharkhand

Salil K Tewari 2008 Farm Forestry: Agro-Forestry, Technical Notes, G. B. Pant University of Agriculture and Technology, Pantnagar, India

Sanchez and Palm CA 1996 Nutrient cycling and agroforestry in Africa, In *Forest Influences, Unasylva* No. 185, An international journal of forestry and forest industries, Vol. 47 - 1996/2, Food and Agriculture Organization of the United Nations

Singh, Gurbachan, Singh NT and Tomar OS 1993, Agroforestry in Salt-Affected Soils, Bulletin No.17, CSSRI, Karnal, 65 page

Smith J 2010 The History of Temperate Agroforestry, The Organic Research Centre, Elm Farm, Berkshire

TEARA 2013 The Encyclopedia of New Zealand, webpage: http://www.teara.govt.nz/en/trees-in-the-rural-landscape/1

Tejwani KG 1994 Agroforestry in India, Oxford & IBH Publication Co. Pvt. Ltd, New Delhi, 233 pages

Tewari DN 1992 A Monograph on Bamboo, International Book Distributors, Dehradun

Tewari DN 1995 Agroforestry for increased productivity, sustainability and poverty alleviation, Dehradun, India: International Book Distributors

Thakur ML 2000 Forest Entomology (Ecology and Management), Chapter XIX: Forest Insect Control, Dehradun: Sai Publishers

TNAU 2009 Reclamation-cum-Development Plan for Wastelands in Tamil Nadu, Centre for Agriculture and Rural Developement Studies, Tamil Nadu Agricultural University, Coimbatore

TNAU Agritech Portal 2013 Forest Disease Management, website http://agritech.tnau.ac.in/forestry/forest_disease_index.html#conclusion

Waters WE and Cowling EB 1976 Integrated Forest Pest Management: A Silvicultural necessity, In Integrated Pest Management (eds) JL Apple and RF smith, Plenum Press, New York

Waters WE and Stark RW 1980 Forest Pest Management: Concept and reality, *Ann. Rev. Entomology* **25** 479-509

Wood PJ and Burley J 1991 A Tree for All Reasons, ICRAF, Nairobi, Kenya

Young A 1989 Agroforestry for Soil Conservation, CAB International, Wallingford, UK.

Zhu Zhaohua, Fu Maoyi and Sastry CB 1997 Agroforestry in China - An Overview, In *Agroforestry Systems in China*, International Development Research Centre, Ottawa, Canada

REFERENCES

AAFC (Agriculture and Agri-Food Canada) 2012 Agro-forestry, Agriculture and Agri-Food Canada

Abdoellah OS 1990 Homegardens in West Java and their future development, In *Tropical Homegardens* (eds) K Landauer and M Brazil, United Nations University Press, Tokyo, 69-79 pages

Abdoellah OS, Takeuchi K, Perikesit GB and Hadikusumah HY 2001 Structure and function of homegarden revisited, In *Proceedings of First Seminar toward Harmonisation between Development and Environmental Conservation in Biological Production*, JSPS-DGHE Core University Program in Applied Biosciences, The University of Tokyo, Tokyo, 167–185 pages

Abebe T, Wiersum KF, Bongers F and Sterck F 2006 Diversity and dynamics in homegardens of southern Ethiopia, In *Tropical homegardens: A time-tested example of sustainable agroforestry* (eds) BM Kumar and Nair PKR, Springer Science, Dordrecht, 123-142 pages

Abel N, Baxter J, Campbell A, Cleugh H, Fargher J, Lambeck R, Prinsley R, Prosser M, Reid R, Revell G, Schmidt C, Stirzaker R and Thorbum P 1997 Design Principles for Farm Forestry: A Guide to Assist Farmers to decide where to plant trees, RIRDC/L.WPRDC, Joint venture agroforestry program

Abrol IP and Bhumbla DR 1979 Crop responses to differential Gypsum application in highly sodic soil and tolerance of several crops to exchangeable sodium under field conditions, *Soil Science* **127** 79-85

AFTA (Association for Temperate Agroforestry) 1997 The status, opportunities, and needs for agroforestry in the United States, Columbia, Missouri: AFTA, University of Missouri

AFTA 2013 Association for Temperate Agroforestry, webpage: http://www.aftaweb.org

Agamuthu PJ and Broughton WJ 1985 Nutrient Cycling within the developing oil palm-legume ecosystem, *Agriculture, Ecosystems and Environment* 13 111-123

Agelet A, Bonet MA and Valles J 2000 Homegardens and their role as a main source of medicinal plants in mountain regions of Catalonia (Iberian peninsula), *Economic Botany* **54** 295-309

Agrios GN 1988 Plant Pathology, New York: Academic Press

Agus F, Watung RL, Suganda H, Tala'ohu SH, Wahyunto, Sutono S, Adi Setiyanto, Mayrowani H, Nurmanaf R and Kundarto M 2003 Assessment of environmental multifunctions of rice paddy farming in Citarum River Basin, West Java, Indonesia, Paper presented in the *3rd Working Group Meeting of the ASEAN-JAPAN Project on Multifunctionality of Paddy Farming and its Effects in ASEAN Member Countries*, Phnom Penh, Cambodia, 27 February to 1 March 2003

Ahmed P 1989 Eucalyptus in agroforestry: its effects on agricultural production and economics, *Agroforestry Systems* **8**: 31-38

Ahmed SH and Singh AK 1999 Integrated aquaculture-cum-fishculture, *Fishing Chimes* **18** 35-45

Ahn PM 1974 Some observations on basic applied research in shifting cultivation, *FAO Soils Bulletin* **24** 123-154

Ahuja LD, Verma CM, Sharma SK and Lamba TR 1978 Range management studies on the contribution of ground storey (grass) in afforested areas in arid regions, *Annals of Arid Zone* **17**: 304-310

AICRPAF 2006 All India Coordinated Research Project on Agroforestry Report, NRCAF, Jhansi

Ajit and Nighat J 2008 Tree Growth Modelling: Indian Experiences, website: http://mirror.iasri.res.in/net/tgm/index.htm

Ajit SPN, Gupta VK and Solanki KR 1999 Sigmoid functions in tree growth modelling: limitation of constant estimation and simulation approach as an alternative, *Indian Journal of Agroforestry* **1**(2) 135-140

Ajit SPN, Gupta VK and Solanki KR 2000 Linear tree growth models: A limitation of negative estimation of size, *Indian Forester* **126**(12) 1336-1341

Akbulut S, Keten A and Stamps WT 2003 Effect of alley cropping on crops and arthropod diversity in Duzce, Turkey, *Journal of Agronomy and Crop Science* **189** 261-267

Akinnifesi FK, Leakey RRB, Ajayi OC, Sileshi G, Tchoundjeu Z, Matakala P and Kwesiga FR (eds) 2008 Indigenous Fruit Trees in the Tropics: Domestication, Utilization and Commercialization, CAB International, Wallingford, UK, 438 pages

Alam MK 1993 Multipurpose Trees and Shrubs in Agroforestry, In *Agroforestry Training Course Module for Bangladesh* (ed) TBS Mahat, Proceedings of the Workshop held at the Bangladesh Agricultural Research Council, Dhaka (27 February - 4 March), Training Support Series-2, Dhaka, Bangladesh

Alam MS, Singh DS and Sehgal VK 2001 Optimization of screening in a water chestnut decorticator, *Journal of Agriculture Engineering* **38** 29-34

Alavalapati JRR and Nair PKR 2001 Socio-economics and institutional perspectives of Agroforestry, In *World Forests, Society and Environment: Markets and Policies* (eds) M Palo and J Uusivuori, Dordrecht, The Netherlands: Kluwer Academic Publishers, 71-81 pages

Albrecht A and Kandji ST 2003 Carbon sequestration in tropical agroforestry systems, *Agriculture, Ecosystems and Environment* **99** 15-27

Albuquerque UP, Andrade LHC and Caballero J 2005 Structure and floristic of homegardens in Northeastern Brazil, *Journal of Arid Environments* **62** 491-506

Alcorn JB 1989a Process as resource: the traditional agricultural ideology of Bora and Huastec resource management and its implications for research, *Advances in Economic Botany* **7** 63-77

Alcorn JB 1989b An economic analysis of Huastec Mayan forest management, In *Fragile Lands of Latin America* (ed) J Browder, Boulder, Colorado: Westview Press

Alcorn PW 1994 The chicle tree (*Manilkara zapota*) in Northwest Belize: natural history, forest floristics, and management, Unpublished M.Sc. Thesis, Botany Department, University of Florida, 136 pages

Altieri MA 1995 Agroecology: The Science of Sustainable Agriculture, Boulder, CO: Westview Press

Altieri MA 1999 The ecological role of biodiversity in agroecosystems, *Agriculture, Ecosystems and Environment* **74** 19-31

Altieri MA and JG Farrell 1984 Traditional farming systems of south central Chile, with special emphasis on agroforestry, *Agroforestry Systems* **2** 3-18

Altieri MA and Nicholls CI 2004 Biodiversity and Pest Management in Agroecosystems, New York: Haworth Press

Altieri MA and Nicholls CI 2008 Ecologically based Pest Management in Agroforestry Systems, In *Ecological Basis of Agroforestry* (eds) DR Batish, RK Kohli, S Jose and HP Singh, CRC Press, Taylor and Francis Group,

Alvim P de T 1977 Cacao, In *Ecophysiology of Tropical Crops* (eds) P de T Alvim and TT Kozlowski, Academic Press, New York, USA, 279-313 pages

Amaro A 1998 Modelling dominant height growth for *Eucalyptus, Forest Science* **4**(1) 37-46

Amend T, Brown J, Kothari A, Phillips A and Stolton S (eds) 2008 Protected Landscapes and Agrobiodiversity Values, Volume 1 in the series, Protected Landscapes and Seascapes, IUCN & GTZ, Kasparek Verlag, Heidelberg

Aminuddin M and Latif MA 1991 Bamboo in Malaysia: Past, present and future research, Proceedings of 4th International Bamboo Workshop: Bamboo in Asia and the Pacific, Chiangmai, Thailand, November 27-30, 349-354 pages

Anderson AB 1990 Extraction and forest management by rural inhabitants in the Amazon estuary, In *Alternatives to Deforestation: Steps Toward Sustainable Use of the Amazon Rain Forest* (ed) AB Anderson, New York: Columbia University Press

Andow DA 1991 Vegetational diversity and arthropod population response, *Annual Review of Entomology* **36** 561-586

Andrevan T 2004 Fodder trees, Agronimca

Anil Kumar Yadava 2010 Biomass production and carbon sequestration in different agroforestry systems in Tarai regaion of Central Himalaya, *Indian Forester* February Issue 234-244

Anjulo A 2009 Component interactions and their influence on the production of apple based agroforestry system in wet temperate zone of Himachal Pradesh, Ph.D. Thesis, Dr. Y S Parmar UHF, Nauni, Solan (H.P.), India

Anonymous 1987 Annual Progress Report of AICRP on Agroforestry, Technical Bulletin, ICAR Research Complex for NEH Region, Shillong, India

Anonymous 1988 Annual Progress Report of AICRP on Agroforestry, Technical Bulletin, ICAR Research Complex for NEH Region, Shillong, India

Anonymous 1988 The state of forest report-1987, Ministry of Environment and Forests, Forests Survey of India, Dehradun

Anonymous 1989 Country report, XIII Commonwealth Forestry Conference, Rotorua, New Zealand

Anonymous 1990 Studies on the production potential of some silvi-horti-pastoral systems, Biennial Report of the AICRPA, ICAR, New Delhi

Anonymous 2003 Annual Report 2002-03, AICRP on Agroforestry, Bhuveneshwar Centre, Odisha, India

Anonymous 2003 Introduction to tropical agroforestry for Indigenous communities, A report for the Joint Venture Agroforestry Program, RIRDC Publication Number: 03/109

Anonymous 2006 Annual report (2005-2006), Research Achievements - Plant genetic resource in Aonla, Central Institute of Arid Horticulture, Bikaner, 7-8 pages

Anonymous 2009 A report of Regional centre, Central Soil &Water Conservation Research & Training Institute, Vasad Dist., Anand

Anonymous 2010 Annual Report, CSWCRTI, Dehardun.

Anonymous 2011 Progress report of All India Co- ordinated Research Project on Agroforestry, TNAU, Mettupalayam, *Annual Report*, 70 pages

Anonymous 2012 Final Report of the Working Group on Agro-Climatic Zonal Planning including Agriculture Development in North-Eastern India for XI Five Year Plan (2007-12) V-I Main Report

Antony Joseph Raj 2012a Timber Measurements and Timber Inventory, In *Forestry: Principles and Applications* (eds) Antony Joseph Raj and SB Lal, Scientific Publishers, Jodhpur, India, 188-213 pages

Antony Joseph Raj 2012b International and National Forest Organizations, In *Forestry: Principles and Applications* (eds) Antony Joseph Raj and SB Lal, Scientific Publishers, Jodhpur, India, 764-786 pages

Arap-Sang FK 1986 Alley cropping under semi-arid conditions in Kenya, In *Alley farming in the humid and subhumid tropics* (eds) BT Kang and L Reynolds, IDRC, Canada

ASA 1987 Proceedings of Symposium on Management of Coastal Ecosystems and Oceanic Resources of the Andamans, Andaman Science Association, Port Blair, India

ASA 1989 Decision reached on sustainable agriculture, American Society of Agronomy, *Agronomy News*. January issue, 15

Asati BS, Tomar JMS and Asati P 2009 Agrihorticulture Development in North Eastern Region, In Winter School on Alternate land use options for resource conservation, emerging market needs and mitigation of climate change in rainfed regions, CRIDA, Hyderabad, AP, India

Asfaw Z and Woldu Z 1997 Crop associations of homegardens in Welayta and Gurage in southern Ethiopia, *Sinet* (*An Ethiopian Journal of Science*) **20** 73-90

Atta-Krah AN and Francis PA 1987 The role of on-farm trials in the evaluation of composite technologies: The case of alley farming in southern Nigeria, *Agricultural Systems* **23**(2): 133-152

Atulya Tankha and Mishra VK 2001 Scope of Participatory Forest Management in Marketing of non-wood forest produce, In *Marketing Monitoring of Tree Products* (eds) GS Rawat and NS Bisht, Forest Research Institute, Dehradun, 140-148 pages

Augusto L, Dupouey JL and Ranger J 2004 Effects of tree species on understory vegetation and environmental conditions in temperate forests, *Annals of Forest Science* **60** 823-831

Austad I and Hauge L 2006 Pollarding in western Norway, In *1er colloque europeen sur led trognes*, 26-28 October, Vendome

Avery ME and Rhodes D 1990 Growth characteristics and total N content of a leucaena/sorghum agroforestry system, *Plant and Soil* **127** 259-267

Avila M 1992 Technical Paper 3: Agroforestry Systems - Concepts and Classification, In *The AFNETA alley farming training manual - Volume 2: Source book for alley farming research* (eds) Tripathl BR and Psychas PJ, Alley Farming Network for Tropical Africa, International Institute of Tropical Agriculture, Ibadan

Avila M and Minae S 1992 Technical Paper-4: Diagnosis and Design Methodology, In *The AFNETA alley farming training manual - Volume 2: Source book for alley farming research* (eds) Tripathl BR and Psychas PJ, Alley Farming Network for Tropical Africa, International Institute of Tropical Agriculture, Ibadan

Awasthi OP, Singh IS and More TA 2009 Performance of intercrops during establishment phase of aonla (*Emblica officinalis*) orchard, *Indian Journal of Agricultural Sciences* **79**(8) 587-91

Azurdia C and Leiva JM 2004 Home Gardens and Agrobiodiversity, In (eds) PB Eyzaguirre and OF Linares, Smithsonian Books, Washington, 168

Babu SC and Rhoe V 2002 Agroforestry Systems for Food and Nutrition Security-Potentials, Pathways and Policy Research Needs, *Journal of Crop Production* **6**(1-2) 177-192

Bagchee K 1952 A review of Indian tree diseases and decay of timber and methods of control, *Indian Forester* **48**(11) 540

Bakshi BK 1973 Forest Pathology - Principles and Practice in Forestry

Bakshi BK and Sujan Singh 1970 Heart rots in trees, *International Review of Forest Research*, Academics Press, New York **3** 197-251

Balaji S 1991 Agroforestry for prosperity, *Forest News*, Tamil Nadu Forest Department, Madras, **1**(3) 9-11

Balee W and Gely A 1989 Managed forest succession in Amazonia: the Ka'apor case, *Advances in Economic Botany* **7** 129-158

Balick MJ, Elisabetsky E and Laird SA 1996 Medicinal resources of the tropical forest, Biodiversity and its importance to human health, Columbia University Press, New York

Ballabh B and Chaurasia OP 2007 *Journal of Ethnopharmacology* **112** 341

Balooni K 2003 Economics of wasteland afforestation in India: A Review, *New Forests* **26** 101-136.

Banda AZ, Maghembe JA, Ngugi DN and Chome VA 1994 Effect of inter-cropping maize and closely spaced Leucaena hedgerows on soil conservation and maize yield on a steep slope at Ntchew, Malawi, *Agroforestry Systems* **27** 31

Barany M, Hammett AL, Sene A and Amichev B 2001 Non-timber forest benefits and HIV/AIDS in sub-Saharan Africa, *Journal of Forestry* **99**(12) 36-41

Bari MS and Rahim MA 2012 Economic Evaluation and Yield Performance of Some Medicinal Plants in Coconut Based Multistoried Agroforestry Systems, *The Agriculturists* **10**(1) 71-80

Barnes RD and Simons AJ 1994 Selection and breeding to conserve and utilize tropical tree germplasm, In *Tropical trees: the potential for domestication and the rebuilding of forest* resources (eds) RRB Leakey and AC Newton, HMSO, London, 84-90 pages

Barradas V and Fanjul L 1986 Microclimatic characterization of shaded and open-grown coffee (*Coffea arabica* L.) plantations in Mexico, *Agriculture, Forestry and Meteorology* **38** 101-120

Barros E, Mathieu J, Tapia-Coral S, Nascimento ARL and Lavelle P 2008 Comunidades da macrofauna do solo da Amazônia Brasileira, In *Biodiversidade do Solo em Ecossistemas Brasileiros* (eds) FMS Moreira, JO Siqueira and L Brussaard, Editora UFLA, Minas Gerais, Brazil, 171-191 pages

Barry Deran 2001 A Bestiary of multi-disciplinary topics that marketing research could address, In *Marketing Monitoring of Tree Products* (eds) GS Rawat and NS Bisht, Forest Research Institute, Dehradun, 39-53 pages

Barton D 1994 Indigenous Agroforestry in Latin America: A Blueprint for Sustainable Agriculture? NRI Socio-economic Series 6, Chatham, UK: Natural Resources Institute

Basavaraju TB1997 Studies on light interception, soil moisture content and yield of finger millet in agroforestry systems in the semi-arid regions of Karnataka State, India, M.Sc. Thesis, Dresden University of Technology, Dresden, Germany

Baumer M 1987 *Agroforesterie et desertification,* Wageningen, Netherlands: Centre Technique de Cooperation Agricole et Rurale, 260 page

Bawa KS et al 2004 Tropical ecosystems into the 21st century, *Science* **306** 227-228

Beedy TL, Ajayi OC, Sileshi GW, Kundhlande G, Chiundu G and Simons AJ 2012 Scaling up Agroforestry to Achieve Food Security and Environmental Protection among Smallholder Farmers in Malawi, Field Actions Science Reports [Online], Special Issue 7, 2012, website: http:// factsreports.revues.org/2082

Beeson CFC 1941 The Ecology and control of Forest Insects of India and the neighboring Countries, Dehradun: Vasantha press, (Reprint in 1961) 761 page

Behari B 2001 Agroforestry Models of bamboo cultivation on degraded agricultural lands, Ph.D. Thesis, Guru Ghasidas University, Bilaspur

Belcher B, Ruíz-Pérez M and Achdiawan R 2005 Global patterns and trends in the use and management of commercial NTFPs: Implications for livelihoods and conservation, *World Development* **33** 1435-1452

Bell S 2000 Agroforestry in the landscape, In *Agroforestry in the UK*, Bulletin 122 (eds) AM Hislop and J Claridge, Forestry Commission: Edinburgh

Bellow J 2004 Fruit-tree-based agroforestry in the western highlands of Guatemala, Dissertation, University of Florida, Gainesville, FL, USA

Bene JG, Beall HW and Cole A 1977 Trees, Food and People, IDRC, Ottawa, Canada

Bentrup G and Leininger T 2002 Agroforestry: Mapping the way with GIS, *Journal of Soil and Water Conservation* **57**(6) 148-153

Bernáth J and Neméth E 1999 Changes in the medicinal plant sector of Hungary since the fall of communism, In *TRAFFIC Europe - Medicinal plant trade in Europe* (ed), Proceedings of the first symposium on the conservation of medicinal plants in trade in Europe, Kew, 55-65 pages

Bertin C, Yang X and Weston LA 2003 The role of root exudates and allelochemicals in the rhizosphere, *Plant and Soil* **256**(1) 67-83

Bhatt BP and Misra LK 2003 Production potential and cost-benefit analysis of agrihorticulture agroforestry systems in Northeast India, *Journal of Sustainable Agriculture* **22** 99-108

Bhatt BP and Todaria NP 1990 Studies on the allelopathic effects of some Agroforestry tree crops of Garhwal Himalaya, *Agroforestry Systems* **12** 251-255

Bhatt BP, Singh R, Misra LK, Tomar JMS, Matvar Singh, Chauhan DS, Dhyani SK, Singh KA, Dhiman KR and Datta M 2001 Agroforestry research and practices: An overview, In *Steps Towards Modernization of Agriculture in NEH Region* (eds) ND Verma and BP Bhatt), ICAR Publication, Meghalaya, India, 365-392 pages

Bhattarai NK 1997 Biodiversity – people interface in Nepal, In *Medicinal plants for forest conservation and health care* (eds) G Bodeker, KKS Bhat, J Burley and P Vantomme, Non-wood Forest Products-11, FAO, Rome, 78-86 pages

Bheemaiah G, Madhusudhan T, Subrahmanyam MVR and Syed Ismail 1998 Effect of Greenleaf manuring and nitrogen application on growth and yield of rainfed castor alley cropped with subabul, *Indian Journal of Agricultural Sciences* **68**(11) 722-725

Bhimaya CP, Kaul RN and Ganguli BN 1961 Sand dune rehabilitation in Western Rajasthan, In *Proceedings of Fifth World Forest Congress*, FAO, Rome, 358-363 pages

Bhumbla DR and Abrol IP 1979 Saline and Sodic Soils, In *Soils and Rice*, International Rice Research Institute, Manila, Philippines, 719-738 pages

Bhumbla DR and *Khare A 1984 Estimates of Wasteland in India,* Society for Promotion of *Wasteland* Development (SPWD), New Delhi

Bhutta MN and Choudhry MR 2000 Biological control of water logging: Bio-drainage and other alternatives to drainage, In 8th ICID International drainage workshop on "Role of Drainage and challenges in 21st century, 31st Jan-4th Feb 2000, New Delhi, 33-45 pages

Bioversity 2009 Annual Report, Bioversity International, Rome

Bisen SS 2001 Marketing of forest based crude drugs and Herbs, In *Marketing Monitoring of Tree Products* (eds) GS Rawat and NS Bisht, Dehradun, 164-173 pages

Blanford HR 1925 Regeneration with assistance of taungya in Burma, Indian Forest Records 2, Old Series (Silv), Also Indian Forest Records, Vol xi, part iii, 37page

Blanford HR 1958 Highlights of one hundred years of forestry in Burma, *Empire Forestry Review* **37**(1) 33-42

Bleasdale JKA 1967 Systematic designs for spacing experiments, *Experimental Agriculture* 3 73-85

Boffa JM 1999 Agroforestry parklands in sub-Saharan Africa, FAO Conservation Guide, No.34, FAO, Rome, Italy

Bolund P and Hunhammar S 1999 Ecosystem services in urban areas, *Ecological Economics* **29** 293-301

Boonkird SA, Fernandes ECM and Nair PKR 1984 Forest villages: an agroforestry approach to rehabilitating forest land degraded by shifting cultivation in Thailand, *Agroforestry Systems* **2** 87-102

Borlaug NE and Dowswell CR 1988 World revolution in agriculture, 1988 Britannica Book of the Year, Encyclopedia Britannica Inc., Chicago, USA, 5-14 pages

Borthakur DN, Prasad RN, Ghosh SP, Singh A, Awasthi RP, Rai RN, Varma A, Datta HH, Sachan JN and Singh MD 1979 Agroforestry based farming system as an alternative to jhuming, Reprinted from Seminar Proceedings on Agro-forestry organised by ICAR at Imphal, Manipur, May 16-18, 1979, 32 pages

Bowonder B, Prasad SSR and Unni NVM 1988 Dynamics of fuelwood prices in India: Policy implications, *World Development* **16** 1213-1229

Boyce JS 1961 Forest pathology, 3rd Editon, New York: Mc Graw-Hill Book Co. Inc., 572 page

Braatz S 1994 Urban forestry in developing countries: status and issues, In *Growing greener communities* (eds) C Kollin, J Mahon and L Frame, Proceedings of the sixth National Urban Forest Conference, Minneapolis, Minnesota, USA, 14-18 September 1993, American Forests, Washington, DC, 85-88 pages

Brandle JR, Hodges L and Zhou X 2004 Windbreaks in sustainable agriculture, *Agroforestry Systems* **61** 65-78

Brenner AJ, Jarvis PG and Van Den Beldt RJ 1995 Windbreak crop interactions in the Sahel: Dependence of shelter on field conditions, *Agriculture and Forest Meteorology* **75** 215-234

Bridge J 1996 Nematode management in sustainable and subsistence agriculture, *Annual Review of Phytopathology* **34** 201-225

Brown IF, Martinelli LA, Thomas WW, Moreira MZ, Ferreira CAC and Victoria RA 1995 Uncertainty in the biomass of Amazonian forests: an example from Rondonia, Brazil, *Forest Ecology and Management* **75** 175-189

Brown S 1997 Estimating biomass and biomass change of tropical forests, a primer, FAO Forestry paper 134, FAO, Rome

Buck LE, Gavin TA, Lee DR, Uphoff NT, Behr DC, Drinkwater LE, Hively WD and Werner FR 2004 Ecoagriculture: A Review and Assessment of its Scientific Foundations, Cornell University, Ithaca, USA

Bulman PA 1991 TREE$PLAN – A model to compare farm tree designs in southern Australia, *Agricultural System and Information Technology Newsletter* **3** 16-18

Burley J and von Carlowitz P (eds) 1984 Multipurpose Tree Germplasm, ICRAF, Nairobi, Kenya, 298 page

Butola JS, Malik AR and Baba JA 2012 Livelihood diversification in cold arid desert of Indian Himalaya: urgent need for transforming traditional agroforestry system with emphasis on adoption of herbal farming, *International Journal of Medicinal and Aromatic Plants* **2**(3) 544-548

Caballero K 1992 Mayan homegardens: past, present and future, *Etnóecolgica* **1** 35-54

Calvet-Mir L, Gómez-Baggethun E and Reyes-García V 2012 Beyond food production: Ecosystem services provided by home gardens: A case study in Vall Fosca, Catalan Pyrenees, Northeastern Spain, *Ecological Economics* **74** 153-160

Campi P, Palumbo AD and Mastrorilli M 2009 Effects of tree windbreak on microclimate and wheat productivity in a Mediterranean environment, *European Journal of Agronomy* **30**(3) 220-227

Cannel M, Van Noordwijk M and Ong C 1996 The central agroforestry hypothesis: the trees must acquire resources that the crop would not otherwise acquire, *Agroforestry Systems* **34** 27-31

Cannell JH 1990 "Apparent" versus "real" competition in plants, In *Perspectives on Plant Competition* (eds) JB Grace and D Tilman, Academic Press, New York, USA, 9-26 pages

Cannell MGR 1983 Plant management in agroforestry: manipulation of trees,. population densities and mixtures of trees and herbaceous crops, In *Plant Research and Agroforestry* (ed) PA Huxley, ICRAF, Nairobi, Kenya, 455-486 pages

Carne RJ 1993 Agroforestry land use: The concept and practice, *Australian Geographical Studies* **31**79-90

Carvalho EF and de Garcia-Torres L 1994 Weed management in agroforestry system in the Amazon, *Agroforesteria en las Americas* **1**(3) 6-9

Castro M 2008 Chapter-6: Silvopastoral Systems in Portugal: Current Status and Future Prospects, In *Agroforestry in Europe: Current Status and Future Prospects* (eds) A Rigueiro-Rodriguez, J McAdam and MR Mosquera-Losada, Springer

Ceccolini L 2002 The homegardens of Soqotra island, Yemen: an example of agroforestry approach to multiple land use in an isolated location, *Agroforestry Systems* **56** 107-115

CGIAR (Consultative Group on International Agricultural Research) 1990 Food Today and Tomorrow, CGIAR, Washington DC, USA

Chadhar SK and Sharma MC 1996 Survival and yield of four medicinal plant species grown under tree plantations of Bhataland, *Vaniki Sandesh* **20**(4) 3-5

Chahal D, Ahmad A and Bhatia JN 2012 Assessment of agroforestry based two tier cropping system in Ambala district of Haryana, In *Proceedings of National seminar on agroforestry: An evergreen agriculture for food security and environmental resilience*, 2-4 February, 9-99 pages

Champion HG and Seth SK 1968 A Revised Survey of Forest Types of India, Govt. of India Publication Division, New Delhi, 404 page

Chander Datt and Niranjan M 2001 Chemical composition of some commonly used weeds as ruminant feeds in Tripura, In *Proceedings of National Seminar on Approaches for Increasing Agricultural Productivity in Hill and Mountain Ecosystem*, ICAR Research Complex for NEH Region, Meghalaya

Chander K, Goyal S, Nandal DP and Kapoor KK 1998 Soil organic matter, microbial biomass and enzyme activities in a tropical agroforestry system, *Biology and Fertility of Soils* **27** 168-172

Chandra JP 2001 Marketing issues and incentives framework for agroforestry, In *Marketing Monitoring of Tree Products* (eds) GS Rawat and NS Bisht, Dehradun, 29-38 pages

Chandrasekharan C 1993 Issues involved in the sustainable development of non-wood forest products, *Proceedings on Expert Consultation on Non-Wood Forest Products for English-Speaking African Countries* (held in Arusha, Tanzania), FAO, Rome

Charreau C and Vidal P 1965 Influence de l'Acacia albida Del sur le sol, nutrition minerale et rendements des mils Pennisetum au Senegal, Agron. Trop. **20** 600-626

Chatterjee PN and Singh P 1968 *Celosterna scarabrator* Fab. (Lamiidae: Coleoptera), a new pest of *Eucalyptus* and its control, *Indian Forester* **94(11)** 826-830

Chaturvedi AN and Sood R 1995 Variation in growth form and volume of clonal and seedling raised plants of *Eucalyptus hybrid, Journal of Tropical Forest Science* **11** 23-29

Chaturvedi OP and Jha MK 1998 Crop production and economics under *Litchi chinesis* Plantation across 1- to 9-year age series in north Bihar, India, *International Tree Crops Journal* **9** 159-168

Chaubey PK, Kabra KN et al 2001 Some aspects of Economics of Timber Markets in India, In *Marketing Monitoring of Tree Products* (eds) GS Rawat and NS Bisht, Dehradun, 1-13 pages

Chaudhury M, Ajayi OC, Hellin J and Neufeldt H 2011 Climate change adaptation and social protection in agroforestry systems: enhancing adaptive capacity and minimizing risk of drought in Zambia and Honduras, ICRAF Working Paper No. 137, Nairobi: World Agroforestry Centre.

Chauhan HS 2000 Performance of Poplar (*Populus deltoides*) based agroforestry system using aromatic crops, *Indian Journal of Agroforestry* **2** 17-21

Chauhan R, Dhillon WS and Chauhan SK 2010 Adoption of lesser known fruit trees: promising livelihood option for farmers, In *Plant Protection in organic farming* (eds) A Sharma and RS Chandel, IBD, Dehradun, 372-401 pages

Chauhan VK, Mishra VK and Khosla PK 1997 Growth and productivity evaluation of agri-horti-silviculture system, *Journal of Tree Science* **16**(2) 69-74

Chen HYN, Klinka K and Kabzens R 1998 Height growth and site-index models for *Populus temuloides* in north British Columbia, *Forest Ecology and Management* **102** 157-165

Chepil WS and Woodruff NP 1965 The Physics of Wind Erosion and Its Control, *Advances in Agronomy* **15** 211-301

Chhabra R and Thakur NP 1998 Lysimeter study on the use of bio-drainage to water logging and secondary salinization in (canal) irrigated arid /semiarid environment, *Irrigation and Drainage System* **12** 265-288

Chifflot V, Rivest D, Olivier A, Cogliastro A and Khasa D 2009 Molecular analysis of arbuscular mycorrhizal community structure and spores distribution in tree-based intercropping and forest systems, *Agriculture, Ecosystems and Environment* **131** 32-39

Chirwa PW, Ong CK, Maghembe JL and Black CR 2006 Soil water dynamics in cropping systems containing *Gliricidia sepium*, pigeonpea and maize in southern Malawi, *Agroforestry Systems* **69** 29-43

Chopra K 1993 The value of non-timber forest products: an estimation for tropical deciduous forests in India, *Economic Botany* **47** 251-257

Chopra N, Chopra NK and Harsh LN 1993 Root trenching of *Acacia tortilis* can increase the yield of intercrops, *Indian Farming* **43**(2) 23-26

Chorley RJ and Barry RG 1971 Introduction to physical hydrology, Routledge Publication (Taylor and Francis), London

Choudhury K and Jansen LJM 1999 Terminology for integrated resources planning and management, Soil Resources, Management and Conservation Service, FAO Land and Water Development Division, Rome

Choudhury SR, Ashwani Kumar, Brahmanand PS, Ghosh S, Mohanty RK, Jena SK, Sahoo N and Panda GC 2011 Application of biodrainage for reclamation of waterlogged situations in deltaic Orissa, ICAR-Research Bulletin-53, Directorate of Water Management, Bhubaneswar, Orissa, India

Christanty L, Abdoellah OL, Marten GG and Iskandar J 1986 Traditional agroforestry in West Java: the Pekaranagan (homegarden) and Kebun-Talun (annual-perennial rotation) cropping systems, In *Traditional Agriculture in South East Asia* (ed) GG Marten, Westview Press, Boulder, CO., 132-158 pages

CISH 2011 Central Institute of Subtropical Horticulture (CISH) Annual Report 2010-2011, Lucknow, Uttar Pradesh, India

Clarke WC and Thaman RR 1993 Agroforestry in the Pacific Islands: Systems for sustainability, United Nations University Press, Tokyo

Cleveland DA and Soleri D 1987 Household gardens as a development strategy, *Human Organization* **46**(3) 259-270

Clifton W, Bruce W and David LH 1986 Basic Windbreak Design Criteria for Farm and Ranch Head Quarters Areas And Large Residential Lots, Technical Note Series No. 190-LI-6, Midwest National Technical Center, Soil Conservation Service, United States Department of Agriculture, USA

Coe R 1994 Through the looking glass: 10 common problems in alley cropping research, *Agroforestry Today* **6** 9-11

Coe R 1998 Participatory on-farm experimentation in agroforestry: experiences and the role of biometrics, In *Proceedings of the International Biometrics Conference*, Invited papers, 13-18 December 1998, Cape Town, South Africa: International Biometrics Association, 1-10 pages

Cohen M and Garrett J 2009 The food price crisis and urban food insecurity, London, UK: IIED

Conklin HC 1957 Hanundo Agriculture, FAO, Rome, Italy

Cook RD and Weisberg S 1990 Confidence curves in nonlinear regression, *Journal of American Statistical Association*, **85** 544-551

Corlett JE, Ong CK and Black CR 1989 Microclimatic modification in intercropping and alley-cropping systems, In *Meteorology and Agroforestry* (eds) WS Reifsnyder and TO Darnhofer, ICRAF, Nairobi, Kenya, 419-430 pages

Correal E, Erena M, Rios S, Robledo A and Vicente M 2008 Chapter-9: Agroforestry Systems in Southeastern Spain, In *Agroforestry in Europe: Current Status and Future Prospects* (eds) A Rigueiro-Rodriguez, J McAdam, MR Mosquera-Losada and M Rosa, Springer

CPREEC 2013 C.P.R. Environmental Education Centre (CPREEC), Centre of Excellence of the Ministry of Environment and Forests (MoEF), Government of India, Website: www.cpreec.org

CRIDA 1995 Annual Reports of Central Research Institute for Dryland Agriculture (CRIDA), Hyderabad, AP, India

CRIDA 1999 Annual Reports of Central Research Institute for Dryland Agriculture (CRIDA), Hyderabad, AP, India

Cromartie WJ 1991 The environmental control of insects using crop diversity, In *CRC Handbook of Pest Management in Agriculture* (eds) D Pimentel and FL Boca Raton, CRC Press, Vol. **1** 223-251

CTFT (Centre Technique Forestier Tropical) 1988 *Faidherbia albida* (Synonym: *Acacia albida*) *Monographie*, Nogent-sur-Marne, France, CTFT, 72 page

Cunningham A, Campbell B and Belcher B (eds) 2005 Carving out a Future: Forests, Livelihoods and the International Woodcarving Trade, Earthscan, London, UK

Cunningham AB 1997 An African-wide overview of medicinal plant harvesting, conservation and health care. In *Medicinal plants for forest conservation and health care* (eds) G Bodeker, KKS Bhat, J Burley and P Vantomme, Non-wood Forest Products-11, FAO, Rome, 116-129 pages

Dadhwal KS, Narain P and Dhyani SK 1989 Agroforestry systems in the Garhwal Himalayas of India, *Agroforestry Systems* **7** 213-225

Dagar JC 1978 Studies on Soil Erosion of Kshipra Ravines, Ph.D. Dissertation, Vikram University, Ujjain, India

Dagar JC 1982 Some ecological aspects of mangrove vegetation of Andaman and Nicobar Islands, *Sylvatrop, the Philippines Forest Research Journal* **7**: 177-216

Dagar JC 1987 Studies on reclamation of Kshipra ravines, *Indian Journal of Forestry* **10**: 83-89

Dagar JC, Sharma HB and Shukla YK 2001 Raised and sunken bed technique for agroforestry on alkali soils of northwest India, *Land Degradation and Development* **12** 107-118

Dagley J 2006 Pollarding in Epping Forest, In *1er colloque europeen sur led trognes*, 26-28 October, Vendome

Dalla Rosa KR 1993 Cattle under coconuts: a practical Pacific tradition, Agroforestry for the Pacific Technologies, Paia, Hawaii, USA, Nitrogen Fixing Tree Association

Damodaram T and Hegde DM 2005 Oilseeds situation: A statistical compendium 2005, Directorate of Oilseeds Research, Hyderabad, 606 pages

Daniel JN and Hegde NG 2007, Tree-Borne Oilseeds in Agroforestry, In *Proceedings of the National Seminar on Changing Global Vegetable Oils Scenario: Issues and Challenges before India* (ed) DM Hegde, Indian Society of Oilseeds Research, Hyderabad, India, 263-276 pages

Danso SKA, Bowen GD and Sanginga N 1992 Biological nitrogen fixation in trees in agro-ecosystems, *Plant and Soil* **141** 177-196

Das B, Singh R and Kumar S 2009 Fruit based agro-forestry system for uplands, In Winter School on Alternate land use options for resource conservation, emerging market needs and mitigation of climate change in rainfed regions, CRIDA, Hyderabad, AP, India

Das DK, Chaturvedi OP, Jha RK and Kumar R 2011 Yield, soil health and economics of aonla (*Emblica officinalis* Gaertn.)-based agri-horticultural systems in eastern India, *Current Science* **101**(6) 786-790

David P and Aaron W 1998 Economic and Environmental Benefits of Agroforestry in Food and Fuelwood Production, In *Agroforestry in Sustainable Agricultural Systems* (eds.) Louise E. Buck, P James, Lassoie and Erick CM Fernandes), CRC Press, USA, 295-312 pages

Day RK, Hill MG, Murphy ST and Nyambo B 1995 Candidate agents for biological control of leucaena psyllid in Africa, In *Leucaena Psyllid: A Threat to Agroforestry in Africa* (eds) WM Ciesla and L Nshubemuki, FAO, Rome, 169-189 pages

Dayal SKN, Grewal SS and Singh SC 1996 Horticulture-based Agroforestry System for Shiwalik Foothills, *Indian Horticulture*

de Moral and Muller R 1969 Fog drip: a mechanism of toxin transport from *Eucalyptus globulus, Bulletin of Torrey Bot. Club* **96** 467-475

Deb Roy R 1991 Agroforestry to meet our food, fodder and fuelwood needs, *Indian Farming* **40**(10) 18-21

Deepa RS and Ramachandra TV 1999 Impact of Urbanization in the Interconnectivity of Wetlands, Paper presented at the National Symposium on Remote Sensing Applications for Natural Resources: Retrospective and Perspective (XIX-XXI 1999), Indian Society of Remote Sensing, Bangalore

DeFries R and Rosenzweig C 2010 Towards a whole landscape approach for sustainable land use in the tropics, *Proceedings of National Academy of Science* **107**(46) 19627-19632

Denevan WM, Treacy JM, Alcorn JB, Padoch C, Denslow J and Paitan SF 1984 Indigenous agroforestry in the Peruvian Amazon: Bora Indian management of swidden fallows, *Interciencia* **9** 346-357

Denis D 2003 The tree behind the forest: ecological and economic importance of traditional agroforestry systems and multiple uses of trees in India, *Tropical Ecology* **44**(1) 63-71

Dery BB, Otsyina R and Ng'atigwa 1999 Indigenous Use of Medicinal Trees and Setting Priorities for Their Domestication in Shinyanga Region, Tanzania, ICRAF, Nairobi, Kenya

Dev S 1997 Ethnotherapeutic and modern drug development: The potential of Ayurveda, *Current Science* **73** 909

Devi Dayal, Bhagirath Ram, Shamsudheen, Swami ML and Patil NV 2009 20 years of CAZRI regional research station Kukma – Bhuj, RRS, CAZRI, Kukma, Gujarat, 35 page

DeWalle DR and Heisler GM 1988 Use of windbreaks for home energy conservation, *Agriculture, Ecosystems and Environment* **22/23** 243-260

Dhar SK and Ahuja G 2001 Marketing strategies of Haryana Forest Development Corporation Ltd. for agroforestry, commercial and community plantations, In *Marketing Monitoring of Tree Products* (eds) GS Rawat and NS Bisht, 86-93 pages

Dhillon GS, Grewal SS and Atwal AS 1979 Developing agri-silvicultural practices: Effect of farm trees (Eucalyptus) on the adjoining crops, *Indian Journal of Ecology* **6** 88-97

Dhir RP 1982 The human factors in ecological history. In *Desertification and Development: Dryland Ecology in Social Perspective*, Academic Press, New York, 311-331 pages

Dhir RP 1997 Characteristics and behavioral aspects of arid and semi-arid zone soils, In *Silvipastoral Systems in Arid and Semi-arid Ecosystems* (eds) MS Yadav, Singh Manjit, SK Sharma, JC Tewari and U Burman, CAZRI, Jodhpur, India, 39-46 pages

Dhruva Narayana VV and Ram Babu 1983 Estimation of soil erosion in India, *Journal of Irrigation and Drainage Engineering* **109** 419-433

Dhyani SK 2011 Alternate Land Use and Agroforestry Systems for Resource Conservation and Enhanced Productivity in Hills, Proceedings of Workshop on Mountain Agriculture in Himalayan Region: Status, Constraints and Potentials, April 2-3, 2011, 107-132 pages

Dhyani SK and Chauhan DS 1989 Evaluation of crops in relation to shade intensities of Khasi pine (*Pinus khasya*), Progress Report, Agroforestry Division, ICAR Research Complex for NEH Region, Shillong, India

Dhyani SK and Sharada VN 2005 Agroforestry systems as rural development options for the Indian Himalayas, *Journal of Tree Science* **24**(1) 1-19

Dhyani SK, Chauhan DS, Singh BP and Prasad RN 1995 Tree effect on yield and quality of pineapple (kew) in a silvi-horti systems of agroforestry in NEH region of India, *Range Management and Agroforestry* **16**(2) 89-93

Dhyani SK, Dadhwal KS and Katiyar VS 1988 Biological measures for rehabilitation of abandoned mine area – preliminary studies, *Indian Farming* **38**(7) **7**-10

Dhyani SK, Kareemulla A, Ajit and Handa AK 2009 Agroforestry potential and scope for development across agro-climatic zones in India, *Indian Journal of Forestry* **32**(2) 181-190

Dhyani SK, Sharda VN and Juyal GP 2007 Conservation forestry and watershed Management, Training Manual, Vol. 3 published by CSWCRTI, Dehra Dun

Diamond J 2002 Evolution, consequences and future of plant and animal domestication, *Nature* **418** 700-707

Dickmann DI 1985 The ideotype concept applied to forest trees, In *Trees as Crop Plants* (eds) MGR Cannell and JE Jackson Rendal, Titus Wilson and Son Ltd., Cumbria, UK

Dimond AE 1955 Pathogenesis in wilt diseases, *A Review of Plant Physiology* **6** 329-350

Divya MP, Chandra Sekaran N, Asokaraja N and Jayarama Soundari R 2008a Development of suitable inter-crops for Pungam based agroforestry system, Paper presented in the "National Symposium on Agroforestry Knowledge for Sustainability, Climate Moderation and Challenges Ahead", NRCAF, Jhansi, 15-17 December, 2008

Divya MP, Chandra Sekaran N, Jayarama Soundari R, Govinda Rao M, Paramathma M and Dhyani SK 2010 A Compendium on 25 years of Agroforestry Research under AICRP on Agroforestry, Tamil Nadu Agricultural University Offset Press, Coimbatore, 105 pages

Divya MP, Jamaludeen V, Parthiban KT and Durairasu P 2011 Casuarina clone based profitable business agroforestry models for Tamil Nadu, Paper presented in the "Short rotation forestry and agroforestry: an exchange of experience between CDM Countries and Europe", Marchesi di Barolo, Barolo, Italy, June 20-22, 2011

Divya MP, Jamaludheen V, Rajalingam GV, Durairasu P and Swaminathan C 2013 Profitable Agroforestry models for Industrial wood species, In *Industrial Agroforestry-Perspectives and Prospective* (eds) P Durairasu, KT Parthiban, R Umarani, S Umesh Kanna, I Sekar, P Rajendran and PS Devanand, Forest College and Research Institute, Tamil Nadu Agricultural University, Mettupalayam, 109-122 pages

Divya MP, Santhi R and Ramesh KR 2006 Evaluation of suitable intercrops for Jatropha based agroforestry system, *Indian Journal of Agroforestry* **8**(1) 1-4

Divya MP, Srinivasan K and Govinda Rao M 2008b Compatible medicinal plants for Simaruba based agroforestry system, Paper presented in the National workshop cum symposium on "Tree Spice cultivation and sustainable development of homegardens for descent livelihood and environment protection", Port Blair, Andaman and Nicobar Islands, 4-6 February 2008

Dixon J, Gulliver A and Gibbon D 2001 Farming systems and poverty improving farmers' livelihoods in a changing world, FAO and World Bank, Rome

Dixon RK 1995 Agroforestry Systems: Sources or Sinks for Greenhouse Gases? *Agroforestry Systems* **31** 99-116

Dixon RK, Andrasko KJ, Sussman FA, Lavinson MA, Trexler MC and Vinson TS 1993 Tropical forests: their past, present and potential future role in the terrestrial carbon budget, *Water Air Soil Pollution* **70** 71-94

Dixon RK, Wimjum JK, Lee JJ and Schroeder PE 1994 Integrated systems: assessing of promising agroforest and alternative land use practices to enhance carbon conservation and sequestration, Climate Change **30** 1-23

DOE 1999 Carbon Sequestration: State of the Science, US Departemnt of Energy (DOE), Washington DC

DoLR and NRSC 2011 Wastelands Atlas of India: Change Analysis based on Temporal Satellite Data of 2005-06 and 2008-09, Department of Land Resources, Ministry of Rural Development, GOI, New Delhi and National Remote Sensing Centre, Hyderabad

Dosskey MG 2001 Toward quantifying water pollution abatement in response to installing buffers on crop land, *Environmental Management* **28**(5) 577-598

Dregne H, Kassas M and Razanov B 1991 A new assessment of the world status of desertification, *Desertification Control Bulletin* (UNEP) **20** 6-18

Dresher AW, Holmer RJ and Iaquinta DL 2006 Urban homegardens and allotment gardens for sustainable livelihoods: management strategies and institutional environment, In *Tropical homegardens: a time-tested example of sustainable agroforestry* (eds) Kumar BM and Nair PKR, Springer, Netherlands, Vol.3, 317-338 pages

Drucker AG, Smale M and Zambrano P 2005 Valuation and sustainable management of crop and livestock biodiversity: a review of applied economics literature, A report published for the CGIAR System-wide Genetic Resources Programme (SGRP) by the IFPRI, IPAGRI, and ILRI

Dudal R and Purnell MF 1986 Land Resources: Salt affected soils, *Reclamation and Revegetation Research* **5** 1-10

Duguma B, Kang BT and Okali DUU 1988 Effect of pruning intensities of three woody green manures produced by alley cropping in West Sumatra, *Agroforestry Systems* **6** 19-35

Dupraz C and Newman SM 1997 Chapter-6: Temperate Agroforestry: The European Way, In *Temperate Agroforestry Systems* (eds) AM Gordon and SM Newman, CAB International, Wallingford, UK

Dutt Vaishnu and Thakur PS 2004 Bio-diesel-economics of cropping systems combining medicinal and aromatic herbs with commercial timber tree species, *Indian Journal of Agroforestry* **6**(1) 1-7

Dwivedi AP 1992 Agroforestry - Principles and Practices, Oxford and Publishing Company Pvt. Ltd., New Delhi, India, 365 pages

Dwivedi AP 2001 Economics of agroforestry systems, In Agroforestry Principles and practices, Oxford & IBH Publishing Co. Pvt. Ltd., 308-320 pages

Dwivedi MK 2001 Apiculture in Bihar and Jharkhand: A study of costs and margins, *Agriculture Marketing* **44** 12-14

Dwivedi RP 2002 Status of extension management perspectives in agroforestry in India, *MANAGE Extension Research Review* **3**(1) 97-103

Dwivedi RP and Shukla SK 1998 Agroforestry for sustainable land management, *Employment News*, New Delhi 5-11 December, 1998, **23**(36) 1&23 pages

East AJ 2008 A future in the past: urban agroforestry systems in future planned urban settlements in Kiribati, a Pacific case study, Ph.D. Thesis, School of Design and Institute for Sustainable Resources, Queensland University of Technology

East AJ and Dawes LA 2009 Homegardening as a panacea: a case study of South Tarawa, *Asia Pacific Viewpoint* **50**(3) 338-352

Eichhorn MP, Paris P, Herzog F, Incoll LD, Liagre F, Mantzanas K, Mayus M, Moreno G, Papanastasis VP, Pilbeam DJ, Pisanelli A and Dupraz C 2006 Silvorable systems in Europe – past, present and future prospects, *Agroforestry Systems* **67** 29-50

Elevitch CR and Wilkinson KM 1998 Mulch from hedgerows of nitrogen fixing trees affects soil nutrient levels in a jackfruit orchard, The Forest, Farm and Community Tree Research Reports Vol.3, FACT-Network, Winrock International, Morrilton, Arkansas

Elevitch CR and Wilkinson KM 2000 Agroforestry Guides for Pacific Islands, Permanent Agricultural Resources, Holualoa, Hawaii,USA, webpage: http://www.agroforestry.net

Elevitch, C.R. (ed.). 2006. Traditional Trees of Pacific Islands: Their Culture, Environment, and Use. Permanent Agriculture Resources, Holualoa, Hawaii, USA, 816p.

Emerton L 1991 Utilization of Kakamega forest reserves by adjacent households: First interim socio-economic report, Kakamega forest reserve, KIFCON series

Eterington DM and Matthews PJ 1983 Approaches to the economic valuation of agroforestry farming systems, *Agroforestry Systems* **1** 347-360

Evans J 1982 Plantation forestry in the tropics, Oxford University Press, New York

Evenson RE and Gollin D 2003 Assessing the impact of the Green Revolution: 1960 to 2000, *Science* **300** 758-762

Ewel JJ 1999 Natural systems as models for the design of sustainable systems of land use, *Agroforestry Systems* 45 1-21

Fanish SA and Priya RS 2012 Review on benefits of agroforestry system, *International Journal of Agricultural Sciences* **2**(3) 80-91

FAO (Food and Agriculture Organisation) 2012 The State of Food Insecurity in the World 2012: Economic growth is necessary but not sufficient to accelerate reduction of hunger and malnutrition, FAO, WFP and IFAD, Rome

FAO (Food and Agriculture Organisation), 1996 FAO AGROSTAT 1995, FAO, Rome, Italy

FAO (Food and Agriculture Organisation), 1997 FAO Production Year book, Rome, Italy

FAO 1969 *Soil and water resources survey of Sokoto Valley, Nigeria*, Final report AGL-UNDP/SF 67/NIR/3. Vol. 5, soil survey and land classification, part 1 of 3, Rome, FAO, 33-34 pages

FAO 1976 A Framework for Land Evaluation, FAO Soils Bulletin-32, FAO, Rome, Italy

FAO 1978 Shifting cultivation, *Forest News for Asia and the Pacific* **2** 1-26

FAO 1980 Forest Resources, *Unasylva* 1985/4. Vol. **37** (ed) Robert Wazeka

FAO 1982 Tropical Forest Resources, FAO, Rome, Italy.

FAO 1984 Institutional aspects of shifting cultivation in Africa, FAO Human Resources, Institutions and Agrarian Reform Division, Based on work by Dr M.Tiffen, Rome

FAO 1986 Tree Growing by Rural People, Forestry Paper-64, Food and Agriculture Organization of the United Nations, Rome

FAO 1995a Improving nutrition through home gardening: a training package for preparing field workers in Southeast Asia, Rome

FAO 1995b Non-wood forest products for rural income and sustainable forestry, FAO Technical Report No-7, Non-wood Forest Products Division, FAO, Rome, 127 pages

FAO 1995c Plantations in tropical and subtropical regions – mixed and pure, FAO, Rome, Italy

FAO 1995d Shifting Cultivation in Bhutan: A Gradual Approach to Modifying Land Use Patterns, A Case Study From Pema Gatshel District, Bhutan, *Community forestry case study* Series **11**

FAO 1996 Report of the International technical Conference on Plant Genetic Resources, Leipzis, Germany, 17-23 June 1996

FAO 1997 Report on the State of the World's Plant Genetic Resources for Food and Agriculture, Food and Agriculture Organization of the UN, Rome, Italy

FAO 1999a Agricultural Biodiversity, Multifunctional Character of Agriculture and Land Conference, Background Paper 1, Maastricht, Netherlands. September 1999

FAO 1999b Women: users, preservers and managers of agrobiodiversity, website: www.fao.org/ FOCUS /E/Women/Biodiv-e.htm

FAO 2000 Global forest resources assessment main report, FAO Forestry Paper 140

FAO 2001 Improving nutrition through home gardening, Nutrition Programmes Service, Food and Nutrition Division, FAO, Rome

FAO 2005 Global forest resources assessment update, Forest Resources Assessment Programme, Working Paper 83/E, Food and Agricultural Organization, Rome

FAO 2007 The State of the World's Animal Genetic Resources for Food and Agriculture, Food and Agriculture Organization of the UN, Rome, Italy

FAO 2009 Second Report on the State of the World's Plant Genetic Resources for Food and Agriculture, Food and Agriculture Organization of the UN, Rome, Italy

FAO 2010a Global forest resources assessment 2010: main report, FAO Forestry Paper 163, Food and Agriculture Organization of the United Nations, Rome

FAO 2010b The second report on the state of the world's plant genetic resources, Food and Agriculture Organization of the United Nations, Rome

FAO 2010c Status of World Report II of Plant Genetic Resources for Food and Agriculture, Rome, Italy

FAO 2011a Forests for improved nutrition and food security, FAO, Rome, website: www.fao.org/forestry /27976-02c09ef000fa99932eefa37c22f76a055.pdf

FAO 2011b State of the World's Forests 2011, FAO, Rome, website: www.fao.org/docrep/ 013/i2000e /i2000e00.htm.

FAO 2013 Non-wood forest products, website: http://www.fao.org/forestry/nwfp/en/

FAO 2013 World Food Situation, website: http://www.fao.org/worldfoodsituation/wfs-home/csdb/en/

FAOStat 2010 Production data relating to food and agriculture, http://faostat.fao.org

Feller C, Albrecht A, Blanchart E, Cabidoche YM, Chevallier T, Hartmann C, Eschenbrenner V, Larre-Larrouy MC and Ndandou C 2001 Soil carbon sequestration in tropical areas: general consideration and analysis of some edaphic determinants for Lesser Antilles soils, *Nutrient Cycle in Agroecosystem* **61** 19-31

Fernandes E and Nair PKR 1986 An evaluation of the structure and function of tropical homegardens, *Agroforestry Systems* **21** 279-310

Fernandes ECM, Oktingati A and Maghembe J 1985 The Chagga home gardens: A multi-storeyed agro-forestry cropping system on Mt. Kilimanjaro, Northern Tanzania, *Food and Nutrition Bulletin* **7**(3) 87

Ferrini F 2006 Pollarding and its effect on tree physiology: a look to mature and senescent tree management in Italy, In *1er colloque europeen sur led trognes*, 26-28 October, Vendome

Filmer D and Pritchett L 2002 Environmental Degradation and the Demand for Children: Searching for the Vicious Circle, *Environmental and Development Economics* **7**(1) 123-146

Fisher JB 1986 Branching pattern and angles in trees, In *On the Economy of Plant Form and Functions* (ed) TJ Givnish, Cambridge University Press, Cambridge

Florence EJM, Sankaran KV and Sharma JK 1987 Two new species of *Phomopsis* from India, *Transactions of the British Mycological Society* **89**(3) 404-407

Florence RG 1986 Cultural problems of eucalypts as exotics, *Commonwealth Forestry Review* **65** 141-163

Follett J 1997 Ginseng production in NZ forests: experiences from Tiketere, New Zealand, *Tree Grower* **18**(3) 19-21

Forde DC 1937 Land and labour in a Cross River village, *Geographical Journal* Vol.**XC** No. 1

Forest/ Fuelwood Research and Development Project 1992 Growing Multipurpose Trees on Small Farms, Bangkok, Thailand: Winrock International, 195+IX page

Forestry Commission 2000 Agroforestry in the UK (eds) Hislop M and Claridge J, *Bulletin* 122, Forestry Commission, UK

Frey R 1995 *Ipomoea carnea* ssp. *fistulosa* (Martius ex Choisy) Austin: taxonomy, biology and ecology reviewed and inquired, *Tropical Ecology* **36**(1) 21-48

FSI (Forest Survey of India) 2011 India State of Forest Report-2011, Dehradun: Forest Survey of India, Ministry of Environment and Forests

FSI 2001 Indian State of Forest Report 2001, Forest Survey of India (MoEF), Dehradun

Fuller, Irvine RA, Devine-Wright P, Warren PH and Gaston KJ 2007 Psychological benefits of green space increase with biodiversity, *Biology Letters* **3** 390-394

Galiana A, Bouillet JP and Ganry F 2004 The importance of biological nitrogen fixation by trees in agroforestry, In *Symbiotic nitrogen fixation: Prospect for enhanced application in tropical agriculture* (ed) Serraj Rachid, Enfield: Science Publishers, 185-199 pages

Gallina S, Mandujano S and Gonzalez-Romero A 1996 Conservation of mammalian biodiversity in coffee plantations of Central Veracruz, Mexico, *Agroforestry Systems* **33** 13-27

Garrett HE and McGraw RL 2000 Alley cropping practices, In *North American agroforestry: an integrated science and practice* (eds) HE Garrett, WJ Rietveld and RF Fisher, ASA, Madison, 149-188 pages

Garrett HE, Buck LE, Gold MH, Hardesty LH, Kurtz WB, Lassoie JP, Pearson HA and Slusher JP 1994 Agroforestry: An integ.ated Land-Use Management System for Production and Farmland Conservation, Resource Conservation Act (RCA) Appraisal of US Agroforestry, USDA Natural Resources Conservation Service, 58 page

Garrett HE, Rietveld WJ and Fisher RF (eds) 2000 North American Agroforestry: an Integrated Science and Practice, American Society of Agronomy, Madison, WI, USA, 402 pages

Garrett HEG 1997 Agroforestry: An integrated land-use management system for production and farmland conservation (USDA-SCS 68-3A7S-3-134), Fort Collins, CO: United States Department of Agriculture-Soil Conservation Science

Garrity D, Amoroso VB, Koffa S, Catacutan D, Buenavista G, Fay P and Dar W 2002 Landcare on the poverty–protection interface in an Asian watershed, *Conservation Ecology* **6**(1) 12

Garrity D, Okono A, Grayson M and Parrott S (eds) 2006 World Agroforestry into the Future, World Agroforestry Centre, Nairobi

Garrity DP 2004 Agroforestry and the achievement of the millennium development goals, *Agroforestry Systems* **61** 5-17

Garrity DP and Mercado Jr AR 1994 Reforestation through agroforestry: smallholder market-driven timber production on the frontier, In *Marketing multipurpose tree species in Asia* (eds) J Raintree and H Fernandez, Bangkok, Winrock International

Gaur VB and Sarkar S 2001 Marketing challenges and the prospective of U.P. Forest Corporation, In *Marketing Monitoring of Tree Products* (eds) GS Rawat and NS Bisht, 94-101 pages

Gautam NC and Narayan LRA 1988 Wastelands in India, Mathura: Pink Publishing House

Gebauer J 2005 Plant Species Diversity of Home Gardens in El Obeid, Central Sudan, *Journal of Agriculture and Rural Development in the Tropics and Subtropics* **106**(2) 97-103

Geisser S 1975 The predictive sample reuse method with application, *Journal of American Statistical Association* **70** 320-328

George MP 1961 Teak plantations of Kerala, *Indian Forester* **50** 644

Gepts 2002 Review and Interpretation: A comparison between Crop Domestication, Classical Plant Breeding and Genetic Engineering, *Crop Science* **42**: 1780-1780

GFU 2008 List of underutilized plants, website: http://www.underutilized-species.org/

Ghosh RC 1997 Handbook of afforestation techniques, Controller of Publications, New Delhi, 418 pages

Ghosh RC and Banerji SP 1974 State of soil aggregation in some forest soils of NW Bengal, *Forestry Conference,* Mimeo

Ghosh SN and Pal PP 2010 Effect of intercropping on plant and soil of Mosambi sweet orange orchard under rainfed conditions, *Indian Journal of Horticulture* **67**(2) 185-190

Gilbert OL 1989 The Ecology of Urban Habitats, Chapman & Hall, London

Gill AS and Ajit 2006 Intercropping of wheat in mango (*Mangifera indica* Linn.), *Indian Journal of Forestry* **29**(4) 417-420

Giller KE and Wilson KF 1991 Nitrogen fixation in tropical cropping systems, Wallingford, UK, CABI

Global Carbon Project 2012 Global Carbon Project (GCP), webpage: http://www.globalcarbonproject.org

Glover N and Beer J 1986 Nutrient cycling in two traditional Central American agroforestry systems, *Agroforestry Systems* **4** 77-87

Gold MA and Garrett HE 2009 Agroforestry nomenclature, concepts and practices In: HE Garrett (ed.) North American Agroforestry: An integrated Science and Practice, 2nd edn. American Society of Agronomy, Madison, 45-55 pages

Gold MA, Rietveld WJ, Garrett HE and Fisher RF 2000 Agroforestry nomenclature, concepts and practices for the USA, In *North American Agroforestry: An Integrated Science and Practice* (eds) HE Garrett et al, 63-77 pages, Madison, WI: American Society of Agronomy, Inc.

Gomez KA and Gomez AA 1984 Statistical procedures for agricultural research, 2nd edition, Johan Wiley & Sons, New York

Gompertz B 1825 On the nature of the function expressive of the law of human mortality and on the new mode of determining the value of life contingencies, *Philosophical Transactions of Royal Society of London* **115** 513-585

Gordon AM, Newman SM and Williams PA 1997 Temperate agroforestry: an overview, In Temperate Agroforestry Systems (eds) AM Gordon and SM Newman, CAB International, Wallingford, UK, 1-8 pages

Gosselink JG and Turner RE 1978 The role of hydrology in fresh water wetland ecosystems, In *Freshwater Wetlands: Ecological Processes and Management Potential* (eds) RE Good, DF Whigham and RL Simpson, Academic Press, New York, 63-78 pages

Goswami SB and Sen H 2001 Swamp taro cultivation in terai climate –a case study, In *International Symposium Abstracts-Tropical roots and tuber crop*, 19-22 Jan 2000, Trivandrum, India, 54 page

Grado and Husak 2005 Economic analyses of a sustainable agroforestry system in the southeastern United States, In *Valuing Agroforestry Systems: Methods and Applications* (eds) Alavalapati JRR and Mercer E, Advances in Agroforestry (Volume-2), Kluwer Academic Publishers, Dordrecht

Grandstaff TB 1980 Shifting cultivation in northern Thailand: Possibilities for development, *Resource Systems Theory and Methodology Series No. 3*, United Nations University, Tokyo, Japan

Greathead DJ 1988 Crop protection without chemicals: pest control in the third world, In *Aspects of Applied Biology*, Wellesbourne, Warwick: The Association of Applied Biologists, Vol. **17** 19-28

Greenland DJ 1974 Evolution and development of different types of shifting cultivation, In *Shifting cultivation and Soil Conservation in Africa, FAO Soils Bulletin* No. **24**

Gregersen HM 1982 Environmental Constraints Versus Economic gain in Tropical forestry

Grewal SS 1992 Eucalyptus and bhabar-grass together conserve rainwater in the Shiwaliks, *Indian Farming* **42** 12-14

Grewal SS and Abrol IP 1986 Agroforestry on alkali soils: effect on some management practices on initial growth, biomass accumulation and chemical composition of selected tree species, *Agroforestry Systems* **4**: 221-232

Grewal SS, Singh, Kehar and Juneja ML 1995 Conservation and production potential of an agroforestry systems integrating grey gum (*Eucalyptus tereticornis*), white popinac (*Leucaena latisiliqua*) and turmeric (*Curcuma longa*), *Indian Journal of Agricultural Sciences* **65**(3) 191-195

Grey GW and Deneke FJ 1986 Urban Forestry, 2nd Edition, John Wiley & Sons, Toronto

Griffin DM 1958 Influence of pH on the incidence of damping-off, *Trans. Br. Mycol. Soc.* **41** 483-490

Grosser D and Liese W 1971 On the anatomy of Asian bamboos, with special reference to their vascular bundles, *Wood Science Technology* **5** 290-312

Gupta AP 1986 Arthropod Immunocytes: Identification, Structure, Functions and Analogies to the Functions of Vertebrate B- and T-lymphocytes, In *Hemocytic and Humoral Immunity in Arthropods* (ed) AP Gupta, John Wiley, New York, 3-59 pages

Gupta JP 1997 Some alternative production systems and their management for sustainability, In *Agroforestry for Sustained Productivity in Arid Regions* (eds) JP Gupta and BM Sharma, Scientific publishers, Jodhpur, India, 31-39 pages

Gupta JP, Kar A and Faroda AS 1997 Desertification in India: problems and possible solutions, Yojna 41 55-59

Gupta LM, Saleem M, Gupta SK and Gupta Meenakshi 2007 Potential of Medicinal and Aromatic Plants in Agroforestry Systems, In *Agroforestry Systems and Practices in India* (eds.) Sunil Puri and Pankaj Panwar, New India Publishing Agency, New Delhi, 519-532 pages

Gupta N, Kukal SS, Bawa SS and Dhaliwal GS 2009 Soil organic carbon and aggregation under poplar based agroforestry system in relation to tree age and soil type, *Agroforestry Systems* **76**(1) 27-35

Gupta RK, Agarwal MC and Kumar N 1996 Growth and phytomass production in *Leucaena leucocephala* under high density plantations on degraded lands of Doon Valley, *Indian Forester* **122** 241-246

Hager N and Otterstedt J 1996 Coope San Juan, a farmers' cooperative, Sustainable use of the natural forest - past and future: a minor field study, Working Paper-International Rural Development Centre, Swedish University of Agricultural Sciences no. 302, Uppsala International Rural Development Centre, Swedish University of Agricultural Sciences, SE, 59 pages

Haggar JP, Briscoe CB and Butterfield RP 1998 Native species: a resource for the diversification of forestry in the lowland humid tropics, *Forest ecology and Management* **106** 195-203

Hailey L 1957 An African Survey, Oxford University Press, Oxford, UKSomarriba E 1992 Revisiting the past: an essay on agroforestry definition, *Agroforestry Systems* **19** 233-240

Hall GG 1963 The application of mathematical thinking, University of Nottingham Press

Hardwood CE and Booth TH 1992 Status of *Grevillea robusta* in forestry and agroforestry, In *Grevillea robusta in agro forestry and foresty: Proceedings of International Workshop* (ed) CE Hardwood, ICRAF, 9-19 pages

Hardy F 1936 Some aspects of tropical soils, Transactions Third International Congress of Soil Science **2** 150-163

Harfouche A, Meilan R, Kirst M, Michele M, Wout B, Maurizio S and Mugnozza GS 2012 Accelerating the domestication of forest trees in a changing world, *Trends in Plant Science* **17**(2): 64-72

Hari Om, Chauhan RS, Malik RK, Singh VP, Singh Dilbag, Lathwal OP, Goyal SP, Yadav SK and Singh Sher 2008 Diversification through Farming System Approach, Technical Bulletin (30), Krishi Vigyan Kendra, Kurukshetra and Directorate of Extension Education, CCS Haryana Agricultural University, Hisar, India, 56 pages

Harlan HR and Martini ML 1936 Problems and results of barley breeding, USDA Yearbook of Agriculture, Washington DC: US Government Printing Office, 303-346 pages

Harlan JR 1975 Crops and Man, American Society of Agronomy/ Crop Science Society of America, Madison, Wisconsin

Harlan JR 1992 Crops and Man, 2nd Edition, American Society of Agronomy, Madison, Wisconsin

Harley IM 2006 Traditional Agroforestry Systems of the Pacific Islands, A Workshop Presentation for the Trees for Improving Profitability, Sustainability and Resource Conservation on Farms and Ranches, Koror, Palau and Hagatna, Guam-June 26-30 pages

Harper JL 1977 Population Biology of Plants, Academic Press, Orlando, Florida, 892 pages

Harsh LN, Tewari JC, Burman U and Sharma SK 1992 Agroforestry in arid regions, *Indian Farming* **45** 32-37

Hartemink AE, Buresh RJ, Jama B and Janssen BH 1996 Soil nitrate and water dynamics in Sesbania fallows, weed fallows and maize, *Soil Science Society of America Journal* **60** 568-574

Hartmans EH 1981 Land development and management in tropical Africa, Ibadan, Nigeria, International Institute of Tropical Agriculture

Harvey CA and Gonzalez-Villalobos JA 2007 Agroforestry systems conserve species-rich but modified assemblages of tropical birds and bats, *Biodiversity and Conservation* **16** 2257-2292

Harvey CA, Schroth G and Zerbock O 2007 Designing agroforestry systems to mitigate climate change, conserve biodiversity and sustain rural livelihoods, website: *http://web.catie.ac.cr/cd_multiestrata/ Poster/session2/ Designing_agroforestry.pdf*

Hawley JG and Dymond JR 1988 How much do trees reduce land sliding? *Journal of Soil and Water Conservation* **43** 495-498

Hazra PK, Banerjee LK and Roy A 1996 Singhara nut/Water chestnut, BSI-Envis News Letter **3** 3-9

He S and Sheng Z 1997 The role of Chinese botanical gardens in conservation of medicinal plants, In *The conservation of medicinal plants* (eds) O Akerele, V Heywood and H Synge, Cambridge University Press, Cambridge, 228-237 pages

Hecht SB and Posey DA 1989 Preliminary results on soil management techniques of the Kayapo Amerindians, *Advances in Economic Botany* **7** 174-188

Hegde R 1995 Nutrient content and decomposition of leaf Litter of *Acacia mangium* Willd. as affected by season and field conditions, M.Sc. (Forestry) Thesis, Kerala Agricultural University, Thrissur, India

Hegde VS et al 2002 In *Agrotechnology for Dryland Farming* (ed) AM Dhopte, Scientific Publishers, Jodhpur, 591-612 pages

Heineman AM, Otieno HJO, Mengich EK and Amadalo BA 1997 Growth and yield of eight agroforestry tree species in line plantings in Western Kenya and their effect on maize yields and soil properties, *Forest Ecology and Management* **1** 103-135

Hervouet JP 1991 Faidherbia albida: A witness of agrarian transformation, In International Workshop on Faidherbia albida: State of the Art and Goals for the Future, 22-26 April 1991, Niamey, the Niger

Herzog F 1998 Streuobst: a traditional agroforestry system as a model for agroforestry development in temperate Europe, *Agroforestry Systems* **42** 61-80

Hicks DL 1995 Control of soil erosion on farm land: A summary of erosion's impact on New Zealand agriculture, and farm management practices which counteract it, MAF Policy Technical Paper 95/4 45 pages

High C and Shackleton CM 2000 The comparative value of wild and domestic plants in homegardens of a South African rural village, *Agroforestry Systems* **48** 141-156

Hildebrand PE (ed) 1986 Perspectives on Farming Systems Research and Extension, Lynne Rienner, Boulder, CO, USA

Hill DB and Buck LE 2000 Forest farming practices, In *North American Agroforestry: An Integrated Science and Practice* (eds) HE Garrett, WJ Rietveld and RF Fisher, American Society of Agronomy: Madison, Wisconsin

Hiraoka M 1989 Agricultural systems on the floodplains of the Peruvian Amazon, In *Fragile Lands of Latin America* (ed) J Browder, Boulder, Colorado: Westview Press

Ho KuenYih and Tsai WeiJe 2011 Adaptations of Casuarina windbreak stands to land subsidence on the southwestern coast of Taiwan, *Taiwan Journal of Forest Science* **26**(1) 71-86

Honda H, Hatta H and Fisher JB 1997 Branch geometry in *Cornus kousa* (Cornaceae): computer simulations, *American Journal of Botany* **84** 745-755

Hoogerbrugge ID and Fresco LO 1993 Homegarden systems: Agricultural characteristics and challenges, International Institute for Environment and Development, Gatekeeper series no. 39, London, 23 page

Hoskins MW 1987 Agroforestry and the social milieu, In *Agroforestry: A Decade of Development* (eds) HA Steppler and PKR Nair, ICRAF, Nairobi, Kenya, 191-203 pages

Howlett DS, Mosquera-Losada MR, Nair PKR, Nair VD and Rigueiro-Rodrigues A 2011 Soil carbon storage in silvopastoral systems and a treeless pasture in northwestern Spain, Journal of Environmental Quality 40(3) 825-832

Huberman 1959 MA Bamboo Silviculture, *Unasylva* **13**(1)

Huel D 1998 Stream bank Stewardship- A Saskatchewan Riparian Project, Saskatchewan Wetlands Conservation Corporation, Canada (Canadian Wetland Inventory, website: http://www.ducks.ca/cwi/)

Huisman O, Rolf AD et al 2009 Principles of Remote sensing and Geographic information system, Enschede

Husch B, Miller CI and Beers TW 1982 *Forest Mensuration*, 3rd edition, The Roland Press Co., New York, 402 pages

Hutchinson JJ, Campbell CA, Desjardins RL 2007 Some perspectives on carbon sequestration in agriculture, *Agriculture and Forest Meteorology* **142** 288-302

Huxley PA 1983 Plant Research and Agroforestry, International Council for Research in Agroforestry, Nairobi, Kenya

Huxley PA 1985a The tree/crop interface or simplifying the biological/ environmental study of mixed cropping agro forestry system, *Agroforestry Systems* **3** 197-207

Huxley PA 1985b Systematic designs for field experimentation with multipurpose trees, *Agroforestry Systems* **3** 251-266

Huxley PA and Greenland PK 1989 Structure and function of traditional agroforestry systems in the western Himalaya. I. Biomass and productivity, *Agroforestry Systems* **9**(1)

Huxley PA and Greenland DJ 1990 Pest management in agroforestry systems, *Agroforestry Abstracts* **2** 37-46

IAASTD 2008 Agriculture at a Crossroads: Global Report, International Assessment of Agricultural Knowledge, Science and Technology for Development, Washington DC

ICAR 1981 Proceedings of the Agroforestry Seminar, Imphal in 1979, Indian Council of Agricultural Research, New Delhi, 268 pages

ICAR 1990 All India Coordinated Research Project on Agroforestry, Biennial Report (1987-89), Indian Council of Agricultural Research, New Delhi, India

ICAR 2013 NAIP- Mass Media Project, DKMA with inputs from RC, CS&WCR&T, Vasad and DMAPR, Anand, website: http://www.icar.org.in/node/5101

ICFRE (Indian Council of Forestry Research and Education) 1990 Indian Forest Utilization Vol I and Vol II, Dehradun: ICFRE, India

ICRAF (World Agroforestry Centre) 1990 Enquête micro-diagnostique, Groupe national de travail du Burkina Faso, ICRAF/SALWA, 40 page

ICRAF (World Agroforestry Centre) 1996 Annual Report-1996, Nairobi, Kenya, 340 page

ICRAF (World Agroforestry Centre) 2008 Transforming Lives and Landscapes: Strategy 2008-2015, World Agroforestry Centre, Nairobi, Kenya

ICRAF (World Agroforestry Centre) 2008 website: http://www.worldagroforestry.org/downloads/publications/PDFS/B15732.pdf.

ICRAF 1992 A selection of useful trees and shrubs for Kenya: Notes on their identification, propagation and management for use by farming and pastoral communities, ICRAF, Nairobi, Kenya

ICRAF 1993 Strategy to the year 2000, ICRAF, Nairobi, Kenya, 78 pages

ICRAF 1995 Annual Report 1995, ICRAF, Nairobi

ICRAF 2008 Medium-Term Plan (2009-2011): Transforming Lives and Landscapes through Agroforestry Science, World Agroforestry Centre, Nairobi, Kenya

IDF 1981 *La réalisation pratique des haies brise-vent et bandes boisées.* Institut pour le Développement Forestier, aris, cited from 'Trees outside forests - Towards a better awareness', FAO CONSERVATION GIUDE No. 35, 129 pages

Igugu GO and Osemeobo GJ 1995 The impact of Forestry in improving small-holder land use for Agricultural production in Nigeria, In Proceedings of the 24 Conference of FAN (ed) Oduwaye, 30 October-4 November 1995, 135-137

India Agronet 2013 India Agronet: Agriculture Resource Centre, Website: www.indiaagronet.com

Intergovernmental Panel on Climate Change (IPCC) 2000 Land use, Land-use Change, and Forestry. A Special Report of the IPCC, Cambridge University Press, Cambridge, UK, 375 page

IPCC 2000 Afforestation, reforestation and deforestation activities, In *Chapter 3: Land use, land use change, and forestry* (eds) R Watson, IR Noble, B Bolin et al, A special report of the IPCC, Cambridge, UK, Cambridge, University Press, 377 pages

ITTO 2000 Tropical Forest Update, Vol. **10**, No. 4: 12

IUCN 1994 Putting plans into action, Report of Metropolitan Open Space Systems (MOSS), International Conference, Durban, South Africa, 9-11 February

IUFRO 2013 Non-wood forest resource management, website: http://www.iufro.org/

Ivory DA 1990 Major characteristics, agronomic features, and nutritional value of shrubs and tree fodders, In *Proceedings of a Workshop on Shrub and Tree Fodders for Farm Animals* (ed) C Devendra, Denpasar, Indonesia, July 1989, IDRC, Ottawa, Canada

Jack WH 1966 The afforestation of blanket bog areas in the British Islands, Proc. VIth World Forestry Congress, Madrid

Jackson G 1970 Vegetation around the City and nearby villages of Zaria, In *Zaria and its region* (ed) MJ Mortimore, Dept. of Geography, Occasional Paper 4, Samaru, Zaria, Nigeria, Ahmadu Bello University, 61-71 pages

Jackson L, Bawa K, Pascual U and Perrings C 2005 Agrobiodiversity: A new science agenda for biodiversity in support of sustainable agro ecosystems, DIVERSITAS Report, 40 pages

Jackson LE, Pascual U and Hodgkin T 2007 Utilizing and conserving agrobiodiversity in agricultural landscapes, *Agriculture, Ecosystems and Environment* **121** 196-210

Jackson NA, Wallace JS and Ong CK 2000 Tree pruning as a means of controlling water use in an agroforestry system in Kenya, *Forest Ecology and Management* **126** 133-148

Jagdish Chander 1998 Poplar - an ideal species for agroforestry, *Indian Farming* January 17-22

Jaggi S** Design and Analysis of agroforestry experiments, I.A.S.R.I, Library Avenue, New Delhi, India

Jaggi S, Gupta VK and Sharma MM 2001 Design and analysis of agroforestry experiments: An overview, *Journal of Indian Society of Agroforestry* **3**(2) 120-129

Jain RC, Tripathi SP, Kishan VS and Kumar S 1996 Volume tables for *Acacia tortilis* plantations based on data collected from kjd abadi plantations of khajuwala range Chattergarh Division, IGNP area, Rajasthan, *Indian Forester* **122** 316-322

Jain RC, Tripathi SP, Singh M and Kumar VKS 1998 Volume tables for *Azadirachta indica* for Gujarat region, *Indian Forester* 124 123-133

Jain SK and Singh P 2000 Economics analysis of industrial agroforestry: poplar (*Populus deltoides*) in Uttar Pradesh (India), *Agroforestry Systems* **49** 225-273

Jaiswal AK, Sharma KK, Kumar KK and Bhattacharya A 2002 Households survey for assessing utilisation of conventional lac host trees for lac cultivation, *New Agriculture 13* 13-17

Jama B, Elias E and Mogotsi K 2006 Role of agroforestry in improving food security and natural resource management in the drylands: a regional overview, *Journal of the Drylands* **1** 206-211

Jama BA, Mutegi JK and Njui AN 2008 Potential of improved fallows to increase household and regional fuelwood supply: evidence from western Kenya, *Agroforestry Systems* **73** 155-166

James R 1994 Agroforestry for the Pacific Technologies, Agroforestry Information Service (AIS) of the Nitrogen Fixing Tree Association (NFTA), Hawaii 96779 USA

Jamnadass R, Leakey RRB, Muriuki J, Dawson I, Kindt R and Beniest J and Simons T 2009 Forest genetic resources and farmers' tree domestication, In *Regional workshop on Learning agrobiodiversity: options for universities in Sub-Saharan Africa*, Nairobi, 21-23 January 2009

Janick J, Schery RW, Woods FW and Ruttan VW 1982 Plant Science, San Francisco, California: Freeman

Janssen PHM and Titus SJ 1995 Calibration of process oriented models, *Ecological Modelling* **83** 55-66

Jeremy SB 2000 Ecophysiology of indigenous trees in agroforestry systems in the semi-arid tropics, BSc. (Hons) thesis, University of Nottingham, 259 pages

Jessica Mott 2001 A framework to consider forest products marketing, In *Marketing Monitoring of Tree Products* (eds) GS Rawat and NS Bisht, 14-28 pages

Jha LK and Lalnunmawia F 2004 Agroforestry with bamboo and ginger to rehabilitate degraded areas in North East India, *Journal of Bamboo and Rattan* **2**(2) 103-109

Jha MN, Pande P and Pathak TC 1979 Studies on the changes in the physico-chemical properties of Tripura soils as a result of *Jhuming*, *Indian Forester* **105** 436-441

Joginder Singh and Grover DK 2001 Marketing of major forest products in Punjab, In *Marketing Monitoring of Tree Products* (eds) GS Rawat and NS Bisht, 76-85 pages

Jose S 2009 Agroforestry for ecosystem services and environmental benefits: an overview, *Agroforestry Systems* **76** 1-10

Jose S 2012 Agroforestry for conserving and enhancing biodiversity, *Agroforestry Systems* **85**(1) 1-8

Jose S, Allen SC and Nair PKR 2008 Tree-crop interactions: lessons from temperate alley cropping systems, In *Ecological basis of Agroforestry* (eds) DR Batish, Kohli RK, Jose S and Singh HP, CRC Press

Jose S, Gillespie AR and Pallardy SG 2004 Interspecific interactions in temperate agroforestry, *Agroforestry Systems* **61** 237-255

Jose S, Gillespie AR, Seifert JR, Mengel DB and Pope PE 2000 Defining competition vectors in a temperate alley cropping system in the midwestern USA: Competition for nitrogen and litter decomposition dynamics, *Agroforestry Systems* **48** 61-77

Jose S, Gold MA and Garrett HE 2012 The future of agroforestry in the United States, In *Agroforestry: The future of global land use* (eds) PKR Nair and D Garrity, Springer, The Netherlands, 217-246 pages

Joshi BC 1996 Poplar cultivation under agroforestry, In *Environment and social issues in Poplar and Willow Cultivation and Utilization*, FAO Proceedings of 20th Session of the International Poplar Commission (ed. Bach, I), 409-423 pages

Joshi PK, Pangare V, Shiferaw B, Wani SP, Bouma J and Scott CA 2004 Socio-economic and policy research on watershed management in India: Synthesis of past experiences and needs for future research, Global Theme on Agro-ecosystem, Report No. 7, International Crops Research Institute for Semi-Arid Tropics, Patancheru, Hydrabad, India

Juo ASR and Lal R 1977 The effect of fallow and continuous cultivation on the chemical and physical properties of an Alfisol in western Nigeria, *Plant and Soil* 47 567-584

Kala CP 2010 Status of an indigenous agroforestry system in changing climate: A case study of the middle Himalayan region of Tehri Garhwal, India, *Journal of Forest Science* 56(8) 373-380

Kalaba KF, Chirwa P, Syampungani S and Ajayi CO 2010 Contribution of agroforestry to biodiversity and livelihoods improvement in rural communities of Southern African regions, In *Tropical Rainforests and Agroforests under Global Change* (eds) T Tscharntke et al, Environmental Science and Engineering, 461-476 pages

Kalaichelvi K and Swaminathan AA 2009 Alternate land use through cultivation of medicinal and aromatic plants - a review, *Agricultural Review* **30**(3) 176-183

Kalloo G 2003 Production technologies for promoting export potential of horticulture, *Indian Horticulture* **48**(1) 6-11

Kamboj VP 2000 *Current Science* **78** 35

Kamugisha JR, Ogutu ZA and Stahl M 1997 Parks and people – Conservation and livelihoods at the crossroads, Technical report No. 17, Nairobi, Kenya

Kanaujia VK, Uttam SK and Kaushal Kumar 2002 Soil moisture content, extraction pattern and splash erosion under Leucaena based cropping system in eroded alluvial soil of central Uttar Pradesh, *Indian Journal of Agroforestry* **4**(2) 113-118

Kang BT 1993 Chapter 2: Potential for sustainable agroforestry and alley farming in tropical Africa, In *Soil tillage in Africa: needs and challenges,* FAO Soils Bulletin 69, Food and Agriculture Organization, Rome

Kang BT and Wilson GF 1987 The development of alley cropping as a promising agroforestry technology, In *Agroforestry: A Decade of Development* (eds) HA Steppler and PKR Nair, ICRAF, Nairobi, 227-243 pages

Kang BT, Reynolds L and Atta-Krah AN 1990 Alley farming, *Advances in Agronomy* 43 315-359

Kang BT, van der Kruijs ACBM and Couper DC 1986 Alley cropping for food crop production in the humid and subhumid tropics, In *Alley farming in the humid and sub-humid tropics* (eds) BT Kang and L Reynolds, IDRC 271e, Ottawa, Canada, 16-36 pages

Kang BT, Wilson GF and Lawson TL 1984 Alley cropping: a stable alternative to shifting tiny cultivation. IITA, Ibadan, Nigeria

Kang BT, Wilson GF and Sipkens L 1981 Alley cropping maize (*Zea mays*) and leucaena (*Leucaena leucocephala*) in southern Nigeria, *Plant and Soil* 63 165-179

Kareemulla K, Dwivedi RP and Rizvi RH 2003 An economic diagnosis of agroforestry systems in western Uttar Pradesh, Completion report of APFC Scheme, NRCAF, Jhansi

Kareemulla K, Rai P, Rao GR and Solanki KR 2002 An economic analysis of Silvipastoral system for degraded lands under rainfed condition, *Indian Forester* **128**(12) 1346-1350

Kareemulla K, Rizvi RH, Singh R and Dwivedi RP 2002 Trees in rainfed agro-ecosystem: A socio-economic investigation in Bundelkhand region, *Indian Journal of Agroforestry* **4**(1) 53-56

Kartasubrata J 1986 Agroforestry in Indonesia with Special Reference to Tumpangsari in Forest Areas, Monsoon Asia Agroforestry Joint Research Team: Comparative Studies on Utilization and Conservation of the Natural Environment by Agroforestry Systems, 29-51

Karur B, Gupta SR and Singh G 2002 Carbon storage and nitrogen cycling in silvopastoral systems on a sodic soil in northwestern India, *Agroforestry Systems* **54** 21-29

Karyono 1990 Homegardens in Java; their structure and function, In *Tropical Homegardens* (eds) K Landauer and M Brazil, United Nations University Press, Tokyo, 138-146 pages

Kaushik N and Kumar V 2003 Khejri based agroforestry systems for arid Haryana-India, *Journal of Arid Environment* **55** 433-450

Kayahara GJ, Klinka K and Marshall PL 1998 Testing site-index site factor relationships for *Pinus contorta* and *Picea engelmannic* x *P.glauca* productivity in central British Columbia, Canada, *Forest Ecology and Management* **110**(1-3) 141-150

Kellogg CF, Baldwin M and Thorp 1960 Soil Classification: A Comprehensive System-"7th Approximation", Soil Conservation Service, U.S. Department of Agriculture, Washington D.C.

Khadse VA and Bharad GM 1996 Performance of annual crops under canopy of *Hardwickia binata* Roxb.(Anjan) in agroforestry system, *Journal of Soils and Crops* **6**(2) 151-153

Khan MA, Tewari JC and Issac VC 1995 Hydrology of small forested catchments in the arid region of Rajasthan, *Annals of Arid Zone* **34**(4) 259-262

Khan N 1990 Reclamation of wastelands in Faizabad district, U.P., In *Utilization of wastelands for sustainable development and their utilization* (Eds.) SC Sharma, RB Chaturvedi and OP Misra, Concept Publ. Co., New Delhi, 419-430 pages

Kho RM 2008 Approaches to tree-environment-crop interactions, In *Ecological basis of Agroforestry* (eds) DR Batish, Kohli RK, Jose S and Singh HP, CRC Press

Khybri ML, Gupta RK, Sewa Ram and Tomer HS 1992 Crop yield of rice and wheat grown in rotation as intercrops with three tree species in outer hills of Himalaya, *Agroforestry Systems* **17** 193-204

Kidd C and Pimentel D 1992 Integrated resources management: agroforestry for development, Academic Press, San Diego, CA

Kiepe P and Rao MR 1994 Management of agroforestry for the conservation and utilization of land and water resources, *Outlook Agriculture* **23**(1) 17-25

Kimble JM, Lal R and Follett RF 2001 Methods for assessing soil C pools, In *Assessment Methods for Soil Carbon* (eds) R Lal, JM Kimble, RF Follett and BA Stewart, Lewis Publishers, Boca Raton, 3-12 pages

King KFS 1968 Agri-Silviculture (the Taungya System), Bulletin No.1, Department of Forestry, University of Ibadan, Ibadan, Nigeria

King KFS 1979a Agroforestry and the utilization of fragile ecosystems, *Forest Ecology and Management* **2** 161-168

King KFS 1979b Agroforestry, Proceedings of the Fiftieth Symposium on Tropical Agriculture, Royal Tropical Institute, Amsterdam, The Netherlands

King KFS 1987 The history of Agroforestry, In *Agroforestry: A Decade of Development* (eds) HA Steppler and PKR Nair, ICRAF, Nairobi, Kenya, 1-11 pages

King KFS and Chandler MT 1978 The wasted lands, ICRAF, Nairobi, Kenya, 85 pages

Kingwell RS and Pannell DJ 1987 MIDAS, a bio-economic model of a dryland farming system (Computer Software), Podoc: Wageningen

Kohli A and Saini BC 2003 Microclimate modification and response of wheat planted under trees in a fan design in northern India, *Agroforestry Systems* **58** 109-118

Kondas S, Sankaramurthy S and Sambasivan V 2012 India and World wood and Wood products Trade, Vision for the Management of wood in Tamil Nadu, Tamil Nadu Forestry Department, 8-21 pages

Korikanthimath VS and Hegde R 1994 Cardamom and arecanut mixed-cropping systems, *Indian Cocoa, Arecanut and Spices Journal* **18**(4) 109-112

Korwar GR 1994 *Hardwickia binata* - A promising MPTS for agroforestry in dryland areas, In *Multipurpose tree species for agroforestry in India*, BAIF Development Research Foundation, Pune, 72-75 pages

Korwar GR and Radder GD 1994 Influence of root pruning and cutting interval of *Leucaena* hedgerows on performance of alley cropped Rabi sorghum, *Agroforestry Systems* **25** 95-109

Kothari A 1999, Agro-Biodiversity: The Future of India's Agriculture Article for MCAER Book; 7 February, 1999

Krankina ON and Dixon RK 1994 Forest management options to conserve and sequester terrestrial carbon in the Russian Federation, *World Resources Review* **6** 88-101

Kricher J 1998 Eastern Forests: A field guide to birds, mammals, trees, flowers, and more, New York, NY: Houghton Mifflin Company

Krishna Kumar N and Buvaneswaran C 2012 Tree Improvement for Enhancing Productivity and Carbon Sink Services of Agroforestry Systems, Proceedings of the National Seminar on 'Agroforestry: An evergreen agriculture for food security and environmental resilience' held on February 2 to 4, 2012 at Navsari Agricultural University, Gujarat (India)

Krishnankutty CN 1990 Demand and supply of wood in Kerala and their future trends, Research Report 67, Kerala Forest Research Institute, Peechi, Kerala, 66 pages

Krishnankutty CN, Thampi KB and Chundamannil M 2008 Trees outside forests (TOF): A case study of the wood production-consumption situation in Kerala, *International Forestry Review* **10**(2) 156-164

Kubicki A, Denby C, Stevens M, Haagensen A and Chatfield J 1993 Determining the long-term costs and benefits of alternative farm plants, In Re-integrating Fragmented Landscapes: Towards Sustainable Production and Nature Conservation (eds) RJ Hobbs and DA Saunders, New York, NY: Spring-Verlag, 245-258 pages

Kuchelmeister G 1998 Urban forestry in the Asia-Pacific Region: status and prospects, Asia-Pacific Forestry Sector Outlook Study Working Paper Series, No. 44, Rome, FAO

Kuchelmeister G and Braatz S 1993 Urban Forestry Revisited, *Unasylva* No.173, **44** 13-18 (Also available as Urban Forestry, Overstory, website: http://agroforestry.net/overstory/ overstory87.html)

Kuiper L and Olenburger J 2005 Biosaline (Agro) forestry: A literature Review, Follow-up BUS Report, Biomassa Upstream Stuurgroep (BUS), Netherlands

Kumar A and Abrol IP 1984 Studies on the reclaiming effect of Karnal grass and Para grass grown in highly sodic soil, *Indian Journal of Agricultural Science* **54**: 189-193

Kumar A and Shukla A 2009 Effect of tree introduction on disease incidence in component crops of agroforestry systems, In *Agroforestry: Natural Resource Sustainability, Livelihood and Climate Moderation* (eds) OP Chaturvedi, A Venkatesh, RS Yadav, Badre Alam, RP Dwivedi, Ramesh Singh and SK Dhyani), Satish Serial Publishing House, Delhi, India, 745-756 pages

Kumar and Gupta 2004 In New Vistas in Agroforestry: A Compendium for 1st World Congress of Agroforestry (eds) PKR Nair, MR Rao and LE Buck, 488 pages

Kumar BM 2006 Agroforestry: the new old paradigm for Asian food security, *Journal of Tropical Agriculture* **44**(1-2) 1-14

Kumar BM 2006 Carbon sequestration potential of tropical homegardens, In *Tropical Homegardens: A Time-Tested Example of Sustainable Agroforestry* (eds) BM Kumar and PKR Nair, Springer, The Netherlands, 185-204 pages

Kumar BM 2008 Assessment of standing stock of thorny bamboo [*Bambusa bambos* (L.) Voss] in the homegardens of Palakkad and Malappuram districts in Kerala, *Journal of Tropical Agriculture* **46** (1-2) 32-37

Kumar BM 2008 Litter dynamics in plantation and agroforestry systems of the tropics: A review of observations and methods, In *Ecological basis of agroforestry* (eds) RD Batish, RK Kohli, S Jose and HP Singh, CRC Press, Florida, USA, 182-216 pages

Kumar BM 2011 Species richness and aboveground carbon stocks in the homegardens of central Kerala, India, *Agriculture, Ecosystems and Environment* **140** 430-440

Kumar BM and Nair PKR (eds) 2006 Tropical Homegardens: A Time-Tested Example of Sustainable Agroforestry, Advances in Agroforestry 3, Springer, Dordrecht, the Netherlands, 380 pages

Kumar BM and Nair PKR (eds) 2011 Carbon Sequestration Potential of Agroforestry Systems: Opportunities and Challenges, *Advances in Agroforestry* **8**, Springer, Springer Science+ Business Media B.V.

Kumar BM and Nair PKR 2004 The enigma of tropical homegardens, *Agroforestry Systems* **61** 135-152

Kumar BM and Nair PKR 2004 The Enigma of Tropical Homegardens, In *New Vistas in Agroforestry: A Compendium for the 1st World Congress of Agroforestry* (eds) PKR Nair, M Rao and L Buck, Kluwer Academic Publishers, Dordrecht, The Netherlands, 700 pages

Kumar BM, George SJ and Chinnamani S 1994 Diversity, structure and standing stock of wood in the homegardens of Kerala in peninsular India, *Agroforestry Systems* **25** 243-262

Kumar BM, Sudheesh KG and Rajesh G 2005 Standing stock of thorny bamboo [*Bambusa bambos* (L.) Voss] in the homegardens of Thrissur, Kerala, *Journal of Tropical Agriculture* **43** 61-66

Kumar M 1995 Tree cop interaction studies in agroforestry systems, productivity, moisture and nutrient dynamics. Ph.D. Thesis, Dr Y.S. Parmar UHF, Solan, H.P India

Kumar P 2003 Development of tree borne oilseeds – Problems and Prospects, In Souvenir of National Seminar on Stress Management in Oilseeds for Attaining Self-Reliance in Vegetable Oils (eds) H Singh and DM Hegde, Indian Society of Oilseeds Research, Hyderabad, January 28-30, 2003, 111-117 pages

Kumar PG, Murkute AA, Gupta S and Singh Shashi Bala 2009 Carbon sequestration with special reference to agroforestry in Cold Desert of Ladakh, *Current Science* **97**(7) 1063-1068

Kumar R, Gupta PK and Gulati A 2004 Viable agroforestry models and their economics in Yamunanagar District of Haryana and Haridwar District of Uttaranchal, *Indian Forester* **130** 131-148

Kumar S 2005 Basics of remote sensing and GIS, Firewall Media, India

Kumar Silpa V 2011 Effect of vermicompost on growth and yield of *Andrographis paniculata* Wall.ex.Ness. and *Stevia rebaudiana* Bert. under Plum based agroforestry system in mid hill zone of Himachal Pradesh, M.Sc.Thesis, Dr Y.S. Parmar UHF, Solan, H.P. India

Kumar SS, Kumar BM, Wahid PA, Kamalam NV and Fisher RF 1999 Root competition for phosphorus between coconut, multipurpose trees and kacholam (*Kaempferia galanga*) in Kerala, India, *Agroforestry Systems* **46** 131-146

Kumar V, Nagawani HC, Kumar D and Singh VP 1996 Propagation and growth studies of *Calligonum polygonoides* (Phog), *Indian Forester* **122** 377-382

Kumari, Anita, Sehgal RN and Shailender Kumar 2008 Traditional agroforestry systems practiced in Lahaul (Lahaul & Spiti) and Kinnaur districts of Himachal Pradesh, *Indian Forester* **134**(8) 1003-1010

Kung'u JB and Temu AB 2003 Role of agroforestry education in sustaining society and environment, In *Rebuilding Africa's capacity for agricultural development: The role of tertiary education* (eds) AB Temu, S Chakeredza, K Mogotsi, D Munthali and R Mulinge, Reviewed papers presented at ANAFE Symposium on Tertiary Education April 2003, Nairobi: ICRAF

Kursten E and Burschel P 1993 CO_2 mitigation by agroforestry, *Water Air Soil Pollution* **70** 533-544

Kwesiga F, Akinnifesi FK, Mafongoya PL, Mcdermott MH and Agumya A 2003 Agroforestry research and development in southern Africa during the 1990s: review and challenges ahead, *Agroforestry Systems* **59** 173-186

Kyuma K and Pairinta C (eds) 1983 Shifting Cultivation-An Experiment at Nam Phrom, Northeast Thailand and its Implication for Upland Farming in the Monsoon Tropics, Ministry of Science, Technology and Energy, Bangkok, Thailand

Lahiri AK 1987 Silvopisciculture as land management system for rural development in mangrove area, *Indian Agriculture* **31**(4) 305-311

Laird SA and Pierce AR 2002 Strategies to Promote Sustainable and Ethical Raw Material Sourcing in the Botanicals Industry: Results from an Industry Survey, Medicinal Plant Conservation, Volume 8(1) June, IUCN Medicinal Plant Specialist Group, 9-14 pages

Lal C 2003 Above and below ground interactions and their implications on the efficiency of agroforestry systems under rainfed conditions, Ph.D. Thesis, Dr.Y.S. Parmar University of Horticulture and Forestry, Nauni-Solan, H.P. India

Lal H, Sharma JS and Tewari RK 2004 Effect of graveliness on peach based agroforestry agri-horti system in Doon valley, *Indian Journal of Agroforestry* **6**(1) 27-32

Lal R 1989a Agroforestry systems and soil surface management of a tropical alfisol-11, Water runoff, soil erosion and nutrient loss, *Agroforestry Systems* **8** 97-111

Lal R 1989b Agroforestry systems and surface management of a tropical alfisols: (1) Soil moisture and crop yields, *Agroforestry Systems* **8** 7-29

Lal R, Kang BT, Moormann FR, Juo ASR and Moomaw JC 1975 Soil management problems and possible solutions in western Nigeria, In *Soil Management in Tropical America* (eds) E Bornemisza and A Alvarado, North Carolina State University, Raleigh, NC, USA, 372-408 pages

Lambert J, Srivastava J and Vietmeyer N 1997 Medicinal plants: Rescuing a global heritage, World Bank Technical Paper 355, World Bank, Washington DC

Lamont SR, Eshbaugh WH and Greenberg AM 1999 Species composition, diversity, and use of homegardens among three Amazonian villages, *Economic Botany* **53** 312-326

Lamprecht H 1989 Silviculture in the tropics, Tropical forest ecosystem and their tree species-possibilities and methods for their long-term utilization, Eschborn

Larsen TH, Williams NM and Kremen C 2005 Extinction order and altered community structure rapidly disrupts ecosystem functioning, *Ecology Letter* **7** 1109-1119

Larson WE and Pierce FJ 1994 The dynamics of soil quality as a measure of sustainable management, In *Declining Soil Quality for a Sustainable Environment* (eds) JW Doran, DC Coleman, DF Bezdicek and BA Stewart, SSSA Special Publication 35 page

Lass RA 1985 Diseases In *Cocoa* (eds) GAR Wood and RA Lass, 4th Edition, London, UK, Longmans, 265-365 pages

Lavelle P, Chauvel A, and Fragoso C 1995 Faunal activity in acid soils, In *Plant Soil Interactions at Low pH* (ed) RA Date, Springer, Berlin, Germany, 201-211 pages

Lawrence JH 1999 Personal Communication, USDA Natural Resources Conservation Service, Guam

Lawson TL and Kang RT 1990 Yield of maize and cowpea in an alley cropping system in relation to available light, *Agriculture and Forest Metereology* **52** 347-357

Le Houerou HN 1991 The role of saltbushes (Artiplex spp.) in arid grazing land rehabilitation in the Mediterranean basin, In *Plant Salinity Research: New Challenges* (ed) Choukr-Allah, IAV Hassan II, Agadir, Morocco, 397-433 pages

Leakey RRB 1996 Definition of agroforestry revisited, *Agroforestry Today* **8**(1) 5-7

Leakey RRB 1998 Tree Domestication, *The Overstory: Agroforestry e-Journal* **31** webpage: http://agroforestry.net/overstory/overstory31.html

Leakey RRB 2010 Agroforestry: a delivery mechansim for Multi-functional Agriculture, In *Handbook on Agroforestry: Management Practices and Environmental Impact* (ed) Lawrence R. Kellimore, Nova Science Publishers, Environmental Science, Engineering and Technology Series, USA

Leakey RRB and Akinnifesi FK 2008 Towards a domestication strategy for indigenous fruit trees in the tropics, In Indigenous Fruit Trees in the Tropics: Domestication, Utilization and Commercialization (eds) FK Akinnifesi, RRB Leakey, OC Ajayi, G Sileshi, Z Tchoundjeu, P Matakala and F Kwesiga, CAB International, Wallingford, UK

Leakey RRB and Jaenicke H 1995 The domestication of indigenous fruit trees: opportunities and challenges for agroforestry, In Proceedings of the 4th International BIO-REFOR Workshop (eds) K Suzuki, S Sakurai, K Ishii and M Norisada, BIO-REFOR, Tokyo, Japan, 15-26 pages

Leakey RRB and Maghembe JA 1994 Domestication of high-value trees for agroforestry: an alternative to slash and burn agriculture, ICRAF Position Paper-1, ICRAF, Nairobi, Kenya

Leakey RRB and Newton AC (eds) 1994a Domestication of Tropical Trees for Timber and Non-timber Products, MAB Digest 17, UNESCO, Paris

Leakey RRB and Newton AC (eds) 1994b *Tropical Trees: The potential for Domestication and the Rebuilding of Forest Resources*, The proceedings of a Conference organized by the Edinburgh Center of Tropical Forests held at Heriot-Watt University, Edinburgh on 23-28 August, 1992

Leakey RRB and Newton AC 1994 Domestication of 'Cinderella' species as the start of a woody-plant revolution, In *Tropical trees: the potential for domestication and the rebuilding of forest resources* (eds) RRB Leakey and AC Newton, HMSO, London 3-5 pages

Leakey RRB and Sanchez PA 1997 How many people use agroforestry products? *Agroforestry Today* **9**(3) 4-5

Leakey RRB and Simons AJ 1998 The domestication and commercialization of indigenous trees in agroforestry for the alleviation of poverty, *Agroforestry Systems* **38** 165-176

Leakey RRB and Tomich TP 1999 Domestication of tropical trees: from biology to economics and policy, In *Agroforestry in Sustainable Ecosystems* (eds) LE Buck, JP Lassoie and ECM Fernandes, CRC Press/Lewis Publishers, New York, USA, 319-338 pages

Leakey RRB, and Schreckenberg K and Tchoundjeu Z 2003 The participatory domestication of West African indigenous fruits, *International Forestry Review* **5**(4) 338-347

Leakey RRB, Atangana AR, Kengni E, Waruhiu AN, Usuro C, Anegbeh PO and Tchoundjeu Z 2002 Domestication of *Dacryodes edulis* in West and Central Africa: characterization of genetic variation, *Forest, Trees and Livelihoods* **12** 57-72

Leakey RRB, Mesen JF, Tchoundjeu Z, Longman KA, Dick J MCP, Newton AC, Matin A, Grace J, Munro RC and Muthoka PN 1990 Low-technology techniques for the vegetative propagation of tropical trees, *Commonwealth Forestry Review* **69** 247-257

Leakey RRB, Tchoundjeu Z, Smith RI, Munro RC, Fondoun JM, Kengue J, Anegbeh PO, Atangana AR, Waruhiu AN, Asaah E, Usoro C and Ukafor V 2004 Evidence that subsistence farmers have domesticated indigenous fruits (Dacryodes edulis and Irvingia gabonenis) in Cameroon and Nigeria, *Agroforestry Systems* **60**: 101-111

Leakey RRB, Temu AB, Melnyk M and Vantomme P (eds) 1996 Domestication and commercialization of non-timber forest products for agroforestry, Non-Wood Forest Products No.9, FAO, Rome Italy, 1-7 pages

Lee AWC, Xuesong B and Perry NP 1994 Selected physical and mechanical properties of giant timber bamboo grown in South Carolina, *Forest Product Journal* **44**(9) 40-46

Leeuw PN and Tuley P 1972 The vegetation, In *The land resources of North East Nigeria* (ed) P Tuley, Vol. 1 Land Resource Study No. 9, London, Land Resources Division, , 121-155 pages

Lehmann J, Weigl D, Droppelmann K, Huwe B and Zech W 1999 Nutrient cycling in an agroforestry system with runoff irrigation in Northern Kenya, *Agroforestry Systems* **43** 49-70

Lejon PH, Chaussod R, Ranger J and Ranjard L 2005 Microbial Community Structure and Density Under Different Tree Species in an Acid Forest Soil (Morvan, France), *Microbial Biology* **50** 614-625

Lely HV 1925 The useful trees of Northern Nigeria, London, Crown Agents for the Colonies, 128 page

Levasseur V and Olivier A 2000 The farming system and traditional agroforestry systems in the Maya community of San Jose, Belize, *Agroforestry Systems* **49** 275-288

Libby WJ 1973 Domestication strategies for forest trees, *Canadian Journal of Forest Research* **3**: 265-276

Lines R and Neustein SA 1966 Afforestation techniques for difficult sites-wetlands, Proceeding of Sixth world Forestry Congress, Madrid, Vol. III, 3750-3756 pages

Loane B 1991 Economic evaluation of farm trees – Methodology and data for FARMTREE model (Computer Software), Melbourne, Australia: Victorian Department of Agriculture and Department of Conservation and Environment

Lorenz C, Van Dijk GM, Van Hattum AGM and Cofino WP 1997 Concepts in river ecology: Implications for indicator development, *Regulated Rivers: Research and Management* **13** 501-506

Lott JE, Khan AAH, Black CR and Ong CK 2003 Water use by trees and crops in a *Grevillea robusta*-based overstorey agroforestry system in semi-arid Kenya, *Forest Ecology and Management* **180** 45-59

Louppe D and Yossi H 1999 Les haies vives défensives en zones sèches et subhumides d'Afrique de l'Ouest, Atelier Jachères, Dakar

Lundgren B and Raintree JB 1982 Agroforestry, presented at the Conference of Directors of National Agroforestry Research Systems in Asia, Jakarta, 12 pages

Lundgren BO 1982 Cited in Editorial: What is Agroforestry? *Agroforestry Systems* **1** 7-12

Lundgren BO 1987 Institutional aspects of agroforestry research and development, In *Agroforestry: A Decade of Development* (eds) HA Steppler and PKR Nair, ICRAF, Nairobi, Kenya 43-51 pages

Lundgren BO and Raintree JB 1982 Sustained Agroforestry, In *Agricultural Research for Develop-ment: Potentials and Challenges in Asia* (ed) B Nestel, ISNAR, The Hague, The Netherlands, 37-49 pages

Maclean RH, Listinger JA, Moody K and Watson AK 1992 The impact of alley cropping *Gliricidia sepium* and *Cassia spectabilis* on upland rice and maize production, *Agroforestry Systems* **20** 213-228

Madhu Goyal and Sharma SK 2009 Traditional wisdom and value addition aspects of arid foods of desert region of North West India, *Indian Journal of Traditional Knowledge* **8**(4) 581-585

Mahapatra AK and Mitchell CP 1999 Biofuel consumption, deforestation and farm level tree growing in rural India, *Biomass Bioenergy* **17** 291-303

Mahapatra R 2003 The multimillion dollar fuel wood trade is the last resort for India's poor, Infochange features, website: www.infochange.org

Maheswarappa HP, Hegde MR and Nanjappa MV 1998 Kacholum (*Kaempferia galanga*) – a potential medicinal-cum-aromatic crop for coconut gardens, *Indian Coconut Journal* **29**(5) 4-5

Maikhuri RK, Semwal RL, Rao KS, Singh K and Saxena KG 2000 Growth and ecological impacts of traditional agroforestry tree species in Central Himalaya, India, *Agroforestry Systems* **48** 257-271

Makundi WR and Sathaye JA 2004 GHG mitigation potential and cost in tropical forestry – relative role for agroforestry, *Environment, Development and Sustainability* **6** 235-260

Malik MS 2004 Promoting intercropping in Bamboo plantations in North-East India, In Agriculture and Environment (eds) MA Khan and MY Zarger, APH Publishing Corporation, New Delhi, India, 221-230 pages

Malik MS 2011 Report of the National Bamboo Mission project on Development Bamboo based Agroforestry system in Jharkhand, Birsa Agricultural University, Ranchi, Jharkhand

Malik RS and Sharma SK 1990 Moisture extraction and crop yield as a function of distance from a row of *Eucalyptus tereticornis*, *Agroforestry Systems* **12** 187-195

Mankelow J 2003 The implementation of the Watershed Development Programme in Zangskar, Ladakh: Irrigation Development, Politics and Society, SOAS, London

Mann HS and Muthana KD 1984 Arid Zone Forestry, CAZRI Monograph No. 23, CAZRI, Jodhpur, 23 page

Mann HS and Saxena SK (eds.) 1980 "Kherji" (*Prosopis cineraria*) in the Indian Desert: Its Role in Agroforestry, CAZRI Monograph No.11, Central Arid Zone Research Institute, Jodhpur, India

Mapa RB and Gunasena NPM 1995 Effect of alley cropping on soil aggregate stability of a tropical Alfisol, *Agroforestry Systems* **32**(3) 237-245

Marsh R 1998 Household food security through home gardening: evidence from Bangladesh and Central America, In *Proceedings of an International Livestock Research Institute (ILRI)-Rockefeller Foundation Workshop*, Ethiopia

Martin FW and Sherman S 1992 Agroforestry Principles Technical Note, ECHO, USA

Martin LK 2012 Agroforestry for biodiversity and ecosystem services - science and practice, Janeza Trdine, Rijeka, Croatia

Massolino F and Pardino A 2009 Results of a project on development of agroforestry systems for food security in Carrefour region, Republic of Haiti, *Journal of Agriculture and Environment for International Development* **103** 201-212

Mathur RS and Sharma KK 1983 Poplars in India, *Indian Forester* **109** 519-628

Mathur RS, Sharma KK and Ansari MV 1984 Economics of *Eucalyptus* plantations under Agroforestry, *Indian Forester* **110** 171-201

Mayer DG and Butler DG 1993 Statistical validation, *Ecological Modelling* **68** 21-32

Mayfield M and Belavadi VV 2008 Pollinator services for sustainable agriculture, Tools for conservation and use of pollination services - Initial survey of good pollination practices, *Global action on pollination services for sustainable agriculture,* Food and Agriculture Organization of the United Nations, Rome

Mburu OM 1981 Agroforestry in Forest Management in Kenya, In *Proceedings of the Kenya National Seminar on Agroforestry* (eds) L Buck, ICRAF, Nairobi, 19-22 pages

McAdam J, Burgess PJ, Graves AR, Rigueiro-Rodríguez A and Mosquera-Losada MR 2008 Chapter 2: Classifications and Functions of Agroforestry Systems in Europe, In *Agroforestry in Europe: Current Status and Future Prospects* (eds) A Rigueiro-Rodríguez, J McAdam, MR Mosquera-Losada and M Rosa, Springer

Mclone RR and Andersons JG 1976 Mathematical modeling, Butter worth group, London (UK)

McNeely JA 2004 Nature vs nurture: managing relationships between forests, agroforestry and wild biodiversity, *Agroforestry Systems* **61** 155-165

McNeely JA and Schroth G 2006 Agroforestry and biodiversity conservation–traditional practices, present dynamics and lessons for the future, *Biodiversity Conservation* **15** 549-554

MEA (Millennium Ecosystem Assessment) 2005a Ecosystems and human well-being: Global assessment reports, Island Press, Washingotn, D.C.

MEA (Millennium Ecosystem Assessment) 2005b Ecosystems and human well-being: Wetlands and water synthesis, Washington, DC, USA: World Resources Institute

Mead R 1967 A mathematical model for the estimation of interplant competition, *Biometrics* **23** 189-205

Mead R, Curnow RN and Hasted AM 1993 Statistical methods in agriculture and experimental botany, 2nd edition, Chapman & Hall, NewYork

Mehta T 1981 A Hand Book of Forest Utilization, Dehradun: International Book Distributors

Mekonnen K 1996 Rooting characteristics and inorganic nitrogen distribution in three land use systems on an Oxisol in western Kenya, M.Sc. Thesis, Moi University, Kenya

Mercer DE 1992 The economics of agroforestry, In *Social science applications in Asian agroforestry* (eds) WR Burch and JK Parker, Winrock International and South Asia Books, New Delhi

Messing I and Noureddine A. Sidi Bouzid 1991 Effects of wind breaks on wind velocity, evapotranspiration and yield of irrigated crops in the arid zone (Rural development studies No. 30), Central Tunisia, 40 pages

Meyer WJ 1985 Concepts of mathematical modeling, McGraw-Hill Book Company, New York

Miah MG, Garrity DP and Aragon ML 1995 Light availability to the understorey annual crops in an agroforestry system, Institute National de la RechercheAgronomie (INRA), (eds), M Sinoquet and P Cruz, 99-107 pages

Michon G and de Foresta H 1999 Agro-forests: Incorporating a Forest Vision in Agroforestry, In *Agroforestry in Sustainable Agricultural Systems* (eds) L Buck, J Lassoie and E Fernandes, Lewis Publishers, Boca Raton, 381-406 pages

Michon G, Bompard J, Hecketsweiler P and Ducatillion C 1983 Tropical forest architectural analysis applied to agroforests in the humid tropics, the example of traditional village agroforestry in West Java, *Agroforestry Systems* **1** 117-129

Mijatovic D, Bordoni P, Eyzaguirre P, Fox E, Hutchinson S, Van Oudenhoven F and Hodgkin T 2011 The use of Agrobiodiversity by indigenous and traditional agricultural communities, In Adapting to climate change, Satoyama Initiative case study, website: *http://satoyama-initiative.org/ en/casestudies-2/group_agriculture-2/the-use-of-agrobiodiversity-byindigenous-and-traditional-agricultural-communities-inadapting-to-climate-change/*

Miller RW 1988 Urban Forestry Planning and Managing Urban Greenspaces, Prentice Hall, New Jersey, USA

Milne G et al 2006 Unlocking Opportunities for Forest-Dependent People in India, Agriculture and Rural Development Sector Unit, South Asia Region, The World Bank/Oxford University Press, New Delhi

Ministry of Agriculture 1983 *Report of Task Force on Shifting Cultivation*, MOA, Govt. of India

Minj AV and Quli SMS 2001 Impact of agroforestry on socioeconomic status of respondents, *Indian Forester* **126** 788-791

Mishra RK and Pandey VK 1998 Inter-cropping of turmeric under different tree species and their planting pattern in agroforestry systems, *Range Management and Agroforestry* **19** 199-202

Mithoefer D and Waibel H 2003 Income and labour productivity of collection and use of indigenous fruit tree products in Zimbabwe, *Agroforestry Systems* **59** 295-305

Mitsch WI and Gosselink IG 1986 Wetlands, Van Nostrand Reinhold, New York

Mittal SP and Singh P 1989 Intercropping field crops between rows of *Leucaena leucocephala* under rainfed conditions in northern India, *Agroforestry Systems* **8**: 165-172

MOA 1994 Annual Report, Ministry of Agriculture, Government of India, New Delhi

Mohan Raj T, Divya MP and Ravi R 2008 Allelopathic effect of *Simarouba glauca* on germination and growth of pulses under pot culture experiment, Paper presented in the National workshop cum symposium on "Tree Spice cultivation and sustainable development of homegardens for descent livelihood and environment protection" Port Blair, Andaman and Nicobar Islands, 4-6 February 2008

Mohanty RK, Jena SK, Ashwani Kumar, Sahoo N, Chowdhury SR and Roy S 2008 Rice-Fish culture: An indigenous Agricultural heritage system, ICAR-Research Bulletin-42, Water Technology Centre for Eastern Region, Bhubaneswar, Orissa, India

Mohindra V 2001 Marketing of Eucalyptus and poplar wood in Punjab – An Appraisal, In *Marketing Monitoring of Tree Products* (eds) GS Rawat and NS Bisht, 124-113 pages

Molina MR, Noguera A, Dary O, Chew F and Valverde C 1993 Principal micronutrient deficiencies in Central America, *Food Nutr. Agric.* **7** 26-33

Montagnini 2006 Homegardens of mesoamerica: biodiversity, food security, and nutrient management, In *Tropical Homegardens: A Time-Tested Example of Sustainable Agroforestry* (eds) BM Kumar and PKR Nair, Springer, The Netherlands, 61-84 pages

Montagnini F and Nair PKR 2004 Carbon sequestration: an underexploited environmental benefit of agroforestry systems, *Agroforestry Systems* **61** 281-295

Montambault JR and Alavalapati JRR 2005 Socioeconomic research in agroforestry: A decade in review, Agroforestry Systems 65 151-161

Montane J (ed) 1994 La Foret Landaise, Editions Privat, Toulouse, France

Monteith JL, Ong CK and Corlett JE 1991 Microclimatic interactions in agroforestry systems, *Forest Ecology and Management* **45** 31- 44

Mopongmetsem PM, Niwah CD, Froumsia M, Kossebe CF and Hamawa Y 2011 Agroforest's potentials for the improvement of the livelihoods and food security in Guinean Highland Savannahs, *International Journal of Social forestry* **4**(2) 180-196

Moran EF 1974 The adaptive system of the Amazonian Caboclo, In *Man in the Amazon* (ed) C Wagley, Florida: Gainesville Press

Moreno G and Pulido F 2008 Chapter-7: The Functioning, Management and Persistence of Dehesas, In Agroforestry in Europe: Current Status and Future Prospects (eds) In *Agroforestry in Europe: Current Status and Future Prospects* (eds) A Rigueiro-Rodriguez, J McAdam, MR Mosquera-Losada and M Rosa, Springer

Morgan 1986 Soil Erosion and Conservation, Harlow: Longman

MOWR 1991 Working group of Ministry of Water resources (only in irrigated commands), Technical papers presented at National workshop on reclamation of waterlogged, saline and alkaline lands and prevention thereof, MOWR, New Delhi

MPSG 2013 IUCN Medicinal plant special group, website: http://mpsg.org/

MPSMFPCF 2013 MFP Resources and their marketing, M.P. State Minor Forest Produce (T & D) Co-op. Fed. Ltd, website: http://www.mfpfederation.com/

Muller CH 1969 Allelopathy as a factor in ecological process, *Vegetatio* **18** 348-357

Munish Kumar and Uttam 2012 Soil and Water Conservation, In *Forestry: Principles and Applications* (eds) Antony Joseph Raj and SB Lal, Scientific Publishers, Jodhpur, India, 671-687 pages

Murphy BC, Rosenheim JA and Granett J 1996 Habitat diversification for improving biological control: abundance of Anagrus epos (Hymenoptera: Mymaridae) in grape vineyards, *Environmental Entomology* **25**(2) 495-504

Muthana KD, Sharma SK and Harsh LN 1985 Study on silvipastoral systems in arid zone, *My Forest* **21**: 23-238

Mwaiko W 1998 Role of *Oecophylla longinoda* (Formicidae) and Amdro (hydramethlynon) ant bait in the integrated pest management of *Pseudotheraptus wayi* (Coreidae) on coconuts in Tanzania, In *Trees for Life – The Key to Development* (eds) CP Topper, PDS Caligari, AK Kullaya, SH Shomari, LJ Kasuga, PAL Masawe and AA Mpunami), BioHybrids International Ltd, Reading, UK, 452-455 pages

Myers N 1994 Protected Areas- protected from a Greater What? *Biodiversity and Conservation* 3 411-418

NAA/ISA 1991 The importance of large tree maintenance in mitigating global climate change, National Arborist Association, Amherst, USA

Nadeau I, Olivier A, Simard RR, Coulombe J and Yelle S 1999 Growing American Ginseng in Maple Forests as an Alternative Land Use System in Quebec, Canada, *Agroforestry Systems* **44** 345-353

Naik GR and Manjunath GTS 2000 Project (UKP) areas through bio-drainage concept, 8th ICID International drainage workshop "Role of Drainage and challenges in 21st century, 31st Jan - 4th Feb 2000, New Delhi

Naik ST 2012 Forest Disease Management, In *Forestry: Principles and Applications* (eds) Antony J. Raj and SB Lal, Scientific Publishers, Jodhpur

Naik ST, Suryanarayana V and Neelannavar TN 1998 Survey and management of forest nursery diseases in Uttar Kannada, Annual Meeting and Symposium on Integrated disease management and crop loss assessment, Indian phytopathological, Society (South Zone) Dec. 10-12, Dept. of plant Pathology, UAS GKVK, Bangalore-65

Nair MA and Sreedharan C 1986 Agroforestry farming systems in the homesteads of Kerala, South India, *Agroforestry Systems* **4** 339-363

Nair GS, Sudhadevi PK and Kurian A 1989 Introduction of medicinal and aromatic plants as intercrops in coconut plantations, In *Recent Advances in Medicinal, Aromatic and Spice Crops* (ed) SP Raychauduri, Vol. 1, Today and Tomorrow's Printers and Publishers, New Delhi, India, 163-165 pages

Nair KSS (ed) 2000 Insect Pests and Diseases in Indonesian Forests: An Assessment of the Major Threats, Research Efforts and Literature, Bogor, Indonesia: Center for International Forestry Research

Nair KSS 1986 Important insect pest problems of forest plantations in tropical India, In *Proceedings of the 18th IUFRO World Congress*, Vol.1. Ljubljana, Yugoslavia: International Union of Forestry Research Organizations, 134-145 pages

Nair KSS 1991 Social, economic and policy aspects of integrated pest management of forest defoliators in India, *Forest Ecology and Management* **39** 283-288

Nair KSS, Sudheendrakumar VV, Varma RV and Chacko KC 1985 Studies on the seasonal incidents of teak defoliators and the effect of defoliation on volume increment of teak, Peechi, Kerela: KFRI Report **No.30**, 85page

Nair MA and Sreedharan C 1986 Agroforestry farming systems in the homesteads of Kerala, South India, *Agroforestry Systems* **4**: 339-363

Nair PKR (ed) 1989 Agroforestry Systems in the Tropics, Kluwer, Dordrecht, The Netherlands, 664 pages

Nair PKR 1979 Intensive multiple cropping with coconuts in India, *Advances in Agro Crop Science* **6** 1-12

Nair PKR 1979 Intensive Multiple Cropping with Coconuts in India: Principles, Programmes and Prospects, Verlag Paul Parey, Berlin

Nair PKR 1983 Agroforestry with coconuts and other tropical plantation crops, In *Plant Research and Agroforestry* (ed) PA Huxley, ICRAF, Nairobi, 79-102 pages

Nair PKR 1984 Soil Productivity Aspects of Agroforestry, ICRAF, Nairobi, Kenya

Nair PKR 1985 Classification of agroforestry systems, *Agroforestry Systems* **3** 97-128

Nair PKR 1986 Alternative and improved land use systems to replace resource depleting shifting cultivation, In *Strategies, approaches and systems in integrated watershed management, FAO Conservation Guide* **14** 61-84, Rome: United Nations Food and Agricultural Organisation

Nair PKR 1987 Agroforestry Systems Inventory, *Agroforestry Systems* **5** 301-317

Nair PKR 1989 Agroforestry Systems in the Tropics, Kluwer Academic Publishers, Dordrecht

Nair PKR 1989a ICRAF's Agroforestry Systems Inventory Project, In Agroforestry Systems in the Tropics (eds) PKR Nair, Kluwer Acdemic publishers, The Netherlands, 21-38 pages

Nair PKR 1989b Classification of agroforestry systems, In *Agroforestry systems in the tropics* (ed.) PKR Nair, Kluwer Academic Publishers in co-operation with ICRAF, Dordrecht, The Netherlands, 39-52 pages

Nair PKR 1990a Classification of agroforestry systems, In *Agroforestry Classification and Management* (eds) KG MacDicken and NT Vergara, John Wiley & Sons, New York, 31-57 pages

Nair PKR 1990b The Prospects for Agroforestry in the Tropics, Technical Paper No.131, World Bank, Washington DC

Nair PKR 1991 State-of-art of agroforestry systems, *Forest Ecology and Management* **45**(1-4) 5-29

Nair PKR 1993 An Introduction to Agrofoestry, Kluwer Academic Publishers, Dordrecht, The Netherlands, 499 pages

Nair PKR 1994 *Agroforestry*: *Encyclopedia of agricultural sciences* (Vol. 1, pp. 13-25), New York: Academic Press

Nair PKR 1997 Directions and tropical agroforestry research: past, present and future, *Agroforestry Systems* **3**(8) 223-245

Nair PKR 2001 Agroforestry in Our Fragile World: Challenges and Opportunities for Sustainable Development, Forerunner to The Encyclopedia of Life Support Systems, Chapter 1.25, vol. I. UNESCO, Paris, France, & EOLSS, UK, 375-393 pages

Nair PKR 2001 Do tropical homegardens elude science, or is it the other way around? *Agroforestry Systems* **53** 239-245

Nair PKR 2007 The coming of age of agroforestry, *Journal of Science Food Agri.* **87** 1613-1619

Nair PKR 2011 Methodological Challenges in Estimating Carbon Sequestration Potential of Agroforestry Systems, In *Carbon Sequestration Potential of Agroforestry Systems: Opportunities and Challenges, Advances in Agroforestry* **8** (eds) BM Kumar and PKR Nair, Springer, Springer Science+Business Media B.V.

Nair PKR 2012 Carbon sequestration studies in agroforestry systems: a reality-check, *Agroforestry Systems* **86** 243-253

Nair PKR and Balakrishnan TK 1977 Ecoclimate of a coconut plus cacao crop combination on the west coast of India, *Agricultural Meteorology* **18** 455-462

Nair PKR and Dagar JC 1991 An approach to developing methodologies for evaluating agroforestry systems in India, *Agroforestry Systems* **16**: 55-81

Nair PKR and Fernandes ECM 1985 Agroforestry as an alternative to shifting cultivation, In *Improved Production Systems as an Alternative to Shifting Cultivation, FAO Soils Bulletin* **53**169-182

Nair PKR and Kumar BM 2006 Introduction, In *Tropical Homegardens: A Time-Tested Example of Sustainable Agroforestry* (eds) BM Kumar and PKR Nair, Springer, The Netherlands, 1-10 pages

Nair PKR and Latt CR 1997 Directions in tropical agroforestry research, *Agroforestry Systems* **38** 1-249

Nair PKR and Merry FD 1995 Development of Non-Wood Forest Products through Agroforestry: Issues and Strategies, In *FAO report of the International Expert Consultation on Non-Wood Forest Products*, FAO Rome

Nair PKR and Nair VD 2003 Carbon storage in North American Agroforestry Systems, In *The Potential of US Forest Soils to Sequester Carbon and Mitigate Greeenhouse Effect* (eds) J Kimble, LS Heath, RA Birdsey and R Lal, Boca Raton, FL: CRC Press LLC

Nair PKR, Bannister ME and Allen SC 2005 Agroforestry today: An analysis of the 750 abstracts presented to the 1st World Congress of Agroforestry - 2004, *Journal of Forest.* **103** 417-421

Nair PKR, Buresh R, Mugendi D and Latt C 1999 Nutrient Cycling in Tropical Agroforestry Systems: Myths and Science, In *Agroforestry in Sustainable Agricultural Systems* (eds) L Buck, J Lassoie and E Fernandes, Lewis Publishers, Boca Raton, 381-406 pages

Nair PKR, Gholz HL, Duryea MI 1990 Recommendations of the International Workshop on Professional Education and Training in Agroforestry, University of Florida, December 1988, In *Agroforestry Systems 12*, Amsterdam: Kluwer Academic Publications, 133-139 pages

Nair PKR, Gordon AM and Mosquera-Losada MR 2008 Agroforestry, In *Ecological Engineering* (eds) SE Jorgensen and BD Fath, Elsevier, Oxford, UK *Encyclopedia of Ecology* **1** 101-110

Nair PKR, Kumar BM and Nair VD 2009 Agroforestry as a strategy for carbon sequestration, *Journal of Plant Nutrition and Soil Science* **172** 10-23

Nair PKR, Nair VD, Kumar BM and Showalter JM 2010 Carbon sequestration in agroforestry systems, *Advances in Agronomy* **108** 237-307

Nair PKR, Rao MR and Buck LE (eds) 2004 New Vistas in Agroforestry: A Compendium for the 1st World Congress of Agroforestry 2004, Adv. Agrofor. 1. Kluwer, Dordrecht, the Netherlands, 480 pages

Nair VD and Graetz DA 2004 Agroforestry as an approach to minimizing nutrient loss from heavily fertilized soils: the Florida experience, *Agroforestry Systems* **61** 269-279

Nambiar EKS, Cossalter C and Tiarks A (eds) 1998 Site management and productivity in tropical plantation forests, CIFOR, Indonesia, 76 pages

Narain P and Tewari JC 2005 Trees on agricultural fields: a unique basis of life support in Thar Desert, In *Multipurpose Trees in the Tropics: Management and Improvement Strategies* (eds) VP Tewari and RL Srivastava, Arid Forest Research Institute, Jodhpur, 516-523 pages

Narain P, Singh G and Joshie P 1992 Technological needs of vegetative land protection measures, Proceeding of 7th ISCO Conference, Sydney, Australia (28-30th September), 638-643 pages

Natarajan M and Willey RW 1980 Sorghum-pigeonpea intercropping and the effects of plant population density, 1.Growth and Yield, 2.Resource Use, *Journal of Agricultural Science and Conservation* **95** 51-65 pages

NCA 1976 Report of the National Commission on Agriculture, Ministry of Agriculture and Irrigation, Govt. of India, New Delhi

Negi MS, Ansari MY et al 2001 Marketing channels of Tree products from agroforestry, In *Marketing Monitoring of Tree Products* (eds) GS Rawat and NS Bisht, 125-133 pages

Negi SS 1986 A handbook of social forestry, International Book Distributers, Dehra Dun (India)

Negi SS 1996 An Introduction to Forest Pathology, Dehradun: IBD

Negi SS 1998 Marketing and trade of forest products, In *Forest Economics, Valuation and projects*, International Book Distributors, Dehradun, Delhi, 126-152 pages

Nelder JA 1962 New kinds of systematic design for spacing experiments, *Biometrics* **18** 283-307

Neter J, Kutner MH, Nachtsheim CJ and Wasserman W 1996 Applied linear statistical models, R.D. Irwin Inc., Chicago

Neufeldt H, Wilkes A, Zomer RJ, Xu J, Nang'ole E, Munster C and Place F 2009 Trees on farms: Tackling the triple challenges of mitigation, adaptation and food security, World Agroforestry Centre Policy Brief 07, World Agroforestry Centre, Nairobi, Kenya

Neumann F and Pietrowicz P 1989 Light and water availability in fields with and without trees: An example from Nyabisindu in Rwanda, In *Meteorology and Agroforestry* (eds) WS Reifsnyder and TO Darnhofer, ICRAF, Nairobi, Kenya, 401-406 pages

Neupane RP and Thapa GB 2001 Impact of Agroforestry Intervention on Soil Fertility and Farm Income under the Subsistence Farming System of the Middle hills, Nepal, *Agriculture, Ecosystems and Environment* 84 157-167

Newman EI 1983 Interactions between plants, In *Physiological Plant Ecology-III, Encyclopedia of Plant Physiology* (eds) OL Lange, PS Nobel, CB Osmond and H Ziegler, Vol-12C, Springer, Berlin, Germany

NFAP (National Forestry Action Programme) 1999 Ministry of Environment and Forests, Government of India, New Delhi, volume 1 & 2

NFNP (New Forest National Park) 2010 Pigs in pannage, website: http://www.newforestnpa. gov.uk/pigs

NFTA 1989 NFT Highlights: A quick guide to useful nitrogen fixing trees from around the world, Winrock, website: http://www.winrock.org/fnrm/factnet/factpub/FACTSH/WhyNFT.htm

NHB 2011 Indian Horticulture Database 2011, National Horticulture Board (NHB), Ministry of Agriculture, Government of India, Gurgaon, India

Nicodemus A, Varghese M and Nagarajan B 2001 Genetic improvement of *Casuarina equisetifolia* through selection and breeding, In *Casuarina: Improvement and Utilization* (eds) K Gurumurthi, A Nicodemus and Siddappa, Institute of Forest Genetics and Tree Breeding, Coimbatore, India, 3-14 pages

Nigh GD and Sit V 1996 Validation of forest height - age models, *Canadian Journal of Forest Research* 26 810-818

Niinemets U and Valladares F 2004 Photosynthetic acclimation to simultaneous and interacting environmental stresses along natural light gradients: optimality and constraints, *Plant Biology* **6** 254-268

Noble IR and Dirzo R 1997 Forests as human-dominated ecosystems, *Science* **277** 522-525

Nokoe S 1992 Technical Paper 7: Statistical and Experimental Design Considerations in Alley Farming, In *The AFNETA alley farming training manual - Volume 2: Source book for alley farming research* (eds) BR Tripathl and PJ Psychas, Alley Farming Network for Tropical Africa, International Institute of Tropical Agriculture, Ibadan

Norberg-Hodge H 2000 Ancient Futures: Learning from Ladakh, London

Nordlund DA 1981 Semiochemicals: a review of the terminology, In *Semiochemicals: Their Role in Pest Control* (eds) DA Nordlund, RL Jones and WJ Lewis, John Wiley & Sons, New York, 13-28 pages

NRCAF 2000 Annual Report, NRCAF, Jhansi, UP, India

NRCAF 2005 Annual Report, National Research Centre for Agroforestry, Jhansi, Uttar Pradesh

NRCAF 2007 Perspective Plan: Vision 2015, National Research Centre for Agroforestry (NRCAF), Jhansi, India

NRCAF 2007 Perspective Plan-Vision 2025: National Research centre on Agroforestry, Jhansi

NRCAF 2011 Annual Report 2010-11, National Research Centre for Agroforestry (NRCAF), Jhansi, Uttar Pradesh, India

NTFP 2013 NTFP Information Resources, website: http://www.ntfp.org

Nye PH and Greenland DJ 1960 The Soil under Shifting Cultivation, Commonwealth Bureau of Soils, Harpenden, UK

O'Kting'ati A, Maghembe JA, Fernandes ECM and Weaver GH 1984 Plant species in the Kilimajaro agroforestry system, *Agroforestry Systems* **2** 177-186

Odoom F 2002 Hardwood plantation in Ghana, Forest plantation working paper FP/24 FAO, Rome, Italy

Oduol P 1989 The Shamba System: An Indigenous System of Food production from Forest Areas in Kenya In *Agroforestry Systems in the Tropics* (eds) Nair PKR, Kluwer Academic Publishers, Dordrecht, 401-410 pages

OECD 1996 Saving Biological Diversity- Economic Incentives, France

OECD 1999 Handbook of Incentives Measures for biodiversity-Design and Implementation

OED 1989 Oxford English Dictionary

Oerke EC, Dehne HW, Schnbeck F and Weber A 1994 Crop Production and Crop Protection: Estimated Losses in Major Food and Cash Crops, Amsterdam: Elsevier

Ogol CKPO and Spence JR 1997 Abundance, population dynamics and impact of the *Leucaena* psyllid *Heterosphylla cubana* Crowford in a maize–*Leucaena* agroforestry system in Kenya, *Insect Science Application* 17 183-192

Ojo GJA 1966 Yoruba Culture, University of Ife and London Press, London, UK

Okali C. and Sumberg JE 1985 Sheep and goats, men and women: Household relations and small ruminant development in southwest Nigeria, *Agricultural Systems* **18** 39-59

Okigbo BN 1981 Alternatives to shifting cultivation, *Ceres* (FAO) **15**(6) 41-45

Okigbo BN 1985 Improved permanent production systems as an alternative to shifting intermittent cultivation, In *Improved Production Systems as an Alternative to Shifting Cultivation*, *FAO Soils Bulletin* **53** 1-100

Olawoye OO 1975 The agri-silvicultural system in Nigeria, *Commonwealth Forestry Review* **54**(3/4) 229-236

Oldeman LR, Van Englen VWP and Pulles JHM 1991 The extent of human-induced soil degradation, In World Map of Status of Human-Induced Soil Degradation: An Explanatory Note (eds) Oldeman LR, Hakkeling RTA and Sombroek WG, Wageningen: International Soil Reference and Information Centre (ISRIC), 27-33 pages

Oldeman RAA 1983 The design of ecologically sound agroforests, In *Plant Research and Agroforestry* (ed) PA Huxley, International Council for Research in Agroforestry (ICRAF), Nairobi, Kenya, 173-207 pages

Ong CK 1994 Alley cropping – ecological pie in the sky? *Agroforestry Systems* **6**(3) 8-10

Ong CK 1995 The 'dark side' of intercropping, In *Proceedings of the Ecophysiology of Intercropping*, Guadeloupe, INRA

Ong CK 1996 A framework for quantifying the effects of various tree crop interactions, In *Tree-crop Interactions - A Physiological Approach* (eds) CK Ong and P Huxley, CAB International, Wallingford, Oxford, UK, 1-23 pages

Ong CK and Swallow BM 2003 Water productivity in forestry and agroforestry, In *Water Productivity in Agriculture: Limits and Opportunities for Improvement* (eds) JW Kijne, R Barker and D Molden, Wallingford, UK: CAB International, 217-228 pages

Ong CK, Black CR and Muthuri CW 2007 Modifying forestry and agroforestry to increase water productivity in the semi-arid tropics, CAB Reviews: Perspectives in Agriculture, Veterinary Science, *Nutrition and Natural Resources* **65** 1-19

Ong CK, Corlett JE, Singh RP and Black CR 1991 Above and below ground interactions in agroforestry systems, *Forest Ecology and Management* **45** 45-57

Ong CK, Odongo CW, Marshall F and Black CR 1992 Water use of agroforestry systems in semiarid India, In *Growth and Water Use of Forest Plantations* (eds) IR Calder, RL Hall and PG Adlard, Wiley, Chichester, UK, 347-358 pages

Oosterbaan A and Kuiters AT 2008 Chapter-16: Agroforestry in Netherlands, In *Agroforestry in Europe: Current Status and Future Prospects* (eds) A Rigueiro-Rodriguez, J McAdam, MR Mosquera-Losada and M Rosa, Springer

Opondo-Mbai ML 1995 Investigations of arthropods associated with agroforestry in Machakos, Ph.D. Thesis, University of Oxford, UK

Osman M and Rao JV 1999 Alternate land use: Hortipastoral systems, In *Fifty years of Dryland Agricultural Research in India* (eds.) HP Singh, YS Ramakrishna, KL Sharma and B Venkateswarlu, 485-495 pages

Otsyina R, Minae S and Cooper P 1996 A never-ending story: rotational woodlots for soil conservation, wood, and Fodder, *Agroforestry Today* **8**(2) April-June 1996

Overton J, Murray W and Ali I 1999 Commodity production and unsustainable production, In *Strategies for sustainable production: experiences from the Pacific* (eds) J Overton and RA Scheyvens, UNSW Press, Sydney, 168-181 pages

Padoch C and de Jong W 1987 Traditional agroforestry practices of native and Ribereno farmers in the lowland Peruvian Amazon, In *Agroforestry: Realities, Possibilities and Potentials* (ed) HE Gholz, Dordrecht: International Centre for Research in Agroforestry and Nijhoff Publishers, 179-194 pages

Padoch C and de Jong W 1991 The housegardens of Santa Rosa: diversity and variability in an Amazonian agricultural system, *Economic Botany* **45** 166-175

Padoch C, Inuma CJ, de Jong W and Unruh J 1985 Amazonian agroforestry: a market-oriented system in Peru, *Agroforestry Systems* **3** 47-58

Palm CA 1995 Contribution of agroforestry trees to nutrient requirements of intercropped plants, *Agroforestry Systems* **30** 105-124

Palmer JR 1994 Designing Ccommercially Promising tropical timber species, In *Tropical trees: The Potential for Domestication and the Rebuilding of Forest Resources* (eds) RRB Leakey and AC Newton

Palsaniya DR and Ram Newaj 2009 Tree-Crop Interactions in Agroforestry, In Course Compen-dium of Training Programme on *Advances in Agroforestry Research and Development* for ICFRE Scientists (eds) OP Chaturvedi, Ajit, Rajendra Prasad, AK Handa and RH Rizvi) held at NRCAF, Jhansi during 24th November to 5th December, 2008, 74-80 pages

Palsaniya DR, Ram Newas and Yadav RS 2009 Tree-Crop Interactions and their management in Agroforestry, In *Agroforestry: Natural Resource Sustainability, Livelihood and Climate Moderation* (eds) OP Chaturvedi, A Venkatesh, RS Yadav, Badre Alam, RP Dwivedi, Ramesh Singh and SK Dhyani), Satish Serial Publishing House, Delhi, India, 241-262 pages

Pandey CB, Lata K, Venkatesh and Medhi RP 2006 Diversity and species structure of home gardens in South Andaman, *Tropical Ecology* **47**(2) 251-258

Pandey DN 1998 Ethnoforestry: Local Knowledge for Sustainable Forestry and Livelihood Security, Himanshu/AFN, New Delhi

Pandey DN 2007 Multifunctional agroforestry systems in India, *Current Science* **92**(4) 455-463

Pandey U and Singh JS 1984 Energy–flow relationships between agro and forest ecosystems in central Himalaya, *Environmental Conservation* **11**(1) 45-53

Panigrahi SK, Reddy AM and Narendra P 2004 Energy security, environmental benefits and carbon trading potential of biofuels in India – with the case study of 1st biodiesel project for CDM, *In Proceedings of International Conference on Biofuels - Perspectives and Prospects*, September 16-17, 2004, New Delhi, 362-368 pages

Panigrahy S, Patel JG, Murthy TVR and Singh TS 2011 Wetlands of India, Brochure under the "National Wetland Inventory and Assessment" project sponsored by Ministry of Environment and Forests, Govt. of India

Panwar P, Charan Singh and Sumit Chakravarthy 2011 *Salix tetrasperma*: Typical phenology a boon to low land agroforestry, *Indian Forester* **137**(3) 395-396

Panwar, Pankaj KD, Bhutia CP, Suresh and Mathew B 2011 Performance of elephant foot yam grown as intercrop in mango orchard in terai zone of West Bengal, *Transient* **1**(1) 148-154

Papanastasis VP, Mantzanas K, Dini-Papanastasi O and Ispikoudis I 2008 Chapter-5: Traditional Agroforestry Systems and Their Evolution in Greece, In *Agroforestry in Europe: Current Status and Future Prospects* (eds) A Rigueiro-Rodriguez, J McAdam, MR Mosquera-Losada and M Rosa, Springer

Paramathma M and Pandey A 2012 Biofuel from Tree Borne Oilseeds, In *Forestry: Principles and Applications* (eds) Antony Joseph Raj and SB Lal, Scientific Publishers, Jodhpur, India

Parera V 1986 The role of *Leucaena leucocephala* in farming systems in Nusa Tenggara Timur, Indonesia, In *Alley farming in the humid and subhumid tropics* (eds) BT Kang and L Reynolds, IDRC, Canada

Parker IM, Lopez I, Peterson J, Anaya N, Cubilla-Rios L and Potter D 2010 Domestication Syndrome in Caimito (*Chrysophyllum cainito*): Fruit and Seed Characteristics, *Economic Botany* **64**: 161-175

Parthiban KT 2010 A Value Chain on Industrial Agroforestry in Tamil Nadu, Baseline Report (2009-2010), NAIP scheme, Forestry College and Research Institute, TNAU, Mettupalayam, 53-54 pages

Parthiban KT, Vennila S and Govinda Rao M 2010 A Value chain on Industrial Agroforestry in Tamil Nadu, Research and World Bank for funding the project (Indian Council Agricultural Research), Forest College and Research Institute, Tamil Nadu Agricultural University, Mettupalayam

Pascual U, Muradian R, Rodríguez LC and Duraiappah A 2010 Exploring the links between equity and efficiency in payments for environmental services: A conceptual approach, *Ecological Economics* **69** 1237-1244

Pathak PS 2000 Agroforestry: A tool for arresting land degradation, *Indian Farming* **49**(11) 15-19

Pathak PS 2002 Common pool degraded lands: technological and institutional options, In *Institutionalizing Common Pool Resources* (ed) DK Marothia, Concept Publishing Co., New Delhi, India, 402-433 pages

Pathak PS, Pateria and Solanki KR 2000 Agroforestry systems in India – A diagnosis & design approach, AICRP on Agroforestry, NRCAF, Jhansi, 223 pages

Pathak RK 2003 Status report on genetic resources of Indian gooseberry-Aonla (*Emblica officinalis*) in South and South east Asia, IPGRI Office for South Asia, NASC, New Delhi, 105 page

Patra SN and Panda GK 1994 Eastern Ghat in Orissa: Environment, Resources and Development, Orissa Environment Society

Pavgi MS and Singh UP 1970 Parasitic fungi from north India – IX, *Sydowia* **24**(1-6) 113-119

Pearson HA, Knowles RL, Middlemiss PG, Baldwin VC and Busby RL 1995 United States agroforestry estate model, *The Compiler* **13** 27-37

Perfecto I, Rice R, Greenberg R and Van der Voort M 1996 Shade coffee: a disappearing refuge for biodiversity, *Bio Science* **46** 598-608

Perrault-Archambault M and Coomes OT 2008 Distribution of Agrobiodiversity in Home Gardens along the Corrientes River, Peruvian Amazon, *Economic Botany* **62**(2) 109-126

Peskin HM, Willem Floor and Douglas F. Barnes 1992 Accounting for Traditional Fuel Production: The Household Energy Sector and Its Implications for the Development Process, Industry and Energy Department, Washington, DC: The World Bank

Peyre A, Guidal A, Wiersum KF and Bongers F 2006 Homegarden dynamics in Kerala, India, In *Tropical Homegardens: A Time-Tested Example of Sustainable Agroforestry* (eds) BM Kumar and PKR Nair, Springer, The Netherlands, 87-103 pages

Pias J 1955 Etude pédologique du bassin alluvionaire du Logone et Chari (nord Cameroun), champagne, Paris, ORSTOM, 1953-54 pages

Pieri CMG 1998 Soil fertility improvement: key connection between sustainable land management and rural well being, 16th World Congress of Soil Science, Montpellier, France

Pinho RC, Miller RP and Alfaia SS 2012 Agroforestry and the Improvement of Soil Fertility: A View from Amazonia, Applied and Environmental Soil Science, Volume 2012, Article ID 616383, 1-11 pages (doi:10.1155/2012/616383)

Pinney A 1991 Farmer augmented designs for participatory agroforestry research, *Agroforestry Systems* **15** 259-274

Place F and Wanjiku J 2006 High Value Tree and Crop Enterprises in Kenya: How Meso and Micro Factors Affect Adoption and Revenues, Paper presented at the International Association of Agricultural Economists Conference, Gold Coast, Australia, August 2006

Planning Commission 2003 Report of the Task Force on Greening India for Livelihood Security and Sustainable Development, Planning Commission, New Delhi, Government of India, 202 & 210 pages

Porro R (ed) 2009 *Alternativa agroflorestal na Amazônia em transformação*, Embrapa Informação Tecnológica, Brasília, Brazil

Posey DA 1985 Indigenous management of tropical forest ecosystems: the case of the Kayapo Amerindians of the Brazilian Amazon, *Agroforestry Systems* **3** 139-158

Poulsen G 1983 Using farm trees for fuelwood (in Africa) *Unasylva* (FAO) 35(141) 26-29

Pradhan JP 1973 Preliminary study of rainfall in interception through leaf litter, *Indian Forester* **99** 440-443

Prajapati ND, Purohit SS, Sharma AK and Kumar T 2003 A Handbook of Medicinal Plants, Agribios Publishers Pvt. Ltd., Jodhpur (India), 553 page

Prasad JVNS et al 2010 Tree row spacing affected agronomic and economic performance of Eucalyptus-based agroforestry in Andhra Pradesh, Southern India, *Agroforestry Systems* **78** 253-267

Prasad R, Pandey DN and Kotwal PC 2000 Trees Outside Forests in India: A National Assessment, Indian Institute of Forest Management, Bhopal, India

Pratap Narain RK, Singh, Sindhwaland NS and Joshi P 1998 Agroforestry for soil and water conservation in the Western Himalayan valley region of India, *Agroforestry Systems* **39** 192-197

Primack RB 1993 Essentials of Conservation Biology, Sinaurer Associate Inc. USA

Prokopy RC 1994 Integration in orchard pest and habitat management: a review, *Agriculture, Ecosystems and Environment* **50** 1-10

Pullan RA 1974 Farmed parkland in West Africa, *Savanna* **3**(2) 119-151

Puri S and Nair PKR 2004 Agroforestry research for development in India: 25 years of experiences of a national program, *Agroforestry Systems* **61** 437-452

Puri S and Panwar P 2006 Agroforestry Systems and Practices, NIPA, New Delhi

Puri S, Anil Kumar and Kumar A 1992 Management and establishment of *Prosopis cineraria* in hot desert of India, In *Integrated land use management for tropical agriculture*, Proceedings of second International symposium, Queensland, 15-25 September 1992, Module 2

Puri S, Bangarwa KS and Singh S 1995 Influence of multipurpose trees on agricultural crops in arid regions of Haryana, *Indian Journal of Arid Environment* **30** 441-451

Puri S, Kumar A and Singh S 1994 Productivity of *Cicer arietinum* (chickpea) under a *Prosopis cineraria* agroforestry system in the arid regions of India, In Special Issue: Acacia and Prosopis, *Journal of Arid Environments* **27**(1) 85-89

Putnam AR and Tang CS 1986 The Science of Allelopathy, Wiley Inter-Science, New York, USA

Quinion A, Chirwa PW, Akinnifesi FK and Ajayi OC 2010 Do agroforestry technologies improve the livelihoods of the resource poor farmers? Evidence from Kasungu and Machinga districts of Malawi, *Agroforestry Systems* **80** 457-465

Quli SMS 2001 Agroforestry for NTFPs conservation and economic upliftment of farmers, *Indian Forester* **127** 1251-1262

Qureshi RH and Barret-Lennard 1998 Saline Agriculture for irrigated land in Pakistan: A handbook, Canberra: Australian Centre for International Agricultural Research, VI, 142 pages

Rabe and Staden JV 1997 *Journal of Ethnopharmacology* **56** 81

Raghavan MS 1960 Genesis and history of the Kumri system of cultivation, In *Proceedings of the Ninth Silviculture Conference*, Forest Research Institute, Dehradun, India

Rai P and Ajit Sharma AK 2007 Tree leaves their production and nutritive value for ruminants: r review, *Animal Nutrition and Feed Technology* **7**(2) 135-159

Rai P, Yadav RS, Solanki KR, Rao GR and Rajendra Singh 2001 Growth and pruned biomass production of multipurpose tree species in silvi-pastoral system on degraded lands in semi-arid region of Uttar Pradesh, India, *Forest Tree and Livelihood* **11** 347-364

Rai SC 1995 Land use options in Sikkim Himalaya: an analysis from the Mamlay Watershed in South District, *Journal of Sustainable Agriculture* **6** 63-79

Rai SN and Chakrabarti SK 2001 Demand and supply of fuelwood and timber in India, *Indian Forester* **127** 263-279

Rai SN and Chauhan KVS 1998 Distribution and growing stock of bamboos in India, *Indian Forester* **118** 89-97

Raintree JB 1980 Conservation farming with soil improving legumes: a farming systems approach to tropical agroforestry, IITA Agric. Econ. Discussion Paper 9/80

Raintree JB 1987a D&D user's manual, ICRAF, Nairobi, Kenya

Raintree JB 1987b State-of-the-art of agroforestry diagnosis and design, *Agroforestry Systems* **5** 219-250

Raintree JB 1993 In Agroforestry Training Course Module for Bangaladesh (ed) Tej BS Mahat, Training Support Series-2, BARC-Winrock International, Agroforestry & Participatory Forestry Research and Training Support Programme, Dhaka, Bangladesh

Raintree JB and Francisco HA 1994 Marketing of multipurpose tree products in Asia, In *Proceedings of an International Workshop*, Baguio City, Philippines, December 1993, Winrock International, Bangkok

Raizada A and Singh C 2010 Silvipastoral systems for wasteland utilization in foothills of the western Himalayas, Technology Brouchere, CSWCRTI, Dehardun, 10 page

Rajiv Ranjan and Upadhyay VP 1999 Ecological problems due to shifting cultivation, *Current Science*, IISC, Bangalore

Rajput MS, Biswas PP, Joshi OP and Srivastava KC 1989 Mango (*Mangifera indica*) based cereal pulse cropping system, *Indian Journal of Agriculture Sciences* **59**(3) 140-153

Rajput MS, Srivastava KC, Joshi OP and Biswas PP 1986 Effect of interplanting fruit crop on Mango, *The Indian Journal of Horticulture* **43**(3&4) 216-220

Rajvanshi AK, Singh V and Nimbkar N 2007 Biofuels – Promise/Prospects, Paper presented at National Oilseeds Conference in Hyderabad, 29-31 January, 2007

Ralhan PK, Singh A and Dhanda RS 1992 Performance of wheat as intercrop under poplar (*Populus deltoides*) plantations in Punjab (India), *Agroforestry Systems* **19** 217-222

Ram J, Dagar JC, Lal K, Singh G, Toky OP, Tanwar VS, Dar SR and Chauhan MK 2011 Biodrainage to combat waterlogging, increase farm productivity and sequester carbon in canal command areas of northwest India, *Current Science* **100**(11) 1673-1680

Ram Newaj, Dar SA, Bhargava MK, Yadav RS and Ajit 2007 Effect of management practices on growth of white siris (*Albizia procera*), grain yield of intercrops, weed population and soil fertility changes in agrisilviculture system in semi-arid India, *Indian Journal of Agricultural Sciences* **77**(7) 403-407

Ramadhani T, Otsyina R and Franzel S 2002 Improving household income and reducing deforestation using rotational woodlots in Tabora district, Tanzania, *Agriculture, Ecosystem and Environment* **89**(3) 229-239

Ramakrishna YS and Shastri ASRAS 1977 Microclimate under *Acacia tortilis* plantation, Annual Progress Report, CAZRI, Jodhpur, 69-70 pages

Ramsay JM and Rose Innes R 1963 Some quantitative observations on the effects of fire on the Guinea savanna vegetation of Northern Ghana over a period of eleven years, *Africa soils* **8** 41-85

Rander HS 1988 Agroforestry systems in Nigeria, Paper presented at the National workshop on Agroforestry, August-1988, Ibadan, Nigeria

Randhawa MS 1980 The history of Indian agriculture, Indian Council of Agricultural Research, New Delhi, Vol. **2**, 414-415 pages

Rands MRW, William MA, Leon B, Stuart HMB, Clements A, Coomes D, Entwistle A, Hodge I, Kapos V, Scharlemann JPW, Sutherland WJ and Vira B 2010 Biodiversity Conservation: Challenges Beyond 2010 *Science* **329** (5997) 1298-1303

Randy Thaman, Craig Elevitch and Kim Wilkinson 2013 Protecting and Expanding Traditional Agroforests in the Pacific The Overstory, Agroforestry e-Journal, No.51, webpage: http://agroforestry.net /overstory/overstory51.html

Rao IVR, Gnanaharan R and Shastry CB 1990 *Current Research*, KFRI, India and IDRC, Canada

Rao JV 2009 Agroforestry – An Approach for Mitigating Adverse Climatic Effects in Rainfed Areas, In Winter School on Alternate land use options for resource conservation, emerging market needs and mitigation of climate change in rainfed regions, CRIDA, Hyderabad, AP, India

Rao M and Sita Ram 1980 Influence of shelter belts on annual crops, *Proceedings of Second Forestry Conference*, FRI, Dehradun, India, 525-529 pages

Rao M, Nair PKR and Ong C 1998 Biophysical interactions in agroforestry systems, *Agroforestry Systems* **38** 3-50

Rao MR, Nair PKR and Ong CK 1998 Biophysical interactions in tropical agroforestry systems, *Agroforestry Systems* **38** 3-50

Rao MR, Ong CK, Pathaka P and Sharma MM 1991 Productivity of annual cropping and agroforestry systems on a shallow alfisol in semi-arid India, *Agroforestry Systems* **15** 51-63

Rao MR, Palada MC and Brian N Becker 2004 Medicinal and aromatic plants in agroforestry systems, *Agroforestry Systems* **61** 107-122

Rao MR, Sharma MM and Ong CK 1990 A study of potential hedgerow intercropping in semi-arid India using a two-way systematic design, *Agroforestry Systems* **11** 243-256

Rao MR, Sharma MM and Ong CK 1991 A tree /crop interface design and its use for evaluating the potential of hedgerow intercropping, *Agroforestry Systems* **13** 143-158

Rao MR, Singh MP and Day R 2000 Insect pest problems in tropical agroforestry systems: contributory factors and strategies for management, *Agroforestry Systems* **50**(3) 243-277

Rao SN 2001 Marketing issues and incentives framework for agroforestry and commercial plantations, In *Marketing Monitoring of Tree Products* (eds) GS Rawat and NS Bisht, Dehradun, 71-75 pages

Ratha Krishnan P, Tewari JC and Bhati TK 2012 A success story and SWOT analysis on increased gum production from *Acacia senegal* using CAZRI gum inducer, Abstract: National symposium on "Managing stress in Dry lands under Climate Change Scenarios" at CAZRI, Jodhpur on December 1-2, 2012, 322-324 pages

Rathore AC, Saroj PL, Sharma NK, Shrimali SS, Jayaprakash J, Chaturvedi OP and Dadhwal KS 2012 Mango based agri-horticultural system for degraded lands of north western Himalaya, Technology Brochure, SWCRTI, Dehardun, 10 pages

Rathore JS 2004 Drought and household coping strategies: A case of Rajasthan, *Indian Journal of Agricultural Economics* **59**(4) 689-708

Ravi R, Divya MP and Mohan Raj T 2008 Allelopathic effect of *Ailanthus excelsa* on germination and growth of fodder crops under pot culture experiment, Paper presented in the National workshop cum symposium on "Tree Spice cultivation and sustainable development of homegardens for descent livelihood and environment protection", Port Blair, Andaman and Nicobar Islands, 4-6 February, 2008

Ravindranath NH and Hall DO 1995 Biomass, Energy and Environment: A Developing Country Perspective from India, Oxford University Press, New York, 376 pages

Ravitchandirane V and Haripriya K 2011 Intercropping with medicinal plants in mango cv Alphanso, *Plant Archives* **11**(1) 413-416

Read H 2006 A brief review of pollards and pollarding in Europe, In *1er colloque europeen sur led trognes*, 26-28 October, Vendome

Rehfuess E, Mehta S, Pruss-Ustan A 2006 Assessing household solid fuel use: Multiple Implications for the Millennium Development Goals, Environmental Health Perspectives, Volume 114, Number 3

Reisner Y, de Filippi R, Herzog F and Palma J 2007 Target regions for silvorable agroforestry in Europe, *Ecological Engineering* **29** 401-418

Richards FJ 1959 A flexible growth function for empirical use, *Journal of Experimental Botany* **10** 290-300

Richards FJ 1969 The quantitative analysis of growth, In *Plant Physiology* (ed) FC Steward, 3-76 pages

Rico-Gray V, Chemas A and Mandujano S 1991 Use of tropical deciduous forest species by the Yucatan Maya, *Agroforestry Systems* **14** 149-161

Rigueiro-Rodriguez A, Fernandez-Nunez E, Gonzalez-Hernandez P, McAdam JH and Mosquera-Losada MR 2009 Agroforestry systems in Europe: Productive, ecological and social perspectives; Current Status and Future Prospects, Springer

Rigueiro-Rodríguez A, McAdam JH and Mosquera Losada MR (eds) 2008 Agroforestry in Europe, Advances in Agroforestry 6, Springer, Dordrecht, The Netherlands, 452 pages

RIRDC (Rural Industries Research and Development Corporation) 2008 Regional opportunities for Agroforestry Systems in Australia, RIRDC Publication No. 08/176 (Project No. CSF-68A)

Rishbeth J 1963 Stump protection against *Fomes annosus* & Inoculation with *Peniophora gigantean, Ann. Appl. Biol.* **52** 63-77

Rocheleau D, Weber F and Field-Juma A 1988 Agroforestry in Dryland Africa, ICRAF, Nairobi, Kenya

Rocheleau DE 1985 Criteria for Re-appraisal and Re-design: Intra-household and Between House-hold Aspects of FSRE in three Kenyan Agroforestry Projects, ICRAF Working Paper 37, ICRAF, Nairobi, Kenya

Rodriguez-Kabana R 1992 Cropping systems for the management of phytonematodes, In *Nematology from Molecules to Ecosystem* (eds) FJ Grommers and PWT Maas, Proceedings of the Second International Nematology Congress, 11-17 August 1990, Veldhcfen, European Society of Nematologists, Dundee, UK, 219-233 pages

Roger JH and Rao MR 1990 Agroforestry field experiments: Discovering the hard facts part I, statistical considerations, *Agroforestry Today* **2**(1) 4-7

Rohilla PP, Rao SS and Jangid BL 2009 Research highlights of CAZRI- RRS, Pali (1959-2009), RRS, CAZRI, Pali, 56 pages

Roonwal ML 1990 Field-ecological observations on the *Ailanthus excelsa* defoliator *Eligma narcissus indica* (Lepidoptera: Noctuidae) in Peninsular India, *Indian Journal Forestry* **13(2)** 81-84

Roothaert RL 2000 The potential of indigenous and naturalized fodder trees and shrubs for intensive use in central Kenya, Doctoral thesis, Wageningen University, The Netherlands

Rosenberg NJ, Blad BL and Verma DB 1983 Microclimate: The Biological Environment, 3rd Edition, Wiley & Sons, New York, USA

Roshetko JM and Gutteridge RC (eds) 1996 Nitrogen fixing trees for fodder production - A Field Manual, Morrilton, Arkansas USA: Forest, Farm, and Community Tree Network (FACT Net), Winrock International- Institute for Agricultural Development

Roshetko M, Delaney M, Hairiah K and Purnomosidhi P 2002 Carbon stocks in Indonesian homegarden systems: Can smallholder systems be targeted for increased carbon storage? *American Journal of Alternative Agriculture* **17** 125-137

Ruark G, Schoeneberger MM and Nair PKR 2003 Agroforestry-helping to achieve sustainable forest management, In: The role of planted forests in sustainable forest management: Reports and Papers of the UNFF Intersessional Expert Meeting (25-27 March), 240-253

Rudebjer, Per G, Temu AB and James Kung'u 2005 Developing Agroforestry Curricula: A Practical Guide for Academic Institutions in Africa and Asia, 45 pages

Rugalema GH, Johnsen FH and Rugambisa J 1994 The homegarden agroforest system to Bukoba district, North-Western Tanzania - 2. Constraints to farm productivity, *Agroforestry Systems* **26** 205-214

Ruthenberg H 1980 *Farming Systems in the Tropics*, 2nd Edition, Oxford University Press, London, UK

SAC (Space Application Centre) 2007 Desertification and Land Degradation Atlas of India, Ahmedabad: Space Application Centre, Indian Space Research Organization

Saha SK, Nair PKR, Nair VD and Kumar BM 2010 Carbon storage in relation to soil size-fractions under tropical tree-based land-use systems, *Plant and Soil* **328**(1) 433-446

Saint Pierre C 1991 Evolution of agroforestry in the Xishuangbanna region of tropical China, *Agroforestry Systems* **13**(2) 159-176

Salick J 1989 Ecological basis of Amuesha agriculture, Peruvian Upper Amazon, *Advances in Economic Botany* **7** 189-212

Salick J and Lundberg M 1990 Variation and change in Amuesha agriculture, Peruvian Upper Amazon, *Advances in Economic Botany* **8** 199-223

Samra JS, Kareemulla K, Marwaha PS and Gena HC 2005 Agroforestry and Livelihood Promotion by Co-operatives, National Research Centre for Agroforestry, Jhansi, India, 104 pages

Samra JS, PratapNarain and Dadhwal KS 1996 Rehabilitation of degraded forest through integrated land use planning with soil and water conservation measures, *Indian Journal of Soil Conservation* **24**(1) 10-18

Samra JS, sahoo N, chowdhury SR, Mohanty RK, Jena SK and Verma HN 2003 Sustainable Integrated Farming System for Waterlogged Areas of Eastern India, ICAR-Research Bulletin No.14, Water Technology Centre for Eastern Region, Bhubaneswar, Orissa, India

Samra JS, Singh and Shiv Charan 2000 Silvipasture systems for soil, water and nutrient conservation on degraded lands of Shiwalik foot hills (subtropical northern India), *Indian Journal of Soil Conservation* **28**(1) 35-42

Samra JS, Vishwanatham MK and Sharma AR 1999 Biomass production of trees and grasses in a silvipasture systems on marginal lands of Doon Valley of north-west India & Performance of grass species, *Agroforestry Systems* **46** 197-212

Samy PR and Ignacimuthu S 1998 *Journal of Ethnopharmacology* **62** 173

Samy PR and Ignacimuthu S 2000 *Journal of Ethnopharmacology* **69** 63

Sanchez and Palm CA 1996 Nutrient cycling and agroforestry in Africa, In *Forest Influences*, *Unasylva* No. 185, An international journal of forestry and forest industries - Vol. 47 - 1996/2, Food and Agriculture Organization of the United Nations

Sanchez P 1995 Science in agroforestry, *Agroforestry Systems* **30** 5-55

Sanchez PA 1976 Properties and Management of Soils in the Tropics, John Wiley, New York, USA

Sanchez PA 1987 Soil productivity and sustainability in agroforestry systems, In *Agroforestry: A Decade of Development* (eds) HA Steppler and PKR Nair, ICRAF, Nairobi, Kenya 205-226 pages

Sanchez PA 1995 Science in agroforestry, *Agroforestry Systems* **30** 5-55

Sanchez PA 2000 Linking climate change research with food security and poverty reduction in the tropics, *Agriculture, Ecosystems and Environment* **82** 371-383

Sanchez PA and Salinas JG 1981 Low-input technology for managing Oxisols and Ultisols in tropical America, *Advances in Agronomy* **34** 279-406

Sanchez PA, Izac AMN, Valencia I and Pieri C 1995 Soil fertility replenishment in Africa: a concept note, ICRAF, Nairobi, Kenya

Sand BK and Baruah RK 2000 Pani-Kachu-A special taro cultivation in southern Assam, *Intensive Agriculture* **38** 26-27

Sandhu HS, Wratten SD and Cullen R 2010 Organic agriculture and ecosystem services, *Environmental Science & Policy* **13** 1-7

Sanginga N, Mulongoy K and Ayanaba A 1986 Inoculation of *Leucaena leucocephala* with *Rhizobium* and its nitrogen contribution to a subsequent maize crop, *Biology, Agriculture and Horticulture* **3** 341-347

Sanjeev K Chauhan, Naveen Gupta, Ritu, Sudhir Yadav and Rajni Chauhan 2009 Biomass and carbon allocation in different agroforestry tree species, *Indian Forester* July Issue 981-992

Sara J. Scherr and Satya Yadav 1996 Land Degradation in the Developing World: Implications for Food, Agriculture and the Environment to 2020, IFPRI Food Agriculture and the Environment Discussion Paper 14, Washington DC: International Food Policy Research Institute

Sargos J 1997 Histoire de la Foret Landaise, Deuxieme edition, L'Horizon chimerique, Bordeaux, France

Satapathy KK 2003 Soil and water conservation measures for sustainable hill agriculture, In *Approaches for Increasing Agricultural Productivity in Hill and Mountain Ecosystem* (eds) BP Bhatt, KM Bujarbaruah, YP Sharma and Patiram, ICAR Research Complex for NEH Region, Umiam, Meghalaya, 3-13 pages

Sathaye JA and Ravindranath NH 1998 Climate change mitigation in the energy and forestry sectors of developing countries, *Annals of Review of Energy Environment* **23** 387-437

Satheesh N and Mohanraj T 2012 Application of Remote Sensing and GIS in Forestry, In *Forestry: Principles and Applications* (eds) Antony Joseph Raj and SB Lal, Scientific Publishers, India, 420-433 pages

Satya Vir and Verma SK 1996 Insect pests of agroforestry leguminous trees in India, *Annals of Arid Zone* **35** 349-359

Saxena KG, Maikhuri RK and Rao KS 2005 Changes in Agricultural Biodiversity: Implications for Sustainable Livelihood in the Himalaya, *Journal of Mountain Science* **2**(1) 23-31

Saxena S, Chandak V, Ghosh SB, Sinha R, Jain N and Gupta AK 2002 Cost of conservation of agrobiodiversity, NBPGR and Indian Institute of Management, New Delhi, India

Scherr SJ 1988a Agroforestry monitoring and evaluation workshop summary and pilot survey of adopted agroforestry practices in the CARE Agroforestry Extension Project, ICRAF Agroforestry Development in Kenya Project Report No.6, ICRAF, Nairobi

Scherr SJ 1988b Current agroforestry practices and extension recommendations of the CARE Agroforestry Project, ICRAF Agroforestry Development in Kenya Project Report No.8, ICRAF, Nairobi

Scherr SJ 1989 Choosing priorities for agroforestry research, In *Social Science Perspectives in Managing Agricultural Technology* (eds) D Groenefeldt and JL Moock, International Irrigation Management Institute, Colombo, Sri Lanka

Scherr SJ 1990 Methods for on-farm agroforestry research: constraints, challenges and opportunities, Paper presented at International Workshop on Methods for Participatory On-farm Agroforestry Research, 19-23 February 1990, ICRAF, Nairobi, Kenya

Schippmann U, Leaman JD and Cunningham AB 2002 Impact of cultivation and gathering of medicinal plants on biodiversity: Global trends and issues, Inter-Departmental Working Group on Biological Diversity for Food and Agriculture, Rome

Schoeneberger MM 2005 Agroforestry - Working Trees for Sequestering carbon on Ag-Lands, In *Moving Agroforestry into Mainstream* (eds) KN Brooks and PF Ffolliot, Proc. 9th North American Agroforestry Conference, Rochester, MN, Dept. of Forest Resources, University of Minnesota

Schroth G 1999 A review of belowground interactions in agroforestry: focusing on mechanisms and management options, In Agroforestry for Sustainable Land Use, Fundamental research and modelling with emphasis on temperate and mediterranean applications (eds) D Auclair and C Dupraz, Kluwer Academic Publishers

Schroth G and Sinclair F 2003 Trees crops and soil fertility: concepts and research methods, CABI, Wallingford, UK, 464 pages

Schroth G, Da Fonseca, Gab, Harvey CA, Gascon C, Vasconcelos H and Izac AN 2004 Agroforestry and biodiversity conservation in tropical landscapes, Island Press

Schroth G, Harvey C and Vincent G 2004 Complex agroforest, In *Agroforestry and Biodiversity conservation in Tropical Landscapes* (eds) G Schroth, GAB da Fonseca, CA Harvey, C Gascon, HL Vasconcelos and AMN Izac, Island Press, Washington DC, 227-260 pages

Schroth G, Krauss U, Gasparotto L, Duarte Aguilar JA and Vohland K 2000 Pest and diseases in agroforestry systems of the humid tropics, *Agroforestry Systems* **50**(3) 199-241

Schroth G, Lehmann J, Rodrigues, Barros M and Macêdo J 2001 Plant-soil interactions in multistrata agroforestry in the humid tropics, *Agroforestry Systems* **53** 85-102

Schultes RE and Hofmann A 1979 Plants of the Gods: Origins of Hallucinogenic Use, New York: McGraw-Hill

Schultz RC, Isenhart TM, Simpkins WW and Colletti JP 2004 Riparian forest buffers in agro ecosystems: lessons learned from the Bear Creek Watershed, central Iowa, USA, *Agroforestry Systems* **61** 35-50

Schulze ED, Kuppers M and Matyssek R 1986 The roles of carbon balance and branching pattern in the growth of woody species, In *On the Economy of Plant Form and Functions* (ed) TJ Givnish, Cambridge University Press, Cambridge

Seiter S, Ingham ER and William RD 1999a Dynamics of soil fungal and bacterial biomass in a temperate climate alley cropping system, *Applied Soil Ecology* **12** 139-147

Seiter S, Ingham ER, William RD and Hibbs DE 1995 Increase in soil microbial biomass and transfer of nitrogen from alder to sweet corn in an alley cropping system, In *Growing a sustainable future* (eds) JH Ehrenreich, DL Ehrenreich and HW Lee, University of Idaho, Boise, ID, 56-158 pages

Seiter S, William RD and Hibbs DE 1999b Crop yield and tree leaf production in three planting patterns of temperate zone alley cropping in Oregon, USA, *Agroforestry Systems* **46** 273-288

Sen-Sarma PK and Thakur ML 1985 Pest management in Indian forestry, *Indian Forester* **111 (11)** 956-964

Seobi T, Anderson SH, Udawatta RP and Gantzer CJ 2005 Influence of grass and agroforestry buffer strips on soil hydraulic properties for an Albaqualf, *Soil Science Society of America Journal* **69**(3) 893-901

Seshadri P 1985 Intercropping of bamboo (D. strictus) with Soybean – An agroforestry study, Ph.D. Thesis, Tamil Nadu Agricultural University, Coimbatore

SFR 2009 The State of Forest Report 2009, Forest Survey of India, Dehradun, India

Shanavas A and Kumar BM 2003 Fuelwood characteristics of tree species in the homegardens of Kerala, India, *Agroforestry Systems* **58** 11-24

Shaner WW, Philip PE and Schmel WR 1982 Farming Systems Research and Development, Westview Press, Boulder, CO, USA

Shankar AV, Gittelsohn J, Pradhan EK, Dhungel C and West KP Jr 1998 Homegardening and access to animals in households with xerophthalmic children in rural Nepal, *Food and Nutrition Bulletin* **19** 34-41

Shankarnarayan KA, Harsh LN and Kathju S 1987 Agroforestry systems in the arid zones of India, *Agroforestry Systems* **5** 69-88

Shankarnarayan KA1984 Influence of *Prosopis cineraria* and *Acacia nilotica* on soil fertility and crop yield, In *Agroforestry in Arid and Semi-arid Zones* (ed) KA Shankarnarayan, Central Arid Zone Research Institute (CAZRI) Publication, Jo.24, Jodhpur (Rajasthan), India, 59-66 pages

Shanmughavel P and Francis K 2001 Intercropping trials of four crops in bamboo plantations *Journal of Bamboo and Rattan* **1**(1) 3-9

Shanmughavel P and Francis K 2002 *Bambusa bambos* – An afforestation trial, XII World Forestry Congress, Quebeck City, Canada, FAO

Sharad Nema 2012 Agroforestry Practices, In *Forestry: Principles and Applications* (eds) Antony J. Raj and SB Lal, Scientific Publishers, Jodhpur, India

Sharda VN, Sikka AK and Juyal GP 2006 Participatory Integrated Watershed Management-A Field Manual, Published by the Director, CSWCRTI, Saraswati Press, Dehra Dun

Sharda VN, Sikka AK and Juyal GP 2012 Participatory Integrated Watershed Management – A Field Manual, CSWCR&TI, Dehradun, 386 page

Sharma AK and Tewari JC 2005 Arid zone forestry with special reference to Indian hot arid zone, In *Forests and Forest Plants*, Encyclopaedia of Life Support Systems (EOLSS), Developed under auspices of the UNESCO, EOLSS publishers, Oxford, UK, website: (http://www.eolss.net)

Sharma AK, Dagar JC and Bandyopadhyay AK 1990 Fodder Resources of Bay Islands, Research Bulletin No. 3, Central Agricultural Research Institute, Port Blair, India

Sharma K, Thakur, Sharma R and Kashyap SD 2008 Production and economics of kinnow cultivation with wheat and gobhi sarson in Himachal Pradesh, *Indian Journal of Soil Conservation* **36**(2) 112-118

Sharma KK and Singh RK 1992 Studies on tree-crop interaction in "*Populus deltoides* G-3" bund plantation under irrigated conditions, Indian Forester **118**(2) 102-108

Sharma KK and Suman S 2001 Marketing channels of agroforestry products in Haryana, In *Marketing Monitoring of Tree Products* (eds) GS Rawat and NS Bisht, Dehradun, 102-113 pages

Sharma LK 1987 Country Report and status of research on bamboos in India, *Indian Forestry Records (New series)* **6** 28 pages

Sharma LR, Bhhati JP and Singh 1999 Emerging farming systems in Himachal Pradesh-Key issues in sustainability, *Indian Journal of Agricultural Economy* **46**(3) 422-427

Sharma M and Parton J 2007 Height–diameter equations for boreal tree species in Ontario using a mixed-effects modeling approach, *Forest Ecology and Management* **249**(3) 187-198

Sharma R, Sharma E and Purohit AN 1994 Dry matter production and nutrient cycling in agroforestry systems of cardamom grown under the *Alnus* and natural forest, *Agroforestry Systems* **27** 293-306

Shastri CM, Bhat DM, Nagaraja BC, Murali KS and Ravindranath NH 2002 Tree species diversity in a village ecosystem in Uttara Kannada district in Western Ghats, Karnataka, *Current Science* **82**(9) 1080-1084

Shaw EB 1940 Geography of mast feeding, *Economic Geography* **16**(3) 233-249

Shigeura GT and McCall WW 1979 Trees and shrubs for windbreaks in Hawaii, University of Hawaii Cooperation Extension Service, Circular No. 447 CTAHR, University of Hawaii, Honolulu, Hawaii

Shiva V 1992 Biodiversity, A Third World Perspective Penang Malaysia: Third World Network, Columbia University Press, New York

Shrivastava TN and Quershi IM 1966 Afforestation of difficult sites, Proceedings of VIth World Forestry Congress, Madrid, 3769-3774 pages

Sikniah B, Sinniah D and Rajeswari B 1983 Epidemiology and control of the human head louse in Malaysia, *Tropical and Geographical Medicine* **35** 337-342

Sileshi G, Akinnifesi FK, Ajayi OC, Chakeredza S, Kaonga M and Matakala PW 2007 Contribution of Agroforestry to ecosystem services in the Miombo eco-region of eastern and southern Africa, *African Journal of Environmental Science and Technology* **4** 68-80

Sileshi G, Maghembe JA, Rao MR, Ogol CKPO and Sithanantham S 2000 Insects feeding on *Sesbania* species in natural stands and agroforestry systems in southern Malawi, *Agroforestry Systems* **49** 41-52

Simmonds NE 1979 Principles of crop improvement, Longman, New York

Simmonds NW 1985 Perspectives on the evolutionary history of tree crops, In *Attributes of trees as crop plants* (eds) MGR Cannell and JE Jackson, Institue of Terrestrial Ecology, Midlothian, U.K.,

Simons AJ 1996 ICRAF's strategy for domestication of indigenous tree species, In *Domestication and Commercialization of Non-timber Forest products in Agroforestry Systems* (eds) RRB Leakey, AB Temu, M Melnyk and P Vantomme, *Non-Wood Forest Products*, FAO, Rome, Italay, **9** 8-22

Simons AJ and Leakey RRB 2004 Tree domestication in tropical agroforestry, *Agroforestry Systems* **61**: 167-181

Sinclair FL 1999 A general classification of agroforestry practice, *Agroforestry Systems* **46** 161-180

Singh A 1987 Studies on some aspects of soil and water in relation to resource management in North Eastern Hills Region, Ph.D. thesis, Bidhan Chandra Agricultural University, West Bengal, India

Singh BG, Rao PV and Hiremath SM 1989 Effect of three varieties of groundnut (*Arachis hypogaea* L.), *Journal of Research APAU* **17** 432-433

Singh C and Chaturvedi OP 2011 Rehabilitation of degraded lands through silvipastoral systems – a balanced approach for alternative land use in the north-western Himalayas, In *Agroforestry Systems for Resources Conservation and Livelihood Security in Lower Himalayas* (eds) P Panwar, AK Tiwari and KS Dadhwal, New India Publishing Agency, New Delhi, 273-296 pages

Singh C and Dadhwal KS 2012 *Grewia optiva* + Hybrid napier based silvipastoral system for degraded lands in north-west Himalayas, Technology Brouchere, CSWCRTI, Dehardun, 8 page

Singh DP, Singh RA, Dabbas MR and Prakash HG 2010 Livelihood security to urban and peri urban farm families through two tiers horticultural based cropping system, *International Journal of Agricultural Sciences* **6**(1) 273-274

Singh G 1990 Effects of irrigation on *Prosopis juliflora* and soil properties of an alkali soil, *International Tree Crops Journal* **6** 81-99

Singh G 2005 Carbon sequestration under an agrisilvicultural system in the arid region, *Indian Forester* **131**(4) 543-552

Singh G, Abrol IP and Cheema SS 1988 Agroforestry on alkali soil: Effect of planting methods and amendments on initial growth, biomass accumulation and chemical composition of mesquite (*Prosopis juliflora*) with inter-space planted with and without Karnal grass (*Deplachne fusca*) *Agroforestry Systems* **7**: 135-160

Singh G, Arora YK, Narain P and Grewal SS 1990 Agroforestry Research in India and Other Countries, Surya Publications, Dehradun, India, 189 page

Singh G, Kapoor IP, Pandey SK, Singh UK and Singh RK 2002 Studies on essential oils - Part 10: Antibacterial activity of volatile oils of some spices, *Phytotheraphy Research* **16** 680-682

Singh G, Ram Babu, Narain P and Abrol IP 1992 Soil erosion rates in India, *Journal of Soil and Water Conservation* **47** 97-99

Singh GB 1987 Agroforestry in Indian subcontinent: past, present and future, In *Agroforestry - a Decade of Development* (eds) HA Steppler and PKR Nair, ICRAF, Nairobi, 117-138 pages

Singh GS, Rao KS and Saxena KG 1997 Energy and economic efficiency of the mountain farming system: a case study in the north-western Himalaya, *Journal of Sustainable Agriculture* **9** 25-49

Singh JS and Singh SP 1992 Forests of Himalaya, Gyanodaya Prakashan, Nainital, 294 pages

Singh JS et al 1986 In Eco-Development Guidelines and Model of Development of the Central Himalaya, Department of Botany, Kumaun University, Nainital, 48 pages

Singh K, Rajeswara Rao BR, Singh CP, Bhattacharya AK and Kaul PN 1998 Production potential of aromatic crops in the alleys of *Eucalyptus citriodora* in semi-arid tropical climate of south India, *Journal of Medicinal and Aromatic Plant Science* 20 749-752

Singh K, Singh V, Hussain A and Kothari SK 1990 Aromatic plants as efficient intercrops under poplars (*Populus deltoids* Bartram ex Marshall), *Indian Forester* **116** 189-193

Singh KA and Pradhan IP 1989 Annual Progress Report of AICRIP on Agroforestry, Technical Bulletin, ICAR Research Complex For NEH Region, Sikkim Centre, Tadong, Sikkim, India

Singh KA, Rai RN, Patiram and Bhutia DT 1989 Large cardamom (*Amormun subulatum* Roxb.) plantation – an age old agroforestry system in eastern Himalayas, *Agroforestry Systems* **9** 241-257

Singh KS and Lal P 1969 Effect of *Prosopis cineraria* and *Acacia arabica* trees on soil fertility and profile characteristics, *Annals of Arid Zone* **8** 33-36

Singh MP, Dayal N and Singh BS 1994 Importance of genetic conservation of tasar host plants in agroforestry programme in Chhotanagpur region of Bihar, *Journal of Palynology* **30** 157-163

Singh P and Bhandari RS 1988 Insect pests of bamboo and their control, *Indian Forester* **114(10)** 670-683

Singh Rathore MP 1995 Insect Pests in Agroforestry, Working Paper No.70, ICRAF, Nairobi, Kenya

Singh RP, Ong CK and Saharan N 1989a Above and below ground interactions in alley cropping in semi-arid India, *Agroforestry Systems* **9** 259-274

Singh RP, Van den Beldt RJ, Hocking D and Korwar GR 1989b Alley Farming in the semiarid regions of India, In *Alley Farming in the Humid and Sub-humid Tropics* (eds) BT Kang and L Reynolds, International Development Research Centre, Ottawa, Canada, 108-122 pages

Singh S and Singh RP 1995 Wastelands and their rehabilitation in India, In *Sustainable development of dryland agriculture in India* (Ed) RP Singh, Scientific Publishers, Jodhpur, 473-496 pages

Singh VK, Ali ZA, Zaidi STH and Siddiqui MK 1996 Ethno-medicinal uses of plants from Gonda district forests of Uttar Pradesh, India, *Fitoterapia* **67** 129-139

Singhal RM and Panwar BPS 1992 Commercial approach to poplar (*Populus deltoides*) based agroforestry system in North- Western Uttar Pradesh, *Van Vigyan* **30**(1) 29-38

Sinniah D, Varghese G, Bhaskaran G and Koo SH 1983 Fungal flora of neem (Azadirachta indica) seeds and neem oil toxicity, *Malaysian Application Biology* 12:1

Siregar AM 1990 Implementation of Agroforestry in Indonesia, In *Agroforestry Systems and Technologies*, BIOTROP Special Publication No. 39. South-east Asian Regional Centre for Tropical Biology, Bogor, Indonesia, 29-36 pages

Sivaprakash M, Manivasakan S and Abhishek Kumar 2013 Alternate Industrial wood species, In *Industrial Agroforestry-Perspectives and Prospective* (eds) P Durairasu, KT Parthiban, R Umarani, S Umesh Kanna, I Sekar, P Rajendran and PS Devanand, Forest College and Research Institute, Tamil Nadu Agricultural University, Mettupalayam, 123-146 pages

Smaling E 1993 An agro-ecological framework for integrating nutrient management, with special reference to Kenya, Ph.D. Thesis, Agricultural University of Wageningen, the Netherlands, 250 page

Smit B and Skinner MW 2002 Adaptation options in agriculture to climate change: A typology, *Mitigation and Adaptation Strategies for Global Change* **7**(1) 85-114

Smith J 2010 The History of Temperate Agroforestry, The Organic Research Centre, Elm Farm, Berkshire

Smith NJH, Williams JT, Placknett DL and Talbot JP 1992 Tropical Forests and their Crops, Comstock Publishing Associates, London

Smith TM, Cramer WP, Dixon RK, Leemans R, Neilson RP and Solomon AM 1993 The global terrestrial carbon cycle, *Water Air Soil Pollution* **70** 19-38

Soares P, Tome M, Skovsgaard and Vanlacy 1995 Evaluating a growth model for forest management using continuous forest inventory data, *Forest Ecology and Management* **71** 251-265

Soderstrom TR and Ellis RP 1987 The woody bamboos a morphological and anatomical study, Smithsonian Contribution, No.72.

Soemarwoto O 1987 Homegardens: A traditional agroforestry system with a promising future, In *Agroforestry: A Decade of Development* (eds) HA Steppler and PKR Nair, ICRAF, Nairobi, 157-170 pages

Somarriba E 1992 Revisiting the past: an essay on agroforestry definition, *Agroforestry Systems* **19** 233-240

Sood KK, Singh B, Sehgal S and Raina NS 2010 Agroforestry Systems in Indian Himalayan Region, In *Himalayan People and Environment* (ed.) S Arora, Soil Society of India, New Delhi, 287-306 pages

Spaulding P 1961 Foreign diseases of forest trees of the world, U.S. Department of Agriculture Handbook 197

Spears J 1987 Agroforestry: A development-bank perspective, In *Agroforestry: A Decade of Development* (eds) HA Steppler and PKR Nair, ICRAF, Nairobi, Kenya, 53-66 pages

Spencer JE 1966 Shifting Cultivation in Southeast Asia, University of California Press, Berkeley, CA, USA

Stamps WT and Linit MJ 1997 Plant diversity and arthropod communities: implications for temperate agroforestry, *Agroforestry Systems* **39**(1) 73-89

Stebbing EP 1914 Indian Forest Insects, (reprinted 1989), Bisen Singh and Mohindra, Dehra Dun

Steppler HA and Nair PKR (eds) 1987 Agroforestry: A Decade of Development, ICRAF (World Agroforestry Centre), Nairobi, Kenya, 335 pages

Stirzaker R and Lefroy EC 1997 Alley Farming in Australia: Current research and future directions, A report for the Rural Industries Research and Development Corporation, CSIRO Centre for Environmental Mechanics, Australia

Stirzaker RJ, Cook FJ and Knight J 1999 Where to plant trees on cropland for control of dryland salinity: some approximate solution, *Journal of Agriculture and Water Management* **39** 115-133

Stockdale EA, Shepherd MA, Fortune S and Cuttle SP 2002 Soil fertility in organic farming systems-Fundamentally different, Soil Use and Management 18 301-308

Story R 1967 Pasture patterns and associated soil water in partially cleared woodland, *Australian Journal of Botany* **15** 175-87

Streiffeler F 1987 Improving urban agriculture in Africa: a social perspective, *Food Nutrition Bulletin* **9**(2) 8-13

Subler S and Uhl C 1990 Japanese agroforestry in Amazonia: a case study in Tome Acu, Brazil, In *Alternatives to Deforestation: Steps Toward Sustainable Use of the Amazon Rain Forest* (ed) AB Anderson, New York: Columbia University Press

Subrahmanyam MVR 2009 Agroforestry - An Alternate Land Use Option for Sustainable Dryland Agriculture, In Winter School on Alternate land use options for resource conservation, emerging market needs and mitigation of climate change in rainfed regions, CRIDA, Hyderabad, AP, India

Sunderland TCH, Tchoundjeu Z and Ngo-Mpeck ML 2000 The exploitation of *Pausinystalia johimbe*, *Medicinal Plant Conservation* **6** 22-23

Sundriyal M and Sundriyal RC 2001 Wild edible plants of the Sikkim Himalaya: Nutritive values of selected species, *Economic Botany* 55 377-390

Suresh KK and Rai RSV 1987 Studies on allelopathic effects of some agroforestry tree crops, *International Tree Crops Journal* **4** 109-115

Swaminathan C 2001 Sustainable tree mixtures: Optimum species combination for a tropical alfisol of southern India, *Biology, Agriculture and Horticulture* **18** 259-268

Swaminathan C, Rai RSV and Suresh KK 1989 Allelopathic proclivities of *Acacia nilotica, Journal of Tropical Forest Science* 2(1) 56-60

Swamy SL, Puri S and Singh AK 2003 Growth, biomass, carbon storage and nutrient distribution in Gmelina arborea stands on red lateritic soils in central India, *Bio-resource Technology* **90** 109-126

Swarup R and Gambhir A 2008 Mass production, certification and field evaluation of bamboo planting stock produced by tissue culture, In *Souvenir of International conference on Improvement of Bamboo productivity and Marketing for Sustainable Livelihood* (eds) ML Choudhary M Pal and K Salam, April 15-17, 2008, New Delhi, India, 1-4 pages

Swift MJ and Anderson JM 1993 Biodiversity and ecosystem function in agricultural systems, In *Biodiversity and Ecosystem Function* (eds) ED Schulze and HA Mooney, Berlin: Springer-Verlag, 15-41 pages

Swinton SM, Lupi F, Robertson GP and Hamilton SK 2007 Ecosystem services and agriculture: Cultivating agricultural ecosystems for diverse benefits, *Ecological Economics* **64** 245-252

Szott LT, Palm CA and Sanchez PA 1991 Agroforestry in acid soils of the humid tropics, *Advances in Agronomy* **45** 275-301

Takimoto A, Nair PKR and Nair VD 2008 Carbon stock and sequestration potential of traditional and improved agroforestry systems in the West African Sahel, *Agriculture, Ecosystems and Environment* **125** (1-4) 159-166

Tamang B, Andreu MG and Rockwood LD 2010 Microclimate patterns on the leeside of single-row tree windbreaks during different weather conditions in Florida farms: implications for improved crop production, *Agroforestry Systems* **79** 111-122

Tan A and Tan AS 1998 Data Collection and Analysis in Turkey – GEF Project on *In Situ* Conservation of Genetic Diversity, In Proceedings of a *Workshop to develop tools and procedures for in situ conservation on farm* (eds) DI Jarvis and T Hodgkins, 31 pages

Tans P 2012 Concentrations of CO_2 in the earth's atmosphere (parts per million) derived from in situ air measurements at the Mauna Loa observatory, Hawaii

Tapia-Coral SC, Luizão FJ and Wandelli EV 1999 Macrofauna da liteira em sistemas agro-florestais sobre pastagens abandonadas na Amazônia Central, *Acta Amazônica* 29(3) 447-495

Taylor CJ 1960 Synecology and silviculture in Ghana, Edinburgh, Thomas Nelson & Sons Ltd (for University College of Ghana), 418 page

Tchoundjeu Z, Duguma B, Fondoun JM and Kengue J 1998 Strategy for domestication of indigenous fruit trees of West Africa: case of *Irvingia gabonensis* in southern Cameroon, *Cameroon Journal of Biology and Biochemical Sciences* **4** 21-28

Teel WS and Buck LE 2002 Between wildcrafting and monocultures: agroforestry options, In *Non-timber forest products in the United States* (eds.) ET Jones, RJ McLain and J Weigand, University of Kansas Press, Lawrence, KS, 199-222

Tejwani KG 1987 Agroforestry practices and research in India, In *Agroforestry: Realities, Possibilities and Potentials* (ed) HL Gholz, Martinus Nijhoff Publishers, Dordrecht, The Netherlands, 109-136 pages

Tejwani KG 1994 Agroforestry in India, Oxford & IBH Publication Co. Pvt. Ltd, New Delhi, 233pages

Tejwani KG, Dhruvanarayan VV and Satyanarayana T 1996 Control of gully erosion in ravine lands of Gujarat, *Journal of Soil and Water Conservation* (India) **8**: 65-74

Temu AB, Rudebjer PG, Kiyiapi J and van Lierop P 2005 Forestry education in sub-Saharan Africa and Southeast Asia: Trends, myths and realities, Rome: FAO

Tewari DD 2008 Management of Non-Timber Forest product Resources of India - An analysis of forest development corporations, International Book distribution Co., Lucknow, India, 152 pages

Tewari JC, Harsh LN and Venkateswarlu J 1989 Agroforestry research in arid regions: a review, In *Agroforestry Systems in India – Research and Development* (eds) RP Singh, IPS Ahlawat and G Saran, Indian Society of Agronomy, New Delhi, 3-17 pages

Thakur ML Sushikumar and Negi A 1989 Chemical control of termites in *Eucalyptus* hybrids, *Indian Forester* **115(10)** 733-743

Thaman RR 1976 Urban gardening in Papua New Guinea and Fiji, In *The Melanesian environmentt* (ed) JM Winslow, Canberra, Australian National University Press

Thaman RR 1977 Urban root crop production in the Southwest Pacific, In *Collected papers: Regional Meeting on Production of Root Crops* (ed) M Lambert, 24-29 October 1975, Suva, Fiji, Technical paper No. 174, Noumea, South Pacific Commission

Thaman RR 1983 Food for urbanizing Polynesian peoples, *Proceedings of the Nutrition Society of New Zealand* **8** 26-37

Thaman RR 1984 Urban agriculture and home gardening in Fiji: a direct road to development and independence, In *Transactions and proceedings of the Fiji Society for the years 1978 to 1980, Vol.* **14** 1-28

Thaman RR 1985 Mixed homegardening in the Pacific Islands: regional paper, In *Proceedings of First International Workshop on Tropical Homegardens*, Bandung, 2-9 December, Bandung, Institute of Ecology, Padjadjaran University

Thaman RR 1987 Urban Agroforestry: The Pacific islands and beyond, In *Urban forestry: Cities, trees and people, Unasylva* No.155, 39(55) 2-13

Thansanga R 1997 New contour farming systems, *Indian Journal of Soil Conservation* **25**(2) 167-169

Thevathasan NV and Gordon AM 2004 Ecology of tree intercropping systems in the North temperate region: Experiences from southern Ontario, Canada, *Agroforestry Systems* **61** 257-268

Thies C and Tscharntke T 2000 Landscape structure and biological control in agroecosystems, *Science* **285** 893-895

Thomas TH 1991 A spreadsheet approach to the evaluation of the economics of temperate agroforestry, *Forest Ecology and Management* **45** 207-235

Thorp J and Smith GD 1949 Higher Categories of Soil Classification into Orders, Suborders and Great Soil Groups, *Soil Science* **67** 117-126

Thrupp LA 1997 Linking biodiversity and agriculture: Challenges and opportunities for sustainable food security, World Resources Institute, USA

Thrupp LA 2000 Linking agricultural biodiversity and food security: The valuable role of agrobiodiversity for sustainable agriculture, *International Affairs* **76**(2) 283-297

Tilman D 1990 Mechanisms for plant competitions for nutrients: the element of predictive theory of competition, In *Perspectives on Plant Competition* (eds) JB Grace and D Toilman, Academic press, New York, USA.

TNAU (Tamil Nadu Agricultural University) 1984 Annual Progress Report of All-India Coordinated Research Project on Dryland Agriculture for North-East Monsoon 1983-84, Kovilpatti Experimental Station, TNAU, Coimbatore, India, 57-59 pages

Toky OP 1997 Poplar an economy booster and eco-friendly agroforestry tree, *Agroforesry News Letter* **9** 2-3

Toky OP, Bist RP and Singh RR 1992 Potential multipurpose trees of Indian forests vegetation: *Parkia javanica*, The Wood Explorer Database, 359-378 pages

Toky OP, Kumar P and Khosla PK 1989 Structure and function of traditional agroforestry systems in the Western Himalaya, *Agroforestry Systems* **9**: 47-70

Tolera M, Asfaw Z, Lemenih M and Karltun E 2008 Woody species diversity in a changing landscape in the south-central highlands of Ethiopia, *Agriculture, Ecosystems and Environment* **128** 52-58

Tomar GS and Shrivastava SK 1986 Preliminary studies of rice cultivation in association with trees, In *Agroforestry Systems - A new challenge* (eds) PK Khosla, S Puri and DK Khurana, ISTS, Solan India, 207-211 pages

Tomar JMS 2006 Ecological interactions between trees and crop components in selected agro-forestry systems of Meghalaya, Ph.D. thesis, North Eastern Hill University, Shillong, India

Tomar JMS and Bhatt BP 2004 Studies on horti-agricultural system in a mid altitude of Meghalaya, *Indian Journal of Agroforestry* **6**(2) 35-39

Tomar JMS, Rathore AC, Kaushal R, Chaturvedi OP and Dadhwal KS 2012 Productive utilization of boulder riverbed lands by cultivation of aromatic grasses, Technical brochure, CSWCRTI, Dehradun, 7 page

Tomar JMS, Satapathy KK and Dhyani SK 2002 Integrated approach of remote sensing/GIS in characterization and evaluation of natural resources for watershed management in Upper shipra watershed, *Indian Journal of Soil and Conservation* **30**(3) 206-213

Tomar JMS, Upadhaya K, Tripathi OP and Pandey HN 2007 Agroforestry Systems and Practices Prevailing in Meghalaya, In *Agroforestry Systems and Practices* (eds) Puri and Panwar, New India Publishing Agency, Pitam Pura, New Delhi, 357-366 pages

Tomar OS and Gupta RK 1985 Performance of some forest-tree species in saline soils under shallow and saline water table conditions, *Plant and Soil* **87**: 329-335

Torquebiau E 1989 Report on the International training Course "Agroforestry Research for Development", May 1989, ICRAF Training and Education reports, Nairobi

Torquebiau E 1992 Are tropical agroforestry homegardens sustainable? *Agriculture, Ecosystems and Environment* **41** 189-207

Tripathi P 2012 Effect of organic manures on yield and biomass production of medicinal and aromatic plants under peach based agroforestry system, Ph.D. Thesis, Dr. Y S Parmar UHF, Nauni, Solan (H.P.) India

Tripathi RS and Barik SK 2003 Shifting cultivation in North East India, In *Proceedings of Approaches for Increasing Agricultural Productivity in Hill and Mountain Ecosystem* (eds) BP Bhatt, KM Bujarbaruah, YP Sharma and Patiram, ICAR Research Complex for NEH Region, Umiam, Meghalaya, 317-322 pages

Tripathi SS, Saini BC, Singh RV and Sharma P 2001 Influence of shisham (*Delbergia sissoo*) on weed management and production potential of wheat under agroforestry system, *Indian Journal of Weed Science* **33**(3&4) 177-179

Tscharntke T, Christoph Leuschner, Edzo Veldkamp, Heiko Faust, Edi Guhardja, Arifuddin Bidin (eds) 2010 Tropical Rainforests and Agroforests under Global Change, In *Environmental Science and Engineering*, Springer-Verlag Berlin, Heidelberg, 461-476

Turnbull JW 2002 Tree Domestication and the History of Plantations, In (ed) VR Squires, The Role of Food, Agricultre, Forestry and Fisheries in Human Nutrition Vol-II

Tyndall J and Colletti J 2007 Mitigating swine odour with strategically designed shelterbelt systems: a review, *Agroforestry Systems* **69** 45-65

Tyndall JC and Wallace DC 2011 Windbreaks: a "fresh" tool to mitigate odors from livestock production facilities, *Agroforestry Notes* **41** 4

Udawatta RP, Gantzer CJ, Anderson SH and Garrett HE 2008 Agroforestry and grass buffer effects on pore characteristics measured by high-resolution X-ray computed tomography, *Journal of Soil Science Society of America* **72** 295-304

Udawatta RP, Krstansky JJ, Henderson GS and Garrett HE 2002 Agroforestry practices, runoff, and nutrient loss: a paired watershed comparison, *Environment Quality* **31** 1214-1225

UMCA 2013 University of Missouri Center for Agroforestry webpage: http://www.centerforagroforestry.org /practices/

UNDP 2003 Human Development Report, United Nations Development Programme, New York

UNEP 1992 Globai biodiversity strategy: guidelines for action to save study and use

UNEP 2006 People and forest: forest destruction fuels east Africa's drought, United Nations Environmental Programme, website: http://www.peopleandplanet.net/doc.php?id=2637

UNEP 2011 Towards a green economy: pathways to sustainable development and poverty eradication, Nairobi, website:www.unep.org/greeneconomy/portals/88/documents/ger/ger_final_dec_2011/ green %20 economyreport_final_dec2011.pdf

UNFCCC 2007 Report of the Conference of Parties on its Thirteenth Session, Bali, Indonesia, United Nations Framework Convention on Climate Change, Geneva

Unger PW, Jones OR and Steiner JL 1988 Principles of crop and soil management procedures for maximising production per unit rainfall, In *Drought Research Priorities for the Dryland Tropics* (eds) FR Bidinger and C Johansen, ICRISAT, India, 97-112 pages

United Nations 2004 Agenda 21, website: www.un.org/esa/sustdev/documents/agenda21/ index.htm

United Nations 2008 World urbanization prospects: the 2007revision, CD-ROM edition, New York, NY: United Nations Department of Economic and Social Affairs, Population Division

USDA-NAC (USDA National Agroforestry Centre) 1999 Alley cropping: An agroforestry practice, AF Note-12, *Agroforestry Notes*, USDA

Vaartaja O 1952 Forest humus quality and light conditions as factors influencing damping off, *Phytopathology* **42** 501-506

Valladares F 1999 Architecture, ecology and evolution of plant crowns, In Handbook of Functional Plant Ecology (eds) FI Pugnaire and F Valladares, Marcel Dekker Inc., New York, 121-194 pages

Valladares F and Niinemets U 2007 The architecture of plant crowns: from design rules to light capture and performance, In *Functional Plant Ecology* (eds) F Pugnaire and F Valladares, Taylor and Francis, New York

van den Bosch R and Telford AD 1964 Environmental modification and biological control, In *Biological Control of Insect Pests and Weeds* (ed) P DeBach, London: Chapman and Hall, 459-488 pages

Van Noordwijk M Spek LY and de Willigen P 1994 Proximal root diameters as predicators of total root system size for fractal branching models, 1.Theory, *Plant and Soil* **164** 101-117

van Noordwijk M, Lawson G, Soumare, Groot JJR and Hariah K 1996 Root distribution of trees and crops: competition and/or complementarity, In *Tree Crop Interactions - A Physiological approach* (eds) CK Ong and P Huxley, CAB International, Wallingford, Oxford, UK, 319-364 pages

van Noordwijk M, Pfund JL, Bakarr M, Jackson L, Schroth G, Hairiah K and Joshi L 2006 Agro-biodiversity and CGIAR tree and forest science: approaches and examples from Sumatra, In Proceedings of Biodiversity Conservation in Agriculture, May 31-June 2, 2006, Dominican Republic

Vandenbeldt RJ 1990 Agroforestry in the semiarid tropics, In *Agroforestry: Classification and Management* (eds) KG MacDicken and NT Vergara, John Wiley, New York, USA, 150-194 pages

Vandenbeldt RJ and Napompeth B 1992 The leucaena psyllid comes to Africa, Agroforestry Today **4(4)** 11-12

Varma RV 1986 Seasonal incidence and possible control of important insect pests in plantations of *Ailanthus triphysa*, Peechi, Kerela: KFRI Research Report No. **39**, 32 page

Vergara NT 1981 Integral agroforestry: a potential strategy for stabilizing shifting cultivation and sustaining productivity of the natural environment, Honolulu, Hawaii, East-West Center, Environment and Policy Institute

Vergara NT 1982 Sustained outputs from legume tree-based agroforestry systems, In *New Directions in Agroforestry: The Potential of Tropical Legume Trees* (ed) NT Vergara

Vergara NT and Nair PKR 1985 Agroforestry in the South Pacific region – an overview, Agroforestry Systems 3(4) 363-379

Verhulst PF 1838 A note on population growth, Correspondence Mathematiques et. Physiques **10** 113-121 (In French)

Verinumbe I and Okali DUU 1985 The influence of coppiced teak *(Tectona grandis)* regrowth and roots on intercropped maize *(Zea mays)*, *Agroforestry Systems* **3** 381-386

Verma KS and Thakur NS 2010 Economic analysis of Ashwagandha (*Withania somnifera* L. Dunal) based agroforestry land use system in mid hill Western Himalayas, *Indian Journal of Agroforestry* **12**(1) 62-70

Verma KS, Bhardwaj DR and Krishan Chand 2006 Agroforestry Systems in Himachal Pradesh, In *Agroforestry Systems and Practices* (eds) Sunil Puri and Pankaj Panwar, NIPA, New Delhi, 67-94 pages

Virginia RA 1986 Soil development under legume tree canopies, *Forest Ecology and Management* **16** 69-79

Vishwanatham MK, Samra JS and Sharma AR 1999 Biomass production of trees and grasses in a silvipasture system on marginal lands of Doon Valley of north-west India, 1- Performance of tree species, *Agroforestry Systems* **46** 181-196

Viswanath S, Dhanya B and Rathore TS 2007 Domestication of *Dendrocalamus brandisii* in upland paddy fields in Coorg, Karnataka, *Journal of Bamboo and Rattan* **6**(3&4) 215-222

Viswanath S, Nair PKR, Kaushik PK and Prakasam U 2000 *Acacia nilotica* trees in rice fields: A traditional agroforestry system in central India, *Agroforesstry Systems* **50** 157-177

Viswanathan S, Kaushik PK, Pandy DK and Amit Sahai 1998 Effect of *Acacia nilotica* (L.) on rainfed rice crop in Chhattishgarh, M.P., *Annals of Forestry* **6** 103-109

Vitousek PM and Sanford RL 1986 Nutrient cycling in moist tropical forests, *Annul Review of Ecological System* **17** 137-167

Von Carlowitz P 1989 The role and place of multipurpose trees and shrubs in agroforestry, Paper presented at the ICRAF/DCO International Agroforestry Course held on 8-26 May 1989 at ICRAF Headquarters, Nairobi, Kenya

Von Hesmer H 1966 Der Kombiniete Land-und Forstwirschaftliche Anbau-I Tropisches Afrika, Ernst Klett Verlag, Stuttgart, Germany

Von Maydell HJ 1995 Agroforestry in central, northern and eastern Europe, *Agroforestry Systems* **31** 133-142

WAC (World Agroforestry Centre) 2013 An introduction to agroforestry, *East Asia Node Heilongtan*, Kunming, 650201 PR China

Wagh R and Rajput JC 1991 Comparative performance of bamboo with the horticultural crops in Konkan, In *Bamboo in Asia and the Pacific*, Proceeding of IV International Bamboo workshop, November 27-30, Chiangmai, Thailand, FORSPA Publication-6, Canada, IDRC and FORSPA-Bangkok, Thailand, 1994, 85-86 pages

Wallace JS 1986 The water balance of mixed tree-crop systems, In *Tree Crop Interactions - A Physiological approach* (eds) CK Ong and P Huxley, CAB International, Wallingford, Oxford, UK, 189-233 pages

Waller DM and Steingraeber DA 1985 Branching and modular growth: Theoretical models and empirical patterns, In *Population Biology and Evolution of Clonal Organisms* (eds) JBC Jackson, LW Buss and RE Cook, Yale University Press, New Haven, London, 225-257 pages

Wanyeki FHM 1981 Agroforestry Practices in High Potential Areas of Kenya, In *Proceedings of the Kenya National Seminar on Agroforestry* (eds) L Buck, ICRAF, Nairobi, 218-309 pages

Warner K 1991 Shifting Cultivators, Local Knowledge and Natural Resource Management in the Humid Tropics. Rome: Food and Agriculture Organization of the United Nations

Waruhiu AN, Kengue J, Atangana AR, Tchoundjeu Z and Leakey RRB 2004 Domestication of *Dacryodes edulis*: pheotypic variation of fruits in 200 trees from four populations in the humid lowlands of Cameroon, *Food, Agriculture and Environemnt* **2**: 340-346

Watson GA 1983 Development of Mixed Tree and Food Crop Systems in the Humid Tropics: A response to Population Pressure and Deforestation, *Experimental Agriculture* **19** 311-332

Watson RT, Noble IR, Bolin B, Ravindranath NH , Vernado DJ, Dokken and Dokken DJ (eds) 2000 Land use, land use change and forestry, IPCC special report, Cambridge University Press, New York

Watters RF 1960 The nature of shifting cultivation, *Pacific Viewpoint* **1** 59-99

Wezel A and Bender S 2003 Plant species diversity of homegardens of Cuba and its significance for household food supply, Agroforestry Systems **57** 39-49

White KJ 1985 Intercrop Strategies and Practices in the Bhabar Terai of Central Nepal (Manual No. 1) Ministry of Forests, Kathmandu, Nepal

Wickens GE 1991 Management issues for development of non-timber forest products *Unasylva* **165**(42) 3-8

Wickramasinghe A 1996 The non-forest wood fuel resources of Sri Lanka, *Wood Energy News* **11** 14-18

Wiersum KF (ed) 1981 Observations on Agroforestry on Java, Forestry Faculty, Gadjah Mada University, Yogyakarta, Indonesia & Department of Forestry Management, Wageningen Agriculture University, Wageningen, The Netherlands, 132 pages

Wiersum KF 1982 Tree Gardening and Taungya on Java: Examples of Agroforestry Techniques in the Humid Tropics, Agroforestry Systems 1 52-70

Wiersum KF 2004 Forest gardens as 'intermediate' land-use system in the nature - Culture continuum: Characteristics and future potential, *Agroforestry Systems* **61** 123-134

Wiersum KF 2006 Diversity and change in homegarden cultivation in Indonesia, In *Tropical Homegardens: A Time-Tested Example of Sustainable Agroforestry* (eds) BM Kumar and PKR Nair, Springer, The Netherlands, 13-24 pages

Wilken GC 1977 Integrating forest and small-scale farm systems in Middle America, *Agroecosystems* **3** 291-302

Wilkinson K and Elevitch C 2013 Improved Fallow, The Overstory No. 42 website: http://agroforestry.net/overstory/overstory42.html

Wilkinson KM and Elevitch CR 2000 Integrating understorey crops with tree crops, IN Agroforestry Guides for Pacific Islands (eds) CR Elevitch and Wilkinson KM, Permanent Agricultural Resources, Holualoa, Hawaii,USA, webpage: http://www.agroforestry.net

Willey RW 1975 The use of shade in coffee, cacao and tea, *Horticultural Abstracts* **45** 791-798

Willey RW and Reddy MS 1981 A field technique for separating above and below ground interactions in intercropping: An experiment with pearl millet/groundnut, *Experimental Agriculture* **17** 257-264

Williams JR 1975 Sediment yield prediction with universal equation using runoff energy factor, In: Present and prospective Technology for predicting sediment yields and sources, ARS-S-40.USDA, Agricultural Research Service, Washington, DC: USDA, 118-124

Williams-Guille NK, Perfecto I and Vandermeer J 2008 Bats limit insects in a tropical agroforestry system, *Science* **320** 70

Wimjum JK, Dixon RK and Schroeder PE 1992 Estimating the global potential of forest and agroforest management practices to sequester carbon, *Water Air Soil Pollution* **64** 213-228

Winrock International 2013 Agroforestry for the Pacific Technologies: Fact Sheets, Forest, Farm, and Community Tree Network, webpage: http://www.winrock.org/fnrm/factnet/factpub/AIS_ list.html

Wischmeier WJ and Smith DD 1978 Predicting Rainfall Erosion loss: A Guide to Conservation Planning, Agri. Hand book No. 537, Washington, DC: USDA

Wood D and Lenne JM 1997 The conservation of agrobiodiversity on-farm: questioning the emerging paradeigm, *Biodiversity and Conservation* **6** 109-129

Wood PJ and Burley J 1991 A Tree for All Reasons, ICRAF, Nairobi, Kenya

World Bank 2008 World Development Report 2008: Agriculture and Development, Washington DC: World Bank Publications

World Bank 2011 Global Partnership for Wealth Accounting and the Valuation of Ecosystem Services (WAVES), Washington, DC, website: www.wavespartnership.org/waves/

Wu Y and Zhu Z 1997 Chapter-5: Temperate Agroforestry in China, In *Temperate Agroforestry Systems* (eds) AM Gordon and SM Newman, CAB International, Wallingford, UK

Xu FY and Wu DX 1989 Control of bamboo scale insects by intercropping rape in the bamboo forest to attract coccinellid beetles, *Chinese Journal of Biological Control* **5** 117-119

Xu JC, Kim KJ and He J 2011 Participatory agroforestry development in DPR Korea, Kumming: World Agroforestry Centre (ICRAF), China and East Asia Node

Yadav JSP 1980 Salt affected soils and their afforestation, *Indian Forester* **106**: 259-272

Yadav JSP and Jai P 1964 Soil suitability for Eucalyptus hybrid plantation in terai region of Uttar Pradesh, *Indian Forester* December Issue

Yeung Y 1987 Examples of urban agriculture in Asia, *Food Nutrition Bulletin* **9**(2) 14-23

Yoshino K and Ando K 1999 Utilization of plant resources in homestead (*bari-bhiti*) in floodplain in Bangladesh, *Japanese Journal of Tropical Agriculture* **43** 306-318

Young A 1987 The Potential of Agroforestry for Soil Conversation and Sustainable Land Use, ICRAF Reprint No. 39. Nairobi, Kenya: ICRAF

Young A 1989a Agroforestry for Soil Conservation, CAB International - International Council for Research in Agroforestry, Wallingford, 246 pages

Young A 1989b Ten hypotheses for soil-agroforestry research, *Agroforestry Today* **1** 13-16

Young A 1997 Agroforestry for soil management, 2nd Edition, CABI Publishing, Wallingford, UK, 320 pages

Yunyin W and Dalmacio RV 1991 Energy balance, water use and wheat yield in a Paulownia-wheat intercropped field, In *Agroforestry systems in China* (eds) Z Zhaohua, C Manfang, W Shiji and J Yonxn,, CAF/IDRC, Singapore

Zang L, Dawes WR, Slavich PG, Meyer WS, Thorburn PJ, Smith DJ and Walker GR 1999 Growth and ground water uptake responses of Lucerne to changes in ground water levels and salinity: lysimeter, isotope and modelling studies, *Journal of Agriculture and Water Management* **39** 265-282

Zhang P 1999 Nutrient inputs from trees via throughfall, stemflow and litterfall in an intercropping system, Thesis presented to the Faculty of Graduate Studies, University of Guelph, Canada

Zhang W, Ricketts TH, Kremen C, Carney K and Swinton SM 2007 Ecosystem services and dis-services to agriculture, *Ecological Economics* **64** 253-260

Zhou SQ 1993 Cultivation of *Amomum villosum* in tropical forests, *Forest Ecology and Management* **60** 157-162

Zhou X 1999 On the three-dimensional aerodynamic structure of shelterbelts, Ph.D. Dissertation, University of Nebraska, Lincoln, NE.

Zomer RJ, Trabucco A, Coe R and Place F 2009. Trees on Farm: Analysis of Global Extent and Geographical Patterns of Agroforestry, ICRAF Working Paper No. 89, Nairobi, Kenya: World Agroforestry Centre.

■■■■